Space Photonic Communications

This book explores the extension of fiber optic communications technology to space optical communications. It presents the specific demands for space communications and examines propagation in a vacuum and also over the turbulence of air-free space.

Space Photonic Communications begins with historical aspects of optical communications in both terrestrial and inter-satellite constellations. It presents the advances in optical communications and integrated photonics that have enabled the technological developments for space photonic communications, especially coherent optical communications to transport Tbps information. The author discusses how the transmission of massive amounts of data over very long distances requires Tera-bps communications in real time for both military and civil applications. The author also explores laser communications with the physics of laser propagation over ultra-long distances without optical amplifications at intermediate locations. He also examines modulation and receiving techniques for space photonic communications.

Intended for communications engineers and professionals, especially those in laser communication technologies, this book could also be used in courses on advanced optical communications, photonic communications, and communication techniques and technologies.

Le Nguyen Binh, B.Eng (Electron. Eng., Hons) and a Ph.D., and a Dr. Eng. (Integrated Photonics and Optical Communications Engineering). He was recently Fellow at Mynaric Laser Communications GmbH, Gilching, Bavaria, Germany and R&D Director of Huawei Technologies European Research Center in Munich, Deutschland. He was awarded five Huawei Technologies Gold Medals for his work on advanced optical/photonics communication technologies. He was previously the chair of Commission D (Electronics and Photonics) of the National Committee for Radio Sciences of the Australian Academy of Sciences and a professorial fellow at Nanyang Technological University, Christian-Albrechts-Universität zu Kiel, and Reader/Professor of Monash University, Australia. Widely published, Dr. Binh is the series editor of *Optics and Photonics* for CRC Press.

Optics and Photonics

Series Editor: Le Nguyen Binh
Huawei Technologies, European Research Center, Munich, Germany

Optical Fiber Communications Systems: Theory and Practice with MATLAB® and Simulink® Models
Le Nguyen Binh

Ultra-Fast Fiber Lasers: Principles and Applications with MATLAB® Models
Le Nguyen Binh and Nam Quoc Ngo

Thin-Film Organic Photonics: Molecular Layer Deposition and Applications
Tetsuzo Yoshimura

Guided Wave Photonics: Fundamentals and Applications with MATLAB®
Le Nguyen Binh

Nonlinear Optical Systems: Principles, Phenomena, and Advanced Signal Processing
Le Nguyen Binh and Dang Van Liet

Wireless and Guided Wave Electromagnetics: Fundamentals and Applications
Le Nguyen Binh

Guided Wave Optics and Photonic Devices
Shyamal Bhadra and Ajoy Ghatak

Digital Processing: Optical Transmission and Coherent Receiving Techniques
Le Nguyen Binh

Photopolymers: Photoresist Materials, Processes, and Applications
Kenichiro Nakamura

Optical Fiber Communication Systems with MATLAB® and Simulink® Models, Second Edition
Le Nguyen Binh

Noises in Optical Communications and Photonic Systems
Le Nguyen Binh

Advanced Digital Optical Communications, Second Edition
Le Nguyen Binh

Optical Modulation: Advanced Techniques and Applications in Transmission Systems and Networks
Le Nguyen Binh

Optical Multi-Bound Solitons
Le Nguyen Binh

Molecular Layer Deposition for Tailored Organic Thin-Film Materials
Tetsuzo Yoshimura

Space Photonic Communications
Le Nguyen Binh

Space Photonic Communications

Le Nguyen Binh

CRC Press
Taylor & Francis Group
Boca Raton London New York

CRC Press is an imprint of the
Taylor & Francis Group, an **informa** business

Designed cover image: Exatronix

First edition published 2025
by CRC Press
2385 NW Executive Center Drive, Suite 320, Boca Raton FL 33431

and by CRC Press
4 Park Square, Milton Park, Abingdon, Oxon, OX14 4RN

CRC Press is an imprint of Taylor & Francis Group, LLC

ISBN: 978-1-032-73101-8 (hbk)
ISBN: 978-1-032-73098-1 (pbk)
ISBN: 978-1-003-42667-7 (ebk)

DOI: 10.1201/9781003426677

Typeset in Times
by codeMantra

Contents

Preface

Optical communication systems and networks have evolved over the last 70 years, from laser invention to laser communications, then to guided wave over optical fibers, integrated optics/photonics, and networking across all continents. However, only about 50% of the global area can access Internet information.

Space photonic communication is the communication technique using photonic modulation of lightwaves generated from finite-linewidth laser sources and coherent reception after long-distance propagation over space, an ideal vacuum, or the atmosphere.

This book gives, first, a historical overview of global laser communications and the diffusion of non-guided and guided terrestrial optical communication systems, especially laser inter-satellite constellations and ground station networking for flexible, dynamic, and wide global area access. Without the advancement of integrated optics and photonics, there would be no ultra-high bit rate transmission. Taking advantage of this advancement, space photonic communications (SPC) can now push interconnects to an extremely high data rate. By photonics, we refer to the coherency of the laser carrier and modulation such that the lightwaves transporting the modulation signal band remain pseudo-asymptotically coherent, enabling coherent reception in such an SPC system.

Thence, the main chapters address (i) technological developments in optical technology and communication, especially photonic communications, focusing on the coherence of the laser and modulated optical beams for long-reach space photonic transmission; (ii) transmission phenomena over vacuum-free space and turbulent propagation in the Earth's tropospheric region, from the optical ground station (OGS) to satellite; (iii) inter-satellite networking via repeaters, optical switching, and all-photonic routing; (iv) optical booster high-power amplification and low noise optical pre-amplification; (v) contrasting two extreme cases: ultra-high-bit rates for ultra-high capacity versus extremely low-intensity signaling for quantum photonic communications over free space media; (vi) appendices 1–7, supplementing basic information relevant to lasers, optical modulation, fiber and integrated optics, and transmission spectral grids.

Furthermore, the data transfer rate, including data collection for AI processing algorithms, can currently be pushed to much higher levels due to artificial intelligence (AI) processes, such as Open AI and ChatGPT. Therefore, there is no doubt that ultra-high-rate transmission can assist in this intelligent networking era via several intensive intelligent searches and combinations of relevant information, including real-time functions, in order to minimize the latency and transmission capacity. Photonic systems and networks must evolve to meet this intelligent information processing. A dynamic ultra-high-capacity network with flexibility within hours would be highly demanded, especially under current geopolitical challenges, enabling the deployment of several groups of people in no time. Therefore, SPC is critical for interconnecting satellite constellations in different orbits and distributing them to any optical ground station on the global ground.

This book presents SPC transmission over the ideal vacuum distance interconnecting satellite optical nodes and to and from OGS nodes and satellite optical nodes via add/drop channels of different wavelength carriers. The reconfigurable optical add-and-drop multiplexers with contention-less, colorless, and directionless features (CDC-ROADM) are in all-optical structures for SPC networking to minimize power consumption and complexity with high fault tolerances. The rotation of the satellite nodes generates frequency shifts and multiple path delays, mitigating pulse dispersion and coherent homodyne or heterodyne by digital signal processing algorithms. We have also adopted advanced optical and photonic integrated technologies developed and matured in terrestrial fiber optic networks for SPC networking. Without these advanced photonic components, especially the integrated optical transceivers, it would be impossible to design and demonstrate SPC multi-wavelength transmission systems at tera-bits/s bit rates.

I acknowledge the information and insights gained via discussions with my colleagues at Mynaric Laser Communications GmbH, Csaba Gal, Branko Miletic, Carsten Fechtmann, Joachim Howarth, Max Koegler, and former colleagues at Huawei Ltd, European Research Centre in Munich, especially Bruce Liu Lei, Prof. Nebojsa Stojanovic, and Xie Chang Song; without these exchanges, I could not have structured the contents of this book.

Finally, I acknowledge the education my parents have provided me over the years, even though their education level is limited. This has helped me overcome difficulties and reach knowledgeable goals. Thi Phuong Nguyen and Lam Le-Nguyen have also accompanied me over the writing period of this book.

Eng. Le Nguyen Binh

Altum discendi, incisionem muniat, eligens ius, uti ea quoque

1 Guided-Wave and Space Technologies

CONTENTS

Optical communication systems and networks have evolved over the last 70 years from laser invention to laser communications, then to guided wave over optical fibers and networking over all continents. However, only 50% of the global area can access the Internet information.

Free space laser communications can now benefit the advancement of guided-wave technologies employed to pushing and accelerating the global interconnection of continental fiber networks at such an extremely high data rate.Furthermore, the data transfer can be pushed to much higher levels due to artificial intelligence (AI) processing, such as OpenAI, ChatGPT, chatAI, and other artificial general intelligence (AGI) data access and processing platforms. Therefore, there is no doubt that ultra-high rate transmission can assist in this intelligent networking. We note that only about 50% of the global area has access to the Internet. It is inevitable that inter-satellite optical networking for universal distribution of information is essential. However, the technology of such satellite laser communication networks is not currently available.[1] A possibility of such ground-to-inter-satellite optical network can be observed in Figure 1.1. Depending on the maximum inter-satellite distance that can be linked by the most advanced optical communication technology, the total number of satellites in a complete round-the-earth can be reduced to 6 or 12. This book aims to reach such maximum transmission distance via the power budget and spatial dispersion budget. Preliminary experimental results can supplement such a degree of confidence for the implementation of such space photonic communication networks. The word 'photonic' is coined here to indicate the coherence in the optical domain of the laser beam under modulation and the reception techniques, thus dealing with the optical beam at the photonic level. More details of the optical multiplexing and demultiplexing techniques to increase the transmission capacity will be described in the chapters.

This chapter presents the following: (i) a brief historical overview of the development of optical communications over the optical communications since the invention of the gas laser in the 1960s, (ii) the initial free space transmission, integrated optics, and fiber-guiding transmission cables for extra-long terrestrial communication systems and networking, and (iii) the current space communication technology, taking advantages of advanced technology for real-time distribution over inter-satellite optical transmission of extra-high capacity information to any earth station on Earth.

DOI: 10.1201/9781003426677-1

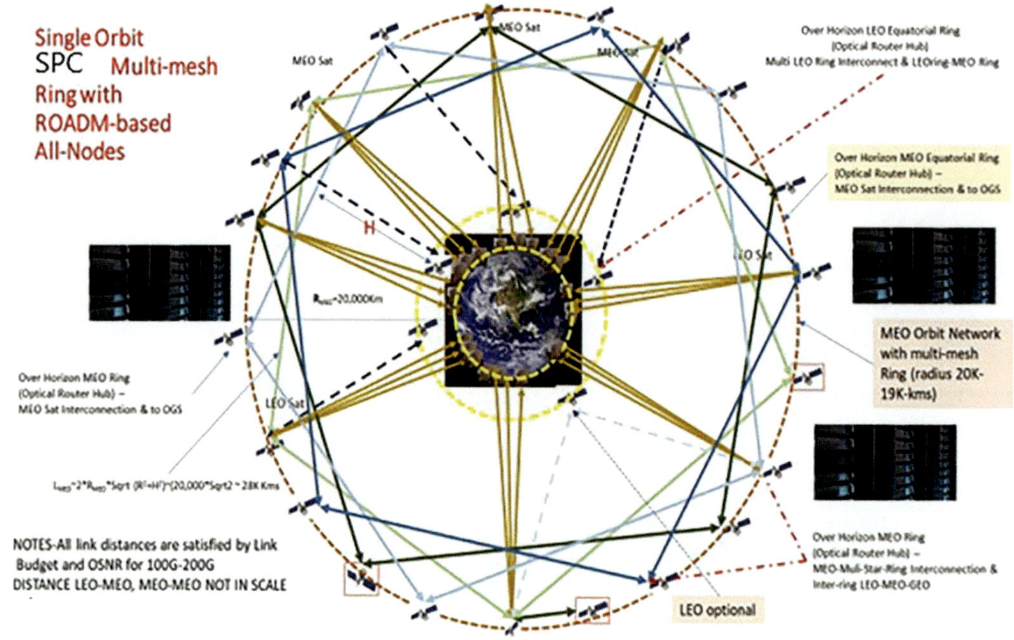

Inter-orbital constellations and optical transmission paths for SOCS networking. Shown are GEO, LEO, and MEO satellite constellations.

FIGURE 1.1 Laser intersatellite network covering the all-round Earth and ground station.

1.1 HISTORICAL ASPECTS OF NON-GUIDED AND GUIDED WAVE OPTICAL TECHNOLOGIES

A general introduction to the different phases of development in space communication and integrated guided line technology is first introduced. Then, further details of these communication techniques and technologies are briefly outlined. Detailed technologies would occupy too much space, so they are provided in appropriate chapters.

1.1.1 General Development Phase

The world of lasers has witnessed numerous breakthroughs over the last 80 years from the invention of the He-Ne gas laser to the current use of embedded lasers in several integrated photonic transmitters and receivers, under intensity modulation and direct detection, to coherent transmission and reception in optical communication systems, including both inland optical lines and space transmission systems and networks.

In the early 1960s, researchers at Bell Labs began exploring the possibility of developing a laser that operated at visible wavelengths. Javan et al.[2] invented and created a practical, efficient, and continuous-wave (CW) laser.

The helium-neon laser's design consists of a gas discharge tube, mirrors, and an external power supply. The gas discharge tube is filled with a mixture of helium and neon gases at specific pressure levels, and electrodes at both ends initiate the electrical discharge. The discharge excites the helium and neon atoms, causing them to emit photons. The photons undergo stimulated emission as they pass through the gain medium, creating a cascade effect of amplified light. One of the mirrors, at the front/output end, is partially reflective, allowing a small portion of the light to escape and form the laser beam, while the other mirror reflects the light into the gain medium to sustain the amplification process.

Since the invention of the He-Ne gas laser in the 1960s, there has been an explosion of inventions of new types in gas, solid-state crystal, semiconductor media, fiber-doped lasers, etc., and their applications have continuously emerged over the last seven decades of the 20th century. Furthermore, the integration of electronic circuits and very large-scale integration (VLSI) circuit technology has permitted the integration of optical devices on the same substrate. This has led to the field of integrated optics or guided wave optics and hence photonics, which deals with optical devices at the photon level, that is, the manipulation of photons in guided wave circuits.[3] A significant development in the photonic fields was accelerated when optical amplification could be implemented using rare-earth metal-doped SiO_2 waveguides, either circular or strip waveguides. This allows for the compensation of propagation loss in guided wave media, enabling much higher integration of optical devices.

In addition, the increase in the sampled electronic processors, reaching up to 256 GSa/s, enables the employment of signal processing in integrated photonic circuits. Therefore, integrated microprocessors and integrated photonics have led to the significant development and invention of several processing systems in telecommunications.

Figure 1.1 depicts typical optical and photonic transmission systems for free space and guided wave transmission line terrestrial systems. Figure 1.2 illustrates a typical space ultra-long-distance satellite photonic communications network that employs multi-wavelength carriers over the transparent spectral windows of space media. The space transmission path can pass through several scenarios, but significantly two main cases are the optical ground station (OGS) to satellite through the air space path in which the turbulence is critical, and hence, the probability or errors can be variable, and the other is transmission via vacuum long path from satellite to satellite of several stages or free space span. It is necessary to multiplex and demultiplex optical channels with different carriers of different frequencies as far as the attenuation of the space transmission medium satisfies the power budget of the transmission system.

In the following, we present the development of several photonic devices and their integrated technology.

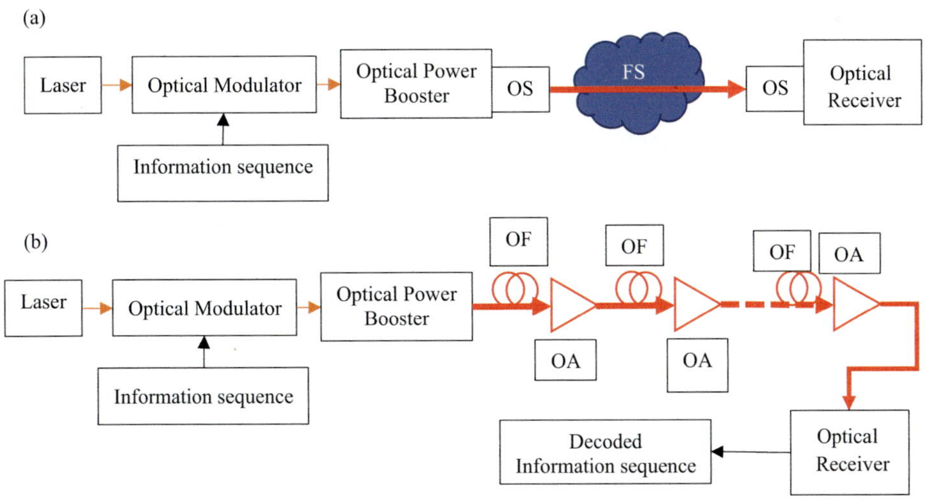

FIGURE 1.2 Generic optical transmission system: (a) laser communication; (b) optical/photonic guided-wave media transmission. OA, optical amplifier; OF, optical fiber; FS, free space; OS, optical system and beam launcher.

1.1.2 Optical Guided Media

The invention of laser has led to several innovative fields of optics applications. One of the significant fields is fiber optics and integrated optics, which is the guiding of optical waves over long distances in optical waveguides of either circular, planar, or channel structures.

Circular optical waveguides, or fibers, are crucial for generating long-haul optical cables. This idea was initially proposed and investigated through dielectric waveguides[4] by Kao and Hockham in 1965, and then demonstrated in circular glass structures by Keck of Corning in 1970.[5] This Corning demonstration proved the guiding phenomenon in circular structures, which consist of a core and a cladding whose refractive indices must differ by a tiny amount. Significant theoretical and practical studies have been conducted, as well as the design and manufacturing of optical fiber structures, leading to the development of various multimode and single-mode types. Subsequently, different transmission techniques were developed and demonstrated for long-haul and long-repeater spacing optical communication systems. This has led to significant developments over many decades with single-mode fibers and coherent optical communications.[6] Furthermore, the guiding of optical waves is required to design and develop guided-wave devices in planar structures, enabling integrated optical devices such as modulators and lasers. This allows for power consumption and high-speed operations, which facilitate fast modulation and efficient operations, leading to transmission of several thousand giga-bits/s of information.

These integrated optical devices surpass their counterparts using bulk structures, which were used in the initial space laser communication systems.[7]

1.1.3 Solid State and Integrated Lasers

Naturally, optical carriers are the main sources for transmitting information over various media, whether free space or guided transmission line. These carriers are lasers whose frequencies can range from a few hundred tera-hertz or the visible wavelength (<870 nm) to the near-infrared (<1700 nm) and far-infrared (2000 nm) spectral regions, and even up to 10 μm.

However, the development and advancement of laser sources that can exhibit a linewidth or optical bandwidth from a few kHz to <10 Hz, such as fiber lasers,[8] have progressed tremendously over the last few decades, especially in the first decade of the 21st century, due to coherent optical communications.[9,10]

There are various types of integrated laser structure whose spectral purity depends on the reflector properties, such as the optical frequencies and bandwidth, to enable the optical amplification via the gain media, as shown in Figure 1.3. This laser is the side-emission type. There is also the vertical-cavity surface-emitting laser (VECSEL), or vertical emission laser type. They are briefly summarized here. It is noted that these integrated lasers must be used to provide a 'single' frequency source in contemporary space laser communication systems, especially when they are multiplexed with other modulated channels.

Advanced integrated lasers can now offer a very narrow linewidth, about 10 Hz, which can be considered single frequency. The modulation of these lasers can lead to a wideband signal band but with a 6 dB peak of the laser. This peak power offers the coherency of the beam to carry signals over long-haul free-space distances.

Figure 1.4a–c shows the structural schematics of the free-space and guided lightwave lasers. Figure 1.4d–f displays the structure and details of optical modulators that use crystal- free space interaction to modulate the laser beam. We can see the difficulties of beam alignment in all these free-space structures and the interaction volume of the electro-, acousto-, and magneto-optic effects. This refers to the interaction between the optical fields of the laser beam and either the electric field of the applied RF voltage, the applied acoustic waves via an interdigital transducer, or the magnetic applied current. The signal voltage or field applied must be very high due to the large distance of the two electrodes deposited on the two faces of crystal block which forms the bulk modulator.

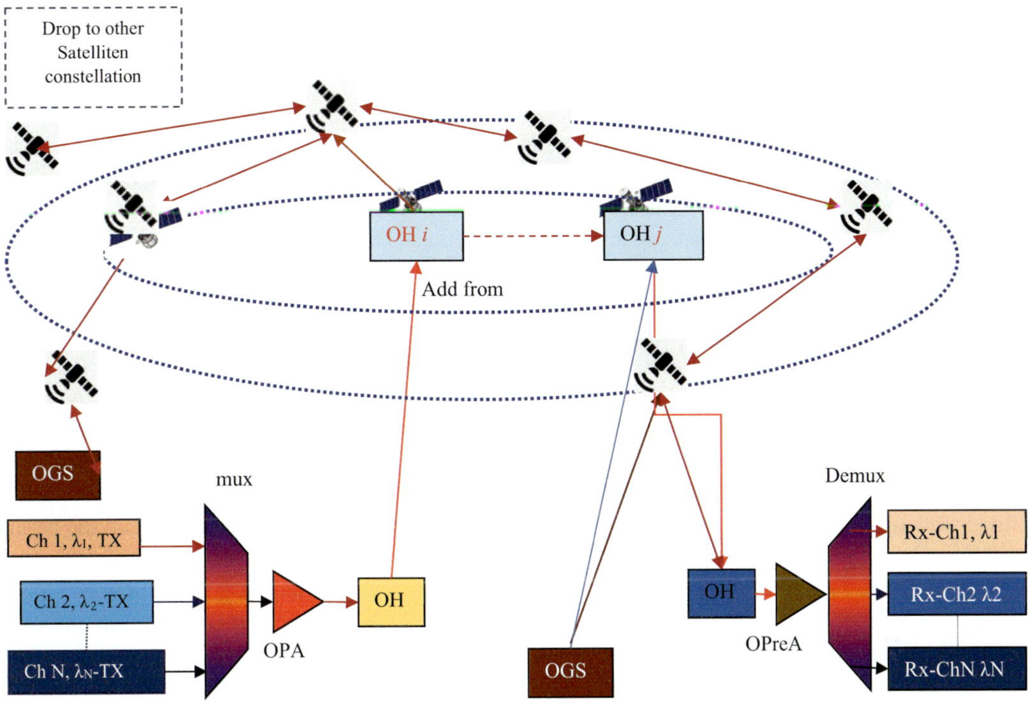

FIGURE 1.3 Space DWDM optical communication channels and inter- and intra-orbital networking. OGS=optical ground station, OH=optical head with pointing control, λ=wavelength, OPreA=optical pre-amplifier, OPA=optical power amplifier, Tx=optical transmitter, Rx=optical receiver, Mux= optical multiplexer, Demux=optical demultiplexer. Ch=channel, Ch N=channel number N.

Therefore, to obtain low voltage-driven modulation, integrated devices or modulators are used. This low-driven voltage allows also high-frequency electronic amplifiers to feed to the electrodes of the modulator. This motivates users to integrate fiber-optic communication systems into space optical communications. So, whenever integrated optical devices can be employed, they are used. The disadvantages are additional coupling losses when optics are input into and output from free space. Provided that these weakened lightwaves can be further amplified, the total loss and gain of the transmission system do not degrade below the critical optical signal-to-noise ratio. These aspects are addressed in the appropriate chapters.

1.1.4 Laser Types

Several basic structures of integrated lasers depend on the types of reflectors, as shown in Figure 1.3, or the reflection frequency and bandwidth of such filters and reflectors. In general, the following types exist.

1.1.4.1 Fabry Perot Resonant Laser (FBL)

The reflectors of the Fabry-Perot (FP) type are just flat mirrors with different reflection coefficients for the front and back. The reflection coefficient has a passband. Thus, there are multiple reflections and, hence, multiple types of laser emission. The gain bandwidth of the main gain medium in the laser cavity also contributes to the number of multiple emitted laser lines. The typical 3 dB bandwidth of this type of laser is a few nanometers, and the frequency spacing between the lines of the spectrum is about 0.01–0.1 nm, depending on the length of the lasing cavity. These spectral ranges depend on the refractive index of the cavity medium and the cavity length, which also affects the uniformity.

1.1.4.2 DFB Laser

Similar to the FP laser, the reflectors incorporate a Bragg diffraction grating so that only the wavelength that matches the Bragg diffraction condition can be reflected, rejecting all other spectral lines. Thus, only the matching wavelength optical waves can be amplified and emitted, ensuring a spectrally pure laser line or single-frequency laser. The typical bandwidth of a Distributed Feedback (DFB) laser is a few MHz.

1.1.4.3 Fiber Laser

Taking advantage of the optical amplification of rare-earth doped fiber medium, this optical amplifier can be incorporated into a total fiber resonator. The fiber laser consists of the fiber gain medium, two Bragg grating reflectors, and a wavelength-division multiplexing (WDM) fiber coupler so that the pump laser can inject the required energy to obtain sufficient optical gain, and consequently, the emitted optical waves. The fiber grating ensures that only the selected spectral lines are produced. Since the length of the fiber resonator can be long, the laser linewidth can be narrow, and the bandwidth can be reduced to a few Hertz.

Furthermore, by doping various rare-earth ions such as erbium, ytterbium, or thulium into pure silica, the emitted laser wavelength ranges can be varied across the near infrared to far infrared spectral regions due to frequency conversion to multiple energy bands of the rare earth ions. The fibre laser reliability is that all-fibre distributed feedback (DFB) structure to ensure the resonant frequency and the line-width of the laser can reach to a few Hz that can be considered to be ‘single frequency’ lasers, that is very close to a single frequency lasing source.

1.1.4.4 Comb Lasers/Ultra-Short Pulse Lasers

This type of laser differs from a CW laser in that the output waves are periodic pulses with very short temporal widths. Thus, its frequency domain equivalent is a series of periodic narrow spectral lines or comb lines. This type of laser is useful when a multiple periodic wavelength laser source is required, such as in dense wavelength optical transmission systems and networks.

Thus, there has been significant development of lasers, especially single-mode types, which allow for all possibilities for coherent and ultra-high bit rate digital optical transmission technologies. These advanced laser sources would not be possible if guided wave optical device technologies had not been developed over the years in integrated optics and optical fiber transmission.

We expect that space optical communication technologies will employ guided optical components whenever possible, launching the optical beams into space at the point of transmission into free space. Similarly, at the free-space reception point, the optical beams are collected and focused into guided wave optical devices and processed in such integrated optic and electronic systems.

Non-guided wave space communication systems and networks are shown in Figure 1.2. In such ultra-long distances, for example, GEO-to-GEO (GEO = Geostationary Orbit) inter-satellite multi-span links, the coherence length of the laser is critical so that the photonic beam can be launched into free space and remain pseudo-coherent. This allows a coherent reception system to be employed in association with polarization multiplexing and DSP (digital signal processing/processor) processing to achieve a high-capacity transmission system.

1.1.5 Optical Modulation

1.1.5.1 Bulk Modulation: Electro-Optic, Acousto-Optic and Magneto-Optic Modulation

Once the laser was invented, several ideas and developments were initiated. Modulation of the lightwaves emitted from the laser is critical for the transmission of information. Thus, in the realm of optics and photonics, the ability to manipulate light is of paramount importance. Optical modulators serve as crucial devices for controlling light intensity, phase, and polarization, enabling a wide range of applications in telecommunications, data transmission, imaging, and sensing.

Bulk crystal optical modulators, also known as electro-optic, magneto-optic, or acousto-optic modulators, utilize the electro-optic, magneto-optic, or acousto-optic effects exhibited by certain crystals in specific directions relative to the crystal structure such as isotropic or anisotropic, to modify the properties of light passing through them. These modulators operate by changing the refractive index of the crystal in response to an applied electric field. The interaction between the electric field and the crystal's optical properties enables precise control over the transmission characteristics of light. Figure 1.4a shows a block diagram of a bulk modulator coupled with a laser via an optical system (OS) and then launching into free space via another OS. We note that the laser beam propagation direction and the electric or acoustic field for electro-optic or acousto-optic interaction must be optimized based on the coefficients of the crystal structure, such as whether it is anisotropic, isotropic, biaxial, or uniaxial. The interaction volume of a bulk modulator is small compared to that of the integrated planar modulator type, shown as in Figure 1.4b, where the interaction length extends along the full length of the electrode.

Optical bulk crystal modulators are devices that can modulate the intensity, phase, or polarization of light using the electro-optic effect in certain types of crystals. These modulators are typically made from materials such as lithium niobate ($LiNbO_3$) or lithium tantalate ($LiTaO_3$), which exhibit a strong electro-optic effect. The crystals are cut and polished to specific orientations to maximize modulation efficiency. When an electric field is applied to the crystal, its refractive index of the crystal changes, allowing modulation of the light passing through it. Optical bulk crystal modulators have been widely used in various applications, including telecommunications, laser systems, and optical signal processing.

However, the distance between the electrodes in these bulk structures, both negative and positive, is large, ranging from a few millimeters to centimeters. This leads to the requirement of high voltage levels so that the electric field across the lightwaves' path is sufficient to change the refractive index of the crystal medium and achieve moderate phase changes or polarization of the lightwaves, and thus amplitude modulation. Furthermore, the high electric field limits the high-frequency operation, microwave and millimeter range, of bulk optical modulators, as there must be wideband and high-power microwave and millimeter-range electric amplifiers.

However, for ultra-high-speed operations, high-speed modulation is required, and the bulk-free-space modulator, as shown in Figure 1.3e, cannot offer high-speed modulation with low average power levels due to the large electrode distance across the crystal faces. Instead, the electrode spacing for an integrated guided wave structure can be as low as a few micrometers. Thus, it is preferred to employ these integrated optic modulators in cooperation with lensed fibers at the input and output to couple to the free-space optic for launching into free-space medium. Another advantage of the integrated optic modulator type is that the laser can be integrated, as well as any processing optical devices, as we will see in later chapters.

This led to the intensive development of the field of integrated optics in the 1970s.[11,12]

An acousto-optic bulk modulator can be used for beam steering using the deflection via the Bragg phase matching condition. Beam steering is essential for long-haul free space to align and minimize coupling loss to the reception subsystem. A bulk modulator or deflector requires a high voltage level to achieve efficient acousto-optic interaction. This is inversely proportional to the electrode separation distance. That is, the optical field and the acoustic or electric field in an integrated acousto-optic deflector can be much more efficient, as demonstrated in the 1980s,[13] with a voltage level of a few hundred mV applied to the integrated interdigital transducer (IDT) electrodes, compared to a few tens of volts at about 450 MHz for the bulk type. In the next section, we describe integrated optic or photonic devices, which are essential for ultra-high capacity information modulation for space communication using optical waves.

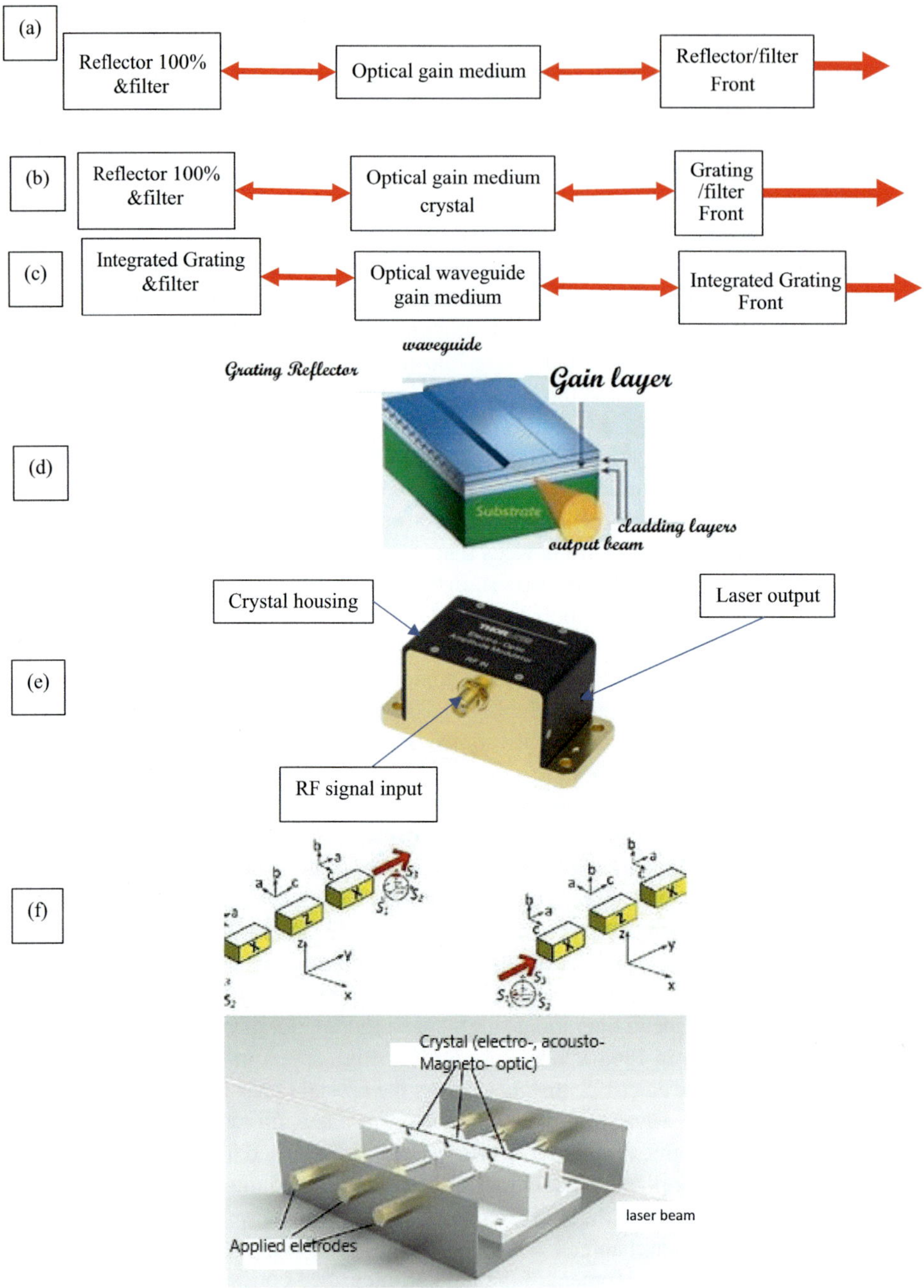

FIGURE 1.4 Generic structure of laser: (a) generic structure, (b) bulk structure, (c) guided resonant structure, (d) integrated semiconductor laser structure, (e) bulk free-space modulator with free-space lightwave input and signal input port, gratings as reflector and filter; (f) internal structures of (e) and crystal orientation for effective interaction between laser beam and electromagnetic or acoustically applied field. Electrodes are designed appropriately for the fields. Note: orientation of crystal axes for efficient interaction.

1.1.5.2 Integrated Optic Modulators

Integrated optics is a field of study that focuses on the design, fabrication, and characterization of optical devices and systems that integrate multiple optical components onto a single substrate. Bell Labs, the research arm of AT&T, has made significant contributions to the field of integrated optics.

Some of their notable work includes the development of planar waveguide technology, which allows for the integration of various optical components such as waveguides, couplers, and modulators onto a single chip. This technology has been widely used in telecommunications. Thus, the photonic integrated circuit (PIC), or integrated optical circuit, was invented. This integration technology in the optical domain, the integrated optic/PIC, is a microchip containing two or more photonic components that form a functioning circuit. This technology detects, generates, transports, and processes light. PICs utilize photons (or particles of light) as opposed to electrons, which are utilized by electronic integrated circuits. The manipulation of photons or guided optical waves in such a planar plane stimulates the fields and the word 'photonics', indicating the coherence of the optical waves. The major difference between the two is that a photonic integrated circuit provides 'functions' via modulation and various optical principles under coherence for information signals imposed on optical wavelengths, typically in the visible spectrum or near-infrared (850–1650 nm) and even in the mid and far-infrared spectral regions.

It is inevitable that modern laser communications employ these integrated photonic circuit technologies as part of the transmitting and receiving systems to increase the transmission and reception efficiencies of the systems. This is why this book title uses the term 'photonic communications'.

1.1.6 Bulk and Large Area Fiber Optical Amplifiers

The advent of optical fiber amplifiers was essential in building the high-speed backbone of the global telecommunications network, which carries our words, pictures, and data around the planet. More recently, in modern optical communication systems and networks, as well as high-power amplifiers and lasers for military applications, fiber lasers have become powerful tools in manufacturing, generating multikilowatt beams that can cut and weld materials ranging from plastics to metals.

Other early solid-state lasers, such as the ruby laser demonstrated by Maiman in 1960,[14,15,16,17,18] were made of bulk materials. The fiber laser uniquely transmits the light it generates along a light-guiding core, concentrating its energy in a small area inside the glass, and making it easy to transfer light from a fiber laser into a passive optical fiber for transmission.

Fiber lasers have also proved exceptionally well-suited for efficiently generating high-quality beams with powers reaching many kilowatts, greatly expanding the applications of lasers in cutting, welding, and other machining of materials. Similarly, the optical fiber amplifier acts on light propagating in a fiber's core. This became particularly important long after Spitzer's work when fiber-optic communications emerged in the 1970s.

The early fiber amplifiers and lasers have led to contemporary fiber amplifiers for NIR and IR regions in silica rare-earth-doped structures.

Optical signals needed to be amplified after passing through tens of kilometers of glass, and initially, this required converting the signals into electronic form for amplification. Innovations by Payne and others[19] led to the development of optical fiber amplifiers, which could boost signals in the important C-spectral window, the silica glass low-loss 1.55-μm telecommunications wavelength region. The advent of optical fiber amplifiers suitable for communications enabled today's broadband fiber networks, which carry signals across continents and oceans and provide the bandwidth necessary to transmit our words, pictures, and data around the planet without interruption.

An optical amplifier amplifies and boosts lightwaves without converting the optical signal to an electrical signal and then vice versa for further transmission. It is an extremely important device that supports long-distance optical communication networks in both fiber-based and free-space scenarios. The major types of optical amplifiers include rare-earth-metal-doped fiber amplifiers (FAs), fiber resonator amplifiers (FRAs), and semiconductor optical amplifier (SOA). OA was initially conceived and developed in the 1960s when lasers were invented, especially in the 633 nm and near infrared (IR) region toward 1100 nm, extending to the IR 1550 nm C, L bands in the last

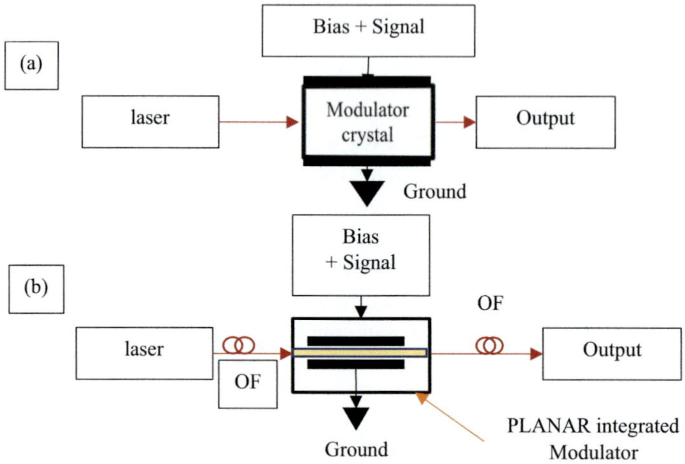

FIGURE 1.5 Bulk (a) and integrated (b) modulator structures.

few decades. However, under the intense interest in optical communications, the essentials of signal amplification in the optical domain, that is, photons in and gain to photons output without undergoing O/E/O (optical-to-electrical-to-optical) conversions became apparent.[20] The early optical amplifiers were also in circular pipe structures, which are not single-mode fiber (SMF) types. The basic structure of an optical amplifier is shown in Figure 1.5a and its use in inter-satellite photonic communication networks (Figure 1.5b).

The rare-earth metal provides a high number of free electrons, which can generate avalanche jumping and accumulation of stimulated emission, leading to transitions of these stimulated electrons from one energy level to another and emitting optical gain at different spectral regions. Thus, a pumping laser to provide energy for optical amplification is essential. For example, for erbium-doped glass, the pump laser wavelengths can be 980 nm or 1480 nm.

In contemporary optical communications, especially in SMF-based transmission, optical fiber amplifiers have attracted intense interest and were invented by Payne et al.[21,22,23]

1.1.7 Lasing Spectrum, Coherence, and Transmission

In both terrestrial and non-terrestrial space transmission systems, especially under coherent reception, the laser linewidth or the bandwidth of the optical frequency emitted by optical waves of the laser sources must be narrow enough so that pseudo-coherency can be approximated. In fiber long-haul and multi-span transmission systems employing coherent DSP-based receivers,[24] the guided waves in long-haul fiber transmission lines are confined in optical waveguides and amplified periodically through optical pre- and post-amplifiers in each span. The laser lines become quasi-coherent can thus be coherently received by the optical receiver. This means that the phases and polarization states of the transmitted waves can be distinguished and processed in the reception system. Similarly, in space transmission systems, optical waves, after being boosted in power and launched into free space over long distances without any optical amplification at intermediate stages, high loss. The demand on the linewidth of the laser is even higher than that for long-haul guided optical systems. We estimate that the laser linewidth should be on the order of a few Hz, or approximately nearly a single frequency at 1550 nm or ~193 THz. Lasers such as fiber lasers, external cavity lasers, and control feedback lasers with temperature stabilization can offer such narrow linewidth lasers.

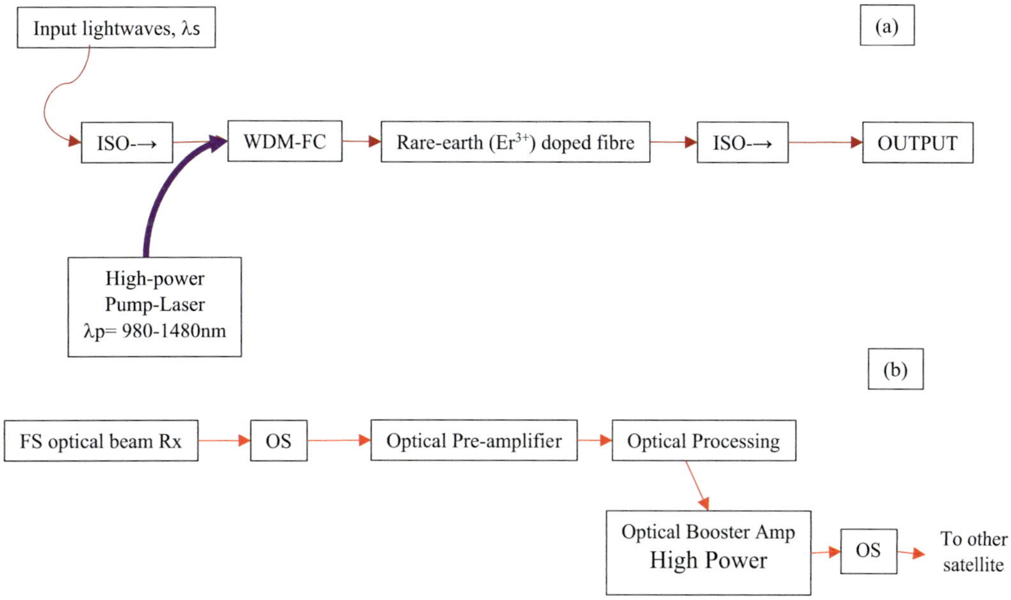

FIGURE 1.6 (a) Optical fiber amplifier basic structures for the 1550 nm operating spectral region. ISO = isolator, WDM = wavelength division multiplexer, FC = fiber coupler; (b) optical amplifier (pre-amp and booster amp) in an intermediate location in an inter-satellite link or OGS (optical ground station) to satellite and vice versa.

We will address and describe these lasers in the appropriate sections of the main chapters.

What happens to the coherency of the lasers when their waves are modulated? It is assumed that the laser output waves are modulated externally, or they are coupled to an integrated optical modulator and modulated with an information digital waveform fed to its electrodes, mostly of the electro-optic type for ultra-high bit rate transmission. The optical spectrum of the modulated waves can consist of an optical carrier on top of a modulated information band, as shown in Figure 1.6. The optical carrier, or laser, can be either suppressed or non-suppressed. In the case of non-suppressed carrier modulation, we can observe that such carrier power can be about 3–6 dB above the modulation band. This power of the carrier above the information band can asymptotically satisfy the pseudo-coherent state of the laser beam for long-distance free-space communication.

If the modulator is external, the CW generated by the laser remains unchanged and undisturbed, so its linewidth remains the same. The double sideband of the modulated spectrum contains the information data.

If the modulated optical waves cannot satisfy the coherency, the beam suffers spatial broadening, unlike the temporal dispersion in guided wave media such as a SMF transmission line.

1.1.8 Integrated Optics and Photonics

As discussed above Section 1.1.8, the coherency of the laser beam is critical for ultra-long-haul free space transmission; hence the need for a narrow linewidth laser and 6 dB power above the signal double sideband spectrum. To obtain such a near-single-frequency laser source, integrated photonic structures are required to obtain short cavity lengths and guided wave structures, ensuring no dispersion and precise reflection of the desired optical line. Such devices fall into integrated photonic and integrated optic technology.

Integrated optics or photonics ensure that the waves are guided in single mode and that polarization is preserved so that the coherency of the lightwaves can be maintained, especially when they are confined and a specific wavelength is selected.

1.1.8.1 Waveguides: Circular and Planar, Channel

Optical waveguides are critical for guiding the lightwaves and form the basis for optical modulators and laser resonators, guiding over extra long distances with minimal dispersion and preservation of polarization modes, guiding, and mixing to generate new sets of lightwaves and processing integrated systems. Figure 1.7 shows three basic waveguiding structures for lightwaves. The refractive indices of the circular guide, or optical fiber, differ by a minute amount of about 0.3%, resulting in minimal chromatic dispersion. This is termed a weakly guiding phenomenon, meaning the lightwaves are gently guided through the circular fiber due to its minute refractive index difference. The guided mode for SMF is linearly polarized, where the polarization remains linear and is preserved over a long distance, although the total polarization vector rotates due to the circular symmetry of the fiber. The power confined in the core region is only about 70%. The main brief explanation is that weakly guiding indicates a minimal number of reflections at the cladding and core interfaces, resulting in longer propagation path before the is reflected to the other side in a spiral fashion. It is noted that for single-mode guiding, the mode field and wave equation must be used to accurately estimate dispersion, and the reflection model can only provide a rough description.

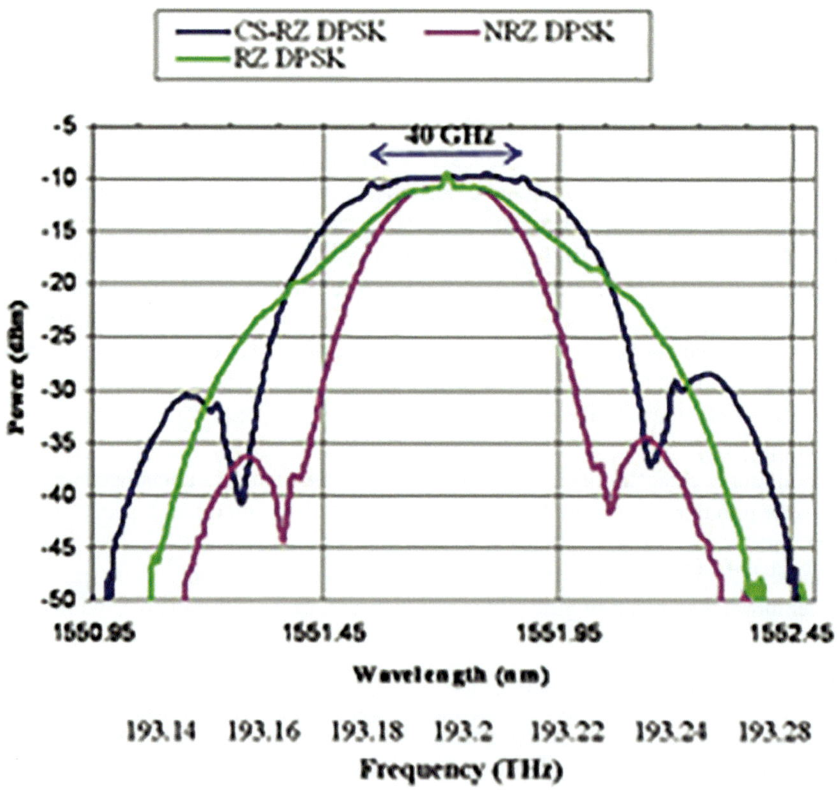

FIGURE 1.7 Optical spectrum of a modulated narrow linewidth laser, zero, modulation formats are NRZ non-return to zero differential shift-keying (DPSK) with carrier suppression (CSRZ-DPSK) – dark curve; RZ return to zero DPSK (RZ-DPSK) – medium dark; and return to zero differential phase shifting keying (RZ-DPSK) – light curve.

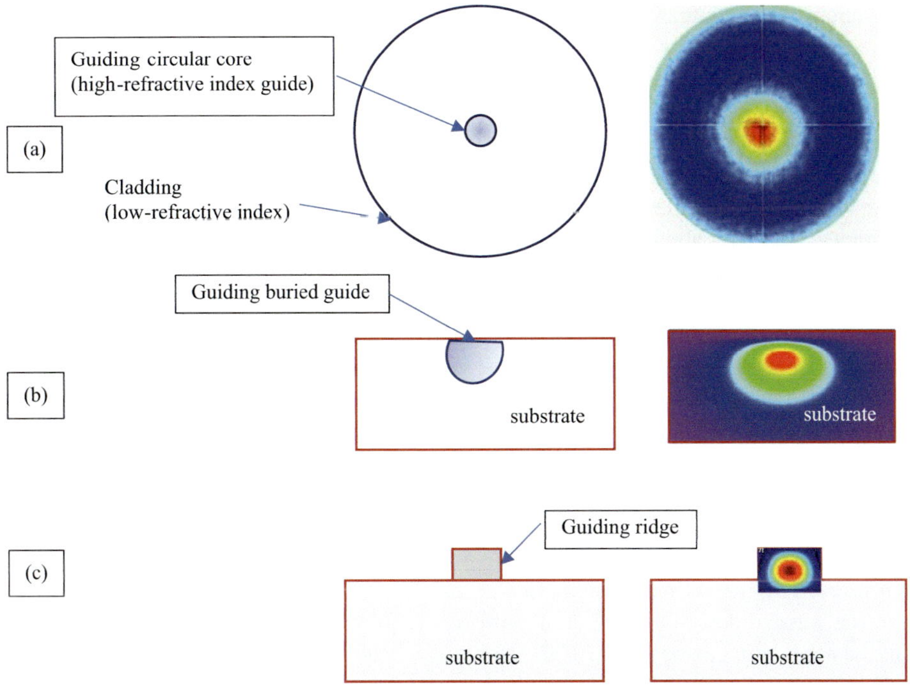

FIGURE 1.8 Cross-sections of optical wave guides: (a) circular optical fiber for long-distance guiding, (b) planar waveguide, (c) planar ridge waveguide. MODE intensity distribution is on the right hand side (RHS) of the structures.

The refractive index difference for planar waveguides is higher than that for circular waveguides, on the order of 10%–20%. Thus, the lightwaves are tightly confined, and the evanescent waves are only slightly distributed in the cladding or substrate area. A very small amount of energy from the guided wave is distributed outside the planar waveguide core. This is the main difference between circular and planar optical waveguides. Planar optical waveguides are used for very short distances, spanning a few centimeters, compared to a few hundred kilometers for circular waveguides.

The mode intensity distributions of the guided modes in circular, diffused, and ridge optical waveguides are shown on the right side of the waveguide structures. The tight confinement of the planar integrated waves is advantageous for laser structures and modulation, as the optical field can be covered by the electric field of the applied voltage for maximum overlap and thus electro-optic interaction (Figure 1.8).

1.1.8.2 Integrated Lasers and Spectra of Lasing: Near IR, Infrared Region and Mid Infrared Region

External modulation using crystal substrate-guided structures for optical modulation and semiconductor modulators have been the major topics in selecting and designing appropriate crystal structures and electro-optic interactions to achieve the most efficient modulation depth on the lightwaves. Integrated structures must be used to minimize the width of the guided lightwaves so that the electrode separation distance is minimized and the high electric field is developed across the optical fields of TE (transverse electric) or TM (transverse magnetic) modes. Thus, the phases of the lightwaves of either polarized field can be modulated. The modulation can be applied to both polarizations, depending on the optical waveguide structures that can support the TE or TM mode.

The integrated structures must have low propagation loss; unlike circular silica guides, the refractive index of the core of the integrated waveguide is much higher than that of the substrate –

normally about 20%, compared to only 0.3% for single-mode optical fiber. Sometimes it is even higher, such as in waveguides for laser structure. The high difference in refractive indices between the guiding core and the substrate leads to a high concentration of the guided field, and thus the curvature can be very sharp, making the integration of components possible.

In space photonic communications (SPC) optical nodes, integrated optical devices, and optical fibers are used to provide compact and efficient manipulation of lightwave channels. For fiber (see Appendix 4 of Chapter 12), the refractive index difference is only 0.3%; thus, the guided mode is very weakly guided, and only about 70% of the energy is concentrated in the guided mode, which also spreads over into the cladding area. This is why we call the guided mode in SMF weakly guiding. The effective area of an SMF is the 1/e value of the maximum peak at the center. This area carries all the channel power and hence the intensity exerted on the silica material. Nonlinearity is generated, leading to cross-talk noises on the transmitting pulse sequences, if such intensity (power over the effective area of the guided mode). The nonlinearity in such short distances of fiber creates nonlinear distortion and cross-talk, as described in Appendix 4 of Chapter 12 and in sections of Chapters 3 and 6.

In modern optical communication systems, optical transceivers are now implemented on SOI structures with $LNbO_3$ and SiN_3 interfaces of optical waveguides. Transceivers can provide 100 G, 200 G, and then 400 G using complex modulation formats like QOSK or 16 QAM, with baud rates of 28 GB, 64 GB, and 128 GB, and electronic amplifiers bandwidth as high as 110 GHz, as along with RF power amplifiers. The designs of the traveling waves or short electrodes depend on the length of the waveguides. SOI integrated platforms require short structures so that the 3 dB bandwidth of the modulator can be up to 70 GHz, making a baud rate of 128 GB possible. Chapters 4 and 5 describe the modulation formats and optical modulators needed to achieve such high-frequency broadband performance. Appendix 1 of Chapter 11 also provides many details about the modulator.

In error analyses of such channels, a digital sampling oscilloscope is required, and Keysight has developed such analyzers.[25]

1.1.9 Reaching Tera-bps Transmission System in Space

As stated above, the limit of modulation and analysis can reach a signal bandwidth of 110 GHz, but that limit may not be possible in practice unless much care is taken in SPC optical nodes. The matured technology of transceivers is at 100–400 G. For 400 G, the receiver sensitivity may not meet the noisy signals transmitted over long-distance interconnecting between satellites in a 43,000 km GEO constellation. Therefore, 100 G Gbps using dual-polarized channels modulated by QPSK at 28 or 56 GB per wavelength is the most likely rate that SPC can practically implement. To reach 1.0 Tbps, we have to combine in parallel at least ten channels, each carrying 100 Gbps, to aggregate to 1.0 Tbps, with a bit extra for forward error coding to lower the acceptable bit-error-rate (BER). This involves the multiplexing and demultiplexing of parallel and different wavelengths separated by 50 or 100 GHz. This section addresses WDM techniques and applications in optical systems.

In optical communications, especially the fiber optic-based systems and networks, WDM is a technology that multiplexes several optical carrier signals onto a single optical transmission path using different wavelengths (i.e., colors or frequencies in the hundreds of tera-hertz range) of laser light. This technique enables optical bidirectional communications over a single strand of fiber (also called wavelength-division duplexing) and multiplication of capacity. If the spacing of two consecutive channels is on the order of <100 GHz, or equivalently 0.4 nm in the 1550 nm C-band, then the term DWDM (dense WDM) is used. If the spacing is more than 1–2 nm, it is called CWDM (coarse WDM). These terms are applicable in SPC. We adapt these terms for SPC in this book.

The term WDM is commonly applied to an optical carrier, which is typically described by its wavelength, whereas frequency-division multiplexing typically applies to a radio carrier, which is more often described by frequency. This distinction is purely conventional because wavelength and frequency communicate the same information. Specifically, frequency (in Hertz, which is cycles per second) multiplied by wavelength (the physical length of one cycle) equals the velocity of the carrier wave. In a vacuum, this is the speed of light. However, frequency is used to determine the exact location of the channel as frequency (in a vacuum) is fixed, while only the wavelength can change as its value depends on the refractive index of the medium and the propagation velocity in that medium, whether guided or non-guided.

For example, the refractive index of silica glass fiber is about 1.448 (and lower if guided in a circular waveguide), so the lightwave velocity is reduced by about 30%. However, its frequency remains unchanged.

1.1.10 Optical and Photonic

The term "optical" indicates that the light may include lightwaves. However, the term "photonic" refers specifically to lightwaves and the art of manipulating them. By light waves, we mean the coherence of the lightwaves or that their phases are synchronized and their polarizations are preserved.

1.1.10.1 Reaching Maximum Capacity by Multi-Wavelength Multiplexing

A WDM system uses a multiplexer at the transmitter to combine several signals in parallel and a demultiplexer at the receiver to split them apart. Lightwaves, in the linear operating region, are parallel and do not interact if the medium is in the linear region. With the right type of optical or photonic system, it is possible to have a device that does both simultaneously and can function as an optical add-drop multiplexer, which enables channels of specific wavelengths to be injected into or dropped out of specified routes, respectively. The optical filtering devices used have conventionally been etalons, which are mostly stable solid-state single-frequency Fabry–Pérot interferometers. Thin-film-coated optical glass must be highly active so that switching out can be done. Alternatively, wavelength-selective switches combine gratings and multilayer filtering to form flexible demultiplexing or multiplexing with selective band filtering to perform such add/drop and through ports. Reconfigurable optical add/drop multiplexing (ROADM) is addressed in Chapters 6 and 9. As there are three different WDM types, one of which is called WDM, the notation xWDM is normally used when discussing the technology, with x defined as dense or coarse.

The concept was first published in 1970 by Delange,[26] and by 1980 WDM systems were being realized in the laboratory. The first WDM systems combined only two signals. Modern systems can handle 160 signals in the C-band and more in the C+-band at 50 GHz spacing and can thus expand a basic 100 Gbit/s system over a single fiber pair to over 25 or 50 tera-bit/s.

WDM systems offer many advantages with telecommunications space networks because they allow us to expand the capacity of the network without installing additional satellite constellations. WDM in SPC is preferred to minimize additional equipment, thereby reducing weight and power consumption on board of satellites. By using WDM and optical amplification modules, several generations of technology development can be accommodated in space optical infrastructure without overhauling the space network. The capacity of a given space link can be expanded simply by upgrading the multiplexers and demultiplexers at each end, using the least absorptive bands of air and vacuum.

These multiplexing and demultiplexing are often performed using O/E/O translation at the very edge of the transport network, thus permitting interoperation with existing vendor equipment under standard optical interfaces. In SPC networks, it is even more important to use standardized interface units to simplify constellation implementation and manufacturing.

Most WDM terrestrial systems operate on SMF optical cables, which have a core diameter of 9 μm. Certain forms of WDM can also be used in multimode fiber cables (also known as premises cables, i.e., for delivery to home or buildings) that have core diameters of 50 or 62.5 μm. However, standardization and a better understanding of the dynamics of WDM systems have made WDM one of the most important technologies for deployment. This is now applicable to SPC as it takes the advanced aspects of fiber optic terrestrial technologies.

Optical receivers, in contrast to laser sources operating in CW mode, tend to be wideband devices. Therefore, the demultiplexer must provide wavelength selectivity for the receiver in the WDM system, including tunable bandwidth and central frequencies.

1.1.10.1.1 Optical Channel Multiplexing and Demultiplexing

WDM systems are divided into three different wavelength patterns: normal (WDM), coarse (CWDM), and dense (DWDM). Normal WDM (sometimes called BWDM) uses the two normal wavelengths, 1310 and 1550 nm, on one fiber. The absorption of air and vacuum is also transparent at these spectral windows.

Coarse WDM provides up to 32 or more channels across multiple transmission windows from the C-band to the L-band, and especially in the O-band (1300 and 60 nm band). Dense WDM (DWDM) uses the C-band (1530–1565 nm) transmission window but with denser channel spacing. Channel plans vary, but a typical DWDM system might use 40 channels at 100 GHz spacing or 80 channels at 50 GHz spacing. Coarse WDM is normally used in the 1300 nm range, but for SPC, coarse WDM can be assigned depending on the low absorption characteristics of the transmission, whether vacuum or air. New amplification options, such as Raman amplification, enable the extension of usable wavelengths into the L-band (1565–1625 nm), doubling the number of channels.

Coarse wavelength-division multiplexing (CWDM), in contrast to DWDM, uses increased channel spacing to allow for less sophisticated and therefore cheaper transceiver designs. To provide 32 channels on a single fiber or space media, CWDM uses the entire frequency band spanning the second and third transmission windows (1310/1550 nm, respectively), including the critical frequencies where OH scattering may occur. Similarly, OH-free air and space or silica fibers are recommended if the wavelengths between the second and third transmission windows are to be used.

WDM, CWDM, and DWDM are based on the same concept of using multiple wavelengths of light on a single fiber but differ in the spacing of the wavelengths, the number of channels, and the ability to amplify the multiplexed signals in the optical space. EDFA provides an efficient wideband amplification for the C-band, while Raman amplification adds a mechanism for amplification in the L-band. For CWDM, wideband optical amplification is not available, limiting the optical spans to several tens kms, now in the range of 80–120 km.

1.1.10.1.1.1 Coarse WDM Originally, the term coarse wavelength-division multiplexing (CWDM) was fairly generic and described several different channel configurations. In general, the choice of channel spacings and frequencies in these configurations precluded the use of erbium-doped fiber amplifiers (EDFAs). Prior to the relatively recent International Telcommunication Union (ITU) standardization of the term, one common definition for CWDM referred to two or more signals multiplexed onto a single strand of circular fiber, with one signal in the 1550 nm band and the other in the 1310 nm band.

In 2002, the ITU standardized a channel spacing grid for CWDM (ITU-T G.694.2) using wavelengths from 1270 to 1610 nm with a channel spacing of 20 nm. ITU G.694.2 was revised in 2003 to shift the channel centers by 1 nm, so, strictly speaking, the center wavelengths are 1271 to 1611 nm.

The main characteristic of the recent ITU CWDM standard is that the signals are not spaced appropriately for amplification by EDFAs. This limits the total CWDM optical span to around 60 km for a 2.5 Gbit/s signal, which is suitable for use in metropolitan applications. For SPC, this distance is only the direct distance between the transmitter and receiver. The combination of the optical booster amplifier at the transmitter site and the low-noise amplifiers at the receiver side is highly critical, ensuring that the optical gain is purely due to the optical gain and minimal amplification stimulated emission noise.

The relaxed optical frequency stabilization requirements allow the associated costs of CWDM to approach those of non-WDM optical components. However, in SPC, frequency stabilization also determines the coherence and phase noise of beam spatial spreading, so a stabilized laser frequency source may eventually be required.

1.1.10.1.1.2 CWDM Applications CWDM is used in cable television networks, where different wavelengths are used for downstream and upstream signals. In these systems, the wavelengths are often widely separated; for example, the downstream signal might be at 1310 nm while the upstream signal is at 1550 nm. Similarly, in SPC, CWDM can be used for bidirectional transmission provided that these wavelengths are within the band of a space bulk optical circulator and its isolation. Thus, separating the unidirectional group of wavelength channels can be much easier. These wavelengths must be in the transparent spectral regions of air and vacuum.

Furthermore, passive CWDM is an implementation of CWDM that uses no electrical power. This is much preferred in SPC, where power is limited. Passive CWDM separates the wavelengths using passive optical components such as bandpass filters and prisms.

1.1.10.1.1.3 Dense WDM Dense wavelength-division multiplexing (DWDM) originally referred to optical signals multiplexed within the 1550 nm band to leverage the capabilities (and cost) of erbium-doped fiber amplifiers (EDFAs), which are effective for wavelengths between approximately 1525–1565 nm (C band) and 1570–1610 nm (L band). EDFAs were developed to replace Synchronous Optical Network/Synchronous Digital Hierarchy (SONET/SDH) optical-electrical-optical (OEO) regenerators, making them practically obsolete. EDFAs can amplify any optical signal within their operating range, regardless of the modulated bit rate. For multi-wavelength signals, so long as the EDFA has enough pump energy, it can amplify as many optical signals as can be multiplexed into its amplification band (though signal densities are limited by the choice of modulation format). Therefore, EDFAs allow a single-channel optical link to be upgraded in bit rate by replacing only the equipment at the ends of the link, while retaining the existing EDFA or series of EDFAs through a long-haul route. Furthermore, single-wavelength links using booster and low-noise EDFAs can similarly be upgraded to WDM links.

1.1.10.1.2 DWDM Systems

At this stage, a basic DWDM system contains several main components: a DWDM terminal multiplexer. The terminal multiplexer contains a wavelength-converting transponder for each data signal, an optical multiplexer, and where necessary, an optical amplifier (EDFA). Each wavelength-converting transponder receives an optical data signal from the client, converts this signal into the electrical domain, and then retransmits it at a specific wavelength using a 1550 nm band laser. These data signals are then combined into a multi-wavelength optical signal using an optical multiplexer for transmission over a single fiber (e.g., SMF-28 fiber). The terminal multiplexer may also include a local transmit EDFA for power booster amplification of the multi-wavelength optical signal and a receiver with low noise and low input optical average power amplification. In the mid-1990s, DWDM systems contained four or eight wavelength-converting transponders; by around 2000, commercial optical systems capable of carrying 128 signals were available.

An intermediate line repeater is placed approximately every 80–100 km to compensate for the loss of optical power as the signal travels along the fiber. The 'multi-wavelength optical signal' is amplified by an EDFA, which usually consists of several amplifier stages.

An intermediate optical terminal, or optical add-drop multiplexer, is a remote amplification site that amplifies the multi-wavelength signal that may have traversed up to 140 km or more before reaching the remote site. Optical diagnostics and telemetry are often extracted or inserted at such a site to allow for the localization of any fiber breaks or signal impairments. In more sophisticated systems (which are no longer point-to-point), several signals from the multi-wavelength optical signal may be removed and dropped locally.

1.1.10.1.3 DWDM Terminal Demultiplexer

At the remote site, the terminal demultiplexer consisting of an optical demultiplexer and one or more wavelength-converting transponders separates the multi-wavelength optical signal back into individual data signals and outputs them on separate fibers for client-layer systems (such as

SONET/SDH). Originally, this demultiplexing was performed entirely passively, except for some telemetry, as most SONET systems can receive 1550 nm signals. However, to allow for transmission to remote client-layer systems (and to allow for digital domain signal integrity determination), such demultiplexed signals are usually sent to O/E/O output transponders prior to being relayed to their client-layer systems. Often, the functionality of the output transponder has been integrated into that of the input transponder, so that most commercial systems have transponders that support bidirectional interfaces on both their 1550 nm (i.e., internal) side and external (i.e., client-facing) side. Transponders in some systems supporting 40 GHz nominal operation may also perform forward error correction (FEC) via digital wrapper technology, as described in the ITU-T G.709 standard.

The optical supervisory channel (OSC) and satellite handshake beat channel (OSSBC) are data channels that use an additional wavelength, usually outside the EDFA amplification band (at 1510, 1620, 1310 nm, or another proprietary wavelength). The OSC carries information about the multi-wavelength optical signal as well as remote conditions at the optical terminal or EDFA site. It is also normally used for remote software upgrades, user (i.e., network operator) network management information, and secret key QKD (Quantum key distribution). ITU standards suggest that the OSC should utilize an OC-3 signal structure, though some vendors have opted to use 100 Mbps Ethernet or another signal format. Unlike the 1550 nm multi-wavelength signal containing client data, the OSC is always terminated at intermediate amplifier sites, where it receives local information before being retransmitted.

The introduction of the ITU-T G.694.1[8] frequency grid in 2002 has made it easier to integrate WDM with older but more standard, SONET/SDH systems. WDM wavelengths are positioned in a grid with exactly 100 GHz (about 0.8 nm) spacing in optical frequency, with a reference frequency fixed at 193.10 THz (or equivalently 1552.52 nm). The main grid is placed inside the optical fiber amplifier bandwidth but can be extended to wider bandwidths. The first commercial deployment of DWDM was in June 1996. Today's DWDM systems use 50 GHz or even 25 GHz channel spacing for up to 160-channel operation.

DWDM systems have to maintain more stable wavelengths or frequencies than those needed for CWDM because of the closer spacing of the wavelengths. Precision temperature control of the laser transmitter is required in DWDM systems to prevent drift from a very narrow frequency window, on the order of a few GHz. In addition, since DWDM provides greater maximum capacity it tends to be used at a higher level in the communications hierarchy than CWDM, such as on the Internet backbone, and is therefore associated with higher modulation rates. This creates a smaller market for high-performance DWDM devices. These factors, smaller volume and higher performance, result in DWDM systems typically being more expensive than CWDM. Refer to Chapter 11 in Appendix 2, Section 11.2.4 about laser autonomous locking for space transmission.

Recent innovations in DWDM transport systems include pluggable and software-tunable transceiver modules capable of operating on 40 or 80 channels. This dramatically reduces the need for discrete spare pluggable modules, as a handful of pluggable devices can handle the full range of wavelengths.

1.1.10.1.4 Wavelength-Converting Transponders

Wavelength-converting transponders originally translated the transmit wavelength of a client-layer signal into one of the DWDM system's internal wavelengths in the 1550 nm band. Note that even external wavelengths in the 1550 nm band will most likely need to be translated, as they almost certainly will not have the required frequency stability tolerances, nor will it have the optical power necessary for the system's EDFA.

In the mid-1990s, however, wavelength-converting transponders rapidly took on the additional function of signal regeneration. Signal regeneration in transponders quickly evolved through 1R to 2R to 3R and into overhead-monitoring multi-bitrate 3R regenerators. These differences are outlined below:

1.1.10.2 Retransmission

Early transponders were "garbage in garbage out" in that their output was nearly an analog "copy" of the received optical signal, with little signal clean-up occurring. This limited the reach of early DWDM systems because the signal had to be handed off to a client-layer receiver (likely from a different vendor) before the signal deteriorated too far. Signal monitoring was basically confined to optical domain parameters such as received power.

1.1.10.2.1 2R

Re-time and re-transmit transponders were not very common and utilized a quasi-digital Schmitt-triggering method for signal clean-up. Some rudimentary signal-quality monitoring was done by these transmitters, which basically looked at analog parameters.

1.1.10.2.2 3R

3R transponders – re-time, re-transmit, re-shape – were fully digital and normally able to view SONET/SDH section layer overhead bytes such as A1 and A2 to determine signal quality health. Many systems offer 2.5 Gbit/s transponders, which usually means the transponder can perform 3R regeneration on OC-3/12/48 signals and possibly gigabit Ethernet, with reporting on signal health by monitoring SONET/SDH section layer overhead bytes. Many transponders will be able to perform full multi-rate 3R in both directions. Some vendors offer 10 Gbit/s transponders, which can perform section layer overhead monitoring for all rates up to and including OC-192.

1.1.10.2.3 Muxponder

The muxponder (short for multiplexed transponder) has different names depending on the vendor. It essentially performs some relatively simple time-division multiplexing of lower-rate signals into a higher-rate carrier within the system (a common example is the ability to accept 4 OC-48s and then output a single OC-192 in the 1550 nm band). More recent muxponder designs have absorbed more and more TDM functionality, in some cases obviating the need for traditional SONET/SDH transport equipment.

1.1.10.2.4 List of DWDM Channels

See Appendix 7 of Chapter 12. For DWDM, the range between C21-C60 is the most common, with Mux/Demux available in 8, 16, 40, or 96 sizes.

1.1.10.2.5 The 100 GHz ITU Channels and 50 GHz ITU Channels

Reconfigurable optical add-drop multiplexer (ROADM)

See Appendix 5 of Chapter 12 or contention-less Directionless Colourless Reconfigurable Optical Add/Drop Multiplexer (CDC-ROADM). As mentioned above, intermediate optical amplification sites in DWDM systems may allow for the dropping and adding of certain wavelength channels. In most systems, add and drop multiplexing is done infrequently because adding or dropping wavelengths requires manually inserting or replacing wavelength-selective cards. However, this process is costly and, in some systems, requires the removal of all active traffic from the DWDM system, as inserting or removing the wavelength-specific cards interrupts the multi-wavelength optical signal.

With a ROADM, network operators can remotely reconfigure the multiplexer by sending soft commands. The architecture of the ROADM is such that dropping or adding wavelengths does not interrupt the pass-through channels. Numerous technological approaches are utilized for various commercial ROADMs, with trade-offs between cost, optical power, and flexibility.

1.1.10.3 Optical Cross Connects (OXCs)

When the network topology is a mesh, where nodes are interconnected by fibers or by optical head (OH) and pre-amplifiers to form an arbitrary signal graph, an additional interconnection device is needed to route the signals from an input port to the desired output port. These devices are called optical cross-connectors (OXCs). Various categories of OXCs include electronic ("opaque"), optical ("transparent"), and wavelength-selective devices can be used, but Wavelength Selective Switch (WSS) is the most flexible and easiest to use, as its loss is minimal.

1.1.10.3.1 Shortwave WDM

Shortwave WDM uses vertical-cavity surface-emitting laser (VCSEL) transceivers with four wavelengths in the 846–953 nm range over single OM5 fiber, or two-fiber connectivity for OM3/OM4 multimode fiber types whose dispersion is optimized.

1.1.10.4 Transceivers and Transponders

1.1.10.4.1 Transceivers

Since communication over a single wavelength is one-way (simplex communication) and most practical communication systems require two-way (duplex communication) communication, two wavelengths will be required if they are on the same transmission optical node. If separate optical nodes are used in a so-called bidirectional node, then the same wavelength can normally be used, and it is not WDM. As a result, at each end, both a transmitter and a receiver will be required. A combination of a transmitter and a receiver is called a transceiver; it converts an electrical signal to and from an optical signal. WDM transceivers designed for single-strand operation require the opposing transmitters to use different wavelengths. WDM transceivers additionally require an optical splitter/combiner to couple the transmitter and receiver paths onto one optical node.

1.1.10.4.2 Transponder

In practice, the signal inputs and outputs will be optical instead of electrical (typically at 1550 nm). This means that wavelength converters are needed, which is exactly what a transponder provides. A transponder can be made up of two transceivers placed sequentially: the first transceiver converts the 1550 nm optical signal to/from an electrical signal, and the second transceiver converts the electrical signal to/from an optical signal at the required wavelength. Transponders that don't use an intermediate electrical signal (all-optical transponders) are thus essential.

1.1.10.5 Optical Amplification

In 1961, Elias Snitzerl4 and colleagues constructed and operated the world's first optical fiber laser in the former American Optical Complex. Three years later this team demonstrated the first optical fiber amplifier. Fiber lasers that can cut and weld steel have since become powerful industrial tools and fiber amplifiers routinely boost signals in the global optical fiber network allowing messages to cross oceans and continents without interruption. However, the emitted wavelengths are in the 1060 nm spectral region with the doping of Neodymium in silica glass.[27] In this window the attenuation of signals is much higher than that in the infrared region of 1550 nm.[28]

The development and innovative works on optical amplification was extended to the 1550nm which is the C-band of low loss transmission of silica single mode fiber, by doping of Erbium rare earth ions[29]. The amplification gains of such Er:doped silica fiber can offer optical gain of about 30dB and a saturated optical power in the 10 to 20 dBm which is too low for the transmission over long haul space system. The overall attenuation of the free-space vacuum medium is about 80 to 100dB if taking the transmission loss and gain of optic system of telescopic range.

Thus, to compensate for this high attenuation, high power optical amplifiers are required to boost the signals at the transmitter, and the optical preamplifiers at the receiver as a low input optical preamplification that can offer a further optical gain of 30-40dB and low signal power reaching -65 dBm.

Typical high-power optical amplifiers as booster can be found, as an example[30]. The amplifiers can offer the followings: (i) Use optical devices such as pump lasers and double-clad active fibers of international famous brands. (ii) Length optimization of Yb3+and Er3+co-doped double-clad fiber; (iii) Residual pump treatment technology; (iv) Reliable optical power output stabilization circuit and laser heating and cooling; (v) Stabilization control circuit ensure the best performance of the whole machine and the long life and stable operation of the laser; (vi) Dual power supply design to ensure safe and stable power supply; and especially (vii) Ultra high-power output: up to 4 W (or 33 to 35 dBm) total output power. This requires minimum input signal average power level in order of 0 dBm to 10 dBm.

1.1.10.6 Space Laser Communication Age and Integration of Integrated Photonics and Optic Systems

SPC systems must take advantage of advancements in terrestrial fiber cable transmission systems and networks developed since the 1970s. These include integrated optical components and fibers, as well as optical amplifiers for booster amplification and low-noise amplifiers for low-level amplification in optical pre-amplifier receivers.

The coherency of the lightwave beam is critical for long-distance transmission in space due to spatial broadening caused by the finite linewidth of the carrier, which is about 6 dB above the signal band. Thus the linewidth of the laser source must be smaller than 10 kHz. In practice, semiconductor laser can offer such narrow linewidths, and even narrower by fiber lasers. These lasers are described in Appendix 2 of Chapter 11. The modulated spectra of modulated lightwave channels are discussed in Chapters 4 and 5.

Spatial broadening leads to higher loss when received by a fixed area of the optical head. This leads to higher optical amplification at the booster and the LNA (low noise amplifier) at the receiver, thus causing higher nonlinear distortion. Therefore, the laser carrier must be not only narrow but also stable to avoid laser phase noises, which can lead to spatial spreading and additional loss of signal quality.

1.1.10.6.1 Coherency of Lasers and Modulated Optical Channels

As required, the laser linewidth must narrow to <10 kHz to obtain the coherence of the modulated laser beam, especially for all multiplexed channels on the same beam.

Over the years, several laser types have become available, such as semiconductor lasers made from III–V and II–VI compounds, InGasAsP, or InP-based materials, which can be fabricated using MOCVD (molecular organic chemical vapor deposition) or MBE (molecular beam epitaxy) techniques. The lasers are pigtailed and coupled to an optical modulator, or they can be directly coupled to an integrated optical modulator on chip.

1.1.10.6.2 Photonic Modulation

Photonics indicates the manipulation of lightwaves with coherence, resulting in a level higher than optical waves and photons in and out, respectively. This photonic modulation is done outside the laser cavity to keep the laser cavity untouched, allowing the laser to operate in CW mode, with its linewidth and frequency kept stable and unchanged by using TEC (thermo-electric control by the Peltier *pn* cooling device).

The modulation is achieved by alternating the phases of the lightwaves, thereby using interference, diffraction, or coupling optical structures to generate variations or switching in the intensity of the lightwaves. Therefore, both phase and amplitude modulation can be performed.

In two-dimensional or complex modulation, intensity modulation can be implemented in both the H_x- and E_y-directions. Thus, the vectorial summation of these two components can be generated. Different amplitudes or intensities can also be generated to create multi-level amplitude, allowing the modulation format m-QAM (quadrature amplitude modulation) to be obtained for lightwaves at the output of the photonic modulators.

Alternatively, the laser current can be modulated, which is known as the internal modulation. This technique is to be avoided, as current modulation of the cavity creates instability and broadening of the laser spectrum.

The photonic modulation described above should be used and is referred to as the external modulation technique for SPC systems.

1.1.10.6.3 Optical System to Launch Optical Beams

In SPC beam booster amplification, passing through a bulk circulator, and then launching into the OH to transmit the beam to the other OH is critical. Any beam misalignment creates distortion in the signals, as the delay differences between the rays at the transmitter and the much longer delay caused by multipath fading effects on the channel signals. This is quite similar to multipath fading in wireless systems. Thus, several digital signal processing algorithms can be used to mitigate such distortion.

A quadrant detector is also placed at the transmitter or receiver to indicate the symmetric beam and provide feedback control for beam alignment, as required to align the beam direction. The frequency is low enough that the electronic amplifier can be of a high-impedance type, and balanced photodetector pairs (PDP) can be used to improve the sensitivity of the alignment receiver.

The schematic of a bidirectional SPC, including the beam alignment, is shown in Figure 1.9. Two optical circulators are used to ensure that the bidirectional receiving and transmitting lightwaves are properly managed. Indeed, such beam circulators are needed to ensure that the specific wavelength for quadrant detection can provide the beam alignment feedback. The quadrant is arranged in such a way that two PDs are connected back-to-back to form a balanced pair. The electric currents coming out of the detection of the two pairs are fed into a differential operational amplifier to provide the difference in the intensities of the beam symmetrically. Only if the output current or voltage of this differential amplifier reaches near zero level is the beam considered to be aligned properly. The OH are mounted on a three-dimensional (3D) micro-manipulator and connected to an optical circulator with three or four ports so that the signal feedback can be fed into the control of a micro-dish that

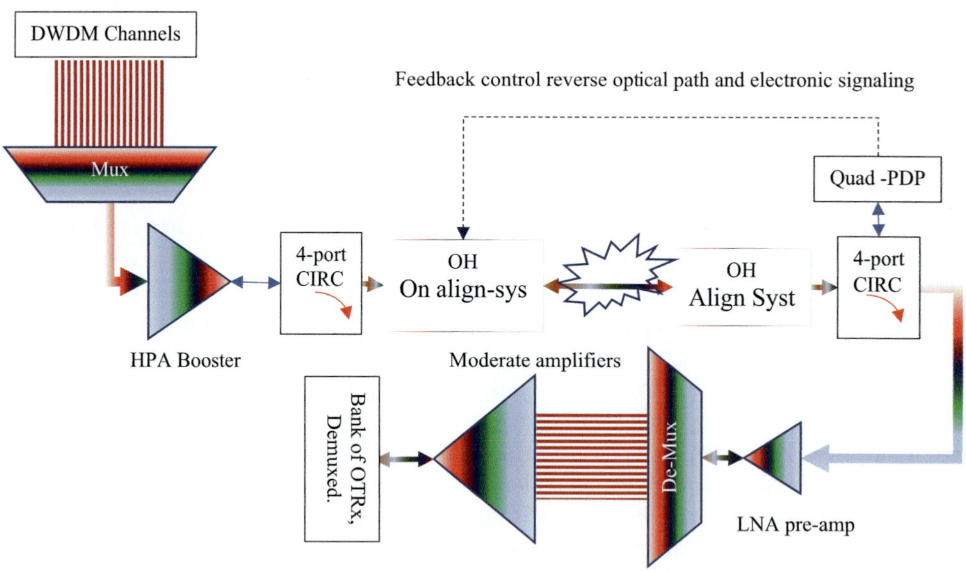

FIGURE 1.9 SPC bi-directional transmission system with quad photodetector for beam alignment feedback and control. PDP=photodetector pair; CRIC=circulator; OH=optical head; DWDM=dense wavelength division multiplexing.

controls the rotation and translation stages of the platform to align the two beams from and to the transmitter and receiver sites and vice versa. The micro-disk must have a resolution of less than 5 micro-rads and can be controlled by a 16-bit processor at low speed.

1.1.10.7 Transmission Media

In SPC, there are two transmission media to pay attention to: atmospheric and vacuum transmission. Both are air, but one without turbulence and transparency, the other is vacuum with a known refrac tive index. The main concern is nuclear radiation affecting the optical and electronic components of the optical transceivers. The radiation test of the transceivers for 100 Gbps is discussed in Chapter 2. The refractive index of turbulence and its probability of errors in transmission are detailed in Chapter 2. Analytical expressions are obtained for the BER or probability of error for different modulation formats. Beam propagation is also modeled using the split-step Fourier method (SSFM) to obtain the conditions for beam transmission without significant beam spreading or decoherent states. The transmittance and absorption coefficients of these media are given in Chapter 2 and are important for selecting spectral windows to transmit as optical carriers of the DWDM parallel channels.

Both integrated planar and channel optical waveguides and circular optical fibers are described in detail in Chapter 12, Appendix 3. Readers should distinguish between these optical waveguides to avoid mixing between the two types. Planar integrated waveguides are used to make optical modulators and integrated optical transceivers, while the optical fibers is used to interconnect the optical components and launch the modulated lightwaves into free space.

1.2 ORGANIZATION OF THE CHAPTERS

The chapters are organized as follows:

Chapter 2 describes in detail the propagation of lightwave beams through the atmospheric layers from the OGS (optical ground station) to satellite optical nodes, and from one optical node to other nodes within a satellite constellation in different orbits, such as LEO, MEO, or GEO. The fundamentals of propagation losses and spatial dispersion of lightwave beams are given. Optical modulation techniques are briefly presented and further details are described in Chapters 3 and 4.

Optical modulation techniques are provided in Chapters 3 and 4. Chapter 3 addresses modulation using only binary levels and whether the optical carriers can be suppressed or not. Duobinary optical modulation is also discussed as a three-level modulation, but with phase alternations under these envelopes in the case of optical duobinary. Electrical duobinary can also be implemented in the electrical domain by filtering. Pulse shaping modulation is also presented to produce a brick-wall-like spectrum, allowing for more effective packing of multiple channels and easier sampling to achieve optimum receiver sensitivity.

Chapter 4 describes the modulation techniques for multiple amplitudes or phases of the lightwaves and a combination of both multiple amplitudes and phases, or effectively the complex modulations. Optical modulation devices are also described.

Chapter 5 details optical reception techniques, in particular the coherent one, as this is the only possible technique to preserve the coherence of the lightwave beams. Noises generated in the reception processes and the additional attenuation factor due to the finite linewidth of the lightwave sources affect the receiver sensitivity. An appendix in Chapter 12 gives an example of a design and noise analysis of electronic amplifiers and quantum shot noise associated with coherent detection.

Chapter 6 then describes the use of digital signal processing in the receiver of digital formats to mitigate impairments and in the carrier phase detection to compensate for the frequency difference between the carrier and the local oscillator laser, allowing perfect homodyne detection to be achieved. The frequency shift of the carrier is due to the Doppler effects between two moving objects, such as the satellites in a constellation or a satellite and an OGS. The test performance of optical transceivers is given on several parameters, especially the Doppler effect on the receiver sensitivity.

Chapter 7 discusses several aspects of optical amplification at the transmitter as a booster amplifier to achieve high power levels in watts and at the receiver as low-noise and low-level optical preamplification. Practical measurements of optical amplifiers are provided, including the biasing technique to keep the booster stable while operating at saturation level. Power budgets for SPC links are given as examples of the design of SPC systems.

Chapter 8 then describes the nonlinear effects in the SPC system, which are mainly contributed by the short length of fiber, about a few meters, connecting from the booster amplifier output to the optical head before launching into space. The high power over the fiber's small area of 50 μm^2 creates ultrahigh intensity on the fiber's effective area, generating nonlinear crosstalk. These are the main issues we address, so designers of booster amplifiers for SPC must be aware that the positions of the wavelength carriers should be placed unevenly to avoid such nonlinear crosstalk.

Chapter 9 addresses the application of photonic communication in the case of power-limited quantum communication over space channels. The QKD (quantum key distribution) is the opposite of Tbps transmission based on the Nyquist capacity theorem. This chapter illustrates how such quantum states can be transmitted in association with the Tbps channels. That is, this quantum channel is used to communicate secret keys so that coding can be done to ensure that the information is completely confidential. The optical routers are provided with space add/drop of channels and multi-dimensional routing of channels in space. DWDM channels are aggregated to form a tera-bits/s bit rate capacity.

Chapter 10 provides a detailed account of transmission over an equivalent one-hop SPC with a multiplexer and demultiplexer, the WSSes, the attenuation equivalent to a one-hop satellite-to-satellite optical link, and the coherent optical transceiver with software and hardware platforms.

Chapter 10 also provides experimental procedures of an SPC transmission platform in a laboratory environment. The step-by-step setup of multiwavelength channels with a multiplexer and demultiplexer and an equivalent space transmission channel is detailed, along with testing under the same space transmission between satellite-to-satellite hop. Then, the multi-hop and all-optical routing are given with a comprehensive power budget.

Chapters 11 and 12 provide the Appendices 1–7. The optical modulators and integrated lasers are given in Chapter 10 as the first two appendices. Chapter 11 includes the remaining five appendices on the fibers or circular waveguides, planar optical waveguides, and the basics of nonlinear effects in silica SMF fibers. Appendix 5 gives the general schematic of reconfigurable OADM and MEMS to form such ROSDM. The final Appendix 7 outlines the wavelength grids of the ITU-T standard for 100 and 50 GHz.

Furthermore, two tables are given at the ends, Table 1 on duality of time and space and guided wave and bulk optical components to contrast their applications in both space and fiber-based communications systems. Then a table of abbreviations and meaning of terms employed in the book chapters, are given.

NOTES

1 European Space Agency, "Developing future optical high-capacity satellite network: HYDRON (high Throughput Optical Network)", https://connectivity.esa.int/developing-future-optical-highcapacity-satellite-networks-hydron-high-throughput-optical-network, access date December 21, 2023.

2 A. Javan, W. R. Bennett, Jr., and D. R. Herriott, "Population inversion and continuous optical maser oscillation in a gas discharge containing a He-Ne mixture", *Physical Review Letters*, 6, 106, 1961.

3 L. N. Binh, *Guided Wave Photonics Fundamentals and Applications with MATLAB*. CRC Press: Boca Raton, FL, 2012.

4 K. C. Kao and G. A. Hockham, "Dielectric-fibre surface waveguides for optical frequencies", *IEE Proceedings*, 113(7), 1151–1158, 1966.

5 Corning Inc., Advanced Optics | Advanced Optical Materials, Components, and Technology Solutions | Corning

6 L. N.Binh, *Advanced Optical Digital Communications*, 2nd Ed. CRC Press: Boca Raton FL, 2015.
7 H. P. Lutz, "Optical communications in space: Twenty years of ESA effort", https://www.esa.int/esapub/bulletin/bullet91/b91lutz.htm, access date November 2023.
8 NKT Photonics, "Koheras single-frequency fibre lasers", https://www.nktphotonics.com/products/single-frequency-fiber-lasers/.
9 M. Rose and H. Hogan, "A history of the laser: 1960 – 2019", Photonic Media. https://www.photonics.com/Articles/A_History_of_the_Laser_1960_-_2019/a42279.
10 L. N. Binh, *Advanced Digital Optical Communications*, 2nd Ed. CRC Press: Boca Raton, FL, 2015.
11 E. Voges, "Integrated optics for optical communications", Part of Springer Proceedings in Physics book series (SPPHY, Vol. 18).
12 L. N. Binh, *Digital Optical Communications (Optics and Photonics)*. CRC Press: Boca Raton, FL, 2008, Chapter 1.
13 L. N. Binh, J. Livingstone, and D. Steven, "A wide-band acousto-optic TE-TM mode converter using a doubly confined structure", *IEEE Journal of Quantum Electronics*, 16(9), 964–971, 1980.
14 IEEE Photonics Society, "IEEE milestone: First optical fibre laser and amplifier, 1961–1964". https://ethw.org/Milestones:First_Optical_Fiber_Laser_and_Amplifier,_1961-1964.
15 E. Snitzer and J. W. Hicks, "Optical wave-guide modes in small glass fibres: (I) theoretical", *Paper TB36, Program of the 1959 Annual Meeting of the Optical Society of America*, vol. 49, p. 1128, November 1959. CHATEAU LAURIER, OTTAWA, CANADA, OCTOBER 8–10, 1959.
16 H. Osterberg, E. Snitzer, M. Polanyi, and R. Hilberg, "Optical wave-guide modes in small glass fibres, II experimental", *Paper TB37, Program of the 1959 Annual Meeting of the Optical Society of America*, vol. 49, p. 1128, November 1959. CHATEAU LAURIER, OTTAWA, CANADA, OCTOBER 8–10, 1959.
17 E. Snitzer, "Optial MASER action of Nd^{+3} in a barium crown glass", *Physical Review Letters*, 7(12), 444–446, 1961.
18 Charles J. Koester and Elias Snitzer, "Amplification in a fibre laser", *Applied Optics*, 3(10), 1182–1186, 1964.
19 a) S. B. Poole, D. N. Payne, and M. E. Fermann, "Fabrication of low-loss optical fibres containing rare-earth ions", *Electronics Letters*, 21, 737–738, 1985; (b) R. J. Mears, L. Reekie, I.M. Jauncey, and D. N. Payne, "Low-noise erbium-doped fibre amplifier operating at 1.54μm", *Electronics Letters*, 23(19), 1026–1028, 1987; (c) E. Desurvire, J. R. Simpson, and P. C. Becker, "High-gain erbium-doped traveling-wave fibre amplifier", *Optics Letters*, 12(1), 888–890, 1987.
20 Bulk optical amplifiers book?
21 S. B. Poole, J. E. Townsend, D. N. Payne, M. E. Fermann, G. J.Cowle, R. I. Laming, and P. R. Morkel, "Characterisation of special fibres and fibre devices", *IEEE Journal of Lightwave Technology*, 7(8), 1242–1255, 1989.
22 R. J. Mears, L. Reekie, S. B. Poole, and D. N. Payne, "Low-threshold tunable CW and Q-switched fibre laser operating at 1.55 μm", *Electronics Letters*, 22(3), 159, 1986.
23 R. J. Mears, L. Reekie, I. M. Jauncey, and D. N. Payne, "Low-noise erbium-doped fibre amplifier operating at 1.54μm", *Electronics Letters*, 23(19), 1026, 1987.
24 L. N. Binh, *Digital Processing Optical Transmission and Coherent Receiving Techniques*. CRC Press, Taylor & Francis: Boca Raton, FL, 2014.
25 Keysight Inc., "Oszilloskop UXR1104A der infiniium UXR-Serie: 110 GHz, 4 Kanäle", access date March 16, 2024.
26 O. E. Delange, "Wideband optical communication systems, part 11: Frequency division multiplexing", *hoc IEEE*, 58, 1683, 1970.
27 E. Snitzer, "Optial MASER Action of Nd+3 in a barium Crown Glass," Physical Review Letters, Volume 7, Number 12, pp. 444-446, December 15, 1961.
28 Peterka, P., Vojtěch, J. (2023). "Optical Amplification", in: Kawanishi, T. (eds) Handbook of Radio and Optical Networks Convergence. Springer, Singapore. https://doi.org/10.1007/978-981-33-4999-5_20-1.
29 Jeff Hecht, SPIE, "Amp it up: How David Payne made fiber optic communications into a bandwidth superhighway", https://spie.org/news/photonics-focus/janfeb-2021/amp-it-up-david-payne#_=_
30 https://www.glhclink.com/index.php?aid=242

2 Space Atmospheric Signal Propagation and Effects on System Performance

CONTENTS

DOI: 10.1201/9781003426677-2

2.1 INTRODUCTORY REMARKS

Various aspects of the propagation of lightwaves in the Earth's atmosphere have been reported, including molecular and aerosol absorption, scattering, rain and snow attenuation, scintillation effects, and beam pointing. These attenuations are key in free-space optical communication systems (FSOCS) due to additional losses and severe effects on the OSNR (optical signal-to-noise ratio). Lightwave propagation through atmospheric and deep space media to satellites suffers high losses due to the extremely long distances and the absence of middle amplification, unlike the case of lightwaves over guided media, such as optical fibers.

Fog seems to be the most penalizing element for free-space optical link operations. The turbulence scintillation layer and its effects on the phase fluctuation of lightwaves are examined, along with the derivation of attenuation and the penalty on the OSNR of the transmitted channels.

Investigating and summarizing such attenuations show that FSO can constitute a reliable broadband alternative to installed fiber networks and lead to better acceptance of free-space communications technology with ultra-high data capacity.

The following attenuations in various atmospheric regions:

i. Free-space propagation loss using the Friis equation, which is assumed under the completely linear "small signal" region and where the laser carrier source is an ideal single-frequency laser, that is, there is no finite linewidth but only one frequency component is considered in the appropriate sections of this book. The finite linewidth of the source leads to spatial spreading and distortion.
ii. Atmospheric propagation loss per 3 km at 100 m height above sea level versus inclination angle.
iii. Average attenuation due to scintillation.
iv. Turbulence coherence length ~ a few tens of meter ≪ the laser coherence length of 10 kHz linewidth laser lightwaves.
v. Pointing loss expected for a 5 μrad error is 1.9 dB.
vi. Depolarization: to be <1.0 μrad; exp results → <5e-5 rad (or 50 μ-rad)– so about 1 dB is allowable for OSNR passing through the turbulence scintillation layer at 20 km.
vii. Beam propagation method for atmospheric transmission of PDM-QPSK: this method involves first performing the FFT over half of the propagation step, then adding the phase evolution in the spatial frequency domain due to phase disturbances by the scintillation layer, followed by taking the IFFT of the two-dimensional signal distribution over the next half of the propagation step, and repeating this process for the rest of the propagation distance.
viii. Doppler shift between satellites is given, and methods for compensating for such shifts in a coherent DSP-based receiver are discussed.
ix. The optical amplification devices (OPA-HPA – high power amplifiers and OPreA-LNA – low noise amplifiers) are operating in the saturated regions, so the Friis formulation

cannot be applied as it was developed and assumed for the small signal or linear region only and a single frequency carrier. Thus, there is no Friis formulation in the Excel file for the power budget.

x. The power budget is evaluated using the 10 $\log_{10}$ (inverse of the total noise-to-signal ratio of the OSNR of the previous stage plus the OSNR referred to as the input of the amplification stage), so that we can derive the OSNR of cascade stages. The Friis formulation cannot be implemented. The finite linewidth of the lightwave carriers leads to spatial broadening or spreading of the beam, resulting in an additional loss added by $10\times\log_{10}$ (beam across section area/area of optical head) dB due to beam spatial spreading. It is noted that the level of input power to the optical amplifier is critical: if this level is much less than or much higher than the optical amplification noise, the OSNR can be high, low, or have only a very minor contribution from the amplification noises. The beam spreading due to the finite linewidth of carriers also adds further complication, including the skew of two polarized beams if polarized channels are used to double the transmission capacity.

The atmospheric propagation is subject to a probability of error (Pe) that follows a fading model similar to the case of fading observed in wireless ray propagation and reflection from various buildings. The probability of error or BER is derived mathematically with the variables dependent on the atmospheric turbulence parameters.

This chapter is organized as follows:

- Section 2.2 provides a brief introduction to the propagation of lightwaves through deep space and the atmospheric air region. Lightwaves demonstrate the coherency of the electromagnetic laser beam.
- Section 2.3 discusses the transmittance and absorption properties of space media.
- Section 2.4 describes the modeling of atmospheric propagation and the dynamics of the lightwave beam, including its spatial spreading.
- Section 2.5 then presents the concept of wave-front errors and synchronization with digital processing, commonly applied to sampled discrete received signals for ultra-fast transmission. The Doppler shift of receiving signals and signal attenuation is addressed, with techniques offered to mitigate these potential signal-degrading effects.
- Optical amplification for photonic communication systems is covered in Section 2.6.
- Section 2.7 provides remarks on atmospheric propagation, possible errors, and expected values of attenuation of space links in both vacuum and atmospheric propagation. The commonly used Friis' relationship for attenuation applies only to the ideal case of absolute single-frequency lightwaves. This equation derives from the radar concept of propagation and gain in the field of antenna and radio propagation and detection; thus, it lacks the modernization of the photonic area, especially in the field of laser and integrated photonics. For finite linewidth carriers and modulated beams, the Friis equation must be modified to account for spatial spreading.
- Finally, Section 2.8 presents a comparison between WFE (wave-front error) in space image processing and EVM (Error Vector Magnitude) in the digital signal processing of complex signals, which are commonly employed to increase transmission capacity and minimize errors.

In brief, this chapter provides an overview of photonic communications in space systems and networks, with a focus on modulated laser beam propagation through space. It details the transmittance and absorption characteristics that influence the probability of errors for advanced modulation formats. The chapter also describes photonic subsystems, optical amplification, and receiving concepts.

Optical coherent transceivers are uniquely required so that further gain in sensitivity can be obtained to reduce the high loss of the long transmission free-space medium. SPC is all-optical and bidirectional, meaning both transmitting and receiving channels are employed to minimize the alignment functionalities of the whole system. They are analyzed and tested here in this chapter, followed by the test and performance of an integrated transceiver of a 100 G DPM-QPSK module in Chapter 9, in which the power-limited case transmission, the quantum SPC is described for secret key transmission.

Detailed techniques in optical modulation, transmission over turbulent free space, attenuation factors, estimation of bit error rate (BER) and probability of error, optical amplification for signal boosting and low-level reception, spatial dispersion due to laser beam finite linewidth, and link budget can be found in the subsequent chapters.

2.2 LIGHTWAVES PROPAGATION THROUGH DEEP FREE SPACE AND ATMOSPHERIC REGION

Free-space optic (FSO) links involve the transmission, absorption, and scattering of light by the Earth's atmosphere. The atmosphere interacts with light due to its composition, which, under normal conditions, consists of a variety of different molecular species and small suspended particles called aerosols. This interaction produces a wide variety of optical phenomena: selective attenuation, scattering, and scintillations.

A selective absorption of radiation that propagates at specific optical wavelengths in the atmosphere results from the interaction between photons and atoms or molecules (N_2, O_2, H_2, H_2O, CO_2, O_3, Ar, etc.,) which leads to the disappearance of the incident photon and an elevation in temperature. The absorption coefficient depends on the type of gas molecules and their concentration. Molecular absorption is a selective phenomenon that results in a spectral transmission of the atmosphere, presenting transparent zones, called atmospheric transmission windows, and opaque zones, called atmospheric blocking windows.

Aerosols are extremely fine solid or liquid particles suspended in the atmosphere with very low fall speeds due to gravity (e.g., ice, dust, smoke). Their size generally lies between a few micrometers and 100 μm. Fog, dust, and maritime spin drift particles are examples of aerosols. Aerosols influence the conditions of atmospheric attenuation due to their chemical nature, sizes, and concentration. In a maritime environment, aerosols are primarily made up of droplets of water (foam, fog, drizzle, rain), salt crystals, and various particles of continental origin. The type and density of continental particles depend on the distance from and characteristics of the neighboring coasts. As the Earth's surface is covered with two-thirds of global oceans, we expect that major turbulence and scattering are due to sea particles.

Atmospheric scattering results from the interaction of part of the light with atoms and/or molecules in the propagation medium. It causes an angular redistribution of this portion of the radiation, with or without modification of the wavelength.

Aerosol scattering occurs when the particle size is of the same order of magnitude as the wavelength of the transmitted wave. In optics, this scattering is mainly due to mist and fog. Attenuation is a function of frequency and the visibility related to the particle size distribution. This phenomenon constitutes the most restrictive factor in the deployment of FSO systems over long distances. Visibility is a concept defined for meteorological purposes. It characterizes the transparency of the atmosphere as estimated by a human observer. It is measured by the Runway Visual Range (RVR), which is the distance a parallel luminous 'beam must travel through the atmosphere until its intensity (or luminous flux) drops to 0.05 times its original value. This is measured using a transmission or diffusion meter.

Under the influence of thermal turbulence inside the propagation medium, randomly distributed cells are formed, which vary in size from 10 cm to 1 km and have different temperatures. These various cells with different refractive indexes *cause scattering, multipath fading, and variations in the arrival angles; the received signal fluctuates at frequencies ranging from 0.01 to 200 Hz.* The *wavefront* varies similarly, causing focusing and defocusing of the beam. Such fluctuations in the signal are called scintillations. The fading effects are quite similar to the fading of wireless signals due to multiple reflections off different buildings. We will see that the statistical properties and the probability of error follow similar trends to RF multipath fading. This fading effect in SPPC is very much like the group delay dispersion in multimode fiber and especially in few-mode fibers, as each mode can propagate in a different path and thus have different effective refractive indices. Therefore, few-mode fibers, whose group delay difference is the reverse of spatial group delay dispersion, can be used as space dispersion compensators to mitigate such spatial spreading. Currently, there are few-mode fibers that can have negative group delay dispersion, but a specially designed few-mode fiber can be created to provide the negative group delay dispersion needed to mitigate the spatial broadening effects.[1]

2.3 TRANSMITTANCE (SIMULATED) AND ATTENUATION VERSUS INCLINATION ANGLE[2]

Total transmission losses depend on the telescope design and beam spatial mode and increase as the *inverse square* of the wavelength (λ). Conversely, atmospheric absorption losses tend to decrease with increasing wavelength, though there are also large spectral bands where the atmosphere is almost entirely opaque. Losses due to atmospheric turbulence also decrease slightly as wavelengths increase. Favorable spectral bands, where relevant laser systems also exist, can be seen in Figure 2.1 and include 665–685 nm, 775–785 nm, 1000–1070 nm, and 1540–1680 nm.

A trade-off when designing telescope optics is the choice between transmissive or reflective telescopes. Reflective mirrors can be larger, although for polarization-based PDM-QPSK 100 G and higher-order modulation, care must be taken to prevent significant depolarization effects. Secondary mirrors in reflector telescopes are often placed within the path of the beam, blocking part of the primary mirror and affecting diffraction spreading. However, the loss is less than 1 dB for a secondary mirror diameter up to 25% of the primary mirror diameter (D). Beam divergence scales with λ/D, so larger apertures produce smaller beam divergences with lower free-space losses. However, due

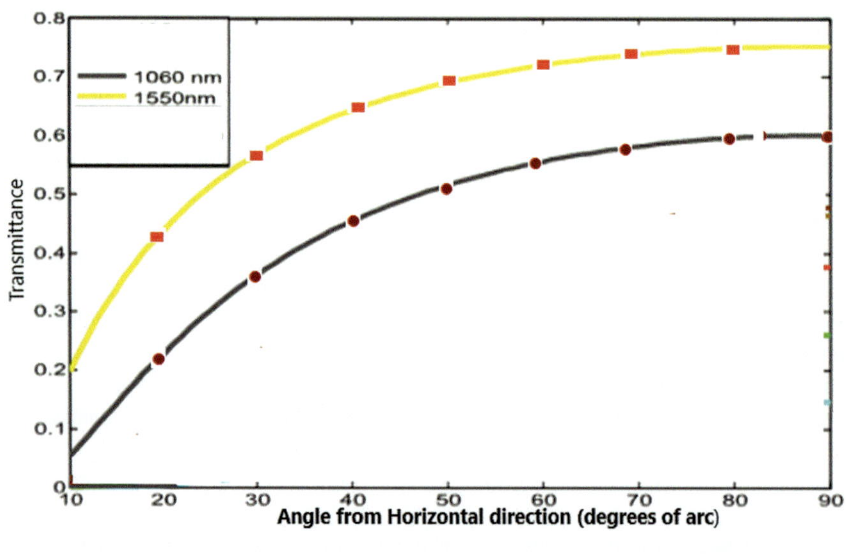

FIGURE 2.1 Simulated transmittance of the atmospheric medium (a—upper graph) at zenith in a typical rural location (b—lower graph) with respect to inclination angle for different wavelengths (nm). (Extracted from Ref.[3])

to atmospheric turbulence, increasing the diameter of ground-based transmitters above a few tens of centimeters makes little improvement on the losses.

A much more accurate and higher resolution spectrum of the transmittance and absorption can be seen in Figure 2.2, with the corresponding absorption characteristics shown in Figure 2.3. This indicates that transmittance can be dynamic, and transparency must be monitored to tune the source carrier wavelength to the appropriate spectral window for transmission. A LIDAR system should be used to monitor such transparent spectra for DWDM transmission. Furthermore, the sensitivity or dynamic range changes based on the transparency of the medium, so high-dynamic-range ORx must be designed to account for these dynamics in the absorption factor.

Another factor affecting ultra-high-speed transmission is the use of polarized channels. Two polarizations of lightwaves are employed to double the capacity of the transmission. Air particles can scatter and depolarize the light waves, so a polarization rotator or DSP compensation technique must be used to counteract such polarization rotation and the resulting skew rate between the two polarized channels. Skew compensation of the two channels ensures synchronization of the polarized channels, thus facilitating the recovery of channel information.

Figure 2.3 shows the absorption coefficients of the air. It is quite transparent from 1549 to 1567 nm, and this spectral window corresponds to the lowest attenuation in silica fibers. Therefore, it is advantageous to use integrated optical devices and fiber optic technology for SPC wherever appropriate, as such technology can offer ultra-high capacity transmission, reaching 400 Gbps with 50 GB and commonly 100 G with PDM-QPSK formats.

The transmittance over the atmospheric medium, as a function of the inclination angle, is simulated and estimated in Figure 2.1a and b, respectively.

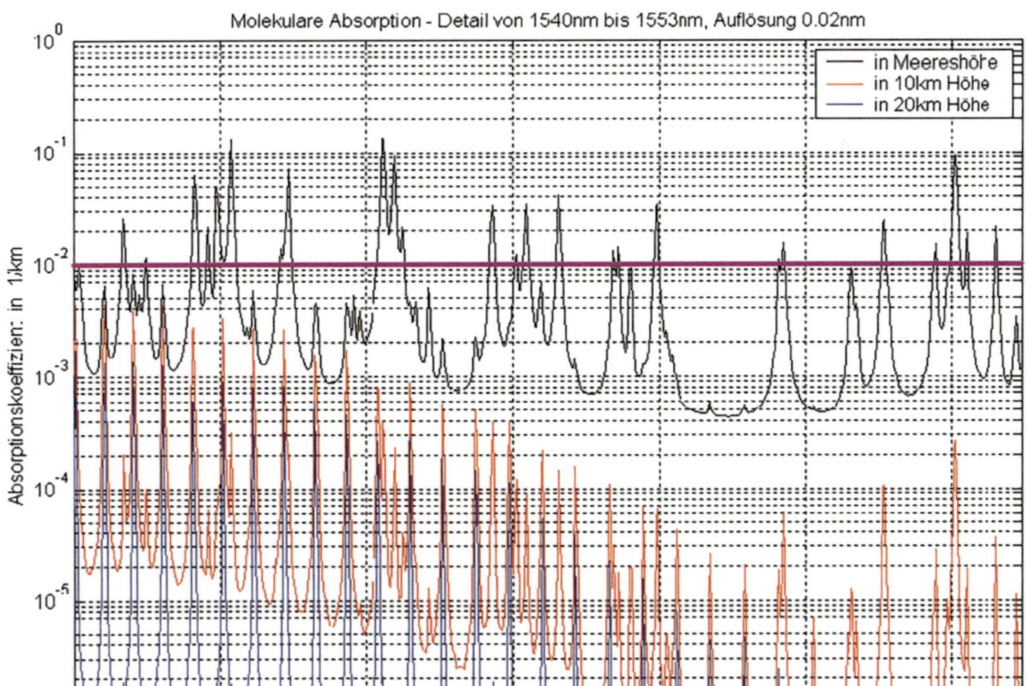

FIGURE 2.2 Detailed absorption spectra over 1.0 km. (With compliments from Mynaric Laser Communications GmbH. Atmospheric absorption coefficient/km versus wavelength (nm).

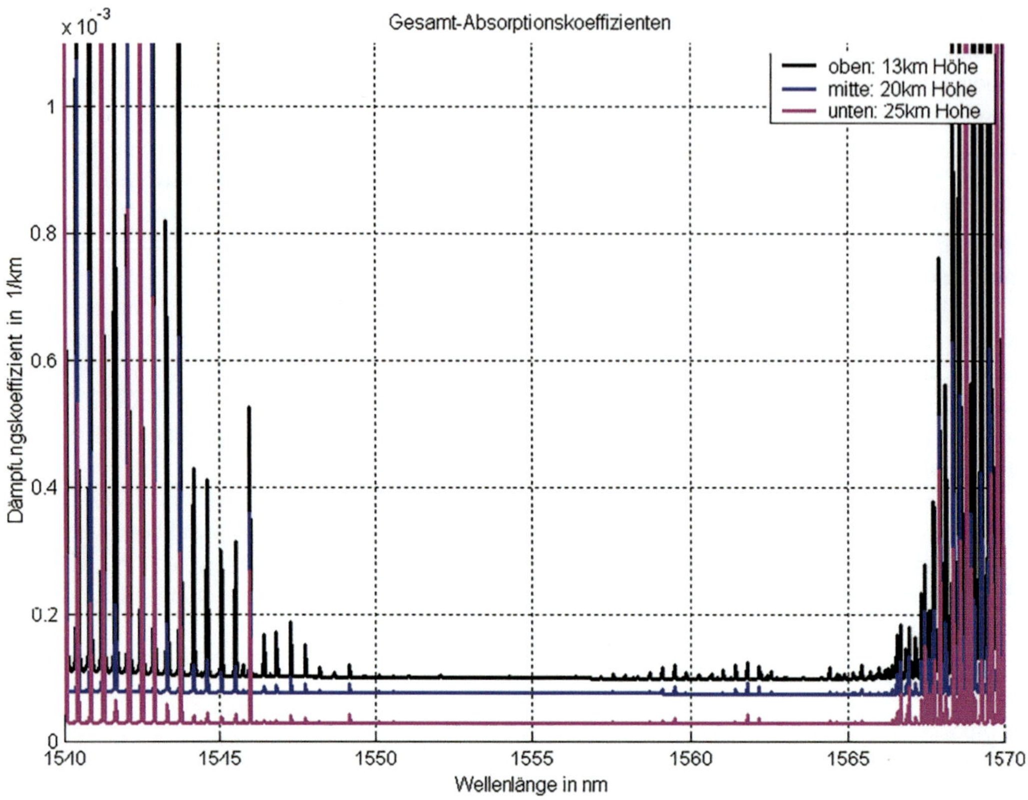

FIGURE 2.3 Measured absorption over 1.0 km. (With compliments from J.H. Howarth of Mynaric Laser Communications GmbH.

2.4 MODELLING OF ATMOSPHERIC REGION AND PROPAGATION

For optical and IR waves, light propagation through the atmosphere is affected by absorption and scattering by air molecules, as well as absorption and scattering by solid or liquid particles suspended in the air. The transmission of light in the atmosphere is described by the Beer-Lambert law:

$$|\tau(\lambda, L)| = \frac{P(\lambda, L)}{P(\lambda, 0)} = e^{-\gamma(\lambda)L} \tag{2.1}$$

where $\tau(\lambda, L)$ is the total transmittance of the atmosphere at wavelength λ; $P(\lambda, L)$ is the signal power at distance L from the transmitter, $P(\lambda, 0)$ is the emitted power at the Tx (origin point), and $\gamma(\lambda)$ is the attenuation or the total extinction coefficient per unit of length.

The extinction coefficient is composed of absorption and scattering terms. Generally, it is the sum of the following terms:

$$\gamma(\lambda) = \alpha_m(\lambda) + \alpha_a(\lambda) + \beta_m(\lambda) + \beta_a(\lambda) \tag{2.2}$$

where $\alpha_m(\lambda)$, $\alpha_a(\lambda)$ are molecular and aerosol absorption coefficients, respectively, and $\beta_m(\lambda)$, $\beta_a(\lambda)$ are molecular and aerosol scattering coefficients, respectively.

In the spectrum region used by free space photonic communications (FSPC), the attenuation coefficient is primarily approximated by the scattering coefficient of the particles present in the atmosphere.

2.4.1 Attenuation Relations

Thus, the attenuation coefficient can be approximated by the following relation (Kruse relation)[4] as

$$\alpha(\lambda) = \frac{3.92}{V_m}\left(\frac{\lambda_{nm}}{550}\right)^{-q} \tag{2.3}$$

where V_m = visibility in meters; λ (wavelength) in nm and q is coefficient dependent on the size of the particles.

The coefficient q depends on the particle size distribution. It is given by[4]:

$$q = \begin{cases} 1.6 & \text{if } V_m > 50\text{ km} \\ 1.3 & \text{if } 6\text{ km} < V_m < 50\text{ km} \\ 0.585\, V_m^{1/3} & \text{if } V_m < 6\text{ km} \end{cases} \tag{2.4}$$

This relation has been widely used in the literature to determine the free-space optical (FSO) equipment link budget. From the last two equations, it is clear that, under any meteorological conditions, as the wavelength increases, the attenuation decreases.

Thus, a recent study proposes an alternative expression for the parameter q. This expression, which has not yet been experimentally checked, is as follows[5]:

$$q = \begin{cases} 1.6 & \text{if } V_m > 50\text{ km} \\ 1.3 & \text{if } 6\text{ km} < V_m < 50\text{ km} \\ 0.16\, V_m + 0.34 & \text{if } 1.0\text{ km} < V_m < 6\text{ km} \\ V_m - 0.5 & \text{if } 0.5\text{ km} < V_m < 1.0\text{ km} \\ 0 & \text{if } V_m < 0.5\text{ km} \end{cases} \tag{2.5}$$

Equation 2.5 implies independence between the atmospheric attenuation values and the wavelength in the presence of dense fog reducing visibility to below 500 m. Beyond 500 m of visibility, this relation respects the conclusion deduced from the previous relations, namely, a smaller attenuation for increasing wavelengths.

2.4.2 Attenuation and Advection Relations

Al Naboulsi et al.[6] developed from FASCOD simple relations allowing the evaluation of attenuation in the 690–1550 nm wavelength range and for visibilities ranging from 50 to 1000 m for two types of fog: advection and convection fog[6]: "The advection fog is generated when the warm, moist air flows over a colder surface. The air in contact with the surface is cooled below its dew point,[7] causing the condensation of water vapor. It appears more particularly in spring when southern displacements of warm, moist air masses move over snow-covered regions."

The attenuation by advection fog can be expressed by the following relation:

$$\sigma_{\text{advection}} = \frac{0.11478\lambda + 3.8367}{V_{\text{m}}} \tag{2.6}$$

where λ is the wavelength (in nano-meters), and V_{m} is the visibility (in meters).

Radiation or convection fog is generated by the radiative cooling of an air mass during the night when meteorological conditions are favorable (very low wind speeds, high humidity, clear skies). It forms when the surface releases the heat accumulated during the day and becomes colder than the air in contact with it, causing the air to cool below the dew point and leading to the condensation of water vapor, which results in the formation of a ground-level cloud. This type of fog occurs more particularly in valleys.

2.4.3 Rain Attenuation

Rain attenuation (dB/km) is generally given by the Carbonneau relation[8]:

$$\text{Att}_{\text{snow}}\left[\frac{\text{dB}}{\text{km}}\right] = \alpha S^{\beta} \text{ in units of dB/km} \tag{2.7}$$

where: Att_{snow} is the space attenuation by snow (dB/km); S is the snowfall rate (mm/h), and α and β are, respectively, coefficients, function of the wavelength, giving by the following relations (λ_{nm} is wavelength in units of nm):

$$\text{Under wet: } \alpha = 0.0001023 \times \lambda_{\text{nm}} + 3.7855466; \quad \beta = 0.72 \tag{2.8}$$

$$\text{Dry snow: } \alpha = 0.0000542 * \lambda_{\text{nm}} + 5.4958776; \quad \beta = 1.38 \tag{2.9}$$

2.4.4 Scintillation Attenuation

Tropospheric scintillation effects are generally studied from the logarithm of the amplitude χ [dB] of the observed signal ("log-amplitude"), defined as the ratio in decibels between its instantaneous amplitude and its average value. The intensity and the speed of the fluctuations (scintillation frequency) increase with the lightwave frequency. For a *plane wave*, low turbulence, and a specific receiver, the scintillation variance σ_x^2 [dB^2] can be expressed by the following relation:

$$\sigma_x^2 = 23.17 \times k^{7/6} \times C_n^2 \times L^{11/6} \text{ in dB}^2 \tag{2.10}$$

where k [m^{-1}] is the lightwave number ($2\pi/\lambda$) in vacuum, L [m] is the length of the link, C_n^2 [$\text{m}^{-2/3}$] is the structural parameter of the refractive index, representing the *degree of the turbulence intensity*, IRT = index of refraction turbulence.

Scintillation peak with peak amplitude is equal to $4\sigma_{\chi}$ and attenuation related to scintillation is equal to $2\sigma_{\chi}$. See later for the average attenuation/loss of the scintillation.

2.4.5 Friis Relation for Attenuation Loss in Ideal Free Space

Friis' original idea behind his transmission formula was to dispense with the use of directivity or gain when describing antenna performance. Instead, he introduced the concept of antenna capture area as one of the two key components of the transmission formula that characterizes the behavior of a free-space radio circuit. This leads to the published form of his transmission formula:

$$\frac{P_R}{P_T} = \left(\frac{A_R A_T}{L^2 \lambda^2}\right) \tag{2.11}$$

where P_T is the power fed into the transmitting optical antenna input terminals; P_R is the power available at the receiving optical antenna output terminals; and A_R *and* A_T are the effective aperture areas of the receiving and transmitting optical head antennas, respectively. These areas refer to the optical head (OH) placed after the fiber to launch into free space and at the receiving end. L=transmission distance between the two optical antennas; λ is the operating wavelength. All parameter units are in SI (International System) units. The main assumption is that the lightwave beam is co-directionally coherent, that is, purely single frequency without any finite linewidth laser source, from the Tx and Rx optical antennas (collimating lenses/telescopes). If the laser carrier and the channel signal spectrum are finite, then the spatial spreading is substantial and creates an additional attenuation factor, which must be added to Equation 2.11.

The modulated signal spectrum must be wide for 100 G 25 GBaud PDM-QPSK (Polarization Division Multiplexed Quadrature phase Shift Keying) and is required to be at least 6 dB below the carrier, whose linewidth is to be <10 kHz in practical properties of laser sources.

Notes: The Friis formula is assumed for cases where all the gain and loss behave in linear regions. Furthermore, the linewidth of the source and modulated signals must be dealt with as if they were absolute single-frequency signals, which is almost impossible. In practice, there is a finite linewidth and signal band. The finite linewidth carrier power would be about 6 dB above the signal band, and the assumption of coherence can be valid if such linewidth is narrow enough, say 1.0 kHz, or much less by using a fiber laser.

In the case of finite linewidth, one must take the summation of all spectral lines and assume that the attenuation factor is equal for all lines. However, at the ultra-high bit rate with a baud rate of 25 GB, the signal bandwidth is only about 25 GHz double sideband, or about 0.5 nm at a 1550 nm wavelength. Such a signal width is narrow enough for the channel to be transmitted without being affected by the transparent spectra of the vacuum or atmospheric spectrum. Due to the vacuum condition, the refractive index remains the same and thus there is no temporal dispersion, but spatial spreading could attenuate the signal, and the reception power is affected by the ratio of the area of the effective beam spot size to that of the optical head (OH).

However, in the power budget incorporating optical amplifiers, the preferred operating region is the saturated region of the gain. Therefore, the Friis formula may result in different estimations due to the power budget estimates using the OSNR step by step, by taking the OSNR from the previous stage plus the noise-to-signal ratio of an amplification stage, and then taking the $\log_{10}(\cdot)$ of the summation inverse to give the OSNR in dB of the resulting stage (output of the amplifier). A saturated signal OSNR can then be obtained.

Assuming the system is operating in the linear region and there is no finite linewidth of the source, then Friis'' antenna effective area should be used to characterize antenna performance over the contemporary use of directivity and gain metrics. Replacing the effective antenna areas with their gain counterparts yields

$$\frac{P_R}{P_T} = G_T G_R \left(\frac{\lambda}{4\pi L}\right)^2 \tag{2.12}$$

alternatively, the propagation loss in dB can be written as

$$\alpha_{\text{prop,dB}} = 20 \times \log_{10}\left(\frac{\lambda}{4\pi L}\right) \text{ in dB}$$

The modified Friis formula, by considering the spatial spreading due to the finite linewidth o the source is

$$\frac{P_R}{P_T} = G_T G_R \cdot \frac{A_{\mathrm{OH}}}{A_{\mathrm{SP}}}$$

where A_{SP}=spreading area of the beam, A_{OH} is OH cross section area.

It is noted that the proportional coefficient of the factor $\left(\frac{\lambda}{4\pi L}\right)^2$ is very small when the distance is in km and the wavelength is in nm. Thus, this relation can be asymptotically approximated by an exponential, as given in the above section.

The attenuation in dB over the length L can then be asymptotically written as

$$\alpha_L = 20 \times \log_{10}\left(\frac{\lambda}{4\pi L}\right) \tag{2.13}$$

2.4.6 Down Link to OGS Attenuation

2.4.6.1 Attenuation Over Atmospheric Layer

Clear-sky (i.e., no clouds) atmospheric signal attenuation down to an OGS roughly depends on elevation angle θ (in degree) and signal wavelength λ. A simple model for OGSs situated 100 m above sea level is given for λ=(850/1064/1550 nm) (outside of atmospheric molecular absorption bands), which is represented as approximation by the following relationship

$$\alpha_{\mathrm{atm}} = \frac{10}{\left(\frac{\lambda}{1550\,\mathrm{nm}}\right)^2 (\in +1)} \ldots \text{ in dB} \tag{2.14}$$

where $\in$=the *inclination angle* in degrees, λ is the operating wavelength in nm. Such attenuation over the atmospheric layer is plotted in Figure 2.4 for a 100 m height with the transmitter located on an airplane at 3 km height. Note that the total attenuation coefficient derived from the plot is 0.5 dB at 20° and reaches 0.1 dB at angles >40°. The calculation for a satellite moving across the pointing angle of 20°–90°. The operation of the beam light of sight (LOS) and the inclination angle is commonly used as 20°–90° of arc. At the spectral window of 1550 nm (C-band), the atmospheric attenuation is approximately 0.45 dB at an inclination angle of 20° and decreases to 0.15 dB for inclination angle >50°.

2.4.6.2 Transmission Loss to Satellite from OGS

This attenuation (same as given in 4.5) can be estimated using Friis formulation[10] as

$$\alpha_L = -20 \times \log_{10}\left(\frac{\lambda}{4\pi L}\right) \text{in dB} \tag{2.15}$$

where L = transmission distance (m); α_L =attenuation in dB; λ = operating wavelength. We note that the factor $\lambda/4\pi$ is very small so such attenuation can be approximated by an exponential decay in an asymptotic approximation sense.

2.4.6.3 IRT-Fading Loss

The classical log-normal fading model of IRT scintillations always implies a certain fraction of signal outages. In Ref.[7] a simplified formula for calculating the IRT losses in dB is given, which requires the received power scintillation index σ_P^2 and the *fading threshold* p_{thr} as input.

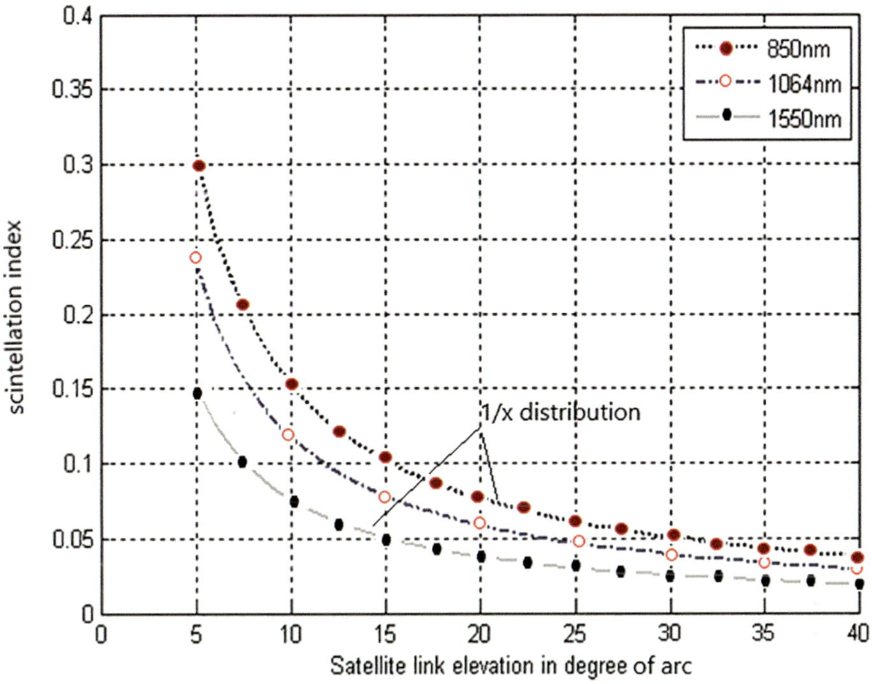

FIGURE 2.4 Attenuation (per 3 km vertical height) over the atmospheric downlink as received at 100 m above sea level, versus the inclination angle, of a transmitter at 3 km on an aircraft. (Extracted from Ref.[9])

$$\text{In dB scale:} \quad \alpha_{\text{sci,dB}} = \left(3.3 - 5.77\sqrt{-\ln\left(p_{\text{thr}}\right)}\right) * \left(\sigma_P^2\right)^{0.4} \text{ dB} \tag{2.16}$$

The variance σ_P^2 depends on the scenario-specific intensity scintillation, together with the size of the averaging effect of the Rx optic head aperture.

The value of σ_P^2 is between 0 and 1 for low to medium turbulence and between 1 and 3 for strong turbulence cases, such as low elevation downlinks. σ_P^2 saturates asymptotically toward a value below 1 for very long distances. Formulae to calculate σ_P^2 can be found in Ref.[11] The parameter p_{thr} states the time-fraction during which the signal may fall below the level required for sound data reception. Values of p_{thr} *are typically chosen* between 2e-1 and 1e-6. This concept requires FEC (forward error coding) specific to this fading channel to recover the data lost during fades, as described in Ref.[12] By combining with such FEC techniques, IRT losses can be reliably compensated.

Estimation of σ_P^2 for an aeronautic downlink can be performed using Equations 2.2 and 2.12, as referenced from Ref.[13] (which are based on the theory presented in Ref.[11]), as shown in Figure 2.5.

2.4.6.4 Depolarization Effect

The Earth's atmosphere can produce changes in the polarization sense, causing depolarization of transmitted lightwaves. This is critical in the case of the modulation format PDM-QAM (polarization multiplexing quadrature amplitude modulation), where each polarization carries a sequence of pulses of in-phase and quadrature components. This effect is referred to as depolarization or cross-polarization in the case of two independent waves, which may interfere with each other due to a depolarizing condition present in the transmission path. This can result in the rotation of polarization and generate delayed differences between two polarized modes of the lightwave beam, leading to skew times between the two synchronized polarized channels.

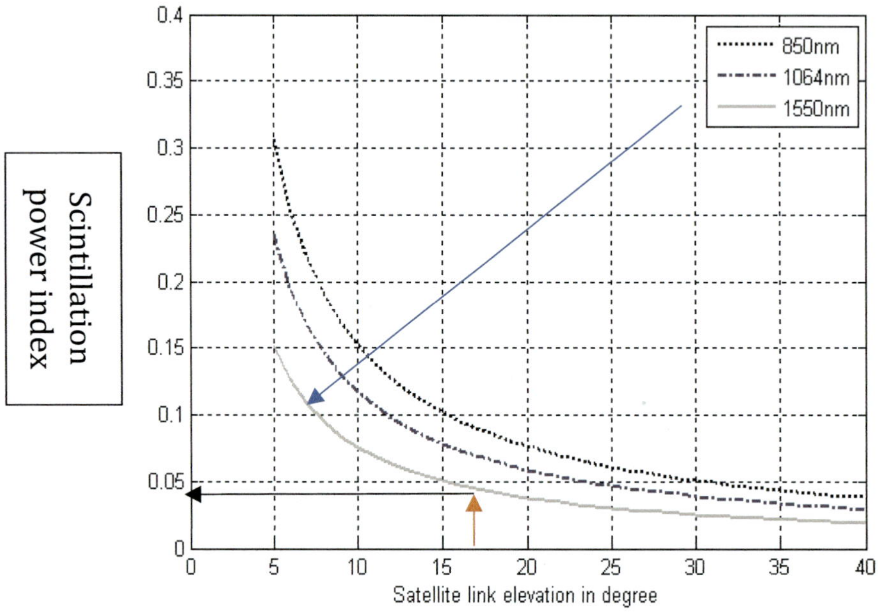

FIGURE 2.5 Satellite downlink IRT-loss approximated from measured data for OGS-OP with its 40 cm Rx-aperture (compare Figure 12 and Ref.[14]). Note: MEO Rx aperture Laser-Light MEO Satellite optic head (OH) diameter = 160 mm or 16 < 40 cm. Left: power scintillation index σ_{P2} versus sat elevation angle; right: IRT-loss with $p_{\text{thr}} = 0.001$. IRT, Index-of-Refraction Turbulence. (Extracted from Ref.[10])

The depolarization of a linearly polarized laser beam can primarily occur over an optical path of 4.5 km. A He–Ne gas laser at 1550 nm can be used with an additional polarizer at the output and a rotating polarization filter assembly in front of the receiver. Values of depolarization can range between 10^{-7} rad and about 5×10^{-5} rad. The lower limit is determined by the quality of the polarizer-analyzer combination used. These values of depolarization are significantly higher than those predicted by theories regarding turbulence-induced depolarization.

This depolarization effect would severely affect the PDM-QPSK modulated light beams when propagating over the SPC SYSTEM link. The atmospheric layer, especially the scintillation turbulence is affected most, and this depolarization is to be controlled such that the polarization rotation should not create any OSNR penalty more than 1 dB or <10 micro-arc-radians as required for Acacia Inc. AC1200 Acacia 100–400 G module.

2.4.6.5 Beam Diffraction Limit and Coupling

The spot size in the receiver focal plane and the divergence of the transmitted beam are critical parameters in the design of coupling a free-space beam into an optical receiving optic, especially in relation to the fiber cross-section and core region. Ideally, both should be as small as possible: the former indicates how well the receiving optical system can focus the received laser signal, and the latter how narrowly the transmitting system can transmit a laser beam.

Figure 2.6 shows the diffraction of the intensity of the optical field after passing through an optical lens of diameter D. It is the spatial Fourier transform of the lightwave beam passing through a circular hole of graded index, equivalent to flat lenses or geometrical concave lens structures. The diffraction limit of the radiation-dependent intensity is then[15]

$$\theta = 1.22\frac{\lambda}{D} \text{ rad.} \tag{2.17}$$

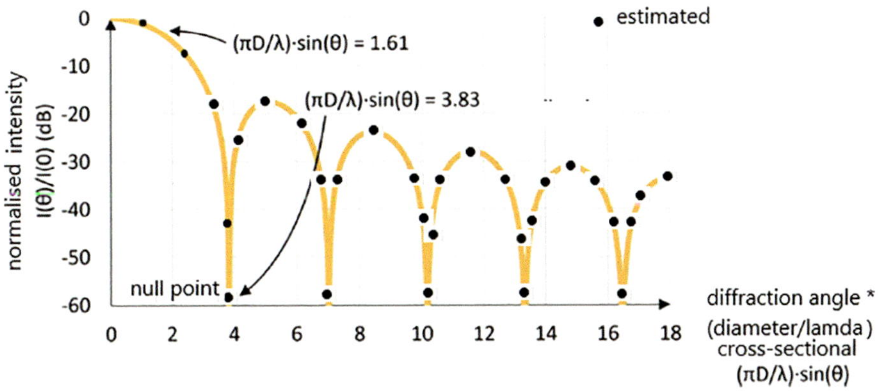

FIGURE 2.6 Normalized intensity of radiation as a function of the wavelength λ, the aperture diameter D and the angular width θ (extracted from Ref.[15]), showing the acceptance angular area.

And, the minimum spot size that a perfect telescope, i.e., diffraction limited can produce at the focal plane (f = focal length)

$$d = 2.44\frac{\lambda f}{D} \tag{2.18}$$

Thus, the larger the diameter of the optics, the smaller the acceptance beam size. So, if the fiber core is used to couple the 1.0 mm optics, then the angle for 1550 nm is 1.892e-6 rad, which is much smaller for a 1632 mm optical head of an MEO receiver. Thus, a GRIN (graded index) or SELFOC (self-focusing)[16] lens should be used in front of a single-mode optical fiber to couple it to the coherent receiver hybrid coupler. The spot-size of a 160 mm telescope requires a short focal length, which makes it impossible to reach the SMF diameter of 8 µm. So, a GRIN lens of 1.0 mm diameter would give a focal length of 42 mm for the telescope. If not, the focal length would need to reach some tens of meters.

Regarding the pointing angle, an allowance can be given as a reference. The point-ahead angle ϕ is approximately 51 µ-rad for 500 km LEO, 18 µ-rad for GEO, and ~32 µ-rad for MEO. This angle can be approximated as

$$\phi = 2\frac{v_t}{c} \tag{2.19}$$

With v_t as the tangential velocity of the satellite and c as the speed of light. For a circular orbit, the worst-case point-ahead angle can be obtained by considering the point where the tangential velocity is the fastest, i.e., the zenith. In this case, the tangential velocity is asymptotically identical to the satellite's orbital velocity v_0, which can be calculated by

$$v_0 = \sqrt{\frac{\mathrm{GM}}{L}} \tag{2.20}$$

where G = gravitational constant; M = mass of the earth; and L = distance from earth's center to the satellite. The pointing attenuation is expected to be –1.39 dB with an accurate pointing error of 5 µrad. This pointing angle tolerance is used as the control feedback parameter for the control system of the beam pointing.

2.5 ATMOSPHERIC EFFECTS AND BEAM PROPAGATION (SIMULATION PLATFORM)

The simulation platform controls the separation between phase screens (propagation step via the split-step Fourier method) such that the step size for consecutive screens only creates weak turbulence while providing a smooth representation of the turbulent stochastic medium along the propagation path. Therefore, to correctly represent the inherently random nature of the turbulent phase screens, it is required that the propagation path be split into at least 10 intermediate steps to achieve a better representation of the final field at the receiver plane (Figure 2.7).

In long-range simulation scenarios, some constraints must be imposed on the propagating optical field to produce physically meaningful results, such as avoiding divergence of the spatially dispersed field while keeping computational efforts and simulation time within reasonable limits. The main constraint is to maintain a minimum spatial transverse sampling interval Δx and Δy through the propagation steps along the transmission path, which can accurately represent the optical traveling field and the atmospheric distortions at each step. On the one hand, the spatial coherence radius indicates the transverse correlation length, or coherence length, of the propagating wave, and it is given by

$$\rho = 1.46k^2C_n^2L^{-3.5} \tag{2.21}$$

where $k = 2\pi/\lambda$ is the wavenumber and λ is the operating optical wavelength in vacuum. The simulation platform automatically and sets the minimum sampling interval in the propagation window to $\Delta z < \rho_0/3$, in order to avoid aliasing effects during step-by-step beam propagation. The dotted lines in Figures 2.3 and 2.4 indicate how the field extent would develop without re-sampling, that is, without an explicit constraint.

2.5.1 Space Channel Loss

Transmission path losses G_L over a distance L of a specific optical carrier can be estimated as

$$G_L = \left(\lambda/4\pi L\right)^2. \tag{2.22}$$

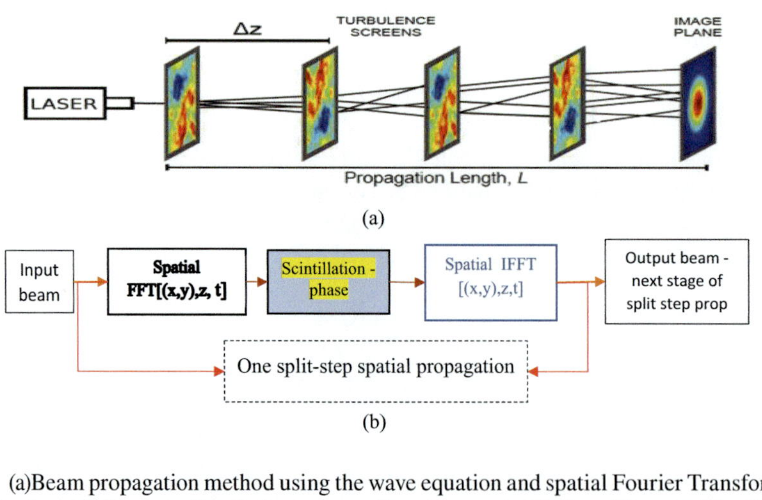

FIGURE 2.7 (a)Beam propagation method using the wave equation and spatial Fourier Transform with propagation step size $\Delta z < 1/3\ R$, where R is the beam spreading due to scintillation effects and the finite linewidth of the laser source. (Extracted from Ref.[21]); (b) split-step-FFT method (SSFFTM), with repeating propagation steps to satisfy 1/3 Kyrov radius (Equation 2.21).

This is valid only for a purely single-frequency laser source. An additional factor of attenuation of beam spatial spreading must be considered as the exponential factor, as indicated by

$$G_T = \left(\lambda/_{4\pi L}\right)^2 * e^{\alpha(x,y)} \tag{2.23}$$

where x, y are the directions of spatial spreading in two dimensions. This represents the modified Friis equation.

In addition, the optical heads of the SPC link at the transmitter and the receiver end offer optical gain. Assuming the beam alignment is perfect with <5 μrad accuracy, the following gains are obtained.

The Tx antenna gain (on-axis) with truncation can be written as[17]

$$G_{\text{Tx}} = \frac{2}{\alpha^2}\left[e^{\alpha^2} - 1\right]^2 \left(\pi D/_{\lambda}\right)^2 \tag{2.24}$$

where $\alpha = D/\omega$ or ratio between aperture diameter and the optical operating frequency, the antenna Tx and reception are denoted as optical heads (OH), which involve a beam pointing and its control for stability of tracking and alignment with other optical OH. The gain given in Equation 2.24 indicates a strong dependence on the aperture diameter and the diffraction limit[18], following a Bessel function distribution both toward and away from the outer region of the aperture. A null degradation of such diffraction occurs at $\left[\pi D/_{\lambda} \sin\theta\right] = 3.83$ with optical aperture diameter D and the angular width $\theta = 1.22\,\lambda/D$. Thus, there is an optimal optical aperture for both receiving and transmitting antennas. Typical diameters are 60, 80, 120, and 162.5 mm, depending on specific applications. In addition to commonly known satellite orbits, there is presently a high demand for optical communications of high capacity between drones and space or ground terminals.

Similarly, the Rx antenna of the reception end is given by

$$G_{\text{RX}} = \left(\pi D/_{\lambda}\right)^2 \tag{2.25}$$

Then the combined gain of Tx and Rx for a free space optical link (FSOL) can be approximated as

$$G_{\text{FSOL}} = \frac{2}{\alpha^2}\left[e^{\alpha^2} - 1\right]^2 \left(\pi D/_{\lambda}\right)^2 \left(\lambda/_{4\pi L}\right)^2 \left(\pi D/_{\lambda}\right)^2 \sim \frac{D^4}{\lambda^2\, L^2} \tag{2.26}$$

This shows a quadratic exponential proportionality with the aperture diameter and is inversely proportional to the quadratic factor of wavelength and the link distance L. The link distance depends on the link scenarios such as LEO-LEO, MEO-MEO, LEO-MEO, LEO-GEO, MEO-GEO, and possibly from these orbital satellites to and from drones.

2.5.2 Index of Refraction of Turbulence

Channel modeling in FSOC is mainly defined by the index of refraction turbulence (IRT) structure parameter C_n^2, which is a parameter that quantifies the atmospheric turbulence strength. Although this parameter is usually given as a function of height above ground level, in this work, C_n^2 will be taken as a constant value, as is the case for horizontal links.

2.5.2.1 Simulation Techniques and IRT

A simulation platform[22] used to simulate the propagation of lightwaves over the atmospheric layer is shown in Figure 2.8. The platform employs the commonly known beam propagation method (BPM – split-step Fourier technique – or split-step FFT) with turbulence applied to generate the phase

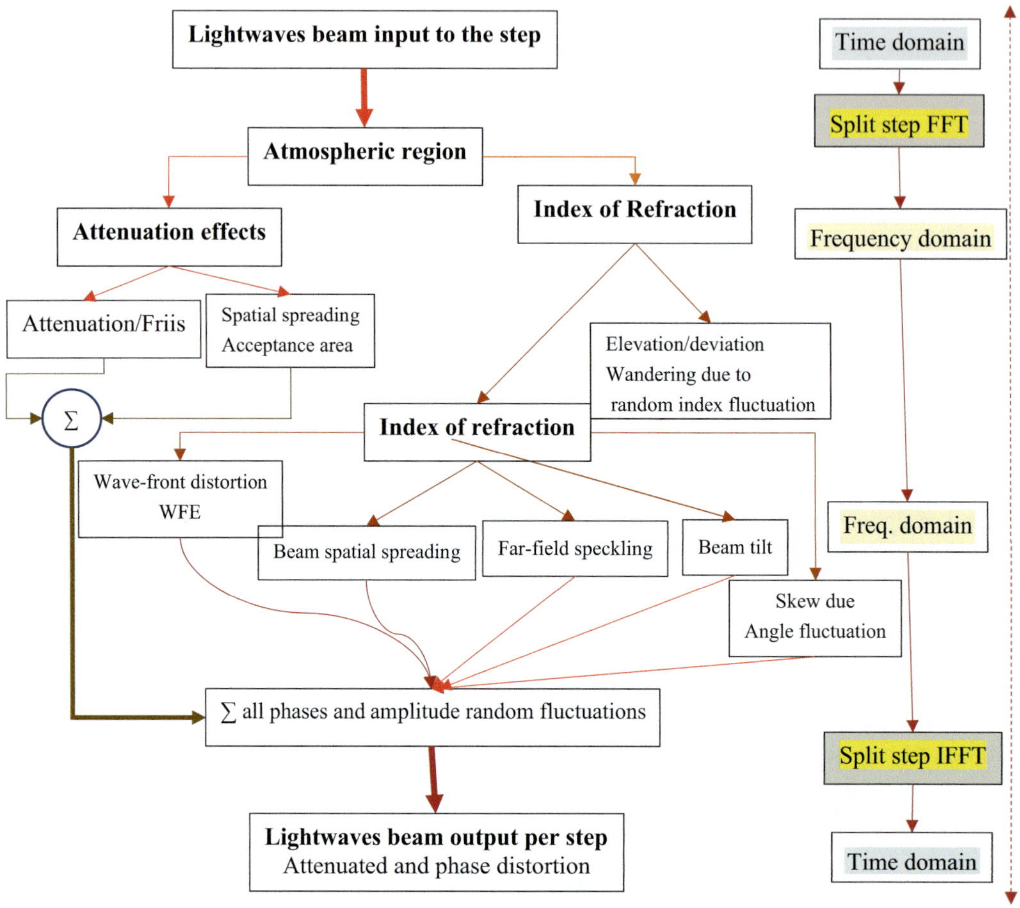

FIGURE 2.8 Effects of optical turbulence on lightwave propagation through free-space optical links. Effects integration in beam propagation, simulated by the split-step method with turbulence effects inserted at the middle of the step.

and spatial turbulence, then taking the IFFT to obtain the output of the propagation step. The split of the step into two halves and injection of the phase and amplitude evolution at the middle allow the spatial broadening and its phases of turbulence and the finite linewidth of the carrier and modulated channel. Either the Helmholtz or Schrödinger equation can be derived from Maxwell's equations and the IRT. The Schrödinger equation offers much more accurate outcomes when nonlinear phase is introduced along the propagation path. In vacuum space links, such nonlinearity seldom occurs. However, over the atmospheric region, the air molecules would face such minute nonlinearity.

As the FSO communication system will operate in desert-like environments, the values chosen for the channel model were extracted from recent literature.[19] In Figures 2.3–2.6, it can be observed that different C_n^2 values are reported for horizontal paths in Neveg Desert, southern Israel, and the area of China Lake in California, USA.[4] It is readily seen how the C_n^2 values directly depend on the time of day, as this parameter is greatly influenced by the energy mixing occurring at the ground-air interface, which in turn is governed by the amount of solar radiation. From the IRT parameter values reported in Refs.[3,4], three values are chosen as representative for the night and daytime conditions. For nighttime, one value is selected, namely $C_n^2 = 5 \cdot$

The atmospheric turbulence-induced fading[20] can be classified, depending on the value of the Rytov variance σ_R^2 from weak to strong turbulence conditions.

- the *weak* fluctuations regime occurs when $\sigma_R^2 \ll 1$,
- the *strong* fluctuations regime is associated with $\sigma_R^2 \gg 1$, and
- *moderate* fluctuations when for $\sigma_R^2 \sim 1$.

Moreover, the saturation regime results when $\sigma_R^2 \to \infty$. The Rytov variance is given by $\rho = \left(1.46k^2C_n^2L\right)^{-3/5}$, as mentioned above. Additionally, the saturation regime results when $\sigma_R^2 \to \infty$, thereby the Rytov variance is given by

$$\sigma_R^2 = 1.23C_n^2k^{7/6}L^{11/6} \tag{2.27}$$

In the platform PILAB report on atmospheric transmission, three-link distances are considered, namely 10, 20, and 30km, which along with the selected C_n^2 values for analysis produce a wide variety of turbulence conditions, spanning from weak turbulence to the saturation regime, as shown in Figure 2.10.

In long range simulation scenarios, some *constraints* must be imposed to the propagating optical field to continue producing physical meaningful results while keeping computational efforts and simulation time to reasonable values. The main constraint that needs to be observed is to maintain a minimum spatial-transverse sampling interval δz through the propagation path, which can accurately represent the optical traveling field and the atmospheric distortions in every step. On the one hand, the spatial coherence radius is the scale that indicates the transversal correlation length of the propagating wave, and it is given by

$$\rho_0 = \left(1.46\,k^2C_n^2L\right)^{-3/5} \tag{2.28}$$

where $k=2\pi/\lambda$ is the wavenumber, and λ is the optical operating wavelength.

Constraint 1: The simulation platform[21] automatically sets the minimum sampling interval in the propagation window to $\Delta x<\rho_0/3$, to avoid aliasing issues during the step-by-step beam propagation. The dotted lines in Figures 2.3 and 2.4 indicate how the field extent would develop without the resampling operation, i.e., without an explicit constraint.

Constraint 2: A second restriction, which is also automatically handled by the simulation platform, tries to guarantee that the *energy of the transmitted wave* is not lost during propagation. To accomplish this, whenever the field is larger than the simulation window after a propagation step is performed, resampling is applied to keep the complete field within the simulation boundaries. Nevertheless, this is a constraint with less precedence than the one explained in the previous paragraph, and if the whole field cannot be kept entirely within the propagation windows to preserve the correct transverse-sampling interval, then the beam is set to continue growing without applying any resampling. Attenuation on the edges of the propagating window needs to be applied so that artificial bouncing of energy is not produced. In this approach, although wave energy is lost, the results for the central portion of the traveling wave are still available. This resampling is illustrated in Figure 2.9.

Constraint 3: Only one polarized mode is considered; that is, the lightwaves coming out of the amplifier are in a polarization monomodal fiber and polarized in a determined direction depending on the rotation of the axis of the polarization fiber. We can add the other polarized mode of the beam and observe the effects mainly, from the skew of the two polarized channels, when the dual-polarized QPSK modulation format is used.

2.5.2.2 Scintillation Index (SI)

A laser beam propagating through the atmosphere will be altered by refractive-index inhomogeneities. At the receiver plane, a random pattern is produced both in time and space. The irradiance

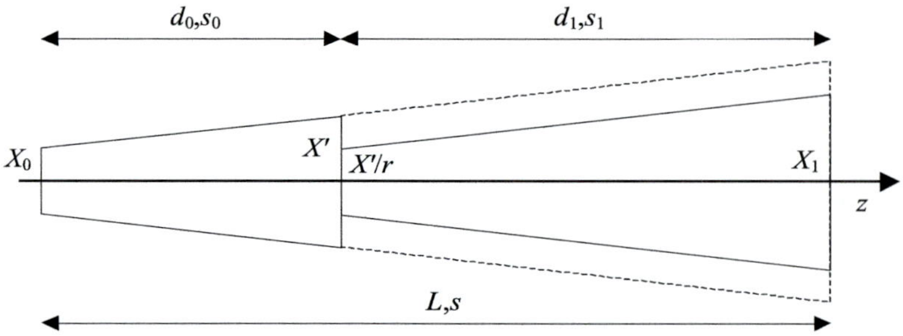

FIGURE 2.9 Re-sampling of the propagating field to maintain different constraints within acceptable ranges for spatial beam propagation (SSFM) under the constraints of spatial coherence radius distance under scintillation, given by Equation 2.27.

fluctuations over the receiver plane resemble the *speckle* phenomenon observed when a laser beam impinges on a rugged surface. The parameter that expresses these irradiance (beam intensity) fluctuations is the scintillation index (SI), given by Ref.[22] (first-order moment of the fluctuation).

$$\sigma_I^2 = \frac{\langle I^2\rangle - \langle I^2\rangle}{\langle I^2\rangle} = \frac{\langle I^2\rangle}{\langle I^2\rangle} - 1 \tag{2.29}$$

where $I=I(0, L)$ is the irradiance after the monochromatic (single mode narrow width laser) lightwave field propagates a distance L as the point of observation. The brackets $\langle\cdot\rangle$ indicate the expected value of the irradiance.

The scintillation index (SI), σ_I^2, of a point-like receiver in a horizontal link is given by

$$\text{For Gaussian point-like Rx}: \sigma_I^2 = \exp\left\{\frac{0.49\sigma_B^2}{\left(1+0.56[1+\Theta]\sigma_B^{12/5}\right)^{7/6}} + \frac{0.5\sigma_B^2}{\left(1+0.69\sigma_B^{12/5}\right)^{5/6}}\right\} - 1 \tag{2.30}$$

where Θ is the parameter describing the *beam phase front radius* at the receiver plane, and σ_B^2 is the Rytov variance for *Gaussian* beams. Their corresponding expressions can be found in Chapter 9 of Ref.[23] Note that the expression given in Equation 2.30 does not include the effects of beam wandering.

For a receiving aperture with *finite aperture* D_{Rx}, the scintillation index can be given by
For finite aperture D_{Rx}

$$\sigma_{\mathrm{nX}}^2(D_{\mathrm{Rx}}) = \frac{0.49\left[(\Omega_G-\Lambda)/(\Omega_G+\Lambda)\right]^2\sigma_B^2}{\left\{\left(1+\dfrac{0.4[2+\Theta](\sigma_B/\sigma_R)^{12/7}}{(\Omega_G+\Lambda)\left(1/3-1/2\bar{\Theta}+1/5\bar{\Theta}^2\right)^{6/7}}\right)+0.5(1+\Theta)\sigma_B^{12/5}\right\}^{7/6}}$$

$$\sigma_{\mathrm{nX}}^2(D_{\mathrm{Rx}}) = \frac{0.49\left[\dfrac{(\Omega_G-\Lambda)}{(\Omega_G+\Lambda)}\right]^2\sigma_B^2}{\left\{\left(1+\dfrac{0.4[2+\Theta]\left(\dfrac{\sigma_B}{\sigma_R}\right)^{12/7}}{(\Omega_G+\Lambda)\left(\dfrac{1}{3}-\dfrac{1}{2\Theta}+\dfrac{1}{2\Theta^2}\right)^{6/7}}\right)+0.56\sigma_B^{12/5}[1+\Theta]\right\}^{7/6}} \tag{2.31}$$

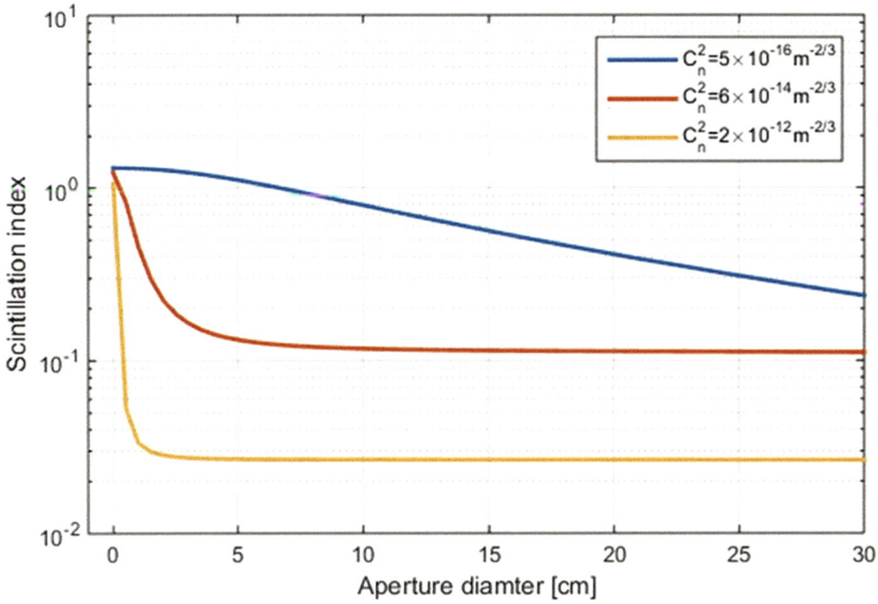

FIGURE 2.10 Scintillation index (SI) vs. receiving OH aperture diameter D_{Rx} for three different refractive-index structure (IRT) constant C_n^2 values. Laser Light MEO constellation aperture dia. = 162 mm, so the steady state of the IRT is reached.

$$\sigma_{nY}^2(D_{Rx}) = \frac{0.51\sigma_B^2(\Omega_G+\Lambda)\left(1+0.69\sigma_B^{12/5}\right)^{-5/6}}{\Omega_G+\Lambda+1.20\left(\sigma_R/\sigma_B\right)^{12/5}+0.83\sigma_R^{12/5}}$$

$$\sigma_{nY}^2(D_{Rx}) = \frac{0.51\sigma_B^2\left[(\Omega_G+\Lambda)\left(1+0.69\sigma_B^{12/5}\right)\right]^{-5/6}}{\Omega_G+\Lambda+1.20\left(\dfrac{\sigma_R}{\sigma_B}\right)^{12/7}+0.83\sigma_R^{12/5}} \tag{2.32}$$

where $\Omega_G = 2L/kW_G$ is a non-dimensional parameter defining the beam radius at the collecting aperture element, and W is related to the receiver aperture size as $W^2 = 8D_{Rx}^2$.

In Figure 2.10, the SI is calculated and plotted against receiving aperture of the optic head for the three different C_n^2 values selected in the previous section for analysis, with the link distance set to $L = 30$ km for the standard scenario. For the special case of $D = 0$ cm, it corresponds to a point receiver.

It is observed from Figure 2.10 that for such long-range propagation, the SI is worse than for the case of weaker turbulence strengths. Furthermore, for the two highest C_n^2 values, the improvement gains in SI by increasing the receiving aperture size have a saturation threshold of about 50 mm, where no enhancement is obtained in terms of SI for apertures larger than this threshold. This effect has also been reported elsewhere. Considering the behavior predicted in Figure 2.11, hereafter only the two lowest C_n^2 values are chosen to perform the simulation analysis, for which the largest SI is expected and therefore more critical for system design.

Figure 2.11 presents the expected behavior of the SI for a varying link distance and two different refractive-index structure parameter values. Moreover, the SI is calculated for a point-like receiver and a 200 mm receiving aperture diameter. It becomes evident that for the lower C_n^2 value, in Figure 2.11a, the SI increases monotonically up to approximately 30 km, and then gets close to the peak value. On the other hand, Figure 2.11b clearly depicts the whole of SI values, peaking

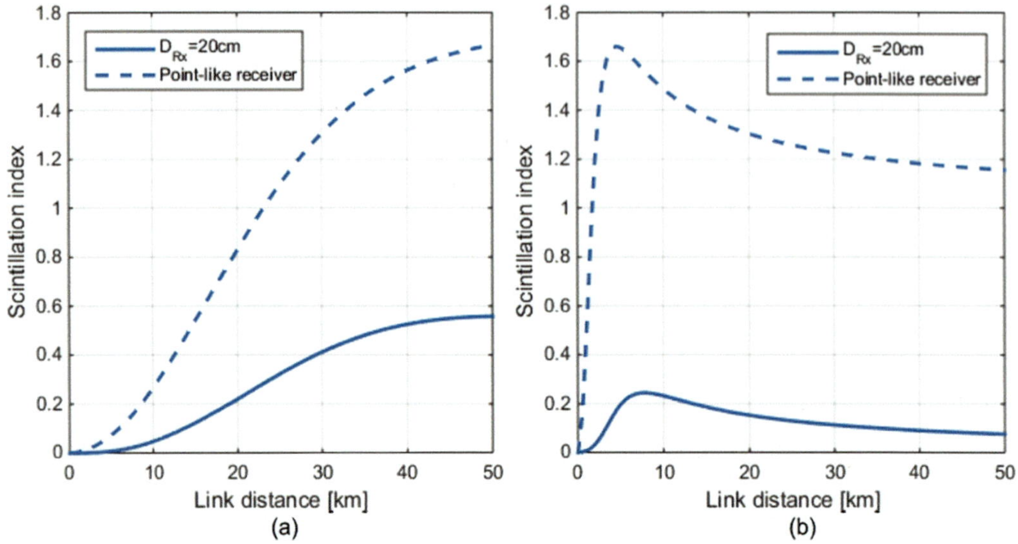

FIGURE 2.11 Scintillation index versus link distance for (a) $C_n^2=5\times10^{-16}\text{m}^{-2/3}$ and (b) $C_n^2=6\times10^{-14}\text{m}^{-2/3}$. Point-like receiver (dashed line) and a receiving aperture diameter of 200 mm (solid line) performances are shown.

at about 5 km and slowly approaching saturation after 40 km. As the standard scenario is set to 30 km, one should expect higher scintillation indices in night-time conditions – when the C_n^2 is small – and lower SI values for daytime conditions when the C_n^2 is high.

2.5.3 Simulation of Beam Propagation[22]

To use the simulation platform, a wave-optics code was prepared to perform simulations of different and selected scenarios. Link distances were set to $L=5$, 10, 30 km, and two refractive-index structure parameter values were selected for the analysis, namely $C_n^2=6\times10^{-14}$ and $5\times10^{-16}\text{m}^{-2/3}$. Moreover, for each combination of link distance and C_n^2 value, the scintillation index (SI) can be obtained for the receiving optical head with aperture diameters ranging from $D=0$ mm (i.e., point receiver) to 300 mm. Due to the stochastic nature of the intensity at the receiver plane, 500 realizations with transverse plane digitized sizes of 512×512, 1024×1024, and 2048×2048, depending on link conditions, are executed for each scenario to estimate the scintillation index.

Figure 2.12 presents all the results of the simulation process, where the scintillation index is plotted versus the receiving aperture diameter. For simulated data, the SI is calculated using Equation 2.31, while the theoretical SI is calculated through Equation 2.32 and related equations. Figure 2.12a–c corresponds to a refractive-index structure parameter $C_n^2=5\times10^{-16}\text{m}^{-2/3}$, and Figure 2.12d–f $C_n^2=6\times10^{-14}\text{m}^{-2/3}$. The arrangement of the plots is such that the results for all scenarios are organized from weak turbulence to the saturation regime, passing through moderate-to-strong turbulence conditions.

The first thing to notice when analyzing Figure 2.12 is that simulation results and theoretical predictions match very well in the weak-to-moderate turbulence regime. Nevertheless, for Rytov variances higher than unity, the theoretical curves predict a somewhat lower scintillation index than the simulations. This is particularly evident in Figure 2.12c and d, well into the strong turbulence regime, where the simulation data shows an SI about two times larger than the theoretical values.

The observed mismatch in the moderate-to-strong turbulence cases, as seen in Figure 2.12c–e, can be explained by the fact that the theory used to obtain Figure 2.13a and b–f only deals with scintillation processes and *does not* take into account additional effects, such as those derived from

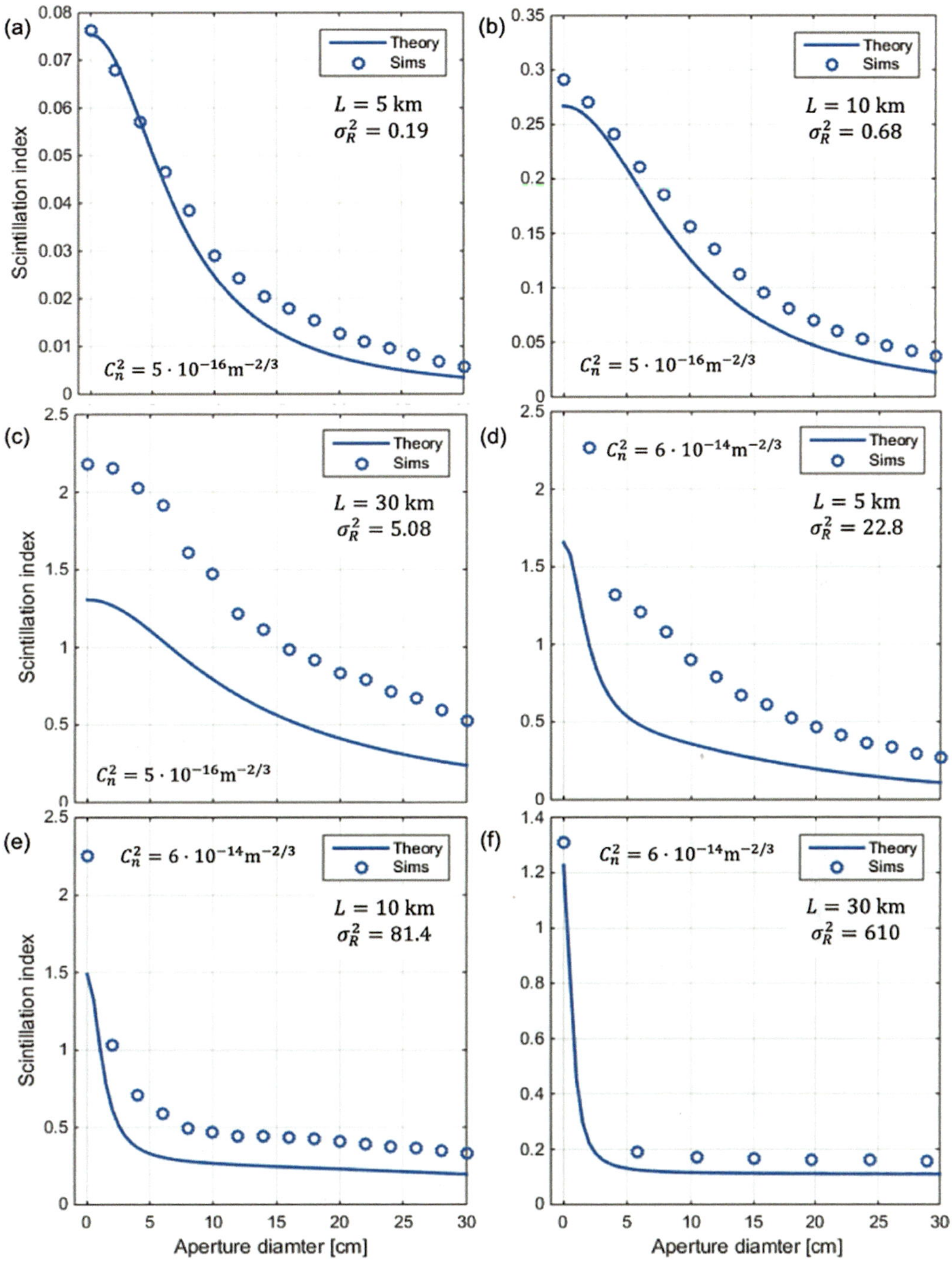

FIGURE 2.12 Scintillation index versus receiving aperture diameter for link distances L=5, 10, 30 km, and refractive-index structure parameter (a)–(c) $C_n^2=5\times10^{-16}\text{m}^{-2/3}$ and (d)–(f) $C_n^2=6\times10^{-14}\text{m}^{-2/3}$. Theoretical predictions are plotted as solid lines and simulation data as circles. (Extracted from Ref.[22]). Continuous line=theoretical; dotted circles: simulated.

beam wander – random displacements at the receiver plane – that would effectively increase the SI value. For such scenarios, the simulation results are more trustworthy, as the optics code includes phase stochastic variations over the traveling wavefront phase, which are responsible not only for purely scintillation but also for beam wander, beam spreading, and angle-of-arrival fluctuations.

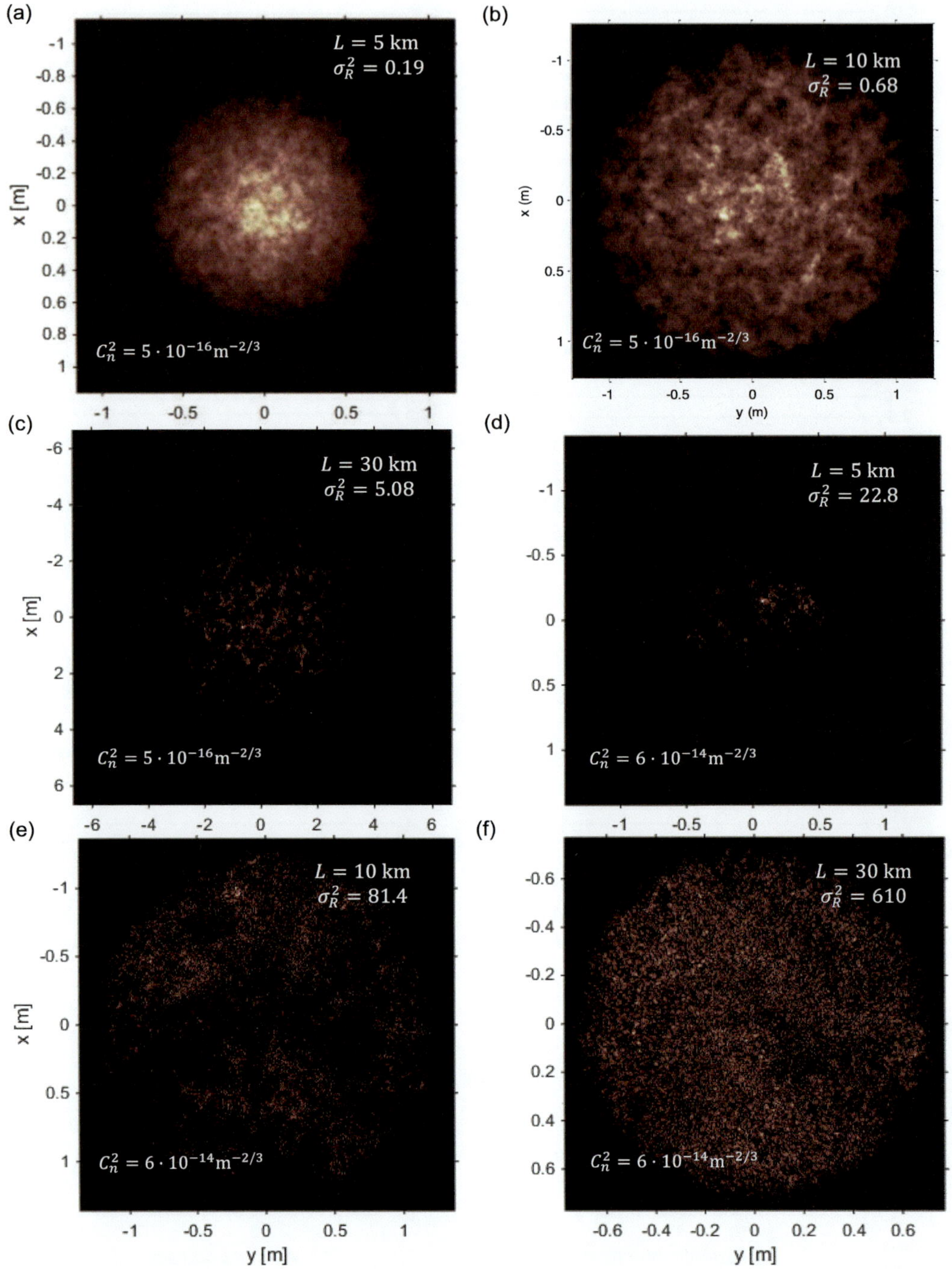

FIGURE 2.13 Intensity spatial distribution for link distances L=5, 10, 30 km, and refractive-index structure parameter (a)–(c) $C_n^2 = 5\times10^{-16}\mathrm{m}^{-2/3}$ and (d)–(f) $C_n^2 = 6\times10^{-14}\mathrm{m}^{-2/3}$. (Extracted from Ref.[22])

Figure 2.13 shows the spatial intensity distribution at the receiver plane for the same scenarios presented in Figure 2.12. It can be readily seen how the intensity profile breaks up into several hot spots across the beam in the moderate-to-strong turbulence regime, as observed in

Figure 2.13c and d. In contrast, under weak turbulence conditions, the profile remains still Gaussian, although scintillated. It is noteworthy that simulation results and theoretical predictions

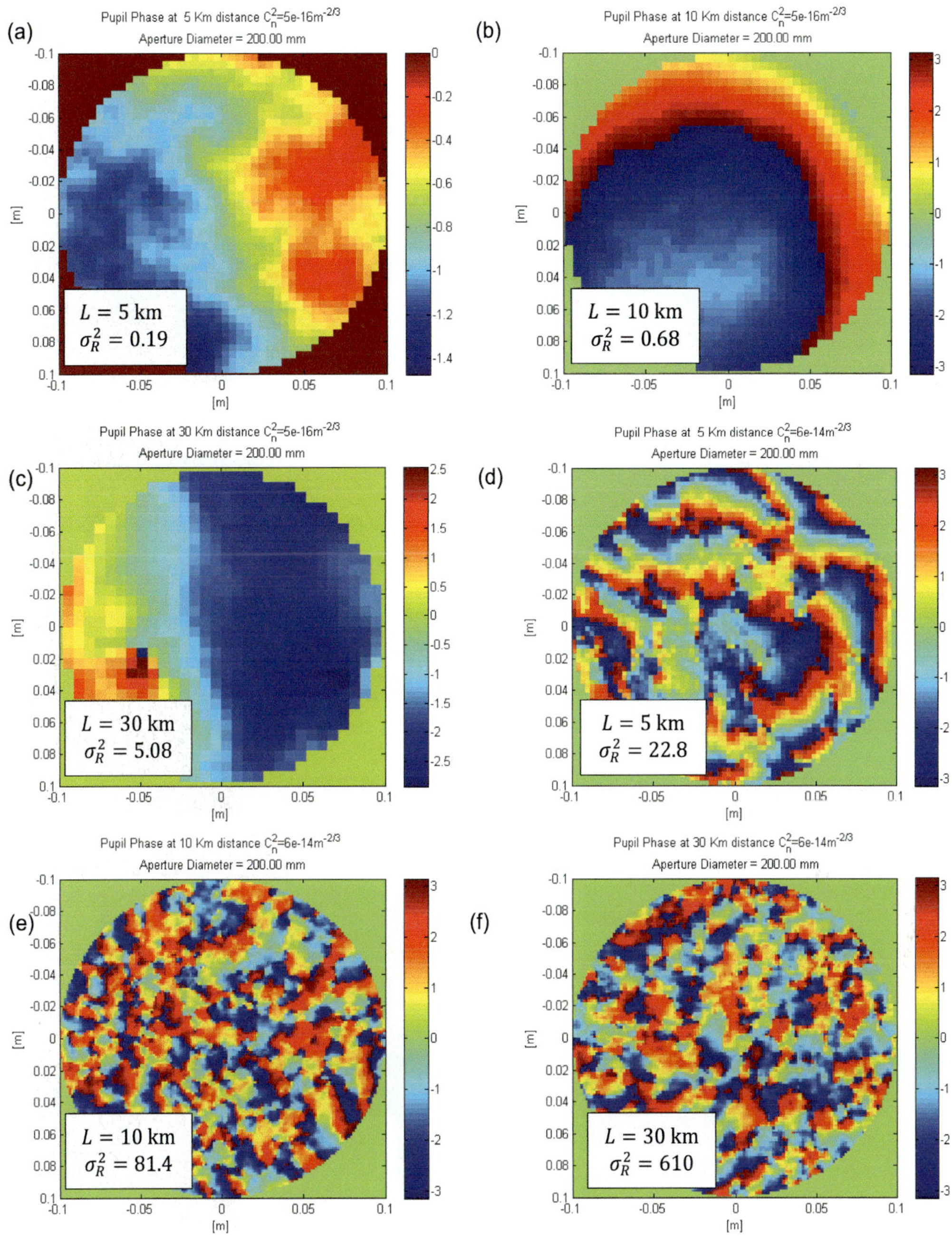

FIGURE 2.14 Pupil phase for a 200 mm Rx aperture at L=5, 10, 30 km and refractive-index structure parameter σ_R^2 (as inserted) with (a)–(c) $C_n^2=5\times10^{-16}\,\mathrm{m}^{-2/3}$ and (d)–(f) $C_n^2=6\times10^{-14}\,\mathrm{m}^{-2/3}$. Note: laser light aperture = 162 mm. (Extracted from Ref.[22])

align again in the saturation regime – refer to Figure 2.12e and f. This phenomenon is easily explained when the beam size at the receiver plane is much larger than the aperture size and the spatial intensity distribution is fairly homogeneous – see Figure 2.13f – making beam wander effects negligible, and the SI is once again driven purely by scintillation processes.

Figures 2.14 and 2.15 illustrate the image of the beam cross-section under the BPM (beam propagation method, i.e., SSFFTM, the Split-Step Fast Fourier Transform Method) with the SI variation

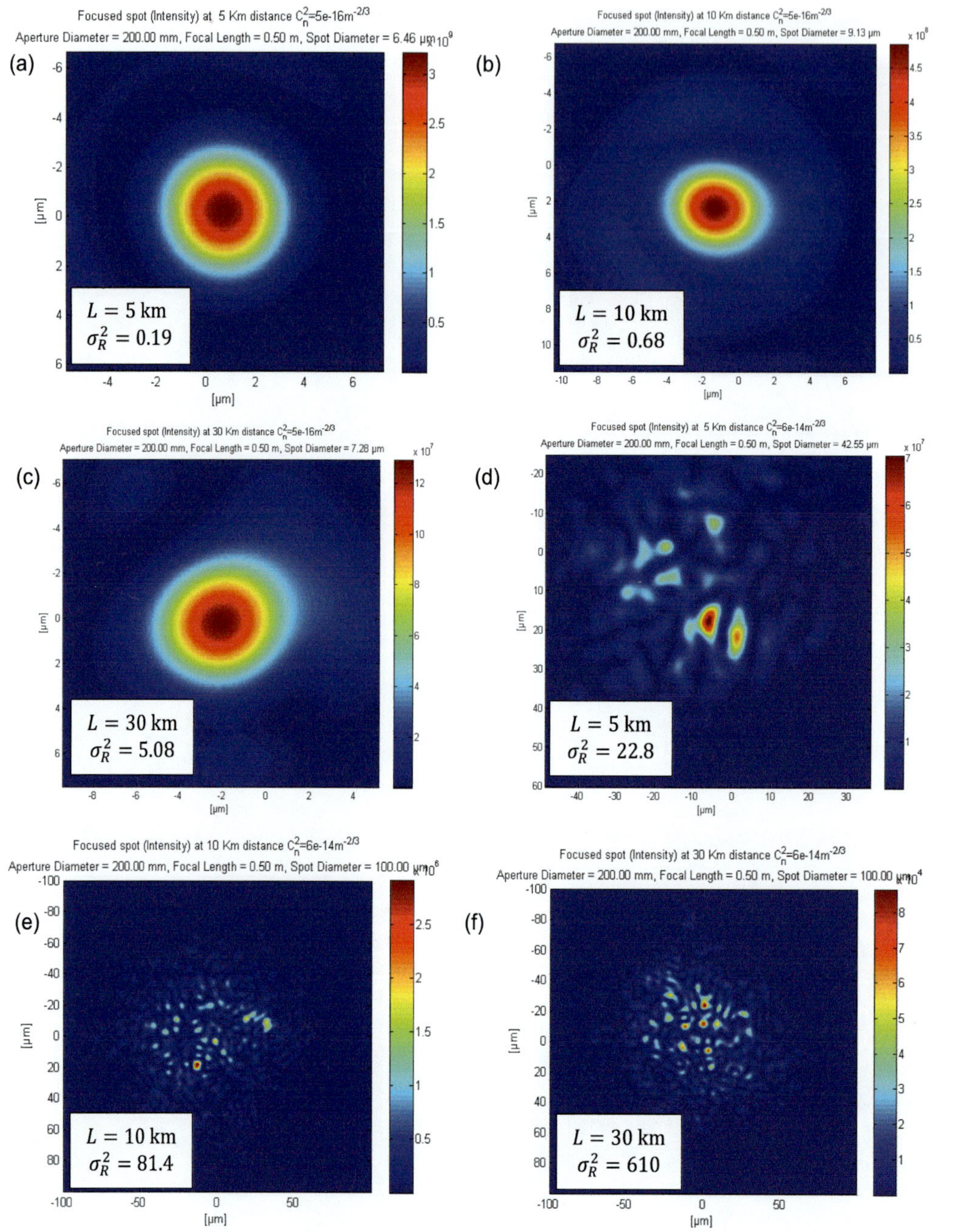

FIGURE 2.15 Focal plane for a 200 mm Rx aperture and an effective Focal Length (EFL) of 500 mm at $L=5$, 10, 30 km, and refractive-index structure parameter (a)–(c) $C_n^2=5\times10^{-16}\,\text{m}^{-2/3}$ and (d)–(f) $C_n^2=6\times10^{-14}\,\text{m}^{-2/3}$. IRT or Scintillation index. (Extracted from Ref.[22])

at different variances and propagation lengths, as indicated in the insets. When the variance is >22.8, the disturbed speckles become seriously affected, and the recovery of such a beam, thus the bit-error-rate (BER), may not be possible to recover, even with FEC.

The variation of the refractive index and the phase evolution is in a small region, that is the variation is much smaller than the signal amplitude and phases. In case the variation is large and approaching the nonlinear region, then the Schroedinger equation must be employed.

It is noted that the intensity distribution (Figures 2.14 and 2.15) of the beam profiles is broadened and rotated when the SI is not too large. This shows that the amplitude and phase of the beam are shifted in the 2D plane. Thus, this broadening and rotation affect the coherent reception of complex modulation such as the M-ary-QAM, when $M=4$, it is QPSK and when $M=16$, then it is QAM16.

2.5.4 Receiver Dynamic Range

The dynamic link budget depends on the receiver sensitivity under a certain bit-error rate and the dynamic behavior of the link under scintillation, pointing error, or beam alignment to set the line of sight of OGS to the MEO optic head and vice versa.

Loss caused by signal fluctuations due to turbulence or system effects can be estimated by p_{thr}, as the probability of fading as

$$\text{Loss} = \frac{F_P^{-1}(P_{\text{thr}})}{P_0} = \frac{P_{\min}}{P_0} \tag{2.33}$$

where $F_P^{-1}(\cdot)$ is the inverse cumulative probability function of the P variable, p_{thr} is the probability that P goes below a certain level P_{min}, and P_0 is the power level under no perturbation. We can reasonably estimate losses at the 1%-probability fade, i.e., by setting $p_{thr}=10^{-2}$.

When the power-scintillation index σ_P^2 of the received optical signal is known, one can calculate a scintillation loss α_{sci} based on the assumption that a certain fraction of time the received power may drop below a threshold value p_{thr}, with p_{thr} typically between 1e-2 and 1e-6. An approximation of the scintillation *attenuation* can be as

$$\alpha_{\text{sci}} = \exp\left(-\text{erf}^{-1}\left(2\ p_{\text{thr}} - 1\right)\sigma_P\sqrt{2} + \sigma_P^2/2\right) \tag{2.34}$$

where erf^{-1} is the inverse of the Gaussian error function or complementary error function. This equation assumes a log-normally probability distribution of the received optical signal and an IM/DD detection system. Nevertheless, the log-normal distribution is no longer valid for stronger turbulences, and a gamma-gamma distribution is usually assumed in the literature.

Because of imperfections in the transmitter and receiver tracking systems, pointing losses occur. Pointing losses have a similar influence on the communication system compared to intensity fluctuations.

If coherent reception is employed, the imperfections result in phase fluctuations, which can then contribute to the spreading of the constellation points. Statistical modulation, as discussed in Chapter 9, can reduce these imperfections and the probability of error or the BER of the system. The phase fluctuation is considered as the product of the field amplitude of the LO laser and that of the carrier over one bit period.

As illustrated in Figures 2.16 and 2.17, a pointing error leads to a misplacement of the beam footprint at the receiver and therefore to *pointing loss*. It is assumed that the standard deviation of the pointing error on one axis, $\sigma_{pointing}$, is normally distributed. Calculating the loss due to pointing errors as a 10^{-2}-probability fade, the pointing loss $L_{pointing}$ can be evaluated according to Ref.[24]

$$L_{\text{pointing}} = 10^{-2\beta^2} \tag{2.35}$$

where $\beta=2\sigma_{pointing}/\theta_{div}$. Thus, for *a given pointing error*, the pointing loss d*ecreases* as the beam divergence increases. On the other hand, increasing the beam divergence *reduces the Tx-antenna gain*, i.e., the free-space loss. Optimizing free-space loss and pointing loss results in an optimum that appears at $\theta_9 \approx 5\sigma_{pointing}$, or equivalently at $\beta \approx 0.4$. However, we have chosen a more restrictive value for the pointing system with $\theta_{div} \approx 10\sigma_{pointing}$. Therefore, a pointing loss *of* $L_{pointing}=-1.39$ dB can be considered.

2.5.5 Estimation of Link Budget under Turbulence

2.5.5.1 Attenuation

The atmospheric attenuation of a collimated light beam means a reduction of intensity in the direction of propagation. It is caused by aerosols (fog, haze, clouds), air pollutants (smoke, pollen, i.e., particles whose diameter is larger than the operating wavelength), Rayleigh scattering from particles whose sizes are smaller than the operating wavelength and molecular absorption lines of the atmosphere's constituents. While the first three factors only vary slowly with wavelength, the molecular absorption lines are highly sensitive to the laser wavelength. In general, the laser wavelength must be chosen to avoid molecular absorption lines. This is the case for wavelengths from 1545 nm through 1565 nm, with some minor absorption from 1540 to 1650 nm.

Figure 2.16 illustrates an overview of the different effects that cause atmospheric attenuation. Due to wavelength-dependent absorption, atmospheric transmission windows exist, i.e., wavelength intervals with possible transmission. As the wavelength is set to minimize within a window of minor atmospheric attenuation, the overall atmospheric attenuation at lower ground level links is determined by scattering from hydrometeors during fog, snow, and heavy rain. This component is slow-varying and determines the overall link availability for ground-to-ground SPC (Space Photonic Communications) links.

2.5.5.2 Atmospheric Attenuation Calculation Based on Meteorological Conditions

The atmospheric attenuation can be expressed as a sum of absorption by atmospheric gases α_{AB}, by scattering on hydrometeors α_{SC}; thereby the power penalty of turbulence α_{SCINT} is given as

$$\alpha_{ATM} = (\alpha_{SC} + \alpha_{AB}) \cdot L_{12} + \alpha_{SCINT} \tag{2.36}$$

where $\alpha_{1,SC}$ and $\alpha_{1},_{AB}$ are the specific attenuation coefficients.

The ***atmospheric turbulence***, α_{SCINT}, can be estimated from the results of the atmospheric simulations as explained above. The atmospheric attenuation due to molecular absorption is negligible for both 850 and 1550 nm spectral windows. The scattering attenuation, which is 0.5 dB/km for the *standard clear* atmosphere, increases substantially during fog, snowfall, and rain. The predominant process for light attenuation by fog is the Mie scattering. Fog density is measured indirectly as the Meteorological Optical Range (MOR) or simply visibility range V_m (in meters).

The relation between A and the visibility range V_m for other wavelengths has been discussed in Refs.[4,5] as it depends on fog density and particle size distribution. Kim's formula[5] was used in this study, as practical observations confirm the wavelength-independence of attenuation.[6]

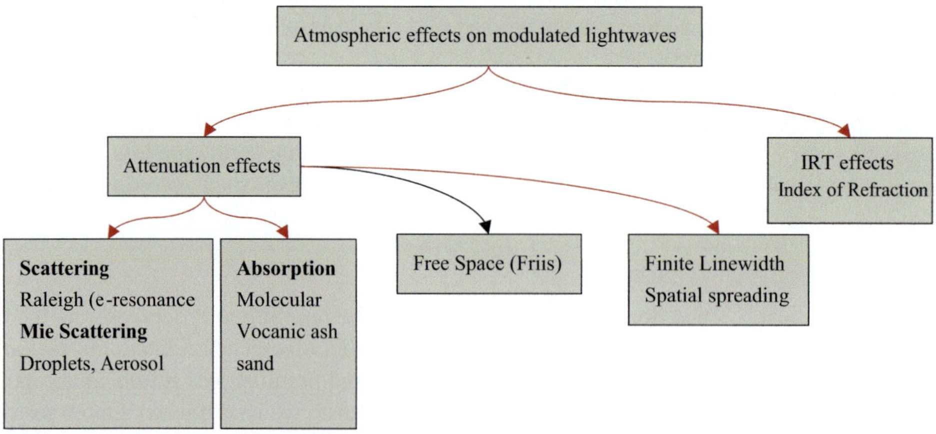

FIGURE 2.16 Estimation categories for atmospheric attenuation.

$$\alpha_{SC.fog} = 10\ln\left(\frac{\lambda}{550}\right)^{-q}\left(\frac{3.912}{V_m}\right) = \frac{4.61}{V_m}\ \text{dB/km} \tag{2.37}$$

where V_m is the meteorological visibility in km for $\eta=0.05$ and λ=operating wavelength. It is noted that this attenuation due to scintillation is taken as the average and absolute value of the random variation of the scintillation as defined by the IRT. This phase variation generates the deduction of the OSNR while the attenuation is evaluated for the overall power budget.

Regarding the atmospheric attenuation α_{ATM} as a random variable, the probability of link unavailability P_{out} can be expressed as

$$P_{out} = P(\alpha_{ATM} > M_1) = E\alpha(M_1) \tag{2.38}$$

where M_1 is the normalized link margin $M_1=M/L_1$, [dB/km] and $E\alpha$ is the cumulative exceedance probability of atmospheric attenuation α_{ATM}. If a fixed value of normalized link margin M_1 is considered, the increase of atmospheric attenuation causes fading events. Therefore, the overall link availability could be expressed as:

$$P_{link}(M_1) = 1 - P_{out} \tag{2.39}$$

2.5.5.3 Pointing and Tracking Estimation

Once the initial uncertainty is resolved, the optical loop is closed for both partners, and the tracking system is in charge of maintaining the pointing error below a certain threshold. This is achieved with the help of optical feedback signals from the optical tracking sensors. The key challenge of this document is to provide a tracking error budget that fulfills the requirements imposed by the communication link budget. Figure 2.17 shows the signal flows for the pointing and tracking error penalty methodology.

As will be described further in this chapter, the tracking sensors chosen for the proposed system design are four-quadrant detectors (4QD), which can detect the intensity of the lightwaves. They must be balanced so that the beam can be centered. As a rough and simple measure, the diodes can be connected back-to-back to offer additional gain. Alternatively, 8QD can be used with four pairs of back-to-back (B2B) connected detectors. The optical system has to be designed together with the tracking system to provide enough resolution for the tracking sensors. On the other hand, this resolution is highly dependent on the amount of optical power received by the sensors. Therefore, a tracking power budget calculation must be performed to ensure that the tracking sensors receive enough signal strength during all possible link scenarios.

Figure 2.17 depicts a diagram of the design methodology. The diagram shows an iterative design process that involves both tracking and optical system designs. A target spot size and field of view must be designed for the tracking sensor to fulfill the tracking requirements. The feasibility of these values has been checked through optical simulations. Figure 2.18 shows a typical ground-to-satellite pointing and control system for tracking and alignment with error <5 µ-rad.

The main output parameter is defined as the ***Noise Equivalent Angle*** **(NEA)**, which determines a boundary for the tracking accuracy of the system and represents the uncertainty of the pointing error estimation due to measurement errors, error vector magnitude (EVM), and phases.

It is noted that using a non-lensed fiber reduces the free space to fiber coupling efficiency. Using a GRIN lensed or SELFOC lensed fiber can be much more efficient. The methodology given in Figure 2.17 is given for non-lensed fiber of single-mode or multi-mode types. An estimated loss due to pointing is **−1.39 dB**.

Figure 2.18 illustrates the demand of pointing between the transmitter and reception point of optical head over a very long distance between satellites in the same or different constellations, or from OGS to satellite orbits LEO/MEO/GEO. The accuracy of only a few micro-radians is required

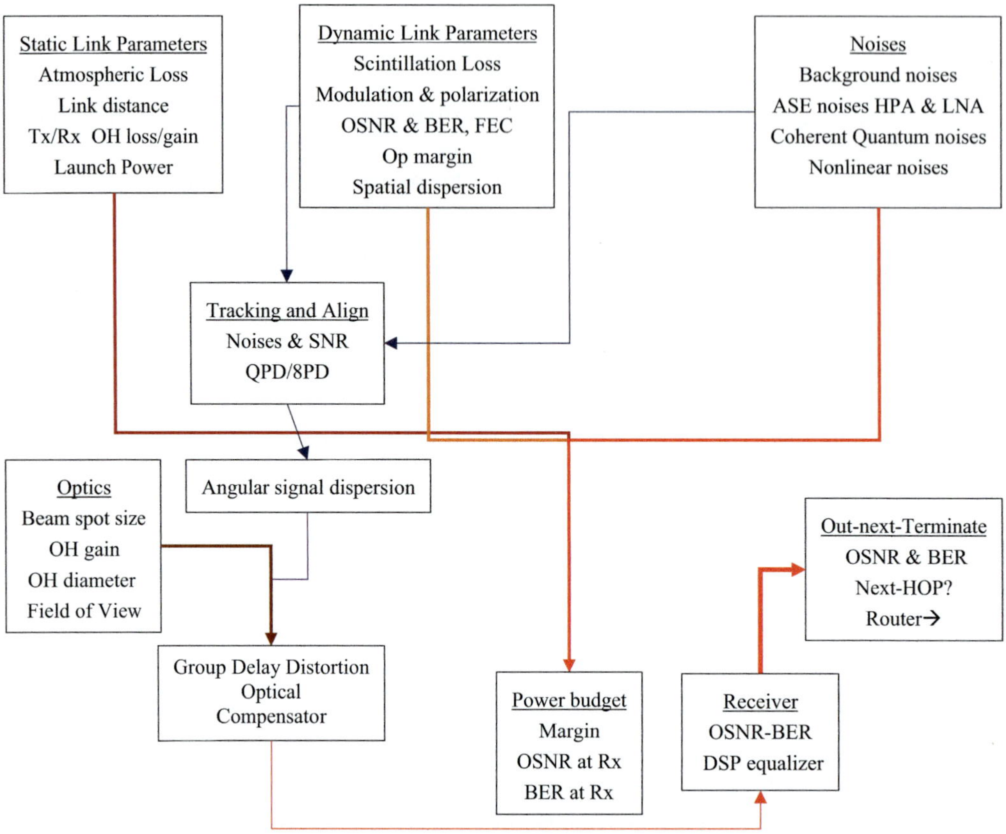

FIGURE 2.17 Tracking and pointing system and link design flow diagram. OH=optical head, OSNR=optical/signal-to-noise ratio, BER=bit-error-rate, FEC=forward error correction, RX=receiver, Tx=transmitter, DSP=digital signal processing, QPD=quad photodetector, 8QPD=octave photodetector.

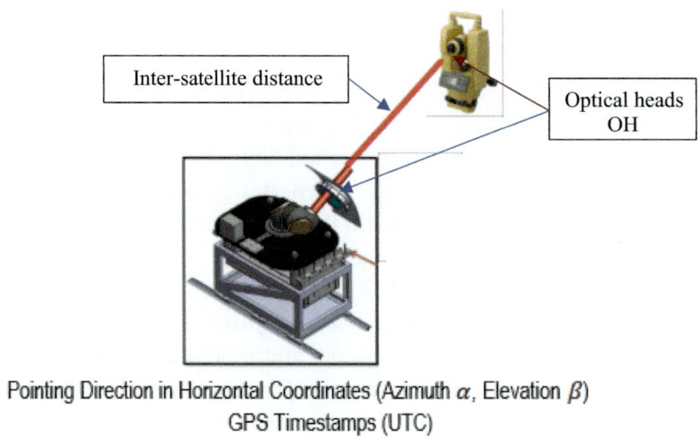

FIGURE 2.18 An arrangement of pointing with coordinate systems and the receiving end satellite positioning and beam alignment control system. GPS=global pointing system, OH=optical head.

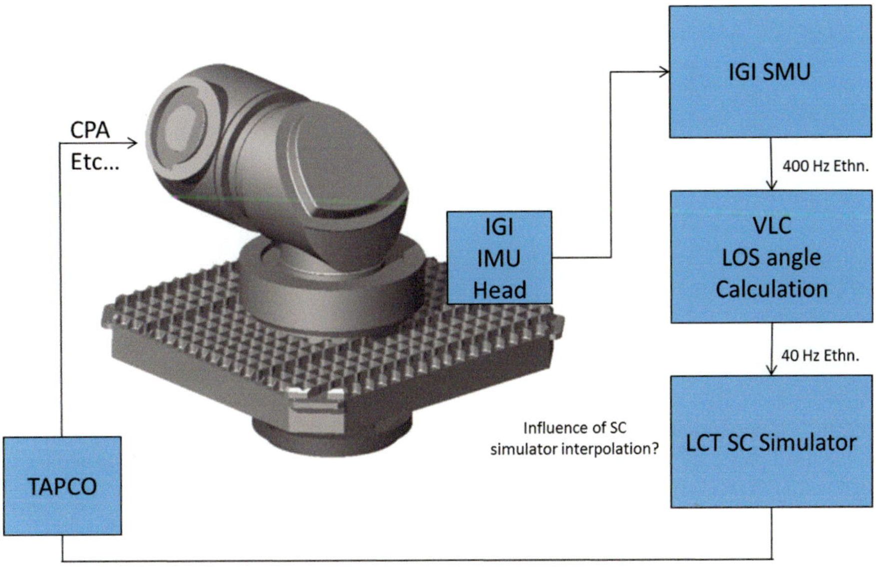

FIGURE 2.19 Schematic diagram of a pointing station of the optic head and control system. FFC= feed-forward control, LOS=lost of sight, IMU=inertial-measurement-unit, IGI=Ingenieur-Gesellschaft für Interfaces GmbH (manufacturer of IMU), CPA=coarse pointing assembly, VLC=a company name ViaLight Co, AVCS=Active Vibration Compensation System.

and thus a control with feedback stability must be achieved as shown in Figure 2.19. Accurate coarse and fine assembly mechanical platform and feedback control have been implemented in steering the beam and align with feedback.

Alternatively, the beam steering can be done using optical steering techniques such as the acousto-optic beam deflection using the Bragg phase matching or electro-optic beam switching, etc. These integrated optical devices can offer a much faster speed of beam steering and control. For high-speed data communication. The control and beam alignment are the critical part of the SPC system. If not aligned, a significant increase in the BER would occur.

The dynamic range of the receiver must then require a dynamic range due to the expected scintillation and pointing errors. A 16dB is expected for the NRZ IM/DD transmission system.[25] For PDM-QPSK 100 G, this dynamic range can be, only in a rigorous approach, about 6dB with the LO power level about 20dB higher than that of the received signal power.

2.5.5.4 Strehl Ratio (SR) and EVM (Error Vector Magnitude)

The Strehl ratio (SR) is defined as

$$\mathrm{SR} = \left|\left\langle e^{j\Delta\varphi}\right\rangle\right| \tag{2.40}$$

with $\Delta\varphi = \dfrac{2\pi\delta}{\lambda}$ being the phase error over the aperture at the operating wavelength λ, while the average (expected value) of the complex quantity inside the brackets is taken over the optic aperture $A(x, y)$. δ=the distortion propagation vector length. This quantity is thus a measure of the distortion of the wavefront of the lightwave beam propagating through an optical head or an atmospheric layer under disturbance or the IRT. In terms of the plane representation of a constellation of modulated lightwaves, the EVM and the wavefront error can be related, as illustrated in Figure 2.19.

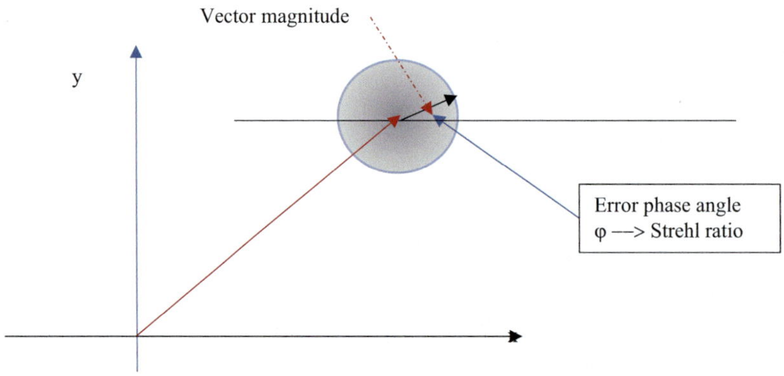

FIGURE 2.20 Representation of the phase error in the complex plane and a constellation point. Phase evolution from the ideal phase vector position in the complex plane constellation. Turbulence generates the spatial spreading, leading to phase front error.

As illustrated in Figure 2.19, the phase angle departed from the original phase vector. So, the SR is the magnitude of the exponential factor of the phase change. Thus, the phase change can be estimated by taking the natural log of the SR, which provides the phase change due to phase front distortion. For example, if SR=0.3 (or a loss=5.2 dB in NRZ) then the phase change –1.1 radians or a rotation of 1.1/3.1416×180°=57° angle of the arc. The non-distorted phase magnitude is determined by the Euclidean distance from the origin to the ideal constellation point. The small vector change shows the contribution of the distorted phase to the original signal constellation under sampling (Figure 2.20).

The SR, in contrast, is analogous to the normalized EVM with respect to the original non-distorted vector, as commonly defined in the signal processing of digital modulation format channels. This allows us to relate the turbulence in the SPC system under the transmission through the turbulence. The penalty in OSNR can be determined by setting a 1 dB penalty, thus setting the margin and the dynamic range of the optical coherent receiver. The SR can be related to the IRT. So, a modified SR is redefined the SR to match that of the EVM in log natural.

The SR can be estimated using only the statistics of the phase deviation, thus equivalent to the average/expected variation of the EVM, as

$$\mathrm{SR} = e^{-\sigma^2} \tag{2.41}$$

where σ is the root mean square (RMS) deviation of the wavefront phase across the optical aperture. This deviation can be represented as

$$\sigma^2 = \left\langle (\varphi - \bar{\varphi})^2 \right\rangle \tag{2.42}$$

where φ and $\bar{\varphi}$ are the instantaneous and average original phase vectors of the wavefront under non-distortion and distortion, respectively.

It is observed that SR=1 or $\sigma^2 = 0$when there is no turbulence, and SR=0 or $\sigma^2 = \infty$ when there is high turbulence.

2.5.5.5 Main Difference between SR and EVM

The SR and EVM are both measures of the quality of an optical wavefront. They are used to quantify how well an optical system can focus light and are commonly used to evaluate the performance of laser systems, telescopes, and other optical systems.

The SR is a *dimensionless* measure of the quality of an optical wavefront. It is defined as the ratio of the peak intensity of the wavefront produced by an optical system to the peak intensity of an ideal wavefront produced by the same system. The SR can be used to evaluate the performance of an optical system by comparing the actual wavefront to the wavefront that would be produced if the system were perfect. A SR of 1 indicates that the wavefront is perfect, while a *lower* value indicates that the wavefront is distorted.

On the other hand, the EVM is a metric that measures the *quality of a modulated signal*, particularly in a free-space optical coherent communication system. It is a measure of the difference between the ideal and the actual signal. In other words, it measures the quality of the signal on the receiver side and describes how closely the received signal matches the ideal signal. EVM is expressed as a percentage, with lower values indicating a better-quality signal.

The SR, in this sense, is indeed an exponential function of the phase fluctuation, while EVM is the expected value of the variation in the phase magnitude relative to the original phase vector. Thus, if the SR ratio is known, the EVM can be derived by taking the natural logarithm of the phase difference, which is derived from the SR. This EVM can then be related to the OSNR penalty. This penalty amount must be accounted for in the power link budget of the SPC system.

In summary, the SR is a measure of the quality of an optical wavefront, comparing the actual wavefront to the ideal wavefront that would be produced by a perfect optical system. The EVM is a metric that quantifies the quality of a modulated signal, specifically in an optical communication system, by measuring the difference between the ideal and the actual signal on the receiver side. Both quantities can be related to the OSNR of an optical communication system.

2.5.5.6 Modified Strehl Ration (MSR)

In order to obtain a parameter that can be related to the EVM, we need to redefine the SR. The SR is defined by the difference in phase between the original or unperturbed phase and the perturbed phase due to turbulence and scattering. The EVM is defined as the ratio between the phase difference and the original or ideal phase, expressed asa percentage, to measure the variation of the spatial phase from the ideal phase position. We therefore can redefine the SR with a modification, referred to as the modified SR (MSR), as

$$\mathrm{SR}_{\mathrm{lin}} = \left| \left\langle e^{j\frac{\Delta\varphi}{\phi}} \right\rangle \right| \rightarrow \frac{\Delta\varphi}{\bar{\varphi}} = \log_e(\mathrm{MSR}) = \mathrm{EVM} \times 100\% \tag{2.43}$$

where $\bar{\varphi}$ is the original phase. Thus, the relative phase ratio can be related to the EVM by taking the natural logarithm of the MSR, in the positive sense.

2.5.6 Determination of Attenuation of Transmission and IRT for Power Budget[26]

2.5.6.1 Transmission Loss

Transmission loss for distances greater than Fraunhofer diffraction:

$$P_{\mathrm{Rx}} = \frac{P_{\mathrm{Tx}} G_{\mathrm{Tx}} A_{\mathrm{Rx,eff}}}{(L\lambda)^2}$$

$$\text{Then} \quad P_{\mathrm{Rx}} = \frac{P_{\mathrm{Tx}} G_{\mathrm{Tx}} \cdot 4\pi\, \lambda^2 A_{\mathrm{Rx,eff}} / (\lambda^2)}{4\pi \cdot (L)^2} \rightarrow \frac{P_{\mathrm{Rx}}}{P_{\mathrm{Tx}}} = G_{\mathrm{Tx}} G_{\mathrm{Rx}} \left(\frac{\lambda}{4\pi L} \right)^2 \tag{2.44}$$

with the receiver antenna optical gain (G_{Rx}) effective area $G_{\mathrm{Rx}} = 4\pi \cdot \lambda^2 A_{\mathrm{Rx,eff}}$.

Therefore, the attenuation $\alpha_{L,\mathrm{lin}} = \left(\frac{\lambda}{4\pi L} \right)^2$ linear scale

Thus, the propagation (spreading) loss in dB (L distance) is

$$\alpha_L = \left(\frac{\lambda}{4\pi L}\right)^2 \rightarrow \alpha_{L,\mathrm{dB}}(\mathrm{dB}) = 20 * \log_{10}\left(\frac{\lambda}{4\pi L}\right) \rightarrow \text{or in per unit length}(m) = 20 * \log_{10}\left(\frac{\lambda}{4\pi}\right)\mathrm{dB} \tag{2.45}$$

Example: For OGS to LEO, the total attenuation can be given as L=8000 km (90°), λ=1550 nm. Then α (dB)=–276 dB loss; at a 20° inclination, the alpha attenuation is –277.1 dB.

2.5.6.2 IRT, SR and Attenuation

Degradation by turbulence SR is given by

$$\mathrm{SR} = e^{-\sigma_\phi^2} \tag{2.46}$$

σ_ϕ^2 is the residual phase variances. Physically, this means that SR measures the magnitude of the square of the magnitude of the phase front fluctuation vector due to random turbulence or the turbulence coefficient C_n^2. Under general turbulence propagation, SR can be rewritten as

$$\mathrm{SR} = \frac{1}{\left(\left[1+\left(D/r_0\right)\right]^{5/3}\right)^{6/5}} \tag{2.47}$$

D=diameter of the optical heads (OH), at both Tx and Rx, and r_0 is the Fried parameter. The coherence length can be defined as

$$r_0 = \left[0.42\sec(\zeta)k^2 \int_0^{L_T} C_n^2(h)\,dh\right]^{-3/5} \tag{2.48}$$

where ζ=the zenith angle, k=wave number in vacuum=$2\pi/\lambda$ (lambda is the operating wavelength in vacuum), h=height above the ground level, L_T is the distance of the turbulence region, $C_n^2(h)$= the IRT coefficient ($C_n^2(h)$ range 10^{-12} to 10^{-13}) and h=integration step. SR is in the range of 10^{-12} to 10^{-13}.

The coherence length r_0 and Fresnel number F_E can be simply approximated as

$$r_0 \sim \frac{\lambda}{B} \quad \text{and} \quad F_E \sim \frac{r_0^2}{Z\cdot\lambda} \tag{2.49}$$

where

B=scattering angle of the sinusoidal fluctuation of the scintillation thin layer, as illustrated below; Z=scintillation layer altitude (0–20 km); OPD=optical path delay, $h_j(x)$ distribution of scintillation along x of transverse plane. B is measured in degrees of arc, typically, a few seconds of arc.[27]

The scintillation layer can be illustrated in [Ref.[27], Appendix C] and Figure 2.21. The coherence length can be approximated by B=1.0 arcsecond, 1=1550 nm, thus, the Rykov radius → r_0=1550e-6/87e-5=1.78 m. The Fresnel number is 454.

For SR=0.32, the loss is –4.9 dB, or an EVM of 32%. The coherence length is inversely proportional to the linewidth of the laser, so the narrower the linewidth, the longer the coherence length and thus the SR reaching unity. Thus, any phase fluctuation must be less than unity to generate much smaller turbulence phase effects on the OSNR.

The coherence length of the lightwaves under *turbulence is less than a few meters*, so the autocorrelation of the lightwaves over the propagation length can be[28] measured and estimated, as shown

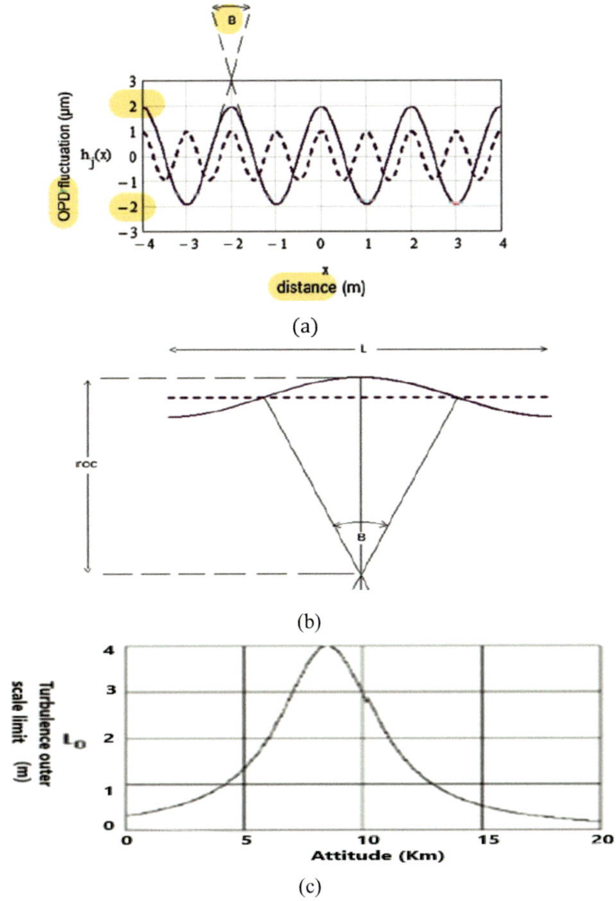

FIGURE 2.21 (a) Scintillation spatial sinusoidal distribution with scattering angle; (b) zoomed overview. The radius of curvature at the center of a sinusoidal wave portion, roc, approximately determines where the central wave portion comes to a focus and where we could expect to see strong scintillation. The distance, roc, scales linearly with the turbulence structure period, L, and inversely with the wave scatter angle, B, (c) turbulence in a thin layer over the distance altitude.

in Figure 2.22. Thus, the laser coherence length, $L_{C,DFB} = \dfrac{c}{\pi \Delta\nu}$, with $\Delta\nu$ as the laser linewidth in Hz and c as the lightwave velocity in vacuum, must be shorter than the turbulence coherence length. The linewidth of the lightwaves carrier <100 kHz to give a coherence length of 30 km, ideally 10 kHz for 300 km, which is also required for coherent modulated lightwaves. The laser coherence length must be much longer than the turbulence coherence length. If not, then much more phase fluctuation is generated, leading to phase noise or the lightwaves.

2.5.7 Doppler Shift

The Doppler shift f_B can be approximated, with the tangential direction of the velocity of the satellite, as

$$f_B = f_0 \frac{\sqrt{1 - \dfrac{\Delta\nu^2}{c^2}}}{1 - \dfrac{\Delta\nu}{c}\cos(\alpha)} \tag{2.50}$$

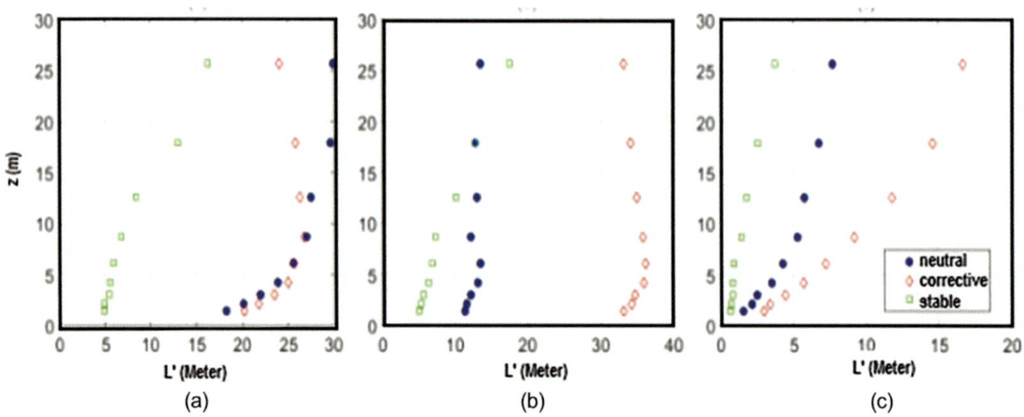

FIGURE 2.22 Turbulence length scales, L_{ix}, of the (a) streamwise u, (b) spanwise v, and (c) vertical w_velocity components, averaged for each of the three different stability conditions in Table 2.1, as a function of height in the TEST atmospheric surface layer (See Refs.[28,28]). Stability markings (Medium dark dot=neutral, light dark circle=corrective, and light circle=stable).

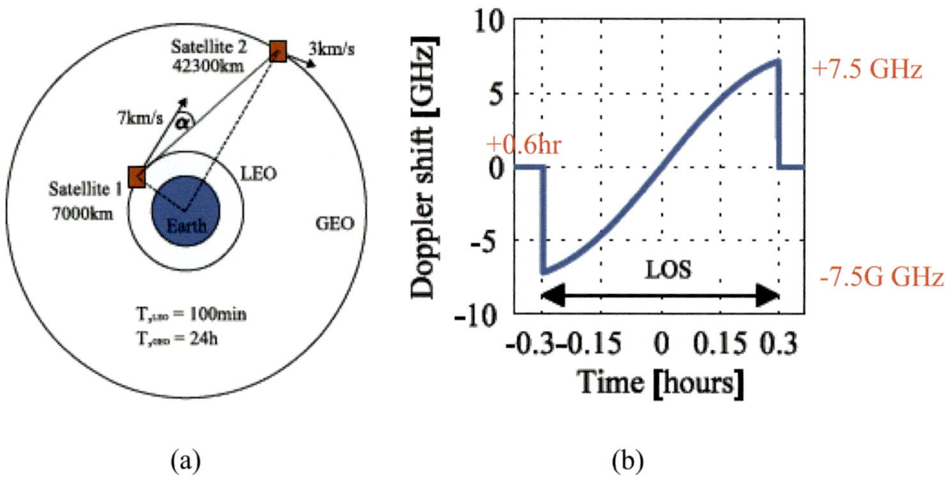

FIGURE 2.23 Illustration of a panel (a) satellite rotating around the Earth and related parameters between LEO and GEO satellites with their tangential velocity and their relative elevation angle. (b) Estimation of Doppler shift in GHz with respect to the time the satellite passes the acceptance area.[29,30] GEO satellite is stationary over 24 hours, and the time LEO satellite passes by the GEO. LOS, line of sight.

where Δv denotes the relative velocity between the satellites, c is the speed of light in vacuum, f_0 is the emitted carrier frequency, and α the angle between LOS of the satellites and the direction of movement of the transmitter satellite. Figure 2.23 illustrates this satellite-to-satellite (S2S) movement, as well as related parameters and the Doppler shift with respect to the time of crossing between the satellites.

Such frequency shift is to be evaluated with the receiver sensitivity and the ability of the DSP-based receiver to shift the coherent local oscillator laser to match the frequency-shifted lightwave carrier.

2.6 OPTICAL AMPLIFICATIONS AND ESTIMATIONS

2.6.1 Optical Amplifications

Transmission of Tbps channels over space links, such as GEO to OGS, MEO, or LEO to OGS, requires optical amplifiers to compensate for the high attenuation factors of such long transmission distances and the spatial broadening of the beam diameter due to the finite linewidth of the source and the modulated bandwidth. The attenuation is 0.01 dB/km in an asymptotically approximation of the Friis loss equation for a zero-linewidth single-frequency source for ISOL (intersatellite optical link), and 0.3 dB/km for scintillation dispersion with random turbulence for links related to atmospheric region of about 10 km. Two main types of optical amplifiers are (i) Booster OPA at the transmitter side to boost up the average optical modulated channels to a level of about 30–40 dBm. (ii) The OPreA (optical pre-amplifier) at the reception side to obtain sufficient sensitivity level required by the coherent receiver, which also requires a local oscillator (LO) to mix and further boost the signal gain and phase. Figure 2.24 shows the typical amplifier structure for the OPA and OPreA. Figure 2.25 shows the amplifiers in a multi-hop inter-satellite transmission network. We note the following:

i. Booster OPA: This high output power amplifier consists of a few amplification stages which are pumped by high power laser at 980 and 975 nm if EYDFA.

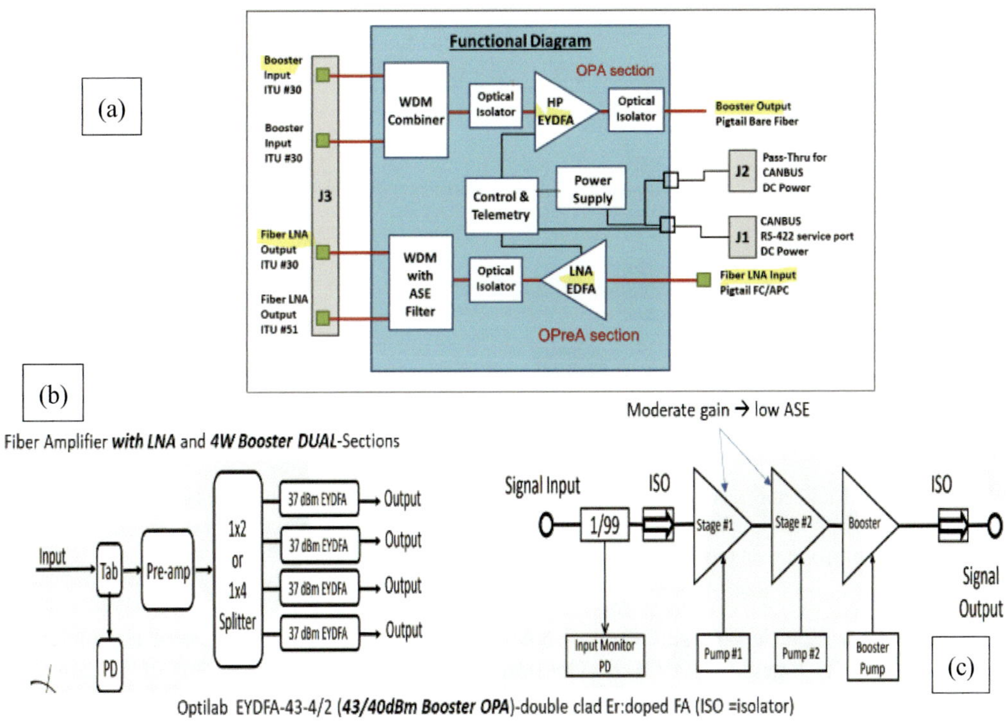

FIGURE 2.24 Optical Amplifications: Optical amplifiers (Booster OPA): Structures and selection. (a) HPA = high power amplifier; (b) optical pre-amplifier LNA and four booster HPAs; (c) three-stage pump EYDFA booster high-power amplifier. LNA = linear amplifier, pre-amp = amplifier of moderate gain, pump = pump laser of different wavelengths and high power to excite electrons in doped ions to upper bands, typically 975 nm and 980 nm or 1480 nm; ISO = isolator, ASE = amplification stimulated emission, EYDFA = Erbium Ytterbium-doped fiber amplifier, LNA (OPreA) = low noise amplifier, with only two outputs of channels 30 and 51 (ITU-T grid of 100 GHz).

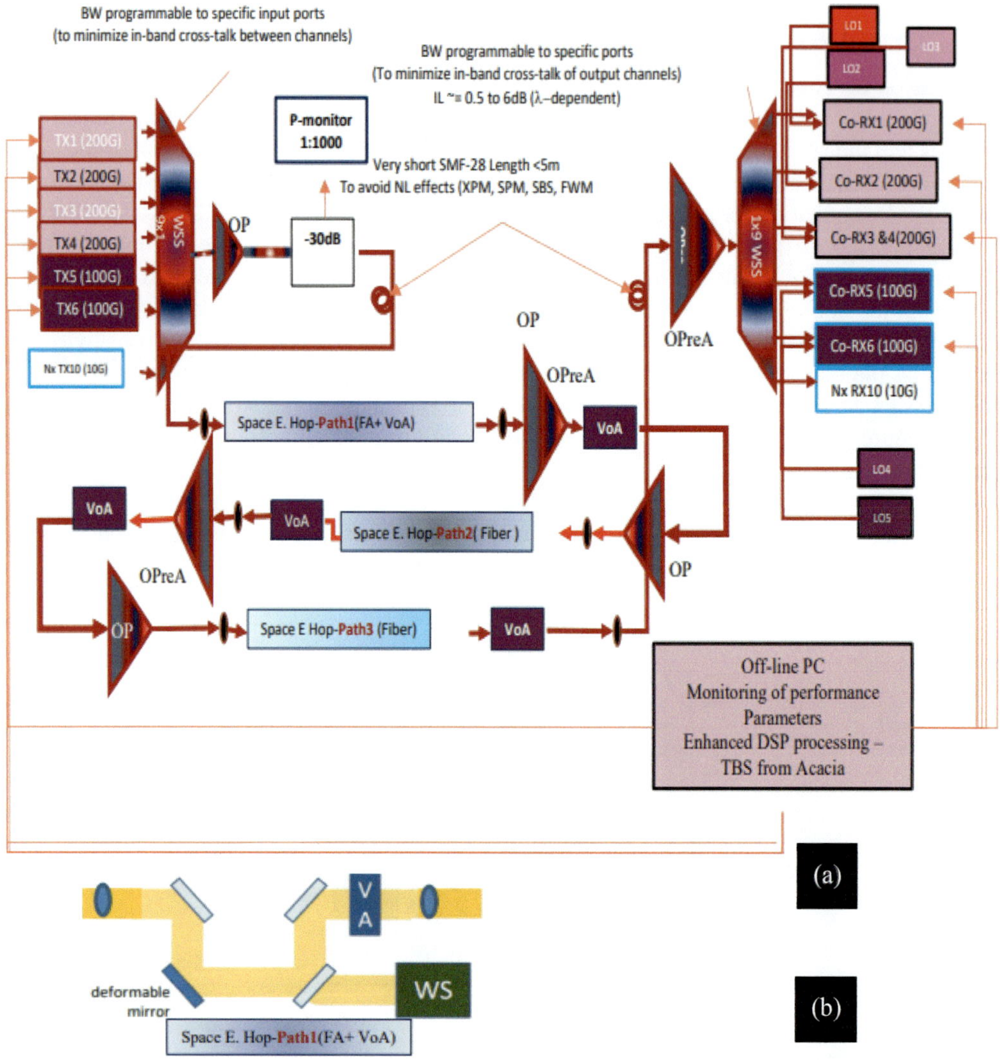

FIGURE 2.25 Multi-hop transmission system: preliminary baseline. E=emulated, LO=local oscillator laser; (a) detailed optical amplifiers in DWDM SPC system; (b) generic block diagram and optical path of SPC through free space with blocks.

ii. LNA Low-input low-noise OPreA: LNA (OPre-A) has only two outputs of channels 30 and 51 (ITU-T grid of 100 GHz). The bottom left shows multiple outputs of 37 dBm by splitting the output of the first stage amplifier and adding further amplifier stages in parallel to provide boosting amplification. The outputs are optically filtered to achieve low noise figure (NF).

iii. Nonlinear distortion and in-band crosstalk: Due to such high-power output and if connected to a single mode fiber of even a few-meter length, nonlinear effects such as SBS, SPM, XPM, SRS may occur, leading to distortion and in-band crosstalk between multiplexed channels.

iv. Super C-band: The C-band can now be extended to the super-C band with an extra 10 nm added to the upper band of the C-band, i.e., from 1565 to 1620 nm. This means 55 nm or 55×125 GHz/50=137 channels with 50 GHz spacing, resulting in a total bandwidth of 68.75 Tera-Hz. This allows the total capacity for the space channel to carry an extra 81,500/50 GHz or 137 channels of 50 GHz spacing, or 13 Tbps (100 Gbps/50 GHz), or 6.5 Tbps if 200 GHz pulse-shaped channels are used. Amplifiers for the C-band can be developed for space transmission.

Figure 2.24b (bottom left) shows multiple outputs of 37 dBm (High Power Amplifier) by splitting the output of the first stage amplifier and further adding amplifier stages in parallel to provide boosting amplification. 1×4 optical splitter splits the input lightwaves into four outputs, each carrying laser power of 37 dBm. The power at the input of the splitter is 43 dBm or 23 W (see Figure 2.25).

2.6.2 Free-Space Optic Equivalent Space Hop

A typical SPC hop in a multi-hop transmission can be structured as shown in Figure 2.26 in which it consists of (i) a booster amplifier that receives modulated signals and boosts its average power into high power output to satisfy the link power budget. In the laboratory platform, the high-power output is split into two paths, one is about –30 to –20 dB down from the other whose power is still high and is fed into an absorber for safety and monitoring of the power. (ii) Then, further attenuation devices can be used to meet the attenuation of the inter-satellite path loss. (ii) The output of this stage is then fed into the input of a low-noise optical amplifier (LNA). The output of this LNA can be further amplified only if the noises do not corrupt the signal levels, that is, satisfying the OSNR required for error-free transceiver module. If it is satisfied, an optical or OEO generator is required to regenerate the optical signals for further transmission. The numerical values inserted in Figure 2.26 indicates the loss of the transmission path. The estimation of the OSNR can be obtained, as described in the next section.

2.6.3 Estimations of OSNR in Cascaded Space Hop

The equation for OSNR of cascade space hop is given by

$$\text{OSNR}_{\text{Hop}\,i}(\text{linear}) = \frac{1}{\dfrac{1}{\text{OSNR}_{\text{stage }(i-1)}} + \dfrac{\text{NF}\cdot \text{h}\cdot \vartheta \cdot \Delta f}{P_{\text{in}}}} \tag{2.51}$$

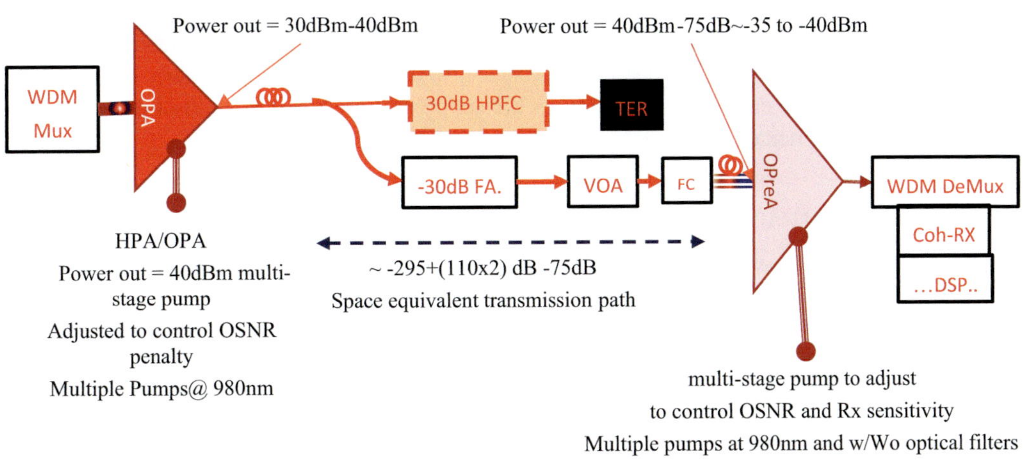

FIGURE 2.26 Emulation of space channel. TER, terminator; FA, fixed attenuator.

The OSNR at the end of a hop, in dB term, can be written as

$$\text{OSNR (dB)} = 10 * \log 10\left[\text{OSNR}_{\text{Hop}\,i}(\text{linear})\right] = 10 * \log 10\left[\frac{1}{\frac{1}{\text{OSNR}_{\text{Hop}\,(i-1)}} + \frac{\text{NF}\cdot h\cdot \vartheta \cdot \Delta f}{P_{\text{in}}}}\right] \tag{2.52}$$

where

- h=Planks constant=6.6260e-34 J s.
- ϑ=optical frequency at 1550 nm in C-band with minute correction=c/λ with c = speed of light in vacuum. λ=operating wavelength.
- Δf=frequency window for measured or defined noise bandwidth, 0.1 nm or 12.5 GHz, which is used as a standard quantity. This BW (bandwidth) is close to the 100Gbps QPSK 25 GBaud channel but about half of that for the 200 Gbps QPSK 50GBaud. The notations of 25 and 50 indicate the baud rate without extra loading bits of FEC (forward Error Correction) that normally about 12%, 30%, or 32% depending on whether FEC encoder is FEC, Soft FEC, or Hard FEC.[31]
- NF (lin.=noise factor, i.e., noise figure in linear scale – its conversion from $\log_{10}$dB scale is given by NF (linear)=$0.5\times 10^{\text{NF (dB)}/10}$). The factor 0.5 is taken to indicate that only one polarization is calculated as the polarization multiplexed channels are demultiplexed, and the electrical SNR is evaluated for one polarized channel.
- P_{in}=input power to the optical amplifier under consideration, with units used is dBm or mW in linear scale.
- $\text{OSNR}_{\text{Hop}\,i}(\text{linear})$=linear ratio of the signal and the noise average power over the defined noise bandwidth → $10\log_{10}\left(\text{OSNR}_{\text{Hop}\,i}(\text{linear})\right)$ → OSNR (dB).
- Therefore, $\text{OSNR}_{\text{Hop}\,(i-1)}(\text{linear})$=OSNR of the previous stage just before stage (ith).

Assumptions: The noise from the booster amplifier contributes minutely to the optical pre-amplifier, as both the signal average power and the noise power are attenuated by the attenuation of the space link. Thus, the noise of the optical pre-amplifier plays a major role.

So, various optical attenuation factors can be combined in the hop, and the optical pre-amplifier's contribution to the OSNR of the considered stage is calculated.

Note: Equations 2.51 and 2.52 can be obtained by referring to Ref.[32]

This equation can then be represented by Friis's formula (see Equations 2.3, 2.4, and 2.53) to yield the same result for OSNR in a cascade in the linear region with a single-frequency carrier, that is, without spatial spreading of the beam in free space. In this calculation, we determine the OSNR for each hop so that we can check and confirm the set-up of each hop in a multi-hop transmission system on the experimental platform.

2.6.4 Gain Equalization

Optical amplification may require to gain equalization when the wavelength carriers are not in the flat band of the amplification, as depicted in Figure 2.27a, to equalize the gain spectrum so that the flatness of the saturation gain can be obtained with an optical fiber length whose attenuation has a loss/attenuation opposite to the gain peaks of the amplifier to obtain equalized gain spectrum, as shown in Figure 2.27b.

This gain equalization is critical when multi-wavelength carriers are used to obtain an aggregate capacity higher than Tbps per SPC hop, so that the OSNR or noise accumulated can be nearly the same for all channels. This ensures that the power budget allows all channels arriving at the

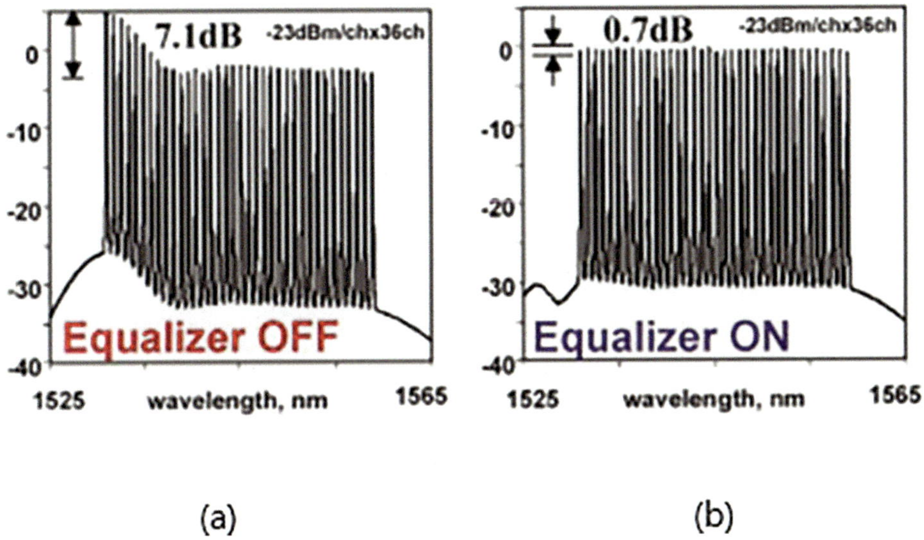

FIGURE 2.27 Gain equalizer for optical amplification and broadband source using integrated broadband filtering (a) non-equalised optical spectrum of amplifier output; and (b) equalized gain spectrum.

receiving ends to have the same OSNR, thus defining a universal design for all photonic components of the SPC system at each hop.

This technique is also commonly used in generating broadband sources by using the ASE noise spectrum and equalization process described here.

2.7 REMARKS ON SPC TRANSMISSION

2.7.1 On Atmospheric Link Attenuation Factors

The various aspects of the propagation of the photons in the Earth's atmosphere were presented (molecular and aerosol absorption, molecular and aerosol scattering, rain and snow attenuation, and scintillation effects). These effects constitute the key to free-space optical communication systems. The fog exerts more penalizing elements to free-space optical links operation.

For a given geographical site, the availability and reliability of a link can be determined according to system parameters such as power, wavelength, system characteristics, and climatic and atmospheric conditions. The link OSNR can be determined for the transmission system, especially the operational margin. The following summarizes the attenuation factors of various effects.

2.7.2 On Expected Values of Attenuation of Atmospheric Link

In conclusion, the following provide the attenuations of various regions:

i. Free-space propagation loss (Friis' equation under a complete linear "small signal" system in which all subsystems are operating in linear region)

$$\alpha_{\text{Prop,dB}} = 20 * \log_{10}\left(\frac{\lambda}{4\pi L}\right)$$

Atmospheric OGS to MEO (Laser light) 8900 km at 20° inclination angle) ~276 dB.

$$\alpha_{\text{Prop,dB}} = 20 * \log_{10}\left(\frac{\lambda}{4\pi L}\right) + 20\log_{10}\left(\frac{A_S}{A_0}\right) \tag{2.53}$$

where A_0=original beam cross-sectional area and A_s=spatial spreading beam cross-sectional area. There is also the differential delay due to the angling of the cross-section area of the receiving optical head. This generates the skew of the polarized beams as well as the differential group delay of the finite linewidth laser source.

ii. Atmospheric propagation loss per (3 km) at 100 m height versus inclination angle

$$\alpha_{\text{atm}} = \frac{10}{\left(\frac{\lambda}{1550\text{nm}}\right)^2 (\in +1)} \text{ in dB} \tag{2.54}$$

So over 10 km attenuation ~0.3 dB × 10/3 = 1 dB at the worst 20° inclination, over 10 km or 2 dB over 20 km of atmospheric layer and on top of this is the turbulence layer, see Figure 2.4, repeated below.

iii. Average attenuation by scintillation:

$$\text{In dB scale} \quad \alpha_{\text{sci,dB}} = \left(3.3 - 5.77\sqrt{-\ln\left(p_{\text{thr}}\right)}\right) * \left(\sigma_P^2\right)^{0.4} \text{ dB} \tag{2.55}$$

$$\alpha_{\text{sci}} = \exp\left(-\text{erf}^{-1}\left(2p_{\text{thr}} - 1\right)\sigma_P\sqrt{2} + \frac{\sigma_P^2}{2}\right) \tag{2.56}$$

~4.5 dB loss over this thin scintillation turbulence layer;

iv. Dispute on this estimation can be as follows: the scintillation and the IRT defined as the average expected values of the exponential of the phase deportation from its original states. The pdf of this phase fluctuation is Gaussian, and the variances are used to derive the attenuation of the average loss to use in the power budget. However, this phase error should be considered as the reduction of the OSNR and equivalent to the EVM in the sampled digital domain. The 2 dB can be assumed for the margin of the OSNR;
   - Turbulence coherence length ~ few tens of meter ≪ laser coherence length of 10 kHz linewidth (~300 to 30 km if 100 kHz);
   - Depolarization: to be <1.0 μrad; exp results → <5e-5 rad – so about 1 dB allowable for OSNR passing through the turbulence scintillation layer at 20 km;
   - Beam propagation method for atmospheric transmission of PDM-QPSK: this method follows the following steps: (i) Step size < 1/3 of coherence distance to ensure no excessive divergence in the spatial domain; (ii) FTT transform of the optical field (two polarized channels) over ½ step size distance; (iii) thus, transverse IFFT over the same ½ step as the distance for spatial IFFT; (iv) disturbance and scintillation – phase evolution properties are applied between steps (ii) and (iii) to perform the phase disturbance effects and any nonlinear phase effects such as nonlinear scattering such as Raman or four-phase mixing; (v) the above steps are repeated till the end of the propagation distance.

Notes:

- Optical amplification gains and saturation: OPA and OPreA are operating in the saturation region, where equal amplitudes are maintained. In this region, the Friis' formulation can no longer be applied and is only applicable in the linear region, which is outside the operating region. Furthermore, the linewidth of the source and modulated signals is zero, or effectively the single-frequency carrier is used, and the signal band is well below the carrier peak. Thus, the modulated channel can be considered a single frequency source to transport the modulated signals into space and transmission without beam spreading.
- Such saturation gains and saturated power output of OPA and OPreA are set in the power budget. Friis' equations are given. Estimations of attenuation can be found for Friis' equation. We note here that Friis equation is applicable for zero linewidth laser, or pure single

frequency laser. In practice suh laser s do not exist but with finite linewidth. So, an additional attenuation factor must be added due to the spatial spreading of the laser beam, thence the acceptance loss of the optic receiving system.

2.8 RELATIONSHIP BETWEEN EVM AND OSNR FOR QAM SIGNALS

2.8.1 EVM's Chamers University Approach

In digital signal processing of complex modulated constellations such as QPSK, M-ary QAM, etc, the error rate and the optical signal-to-noise ratio are important to gauge the performance of a transmission system. In the case of SPC, the signals operate at an ultra-high rate and do not involve in massive parallel resolution as in spatial image processing with wavefront sensors (WFS). In such scenarios, beam's cross-sectional area of rays passing through the detection area is evaluated by using a matrix of a minute optical detector, with which both the amplitudes and the displacement of the detected area can be evaluated. Hence, this aids in assessing the wavefront error compared to non-distorted matrix pattern.[33]

Error by WFS (Figure 2.28).

The MLSE (mean square logarithmic error) of a vector error sensor is used to evaluate the phase error and the spatial spreading. EVM and MLSE can be quite comparable, and this section addresses this issue. Understanding the definition and compatibility of the two systems can be useful in adaptive optics and adaptive optics with DSP for ultra-high-speed SPC systems.

The definition of the EVM in DSP-based optical coherent receiver can be used to evaluate the BER. The MSLE, as defined by a group at Chalmers' University, is similar to the error function. The MSLE between the true EVM and the estimated EVM can be expressed as:

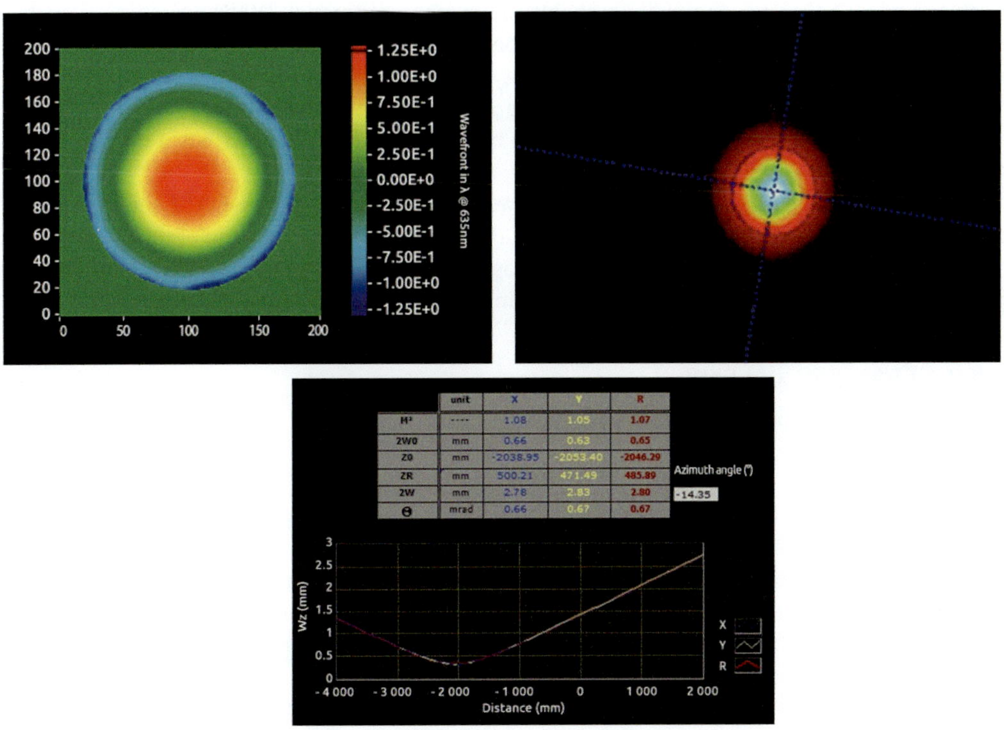

FIGURE 2.28 Image and WFS measurement. (a) Original undistorted beam cross-section, (b) wave front distorted beam, (c) measurement of the wavefront, hence WFE.[33] WFE, wavefront error.

$$\text{MSLE} =_n \sum_{i=1} \left(\log\left(\text{EVM}_{ti} + 1\right) - \log\left(\text{EVM}_{ei} + 1\right)\right)^2 \tag{2.57}$$

$$\text{or} \quad \frac{\text{EVM}_t + 1}{\text{EVM}_e + 1} = e^{-0.1*\text{MSLE}} = e^{-0.1*\text{OSNR(dB)}}$$

thus, for a 0.1 (10%) change in EVM, the OSNR $\rightarrow \text{EVM}(\%) = e^{2.3(\text{OSNR-dB})}$

Thus, under MSLE, the OSNR in power term is plotted in Figure 2.29, with n representing the total number of samples.

By comparing this with Equations 2.41 and 2.42, we observe the equivalence between the SR and EVM differential ratio, as well as the phase front change with the OSNR in a linear scale. The main affected point is the magnitude of the phase vector change, which is empirically significant for QPSK (coherent) and hence for PDM-QPSK.

The relationship between EVM and SNR, also known as MER, is relatively straightforward. EVM is a voltage ratio of the error signal to the peak state voltage, while SNR is a ratio of the *signal power* to the *error* power. The equation is as follows:

$$\text{SNR} = -\left[3.7 + 20 \times \log_{10}\left(\text{EVM}/100\%\right)\right] \tag{2.58}$$

Hence, the optical SNR (OSNR) is the square of the SNR (due to square-law detection) → or $10 \times \log_{10}(\text{SNR}) = 20 \times \log_{10}(\text{OSNR})$. So $2 \times \text{SNR (dB)} = \text{OSNR (dB)} \rightarrow \text{OSNR} = \text{SNR}/2$ in dB

$$\text{OSNR} = \text{SNR}/2 = -0.5 \times \left[3.7 + 20 \times \log_{10}\left(\text{EVM}/100\%\right)\right] \tag{2.59}$$

Equations 2.58 and 2.59 are applicable only to 64 QAM, and are accurate only when averaging is disabled and all 64 states are equally probable in the measured signal. When averaging is enabled, the measured SNR and the SNR calculated from EVM will be different as the SNR number is biased because of the averaging of a $\log_{10}$ quantity.

As EVM is a ratio of voltages, we use 20 $\log_{10}$ to convert it into a power ratio. The minus sign is necessary because SNR is a ratio of signal to noise whereas EVM represents a ratio of noise to signal.

The EVM represents the peak-to-average power ratio (in dB) of the 64-QAM signal. This ratio can be easily computed from the constellation diagram by computing the ratio of the average power for all 64 states (which is the same as the average power in one quadrant) and dividing it by the power of a corner state (the corner state has the highest power).

2.8.2 EVM, BER and SNR

According to the derivation of Ref.[35], the relationship is given by

$$\text{EVM}_{\text{rms}} = \sqrt{\frac{1}{N}\sum_{k=1}^{N}\left|r_k^{\,2} - t_k^2\right|^2 \Big/ P_a} \tag{2.60}$$

where P_a is the average received power for a particular modulated signal; r_k and t_k are the received and transmitted vectors, distorted/attenuated and non-distorted vectorial representation of the signal at a sampled point; N=number of symbols over which the EVM is estimated. Thus, EVM is a noise-over-signal ratio in power terms.

The vector t_k can either be a known reference pattern for data-aided EVM calculation or can be computed relative to the closest symbol in the constellation for non-data-aided EVM calculation. EVM is related to the signal-to-noise ratio (SNR) as follows:[36]

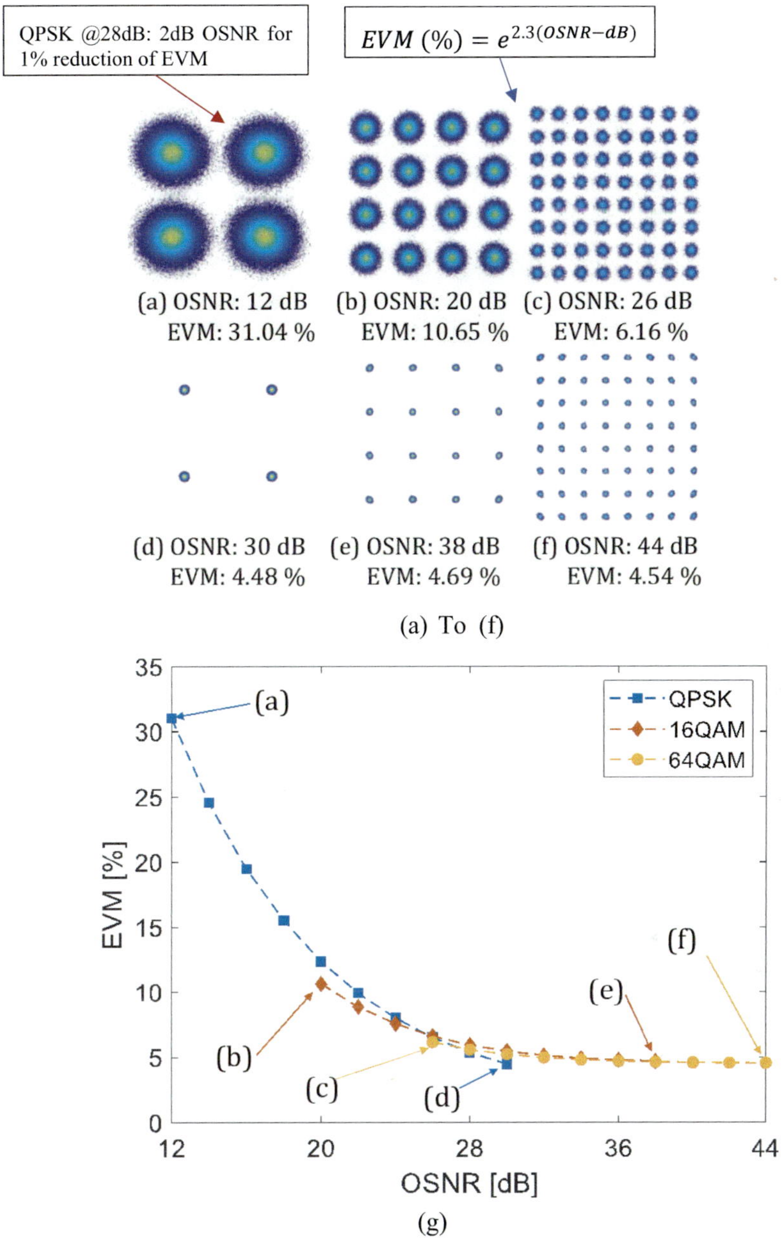

FIGURE 2.29 (a) The true EVM values of simulated QPSK, 16 QAM, and 64 QAM signals with respect to the OSNR. Panels (a–f) are the corresponding constellation diagrams for the end-points of the considered OSNR ranges.[34] Thus, the phase fluctuation due to turbulence <12.5% so that for 1 dB OSNR penalty, the phase fluctuation <1%. (g) EVM versus OSNR for modulation formats QPSK, 16 and 64 QAM. EVM large to small from (a) to (f).

$$\mathrm{EVM_{rms}} = \sqrt{\frac{1}{\mathrm{SNR}}} \tag{2.61}$$

The error analysis and EVM estimated values for QPSK are plotted in Figure 2.30, showing <0.1 dB difference when using Equation 2.60 to obtain the SNR required for QPSK and higher orders of QAM modulation.

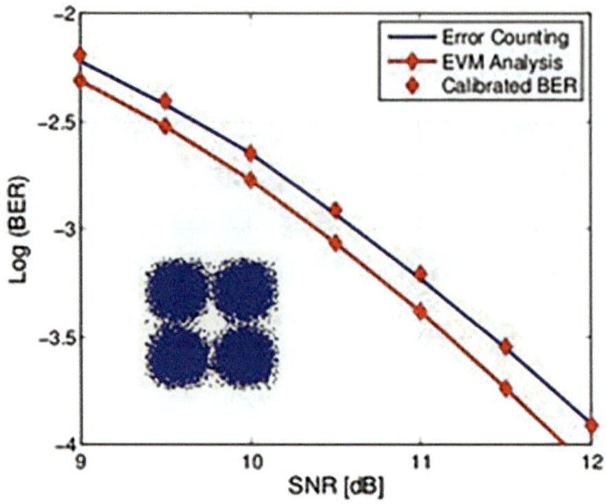

FIGURE 2.30 (Figure 5 of Ref.[36]): Performance of calibrated BER compared to the BER estimated from EVM analysis and error counting. The inset shows the recovered QPSK constellation at an SNR of 10.5 dB ($\Delta\upsilon{\cdot}Ts=10^{-4}$ and $\Delta P=0.245$ dB).

For data-aided calculation in the QPSK modulation format and differential decoding, where the EVM is caused only by amplified spontaneous emission (ASE) noise, the BER can be estimated using

$$\mathrm{BER} = \mathrm{erf}c\left\{\sqrt{\frac{1}{2 * \mathrm{EVM}_{\mathrm{rms}}^{2}}}\right\} \tag{2.62}$$

where factor 2 is included for the differential coding penalty, and erfc is the complementary error function. Although the relationship (3) is valid for data-aided EVM calculation, it can also be applied to non-data-aided EVM calculation for $\mathrm{BER} < 10^{-2}$ as discussed in Ref.[37] The performance of (3) with carrier phase recovery is discussed in the next section.

2.8.2.1 BER Estimation with Carrier Phase Recovery

The feed-forward Viterbi & Viterbi algorithm (VVA) is the widely adopted scheme for the QPSK modulation format in digital coherent receivers for carrier phase recovery. The estimated phase, ϕ, using the VVA can be obtained using[38]

$$\phi = \frac{1}{4}\arg\left(\sum_{k=1}^{N_b} r_k^4\right) \tag{2.63}$$

where N_b is the averaging block size, the VVA is typically followed by differential decoding to resolve the fourfold phase ambiguity and avoid cycle slips. The analytical expression for the BER of a differentially encoded QPSK signal using the VVA with block averaging can be approximated as:

$$P_e = \int_{-\frac{\pi}{4}}^{\frac{\pi}{4}} \mathrm{erfc}\left(\sqrt{\gamma_s/2}\right)\left(\cos(\varepsilon) - \sin(\varepsilon)\right) P_{\Delta\phi}(\varepsilon)\, d\varepsilon \tag{2.64}$$

where $\gamma_s = \mathrm{SNR}$; $P_{\Delta\phi}(\varepsilon)$ is the Gaussian probability density function (pdf) of the phase estimation error with zero mean and variance.[39]

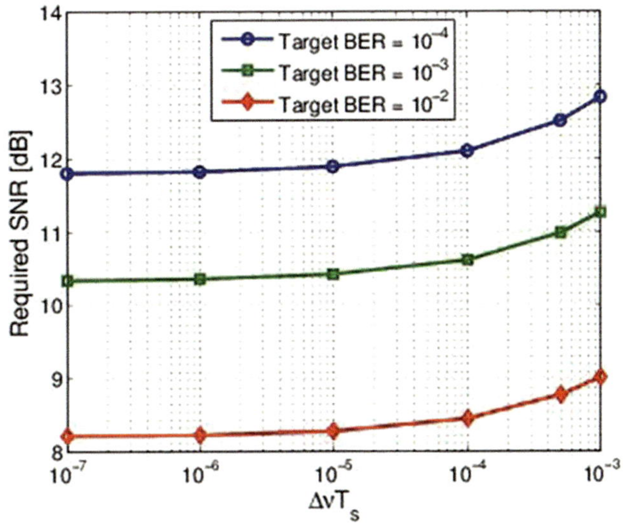

FIGURE 2.31 Required SNR to achieve target BER of 10^{-4}, 10^{-3}, and 10^{-2} using analytical approximation (5) with the optimum block size for the Viterbi and Viterbi algorithm. For optical power, the SNR is to be raised by 3 dB.

From Equations 2.61 and 2.62, it can be observed that the calibration can be achieved by scaling the estimated SNR from EVM analysis when calculating the BER as follows:

$$\text{BER} = \text{erfc}\left\{\sqrt{\frac{10^{-\Delta P/10}}{2 * \text{EVM}_{\text{rms}}^2}}\right\} \tag{2.65}$$

where ΔP is the implementation penalty in dB associated with the carrier phase recovery algorithm at the calibration point (Figure 2.31).

2.8.3 EVM, BER AND OSNR

2.8.3.1 BER versus OSNR

The relationship between BER or Pe (probability of error) and OSNR, and BER versus EVM can be written and derived as[36]

$$\text{BER} = P_e = \frac{2\left[1-\frac{1}{L}\right]}{\log_2(L)} Q\left\{\sqrt{\left[\frac{2\log_2 L}{L^2-1}\right]\frac{2E_b}{N_0}}\right\} \tag{2.66}$$

where L is the number of levels in each dimension of the M-ary modulation scheme; E_b is the energy per bit, and $N_0/2$ is the noise power spectral density; and $Q[\cdot]$ is the Gaussian co-error function given by $Q(x) = \int_x^\infty \frac{1}{\sqrt{2\pi}} e^{-0.5y^2} dy$.

Assuming raised cosine-shaped pulses with sampling at a data rate, Equation 2.66 also gives the BER in terms of SNR or in the case of optics (OSNR) as

$$\text{BER} = P_e = \frac{2\left[1-\frac{1}{L}\right]}{\log_2(L)} Q\left\{\sqrt{\left[\frac{3\log_2 L}{L^2-1}\right]\frac{2E_s}{N_0 \log_2 M}}\right\} \tag{2.67}$$

where $\frac{E_S}{N_0}$ = SNR or OSNR for M-ary modulation system and raise cosine at the data rate, the baud rate.

This equation defines the relationship between BER and OSNR.

2.8.3.2 BER versus EVM for QAM Modulated Signals

EVM measurements are often performed on vector signal analyzers (VSAs), real-time analyzers, or other instruments that capture a time record and internally perform a Fast Fourier Transform (FFT) to enable frequency domain analysis. Signals are down-converted before EVM calculations are made. The IEEE802.11a-1999 specification describes a set of different schemes that are used adaptively: binary phase-shift keying (BPSK), 4 quadrature amplitude modulation (4-QAM or QPSK), 16 quadrature amplitude modulation (16-QAM), 64 quadrature amplitude modulation (64-QAM), etc. These modulation schemes are normalized to facilitate the calculation of EVM.

The normalization is derived such that the mean square amplitude of all possible symbols in the constellation of any modulation scheme is one. Thus, EVM is defined as the root-mean-square (RMS) value of the difference between a collection of measured symbols and ideal symbols. These differences are averaged over a given, typically large number of symbols and are often shown as a percent of the average power per symbol of the constellation. As such, EVM can be mathematically given as

$$\mathrm{EVM}_{\mathrm{rms}} = \frac{\frac{1}{N}\sum_{n=1}^{N}\left|S_n - S_{0,n}\right|^2}{\frac{1}{N}\sum_{n=1}^{N}\left|S_{0,n}\right|^2} \tag{2.68}$$

where S_n is the normalized nth symbol in the stream of measured symbols, $S_{0,n}$ which is ideal in the normalized constellation points of the nth symbol; N is the number of unique symbols in the constellation.

The expression in Equation 2.68 cannot be replaced by its unnormalized value since the normalization constants for the measured constellation and the ideal constellation are not the same. The normalization scaling factor for ideal symbols is given by

$$|A| = \sqrt{\frac{1}{\frac{P_v}{T}}} = \sqrt{\frac{T}{P_v}} \tag{2.69}$$

where P_v is the total power of the measured constellation of T symbols. For RMS voltage levels of in-phase and quadrature components, V_I and V_Q, and for $T \gg N$, it can be shown that Pv can be expressed as

$$P_v = \sum_{t=1}^{T}\left[V_{I,t}^2 + V_{Q,t}^2\right] \tag{2.70}$$

The normalization factor for an ideal case can be directly measured from N unique ideal constellation points and is given by

$$|A_0| = \sqrt{\frac{N}{\sum_{n=1}^{N}\left[V_{I,t}^2 + V_{Q,t}^2\right]}} \tag{2.71}$$

Hence, Equation 2.71 can be further extended using normalization factors given in Equations 2.69 and 2.70 as

$$\mathrm{EVM}_{\mathrm{rms}} = \left\{\frac{\frac{1}{T}\sum_{t=1}^{T}\left|I_t - I_{0,t}\right|^2 + \left|Q_t - Q_{0,t}\right|^2}{\frac{1}{N}\sum_{n=1}^{N}\left|\left|I_{0,n} - Q_{0,n}\right|^2\right|}\right\}^{\frac{1}{2}} \tag{2.72}$$

where $I_t = (VI_t)|A|$ is the normalized in-phase voltage for measured symbols, and $I_{0,t} = (VI_{0,t})|A_0|$ is the *normalized* in-phase voltage for ideal symbols in the constellation. $Q_t = (VQ_t)|A|$ is the normalized quadrature voltage for measured symbols, and $Q_{0,t} = (VQ_{0,t})|A_0|$ is the normalized quadrature voltage for ideal symbols in the constellation.

This definition is currently used as the standard definition of the EVM in IEEE 802.11a – 1999TM.[40]

It is evident that the EVM is essentially the normalized error magnitude between the 'measured' constellation and the 'ideal' constellation. For the Gaussian noise model, Equation 2.72 can be simplified in terms of the noise in-phase component, $n_{I,t}$ and quadrature component, $n_{Q,t}$ as

$$\mathrm{EVM}_{\mathrm{rms}} = \left\{ \frac{\frac{1}{T}\sum_{t=1}^{T} \left|n_{I,t}\right|^2 + \left|n_{Q,t}\right|^2}{P_0} \right\}^{\frac{1}{2}} \tag{2.73}$$

where P_0 is the power of the normalized ideal constellation or the transmitted constellation. The numerator of Equation 2.73 sets up the normalized noise power. However, for $T \gg N$, the ratio of normalized noise power to the normalized power of ideal constellation can be replaced by their unnormalized quantities, i.e. the Equation 2.73 rewritten as

$$\mathrm{EVM}_{\mathrm{rms}} = \left\{ \frac{1}{\mathrm{SNR}\ (\mathrm{or\ OSNR})} \right\}^{\frac{1}{2}} = \left\{ \frac{N_0}{E_S} \right\}^{\frac{1}{2}} \tag{2.74}$$

In order to establish relationship between BER and EVM, SNR in Equation 2.74 can be expressed in terms of EVM as

$$\mathrm{OSNR} = \left\{ \frac{1}{\mathrm{EVM}^2} \right\}^{\frac{1}{2}} \tag{2.75}$$

It is noted here that the EVM is in unit or normalized voltage vector after detection and sampling in the linear domain, while the OSNR is in the power domain, so the square root term appears on the EVM inverse in this equation.

Now, combining Equation 2.75 and the BER expression given in Equation 2.67, we can obtain the relationship between BER and EVM as follows:

$$\mathrm{BER} = P_e \cong \frac{2\left[1 - \frac{1}{L}\right]}{\log_2(L)} Q\left\{ \sqrt{\left[\frac{3\log_2 L}{L^2 - 1}\right] \frac{2}{\mathrm{EVM}_{\mathrm{rms}}^2 \log_2 M}} \right\} \tag{2.76}$$

The BER is plotted against OSNR for different modulation formats for a single polarization of the PDM-M-ary QAM, as shown in Figure 2.32, and BER versus EVM is shown in Figure 2.33.

At the SDFEC level ($\mathrm{BER} = 2 \times 10^{-2}$), as indicated by the arrows, and comparing the BER versus OSNR in Figure 2.32, calculated using Equation 2.67, with the BER versus E_b/N_0 (total signal energy/bit and noise per bit) in Figure 2.34 (by MATLAB's bertool.m), we can see a difference of about 3 dB when taking the energy of the raised cosine-shaped pulse sequence of QPSK for a single polarized channel. A typical practical OSNR required by the Acacia 1200 module is 9.8 dB for PDM-QPSK. This extra 3 dB is contributed by the wide noise bandwidth in the model, which can be used for double wide bandwidth (i.e., 56 and 25 GSa/s). The specified OSNR by Module AC1200 is approximately equal to that for 1e-6 in the curve shown in Figure 2.34.

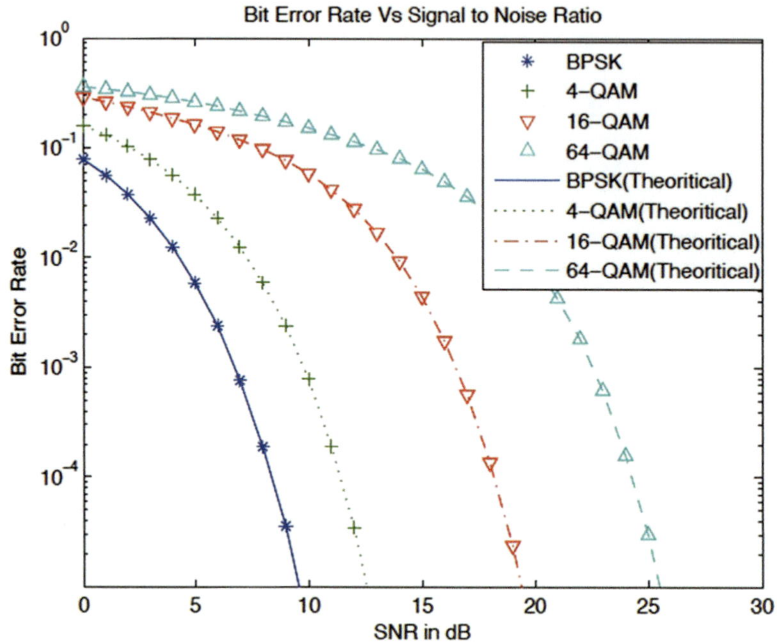

FIGURE 2.32 BER (in log scale) plotted against (SNR) OSNR from (page 65) Equation 2.76. Single polarization is considered here. For a QPSK, the OSNR is the same for each polarized channel plus the OSNR (dB) required for polarization dispersion loss. Non-dotted are simulations, *, +, -- are simulated; dotted lines are theoretical.

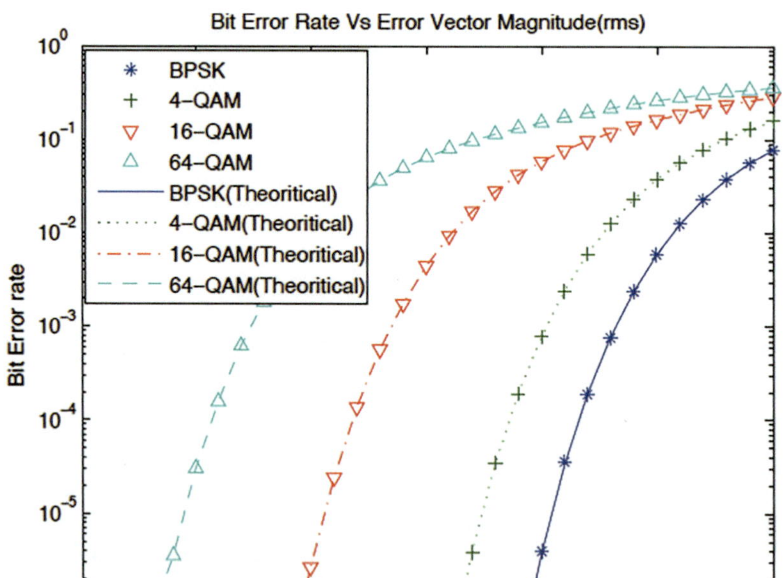

FIGURE 2.33 BER plotted against EVM in dB scale for different levels of M-ary QAM systems. For 20% or –7 dB EVM, the required OSNR = and BER = 2×10^{-2}. While for 31% (or –5 dB), the BER increases to 3×10^{-2} under SDFEC.

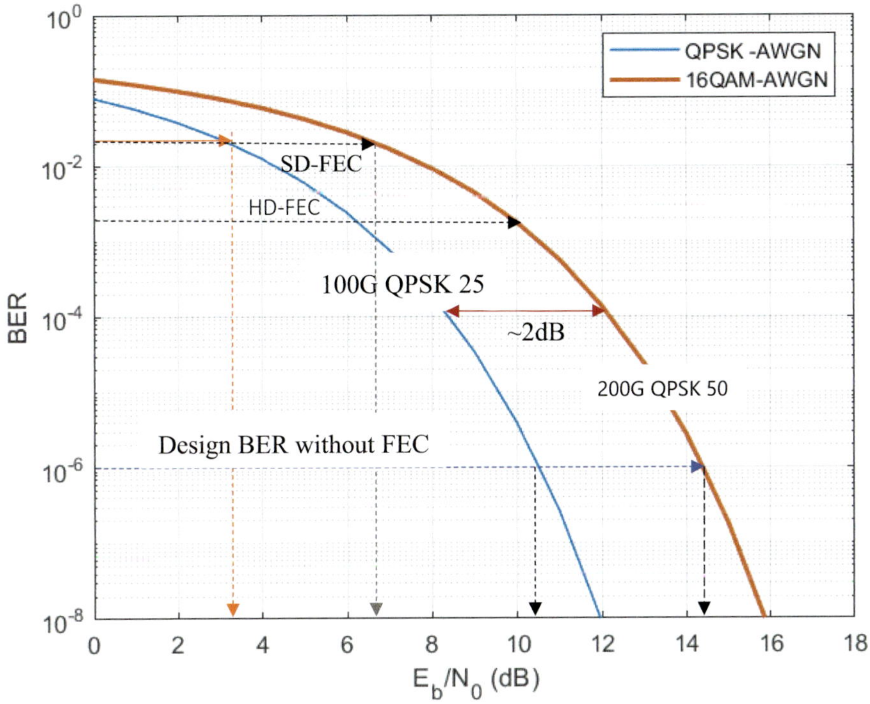

FIGURE 2.34 Baseline key performance of polarized division multiplexing (PDM)-QPSK and PDM-16 QAM (100 and 200 G respectively) plotted by BER versus optical signal-to-noise ratio (OSNR) in cases of no FEC, and pre-FEC of HD-FEC or soft-FEC. SD-FEC=software-defined forward error correction, AWGN=additive white Gaussian noise, Eb=energy of signal in a bit period, No=noise energy contained in the bit period, BER=bit error rate, QPSK=quadrature phase shift-keying.

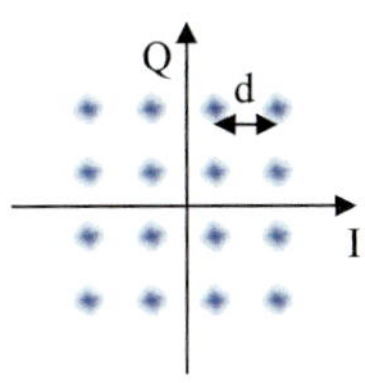

FIGURE 2.35 Constellation of 16 QAM on IQ plane.

The performance of 200 G PDM-QPSK is similar to that of PDM-16QAM due to the wider bandwidth and hence the total noise contribution. It is noted that the symbol rate of 200 G PDM-QPSK is twice that of 100 G PDM-QPSK, resulting in a doubling of the channel bandwidth.

2.8.3.3 Approximation Method for Estimating Penalty of OSNR in Multi-Level Modulation Formats[41]

The electrical field vector of an amplitude and/or phase-modulated optical signal can be represented in an *IQ* (in-phase and quadrature) constellation diagram in the complex plane, as shown in Figure 2.1 for the example of 16-QAM. Using M constellation points, $b = \ln(M)$ bits can be encoded into one symbol. The value b can be a non-integer number.[2] During propagation in the transmission system, optical noise is added to the signal by optical amplifiers. Here, we assume white noise with a Gaussian amplitude distribution and uniformly distributed phase (noise amplitude N_0 in Figure 2.35).

The minimum Euclidean distance d to the next constellation neighbor determines the sensitivity of the modulation format towards noise impairment. Assuming linear detection, like in a coherent receiver, results are approximately valid for both coherent and direct detection receivers.

The OSNR is defined as the ratio of the average signal power P_{av}, the square of the normalized constellation vectors, to the optical noise power integrated over a predetermined optical bandwidth, for example, 0.1 nm for 10 G. The OSNR, SNR, and the OSNR penalty can then be written as

$$\text{OSNR} = \frac{P_{av}}{N_0^2}$$

$$\text{SNR} = \frac{d^2}{N_0^2 b_{rel}} \tag{2.77}$$

$$\text{OSNR}_{penalty} = \frac{4P_{av}b_{rel}}{d^2}$$

In which a bandwidth reduction factor b_{rel} due to high-level formats with respect to binary signaling is introduced. The SNR is given after filtering. To compare different modulation formats, the required OSNR is related to a fixed SNR, which is determined by the constellation distance d. If one normalizes to the required OSNR for the BPSK format, then a penalty relative to BPSK for an arbitrary multi-level format is given by the simple approximation in Equation 2.77. Verification of this approximation was carried out in Ref.,[42] where numerical Monte Carlo simulations were performed. Gray-coded M-ary symbols with 50% return-to-zero (RZ) pulse shaping ere transmitted using coherent homodyne detection. Ideal synchronization of the carrier laser and the local oscillator (LO) at the receiver, and negligible laser linewidths were assumed. Optical and electrical filter bandwidths at Rx (third-order Gaussian, optical Bessel fifth-order electrical) were optimized. The OSNR penalty compared to BPSK is calculated for BERs of 10^{-3} and 10^{-5}.

Table 2.1 shows the values of normalized average power (P_{avg}), normalized constellation point distance (d), and relative bandwidth factor (b_{rel}) for various multi-level modulation formats. The resulting OSNR penalty is evaluated using Equation 2.77. In addition, the general case of M modulation levels is shown. Based on these equations, the OSNR penalties are shown in Figure 2.36. It is found that optimum OSNR tolerance can be obtained with a 3-PSK modulation, where the required OSNR is approximately 0.8 dB lower than that for BPSK (binary phase shift keying) or 4QAM (4-level quadrature amplitude modulation). Using this modulation format, >1.5 bits can be encoded into one symbol, or three bits can be encoded in two symbols.

TABLE 2.1
Mary QAM Modulation Formats and Associated OSNR Penalty as Compared to BPSK

Parameters	BPSK	Square Mary-QAM $M=4$	$M=16$	M
P_{av}	1	2	1.111	$\frac{2(M-1)}{3(\sqrt{M}-1)^2}$
d	2	2	0.67	$\frac{2}{(\sqrt{M}-1)}$
b_{rel}	1	1/2	1/4	$\frac{1}{(M)^2}$
OSNR penalty (linear scale)	1	1	2.5/3.9 dB	$\frac{2(M-1)}{3M^2}$

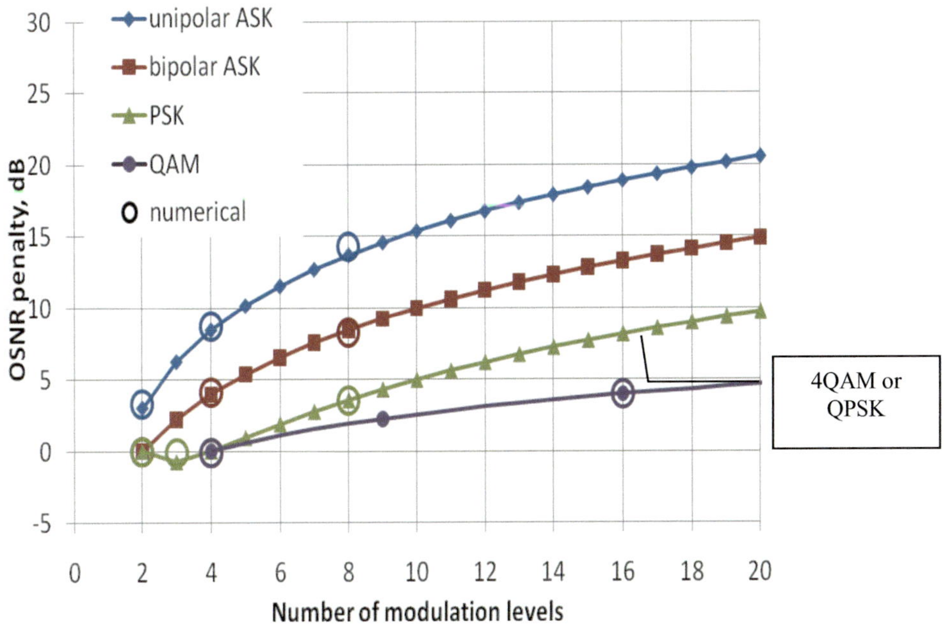

FIGURE 2.36 OSNR penalty versus level of modulation (extracted from Ref.[1]) estimated by simulation for BER = 1e-3 and 1e-5 Monte Carlo.

2.8.4 Spatial Distortion and Compensation

2.8.4.1 Dispersion Due to Spatial Spreading

The transmission of a laser beam over long-reach space link distances is spread due to the finite bandwidth of the laser carrier. Furthermore, the modulation of the signal information also adds to the signal bandwidth. As far as the carrier line peak is more than 6 dB above the signal double sideband, the beam will remain pseudo-coherent, allowing it to carry the signal information over the vacuum space distance.

The spatial spreading of the beam of finite radius will generate differences in propagation time for each ray of the lightwave beam. Thus, there are several delay differences between the central ray and the other rays of the beam, whose profile is assumed to be Gaussian. The pulse spreading in the time domain is due to the ray propagation time difference, which is related to group spatial delay and dispersion. This spreading varies several times a day when the satellite's tangential link varies during the satellite's flying and rotation speed.

On top of this group delay dispersion, there is the Doppler shift of the signal and carrier. This shift affects the coherent reception of the receiver, shifting from intra-coherent to heterodyne reception. However, digital signal processing can mitigate such Doppler shifts to restore the reception to true homodyne reception.

Similarly, the DSP technique can compensate for the group delay dispersion. The adaptive optics technique can be used as part of the compensation as this is an analog technique. This adaptive optics for ultra-fast signaling is quite different from adaptive optics for image distortion in spatial imaging, which involves compensating for the phase error distortion by measuring the wavefront phase error with spatial detection cells and then correcting the optics to reduce WFE to zero. The WFS (wavefront sensor) is too slow for compensating spatial group delay dispersion of ultra-fast signals; 40 ps pulse width for 100 Gbps DP-QPSK and 20 ps for DP-QPSK 400 Gbps.

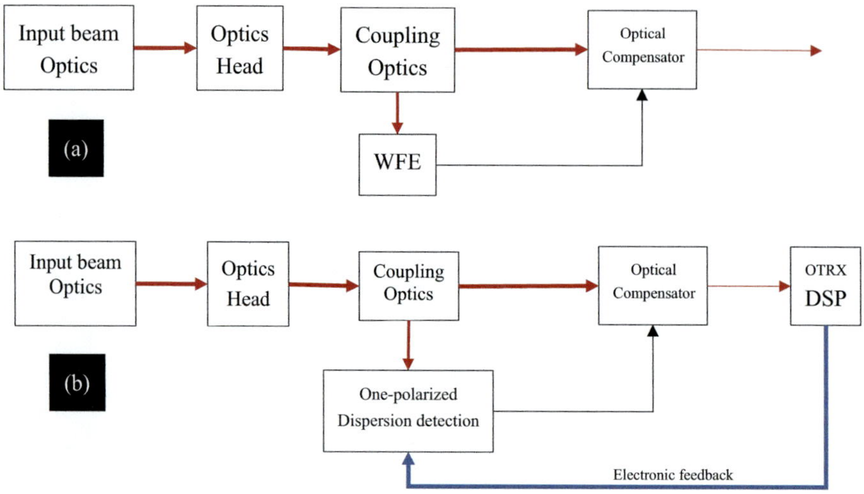

FIGURE 2.37 Spatial dispersion compensation systems (a) pure wavefront sensor and compensation; (b) optical compensation and DSP correction. WFE, wavefront error.

2.8.4.2 WFS and Adaptive Optics for SPC

Figure 2.37a shows the schematic system using optical compensation with a wavefront error monitor and measuring device to determine the spatial phase distortion of beam propagation in space. The feed signals to change the optical device as a compensator, such as a multiple array of optical reflectors to reverse the distortion of the optics phase of the signal beam. However, the speed of the WFE device is slow, and the ultra-fast signals with periods of 40 or 20 ps are too fast for such WFE adaptive optics.

2.8.4.3 DSP Compensation by Digital Technique

On the other hand, a hybrid structure using both optical and DSP signal processing, as shown in Figure 2.37b, combines an optical compensator as a coarse equalizer with the digital signal processing to correct signal distortion at the receiver. A signal feedback loop can be created to provide tuning of the optic compensation. Such a fiber optic equalizer can be a few-mode fiber,[43] whose group delay difference can have the opposite sign of the SPC spatial group delay. Furthermore, a digital compensating algorithm can be used to correct the distortion. The optical distortion compensator can be a short length of multimode fiber, whose group delay difference is similar but opposite to the propagation delay difference. Minute over- or under-compensation due to variation in the entry angle and non-perpendicular alignment of the beams can be compensated by the DSP algorithm.

2.9 MULTI-WAVELENGTH MULTI-HOP SPACE PHOTONIC COMMUNICATIONS NETWORKS

A multi-hop photonic transmission link in space is shown in Equation 2.36, consisting of three optically amplified hops as a model. A higher number of hops can be employed, provided that the required OSNR at the reception end satisfies the noise and signal levels such that the signals are not corrupted by the accumulated noise due to optical booster amplification and low-noise amplifications of simulated emission processes.

Several optical channels transport a number of wavelength carriers and their modulated signal bands. The equivalent space loss is due to transmission, which includes space propagation losses and the gain of the two optical heads. There is then an optical feedback path to extend the number

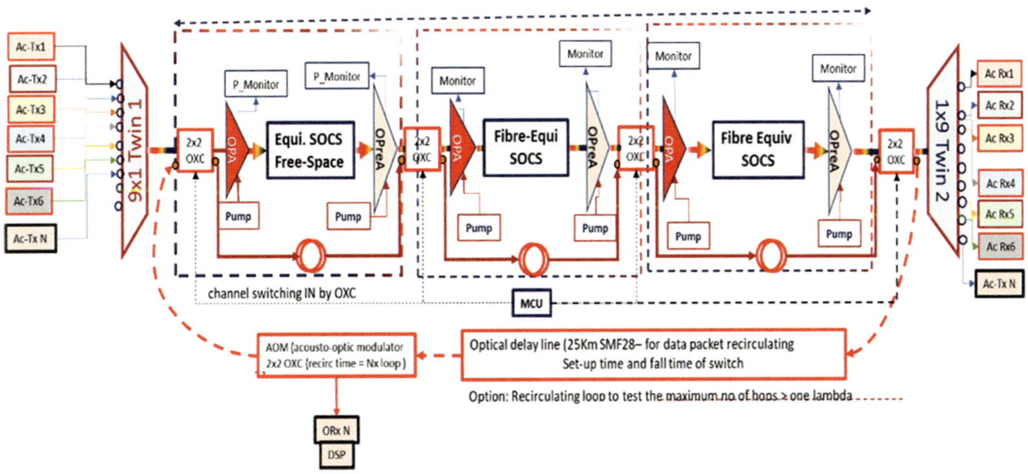

FIGURE 2.38 Prelim baseline design: Demonstration platform 1: SPC SYSTEM 3-Hop link with B2B test without ROADM and reconfigurable using 2×2 optical switches (OXC). This is to evaluate the accumulated noise of the amplified space spans.

of hops if required, increasing the accumulated noise level until signal corruption occurs. This feedback extension is implemented through an optical switch, which switches the link hops in and out.

To demonstrate the multi-hop transmission, a laboratory platform of the SPC multi-hop SPC system is shown in Figure 2.39. The feedback of the optical path is through one hop only, passing through the same amplifiers, along with booster and preamplifiers, to ensure that the noise contributions are the same to the hop. The input level and gain of the preamplifier is adjustable so that the gain can be increased or decreased based on the signal level. The ASE noise is also adjusted by the preamplifier. The noise equation used to estimate the total noise contribution to the link can be referred to in Chapter 9.

An OXC is used to connect the number of hops, and this can also reconfigure the number of hops in the multi-hop structure. This is done to evaluate the maximum number of hops such that the required Optical Signal to Noise Ratio (rOSNR) can be met. R = required

A hop link in Figure 2.38 is represented by a block, for example, GEO-GEO, LEO-GEO, MEO-MEO, etc. in Figure 2.39.

The power budget is given in Section 2.13, outlining the parameters and performance with noises for the multi-hop space link. An attenuation factor is hidden,but reconfigurable, as a parameter can be changed to include in the configurability of the OXC.

The performance of the link of one hop is described and analyzed in Section 2.10.

2.10 TBPS SPC SYSTEM SET-UP AND POWER BUDGET SHEETS

Figure 2.40 shows a typical single stage of space transmission from one MEO satellite to another MEO satellite. The OSNR is evaluated using the OSNR equations represented in Equations 2.51 and 2.52. Negative signs indicate losses, while positive signs indicate gain. The inter-satellite hop distance is 8000 km. At each hop, the required OSNR must be met. If the OSNR at the end of the hop is lower than the required OSNR, no further transmission is possible, and repeating and regeneration of channels are to be done for further transmission. The OSNR for different modulation formats can be referred to in Section 2.8.

2.10.1 Atmospheric Channels and Impact on Modulated Channel

FSO links over atmospheric regions, e.g. OGS to LEO/MEO/GEO satellites suffer from random changes of refractive index caused by the variation of air temperature and pressure.[44,45]

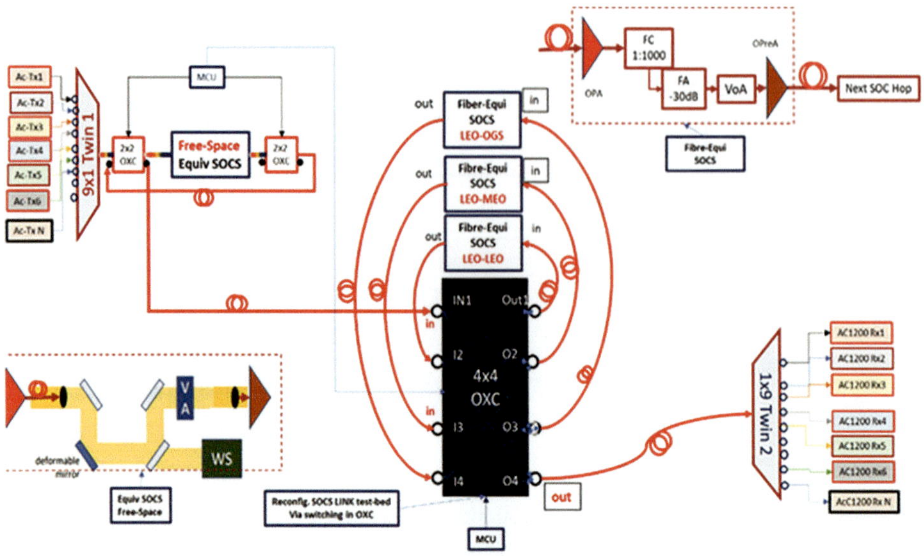

FIGURE 2.39 Schematic diagram of the proposed setup for Tbps transmission using equivalent OISL in a laboratory platform for demonstration, Demo 2, which is reconfigurable by using a 4×4 OXC. Reconfigurable numbers of OISL hops can be implemented through connections or interconnections via OXC to Fiber-equi SOCS link LEO-OGS, LEO-MEO, or LEO-LEO. SOCS = space optical communication system. The insert on the left shows the free-space representation of the SPC link.

OISL Type (Description) (aperture size)	L(kms)	OPA ouput power (Multiplexed channels)	Multiplexed channel	OPA output per channel	OSNRo(dB) - Tx AC1200 can offer to 32 dB	OTRX average modulated power (dBm)	WSS (Lamda Mux/Demux+Filter)			OPA Gain/channel (all 50dB -> -10*Log10(6))	Launch-P(dBm) 0dBm input into OPA (--> OSNR10) per channel	Power launched into OGS-MEO per mux channel	OSNR10 after OPA launching into Hop1
(Start-up) OGS-MEO-OISL-MEO-OGS(down) aperture (mm)	Link Distance	OPA power OUT (all channels)	channel number (total)	ouput power /channel (dBm)	OSNR0 Initial IN OSNR	Input Ave. Power (Pin)	WSS_IL (dB)	NF(dB) OPA	F (OPA) - noise	Gain (dB)/channel	Launch Power (OPA out-dBm))	OPA out/Mux channel (dBm) to MEO	OSNR (dB)
MEO-MEO Laser Light Constellation (163mm) input from UPLINK output 100G-PDM-QPSK 28GSy./s contellation 12 satellites					start OSNR0 from Uplink						channel Demuxed power (dBm) 43dBm-10*Log10(6)		OSNR10 OSNR10
N3_MEO-MEO Laserlight (163mm)	8000				23.86		-4.5	5.9	3.890	42.22	35.22	35.2185	23.8501

FIGURE 2.40 A typical stage of the power budget for one link of a satellite MEO-MEO constellation.

This impairment is shown as a temporal and spatial variation in light intensity,[46] which is called scintillation, similar to the fading effect in wireless communication.[47]

There are mainly three statistical models to describe the atmospheric turbulence channel, namely, the Log-normal distributed channel model,[20,48,49] the K-distributed channel model,[50,51] and the Gamma–Gamma distributed channel model.[52,53] The Gamma–Gamma distributed channel model is found to be most suitable for modeling the irradiance of FSO channels in all the turbulent scenarios, from weak to strong,[54] and hence, this model can be considered to offer close inspection of atmospheric effects on the laser communication system.

The optical field has three physical attributes (intensity, phase, and polarization) that can be used to transmit information. Considering the atmospheric turbulence mainly affects the light intensity, pulse-position modulation (PPM) is commonly used in FSO communication. Since fiber-optic technologies have been well-developed and fiber-optic networks have been widely deployed from the local access networks to the long-haul intercontinental networks, some commonly used modulation formats in fiber-optic transmission systems, including on–off keying (OOK), and differential phase-shift keying (DPSK), have also been investigated in FSO systems.

Simplicity is the advantage of OOK, while the DPSK format, which encodes information on its phase, can mitigate the severe effect of scintillation to some extent. Compared with the binary formats such as OOK and DPSK, the differential quadrature phase-shift keying (DQPSK) format doubles spectral efficiency by taking advantage of the two-signal quadrature of an optical carrier. Thus, two information bits are transmitted per symbol, being represented by four possible optical phase variations between successive symbol periods.

This section investigates the BER performance of various commonly used modulation formats, along with SDRT in the FSO link. We derive an analytical BER result for the DPSK signal in the Gamma–Gamma distributed channel using two receivers with SDRT and carry out a comprehensive transmission performance comparison of OOK, DPSK, and DQPSK formats. It is found that, in the strong turbulence scenario, the OOK and DPSK formats can gain SDRT as large as 19.5 and 20.3 dB at a BER of 103, respectively. Using SDRT, the modulation gains of the DPSK format over the OOK format are 3.2 dB in the strong turbulence scenario and 4.5 dB in the weak turbulence scenarios, respectively. In addition, in the moderate and strong turbulent scenario, it is found that the DPSK and DQPSK formats have almost the same BER performance.

2.10.2 Gamma–Gamma Distributed Channel Model[55]

In the Gamma–Gamma distributed atmospheric channel model,[56] the normalized light intensity can be partitioned into large-scale and small-scale atmospheric effects. Both the large-scale and small-scale intensity fluctuations follow a gamma distribution. Hence, the probability density function (pdf) of light intensity I is given by Ref.[57]

$$f(I)=\frac{2(\alpha\beta)^{\frac{\alpha+\beta}{2}}}{\Gamma(\alpha)\Gamma(\beta)}I^{\frac{\alpha+\beta-2}{2}}K_{\alpha-\beta}\left(2\sqrt{\alpha\beta I}\right);I>0 \tag{2.78}$$

where K_n (·) is the modified Bessel function of the second kind of order n, and $\Gamma(\cdot)$ is the gamma function. In Equation 2.78, the positive parameters and α and β represent the large-scale and small-scale optical wave intensity fluctuation, which are given by

$$\alpha=\left\{\exp\left[\frac{0.49\sigma_R^2}{\left(1+1.11\sigma_R^{12/5}\right)^{7/6}}\right]-1\right\}^{-1} \tag{2.79}$$

$$\beta=\left\{\exp\left[\frac{0.49\sigma_R^2}{\left(1+1.11\sigma_R^{12/5}\right)^{5/6}}\right]-1\right\}^{-1} \tag{2.80}$$

where $\sigma_R^2=1.23C_n^2k_r^{7/6}L^{11/6}$ is the Rytov variance, representing the variance of log-intensity fluctuation,[58] in which C_n^2 is the refractive index structure parameter, k is the wavenumber in vacuum, and L=distance between Tx and Rx.

The Rytov parameter is given by

$$\sigma_R^2 = 1.23 C_n^2 k^{7/6} L^{11/6} \tag{2.81}$$

L=communication distance. C_n^2 =refractive index structure parameter, which is determined by wind speed and altitude according to the H-V turbulence model. When $\sigma_R^2 < 1$, it means the light intensity fluctuation is weak, and when $\sigma_R^2 > 1$, it means strong intensity fluctuation. Since the Gamma-Gamma model covers all possible levels of turbulence, the performance of BER under the Gamma-Gamma model is used in this document.

According to the Hufnagel–Valley (H–V) turbulence model,[58] the refractive-index structure parameter C_n^2 is determined by wind speed (w) and altitude (h), which is given by

$$C_n^2(h) = 0.00594\left(\frac{w}{27}\right)^2 \left(10^{-5}h\right)^{10} e^{\left(-\frac{h}{100}\right)} + 2.7\times10^{-16} e^{-\frac{h}{1500}} + C_n^2(0)\cdot e^{-\frac{h}{1000}} \tag{2.82}$$

where $C_n^2(0)$ is the value of the structure constant on the ground, which is usually $1.7\times10^{-14}\text{m}^{-2/3}$. Since this Gamma–Gamma channel model can cover all turbulence scenarios from weak to strong,[58,54] it is employed in our study on the BER performance of various widely used modulation formats. When $\sigma_R^2 < 1$, the light intensity fluctuation is weak. The moderate intensity fluctuation is defined by $\sigma_R^2 \triangleq 1$. The strong intensity fluctuation emerges when $\sigma_R^2 > 1$. The saturation regime is associated with $\sigma_R^2 \to \infty$.

2.10.3 Performance without SDRT (Space Diversity Reception Technique)

In the Gamma–Gamma distributed channel, the transmitted signal is assumed to experience independent and identically distributed intensity fading. The received signal can be represented as $y = hx + n = \frac{\eta}{x} + n$, where parameter h is the Gamma–Gamma channel gain, I is the normalized light intensity/power, x is the transmitted signal (being logical "0" or "1"), n is the additive white Gaussian noise with zero mean and variance N_0=2, and η is the photo-current conversion ratio, and PD quantum efficiency. The BER of an optical signal transmitted in the Gamma–Gamma-distributed channel can be expressed by

$$p_e = \int_0^\infty p\left(I^2\gamma\right) f(I)\, dI \tag{2.83}$$

where $p\left(I^2\gamma\right)$ represents the BER of an optical signal transmitted in the additive white Gaussian noise (AWGN); γ =average SNR.

The BER of OOK, DPSK, and DQPSK, QPSK modulation formats via an AWGN channel can then be written by Refs.[59,60,61] respectively.[62,63,64]

$$p_{\text{OOK}} = Q\left(\sqrt{I^2\gamma/2}\right) \tag{2.84}$$

$$p_{\text{DPSK}} = \exp\left(-I^2\gamma\right) \tag{2.85}$$

$$p_{\text{DQPSK}} = \exp\left(-I^2\gamma\right)\left\{1 + \sum_{n=0}^{\infty}\left(\sqrt{2}-1\right)^n I_n\left(\sqrt{2}I^2\gamma\right) - \frac{1}{2} I_0\left(\sqrt{2}I^2\gamma\right)\right\} \tag{2.86}$$

where $Q(x)=\frac{1}{2}\text{erfc}\left(\frac{x}{\sqrt{2}}\right)$ and $\text{erfc}(x)$ are complementary error functions. By substituting Equations 2.84 and 2.86 into Equation 2.83, we obtain the BER of OOK, DPSK, and DQPSK signals in the Gamma-Gamma distributed turbulence channel for free-space link.

For BPSK under turbulence and detector noise with balanced detection, the system BER can be given as the summation of P('1'/'0')+P('0'/'1'). The probability of error, which is the BER, can be written as

$$P_e(I)=\int_0^\infty \frac{1}{2}\text{erfc}\left(\sqrt{\text{SNR}}\right)f(I)\,dI \tag{2.87}$$

$$P_e(I)=\int_0^\infty \frac{1}{2}\text{erfc}\left(\frac{hI}{\sqrt{N_0}}\right)\frac{2(\alpha\beta)^{(\alpha+\beta)/2}}{\Gamma(\alpha)\Gamma(\beta)}I^{(\alpha+\beta)/2-1}K_{\alpha-\beta}\left(2\sqrt{\alpha\beta I}\cdot dI\right) \tag{2.88}$$

We can use the Meijer G-function defined for the erfc(·) and $K_v(\cdot)$[65,66] as

$$\text{erfc}(x)=\frac{1}{\sqrt{\pi}}G_{1,2}^{2,0}\left[x\middle|\begin{matrix}1\\0,\frac{1}{2}\end{matrix}\right];\quad K_v(x)=\frac{1}{2}G_{1,2}^{2,0}\left[\frac{x^2}{4}\middle|\begin{matrix}-\\\frac{v}{2},\frac{v}{2}\end{matrix}\right] \tag{2.89}$$

If we define $\delta=\frac{h}{\sqrt{N_0}}$, then the average BER for the OOK modulation format can be expressed as

$$P_{\text{e-OOK}}=\frac{1}{2\sqrt{\pi}\Gamma(\alpha)\Gamma(\beta)}G_{3,2}^{2,2}\left[\begin{matrix}1-\beta,1-\alpha,1\\0.1-\frac{\alpha+\beta}{2},1/2\end{matrix}\right] \tag{2.90}$$

In terms of the Meijer G- function, the BER for the QPSK format can be expressed in the following[67]:

For BPSK and QPSK (twice of BPSK)

$$P_{\text{e-BPSK}}=\frac{(\alpha\beta)^{(\alpha+\beta)/2}}{2\sqrt{\pi}\Gamma(\alpha)\Gamma(\beta)}\delta^{(\alpha\beta)^{-(\alpha+\beta)/2}}G_{2,3}^{2,2}\left[\frac{\alpha\beta}{\delta}\middle|\begin{matrix}1-(\alpha+\beta)/2,1-(\alpha+\beta)/2\\\frac{\alpha-\beta}{2},-\frac{\alpha+\beta}{2},\frac{-\alpha+\beta}{2}\end{matrix}\right] \tag{2.91}$$

Hence, QPSK is doubling that of BPSK and given by Ref.[67]

$$P_{\text{e-QPSK}}=\frac{(\alpha\beta)^{(\alpha+\beta)/2}}{\sqrt{\pi}\Gamma(\alpha)\Gamma(\beta)}\delta^{(\alpha\beta)^{-(\alpha+\beta)/2}}G_{2,3}^{2,2}\left[\frac{\alpha\beta}{\delta}\middle|\begin{matrix}1-(\alpha+\beta)/2,\ 1-(\alpha+\beta)/2\\\frac{\alpha-\beta}{2},-\frac{\alpha+\beta}{2},\frac{-\alpha+\beta}{2}\end{matrix}\right] \tag{2.92}$$

where $\Gamma(\cdot)$ is the Gamma function, and $G(\cdot)$ is the Meijer-G function, as defined above for erfc(·) and $\Gamma(\cdot)$ function.

2.10.4 BER versus SNR under Moderate Turbulence

NOTE: BER versus SNR for PSK and QPSK modulated channels propagating under turbulence: a tentative estimation. Turbulence is moderate, i.e., parameters. For a pre-FEC BER of 2e-2, the SNR is ~17.5 dB. Scintillation and turbulence conditions are moderate. Note that for pre-FEC BER of 2e-2, SNR required is 17.5 dB or about 7.5 dB above the OSNR required for an atmospheric non-turbulent propagation link (Figures 2.41 and 2.42).

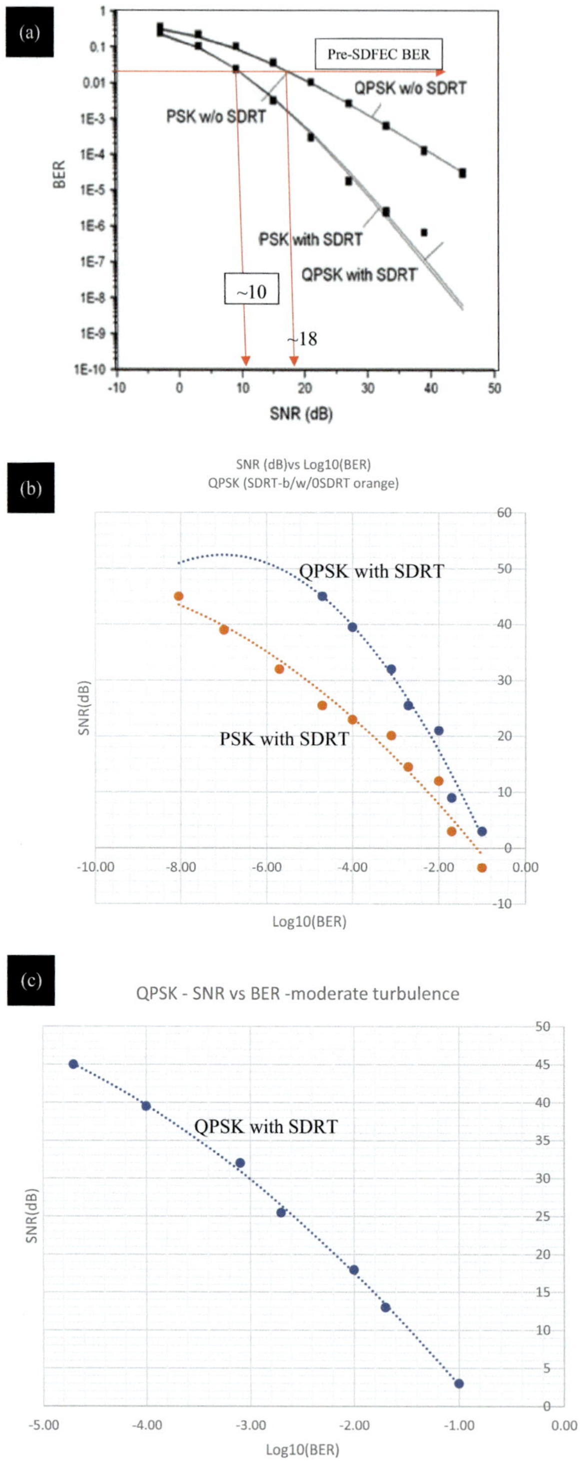

FIGURE 2.41 BER versus SNR for PSK and QPSK modulated channels propagating under turbulence. A tentative estimation. Turbulence is moderate, i.e., parameters: SDRT = space diversity reception technique. For a pre-FEC BER of 2e-2, the SNR is 18 and 10 dB without SDRT. Scintillation and turbulence condition: moderate. (a) P_e (BER) versus SNR (power); (b) SNR in dB) versus $\log_{10}$ (BER) for QPSK and PSK with SDRT; and (c) SNR versus $\log_{10}$ (BER) for QPSK modulation format.

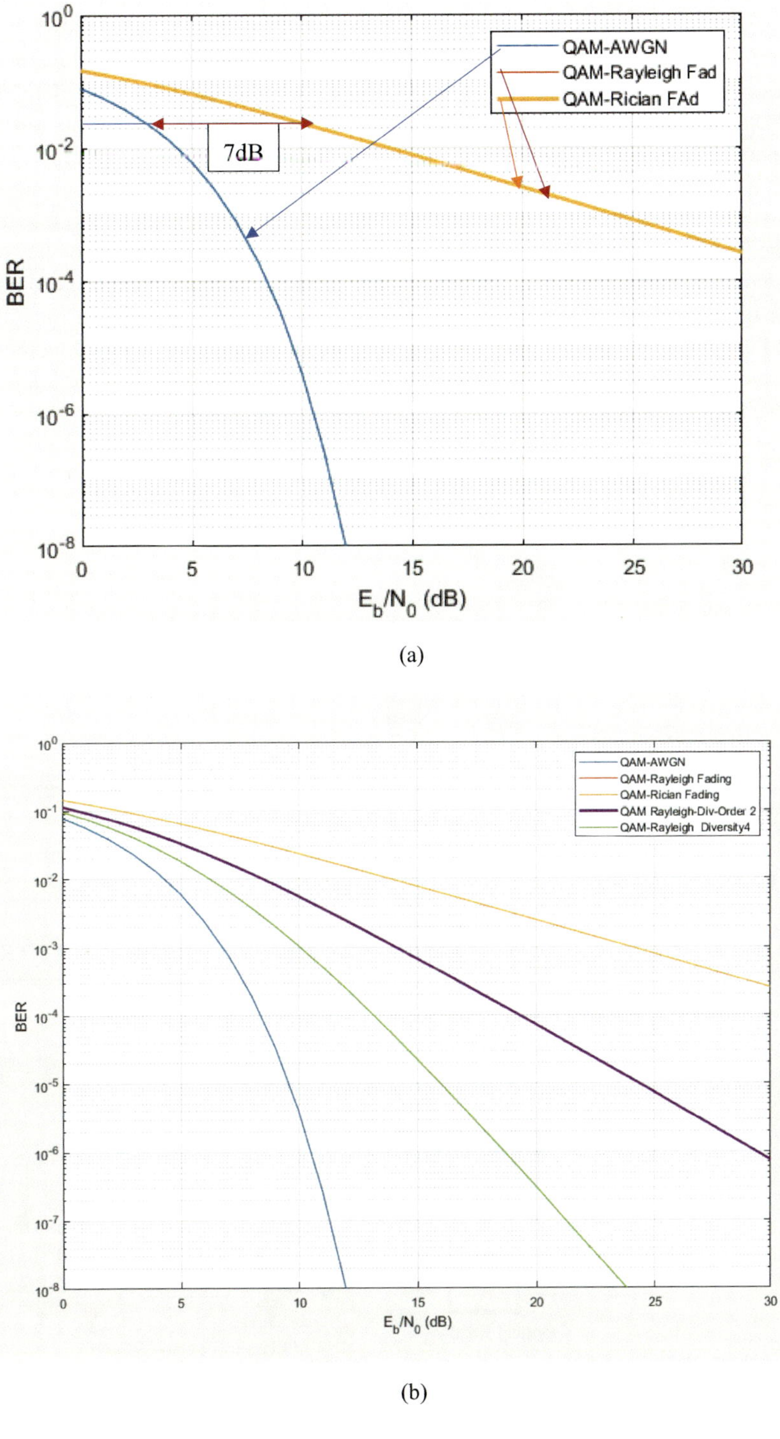

FIGURE 2.42 BER versus Eb/No for QAM under AWGN and Rayleigh fading channels. The Rician channel gives the same (identical) results as Rayleigh fading. Estimation using bertool.m. (a) Without space diversity reception detection (SDRT); (b) with SDRT in fading channels of order 2 and 4 (4 diversity dimensions) and without diversity (SDRT technique – space diversity reception technique). Rician and Rayleigh fading curves as indicated.

2.11 ADAPTIVE OPTICS FOR TBPS SPC TRANSMISSION

Figure 2.43 shows the schematic of an adaptive optics for optical mage of a telescope. The WFE of the image are obtained by the WFS,[68,69,70] which produces a matrix of spatial elements, and the phase distortion is obtained. Indeed, the micro-CCD (charge-coupled devices) elements are integrated with 2D arrays to measure the intensity distribution and the shift of their position in a matrix to give the spatial dispersion or broadening. An adaptive optical system is designed such that the phase error of a matrix is inverse of the incoming matrix of the element, or the two matrices of the elements are Hermitian. The response of the WFS is very slow compared to the few tens of picoseconds of the 25 GB rate, due to the image processing of the micro elements of the sensor. Such a WFS is only appropriate for optical systems in image processing in astronomy to correct beam alignment for imaging. In SPC, the adaptive optical system is quite different; it must be real time to correct and equalize the optical pulse sequence. This means that the control feedback must be within a few pico-seconds to avoid distortion of the short pulse period. Therefore, it is impossible for *p*s response for a bit period.

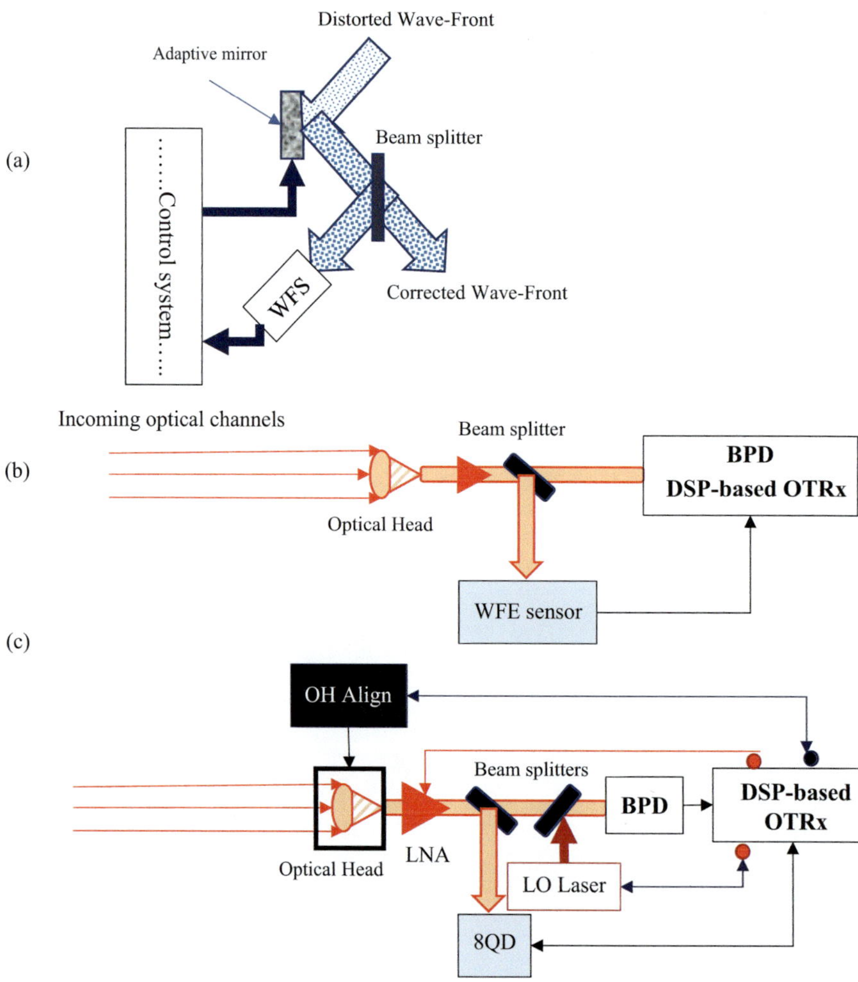

FIGURE 2.43 Adaptive optics or image correction for telescopes. Parallel beam and spatial phase front error. (a) Concept of wavefront sensor (WFS) and adaptive optics – parallel spatial distortion correction; (b) adaptive optics and SPC-based ORx; (c) Real-time adaptive optics ORx for Tbps SPC transmission, processing, and feedback control via DSP of BER and alignment control, and control of LNA gain. WFS=wavefront sensor.

In SPC, such an adaptive optics matrix is valid for parallel beam operation at slow speeds; this is not applicable for SPC at 100 Gbps per channel. The baud rate is 25 GB or 50 GB, with bit periods of 25 and 12.5 ps respectively. The spatial distortion comes from (i) the phase rotation of the beam cross-section due to air disturbance, with the polarization directions also rotating, and (ii) Doppler shift of the carrier frequencies. Such shifts can happen within an hour or, twice a day depending on the relative velocity of the satellites in the orbit and other receiving stations or photonic router.

Two adaptive optical systems are shown in Figure 2.43b and c, using the DSP processing algorithm to evaluate the time-domain distortion and feedback signal to correct the wavefront. Once the wavefront is corrected, the action of the phase correction can be performed once every period, which is much longer than the pulse period of the high-speed modulated lightwaves.

Alternatively, to the phase front corrector, a length of a few modes of fiber can be used to counteract the dispersion of the pulse sequence due to group delay time, opposite to that of the phase delay dispersion caused by the angle of arrival of the rays of the lightwave beam.

2.12 COHERENT COMPLEX INTEGRATED PLUGGABLE MODULE FOR SPC

Pluggable modules with small and compact forms offer much advantages for terrestrial fiber networks, data centers, space communication systems, and optical routers in space laser communication networking. In SPC, the compact modules are integrated into the transceiver section for transmitting and launching in association with optical amplifiers as a power booster and receiving low level with low noise amplifiers in the optical domain. Thus, these pluggable modules offer compact, low-noise, and low-energy components for SPC.

2.12.1 Mechanical Form Factor Evolution

When the coherent 40 Gb/s transponder was first introduced in 2008, it was a line card with all components spread out on multiple printed circuit boards (PCBs), as shown in Figure 2.1. To increase the number of interoperable vendors, the Optical Inter-Networking Forum (OIF) developed Multi-Source Agreements (MSA) for 100 Gb/s long-haul transponder form factors, as shown in the left two pictures of Figure 2.2. These MSA modules are not hot-pluggable but are fastened through screws and electrical connectors to the host PCB during line card assembly. Over time, many proprietary embedded modules, similar to the MSA modules, have been developed for advanced system performance, including those used in today's 90^{+}-GBaud-based (112 Gbaud) 800 Gb/s line cards.

2.12.2 Coherent Complex Pluggable Modules

Compact modules have been developed to take advantage of their low power, low cost, field-replaceable, and pay-as-you-grow features. The first such coherent pluggable module was the CFP, introduced around 2016, which evolved into the smaller CFP2-DCO and finally to today's QSFP, QSFP-DD, and OSFP modules, as shown in Figure 2.2. The next-generation OSFP-XD and QSFP-DD can accommodate 30+W of power and are preparing for 1.6 T per module. The significant reduction in size and power consumption of these hot-pluggable modules drastically improves the port density on switches, routers, and DWDM transport equipment.

Chapters 4–9, which follow this chapter, will address the technological essentials for space photonic communication systems and networks, including:

Chapter 4 on optical modulation in intensity binary format and complex but self-coherent detection with the NRZ or RZ formats, with or without suppression of the optical carrier. Self-coherent indicates that the lightwave beam is split into two paths, one is delayed by a one-bit period, and then combined to detect the differential phase and amplitudes of the two paths by interference.

Chapter 5 on advanced complex modulation, mainly the Mary-QAM, ranging from four-point constellation to 2^n (n=2, 4, 8, etc. constellation points such as QPSK, 16 or 64 QAM, etc. The reception subsystem by photonic circuit and digital processing to mitigate the dispersion or spatial spreading of the SPC system.

Chapter 6 on optical reception at ultra-high bit rate with Tbps capacity and integrated transceiver, especially for SPC inter-satellite constellation.

Chapter 7 on optical amplifications in different types for SPC (Space Photonic Communications), including HPA, LNA with low-level sensitivity input, noise suppression via optical filtering, and high gain to ensure that the output signal average can meet the OSNR requirements for the next stage transmission and routing.

Chapter 8 on all optical routing techniques and technologies for multi-directional add and drop or through transmission in the optical domain.

Chapter 9 on networking aspects and the employment of multi-wavelength sources for ultra-dense optical transmission ad routing in space inter-constellation networks.

Chapter 10, finally makes remarks and states the potential photonic systems and networking for space multi-orbital networking (Figures 2.44 and 2.45).

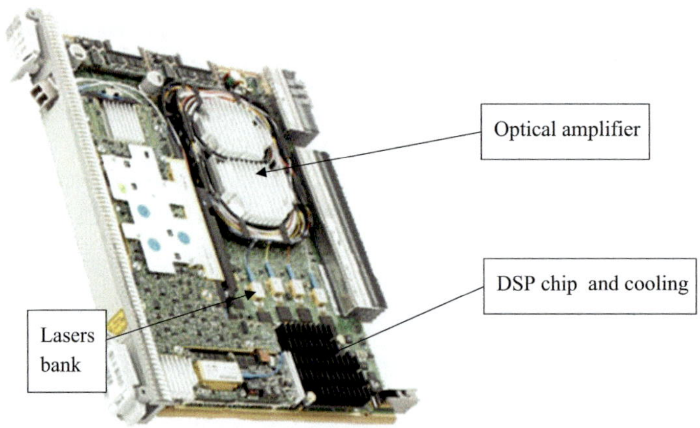

FIGURE 2.44 A 100 Gb/s line card using a discrete integration module, not pluggable. Now it is wholly integrated as given in Chapter 10 as QSFP (Quad Special Form Factor) for coherent reception.

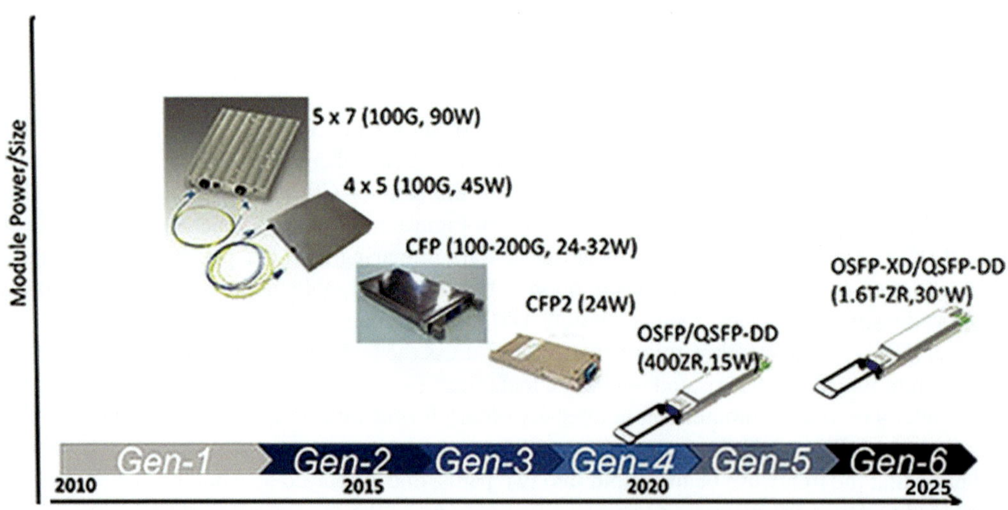

FIGURE 2.45 Module size and power consumption evolution from MSA to hot-pluggable modules from 2011 to 2024. No fiber amplifier integrated, but soon with Er: doped silica in planar structures. OSFP-XD=extra-dense op.tical special form package, CFP=compact form package, QSFP-DD=direct detection quad special form package, Gen=generation.

2.12.3 Optical System Performance Evolution[71,72,73]

Figure 2.46 shows the rOSNR for different modules at operating bit rates of 100–400 Gbps at different baud rates of 28–112 GBaud, respectively. It is noted that the higher the baud rate, the shorter the transmission distance in fiber due to chromatic dispersion because CD depends on the bandwidth of the signals. Thus, the wider the bandwidth the more chromatic delay times with respect to fiber length, leading to the broadening of the pulse sequence, and hence a limited length of transmission.

However, in space transmission, CD is not valid, but the spatial spreading of the beam cross-sectional area and the power boosting through HPA, and the low-noise sensitivity of the LNA or OpreA are important. Severe tests must be conducted to ensure the 400 GbBps of 112 GBaud in test transmission. The wider the bandwidth of the signals, the more sensitive the receiver must be and the bandwidth must be available to obtain acceptable BER under the same Soft-FEC for 100 Gbps module.

The 400 Gbps is mainly developed for data center networking. A data center in space is possible with the data center located in one of the OGS networks due to ultra-high-power consumption. Thus, space photonic networking can only contain the main communication and routing in space within the satellite constellations.

Therefore, this chapter describes the technological aspects of transmission, routing, and regeneration in space photonic communications.

The difference between hot-pluggable and embedded modules. Since embedded modules are large and can tolerate very high power consumption of up to ~100 W, the general trend is that embedded modules were first developed to provide the best system performance by using high-power DSPs and best-in-class optoelectronic components. Then, 3~4 years later, they were followed by hot-pluggable modules, which use next-generation DSP chips and more power-efficient, high-sensitive optoelectronic components, along low-noise amplifications, to achieve similar system performance (aiming for metro or long-haul applications) but at much lower power consumption.

As shown in Figure 2.46, the rOSNR is compared for coherent pluggable and embedded modules for 100, 200, and 400 Gb/s compound data rates. For each data rate, although the rOSNR numbers represent different generations of DSP, baud rate, and FEC algorithms, we can see that the rOSNR performance for pluggable and embedded modules is essentially the same for 100 and 200 Gb/s.

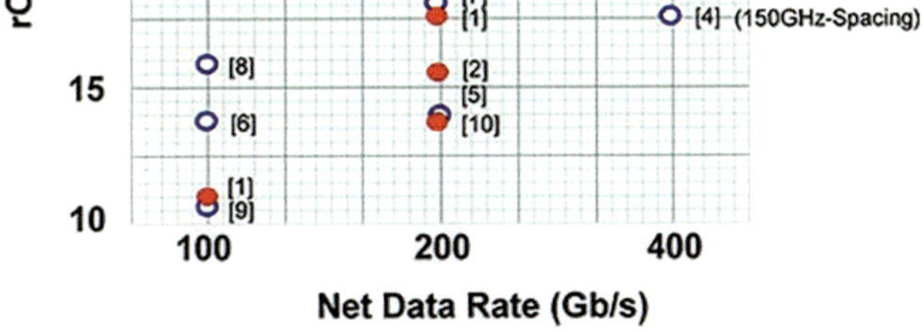

FIGURE 2.46 Required optical signal-to-noise ratio (rOSNR) of various pluggable and embedded modules at 100, 200, and 400 Gb/s. OFEC=optical fiber engineers conference.

In the case of 400 Gb/s, the best rOSNR for a state-of-the-art pluggable module is about 2.5 dB worse than that of the state-of-the-art embedded module. This is because the pluggable module uses 69 Gbaud (requiring an efficient 75 GHz-spacing DWDM multiplexer and demultiplexer). In comparison, the embedded module uses 90–95 Gbaud (requiring a twice-as-wide 150 GHz-spacing DWDM multiplexer and demultiplexer, and consequently a lower spectral efficiency, to accommodate the signal bandwidth and the roll-off the optical filters). Thus, higher sensitivity of the LNA in the space transceiver must be available. This can be provided by an integrated planar optical amplifier placed in front of the hybrid coupler as optical pre-processing ahead of the balanced photodetector pair, before the balanced photodetector and amplifiers. See Chapter 7 on transceiver systems or receiver structures using polarization-multiplexed complex IQ modulation.

2.13 TYPICAL POWER BUDGET

Figure 2.47 shows the typical power budget and power consumption of an LCT on the platform of an LEO constellation satellite, while Figure 2.48 illustrates the same for MEO, and Figure 2.49 for a GEO platform. Figure 2.47 lists the power budget for a three-hop optically amplified SPC link.

2.13.1 Link Budget for LEO Platform

LCT 163	LEO-GEO OISL			
Data rate	10 Gb/s	10 Gb/s	100 Gb/s	100 Gb/s
Link distance	45000 km			
Wavelength	1064 nm	1550 nm	1064 nm	1550 nm
Wavefront error RMSt	110 nm			
Modulation Format	OOK		DP-QPSK	
Telescope Aperture	163 mm	163 mm	163 mm	163 mm
Transmit telescope gain	**112,7 dB**	**109,5 dB**	**112,7 dB**	**109,5 dB**
Transmit terminal efficiency	**-2,8 dB**	**-1,8 dB**	**-2,8 dB**	**-1,8 dB**
TX Pointing loss @ 2 µrad	**-3,9 dB**	**-2,4 dB**	**-3,9 dB**	**-2,4 dB**
Free Space Loss	**-294,5 dB**	**-291,2 dB**	**-294,5 dB**	**-291,2 dB**
Telescope Aperture	163 mm	163 mm	163 mm	163 mm
Receiver telescope gain	**113,6 dB**	**110,4 dB**	**113,6 dB**	**110,4 dB**
Receive terminal efficiency	**-2,8 dB**	**-1,8 dB**	**-2,8 dB**	**-1,8 dB**
Fiber coupling / tracking loss	**-3,0 dB**	**-3,0 dB**	**-3,0 dB**	**-3,0 dB**
Total receive loss	-80,7 dB	-80,5 dB	-80,7 dB	-80,5 dB
Required receiver power	-39,3 dBm	-40,9 dBm	-39,7 dBm	-41,4 dBm
Optical power	13,7 W	9,0 W	12,4 W	8,2 W
Electrical power	68,3 W	50,2 W	62,1 W	45,6 W
Delta power (1064 - 1550 nm)	18,1 W		16,4 W	

(a)

LCT 163	LEO-Ground			
Data rate	10 Gb/s	10 Gb/s	100 Gb/s	100 Gb/s
Link distance	45000 km			
Wavelength	1064 nm	1550 nm	1064 nm	1550 nm
Wavefront error RMSt	110 nm			
Modulation Format	OOK		DP-QPSK	
Telescope Aperture	163 mm	163 mm	163 mm	163 mm
Transmit telescope gain	**112,7 dB**	**109,5 dB**	**112,7 dB**	**109,5 dB**
Transmit terminal efficiency	**-2,8 dB**	**-2,8 dB**	**-1,8 dB**	**-1,8 dB**
TX Pointing loss @ µrad	**-8,3 dB**	**-5,1 dB**	**-3,5 dB**	**-2,2 dB**
Free Space Loss	**-294,5 dB**	**-291,2 dB**	**-294,5 dB**	**-291,2 dB**
Telescope Aperture	800 mm	800 mm	800 mm	800 mm
Receiver telescope gain	**127,5 dB**	**124,2 dB**	**127,5 dB**	**124,2 dB**
Receive terminal efficiency	**-5,3 dB**	**-3,5 dB**	**-5,3 dB**	**-3,5 dB**
Atmospheric effects after AO	**-10,0 dB**	**-10,0 dB**	**-10,0 dB**	**-10,0 dB**
Total receive loss	-80,7 dB	-79,0 dB	-74,9 dB	-75,1 dB
Required receiver power	-40,2 dBm	-41,8 dBm	-39,7 dBm	-41,4 dBm
Optical power	11,2 W	5,3 W	3,3 W	2,4 W
Electrical power	56,1 W	26,3 W	16,6 W	11,9 W
Delta power (1064 - 1550 nm)	29,8 W		4,7 W	

(b)

FIGURE 2.47 (a) For an LCT (Light Comm Terminal) with a 162.5 mm optical head (OH) aperture; and (b) towards an 800 mm ground station with adaptive optics from LEO satellites.

2.13.2 Link Budget for MEO Platform

LCT 400	GEO-GEO OISL			
Data rate	10 Gb/s	10 Gb/s	1000 Gb/s	1000 Gb/s
Link distance	45000 km			
Wavelength	1064 nm	1550 nm	1064 nm	1550 nm
Wavefront error RMSt	110 nm			
Modulation Format	OOK		DP-QPSK	
Telescope Aperture	400 mm	400 mm	400 mm	400 mm
Transmit telescope gain	**120,6 dB**	**117,3 dB**	**120,6 dB**	**117,3 dB**
Transmit terminal efficiency	**-2,8 dB**	**-1,8 dB**	**-2,8 dB**	**-1,8 dB**
TX Pointing loss @ 2 µrad	**-3,9 dB**	**-2,4 dB**	**-3,9 dB**	**-2,4 dB**
Free Space Loss	**-294,5 dB**	**-291,2 dB**	**-294,5 dB**	**-291,2 dB**
Telescope Aperture	400 mm	400 mm	400 mm	400 mm
Receiver telescope gain	**121,4 dB**	**118,2 dB**	**121,4 dB**	**118,2 dB**
Receive terminal efficiency	**-2,8 dB**	**-1,8 dB**	**-2,8 dB**	**-1,8 dB**
Fiber coupling / tracking loss	**-3,0 dB**	**-3,0 dB**	**-3,0 dB**	**-3,0 dB**
Total receive loss	-65,0 dB	-64,8 dB	-65,0 dB	-64,8 dB
Required receiver power	-39,3 dBm	-40,9 dBm	-29,7 dBm	-31,4 dBm
Optical power	0,4 W	0,2 W	3,4 W	2,2 W
Electrical power	1,9 W	1,4 W	16,9 W	12,4 W
Delta power (1064 - 1550 nm)	0,5 W		4,5 W	

(a)

LCT 400	GEO-Ground			
Data rate	10 Gb/s	10 Gb/s	1000 Gb/s	1000 Gb/s
Link distance	45000 km			
Wavelength	1064 nm	1550 nm	1064 nm	1550 nm
Wavefront error RMSt	110 nm			
Modulation Format	OOK		DP-QPSK	
Telescope Aperture	400 mm	400 mm	400 mm	400 mm
Transmit telescope gain	**120,6 dB**	**117,3 dB**	**120,6 dB**	**117,3 dB**
Transmit terminal efficiency	**-2,8 dB**	**-2,8 dB**	**-1,8 dB**	**-1,8 dB**
TX Pointing loss @ µrad	**-8,3 dB**	**-5,1 dB**	**-3,5 dB**	**-2,2 dB**
Free Space Loss	**-294,5 dB**	**-291,2 dB**	**-294,5 dB**	**-291,2 dB**
Telescope Aperture	800 mm	800 mm	800 mm	800 mm
Receiver telescope gain	**127,5 dB**	**124,2 dB**	**127,5 dB**	**124,2 dB**
Receive terminal efficiency	**-5,3 dB**	**-3,5 dB**	**-5,3 dB**	**-3,5 dB**
Atmospheric effects after AO	**-10,0 dB**	**-10,0 dB**	**-10,0 dB**	**-10,0 dB**
Total receive loss	-72,8 dB	-71,2 dB	-67,1 dB	-67,3 dB
Required receiver power	-40,2 dBm	-41,8 dBm	-29,7 dBm	-31,4 dBm
Optical power	1,9 W	0,9 W	5,5 W	3,9 W
Electrical power	9,3 W	4,3 W	27,3 W	19,6 W
Delta power (1064 - 1550 nm)	4,9 W		7,7 W	

(b)

FIGURE 2.48 (a) For an LCT (Light comm terminal) with a 162.5 mm aperture OH, and (b) towards an 800 mm ground station with adaptive optics from GEO satellites.

2.13.3 Link Budgets for GEO Constellation Platform

LCT 400	GEO-GEO OISL			
Data rate	10 Gb/s	10 Gb/s	1000 Gb/s	1000 Gb/s
Link distance	45000 km			
Wavelength	1064 nm	1550 nm	1064 nm	1550 nm
Wavefront error RMSt	110 nm			
Modulation Format	OOK		DP-QPSK	
Telescope Aperture	400 mm	400 mm	400 mm	400 mm
Transmit telescope gain	120,6 dB	117,3 dB	120,6 dB	117,3 dB
Transmit terminal efficiency	-2,8 dB	-1,8 dB	-2,8 dB	-1,8 dB
TX Pointing loss @ 2 µrad	-3,9 dB	-2,4 dB	-3,9 dB	-2,4 dB
Free Space Loss	-294,5 dB	-291,2 dB	-294,5 dB	-291,2 dB
Telescope Aperture	400 mm	400 mm	400 mm	400 mm
Receiver telescope gain	121,4 dB	118,2 dB	121,4 dB	118,2 dB
Receive terminal efficiency	-2,8 dB	-1,8 dB	-2,8 dB	-1,8 dB
Fiber coupling / tracking loss	-3,0 dB	-3,0 dB	-3,0 dB	-3,0 dB
Total receive loss	-65,0 dB	-64,8 dB	-65,0 dB	-64,8 dB
Required receiver power	-39,3 dBm	-40,9 dBm	-29,7 dBm	-31,4 dBm
Optical power	0,4 W	0,2 W	3,4 W	2,2 W
Electrical power	1,9 W	1,4 W	16,9 W	12,4 W
Delta power (1064 - 1550 nm)	0,5 W		4,5 W	

(a)

LCT 400	GEO-Ground			
Data rate	10 Gb/s	10 Gb/s	1000 Gb/s	1000 Gb/s
Link distance	45000 km			
Wavelength	1064 nm	1550 nm	1064 nm	1550 nm
Wavefront error RMSt	110 nm			
Modulation Format	OOK		DP-QPSK	
Telescope Aperture	400 mm	400 mm	400 mm	400 mm
Transmit telescope gain	120,6 dB	117,3 dB	120,6 dB	117,3 dB
Transmit terminal efficiency	-2,8 dB	-2,8 dB	-1,8 dB	-1,8 dB
TX Pointing loss @ µrad	-8,3 dB	-5,1 dB	-3,5 dB	-2,2 dB
Free Space Loss	-294,5 dB	-291,2 dB	-294,5 dB	-291,2 dB
Telescope Aperture	800 mm	800 mm	800 mm	800 mm
Receiver telescope gain	127,5 dB	124,2 dB	127,5 dB	124,2 dB
Receive terminal efficiency	-5,3 dB	-3,5 dB	-5,3 dB	-3,5 dB
Atmospheric effects after AO	-10,0 dB	-10,0 dB	-10,0 dB	-10,0 dB
Total receive loss	-72,8 dB	-71,2 dB	-67,1 dB	-67,3 dB
Required receiver power	-40,2 dBm	-41,8 dBm	-29,7 dBm	-31,4 dBm
Optical power	1,9 W	0,9 W	5,5 W	3,9 W
Electrical power	9,3 W	4,3 W	27,3 W	19,6 W
Delta power (1064 - 1550 nm)	4,9 W		7,7 W	

(b)

FIGURE 2.49 (a) For an LCT with a 400 mm aperture, and (b) towards an 800 mm ground station with adaptive optics.

2.13.4 Link Budget for Three-Hop SPC Link

For an overview of the power budget for the three-hop optically amplified SPC link, see Tables 2.2–2.5 sequentially: **OSNR0-OSNR1**, **OSNR1-OSNR2**, **OSNR2-OSNR3**, and finally **OSNR3-OSNR4** and **receiver performance**. The OSNR at the end of each hop must satisfy the rOSNR for launching into the next hop till it can no longer meet the ratio. At this point, the channels must be repeated, or the signals are converted to the electrical domain and then used to modulate the next transmitter, which is amplified to satisfy the signal-to-noise ratio.

2.14 SUMMARY

In summary, this chapter provides a brief overview of photonic communications in space systems and networks, focusing on the details of modulated laser beam propagation through the space medium, including its transmittance and absorption characteristics so as to drive the probability of errors for advanced modulation formats. Photonic subsystems, optical amplification, and receiving concepts are also described.

Optical coherent transceivers are required in SPC, as the photonic system is bidirectional, meaning both transmitting and receiving channels are required to minimize the alignment functionalities of the whole system. They are analyzed and tested in this chapter, followed by the testing and performance of an integrated transceiver of 100 Gbps DPM-QPSK module in Chapter 9, where the power-limited case of transmission and the quantum SPC for secret key transmission is described.

The rest of the chapters of the book will give detailed description of the components of SPC transmission system, phus modulation techniques and modulators as well as the receiver technological features and digital signal processing to mitigate distortion or onlinear effects on hannel signals and transmission performances. Advanced optical routing techniques will aloso be address for intersatellite optical networking.

TABLE 2.2
OSNR0-OSNR1

ISCL Type (aperture) - Original	*L* (km)	OSNRo (dB)	Launch-P (dBm)	TxAnt G (dB)	ISL IL (dB)	Tx_ OH (dB)	Rx_ OH-G (dB)	OPreAo (-60-10 dBm)	OPreA-G1 (dB)	NF1 (0.1 nm)-dB	*F* (lin.)	OSNR1 (dB)
IESA Format Terms	**Link Distance**	**OSNR0**	**Launch Power**	**TX Ant Gain (dB)**	**ISL IL (dB)**	**TX OH (dB)**	**RX Ant Gain (dB)**	**Power into OPreAmp (dBm)**	**Gain (dB) OPreAmp**	**Noise Figure NF1 (dB)**	**Noise Factor F1 (Linear)**	**OSNR1 (dB)**
MEO_MEO												
MEO_ to MEO	**100 GQPSK28**							**HOP 1-HOP2 (50 dB OpreA)**				
N3_MEO-MEO Myn	**28,000**	**15**	**40**	109	−28,778	−4.51	110	−33.29	35	5	1.581139	**14.31**
N9_MynMEO-MynMEO **(162 mm)**	***28,700***	**15**	40	110	−287	−3.72	110	−30.72	35	5	1.58	**14.61**
N10_MynMEO to MynMEO **(162 mm)**	***35,000***	**15**	40	110	−289	−3.72	110	−32.72	35	5	1.58	**14.39**
LEO-MEO	**100 GQPSK28**							**HOP 1 (GA) → HOP2 (35 dB OPreA)**				
N14_SdaLEO –MynMEO **(80 mm)**	***23,400***	**15**	40	104	−285.56	−2.1	110.34	−33.32	35	5	1.58	**14.31**
N15_MynMEO to SdaLEO **(80 mm)**	***45,000***	**15**	40	104	−291.24	−2.1	110.35	−38.99	40	5	1.58	**12.87**
N19_CndorMk3-Myn-MEO **(80 mm)**	***45,000***	**15**	40	104.2	−291.24	−2.1	110.34	−38.8	40	5	1.58	**12.94**
MEO-GEO	**100 GQPSK28**							**HOP 1 (GA) → HOP 2 (50 dB OPreA)**				
N22_CndorMk3 to Myn_GEO **(80 mm)**	***45,000***	**20**	43	104.2	−291.24	−2.84	**104**	−42.88	40	5	1.58	**12.28**
N23_CndorMk3 to SpaceCable**(80 mm)**	***8000***	**15**	40	104.2	−276.**24**	−3.64	104.2	−31.48	40	5	1.58	**14.54**
N24_CndorMEO to SpaceCable **(162.5 mm)**	***8000***	**15**	40	110.35	−276.24	−3.64	110.35	−19.18	40	5	1.58	**14.97**
N25_CndorMEO to Space-Cab **(162.5 mm)**	***16,000***	**15**	40	110.35	−282.26	−3.64	110.35	−25.2	35	5	1.58	**14.89**
N26_CndorMEO to OriGEO	***384,400***	**20**	**50**	110.35	−309	−3.64	110.35	−41.94	**35**	5	1.58	**13.05**
OGS-MEO	**100 GQPSK28**							**HOP 1 (GA) → HOP2 (50 dB OPreA)**				
N29_CondorMEO to OGS (162.5 mm)	***18,000***	**15**	40	110.35	−283.28	−3.64	110.04	−26.53	40	5	1.58	**14.85**
N30_OGS -CndorMEO	***18,000***	**15**	40	113.74	−283.28	−3.64	110.35	−22.83	40	5	1.58	**14.93**

IL = insertion loss, OPreA = optical pre-amplifier, OSNRo = OSNR at the origin, ISCL = intersatellite Communication Link, Rx OH = optical head at the reception site, TxAnt_G = antenna gain at the transmitter site (optical telescope optics), NF = noise figure, lin. = linear scale.

TABLE 2.3
OSNR1 –OSNR2

OSNR1 (dB)	OPreA-out (dBm)	Margin (dB)	RxOH-IL (dB)	OA2 preA	Pin (PreA)	NF2 (lin.-5 dB	P2o-MidOA2 (dBm) → WSS1	Sat-P2o (dBm)	OSNR2 (dB)
OSNR1 (dB)	Power Out OPreAmp (dBm)	Margin (dB)	Rx -OH (dB)	OPAmp 2(dB)	Power In (dBm)	Noise Figure 2 (linear)	Power In - Mid OA to go to WSS (dBm)	Amp. Saturated Power (dBm)	OSNR2 (dB)
				HOP2 → HOP3 (50 dB OPreA)					
14.31	1.71	−6	−4.43	50	−33.29	1.58	−32.01	17.99	**13.72**
14.61	4.28	−6	−4.43	50	−30.72	1.58	−26.87	23.13	**14.25**
14.39	2.28	−6	−4.43	50	−32.72	1.58	−30.87	19.13	**13.86**
				HOP2 → HOP3 (50 dB OPreA)					
14.31	1.68	−6	−4.43	50	−33.32	1.58	−32.07	10	**13.72**
12.87	1.01	−7	−4.43	50	−38.99	1.58	−39.41	10	**12.87**
12.94	1.2	−5	−4.43	50	−38.8	1.58	−37.03	10	**12.94**
12.28	−2.88	−3	−4.43	50	−42.88	1.58	−46.19	10	**9.65**
14.54	8.52	−5	−4.43	50	−31.48	1.58	−22.39	10	**14.12**
14.97	20.82	−6	−4.43	50	−19.18	1.58	1.21	10	**14.97**
14.89	9.8	−6	−4.43	50	−25.2	1.58	−15.83	10	**14.89**
13.05	−6.94	−6	−4.43	50	−41.94	1.58	−59.31	10	**13.05**
				HOP2 → HOP3 (50 dB OPreA)					
14.85	13.47	−10	−4.43	50	−26.53	1.58	−17.49	10	**14.70**
14.93	17.17	−10	−4.43	50	−22.83	1.58	−10.09	10	**14.9**

Pin = input power, OH = optical head, RxOH = optical head at reception site.

TABLE 2.4
OSNR2 -OSNR3

OSNR2 (dB)	WSS (IL)	Inter-WSS1 G (dB)	NF3(dB)	NF3 (lin)	Power P3 (dBm)	WSS2 (IL)	Margin (dB)	Tx Optics (IL)	PinOPA (dBm)/Orx-E	OPA-OISL G	NF (dB)	NF (lin)	out-OISL (dBm)	OSNR3 (dB) before AF4(OPA)
OSNR2 (dB)	WSS1-IL (dB)		No WSS			WSS2-IL (dB)	Margin (dB)	TX-OH IL (dB)	Power into OPA	OPA Gain (dB)	OPA Noise Figure (dB)	Noise Factor F (lin.	OPA Power Out (dBm)	OSNR3
WSS → HOP 3- Assuming NF Same as in Previous (OpreAmp)														
13.72	−6.4					−6.4	−6	−4.43	−5.24	40	6	1.99	34.76	**13.72**
14.25	−6.4					−6.4	−6	−4.43	−0.1	40	6	1.99	39.9	**14.2475**
13.86	−6.4					−6.4	−6	−4.43	−4.1	40	6	1.99	35.9	**13.86**
WSS → HOP 3														
13.72	−6.4					−6.4	−6	−4.43	−13.23	40	6	1.99	26.77	**13.71**
12.87	−6.4					−6.4	−6	−4.43	−13.23	40	6	1.99	26.77	**12.86**
12.94	−6.4					−6.4	−6	−4.23	−13.03	30	6	1.99	43.97	**12.93**
WSS → HOP 3														
9.65	−6.4					−6.4	−6	13.72233	4.92233	40	6	1.99	42.09	**9.65**
14.12	−6.4					−6.4	−6	−4.23	−13.03	40	6	1.99	51.29	**14.11**
14.97	−6.4					−6.4	−6	−4.23	−13.03	40	6	1.99	63.59	**14.96**
14.89	−6.4					−6.4	−6	−4.23	−13.03	40	6	1.99	57.57	**14.88**
13.05	−6.4					−6.4	−6	−4.23	−13.03	40	6	1.99	51.83	**13.04**
WSS → HOP 3														
14.70	−6.4					−6.4	−6	−4.23	−13.03	40	6	1.99	69.24	**14.69**
14.9	−6.4					−6.4	−6	−4.23	−13.03	40	6	1.99	68.94	**14.86**

TABLE 2.5
OSNR3 -OSNR4 & Performance2

OSNR3 (dB) before (OPA)	P4o-noAmp3(dBm)	Satd-OPA Out (dBm)	Link 4 (loss)+Marg (10 dB)	Input to OPreAmp (dBm)	low-in Pre-amp G	OSNR4 (fin) dB	Fin P-Rx(dBm)	100G/200G QPSK25/50	Aca Co-QPSK(dBm)	ReqRx(QPSK)-coh Aca 200G
OSNR3	Power Out (Po4) dBm	OPA Saturated Power (dBm)	Hop4 IL (dB)+Margin (10 dB)	Power in OPreAmp (dBm)	OPreAmp Gain (dB)	OSNR4	Power into Orx (dBm)	TRx 100 G/200 G type	100 G ORx Sensitivity (FEC) min. (dBm)	200 G ORX Sensitivity (dBm)
	HOP3 → HOP4 (Terminus-RX Performance)									
13.72	30.33	35	−77	−42	35	**10.32**	−7.00	OK for AC1200	−42	−21.00
14.2475	35.47	35	−77.4	−42.4	35	**10.31**	−7.40	OK for AC1200	−42	−21.00
13.86	31.47	35	−78.2	−43.2	35	**9.68**	−8.20	OK for AC1200	−42	−21.00
13.71	22.34	35	−77.2	−42.2	35	**10.20**	−7.20	OK for AC1200	−42	−21.00
12.86	22.34	35	−78.2	−43.2	35	**9.27**	−8.20	OK for AC1200	−42	−21.00
12.93	12.74	40	−78.2	−38.2	35	**11.43**	−3.20	OK for AC1200	−42	−21.00
9.65	58.64	40	−77	−37	35	**9.06**	−2.00	OK for AC1200	−42	−21.00
14.11	22.74	40	−78.2	−38.2	35	**12.23**	−3.20	OK for AC1200	−42	−21.00
14.96	22.74	41	−77.6	−36.6	35	**13.33**	−1.60	OK for AC1200	−42	−21.00
14.88	22.74	40	−77.7	−37.7	35	**12.90**	−2.70	OK for AC1200	−42	−21.00
13.04	22.74	42	−92.2	−50.2	35	**4.16**	−15.20	OK for AC1200	−42	−21.00
	HOP3 → HOP4 (Terminus-RX Performance)									
14.69	22.74	40	−76.9	−36.9	35	**13.05**	−1.90	OK for AC1200	−42	−21.00
14.86	22.74	41	−76.8	−35.8	35	**13.49**	−0.80	OK for AC1200	−42	−21.00

NOTES

1 L. N. Binh, *Advanced Digital Optical Communications*, 2nd Ed. CRC Press, Taylor & Francis: Boca Raton, FL, 2015.
2 R. Bedington, J. M. Arrazola, and A. Ling, "Progress in satellite quantum key distribution", arXiv:1707.03613v2 [quant-ph] 25.
3 J.-P. Bourgoin, et al., "A comprehensive design and performance analysis of low Earth orbit satellite quantum communication", *New Journal of Physics*, 15, 023006, 2013.
4 P. W. Kruse, et al., *Elements of Infrared Technology: Generation, Transmission and Detection.* John Wiley & Sons: New York, 1962.
5 I. I. Kim, B. McArthur, and E. Korevaar, "Comparison of laser beam propagation at 785 nm and 1550 nm in fog and haze for optical wireless communications", *Proceeings of SPIE*, 4214, 26–37, 2001.
6 M. Al Naboulsi, H. Sizun, and F. de Fornel, "Fog attenuation prediction for optical and infrared waves", *Journal SPIE (International Society for Optical Engineering*, 43(2), 319–329, 2004.
7 Dewpoint temperature, Td, is the temperature at which the water vapor in air at constant barometric pressure condenses into liquid water at the same rate at which it evaporates.
8 T. H. Carbonneau and D. R. Wisely, "Opportunities and challenges for optical wireless; the competitive advantage of free space telecommunications links in today's crowded market place", SPIE Conference on Optical Wireless Communications, Boston, MA, vol. 3232, 1998.
9 D. Giggenbach, et al, "Space communications protocols for future optical satellite-down links", *The 62nd International Astronautical Congress*, Cape Town, SA, 2011.
10 D. Giggenbach, "Mobile optical high-speed data links with small terminals", *Proceedings of SPIE - The International Society for Optical Engineering*, Berlin, September 2009.
11 L. C. Andrews and R. L. Phillips, *Laser Beam Propagation through Random Media*, 2nd Ed. SPIE-Press: Bellingham, WA, 2005.
12 H. Henniger and A. Gonzalez, "Transmission scheme and error protection for simplex long-distance atmospheric FSO systems", *Mediterranean Journal of Electronics and Communications*, 2(3), 118–126, 2006.
13 D. Giggenbach and H. Henniger, "Fading-loss assessment in atmospheric free-space optical communication links with on-off keying", *Optical Engineering*, 47(4), 046001, 2008.
14 T. Jono, et al., "Report on DLR-JAXA joint experiment: The Kirari Optical Downlink to Oberpfaffenhofen (KIODO)", JAXA, 2007. ISSN 1349-1121.
15 A. Carrasco-Casado and R. Mata-Calvo "Free-space optical links for space communication networks". In B. Mukherjee, I. Tomkos, M. Tornatore, P. Winzer, and Yongli Zhao (eds., *Springer Handbook of Optical Networks.* Springer Nature, 2020. https://doi.org/10.1007/978-3-030-16250-4_34
16 Thorlab GRIN lense collimator, https://www.thorlabs.com/newgrouppage9.cfm?objectgroup_ID=1209
17 B. J. Klein and J. J. Degnan, "Optical antenna gain. 1: Transmitting antennas", *Applied Optics*, 13(9), 2134–2141, 1974.
18 A. Carrasco and R. Mata-Calvo, "Free space optical links for space communications", https://doi.org/10.1007/978-3-030-16250-4_34.
19 D. Giggenbach, "Optimierung der optischen Freiraumkommunikation durch die Atmosphäre: Focal array receiver", Ph.D. dissertation, Universität der Bundeswehr München, Munich, Germany, 2004.
20 D. Giggenbach and H. Henniger, "Fading-loss assessment in atmospheric free-space optical communication links with on-off keying", *Optical Engineering*, 47(4), 046001, 2008.
21 PILab GmbH, simulation package.
22 ViaLight GmbH, "System design of a laser terminal for tower-to-tower applications", January 2015.
23 N. Perlot, "Characterization of signal fluctuations in optical communications with intensity modulation and direct detection through the turbulent atmospheric channel", PhD thesis, 2005.
24 K. Kiasaleh, "On the probability density function of signal intensity in free-space optical communications systems impaired by pointing jitter and turbulence", *SPIE Optical Engineering*, 33, 3748–3757, 1994.
25 G. G. Ortiz, S. Lee, S. Monacos, M. Wright, and A. Biswas, "Design and development of a robust ATP subsystem for the Altair UAV-to-Ground Lasercomm 2.5 Gbps Demonstration", *Proceedings of SPIE: The International Society for Optical Engineering*, 4975, 103–114, 2003.
26 L. B. Stotts, et al., "Free-space optical communications link budget estimation", *Applied Optics*, 49(28), 5333, 2010.
27 T. S. McKechnie, *General Theory of Light Propagation and Imaging through the Atmosphere*, Springer Series of Optics Sciences, vol. 196. Springer, pp. 120 & 123, 2016, DOI: 10.1007/978-3-319-18209-4.

28 M. J. Emes, et al., "Estimating the turbulence length scales from cross-correlation measurements in the atmospheric surface layer", *21st Australasian Fluid Mechanics Conference*, Adelaide, 10 August–13 December, 2018.
29 S. Schaefer, M. Gregory, and W. Rosenkranz, "Coherent receiver design based on digital signal processing in optical high-speed intersatellite links with M-phase-shift keying", *Optical Engineering*, 55(11), 111614, 2016.
30 S. Schaefer, et al., "Numerical investigation of a free-space optical coherent communication system based on optical phase-locked loop techniques for high-speed intersatellite data transmission", *Conference Paper*, 2015, https://www.nt.tf.uni-kiel.de/en/publications/conferences-and-journals-papers/schaefer_itg2015d.pdf
31 See L. Khermosh, "Functions of computing FEC overhead", IEEE802.org, Access date 10.04.2024.
32 G. P. Agrawal, *Fiber Optic Communication Systems*, Chapter 7, Section 7.2, page 305, 4th Edition. Academic Press: Cambridge, MA, 2010 (Eq. 7.2.8 &7.2.9) of page 305.
33 "Laser beam testing", https://www.phasics.com/en/application-areas/laser-measurement-beam-profiling-adaptive-optics/wavefront-sensor-for-laser-beam-qualification/
34 "Fast signal quality monitoring for coherent communications enabled by CNN-based EVM estimation", https://research.chalmers.se, 2023-01-01 22:06 UTC.
35 I. Fatadin, "Calibration of estimated ber from error vector magnitude with carrier phase recovery", *Advances in Optics and Photonics*, 1(1), 1–6, 2017.
36 R. A. Shafik, et al., "On the extended relationships among EVM, BER and SNR as performance metrics", *International Conference on Electrical and Computer Engineering*, Bangladesh, pp. 408–411, 2006.
37 R. Schmogrow, et al., "Error vector magnitude as a performance measure for advanced modulation formats", *IEEE Photon Technology Letters*, 24, 61–63, 2012.
38 L.-G. Dany, et al. "Coherent detection of optical quadrature phase-shift keying signals with carrier phase estimation", *Journal of Lightwave Technology*, 24, 12–21, 2006.
39 G. Goldfarb and G. Li, "BER estimation of QPSK homodyne detection with carrier phase estimation using digital signal processing", *Optics Express*, 14, 8043–8053, 2006.
40 (a) IEEE, IEEE Standard 802.11b-1999. IEEE Standard for Wireless LAN Medium Access Control (MAC) and Physical Layer (PHY) Specifications: High Speed Physical Layer; (b) IEEE, IEEE Standard 802.11a-1999. IEEE Standard for Wireless LAN Medium Access Control (MAC) and Physical Layer (PHY) Specifications: High Speed Physical Layer in the 5GHz Band.
41 Le Binh personal notes only – Huawei confidential.
42 M. Eiselt, A. Dochhan, and W. Rosenkranz, "OSNR sensitivity of multi-level modulation formats", OFC, 2010.
43 A few-mode fiber is a fiber at the operating wavelength, the number of guided modes is higher than unity and up to 16. That is the single mode fiber at the operating wavelength 1550 nm, can have its diameter large and/or the refractive index difference higher than 0.3%, the common requirement for single mode operation at 1550 nm. See Appendix 3 of Chapter 12 for the design of few-mode fibers.
44 W. Gappmair and M. Flohberger, "Error performance of coded FSO links in turbulent atmosphere modelled by Gamma- Gamma distribution", *IEEE Transactions on Wireless Communications*, 8(5), 2209–2213, 2009.
45 N. Letzepis, I. Holland, and W. Cowley, "The Gaussian free space optical MIMO channel with Q-ary pulse position modulation", *IEEE Transactions on Wireless Communications*, 7(5), 1744–1753, 2008.
46 J. Li, J. Q. Liu, and D. P. Taylor, "Optical communication using subcarrier PSK intensity modulation through atmospheric turbulence channels", *IEEE Transactions on Communications*, 55(8), 1598–1606, 2007.
47 M. K. Simon and M.-S. Alouini, *Digital Communication over Fading Channels: A Unified Approach to Performance Analysis*. John Wiley & Sons, Inc.: New York, 2000.
48 X. Zhu and J. M. Kahn, "Free-space optical communication through atmospheric turbulence channels", *IEEE Transactions on Communications*, 50(8), 1293–1300, 2002.
49 R. L. Phillips and L. C. Andrews, "Measured statistics of laser-light scattering in atmospheric turbulence", *Journal of the Optical Society of America*, 71(12), 1440–1445, 1981.
50 L. C. Andrews and R. L. Phillips, *Laser Beam Propagation through Random Media*. SPIE: Bellingham, WA, 2005.
51 G. Parry and P. N. Pusey, "K distribution in atmospheric propagation of laser light", *Journal of the Optical Society of America A*, 69(5), 796–798, 1979.
52 H. E. Nistazakis, E. A. Karagianni, A. D. Tsigopoulos, M. E. Fafalios, and G. S. Tombras, "Average capacity of optical wireless communication systems over atmospheric turbulence channels", *Journal of Lightwave Technology*, 27(8), 974–979, 2009.

53 M. A. Al-Habash and L. C. Andrew, "Mathematical model for the irradiance probability density function of a laser beam propagating through turbulent media", *Optical Engineering*, 40(8), 1554–1562, 2001.

54 W. O. Popoola and Z. Ghassemlooy, "BPSK subcarrier intensity modulated free-space optical communications in atmospheric turbulence", *IEEE Journal of Lightwave Technology*, 27(8), 967–973, 2009.

55 Z. X.Wang, W.-D. Zhong, S. Fu, and C. Lin, "Performance comparison of different modulation Formats Over Free-Space Optical (FSO) turbulence links with space diversity reception technique", *IEEE Photonics Journal*, 1(6), 277–285, 2010.

56 M. K. Simon, M.-S. Alouini, *Digital Communication over Fading Channels: A Unified Approach to Performance Analysis*. John Wiley & Sons, Inc: New York, 2000, Print ISBN 0-471-31779-9; Electronic ISBN 0-471-20069-7.

57 X. Zhu and J. M. Kahn, "Free-space optical communication through atmospheric turbulence channels", *IEEE Transactions on Communications*, 50(8), 1293–1300, 2002.

58 L. C. Andrews and R. L. Phillips, *Laser Beam Propagation through Random Media*. SPIE: Bellingham, WA, 2005.

59 X. Zhu and J. M. Kahn, "Free-space optical communication through atmospheric turbulence channels", *IEEE Transactions on Communications*, 50(8), 1293–1300, 2002.

60 L. C. Andrews and R. L. Phillips, *Laser Beam Propagation through Random Media*. SPIE: Bellingham, WA, 2005.

61 H. E. Nistazakis, E. A. Karagianni, A. D. Tsigopoulos, M. E. Fafalios, and G. S. Tombras, "Average capacity of optical wireless communication systems over atmospheric turbulence channels", *Journal of Lightwave Technology*, 27(8), 974– 979, 2009.

62 W. Gappmair and M. Flohberger, "Error performance of coded FSO links in turbulent atmosphere modelled by Gamma-Gamma distribution", *IEEE Transactions on Wireless Communications*, 8(5), 2209–2213, 2009.

63 B. Sklar, *Digital Communications: Fundamentals and Applications*. Pearson Education: Patparganj, India: 2001.

64 D. Lim and Y. Chong, "The effect of mismatched filtering on the BER performance of pi/4 shift DQPSK", In *Proceedings of APCC/MDMC*, Beijing, China, vol. 2, pp. 779–781, 2004.

65 R. Beals and J. Szmigielski, "Meijer G–functions: A gentle introduction", *Notices of the AMS*, 60(7), 866, 2013.

66 https://en.wikipedia.org/wiki/Meijer_G-function.

67 H. Zhang, H. Li, Xiao D., and C. Chao, "Performance analysis of different modulation techniques for free-space optical communication system", *Telkomnika*, 13(3), 880–888, 2015.

68 SHSLab; WFS, Wellenfrontsensoren, https://www.optocraft.de/shslab/?gclid=EAIaIQobChMIiNKA5YrQhAMVrBoGAB13wwblEAAYASAAEgJ7O_D_BwE

69 Wavefront Sensor Applications, https://www.axiomoptics.com/application/wavefront-sensing-aaplications/

70 P. M. Palomo, A. Zepp, and S. Gladysz, "Characterization of the digital holographic wavefront sensor", https://www.ptb.de/cms/en/ptb/fachabteilungen/abt4/fb-42/ag-422/characterizing-and-calibrating-optical-wavefront-sensors.html

71 Y. Loussouarn, et al., OFC, MIE.5, 2018; Y. Loussouarn, et al., OFC, M2D.2, 2020; Y. Chen, et al., OFC, Tu6D.3, 2021.

72 H. Sun, JLT, p. 4744, 2020; J. Li, et al., ECOC, Th 1.A.2, 2019; C. Zhang, et al., OECC, p. 210, 2011.

73 J. Li, et al., OECC, 2015; L. E. Nelson, et al., NWA2, OFC2011; F. Buchali, et al., Th2A.16, 2015.

3 Digital Binary Photonic Modulation for SPC

CONTENTS

Modulation of lightwaves for SPC (Space Photonic Communications) is very critical for long-haul transmission over the inter- or intra-constellation of satellite photonic nodes. Photonics is the art of manipulating or modulating lightwaves using several effects which alter the refractive index of the guided waveguides, changing the phase of the lightwaves. Electro-optic interaction occurs via an applied electric field across the guided propagation path of the lightwaves. Therefore, phase modulation can be effectively implemented over an integrated optical waveguide. Alternatively, this phase change can be incorporated with an optical interferometer so that the interfered lightwaves at the output can be intensity switched from high to low, allowing amplitude modulation to be achieved.

DOI: 10.1201/9781003426677-3

The binary photonic modulation technique is mainly addressed in this chapter. Both single-dimensional and orthogonal two-dimensional complex signal constellations are presented.

Further, amplitude modulation can be formed in two-dimensional coordinates to generate quadrature amplitude modulation (QAM), which involves an amplitude modulation in the x-direction or real part and the y-direction or imaginary part to form complex modulation schemes; hence achieving higher order to form M-ary QAM (or mQAM). This chapter introduces complex modulation techniques for SPC. The next chapter, Chapter 4, describes the binary modulation and extending to multi-level complex digital modulation, that is combining two uni-directinal modulation schems onto 2-dimensional complex plane with to create mQAM (or M-ary QAM – quadrature amplitude modulation of m constellation points).

Therefore, the modulator must be integrated with another set of amplitude modulation device, meaning two independent sets of amplitude modulation are required for mQAM modulation frmat scheme.

In addition, if another polarization of the lightwaves can be used, then we would have PDM-mQAM (polarization division multiplexing mQAM) channels. This chapter presents only binary modulation techniques.

Coherent reception for complex modulated channels, with polarization diversity of degree two, has matured with integrated photonic transceivers incorporating digital signal processing (DSP) algorithms to mitigate several impairments, in particular the turbulence and spatial spreading of lightwave channels.

Chapter 5 will describe complex and multi-level photonic modulation techniques for SPC transmission systems. Photonic modulators in details are supplemented in Appendix 1 of Chapter 11.

3.1 INTRODUCTORY REMARKS

Since the invention of optical fibers, the modulation of lightwaves for transmission in optical communication systems, both in free space and guided media, has mainly been based on on-off switching of the intensity of the waves, rather than manipulating the phase or frequency of the field of the waves. In the 1980s, interest in extending the repeaterless distance from 40 km to longer distances, such as 60 km, in coherent optical communications led to the use of external modulation with integrated optic devices for the carrier wave. This concept of external modulation was briefly introduced in Chapter 1. Integrated photonic modulators are further described in Appendix 1 of Chapter 11.

Since the invention of the laser, ideas for optical communications appeared in which bulk modulators were used to encode input electrical data into optical pulse sequences. The bulk devices cannot be practical in systems, particularly because the operation speed is limited caused by the wide distance between the electrodes, which in turn limits the production of the electromagnetic field. Thus, integrated modulators must be developed. A typical diffused waveguide structure is shown in Figure 3.1, with refractive index displayed for both vertical and lateral directions.

An integrated optic device must include (i) an optical waveguide in planar form, which is a guide whose refractive index is higher than that of a substrate, normally about 0.2 to 0.002. This large refractive index difference ensures strong confinement of the lightwaves in the guide, thus requiring smaller driving voltage, especially at high frequencies. (ii) An electrical driving electrode pair is placed across the optical waves to provide the electric field to modulate the refractive index of the optical waves via the change in its refractive index using its electro-optic coefficient. Depending on the substrate's crystal structure, the electrodes are arranged accordingly. (iii) The electrodes for high frequency must be a traveling type so that the interaction between the optical field and the electrical field must be maximized, normally to match their phase propagation velocity. (iv) The waveguide is coupled with the laser output, so a coupling structure must be available at the

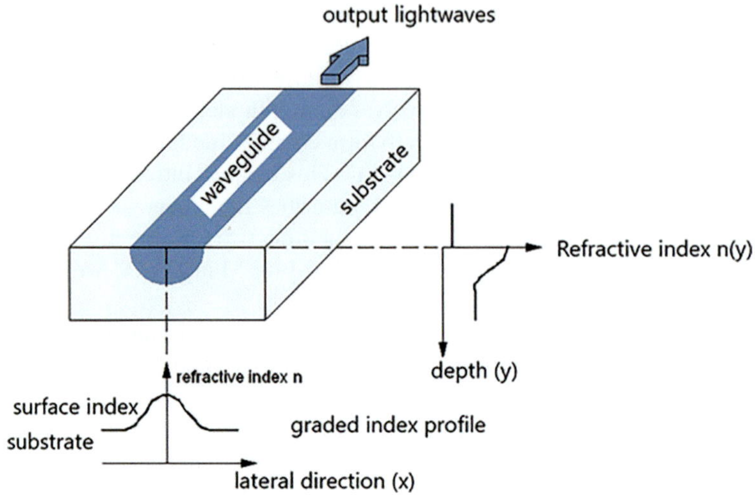

FIGURE 3.1 Optical waveguide structures and refractive index profiles in vertical and lateral directions.

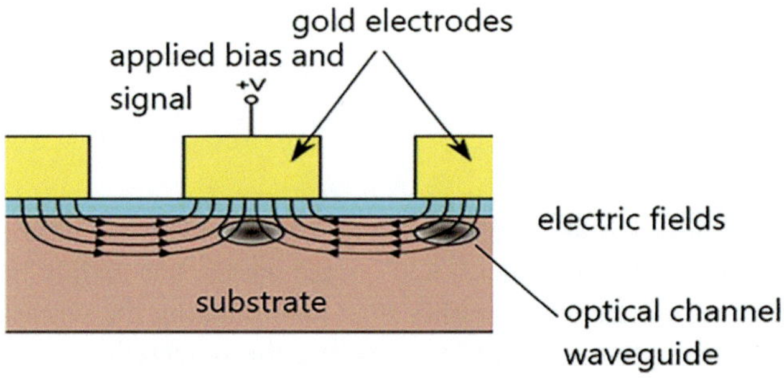

FIGURE 3.2 A typical cross-section of an integrated optical modulator using electro-optic interaction, where the electric field creates changes in the refractive index of the waveguide, hence changing the phase of guided lightwaves.

modulator's input and output. Depending on the optical waveguide type, the coupler can vary. For example, in the case of the SOI (Silicon on Insulator – Oxide) modulator, an optical grating may be more effective. Several coupling structures are available, but they are not the main topic of this chapter or this book (Figure 3.2).

The integrated optic structures allow the modulation electrodes of electro-optic modulator types to be as narrow as possible for low driving voltage, thereby achieving broadband characteristics so that high data rates can be implemented.

Furthermore, there was a lack of sufficiently narrow linewidth lasers and integrated optical components that would facilitate the development of such techniques. The invention of optical amplification devices diverts the attention of coherent communications to ultra-long-haul SPC and terrestrial networks. However, with bit rates now extended to 40 Gbps, 100 Gbps, 400 Gbps, and even 1.0 Tbps, the necessity of narrowing the signal bandwidth to minimize dispersion effects has led to the successful demonstration of modulation techniques for direct detection systems. Hence, several modulation techniques have been studied to maximize transmission bit rates and reception sensitivity.

Coherent communications have been revived in this trend for terrestrial optical systems, in association with high-sampling-rate electronic DPS and processors for the estimation of the phase of the demodulated signals, compensation for frequency offsets and compensation of dispersion and nonlinear impairments in guided media.

In space optical communications, the coherency of the laser and modulated signals is even more critical for long-distance free-space transmission to minimize spatial beam spreading, which leads to a reduction in receiver sensitivity and beam alignment.

Suppressing the carrier can reduce nonlinearity due to its power level being above the signal band. This approach is acceptable for terrestrial distances, but not realizable for non-terrestrial space transmission, as the beam coherency leads to significant beam spreading and thus a significant reduction in reception power. Furthermore, nonlinearity in the fiber over short distances of about 1–5 m, that is employed in optical fiber amplifiers for power boosting before launching into free space, is not insignificant.

In long-haul space photonic transmission, the linewidth of the laser is critical as its coherence can provide a pseudo/coherent lightwave beam. The modulation is implemented with its optical carrier being about 6 dB above the optical signal band.

One of the main aims of this chapter is to examine and try to provide an answer for the best modulation format that can allow the propagation of such modulated photonic beams over ultra-long distances in both free space, with/without turbulence, from OGS to satellite orbits such as GEO, MEO, or LEO (Geostationary Earth orbit/Medium Earth Orbit/Low Earth Orbit), and satellite to satellite. The addition and drop of these optical channels via switching and multiplexing are considered in another chapter.

This chapter gives an account of optical modulation of binary signals in pulse shaping, including return-to-zero (RZ) or non-return to zero (NRZ), and the manipulation of either the amplitude or phases (discrete or continuous) of the lightwave carrier. Furthermore, in Chapter 5, complex modulation and single-sideband (SSB) modulation techniques are described and illustrated. The requirements for the performance of the laser and modulator are stated with reference to their brief descriptions in the appendices. It should be noted that pulse shaping of the modulated optical channels, such as Nyquist filtering, is important for spectral-efficient modulation. This technique is not covered here but in a later separate chapter, Chapter 6.

Two Appendices 1 and 2 of Chapter 11 are provided and referred to for a bried outline of the optical modulators and types of lasers that are suited for coherence in space photonic transmission.

3.1.1 Photonic Modulation and Challenging Issues

Beginning with the first research phase around the early 1970s, fiber-optic communications have progressed tremendously over the last three decades. Free space optical communications also began with visible lasers at 633 nm and far-infrared lasers at 10.6 μm due to the transparency of the medium at these spectral windows and the availability of the lightwave sources.

However, the modulation of the laser beam can only be conducted via bulk crystal type, and the electrodes are deposited at the two parallel faces of the crystal, much wider than a few millimeters. Thus, due to small electro-optic coefficients, the required phase changes, leadng to high-voltage amplitude, therefore, limiting the speed and power of the RF amplifiers.

The first-generation lightwave systems were commercially deployed in 1983 and operated in the first wavelength window of 800 nm at transmission bit rates of up to 45 Mb/s.[1,2] After the introduction of ITU-G652 Standard Single Mode Fibers (SSMF) in the 1970s,[2,3] the second generation of lightwave transmission systems became available in the early 1980s.[4,5] The operating wavelengths were shifted to the second window of 1300 nm, and this window offers much lower attenuation for the optical fiber compared to the previous 800 nm region.

The optical fiber communications technologies and fast developments of integrated optic lasers and modulators have made it possible for high-speed modulations and coherency of the beams.

These developments have opened many possibilities for the implementation of ultra-high speed, ultra-high bit rate, and high capacity via wavelength multiplexing for ultra-long space photonic communications.

For free-space optical transmission, the transparent spectrum has a number of windows depending on the transmissivity of free space, whether in the air, vacuum regions, or clear atmospheric conditions in sea areas or mountainous regions. The transmittance of the air medium is discussed in Chapter 2. However, the transmittance would be best over the 1550 nm regions, which would also match those of silica-doped core fibers. Thus, the modulation and lightwave sources can be employed for ultra-high-rate optical communication over free-space media.

These systems could operate at transmission rates of up to 1.7 Gbps and had a repeater-less transmission distance of about 50 km. Optical communications then evolved to third-generation transmission systems, which utilized the lowest attenuation 1550 nm wavelength window and operated at a 2.5 Gbps bit rate.[6] These systems were commercially available in 1990,[7] with a repeater spacing of around 60–70 km, and then increased to 80–120 km as the attenuation of silica single-mode fiber was reduced to 0.15 dB/km in the spectral range from 1545 to 15,620 nm.

Fortunately, the attenuation or absorption factor of free space is minimal in such spectral regions, with few peaks in the S, C, and L bands of the silica fiber. Therefore, integrated photonic components can be used in SPC, especially optical transceivers and optical amplifiers.

At this stage, the generation of optical signals was mainly based on the direct modulation of the semiconductor laser source. Then, since the invention of Erbium-Doped Fiber Amplifiers (EDFA) in the early 1990s,[8,9] lightwave systems have rapidly evolved to Tbps and long repeater distances with pre-amplification (power booster) and post-amplification (low-level detection), thence cascading of optically amplified spans. (ii) Increasing capacity by multiplexing in the spectral domain, that is, Wavelength Division Multiplexing (WDM) and then Dense WDM (DWDM with channel carrier spacing <50 GHz), optically amplified transmission systems can transmit multiple 10 Gbps channels, thence $N \times 100$ Gbps channels. These high-speed, long-haul systems employ external modulation in their optical transmitters to preserve laser coherencies and the linewidth of the laser. The present fiber-optic transmission systems are considered to be in the fifth generation, with a transmission capacity of a few Tbps. These lasers and photonic modulation techniques can be much applicable to space photonic communications, with the only difference beings the long space distance without any optical amplification in the middle, and only pre- and post-amplification subsystems.

Coherent detection was the focus of extensive research and development during the 1980s and early 1990s,[10,11,12,13,14,15] and was the main detection technique in the first three generations of lightwave transmission systems. At that time, the main motivation for developing coherent optical systems was to improve receiver sensitivity, commonly by 3–6 dB.[11,14] The repeater-less transmission distance was thus extended to more than 60 km of SSMF (with a 0.2 dB/km fiber attenuation factor). The free-space ideal vacuum transmission loss is inversely proportional to the square of the distance. A 36,000 km distance in an optical system would result in a total effective loss of about 72 dB, but with the gain of the optical antenna of about 220 dB and the intrinsic loss of propagation of 292 dB (according to Friis's transmission formula for gain/loss[16]).

Where D_t is the directivity of the transmitting optical launching system, D_r is the directivity of the receiving optical system, λ is the signal operating wavelength, and d is the distance between the optical transmitting and receiving systems, P_t is the transmitting power, and P_r is the receiving power. We note that the Friis formula is an ideal equation and assumes that the laser is a pure single line, with no linewidth. In practice, the linewidth of the laser is finite, and about <10 MHz for a semiconductor laser and a few Hz for a fiber laser. The finite linewidth of the laser and modulated spectrum will lead to finite spreading of the laser beam, and hence, depending on the reception area of the optical system and photodetector, there is additional loss. The directivity of the transmitting and receiving optical systems can be assumed to be a straight line and under control feedback so that the moving satellite can be assumed to be stationary.

However, coherent optical systems suffer severe performance degradation due to fiber dispersion impairments. In addition, maintaining phase coherence between the lightwave carriers of the laser source and the local laser oscillator was very difficult. On the contrary, incoherent detection technique minimizes the linewidth obstacles of the laser source as well as the local laser oscillator, and relaxes the requirement for phase coherence. Moreover, incoherent detection mitigates the problem of polarization control for the mixing of the transmitted lightwaves and the local laser oscillator at multi-THz optical frequency ranges. The invention of Erbium-doped optical amplifiers (EDFAs),[17] capable of producing optical gains of 25 dB and above, has also greatly contributed to further progress in incoherent optical transmission systems. Therefore, since the mid-1990s, incoherent detection has become the preferred choice for digital photonic transmission systems.

Appendix 1 of Chapter 11 is provided for optical modulators. The phasor diagram is recommended for understanding the behavior of the phases of the optical waves and thence interferences.

3.1.1.1 Challenging Issues

Recent years have seen enormous increases in demand for broadband communications, driven mainly by the rapid growth of multimedia services, peer-to-peer networks, and IP streaming services, in particular IPTV (Internet Protocol television) and the emerging Internet of Things (IoT) and 5G/6G (fifth and sixth generations) mobile networking, as well as cloud networks. It is most likely that such tremendous growth will continue in the coming years. This is the main driving force for local and global telecommunications service carriers to develop high-performance and high-capacity next-generation optical networks. The overall capacity of WDM or DWDM optical systems can be boosted by either increasing the base transmission bit rate of each optical channel, multiplexing more channels in a DWDM system, or, preferably, by combining both of these schemes. While implementing these schemes, optical transmission systems encounter a number of challenging issues which are outlined in the following paragraphs.

Current 10 Gbps transmission systems employ Intensity Modulation (IM), also known as On-Off Keying (OOK), and utilize Non-Return-to-Zero (NRZ) pulse shapes. The term OOK can be used interchangeably with Amplitude Shift Keying (ASK) when no coherent reception is employed. It is noted here that the OOK format simply implies the on-off states of the optical lightwaves, where only the manipulation of intensity is considered. On the other hand, the ASK format considers not only the amplitude but also the phase in the constellation diagram.

3.1.1.2 Moving towards High Bit Rate Transmission

In free space communications using lasers, the most common modulation format is pulse position modulation, where the pulse width of the pulse located in the bit period can be much shorter and positioned at different slots. Thus, it requires pulse widths much smaller than the bit period and so high frequency must be employed. This limits the operation speed, especially imposing constraints on bulk modulators. Current integrated optical modulators can offer such PPM (pulse position modulation) but to reach the highest operation bit rates. PPM is not preferred due to the significant increase in the operational speed required by the modulation format. Complex modulation techniques that use both the real and imaginary components of the modulation, and corresponding DSP-based optical reception using coherent techniques, can offer much higher receiver sensitivities. Therefore, complex QAM at single or multiple levels, which offers much higher capacity, is selected in long-haul terrestrial fiber networks. The next chapter can also prove that such complex modulation and coherent reception can be most appropriate for SPC systems in long-reach transmission. The complex modulation for optical beams in SPC will be discussed in Chapter 4. Coherent reception can also be employed in binary modulation schemes.

The performance of OOK photonic systems is severely degraded due to free-space turbulence and fiber impairments, including fiber dispersion and fiber nonlinearities. Fiber dispersion is classified into Chromatic Dispersion (CD) and Polarization Mode Dispersion (PMD), causing the Inter-Symbol Interference (ISI) problems. On the other hand, severe deterioration in the performance due to fiber nonlinearities results from the high-power spectral components at the carrier and signal frequencies of OOK-modulated optical signals. It is also a concern that existing transmission networks comprise millions of kilometers of SSMF, which have been installed for approximately two decades. The properties of these fibers are not only lower in performance compared to the state-of-the-art fibers used in recent laboratory 'hero' experiments but have also degraded after many years of use.

The total transmission capacity can also be enlarged by increasing the number of multiplexed DWDM optical channels. This can be carried out by reducing the spacing between these channels, for instance from 100 GHz down to 50 GHz, and then to 25 GHz and 12.5 GHz spacing, which has been demonstrated.[18,19] The reduction of channel spacing also results in narrower bandwidths for the optical multiplexers (mux) and demultiplexers (demux). Passing through these narrowband optical filters, signal waveforms are distorted, and optical channels suffer from inter-channel crosstalks. These narrowband filtering problems are getting more severe especially at high transmission bit rates, such as 40 Gbps, thus significantly degrading the system performance.

Together with the demand for boosting the total system capacity, another challenge for service carriers is to find cost-effective solutions for the upgrading process. These cost-effective solutions should require minimal renovation to the existing photonic and electronic sub-systems, i.e. the upgrading only takes place at the transmitter and receiver ends of an optical transmission link. In short, next-generation optical networks are required to meet the desired high performance while also considering the cost of system upgrading.

The main differences between fiber-terrestrial and space-non-terrestrial systems are that (i) optical amplifiers, post- and pre-amplifications can be inserted at the end of a fiber span and the beginning of the next consecutive span, while in space transmission no amplifiers can be inserted between the transmitting and receiving optical ends; (ii) the coherency of the modulated signal spectra is critical for both systems as the transmission length is long, and pseudo-coherency must be assumed. In particular, for the space system, no intermediate amplification is possible, and coherency over a great distance is required; (iii) any optical routing in a space transmission system must be done after the O/E/O (optical to electrical and back to optical) regeneration, meaning the optical signals are detected and amplified in the electronic domain to provide signals to drive and generate optical channels before further switching, routing, and transmitting to the next inter-satellite spans.

Further, the probability of error distribution in space transmission due to disturbances is quite different from the typical Gaussian distribution for fiber-optic transmission. This is applied to all modulation formats for space transmission. The noises contributing to space transmission lead to considerations of the sensitivity of the reception system in phase or amplitude or a combination of amplitude and phase of the receiving modulated lightwaves. Both fiber-optic and space transmission systems must employ digital signal processing systems to recover and make intelligent decisions on the constellation points of the complex coherent reception.

In fiber optic transmission the main distortion mechanism in the linear operating region of the fiber is chromatic dispersion (CD). This CD is a natural phenomenon due to (i) the propagation constant of the guided mode of the single-mode waveguide varying with respect to optical frequency and the λ^{-4} dependence of the impurity-doped silica refractive index, which affects the propagation velocity variation.

The distortion of lightwave beams in space photonic communications comes from spatial spreading, which is dependent on the linewidth of the laser and the modulated bandwidth of the information channel. This spreading reduces the reception of optical power over the detected area of the photoreceiver and the receiving optics. Furthermore, the disturbance of the air layer in the isospheric region eads to multi-path delays and hence differential delays and spreading of the pulses, thus errors and reduction of transmission performance.

The modulation formats of the lightwaves must be considered, mainly the modulation of the phase and/or amplitude of the lightwaves and their responses to the propagation distortion of the medium.

The nonlinearity of the fiber optic medium in optical transmission lines and high-power pump over short lengths of fiber in beam launching in space photonic communication systems will be examined in the optical amplifications in Chapter 6.

All of the above-described problems facing contemporary fiber-optic and space photonic communications can be overcome with spectrally efficient transmission schemes. The main motivation of this research is to find effective solutions to upgrade the aggregate transmission capacity while minimizing the effects of fiber shortfalls, particularly dispersion impairments. The enabling technologies presented in this research aim to overcome the challenging issues addressed above.

3.1.1.3 Enabling Technologies

Spectrally efficient transmission schemes can be achieved by utilizing advanced modulation formats. A number of modulation formats have recently been proposed as alternatives to the OOK (on-off keying) format. These include return-to-zero (RZ) pulses in OOK or ASK (amplitude shift keying) systems,[20,21,22] differential phase shift keying (DPSK),[18,23,24,25] and more recently, minimum-shift keying (MSK).[26,27,28,29 30,31,32,33] The modulation characteristics are based on the phase or amplitude of the lightwave carrier under the signal envelope. If the phase is modulated, the lightwaves may be continuous or switch to another phase. MSK requires that the phase be continuous at the boundary of the baud period.

These formats are adopted into photonic communications based on the knowledge from wireline and wireless communications.

DPSK has received much attention in the last few years, particularly when combined with RZ pulses. The main advantages of this format are: (i) a 3-DB improvement over OOK in receiver sensitivity by using an optical balanced receiver,[21,34,35] and (ii) high tolerance to fiber nonlinearities,[20,34,36] such as intra-channel self phase modulation (SPM) and inter-channel cross phase modulation (XPM) effects. Several experimental demonstrations of DPSK long-haul DWDM transmission systems[37] for 10 Gbps, 40 Gbps, and higher bit rates have been reported recently.[18,23,38,39,40] However, there are few practical experiments addressing the performance of cost-effective 40 Gbps DPSK and 10 Gbps OOK hybrid systems on the existing old-vintage SSMF infrastructure.[41,42] In addition, the performance of 40 Gbps DPSK in this hybrid transmission scheme has not been thoroughly studied. Therefore, one of the main contributions of this chapter and others in this book is to present the feasibility of overlaying 40 Gbps DPSK channels on the existing 10 Gbps OOK channels for implementing hybrid systems.

The MSK format offers a spectrally efficient modulation scheme compared to the DPSK counterpart at the same bit rate. As a subset of Continuous Phase Frequency Shift Keying (CPFSK), MSK possesses the spectrally efficient attributes of the CPFSK family. The frequency deviation of MSK is equal to a quarter of the bit rate, and this frequency deviation is also the minimum spacing required to maintain orthogonality between two FSK-modulated frequencies. On the other hand, it can also be considered a particular case of Offset Differential Quadrature Phase Shift Keying (ODQPSK),[43,44,45,46,47] which enables MSK to be represented by I and Q components on the constellation. The advantageous characteristics of the optical MSK format can be summarized as follows: (i) Compact spectrum, which is of particular interest for spectrally efficient high-speed transmission systems. In addition, this provides robustness to the problem of tight optical filtering; (ii) High suppression of the side lobes in the optical power spectrum compared to the DPSK. The roll-off factor follows f^{-4} rather than f^{-2} as in the case of DPSK; (iii) Constant envelope property, which eases the measurement of the average optical power; (iv) No high-power spikes in the power spectrum and high energy concentration in the main lobe offer the advantage of lower peak input powers into the fiber, thus reducing fiber nonlinear effects.

Being a subset of either CPFSK or ODQPSK, MSK can be detected either incoherently based on the phase or the frequency of the lightwave carrier or coherently based on the popular I-Q detection structure.

In addition, if MSK can be combined with multi-level modulation schemes, the baud rate would be reduced, and the spectral efficiency can be even more effective. This is of great interest for long-haul, metropolitan, and free-space inter-satellite optical networks and provides the main motivation for investigating the performance of the proposed dual-level MSK modulation format.

Several studies have been conducted[26–29] focusing on the generation and direct detection of externally modulated optical MSK, as well as the format's CD tolerance. However, there are only a few research works investigating the potential of the externally modulated MSK format for digital photonic transmission systems, particularly at high bit rates such as 40 Gbps.[26,48,49] Therefore, another key contribution of this research is to provide comprehensive studies on the performance of MSK-based modulation formats for long-haul and metropolitan optical transmission systems.

3.1.2 Coherent and Self-Coherent Optical Receivers

Self-coherence indicates that the signals can be used as the local oscillator and can be mixed to produce the beating signals. The advantages are that the frequency of the local oscillator and the signal carrier is the same, thus avoiding intradyne frequency. The disadvantage is that the power of the LO is low, and not much gain can be obtained from the resulting product.

The modulation formats described in this section, the optical DPSK and MSK-based formats, can be demodulated incoherently using an optical balanced receiver that employs a Mach-Zehnder Delay Interferometer (MZDI). In the incoherent detection of optical DPSK, the MZDI is used to detect the differentially coded phase information between every two consecutive symbols.[34,35,50] This detection is carried out in the photonic domain as it is beyond the speed of the electrical domain, especially at very high bit rates of 40 Gbps or above. The MZDI balanced receiver is also used for the incoherent detection of optical MSK signals.[26,28,29] However, using the MZDI-based detection scheme, optical MSK does not significantly improve the CD or spatial spreading tolerance compared to its DPSK counterpart.[28,30]

To fully exploit the advantages of optical MSK spectral characteristics, MSK-modulated lightwaves can be incoherently detected based on the principles of optical frequency discrimination. Thus, an Optical Frequency Discrimination Receiver (OFDR) is proposed in this research, which employs dual narrowband optical filters and an optical delay line (ODL). This receiver scheme effectively mitigates the CD-induced ISI effects and enables breakthrough CD tolerances for optical MSK transmission systems. These results have been reported in Refs.[31,32] The feasibility of this novel optical receiver is based on recent advances in the design of optical filters, in particular the micro-ring resonator filters. Such optical filters have very narrow bandwidths, e.g., <2 GHz (3-dB bandwidth), and thy have been commercially realized by Little Optics.[48,49,50,51] This research provides a comprehensive study not only on this OFDR scheme, from the receiver operational principles to the analysis of the receiver design, but also on the performance of OFDR-based optical MSK systems.

In phase modulation techniques, the phase transition between two consecutive bit slots can be either non-continuous or continuous. As a non-continuous format, DPSK phase, in particular RZ DPSK, has received considerable research attention as RZ DPSK displays advantages, such as 3-dB improvement in the receiver sensitivity with the use of balanced receivers and robustness to fiber nonlinear effects.

Spectral efficiency is a critical requirement for modern high-performance digital photonic transmission systems. Thus, the fact that non-continuous phase modulation schemes, such as DPSK, are not spectrally efficient motivates the exploration of Continuous Phase Modulation (CPM), especially CPFSK formats, which are popularly known for their compact power spectrum.[9,10] Among CPFSK formats, MSK is the most widely used in bandwidth-limited digital wireless and satellite communications. However, the potential of MSK using external modulation techniques in non-coherent optical communications has been approached. In MSK, the frequency deviation from the lightwave carrier is equal to a quarter of the bit rate, which is also the minimum spacing to maintain the orthogonality attribute between the two modulated frequencies.[52] Additionally, MSK is a special case of ODQPSK in which the I and Q components are interleaved by one bit period.[9,18,19] The two approaches, CPFSK and ODQPSK, enable two possible schemes for generating optical MSK signals.

If a multi-level scheme can be incorporated with the MSK format, the transmission baud rate would be reduced, in addition to increasing spectral efficiency. This effectively mitigates fiber impairments for optically amplified long-haul (LH) and ultra-long-haul (UHL) transmission systems and relaxes the requirement for high-speed processing electronics. This demand is the principal motivation for proposing the dual-level MSK modulation scheme in fiber-optic communications, which utilizes two binary bits for each modulated symbol. The transmission symbol rate is thus halved compared to the data rate from a bit pattern generator.

The advanced modulation formats described so far can be detected either coherently or incoherently. Incoherent detection is preferred as it overcomes the laser's linewidth obstacles faced by homodyne or heterodyne coherent detection systems. Furthermore, the incoherent detection scheme can be easily implemented using a Mach-Zehnder delay interferometer (MZDI). This structure reflects the main advantages of the optical domain over the electrical domain for systems operating at ultrahigh bit rates, such as 40 Gbps. The MZDI is normally utilized in balanced receivers for incoherent detection of both DPSK and MSK optical signals. Additionally, optical MSK signals can also be incoherently detected using an optical frequency discrimination receiver (OFDR) that employs dual optical narrowband filters and an optical delay line (ODL).

This chapter is organized as follows: Section 3.1 presents principles for generating several types of RZ optical pulses, with and without the suppression of the lightwave carrier. The modulation process of these RZ pulses can be clearly explained using a phasor representation. Section 3.3 then reviews optical transmitter and receiver configurations for the DPSK format. In Section 3.5, two MSK optical transmitter schemes are presented. These two schemes are derived from the knowledge that MSK is either a subset of CPFSK or a special case of ODQPSK. Section 3.5.2, and section 3.5.3 also describes incoherent detection techniques for optical MSK systems, which use either an MZDI balanced receiver or an OFDR scheme. In Section 3.3.6, the novel dual-level MSK modulation format is proposed for optical communications, along with its generation and incoherent detection schemes. Finally, Section 3.6 compares key spectral characteristics of all modulation formats under the study before providing a summary of the chapter.

For coherent reception, the channel signal is not split into two parallel arms and delayed as in the self-coherent case; instead, an additional local oscillator (LO) laser with a narrow linewidth is employed to interfere with the channel lightwaves. The mixed signal between the channel lightwaves and the LO laser is imposed on the photodetector and mixed to generate the beating components, which are proportional to the amplitudes of the LO laser and the channel lightwaves. When the frequencies of both lightwaves are the same, it is homodyne coherent reception; if there is a frequency difference, it is heterodyne. If the frequency difference is smaller than the bandwidth of the channel, it is an intradyne type. In practice, frequencies of the channel carrier and the LO frequencies are never identical, but there is a small difference. This frequency difference can be matched by a phase recovery algorithm via the digital signal processor. Thus, it can be equivalent to an absolute homodyne detection.

Further coherenct reception techniques are discussed in Chapter 6.

3.2 RETURN-TO-ZERO BINARY ASK MODULATION

Why RZ format? In binary modulation, RZ means that the pulse rises from low to high and then drops from high to low within a pulse period. This indicates that the speed of the sequence is higher than that of a, NRZ modulation sequence. The advantage of the RZ sequence is that the total average power is lower than that of the NRZ sequence. This helps avoid the nonlinear effects, thus preventing distortion in both phase and amplitude (see Appendix 5 of Chapter 12) in the transmitted sequence. The lower average power leads to higher receiver sensitivity if the receiver can afford to use her gain and lower noise of the optical preamplifiers. The SPC systems use both a booster at the transmitter and low noise at the amplifiers.

Furthermore, the RZ can be done with suppression of the optical carrier so one can suppress the nonlinearity effects in the nonlinear-sensitive medium for transmission. However, in SPC, this suppression of the lightwave carrier means that the coherency of the lightwave beam can be severely affected, leading to larger spreading of the beam spot size, higher group delay times, and thus higher distortion. So, no recommendation is made for the use of the RZ format for modulation of data sequences in the SPC transmission system.

3.2.1 RZ Generation

A conventional return-to-zero (RZ) RZ-ASK (amplitude shift keying) optical transmitter can be constructed using two optical Mach-Zehnder Interferometric Modulators (MZIM), as shown in Figure 3.3. One MZIM, on the left or at the input, is used to carve the laser into two optical pulses, whose width is shortened while the frequency is doubled, as shown in the pulse sequence on the far right of Figure 3.3. The second MZIM is fed with the data sequence to turn the periodic optical pulses at the output of MZIM 1 on or off. Therefore, we can obtain an optical data sequence with a shortened pulse width and a doubled pulse sequence or data rate, requiring higher detection speed from the receiver.

The biasing of such an RZ-ASK pulse sequence can suppress the carrier so that the total optical power of the data sequence is reduced to avoid the nonlinearity of the optical pulses passing through the optical power booster.

The biases and evolution of the transfer function of the two MZIMs are shown in Figure 3.4. The main concern now, for a space photonic communication system, is the coherency of such a carrier-suppressed optical sequence. Although the carrier can be suppressed, the carrier of the optical pulse

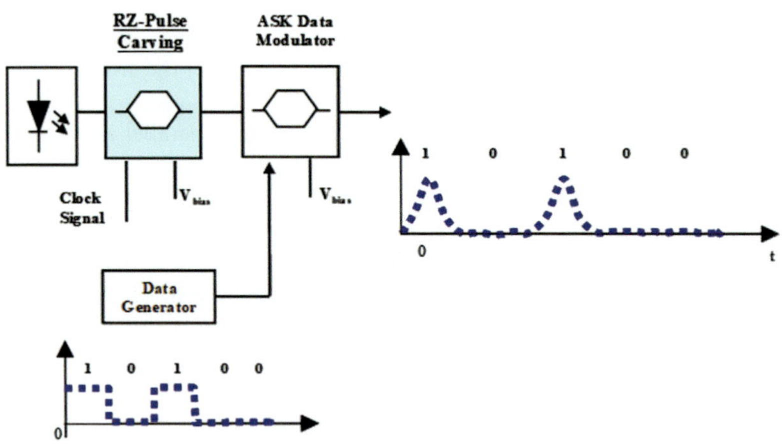

FIGURE 3.3 Conventional structure of an OOK/ASK optical transmitter utilizing two MZIMs, one for carving pulses into a return-to-zero sequence and the other for switching on and off based on data sequence.

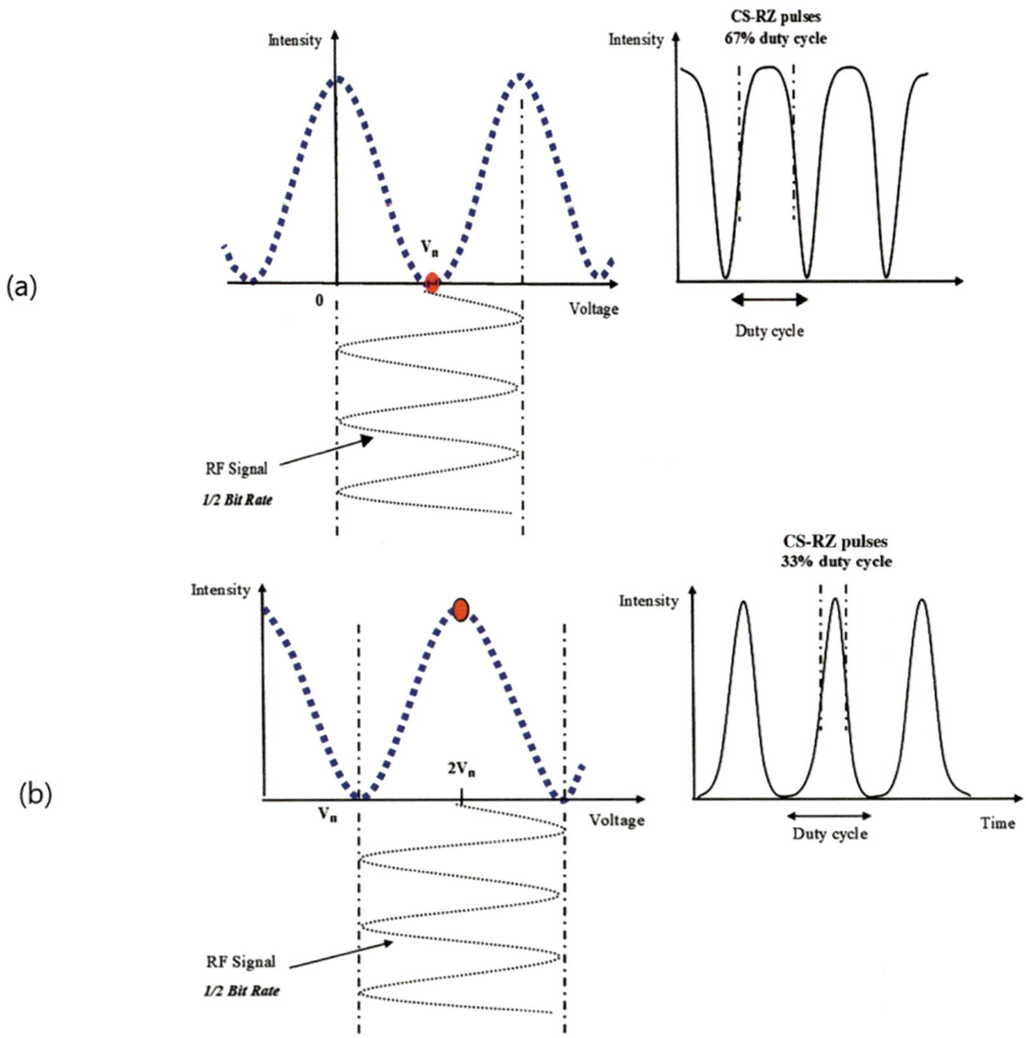

FIGURE 3.4 Bias positions (medium dark dot) on the optical-voltage transfer characteristics and superimposed driving RF signals (sinusoidal for $MZIM_1$ and pulse sequence for $MZIM_2$) for generation of (a) CS-RZ (67% pulse width duty cycle) and (b) RZ33 (33% pulse width duty cycle) pulses. Intensity (square of optical field) of the generated pulse sequence.

sequence can still shift left and right with reduced power and become superimposed on the spectra of the pulse sequence, both to the left and right of the main suppressed carrier.

In SPC transmission, the carrier is important for transporting the signal sidebands through free space. A linewidth of 10 kHz or less can maintain a pseudo-coherent state as far as the peak power is at least 6 dB above the signal mid-band flatness. Thus, while the central carrier can be suppressed, the frequency-shifted carrier on both sides of the central one must also meet this 6 dB condition, which is not commonly satisfied. Therefore, the CS-RZ is not preferred for free-space transmission. The reason the CS-RZ format is used in fiber optic transmission systems is to reduce the nonlinear effects caused by the carrier's peak power. The nonlinearity of the fiber can be reduced if the carrier is suppressed.

The pulse width of the CS pulse sequence can be controlled by varying the biasing point as indicated in Figure 3.4. The 33% and 67% of the original non-carrier-suppressed pulses can be generated. The shortened pulse width offers higher energy concentration in the pulses, enabling the peak power to support longer transmission distance. Current receiver operating speeds can detect frequencies up to 100 GHz. Therefore, this type of modulation offers advantages for space photonic communication systems.

It is noted that one can generate these CS-RZ optical pulse sequences by using only one MZIM with two electrodes dual-drive modulators, as shown in the part following the driving circuit section, instead of using two separate MZIMs in this section.

The first MZIM serves as a pulse carver to generate a periodic pulse train of a specific RZ pulse shape format, and it is usually implemented with a chirp-free X-cut single-drive MZIM. Different types of RZ pulses can be generated depending on the amplitude of sinusoidal electrical driving voltages and the biasing schemes of the MZIM. Suppression of the lightwave carrier can be carried out at this stage to form carrier-suppressed-RZ (CS-RZ) pulses. Compared to other RZ types, CS-RZ is found to have interesting attributes for long-haul transmission including the π phase difference between adjacent modulated bits, the suppression of a high-power carrier, and a narrower spectral width. Another type of RZ pulse, RZ33, also has several advantages due to its soliton-like properties. The RZ33 pulse has full-width at half maximum (FWHM) positions occupying 33% of the duty circle. However, a FWHM of 67% is obtained for the case of the case of CS-RZ modulation format. These optical RZ pulses are represented in the optical field $E(t)$as follows[20]:

$$E(t)=\begin{cases}\sqrt{\dfrac{E_b}{T}}\sin\left[\dfrac{\pi}{2}\cos\left(\dfrac{\pi t}{T}\right)\right] & 67\%\,\text{duty-ratio RZ pulses or CS-RZ}\\[2ex] \sqrt{\dfrac{E_b}{T}}\sin\left[\dfrac{\pi}{2}\left(1+\sin\left(\dfrac{\pi t}{T}\right)\right)\right] & 33\%\,\text{duty-ratio RZ pulses or RZ33}\end{cases} \tag{3.1}$$

where E_b is the pulse energy per transmitted bit and T is the bit period.

The art of generating these two RZ pulse types relies on the biasing position on the transmittance characteristic curve of the MZIM. The pulse carver and the amplitude of the sinusoidal driving signal can be obtained by locating the minimum transmission point and the amplitude equal to the V_π as shown in Figure 3.4a and b. The bias point is located at the minimum or maximum point of the P-V transfer characteristic curve for the case of CS-RZ or RZ, respectively. The peak-to-peak amplitude of the RF driving signal for these two RZ pulses is similar and equal to $2V_\pi$ where V_π is the required driving voltage to obtain a π phase shift on the lightwave carrier at the output of the MZIM. It should be emphasized that the electrical driving signal is operating at only half of the bit rate; hence, the pulse carver is actually implementing frequency doubling. Figure 3.4a and b demonstrates all the above-described settings as well as the pulse shapes of the generated CS-RZ and RZ33 optical pulse trains.

The pulse carver for generating RZ-ASK optical pulses can also utilize a dual-drive MZIM, which is driven by two complementary sinusoidal RF signals. The settings of the biasing scheme and driving signal, as well as the modulation process in these dual-drive MZIMs, are explained in the next section using the phasor representation.

Appendix 1 of Chapter 11 provides more details on photonic modulators using Mach-Zehnder interferometric principles.

3.2.2 Phasor Representation of Modulated Lightwaves

Phasor representation of optical fields of optical pulses can offer the evolution of the phase of the optical carrier under the pulse envelope. That is, the carrier phase at optical frequency is much higher than the pulse sequence speed; therefore, the envelope can be considered stationary compared to the frequency of the lightwaves. The phase evolution can then be considered, and detection can occur in the complex plane.

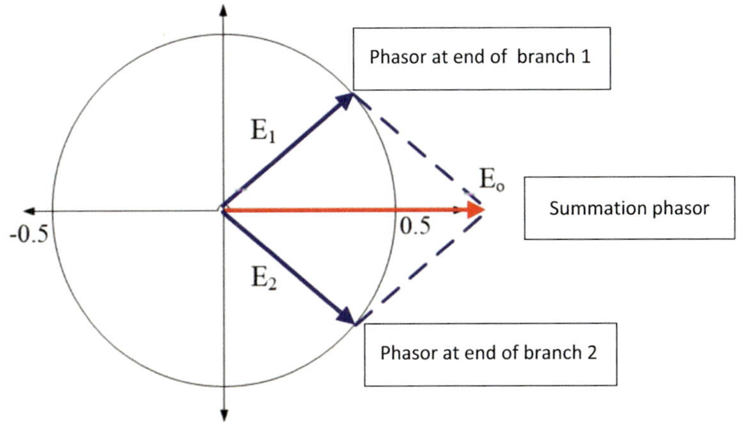

FIGURE 3.5 Phasor representation of the generation of optical signals using a dual-drive MZIM. Vectorial representation of the states of lightwave vector in each path E_1 and E_2 propagating through two arms of the MZ. Normalized amplitude.

Therefore, under the coherent reception technique, the phase states of the lightwaves under the envelope of the pulse sequences can be recovered after the mixing of the signals and the local oscillator laser in the photodetector. Thus, the sampling and processing of such an envelope-by-envelope detection in each dimension of the phasor can lead to the recovery of the pulses in the DSP system of the photonic receiver.

So, the phases of lightwaves under the envelope of the modulated pulse sequence can be represented on the phase plane as shown in Figure 3.5.

Let the optical fields of the laser beam be the input coupled into the guided modulator. The beam is then split into two equal optical paths of the MZ interferometer. These two split beams are modulated electro-optically via the applied electric fields developed across the optical waveguides, resulting in the optical fields E_1 and E_2. The voltage applied to the electrodes consists of a DC bias voltage superimposed by an RF data pulse sequence via a T-coupler.

The analysis of the modulation by phasor representation can be as follows.

The output optical field of a dual-drive MZIM is given as:

$$E_o = \frac{E_i}{2}\left[e^{j\varphi_1(t)} + e^{j\varphi_2(t)}\right] = \frac{E_i}{2}\left[e^{j\pi V_1(t)/V_\pi} + e^{j\pi V_2(t)/V_\pi}\right] = E_1 + E_2 \tag{3.2}$$

This equation suggests that a phasor diagram (see Figure 3.5), which involves vector addition and simple trigonometric calculus, can provide a clear view of the superposition of optical fields at the coupling output of a dual-drive MZIM. The RF driving signals follow a push-pull configuration, i.e., the data $V_1(t)$ and the inverse data $(\overline{\text{data}})$ $V_2(t)$ $(V_2(t)=-V_1(t))$ are applied to each arm of the MZIM, respectively.

It should be noted that the widths of RZ33 and CS-RZ pulses are measured at the FWHM positions that are defined for intensity pulses, whereas the phasor diagram considers optical field vectors. Thus, the normalized output optical field (E_o) has a magnitude value of $1/\sqrt{2}$ at FWHM positions. In addition, the time interval between these FWHM points defines time ratios such as 33% or 67%, and these are normalized with respect to the duty cycle.

3.2.2.1 Phasor Representation for CS-RZ Pulses

Figure 3.6a shows key parameters in the generation scheme of CS-RZ pulses, including the biasing voltage V_{bias} and the amplitude of RF driving signals. Accordingly, its initial phasor representation

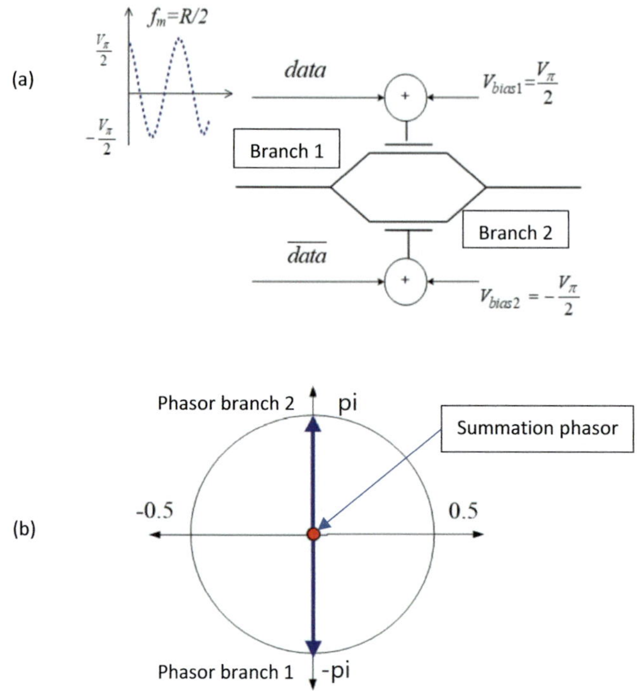

FIGURE 3.6 Initialized stage for generation of CS-RZ optical pulses using *dual-drive* MZIM modulator: (a) RF driving signal and biasing voltages; (b) initial phasor representation at π/2 and −π/2 states.

is demonstrated in Figure 3.6b, where the red dot indicates the sum of the two initialized field vectors of the two arms. We note again that the modulator is of the dual-drive Mach-Zehnder type and the optical phase is manupuated by the applied voltages on the two branches of the MZ interferometer.

Values of these key parameters are outlined as follows: (i) Biasing: $V_{\text{bias}} = \pm\frac{V_\pi}{2}$, that is the dual drive MZIM is driven at the minimum transmission point, and thus frequency doubling of any sinusoidal RF signal is superimposed on this biasing voltage. (ii) Pulse carving: The optical waves driven by the RF driving signal superimposed on each arm have an amplitude of $\frac{V_\pi}{2}$ (i.e. $V_{\text{p-p}}=V_\pi$); it swings in opposite directions and operates at half of the bit rate ($R/2$). (iii) Thus, the generated 'clock' optical pulse sequence is at the bit rate R. (iv) This operation allows for the suppression of the carrier due to the biasing point, with two optical branches of the MZIM at the input always at π phase difference, thus depleting the carrier at this point.

Considering the generation of 40 Gbps CS-RZ pulses (R=40 Gbps), the modulating frequency is $f_m=R/2=20\,\text{GHz}$. At the FWHM positions of the optical intensity pulse, $E_o = \pm 1/\sqrt{2}$ and the field vectors E_1 and E_2 form a phase of π/4 with the vertical axis, as shown in Figure 3.7.

The conditions shown in Figure 3.7 lead to

$$\frac{\pi}{2}\sin(2\pi f_m) = \frac{\pi}{4} \Rightarrow \sin 2\pi f_m = \frac{1}{2} \Rightarrow 2\pi f_m = \left(\frac{\pi}{6}, \frac{5\pi}{6}\right) + 2n\pi \tag{3.3}$$

Thus, the normalized FWHM (with reference to the duty cycle) of a CS-RZ optical pulse can be obtained as follows:

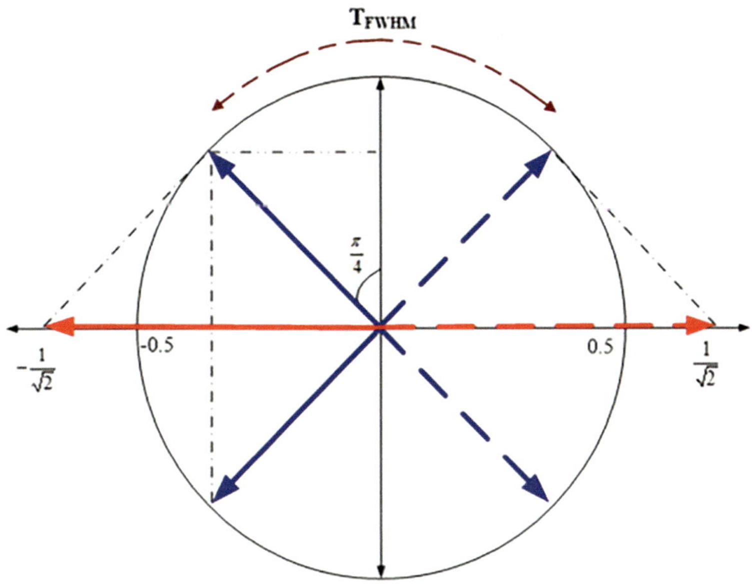

FIGURE 3.7 Phasor representation of CS-RZ generation at FWHM positions of the pulse sequence using a dual-drive MZIM.

$$T_{\text{FWHM}} = \left(\frac{5\pi}{6} - \frac{\pi}{6}\right)\frac{1}{2\pi R} = \frac{1}{3}\pi \times \frac{1}{R} \Rightarrow \frac{T_{\text{FWHM}}}{T} = \frac{1.66 \times 10^{-4}}{2.5 \times 10^{-11}} = 66.67\% \tag{3.4}$$

Thus, the generation of CS-RZ pulses is clearly explained via the phasor representation.

3.2.2.2 Phasor Representation for RZ33 Pulses

Key parameters and initial phasor representation for the generation of RZ33 pulses are shown in Figure 3.8a and b. The red vector is the sum vector of the two initialized field vectors.

Values of the key parameters are as follows: $V_{\text{bias}} = V_\pi$ for both arms; the RF driving signal on each arm has an amplitude of $\frac{V_\pi}{2}$ (i.e. $V_{\text{p-p}} = V_\pi$), swings in opposite directions, and operates at half of the bit rate ($R/2$); and the modulating frequency is $f_m = 20\,\text{GHz}$.

At FWHM (full-width half mark) positions of the optical intensity pulse, $E_o = \pm 1/\sqrt{2}$ and the field component vectors E_1 and E_2 form a phase of $\pi/4$ with the horizontal axis, as displayed in Figure 3.9.

The conditions in Figure 3.9, give the following characteristics:

$$\frac{\pi}{2}\cos(2\pi f_m t) = \frac{\pi}{4} \Rightarrow t_1 = \frac{1}{6f_m} \tag{3.5}$$

$$\frac{\pi}{2}\cos(2\pi f_m t) = -\frac{\pi}{4} \Rightarrow t_2 = \frac{1}{3f_m} \tag{3.6}$$

Thus, the normalized FWHM of RZ33 optical pulses can thence be calculated:

$$T_{\text{FWHM}} = \frac{1}{3f_m} - \frac{1}{6f_m} = \frac{1}{6f_m} \therefore \frac{T_{\text{FWHM}}}{T} = \frac{1/6f_m}{1/2f_m} = 33\% \tag{3.7}$$

Equation 3.7 verifies the generation of RZ33 optical pulses derived from the phasor representation.

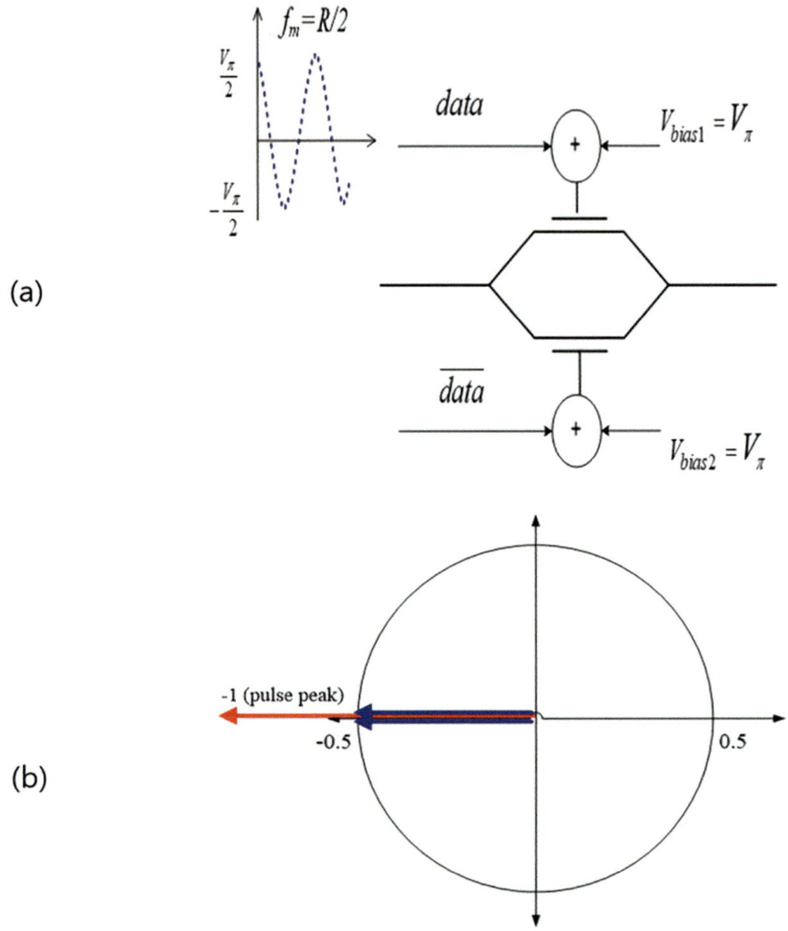

FIGURE 3.8 Initialized stage for generation of RZ33 pulses: (a) RF driving signals and the biasing voltages; (b) initial phasor representation of lightwaves after passing through a DD (dual-drive) MZIM.

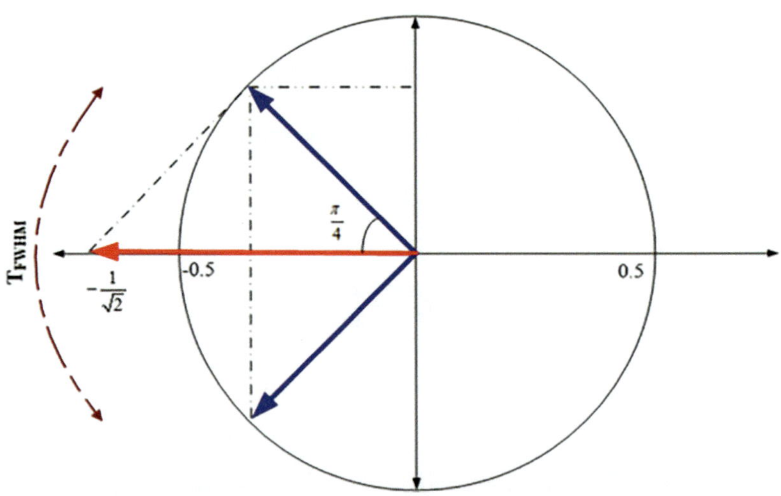

FIGURE 3.9 Phasor representation of RZ33 generation at FWHM positions using dual-drive MZIM.

3.3 DIFFERENTIAL PHASE SHIFT KEYING (DPSK)

3.3.1 Background

Digital encoding of data information by modulating the phase of a lightwave carrier is referred to as optical phase shift keying (PSK). Optical PSK has been extensively researched for coherent photonic transmission systems.[21–24] These systems require strict coherence between the phase of the lightwave carrier and the phase of the laser local oscillator. This phase coherence was difficult to be maintained because of problems of the phase noise, broad linewidth, and chirping effects of the laser source. In contrast, the DPSK format minimizes these problems. The detection of DPSK-modulated optical signals is conducted incoherently by comparing differentially coded phases between every two consecutive optical pulses. The existing incoherent receiver scheme for the optical DPSK format utilizes the MZDI balanced receiver.

In DPSK, a binary '1' is encoded if the present and the past encoded bits have opposite logics, whereas a binary '0' is encoded if the logics are similar. This operation is equivalent to an XOR logic operation. Thus, an XOR gate is employed as a differential pre-coder. In addition, the XOR gate can be replaced by a NOR gate (see Figure 3.10a. In DPSK, a data '1' indicates a π phase change on the optical carrier between consecutive data bits, while a binary '0' implies no phase change between them. Thus, two discrete points, which are located exactly at π phase difference with respect to each other in a signal constellation diagram, are utilized to represent this encoding scheme (refer to Figure 3.10b). In the case of a continuous phase shift keying format, such as MSK, the signal phase evolves continuously from one state to the other, as indicated by the inner bold circle in Figure 3.10b.

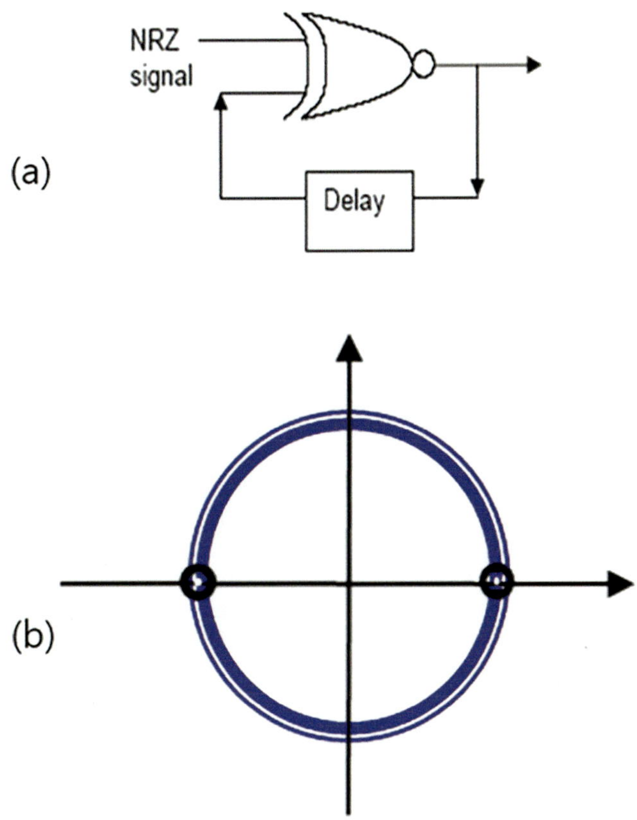

FIGURE 3.10 (a) DPSK pre-coder in electrical domain before modulating lightwaves and (b) DPSK signal constellation diagram by phasor representation.

3.3.2 Optical DPSK Transmitter

The conventional structure of a DPSK optical transmitter employs two external $LiNbO_3$ optical MZIMs (see Figure 3.11). The first MZIM is an optical pulse carver, as described in the previous section. The generated RZ optical pulses are then fed into the second MZIM, where these RZ pulses are modulated with pre-coded binary data to generate RZ DPSK optical signals.

Binary electrical data is differentially encoded in a pre-coder using an XOR or a NOR operation. Without the pulse carver, the structure shown in Figure 3.11 becomes an NRZ DPSK optical transmitter. In DPSK data modulation, the second MZIM is biased at the minimum transmission point. The pre-coded RF driving signal has a peak-to-peak amplitude equal to $2V_\pi$ and operates at the transmission bit rate. The data modulation of NRZ DPSK optical signals is demonstrated in Figure 3.12. See also Appendix 1 of Chapter 11 for optical modulators. The phasor diagram is recommended to understand the behavior of the phases of the optical waves and their resulting interferences.

In addition, an EOPM might also be used for the generation of NRZ DPSK optical signals. However, an EOPM produces chirps, whereas an MZIM, in particular the X-cut type, generates chirp-free optical signals. In practice, a small amount of chirp might be desirable in transmission.[25]

3.3.3 Self-Coherent Detection of Optical DPSK

An optically delayed balanced receiver is well known for the self-coherent detection of DPSK-modulated lightwaves. It consists of a photonic MZDI followed by dual photo-detectors connected in a balanced structure, as shown in Figure 3.13a.53 At the input of the MZDI, the lightwaves are split into two guided paths, and one path is delayed with respect to the other by a physical length that equals a propagation time delay of one-bit interval.

MZDI can be effectively seen as an optical self-heterodyne structure that performs the differential phase demodulation in the photonic domain. This photonic processing overcomes the bandwidth and speed limitations of the electrical domain. The MZDI has constructive and destructive ports that are inputs to the balanced-structure photo-detectors. Hence, a push-pull eye diagram is produced (see Figure 3.13b). This detection scheme offers a 3-dB improvement over direct detection using a single photo-detector.[53] In addition, the DPSK MZDI balanced receiver is inherently

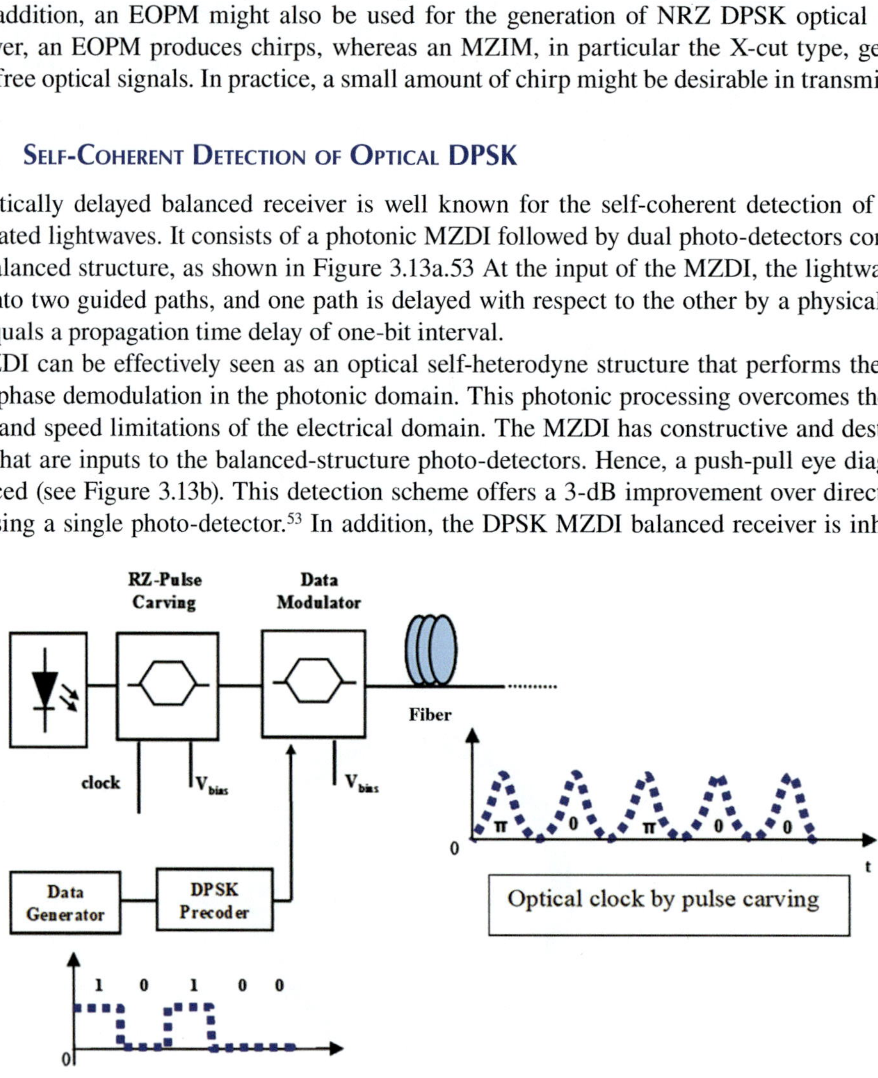

FIGURE 3.11 DPSK optical transmitter with RZ pulse carver, the pulse carving pulse sequence. Phase differential by data switching on and off after carving.

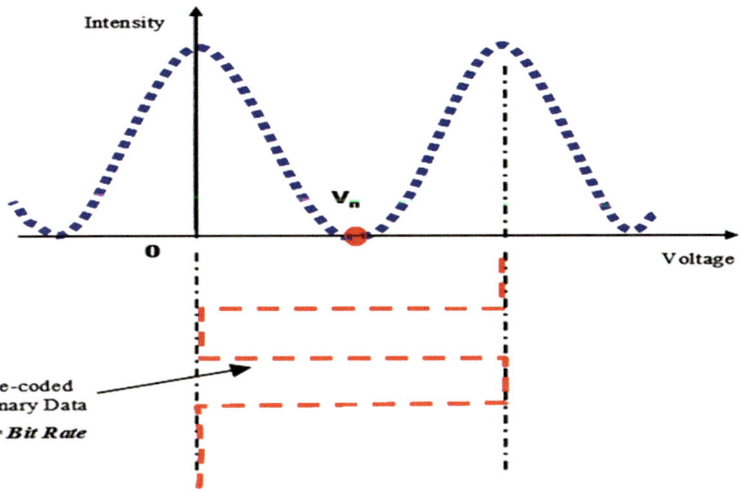

FIGURE 3.12 Bias point on the optical transfer characteristics (upper curve) and the superimposed RF driving signal (medium dark point) swinging from '0' to π for generation of optical NRZ DPSK signals. Lightwaves under the pulse envelope are optical phases varying from 0 to π. The optical carrier is suppressed except for the swing states.

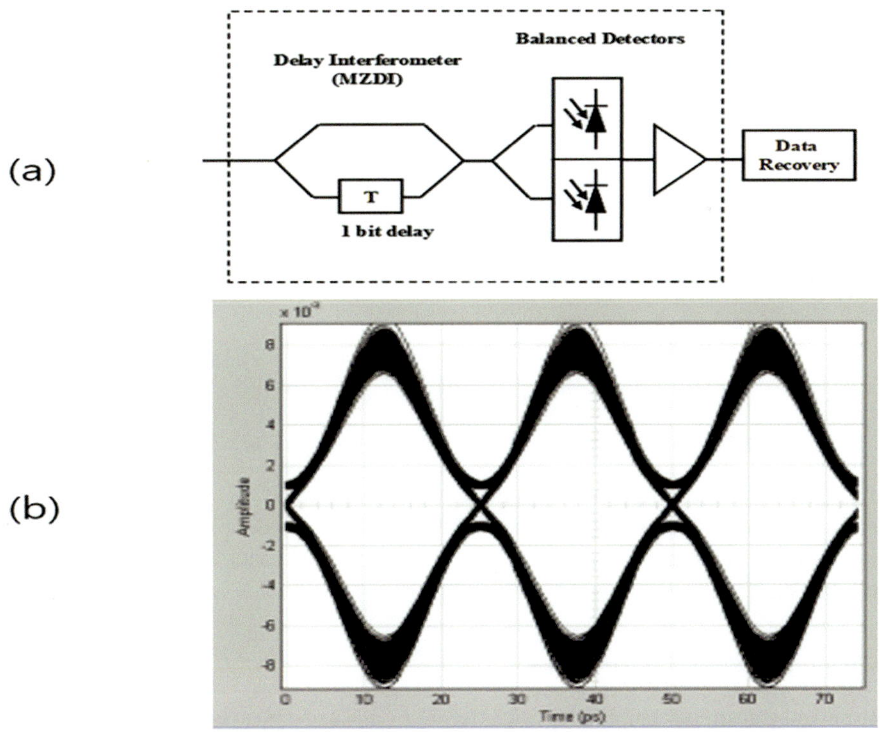

FIGURE 3.13 (a) MZDI balanced receiver for incoherent detection of optical DPSK signals and (b) detected electrical eye diagrams.

insensitive to the absolute phase of the lightwave carrier. The received signal current at the output of the push-pull photo-detectors can be expressed as[54]:

$$\begin{aligned} i(t) &= \left|E(t)+E(t-T)\right|^2 - \left|E(t)-E(t-T)\right|^2 \\ &= 4\Re\left\{E(t)E^*(t-T)\right\} = 4\left|E(t)\right|^2\cos\left(\Delta\phi+\varsigma\right) \end{aligned} \quad (3.8)$$

where $E(t)$ and $E(t-T)$ are the current and the one-bit-delay version of the optical field respectively; $\Delta\phi$ represents the differential phase, and ς indicates the phase noise caused by either the nonlinear phase noise or the imperfection of waveguide path lengths in the MZDI. This MZDI imperfection can be overcome by thermally fine-tuning the thin-film heater integrated into the waveguide of the MZDI.[53,55]

3.4 CPFSK MODULATION FORMAT

3.4.1 Theoretical Background

In CPFSK modulation, the binary data sequence is modulated such that either upper side band (USB) frequency f_1, the right side of the carrier, or lower side band (LSB) frequency f_2 such that $f_1 = f_c + f_d$ and $f_2 = f_c - f_d$, where f_c is the lightwave carrier frequency and f_d is the frequency deviation from f_c. In the case of MSK format, $f_d = {}^1\!/_{4T}$, and the frequency modulation index, defined as $h = 2f_dT$, is equal to 1/2 (h=1/2). It is emphasized that the phase of the optical MSK carrier is continuous at symbol transitions as the MSK meets the condition of the frequency spacing between USB and LSB to be a multiple of ${}^1\!/_{2T}$ to obtain the phase continuity.[19,28,66,67] In addition, this frequency spacing (h=1/2) is also the minimum value to maintain the orthogonality between the two frequencies/modulated signals.[66,67] The baseband representation of an optical MSK pulse train $x(t)$ at the nth symbol interval can be expressed as[66]:

$$x(t) = \sqrt{\frac{E_b}{T}}e^{j\Phi_n(t,a)} \qquad nT - T/2 \le t \le nT + T/2 \quad (3.9)$$

where E_b is the bit energy. The phase term, $\Phi_n(t,a)$, is given by:

$$\begin{aligned} \Phi_n(t,a) &= \left[\pi\Big/\sum_{k=-\infty}^{n-1} a_k\right] + 2\pi f_d\left(t+T/2-nT\right)a_n \\ &= \theta_{n-1}(t,a) + 2\pi f_d\left(t+T/2-nT\right)a_n \end{aligned} \quad (3.10)$$

where $a_n \in \{+1,-1\}$ is the data logic transmitted at the nth interval.

Equation 3.10 shows that the MSK phase evolution follows a linear phase trellis (see Figure 3.14). It increases linearly up to π/2 within a bit slot with a transmitted data '1' and linearly decreases by π/2 with a transmitted data '0'.

3.4.2 Proposed Generation Scheme

An MSK optical transmitter employing two external EOPMs can be arranged in a cascaded configuration, as shown in Figure 3.15. The first EOPM enables the frequency modulation of data logic into either USB or LSB frequencies. The RF driving voltage $V_d(t)$ requires a periodic ramp signal source with a duty cycle of $4T$. It should be noted that differential phase pre-coder and bias voltages are not required in this configuration, reducing the complexity of this MSK optical transmitter scheme.

The second external EOPM enforces the phase continuity of the lightwave carrier at transitions between every two consecutive symbols. The driving voltage into the second EOPM (V_{prep})

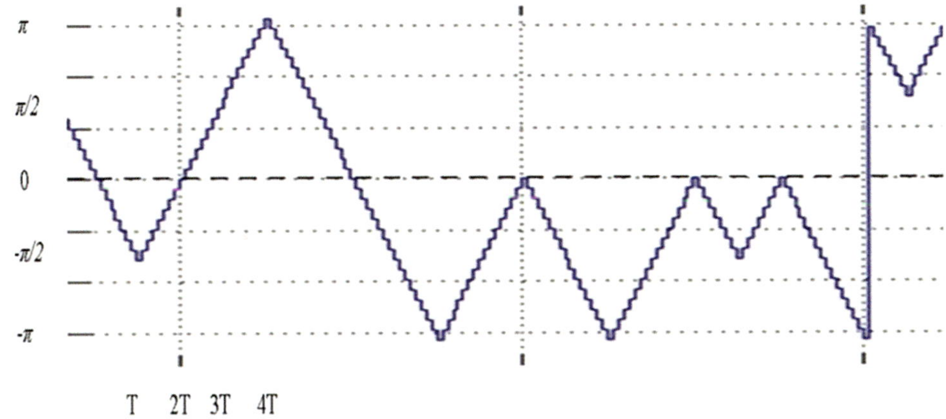

FIGURE 3.14 Plot of the phase trellis, showing the evolution of the optical carrier phase with respect to time, for MSK signals.

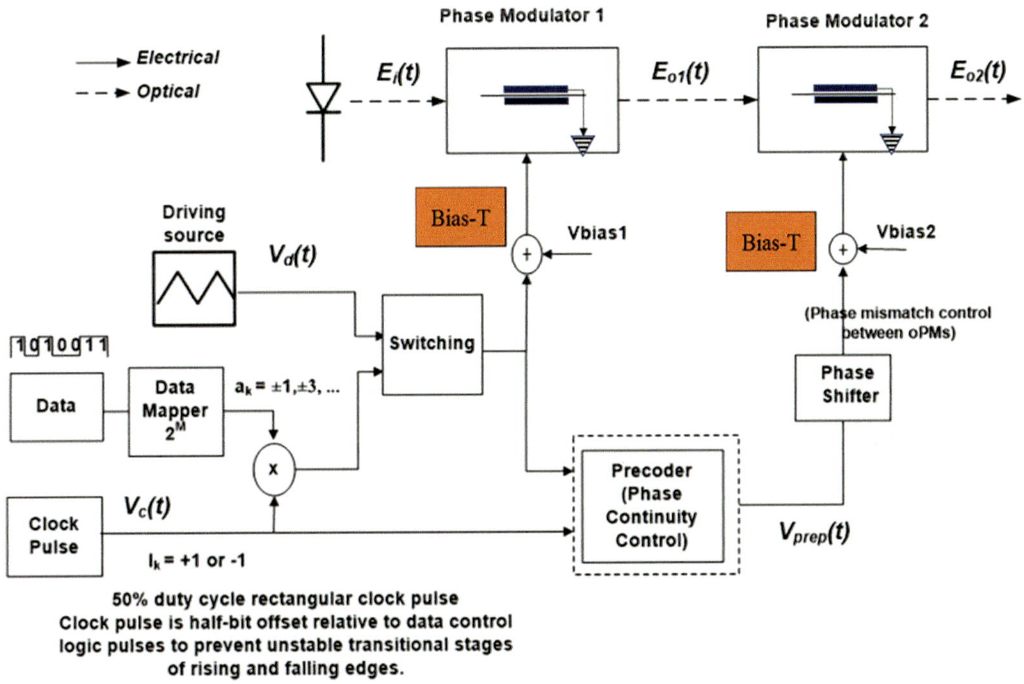

FIGURE 3.15 Schematic block diagram of MSK optical transmitter employing two cascaded external optical phase modulators. Optical paths in dashed lines; Electrical paths: dark solid lines.

is pre-computed to fully compensate for transitional phase discrepancy at the output $E_{01}(t)$ of the first EOPM. An electrical phase shifter is usually used to control the synchronization between the EOPMs. The calculation of V_{prep} at the nth instance is given by the following algorithm:

$$V_{\text{prep}_n}(t,a) = \frac{V_\pi}{2}\left(\sum_{k=0}^{n-1} a_k - a_n I_n \sum_{k=0}^{n-1} I_k\right) \tag{3.11}$$

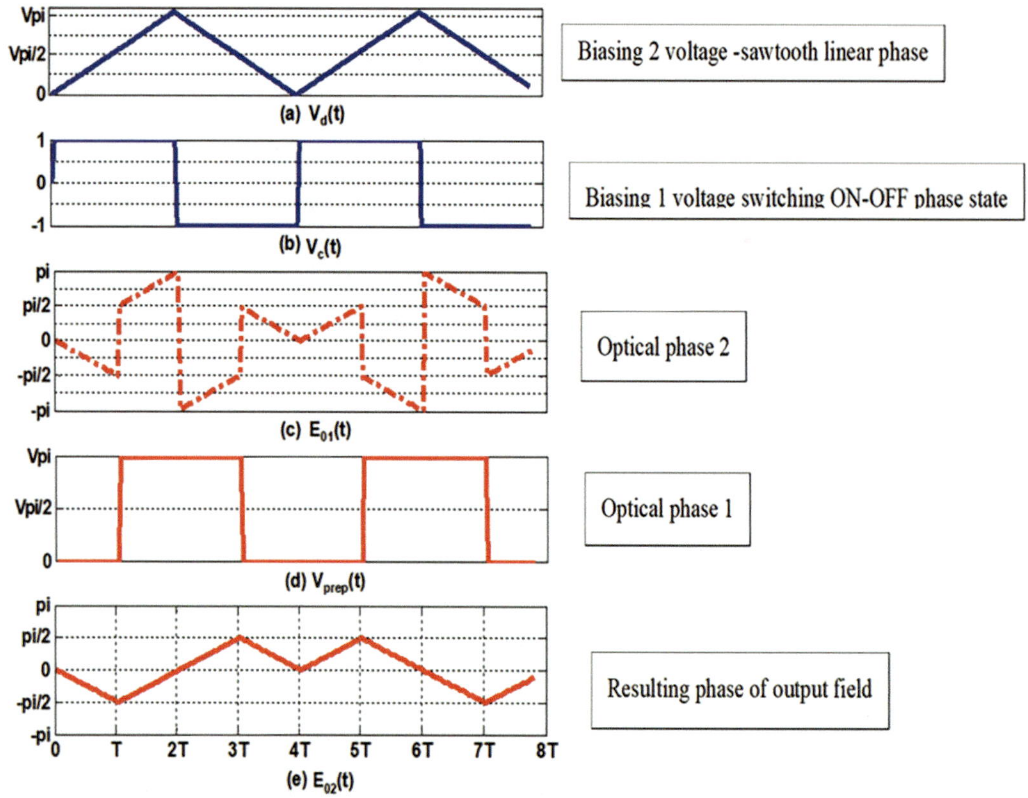

FIGURE 3.16 Evolution of time-domain phase trellis of an optical MSK sequence [-1 1 1-1 1-1-1 1] at different stages of the MSK optical transmitter.

where $a_n = \pm 1$ are the logic levels; $I_n = \pm 1$ represents two stages of a clock pulse $V_c(t)$ whose duty cycle is equal to $4T$. It is emphasized that the values of V_{prep} are either 0 or V_π.

The time-domain signal phase evolution during the modulation process for an optical MSK sequence [-1 1 1-1 1-1-1 1] is demonstrated step by step in Figure 3.15. The key points of Figure 3.16 are as follows: (i) Shows the periodic ramp signal $V_d(t)$ with a duty cycle of $4T$. (ii) Illustrates the clock pulse $V_c(t)$ with a duty cycle of $4T$. To mitigate the overshooting effect at rising and falling edges of electronic circuits, the clock pulse $V_c(t)$ is delayed, e.g., by $2T$, compared to the driving signal $V_d(t)$. (iii) Outlines the phase of the carrier at the output of the first EOPM $E_{01}(t)$. It can be observed that optical signals at this stage are of FSK type and thus, the phase at transitions of consecutive symbols is not continuity. (iv) The transitional phase discrepancy is fully compensated by the pre-computed driving signal $V_{\text{prep}}(t)$, which is calculated from Equation 3.11. (v) Demonstrates the continuous phase trellis of generated optical MSK sequence [-1 1 1-1 1-1-1 1] at the output of the second EOPM.

The proposed generation scheme is realizable with the advance in the design of EOPM. EOPMs operating up to 40 Gbps using resonant-type electrodes have been reported in Refs.[56,57] In addition, high-speed electronic driving circuits evolved with the ASIC (application-specific integrated circuits) technology using 0.1 μm GaAs P-HEMT or InP HEMTs have recently been realized.[58] Furthermore, a simplified version of the proposed optical MSK transmitter can be implemented by alternating the ramp driving signals $V_d(t)$, with a sinusoidal driving source. Finally, this transmitter scheme is also capable of generating M-ary FSK optical signals. The next section describes the optical transmitter configuration for generating optical MSK signals based on the I-Q approach of ODQPSK format.

3.5 MINIMUM SHIFT KEYING MODULATION

3.5.1 Background

MSK modulation is a frequency modulation format such that the phase at the bit transition is continuous. This continuity gives rise to the chirping of the frequency of the carrier, the lightwaves, under the information envelope, and results in a sharper frequency passband of the channel. Coherent reception of such lightwave channels can be easier, and popular integrated optical transceivers can be used with different DSP algorithms.

The MSK modulation format can be formulated in the form of DQPSK by using both in-phase I and quadrature-phase Q components,[56,66] which can be written as:

$$s(t) = \pm\sqrt{\frac{E_b}{T}}\cos\left(\frac{\pi t}{2T}\right)\cos\left(2\pi f_c t\right) \pm \sqrt{\frac{E_b}{T}}\sin\left(\frac{\pi t}{2T}\right)\sin\left(2\pi f_c t\right) \tag{3.12}$$

The I and the Q components consist of a half-cycle cosine and sine waveforms respectively, expressed as:

$$s_I(t) = \pm\sqrt{\frac{E_b}{T}}\cos\left(\frac{\pi t}{2T}\right) \qquad -T < t < T \tag{3.13}$$

$$s_Q(t) = \pm\sqrt{\frac{E_b}{T}}\sin\left(\frac{\pi t}{2T}\right) \qquad 0 < t < 2T \tag{3.14}$$

During the even bit intervals, the I-component consists of positive cosine waveforms for the zero phase and negative cosine waveforms for the π phase. On the other hand, during odd bit intervals, the Q component consists of positive sine waveforms for the phase of π/2, and negative sine waveforms for the phase of −π/2. Hence, any of the four states '0', 'π/2', '−π/2', 'π' can arise. However, only state 0 or π can occur during any even bit intervals and only state π/2 or −π/2 can occur during any odd bit intervals. The output signal is the superposition of the I and Q signal components.

Two important characteristics in generating optical MSK signals using the I-Q approach of the ODQPSK format are that the I and Q driving signals operate at half of the bit rate and are interleaved by a one-bit period. Therefore, only I or Q component can change at a time, i.e., when one is at zero-crossing, the other is at its maximum peak.

3.5.2 MSK Modulation Scheme

The MSK optical transmitter shown in Figure 3.17 consists of two dual-drive MZIMs in parallel, reflecting I-Q structure. The binary logic data is pre-coded and de-interleaved into even and odd symbol streams, staggered by one-bit duration. These pre-coded I and Q data operate at half the data bit rate. In this generation scheme, phase-shaping filters can also reduce the driving signals' spectral width before modulating the dual-drive MZIMs. In addition, the two arms of the dual-drive MZIMs are biased at $V_\pi/2$ and $-V_\pi/2$ and driven by push-pull driving voltages, i.e., data and its complementary $\overline{\text{data}}$.

The phase-shaping driving source can be either *a periodic ramp source* in the case of linear MSK or, more preferably, a sinusoidal source for nonlinear optical MSK. By using sinusoidal waveforms, the hardware complexity for implementing this I-Q MSK optical transmitter scheme is significantly reduced. This is the main motivation for *generating weakly nonlinear* and *strongly nonlinear* types of optical MSK signals. The waveform amplitude of nonlinear MSK optical signals fluctuates, and this ripple level depends on the magnitude of the electrical sinusoidal driving source. The main properties in the generation of linear and nonlinear optical MSK signals are described below.

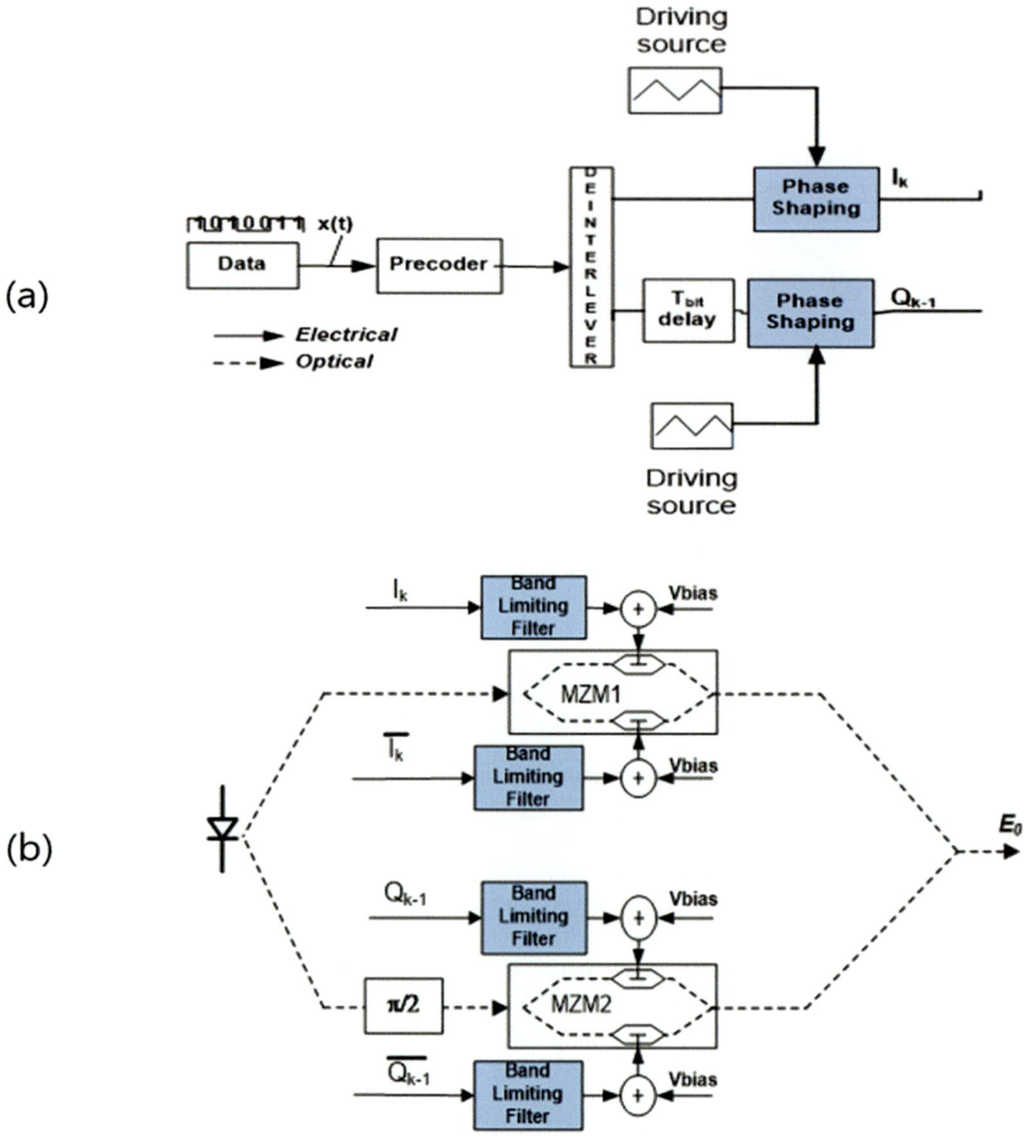

FIGURE 3.17 Block diagram of the I-Q MSK optical transmitter employing two dual-drive MZIMs (a) general block structure (b) device and signals.

3.5.3 MSK Pre-Coding

The logic operations of a MSK pre-coder are constructed based on a state diagram as shown in Figure 3.18a.

It can be observed that the current state of the signal is dependent on the previous state. Therefore, memory is needed to store the previous state. The state diagram in Figure 3.18a is then developed into a logic state diagram, demonstrated in Figure 3.18b. In this case, S_0, S_1 and S_0', S_1' represent the current and previous states, respectively. The pre-coding logic block computes the current state and the

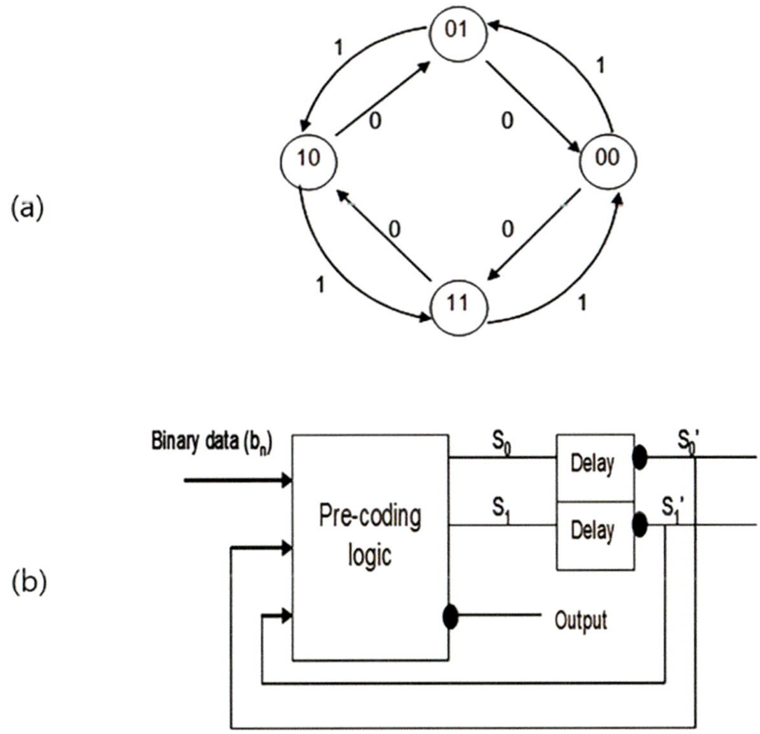

FIGURE 3.18 (a) State diagram for MSK (the arrows indicate continuous increment or decrement of the phase of the carrier); (b) combinational logic for constructing the truth table of the pre-coder.

output based on the feedback state (previous state) and binary data (b_n) from the bit pattern generator. Hence, a Karnaugh map is constructed to derive the logic gates within the pre-coding logic block:

$$S_0 = \overline{b_n}\,\overline{S_0'}\,\overline{S_1'} + b_n\,\overline{S_0'}S_1' + \overline{b_n}S_0'S_1' + b_nS_0'\,\overline{S_1'} \tag{3.15}$$

$$S_1 = \overline{S_1'} = \overline{b_n}\,\overline{S_1'} + b_n\,\overline{S_1'}$$

$$\text{Output} = \overline{S_0}$$

As a result, the truth table (Table 3.1) is constructed. For a positive half-cycle cosine wave and a positive half-cycle sine wave, the output is ‘1’, whereas, for a negative half-cycle cosine wave and a negative half-cycle sine wave, the output is ‘0’.

3.5.4 Incoherent Detection of MSK

3.5.4.1 MZDI Balanced Receiver

Similar to optical DPSK, an MZDI balanced receiver is used for the direct detection of both O-QPSK and MSK-modulated optical signals, as shown in Figure 3.19. However, the main difference in this MZDI configuration, compared to that of the DPSK counterpart, is an additional π/2 phase shift inserted on one arm of the MZDI.[59,60,61,62] This π/2 phase shift can be located on either arm of the MZDI.

TABLE 3.1
Truth Table Based on MSK State Diagram

$b_n S_0' S_1'$	$S_0 S_1$	Output
100	01	1
001	00	1
010	01	1
101	10	0
110	11	0
111	00	1
000	11	0
011	10	0

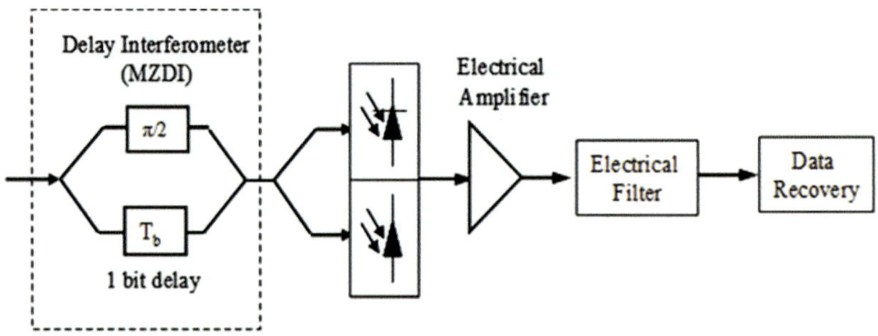

FIGURE 3.19 MZDI balanced receiver for incoherent detection of optical MSK signals.

3.5.4.2 Optical Frequency Discrimination Receiver

The detection of optical MSK signals using frequency discrimination principles was studied in the early 1990s, however, mainly for heterodyne coherent receivers.[23,34–37] The frequency discrimination was conducted in the electrical domain, at the Intermediate Frequency (IF) range. Non-coherent frequency discrimination in the photonic domain for the direct detection of MSK lightwaves could not be effectively implemented at that time due to the unavailability of optical filters with very narrow bandwidths (as a fraction of the signal bandwidth).

However, with recent advances in the design and fabrication of optical filters, in particular using micro-ring resonator type, the 3-dB bandwidth of optical filters can be narrowed down to around 2 GHz.[63,64,65] The availability of such narrowband optical filters thus enables the feasibility of non-coherent optical frequency discrimination for detecting modulated frequencies of optical MSK signals. An implementable receiver scheme employing dual narrowband optical filters and an optical delay line is proposed in this research for optical MSK transmission systems. Detailed descriptions of the design and performance characteristics of this receiver are provided in Chapter 6.

3.5.5 Dual-Level MSK

3.5.5.1 Constellation of Binary MSK

MSK can be formulated from either the continuous phase frequency shift keying (CPFSK) or offset DQPSK (OQPSK) format.[66,67] The signal trajectory of optical MSK signals, as shown in Figure 3.20, continuously evolves. This is unlike the discrete DPSK format, where encoded bits are located *exactly* at the '0' and 'π' phase positions. Two novel schemes are proposed for generating optical MSK signals.

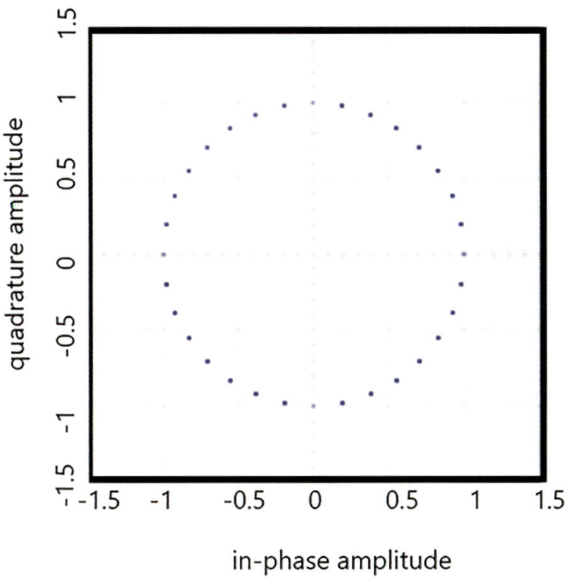

FIGURE 3.20 Scatter plots of the signal phase trajectory, in-phase amplitude against quadrature amplitude, of optical MSK modulated signals.

3.5.5.2 Constellation of Dual Level MSK Signal

Dual-level MSK is a multi-level scheme that encodes two data bits into a modulated symbol. One of the information bits is for coding amplitude levels, whereas the other is for the linear phase trellis coding of $\pm\pi/2$ phase change within one symbol period. Therefore, the transmission baud rate is halved compared to the bit rate from the data pattern generator. This clearly offers transmission bandwidth efficiency and significantly reduces the requirement for high-speed electronic processing. Figure 3.21 illustrates the signal constellation of the dual-level MSK format. The mapping scheme for data information a and b in dual-level MSK is shown in Table 3.2.

The dual-level MSK modulation format can be generated from the superposition of two MSK signals having different amplitudes, as given by Ref.[66]:

$$s(t) = A\cos\left(2\pi f_c t + \phi_n(t,a)\right) + B\cos\left(2\pi f_c t + \phi_n(t,b)\right) \tag{3.16}$$

where f_c is the optical carrier frequency, A and B are the signal amplitude levels. At the transition of the nth symbol interval, these two amplitudes are either in phase or π phase-shift with respect to each other. The phase terms $\phi_n(t,a)$ and $\phi_n(t,b)$ are defined as:

$$\phi_n(t,a) = \pi h \frac{a_n(t-nT)}{T} + \pi h \sum_{k=-\infty}^{n-1} a_k \qquad nT \le t \le (n+1)T \tag{3.17}$$

$$\phi_n(t,b) = \pi h \frac{b_n(t-nT)}{T} + \pi h \sum_{k=-\infty}^{n-1} b_k \qquad nT \le t \le (n+1)T \tag{3.18}$$

The logic values a_n and b_n are statistically independent and are taken from the set $\{\pm 1\}$. The frequency modulation index h of the dual-level MSK is equal to 1/2.

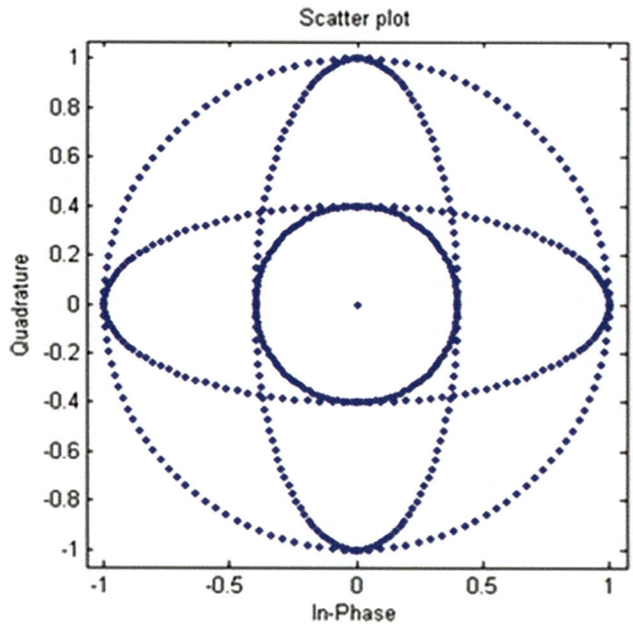

FIGURE 3.21 Signal constellation of dual-level MSK.

TABLE 3.2
Mapping of Data Information in Dual-Level MSK Format

a	b	Remarks on Signal Constellation
1	1	Amplitude unchanged, phase increased
1	−1	Amplitude unchanged, phase decreased
−1	1	Amplitude changed, phase increased
−1	−1	Amplitude changed, phase decreased

3.5.5.3 Multi-Level MSK Pre-Coder

A pre-coder is compulsory for the generation of dual-level MSK signals. At the *n*th instance, the logics of a_n and b_n taken from {±1} are pre-coded from the binary logics *d* and *d*' as follows:

$$a_n = 2d_n - 1; \qquad \text{and} \qquad b_n = a_n\left(1 - \frac{d'_n}{h}\right) \tag{3.19}$$

where the logics *d* and *d*' are statistically independent and taken from {0,1}.

3.5.5.4 Dual-Level Modulation Scheme

Any configuration of MSK optical transmitters presented in Section 3.4 or reported in previous research,[33,68,69,70] can be utilized for the proposed generation scheme of dual-level MSK optical signals.

Figure 3.22 shows the block diagram of a dual-level MSK optical transmitter in which two MSK optical transmitters are arranged in a parallel configuration.

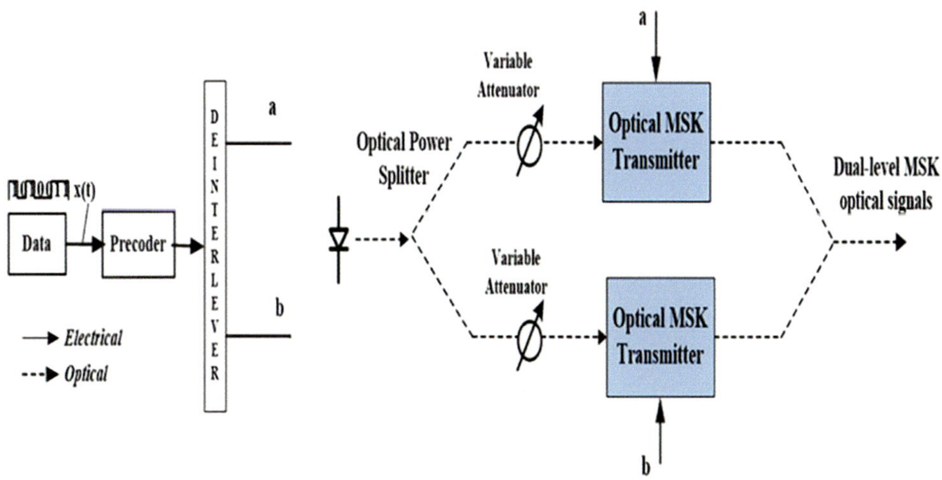

FIGURE 3.22 Block diagram of the proposed dual-level MSK optical transmitter.

It can be seen from Figure 3.22 as well as from Equation 3.16 that the amplitude levels *A* and *B*, which are input into the two MSK optical transmitters, play significant roles in the proposed generation scheme as they determine the amplitude levels of the transmitted dual-level MSK optical signals. In practice, optical components are normally specified by optical intensity rather than the magnitude of the optical field. Thus, we consider the optical powers input into the two MSK optical transmitters rather than the amplitude levels *A* and *B*. Moreover, the splitting ratio between these two optical powers is more important than their individual values. This intensity-splitting ratio can be obtained by using a high-precision optical power splitter or simply by using a 3-dB or 50:50 optical power splitter followed by a variable optical attenuator on each path.

3.5.6 Linear and Nonlinear MSK

3.5.6.1 Linear MSK

The phase-shaping signal used to drive the dual-drive MZIMs is a periodic ramp signal with a duty cycle of $4T$. The Q driving path is delayed by one-bit period with respect to the I path. Optical MSK signals are obtained at the output of the two MZIMs as the superposition of I and Q modulated signals, and the generated MSK waveforms clearly display the constant amplitude characteristic of the CPM format. In addition, the signal trajectory forms a perfect circle. These characteristics are illustrated in Figure 3.23a.

3.5.6.2 Weakly-Nonlinear MSK

The generation of a weakly-nonlinear MSK format requires a sinusoidal driving signal with an amplitude of about $V_\pi/4$ and operating at half the data bit rate. The in-phase pulse shaper uses a cosine waveform, while the quadrature counterpart uses a sine waveform. The time-domain variation of the carrier phase is thus nonlinear. As a result, this causes a mismatch between the two MZIMs when I- and Q-waveforms are superimposed, resulting in ripples on the generated optical MSK waveforms. The signal trajectory is thus not a perfect circle. The magnitude ripples of weakly-nonlinear optical MSK waveforms are <10% of the push-pull signal levels. All of these characteristics are illustrated in Figure 3.23b.

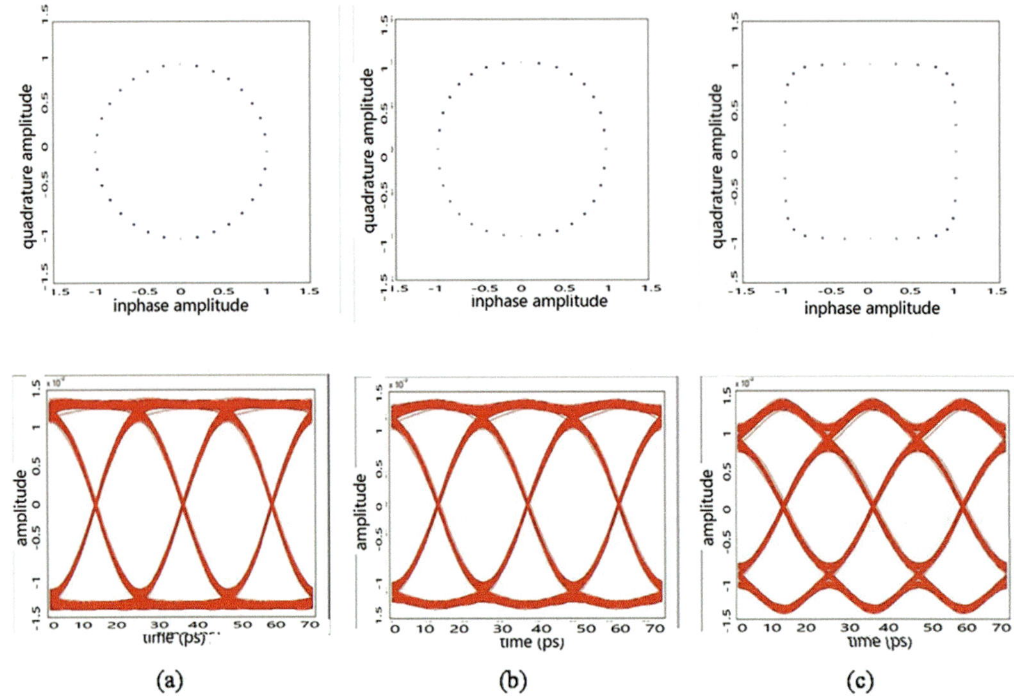

FIGURE 3.23 Signal constellation (above) and eye diagrams (in medium dark dots) of MSK optical signals: (a) linear, (b) weakly nonlinear, and (c) strongly nonlinear types.

3.5.6.3 Strongly-Nonlinear MSK

Sinusoidal phase-shaping waveforms are also used for generating strongly nonlinear MSK optical signals. However, the magnitude of this driving signal is close to V_π compared to $V_\pi/4$ as in the case of weakly nonlinear MSK. Strongly nonlinear optical MSK signals have amplitude ripples of above 10% of the push-pull signal levels. This fluctuation is also caused by the mismatch of the waveforms on two arms at the output of the MZIMs. Figure 3.23c illustrates the signal trajectory and the ripples of strongly nonlinear optical MSK signals.

3.5.7 Self-Coherent Detection of Optical Dual-Level MSK

As a combination of optical MSK and multi-level modulation formats, the demodulation of dual-level MSK optical signals requires the incoherent detection of both amplitude and phase components, as shown in Figure 3.24.

Similar to optical MSK, an MZDI balanced receiver is implemented for phase detection, with an additional $\pi/2$ phase shift inserted on one arm of the MZDI. On the other hand, the lightwave amplitude is detected by a single photodetector, which is used for decision-making on the amplitude level of the MSK multilevel. It should be noted that a differential decoder is required to detect amplitude changes between two consecutive bits. The detected eye diagrams for amplitude and phase detections of dual-level MSK received signals are demonstrated in Figure 3.25a and b. In these figures, the dotted line represents the decision threshold voltage.

It can be observed that the phase detection of dual-level MSK gives a push-pull eye diagram having multiple levels. However, the focus is only on either the positive or the negative attribute of these levels, which corresponds to either zero phase or π phase change respectively. Thus, the threshold level for phase detection is set at zero, whereas the amplitude threshold is a non-zero value to distinguish between the two modulated signal levels.

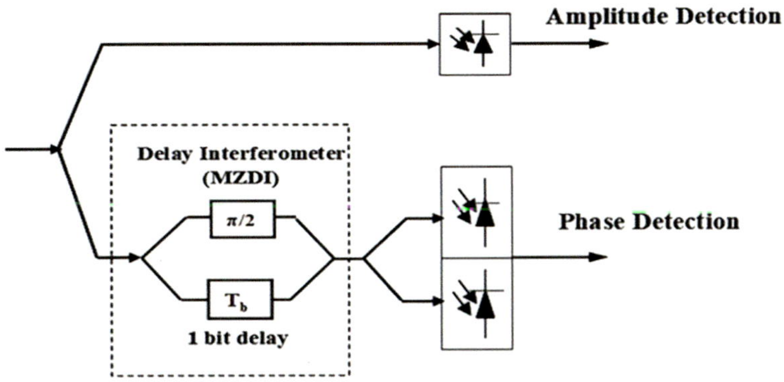

FIGURE 3.24 Incoherent and self-coherent detection for both amplitude and phase of dual-level MSK optical signals.

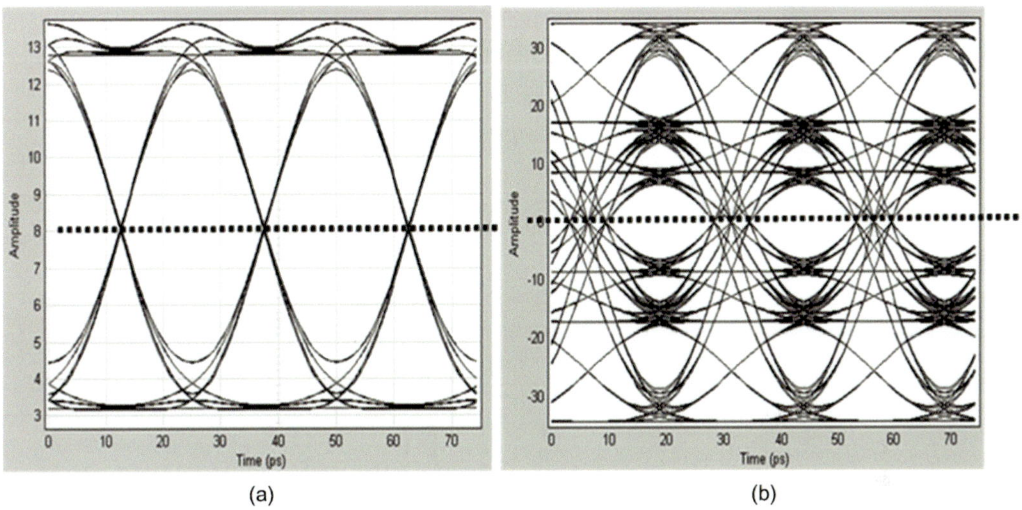

FIGURE 3.25 Eye diagrams for (a) amplitude and (b) phase detection of dual-level MSK received signals. Decision threshold in broken-line.

3.6 SPECTRAL CHARACTERISTICS OF BINARY MODULATION

Optical power spectra of 40 Gbps NRZ/RZ33/CS-RZ DPSK and OOK optical signals are shown in Figures 3.26 and 3.27. It is noted that the carrier amplitudes cannot be seen in full depending on the resolution of the optical spectrum analyzer (OSA), which has a minimum resolution of about 5 pm with a wide spectral span of 1.0 nm. With this 5 pm resolution, it is not possible to observe the carrier linewidth of 10 kHz. It is possible to use a coherent optical spectrum analyzer, but the spectral range is very limited.

Several main points observed in Figures 3.26 and 3.27 are (i) The optical power spectra of OOK formats have high power components at the carrier and signal frequencies, which cause fiber nonlinear effects. On the contrary, DPSK spectra do not contain high-power frequency components, making optical DPSK formats more resilient to fiber nonlinear effects; (ii) RZ pulses are more susceptible to fiber CD or spatial broadening in free space due to their broader spectra compared to NRZ pulses. In particular, the RZ33 type has the broadest spectrum (at 20 dB level down from the peak amplitude) and thus, RZ33 is least tolerant to fiber CD and spatial spreading in free space;

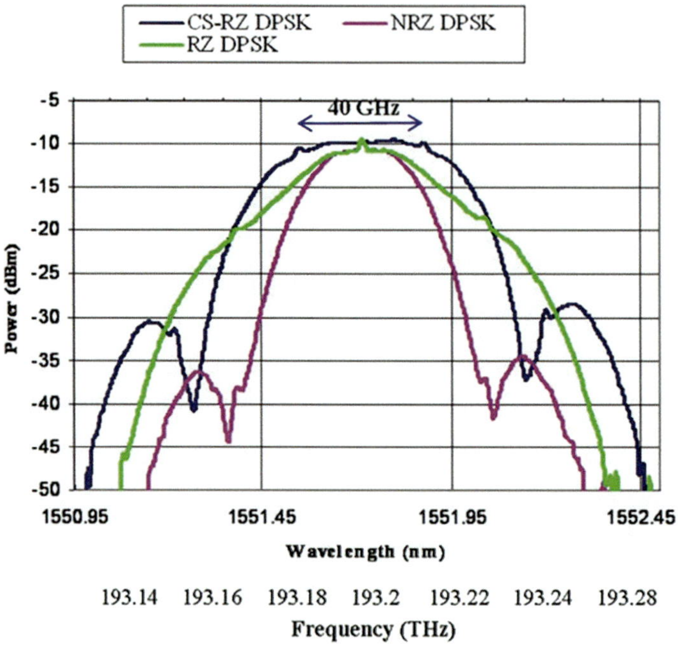

FIGURE 3.26 Spectra of 40 Gbps CS-RZ/RZ33/NRZ DPSK optical signals.

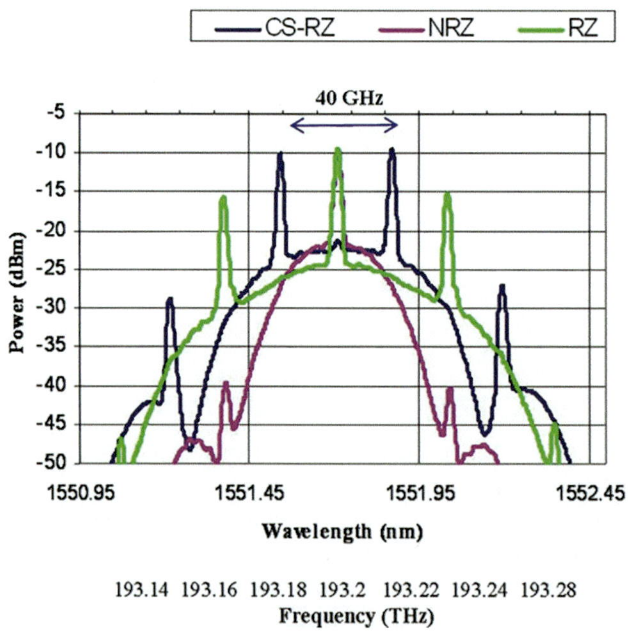

FIGURE 3.27 Spectra of 40 Gbps CS-RZ/RZ33/NRZ OOK optical signals.

(iii) However, the susceptibility of RZ33 and CS-RZ pulses to fiber CD provides a trade-off in the improvement of the resilience to fiber nonlinearities. This is because time-domain peaks of these fast-spreading optical pulses decrease rapidly when propagating along the fiber, hence lowering the effects of fiber nonlinearities.

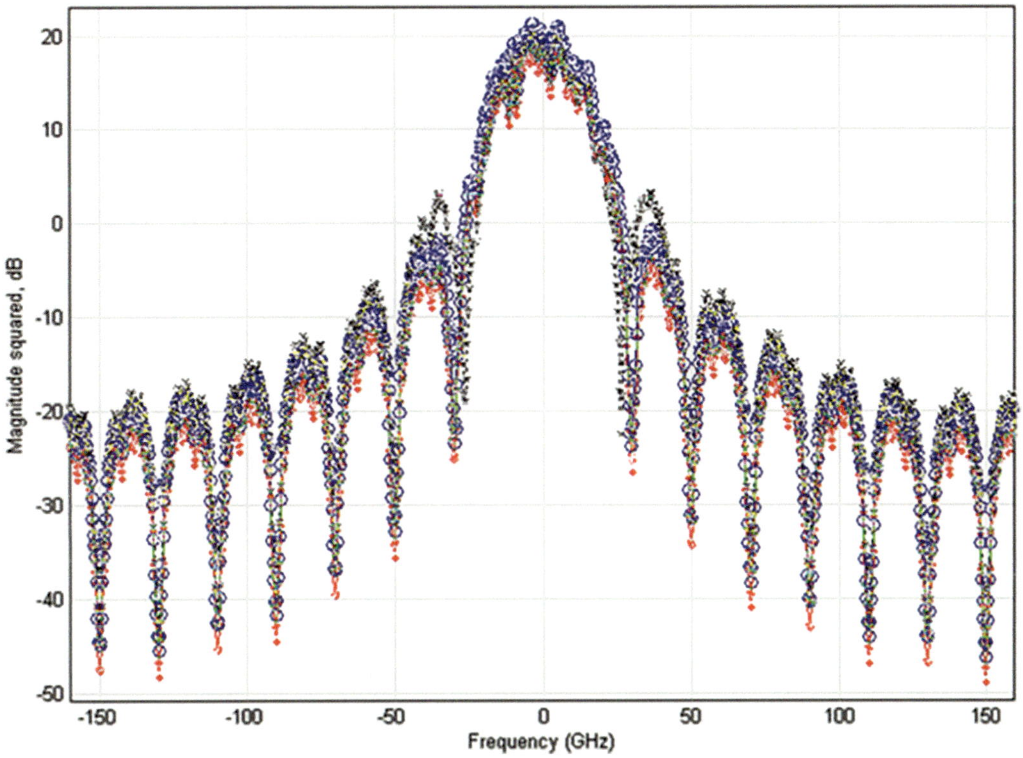

FIGURE 3.28 Optical power spectra of three 40 Gbps optical MSK formats: linear (medium dark line), weakly nonlinear (light dark line), and strongly nonlinear (strong dark line).

The carrier power must be about 6 dB above the signal band, which is not observed in the spectrum analyzer due to resolution, the same is in practice when using OSA.

Optical power spectra of 40 Gbps linear, weakly nonlinear, and strongly nonlinear MSK formats are compared in Figure 3.28. It can be observed that these formats' power spectra are similar, with the exception being that sidelobes in the power spectrum of strongly nonlinear MSK are not as highly suppressed as in the case of linear and weakly nonlinear MSK. All three formats offer better spectral efficiency compared to the DPSK counterpart.

Figure 3.29 compares the power spectra of three modulation formats: 80 Gbps dual-level MSK, 40 Gbps linear MSK, and 40 Gbps NRZ DPSK. The intensity-splitting ratio for the optical dual-level MSK format is set at '0.8/0.2'.

Several key points can be observed: (i) The signal power spectrum of 80 Gbps optical dual-level MSK has characteristics identical to that of the 40 Gbps MSK format. The main lobe spectral widths of these two formats are narrower than that of NRZ DPSK. More specifically, the base-width takes a value of approximately ±32 GHz on either side compared to ±40 GHz in the case of DPSK. Hence, it is expected that these MSK-based formats are more tolerant to fiber CD and more robust to tight optical filtering than NRZ DPSK. (ii) The high suppression of side lobes and the confinement of signal energy in the main spectral lobe of 80 Gbps dual-level MSK and 40 Gbps MSK optical power spectra significantly mitigate inter-channel cross-talk between DWDM channels.

3.7 PARTIAL RESPONSE MODULATION

Photonic communication systems have continuously evolved over the last seven decades in both fiber-guided transmission lines and space inter-satellite links. The increasing demand for higher transmission capacity has driven the development of communication systems with ultra-high

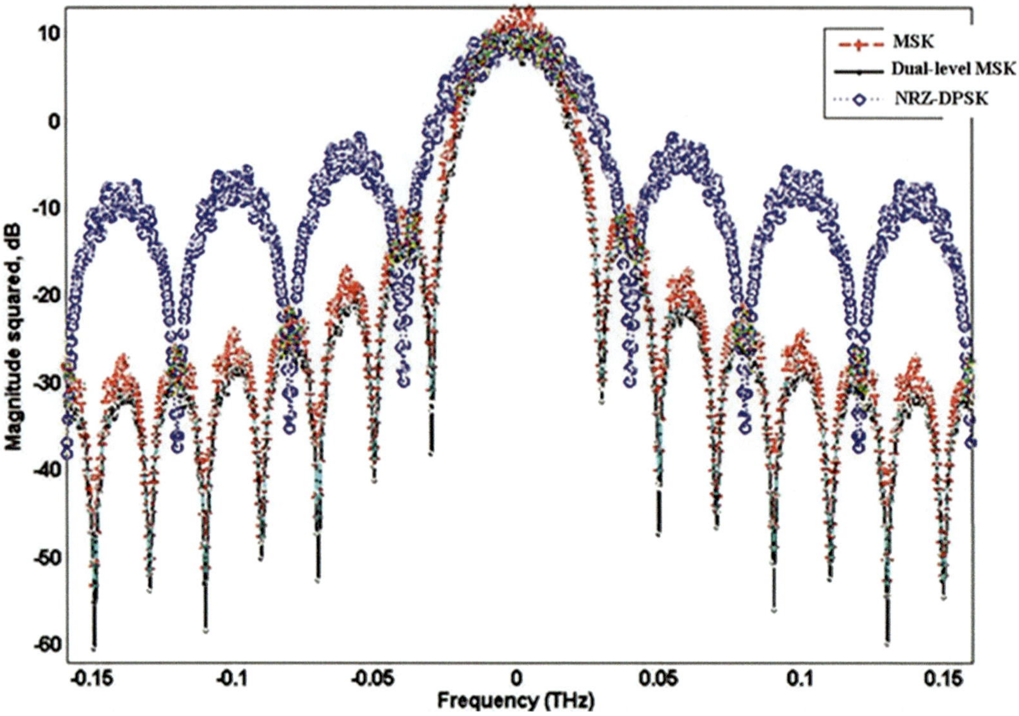

FIGURE 3.29 Spectral properties of three modulation formats: 40 Gbps MSK (medium dark dash), 80 Gbps dual-level MSK (dark black solid dot), and 40 Gbps NRZ DPSK (light dark dot).

capacity and ultra-high bit rates. Photonic transmission systems operating at 40 Gbps, 100 Gbps, and higher have extended transmission reach and improved capacity, mainly in fiber terrestrial lines. Over the last several years, they have become an attractive alternative to 10 Gbps communication systems. These modulation formats generate channel signals with the carrier located in the middle of a double sideband (DSB) signal in the frequency spectra.

In space photonic communications, spatial broadening leads to a reduction in the receiving power by the photodetector (PD) via the optic system. The bandwidth of the modulated sequence also contributes to this broadening, thus reducing the channel bandwidth without sacrificing its bit rate is desired. Partial response modulation, a method to generate a single sideband of the channel with an equivalent bit rate, can offer such advantages. All the modulation formats described in the above sections typically offer DBS spectra.

System performance is further enhanced by employing various advanced modulation formats, such as Duobinary (DB) in optical multi-phase or by electrical filtering, RZ-DPSK, and vestigial NRZ-DPSK. Investigations and developments have been carried out to determine the most appropriate and efficient formats that meet current as well as future demands. The partial response modulation formats aim to reduce the wideband of the double sideband around the carrier to a single sideband, reducing the fiber chromatic dispersion in optically guided media and spatial broadening in space photonic transmission, thus reducing the sensitivity in space communications and pulse temporal broadening in the fiber transmission line.

This section examines and describes the partial response modulation and their performance over the optical transmission systems, in particular the duo-binary modulation format.

Duo-binary modulation (DBM) format, in the photonic domain, offers three-level coding, but unlike electronic or wireless communication systems, the "–1" and "+1" are coded using the phase of the lightwave carrier, that is either "0" or "π". This coding would overcome the dispersion due to its single sideband property; the detection scheme is much simpler as a direct detection technique can be employed. The single sideband can also be implemented using the vestigial sideband modulation technique, which is an optical filter that can be inserted after the optical modulator to filter half of the band of the spectrum. Alternatively, the modulators can be conditioned with two Hilbert transform signals, hence the suppression of half of the band.

The first part of this section presents a comprehensive modeling platform for a duo-binary modulation format optical fiber transmission system. The system modeling is developed using MATLAB Simulink 7.0 or higher. The Simulink model has been formed with component block systems, such as communication block sets and signal processing block sets. More comprehensive Simulink models can be found in Ref.[71] Furthermore, the transmission of 40 Gbps alternating phase 0 and π with duo-binary modulation formats of 33% and 50% width RZ pulse sequences can be demonstrated with 850 ps/nm dispersion tolerance. At least a 2-dB receiver sensitivity improvement can be achieved compared to carrier-suppressed DPSK.

The second part of this section presents the transmission of optical multiplexed channels of 40 Gbps using vestigial single-side band modulation format over a long-reach optical fiber transmission system. Thus, an optical filter must be designed to follow the optical modulator to filter half of the signal band. The effects on the *Q*-factor of fiber dispersion, the passband and roll-off frequency of the optical filters, and the channel spacing are described. The performance of optical transmission using low and non-zero dispersion fibers or/and dispersion compensation is analyzed. It has been demonstrated that a BER of 10–12 or better can be achieved across all channels, with minimum degradation of the channels under this modulation format. Optical filters are designed with asymmetric roll-off bands. Simulations of the transmission system are also given and compared for channel spacings of 20, 30, and 40 GHz. It is shown that a passband of 28 GHz and a 20 dB cut-off band perform best for 40 GHz channel spacing.

3.7.1 Optical Phase Duo-Binary Modulation

3.7.1.1 Introductory Remarks

The demand for high-capacity long-haul telecommunication systems has increased in recent years. To achieve high throughput with minimal errors, different advanced modulation formats, such as ASK, PSK (coherent and differential incoherent), and FSK, have been proposed, and comparisons are made to determine which modulation format offers the best transmission performance. To counter performance degradation, modulation formats aim to narrow down the optical spectrum to enable closer channel spacing in the network. They increase symbol duration so that more uncompensated dispersion accumulates before ISI becomes significant. Furthermore, this format is more resilient to fiber non-linearities and optical signal distortion.

The modulation format is important in determining the performance of 40 Gbps optical fiber communication systems. Duo-binary (DB or DuoB) and continuous phase modulation (CPM) DB are shown to offer high spectral efficiency.[72] DBM formats minimize inter-symbol interference (ISI) impairments in a controlled way instead of trying to eliminate them. It is possible to achieve a signaling rate equal to the Nyquist rate of 2W symbols per second in a channel with a bandwidth of W Hz. The Optical DuoB technique has received much attention due to its high dispersion tolerance and high frequency-utilization efficiency by means of spectral narrowing. The DBM format is similar to the NRZ format, with the inclusion of phase coding. The phase characteristics of DBM signals compensate for the group velocity dispersion by reducing the spectral components found in conventional NRZ modulation. ISI is reduced because bit patterns such as 101 are transmitted with the ones carrying the opposite phase. Therefore, if pulses spread into the zero-time slot due to dispersion, they tend to cancel each other out. The recovery of signals at the receiver is relatively simple. Furthermore, a conventional direct detection receiver is applicable, resulting in a simple receiver structure. There are two types of DBM schemes: constant phase and alternating phase in blocks of logical “1s”.

This section presents the models for photonic transmission with optical channels operating under the DBM format. This includes the development and implementation of the photonic transmitter, optical fiber propagation, and the opto-electronic receiver. DBM encodes a two-level electrical signal into a three-level electrical signal before modulating the lightwave carrier. The transmitter in the MATLAB Simulink© model consists of a DuoB encoder and a dual-drive MZIM. A baseband modulation is first implemented in the DuoB encoder, which encodes the binary signal into three levels signals: "1", "0", and "–1". The MZIM is an electro-optic modulator that converts the electrical signal into an optical signal.

The DB or phase-shape binary modulation (DBM) formats can be generated by modulating a dual-drive MZIM. Recent works have shown that the driving voltages for the modulator can be reduced to generate variable pulse-width DB optical signals. However, the pulse width of the DB DPSK has not been thoroughly investigated under the alternating phase of the "1" coded bits. This means the "0" "π" "0" "π" phases of consecutive "1" in contrast to conventional duo-binary formats. We also present the modeling performance of alternating-phase DBM with a full-width at half-mark (FWHM) ratio with respect to the bit period of 100%, 50%, and 33%, and compare it with the experimental transmission of carrier-suppressed DPSK over 50 km of SSMF and dispersion compensation. For the DB case, transmission without dispersion compensation over the same SSMF length offers better performance for 50% FWHM DBM and slightly worse for 33%. The transmission performance, the bit-error rate versus receiver sensitivity, of these DBM formats is compared with the carrier-suppressed DPSK experimental transmission.

Section 3.7.1.2 describes the fundamental aspects of the DBM format. It is followed by the description of each component in the 40 Gbps DBM photonic transmission systems in Section 3.7.2. Section 3.7.4 describes the alternative implementation of duo-binary via electrical filtering. Lastly, Sections 3.7.5 and 3.7.4 briefly outline the partial responses with single sideband and the performance in transmission.

3.7.1.2 DBM System

DuoB Modulation (DBM) format aims to modulate one or more field properties to suit system needs. There are four types of field properties: intensity, phase, polarization, and frequency. Symbols are constellated in one or more dimensions, to carry more information and to travel a greater distance. The data modulation format is the information-carrying property of the optical field.

DBM format has become an attractive modulation format in recent years compared to other formats, such as NRZ and RZ. This is because it can overcome fiber chromatic dispersion in high-capacity transmissions. It is the characteristics of the DBM format that make it a preferred format.

DBM schemes can be described as correlative-level coding or partial-response signaling schemes. Correlative-level coding schemes add ISI to the transmitted signal in a controlled manner, and a signaling rate equal to the Nyquist rate of 2W symbols per second in a channel of bandwidth W Hz can be achieved. The "D" in DB indicates the doubling of the transmission capacity of a conventional binary system. DBM format is, in fact, NRZ modulation with the inclusion of phase coding. The one bit in the data input is phase-modulated. For instance, for a bit pattern of "101", this data will be transmitted with the ones carrying opposite phases, 0 and π. If the pulse of the one-bit spreads out to the zero time-slot in between, they can cancel each other. This effect increases dispersion tolerance and allows the signal to be transmitted over a longer distance.

DuoB coding converts a two-level binary signal of 0's and 1's into a three-level signal of "–1", "0" and "+1". This is done by first applying the binary sequence to a pulse-amplitude modulator to produce two-level short pulses with amplitudes of –1 and +1, where –1 corresponds to 0 and +1 corresponds to 1. This sequence is then, applied to a DuoB encoder to produce a three-level output of "–2", "0" and "2".

As shown in Figure 3.30, the input sequence $\{a_k\}$ of uncorrelated two-level pulses is transformed into $\{c_k\}$, a sequence of correlated three-level pulses. The correlation between adjacent pulses is equivalent to introducing ISI into a transmitted signal in an artificial manner. The DuoB encoder is simply

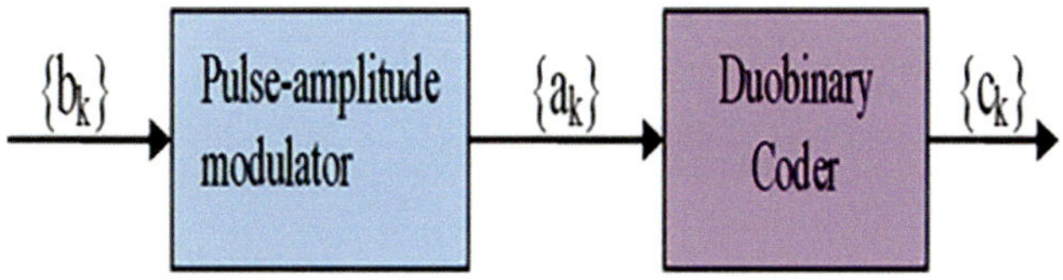

FIGURE 3.30 Brief overview of DB signaling.

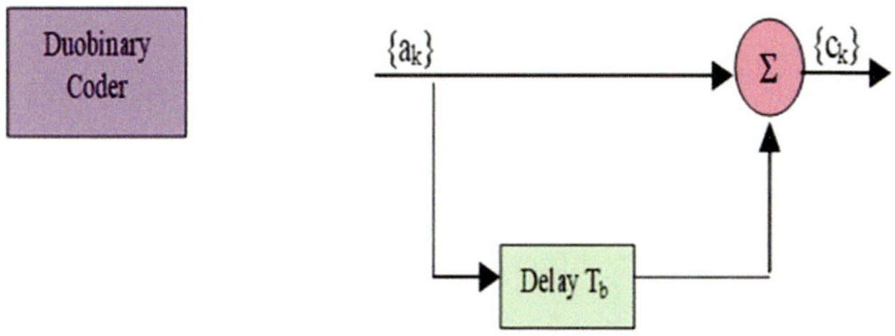

FIGURE 3.31 DuoB encoder – the block on the left is represented by the signal flow diagram shown on the right.

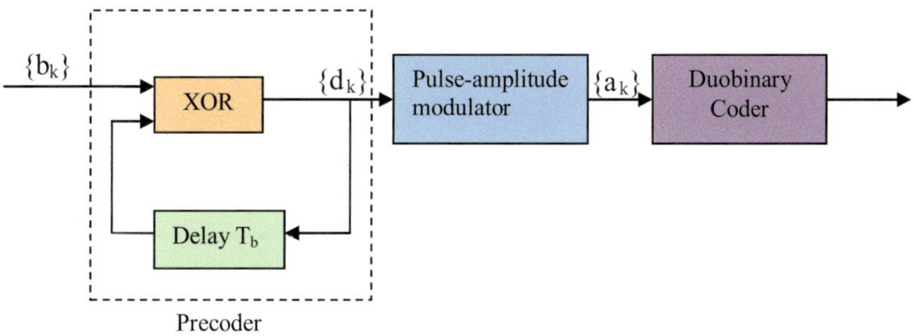

FIGURE 3.32 DBM scheme with precoder.

a filter involving a single delay element and a summer, as shown in Figure 3.31. However, once errors are made, they tend to propagate through the output. This is because a decision made on the current input a_k depends on the decision made on the previous input a_{k-1}. Therefore, precoding is needed to avoid this error propagation phenomenon. The binary sequence $\{b_k\}$ is converted into another binary sequence $\{d_k\}$ by modulo-two addition, exclusive OR of b_k and d_{k-1}, as shown in Figure 3.32.

$$c_k = a_k + a_{k-1} \tag{3.20}$$

$$d_k = b_k \oplus d_{k-1} \tag{3.21}$$

The three-level DB output $\{c_k\}$ is then modulated into a two-level optical signal by an external modulator. The most commonly used external modulator is the MZIM. The optical DuoB signal has two intensity levels, "on" and "off". The "on" state signal can have one of two optical phases, 0 and π. The two "on" states correspond to the logic states "1" and "–1" of the DB encoded signal, $\{a_k\}$, and the "off" state corresponds to the logic state "0". Figure 3.33a shows an example of the original binary signal, the DuoB encoded signal, the optical DuoB signal, and a summary of the coding rules in Figure 3.33b.

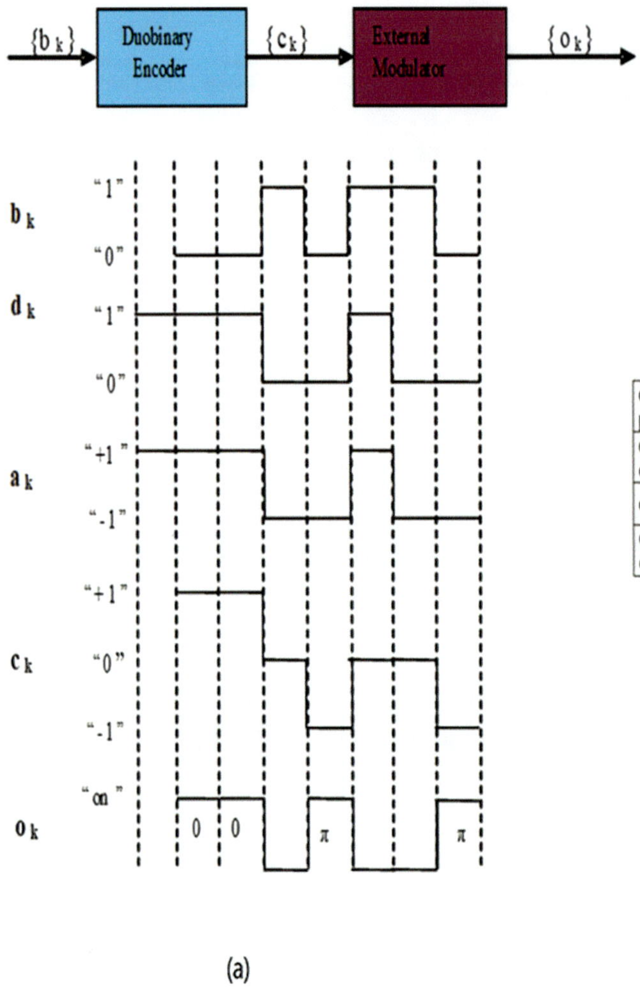

duobinary precoding	$b(k)=\overline{d(k)}\oplus b(k-1)$
duobinary encoding	$c(k)=b(k)+b(k-1)$
offset	$c_{AC}(k)=c(k)-1$
duobinary decoding	$\hat{d}(k)=\lvert c_{AC}(k)\rvert$

(b)

FIGURE 3.33 (a) Example of original binary signal (b_k), pre-coded signal (d_k), DuoB encoded signal (c_k) and optical DB signal (o_k). (b) Summary of coding rules.

3.7.2 Photonic Phase Duobinary Modulation

Ultra-long terrestrial networks transmitting signals at a bit rate of 40 Gbps have matured in recent years. Various advanced modulation schemes, such as RZ, NRZ, NRZ-DPSK, and RZ-DPSK, have been proposed to achieve extended reach and improved capacity in communication systems.

Figure 3.34 shows a typical DWDM optical fiber communication system. Signals are modulated at the transmitters and are multiplexed together at the wavelength multiplexer before transmitting them into the fiber. The fiber link is divided into several spans, aach consisting of an SSMF and DCF. The EDFA is used to compensate for optical power loss in the transmission span. At the end of the fiber, the signals are demultiplexed and detected at the receivers.

The DBM format has become popular compared to other modulation formats because it extends the transmission distance limited by fiber loss, without the need for regenerative repeaters. It also extends the dispersion limit without requiring additional optical components, such as dispersion-compensating fiber (DCF). Chromatic dispersion has become a main factor limiting the transmission distance. The optical three-level transmission can overcome this limitation, as a narrowband signal has higher tolerance to chromatic dispersion compared to a broadband signal. Furthermore, the DuoB optical fiber communication system can suppress stimulated Brillouin scattering (SBS).

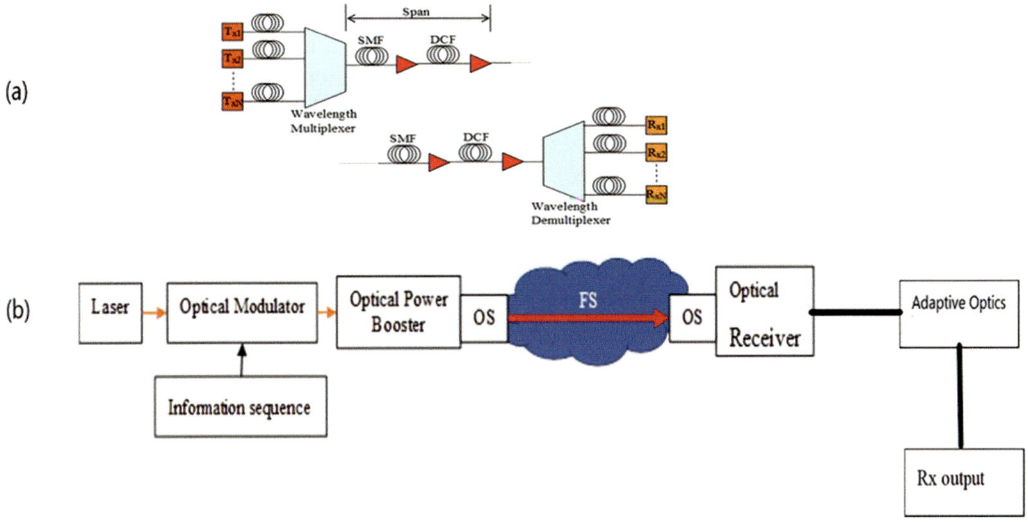

FIGURE 3.34 Ultra long-haul fiber transmission. (a) Fiber terrestrial system with dispersion compensation. (b) Free-space optical communications system employing an adaptive optics feedback system plus spatial distortion compensation using a first-order finite impulse (FIR) response filter. In optical technology this can be implemented the 1st order FIR by single MZDI (Mach-Zehnder delay interferometer) incorporating a tunable optical delay line. AO=adaptive optics, OS=optic system, FS=Free-space; Rx=reception system, DCF=dispersion compensation fiber, SSMF=standard single-mode fiber.

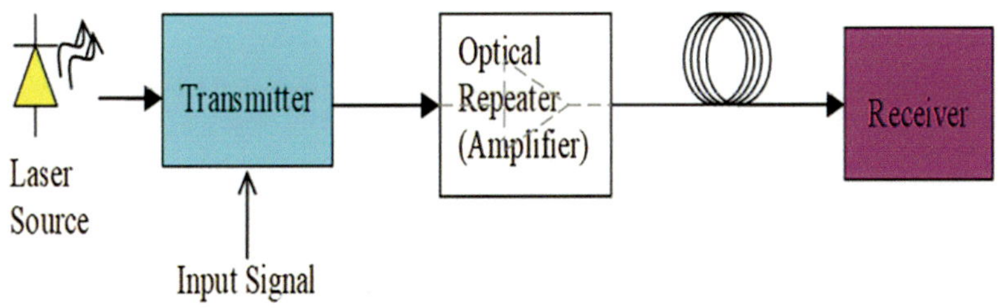

FIGURE 3.35 Main modules of a Duo-binary modulation optical communication system (a) coder and decoder and (b) generic transmission system.

The main modules of the communication system are the transmitter, optical fiber and the opto-electronic receiver, as shown in Figure 3.35. The transmitter consists of the DuoB encoder and the MZIM. A series of 0's and 1's is modulated using the DBM scheme. This three-level electrical data is then used to modulate the laser source, producing a two-level optical signal. This modulated signal is transmitted along an optical fiber transmission link toward the electro-optic receiver. The signal is detected using a photodetector, which converts the two-level optical signal back into an electrical signal. Optical amplification can be done at some points along the transmission link to minimize the effect of fiber loss.

3.7.3 Photonic Modulation

The transmitter modulates and converts the incoming electrical signal into the optical domain. Depending on the nature of the signal, the resulting modulated light may be turned on and off, or may vary linearly in intensity between two levels. The output of the DB transmitter is the modulated lightwaves switched on and off at the transitional instances of the input electrical signal. In general, a DBM transmitter, as shown in Figure 3.36, consists of a monochromatic laser source, a coder, and a photonic modulator. Binary data is encoded by the DB encoder, as shown in Figure 3.37. This resulting three-level electrical signal is converted into a two-level optical signal using the folding characteristic of an optical MZIM, which is then transmitted through the optical fiber.

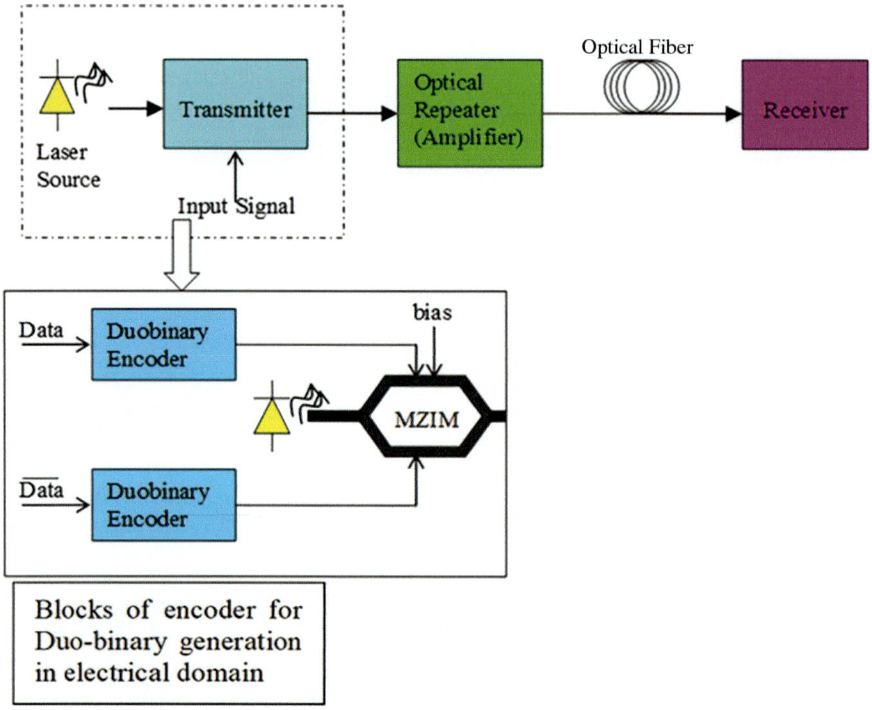

FIGURE 3.36 Transmitter of the 40 Gbps DBM photonic transmission system.

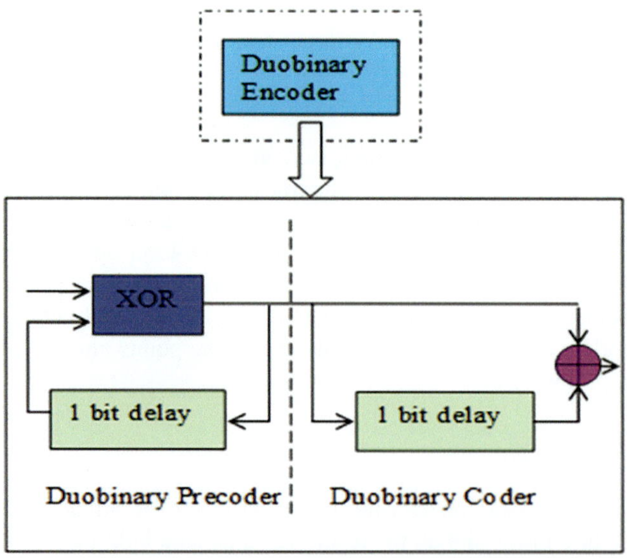

FIGURE 3.37 The DBM encoder of the 40 Gbps DBM photonic transmission system.

More on the physical fundamentals of optical waveguides and optical modulation as external modulators for coherent photonic communication technology can be found in Appendix 1 of Chapter 11. The laser is kept excited in continuous mode so that the linewidth of the laser remains unchanged and the lasing resonant cavity undisturbed. Modulation is achieved via the applied voltage to the electrode so that the modulation of the refractive index change can alter the phases of the guided lightwaves.

There are two types of DuoB transmitters. The conventional DuoB transmitter, as previously mentioned, includes a dual-drive MZIM driven by three-level electrical signals in a push-pull configuration. Since the MZIM is normally driven by a two-level signal, the effect of driving it with a three-level signal has its uncertainties. It is proposed that three-level signals may experience significant distortion in electrical amplifiers operating in saturation, leading to penalties for long word lengths. This may also cause degradation of receiver sensitivity. For these reasons, a second type of DuoB transmitter has been proposed. This type of transmitter has the MZ modulator driven by only two-level electrical signals. The optical DuoB signal generated is the same as type one of DB transmitter, with a constant phase in blocks of 1's.

3.7.3.1 DB Encoder

The DB encoder encodes binary signals, which are sequences of 0's to a three-level electrical signal. The DB signal is a fundamental correlative coding in partial response signaling. A DuoB encoder consists of a precoder and a DuoB coder. A precoder is used before DB coding to allow for easier recovery of binary data at the receiver and to avoid error propagation. The precoder is a simple binary digital circuit that consists of an exclusive OR (XOR) and a 1-bit delay feedback. The DB coder is a filter that consists of a single delay element and a summer.

Binary data input is pre-coded, with the initialization of the 1-bit delay to 0. The output of the DuoB precoder is then modulated by a pulse-amplitude modulator to produce a two-level electrical signal with amplitudes of –1 and 1. The DuoB signal is produced by adding data delayed by 1-bit period to the present data. This DB signal is a three-level electrical signal of –2, 0, and 2. Finally, it is converted to levels of –1, 0, and 1. The three-level signal is mapped into the optical domain by modulating both amplitude and phase. The "+1" and "–1" levels have the same optical intensity but opposite optical phases.

3.7.3.2 Optical External Modulator for DBM

Although the electro-optic modulator is detailed in Chapter 11, Appendix 1, it is essential to revisit the operation of the MZIM for DBM operation here. In an MZIM, the input optical carrier is split into two paths via a Y junction. This Y junction splits the input signal into $E_i/\sqrt{2}$ each. The resultant signal is

$$E_o = \frac{E_i}{2}\left[1+\exp\left(j\pi\frac{V(t)}{V_\pi}\right)\right] \tag{3.22}$$

where V_π is the voltage required to provide a π phase shift of each phase modulator, and $V(t)$ is the driving voltage. The input and output relationship of this MZIM is shown in Figure 3.38. It is accompanied by a phase modulation of exp $(j\varphi(t))$ with $\varphi(t)=\pi V(t)$. For $V(t)$ ranging from 0 to V_π, E_o and E_i have the same phase, whereas for $V(t)$ ranging from V_π to $2V_\pi$, E_o and E_i have different phases.

MZIMs can be either single-drive or dual-drive, that is the signals are applied to either one electrode or two electrodes with respect to a common ground electrode, in push-pull mode. A single-drive x-cut $LiNbO_3$ MZM can generate a phase modulation by the strength of the amplitude of the electrical signal. The transfer characteristics given in Figure 3.38 illustrates the optical output power against the applied voltage amplitude. Thus one can bias to create an initial phase at DC level, that is, then supperimposing on thos bias poit an RF (radio frequency) signal to swing the ptical phase of the output ligtwaves.

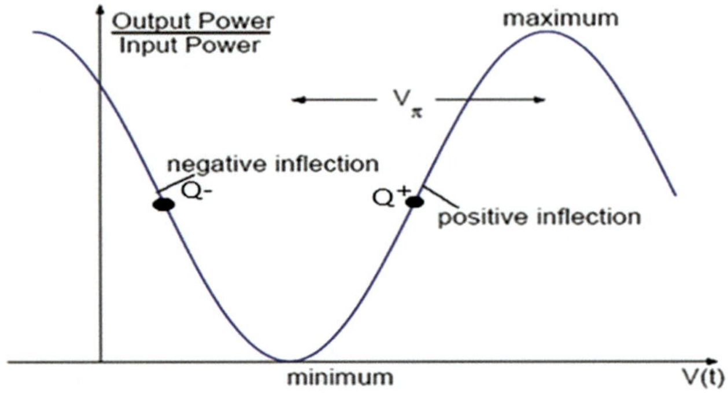

FIGURE 3.38 Practical input-output power versus applied voltage transfer characteristics of the MZIM. Q=quadrature point: $Q+$ on the potitive slope and $Q-$ on negative slope. Note that no maximum at zero applied voltage

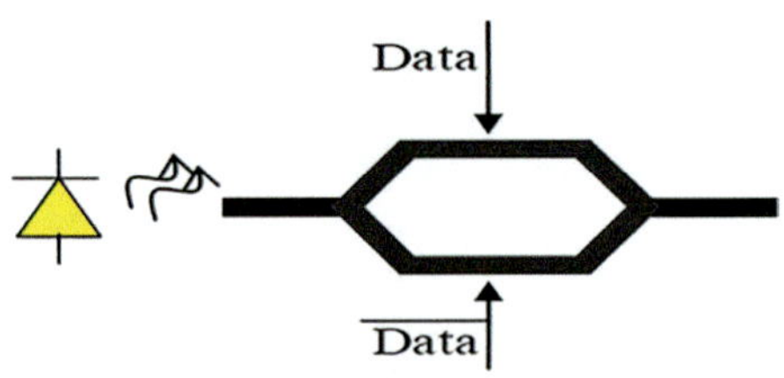

FIGURE 3.39 Dual-drive MZIM.

The dual-drive X-cut $LiNbO_3$ MZIM, on the other hand, has two paths phase-modulated with opposite phase shifts in a push-pull operation. The V_π in Figure 3.11 is reduced by half in this case. For a dual-drive y-cut $LiNbO_3$ MZIM, the two paths are driven by complementary signals, with V_1 equal to $-V_2$. The output electric field of a dual-drive MZIM is

$$E_o = \frac{E_i}{2}\left[\exp\left(j\pi\frac{V_1}{V_\pi}\right) + \exp\left(j\pi\frac{V_2}{V_\pi}\right)\right] \tag{3.23}$$

DuoB optical signal can be generated by driving a dual-drive MZIM with a push-pull operation, as shown in Figure 3.39.

One arm is driven by the DB signal, and the second arm is driven by the inverted DB signal. Figure 3.40 shows the operation of the MZIM. The output electric field $E_o(t)$ can be expressed as

$$E_o(t) = E_i \cos\frac{\Delta\phi(t)}{2} \cdot \exp\left(-j \cdot \frac{\phi_0}{2}\right) \tag{3.24}$$

where E_i is the input electric field, $\Delta\varphi(t)$ is the phase difference between the lightwaves propagating in the two optical waveguides, and φ_0 is a constant when the MZIM is driven in a push-pull operation. At point B in Figure 3.40, the phase of the output optical signal is inverted. The optical DB signal depends on the biasing point of the driving signal, which is the electrical DB signal. By biasing at point B in Figure 3.13, the "–1" and "+1" levels of the electrical DB signal correspond to the "on" state of the optical signal, while the "0" level corresponds to the "off" state. To achieve carrier suppression, there must be a π phase difference between the two arms.

The transmitter model generally consists of the DuoB coder and the MZIM. The DuoB coder encodes the incoming binary sequence of 0's and 1's into a DB electrical signal. This signal is, then, used to drive the arms of the dual-drive MZIM. One arm is driven by the DuoB signal, and

(a)

(b)

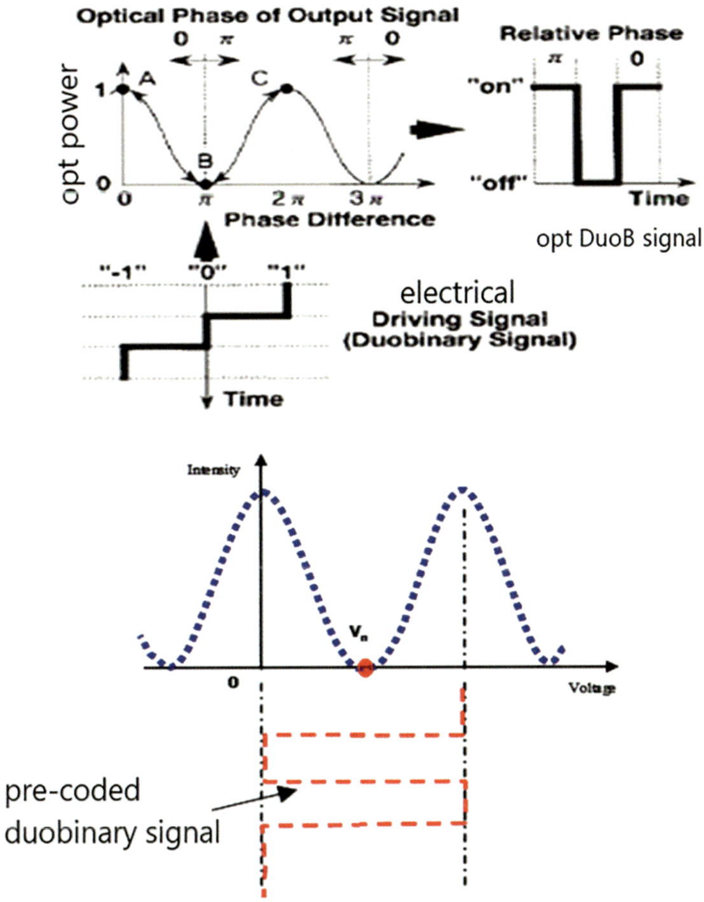

FIGURE 3.40 Driving operation of dual-drive MZM[73] (a) DuoB signaling, (b) biasing to generate optical DuoB signals with phase states; upper curve is power versus applied voltage of the MZIM.

the other arm is driven by the inverted DuoB signal. The Bernoulli binary generator produces a random sequence of binary electrical signals at a rate of 40 Gbps. This signal is encoded by the DuoB encoder, which consists of a DuoB precoder and a DB coder. The first output of the encoder is shifted up by 1 to produce levels of "0", "1" and "2". This electrical DuoB signal is sent to the phase shift block, as shown in Figure 3.41, to represent these levels with a certain phase. This, in fact, represents the biasing point on the transmittance curve. For the dual-drive MZIM, the driving signal is biased at $V_\pi/2$. The second output of the DB encoder is the inversion of output 1. This output is used to modulate the second arm of the MZIM. The output 2 signal is shifted down by –1, to bias at the point $-V_\pi/2$ of the transmittance curve. MZIM is an amplitude modulator, accompanied by a shift of phase. This modulation is also called AM-PSK. The lightwave, which is a sine wave produced by the sine wave function, is modulated by the DuoB signal through the complex phase shift block. The input sine wave is shifted by the amount specified at the input.

The DB coder consists of a precoder and a coder, as shown in Figure 3.42. The precoder is a differential coder, with an exclusive OR (XOR) gate and a 1-bit delay feedback path. The addition of –0.5 and division by 2 functions as the amplitude modulator, which shifts levels of the signal ranging from "0" and "1" to "–1" and "+1". The signal is then added to its one-bit delay to produce a three-level DB signal with levels of "–2", "0", and "+2", followed by a conversion to a level of "–1", "0" and "+1". The summation of the signal with its one-bit delay is the DuoB coder. For the second output, the output of the differential coder is inverted, before going through the same operation as Out1.

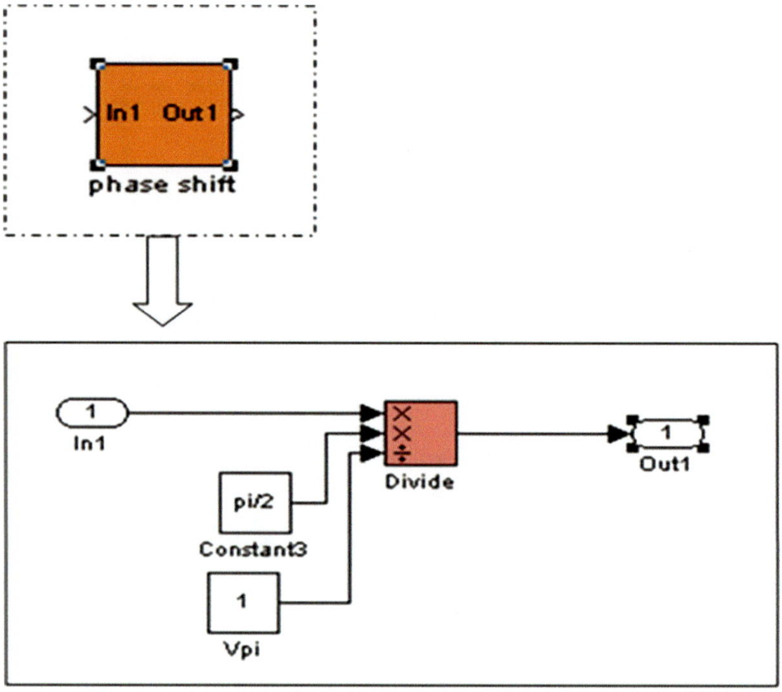

FIGURE 3.41 Phase shift block of the transmitter model.

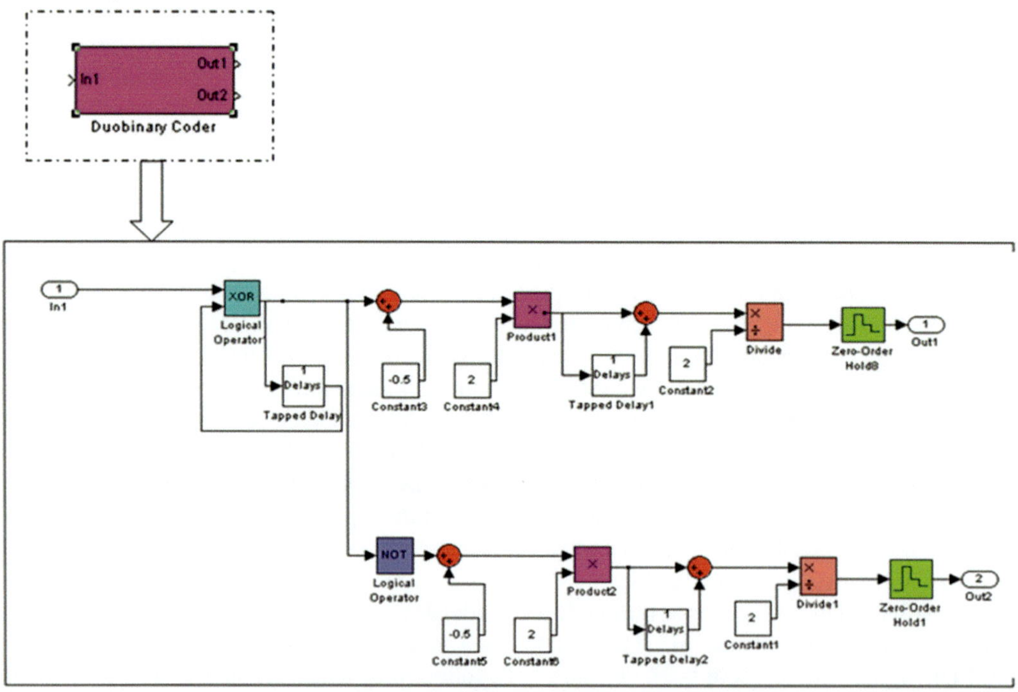

FIGURE 3.42 DB coder of the transmitter model.

A zero-order hold is placed before the production of the DB encoder functions to discretize the signals to have a fast-to-slow transition of signals. It holds and samples the signal before transmitting it out. If the signals are transmitted without the zero-order hold, the transition to the "0" level will be overseen. The signal will only have two levels, "–1" and "+1".

3.7.4 Alternative Optical Phase DBM Transmitter

Two types of DBM transmitter models are proposed. The conventional DBM transmitter, as mentioned previously, uses a dual-drive Mach-Zehnder modulator driven by three-level DB electrical signals. The MZIM shown in Figure 3.43 (and Figure 3.40a) is usually driven by a two-level electrical signal. In some cases, this three-level driving signal may experience significant distortion in electrical amplifiers operating at saturation, leading to penalties for long word lengths. It may also occur that there is a degradation of receiver sensitivity. Due to these uncertainties, an alternative DB transmitter, as shown in Figure 3.43, is simulated based on MATLAB Simulink. This second type of DBM transmitter has the MZIM driven by two-level electrical signals. It consists of a differential encoder, a one-bit-period electrical time delay, and a MZ modulator. One arm of the dual-drive MZ modulator is driven by the signal from the, whereas the other arm is driven by the inverted signal, delayed by a one-bit period. Both DB transmitters produce the same result, which is constant phase in blocks of 1's.

The whole duobinary modulated SPC system model can be seen in Figure 3.44 in which the models are based on MATLAB Simulink.[37]

3.7.5 Single-Side-Band Modulation

Why SSB for SPC? As we already know, the modulation in the optical domain generates the channel in the frequency of DBSs on both sides of the lightwave carriers. However, only one sideband is used in the electronic domain after detection by the photodetector, either by a coherent or non-coherent receiver.

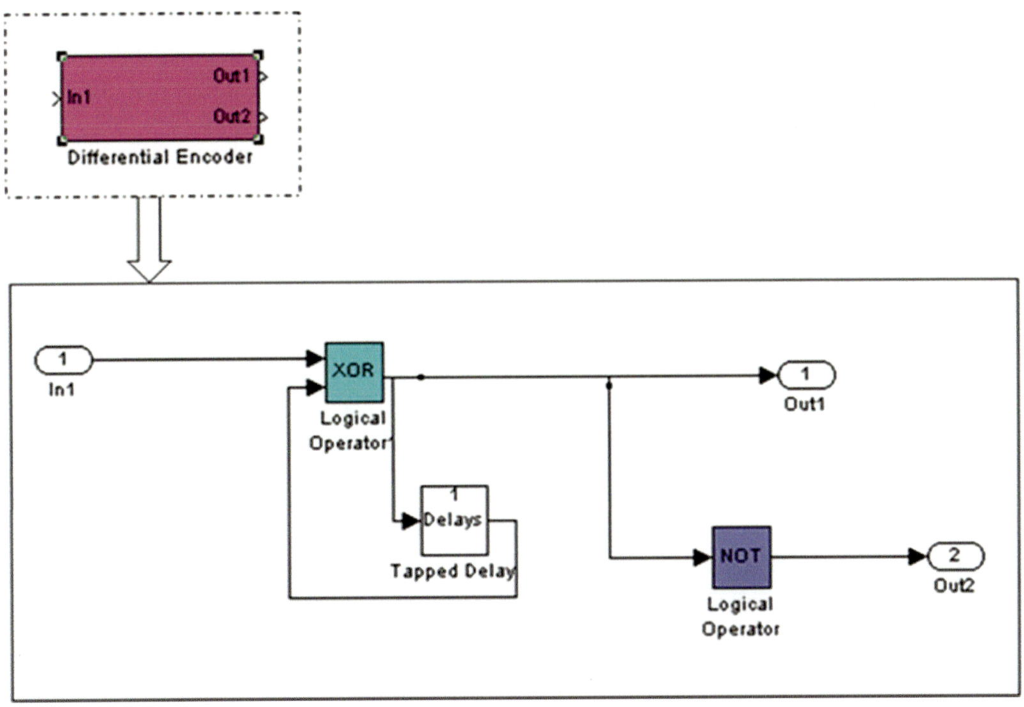

FIGURE 3.43 Simulink model of MZIM.

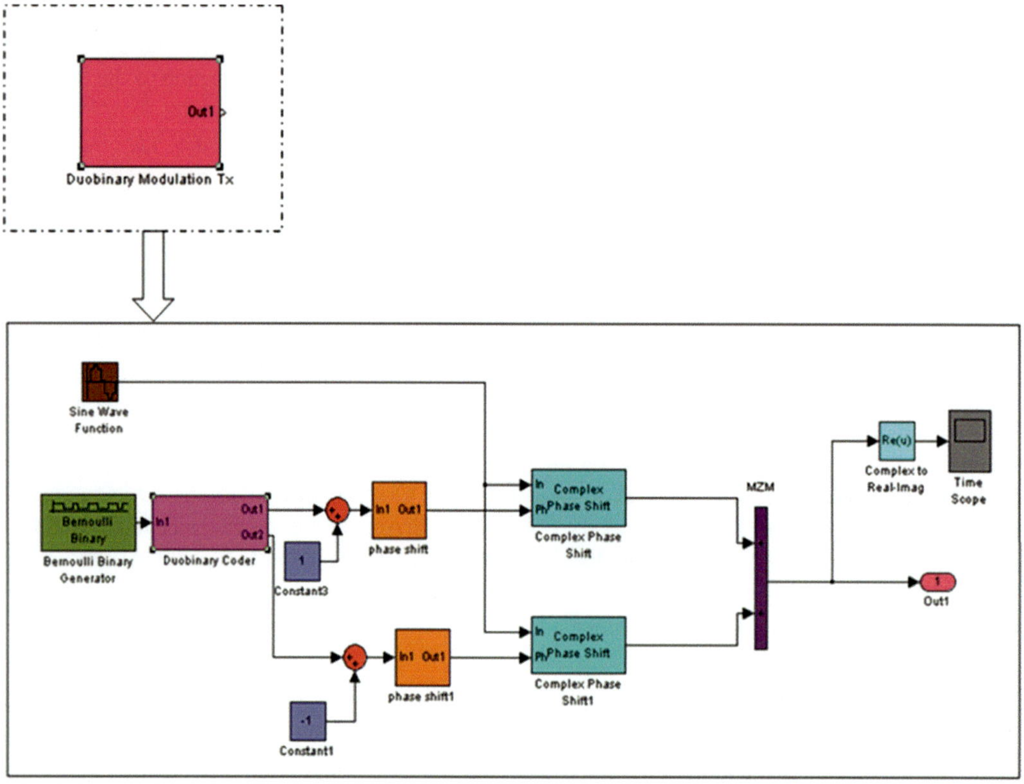

FIGURE 3.44 Conventional transmitter model of the 40 Gbps Duobinary modulation SPC and fiber system.

If a DBS is used, then the noise contributed to the receiver doubles. So, a SSB of either the upper or lower band, is sufficient to recover the data sequence. This section thus first presents the principles of generating an SSB-only channel via the Hilbert transform and then by a compatible implementation for photonic modulation. Then, the demodulation of such SSB lightwaves is discussed using direct detection or coherent reception. The noise contribution in SPC comes from the HPA (high power booster amplifier) at the transmitter before launching into space and from the low noise amplifier with low input signal average power, which acts as the photonic pre-amplifier of the receiver. This noise would be doubled or at least 3dB above the SSB. The main difficulty is ensuring that the optical filter at the optical receiver minimizes noise contribution and avoids channel distortion if the filter band is shifted with respect to the carrier's referenced frequency.

Thus, the SSB channel would allow us to minimize noise, leading to a higher OSNR when the optical filter of the optical pre-amplifier is stable and narrow enough to cover the SSB channel spectrum.

3.7.5.1 Hilbert Transform SSB MZ Modulator Simulation

The Hilbert transform of signal $m(t)$ is defined to be the RF signal whose frequency components are all phase-shifted by $\pi/2$ radians.[74] Thus, we have,

$$\hat{m}(t) = H\{m(t)\} \rightarrow m(t) = A\cos 2\pi f_o t \quad \text{then} \quad \hat{m}(t) = A\cos\left(2\pi f_o t - \frac{\pi}{2}\right) \tag{3.25}$$

Taking the Fourier transform, we have

$$\begin{aligned} M(f) &= -j\,\text{sgn}(f)\frac{A}{2}\left[\delta(f+f_o)+\delta(f-f_o)\right] \\ &= \frac{A}{j2}\left[-\delta(f+f_o)+\delta(f-f_o)\right] \end{aligned} \tag{3.26}$$

$$\hat{m}(t) = A\cos\left(2\pi f_o t - \frac{\pi}{2}\right) = A\sin 2\pi f_o t \tag{3.27}$$

3.7.5.2 SSB Demodulator Simulation

The modulated signal is:

$$u(t) = A_c/2\left\{\cos\left(\omega_c t + \gamma\pi + \alpha\pi\cos\omega_{rf}t\right) + \cos\left(\omega_c t + \alpha\pi\cos\left(\omega_{rf}t + \theta\right)\right)\right\} \tag{3.28}$$

The demodulated signal is:

$$s(t) = A_c\cos 2\pi f_c t * u(t)a \tag{3.29}$$

The low-pass filter (LPF) is used to filter out the high-frequency components of the signal. The only component left is the modulating signal (10 Gbps [binary * cosine signal]) after demodulation. It is required to multiply the demodulated signal by an in-phase cosine signal. Since there is also a high-frequency signal, the LPF is required to filter the high-frequency components in the signal.

3.7.5.3 SSB for Optical Multi-Carrier Comb-Generation

As an example of such SSB modulation, we illustrate an application of the SSB technique to create a multi-carrier generator (comb-generator) as follows (see Figure 3.45):

A continuous-wave (CW) external cavity laser (ECL) or a multi-laser bank of ECLs, witha sufficiently narrow linewidth, possibly in order of <100 kHz, can be employed as the original lightwave carrier/carriers.

An MZM for pulse shaping (CSRZ) will make the output spectrum of the super-channel flat: (i) An MZM is used as a phase modulator, with the phase control signal coming from an RF signal generator; (ii) An RF generator, where its frequency is the spacing of the sub-channels; (iii) A Sinusoidal signal generator, whose output is split into two branches: one branch is applied to the amplitude modulator, and the other is phase-shifted by $\pi/2$ radians and applied to the other branch of the optical modulator; (iv) 90° phase shifter is used as a Hilbert transformer to generate an SSB carrier, creating a the spectrum to be only a single carrier shifted in the frequency domain, i.e., the SSB carrier.

Figure 3.46 shows a laboratory bench setup for the SSB modulation of a laser carrier, which is then recirculated to continually shift the carriers, resulting in multiple carrier outputs. The spectra of the multiple-carrier laser are shown in Figure 3.47, showing the shift to the right and left by a positive or negative phase shift of $\pi/2$ rad, respectively. This phase shift is applied to the optical modulator for single sideband frequency shifting.

3.7.6 Discrete Multi-Tone Modulation

In SPC systems where the distance is moderate such as inter-LEO inter-satellite optical networks or inter-micro-satellite networking, the communication link between satellite is shorter and has become very critical and in high demand. For these applications, direct detection (DD) may be

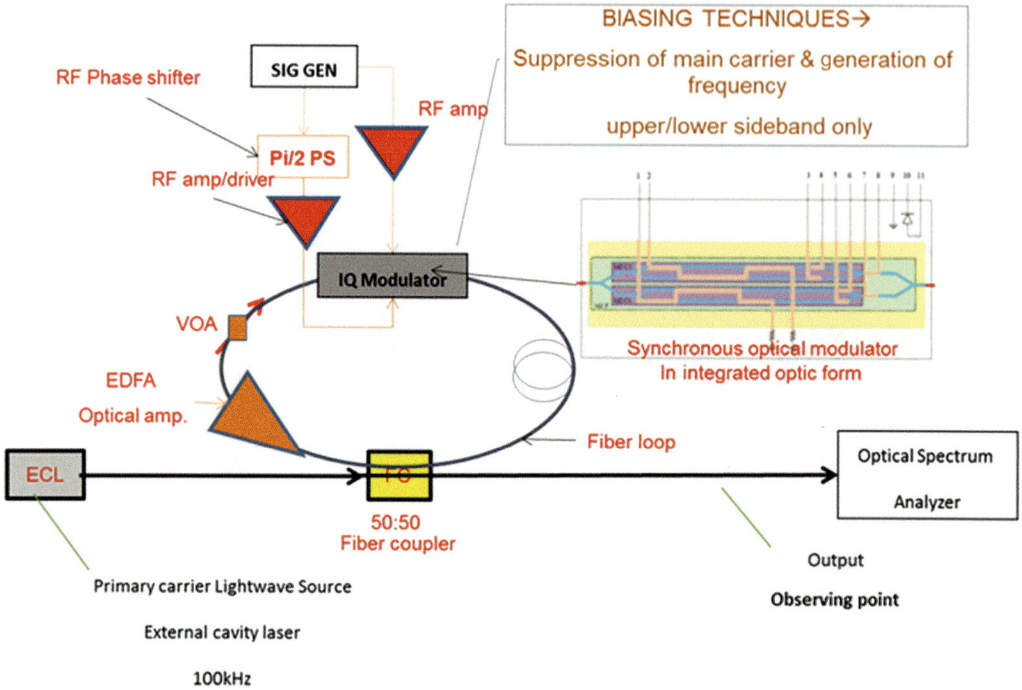

FIGURE 3.45 Schematic of the experimental setup of the comb generator using re-circulating SSB frequency shifting. Insert: plane view structure of a dual-drive optical modulator.

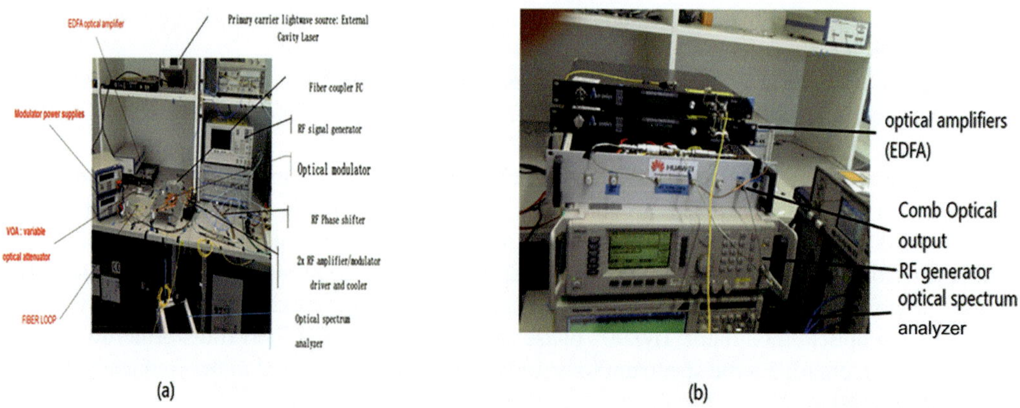

FIGURE 3.46 Laboratory setup of a comb-generator prototype employing the re-circulating frequency shifting technique (a) comb generator under development, (b) packaged comb generator with external EDFA and signal synthesizer.

more attractive due to its lower system cost and power consumption. In a 100-Gbps discrete multitone (DMT) system at 1550 nm, the main challenge is the frequency-related power fading induced by chromatic dispersion (CD) in fiber or spatial beam spreading after direct detection and reduced sensitivity if the receiving optics and receiver of SPC. Although bit-loading and power-loading can be used to optimize the bandwidth utilization of DMT sub-carriers, power fading still reduces the available bandwidth and capacity, especially after 80-km of fiber-optic transmission or inter-LEO satellite transmission.

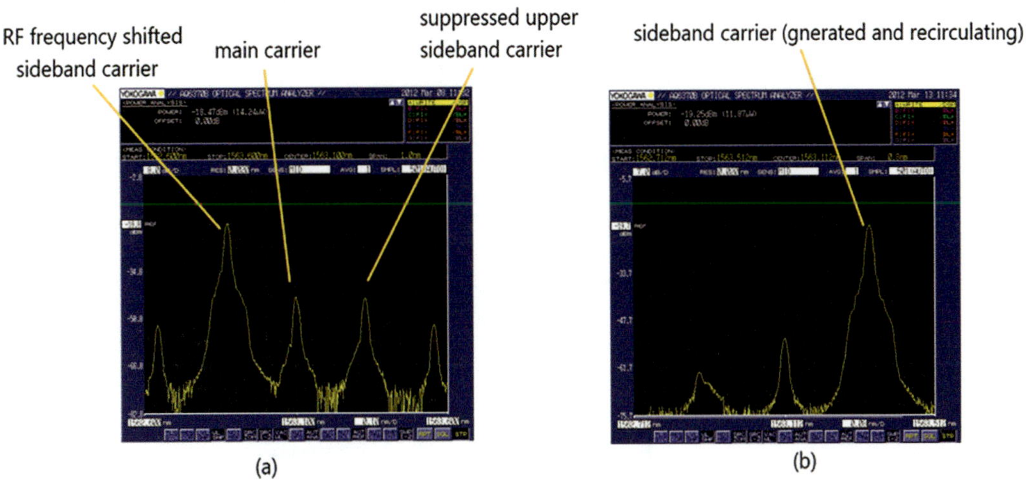

FIGURE 3.47 Spectrum of the SSB lightwave at the output of the optical modulator (a) shifting right (+π/2 bias phase shift applied) in the frequency scale, (b) shifting right in frequency (−π/2 bias phase shift applied).

SSB-DMT can be used to resolve the power-fading issue and the noise contribution as already mentioned in Section 3.7.5. In theory, the average signal-to-noise (SNR) ratio of SSB-DMT exhibits a 3-dB penalty compared with DSB-DMT, since the spectrum of half of the data sequence, either left sideband or right sideband, is suppressed.

In this section, enabled by digital distortion pre-compensation, a C-band 100-Gbps DSB-DMT transmission over 80-km SMF and/or spatial distortion of an inter-satellite link of more than 32,000 km with a laser linewidth of <10 kHz can be experimentally demonstrated[75] using an integrated 10 GHz tunable transmitter assembly (TTA), consisting of a C-band tunable DS-DBR (double-sided distributed Bragg reflector) laser and an InP DDMZM.[76,77,78,79] Maximum capacities of up to 125, 106, and 103 Gb/s could be achieved for back-to-back (BTB), 40-, and 80-km single-mode fiber (SMF) transmissions, respectively. Therefore, the space transmission in an ideal vacuum can reach a distance with a low tolerance of 260–300 km, and thence 32,000 km with a narrower linewidth laser of <10 kHz. Thus, compared with the SSB-DMT, better OSNR performance and higher capacity can be realized.

A twin SSB-DMT is shown in the schematic principles of Figure 3.48. Two parallel sections of the transmitter and receivers are shown as mirroring each other and orthogonal, meaning they are in Hilbert transform of each other. This orthogonality permits detection and processing without interference between channels. The filtered SSB-DMT spectra of the two orthogonal channels can exhibit high carrier power. The biasing of the two branches of the DD-MZIM allows the generation of two complex conjugates of the input lightwaves that carry the information, as indicated in Figure 3.48. This ensures the coherency of the channel for long-haul transmission over the space link. If a single-frequency laser source, such as the 10 kHz linewidth laser, is used along with external modulation of the DD-MZIM, the long-haul space link can be established.

An SSB-filtered left and right can be achieved by using a tunable optical filter, whose spectrum is shown in Figure 3.49d. Such DMT can be termed a twin DMT modulated system.

The reception sub-system is quite simple, involving direct detection and sampling with DSP processes, as listed in Figure 3.49. The blocks for post-processing and pre-distortion are given in the receiver and transmitter sections, respectively. A and B and their conjugates are the information to be transmitted.

The dual conjugated channels can be transmitted over space photonic systems. A dual receiver, associated with an integrated DSP processor, can recover and process to combine the conjugated information to eliminate any distortion of the digital sequences A and B.

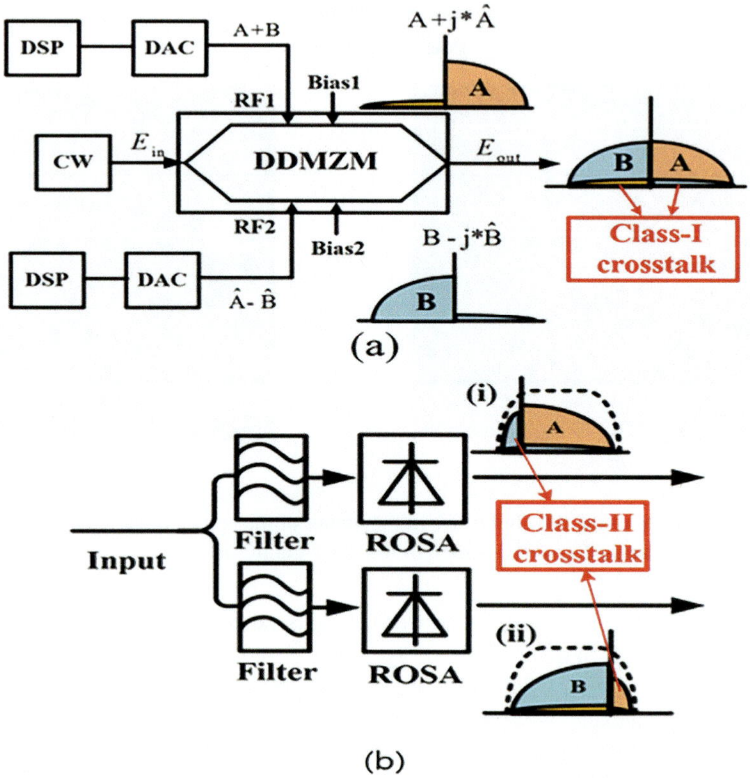

FIGURE 3.48 Schematic principles of dual orthogonal SSB-DMT modulation transmitter (a) and receiver, (b) with direct detection and signal recovery.

The constellation of two orthogonal dual channels under SSB-DMT modulation is shown in Figure 3.50b without MIMO and in Figure 3.50c with MIMO. MIMO (multiple input multiple output) is a processing algorithm most applicable when there are multiple inputs and multiple outputs. In the case of dual orthogonal channels SSB-DMT, the two channels are orthogonally used as the inputs, and the two output channels can be complementary to each other.

The BER is plotted versus the transmission rate, under MIMO and without MIMO DSP algorithms, in Gbps, as shown in Figure 3.51. For a certain BER of 2e-2 in association with hard FEC (forward error coding), HD-FEC, the bit rate can be increased by an additional 50 Gbps or equivalently to about 25 dB in terms of loss, or more than 100 km in ideal vacuum transmission distance (under Friis' formula, ideal single-frequency lightwave source).

The dual SSB-DMT can thus offer many advantages in space transmission, as coherency can be kept reserved for long-hail transmission, and the conjugation allows different information capacities to be transmitted without loss of information, thence improving the BER of the transmission channels. Therefore, depending on the transmission space distance, the SSB-DMT can be used, and higher-gain optical booster amplifiers are available. The SSB with carriers can be seen in Figure 3.49. The carrier power peaks can be observed, and coherent reception can be used for additional gain of 20–30 dB.

As we can observe from the optical spectra of the dual orthogonal SSB-DMT given in Figure 3.47, the two spectra can overlap, and filtering can eliminate one carrier from one spectrum. This referenced carrier may be lost, leading to issues in processing. So, to avoid this problem, one can shift the carrier in one branch of the SSB-DMT by superimposing two RF carriers onto the optical carrier, as shown in Figure 3.52. It shows the band $f_{RF1}+f_{RF2}$ is the spectral guard for filtering safety, ensuring that no distortion is created by filtering.

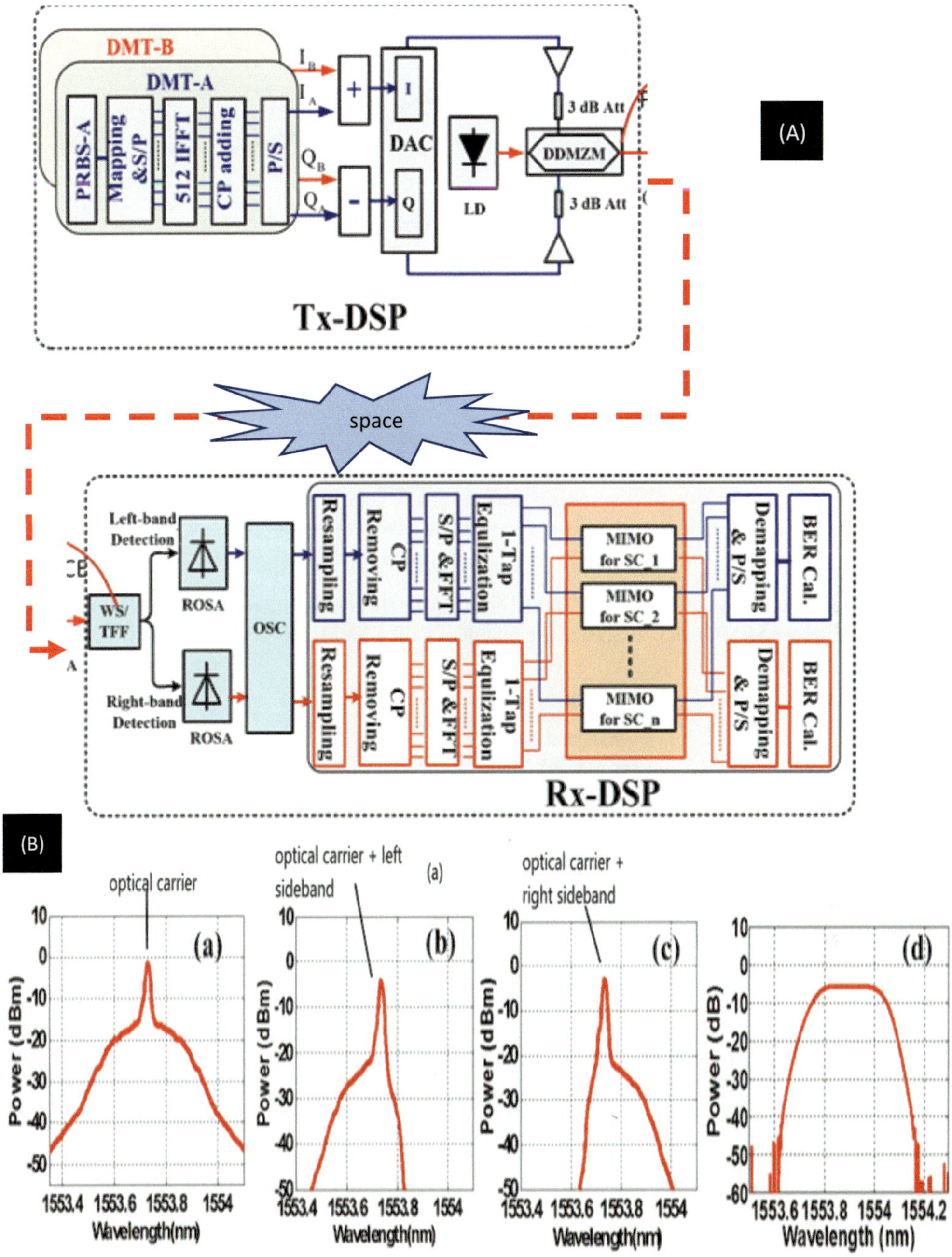

FIGURE 3.49 Twin SSB-DMT optical transmission system using either fiber-optic or photonic space transmission. (A) System schematic, (B) optical spectra at different stage points: (a) DSB, (b) shift right SSB (c) shift left SSB, and (d) optical filtering with passband and cut-off band. TOSA = transmitter optical sub-assembly, ROSA = receiver optical sub-assembly, DDMZIM = dual-drive Mach-Zehnder modulator, I = in-phase, Q = quadrature phase, PRBS = pseudo-random bit sequence, WS-TFF = wavelength shifter/tunable filter, OSC.

The schematic of the transmission system employing RF-shift dual orthogonal SSB-DMT is shown in Figure 3.53. RF_1 and RF_2 must differ by half the signal bandwidth of the channels so that the filtering can be distinguished. These RF sources are added to the digital channels at the input to the modulator port or that can be integrated into the FFT shift if the sampling rate is

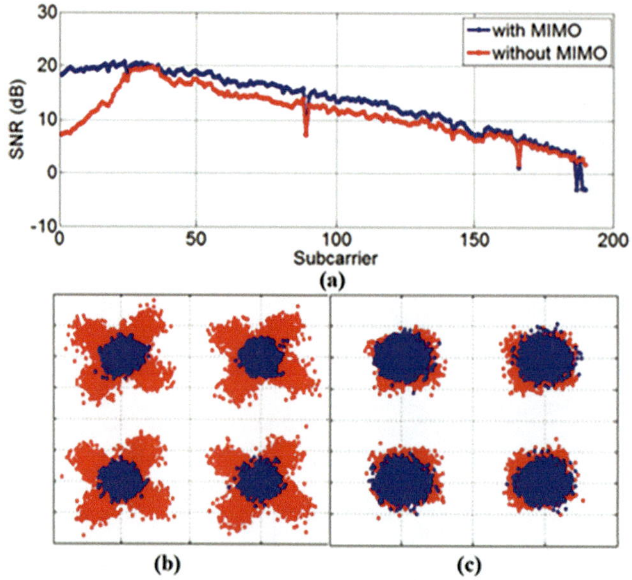

FIGURE 3.50 Transmission performance of dual SSB-DMT: (b) constellation under no MIMO processing, (c) constellation under MIMO, and (a) SNR at the output of DSP with MIMO and without MIMO processing.

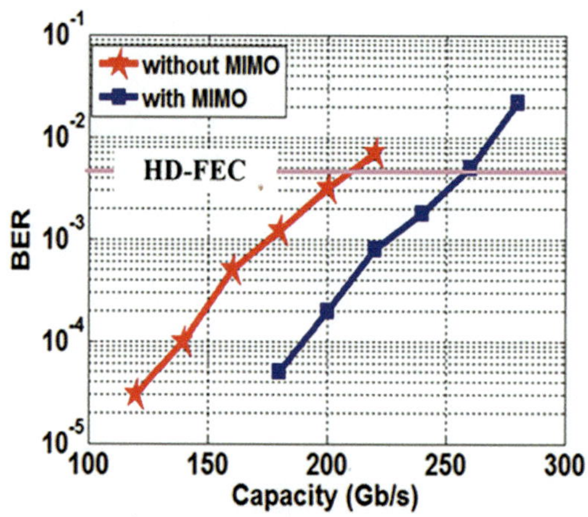

FIGURE 3.51 BER versus transmission capacity of dual orthogonal channels SSB-DMT modulation and direct detection, detection, and DSP algorithm using 2×2 MIMO and without MIMO (multiple input/multiple output).

high enough. The steps of predistortion and post-processing using the MIMO algorithm, along with compensation for distortion, are also included in the diagram. Direct detection is employed. If coherent detection is used, the improvement in OSNR can be at least 20 dB when the power of the local oscillator is about 30 dB higher than that of the signal power. The spectra of the two SSB channels, SSB-DMT(A) and SSB-DMT (B) orthogonal to A are shown in Figure 3.54. The fast roll-off of the spectra for channels A and B can be observed in Figure 3.54b and c. Therefore, the sharp roll-off of the optical filters is critical to isolating any crosstalk from one channel to the other. In addition, the orthogonality of the channels can ensure that the DSP processing can distinguish the channels, resulting in higher BER due to reduced noise contribution from any crosstalk.

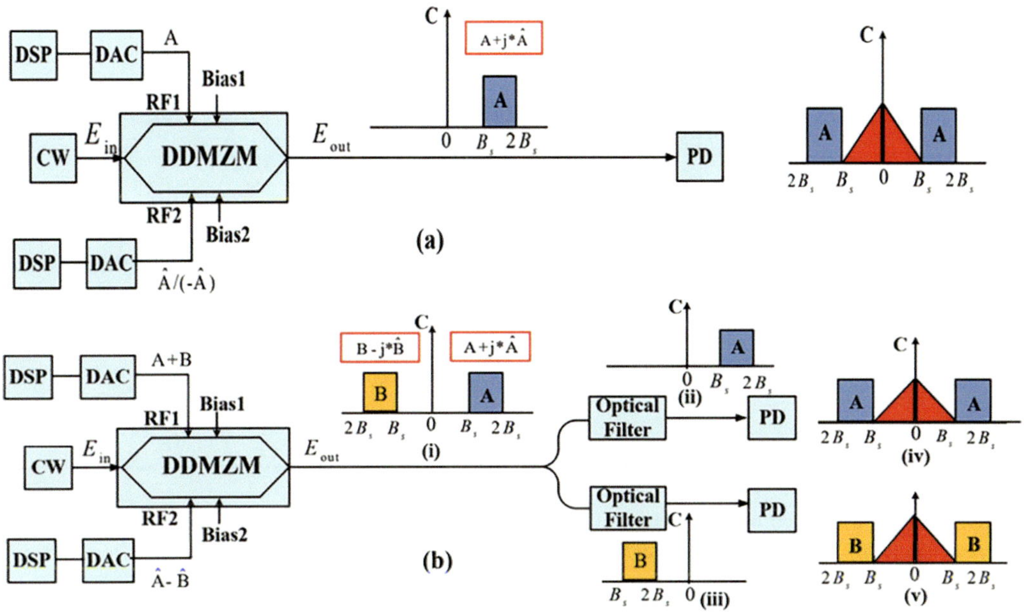

FIGURE 3.52 RF-shifted optical carriers in dual orthogonal SSB-DMT: (a) schematics of modulation, biasing, and RF shifting of carriers for SSB-DMT, original channel shifted by RF1, and spectrum on far right; (b) conjugated channel shifted by RF2, and shifted spectrum on the far right. Legends are the same as in Figure 3.49.

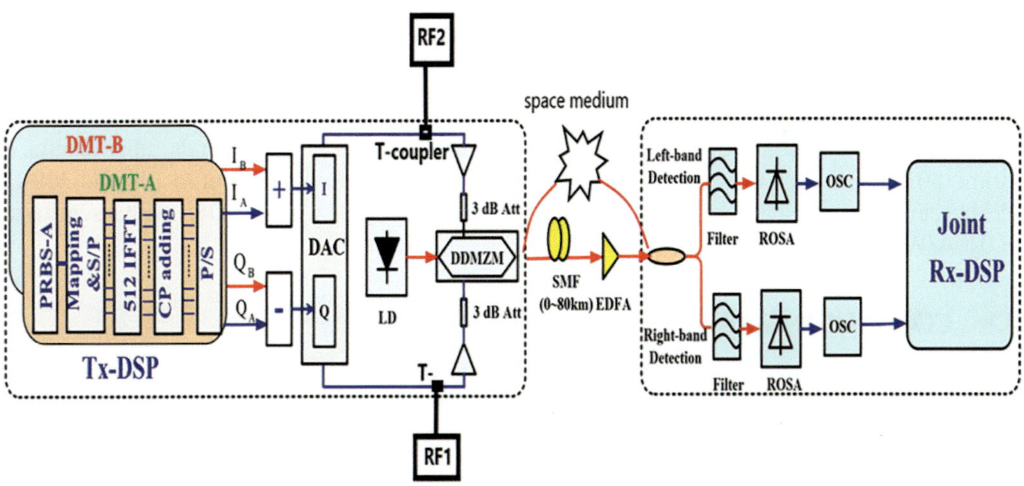

FIGURE 3.53 Transmission link using dual-orthogonal SSB-DMT with RF shifting via external RF sources superimposed on the data channel via the T-bias DC-RF couplers; alternatively, the RF shift can be generated in the FFT processing of the DSP.

At the Tx side, the pseudo random binary pattern (PRBS) can be generated from the DSP processor which is integrated in the transceiver. This serial sequence is then is mapped to parallel data patterns. They are then passed through a 512 inverse FFT transformer, using the FFT generator. Only the the real part which are obtained from the full FFT processor. This the discrete (cosine) tone DMT modulation format. The bit padding is filled to the circular FFT to ensure no distortion. The serial sequence is then converted to parallel mode with the I (in-phase) and Q (quadrature)

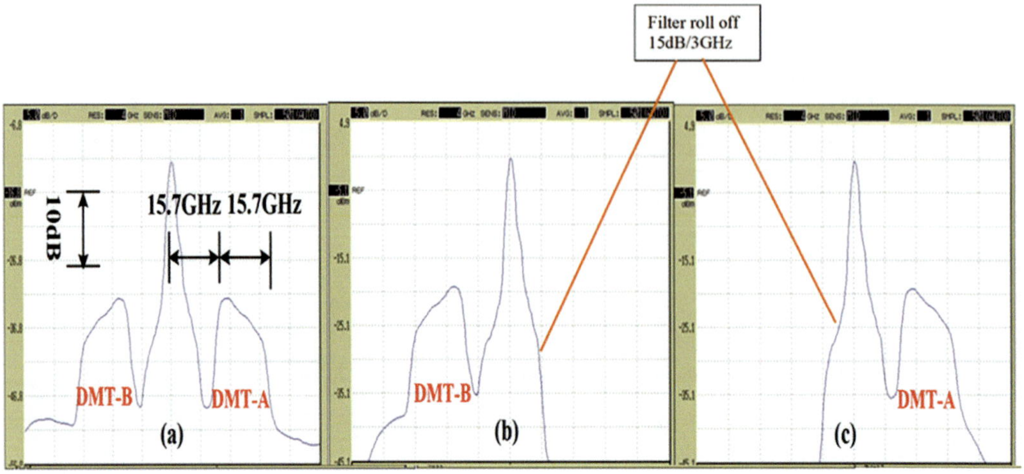

FIGURE 3.54 Measured optical spectra of panel (a) total dual orthogonal channels without filtering; (b) filtered channel B, SSB-DMT (b), and (c) filtered channel A- SSB-DMT.

components, which are then fed into each arm of the DD-MZIM. Therefore, two orthogonal modulated lightwaves are generated and then combined at the output of the MZIMs: The two orthogonally lightwaves are then launched into the transmission medium, whether fiber optic or free space channels.

The OSNRs for the right and left channels differ by 1 dB due to the tuning of the optical filter and their different roll-off characteristics. As required OSNR of 30 dB is required for the HD-FEC 5e-2 for such transmission distances. This can be achieved using optical pre-amplifiers, low-noise and high-gain optical amplifiers in the space photonic communications systems.

The transmission performance of the dual orthogonal RF-shifted SSB-DMT is depicted in Figure 3.55a and b for the right and left filtered, respectively, with the BER against OSNR. The hard FEC level of 5e-2 indicates that an error-free performance can be achieved. Under fiber lengths of 40 km, 80 km, and back-to-back (BTB) link conditions, this is also equivalent to 100 and 200 km ideal vacuum space distances. The linewidth of the laser is <100 kHz, and a ROSA with a sensitivity of 0.8 A/W for direct detection is used.

3.8 CONCLUDING REMARKS

This chapter has addressed the challenging issues of optical modulation whose format follows binary format in either amplitude or phases of the lightwaves. The phases of the lightwaves can be either discrete or continuous. Both can be modulated by optical modulators, mainly the high-speed electro-optic type in which both the RF signals and electrical voltage DC bias can be applied via a T-coupler. For amplitude shift keying (ASK), the carrier can be suppressed or non-suppressed, and the RZ formats can also be generated by different biasing techniques applied to the modulator transfer characteristics. Furthermore, the $Q+$ or $Q-$ quadrature points of the transfer curve of MZIM can be used to generate linear optical signals as long as the magnitude of the driving signal voltage does not exceed the linear region of a $\cos^2(\cdot)$ curve.

The continuous phase modulation, where the phases of the lightwaves can be changed by up chirp or down chirp but remain continuous at the transition point between two consecutive bits, offers many advantages in SPC transmission and propagation of the lightwave beam through turbulence

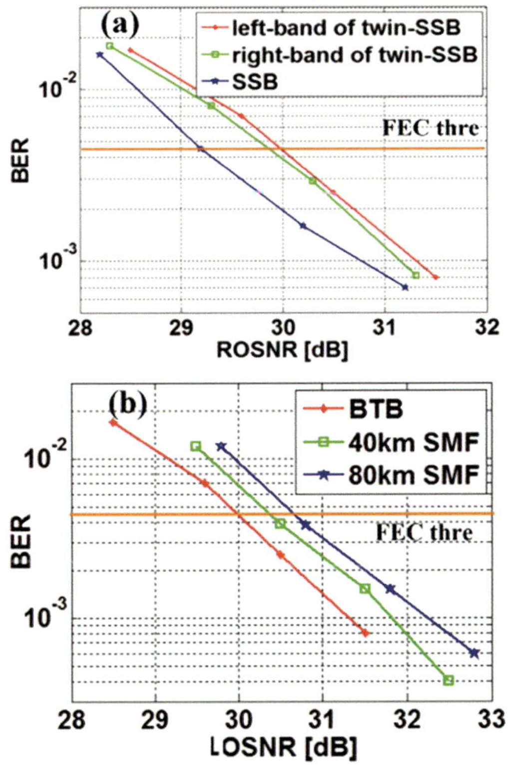

FIGURE 3.55 DMT system transmission performance: BER versus OSNR. BTB = back-to-back test platform, SMF = single-mode fiber, coherent but direct detection, FEC = forward error correction, ROSNR = Right sideband OSNR, LOSNR = Left sideband OSNR.

and vacuum ideal long reach space distance. MSK is the strongest phase continuity modulation and offers the minimum bandwidth for transmission. However, nonlinearity can be a limiting factor.

The high power of the beam to be boosted for launch into space creates some nonlinear crosstalk and requires careful attention. Thus, modulation formats such as RZ formats can suppress the carrier power level to a minimum to avoid such nonlinear impairment. However, the carrier power and its narrow linewidth are critical for providing coherency to the beam, and thus there are no advantages in suppressing the carrier. RZ formats without carrier suppression provide two carriers above the two sidelobes of the signal spectrum, offering significant advantages in coherent transmission. Additionally, the carrier phase can be recovered for determining the impairments due to the space turbulent layer.

SSSB modulation formats, such as the DMT, offer bandwidth savings in limited availability of transparent spectral window of atmospheric and vacuum ideal media. Although direct detection can be used, coherent reception offers an additional 20 dB. DSP is important for generating the homodyne coherence type, which is important for SPC.

Having described the binary modulation formats in both phase and amplitude, higher modulation formats in multiple amplitude and multiple phases can then be used to further increase the system capacity, especially to reach tTera-bits/s per single carrier wavelength or through multiplexing. These topics will be covered in Chapter 5.

NOTES

1 I. P. Kaminow and T. Li, *Optical Fiber Communications*, Volume IVB. Elsevier Science: Philadelphia, PA, 2002.
2 R. S. Sanferrare, "Terrestrial lightwave systems", *AT&T Technology Journal*, 66, 95–107, 1987.
3 C. Lin, H. Kogelnik, and L. G. Cohen, "Optical pulse equalization and low dispersion transmission in single-mode fibers in the 1.3–1.7 mm spectral region", *Optics Letters*, 5, 476–478, 1980.
4 A. H. Gnauck, S. K. Korotky, B. L. Kasper, J. C. Campbell, J. R. Talman, J. J. Veselka, and A. R. McCormick, "Information bandwidth limited transmission at 8 Gbps over 68.3 km of single mode optical fiber", In *Proceedings of OFC'86, Paper PDP6*, Atlanta, GA, 1986.
5 H. Kogelnik, "High-speed lightwave transmission in optical fibers", *Science*, 228, 1043–1048, 1985.
6 A. R. Chraplyvy, A. H. Gnauck, R. W. Tkach, and R. M. Derosier, "8x10 Gbps transmission through 280 km of dispersion-managed fiber", *IEEE Photonics Technology Letters*, 5, 1233–1235, 1993.
7 H. Kogelnik, "High-capacity optical communications: Personal recollections", *IEEE Journal on Selected Topics in Quantum Electronics*, 6(6), 1279–1286, November/December 2000.
8 C. R. Giles and E. Desurvire, "Propagation of signal and noise in concatenated erbium-doped fiber amplifiers", *IEEE Journal of Lightwave Technology*, 9(2), 147–154, 1991.
9 P. C. Becker, N. A. Olsson, and J. R. Simpson, *Erbium-Doped Fiber Amplifiers, Fundamentals and Technology*. Academic Press: San Diego, CA, 1999.
10 T. Okoshi, "Heterodyne and coherent optical fiber communications: Recent progress", *IEEE Transactions on Microwave Theory and Techniques*, 82(8), 1138–1149, 1982.
11 T. Okoshi, "Recent advances in coherent optical fiber communication systems", *IEEE Journal of Lightwave Technology*, 5(1), 44–52, 1987.
12 J. Salz, "Modulation and detection for coherent lightwave communications", *IEEE Communications Magazine*, 24(6), 38–49, 1986.
13 T. Okoshi, "Ultimate performance of heterodyne/coherent optical fiber communications", *IEEE Journal of Lightwave Technology*, 4(10), 1556–1562, 1986.
14 P. S. Henry, *Coherent Lightwave Communications*. IEEE Press: New York, 1990.
15 A. F. Elrefaie, R. E. Wagner, D. A. Atlas, and A. D. Daut, "Chromatic dispersion limitation in coherent lightwave systems", *IEEE Journal of Lightwave Technology*, 6(5), 704–710, 1988.
16 Free space path loss formulae.
17 M. C. Farries, P. R. Morkel, R. I. Laming, T. A. Birks, D. N. Payne, and E. J. Tarbox, "Operation of erbium-doped fiber amplifiers and lasers pumped with frequency-doubled Nd:YAG lasers", *IEEE Journal of Lightwave Technology*, 7(10), 1473–1477, 1989.
18 G. Charlet, E. Corbel, J. Lazaro, A. Klekamp, W. Idler, R. Dischler, S. Bigo, P. Tran, T. Lopez, H. Mardoyan, A. Konczykowska, and J.-P. Thierry, "Comparison of system performance at 50, 62.5 and 100 GHz channel spacing over transoceanic distances at 40 Gbit/s channel rate using RZ-DPSK", *Electronics Letters*, 41(3), 145–146, 2005.
19 P. S. Cho, V. S. Grigoryan, Y. A. Godin, A. Salamon, and Y. Achiam, "Transmission of 25-Gbps RZ-DQPSK signals with 25-GHz channel spacing over 1000 km of SMF-28 fiber", *IEEE Photonics Technology Letters*, 15(3), 473–475, 2003.
20 K. Ishida, T. Kobayashi, J. Abe, K. Kinjo, S. Kuroda, and T. Mizuochi, "A comparative study of 10 Gbps RZ-DPSK and RZ-ASK WDM transmission over transoceanic distances", In *Proceedings of OFC'03*, vol. 2, pp. 451–453, Atlanta, GA, USA, 2003.
21 W. A. Atia and R. S. Bondurant, "Demonstration of return-to-zero signaling in both OOK and DPSK formats to improve receiver sensitivity in an optically preamplified receiver", In *Proceedings of IEEE LEOS'99*, vol. 1, pp. 226–227, San Francisco, CA, USA, 1999.
22 G. Bosco, A. Carena, V. Curri, R. Gaudino, and P. Poggiolini, "On the use of NRZ, RZ, and CSRZ modulation at 40 Gbps with narrow DWDM channel spacing", *Journal of Lightwave Technology*, 20(9), 1694–1704, 2002.
23 A. H. Gnauck, G. Raybon, P. G. Bernasconi, J. Leuthold, C. R. Doerr, and L. W. Stulz, "1-Tb/s (6/spl times/170.6 Gbps) transmission over 2000-km NZDF using OTDM and RZ-DPSK format", *IEEE Photonics Technology Letters*, 15(11), 1618–1620, 2003.
24 B. Zhu, L. E. Nelson, S. Stulz, A. H. Gnauck, C. Doerr, J. Leuthold, L. Gruner-Nielsen, M. O. Pedersen, J. Kim, and R. L. Lingle, Jr., "High spectral density long-haul 40-Gbps transmission using CSRZ-DPSK format", *IEEE Journal of Lightwave Technology*, 22(1), 208–214, 2004.
25 A. Hirano, Y. Miyamoto, and S. Kuwahara, "Performances of CSRZ-DPSK and RZ-DPSK in 43-Gbit/s/ch DWDM G.652 single-mode-fiber transmission", In *Proceedings of OFC'03*, Anaheim, CA, vol. 2, pp. 454–456, 2003.

26 L. N. Binh and T. L. Huynh, "Linear and nonlinear distortion effects in direct detection 40Gbps MSK modulation formats multi-span optically amplified transmission", *Optics Communications*, 237(2), 352–361, May 2007.

27 J. Mo, D. Yi, Y. Wen, S. Takahashi, Y. Wang, and C. Lu, "Optical minimum-shift keying modulator for high spectral efficiency WDM systems", In *Proceedings of ECOC'05*, vol. 4, pp. 781–782, Glasgow, U.K., 2005.

28 J. Mo, Y. J. Wen, Y. Dong, Y. Wang, and C. Lu, "Optical minimum-shift keying format and its dispersion tolerance", In *Proceedings of OFC'05*, paper JThB12, Anaheim, CA, 2005.

29 M. Ohm and J. Speidel, "Optical minimum-shift keying with direct detection (MSK/DD)", *Proceedings of SPIE on Optical Transmission*, Switching and Systems, vol. 5281, pp. 150–161, 2004.

30 T. L. Huynh, T. Sivahumaran, L. N. Binh, and K. K. Pang, "Narrowband frequency discrimination receiver for high dispersion tolerance optical MSK systems", In *Proceedings of ACOFT'07*, paper TuA1-3, Melbourne, Australia, June 2007.

31 T. L. Huynh, T. Sivahumaran, L. N. Binh, and K. K. Pang, "Sensitivity improvement with offset filtering in optical MSK narrowband frequency discrimination receiver", In *Proceedings of ACOFT-2007*, paper TuA1-5, Melbourne, June 2007.

32 T. Sivahumaran, T. L. Huynh, K. K. Pang, and L. N. Binh, "Non-linear equalizers in narrowband filter receiver achieving 950 ps/nm residual dispersion tolerance for 40Gbps optical MSK transmission systems", In *Proceedings of OFC'07*, paper OThK3, Anaheim, CA, 2007.

33 T. Sakamoto, T. Kawanishi, and M. Izutsu, "Optical minimum-shift keying with external modulation scheme", *Optics Express*, 13, 7741–7747, 2005.

34 A. H. Gnauck and P. J. Winzer, "Optical phase-shift-keyed transmission", *IEEE Journal of Lightwave Technology*, 23(1), 115–130, 2005.

35 J. A. Lazaro, W. Idler, R. Dischler, and A. Klekamp, "BER depending tolerances of DPSK balanced receiver at 43Gbps", In *Proceedings of IEEE/LEOS Workshop on Advanced Modulation Formats 2004*, pp. 15–16, San Francisco, CA, USA, 2004.

36 H. Kim and A. H. Gnauck, "Experimental investigation of the performance limitation of DPSK systems due to nonlinear phase noise", *IEEE Photonics Technology Letters*, 15(2), 320–322, 2003.

37 L. N. Binh, *Advance Optical Communication Systems*, 2nd Ed. CRC Press, Taylor & Francis: Boca Raton, FL, 2014.

38 S. Bhandare, D. Sandel, A. F. Abas, B. Milivojevic, A. Hidayat, R. Noe, M. Guy, and M. Lapointe, "2/spl times/40 Gbit/s RZ-DQPSK transmission with tunable chromatic dispersion compensation in 263 km fibre link", *Electronics Letters*, 40(13), 821–822, 2004.

39 T. Mizuochi, K. Ishida, T. Kobayashi, J. Abe, K. Kinjo, K. Motoshima, and K. Kasahara, "A comparative study of DPSK and OOK WDM transmission over transoceanic distances and their performance degradations due to nonlinear phase noise", *Journal of Lightwave Technology*, 21(9), 1933–1943, 2003.

40 C. Xu, X. Liu, L. F. Mollenauer, and X. Wei, "Comparison of return-to-zero differential phase-shift keying and ON-OFF keying in long-haul dispersion managed transmission", *IEEE Photonics Technology Letters*, 15(4), 617–619, 2003.

41 L. N. Binh and T. L. Huynh, "Phase-modulated hybrid 40Gbps and 10Gbps DPSK DWDM long-haul optical transmission", In *Proceedings of OFC'07*, paper JWA94, Anaheim, CA, 2007.

42 T. Ito, K. Sekiya, and T. Ono, "Study of 10 G/40 G hybrid ultra-long-haul transmission systems with reconfigurable OADM's for efficient wavelength usage", In *Proceedsings of ECOC'02*, paper 1.1.4, Copenhagen, Denmark, 2002.

43 K. Iwashita and N. Takachio, "Experimental evaluation of chromatic dispersion distortion in optical CPFSK transmission systems", *IEEE Journal of Lightwave Technology*, 7(10), 1484–1487, 1989.

44 J. G. Proakis, *Digital Communications*, 4th Ed. McGraw-Hill: New York, 2001.

45 J. G. Proakis and M. Salehi, *Communication Systems Engineering*, 2nd Ed. Prentice Hall, Inc: Hoboken, NJ, pp. 522–524, 2002.

46 K. K. Pang, *Digital Transmission*, Mi-Tec Media Pty. Ltd.: Melbourne, Australia, 2005.

47 K. Iwashita and T. Matsumoto, "Modulation and detection characteristics of optical continuous phase FSK transmission system", *IEEE Journal of Lightwave Technology*, 5(4), 452–460, 1989.

48 B. E. Little, "Advances in micro-ring resonator", *Integrated Photonics Research Conference 2003*, Invited Paper, Marseille, France, 2003.

49 V. Van, B. E. Little, S. T. Chu, and J. V. Hryniewicz, "Micro-ring resonator filters", In *Proceedsings of LEOS'04*, vol. 2, pp. 571–572, Boulder, CO, USA, 2004.

50 P. P. Absil, S. T. Chu, D. Gill, J. V. Hryniewicz, F. Johnson, O. King, B. E. Little, F. Seiferth, and V. Van, "Very high order integrated optical filters", In *Proceedsings of OFC'04*, Anaheim, CA, vol. 1, 2004.

51 L. Brent, C. Sai, C. Wei, C. Wenlu, H. John, G. Dave, K. Oliver, J. Fred, D. Roy, D. Kevin, and G. John, "Advanced ring resonator-based PLCs", *IEEE Lasers and Electro-Optics Society 2006*, pp. 751–752, Montreal, QC, Canada, 2006.
52 L.N.Binh " Advanced Digital Optica Communications", CRC Press, Taylor & Francis, Chapter 8, Boca Raton, FL, USA, 2015.
53 A. H. Gnauck and P. J. Winzer, "Optical phase-shift-keyed transmission", *IEEE Journal of Lightwave Technology*, 23(1), 115–130, 2005.
54 U.-V. Koc and X. Wei, "Combined effect of polarization-mode dispersion and chromatic dispersion on strongly filtered /spl pi//2-DPSK and conventional DPSK", *IEEE Photonics Technology Letters*, 16(6), 1588–1590, 2004.
55 H. Kim and P. J. Winzer, "Robustness to laser frequency offset in direct-detection DPSK and DQPSK systems", *IEEE Journal of Lightwave Technology*, 21(9), 1887–1891, 2003.
56 T. Kawanishi, S. Shinada, T. Sakamoto, S. Oikawa, K. Yoshiara, and M. Izutsu, "Reciprocating optical modulator with resonant modulating electrode", *Electronics Letters*, 41(5), 271–272, 2005.
57 R. Krahenbuhl, J. H. Cole, R. P. Moeller, and M. M. Howerton, "High-speed optical modulator in $LiNbO_3$ with cascaded resonant-type electrodes", *IEEE Journal of Lightwave Technology*, 24(5), 2184–2189, 2006.
58 I. P. Kaminow and T. Li, *Optical Fiber Communications*, vol. IV-A. Elsevier Science: Amsterdam, The Netherlands, Chapter 16, 2002.
59 M. Ohm and J. Speidel, "Optical minimum-shift keying with direct detection (MSK/DD)", In *Proceedsings of SPIE on Optical Transmission, Switching and Systems*, vol. 5281, pp. 150–161, Wuhan China, `2004.
60 T. Sakamoto, T. Kawanishi, and M. Izutsu, "Optical minimum-shift keying with external modulation scheme", *Optics Express*, 13, 7741–7747, 2005.
61 J. Mo, D. Yi, Y. Wen, S. Takahashi, Y. Wang, and C. Lu, "Optical minimum-shift keying modulator for high spectral efficiency WDM systems", In *Proceedings of ECOC'05*, vol. 4, pp. 781–782, Glasgow, U.K., 2005.
62 T. L. Huynh, L. N. Binh, K. K. Pang, and L. Chan, "Photonic MSK transmitter models using linear and non-linear phase shaping for non-coherent long-haul optical transmission", In *Proceedings of SPIE APOC'06*, paper 6353-85, Gwangju, Korea, September 2006.
63 B. E. Little, "Advances in microring resonator", *Integrated Photonics Research Conference 2003*, Invited Paper, 2003.
64 L. Brent, C. Sai, C. Wei, C. Wenlu, H. John, G. Dave, K. Oliver, J. Fred, D. Roy, D. Kevin, and G. John, "Advanced ring resonator-based PLCs", IEEE Lasers & Electro-Optics Society, 751–752, Montreal, QC, Canada, 2006.
65 V. Van, B. E. Little, S. T. Chu, and J. V. Hryniewicz, "Micro-ring resonator filters", In *Proceedings of LEOS'04*, vol. 2, pp. 571–572, Boulder, CO, USA, 2004.
66 J. G. Proakis, *Digital Communications*, 4th Ed. McGraw-Hill: New York, 2001.
67 F. Xiong, *Digital Modulation Techniques*. Artech House Inc: Boston, London, 2000.
68 J. Mo, Y. J. Wen, Y. Dong, Y. Wang, and C. Lu, "Optical minimum-shift keying format and its dispersion tolerance", In *Proceedings of OFC'05*, paper JThB12, Anaheim, CA, 2005.
69 M. Ohm and J. Speidel, "Optical minimum-shift keying with direct detection (MSK/DD)", In *Proceedinsg of SPIE on Optical Transmission, Switching and Systems*, vol. 5281, pp. 150–161, Wuhan, China, 2004.
70 R. Krahenbuhl, J. H. Cole, R. P. Moeller, and M. M. Howerton, "High-speed optical modulator in $LiNbO_3$ with cascaded resonant-type electrodes", *IEEE Journal of Lightwave Technology*, 24(5), 2184–2189, 2006.
71 L. N. Binh, *Advanced Digital Optical Communications*. CRC Press, Taylor & Francis: Boca Raton, FL, 2015.
72 (a) D. Boivin, M. Haris, M. Hanna, J. Yu, G.-K. Chang, and J. R. Barry, "Experimental measurement of optical phase variance in RZ-DPSK systems using direct detection after demodulation by an MZDI", *IEEE Photonics Technology Letters*, 18(19), 1990–1992, 2006; (b) D. Bolvin, et al., IEEE Photonics Technology Letters, 14(6), 843, 2002.
73 T. Frank, P. B. Hansen, T. N. Nielsem, and L. Eskildson, "DB transmitter with lower inter-symbol interference", *IEEE Photonics Technology Letters*, 10(4), 597 – 599, April 1998.
74 MATLAB Inc., Hlp Centre, "Single sideband modulation via the Hilbert transform", https://www.mathworks.com/help/signal/ug/single-sideband-modulation-via-the-hilbert-transform.html, Access date: May 2024.

75 L. Zhang et al., "Single wavelength 248-Gb/s transmission over 80-km SMF based on twin-SSB-DMT and direct detection", *Proceedings of ECOC 2016, 42nd European Conference and Exhibition on Optical Communications*, Düsseldorf, Germany, September 18–22, 2016.
76 M. S. Erkilinc, et al., "Spectrally-efficient single sideband 16-QAM Nyquist-Subcarrier modulation-based WDM transmission using an InP dual-drive mach-zehnder modulator and direct-detection," In *Proceeedings of ECOC'15*, paper 0056, Valencia, Spain, 2015.
77 J. Wei, et al., "First demonstration of real-time end-to-end 40 Gb/s PAM-4 system using 10-G transmitter for next generation access applications," *Proceedings of ECOC'15*, PDP4.4r, Valencia, Spain, 2015.
78 S. Chandrasekhar, et al., "Compact All-Inp laser-vector-modulator for generation and transmission of 100-Gb/s PDM-QPSK and 200-Gb/s PDM-16-QAM", *Journal of Lightwave Technology*, 32(4), 736, 2014.
79 T. Frank, P. B. Hansen, T. N. Nielsem, and L. Eskildson, "DB transmitter with lower inter-symbol interference", *IEEE Photonics Technology Letters*, 10(4), 597 – 599, April 1998.

4 Multi-Level Complex Photonic Modulation

CONTENTS

Modulation of lightwaves for SPC is crucial for long-haul transmission over the inter- or intra-constellation of satellite photonic nodes. Photonics is the art of manipulating or modulating lightwaves using several effects that alter the refractive index of guided waveguides, thereby changing the phase of the lightwaves. Electro-optic interaction occurs via an applied electric field across the propagation path of the lightwaves. Therefore, phase modulation can be implemented. Alternatively, this phase change can be incorporated into an optical interferometer so that the interfered lightwaves at the output can be intensity-switched from high to low, achieving amplitude modulation.

DOI: 10.1201/9781003426677-4

Furthermore, amplitude modulation can be formed in a two-dimensional coordinate to generate quadrature amplitude modulation (QAM), and then to a higher order to form M-ary QAM (or mQAM). This chapter is the second part introducing the complex modulation techniques for SPC. The previous chapter, Chapter 3, described the binary modulation technique. Binary modulation is extended to multi-level digital modulation and combined with another dimension to create mQAM.

Therefore, the modulator must be integrated with another set of amplitude modulation devices, resulting in two independent sets of amplitude modulation.

In addition, if another polarization of the lightwaves can be used, we would have PDM-mQAM (polarization division multiplexing mQAM) channels. This chapter describes these complex modulation techniques.

4.1 INTRODUCTION

This chapter addresses the binary modulation format, where the amplitude varies along the real axis or in a single dimension, taking real amplitude values. On the other hand, if two orthogonal axes are taken as the variation of the amplitudes, the vector summation of these two orthogonal components generates a complex combination of the amplitude and phase in an angular coordinate system. Using such complex or two-dimensional (2D) modulation offers the advantage of doubling the information capacity. Both the real and complex axes are employed and the derivation of the phase and magnitude of the complex vector is shown in Figure 4.1. The 2D modulation increases or doubles the total capacity of the transmission system with both the amplitude and phase representation. Further increases can be implemented by using multi-dimensional levels of modulation such as m-QAM (quadrature amplitude modulation). These 2D modulation and reception techniques require coherent receivers and transmitters, which are also critical for ULH non-terrestrial transmission, as described in Chapter 1. Multi-level modulation requires low noise contribution in the optical amplification and optoelectronic receivers so that higher resolution, and thus the eye-opening of each level, can meet the required sensitivity in the free-space system. Furthermore, the use of polarizations of the guided wave or coherent beams in space transmission offers another capacity doubling with the two orthogonal polarized modes.

This section, therefore, presents modulation formats for generating complex optical signaling that combine the amplitude modulation and differential phase modulation schemes, as well as the multi-level amplitude and phase shift keying. Comparisons between multi-level and binary modulation are made. Critical issues of transmission performance for multi-level modulation are identified. A simulation platform based on MATLAB Simulink[1,2] can be used to briefly describe and

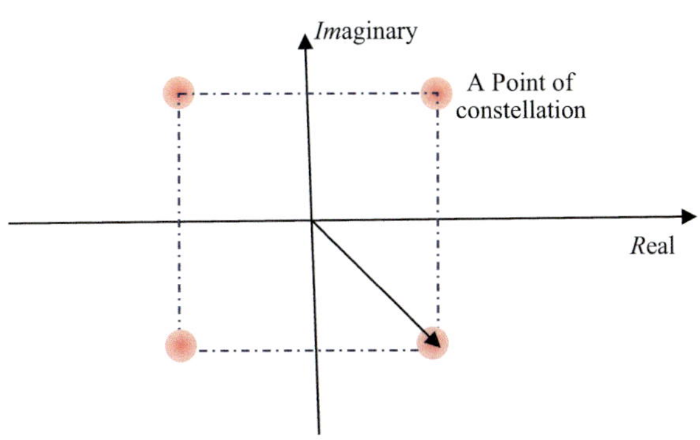

FIGURE 4.1 Complex modulation representation of modulated points in a 2D constellation.

implement the complex modulation. Furthermore, to achieve the 100 Gbps Ethernet, several multi-level modulations, such as PSK and OFDM, are proposed. The optical modulators that employ both amplitude and phase in integrated optic device structures are described.

Multi-level multi-channels for SPC are presented in the last section, along with the experimental demonstration of Tbps using five or ten channels with 200 and 100 Gbps, respectively. A more detailed experimental platform for one SPC hop with WSS as mux and demux, optical booster amplifier, and optical preamplifier is presented in Chapter 10.

4.2 AMPLITUDE AND DIFFERENTIAL PHASE PHOTONIC MODULATION (ADPM)

4.2.1 Remarks

Under the conventional on-off keying (ASK) modulation format, the transmission bit rate beyond 40 Gbps per optical channel is costly because the electronic signal processing technology may have reached its fundamental speed limit. It is expected that advanced photonic modulation formats, such as M-ary amplitude and differential phase shift keying, will replace ASK in the near future. These advanced formats would offer more efficient spectral properties, thus enabling an increase in the transmission rates without placing stringent requirements on high-speed electronics, while using the same existing photonic communication infrastructure.

In SPC, the demands for optical amplification for achieving higher gain can be obtained by exploiting the current coherent at the transmitter site, and low noise low-level preamplifiers are compulsory to compensate for the high-lost over the transmission path and exploring the transceiver technologies currently available for long hail terrestrial fiber optic networks. Therefore, noises are accumulated, corrupting the digital constellation points. Coherent reception is preferred for the SPC reception subsystem so that higher gain can be obtained; at least 6 dB from balanced receiver and 20–30 dB from coherent mixing with a LO laser. Furthermore, the constellation distance of complex points provides additional Euclidean distance for better noise tolerance.

Coherent communications, developed in the mid-1980s, have extensively exploited different modulation techniques to improve the 'optical' signal-to-noise ratio (OSNR). Coherent reception is by mixig the coming signals with a local oscillator (LO) laser, so that the resulting RF signals can be derive at frequency of the difference of the LO and that of the signal carrier. The mixing of a high magnitude LO laser and the weak signal amplitude in the square-law detector, leads to a resulting product of these magnitudes, thus boosting the weak signal amplitude. If the LO lasEr frequency is the same as that of the signal carrier then the reception is a homodyne time and the information band is located in the baseband.

In these early years, the coherent detection faced considerable difficulties due to the stability of the laser source spectrum and its wide linewidth, leading to the laser phase noises and frequency shifting, which caused difficulties in wider bandwidth of the receiver. Furthermore, the advances in optical modulators, that allows external modulation of the lightwaves of the laser without disturbing the laser cavity, have kept the laser line stable and its linewidth sufficient narrow for near-ideal coherent receptions.

The shift of the frequency difference leads to heterodyne from the ideal homodyne detection. The receiver sensitivity can be of 3 dB for heterodyne case, and 6 dB for homodyne detection. Under current technology on laser technology and digital signal processing, the frequency shift can be compensated by a soft algorithm in the DSP to mitigate this shift, and an ideal homodyne can be achieved. These techniques are further presented in Chapter 6.

Optical amplifiers, which process photon signal input and output photonic signals without going through the O/E/O (Optical-Electronic-Optical) conversion process, are required for space photonic transmission systems. The power booster amplifiers and optical pre-amplifiers are also required to boost the optical power launched into space medium and to detect low-power from high attenuation in long-haul free-space medium.

The high attenuation requires low-level amplification, which leads to a high noise contribution from the ASE (amplification stimulated emission), an amplitude input dependent. Therefore, for further transmission of channels transmitted from OGS, the photonic signals are normally received and converted back to the electrical domain, followed by modulation of a new transmitter. This repeater process ensures the new path transmission signal power to have sufficient OSNR for the next ultra-long-haul (ULH) hop transmission and/or routing to other satellites in the constellation.

The invention of optical amplifiers (OA) in the early 1990s overcame the fiber attenuation limit and thus offers a significant improvement in optical transmission technology. As a result, ultra-long haul and ultra-high-capacity optical transmission systems have been deployed widely around the world in the last decade. The technology has matured, with ASK modulation reaching 40 Gbps per optical channel, a total channel count in the hundreds, and 100/50 GHz channel spacing.

The low-level optical amplifiers will suffer an increase in the quantum amplification stimulated noise, leading to a reduction in the sensitivity of the receiver. This would require careful design of system budget and amplification of the implemented transmission systems.

Based on the proven efficient spectra and transmission technology, especially the controllable total dispersion of the transmission and compensating fibers, it is much more advantageous to use these spectral regions efficiently. Therefore, the contribution of advanced modulation techniques and formats would offer higher spectral efficiency for photonic transmission.

Further, digital modulation techniques have been well established over the last half-century with amplitude, frequency, or phase modulations. These techniques, especially phase modulation, rely principally on the detection schemes – on whether it is coherent or pseudo-coherent differential detection – have been heavily exploited in wireless communication networks. In the photonic domain, for a long time, technological difficulties associated with manufacturing narrow linewidth lasers have prevented the use of coherent and differential phase modulation. Only over the last several years, due to advancements in laser fabrication and structural technologies, particularly distributed feedback (DFB) lasers, external cavity lasers, and double cavity resonant feedback control lasers, has the laser linewidth become much smaller than the signal modulation bandwidth. The coherence of the sources is now sufficient for differential phase modulation and detection applications, which require the phase of the sources to remain stable over at least two consecutive symbol periods. Furthermore, such narrow linewidth provides sufficient coherency for long-haul space photonic communications. By 'photonic', we indicate the photon wave and its operation, which can be examined and manipulated at the photon level. Hence, the title of this book, 'Photonic Communication', pertains to space transmission mediums.

Recently, advanced modulation techniques have attracted significant interest from the photonic transmission research and system engineering community. Several modulation schemes and formats, such as binary differential phase shift keying (BDPSK), differential-quadrature phase shift keying (DQPSK), Duo-binary ASK associated with non-return-to-zero (NRZ), return-to-zero (RZ), and carrier-suppress return-to-zero (CSRZ) formats, have been widely studied. However, what have not been widely explored are the multi-level modulation schemes. Which modulation formats are most suitable for SPC systems and multi-hop space transmission?

Although multi-level schemes have been intensively exploited in wireless communications, there are minimum works to date that incorporate the in-coherent multi-level optical amplitude-phase shift keying modulation schemes.[3] The fading effects in wireless transmission are much similar to the turbulence effects of free space beam propagation on coherent mixed light rays.

The multi-level concentric circular phase distribution can offer the following advantages: (i) a lower symbol rate, hence for the same available spectral region, a multi-level modulation scheme would offer higher transmission capacity than binary modulation counterparts; (ii) efficient bandwidth utilization, as the photonic transmission of these multi-level signals could be implemented over the existing SPC infrastructure without significant alteration of the system architecture, thus saving on CAPEX (capital expenditure) and easing system management; (iii) the complexity of the coder and demodulation subsystems falls within the technological capabilities of current microwave and photonic technologies.

The principal objectives of this section are: (i) to evaluate different modulation and coding techniques and signal pulse formats in the complex plane, especially complex modulation techniques for long-haul ultra-high and ultra-long distance capacity transmission over space non-terrestrial mediums; hence, the determination of novel modulation schemes, such as multi-level amplitude-phase shift keying (MAPSK); (ii) to briefly present some simulation and experimental test results to demonstrate the uniqueness and superiority of these novel schemes, with particular emphasis on space photonic communications; (iii) a comparative study of the modulation formats so as to unveil the principal directions for photonic modulation and transmission technologies for the next transmission generation; (iv) the fabrication and practical structures of the modulators and photonic transmitters for space photonic systems.

Thus, a conceptual photonic transmission system is proposed based on a hybrid technique that combines phase and amplitude modulation, the multi-level amplitude-differential phase shift keying (MADPSK) format. This technique combines two modulation formats: the well-known M-ary ASK and the M-ary DPSK, leveraging the high receiver sensitivity and dispersion tolerance (DPSK), and the enhancement of total transmission capacity (M-ASK), compared to the traditional ASK format.

The models of MADPSK transmitter and receiver are structured for MADPSK signaling. A simulation model based on the MATLAB Simulink platform is developed for the proof-of-concept. The system performance is evaluated for back-to-back (B2B) and long-haul space transmission. Analytical and simulation results of the transmission configurations are provided.

The following topics are presented: (i) Noise mechanisms, such as quantum shot noises, quantum phase noises, optically amplified noises, noise statistics, and nonlinear phase noises; hence the design of an optimum detection and decision-level schemes for MADPSK; (ii) Linear and nonlinear and polarization dispersion impairments and their impacts on MADPSK system performance; (iii) Matched filter design for optimum MADPSK signal detection; (iv) Offset MADPSK (O-MADPSK) modulation schemes; (v) Multi-level amplitude-minimum shift keying (MAMSK) modulation, (vi) MADPSK modulation for applications in subcarrier transmission systems, especially for metropolitan-wide area multi-add/drop networks. (vii) Other issues or additional modulation formats suitable for MADPSK.

This section can be organized as follows: Section 4.1 provides a brief review of several advanced photonic modulation formats. Section 4.2 reviews and compares different modulator structures designed for generating advanced photonic modulation signals and emphasizes the advantages of dual-drive MZIM as a modulator for generating MADPSK signals. In Section 4.4, a novel photonic transmission system with MADPSK modulation format is proposed. Section 4.5 summarizes the results. Section 4.6 identifies numerous critical issues and alternative multi-level signaling for optical systems.

4.2.2 Two-Dimensional ASK and Complex Modulation

4.2.2.1 NRZ/RZ-ASK Modulation

ASK has been the dominant modulation technique since the early 1970s in optical communications. The main advantage of this modulation is that the ASK signal is not sensitive to phase noise. ASK

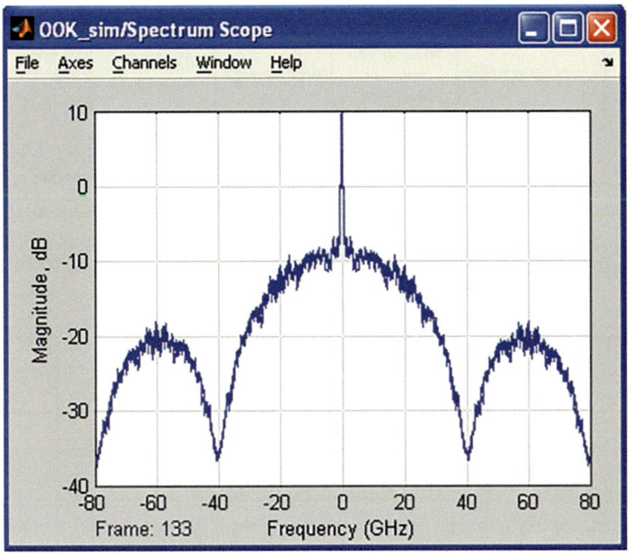

FIGURE 4.2 Spectrum of a 40 Gbps NRZ-ASK (OOK) signal.

modulation can take two principal formats: the first is called NRZ-ASK, where the "1" optical bit occupies the whole bit period; the second is RZ-ASK, where "1" bit is present only during the first half of the bit period.

Figure 4.2 depicts the spectrum of a 40 Gbps NRZ-ASK signal, with the carrier seen at the highest peak and the 3 dB bandwidth reaching the bit rate. The main advantage of the NRZ-ASK signal is that its spectrum is generally the most compact compared to other formats such as RZ-ASK, and CSRZ-ASK. On the other hand, however, the NRZ-ASK signal is affected by fiber chromatic dispersion and is more sensitive to fiber nonlinear effects compared to its RZ- and CSRZ-ASK counterparts.

4.2.2.2 RZ-ASK Modulation

A RZ-ASK signal, as shown in Figure 4.3, is similar to NRZ-ASK, except for the "1" bit occupying only the first half of the bit period. This signal can be generated by a transmitter shown in the same figure in which an NRZ-ASK transmitter is followed by a pulse caver driven by a pulse train synchronized with the data source. The pulse train has a frequency equal to the data rate. The RZ-ASK pulse width can take the form of 33%, 50%, and 66% duty-ratio. Due to its narrower pulse width, the spectrum of the RZ-ASK signal, as shown in Figure 4.4, is larger than that of the NRZ-ASK signal, leading to less spectrum efficiency. In this spectrum, the carrier is seen as the highest peak, the two side peaks are RF-modulating signals positioned at 80 GHz apart.

4.2.2.3 CSRZ-ASK Modulation

CSRZ-ASK modulation format is similar to standard RZ-ASK, except that the neighboring optical pulses have π-phase difference. The carrier in neighboring time slots is thus cancelled and effectively excluded from the signal spectrum. The CSRZ-ASK signal can be generated by a transmitter with scheme, as shown in Figure 4.5. In this scheme, the first MZIM modulates the intensity of optical signal coming from a laser source, while the second MZIM, driven by a clock signal at the haft data rate, caves the NRZ pulses into RZ. Because the second MZIM is biased at the minimum-intensity point, it provides an RZ pulse train at the data rate with alternating phase 0 and π for neighboring time slots. The CSRZ-ASK signal can also be detected by a direct-detection receiver as it would not be phase sensitive.

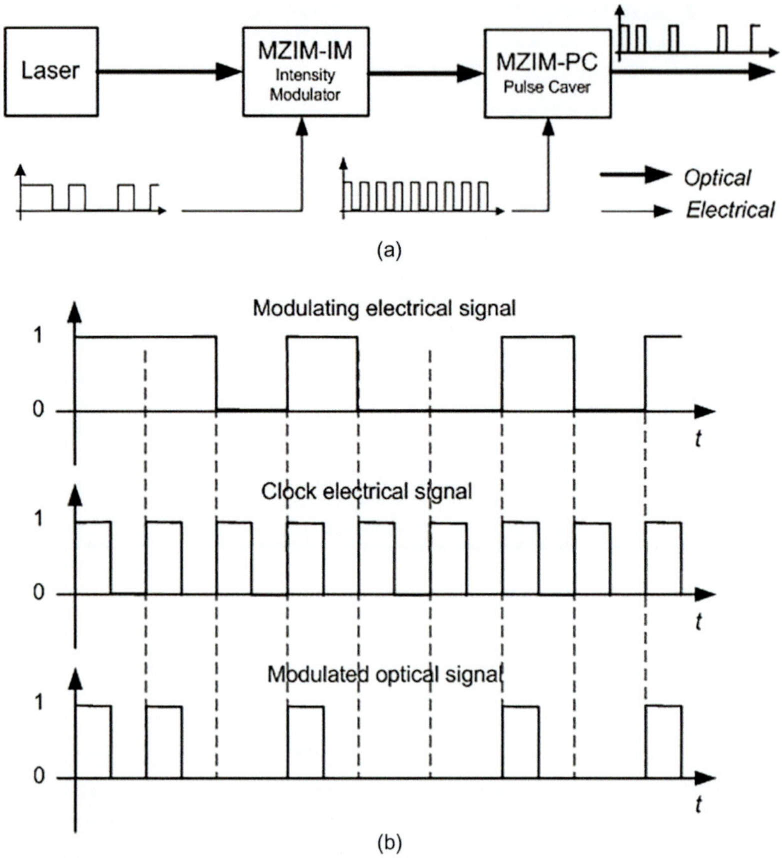

FIGURE 4.3 Schematic structure of RZ-ASK transmitter (a) and electrical pulse sequence for generation of the optical signal (b).

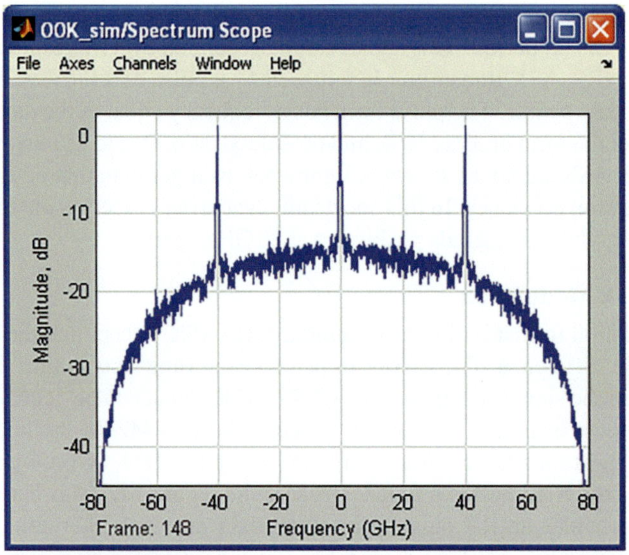

FIGURE 4.4 Spectrum of 40 Gbps with 50% pulse width of RZ-ASK signal.

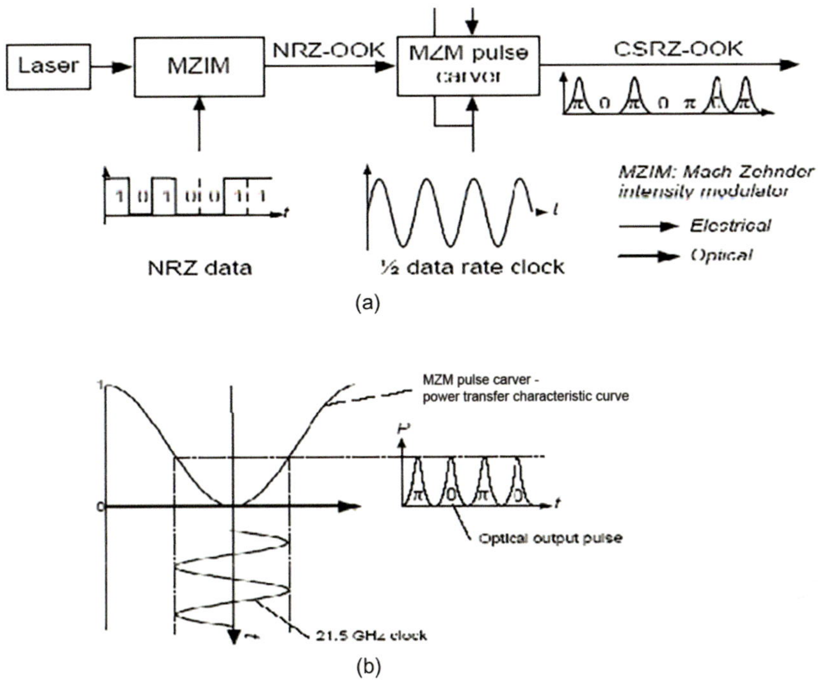

FIGURE 4.5 (a) Block diagrams of CSRZ-ASK transmitter, (b) generation of optical pulse with alternative phase using biasing control and amplitude.

The main advantages of CSRZ-ASK include narrower spectrum, higher tolerance to dispersion, and stronger robustness against fiber nonlinear effects compared to the standard RZ-ASK. Since its peak optical power is much lower than that of other formats, it is less affected by both self-phase modulation (SPM) and cross-phase modulation (XPM). Figure 4.6 shows the spectrum of 40 Gbps CSRZ-ASK signal with very low carrier power level.

ASK is a modulation technique that generates a signal $s(t)$ by multiplying a digital signal $m(t)$ by a carrier f_c

$$s(t) = A \cdot m(t)\cos 2\pi f_c t; \quad \text{for} \quad 0 < t < T \tag{4.1}$$

where A is amplitude envelope; the digital signal $m(t)$ may take one of M levels $\left[b_0, b_1, ..., b_M\right]$. When $M=2$, $s(t)$ is a binary ASK signal, with ASK being a special case. ASK is also implemented in NRZ, RZ, and CSRZ formats, whose spectra are shown in Figure 4.7 on the same graph for the purpose of comparison. Like their ASK analogues, NRZ-ASK has the most compact spectrum, while RZ-ASK has the broadest. In terms of energy, CSRZ-ASK has the lowest peak power because the carrier signal has been effectively removed.

4.2.2.4 QPSK Modulation/Modulator as a Summation of X- and Y-ASK

Two ASK signals, whose amplitude distribution is based on the two orthogonal axes, the real and imaginary, can form the complex QPSK as shown in Figure 4.1. In hardware implementation, a QPSK optical modulator can be formed using two MZIM modulators, with their phases shifted by $\pi/2$ relative to each other by an optical phase shifter, as shown in Figure 4.8. Synchronization of the two data sequences to generate the in-phase and quadrature-phase components is critical, and the delay time difference must be very small. Similarly, at the receiver, their synchronization is also important to control. The bit-error rate, or probability of errors of the QPSK detection, can

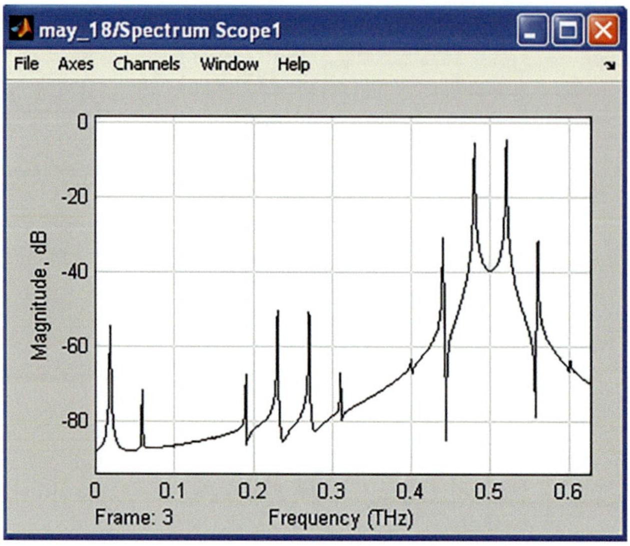

FIGURE 4.6 Spectrum of 40 Gbps with CSRZ-ASK modulation format. Note: two carriers are spaced on either side of the central carrier.

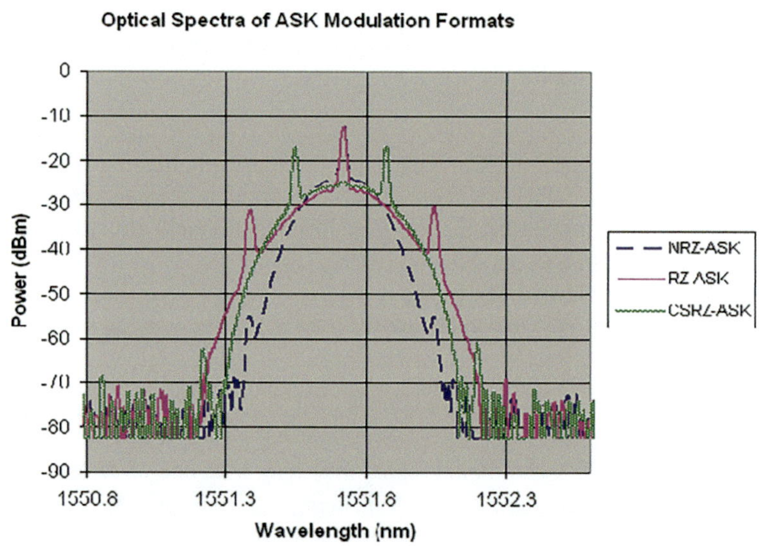

FIGURE 4.7 Spectra of NRZ-ASK, RZ-ASK, and CSRZ-ASK signals. Note: modulation formats are coded in different intensity of lines. Note: the carrier peak power is above the signal spectra.

be viewed as the summation of the detection of the ASK sequences in the X- and Y-directions. The laser is turned on at all times, operating in the continuous wave (CW) mode.

If two orthogonal polarizations are used to carry another set of data sequences, then a polarization beam splitter in integrated waveguide is used to split the polarization from the laser lightwaves at the input. The modulator for the other orthogonal channel can be formed by another two MZIM modulators in parallel with the IQ modulator shown in Figure 4.8. The details of this dual-polarized IQ modulator are provided in Chapters 5, 6, and 10.

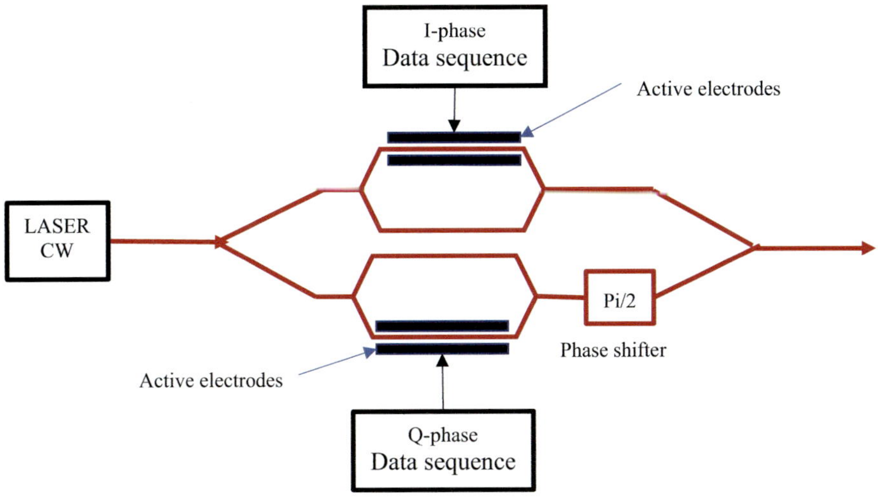

FIGURE 4.8 Schematic diagram of an integrated optic QPSK modulator with MZIM or real (in phase X) and quadrature (Y-component) components. CW = continuous wave.

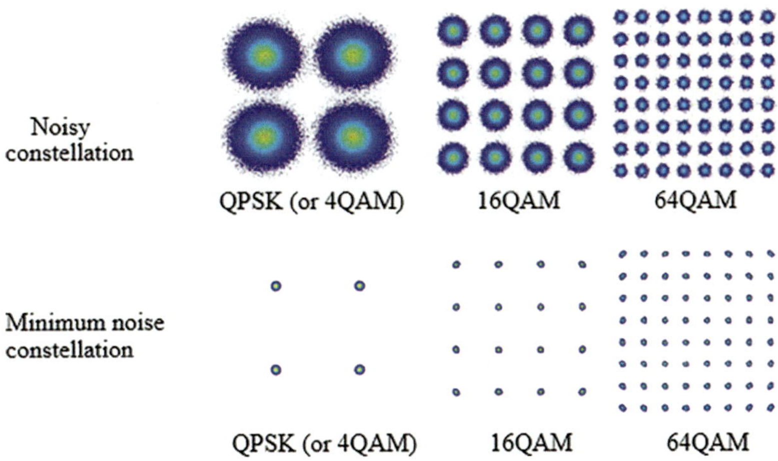

FIGURE 4.9 Constellations of mQAM modulated signals of 4 QAM (QPSK), 16 QAM, and 64 QAM under minimum noise and noisy scenarios.

The constellation for QPSK and 16 and 64 QAM under minimum noise and noisy scenarios can be seen in Figure 4.9. For 16 and 64 QAM, the sequence applied to the electrodes of IQ modulator has four level ASK modulation. Thus, we have a higher degree of probability of error. This is possible if the optical sequence is amplified by optical amplifiers and higher receiver sensitivity.

Nowadays, most mQAM transceiver consists of (i) integrated dual-polarized IQ modulator; (ii) DSP for each transmitter section and one DSP processor for the receiver, two lasers, one for lightwaves transport of data through the transmitter and one as a local oscillator for coherent reception; (iv) a hybrid coupler with optical phase shifter to extract the *I*- and *Q*-components; and (v) phase shifter integrated with the dual-polarized modulator for orthogonality generation of the *I*- and *Q*-optical sequence.

4.2.3 Multi-Amplitude Differential Phase Shift Keying

Differential phase coding and modulation offer better performance for SPC, as the turbulence or fading effects in a turbulent layer can be eliminated by comparing and detecting the phase difference over two consecutive pulse periods. During the time of differential code, one-bit period, the laser phase does not change; hence, it is better than the absolute phase. The main reason for the differential coding is that the phase shift between consecutive bits can be differentiated, while the carrier frequency remains unchanged during this short bit period. Thus, the effects of the frequency fluctuation of the laser are minimized, especially when coherent detection is used. This section presents the differential phase techniques in association with binary and multi-level amplitude modulation.

Under ASK/ASK modulation schemes with the associated NRZ, RZ, and CSRZ formats, the amplitude of the optical carrier varies accordingly. Phase modulation, on the other hand, modulates carrier phase, thus facilitating the use of bi-polar signals "±1". This distinguished feature means that the phase modulation offers significant improvement in receiver sensitivity compared to the ASK modulation. With the recent advancement in photonic lightwave technology, especially integrated optic delay interferometer, differential phase modulation, as well as demodulation and balanced receiver, has become realizable. This section provides an overview of the differential modulation techniques and their implementations in the photonic domain, especially the MADPSK (multi-amplitude differential phase shift keying). This modulation format is further expanded in Section 4.2.5.

The term NRZ-BPSK, or traditionally NRZ-DPSK, is commonly used for denoting a modulation technique in which the optical carrier is always present with constant power, and only its phase is alternated between 0 and π radians. The modulation rule is as follows: (i) At the transmitter, initially, a reference "0" bit is entered as the present encoded bit. Then, the next data bit is compared with that current encoded bit. If they are different, the next encoded bit is "1" for which a phase change of π occurs, else the next encoded bit is "0", which causes no (or 0) phase change. (ii) At the receiver, the phase of the carrier at the current bit slot is compared with that of the previous. If the phase difference is π, then the data is decoded as "1", otherwise the data is "0" when the phase difference is 0.

One of the NRZ-DPSK transmitter structures is shown in Figure 4.10. User data are first encoded by a differential encoder into the driving voltage that then alternates the phase of the carrier signal between 0 and π. In detecting an NRZ-DPSK signal, a delay Mach-Zehnder Interferometer (MZI) in combination with a balanced optoelectronic receiver can be used. The interferometer acts as the phase comparator with constructively and destructively interfered outputs. As shown in Figure 4.10, the received optical signal is split into two arms of a MZI, with one arm passing through a one-bit optical delay line. The two branches of the MZI combines at its output, and via its interference pattern, the phase of each bit is compared with the phase of its previous bit. The photodetector converts the optical phase difference to the intensity, and hence the current and voltage at its output voltage level.

When there is 'no' phase shift between two bits, the two optical path lightwaves are added constructively, and the receiver output gives a maximum level, otherwise they cancel destructive interference when their phase shift equals *to* π.

If the differential phase shift is $\Delta\phi$, then the differential current at the output of the balanced photodetector can be written as:

$$i = A^2 \cos \Delta\varphi \tag{4.2}$$

If under coherent reception, the amplitude is the product of the signal's amplitude and that of the local oscillator (LO) laser, along with another term whose frequency is the difference between the optical frequency of the LO laser and that of the carrier. This determines whether it is the

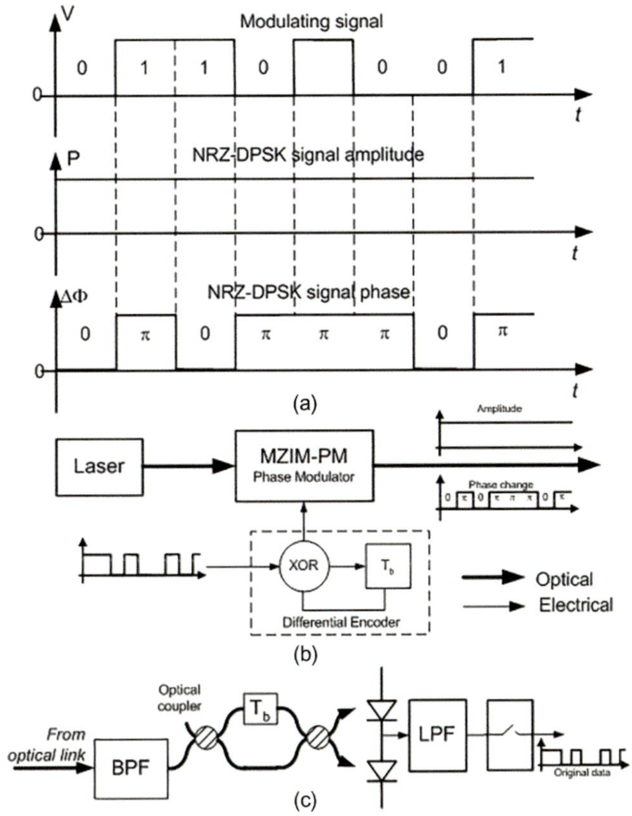

FIGURE 4.10 (a) NRZ-DPSK signal, (b) transmitter, and (c) receiver using a phase delay interferometer for detection of dual-phase binary modulated signals. Bold lines=optical path, thin black line: electrical path.

heterodyne or homodyne or intradyne type, based on the shift of the intermediate frequency (IF) of the received product.

Because the balanced receiver uses both constructive and destructive ports of the MZI, the detected signal level can swing from "1" to "–1". Compared with ASK under the detection of an unbalanced receiver, the balanced signal amplitude is limited between "1" and "0", and DPSK can offer a 3 dB improvement in the receiver sensitivity.

Due to its constant envelope, the NRZ-DPSK signal sequence is less sensitive to the power modulation-related non-linearity, especially when it is boosted by the high-power booster optical amplifiers (HPA). The HPA is integrated at the last stage of the optical transmitter before launching the beam into the space. The nonlinear effects can be SPM, SBS, SRS, and XPM. The signal modulated by NRZ-DPSK, suffer nonlinear effects much less than NRZ-ASK counterpart.

On the other hand, however, SPC long-haul DPSK systems, including both NRZ and RZ, with high power optical amplifier are affected by the non-linear phase noise. Amplified spontaneous emission (ASE) noise from optical amplifiers is converted into the phase noise, leading to the waveform distortion and, consequently, signal degradation. The spectrum of NRZ-DPSK signal is shown in Figure 4.11, along with other DPSK formats. It can be seen that the NRZ-DPSK signal has the most compact spectrum compared to that of other DPSK formats. This can be explained by the fact that the NRZ-DPSK signal amplitude remains constant, regardless of whether bit "1" or bit "0" is transmitted, allowing for more equal energy distribution compared to RZ- and CSRZ-DPSK signals.

Similarly, RZ-DPSK optical data sequence is quite similar to the NRZ-DPSK format, with only difference being that, instead of constant optical power, the pulse width is narrower than bit period

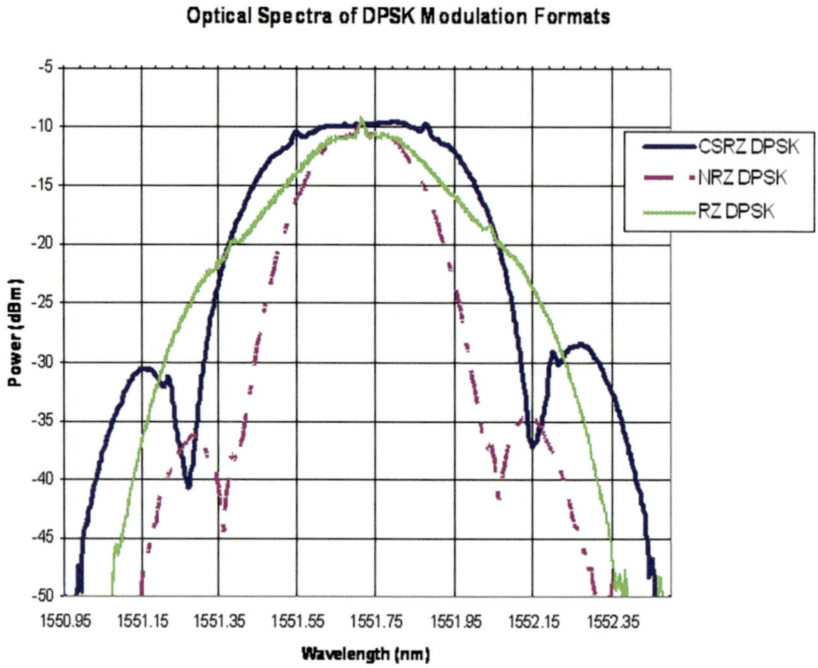

FIGURE 4.11 Experimentally measured spectra of NRZ-DPSK, RZ-DPSK, and CSRZ-DPSK signals.

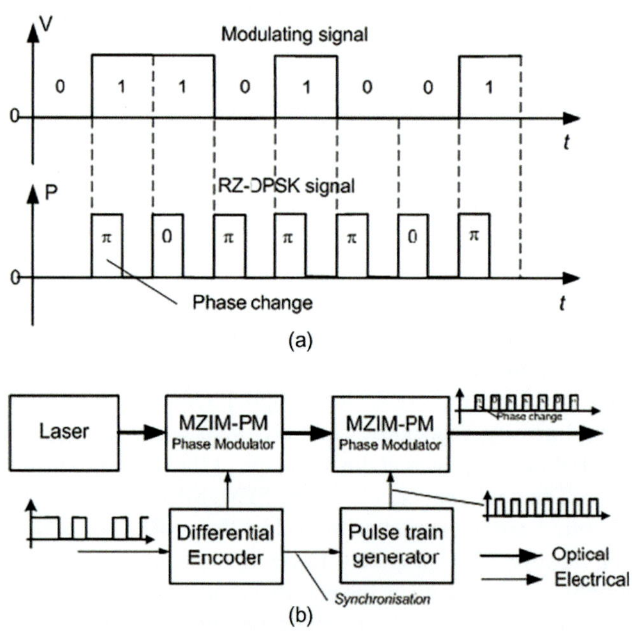

FIGURE 4.12 (a) RZ-DPSK signal sequence, (b) transmitter structure with pulse sequences at input electrical port and output optical port.

appears in each bit slot, as shown in Figure 4.12. The RZ-DPSK transmitter, however, resembles an RZ-ASK transmitter with the optical phase modulator (OPM) replacing the intensity modulator (IM). The RZ-DPSK signal can also be detected by the same receiver used for the NRZ-DPSK signal. Due to its narrow pulse, the RZ-DPSK format is expected to minimize the effects of inter-symbol interference (ISI) and thus achieve a longer transmission distance. A narrow pulse, however, spreads the spectrum of the RZ-DPSK signal wider than that of NRZ-DPSK, making RZ-DPSK systems more susceptible to chromatic dispersion (CD) in fiber transmission line but not severely affected in space transmission distance. To reduce the effect of this impairment, CD compensation devices are used in fiber optic terrestrial system.

RZ-DPSK signal energy does not distribute equally as in the case of NRZ-DPSK. Most of it concentrates on only a fraction of the bit duration, while reduces nearly to zero for the rest of the bit period. This large energy fluctuation makes the signal more susceptible to fiber non-linearity and signal detection more difficult.

The carrier suppression (CS) technique can also be used in conjunction with RZ-DPSK modulation to produce the CSRZ-DPSK signal, which can be demonstrated as one of the most attractive modulation formats in high spectral efficiency wavelength division multiplexing (WDM) and dense WDM (DWDM) systems.

It is due to CSRZ-DPSK's higher energy and spectral efficiency, which increases tolerance to fiber non-linearity-induced impairments, CD, and polarization-mode dispersion (PMD) as compared to its RZ-DPSK counterpart. The CSRZ-DPSK, RZ-DPSK, and NRZ-DPSK spectra are shown together in Figure 4.11 for comparison.

The CSRZ-DPSK signal can be generated by a transmitter with a scheme similar to that of CSRZ-ASK, as shown in Figure 4.13. The main difference is that, in the CSRZ-DPSK transmitter,

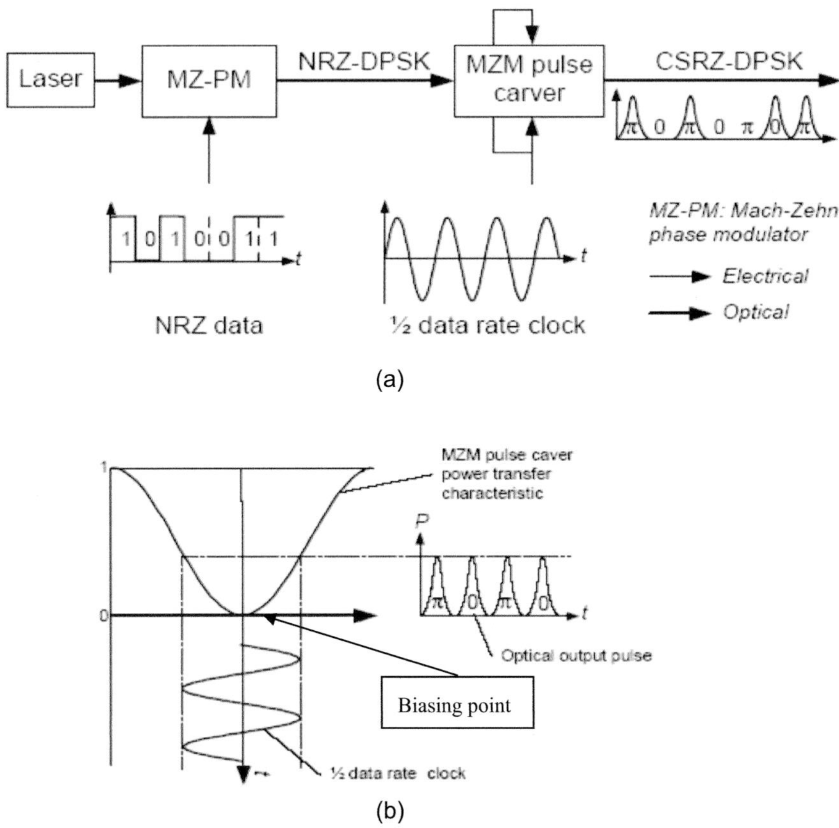

FIGURE 4.13 (a) Block diagrams of CSRZ-DPSK transmitter, (b) optical transfer characteristics of optical MZIM modulator and generation of optical pulses with alternative optical phases by driving the dual-drive MZIM with a 2 V_π voltage swing and biasing at the minimum transmission bias point.

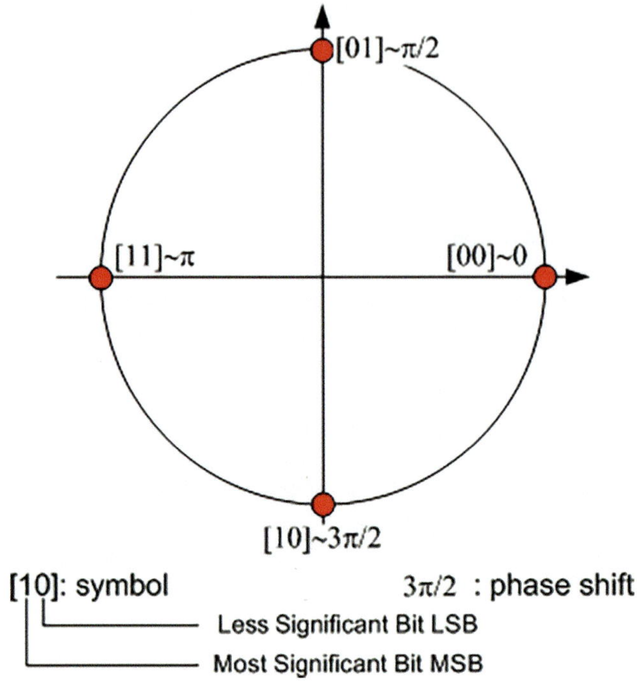

FIGURE 4.14 DQPSK signal constellation.

a PM replaces the IM used in the CSRZ-ASK transmitter. The receiver for CSRZ-DPSK has the same structure as that of the NRZ-DPSK scheme.

To increase transmission bit rate without suffering bandwidth requirements, one can code more than one bit into a data symbol. DQPSK modulation is the first step in the realization of this idea.

A signal constellation or signal space (see Figure 4.14) is the best way to represent a DQPSK signal, where the points representing phase-modulated signals are located along two orthogonal axes called I and Q (for in-phase and quadrature components, respectively). Two data bits $\left[D_0 D_1\right]$ are first pre-encoded into a symbol, which is then is encoded into phase shift that can take one of four values $\left[0, \pi/2, \pi, 3\pi/2\right]$ depending on the bit combination it represents. The DQPSK symbol rate is thus equal to only half of the bit rate. Intuitively, one can say that with the same bandwidth available, DQPSK can offer twice the transmission capacity compared to ASK and binary DPSK counterparts.

The DQPSK signal can be generated by an optical transmitter, as shown in Figure 4.15. This structure consists of two MZMs connected in parallel. A $+\pi/2$ phase shift is introduced in one of these MZM, making optical signals in two paths orthogonal to each other. A pre-coder encodes user data in accordance to the differential rule to generate I and Q driving voltages, which then modulate the carrier's phase in the two optical paths. Modulated carrier components are then combined at the output of the MZI. If the two normalized driving signals are denoted by I and Q, respectively, then the output signal is:

$$E_{\text{output}} = I \cos 2\pi f_c t + Q \sin 2\pi f_c t \tag{4.3}$$

where f_c is the frequency of the optical carrier. The coding and mapping bits $\left[D_1, D_0\right]$ into I and Q and signal constellation points follow the rule (Table 4.1):

DQPSK receiver uses two sets of MZ delay interferometers (DI) and balance receivers to detect in-phase (I) and quadrature-phase (Q) components of the received signal. Each set is similar to the one used in the NRZ-DPSK receiver. There are, however, two main differences: first, the delay introduced in the first branches of interferometers is now replaced by the symbol duration T_s;

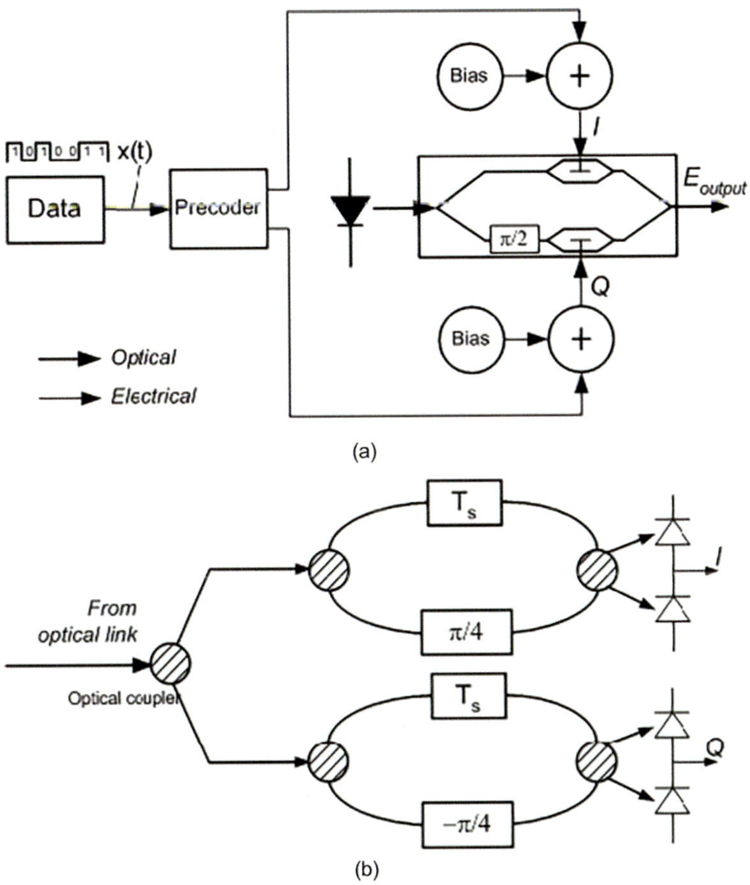

FIGURE 4.15 Parallel structure of a DQPSK transmitter, Ts=symbol duration (a) optical transmitter employing two pparallel MZMs plus one optical phase shifted inline with MZM in one path (QPSK modulator); (b) Reception of modulated lightwaves generated and transmitted of otput of transmitter given in (a).

TABLE 4.1
DQPSK Signal Bit-Phase Mapping

D_1	D_0	I	Q	Shifted Phase (rad)
0	0	0	0	0
0	1	0	1	$\pi/2$
1	1	1	1	π
1	0	1	0	$3\pi/2$

second, the phases of the signal in the second branches are shifted by $+\pi/4$ and $-\pi/4$ for I and Q components, respectively. These additional phase shifts are needed to separate two orthogonal phase components I and Q.

Figure 4.16a shows the spectrum of 40 Gbps NRZ-DQPSK signal with the single-sided bandwidth of the main lobe equal to 20 GHz, which is only half of the transmitted bit rate. The spectra of the RZ-DQPSK signal, as shown in Figure 4.16b, are much broader with strong harmonics beside the main lobe.

Despite numerous advancements in optical modulation techniques, the number of levels encoded in a signal symbol falls far behind the 256 or 1024 achieved in microwave modulation schemes.

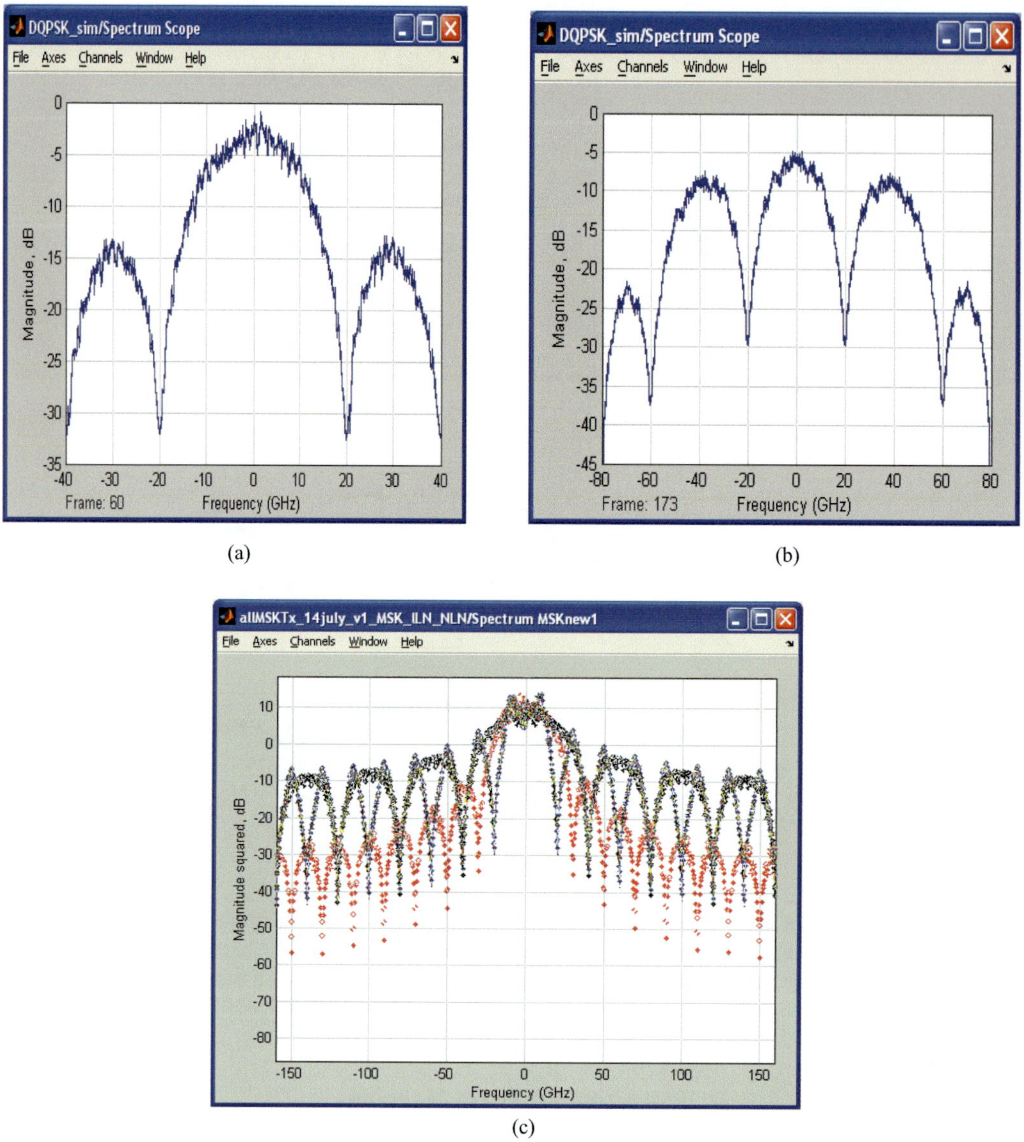

FIGURE 4.16 Optical spectra of 40 Gbps (a) NRZ-DQPSK, (b) 50% RZ-DQPSK, (c) DPSK as compared with MSK (light color curve).

The phase noise associated with optical sources and OAs has hindered the use of phase-related modulation schemes to current fluctuations in the photodetection (PD), hence the degradation of the BER. The differential phase demodulation process based on the phase comparison of two consecutive symbols requires the phase to remain stable during two symbol periods. Thus, narrow linewidth lasers are critical for phase-modulated systems. It has been shown that to achieve a power penalty <1 dB, $\Delta v / B < 1\%$ with Δv and B, the laser linewidth and system bit rate, respectively. In optical transmission systems where OAs are used, the ASE noises intermingle with the fiber nonlinear phase effect, thus enhancing the non-linear phase noise. While SPM-induced nonlinear phase noise is the dominant phase noise in single optical channel systems, XPM-induced phase noise is the main phase noise for multi-channel (WDM) systems.

Significant phase noises caused by optical sources and OA have prevented optical DPSK schemes from having many levels in each symbol. To increase the number of levels in the signal space, and thus the number of bits per symbol higher than four, one of the most preferred solutions is a combination of DQPSK and ASK.

Hayase et al.[3] have demonstrated experimentally a 30 Gbps, 8-states-per-symbol optical modulation system using a combined ASK and DQPSK modulation scheme, as shown in Figure 4.17. This system maps three bits into a symbol, creating a transmission bit rate three times higher than the symbol rate. The transmitter consists of two cascaded PMs and an amplitude modulator (AM). The first PM, driven by data bit D_0, creates 0 and π phase shifts, while the second, driven by D_1, forcing two further phase shifts 0 and $\pi/2$, the quadrature phase to generate four distinct phases of DQPSK signal. The AM, driven by D_2 bit, shifts the four phases between two amplitudes to create a total of eight signal points.

At the receiver side, optical signals are detected in both amplitude and differential phase. An ASK demodulator detects the D_2 bit, while a DQPSK demodulator recovers D_1 and D_0 bits. Sekine et al.[4] reported experimentally a similar scheme, but with 4-bits $\left[D_3, D_2, D_1, D_0\right]$ mapped into a symbol: $\left[D_1, D_0\right]$ bits are used to generate a "normalized" DQPSK signal, while $\left[D_3, D_2\right]$ bits manipulate the amplitude of this DQPSK signal between four concentric circles. Thus, a 16-ary MADPSK signal can be generated, offering 40 Gbps bit rate with a symbol rate of only 10 Gbauds.

4.2.4 Comparison of Different Optical Modulation Formats

Different amplitude and phase optical modulation formats are summarized in Table 4.2. In most cases, NRZ-ASK parameters are used as references. From the comparison, it can be concluded that MADPSK takes advantage of other modulation formats in terms of spectral efficiency and ability to significantly increase transmission bit rate, which are very, if not most, important parameters for an optical transmission system. It is also expected that MADPSK inherits good properties (and of course, the bad ones, if any) from two basic ASK and DPSK modulation formats. The spectra of different modulation formats, such as NRZ-DQPSK, 67% CSRZ-DQPSK, 100 Gbps CSRZ 16ADPSK ($M=16$), are shown in Figure 4.18 indicating their spectral efficiency. The RZ features of the modulation format result in a wider bandwidth of the signal spectrum. The 16 ADPSK offers higher efficiency, and filtering is required to minimize the noise and optimize the distinct levels. The bandwidth remains the same as for the CSRZ-DQPSK format.

Table 4.2 outlines the pros and cons of different modulation formats and the suppression or non-suppression of the carrier for SPC transmission systems.

4.2.5 Multilevel Optical Transmitter

In this section, several optical transmitter structures used to generate the DQPSK signal are described. It is necessary because a novel optical transmission system will be developed based on the DQPSK modulation format. All these structures have MZIM as their base component, which can be single- or dual-electrode structures.

Unlike single-drive MZIM, a dual-drive electrode structure with two traveling wave RF electrodes can modulate the phase of optical signals in both branches, enabling push-pull operation. Interference at the output of dual-drive MZIM will produce phase-modulated signal. However, when the effects of phase modulation in the two branches are exactly equal but opposite in sign, the output signal becomes intensity-modulated. In this manner, dual-drive can be used for both phase and intensity modulation. The relationship between input and output signals of a dual-drive MZIM can be described by:

$$E_{\text{output}} = \frac{E_{\text{input}}}{2}\left[\exp\left(j\pi\frac{V_1(t)}{V_\pi}\right) + \exp\left(j\pi\frac{V_2(t)}{V_\pi}\right)\right] \tag{4.4}$$

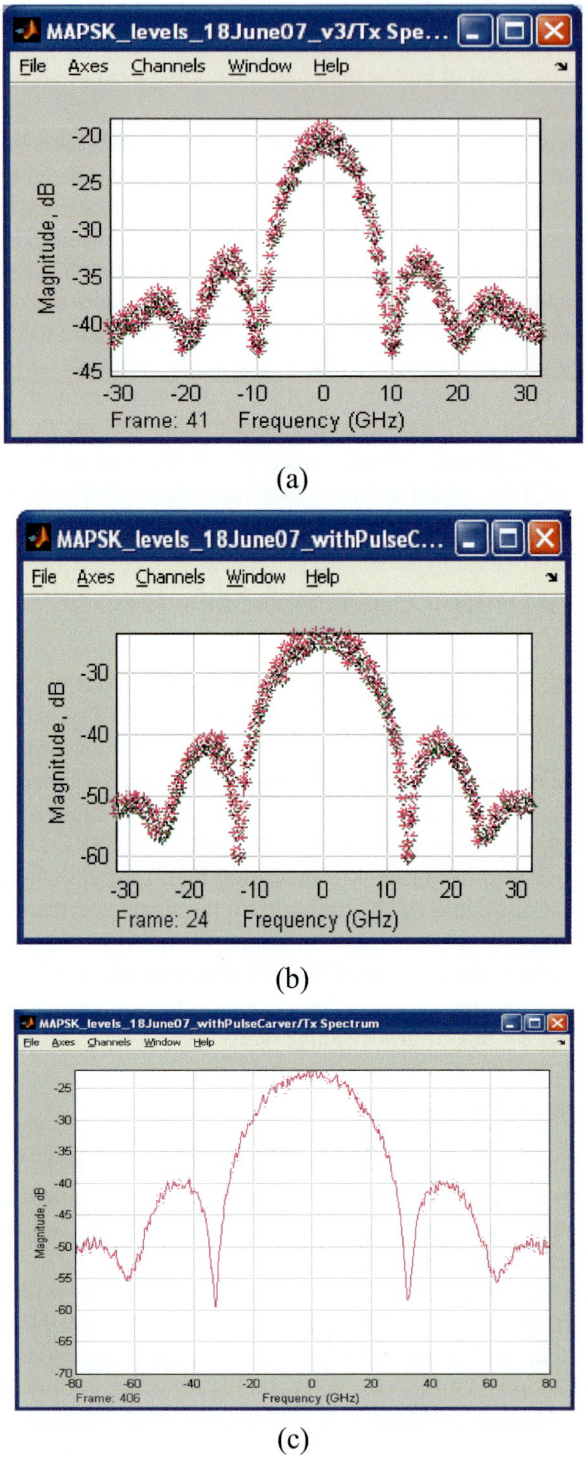

FIGURE 4.17 The 8-ary APSK modulation experimental configuration (extracted from 0), 10 GHz clock assign synchronization of symbol rate, data modulator, and quadrature phase shift in optical domain using the PM, two balanced receivers for differential phase shift detection and direct detection for amplitude detection. (a) The 8-ary ASK-DPSK signal, (b) transmitter configuration using multiple cascaded MZIM and phase modulator, and (c) receiver configuration employing dual MZDI for phase detection, with balanced photodetector and a PD for amplitude detection.

TABLE 4.2
Comparison of Different Optical Modulation Formats for SPC Transmission

Mod. Format	Spectral Width	Receiver Sensitivity	Resilience to Dispersion	Resilience to SPM	Resilience to XPM	Current Transmission Bit Rate Limits
NRZ-ASK	Narrowest	Lowest	Worst	High	High	112 Gbps
RZ-ASK	2xRZ-ASK (at 50% duty ratio)	Higher NRZ-ASK	Higher NRZ-ASK	High	High	224 Gbps
CSRZ-ASK	Same as RZ-ASK	Higher NRZ-ASK	Higher NRZ-ASK	High	High	40 Gbps
NRZ-DPSK	Same as NRZ-ASK	3 dB better than NRZ-ASK	Higher NRZ-ASK	Worse than ASK	Worse than ASK	112 Gbps
RZ-DPSK	Same as RZ-ASK	3 dB better than RZ-ASK	Higher NRZ-ASK	Worse than ASK	Worse than ASK	224 Gbps
CSRZ-DPSK	Same as CSRZ-ASK	3 dB better than CSRZ-ASK	Higher CSRZ-ASK	Higher than ASK	Higher than ASK	112 Gbps
DQPSK	1/2 DPSK	1.5 dB better than ASK	Higher*	Worse than ASK	Worse than ASK	448 QPSK
MADPSK	1/M DPSK					$(\log_2 M)\times$DPSK - expected

where $V_1(t)$ and $V_2(t)$ are driving voltages applied to the modulator, V_π is the voltage required to provide a π phase shift of the carrier in each branch of MZIM. It is also noted that unlike in single-drive MZIMs, the chirp effect does not exist in dual-drive MZIMs.

The transmitter structure, as shown in Figure 4.15, is called parallel type. It is only one of the several structures, which can be used specifically for generating DQPSK signal, namely parallel structure, serial structure, single PM structure, and dual-drive MZIM structure. These terms indicate the structuring of MZIMs whether they are connected in tandem, parallel, or just a pure PM with a single electrical drive port. Another possible modulation structure for generating optical multi-level amplitude and phase can be shown in Figure 4.19. It is an electro-optic transmitter of the serial type, an MZIM-generating in-phase component, and a PM-generating quadrature component are connected in tandem. Pre-encoded data generate two signals: one is used for driving the MZIM and the other for driving the PM. Usually, the square shape of the pre-encoded waveforms is replaced by the raised cosine before being fed into the modulators. Furthermore, the biasing conditions and the amplitude of the modulators can be used to generate 33%–67% pulse width RZ formats. It is also noted that the pulse shape would also follow a $\cos^2$ profile due to the property of the intensity modulator. This transmitter would suffer the chirping effects due to the rise time of the electrical driving signals, contributing to the distortion of the lightwave signals, in particular when switching between the lowest level to the highest level.

A typical multi-level eye diagram of a multi-level modulation scheme is shown in Figure 4.20. Accuracy and noise levels are critical to ensuring the degree of eye-opening and the distance of the eye levels. Due to the threshold of the nonlinear effects in the transmission fibers, the maximum level of a multi-level signal must not exceed a certain limit to maintain a maximum eye-opening in multi-level amplitude and phase. Thus, minimum noise contributions due to electronics and optical energy at the reception subsystems as well as at the transmitter must be minimized.

The single PM structure (Figure 4.21) uses only one MZIM as a phase modulator. Pre-encoded data are added to create a single driving voltage. One of the two pre-encoded data is amplified, and along with the other signal represents four positions of the DQPSK signal.

The dual-drive MZIM structure (Figure 4.22) uses two driving voltages for modulating the optical carrier phase in two branches of an MZIM. Data are first pre-encoded following the differential rule and then used to create driving voltages $V_1(t)$ and $V_2(t)$ corresponding to the signal constellation points.

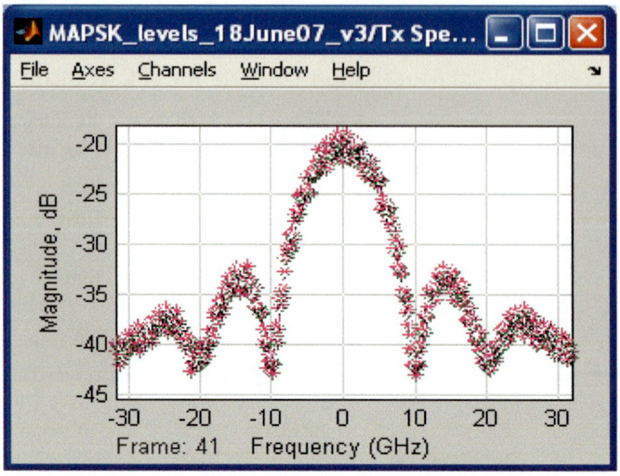

(a)

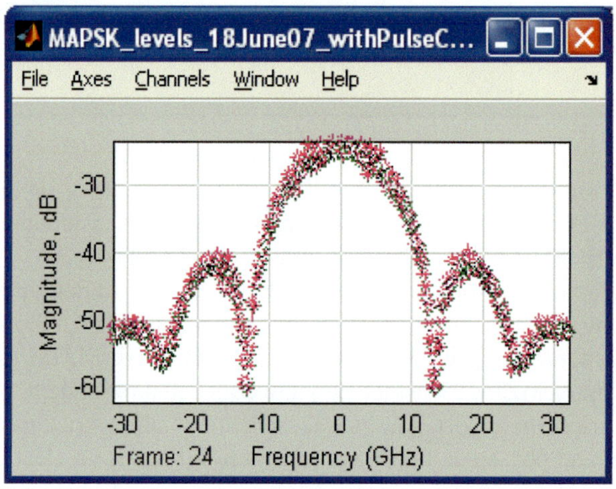

(b)

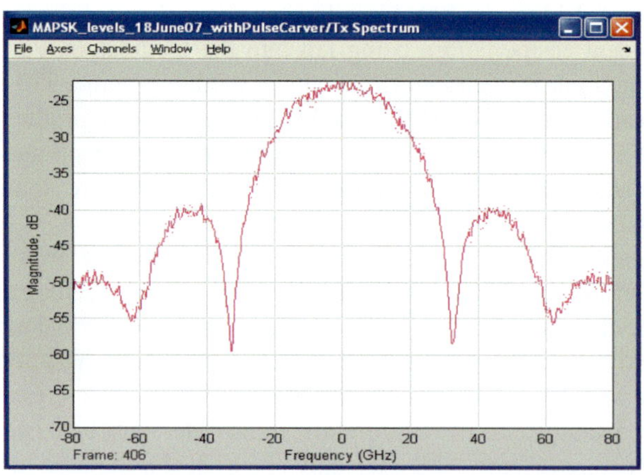

(c)

FIGURE 4.18 Optical spectra of 40 Gbps (a) NRZ-DQPSK, (b) 67% CSRZ-DQPSK, (c) 100 Gbps CSRZ-16ADPSK with an optical filter to minimize noises.

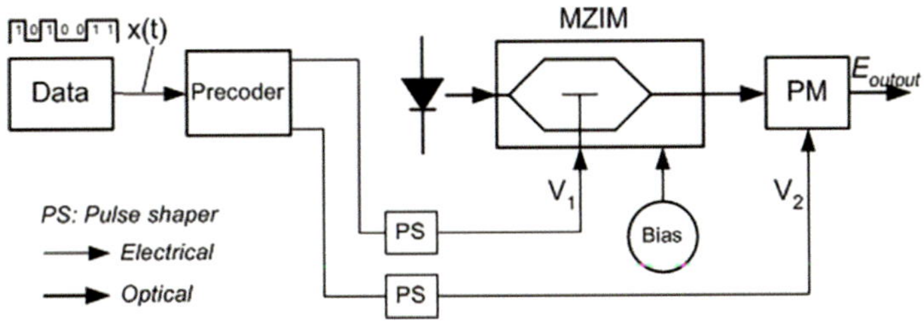

FIGURE 4.19 Cascade PM and MZIM for DQPSK signal generation with laser in CW mode as an external source.

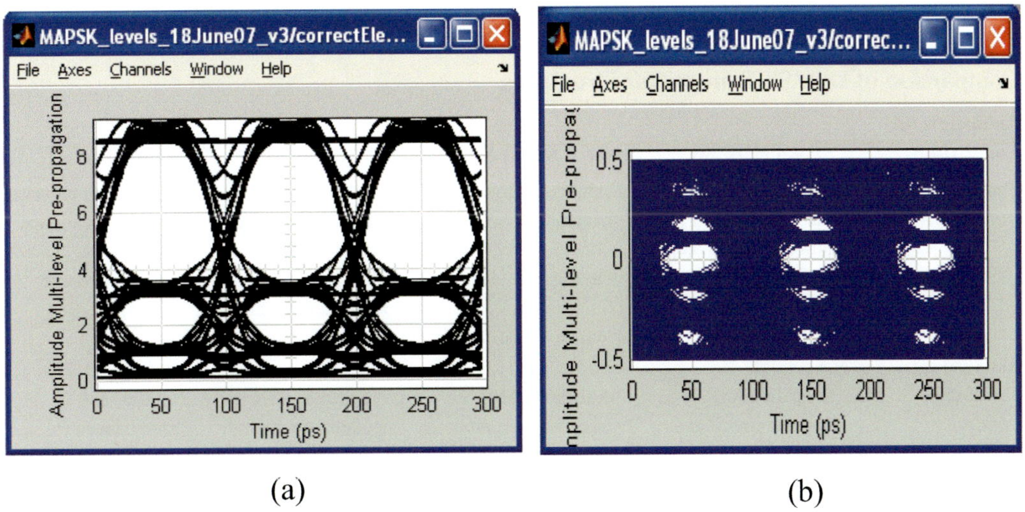

FIGURE 4.20 Eye diagram – amplitude detection section of 40 Gbps after transmission of panel (a) 5 km SSMF transmission under direct detection; and (b) quadrature phase – coherent detection without digital signal processing.

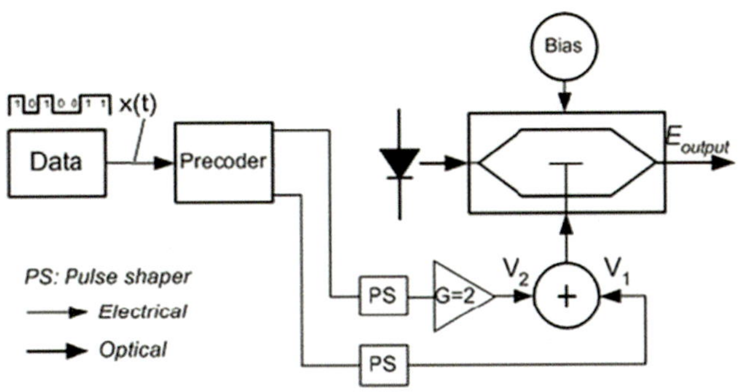

FIGURE 4.21 Single drive PM structure using MZIM.

In the four-transmitter structures described above, the parallel and serial structures are the most complex and difficult to implement because they have discrete devices connected together. The dual-drive MZIM and single PM structures, on the other hand, are much simpler because they require fewer discrete devices. Furthermore, as will be discussed in the next section, dual-drive MZIM can be configured to work as both phase and amplitude modulators at the same time; it can easily generate not only DQPSK but also MADPSK signals. Thus, a dual-drive MZIM is the

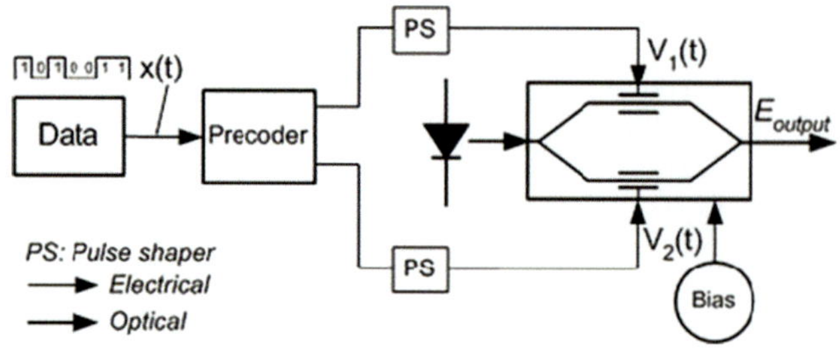

FIGURE 4.22 Dual-drive MZIM structure.

TABLE 4.3
Comparison of DQPSK Transmitter Structures

Parameters for Comparison	Parallel MZIM	Serial MZIM & PM	Single PM	Dual-Drive MZIM
Complexity of circuit design	Complicated in the matching of ultra-high frequency electrical paths; high insertion loss. Flexible in biasing	Complicated in the matching of ultra-high frequency electrical paths; high insertion loss Flexible in biasing	Simple in photonic but complicated in the realization of ultra-high frequency signal connections	Simplest but required multi-level voltage switching at symbol rate (microwave speed)
Ability to create MADPSK signal	Not possible. A separate ASK modulator required.	Not possible. A separate ASK modulator required	Impossible. A separate ASK modulator required	Dual-drive MZIM acts as ASK and DPSK simultaneously

principal part of the MAPSK transmission system. Table 4.3 provides a comparison between different transmitter structures; the dual drive modulator is outstanding as a combined amplitude and phase switching between the states of a multi-circular constellation.

4.2.5.1 MADPSK Photonic Transmission

4.2.5.1.1 MADPSK Transmitter

In general, the structures of the MADPSK can be given as shown in Figure 4.23. A model has been constructed for investigating the performance of systems based on the MADPSK modulation format. It consists of a signal coding model, transmitter model, receiver model, and transmission and dispersion compensation fiber models.

The 16-ary MADPSK signal model described in Section 4.2.6.1 will be used. To balance the ASK and DQPSK sensitivities, ASK signal levels are preliminary adjusted to the ratio $I_3/I_2/I_1/I_0 = 3/2/1.5/1$, as shown in Figure 4.24. These level ratios can be determined from the signal-to-noise ratio at each separation distance of the eye diagram or q-factor. The noise is assumed to be dominated by the beat noise between the signal level and the ASE noise.

The transmitter model shown in Figures 4.23 and 4.24 is used to produce the 16-ary MADPSK signal. It consists of a DFB[5] (distributed feedback Bragg) laser source generating CW light (carrier), whose lightwaves are fed into a cascade of two optical modulators, one dual drive MZIM amplitude modulator and one single drive MZM phase modulator. The modulated lightwaves vary in both phase and amplitude, thus realizing complex modulation. Each four bits of user data $[D_3D_2D_1D_0]$ are first grouped into a symbol and then encoded to generate two electrical driving signals $V_1(t)$ and $V_2(t)$ under which the amplitude and phase of the carrier in two optical paths of the dual-drive MZIM will be modulated to produce the NRZ 16-ary MADPSK signal. The following RZM-PC converts the NRZ pulse train into RZ one to minimize the effects of ISI.

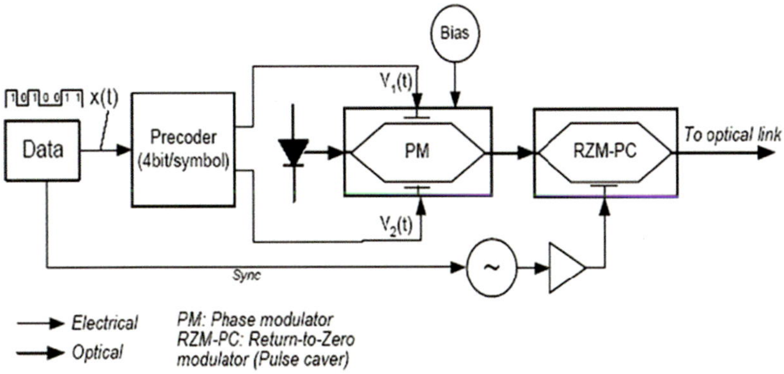

FIGURE 4.23 Schematic of MADPSK transmitter. RZM=return to zero modulator, PM=phase modulator, electrical optical path distinguished by light dark and dark lines.

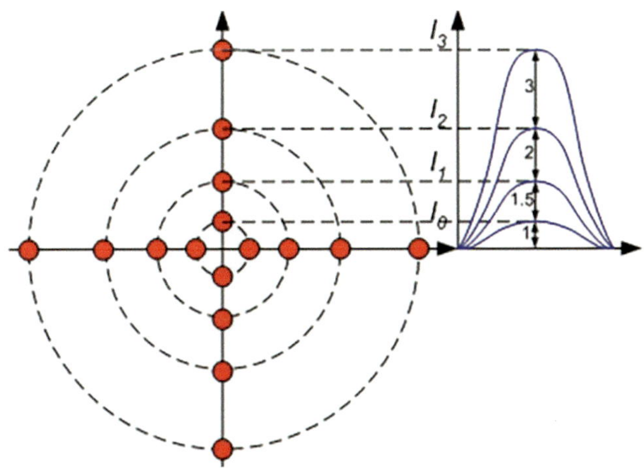

FIGURE 4.24 ASK inter-level spacing in MADPSK.

4.2.5.1.2 MADPSK Receiver

The complex vector signals can be received by the receiver model shown in Figure 4.25. The complex signal constellation is shown in Figure 4.24. It consists of two-phase demodulators, an amplitude demodulator, and a data multiplexer MUX. Two-phase demodulators are used for extracting $[D_1D_0]$ bits and work exactly in the same way as ones in the DQPSK receiver described in Section 4.2.5. The amplitude demodulator (AD) is used for detecting four amplitude levels of the MADPSK signal. It is a well-known direct-detection scheme consisting of a photodiode followed by an electronic receiver. The amplitude-modulated signal is then threshold detected in association with a clock recovery circuit to recover two bits $[D_3D_2]$. These two bits $[D_3D_2]$ are interleaved with two bits $[D_1D_0]$ by the MUX to reconstruct the original binary data stream. It is noted that the demodulators using optical modulation structure indicate the demodulation in the optical domain, which is much easier than detected in the electronic domain and demodulating in the electrical domain due to the limitations in the electronic bandwidth of devices. So, the optical demodulation of optically modulated lightwaves can be demodulated to optical sequences, which are then received by photodetector and processed in electronic DSP subsystems.

The detected amplitudes are derived from the amplitude of the complex vectors, amplitude and phase, which can be resolved from the vertical and horizontal amplitudes. The right side of Figure 4.24

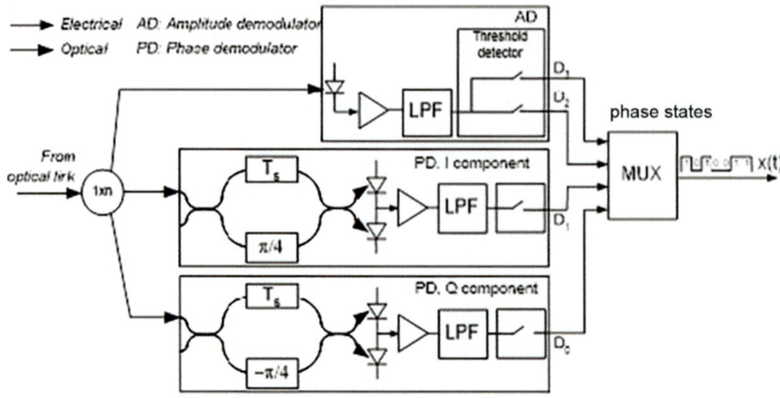

FIGURE 4.25 Amplitude direct detection and photonic phase comparator with balance receiver for MADPSK demodulation. Phase shifter in the MZDI of π/4 and −π/4 generating two X-Y orthogonal components. LPF=low pass filter, Mux=multiplexer, AD=amplitude detection by amplitude comparator, D1, D2: digital bit 1 and 2, T_s=sampling interval.

depicts such amplitude distribution at different amplitudes of the constellation points displayed on the left-hand side of the figure. This is used to obtain the amplitude distribution of only the vertical amplitude of the received constellation point. The variation of this directional amplitude varies according to the amplitude variation with respect to time. Similarly, the amplitude and its variation concerning the amplitude direction can be obtained. The combined vertical and horizontal amplitudes generate a vector with the amplitude and phase of the complex constellation point.

Phase shifters in the MZDI of π/4 and −π/4, incorporated in the receiver, are used to generate two X-Y orthogonal components. The delay $T_s/2$ is used to differentiate the phases of consecutive bits of the orthogonal components. Without this delay, the detection is for QPSK and not in differential mode.

The complexed vectorial receiver can be structured as shown in Figure 4.25, which consists of a pair of optical hybrid couplers in Mach-Zehnder delay structure with a π/4 phase and a unit sampling time for differential phase. A balanced photodetector is used to obtain 3 dB improvement and minimize phase noise. A section for amplitude detection can also be shown in the upper most branch. Thus, both amplitude and differential phase can be detected for DQPSK-modulated signals.

Multi-level amplitude can be processed in the digital domain by the DSP. Amplitude discrimination can be easily implemented in the DSP (digital signal processor) using the digital amplitude and the amplitude distribution at sampling intervals.

4.2.5.1.3 MADPSK Single Dual-Drive MZIM Transmitter

The main reason for explaining why the dual-drive MZIM structure has attracted our attention in this chapter is that it can play the role of both amplitude and phase modulators simultaneously, which is impossible with other transmitter structures. This means that to generate a MADPSK signal, employing separate phases and amplitude modulators is not required as it has been implemented in the works of Seikine et al.[4] 0 and Hayase et al.[3] This section describes a method for generating a 16-ary MADPSK signal using this dual-drive MZIM structure.

The 16-ary MADPSK signal constellation of interest is shown in Figure 4.26. It is a combination of a 4-ary ASK and a DQPSK signal, with four bits $[D_3, D_2, D_1, D_0]$ mapped into a symbol, among them two bits $[D_1D_0]$ are coded into four phases $[0, \pi/2, \pi, 3\pi/2]$and two bits $[D_3D_2]$ coded into four amplitude levels $[I_3, I_2, I_1, I_0]$. With the use of a balanced receiver and DI, solely available today in practical optical phase demodulator, the MADPSK signal produces clear DQPSK eye patterns centered at zero-voltage decision level only when constellation points are positioned in a radial pattern.

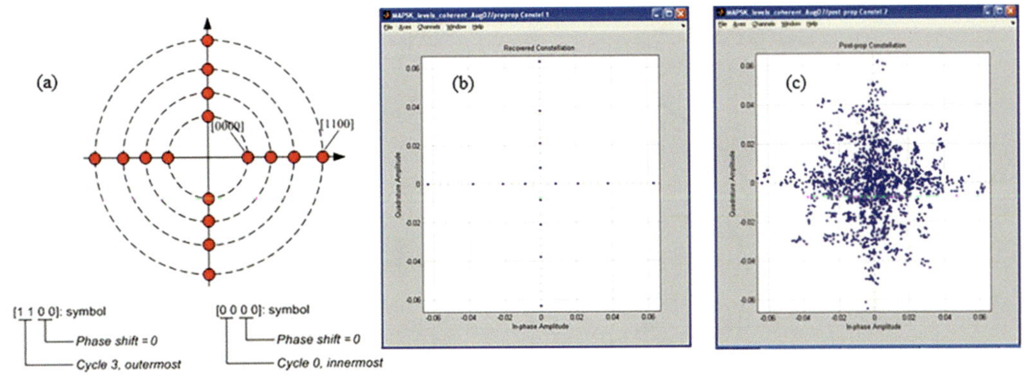

FIGURE 4.26 The 16-ary MADPSK signal bit-phase mapping (a) design, (b) Simulink scattering plot before transmission, (c) after 200 km transmission with 2 km mismatch of fiber dispersion factors.

Recall that the signal at the output of the dual-drive MZIM can be represented as:

$$E_o = \frac{E_i}{2} e^{j\phi_1} + \frac{E_i}{2} e^{j\phi_2} \tag{4.5}$$

with $\phi_1 = \pi \frac{V_1(t)}{V_\pi}$, $\phi_2 = \pi \frac{V_2(t)}{V_\pi}$, where E_i and E_o are electrical fields of the input and output optical signal, respectively; $V_1(t)$, $V_2(t)$ are driving voltages applied to the modulator, and V_π is the voltage required to provide a π phase shift for the carrier in each MZIM branch.

Equation 4.7 suggests that with a properly chosen input signal E_{input} and driving voltages, $V_1(t)$, $V_2(t)$, all signal points of the constellation in Figure 4.26 can be constructed from two phasor signals $\frac{E_i}{2} e^{j\phi_1}$ and $\frac{E_i}{2} e^{j\phi_2}$. Indeed, if E_i is chosen to be equal to the electrical field corresponding to signal points in the largest circle of the constellation, then a constellation signal point E_{output} with the phase shift θ_i in the circle n can be found as a sum of two vectors $\frac{E_i}{2} e^{j\phi_{ni1}}$ and $\frac{E_i}{2} e^{j\phi_{ni2}}$, where $\phi_{ni1} = \theta_i + \text{arc}\cos\left(E_n/E_{\text{input}}\right)$, $\phi_{ni2} = \theta_i - \cos^{-1}\left(E_n/E_i\right)$. The subscripts i and n denote the phase position and the order of the circle of interest. Figure 4.27 illustrates the relationship between E_i, E_0, ϕ_{ni1} and ϕ_{ni2}. For simplicity the signal point is chosen with $\theta_i = 0$.

By substituting ϕ_1 and ϕ_2 into Equation 4.7, the driving voltages for this point can be obtained (Figure 4.28)

$$V_{ni1}(t) = \frac{V_\pi}{\pi}\left[\theta_i + \cos^{-1}\left(E_n/E_i\right)\right]; \quad V_{ni2}(t) = \frac{V_\pi}{\pi}\left[\theta_i - \cos^{-1}\left(E_n/E_i\right)\right] \tag{4.6}$$

4.2.5.2 MADPSK Transmission Performance

Under performance evaluation, the following main parameters are investigated: (i) System bit error rate BER versus SNR: an analytical solution for system BER will be determined, and BER will be computed against different SNR values and bit rates. System BER versus SNR will also be obtained through system simulation and cross-checked with results obtained analytically. Graphs of BER versus SNR will be plotted; (ii) System BER versus receiver sensitivity: BER versus receiver sensitivity will be obtained analytically and through simulation, with results cross-checked. Graphs of BER versus receiver sensitivity will be plotted; (iii) Dispersion tolerance: transmission over fibers of types ITU-G652, ITU-G.655, and LEAF, with their corresponding dispersion factors, will be considered. Graphs of the power penalty due to the dispersion as compared with back-to-back transmission will be plotted against the dispersion factor in ps/nm; (iv) Tolerance to other system

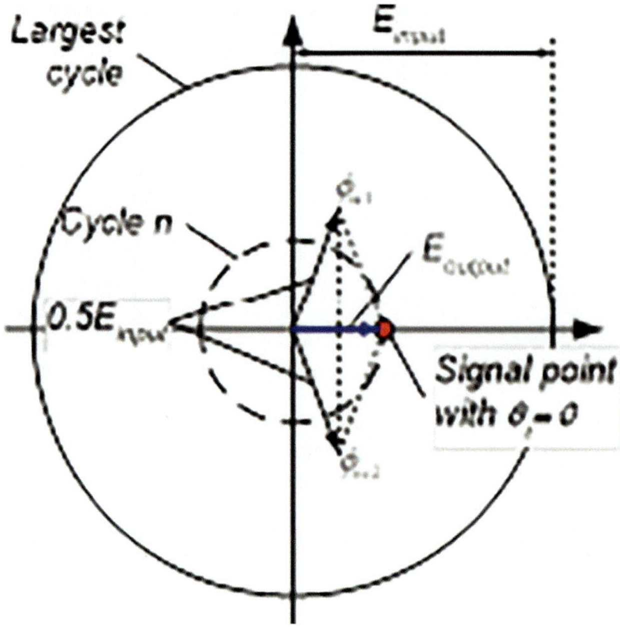

FIGURE 4.27 Relationship between E_i, E_o, ϕ_{ni1} and ϕ_{ni2} using phasor representation.

impairments: similar to dispersion tolerance, the power penalty due to other impairments such as laser and OA phase noise, and receiver phase error will be investigated. Corresponding graphs will be plotted.

Performance evaluation is conducted under the effects of the following conditions or contexts: (i) Different pulse shapes: raised cosine, rectangular, Gaussian; (ii) Modulation formats: NRZ, RZ, CSRZ; (iii) ASE noise of optical amplifiers; (iv) Transmitter impairments: laser noise; (v) Receiver impairments: phase error of DI-based phase demodulators; (vi) Change of ASK inter-level spacing; (vii) Optical and electrical filtering; and (viii) Multi-channel environment: system performance in combination with DWDM technology can be further investigated.

4.2.6 A System Modeling by MATLAB Simulink©

The following simulation models have been built on the MatLab Simulink platform to prove the working principles and investigate the performance of systems using the optical MADPSK modulation. A transmitter is simulated to generate a 16-ary MADPSK signal, and a receiver is to reconstruct the original binary signal. These models run over a simulated single-mode optical fiber. Laser chirp, OA phase noise, nonlinearities, CD, PMD, and other impairments will be involved in later stages to evaluate different performance characteristics of the modulation format: system BER, receiver sensitivity, and power penalties due to several impairments.

The phases and the driving voltages for creating signal points of the 16-ary MADPSK constellation can be computed and tabulated in Table 4.4.

4.2.6.1 MATLAB Simulink™ Transmitter Model

The MatLab Simulink model of the system is shown in Figure 4.29. The transmitter model using the dual-drive MZIM structure is shown in Figure 4.30. The purpose of blocks is as follows: (i) "User data and ADPSK pre-coder" block generates a pseudo-random data sequence to simulate the user data

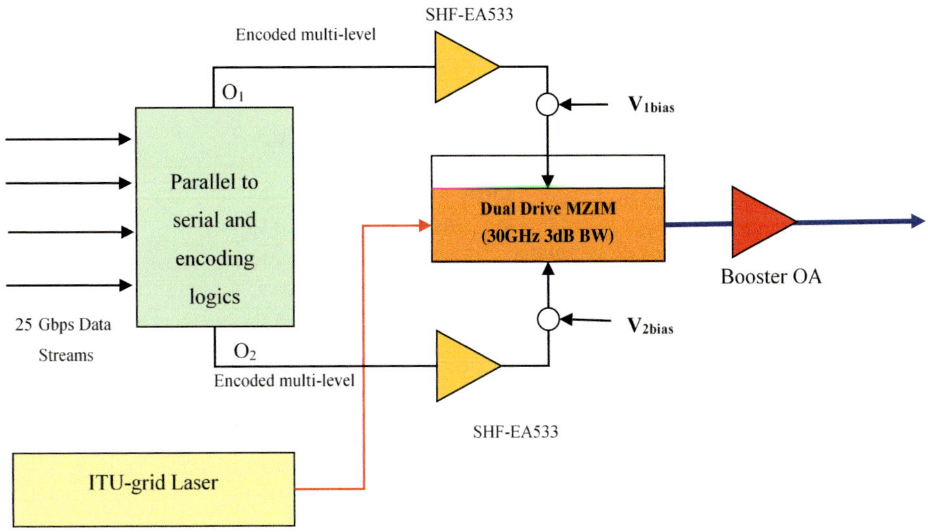

(a)

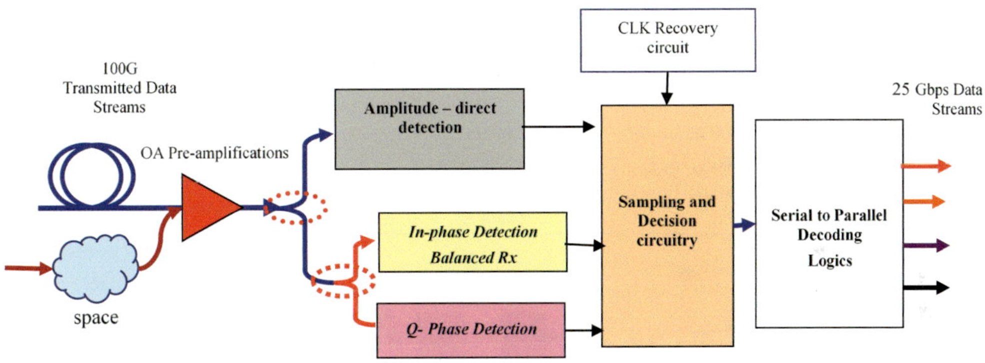

(b)

FIGURE 4.28 Schematic diagram of the photonic transmitters and receivers for the 16ADQPSK transmission scheme (a) transmitter, (b) receivers with branches for detection of amplitude, in-phase and quadrature-phase components (a) photonic transmitter; (b) amplitude and phase detection of in-phase and quadrature components of QPSK modulated optical sequence.

stream and encodes each group of 4 data bits into a symbol. (ii) "Voltage driver 1" and "Voltage driver 2" blocks map pre-encoded data into driving voltages for modulating the amplitude and phase of the carrier in the dual-drive MZIM. (iii) Two "complex phase shift" blocks simulate the variation of optical phases of two lightwave paths of the dual-drive MZIM. (iv) The "Sum block" represents the combiner of the two Mach-Zehnder paths to give the output of MZIM. (v) "Gaussian noise generator" block simulates noise source. (vi) "Amplifier" block simulates optical amplifier.

The parameters of these blocks are stored in the program initialization in MATLAB and are associated with the running procedures of the Simulink platform.[6]

TABLE 4.4
Phase and Driving Voltages for 16-ary MADPSK Constellation

Positions	θ_i	V_{i1}	V_{i2}
		Circle 3	
1100	0°	0.0 V_π	0.0 V_π
1101	90°	0.5 V_π	0.5 V_π
1111	180°	1.0 V_π	1.0 V_π
1110	270°	1.5 V_π	1.5 V_π
		Circle 2	
1000	0°	0.2952 V_π	–0.2952 V_π
1001	90°	0.7949 V_π	0.2046 V_π
1011	180°	1.2947 V_π	0.7043 V_π
1010	270°	1.7944 V_π	1.2041 V_π
		Circle 1	
0100	0°	0.3919 V_π	–0.3919 V_π
0101	90°	0.8917 V_π	0.1078 V_π
0111	180°	1.3914 V_π	0.6076 V_π
0110	270°	1.8912 V_π	1.1073 V_π
		Circle 0	
0000	0°	0.4575 V_π	–0.4575 V_π
0001	90°	0.9573 V_π	0.0422 V_π
0011	180°	1.4570 V_π	0.5420 V_π
0010	270°	1.9568 V_π	1.0417 V_π

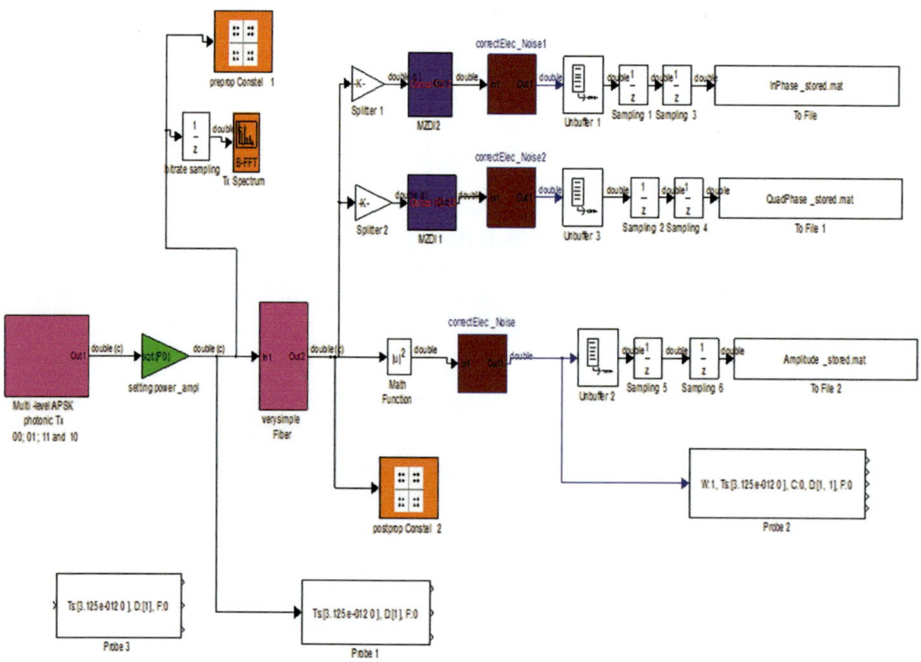

FIGURE 4.29 MATLAB Simulink photonic communication system model.

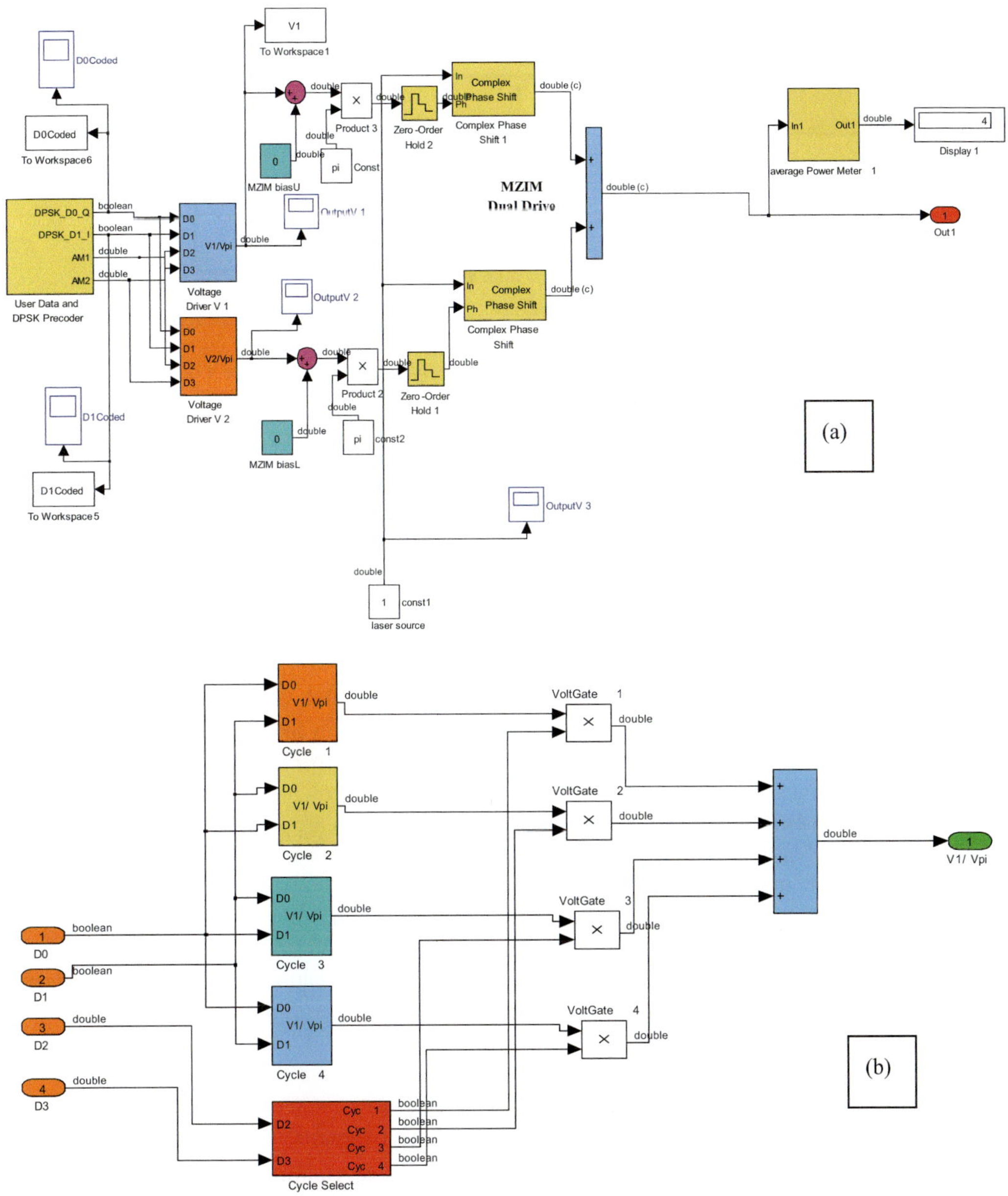

FIGURE 4.30 MATLAB™ Simulink model of MADPSK (a) transmitter and (b) logic precoder.[7]

4.2.6.2 Simulink Receiver Model

The receiver structure is shown in Figure 4.31. The functions of the blocks are as follows: (i) Each DI is simulated by a set of two "Magnitude-Angle" blocks, a "Delay block", and a "Sum" block. The "Delay block" stores the phase of the previous symbol, and the "Magnitude-Angle" blocks extract the phase and amplitude of present and the previous symbols, which will be used in the following different phase demodulation and detection operations. (ii) "Constant $\pi/4$" and "Constant $-\pi/4$", along with the next two "Sum" blocks, simulate extra phase delay in each branch of DI. (iii) Two "Cos" blocks and two "Product" blocks simulate two balanced receivers. (iv) "Amplitude

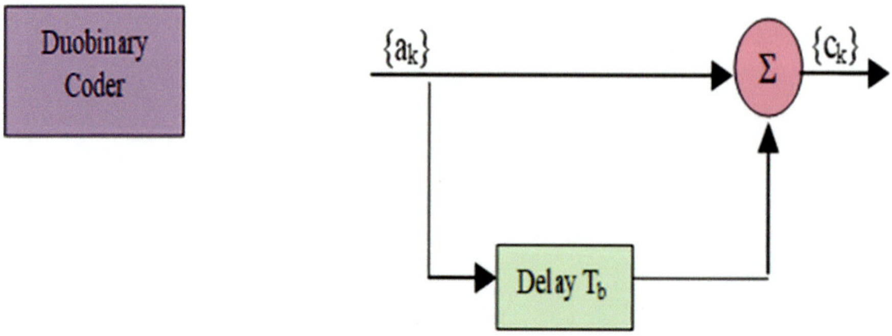

FIGURE 4.31 MATLAB Simulink model of MADPSK receiver (a) amplitude direct detection and (b) balanced receiver detection – in-phase and quadrature.

Detector D2 and D3" block simulates ASK detector for D_2 and D_3 bits. (v) Three "Analog Filter Design" blocks simulate electrical low pass filters. (vi) "Phase Detector D0_I" and "Phase Detector D1_Q" blocks simulate the threshold detectors for D_0 and D_1 bits (I and Q components of a DQPSK signal), respectively.

4.2.6.3 Dispersion-Compensating Optical Transmission

The propagation of optical signal in a fiber medium, which is both dispersive and nonlinear, is best described by the non-linear Schrödinger equation (NLSE).[2] For SPPC over free space, including vacuum and atmospheric turbulence layers, SFFTM (Split-step FFT method) is presented in Chapter 3. The compensation of the spatial dispersion through beam broadening is achieved using adaptive optics and DSP processing algorithms. Other parameters are explained below. The transmission fiber model, as shown in Figure 4.32, is used to simulate the propagation of optical signals and the impairments impacting system performance.

All characteristic parameters of the fiber medium along with optical input signals are taken by the "Matrix Concatenation" block and then processed by a MATLAB function that solves the NLSE using the split-step Fourier method.[8] The dispersion compensation fiber model has the same structure as the propagation fiber model, except that the sign of the propagation constant β_2 in the two models are opposite.

4.2.6.4 Transmission Performance

4.2.6.4.1 Signal Spectrum, Signal Constellation, and Eye Diagram

The spectrum of 40 Gbps 16-ary MDAPSK signal obtained by running the transmitter model is represented in Figure 4.33. As seen in the graph, the single-sided bandwidth of the main lobe equals 10 GHz. Numerically, it amounts to only one-fourth of the transmission bit rate, and from that it can be concluded that MADPSK is a high bandwidth-efficient modulation format.

Figure 4.34 exhibits the signal constellation recovered at the receiving end. Noise and nonlinear properties of fiber cause amplitude and phase fluctuations and scatter signal points around some mean value. The MADPSK eye diagram is shown in Figure 4.35 for the I component (the Q component should have a similar diagram). This eye diagram shows four amplitude levels associated with two-phase shifts 0 and π.

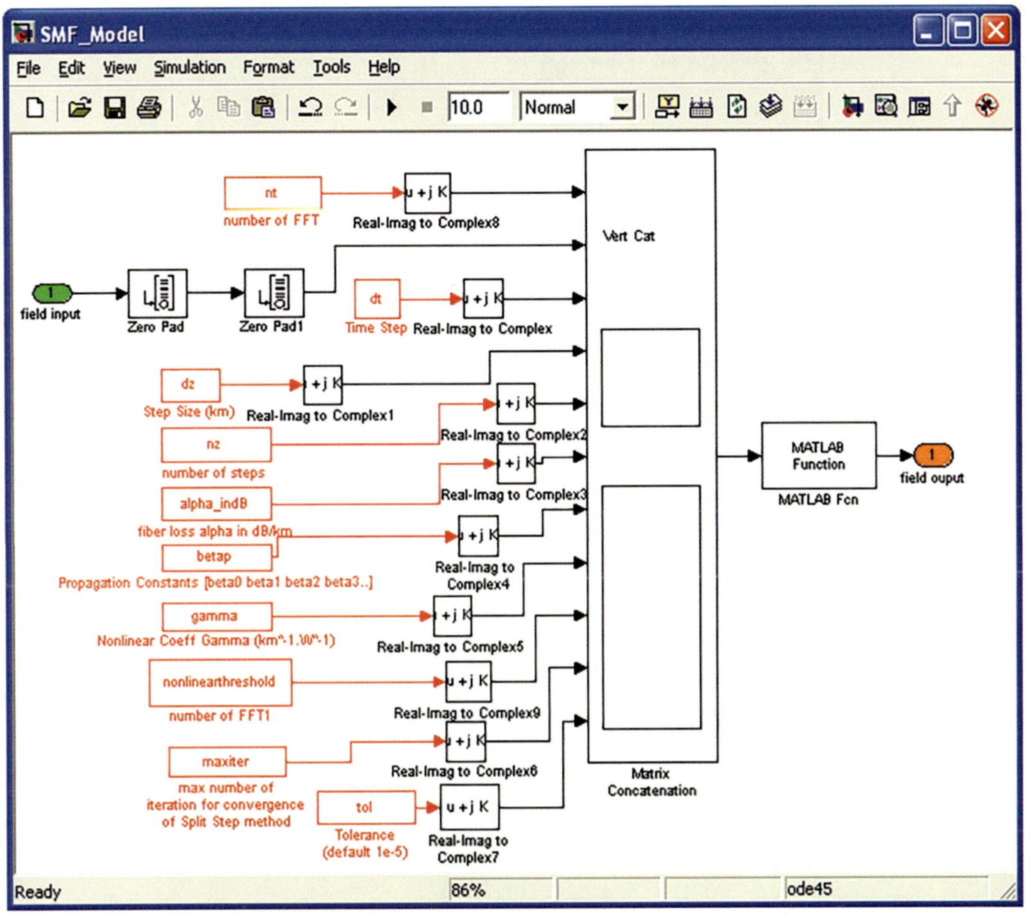

FIGURE 4.32 Input parameters for single mode fiber SSMF model.

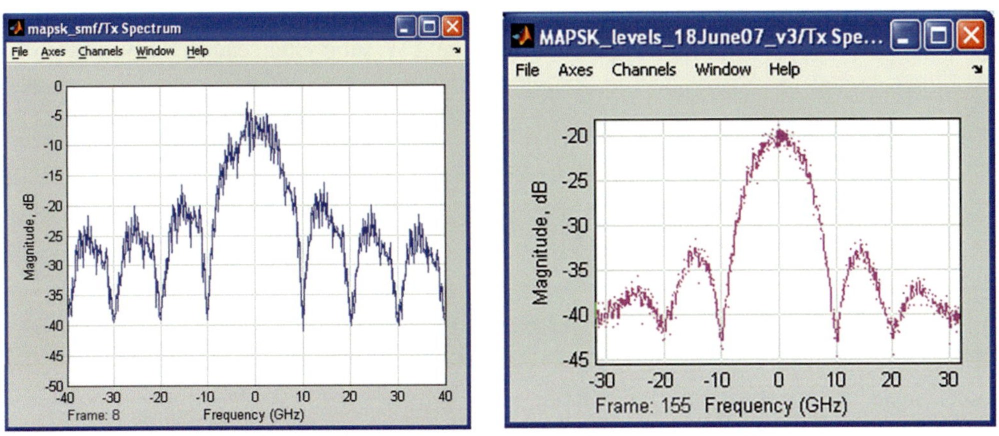

FIGURE 4.33 The 40 Gbps MADPSK spectrum.

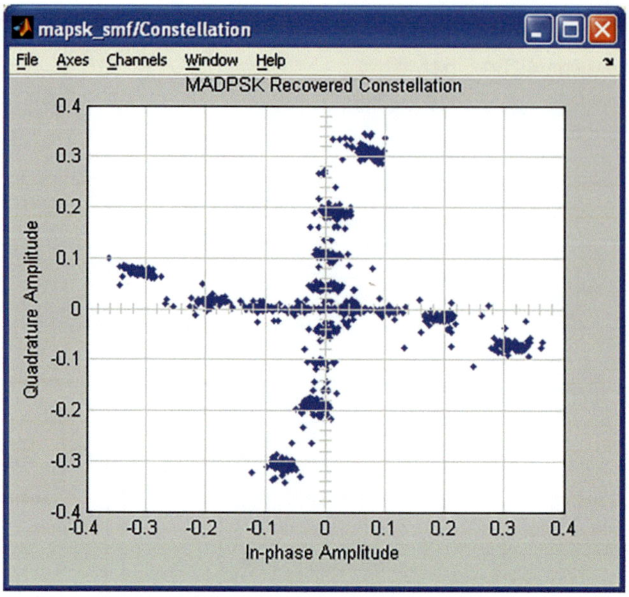

FIGURE 4.34 The 40 Gbps MADPSK constellation recovered at the receiver.

4.2.6.4.2 BER Evaluation

The MADPSK system can be considered as consisting of two subsystems, ASK and DQPSK. Its error probability can be evaluated as a joint error probability of the two:

$$P_{\text{ADPSK}} = \left[\frac{1}{2}P_{\text{ASK}} + \frac{1}{2}P_{\text{DPSK}} - \frac{1}{2}P_{\text{ASK}} \cdot \frac{1}{2}P_{\text{DPSK}}\right] = \frac{1}{2}\left[P_{\text{ASK}} + P_{\text{DPSK}} - P_{\text{ASK}} \cdot P_{\text{DPSK}}\right] \quad (4.7)$$

where P_{ASK} and P_{DPSK} are the error probabilities of ASK and DQPSK subsystems, respectively.

For ASK sub-system the error probability can be shown in Figure 4.36 which depicts four ASK signal levels b_0, b_1, b_2, b_3, three decision levels d_1, d_2, d_3, and standard deviation of noise at different signal levels σ_0, $\sigma_1, \sigma_2, \sigma_3$. The error probability of the ASK sub-system can be derived as

$$P_{\text{ASK}} = \frac{2}{M+1}\sum_{1}^{M} Q\left(\frac{b_i - d_i}{\sigma_i}\right) = \frac{2}{3+1}\begin{bmatrix} Q\left(\frac{b_1 - d_1}{\sigma_1}\right) \\ +Q\left(\frac{b_2 - d_2}{\sigma_2}\right) \\ +Q\left(\frac{b_3 - d_3}{\sigma_3}\right) \end{bmatrix} \quad (4.8)$$

For example, in our system: (i) $b_1 = 8.08e-2, b_2 = 1.45e-1, b_3 = 2.42e-1$; (ii) $d_1 = 5.11e-2$, $d_2 = 1.08e-1, d_3 = 1.88e-1$; and (iii) $\sigma_1 = 5.00e-3, \sigma_2 = 6.70e-3, \sigma_2 = 8.65e-3$ at an OSNR=20 dB

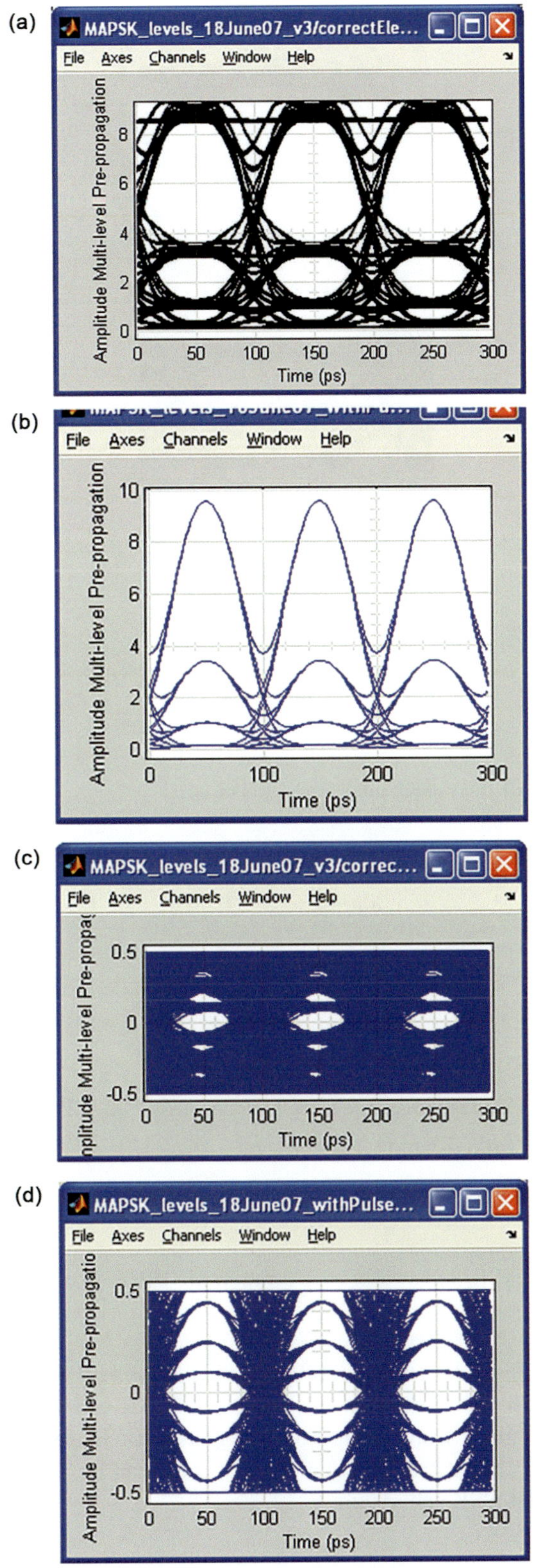

FIGURE 4.35 The 40 Gbps MADPSK eye diagram at OSNR = 20 dB (a) NRZ amplitude, (b) CSRZ amplitude, (c) NRZ in-phase, and (d) CSRZ in-phase.

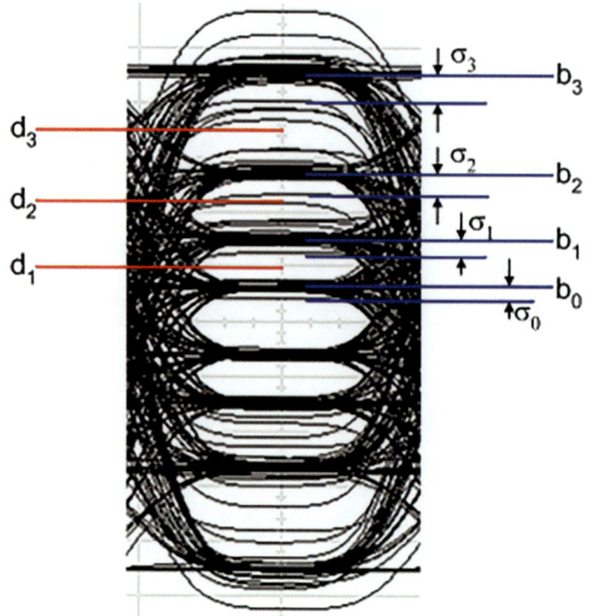

FIGURE 4.36 MADPSK eye diagram: signal levels, decision levels, and standard deviation of noise.

The error probability of the ASK subsystem thus equals:

$$P_{\text{ASK}} = \frac{1}{2}\left[\begin{matrix} Q\left(\dfrac{(8.08e-2)-(5.11e-2)}{5.00e-3}\right)+Q\left(\dfrac{(1.45e-1)-(1.08e-1)}{6.70e-3}\right) \\ +Q\left(\dfrac{(2.42e-1)-(1.88e-1)}{8.65e-3}\right) \end{matrix}\right]$$

$$\begin{aligned} P_{\text{ASK}} &= \frac{1}{2}\left[Q(5.94)+Q(5.52)+Q(6.24)\right] \\ &= \frac{1}{2}\left[(1.47e-9)+(1.73e-8)+Q(2.26e-10)\right] \\ &= 9.49e-9 \end{aligned}$$

The error probability of ASK subsystem over a range of OSNR from 6 to 24 dB is evaluated and shown in Figure 4.37.

4.2.6.4.2.1 DQPSK Sub-System Error Probability Evaluation In terms of the DPSK modulation, the system can be divided into four independent DQPSK subsystems corresponding to four circles, circle 0, circle 1, circle 2, and circle 3 of the signal constellation. The error probability BER of each subsystem can be evaluated, and then their average can be taken to obtain the BER error probability for the 16A-DQPSK subsystem.

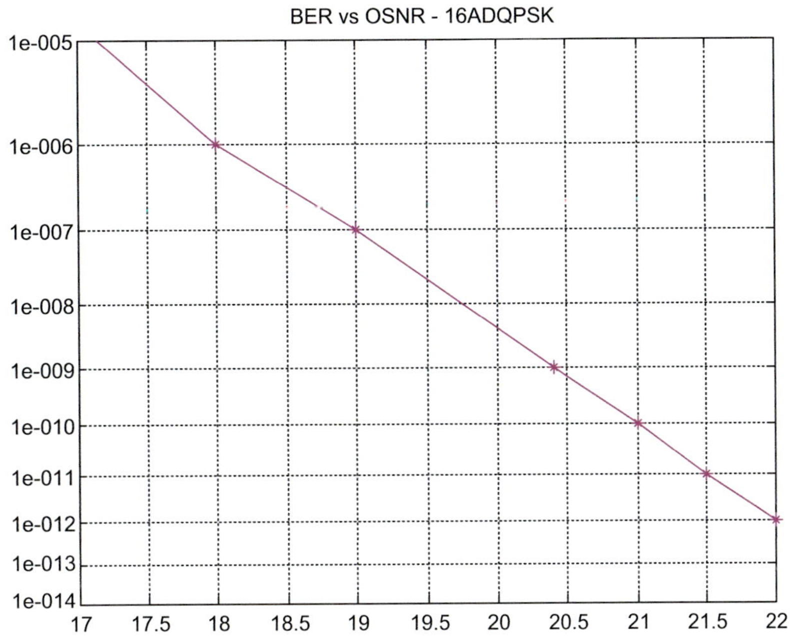

FIGURE 4.37 Error probability BER of ASK sub-system versus OSNR (in dB).

Each DQPSK subsystem, in turn, can be assumed as consisting of two 2-ary DPSK subsystems. The error probability of each 2-ary DPSK subsystem can be evaluated, and their average can be taken to get the error probability of the DQPSK subsystem as

$$P_{\text{DQPSK}} = 1-\left(1-P_{\text{DPSK}\_I}\right)\left(1-P_{\text{DPSK}\_Q}\right) = P_{\text{DPSK}\_I} + P_{\text{DPSK}\_Q} - P_{\text{DPSK}\_I}\cdot P_{\text{DPSK}\_Q} \tag{4.9}$$

where P_{DPSK_I} and P_{DPSK_Q} are error probability of in-phase (I) and quadrature-phase (Q) components of each DPSK subsystem (circle). Because the imaginary part I is coded by bit D_0, the quadrature component Q is coded by bit D_1, thus detecting both the amplitude of the I and Q components in the same way. When the D_0 and D_1 can be considered equally probable, then Equation 4.20 becomes:

$$P_{\text{DQPSK}} = 2P_{\text{DPSK}\_I} - P^2_{\text{DPSK}\_I} \; = 2P_{\text{DPSK}\_Q} - P^2_{\text{DPSK}\_Q} \tag{4.10}$$

P_{DPSK_I} is evaluated based on the δ-factor[8] as

$$P_{\text{DPSK}\_I} = \frac{1}{2}\left(\frac{\delta}{\sqrt{2}}\right) \approx \frac{\exp\left(-\delta^2/2\right)}{\delta\sqrt{2\pi}} \tag{4.11}$$

where $Q = \dfrac{i_H - i_L}{\sigma_H + \sigma_L}$, i_H, i_L and σ_H, σ_L are mean value and standard deviation of signal currents at high and low levels at the input of the receiver, respectively. For example, the transmission parameters can be set as follows: $i_H = 3.23e-02,\; i_L = (-3.23e-02)$, at OSNR=20 dB $\sigma_H = \sigma_H = 3.16e-3$. The δ-factor for a single DQPSK sub-system of circle 0 thus equals, and the corresponding the error probability is

$$P_{\text{DPSK\_I\_CYCLE}_0} = \frac{1}{2}\text{erfc}\left(\frac{10}{\sqrt{2}}\right) \approx 7.7e-24$$

The BER error probability of circle 0 (the inner most circle) is $P_{\text{DQPSK\_CYCLE}_0} = 2*(7.7e-24)-(7.7e-24)^2 = 1.54*10e-23$. Thus, the error probability of all four circles is:

$$P_{\text{DQPSK}} = \frac{1}{4}[P_{\text{DQPSK\_CYCLE}_0} + P_{\text{DQPSK\_CYCLE}_1} + P_{\text{DQPSK\_CYCLE}_2} + P_{\text{DQPSK\_CYCLE}_3}] \tag{4.12}$$

P_{DQPSK} over a range of OSNR from 6 to 24 dB is evaluated and shown in Figure 4.38.

4.2.6.4.2.2 MADPSK System BER Evaluation The error probability of the modulation format MADPSK transmission system can be evaluated based on Equation 4.17. Figure 4.39 shows the graphs of error probability versus the optical OSNR for the ASK sub-system, DQPSK subsystem, and MADPSK system. They are plotted on the same coordinates for comparison purposes. The noises are estimated based on the optical and electronic noises under coherent reception. As it can be observed from Figure 4.39,

i. at a 16 dB OSNR lead to an error probability of 1e-9 for DPSK and 25 dB for both MADPSK and ASK. This shows that the MADPSK and ASK are more tolerant to the nonlinear effects in the transmission medium, the fiber in terrestrial optical lines, and the vacuum air in space. The nonlinearity in the space system mainly comes from the short fiber at the output of the high-power booster amplifier before launching into space. Therefore, the DPSK modulation offers better resilient to nonlinearity compared to the ASK and MADPSK. Amplitude noise contributes to the worst BER than that for DPSK. The differential phase between consecutive bits improves detection and resilient to noises. This is very important implication for space photonic transmission, especially when noise is caused by air disturbances via the air layer from the Optical ground station (OGS) to the satellite link. Thus,

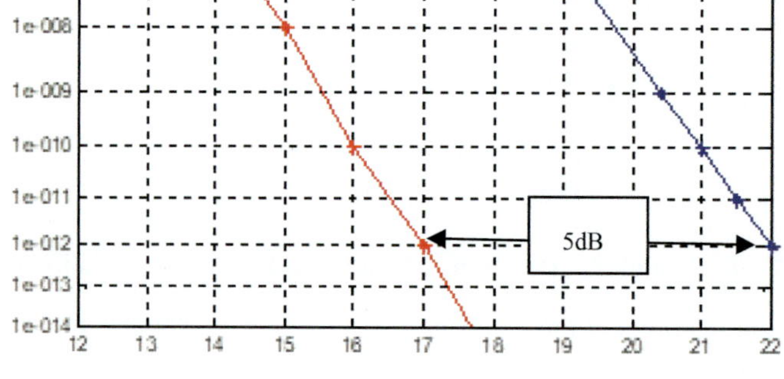

FIGURE 4.38 Error probability of DQPSK (M=1 left-curve and M=16-right curve) subsystem versus OSNR.

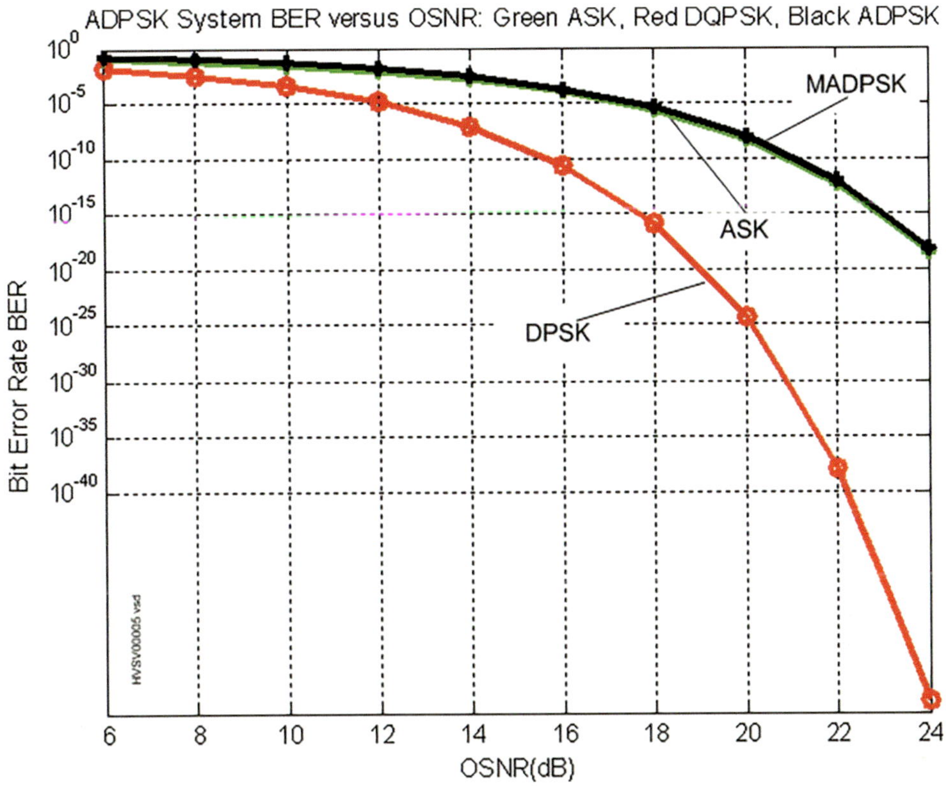

FIGURE 4.39 Error probability of MADPSK (black dark) system versus OSNR, logarithm scale. Error probability of ASK (light adark and upper line) and MADPSK (Dark and upper line lines) are nearly coincided.

a 2.5 dB better in OSNR at the same BER is required for the differential phase format compared to amplitude ASK or MADPSK.

ii. With an OSNR of 24 dB, the MADPSK leads to a complete error-free 1e-40. For the same value of OSNR, especially when it is high, the DQPSK subsystem outperforms the ASK counterpart, and the overall performance of the MADPSK system is dominated by the ASK subsystem performance. Thus, the spaces between ASK levels could be adjusted for a better balance between BER ASK and BER DQPSK to achieve a better overall MADPSK BER performance. This is probably due to the ISI during the transition of different levels.

Figure 4.36 shows simulation results for 16 ADPSK at 100 Gbps transmission (extreme left graph), compared with other modulation formats such as Duo-binary 50, Duo-binary 67, and experimental results of CSRZ DPSK. The bit rates of these other transmission results operate at 40 Gbps. It is observed that for 16MADPSK, the receiver sensitivity is close to the –28 dBm performance standard used in 10 Gbps NRZ transmission and performs better at 100 Gbps compared to other modulations operating at the lower rate of 40 Gbps. However, this superior performance at 100 Gbps incurs a penalty of approximately 3 dB compared with 10 Gbps transmission systems. Fortunately, this penalty can easily be compensated for using a low-noise optical pre-amplifier at the receiver end. For example, a 15 dB gain optical pre-amplifier with a 3 dB noise figure would adequately resolve the issue. The BER versus the receiver sensitivity for 16 ADPSK, duo-binary formats, and ASK is shown in Figure 4.40. It indicates a 2–3 dB improvement of the MADPSK.

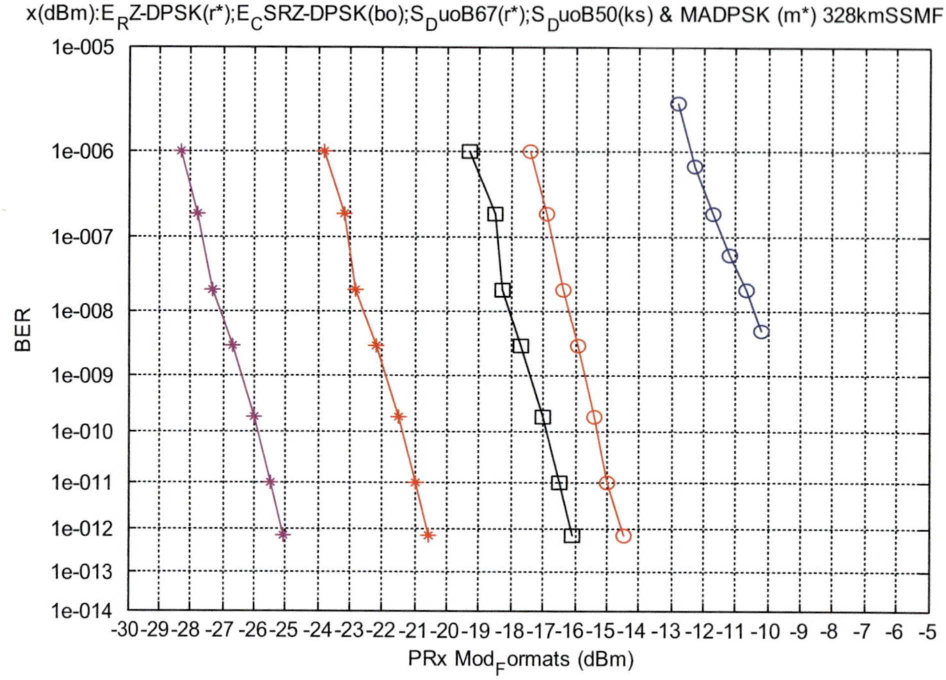

FIGURE 4.40 BER versus receiver sensitivity for MADPSK format and other duo-binary and ASK (simulation) and CSZ and CSRZ DPSK (experimental). Lightest * is the MDAPSK.

The detection of the lowest level may have been affected by the noise level of the optical preamplifier when only amplitude information is used. This can be improved significantly if both the phase detection and amplitude detection are used, as observed in Figure 4.35c and d.

4.2.6.4.2.3 Chromatic Dispersion Tolerance The residual chromatic dispersion of the optical link is characterized by the *DL* product defined as the product of the dispersion coefficient *D* and the total fiber length *L*. Figure 4.41 illustrates the signal phase evolution under the effect of chromatic dispersion. It can be observed that with a predetermined $DL=50$ ps/nm, all signal points are rotated around the [0, 0] origin by the same angle of approximately 0.125 rad. This confirms the parabolic phase shift due to the chromatic dispersion. This phenomenon is called linear phase distortion contrary to the non-linear phase distortion caused by the fiber non-linearity.

Figure 4.42 shows the BER penalty versus different values of *DL* product. The BER performance of the NRZ format is severely affected by the fiber dispersion. When *DL* increases from 0 ps/nm (fully CD compensated) to 35 ps/nm, its BER performance is improved by 1.5 dB, but sharply degraded by 28 dB penalty at $DL=50$ ps/nm, and should be worse for higher value of dispersion. This leads to the conclusion that it is undesirable to use the NRZ format in MADPSK systems because the optical link residual dispersion usually cannot be compensated to a small amount, and ineffective dispersion management and control plan could lead to a very high BER.

The 66%-RZ format, on the other hand, can tolerate a much higher degree of chromatic dispersion. Its BER performance is even slightly improved at $DL\approx 50$ ps/nm, and the BER penalty is just <1 dB at $DL=100$ ps/nm. That is equivalent to the transmission of over a 6 km of uncompensated standard SMF fiber without significantly giving up the BER performance.

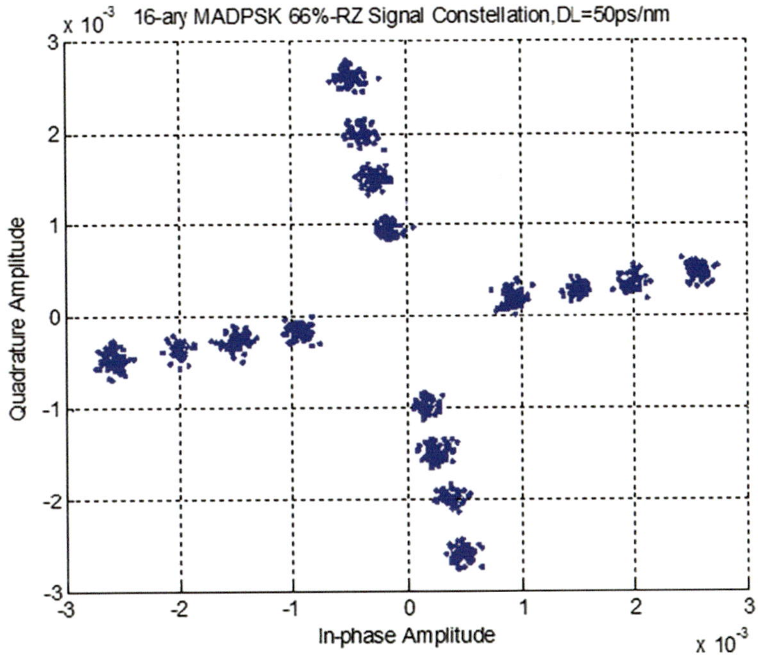

FIGURE 4.41 Evolution of the phase scattering of the MADPSK signal constellation under chromatic dispersion effects.

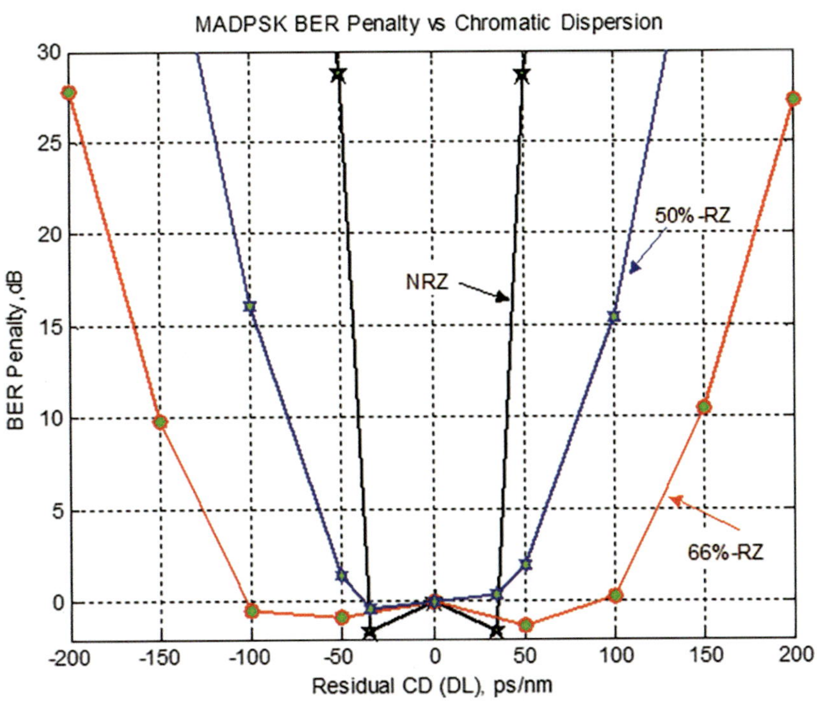

FIGURE 4.42 Error probability of MADPSK system versus the dispersion-length DL (length dispersion product) product.

The MAPSK offers a lower symbol rate and higher channel capacity, allowing for upgrades to higher rates merging with a low-bit-rate optical fiber transmission system without modifying the photonic infrastructure of the optical networks.

4.2.6.5 Critical Issues

This section outlines the critical issues to evaluate the performance of MADPSK systems.

4.2.6.5.1 Noise Mechanism and Noise Effect on MADPSK

Although receiver noises in multi-level amplitude modulation were investigated intensively in the 1980s, little has been reported on multi-level phase and differential phase modulation. One of the principal goals in the system design, especially for long-haul transmission systems, is to achieve high receiver sensitivity. At a given optical power, the error probability depends on the noise power, which affects the receiver sensitivity.

The quantum shot noise (QSN) is the fundamental noise mechanism in the photodiode, which leads to the fluctuation in the detected electrical current even when the incident optical signal has a constant or varied power. Thus, it is signal-dependent. Furthermore, the beating of the currents of the signal and the optical phase noise would generate an amplitude-dependent noise at different level signals of the MADPSK. It is caused by a random generation of electrons contributing to the photoelectric current, a random variable. All photodiodes generate some current even in the absence of an optical signal due to the stray light and/or thermal generation of electron-hole pairs, known as the dark current.

In long-distance free space links, both booster and pre-amp amplifiers are the main components to compensate for the losses, their ASE accumulated noises are critical and must be optimized.

In MADPSK, the signal amplitude of the outermost circle of the constellation is affected by the quantum shot noise, which is strongly dependent on the signal amplitude, especially when an optical preamplifier is used. On the other hand, it is desirable for the innermost constellation to have the largest magnitude to maximize the optical signal energy for long-haul transmission. Therefore, an optimum receiving scheme must be developed, both analytically and through modeling, eventually by experimental demonstration.

However, the amplitude of the outermost constellations is limited by the nonlinear self-phase modulation effects, which will be further explained in the next few sections. Thus, the lower and upper limits of the MADPSK amplitude will be extensively investigated in the next phases of the research.

The electronic equivalent noise as seen from the input of the electronic pre-amplifier following the PD (photodetector) can be measured and considered the total noise process caused by thermal noises of the input impedance, the biasing current shot noise, and the noises at the output of the electronic preamplifier. These noises are combined with the signal-dependent quantum shot noises to gauge their contribution to the MADPSK receiver. Thus, we may consider new structures of electronic amplifiers or matched filters at the input of the receiver to achieve the optimum MADPSK receiver structure.

For long-haul transmission systems, accumulation of ASE noises of cascaded-span optical amplifiers or free-space hops is probably the important noise mechanism. In optical amplifiers, even in the absence of an input optical signal, spontaneous emission always occurs stochastically when electron-hole pairs recombine and release energy in the form of random photons. This spontaneous emission creates noise and is further amplified by the optical amplifiers, along with the optical signal, and accumulated along an optical transmission link.

Noise reduces the SNR (signal-to-noise ratio), hence the system BER and receiver sensitivity. Noise models also affect the design of optimum detection schemes such as decision thresholds. The noise mechanism and their impacts on multi-level signaling have not been investigated for SPC systems. These noises are used to estimate the optimum decision level of the amplitude detection of the multilevel eye diagram, especially when ASE noises are dominant.

4.2.6.5.2 Transmission Impairments

For broadband optical signals, the transmission medium is an optical fiber or over a free-space hop, associated by OAs, dispersion compensation devices, multi-path fading, or a leased wavelength running on top on a DWDM system. Impairments are always part of the transmission medium; among the most critical are CD, PMD, multi-path fading, spatial spreading, and nonlinearity in free space.

When an optical pulse propagates along a fiber or a free-space long distance, its spectral components disperse due to the differential group delay (DGD), and the output pulse will be broadened. Furthermore, for free space, the spatial spreading is due to the finite linewidth of the lightwave carrier. Chromatic Dispersion in SMF (single mode fiber) is proportional to the fiber length and the laser linewidth, especially the spectrum of the lightwave modulate signals. The CD causes optical pulses to overlap with each other, thus leading to ISI and increased system BER, especially for ASK systems. DPSK systems, on the other hand, are more CD tolerant and to the effects of laser phase noises on differential phase detection of two consecutive bits. For MADPSK (multi-amplitude differential phase shift keying) systems, the phase constellation, as shown in Figure 4.43, is rotating from its ideal original constellation, as shown in Figure 4.44, when the MADPSK is under a linear chromatic dispersion effect or multi-path fading. It is also well-known and developed in our model that this CD can be compensated by dispersion-compensating fiber modules. However, the mismatch between the dispersion slopes of the transmission and compensating fibers is very critical for multi-channel, multi-level modulation schemes.

Optical pulse is also broadened by PMD (polarization mode dispersion), which results from the time mismatch between two orthogonal polarizations channels of lightwaves as they traverse along a circular fiber and the free-space medium under turbulence or radiation.

In an ideal optical fiber having truly homogeneous glass and truly coaxial geometry of the core, the two optical polarizations will propagate with the same velocity. However, this is not the practice case. The core circularity is more elliptical at random, causing the lightwaves to become randomly polarized and hence changing their polarized-mode propagation velocity. Thus, there are differences in arrival times of the polarized channels, that leads to the polarization dispersion of the pulse sequence.

In the free-space link of long-distance, the angle of alignment between the transmitter and the receiver can vary due to their rotation speeds. Thus, there is a geometrical difference between the channel light rays and the random rotation of polarization due to the air refractive index of

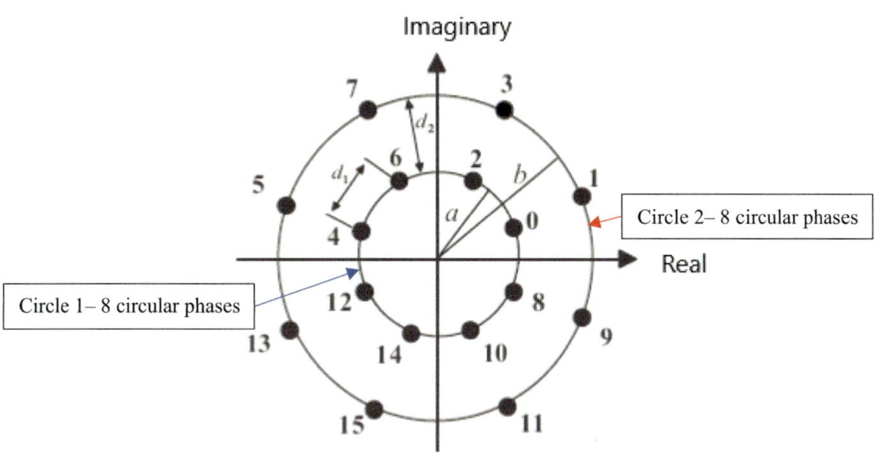

FIGURE 4.43 The 40 Gbps MADPSK eye diagrams of the I (a) and Q (b) components (A) 0 km – back-to-back; (B) 2 km SSMF mismatch over three 100 km SSMF transmission spans.

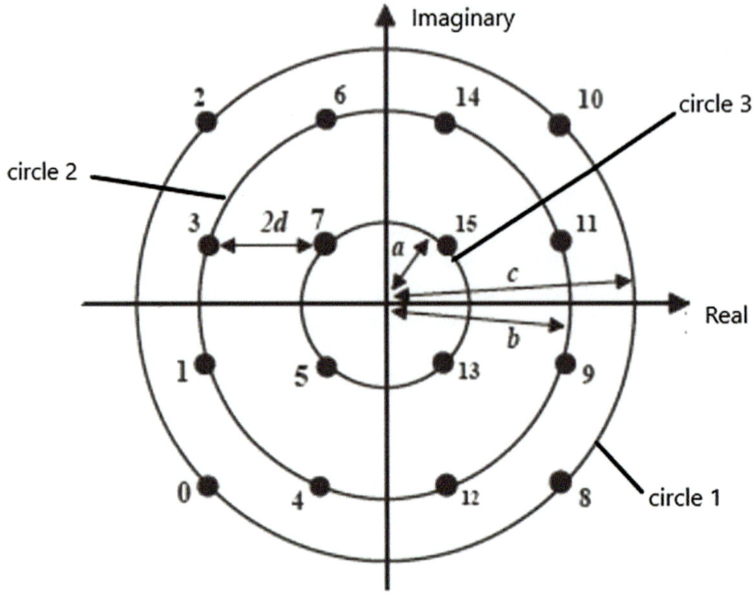

FIGURE 4.44 Constellation of 16-square QAM after two optically amplified spans and 2 km SSMF dispersion or equivalently 40,000 km vacuum space distance with 40° of arc angle alignment (a) pre-transmission and (b) post-transmission.

turbulence. Furthermore, the spatial broadening of the lightwave beam also creates additional randomness of fading and polarization dispersion. For ease of notation, we term these effects as chromatic spatial dispersion (CSD) and spatial polarization mode dispersion (SPMD).

Similar to the CD/CSD effects, the PMD/SPMD can cause pulse overlapping, worsening the BER system. However, unlike CD/CSD, which is practically constant over time and can be in a large-scale compensation, PMD/SPMD is a stochastic process and cannot be managed easily. It has been well-known that PMD/SPMD has the Maxwellian probability density function with a mean value $\langle \text{PMD} \rangle = K_{\text{PMD}} \cdot \sqrt{L}$, where K_{PMD} is defined as PMD/SPMD coefficient or spatial disturbance coefficient, whose measured values vary from fiber to fiber or space to space disturbance in the range $\left[0.01-1 \text{ ps}/\sqrt{\text{km}}\right]$, and L is fiber length.

Under the MADPSK, the signal-space of the constellations are affected either by the magnitude or phases due to PMD/SPMD but expected to be dominated by the phase distortion. It is well-known that in fiber optics, the first and second effects of PMD are critical for ASK modulation. For space medium, the SPMD is more dependent on the first order, that is, its wavelength dependence is much smaller than that caused by the first-order dependence on wavelength.

For DPSK, the principal axes of the polarization modes propagating through the fiber or space medium are expected to be minimally affected. Thus, under a combined amplitude-phase modulation scheme, there are several issues to be resolved. In the MADPSK scheme, the delay difference of the polarization channels can generate the phase difference or phase distortion on I- and Q- components, hence enhancing the distortion effects of the ISI. The amplitude distortion would then be increased but considered to be a secondary effect.

4.2.6.5.3 *Nonlinear Effects on MADPSK*

Nonlinear effects in fiber optic transmission lines and SPC systems occur due to the nonlinear response of the fiber glass to the applied optical power and the vibration of the bonding of Silicate (SiO_2). This leads to the generation of acoustic waves propagating along the guided medium, generating the SBS, which modulates the amplitude of the bit sequence and causes the Brillouin shift. Fiber nonlinearity can be classified into stimulated scattering effects – SBS and stimulated Raman scattering (SRS) – as well as the Kerr effect and stimulated Brillouin scattering. Among several stimulated scattering effects, SRS, caused by interaction between light and acoustical vibration modes in the fiber glass, is the most critical. Under this mechanism, the optical signal is reflected to the transmitter, and in WDM systems, its power is also transferred from shorter to longer wavelengths, thus attenuating the signal and causing crosstalk. The Kerr effect is the root of the intensity-dependent phase shift of the optical field.

The nonlinear effects that generate distortions and crosstalks to other channels are self-phase modulation (SPM), cross-phase modulation (XPM), and four-wave mixing (FWM), provided the phase matching is met. The details of the nonlinearity effects on SPC are discussed in Chapters 3 and 7. Both fiber optic transmission and SPC systems experience the same nonlinearities, as these effects occur in the single-mode fiber, across terrestrial lines in cascaded fiber spans, and in space systems, only in the short length of SMF connected to HPA booster amplifier and the optical head before launching into space.

SPM in fiber is usually the dominant effect in a single-channel DPSK systems. The changes in instantaneous power of optical pulses, along with the ASE from associated OAs, lead to intensity-dependent changes in the guided medium refractive index, known as the Kerr effect. This modifies the effective index of the guided mode, converting into the phase shifts or phase noise of the lightwave carriers. At the receiver, the phase noise is transferred back to intensity noise, which degrades BER. As mentioned above, the noise's contribution to different levels of the MADPSK scheme is critical to determining optimum decision thresholds. These additional nonlinear noises further complicate the prediction of the systems BER and OSNR. Especially, the nonlinear phase noises (NLPN) are usually contributed by the SPM due to the outermost constellation. These NLPN effects from the outermost constellation to other inner circle signal spaces have not been investigated.

The best technique is to provide sufficient margin of system operation at different stages of the subsystem, importantly focusing on the dynamic range of the optical transceiver.

In WDM systems, XPM becomes the most critical nonlinearity. where, the phase shifts (noise) in one channel result from refractive index fluctuations caused by power changes in other channels. In the SPC link, this fluctuation is quite random, as channels pass through the turbulence layer. Such XPM becomes more pronounced when neighboring channels have equal spaced and equal bit rate. Four-wave mixing (FWM) is basically a crosstalk phenomenon in DWDM systems. When three wavelengths with frequencies, ω_1, ω_2, and ω_3 propagate in a nonlinear fiber medium at which there is no dispersion, they combine and create a degenerate fourth wavelength. This new wavelength will fall on the location of an active wavelength channel. This is also the characteristics of the vacuum in vacuum free-space links between satellite points in a constellation.

If these parametric wavelengths fall into other channels, they cause crosstalk and degrade system performance. Although FWM is expected to reduce the receiver sensitivity, the proposed system should minimize the effects of fiber nonlinearity. The maximum power of the optical signal should not be set higher than a certain threshold. This maximum power dictates the amplitude of signal points in the outermost circle (circle 3), and hence influences other circles of the signal space. Thus, optimization of the signal amplitude levels for MADPSK is very critical.

Therefore, it is highly critical for SPC systems that the DWDM wavelength carriers of the channels making up Tera-bps capacity be selected in unequal spacing and within the transparent spectral window whose spectrum can be inspected as given in Chapters 2 (Figures 2.2 and 2.3).

4.2.6.6 Offset Detection

The 16-ary MADPSK signal model described in Section 4.2.6.1 can be modified. To balance the ASK and DQPSK sensitivities, ASK signal levels are preliminary adjusted to the ratio $I_3/I_2/I_1/I_0 = 3/2/1.5/1$ and rotated by $\pi/4$, as shown in Figure 4.45. These level ratios can be determined from the signal-to-noise ratio at each separation distance of the eye diagram or the Q-factor. The noise is assumed dominated by the beat noise between the signal level and that of the ASE noise. The eye-opening is expected to improve significantly, as shown in Figure 4.46.

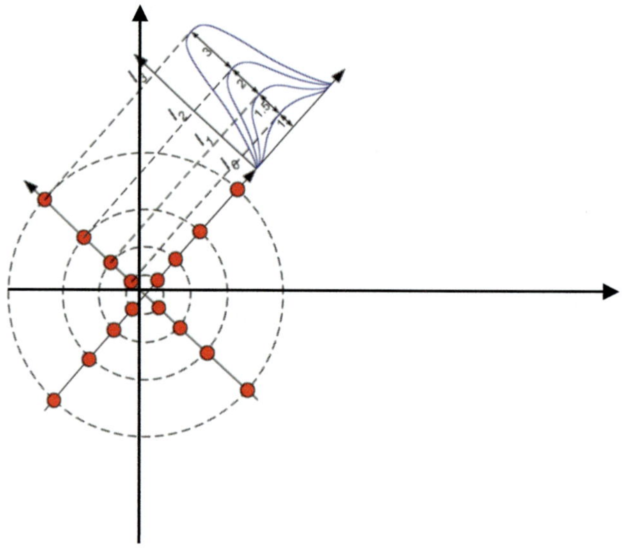

FIGURE 4.45 ASK inter-level spacing and offset modulation and detection line.

(a) 0 km mismatch SSMF

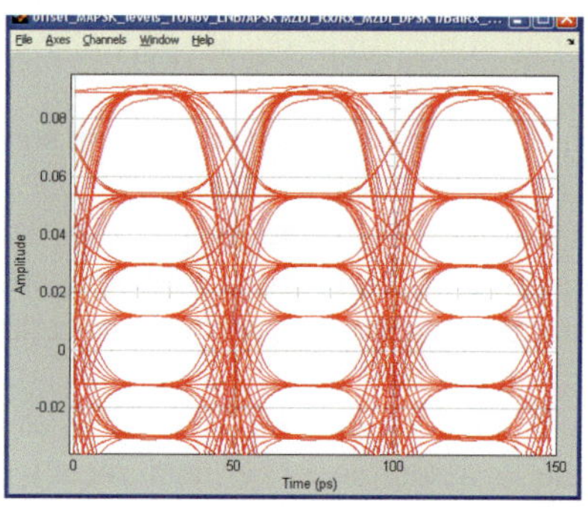

(a1)

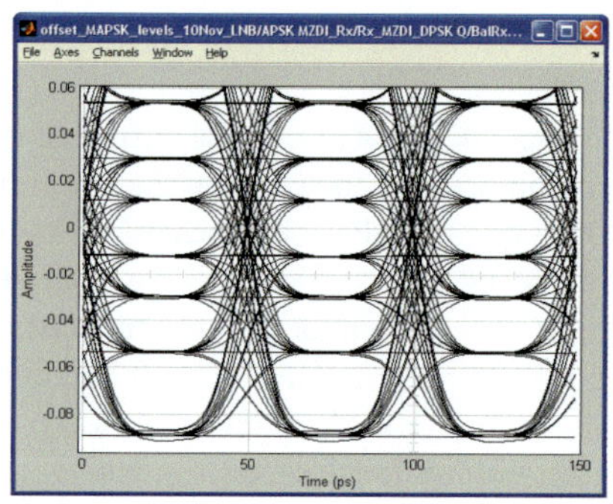

(a2)

(b) 2 km mismatch dispersion SSMF.

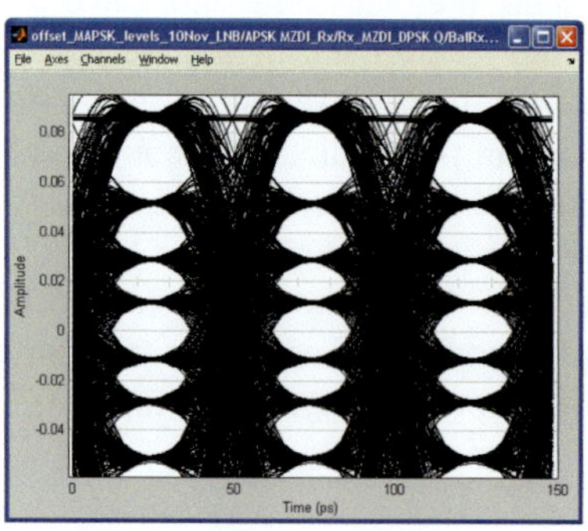

(b1)

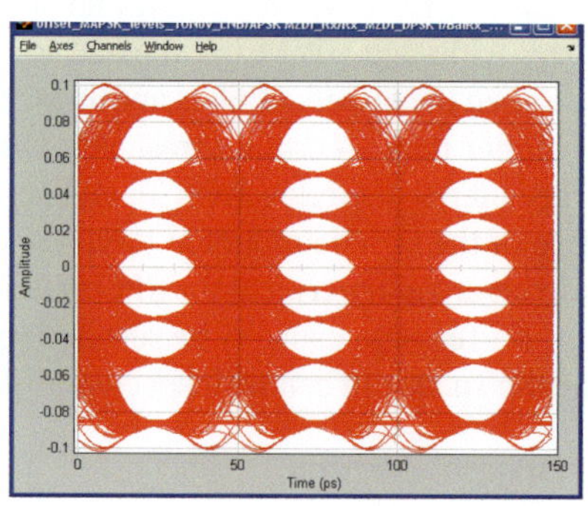

(b2)

FIGURE 4.46 Modulation using multi-phase equal amplitude constellation for BACP 16-QAM dual-amplitude levels and eight circular phases per circle.

4.3 CONSTANT AMPLITUDE PHASE CIRCULAR MODULATION

This section provides a briefing on the simulation of the transmission performances of SPC transmission systems over N-hop space transmission systems. The modulation format used is an N-phase circular constellation points with a number of concentric N-phase state states. Optical transmitters and coherent receivers, as the main subsystems components, are described. Simulation results have shown that it is possible to transmit and detect the data symbols for 43 Gbps with the possibility of scaling to 107 Gbps without much difficulty. The OSNR, with 0.1 nm 3 dB bandwidth optical filters, can be achieved with 18 and 23 dB for back-to-back (B2B) and long-haul transmission cases of 42,500 km inter-satellite GEO constellation distance.

4.3.1 Introductory Remarks

To increase the channel capacity and bandwidth efficiency in optical transmission, multi-level modulation formats like QAM formats can be of much interest. In the digital transmission with multi-level (M-levels) modulation, m bits are collected and mapped onto a complex symbol from an alphabet with $M=2$m possibilities at the transmitter side. Furthermore, in SPC, the attenuation factor and non-optical amplification between the long-distance alignment are critical, offering phase modulation many advantages. Phase modulation can be detected by IQ coherent reception sub-system because, once the I and Q-components of a constellation point can be found, its phase and amplitude in the angular coordinate can be derived, and hence the phase of the constellation points.

The symbol duration is $T_s=mT_B$ with T_B as the bit duration and the symbol rate is $f_s=f_B/m$ with $f_B=1/T_B$ as the bit rate. This shows that for a given bit rate, the symbol rate decreases if the modulation level increases. This implies that higher bandwidth efficiency can be achieved through a higher order modulation format. For the 16-QAM format, $m=4$ bits are collected and mapped to one symbol from an alphabet with $M=16$ possibilities. In comparison to the case of binary modulation format, only $m=1$ bit is mapped to one symbol from an alphabet with $M=2$ possibilities. With the 16-QAM format and a data source with a bit rate of $f_B=40$ Gbps, only a symbol rate of f_s of 10 Gbaud/s is necessary. From a commercial point of view, it means a 40 Gbps data rate can be transmitted with 10 Gbps transmission devices. In the case of binary transmission, the transmitter needs a symbol rate of f_s of 40 Gbaud/s. It means that the 16-QAM transmission requires four times slower transmission devices than that of the binary transmission. It is noted here that a 10.7 GSy/s is used as the symbol rate to compare the simulation results with the well-known 10.7 Gbps modulation schemes, such as DPSK, CSRZ, DPSK, etc.

This section provides a general approach to the design and simulation of Star 16-QAM, with two amplitude levels and eight phase states, forming two-circular phase circles. We term this BACP (bi-amplitude constant phase) 16-QAM as 2A-8P Star 16-QAM, or two-amplitude-level and eight-phase states. The transmission format can be discussed with theoretical estimates and simulation results to determine the transmission performance. The optimum Euclidean distance is defined for the design of a BACP 16-QAM. In the second section, the two detection schemes, namely, the direct detection and the coherent detection, for star-QAM constellations are discussed.

4.3.2 The 16-QAM Signal Constellation

There are many possibilities for designing a 16-QAM signal constellation. Three of the most popular constellations can be introduced. For 16-QAM modulation schemes, there can be (i) Star 16-QAM (ii) Square 16-QAM; and (iii) Shifted-square 16-QAM. The first two of these constellations can be well-known. However, only the BACP 16-QAM with two amplitudes and eight circular phases per amplitude level are investigated.

TABLE 4.5
Symbol Mapping and Coding for Bi-Amplitude Circular-Phase 16-QAM

Mapping for Bi-Amplitude Circular-Phase 16QAM			
$0 \to 1$	$4 \to 9$	$8 \to 13$	$12 \to 5$
$1 \to 0$	$5 \to 8$	$9 \to 12$	$13 \to 4$
$2 \to 3$	$6 \to 11$	$10 \to 15$	$14 \to 7$
$3 \to 2$	$7 \to 10$	$11 \to 14$	$15 \to 6$

4.3.2.1 Signal Constellation

The signal constellation for BACP 16-QAM with Gray coding is shown in Figure 4.46. The binary representation of the symbols in the figure is shown in Mapping Table 4.5. As seen from this figure, the symbols are evenly distributed on two rings, and the phase difference between the neighboring symbols on the same ring is equal (π/4). In order to detect a received symbol, its phase and amplitude must be determined. In other words, between the two amplitude levels of the rings and among eight phase possibilities, there are several ways to form this constellation.

The ring ratio (RR) for this constellation is defined as: RR=b/a, where a and b are the ring radii, as shown in Figure 4.46. The RR can be set to different values to optimize transmission performance. It is possible to use the same coding scheme, as described above, but with the codes randomly distributed so that adjacent codes, which result in the worst error rate, are avoided. Thus, an improvement of the BER can be achieved. The improvement in OSNR for the same BER can be 1–2 dB, which is quite substantial.

4.3.2.2 Optimum Ring Ratio for BACP Constellation

From Figure 4.46, it can be seen that there are many possibilities to choose the RR for the star 16-QAM constellation. Here, the theoretical best RR is defined to minimize the error probability in an AWGN channel by maximizing the minimum distance $d_{\min}$ between the neighboring symbols. The results for the AWGN channel can be used approximately for optical transmission. For star 16-QAM, the minimum distance $d_{\min}$ is maximized when

$$d_1 = d_2 = b - a = d_{\min} \tag{4.13}$$

With some geometrical calculations, it can be obtained that

$$d_{\min} = 2a \cdot \sin(22.5°) \tag{4.14}$$

which leads to an optimal ring ratio of

$$\mathrm{RR}_{\mathrm{opt}} = b/a = (d_{\min} + a)/a = (2 \cdot a \cdot \sin 22.5 + a)/a \approx 1.77 \tag{4.15}$$

The average power of the bi-amplitude circular phase 16-QAM constellation can be determined as

$$P_0 = (8a^2 + 8b^2)/16 = (a^2 + b^2)/2 \tag{4.16}$$

Thus, we have the relationship between the average optical power and the minimum distance between the two rings of the two amplitude levels as

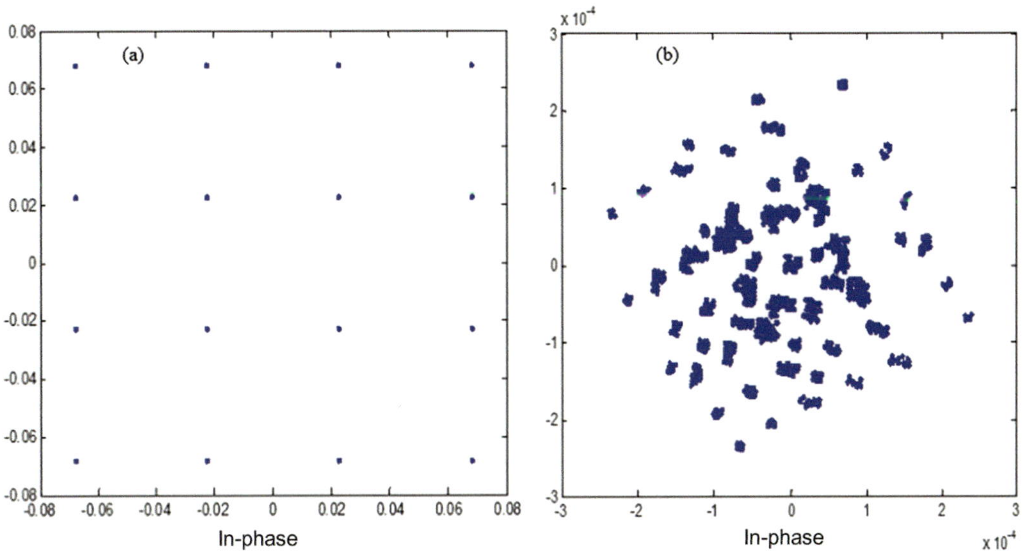

FIGURE 4.47 Square 16-QAM or 3ASK CP-16QPSK signal constellation represented in three-level (Tri-) concentric amplitude (TACP) circular levels.

$$d_{\min} \approx 0.53(P_0)^{1/2} \tag{4.17}$$

The obtained $RR_{opt}=1.77$ does not depend on P_0 and is constant for each P_0 value. For an average power of 5 dBm (3.16 mW), $d_{min}=2.98\cdot10^{-2}\sqrt{W}$, $a=3.89\cdot10^{-2}\sqrt{W}$, and $b=6.87\cdot10^{-2\sqrt{W}}$ can be obtained.

4.3.2.2.1 Square 16-QAM

The signal constellation of the square 16-QAM with Gray coding is shown in Figure 4.47.

The square 16QAM can be generated by combining two rings and a QPSK in the outermost ring, which can also be a ring. Thus, the circular phasor generation scheme can be used to form signals with amplitude and phases to be applied to the optical modulators. The binary representation of the symbols in the figure is shown in mapping Table 4.5. In the constellation of the square 16-QAM, the 16 symbols have equal distance to their direct neighbors and a total of 12 different phases, i.e., three phases per quarter, distributed on three rings. The phase differences between neighboring symbols on the inner and outer rings are equal ($\pi/2$), but the phase differences between neighboring symbols on the middle ring are different (37° or 53°). If the distance between direct neighbors in the square 16-QAM is rotated as $2d$, the average symbol power (P_0) of the constellation is:

$$P_0 = 10 \cdot d_2 \tag{4.18}$$

For an average power of 5 dBm (3.16 mW), it can be computed that $d=1.77\cdot10^{-2}\sqrt{W}$ and thus: $a=2.5\cdot10^{-2}\sqrt{W}$, $b=5.6\cdot10^{-2}\sqrt{W}$, and $c=7.5\cdot10^{-2}\sqrt{W}$. In comparison with BACP 16-QAM, here the distances between the middle ring and the outer ring are much smaller. This means that to achieve the same BER, square 16-QAM needs a higher average power than star 16-QAM. The decision method for square 16-QAM is more complicated than that for BACP 16-QAM. Firstly, the decision between three amplitude possibilities for each ring should be made; then, depending on the ring level, the decision is made between four or eight phase possibilities.

4.3.2.2.2 Offset-Square 16-QAM

Square QAM is the most common format for coding and reception, thus it is given here for comparison with the dual amplitude circular phase constellation format. To optimize the phase detection of the middle ring, it is envisaged that the phase differences between neighboring symbols on the middle ring in square 16-QAM should be equal. Thus, the shifted-square 16-QAM is introduced by shifting (rotation of) symbols on the middle ring to obtain equal phase differences between all neighboring symbols, as shown in Figure 4.52. After shifting the symbols on the middle ring, the distances between all direct neighbors are not necessarily equal. In comparison to square 16-QAM, this constellation may offer more robust detection against phase distortions according to the amplitude and phase detection method introduced in the later section.

4.3.2.3 Detection Methods

In the case of differential encoding for the 16-QAM format, there are two different detection methods to demodulate and recover the data in the receiver: (i) Direct detection, and (ii) Coherent detection.

In this section, direct detection means detection with an MZDI or (2×4) 90° Hybrid, while coherent detection is similar except that a local oscillator, a very narrow linewidth laser, is used to mix the signal and its lightwaves to generate the IF or base-band signals while preserving the modulated phase states. Each of these two receiving methods has different implementations, which can be introduced as follows.

4.3.2.3.1 Direct Detection

In contrast to coherent detection, differential decoding is done for direct detection in the optical domain. Indeed, this is equivalent to a self-homodyne coherent detection. This has the disadvantage that the transmitted absolute phase is lost after differential decoding. However, the relative phase (the phase of the differential decoded signal) remains in the electrical domain, making the electrical equalization still possible. Equalization with relative phases is more difficult and yields worse results than equalization with absolute phases. The advantage of direct detection compared to coherent detection is that synchronization of the local laser with the signal light wave is omitted. There are two methods to implement direct detection: one is with MZDI, and the other is with a two input, four output 90° hybrid coupler, which has two optical inputs and four optical outputs.

4.3.2.3.2 Coherent Detection

In a coherent receiver a local oscillator (LO) is used to mix its signal with the incoming signal light wave for demodulation. As a result, it makes it possible to preserve the phase in the electrical domain, which makes electrical equalization very effective in coherent detectors. For coherent detectors, the differential decoding is done in the electrical domain. Depending on the intermediate frequency (f_{IF}) defined as $f_{IF} = f_s - f_{LO,}$ three different coherent methods can be distinguished: (i) Homodyne receiver, (ii) Heterodyne receiver, and (iii) Intradyne receiver. Only the homodyne receiver is included in this section, and the other two are only briefly mentioned.

4.3.2.3.2.1 Homodyne Receiver A receiver is called homodyne when the carrier frequency (f_s) and the local oscillator frequency (f_{LO}) are the same. It means

$$f_{IF} = f_s - f_{LO} = 0 \tag{4.19}$$

In practice, because of the laser linewidth, carrier synchronization must be implemented to set the center frequency and the phase of the LO to the same values as those in the incoming signal. For homodyne receivers, carrier synchronization can be implemented in the optical domain via an optical phase-locked loop (OPLL). Carrier synchronization failure causes degradation in the receiver's performance, but in this document, this effect is not considered, and perfect synchronization in the receiver (a perfect single-spectrum line) is assumed. Alternatively, as mentioned later, a heterodyne receiver using only one 90° hybrid coupler with associated electronic demodulation circuitry can be used to simplify the receiver configuration for coherent detection. Polarization control is another critical difficulty in all coherent receivers, which is also not included in this chapter.

4.3.2.3.2.2 Heterodyne Receiver For this type of reception subsystem, the difference in frequency between the carrier and the LO gives an IF given as:

$$f_{\mathrm{IF}} = f_s - f_{\mathrm{LO}} \neq 0 > B_{\mathrm{opt}} \tag{4.20}$$

where B_{opt} is the optical bandwidth of the channel signal. The IF can be mixed in the electrical domain using synchronous or asynchronous methods to produce channel signals in the low-pass baseband region. In the case of synchronous demodulation, phase synchronization can be done in the electrical domain. The implementation complexity of heterodyne receivers in the optical domain is less than that of homodyne receivers.

4.3.2.3.2.3 Intradyne Receiver Intradyne receiver requires

$$f_{\mathrm{IF}} = f_s - f_{\mathrm{LO}} \neq 0 < B_{\mathrm{opt}} \tag{4.21}$$

The phase synchronization in the intradyne receiver can be done in the digital domain, which makes the intradyne receiver less complex in the optical domain than the homodyne receiver. As compared to heterodyne type, intradyne receiver has the advantage of a smaller processing bandwidth. The disadvantage of the intradyne receiver is the higher requirement for the carrier-laser linewidth. Current technology can provide linewidth laser <10 kHz, which is more than sufficient for SPC transmission systems.

4.3.3 Dual-Amplitude Circular Phase (BACP) 16-QAM

In this section, the transmission performance of the BACP 16-QAM modulation format is evaluated by simulations and compared with square QAM. The implementation of the transmitter for BACP 16-QAM was introduced in Section 4.3.2.2. Both direct and coherent homodyne receivers for BACP 16-QAM are described in Section 4.3.2. In the case of the homodyne receiver (without laser phase noise and ideal frequency locking), two different receiver models can be realized: one without phase estimation and the other with phase estimation. The simulation results of all three receivers are shown in the following sections.

4.3.3.1 Transmitter Design

There are many possibilities to implement the transmitter for BACP-16-QAM, as described in Section 4.3.2. For the simulations, the parallel transmitter shown in Figure 4.48 is implemented. The bit stream enters the differential encoder module after serial-to-parallel conversion.

Referring to Figure 4.48, the operation of the photonic transmitter works as follows:

- The data sequence is converted from serial to parallel by an encoder. The bit sequence is then formed and fed to RF amplifiers to drive the optical modulators. The OPM is a photonic optical phase modulator that provides the variation of the phases of the lightwaves, which are the circular distribution of the phases as shown in Figures 4.46, 4.47, 4.49, and 4.50.
- Bit numbers 1 and 2 are used to set the amplitudes of the constellation points, which are the binary or quad levels for 16 constellations in one level or two levels. These two-bit sequences are fed into RF amplifiers and drive two MZM modulators (MZM1 and MZM2) so that amplitude modulation of lightwaves can be achieved.
- Bit number 4 is used for the third amplitude ring of QAM, equivalent to rings as shown in Figure 4.47.

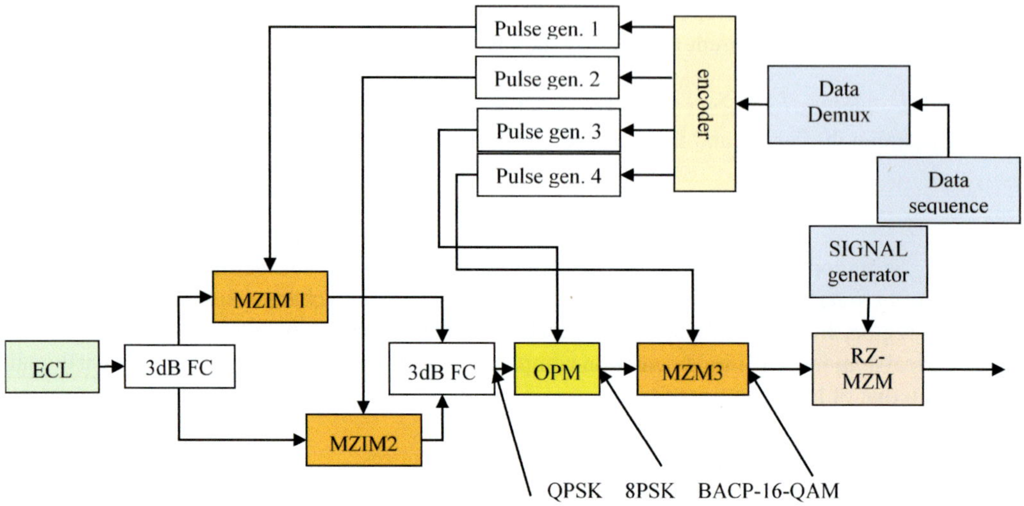

FIGURE 4.48 Schematic diagram of the optical transmitter for Start 16-QAM. Sig gen. = signal generator, pulse gen = pulse generator, ECL = external cavity laser linewidth <10 kHz, MZIM = MZ intensity modulators, OPM = optical phase modulator, Gen. = generator/generation.

- All rings and their constellation points must be synchronized. This is very critical when operating at high speeds, as the electrical paths may not provide the same delay time.
- Finally, the RZ-MZM photonic modulator can be used to generate only the RZ format of pulse sequences. This modulator operates on principles similar to the RZ modulator described in Chapter 4. The RZ format offers a lower average power of the pulse sequences, resulting in fewer nonlinear effects or higher resilience to nonlinearity in fiber. However, the speed is higher as the pulses return to the zero level, thus the pulse width must be less than the bit period, requiring higher bandwidth.

In the differential encoder, the following processes are provided: (i) The four parallel bits arriving at the module are mapped by Gray coding into symbols. The pre-coded symbols are differentially encoded; and (iii) The differentially encoded symbols are mapped again to other symbols and then fed into RF amplifiers to drive the MZMs according to the mapping Table 4.5.

Statistical modulation can be implemented at this stage to ensure the maximum Euclidean distance between consecutive constellation points. This ensures the optimum improvement of the BER and, hence, the receiver sensitivity of BACP 16-QAM or M-ary-QAM.

Each symbol at the output of the differential encoder module is represented by four bits. The bits are sent to pulse formers. The first two bits drive the first two Mach-Zehnder modulators (MZMs) coupled with CW (continuous wave) lightwaves from an external cavity laser. If the input bit is equal to '1', then the output of MZM is '–1'; otherwise, the output of MZM is '1' (after sampling). After combining the output signals of these two MZMs and with an optical phase of $\pi/2$ by an optical delay path in one arm, we obtain the QPSK signal shown in Figure 4.48.

The third bit from the differential encoder output drives an Optical Phase Modulator (OPM) to obtain the 8-PSK signal constellation from the QPSK signal. If this bit is equal to '1', then the QPSK symbol will rotate by $\pi/4$. The 8-PSK signal constellation is shown in Figure 4.49. To achieve the two-level star 16-QAM signal constellation, another MZM is used to generate the second amplitude. If the fourth bit of the differential encoder output is '1', then this output symbol is set on the outer ring of the constellation; otherwise, it is set on the inner ring. This MZM sets the RR of the constellation. The signal constellation after MZM3 is shown in Figure 4.50.

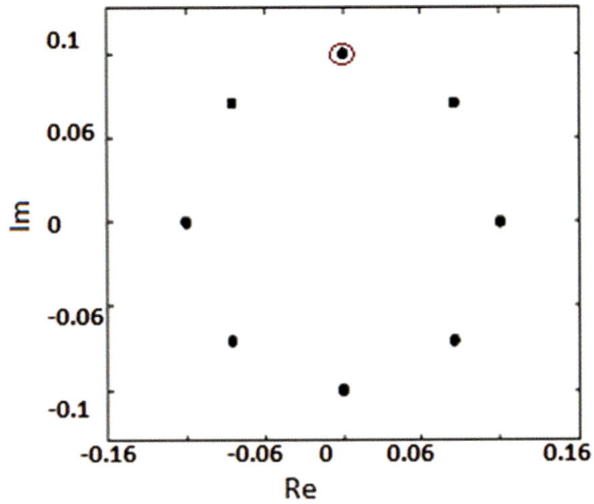

FIGURE 4.49 Constellation of the first amplitude level generated from the optical transmitter for single amplitude and circular phase 8-QAM.

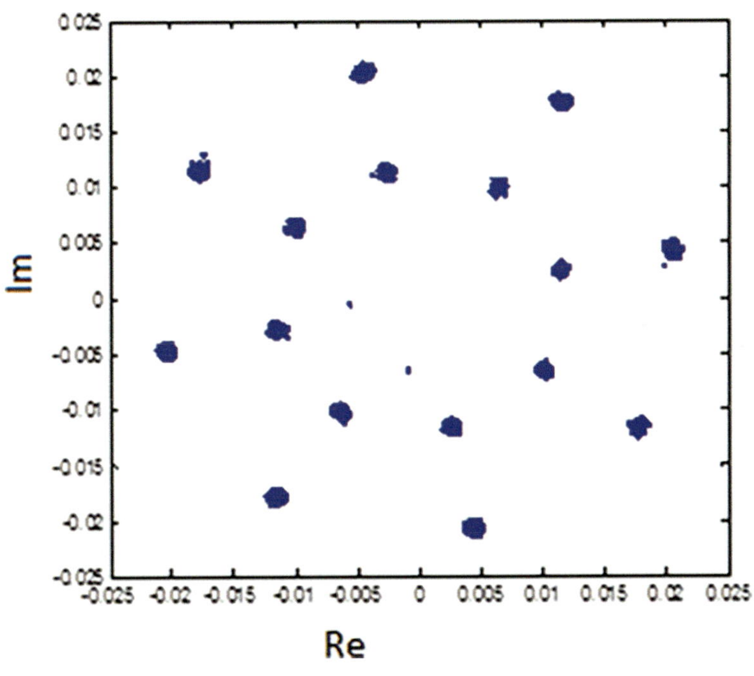

FIGURE 4.50 A constellation of the first and second amplitude level at the receiver of BACP 16-QAM after N-optically amplified spans with dispersion compensated standard fiber lines or multi-haul inter-satellite spatial broadening space photonic communication hops. Im=imaginary, Re=real parts.

The signal constellation in Figures 4.49 and 4.50 can be constructed from the whole constellation in Figure 4.46 with a rotation of $\pi/6$. The advantage of this rotation is that on the real and imaginary axes of the constellation, only eight different amplitude levels instead of nine levels exist. Thus, another OPM can be used between MZM3 and MZM-RZ to rotate the constellation by $\pi/6$ radians. This additional OPM is not shown in Figure 4.48. To increase receiver sensitivity and reduce signal chirp, RZ pulse curving with a duty cycle of 50% can be implemented at the end of the transmitter with an MZM driven by a sinus signal generator (SG). In our simulations, the MZMs shown in Figure 4.48 work in a push-pull mode and are biased in such a way that they act as phase modulators. The received constellation of this BACP QAM after transmission through N-spans/hops of SPC with optical boosters and low-noise amplifications is shown in Figure 4.50, using coherent reception and DSP to recover the original constellation.

4.3.3.2 Receiver for BACP 16-QAM

The implementation of direct detection and coherent receivers for BACP 16-QAM can be described as follows. For coherent detection, there are many possibilities in the digital domain of the receiver to recover the data. The two methods implemented in this study detect the symbol before realizing the differential decoding. The difference is that one method detects the symbol directly using the one described in Section 4.3.2, while the other employs a phase estimation algorithm as described in Section 4.2.2.4, prior to symbol detection to eliminate the phase distortions (phase synchronization between the LO and the received signal). Another possibility, that is not implemented in this work, is to first perform the differential decoding of the incoming signal and followed by symbol detection.

4.3.3.3 Optically Amplified Coherent Reception without Phase Estimation

The structure of the coherent detection receiver is shown in Figure 4.51a. After transmission over fiber, the signal is amplified by an EDFA. The input power of the EDFA can be changed via an

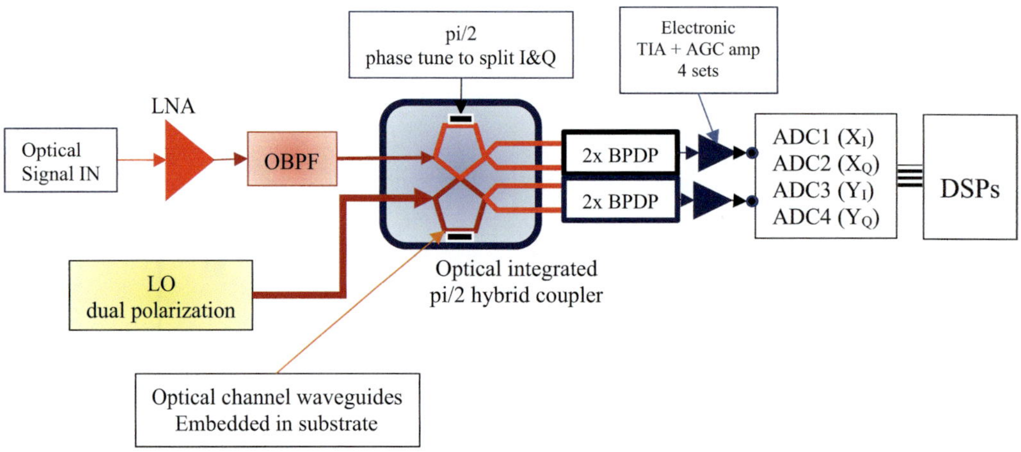

(a)

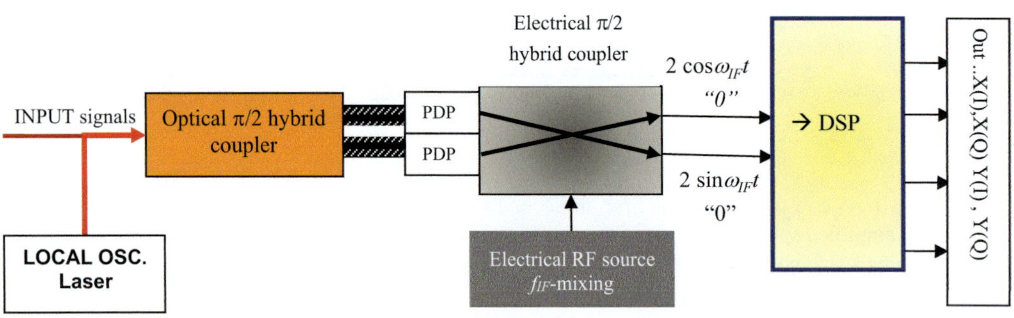

(b)

FIGURE 4.51 Coherent receiver for BACP 16-QAM (a) homodyne and heterodyne I and Q detection model, (b) heterodyne model with optical and electrical π/2 hybrid couplers- electrical detection of I and Q components. PDP = photodetector pair (balanced), X = polarization horizontal, Y = vertical polarization, DSP = digital signal processor, RF = radio frequency.

attenuator to set the OSNR. The output signal of the EDFA is then sent to a band-pass filter (BPF) to reduce the noise bandwidth.

An attenuator is used to set the OSNR as required. The output signal of the EDFA is sent to a band pass (BP) filter to reduce the noise bandwidth and the noise strength after integrating its density over the system bandwidth, as given in the appendices of Chapters 6 and 7. The signal from the LO and the output of BPF are sent to a (2×4) π/2-hybrid and, subsequently, to two balanced detectors. The (2×4) 90°- optical hybrid coupler and the balanced detectors can then demodulate and separate the received signal into in-phase (*I*) and quadrature (*Q*) components via the applied phase shift on the optical phase shifter embedded in the optical path to create a π/2 phase shift or stretching the optical waveguide by a $\lambda/2$ amount or 1550/2 (i.e., 750 nm or there abouts). This can be achieved through the pyroelectric effect by thermal heating of the guided waveguide section. The structure of (2×4) π/2-hybrid coupler, the balanced detectors, and their mathematical descriptions can be found in my published works for coherent optical communication technology.[9] This coherent detector can be further simplified if heterodyne detection is used, as shown in Figure 4.51b.10 However, the electrical signals must be demodulated into the *I*- and *Q*-components, as described above.

Furthermore, an amplitude direct detection in the electronic digital processor must be able to process the magnitude of the vector formed by the *I*- and *Q*-components to decide the amplitude and phase of the received signals, their corresponding position on the constellation, and thus the decoding of the data symbols.

Following the low-pass (LP) filtering and sampling of the *I* and *Q* components, the samples are sent to the symbol detection and differential decoding module. The sampling is done at the center of the eye diagram. In the symbol detection and differential decoding module, we first recover the symbols from the incoming samples and then make the differential decoding of the symbols. To recover the symbols, the *I* and *Q* components are added to a complex signal. According to the original signal constellation, a decision must be made about which symbol our complex sample must be mapped to. This decision has two parts: first, an amplitude decision determines which ring the sample belongs to, and then a phase decision determines which of the eight possible symbols on that ring the sample belongs to.

For amplitude decision, a known bit sequence for the receiver (training sequence) is used and an amplitude threshold a_{th} is defined according to:

$$a_{th} = \left\{\max 1 \le k \le n\left|s_1(k)\right| + \min 1 \le k \le m\left|s_2(k)\right|\right\}/2 \tag{4.22}$$

where $s_1(k)$ is the kth complex received sample of symbols on the inner ring, and n is the whole number of symbols on the inner ring. $s_2(k)$ is the kth received sample of complex symbols on the outer ring, and m is the whole number of symbols on the outer ring. If the amplitude of one sample is larger than the threshold, the symbol is decided to be on the upper ring; otherwise, it is on the inner ring. The EVM can be used to determine the average of the circular lines of the constellation points of the circular phase plane. Then, the BER can be calculated using the EVM technique, as presented in Chapters 3 and 9.

For the phase decision, the complex plane is divided into eight equi-phase intervals. Depending on which interval the phase of the sample falls into, an index from 0 to 7 is assigned to this sample[11]. The next steps are differential decoding and mapping. Amplitude differential decoding and phase differential decoding are done separately. From their results, symbol detection and, subsequently, symbol-to-bit mapping is performed in the inverse order of the encoding in the transmitter.

Alternatively, reception can be conducted with a heterodyne receiver that uses only one single π/2 hybrid optical coupler. This is then detected by two photodiodes and coupled through a π/2 electrical hybrid coupler to detect the *I* and *Q* components for the phase and amplitude reconstruction of the received signals, as shown in Figure 4.51b. The phase estimation can then be estimated by processing *I* and *Q* signals in the electronic domain, as described in the next section.

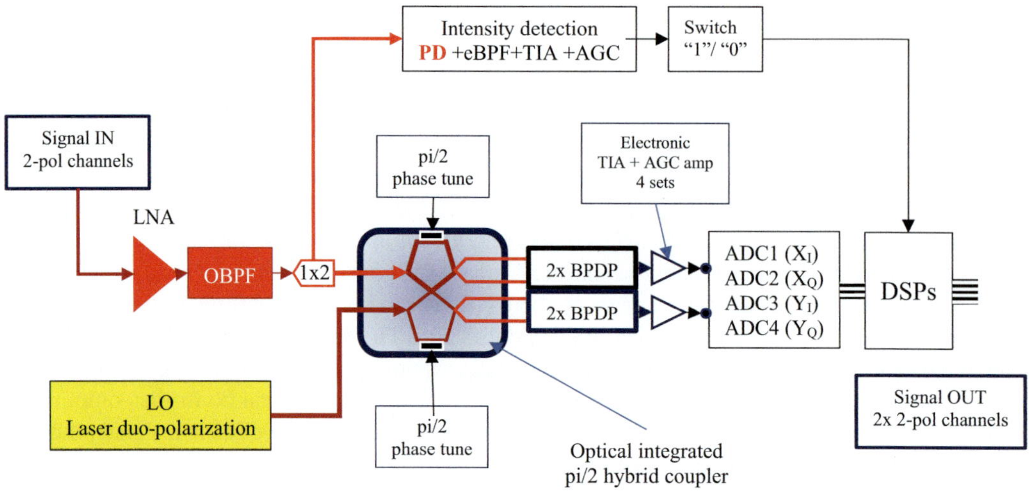

FIGURE 4.52 Direct detection receiver for BACP 16-QAM. ADC = analog to digital converter, DSP = digital signal processor, TIA = transimpedance amplifier, AGC = automatic gain control, OBPF = optical bandpass filter, LO = local oscillator, LNA = low noise amplifier, BPDP = balanced photodetector pair, eBPF = electrical bandpass filter.

Due to the two levels of the BACP 16-QAM, there must be an amplitude detection sub-system which can be implemented using a single photo-detector followed by an electronic preamplifier (TIA and automatic gain control), as shown in Figure 4.52. An electronic processor would be able to determine the position of the received signals on the constellation, and hence decoding can be implemented without any problem. The transmission performance presented in this article would not be affected. Only the technological implementation would be affected, and hence the contribution of electronic noise or optical noise to the receiver can be considered.

4.3.3.4 Coherent Detection Receiver with Phase Estimation

The method and structure of this receiver are almost the same as for the previous receiver shown in Figure 4.51. The difference is that here a phase estimation is done before the phase decision in the symbol detection and differential decoding module. The dispersion of the single-mode optical fiber involves purely phase effects, causing phase rotation and resulting in phase decision errors. The effect of dispersion on the star 16-QAM format is shown in Figure 4.53.

The left plot (a) in the figure shows the sampled input signal into the fiber, while the plot on the right side shows the sampled output of the fiber with a chromatic dispersion of 300 ps/nm. The fiber is considered linear in this simulation. A comparison of point A in both figures shows that this point is spread and rotated due to dispersion. The spatial spreading causes both phase and amplitude distortion, while rotation causes only phase distortion. To solve the problem of phase rotation, the following phase estimation method is implemented. Generally, this phase estimation method is used for phase synchronization between the LO with linewidth and the signal, to replace the optical phase lock loop (OPLL).

The incoming signal after sampling can be described as:

$$c(k) = Ae^{j(\Phi'\text{tot}(k) \pm \text{mod}(k))} \tag{4.23}$$

where Φ'_{tot} is the phase distortion due to dispersion and noise, and ϕ_{mod} is the signal phase that must be recovered. Now ϕ_{tot} must be eliminated from $c(k)$. Φ_{mod} values are $\pi/8$, $3\pi/8$, $5\pi/8$, $\pi/8$, $9\pi/8$, $11\pi/8$, $13\pi/8$ and $15\pi/8$. If these phase values are multiplied by 8, then:

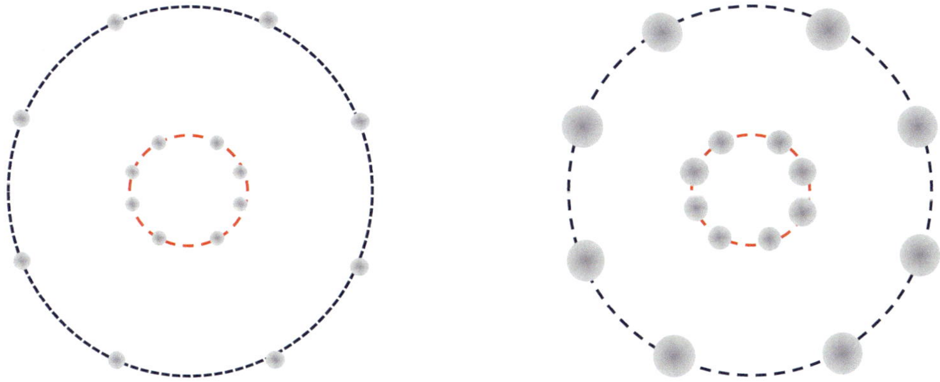

(a) (b)

FIGURE 4.53 Signal constellation (imaginary versus real electrical components) before (at the transmitter) (a) and (b) after (at receiver end), propagation through the optical fiber link or free space multi inter-satellite optical pre-amp and boosting amplifiers in space link with phase rotation or spatial spreading. Arbitrary units in both directions.

$$c_8(k)=c^8(k)=A^8e^{j(8'_{tot}(k)+8\_mod(k))}=A^8e^{j(8'_{tot}(k)+\_)}=-A^8e^{j(8'_{tot}(k))} \tag{4.24}$$

and from this, we obtain:

$$\Phi'_{tot}(k)=1/8\arg(-c_8(k)) \tag{4.25}$$

$\Phi'_{tot}(k)$ is the estimated phase for $\phi'_{tot}(k)$. In the simulations in this work, $\arg(c_8(k))$ is filtered; the filter makes an average between 20 neighbor symbols) to avoid the phase jumps from symbol to symbol. The filter order of 20 is not optimized for each CD or SMD.

Now, the signal phase $\phi_{mod}(k)$ can be estimated as:

$$\phi_{mod}(k)=\arg(c(k))-\phi'_{tot}(k)=\Phi_{mod}(k)+\phi'_{tot}(k)-\phi'_{tot}(k). \tag{4.26}$$

Following this phase estimation, the signal decision takes place with the same method as for the amplitude decision case.

4.3.3.5 Direct Detection Receiver

The block diagram of the direct detection receiver is shown in Figure 4.52. After the optical filter, the signal is split into two branches via a 3dB coupler. We name these two branches: the intensity branch and the phase branch. In the phase branch, the phase differential demodulation is done in the optical domain. The signal and the delayed signal at T_s (symbol duration) are sent into the (2×4) 90°-hybrid and then again into balanced detectors. At the output of the balanced detectors, the in-phase and quadrature components of the demodulated and differentially decoded received signal can be derived. After electrical filtering, the signal is sampled and then sent into the symbol detection module. In the amplitude branch, the amplitude is determined and differentially decoded. After the photodiode, the signal is low-pass filtered, sampled, and then fed into the amplitude detection and differential decoding module. A well-known optical coherent structure can be used to accomplish the amplitude decision and differential decoding. Finally, the in-phase and quadrature components and the amplitude branch are sent to the symbol detection module for further processing as symbol detection and symbol-to-bit mapping.

4.3.3.6 Coherent Receiver without Phase Estimation

In this section, the required OSNR at 10^{-4} BER (using Monte Carlo simulation) is determined for the coherent and incoherent direct detection receivers. For each detection method, the optimum RR is obtained to minimize the OSNR at BER = 10^{-4} (the BER is determined via Monte Carlo simulations). After that, the dispersion tolerance (DT) at a 2dB OSNR penalty at BER = 10^{-4} is determined. The OSNR penalty is only 2dB in this work but can have other values. The OSNR penalty is defined as the OSNR difference in dB between the OSNR of back-to-back case and the OSNR of the other CD values. DT is the CD interval that can be achieved with a certain OSNR penalty. In practice, DT describes how much dispersion (residual dispersion) a system can tolerate with an OSNR penalty smaller than 2dB. The simulations in this work are performed using the simulation tool. The simulations are done for both the linear and the nonlinear channels. The average input power of the nonlinear fiber in this section is set at 5 dBm.

4.3.3.6.1 Linear Channel

It is observed that the optimum RR (RR_{opt}) can be determined for the effects of CD or SMD to minimize the OSNR for a given CD or SMD. The optimum RR changes nearly linearly with the dispersion and can be expressed as

$$RR_{opt} = -0.002|CD| + 1.92 \quad \text{for} \quad \frac{50\,\text{ps}}{\text{nm}} \le |CD| \le \frac{300\,\text{ps}}{\text{nm}} \quad \text{or} \quad 0.3\,\text{nm}/10^3\,\text{km} \tag{4.27}$$

RR_{opt} increases with CD/SMD because an increase in CD or SMD means an increase in phase rotation due to dispersion. This causes more phase detection errors and thus a higher OSNR requirement. Increasing RR reduces the phase error probability but increases the amplitude error probability. RR_{opt} is the best trade-off between phase errors and amplitude errors for each CD or SMD value.

In the B2B case, RR_{opt} is found to be approximately 1.87. The theoretical value for RR_{opt} obtained is 1.77. The difference is because, in the introduced coherent receiver, symbol detection is performed first and then the differential decoding. In the case of differential decoding before symbol detection, the optimum RR is around 1.77 as expected. To determine the RR_{opt} for each CD, RR is changed for each CD. The RR value that yields the smallest OSNR is RR_{opt}. The RR step (which determines the RR accuracy) in simulations is 0.05. To compare the OSNRs, the reference in this work is the OSNR from the back-to-back case. When considering residual dispersion, it is important to see how the system performance changes with changes in RR. The simulation results for three different RR can be seen in Figure 4.57. As shown in Figure 4.57a, the OSNR performance for the B2B case decreases if (in simulated interval) the RR from 1.87 decreases. A degradation of 6.5dB is determined if the RR from 1.87 decreases to 1.32. On the other hand, the (Temporal or spatial) Dispersion Tolerance (DT) at a 2dB OSNR penalty can be increased. The DT for RR =1.87 is 220 ps/nm, while for RR = 1.32 it is 460 ps/nm. To understand the reason for this behavior: for each CD, if the RR decreases from the RR_{opt} value, the OSNR increases rapidly. This means that RR = 1.32 for back-to-back case increases the OSNR, but CD = 300 ps/nm has the minimum OSNR for this RR. The result is that the OSNR for the back-to-back case increases, but for a linear channel without phase estimation, CD = 300 ps/nm decreases. This effect causes a larger DT at a certain OSNR penalty. In a practical system, depending on the higher requirements for OSNR or DT, the RR can be chosen (Figure 4.54).

It can be estimated and observed that the required OSNR for |CD| = 300 ps/nm or 0.3nm per 10^3km of SMD shows a significant increase compared to other values of CDs/SMDs. With |CD| = 350 ps/nm, it is not possible to reach a BER of 10^{-4} pre-FEC. The reason is that |CD| = 350 ps/nm is the limit of the system. For this CD/SMD, it is not possible to transmit error-free even without noise due to phase rotations caused by dispersion (phase detection errors). The signal constellation of the received signal after sampling with CD = 350 ps/nm or equivalent SMD can be seen in Figure 4.55; for example, some of the received symbols of A are over the phase threshold line, and they generate

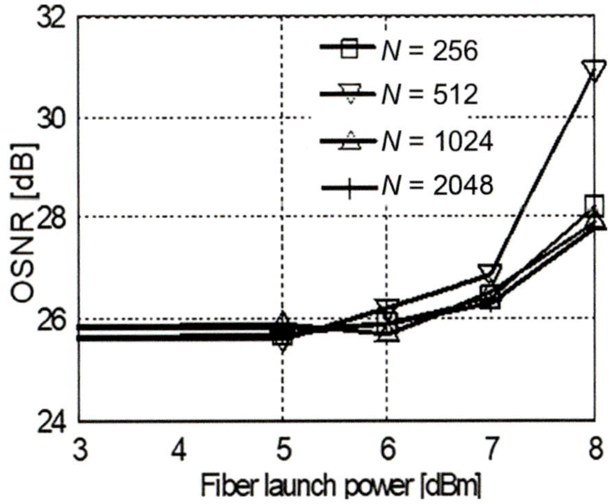

FIGURE 4.54 Simulation result for 42.7 Gbps oOFDM transmission over 640 km of standard single-mode fiber; OSNR (in dB) required for BER = 10^{-3} (Monte Carlo simulation and FEC) as function of launch power (Ref.[12]). N = number of subcarriers. Booster amplifiers in SPC contribute to 3–5 dB noise figure.

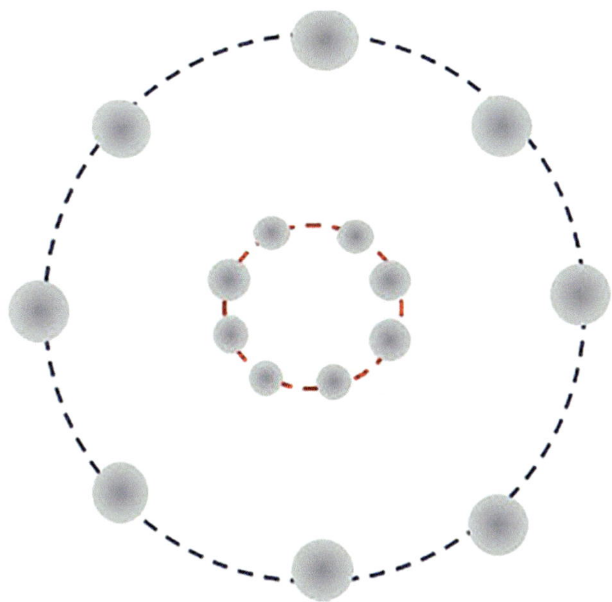

FIGURE 4.55 Received signal constellation in the amplitude of the vertical versus horizontal components at the output of DSP with CD or SMD without noise for a coherent receiver in spatial spreading due to linewidth of the laser and modulated bandwidth of SSB or DSB plus the amplified noises in multi-distance intersatellite amplified link. Pure phase rotation can be detected with DSP rotation of sampled constellation points. SSB = single sideband, DSB = double sideband.

the detection errors. Typical eye diagram at the output of the coherent receiver at the limit of the distortion is shown in Figures 4.56 and 4.57. Phase estimation can be implemented in the digital signal processor. It is noted that the signal constellation is rotated uniformly due to the property of the single-mode fiber that is purely phase distortion. Thus, in the processing of the constellation, it is best if the reference frame of the phasor diagram is rotated to align with the constellation, simplifying the phase estimation process at $\pi/8$ and its multiple values.

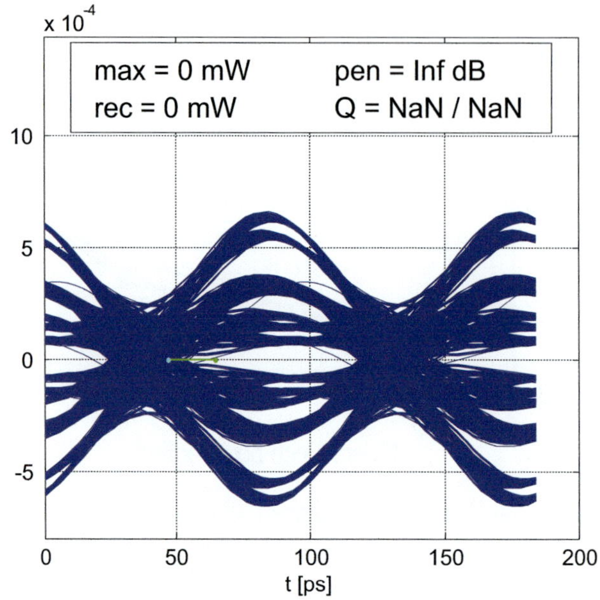

FIGURE 4.56 Typical eye diagram of multi-level N-ACP M-ary-QAM at the output of the coherent receiver under significant distortion limit of the 2A-8P BACP 16-QAM.

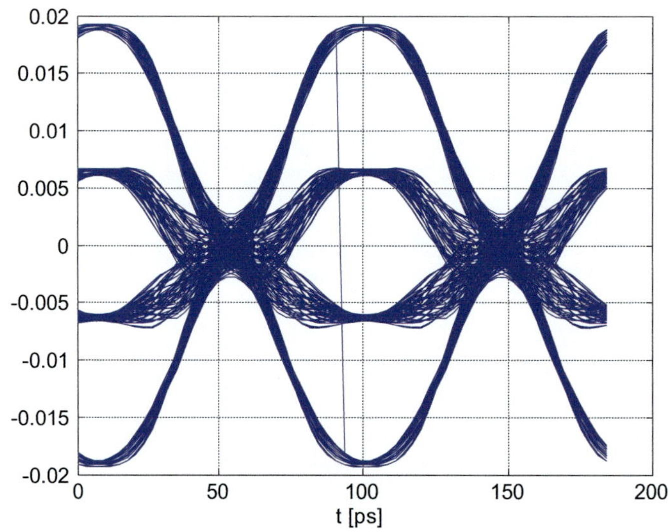

FIGURE 4.57 Eye diagram of the 16 QAM detected at the output of a balanced receiver. A bit rate of 40 Gbps with a baud rate of 10 GB.

4.3.3.6.2 Non-Linear Effects

The optimum RR in the case of non-linear channels for different CDs can be obtained. As mentioned for a linear channel, the RR_{opt} also decreases with an increase of CD or SMD. The difference is that the constellation diagram will not be symmetric. The reason for this is the interaction between dispersion and SPM. For a positive CD or SMD, the dispersion and SPM have a constructive interaction, which results in a better OSNR performance. For a negative CD, the dispersion and SPM have a destructive interaction, resulting in a much poor OSNR performance. It indicates that the phase distortion is higher in the nonlinear channel. It is observed that, for the positive CD/SMD, the slope in nonlinear channel is lower, which means the phase distortion is less. Here, it can be seen again that the required OSNR for the B2B case increases with a decrease in RR. Similarly, for the same reasons as in the linear case, the dispersion tolerance at a 2 dB OSNR penalty increases as well.

4.3.3.7 Remarks

The design of a BACP 16-QAM modulation scheme is described for coherent detection in ultra-high-speed ultra-high capacity optical fiber communication schemes. Two amplitude levels and eight phases (BACP 16-QAM) are considered to offer simple transmitter and receiver configurations while providing optimal receiver sensitivity at the receiver. An OSNR of about 22 dB is required for the transmission of BACP 16-QAM over an optically amplified transmission, dispersion-compensated link. A dispersion tolerance of 300 ps/nm is possible with a 1.0 dB penalty of the eye-opening at 40 Gbps bit rate or 10 Giga-symbols/s (GSy./s), with an OSNR of 18 dB. An OSNR of about 22 dB could be achieved for a 107 Gbps bit rate and a symbol rate of 26.75 G symbols/s. The transmission link consists of several spans totaling 1000 km, with dispersion-compensated optically amplified transmission link. The optical gain of the in-line optical amplifiers is set to compensate for the attenuation of the transmission and compensating fibers, with a noise figure (NF) of 3 dB.

The optical transmitter and receivers incorporating commercially available coherent receivers are structured and sufficient for engineering the optical transmission terminal equipment for a bit rate of 107 Gbps and a symbol rate of 26.3 GSy/s (Sy=Symbol or Baud=Bd). Furthermore, electronic equalization of the receiver phase shift keying signals can be done using blind equalization, which would improve dispersion tolerance much further. For a symbol rate of 10.7 Gbps, this dispersion tolerance for a 1 dB penalty would reach 300 km of standard SMF. This electronic equalization can be implemented without any difficulty at 2 x 10.7 GSy/s.[13] For 107 Gbps bit-rate, similar improvement in dispersion tolerance can be at 2 x 26.5 GSy/s, provided that the electronic sampler can offer more than a 50 GSy/s sampling rate.

4.4 OTHER MULTILEVEL AND MULTI-SUBCARRIER MODULATION FORMATS FOR 100 GBPS TRANSMISSION

Numerous technologies have been introduced in recent years to cope with the ever-growing demand for transmission capacity in optical communications. Although the optical single-mode fiber offers enormous bandwidth in the order of magnitude of 10 THz, efficient bandwidth exploitation has started to become an issue a couple of years ago. Moreover, the limited speed of electronics and electro-optic devices such as modulators and photo receivers are considered bottlenecks for further increase of data rate based on binary modulation.

Similarly, in SPC, we need to increase the transmission capacity by multi-dimensional signals. Our motivation here is different from that of terrestrial fiber optic due to the absorption spectrum of air and vacuum in which there are only some periodic transparent slots. So, we need to pack the channel spectra tightly to achieve the Tbps aggregate bit rate. Furthermore, advanced modulation formats and multi-level signaling can be used to increase the bit rate while occupying the same bandwidth. Naturally, the noise and distinct levels of the eye diagram would be reduced.

For all these reasons, optical modulation formats offering a high ratio of bits per symbol are essential technologies for the next generation of high-speed data transmission. This way, data throughput can be increased while the required bandwidth in the optical domain as well as for electronics is kept at a lower level.

Based on the demand for transmission technologies offering a high ratio of bits per symbol, two promising candidates to achieve a data rate of 100 Gbps per optical carrier are discussed: optical orthogonal frequency division multiplexing and 16-ary multilevel modulation.

In this section, these promising candidates are introduced, and their performance is analyzed by means of numerical simulation and experimentation.

4.4.1 Multi-Level Amplitude-Phase Modulation

Optical modulation formats incorporating four or eight bits per symbol have been investigated in numerous contributions over the last couple of years (e.g. DQPSK[4, 8] and 8-DPSK[4, 9]). However, to carry out the steps from 10 to 40 Gbps data rate using devices designed for 10 Gsymb/s (or GSy/s), a number of 16 bits are required per symbol. The main challenge is to find the optimal combination of ASK and DPSK formats. Several approaches, reviewed in Refs.[3,10], are possible.

The simplest structure can be an extension of a 30 Gbps 8-DPSK by adding phase modulator, resulting in a 40 Gbps 16-DPSK system. In the complex plane, 16 symbols are placed on a unit circle, as shown in Figure 4.58a. Depending on the current bit at the data input of the additional phase modulator, the 8-DPSK symbol is shifted by $\pi/8$ in the case of a '1', while the incoming phase of the symbol is preserved in case of a '0'. Although it seems simple, the experimental implementations have shown that the phase stability of the modulators and corresponding demodulators is very critical. Thus, experimental setup must be stabilized. Moreover, 16-DPSK is sub-optimum regarding the exploitation of the full area that the complex plane offers. That is the ratio of symbol distance and signal power is low, resulting in poor receiver sensitivity.

Similar behavior can be observed in the extreme case of 16-ASK, as shown in Figure 4.58b. Here, the 16 symbols are placed on the positive real axis. The symbol distance is extremely narrow, resulting in poor sensitivity as well.

Much improved performance can be achieved by combining the amplitude and phase shift keying modulation of ASK and DPSK, forming the M-ary ADPSK as described above. There are several combinations of M-ary ADPSK. One approach is the extension of an 8-DPSK by two rings of ASK levels. Thus the 16 symbols appear as two rings in the complex plane with 8 symbols per ring, as shown in Figure 4.58c. Alternatively, this topology is known as BACP-16 QAM. A second structure

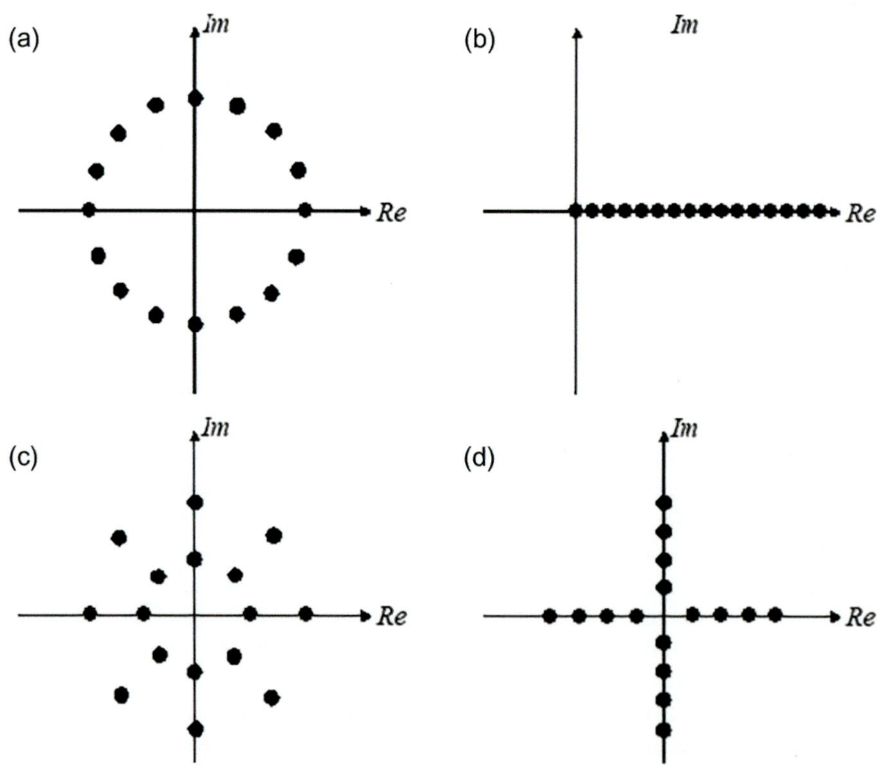

FIGURE 4.58 Constellation of symbols in the complex plane for (a) 16 CP-DPSK, (b) 16-ASK, (c) BACP 16-QAM, (d) 16-ADPSK. Im=imaginary, Re=real.

is formed by combining DQPSK with four-level ASK (or equivalently M-ary ADPSK), resulting in four rings with four symbols, each as analyzed in Section 4.3 and shown in Figure 4.58d. Both structures effectively utilize the complex plane. However, both structures require that the sensitivity or the magnitude (diameters) of the rings be optimized to balance the sensitivity performance of DPSK and the ASK geometrical distribution. Especially, the inner ring has to be of sufficient size to enable distinction of the different phases of the symbols on this ring. They are limited by the nonlinear effects of the transmission fiber and the noise contributed by the receiver and the in-line optical amplifiers. Hence, the distance between the constellations is limited by these two factors.

A strategy to mitigate this trade-off was introduced in[4, 8] by using a special pulse shaping called inverted RZ. For binary ASK in conjunction with inverse RZ, a '0' is encoded as a temporary breakdown of the optical power, while a '1' is represented the optical power remaining at a high level. Using this pulse shaping for the M-ary ADPSK format, the four levels of the QASK part are transmitted by means of four different values for temporary decay of optical power. The DQPSK part, however, is transmitted by modulating the phase of the signal in the time slot between, which implies that in the transmitter the phase of the signal can be detected while the signal has maximum power.

Measurement results for this modulation format are depicted in Figure 4.59, where the BER is plotted as a function of the OSNR measured within a 0.1 nm optical filter bandwidth. The main outcome is that the DQPSK part is insignificantly disturbed by the additional QASK part. Moreover, even after transmission over 75 km of standard single-mode fiber, the DPSK part shows a very low penalty. In contrast, the QASK component inherently shows low performance due to the low symbol Euclidean distance. Improvement might be achievable by optimizing the duty cycle of inverse RZ

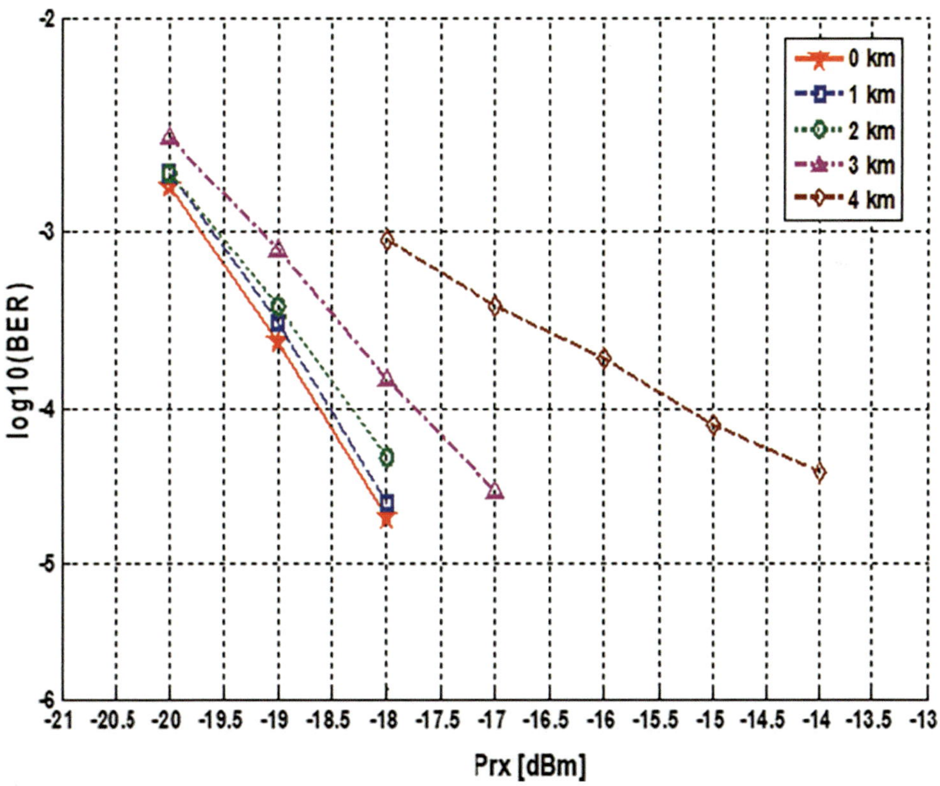

FIGURE 4.59 Plot of BER versus receiver sensitivity (back-to-back) and temporal/spatial dispersion tolerance for the 2ASK_8-DPSK modulation scheme. Prx (average power received by the receiver), BER = bit error rate (dimensionless). The parameter in equivalent km of dispersion of spatial broadening/spreading.

and the ring ratio. A simulated eye diagram of the 16-square QAM is shown, with the constellation before and after the optical transmission link of two optical spans with a 2 km SSMF mismatch, as shown in Figure 4.43.

4.4.2 Optical Orthogonal Frequency Division Multiplexing (oOFDM)

Orthogonal frequency division multiplexing (OFDM) is a transmission technology primarily known from wireless communications and wired transmission over copper cables. It is a special case of the widely known frequency division multiplexing (FDM) technique, in which digital or analog data is modulated onto a certain number of carriers and transmitted in parallel over the same transmission medium. The main motivation for using FDM is that, due to parallel data transmission in the frequency domain, each channel occupies only a small frequency band. Signal distortions originating from frequency-selective transmission channels, such as fiber chromatic dispersion, can be minimized. The special property of OFDM is characterized by its very high spectral efficiency. While conventional FDM has limited spectral efficiency due to the selectivity of the bandpass filters required for demodulation, OFDM is designed so that the different carriers are pair wise orthogonal. As a result, at the sampling point, inter-carrier interference (ICI) is suppressed, though the channels are allowed to overlap spectrally.

Orthogonality is achieved by placing the different RF carriers onto a fixed frequency grid and assuming rectangular pulse shaping. It can be shown that in this special case, the OFDM signal can be described as the output of a discrete inverse Fourier transform with the parallel complex data symbols as input. This property has been one of the main driving aspects of OFDM in the past, as modulation and demodulation of a high number of carriers can be realized by simple digital signal processing (DSP) instead of using many local oscillators in the transmitter and receiver. Recently, OFDM has become an attractive topic for digital optical communications. It is just another example of the current tendency in optical communications to consider technologies originally known from classical digital communications. Using OFDM appears to be very attractive, as the low bandwidth occupied by a single OFDM channel increases the robustness towards fiber dispersion drastically, allowing the transmission of high data rates of 40 Gbps and higher over hundreds of kilometers without the need for dispersion compensation. Similarly, over SPC inter-satellite constellation hops of space distance, the coherency and the spatial broadening create the group delay difference. The multi-frequency components with orthogonality make the interaction between these components self-coherent, thus minimizing the broadening of the spatial beam over space distance.

In the same way as with modulation formats like DPSK or DQPSK introduced in recent years, the challenge for optical system engineers with OFDM is to adapt a classical technology to the special properties of the optical channel and the requirements of optical transmitters and receivers.

Thus, two approaches have been reported recently. An intuitive approach introduced by Lorente et al.[14] makes use of the fact that the wavelength-division multiplexing (WDM) technique already realizes data transmission over a certain number of different carriers. By means of special pulse shaping and carrier wavelength selection, orthogonality between the different wavelength channels can be achieved, resulting in the so-called orthogonal WDM technique (OWDM). However, in this approach, the option of simple modulation and demodulation by means of discrete Fourier transforms (DFT) cannot be utilized, as this kind of digital signal processing is not available in the optical domain.

As an alternative and popular method, an electrical OFDM signal is generated through electrical signal processing and then modulated onto a single lightwave carrier. This approach is known as optical OFDM (oOFDM). The modulation is a two-step process: first, the electrical OFDM signal is already a broadband bandpass signal, which is then modulated onto the optical carrier. Second, to increase data throughput, oOFDM can be combined with WDM, resulting in a multi-Tb/s transmission system, as shown in Figure 4.60. Nevertheless, oOFDM itself offers different implementation options. One important issue is optical demodulation, which can be realized either of direct

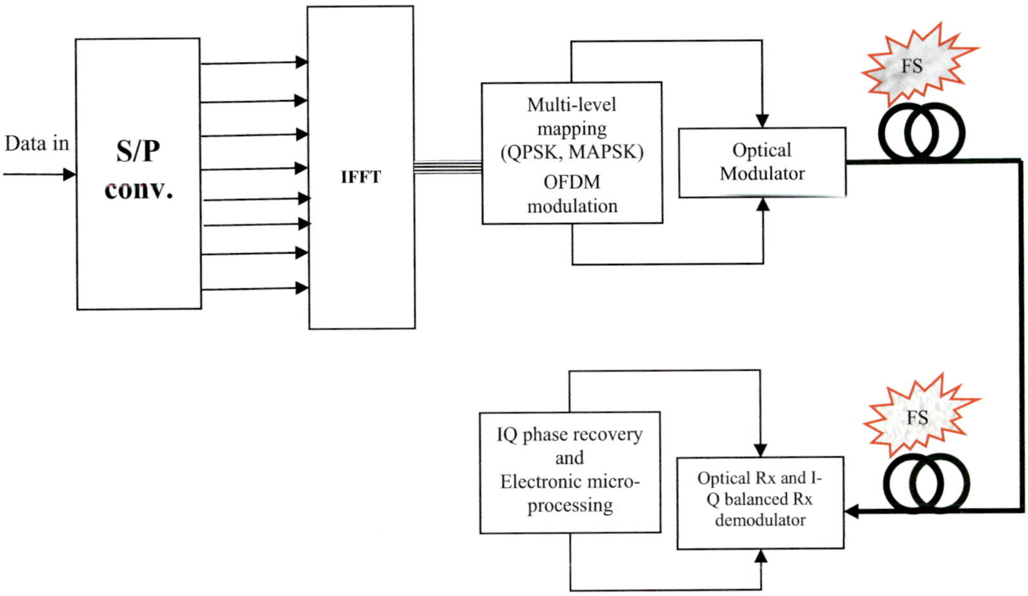

FIGURE 4.60 Schematic diagram of optical OFDM transmitter of a long-haul optical transmission system using multi-level modulation formats. S/P=serial to parallel converter. FS = free space.

detection (DD) or coherent detection (CD) using a local oscillator. DD is preferable due to its simplicity. However, for DD, the optical intensity must be modulated. Since the electrical OFDM signal is quasi-analog with zero mean and a high peak-to-average ratio, the majority of the optical power has to be wasted on the optical carrier (i.e., an additional DC value of the complex baseband signal), resulting in low receiver sensitivity. For CD, the bandwidth efficiency is twice as high as for DD, as pure intensity modulation inherently generates a double-sideband signal. With CD, a complex optical I-Q modulator, composed of two real modulators in parallel followed by superposition with a $\pi/2$ phase shift, allows the transmission of twice as much data within the same bandwidth. For intensity modulation, bandwidth efficiency may be increased by suppressing one of the redundant sidebands, resulting in optical OFDM with single-sideband (SSB) transmission. First, the serial data can be converted to parallel streams. These parallel data sequences are then mapped to QAM constellation in the frequency domain, then using IFFT, converted it back to the time domain. The time-domain signals are in the *I*- and *Q*-components, which are then fed into *I* and *Q* optical modulators. This optical modulation can be DQPSK or any other multilevel modulation subsystem. At the end of the optical fiber transmission, the *I*- and *Q*-components are detected either by direct detection or coherent detection. For coherent detection, a 2×4 90-degree hybrid coupler is used to mix the polarizations of the local oscillator and the received signals. The outputs of the couplers are then fed into the balanced optical receivers. The mixing of the local laser source with the signals preserves the phase of the signals, which are then processed by a high-speed electronic processor. For direct detection, the *I* and *Q* components are detected differentially; the amplitude and phase detection are then compared and processed similarly to the coherent case.

To show the robustness of oOFDM towards fiber dispersion and fiber nonlinearity, numerical simulations are carried out for a data stream with a 42.7 Gbps data rate. The number of OFDM channels varies between $N_{min}=256$ and $N_{max}=2048$. A guard interval of 12 ns can be inserted, a strategy inherent to OFDM technology that ensures orthogonality of the different channels in the presence of a transmission channel with memory. For optical modulation, intensity modulation using a single Mach-Zehnder modulator in conjunction with SSB filtering and direct detection was implemented. The nonlinear optical transmission channel consisted of eight 80 km non-DCF spans of SSMF. In SPC, the nonlinearity arises from the short length and the high-power output of the HPA booster amplifier at the transmitter before launching into space. As a criterion for performance, the required OSNR for a BER of 10^{-3} (Monte Carlo) is measured. Using FEC (Forward Error Correcton), this BER can reach 10^{-9} or error free.

Figure 4.54 shows the required OSNR as a function of fiber launch power for different values of N. The most important result is that transmission is possible over 640 km of SSMF without any dispersion compensation. This can be explained by the fact that even for the lowest value of $N_{min=}256$, each sub-channel occupies a bandwidth of approximately 42.7 GHz/256=177 MHz resulting in high robustness towards fiber dispersion or spatial mode dispersion. This can be equal to an equivalent distance of thousands of kilometers in free vacuum.

The principal difficulties of optical OFDM are that the pure delay due to the variation of the refractive index of the fiber with respect to the optical frequency leads to bunching of the sub-channels and hence an increase in the optical power. Thus, unexpected SPM may occur in a random manner.

4.4.3 The 100 Gbps 8-DPSK_2-ASK Circular Phase 16-QAM

4.4.3.1 Introduction

The multi-level modulation scheme enables the transmission baud rate to be reduced, thus, obtaining better spectral efficiency. Another significant advantage of this modulation scheme is that it reduces the requirement for high-speed processing electronics. This is of particular interest for high-speed optical transmission systems.

This part of the report investigates a multilevel modulation scheme, which has eight phase and two amplitude levels. This scheme, which is named in short 8-DPSK_2-ASK, effectively utilizes 4 bits per symbol for transmission, with the first 3 bits coding phase information and the coding of the amplitude levels is implemented with the fourth bit. As a result, the transmission baud rate is equivalently a quarter of the bit rate from the bit pattern generator.

This section is organized as follows: Section 4.4.3.2 presents a detailed description of the optical transmitter for generating 8-DPSK_2-ASK signals. In Section 4.4.3.3, the detailed configuration of the receiver is provided. The configurations of the optical transmitter and receiver are referenced from those reported by Djordjevic and Vasic.[15] Section 4.4.3.4 provides study on dispersion tolerance and transmission performance of the 8-DPSK_2-ASK scheme. Finally, a short summary of the report is provided.

4.4.3.2 Configuration of 8-DPSK_2-ASK Optical Transmitter

There have been several configurations of optical transmitters for generating multi-phase/level optical signals using amplitude or phase modulators arranged in either serial or parallel configurations. However, the optical transmitters reported in Refs.[16,17,18] require a pre-coder with high complexity. On the contrary, the configuration reported by Ivan, et al.[14] utilizes the Gray mapping technique to differentially encode the phase information, and this significantly reduces the complexity of the optical transmitter. In addition, as elaborated in more detail in Section 4.3, this pre-coding technique enables a detection scheme using I-Q demodulation techniques, as equivalently used in coherent transmission systems.

The optical transmitter of the 8-DPSK_2-ASK scheme employs the I-Q modulation technique with two Mach-Zehnder Modulators (MZMs) in parallel and a $\pi/2$ optical phase modulator, as shown in Figure 4.61. At each kth instance, the absolute phase of transmitted lightwaves θ_k is expressed as: $\theta_k=\theta_{k-1}+\Delta\theta_k$ where θ_{k-1} is the phase at the $(k-1)$th instance and $\Delta\theta_k$ is the differentially coded phase information. The encoding of this $\Delta\theta_k$ for generating 8-DPSK_2-ASK modulated optical signals (4 bits per one transmitted symbol) follows the well-known Gray mapping rules. This Gray mapping phasor diagram is shown in Figure 4.62. The phasor is normalized with the maximum energy on each branch, i.e. $E_1/2$.

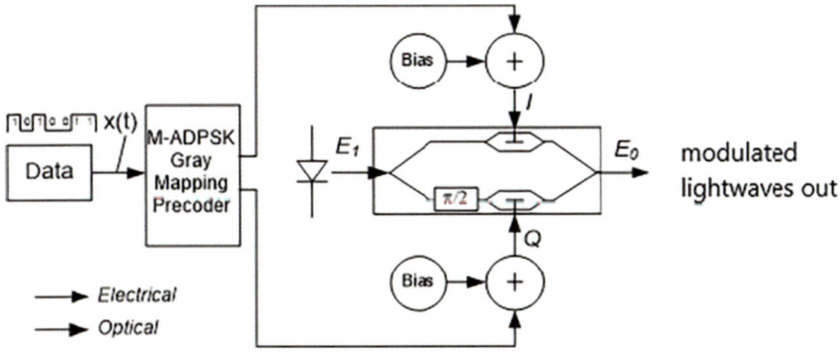

FIGURE 4.61 Optical transmitter configuration of the 8-DPSK_2-ASK modulation scheme.

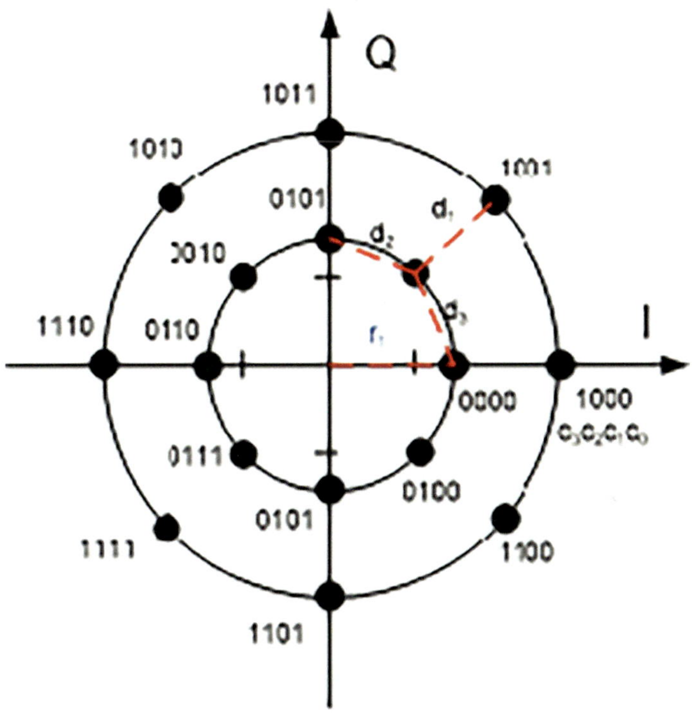

FIGURE 4.62 Gray mapping for optimal 8-DPSK_2-ASK modulation scheme.

The amplitude levels are optimized so that the Euclidean distances d_1, d_2, and d_3 are equal, i.e. $d_1 = d_2 = d_3$. After derivation, we obtain $r_1 = 0.5664$. The I and Q field vectors corresponding to Gray mapping rules from the M-ADPSK pre-coder (see Figure 4.61) are provided in Table 4.6.

The above-described transmitter configuration can be replaced with a dual-drive MZIM. The explanation and derivation for generating 8-DPSK_2-ASK optical signals are also based on the phasor diagram of Figure 4.62. In this case, the output field vector is the summation of two component field vectors, each of which is determined not only by the amplitude but also by the initial bias phases.

4.4.3.3 Configuration of 8-DPSK_2-ASK Detection Scheme

The detection of 8-DPSK_2-ASK optical signals is implemented using two Mach- Zehnder Delay Interferometric (MZDI) balanced receivers (see Figure 4.63). Several key points in this detection

TABLE 4.6
***I* and *Q* Field Vectors in 8-DPSK_2-ASK Modulation Scheme Using Two MZIMs in Parallel**

Binary Sequence	$\Delta\theta_k$, Amplitude	I_k	Q_k
1000	(0, 1)	1	0
1001	(π/4, 1)	$\sqrt{2}/2$	$\sqrt{2}/2$
1011	(π/2, 1)	0	1
1010	(3π/4, 1)	$-\sqrt{2}/2$	$\sqrt{2}/2$
1110	(π, 1)	−1	0
1111	(−3π/4, 1)	$-\sqrt{2}/2$	$-\sqrt{2}/2$
1101	(−π/2, 1)	0	−1
1100	(−π/4, 1)	$\sqrt{2}/2$	$-\sqrt{2}/2$
0000	(0, 0.5664)	1 0.5664	0
0001	(π/4, 0.5664)	$\sqrt{2}/2*0.5664$	$\sqrt{2}/2*0.5664$
0011	(π/2, 0.5664)	0	1*0.5664
0010	(3π/4, 0.5664)	$-\sqrt{2}/2*0.5664$	$\sqrt{2}/2*0.5664$
0110	(π, 0.5664)	−1*0.5664	0
0111	(−3π/4, 0.5664)	$-\sqrt{2}/2*0.5664$	$-\sqrt{2}/2*0.5664$
0101	(−π/2, 0.5664)	0	−1*0.5664
0100	(−π/4, 0.5664)	$\sqrt{2}/2*0.5664$	$-\sqrt{2}/2*0.5664$

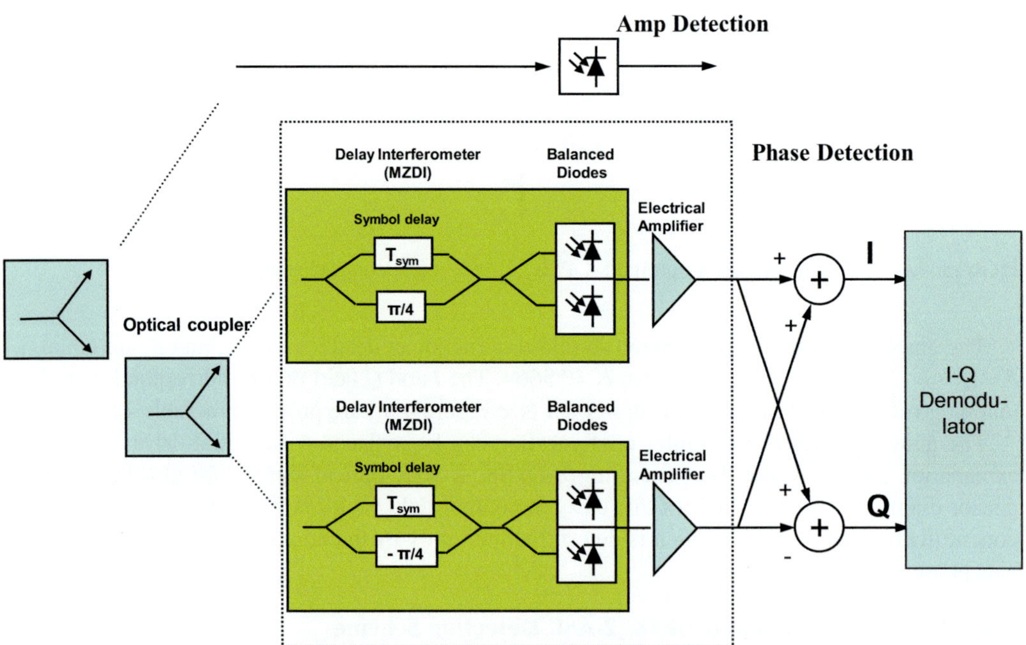

FIGURE 4.63 Detection configuration for the 8-DPSK_2-ASK modulation scheme.

structure can be stated as (i) The MZDI introduces a delay corresponding to the baud rate. (ii) One arm of the MZDI has a $\pi/4$ optical phase shifter while the other arm has an optical phase shift of $-\pi/4$. (iii) The outputs from the two balanced receivers are superimposed positively and negatively, which leads to the I and Q detected signals respectively. The I and Q detected components are expressed as $I = \mathrm{Re}\left\{E_k E_{k-1}^*\right\}$ and $Q = \mathrm{Im}\left\{E_k E_{k-1}^*\right\}$; (iv) I-Q detected components are demodulated using the popular I-Q demodulator in the electrical domain, which involves a $\pi/2$ phase shift in the electrical signal. Alternatively, this $\pi/2$ phase shift can be used by the $\pi/2$ hybrid coupler given in Chapter 4 for coherent QPSK complex signal detection.

These detected signals are then sampled and represented as shown in the signal constellation.

4.4.3.4 Transmission Performance of 100 Gbps 8-DPSK_2-ASK Scheme

The performance characteristics of the 8-DPSK_2-ASK scheme operating at a 100 Gbps bit rate are studied in terms of receiver sensitivity, dispersion tolerance, and feasibility for long-haul transmission. Bit error rates (BER) are the pre-Forward Error Correction (FEC) BERs, and the pre-FEC limit is conventionally referenced at 2e-3. In addition, the BERs are evaluated using the Monte Carlo method.

4.4.3.4.1 Power Spectrum

The power spectrum of 8-DPSK_2-ASK optical signals is shown in Figure 4.64. It can be observed that the main lobe spectral width is about 25 GHz as the symbol baud rate of this modulation scheme is equal to a quarter of the bit rate from the bit pattern generator. The harmonics are not highly suppressed, thus requiring the bandwidth of the optical filter to be necessarily large in order not to severely distort signals.

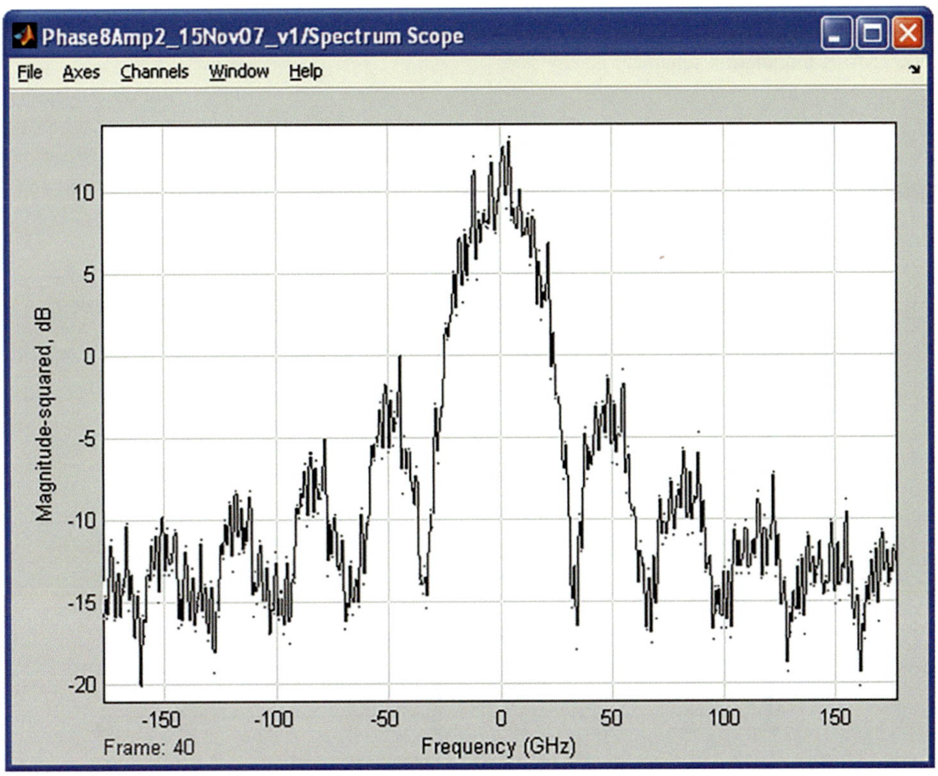

FIGURE 4.64 Power spectrum of 8-DPSK_2-ASK signals.

4.4.3.4.2 Receiver Sensitivity and Dispersion Tolerance

The receiver sensitivity is studied by connecting the optical transmitter of the 8-DPSK_2-ASK scheme directly to the receiver in a back-to-back setup (see Figure 4.65). On the other hand, the dispersion tolerance is studied by varying the length of SSMF from 0 to 5 km ($|D| = 17$ ps/(nm km)). Received powers are varied using an optical attenuator. The optical Gaussian filter has BT = 3 (B is approximately 75 GHz). Modeling of receiver noise sources comprises shot noise, an equivalent noise current density of $20\,\text{pA}/\sqrt{\text{Hz}}$ at the input of the trans-impedance electrical amplifier, and a dark current of 10 nA for each of the two photodiodes in the balanced structure. A fifth order Bessel electrical filter with a bandwidth of BT = 0.8 is used.

The numerical BER curves for the receiver sensitivity for cases of 0–5 km SSMF are shown in Figure 4.59. The receiver sensitivity of the 8-DPSK 2-ASK scheme is approximately –18.5 dBm at BER = le-4. The receiver sensitivity at BER = le-9 can be obtained by extrapolating the BER curve for the 0 km case. The power penalty versus residual dispersion results are then obtained and plotted in Figure 4.66. It is realized that a 2-dB penalty occurs for a residual dispersion of approximately 60 ps/nm, which is equivalent to 3.5 km of SSMF.

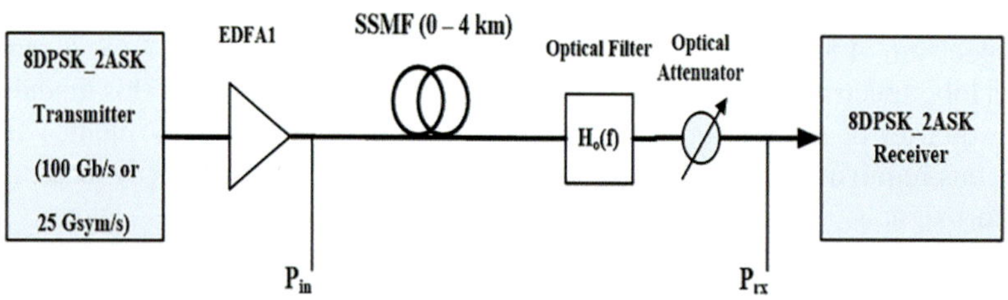

FIGURE 4.65 Set up for the study of receiver sensitivity (back-to-back) and dispersion tolerance for the 8-DPSK_2-ASK modulation scheme.

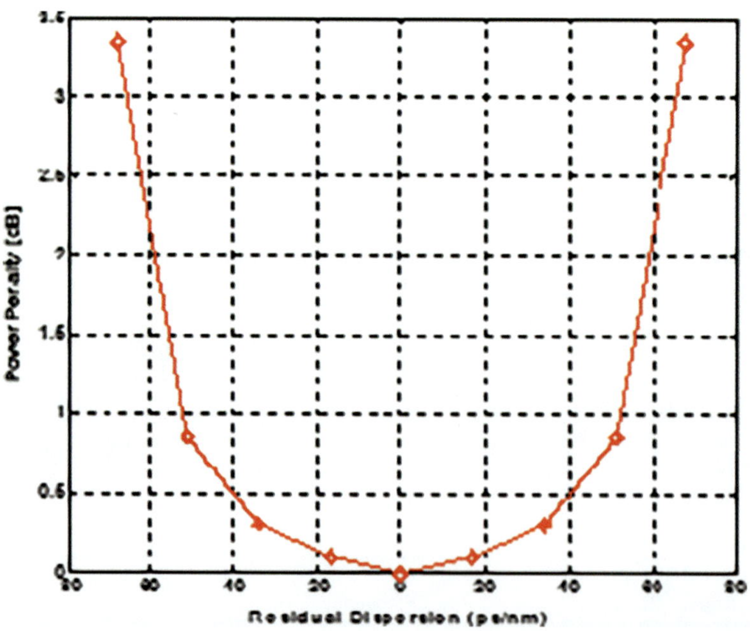

FIGURE 4.66 Power penalty due to residual dispersions for the 2-ASK_8-DPSK modulation scheme.

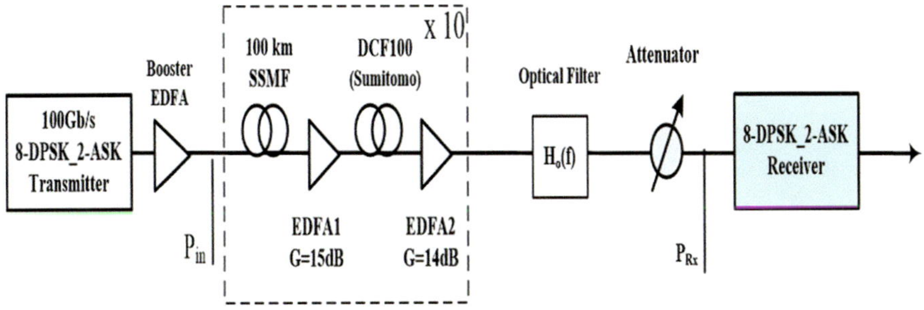

FIGURE 4.67 Transmission set up of 1100 km optically amplified and fully compensated multi-span fiber link, that is, attenuation only or equivalent to space photonic transmission of a loss of 25 dB.

4.4.3.4.3 Long-Haul Transmission

The long-haul transmission performance of this modulation format is conducted over 10 optically amplified and fully compensated spans, with each span composed of 100 km SSMF and 10 km DCF100 (Sumitomo fiber). As a result, the length of the transmission fiber link is 1100 km. This long-haul range is selected to reflect the distance between Melbourne and Sydney. The wavelength of interest is 1550 nm, and the dispersion at the end of the transmission link is fully compensated. The simulation setup is shown in Figure 4.69. Additionally, fiber attenuation due to SSMF and DCF is also fully compensated by using two EDFAs with optical gains, as depicted in Figure 4.67. These EDFAs have a Noise Figure (NF) set at 5 dB.

Numerical transmission BERs are plotted against received powers in Figure 4.68 and compared to the back-to-back BER curve. It can be observed that the BER curve for 1100 km follows a linear trend and feasibly reaches 1e-9 if extrapolated, as shown in Figure 4.68. It should be noted that this transmission performance can be significantly improved with the use of a high-performance FEC scheme.

4.4.3.4.4 Remarks

The ever-increasing bandwidth demand in telecommunication networks based mainly on optical fiber communication technology indicates that low bandwidth-efficient modulation formats such as ASK will no longer satisfy transmission capacity demands, and new advanced optical modulation schemes should replace ASK in the near future. Advanced optical modulation schemes, especially the multi-level amplitude and phase schemes presented above, are able to: (i) provide long reach, error-free, and high transmission capacity; (ii) have high bandwidth efficiency (number of bits/Hz parameter); (iii) push the bit rate well above what could be offered by electronic technology, e.g., 100 Gbps with detection at the symbol rate; (iv) tolerate dispersion and nonlinearity; (v) maximize the utilization of the existing optical network infrastructure.

Current developments in photonic technology have enabled the use of differential phase modulation and demodulation in the optical domain. Presently, BDPSK and DQPSK formats have received great attention due to their improvement in receiver sensitivity compared with ASK. Furthermore, the RZ and CSRZ formats assist in combating nonlinear effects. However, as far as transmission capacity and bandwidth efficiency are concerned, MADPSK modulation formats offer better performance in trading off for its complexity in receiver structures.

Alternatively, there are other possible multi-level modulation schemes that could further improve optical transmission performance.[19]

4.4.4 Offset MADPSK Modulation

The binary DPSK and quaternary DPSK (DQPSK) are special cases of a more general class of differential phase modulation formats that map data bits into the phase difference between neighboring symbols. This phase difference $\Delta\phi_i$ can be described as:

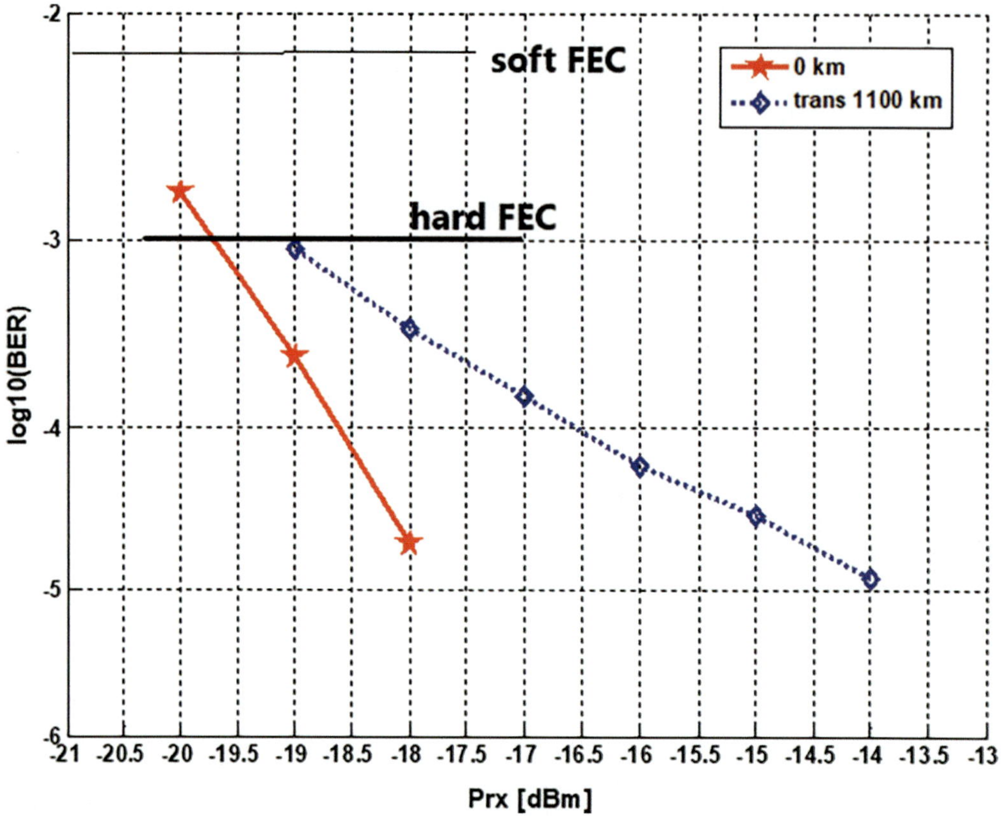

FIGURE 4.68 Plot of $\log_{10}$ (BER) versus receiver sensitivity (average received power in dBm) for 8DPSK 2ASK modulation format transmission. These are pre-FEC BER. For Soft FEC, the BER=2e-2 and hard FEC at BER=2e-3 as pre-FEC BER. Note: 1000 km fiber terrestrial distance – equivalent to a gray delay of 450,000 km space distance. 0 km=back-to-back transmission.

$$\Delta\phi_i = \theta + \frac{2\pi(i-1)}{M}, \quad i = 1, 2, \ldots, M-1 \tag{4.28}$$

where θ is the initial phase, M is total number of phase levels. This class of formats is called offset DPSK (O-DPSK) and is denoted as θ-M-DPSK. Specifically, with $\theta = 0$, M=2, or M=4, 0-M-DPSK become 0-2-DPSK or 0-4-DPSK, which are the conventional binary DPSK and DQPSK mentioned above.

O-DPSK has been used to transmit over satellite non-linear channels because its phase transition is smooth and can avoid 180° phase jumps. As fiber mediums also exhibit nonlinearity, this modulation format attracts our attention as a candidate, along with ASK, for creating a new multi-level modulation format, possibly termed Offset MADPSK.

4.4.5 Multi-Level Amplitude-Minimum Shift Keying (mASK-MSK) Modulation

Section 4.4 of Chapter 4 addresses the principles of the dual-level MSK modulation format. It features continuous phase along the bit period, and the phase of the carrier is continuous at the transition point between the two consecutive bits. Thus, minimum shift keying (MSK) is a form of OQPSK with sinusoidal pulse weighting. In MSK, data bits are first coded into bipolar signals ±1, which are then separated into $V_I(t)$ and $V_Q(t)$ streams consisting of even and odd bits, respectively. In the next stage, $V_I(t)$ and $V_Q(t)$ are used to modulate a carrier f_c to create an MSK signal $s(t)$, which can be presented as

$$s(t) = V_I(t)\cos\left(\frac{\pi t}{2T}\right)\cos\left(2\pi f_c t\right) + V_Q(t)\cos\left(\frac{\pi t}{2T}\right)\sin\left(2\pi f_c t\right) \quad (4.29)$$

MSK is a well-known modulation format in wireless communications with efficient spectral characteristics due to the high compactness of its main lobe compared with DPSK and the high suppression of the side lobes. These characteristics indicate that the MSK signal is highly dispersion-tolerant. A combination of MSK (minimum shift keying) and ASK into a multi-level amplitude-minimum shift keying (mASK-MSK) modulation would further improve transmission performance without increasing the complexity of the detection scheme.

Multi-level techniques for 100 Gbps are given; in particular, OFDM with multi-carrier and multilevel amplitude modulation with orthogonality between adjacent channels have been proven to be cost-effective and appropriate for current electronic technologies. Two interesting approaches to achieve data transmission of 40 Gbps and beyond (e.g. 100 Gbps Ethernet) based on low symbol rate are discussed. On the one hand, oOFDM can combine a large number of parallel data streams into one broadband data stream with high spectral efficiency. Simulation results are shown for different values of the number of parallel data streams in a nonlinear environment. On the other hand, 16-ary modulation formats enable 40 Gbps transmission with 10 GSym/s (i.e., 100 Gbps with 25 GSym/s). For a special case, namely inverse RZ Single ACP-QPSK, measurement results are demonstrated.

4.4.6 N-Amplitude Circular Phase QAM and Coherent Detection

This section describes two optical transmission schemes: BACP 8DPSK and BACP 16-QAM for 100 Gbps or 25 G symbols/s using both coherent and incoherent transmission and detection techniques.

Firstly, the design of a BACP 16-QAM modulation scheme is proposed for coherent detection in ultra-high-speed, ultra-high-capacity optical fiber communications schemes. Two amplitude levels and eight phases (BACP 16-QAM) are considered to offer significant simple transmitter and receiver configurations and at the same time the best receiver sensitivity at the receiver. An optical SNR of about 22 dB is required for the transmission of BACP 16-QAM over an optically amplified, dispersion-compensated link. Dispersion tolerance of 300 ps/nm is possible with a 1 dB penalty in eye-opening at a 40 Gbps bit rate or 10 Giga-symbols/s. with an OSNR of 18 dB. An OSNR could be about 22 dB for a 107 Gbps bit rate and a symbol rate of 26.75 G symbols/s. The transmission link consists of several spans of a total 1000 km of dispersion-compensated, optically amplified transmission link. The optical gain of the in-line optical amplifiers is set to compensate for the attenuation of the transmission and compensating fibers, with an NF of 3 dB. The optical transmitter and receivers, incorporating commercially available coherent receivers, are structured and sufficient for engineering of the optical transmission terminal equipment for a bit rate of 107 Gbps and a symbol rate of 26.3 Gbauds/s.

Furthermore, electronic equalization of the receiver phase shift keying signals can be done using blind equalization, which would improve the dispersion tolerance much further. For a symbol rate of 10.7 Gbps this dispersion tolerance for a 1 dB penalty would reach 300 km of standard SMF. This electronic equalization can be implemented without any difficulty at 10.7 GB. For a 107 Gbps bit rate, similar improvement in dispersion tolerance can be achieved at 26.5 Giga-Bauds/s (GB/s), provided that the electronic sampler can offer more than 50 Giga-sampling rates.

Secondly, the transmitter and receiver configurations for generating 8-DPSK_2-ASK optical signals and direct detection are described, as well as the transmission performances. In addition, performance characteristics of this modulation format at 100 Gbps (equivalently 25 GB) have also been investigated in terms of receiver sensitivity, dispersion tolerance, and the long-haul transmission performance. The simulation results show that 8-DPSK_2-ASK, which involve two levels with 8 states of DPSK in differential coding, is a promising modulation for very high-speed (100 Gbps) and long-haul optical communications as shown in Figure 4.49. Each ring of phases has eight constellation points. The Euclidean distance can be designed to maximize the distance between the constellation points and their neighboring points.

4.5 PULSE SHAPING MODULATION

4.5.1 Requirements and Specifications

Transmission distance: As the next generation of backbone transport, the transmission distance should be comparable to the previous generation, namely the 100 Gbps transmission system. As the most important requirement, we require that the 1 Tbps transmission for the long haul should be 1500~2000 km, and for metro application ~300 km.

CD (chromatic dispersion) tolerance: As SSMF (standard single mode fiber) fiber CD factor/coefficient 16.8 ps/nm is the largest among the current deployed fibers, CD tolerance should be up to 30,000 ps/nm at the central channel whose wavelength is approximated at 1550 nm. At the edge of C-band, this factor is expected to increase by about 0.092 ps/(nm^2·km) or about 32,760 ps/nm at 1560 nm and 26,400 ps/nm at 1530 nm.[20]

PMD (Polarization mode dispersion) tolerance: The worst case of deployed fiber with 2000 km would have a DGD (differential group delay) of 75 ps. So, the PMD (mean all-order DGD) tolerance is 25 ps.

SOP rotation speed: According to the 100 Gbps experiments, SOP rotation can be up to 10 kHz, we take the same spec as the 100 G system.

Modulation format: PDM-QPSK for long-haul transmission; PDM-16 QAM for metro application.

Spectral efficiency: Compared to the 100 G system, there is an increase by a factor 2. Both Nyquist-WDM and CO-OFDM (coherent optical OFDM) can fulfill this. However, it depends on technological and economic requirements that will determine the suitability of the technology for optical network deployment.

4.5.2 Comparison of Nyquist_*N*_Tbps Transmission System

Table 4.7 provides a comparison of two different Tera-bps superchannels using different comb generation techniques, incorporating pulse shaping with variable roll-off coefficients and requirements for the digital processing systems, as well as possible applications. Various parameters of the transmission platform are also specified.

4.5.3 Superchannels by Pulse Shaping Modulation for Tera-bps Transmission

Single-channel 1 Tbps transmission is impossible to realize using only one carrier or sub-carrier and modulation. If we keep the non-FEC baud rate close to 25 GBaud, we need PDM-4096QAM modulation scheme to achieve 1 Tbps per single lightwave carrier. This is impossible under the constraint of maximum laser amplitude power available to date. Furthermore, as the OSNR requirement increases exponentially with the modulation format level, it is impossible to reach the transmission distance with limited laser power and the threshold of the nonlinear level of the transmission fiber. Even PDM-16QAM needs, theoretically, 7 dB more OSNR than PDM-QPSK to reach the same distance. To solve this problem, one can either increase the baud rate or employ more sub-carriers, that is, a super-channel with multiple sub-carriers. Increasing the baud rate leads to higher demand on the bandwidth for O/E components, thus challenging technological advances and cost. So employing a high spectral efficiency super-channel with sub-channel bandwidth of 50 GHz or below seems to be the most favorable choice.

Considering the overhead of the OTN framer and FEC of either 7% or 20%, the total bit-rate can thus be 28 G×4×10 to obtain 1.12 Tbps or 32 G×4×10 to obtain 1.28 Tbps

For multiple sub-carrier transmission, we would need multi-carrier modulation techniques and corresponding multi-carrier demodulation and coherent detection. Currently, there are a few methods and apparatus to generate multi-carriers. The commonly used method involves a single ECL laser source, a recirculating shift loop, and an MZM modulated by an RF signal. The frequency of

TABLE 4.7
The 1 Tbps Off-Line System Specification

Parameter	Channel	Channel	Technical Specs	Remarks
Technique	Comb generator	Nonlinear comb gen		
Bit rate (Aggregate)	1, 2, …, *N* Tbps (whole C-band)	1, 2, …, *N* Tbps	~1.28 Tbps @28–32 GB	20% OH for OTN, FEC
Number of ECLs	1	$N \times 2$		
Nyquist roll-off α	0.1 or less	0.1 or less		DAC pre-equalization required
Baud rate (GBauds)	28–32	28–32	28, 30 or 31.5 GBaud	Pending on FEC coding allowance
Transmission distance	2500	2500	1200 km (16 span) ~ 2000 km (25 spans) 2500 km (30 spans)	20% FEC req. for long-haul application
			500 km	Metro application
Modulation format	QPSK/16 QAM	QPSK/16 QAM	Multi-carrier Nyquist WDM PDM-DQPSK/QAM	For long haul
			Multi-carrier Nyquist WDM PDM-16 QAM	For metro
Channel spacing			4×50 GHz	For long haul
			2×50 GHz	For long haul
Launch power	≪ 0 dBm if 20 Tbps used		~-3 to 1 dBm lower if $N>2$	Depending on QPSK/16 QAM and long haul/metro can be different
B2B ROSNR @2e-2 (BOL) (dB)	14.5	14.5	15 dB for DQPSK 22 dB for 16 QAM	1 dB hardware penalty 1 dB narrow filtering penalty
Space transmission distance	GEO -GEO	GEO-GEO	GEO-GEO	
Span loss	72–78	72–78	72–78	
Amplifiers (Booster)	EDFA ($G>22$ dB); NF<5 dB	30 dB 40 W	EDFA (HPA) High power – multi-stage pumping	
Amplifier (optical pre-amp)	EDFA ($G>40$) NF<4 dB	30–40 dB –65 dBm	EDFA -LNA (low noise, high sensitivity)	Receiving sensitivity –65 dBm, NF< 4 dB
BER	2e-3	2e-3	Pre-FEC 2e-2 (20%) or 1e-3 classic FEC (7%)	
CD penalty (dB)			0 dB@ ±3000 ps/nm <0.3 dB@±30,000 ps/nm	16.8 ps/nm/km and 0.092 ps/(nm · km)
PMD penalty (DGD)			0.5 dB@75 ps, 2.5 symbol periods	
SOP rotation speed	10 kHz	10 kHz	10 kHz	
Filters cascaded penalty			<1 dB @12 pcs WSS	
Driver linearity	Required	Required	THD<3%	16 QAM even more strict

THD = third harmonic distortion, FECC = forward error coding, PMD = polarization mode dispersion.

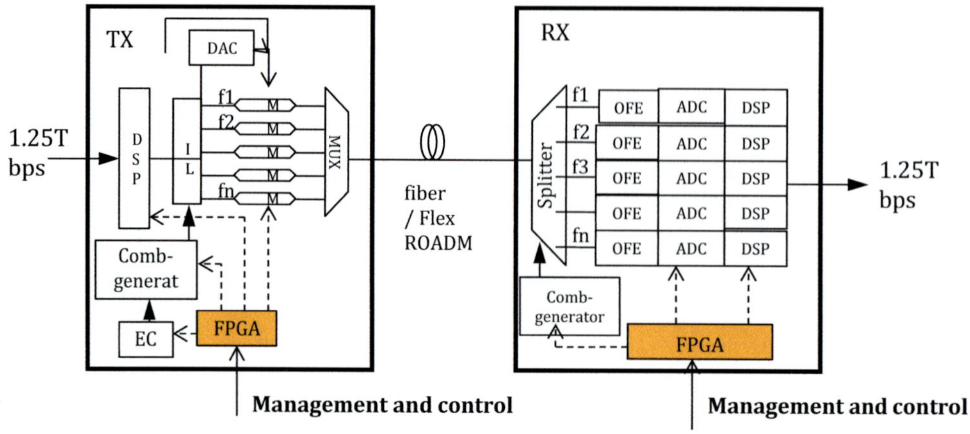

FIGURE 4.69 Architecture of a super-channel DWDM 1 Tbps transmission platform using DSP-based coherent reception systems and pulse shaping modulation. M=modulator, ADC=analog to digital converter, OFE=optical frequency equalizer, DSP=digital signal processing, DC=digital to analog converter, FPGA=field programmable gate array, ROADM=reconfigurable optical add/drop multiplexer.

the RF signal determines the spacing of the sub-channels. Tx-DSP and DAC are needed for shaping the signal pulse, for example, Nyquist pulse shaping, and pulse shaping to compensate for non-ideal O/E components transfer function and CD pre-compensation, among other factors (Figure 4.69).

An interleaver can be used to separate the sub-carriers for individual sub-channel modulation. At the output, all the signals of the sub-channels are multiplexed together to form a super-channel 1.12 Tbps optical signal (Figure 4.70).

The principal question is how we can create superchannels under laboratory conditions to achieve proof of concept for Tbps transmission, specifically how to generate superchannels, transmit and detect/demodulate at the receiving end. Thus, one of the main principles of this chapter is to describe techniques employed to generate superchannels and probe channels to evaluate the performance of transmission systems for Tbps superchannels.

4.5.4 Multi-Carrier Nyquist Pulse Shaping Modulation under Transmission System

In Nyquist WDM the goal is to place different optical sub-channels as close as possible to their baud rate, the Nyquist rate, while minimizing channel crosstalk through the formation of approximately close to rectangular spectrum shape of each channel by either optical filtering or electrical filtering and equalization.

4.5.4.1 Nyquist Signal Generation Using DAC by Equalization in Frequency Domain

The rectangular spectrum has a $\sin c\,\text{i.e.}\,\sin x/x$ time domain impulse response. At the sampling instants, $t = kT$ $(k = (1,2,\ldots,N)$ is a nonzero integer), its amplitudes reach zero. This implies that at the ideal sampling instants, the ISI from neighboring symbols is negligible, or free of ISI. Figure 4.71 depicts such a Nyquist pulse and its spectrum for either a single channel or multiple channels. Note that the maximum of the next pulse rise is the minimum of the previous impulse of the consecutive Nyquist channel.

Considering one sub-channel with a carrier 25 GBaud PDM-DQPSK signal, the resulting capacity is 100 Gbps for a sub-channel, hence to reach 1 Tbps, ten sub-channels would be required. To increase spectral efficiency, the bandwidth of these ten sub-channels must be packed densely together. The most likely technique for packing the channels as close as possible in frequency with minimum ISI is Nyquist pulse shaping, which is described later in this section. Thus, the term

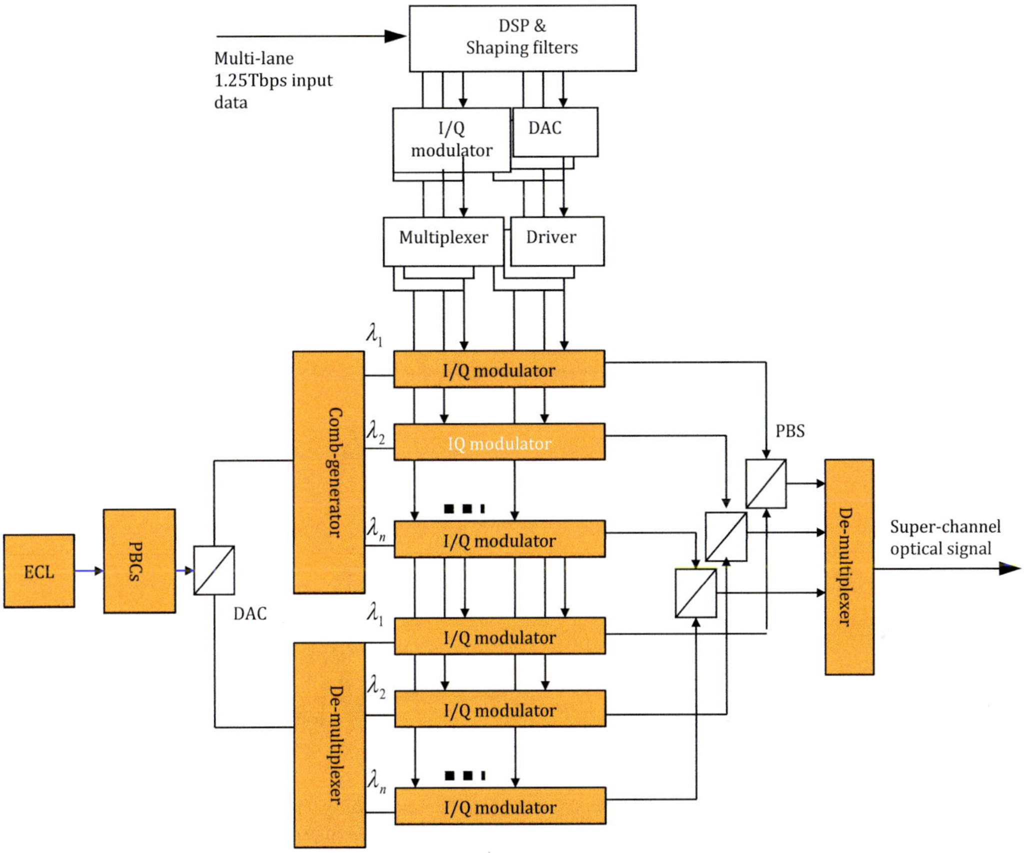

FIGURE 4.70 General architecture of a super-channel transmitter for coherent reception systems. ECL=external cavity laser, PBC=polarization beam coupler, DAC=digital to analog converter, PBS=polarization beam splitter, I/Q=in-phase and quadrature, PBC=polarization beam combiner, DSP=digital signal processor.

"Nyquist-WDM system" is coined. However, in practice, such a "brick wall" like spectrum shown in Figure 4.71 is impossible to obtain and hence a non-ideal solution for non-ISI pulse shape should be found, such as the raised cosine pulse with some roll-off property condition.

The raised-cosine filter is an implementation of a low-pass Nyquist filter, i.e., one that has the property of vestigial symmetry. This means that its spectrum exhibits odd symmetry about $1/2T_s$, where T_s is the symbol-period. Its frequency-domain representation is a 'brick-wall-like' function, given by:

$$H(f)=\begin{cases} T_s & |f|\leq\dfrac{1-\beta}{2T_s} \\ \dfrac{T_s}{2}\left[1+\cos\left(\dfrac{\pi T_s}{\beta}\left\{|f|-\dfrac{1-\beta}{2T_s}\right\}\right)\right] & \dfrac{1-\beta}{2T_s}<|f|\leq\dfrac{1+\beta}{2T_s} \\ 0 & \text{otherwise} \end{cases} \quad (4.30)$$

$$\text{with} \qquad 0\leq\beta\leq 1$$

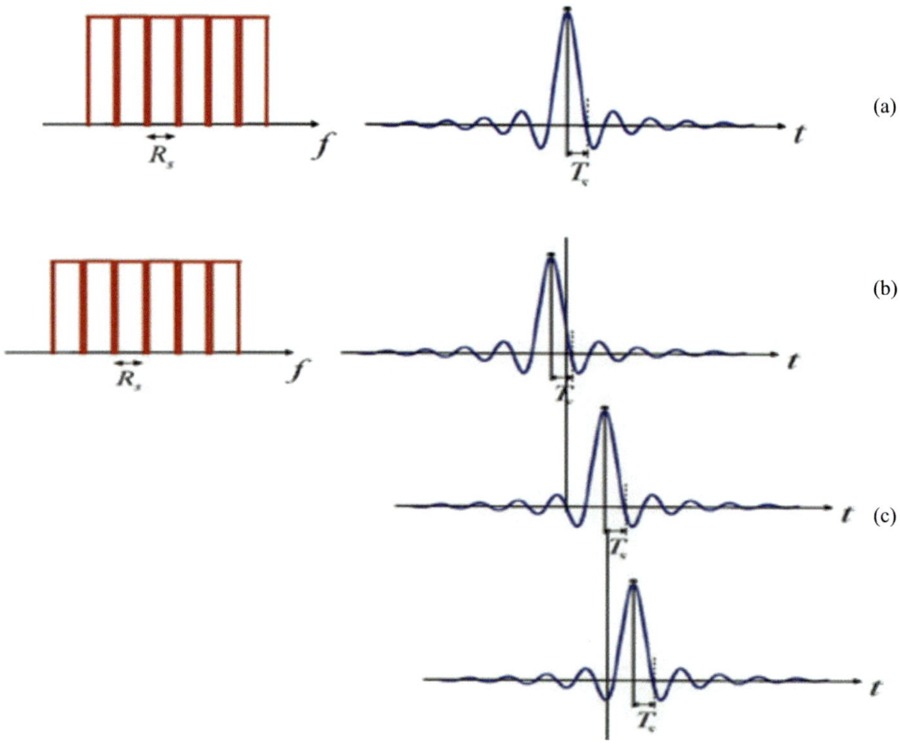

FIGURE 4.71 A super-channel Nyquist spectrum and its corresponding "impulse" response (a) spectrum, (b) impulse response in the time domain of a single channel, (c) sequence of pulse to obtain consecutive rectangular spectra. A superposition of these pulse sequences would form a rectangular 'brick wall-like' spectrum.

This frequency response is characterized by two values: β, the *roll-off factor*, and T_s, the reciprocal of the symbol rate in Sym/s, that is $1/2T_s$ is the half bandwidth of the filter. The impulse response of such a filter can be obtained by analytically taking the inverse Fourier transformation of Equation 4.30, in terms of the normalized *sinc* function, as

$$h(t) = \sin c\left(\frac{t}{T_s}\right)\frac{\cos\left(\frac{\pi\beta t}{T_s}\right)}{1-\left(2\frac{\pi\beta t}{T_s}\right)^2} \tag{4.31}$$

where the roll-off factor, β, is a measure of the *excess bandwidth* of the filter, i.e., the bandwidth occupied beyond the Nyquist bandwidth as from the amplitude at $1/2T$. Figure 4.72 depicts the frequency spectra of raise cosine pulse with various roll-off factors. Their corresponding time domain pulse shapes are given in Figure 4.72b.

When used to filter a symbol stream, a Nyquist filter has the property of eliminating ISI, as its impulse response is zero at all nT (where n is an integer), except when $n = 0$. Therefore, if the transmitted waveform is correctly sampled at the receiver, the original symbol values can be recovered completely. However, in many practical communications systems, a matched filter is used at the receiver to minimize the effects of noises. For zero ISI, the net response of the product of the transmitting and receiving filters must equate to $H(f)$, thus we can write:

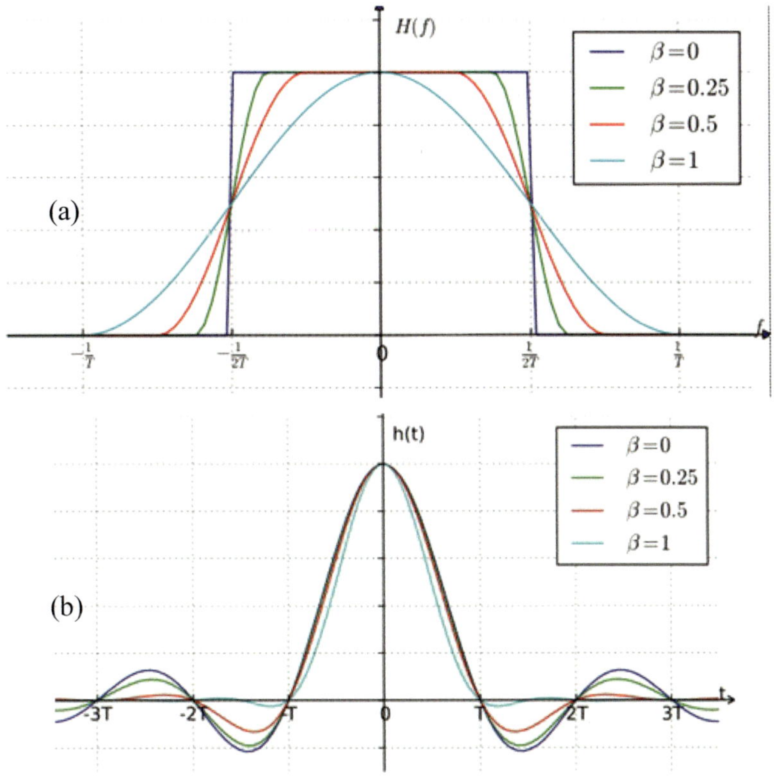

FIGURE 4.72 (a) Frequency response of raised-cosine filter with various values of the roll-off factor β; (b) impulse response of raised-cosine filter with the roll-off factor β as a parameter.

$$H_R(f)H_T(f) = H(f) \tag{4.32}$$

Alternatively, we can rewrite that

$$|H_R(f)| = |H_T(f)| = \sqrt{|H(f)|} \tag{4.33}$$

The filters that can satisfy the conditions of Equation 4.31 are root-raised-cosine filters. The main problem with root-raised-cosine filters is that they occupy a larger frequency band than that of the Nyquist sin*c* pulse sequence. Thus, for the transmission system we can split the overall raised cosine filter into root-raised-cosine filters at both the transmitting and receiving ends, provided the system is linear. This linearity needs to be specified accordingly. An optical fiber transmission system can be considered linear if the total power of all channels is below the nonlinear SPM threshold limit. When it is over this threshold, a weakly linear approximation can be used.

The design of a Nyquist filter influences the performance of the overall transmission system. Oversampling factor, selection of roll-off factor for different modulation formats, and FIR Nyquist filter design are key parameters to be determined. If considering the transfer functions of the overall transmission channel including fiber, WSS, and the cascade of the transfer functions of all O/E components, the total channel transfer function is more Gaussian-like. To compensate for this effect in the Tx-DSP, one would thus need a special Nyquist filter to achieve the overall frequency response equivalent to that of the rectangular or raised cosine with roll-off factor shown in Figures 4.73 and 4.74.

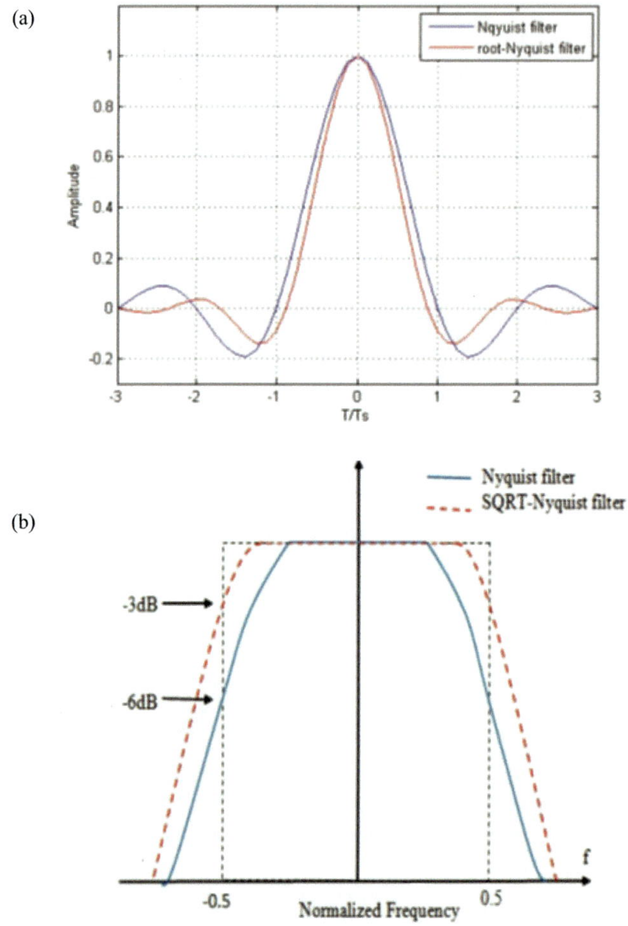

FIGURE 4.73 (a) Impulse and (b) corresponding frequency response of *sinc* Nyquist pulse shape or root raised cosine (RRC) Nyquist filters.

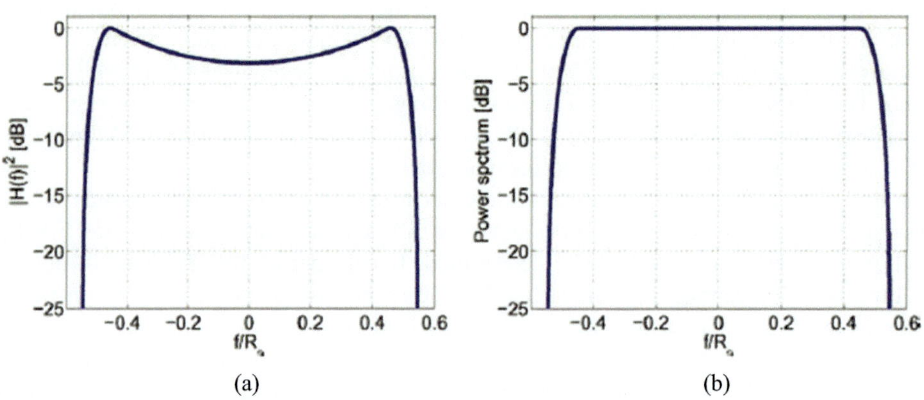

FIGURE 4.74 (a) Desired Nyquist filter for spectral equalization; (b) output spectrum of the Nyquist filtered QPSK signal.

4.5.4.2 Functional Modules of Nyquist Pulse Shaper

A generic schematic of the functional modules of the Nyquist-WDM system is described in Figure 4.2. Two special features should be considered. At the Tx side, pulse shaping is so defined that the overall transfer function from the driver to the ADC on the Rx side should ideally be rectangular or NRZ-form. The implementation can be either using a look-up-table or Nyquist low-pass filter in the frequency domain or time domain. Alternatively, the optical spectrum of the modulated lightwaves at the output of the optical modulator can be obtained via the port of the optical spectrum analyzer (OSA). This is then equalized to achieve a "brick wall" like spectrum. Thus, the required spectrum on the DAC to achieve this equalization is known and used to modify the DAC driving output voltage levels at (H_I^+, H_I^-) and (H_Q^+, H_Q^-) and (V_I^+, V_I^-) and (V_Q^+, V_Q^-), where the equalized optical spectra can be obtained for the two polarized modes of the linearly polarized mode to be launched into the SSMF optically amplified transmission lines. The subscripts *I* and *Q* indicate the in-phase and quadrature phase respectively.

For each sub-channel, 4x DACs are needed to convert the discrete digital signal to an analog signal (X_I, X_Q, Y_I, Y_Q). To match the voltage amplitude (power) requirement of the I-Q modulator, 4x RF broadband drivers are required to provide appropriate RF signal amplitude for driving the MZM modulator to obtain the phase difference between the in-phase and quadrature-phase constellation points. At the output of the transmitter, all sub-channels are multiplexed together to form a superchannel (Figure 4.75).

There is another way to generate the Nyquist signal, which is suitable when Tx-DSP is not available. The Nyquist spectrum can be generated by an optical rectangular (or raised-cosine roll-off) filter, as shown in Figure 4.76. However, in practice, it is very difficult to achieve a sharp filter roll-off without much phase fluctuations at the edges of the filter.

At the receiver, multi-carrier parallel demodulation is used. First, the super-channel will be separated into individual sub-channels. Each sub-channel will be demodulated as if it were a single-channel reception: OFE down-converts the optical spectrum to the baseband (homodyne coherent demodulation), the 4-lane analog signals are converted into the digital domain using four ADCs. The digitized signals are processed by the ASIC containing the software processing modules for CD compensation, timing recovery, MIMO FIR filtering, carrier recovery, and so on (Figure 4.77).

4.5.4.3 DSP Architecture

Single sub-channel DSP can use either the traditional FDEQ+TDEQ architecture or a pure FDEQ architecture (Figure 4.78).

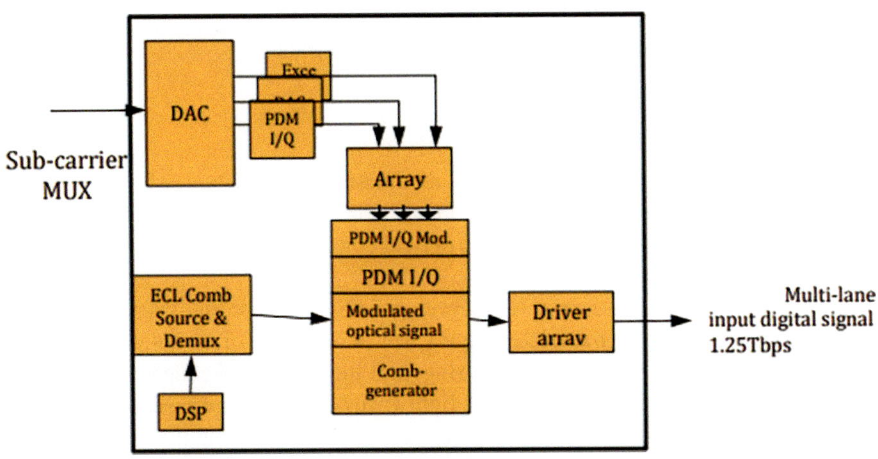

FIGURE 4.75 Transmitter architecture of a Nyquist-WDM system with an electrical Nyquist filter.

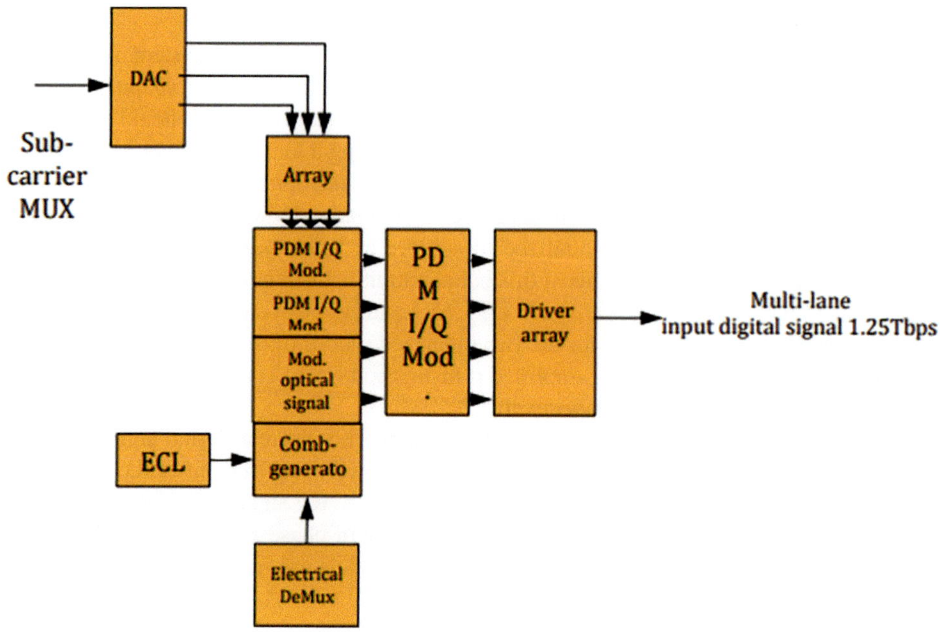

FIGURE 4.76 Schematic of the Nyquist optical transmitter architecture with optical Nyquist filter.

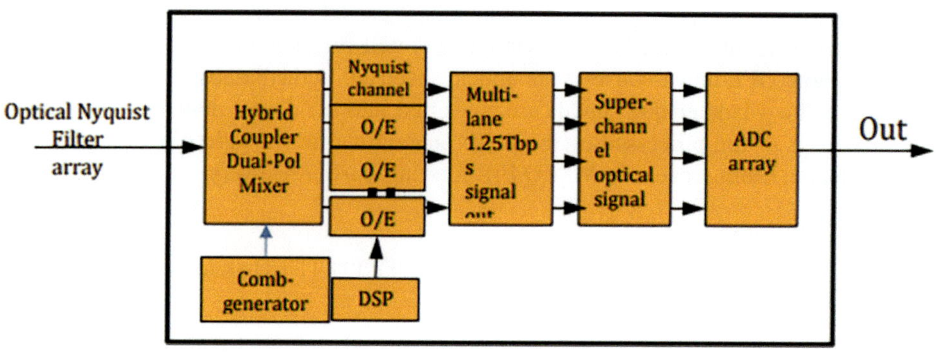

FIGURE 4.77 Receiver architecture of Nyquist-WDM system. O/E=optical to electrical, ADC=analog to digital converter, Pol=polarization.

4.5.4.4 Superchannel Spectrum

To demonstrate the multi-carrier generation and modulation of 100 Gbps or 200 Gbps, we conduct experimental platform, where the optical components and the data modulated signals are generated by the DAC. A special optical component is the optical comb laser, which can be formed by various techniques as shown in references.[21,22]

Figure 4.79a shows the multi-carrier lightwaves generated by a comb laser, whose structure is formed by frequency shifting of modulated lightwaves and the use of an optical amplifier to compensate for ring loss. The resonant cavity is a ring, and the periodic frequency shifting allows the formation of a comb laser with several periodic lightwaves of different wavelengths. The spacing of these lightwave lines is determined by the RF frequency shift, ensuring that the signal bandwidth can accommodate the bandwidth of the modulated signals. Figure 4.79b shows two adjacent carriers with very narrow linewidths of about 10 kHz. This linewidth is based on the rise time and fall

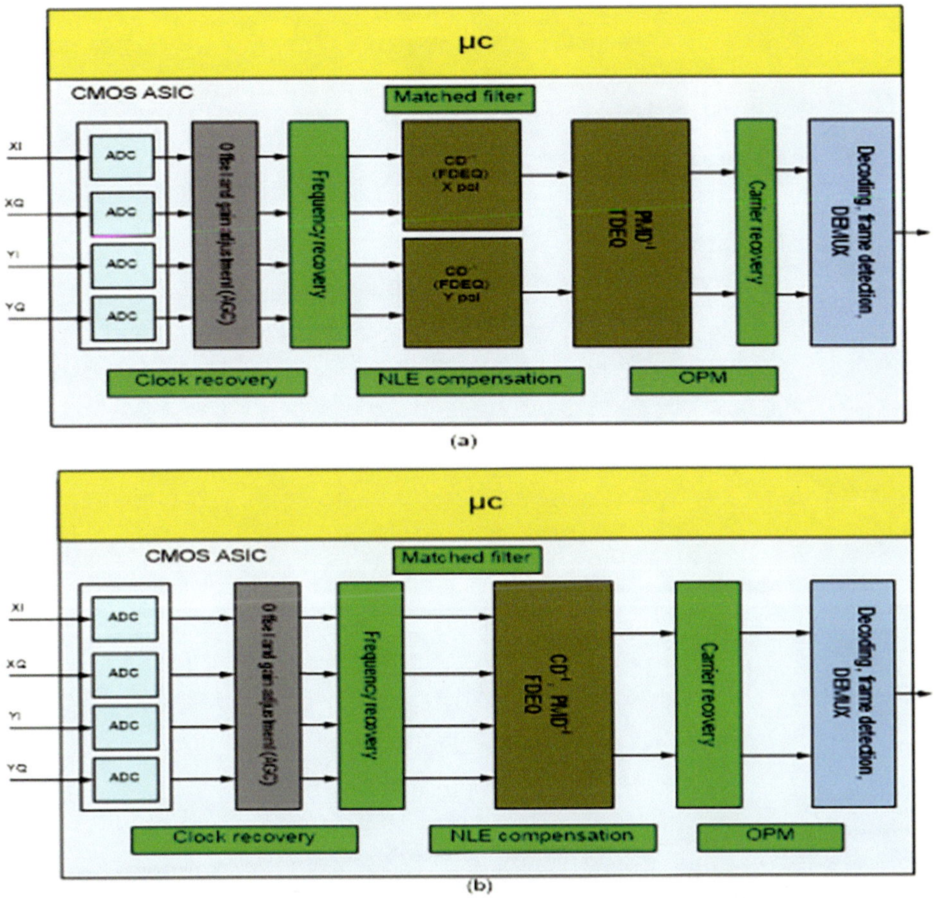

FIGURE 4.78 (a) DSP architecture, FDEQ+TDEQ (frequency domain equalizer+time domain equalizer); and (b) pure FDEQ.

time of the RF signals that modulate the optical modulator. The rise and fall times, or bandwidth, of optical modulator, affect this linewidth. This linewidth can be considered sufficient for coherent SPC transmission systems.

Figure 4.80a shows the spectra of several multiplexed channels without pulse shaping. The carrier-to-noise ratio reaches 32 dB, which is quite substantial and meets the requirements for long-haul transmission in space. Figure 4.80b shows the spectra of 14 carriers modulated with pulse shaping. This shows clearly the wall-like spectral shape of the multiplexed channels. Figure 4.80c and d shows the spectrum of multiplexed channels under optical filtering and without filtering to suppress unequal channels in power. A total number of 14 channels can be generated. Figure 4.81 shows the modulated channels with sharper pulse shaping, which clearly shows the brick-wall spectral shape. Observing the spectrum of the pulse-shaped channels given in Figure 4.79, the optical filter used in the LNA optical preamplifier and the WSS as a demultiplexer can suppress noise to give the OSNR for the channels that can be higher than 40 dB.

The single sideband generation is achieved by adapting the technique presented in Section 3.7.5.3 of Chapter 3.

The technique of generating modulated channels using only one comb laser source allows more compact channels reaching Tbps with the baud rate of 28 or 56 GBaud and the modulation format

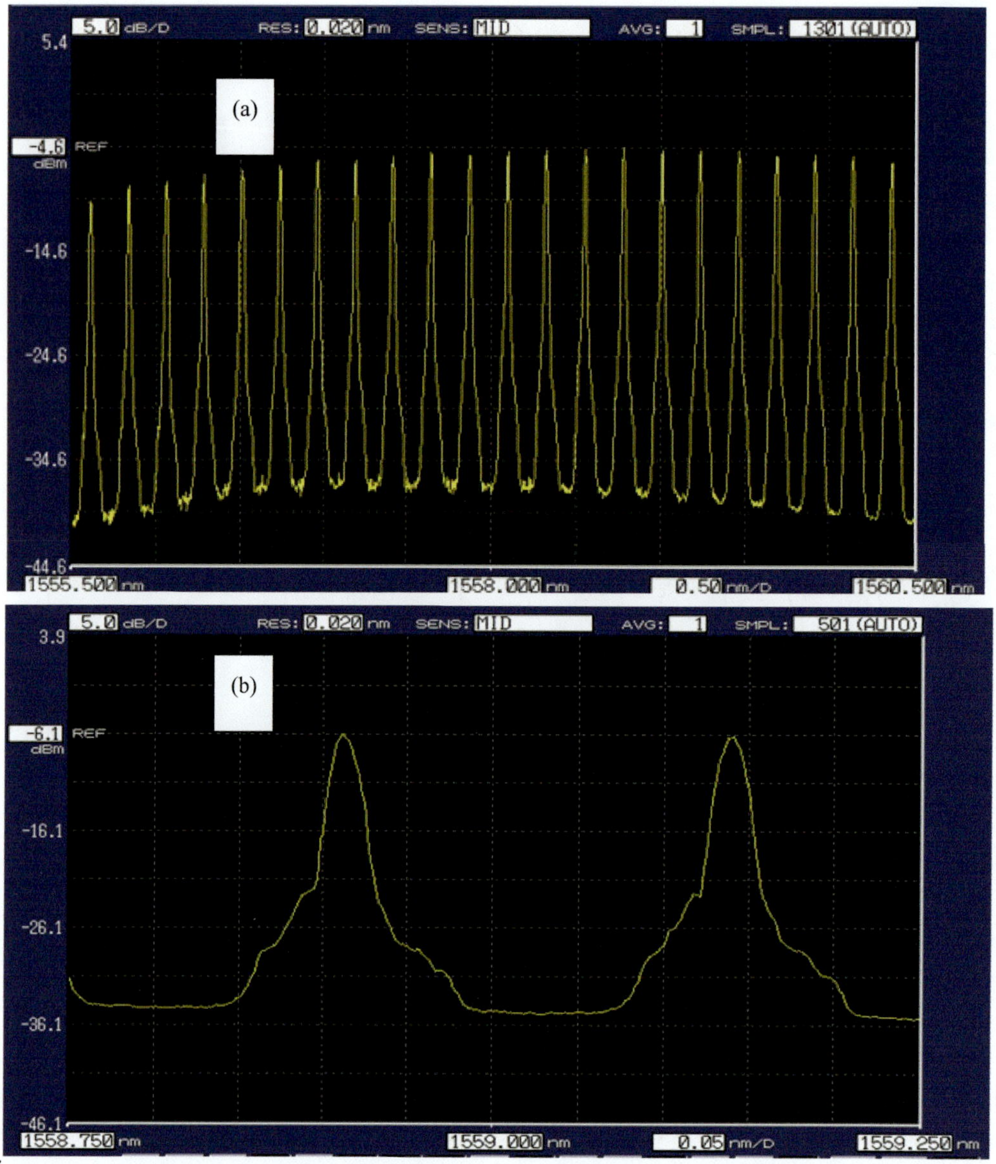

FIGURE 4.79 Spectrum of multi-subcarriers generated by a comb generator of a recirculating frequency shift loop (a) frequency-shifted carrier spectrum; (b) close up spectrum of two carriers.

of QPSK or 16 QAM with ten or five carriers, respectively. This generation and modulation are quite suitable in the SPC if WSS (wavelength selective switch) is used as a multiplexer or demultiplexer, as described in Chapters 3 and 7.

The transmission and performance of 100 Gbps dual polarization QPSK and 16 QAM systems over SPC are reported and presented in Chapter 10.

Chapters 5 and 6 will present the technological development and design of optical coherent receivers and DSP processing techniques for digital DWDM channels. Chapter 10 uses these techniques for SPC over single and multi-hop transmission through inter-satellite constellations with optical routing.

FIGURE 4.80 A spectrum of carrier comb generator 1×5 sub-carriers (a) sub-carriers at 30 GHz spacing, (b) pulse shaping modulated sub-channels. Note: the resolution of optical spectrum analyzer and narrower linewidth of a carrier (10 kHz); and OSNR (signal = carrier in this case) = ~19 dB.

4.6 STATISTICAL MODULATION

Statistical modulation is a technique that uses standard modulation, such as QPSK, M-ary QAM, with an additional distribution of the constellation points so that their Euclidean distance is maximized. This means that the distance can be larger than the minimum distance between adjacent constellation points. Figure 4.82 shows the 16 QAM modulation with statistical modulation at the receiver and transmitter, and the corresponding number of bits per symbol which is less than four for 16 QAM, that is the ratio is less than unity.

Figure 4.83 shows the optical spectrum of 64 QAM with statistical distribution and the constellation points with the distribution. The carrier is not suppressed, as seen from the spectrum of the channels. The effective ratio of the number of bits per symbol is 0.93,[23] and dual polarizations are used with an entropy of 5.5 bits/symbol (Baud). The resultant bit rate is 105 Gbps. Pulse shaping (see Section 4.5) is also used, as evidenced by the sharp roll-off of the spectrum. The higher the constellation points, the darker the points.

4.7 CONCLUDING REMARKS

Further to the binary modulation of lightwaves presented in Chapter 4, this chapter describes the techniques to modulate lightwaves with higher levels in amplitude and phases so as to increase the transport system capacity. The SPC system must operate with coherence in both transmission and reception. Therefore, preserving the narrow linewidth of modulated carriers and the channel bandwidth is highly critical.

Multi-level in amplitude and phase of the lightwaves, along with statistical modulation, can assist in the sensitivity and capacity of space transmission over long distances.

In SPC the optical booster amplifier at the transmitter and the low-noise low-power optical preamplifier at the receiver are critical to compensate for the long-distance transmission losses. The noise generated by these amplifiers superimposes on the constellation points, increasing EVM and

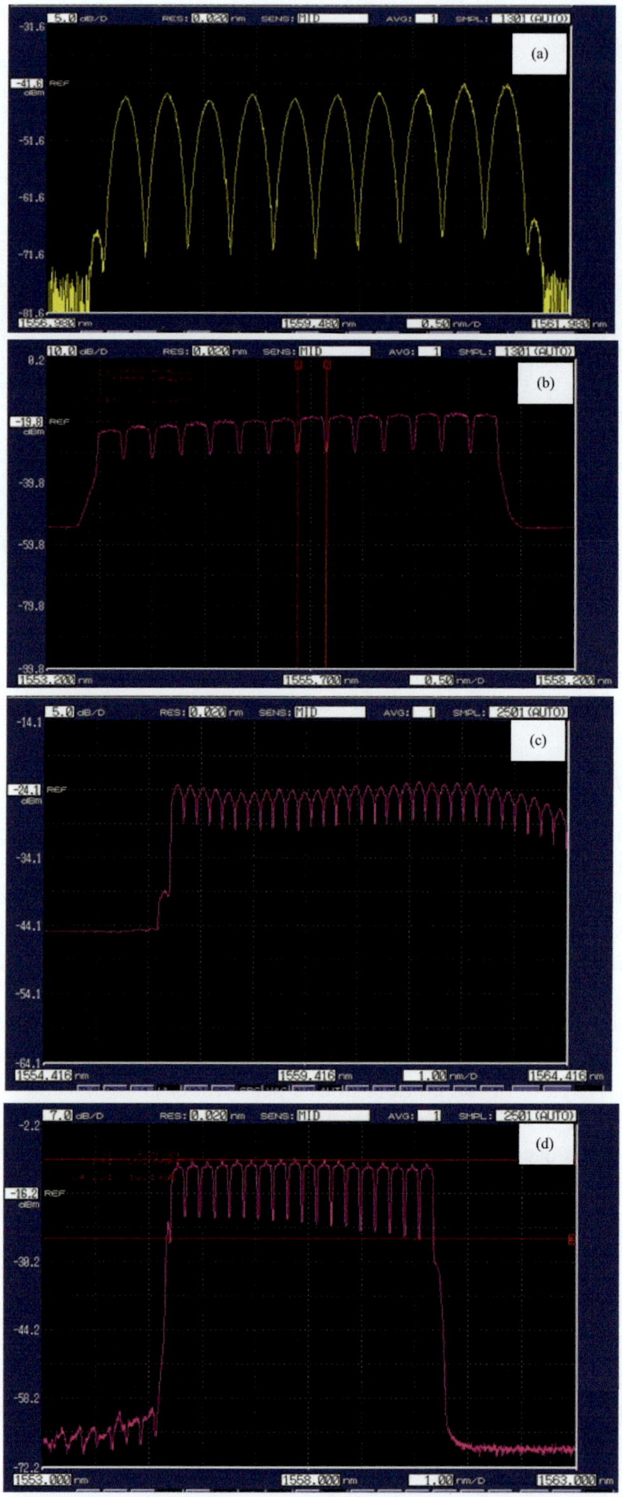

FIGURE 4.81 Spectrum of 10×25 GBaud QPSK super-channel modulation (a) under QPSK with 50 GHz spacing without pulse shaping modulation; (b) 14 channels under Nyquist QPSK after transmission over 2000 km optically amplified SSMF spans with pulse shaping modulation; (c) higher than 20 sub-carriers under Nyquist QPSK modulation with pulse shaping; and (d) 20 modulated lightwave super-channels with pulse shaping modulation, after optical filtering.

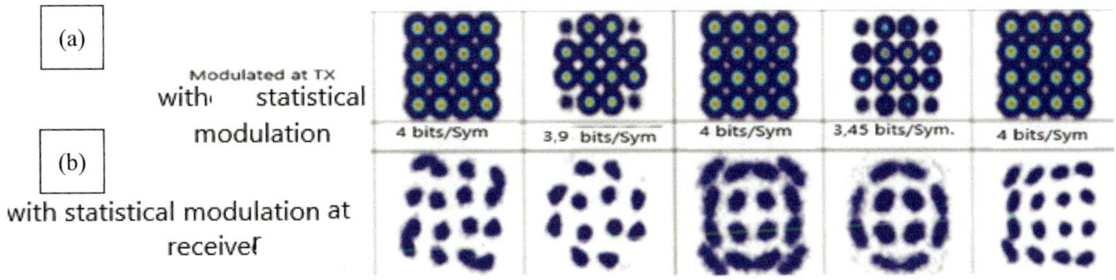

FIGURE 4.82 The 16 QAM under modulation (a) with transmitter and (b) with statistical modulation at receiver and effective bits/symbol transmission effects on the constellation with purely phase rotation (frequency carrier shift), amplitude and phase distortion, etc.

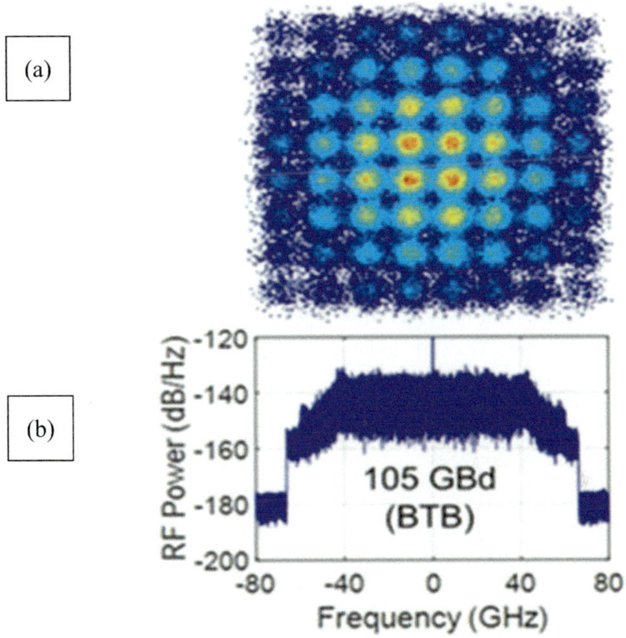

FIGURE 4.83 (a) Constellation of statistical distribution of 64 QAM and spectrum of the modulated channel with statistical modulation, back-to-back (BTB), (b) spectrum of the channel is shown below the constellation. Of 105 GBaud the resultant bot rate is 105 GBaud * 6 bit per Baud=630 Gbps.

thereby lowering the BER. Thus, the design of the multi-level modulation format of the SPC system must be optimized.

Chapter 7 describes optical amplifications and addresses these issues, along with the optical router for intersatellite constellation networking.

Although we have considered various modulation formats for SPC ultra-long transmission systems, the practicality of the transmitter and receiver must also be considered. Currently, many integrated coherent photonic transceivers are commercially available in DP-QPSK but not in other formats. If other formats are required, extensive development time is required. QPSK offers the main advantage that both real and imaginary components can be generated by ASK modulation, and the phase of a bit period remains the same over the whole baud period. For phase and frequency shift keying, the phase of the carrier changes at the transition state, making it difficult to detect practically. MSK can offer better detection as the phase of the lightwave carrier is continuous over the baud period, allowing for differential detection. Integrated coherent photonic transceivers can be modified and programmed using the DSP embedded in the transmitter section to implement this

MSK modulation format. Chapter 6 addresses such DSP techniques for both the transmitter and receiver sections.

Statistical modulation can be superimposed on one-dimensional and multi-dimensional complex modulations.

NOTES

1 L. N. Binh, "Monash optical communication system simulator. Part I: Ultra-long ultra-high-speed optical fiber communication systems", Manual for Technical Development, Monash University, Melbourne, Australia, 2004.
2 L. N. Binh, "Monash optical communication system simulator. Part II: Ultra-long ultra-high-speed optical fiber communication systems", Manual for Technical Development, Monash University, Melbourne, Australia, 2004.
3 S. Hayase, et al., "Proposal of 8-state per symbol (binary ASK and QPSK) 30-Gb/optical modulation demodulation scheme", *Proceedings of ECOC2003, European Conference on Optical Communication*, Rimini, Italy, paper Th2.6.4, pp. 1008–1009, September 2003.
4 K. Sekine, et al., "40 Gbps 16-ary (4 bit/symbol) optical modulation/demodulation scheme", *IEEE Electronics Letters*, 41(7), 1–2, March 2005.
5 DFB laser is a guided waveguide with a gain section and two DFB structures at the output and input of the waveguides. The DFB is designed with desired wavelength reflected feed back into the cavity, thus generating a resonant cavity for lasing the desired-wavelength single frequency source. Typical linewidth is about 2 MHz.
6 L. N. Binh, *Optical Fiber Communication Systems with MATLAB and Simulink Models (Optics and Photonics)*. CRC Press, Taylor & Francis: Boca Raton, FL, 2014.
7 L.N. Binh, *Optical Modulation: Advanced Techniques and Applications in Transmission Systems and Networks*. CRC Press, Taylor & Francis: Boca Raton, FL, 2018.
8 L. N. Binh, *Advanced Digital Optical Communications*, 2nd Ed. CRC Press: Boca Raton, FL, 2015.
9 L. N. Binh, *Digital Processing for Optical Transmission and Coherent Receiving Technique*. CRC Press, Taylor & Francis: Boca Raton, FL, Chapter 4, 2013.
10 C. Wree, et al., "Coherent receivers for phase-shift keying transmission", *Optical Fiber Communication Conference*, Anaheim, CA, 2007.
11 L.N.Binh, *Advanced Digital Optical Communications*, 2nd Ed. CRC Press, Taylor & Francis: Boca Raton, FL, Chapter 7, 2015.
12 C. Wree, J. Leibrich, and W. Rosenkranz, "RZ-DQPSK format with high spectral efficiency and high robustness towards fiber nonlinearities", *Proceedings of ECOC 2002*, Copenhagen, Denmark, paper 9.6.6, September 2002.
13 Nyquist sampling rate
14 R. Llorente, J. H. Lee, R. Clavero, M. Ibsen, and J. Martí, "Orthogonal wavelength-division-multiplexing technique feasibility evaluation", *Journal of Lightwave Technology*, 23, 1145–1151, March 2005.
15 I. B. Djordjevic and B. Vasic, "Multilevel coding in M-ary DPSK/differential QAM high-speed optical transmission with direct detection," *IEEE Journal of Lightwave Technology*, 24, 420–428, 2006.
16 M. Serbay, C. Wree, and W. Rosenkranz, "Experimental investigation of RZ-8DPSK at 3x 10.7Gbps," Proceedings of *LEOS'05*, Sydney, NSW, Australia, 2005.
17 M. Seimetz, M. Noelle, and E. Patzak, "Optical systems with high-order DPSK and star QAM modulation based on interferometric direct detection," *IEEE Journal of Lightwave Technology*, 25, 1515–1530, 2007.
18 H. Yoon, D. Lee, and N. Park, "Performance comparison of optical 8-ary differential phase-shift keying systems with different electrical decision schemes," *Optic Express*, 13, 371–376, 2005.
19 S. Walklin and J. Conradi, "Multilevel signaling for increasing the reach of 10Gbps lightwave systems", *IEEE Journal of Lightwave Technology*, 17(11), 2235–2248, November 1999.
20 See Appendix for technical specification of Corning fiber G.652 SSMF.
21 L. N. Binh, *Advanced Digital Optical Communications*. CRC Press, Taylor & Francis Group: Boca Raton, FL, 2015.
22 L. N. Binh, *Optical Modulation: Advanced Techniques and Applications in Transmission Systems and Networks*. CRC Press, Taylor & Francis Group: Boca Raton, FL, Chapter 10, 2017.
23 H. Chien, et al., "Approaching terabits per carrier metro-regional transmission using beyond-100GBd coherent optics with probabilistically shaped DP-64QAM modulation", *Journal of Lightwave Technology*, 37(8), 1751–1755, 2019.

5 DSP-Based Optical Coherent Receiver for SPC

Physical

CONTENTS

Detection of optical signals can be carried out at the optical receiver by direct conversion of optical signal power to electronic current in the photodiode and then electronic amplification. On the other hand, a coherent reception system would offer at least 20–30 dB higher gain in receiver sensitivity due to the mixing of the weak signals with that of the high-power local oscillator laser. In SPC, such gain is highly significant and desired to compensate for the attenuation factor of long-distance space transmission. Thus, coherent reception is the main reception technology to be considered for SPC systems.

This chapter provides a fundamental understanding of the coherent detection of optical signals, which requires the mixing of the optical fields of the optical signals with those of the local oscillator, a high-power laser, so that its beating product results in modulated signals preserving both their phase and amplitude characteristics in the electronic domain. Optical pre-amplification in coherent

DOI: 10.1201/9781003426677-5

detection can also be integrated at the front end of the optical receiver to further improve the sensitivity of the receiver. The LNA presented in Chapter 7 offers high gain, low power, and low noise sensitivity for SPC systems.

5.1 INTRODUCTION

With the exponential increase in data traffic, especially due to the demand for ultra-broad bandwidth driven by multimedia applications, cost-effective ultra-high-speed optical networks have become highly desired. It is expected that Ethernet technology will not only dominate in access networks but also become the key transport technology of next-generation metro/core networks. The 100 Gigabit Ethernet (100 GbE) is currently considered the next logical evolution step after 10 Gigabit Ethernet (10 GbE). Based on the anticipated 100 GbE requirements, a 100-Gbit/s data rate of serial data transmission per wavelength is required. To achieve this data rate while complying with current system design specifications such as channel spacing, chromatic dispersion, and polarization mode dispersion (PMD) tolerance, coherent optical communication systems with multilevel modulation formats will be desired, as they can provide high spectral efficiency, high receiver sensitivity, and potentially high tolerance to fiber dispersion effects.[1] Compared to conventional direct detection in intensity-modulation/direct-detection (IMDD) systems, which only detects the intensity of the light of the signal, coherent detection can retrieve the phase information of the light and, therefore, can significantly improve receiver sensitivity.

Coherent optical receivers are important components in long haul optical fiber communication systems and networks in order to improve the receiver sensitivity and thus extra transmission distance. Coherent techniques were considered for optical transmission systems in the 1980s when the extension of repeater distance between spans was pushed to 60 km instead of 40 km for single-mode optical fiber at a bit rate of 140 Gb/s. However, in the late 1980s, the invention of optical fiber amplifiers overcame this attempt. Recently, interest in coherent optical communications has attracted significant research activities for ultra-bit rate DWDM optical systems and networks. The motivation has been due to (i) the use of optical amplifiers in cascaded fiber spans adding significant noise and thus limiting the transmission distance, (ii) advances in digital signal processors whose sampling rates can reach a few tens of giga-samples per second,econds allowing the processing of beating signals to recover the phase or phase estimation, (iii) the availability of advanced signal processing algorithms such as Viterbi and Turbo algorithms, and (iv) the fact that differential coding, modulation, and detection of such signals may not require an optical phase lock loop, hence enabling self-coherent and digital signal processing to recover transmitted signals.

As is well known, the typical arrangement of an optical receiver involves detecting optical signals with a photodiode (a PIN diode, APD, or photon counting device). The electrons generated in the photodetector are then electronically amplified through a front-end electronic amplifier. The electronic signals are then decoded to recover the original format. However, when the fields of incoming optical signals are mixed with those of a local oscillator, whose frequency can be identical to or different from that of the carrier, the phase and frequency properties of the resultant signals reflect those of the original signals. Coherent optical communication systems have also been reviving dramatically due to advances in electronic processing and the availability of stable, narrow linewidth lasers.

This chapter deals with the analysis and design of coherent receivers with optical phase lock loops and the mixing of optical signals with those of a local oscillator in the optical domain, followed by detection by opto-electronic receivers after this mixing. Thus, both the optical mixing and photodetection devices act as fundamental elements of a coherent optical receiver. Depending on the frequency difference between the lightwave carrier of the optical signals and that of the local oscillator, coherent detection can be termed heterodyne or homodyne detection.

In heterodyne detection, there is a frequency difference, and thus the beating signal falls into a passband region in the electronic domain. Therefore, all electronic processing at the front end must be in this passband region. In homodyne detection, there is no frequency difference, and thus the detection occurs in the baseband of the electronic signal. Both cases require locking the local oscillator to the carrier of the signals. An optical phase lock loop (OPLL) is thus discussed in this chapter.

This chapter is organized as follows: Section 5.2 provides an account of the components of coherent receivers, Section 5.3 outlines the principles of optical coherent detection under heterodyne, homodyne, or intradyne techniques, and Section 5.4 details the optical phase lock loop, which is a crucial development for modern optical coherent detection.

5.2 COHERENT RECEIVER COMPONENTS

The design of an optical receiver depends on the modulation format of the signals and thus is transmitted through the transmitter. The modulation of the optical carrier can be in the form of amplitude, phase, or frequency. Furthermore, phase shaping also plays a critical role in the detection and the bit error rate of the receiver and, hence, the transmission systems. In particular, it is dependent on the modulation in analog or digital, Gaussian or exponential pulse shape, on-off keying or multiple levels, etc.

Figure 5.1 shows the schematic diagram of a digital coherent optical receiver, which is similar to the direct detection receiver but with an optical mixer at the front end. However, the phase of the signals at base or passband of the detected signals in the electrical domain remains in the generated electronic current and voltages at the output of the electronic preamplifier. An optical front end is an optical mixer combining the fields of the optical waves of the local laser and the optical signals, so that the envelope of the optical signals can beat with each other to produce a summation and a difference of the frequencies of the light waves. Only the lower frequency term, which falls within the absorption range of the photodetector, is converted into electronic current, preserving both the phase and amplitude of the modulated signals (Figure 5.2).

Thus, an optical receiver front end, very similar to that of direct detection, is connected following the optical processing front end, which consists of a photodetector for converting lightwave energy into electronic currents, an electronic preamplifier for further amplification of the generated electronic current, and an electronic equalizer for bandwidth extension, usually in voltage form. This is followed by a main amplifier for further voltage amplification, clock recovery circuitry for

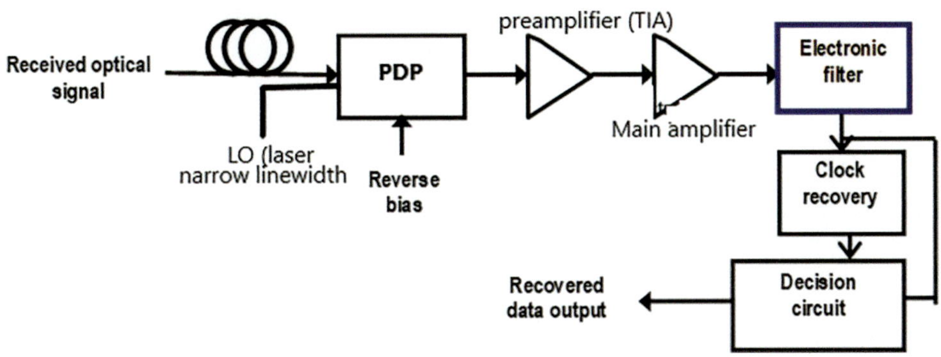

FIGURE 5.1 Schematic diagram of a digital optical coherent receiver with a local oscillator laser mixing in the photodetector with the received optical signals and then amplified by an electronic amplification system. PDP=photodetector pair for balanced detection, LO=local oscillator.

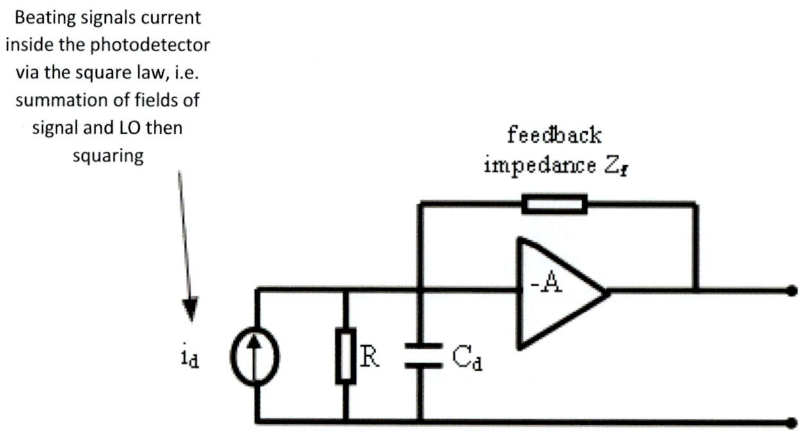

FIGURE 5.2 Schematic diagram of an electronic pre-amplifier in an optical receiver with a transimpedance electronic amplifier at the front end. The current source represents the electronic current generated in the photodetector due to the beating of the local oscillator and the optical signals. C_d=photodiode capacitance.

regenerating the timing sequence, and a voltage-level decision circuit for sampling the waveform for the final recovery of the transmitted and received digital sequence. Therefore, the opto-electronic preamplifier is followed by a main amplifier with automatic control to regulate the electronic signal voltage, which is then filtered and sampled by a decision circuit synchronized by the clock recovery circuitry.

An in-line fiber optical amplifier can be incorporated in front of the photodetector to form an optical receiver with an optical amplifier front end, improving its receiving sensitivity. This optical amplification at the front end of an optical receiver will be discussed in the chapter dealing with optical amplification processes.

The structure of the receiver thus consists of four parts: the optical mixing front end, the front end section, the linear channel of the main amplifier and AGC (if necessary), and the data recovery section. The optical mixing front end performs the summation of the optical fields of the local oscillator and the optical signals. Polarization orientation between these light waves is critical for maximizing the beating of the additive field in the photodiode. Depending on whether the frequency difference between these fields is finite or null, the resulting electronic signals derived from the detector can be in the baseband or the passband. The detection technique is termed heterodyne or homodyne, respectively (Figure 5.3).

5.3 COHERENT DETECTION TECHNIQUES

Optical coherent detection can be distinguished by the "demodulation" scheme in communication techniques according to the following definitions: (i) Coherent detection involves mixing two lightwaves or optical carriers – one carrying the information and the other being a local oscillator with an average energy much larger than that of the signals; and (ii) Demodulation refers to the recovery of baseband signals from the electrical signals.

A typical schematic diagram of coherent optical communications employing a guided wave medium and components is shown in Figure 5.1. In this setup, a narrow-band laser incorporating an optical isolator, cascaded with an external modulator, usually serves as the optical transmitter.

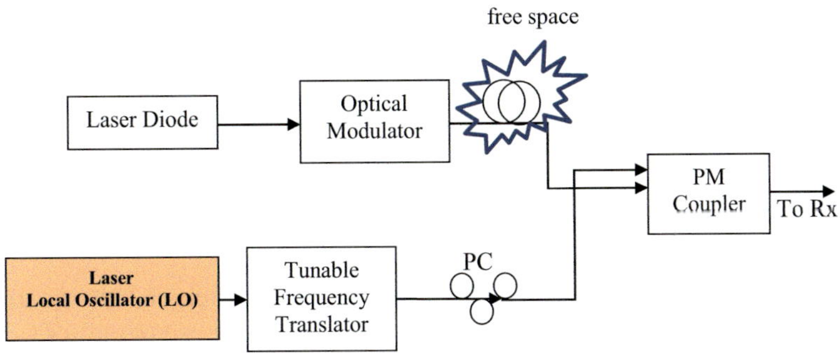

FIGURE 5.3 Typical arrangement of coherent optical communication systems[44,45] LD/LC is a very narrow linewidth laser diode used as a local oscillator without any phase locking to the signal optical carrier. PM coupler is a polarization-maintaining fiber-coupled device. PC=polarization controller, PM=polarization maintaining, coupler or fiber.

Information is fed via a microwave power amplifier to an integrated optic modulator, commonly of $LiNbO_3$ or EA types. Coherent detection is a principal feature of coherent optical communications and can be further distinguished by heterodyne and homodyne techniques, depending on whether there is a frequency difference between the local oscillator and the carrier of the signals. A local oscillator is a laser source with a tunable frequency, approximately equivalent to a monochromatic source. A polarization controller is also used to match its polarization with that of the information-bearing carrier. The local oscillator and the transmitted signal are mixed via a polarization-maintaining coupler and then detected by a coherent optical receiver. Most previous coherent detection schemes are implemented in a mix of the photonic domain and the electronic/microwave domain.

Coherent optical transmission has returned to the focus of research. One significant advantage is the preservation of all the information in the optical field during detection, leading to enhanced possibilities for optical multi-level modulation. This section investigates the generation of optical multi-level modulation signals. Several possible structures of optical M-ary-PSK and M-ary-QAM transmitters are shown and theoretically analyzed. Differences in the optical transmitter configuration and electrical driving lead to different properties of the optical multi-level modulation signals. This is demonstrated by deriving general expressions applicable to every M-ary-PSK and M-ary-QAM modulation format and exemplified for Square-16-QAM modulation.

Coherent receivers are distinguished between synchronous and asynchronous types. Synchronous detection requires an optical phase lock loop (OPLL), which recovers the phase and frequency of the received signals to lock the local oscillator (LO) to that of the signal, allowing the measurement of the absolute phase and frequency of the signals relative to the LO. Thus, synchronous receivers enable the direct mixing of the bandpass signals and the baseband, a technique termed homodyne reception. For asynchronous receivers, the frequency of the LO is approximately the same as that of the received signals, and no OPLL is required. In general, the optical signals are first mixed with an intermediate frequency (IF) oscillator, which is about two to three times the 3 dB passband. The electronic signals can then be recovered using an electrical PLL at a lower carrier frequency in the electrical domain. The mixing of the signals and a local oscillator of an IF frequency is referred to as heterodyne detection.

If no LO is used for demodulating digital optical signals, then differential or self-homodyne reception may be utilized, or classically termed as autocorrelation reception process or self-heterodyne detection. Coherent communications were an important technique in the 1980s and early 1990s, but research came to an interruption with the advent of optical amplifiers in the late 1990s, which offered up to 20 dB gain without difficulty. Nowadays, however, coherent systems are emerging again into the focus of interest due to the availability of digital signal processing and low-priced components, the partly relaxed receiver requirements at high data rates, and several advantages that coherent detection provides. The preservation of the temporal phase of coherent detection enables new methods for adaptive electronic compensation of chromatic dispersion. When considering WDM systems, coherent receivers offer tunability and allow channel separation via steep electrical filtering.

Furthermore, only the use of coherent detection permits convergence to the ultimate limits of spectral efficiency. To achieve higher spectral efficiencies, the use of multi-level modulation is required. In this regard, coherent systems are also beneficial because all the information of the optical field is available in the electrical domain. This allows complex optical demodulation with interferometric detection – which must be used in direct detection systems – to be avoided, and the complexity is transferred from the optical to the electrical domain. Several different modulation formats based on the modulation of all four quadratures of the optical field were proposed in the early 1990s, describing the possible transmitter and receiver structures and calculating the theoretical BER performance. However, a more detailed and practical investigation of multi-level modulation coherent optical systems for today's networks and data rates is still missing.

Currently, coherent reception has attracted significant interest due to the following reasons: (i) the received signals of coherent optical receivers are in the electrical domain, which is proportional to those in the optical domain. This, in contrast to direct detection receivers, allows for exact electrical equalization or precise phase estimation of the optical signals. (ii) Using heterodyne receivers, DWDM channels can be separated in the electrical domain by using electrical filters with a sharp roll-off from the passband to the cutoff band. Presently, the availability of ultra-high sampling rate digital signal processors (DSPs) allows users to conduct filtering in the DSP, where the filtering can be easily adjusted.

However, there are disadvantages that coherent receivers may face: (i) Coherent receivers are polarization sensitive and require polarization tracking at the front end of the receiver. (ii) Homodyne receivers require OPLL and electrical PLL for heterodyne detection, which necessitates control and feedback circuitry – either optical or electrical – that may be complex. (iii) For differential detection, compensation may be complicated due to the nature of differential reception.

In a later chapter, when advanced modulation formats are presented for optically amplified transmission systems, the use of photonic components is extensively exploited to take advantage of the advanced technology of integrated optics and planar lightwave circuits. Modulation formats of signals depend on whether the amplitude, phase, or frequency of the carrier is manipulated, as mentioned in Chapter 2. In this chapter, the detection is coherently converted to the intermediate frequency range in the electrical domain and the signal envelope. The downconverted carrier signals are detected and then recovered. Both binary level and multilevel modulation schemes employing amplitude, phase, and frequency shift keying modulation are described in this chapter.

Thus, coherent detection can be distinguished by the difference between the central frequency of the optical channel and that of the local oscillator. Three types can be classified: (i) heterodyne, when the difference is greater than the 3 dB bandwidth of the baseband signal; (ii) homodyne, when the difference is zero; and (iii) intradyne, when the frequency difference falls within the baseband of the signal.

It is noted that to maximize the beating signals at the output of the photodetector, the polarizations of the LO and the signals must be aligned. In practice, this can be best implemented using the polarization diversity technique (Figure 5.3).

5.3.1 Optical Heterodyne Detection

The basic configuration of optical heterodyne detection is shown in Figure 5.4. The local oscillator, whose frequency can be higher or lower than that of the carrier, is mixed with the information-bearing carrier, allowing for down or up conversion of the information signals to the intermediate frequency range. The down-converted electrical carrier and signal envelope are received by the photodetector. This combined lightwave is converted by the photodetector into electronic current signals, which are then filtered by an electrical bandpass filter (BPF) and demodulated by a demodulator. A low-pass filter is also used to remove higher-order harmonics of the nonlinear detection process, specifically square-law detection. With the use of an envelope detector, the process is asynchronous, hence the term asynchronous detection. If the down-converted carrier is recovered and then mixed with the IF signals, this constitutes synchronous detection. It is noted that detection is conducted in the intermediate frequency range in the electrical domain, hence the need for controlling the stability of the frequency spacing between the signal carrier and the local oscillator. This means that mixing these carriers would result in an IF carrier in the electrical domain prior to the mixing process or envelope detection to recover the signals.

Coherent detection relies on the electric field components of the signal and the local oscillator. The polarization alignment of these fields is critical for optimum detection. The electric field of the optical signals and the local oscillator can be expressed as

$$E_s(t) = \sqrt{2P_s(t)}\cos\{\omega_s t + \phi_s + \varphi(t)\} \tag{5.1}$$

$$E_{LO} = \sqrt{2P_L}\cos\{\omega_{LO} t + \phi_{LO}\} \tag{5.2}$$

where $P_s(t)$ and P_{LO} are the instantaneous signal power and average power of the signals and local oscillator, respectively, $\omega_s(t)$ and ω_{LO} are the signal and local oscillator angular frequencies, ϕ_s and

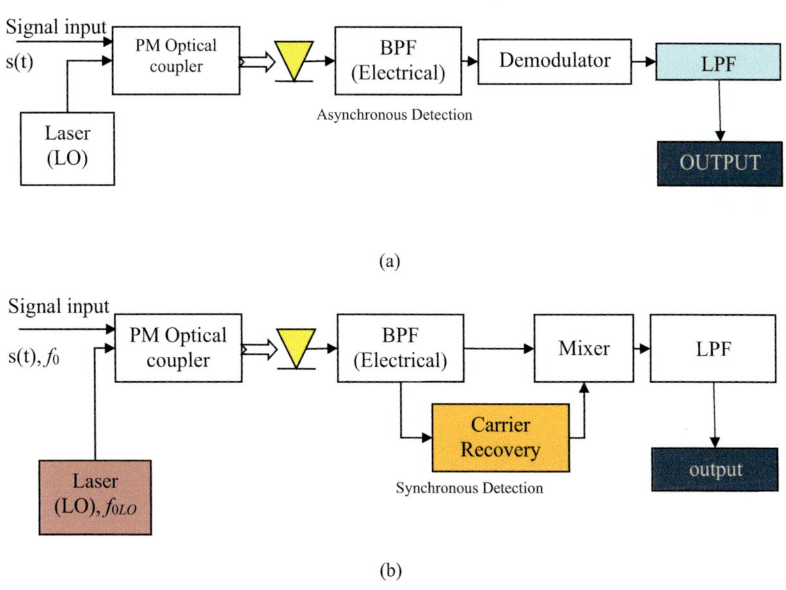

FIGURE 5.4 Schematic diagram of optical heterodyne detection – (a) asynchronous and (b) synchronous, receiver structures. LPF is a low-pass filter. BPF=bandpass filter, LPF=low-pass filter, PD=photodiode.

ϕ_{LO} are the phase including any phase noise of the signal and the local oscillator, and $\psi(t)$ is the modulation phase. The modulation can be amplitude with the switching on and off (amplitude shift keying – ASK) of the optical power or phase or frequency with the discrete or continuous variation of the time-dependent phase term. For discrete phase, it can be phase shift keying (PSK), differential PSK (DPSK), or differential quadrature PSK – DQPSK, and when the variation of the phase is continuous, we have frequency shift keying if the rate of variation is different for the bit "1" and bit "0", as given in Chapter 4.

Under an ideal alignment of the two fields, the photodetection current can be expressed by

$$i(t) = \frac{\eta q}{h\upsilon}\left[P_S + P_{LO} + 2\sqrt{P_S P_{LO}}\cos\left\{(\omega_S - \omega_{LO})t + \phi_S - \phi_{LO} + \varphi(t)\right\}\right] \tag{5.3}$$

where the higher frequency term (the sum) is eliminated by the photodetector frequency response, η is the quantum efficiency, q is the electronic charge, h is Plank's constant, and v is the optical frequency.

Thus, the power of the local oscillator dominates the shot noise process and, at the same time, boosts the signal level, enhancing the signal-to-noise ratio (SNR). The oscillating term is the beating between the local oscillator and the signal and is proportional to the signal with the amplitude being the square root of the product of the power of the local oscillator and the signal.

The electronic signal power S and shot noise N_s can be expressed as

$$\begin{aligned} S &= 2\Re^2 P_s P_{LO} \\ N_s &= 2q\Re\left(P_s + P_{LO}\right)B \\ \Re &= \frac{\eta q}{hv} = \text{responsivity} \end{aligned} \tag{5.4}$$

where B is the 3 dB bandwidth of the electronic receiver. Thus, the optical signal-to-noise ratio (OSNR) can be written as

$$\text{OSNR} = \frac{2\Re^2 P_s P_{LO}}{2q\Re\left(P_s + P_{LO}\right)B + N_{eq}} \tag{5.5}$$

where N_{eq} is the total electronic noise equivalent power at the input to the electronic preamplifier of the receiver. From this equation, we can observe that if the power of the local oscillator is significantly increased so that the shot noise dominates over the equivalent noise, at the same time the SNR increases. The sensitivity of the coherent receiver can only be limited by the quantum noise inherent in the photodetection process. Under this quantum limit, the OSNR_{QL} is given by

$$\text{OSNR}_{QL} = \frac{\Re P_s}{qB} \tag{5.6}$$

5.3.1.1 ASK Coherent System

Under the ASK modulation scheme, the demodulator inFigure 5.4 is an envelope detector (in lieu of the demodulator) followed by decision circuitry. The eye diagram is obtained, and a sampling instant is established with a clock recovery circuit. While synchronous detection would require locking between the carrier frequencies, it is then tuned to the local oscillator frequency to track the frequency component of the signal. The amplitude-demodulated envelope can be expressed as

$$\begin{aligned} r(t) &= 2\Re\sqrt{P_s P_{\mathrm{LO}}}\cos(\omega_{\mathrm{IF}})t + n_x\cos(\omega_{\mathrm{IF}})t + n_y\sin(\omega_{\mathrm{IF}})t \\ \omega_{\mathrm{IF}} &= \omega_s - \omega_{\mathrm{LO}} \end{aligned} \tag{5.7}$$

The IF frequency ω_{IF} is the difference between the LO and the signal carrier, n_x and n_y are the expected values of the orthogonal noise power components, which are random variables.

$$r(t) = \sqrt{\left[2\Re P_s P_{\mathrm{LO}} + n_x\right]^2 + n_y^2}\cos(\omega_{\mathrm{IF}}t + \Phi)t \qquad \text{with...}\Phi = \tan^{-1}\frac{n_y}{2\Re P_s P_{\mathrm{LO}} + n_x} \tag{5.8}$$

5.3.1.1.1 Envelop Detection

The noise power terms can be assumed to follow a Gaussian probability distribution and are independent of each other, with a zero mean and a variance σ. The probability density function (PDF) can thus be given as

$$p(n_x, n_y) = \frac{1}{2\pi\sigma^2} e^{\frac{-(n_x^2 + n_y^2)}{2\sigma^2}} \tag{5.9}$$

with respect to the phase and amplitude, this equation can be written as[2]

$$p(\rho, \phi) = \frac{\rho}{2\pi\sigma^2} e^{\frac{-(\rho^2 + A^2 - 2A\rho\cos\phi)}{2\sigma^2}} \tag{5.10}$$

where

$$\begin{aligned} \rho &= \sqrt{\left[2\Re\sqrt{P_s(t)P_{\mathrm{LO}}} + n_s(t)\right]^2 + n_y^2(t)} \\ A &= 2\Re\sqrt{P_s(t)P_{\mathrm{LO}}} \end{aligned} \tag{5.11}$$

The PDF of the amplitude can be obtained by integrating the phase amplitude PDF over the range of 0 to 2π and given as

$$p(\rho) = \frac{\rho}{\sigma^2} e^{\frac{-(\rho^2 + A^2)}{2\sigma^2}} I_0\left\{\frac{A\rho}{\sigma^2}\right\} \tag{5.12}$$

where I_0 is the modified Bessel function. If a decision level is set to determine the "1" and "0" levels, then the probability of error, and then the bit error rate (BER), can be obtained assuming equal probability of error between the "1s" and "0s", as:

$$BER = \frac{1}{2}P_e^1 + \frac{1}{2}P_e^0 = \frac{1}{2}\left[1 - Q\left(\sqrt{2\delta}, d\right) + e^{\frac{d^2}{2}}\right] \tag{5.13}$$

where Q is the Magnum function given in Chapter 7, and δ is given by

$$\delta = \frac{A^2}{2\sigma^2} = \frac{2\Re^2 P_s P_{\mathrm{LO}}}{2q\Re(P_s + P_{\mathrm{LO}})B + i_{\mathrm{N_{eq}}}^2} \tag{5.14}$$

When the power of the local oscillator is much larger than that of the signal and the equivalent noise current power, then this SNR becomes

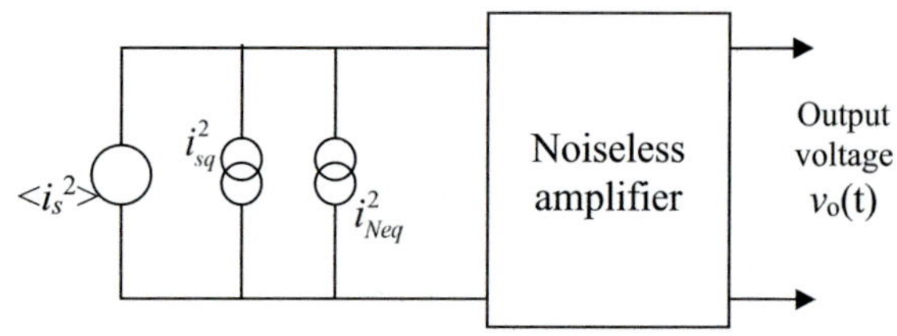

FIGURE 5.5 Equivalent current model at the input of the optical receiver, average signal current, and equivalent noise current of the electronic preamplifier as seen from its input port.

$$\delta = \frac{\Re P_s}{qB} \tag{5.15}$$

The physical representation of the detected current and the noise current due to quantum shot noise and the noise equivalent of the electronic pre-amplification can be seen in Figure 5.5, where the signal current is general and derived from the output of the detection scheme, whether from a photodetector or a back-to-back pair of photodetectors in a balanced receiver for detecting the phase difference of DPSK, DQPSK, or CPFSK signals and converting them to amplitudes.

The BER is optimum when setting its differentiation with respect to the decision level δ. An approximate value of the decision level can be obtained as:

$$d_{\text{opt}} \cong \sqrt{2+\frac{\delta}{2}} \Rightarrow \text{BER}_{\text{ASK-e}} \cong \frac{1}{2} e^{-\frac{\delta}{4}} \tag{5.16}$$

5.3.1.1.2 Synchronous Detection

ASK can be detected using synchronous detection,[3] and the BER is given by

$$\text{BER}_{\text{ASK-S}} \cong \frac{1}{2} \text{erfc}^{\frac{\sqrt{\delta}}{2}} \tag{5.17}$$

5.3.1.2 PSK Coherent System

Under the phase shift keying modulation format, the detection is similar to that of Figure 5.4 for heterodyne detection, but after the BPF, an electrical mixer is used to track the phase of the detected signal (Figure 5.6). The received signal is given by

$$r(t) = 2\Re\sqrt{P_s P_{\text{LO}}} \cos\left[(\omega_{\text{IF}})t + \varphi(t)\right] + n_x \cos(\omega_{\text{IF}})t + n_y \sin(\omega_{\text{IF}})t \tag{5.18}$$

The information is contained in the time-dependent phase term $\varphi(t)$.

When the phase and frequency of the voltage control oscillator (VCO) are matched with those of the signal carrier, then the received electrical signal can be simplified to:

$$\begin{aligned} r(t) &= 2\Re\sqrt{P_s P_{\text{LO}}}\, a_n(t) + n_x \\ a_n(t) &= \pm 1 \end{aligned} \tag{5.19}$$

Under Gaussian statistical assumption, the probability of the received signal of a "1" is given by

$$p(r) = \frac{1}{\sqrt{2\pi\sigma^2}} e^{-\frac{(r-u)^2}{2\sigma^2}} \tag{5.20}$$

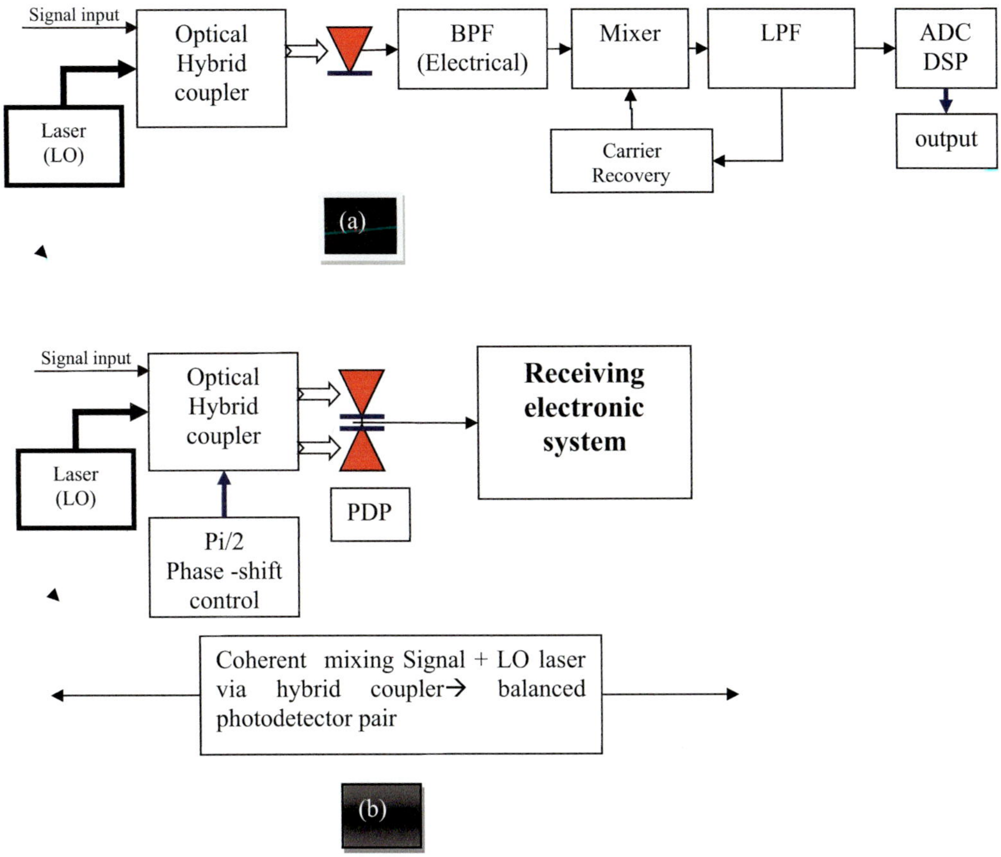

FIGURE 5.6 Schematic diagram of optical heterodyne detection for PSK format (a) single PD detector and coherent reception via hybrid coupler for mixing with LO laser; (b) balanced photodetector pair (PDP).

Furthermore, the probability of the "0" and "1" are assumed to be equal. We can obtain the BER as the total probability of the received "1" and "0" as

$$\text{BER}_{\text{PSK}} = \frac{1}{2}\text{erfc}(\delta) \tag{5.21}$$

5.3.1.3 Differential Detection

As observed in synchronous detection, there is a need for carrier recovery circuitry, usually implemented using a PLL, which complicates the overall receiver structure. It is possible to detect the signal through a self-homodyne process by beating the carrier of one bit period with that of the next consecutive bit; this is called differential detection. The detection process can be modified as shown in Figure 5.7, where the phase of the IF carrier of one bit is compared with that of the next bit, and a difference is recovered to represent a bit as "1" or "0." This requires differential coding at the transmitter and an additional phase comparator for the recovery process. In a later chapter on differential PSK, the differential decoding is implemented in the photonic domain via a photonic phase comparator in the form of an MZ delay interferometer (MZDI) with a thermal section for tuning the delay time of the optical delay line. The BER can be expressed as

$$\text{BER}_{\text{DPSK-e}} \cong \frac{1}{2}e^{-\delta} \tag{5.22}$$

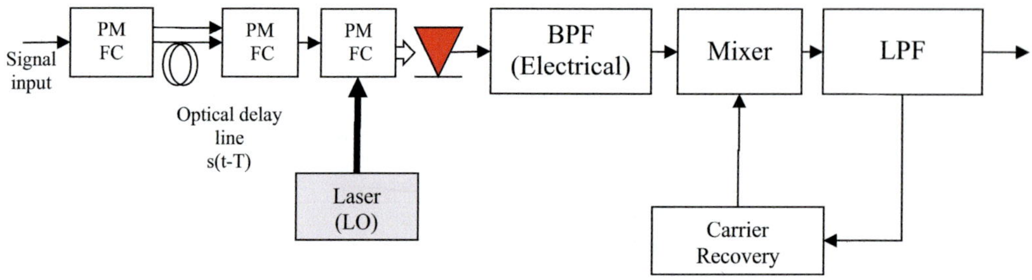

FIGURE 5.7 Schematic diagram of optical heterodyne and differential detection for PSK format.

$$r(t) = 2\Re\sqrt{P_s P_{LO}}\cos\left[\pi A_k s(t)\right] \tag{5.23}$$

where $s(t)$ is the modulating waveform, and A_k represents the bit "1" or "0". This is equivalent to the baseband signal, and the ultimate limit is the BER of the baseband signal.

The noise is dominated by the quantum shot noise of the local oscillator with its square noise current given by

$$i_{N\text{-sh}}^2 = 2q\Re\left(P_s + P_{LO}\right)\int_0^\infty \left|H(j\omega)\right|^2 d\omega \tag{5.24}$$

where $H(j\omega)$ is the transfer function of the receiver system, normally a transimpedance of the electronic preamp and that of a matched filter. As the power of the local oscillator is much larger than the signal and integrates over the dB bandwidth of the transfer function, this current can be approximated by

$$i_{N\text{-sh}}^2 \simeq 2q\Re P_{LO} B \tag{5.25}$$

Hence, the SNR (power) is given by

$$\text{SNR} \equiv \delta \simeq \frac{2\Re P_s}{qB} \tag{5.26}$$

The BER is the same as that of a synchronous detection and is given by

$$\text{BER}_{\text{homodyne}} \cong \frac{1}{2}\text{erfc}\sqrt{\delta} \tag{5.27}$$

The sensitivity of the homodyne process is at least 3 dB better than that of the heterodyne and the bandwidth of the detection is half that of its counterpart due to the double sideband nature of heterodyne detection.

5.3.1.4 FSK Coherent System

5.3.1.4.1 Reception System

The nature of FSK is based on two frequency components that determine the bits "1" and "0." There are several formats related to FSK, depending on whether the change in frequencies representing the bits is continuous or non-continuous, such as FSK or CPFSK modulation formats. For non-continuous FSK, detection is usually performed using a dual frequency discrimination structure, as shown in Figure 5.8, where two narrowband filters are used to extract the signals. For CPFSK, both frequency discriminators and balanced receivers for PSK detection can be used. Frequency

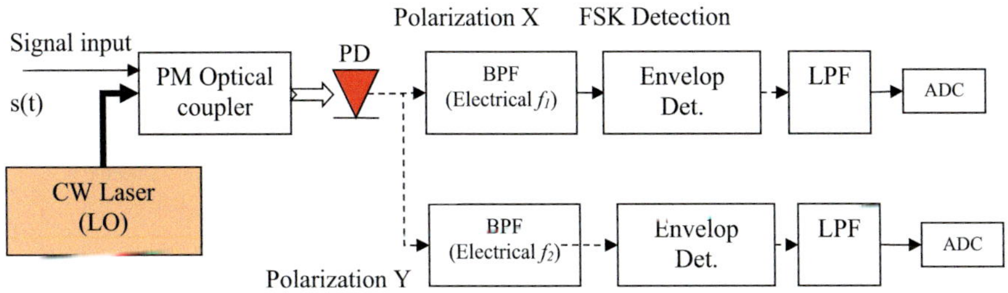

FIGURE 5.8 Schematic diagram of optical homodyne detection of FSK format. CW=continuous wave, LPF=low pass filter, BPF=band pass filter, PM=polarization maintaining coupler, PD=photodetector, LO=local oscillator, FSK=frequency shift keying, ADC=analog-to-digital converter.

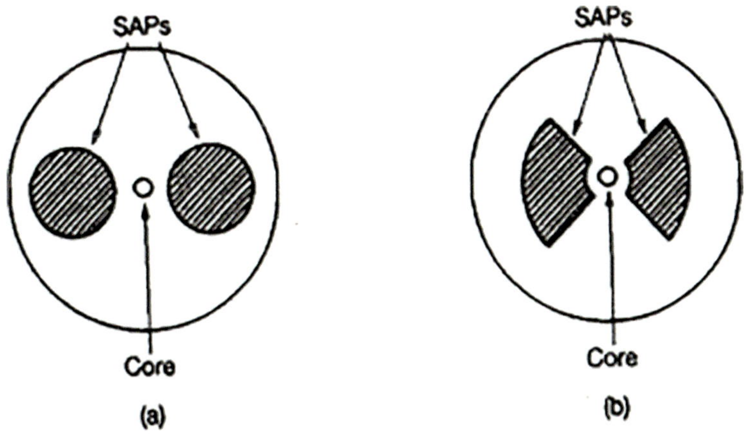

FIGURE 5.9 Polarization maintaining (PM) fibers (a) PANDA type and (b) Bow tie type.

discrimination is preferred over balanced receiving structures because it eliminates the phase contribution from the local oscillator or optical amplifiers, which may be used as optical preamps.

When the frequency difference between "1" and "0" equals a quarter of the bit rate, the FSK can be termed the minimum shift keying (MSK) modulation scheme. At this frequency spacing the phase is continuous between these states.

5.3.1.4.2 Polarization Handling

Mixing the LO and signal dual-polarized channels is critical due to their polarization alignment. Therefore, the hybrid coupler must be polarized in such a way that both the polarizations of the LO laser and the incoming signals are aligned. This can be achieved using a polarization-maintaining coupler and splitter, as shown in Figure 5.9.

5.3.2 Optical Homodyne Detection

Optical homodyne detection matches the transmitted signal phases to that of the local oscillator phase signal. A schematic of the optical receiver is shown in Figure 5.10. The field of the incoming optical signals is mixed with the local oscillator, whose frequency and phase are locked to that of the signal carrier waves via a phase-locked loop. The resultant electrical signal is then filtered and sent to a decision circuitry.

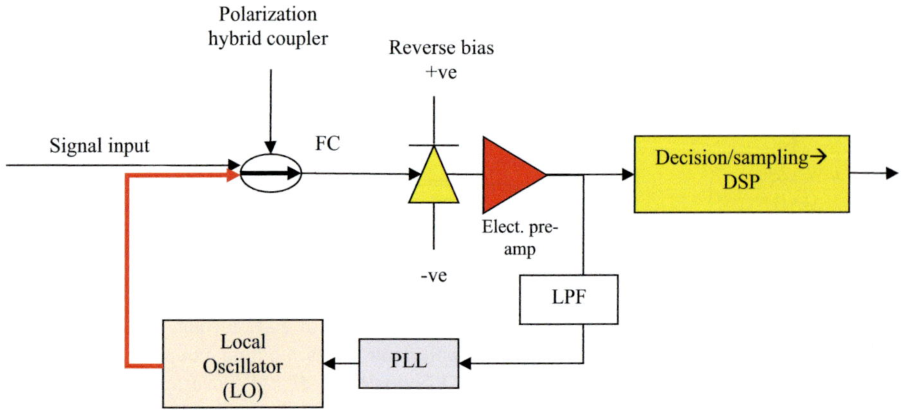

FIGURE 5.10 The general structure of an optical homodyne detection system with phase locking to match frequencies of signal carrier and LO laser. FC=fiber coupler, LPF=low pass filter, PLL=phase lock loop, PHC=polarization hybrid coupler.

5.3.2.1 Detection and Optical PLL

Optical homodyne detection requires the phase matching of the signal carrier frequency and that of the local oscillator. This type of detection would give a very sensitive result, in principle, of 9 photons/bit. Implementation of such a system would normally require an optical PLL, whose structure in a recent development[4] is as shown in Figure 5.11. The local oscillator frequency is locked to the carrier frequency of the signal by shifting it to the modulated sideband component via the use of the optical modulator. A single-sideband optical modulator is preferred; however, a double-sideband may also be used. This modulator is excited by the output signal of a voltage-controlled oscillator, whose frequency is determined by the voltage level of the output of an electronic bandpass filter conditioned to meet the required voltage level for driving the electrode of the modulator. The frequency of the local oscillator is normally tuned to the region such that the frequency difference with respect to the signal carrier falls within the passband of the electronic filter. When the frequency difference is zero, there is no voltage level at the output of the filter, and thus the optical PLL has reached the final stage of locking. The bandwidth of the optical modulator is important so that it can extend the locking range between the two optical carriers.

Any difference in frequency between the LO and the carrier is detected and noise-filtered by the low-pass filter (LPF). This voltage level is then fed to a voltage-controlled oscillator to generate a sinusoidal wave, which is then used to modulate an intensity modulator that modulates the light waves of the LO. The output spectrum of the modulator would exhibit two sidebands and the LO light wave (see Figure 5.11b). One of these components would then be locked to the carrier. A closed loop would ensure stable locking. If the intensity modulator is biased at the minimum transmission point and the voltage level at the output of the VCO is adjusted to $2V_{\pi}$ with driven signals of $\pi/2$ phase shift with each other, then we would have carrier suppression and sideband suppression. This eases the confusion of the closed-loop locking.

For dual-polarized channels modulated by complex formats, including the *I*- and *Q*-components, the coherent receiver schematic is given in Figure 5.11c. Polarization splitters are placed at the input of the dual-polarized optical channels and for the dual-polarized LO laser. The integrated polarization beam splitter splits the two polarized channels and those of the LO laser, and the mixing of the appropriate polarized components applies an optical phase shift of $\pi/2$ before beating each other in the PDP (photodetector pair). The coherent received signals are converted to electronic current, then amplified in a transimpedance amplifier (TIA) and an automatic gain control (AGC) amplifier before being sent to the ADC (analog-to-digital converter) for digital processing. The AGC is required for the SPC system because of the high dynamic range or variation of the

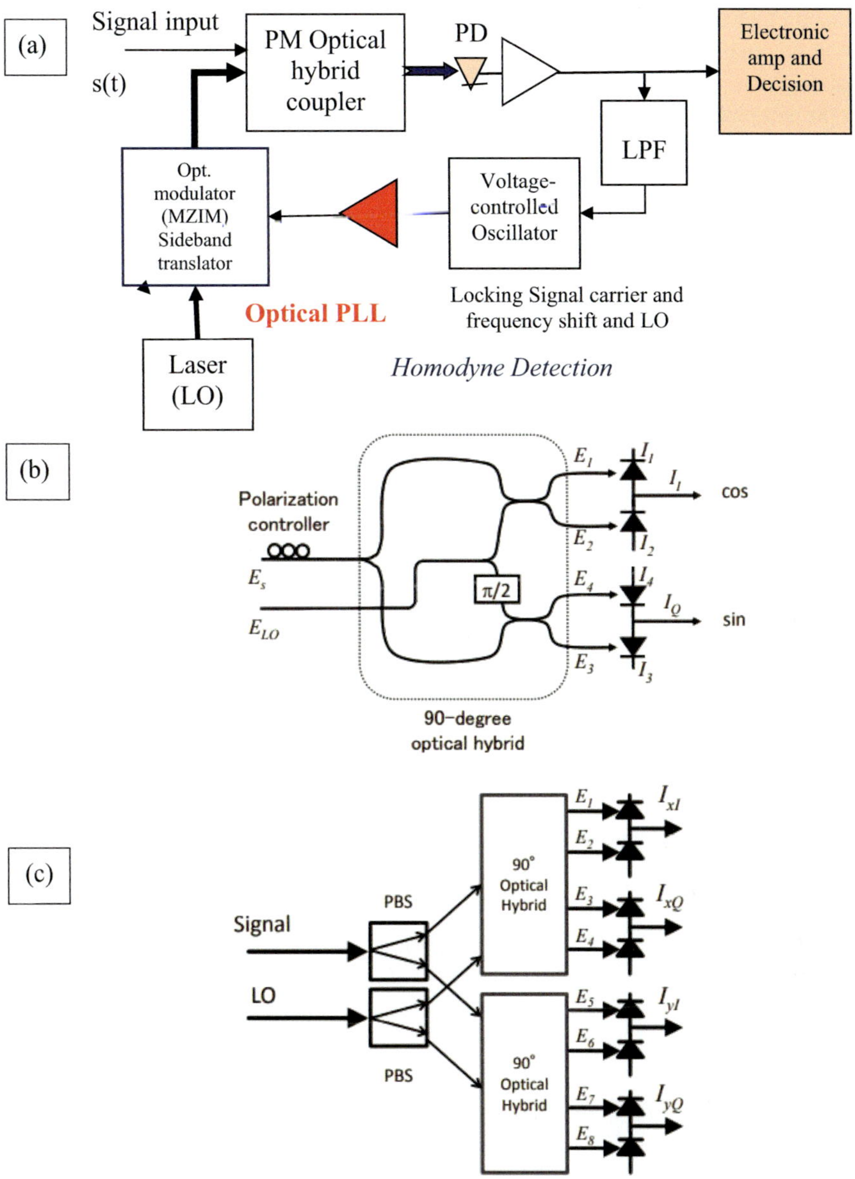

FIGURE 5.11 (a) Schematic diagram of optical homodyne detection --- electrical line (dashed) optical line (continuous and solid) using an optical phase lock loop; (b) schematic of a π/2 optical hybrid coupler for complex modulated signals with balanced photo-detector pair and optical waveguides to couple input signals and LO laser with π/2 phase shifter; (c) dual-polarized channels and dual π/2 optical hybrid coupler. LO=local oscillator, E=electric field of lightwaves, *I*=*I* phase, *Q*=quadrature, sin=sinusoidal, cos=cosinusoidal, PM=polarization maintaining, PBS=polarization beam splitter, x=horizontal direction, y=vertical direction.

signal's average power due to the fluctuation of the turbulence layer and spatial broadening of the beam under alignment. The phase shifter of π/2 is implemented by a thermal heater, using the pyroelectric effects on the planar waveguide made of silica or Si_3N_4 substrate. This shifter splits the in-phase and quadrature components, which are then detected separately. Thus, we can have four balanced PDPs to produce the *I*- and *Q*-components of the *X* and *Y* polarization channels. Dual polarization channels can double the total transmission capacity, but the skew of these channels must be managed to ensure synchronization at the sampling instants of the sequence.

Under perfect phase matching, the received signal is given by

$$i_s(t) = 2\Re\sqrt{p_s P_{LO}}\cos\{\pi a_k s(t)\} \tag{5.28}$$

where a_k takes the value ±1, *s*(*t*) is the modulating waveform. This is a baseband signal, thus the error rate is the same as that of the baseband system.

The shot noise power induced by the local oscillator and the signal power can be expressed as

$$i_{NS}^2 = 2q\Re\left(p_s + P_{LO}\right)\int_0^{\infty}\left|H(j\omega)\right|d\omega \tag{5.29}$$

where $\left|H\left(j\omega\right)\right|$ is the transfer function (TF) of the receiver whose expression, if under a matched filtering, can be

$$\left|H\left(j\omega\right)\right|^2 = \left[\frac{\sin\left(\omega T/2\right)}{\omega T/2}\right]^2 \tag{5.30}$$

with *T* being the bit period. Then the noise power becomes:

$$i_{NS}^2 = q\Re\left(p_s + P_{LO}\right)\frac{1}{T} \simeq \frac{q\Re P_{LO}}{T} \quad \text{when...} p_s \ll P_{LO} \tag{5.31}$$

Thus, the SNR is:

$$\text{SNR} = \frac{2\Re p_s P_{LO}}{q\Re P_{LO}/T} = \frac{2p_s T}{q} \tag{5.32}$$

Hence, the bit error rate is

$$P_E = \frac{1}{2}\text{erfc}\left(\sqrt{\text{SNR}}\right) \rightarrow \text{BER} = \text{erfc}\left(\sqrt{\text{SNR}}\right) \tag{5.33}$$

5.3.2.2 Quantum Limit Detection

For homodyne detection, a superquantum limit can be achieved. In this case, the local oscillator is used in a very specific way that matches the incoming signal field in polarization, amplitude, and frequency and is assumed to be phase-locked to the signal. Assuming that the phase signal is perfectly modulated such that it acts in-phase or counter-phase with the local oscillator, the homodyne detection would give a normalized signal current of

$$i_{sC} = \frac{1}{2T}\left[\mp\sqrt{2n_p} + \sqrt{2n_{LO}}\right]^2 \ldots\text{for}\ldots 0 \le t \le T \tag{5.34}$$

Assuming further that $n_p = n_{LO}$ the number of photons for the LO for generation of detected signals then the current can be replaced with $4n_p$ for the detection of a "1" and nothing for a "0" symbol.

5.3.2.3 Linewidth Influences

5.3.2.3.1 Heterodyne Phase Detection

When the linewidth of the light sources is significant the intermediate frequency deviates due to a phase fluctuation, and the probability density function is related to this linewidth conditioned on the deviation $\delta\omega$ of the intermediate frequency. For a signal power of p_s we have the total probability of error is given as

$$P_E = \int_{-\infty}^{\infty} P_C\left(p_s, \partial\omega\right) p_{\mathrm{IF}}\left(\partial\omega\right) \partial\omega \tag{5.35}$$

The probability density function of the intermediate frequency under a frequency deviation can be written as[5]

$$p_{\mathrm{IF}}(\partial\omega) = \frac{1}{\sqrt{\Delta\upsilon BT}} e^{-\frac{\partial\omega^2}{4\pi\Delta\upsilon B}} \tag{5.36}$$

where Δv is the full IF linewidth at FWHM of the power spectral density, and T is the bit period.

5.3.2.3.2 Differential Phase Detection with Local Oscillator

5.3.2.3.2.1 DPSK Systems The DPSK detection requires an MZDI and a balanced receiver in either the optical domain or the electrical domain. If in the electrical domain, the beating signals in the PD between the incoming signals and the LO would give the beating electronic current, which is then split and one branch is delayed by a one-bit period and then summed up. The heterodyne signal current can be expressed as[6]

$$i_s(t) = 2\Re\sqrt{P_{\mathrm{LO}} p_s}\cos\left(\omega_{\mathrm{IF}} t + \phi_s(t)\right) + n_x(t)\cos\omega_{\mathrm{IF}} t - n_y(t)\sin\omega_{\mathrm{IF}} t \tag{5.37}$$

The phase $\phi_s(t)$ is expressed by

$$\phi_s(t) = \varphi_s(t) + \left\{\varphi_N(t) - \varphi_N(t+T)\right\} - \left\{\varphi_{pS}(t) - \varphi_{pS}(t+T)\right\} - \left\{\varphi_{pL}(t) - \varphi_{pL}(t+T)\right\} \tag{5.38}$$

The first term is the phase of the data and takes the value 0 or π. The second term represents the phase noise due to the shot noise of the generated current and the third and fourth terms are the quantum shot noise due to the local oscillator and the signals. The probability of error is given by

$$P_E = \int_{-\pi/2}^{\pi/2} \int_{-\infty}^{\infty} p_n\left(\phi_1 - \phi_2\right) p_q\left(\phi_1\right) \partial\phi_1 \, \partial\phi_1 \tag{5.39}$$

where $p_n(\cdot)$ is the probability density function of the phase noise due to the shot noise, and $p_q(\cdot)$ is for the quantum phase noise generated from the transmitter and the local oscillator.[7]

The probability of error can be written as

$$p_N\left(\phi_1 - \phi_2\right) = \frac{1}{2\pi} + \frac{\rho e^{-\rho}}{\pi} \sum_{m=1}^{\infty} a_m \cos\left(m\left(\phi_1 - \phi_2\right)\right)$$
$$a_m \sim \left\{ \frac{2^{m-1}\Gamma\left[\frac{m+1}{2}\right]\Gamma\left[\frac{m}{2}+1\right]}{\Gamma[m+1]} \left[I_{m-1/2}\frac{\rho}{2} + I_{(m+1)/2}\frac{\rho}{2} \right] \right\}^2 \tag{5.40}$$

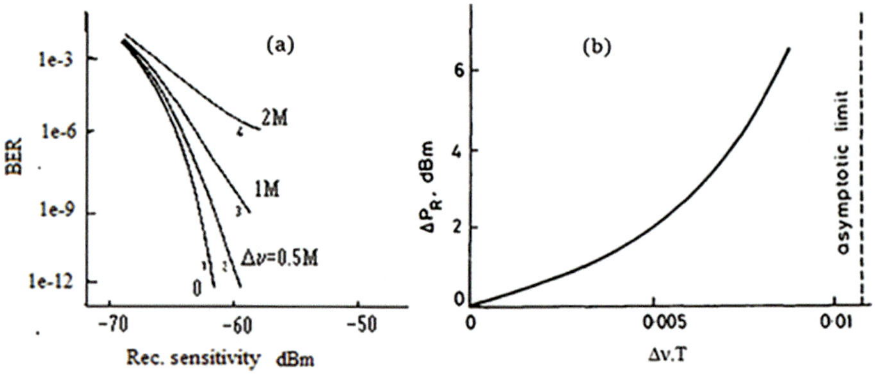

FIGURE 5.12 (a) Probability of error versus receiver sensitivity with linewidth as a parameter in MHz. (b) Degradation of optical receiver sensitivity at BER = 1e-9 for DPSK systems as a function of the linewidth and bit period – bit rate = 140 Mb/s. (Extracted from Nicholson (1984)[46].)

where $\Gamma(\cdot)$ is the gamma function and is the modified Bessel function of the first kind. The PDFof the quantum phase noise can be given as[8]

$$p_q(\phi_1) = \frac{1}{\sqrt{2\pi D\tau}} e^{\frac{\phi_1^2}{2D\tau}} \tag{5.41}$$

with D as the phase diffusion constant, the standard deviation from the central frequency is given as

$$\Delta\upsilon = \Delta\upsilon_R + \Delta\upsilon_L = \frac{D}{2\pi} \tag{5.42}$$

is the sum of the transmitter and the local oscillator FWHM linewidth. Substituting Equations 5.41 and 5.40 into 5.39, we obtain

$$P_E = \frac{1}{2} + \frac{\rho e^{-\rho}}{2} \sum_{n=0}^{\infty} \frac{(-1)^n}{2n+1} e^{-(2n+1)^2 \pi \Delta\upsilon T} \left\{ I_{n-1/2} \frac{\rho}{2} + I_{(n+1)/2} \frac{\rho}{2} \right\}^2 \tag{5.43}$$

This equation gives the probability of error as a function of the received power. The probability of error is plotted against the receiver sensitivity, and the product of the linewidth with the bit rate (or the relative bandwidth of the laser linewidth and the bit rate) is shown in Figure 5.12 for the DPSK modulation format at a 140 Mb/s bit rate, with the variation of the laser linewidth from 0 to 2 MHz.

5.3.2.3.3 Differential Phase Detection under Self-Coherence

Recently, the laser linewidth requirement for DQPSK modulation and differential detection for DQPSK has also been studied. No local oscillator is used, meaning self-coherent detection. It has been shown that a transmitter laser linewidth of up to 3 MHz would not significantly influence the probability of error, as shown in Figures 5.13 and 5.14.[10]

5.3.2.3.4 Differential Phase Coherent Detection of Continuous Phase FSK Modulation Format

The probability of error for CPFSK can be derived by taking into consideration the delay line of the differential detector, the frequency deviation, and phase noise.[9] Similar to Figure 5.8, the differential detector configuration is shown in Figure 5.15a, and the conversion of the frequency-to-voltage

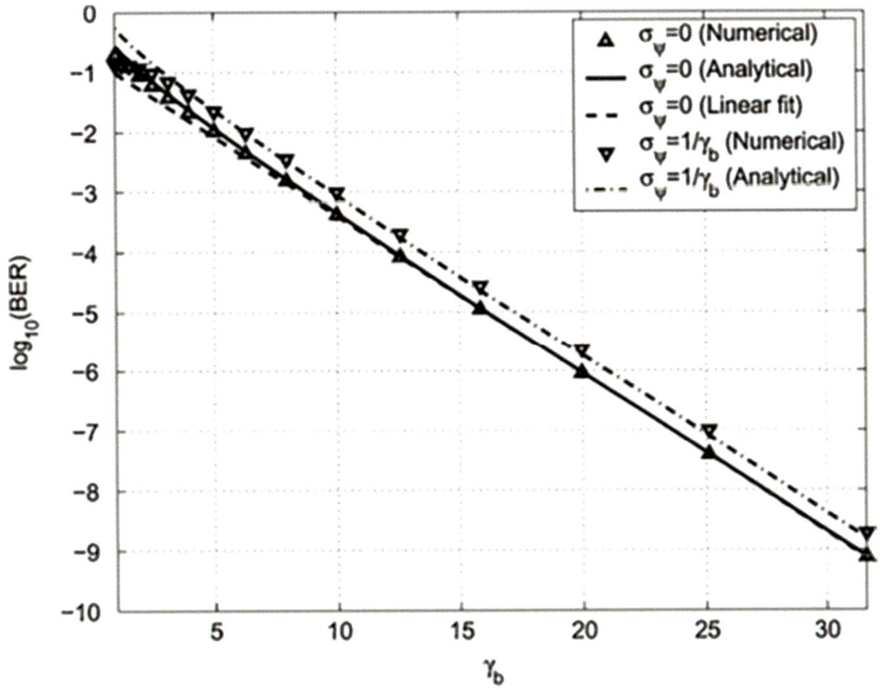

FIGURE 5.13 Analytical approximation (solid line) and numerical evaluation (triangles) of the BER for the cases of zero linewidth and that required to double the BER. The dashed line is the linear fit for zero linewidth. Bit rate 10 Gb/s per channel. (Extracted from Ref.[47])

relationship is shown in Figure 5.15b. If heterodyne detection is employed, a bandpass filter is used to bring the signals back to the electrical domain.

The detected signal phase at the shot noise limit at the output of the low-pass filter (LPF) can be expressed as:

$$\phi(t) = \omega_c t + a_n \frac{\Delta\omega}{2}\tau + \varphi(t) + \varphi_n(t)$$
$$\text{with}\ldots\omega_c = 2\pi f_c = (2n+1)\frac{\pi}{2\tau} \tag{5.44}$$

And τ is the differential detection delay time, $\Delta\omega$ is the deviation of the angular frequency of the carrier for the "1" or "0" symbol; φ(t) is the phase noise due to shot noise; $n(t)$ is the phase noise due to the transmitter and the local oscillator quantum shot noise; an takes the values of ±1, the binary data symbol.

Thus, by integrating the detected phase from $-\frac{\Delta\omega}{2}\tau \to \pi - \frac{\Delta\omega}{2}\tau$, we obtain the probability of error as

$$P_E = \int_{-\frac{\Delta\omega}{2}\tau}^{\pi-\frac{\Delta\omega}{2}\tau/2} \int_{-\infty}^{\infty} p_n(\phi_1 - \phi_2)\, p_q(\phi_1)\, \partial\phi_1\, \partial\phi_1 \tag{5.45}$$

Similar to the case of DPSK system, substituting Equations 5.40 and 5.41 into 5.45, we obtain

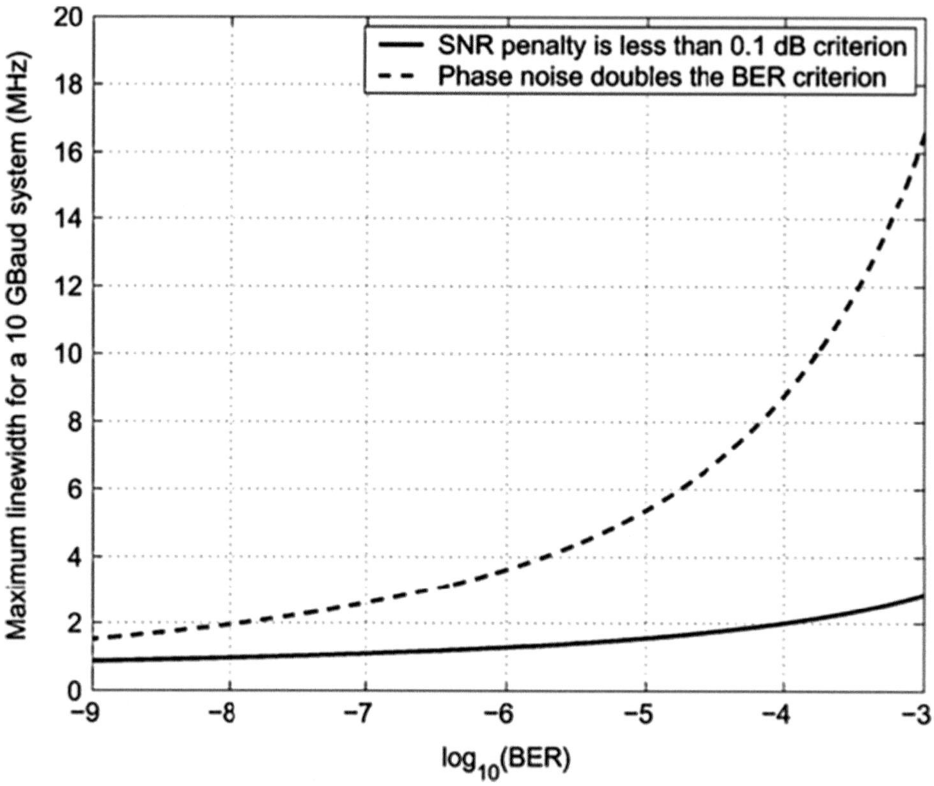

FIGURE 5.14 Criteria for neglecting linewidth in a 10-GSymbols/s system. The loose bound is to neglect linewidth if the impact is to double the BER, while the tighter bound is to neglect linewidth if the impact is a 0.1-dB SNR penalty. Bit rate 10 GSymbols/s. (Extracted from Ref.[10])

$$P_E = \frac{1}{2}\frac{\rho e^{-\rho}}{2}\sum_{n=0}^{\infty}\frac{(-1)^n}{2n=1}e^{-(2n+1)^2\pi\Delta\upsilon\tau}\left\{I_{n-1/2}\frac{\rho}{2}+I_{(n+1)/2}\frac{\rho}{2}\right\}^2 e^{-(2n+1)^2\pi\Delta\upsilon\tau}\cos\{(2n+1)\alpha\} \tag{5.46}$$

$$\alpha = \frac{\pi(1-\beta)}{2}\ldots\text{and}\ldots\beta = \frac{\Delta\omega}{\omega_m} = 2\pi\tau/T_0$$

ω_m is the deviation of the angular frequency, where m is the modulation index, and T_0 is the pulse period or bit period. The modulation index parameter β is defined as the ratio of the actual frequency deviation to the maximum frequency (Figure 5.16).

5.3.3 Optical Intra-Dyne Detection

Optical phase diversity receivers combine the advantages of homodyne detection, which requires minimum signal processing bandwidth, with heterodyne reception, which requires no optical phase locking. The term diversity is well known in radio transmission links and describes transmission over more than one path. In optical receivers, the optical path is considered in terms of different polarization and phase paths. In intradyne detection, the frequency difference, the intermediate frequency or local oscillator frequency offset (LOFO), between the LO and the central carrier is non-zero and lies within the signal bandwidth of the baseband signal, as illustrated in Figure 5.17.[10] Naturally, the

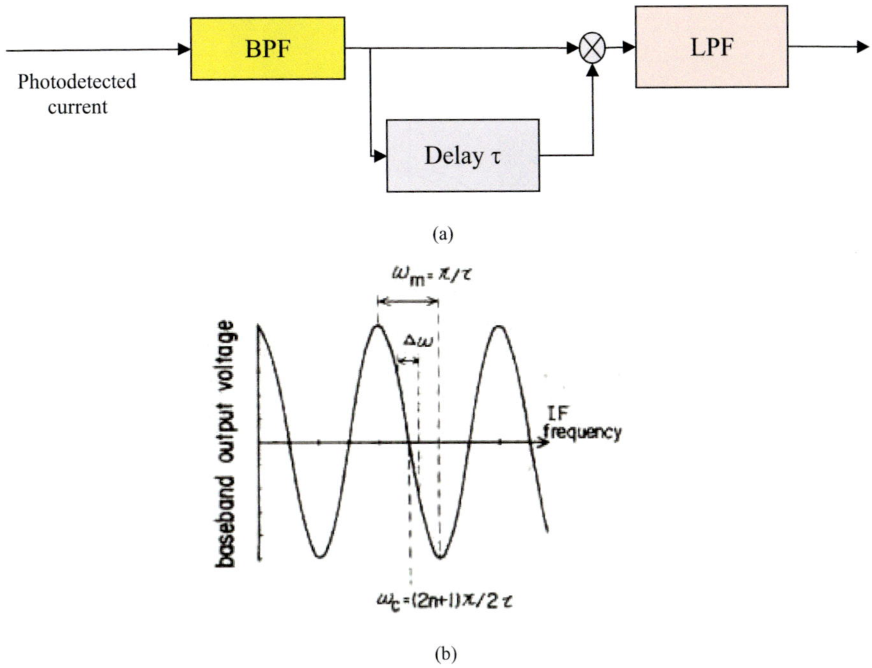

FIGURE 5.15 (a) Configuration of a CPFSK differential detection, (b) frequency-to-voltage conversion relationship of FSK differential detection. (Extracted from Ref.[48])

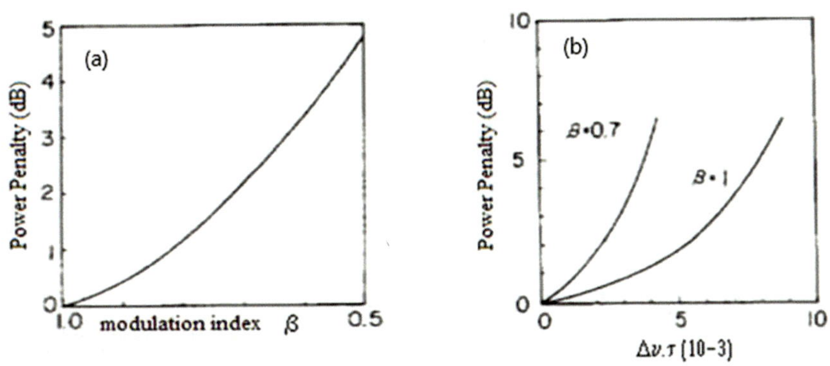

FIGURE 5.16 (a) Dependence of receiver power penalty at a BER of 1e-9 on modulation index β (ratio between frequency deviation and maximum frequency spacing between f_1 and f_2). (b) Receiver power penalty at BER 1e-9 as a function of the product of beat bandwidth and bit delay time – effects, excluding LD phase noise. (Extracted from Ref.[49])

control and locking of the carrier and the local oscillator cannot be exact, sometimes due to jitter of the source. Most of the time, the laser frequency is locked stably by adjusting the reflection mirror; hence, the central frequency varies by a few hundreds of kHz. Thus, intradyne coherent detection is more realistic. Furthermore the digital signal processor in modern coherent reception systems can extract this difference without much difficulty in the digital domain.[11] Obviously, heterodyne detection requires a large frequency range of operation for electronic devices, while homodyne and intradyne reception require simpler electronics. Either differential or non-differential formats can

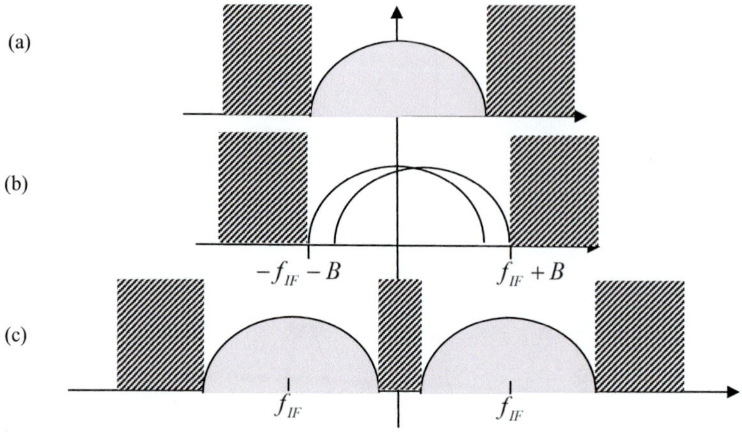

FIGURE 5.17 Spectrum of coherent detection (a) homodyne, (b) intradyne, and (c) heterodyne.

be used in DSP-based coherent reception. For differential-based reception, differential decoding gains an advantage when there are slips in the cycles of bits due to walk-off of the pulse sequences over very long transmission lines with non-compensating fiber.

The diversity in phase and polarization can be achieved by using a $\pi/2$ hybrid coupler that split the polarization of the LO and the received channels and mixes them with $\pi/2$ optical phase shift. The mixed signals are then detected by balanced photodetectors. This diversity detection is described in the next few sections (see also Figure 5.24).

5.4 SELF-COHERENT DETECTION AND ELECTRONIC DSP

The coherent techniques described above offer significant improvements but face setbacks due to the availability of a stable local oscillator and an optical phase lock loop for locking the frequency of the local oscillator and that of the signal carrier.

Digital signal processors (DSP) have been widely used in wireless communications and play key roles in the implementation of DSP-based coherent optical communication systems. Applying DSP techniques to coherent optical communication systems helps overcome the difficulties associated with optical phase-locking loops (PLL) and also improves the performance of transmission systems in the presence of fiber-degrading effects, including chromatic dispersion, PMD, and fiber nonlinearities.

Coherent optical receivers have the following advantages: (i) the shot-noise limited receiver sensitivity can be achieved with sufficient local oscillator (LO) power; (ii) closely spaced WDM channels can be separated with electrical filters having sharp roll-off characteristics; and (iii) the ability of phase detection can improve receiver sensitivity compared with IMDD systems.[12] In addition, any kind of multi-level phase modulation format can be introduced using the coherent receiver. While the spectral efficiency of binary modulation formats is limited to 1 bit/s/Hz/polarization (known as the Nyquist limit), multilevel modulation formats with N bits of information per symbol can achieve up to N bit/s/Hz/polarization in spectral efficiency. Recent research has focused on M-ary phase-shift keying (M-ary PSK) and even quadrature amplitude modulation (QAM) with coherent detection, which can increase spectral efficiency by a factor of $\log_2 M$.[13,14,15] Moreover, for the same bit rate, since the symbol rate is reduced, the system can have higher tolerance to chromatic dispersion and PMD.

However, one of the major challenges in coherent detection is overcoming carrier phase noise when using a local oscillator (LO) to beat with the received signals to retrieve the modulated phase information. Phase noise can result from lasers, which will cause a power penalty to receiver sensitivity. A self-coherent multi-symbol detection of optical differential M-ary PSK is introduced to improve system performance; however, it requires higher analog-to-digital conversion resolution and more digital signal processing power compared to a digital coherent receiver.[16] Furthermore, differential encoding is also necessary in this scheme. For the coherent receiver, an optical PLL is initially an option to track the carrier phase with respect to the LO carrier in homodyne detection. However, an optical PLL operating at optical wavelengths in combination with distributed feedback (DFB) lasers may be quite difficult to implement because the product of laser linewidth and loop delay is too large.[17] Another option is to use an electrical PLL to track the carrier phase after down-converting the optical signal to an intermediate frequency (IF) electrical signal in a heterodyne detection receiver, as mentioned above. Compared to heterodyne detection, homodyne detection offers better sensitivity and requires a smaller receiver bandwidth.[18] On the other hand, coherent receivers employing high-speed analog-to-digital converters (ADCs) and high-speed baseband digital signal processing (DSP) units are becoming increasingly attractive compared to using an optical PLL for demodulation. A conventional block *M*th power phase estimation (PE) scheme is proposed in Refs.[19,20] to raise the received M-ary PSK signals to the *M*th power to estimate the phase reference in conjunction with a coherent optical receiver. However, this scheme requires nonlinear operations, such as taking the *M*th power and the $\tan^{-1}(\cdot)$, and resolving the $\pm 2\pi/M$ phase ambiguity, which incurs a large latency in the system. Such nonlinear operations limit the potential for real-time processing of the scheme. In addition, nonlinear phase noises always exist in long-haul systems due to the Gordon -Mollenauer effect,[21] which severely affects the performance of a phase-modulated optical system.[22] The results in Ref.[23] show that such *M*th power PE techniques may not effectively deal with nonlinear phase noise.

The maximum likelihood (ML) carrier phase estimator derived in Ref.[24] can be used to approximate ideal synchronous coherent detection in optical PSK systems. The ML phase estimator requires only linear computations, making it more feasible for online processing in real systems. Intuitively, one can show that the ML estimation receiver outperforms the *M*th power block phase estimator and conventional differential detection, especially when nonlinear phase noise is dominant. This leads to significantly improved receiver sensitivity and tolerance to nonlinear phase noise. The algorithm of ML phase estimator is expected to improve the performance of coherent optical communication systems using different M-ary PSK and Quadrature Amplitude Modulation (QAM)formats. The improvement by DSP at the receiver end can be significant for transmission systems in the presence of fiber-degrading effects, including chromatic dispersion, PMD, and nonlinearities for both single-channel and also DWDM systems.

5.5 OPTICAL PRE-AMPLIFIER RECEIVER

In optical communication systems, especially the SPC system, a pre-amp Erbium-Doped Fiber Amplifier (EDFA) with low noise and low input level is required and used at the input of the optical receiver to increase receiver sensitivity and to compensate for long-haul space transmission losses by amplifying the photon detector's incoming optical signal. In this text, the pre-amp EDFA noise performance must first be characterized at the pre-amp level through modeling. Then, the pre-amp noise performance can be characterized experimentally at the optical receiver level. The optimization of the pre-amp EDFA performance can be achieved. The pre-amp EDFA noise performance characterization and analysis at the optical receiver level can be done. This is the ultimate performance characterization method for the pre-amp EDFA, and it was performed by testing the optical receiver's transmission performance under different pre-amp operating conditions (Figure 5.18).

The LNA is employed as a pre-amplification device at the input of the receiver site. The gain of such an LNA operating at the saturation level and linear gain range is given in Chapter 7. Figure 5.19 shows a typical receiver sensitivity versus the optical gain factor for an LNA with a noise figure (NF) of about 3–4 dB. The gain and receiver sensitivity vary linearly until the saturation level, at which the gain is saturated at a constant level and thus the sensitivity cannot be lowered.

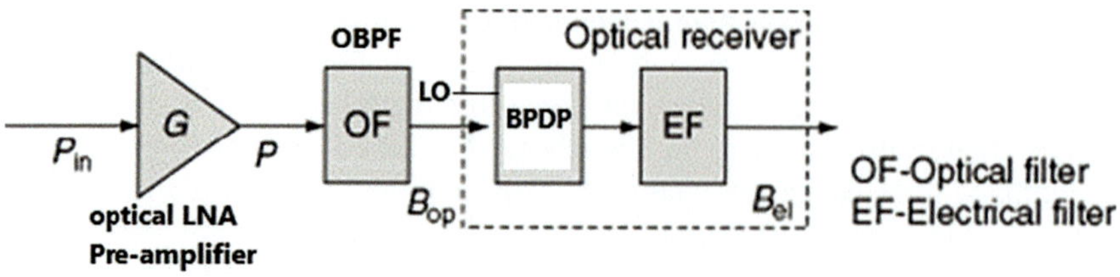

FIGURE 5.18 Schematic of an optical pre-amplifier receiver with coherent reception. LO=local oscillator laser, Pin=average power of the input channel, G=optical gain of LNA, OF=optical filter, usually OBPF=optical bandpass filter, BPDP=balanced photodetector pair, EF=electric filter if heterodyne.

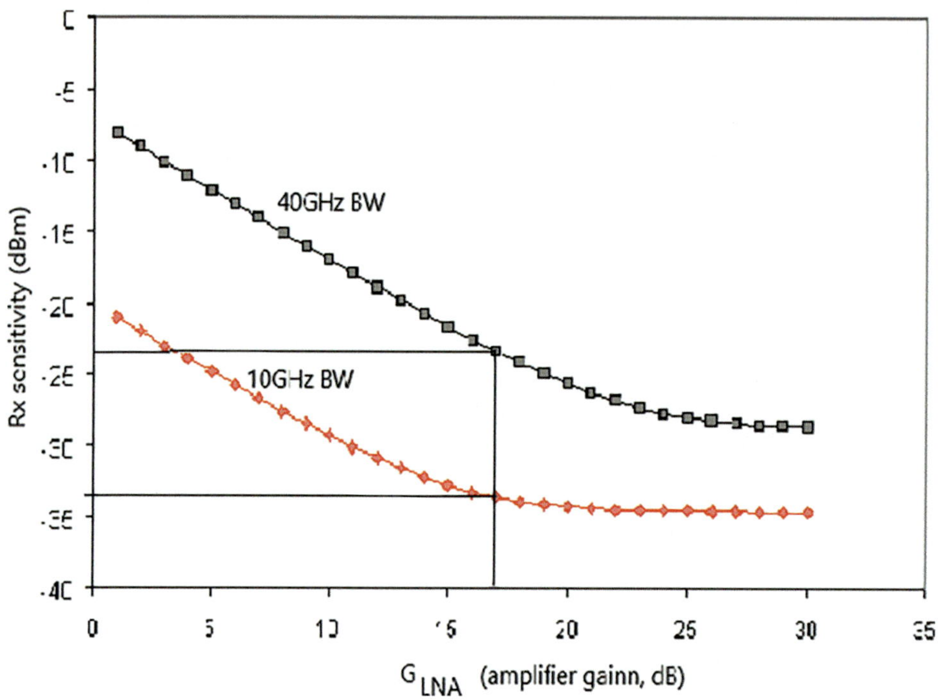

FIGURE 5.19 Receiver sensitivity in dBm versus the low noise amplifier gain for bandwidths of 10 and 40 GHz in receiver system, including all noises. Slope –30/17 dB or –1.76 factor.

The sensitivity improves with respect to the gain of –1.76 or as the noise figure of the amplifier increases with the gain, which reduces the sensitivity. This factor becomes zero when the LNA reaches the saturation level, with no further improvement in receiver sensitivity at $G=17$ dB for one-stage pump or at –47 dB for a two-stage bidirectional pump pre-amplifier with an OBPF inserted in between. It is the sharp roll-off of the OBPF that generates the low noise level so that the signal average power can be lowered to operate.

In summary, the optical preamplifier operates to offer optical gain and at the same time contribute to the optical amplification noise. The OBPF is the main component that filters the out-of-band noises, allowing low-level signal average power to remain uncorrupted by the noise from the preamplifier and the system. The optical noise of the booster amplifier is also attenuated by the same amount as that of the signals, so the OSNR at the input of the preamplifier is the same as at the output of the optical head. The ASE noise of the preamp is significant but filtered by the OBPF to a low level, allowing the signal to operate without corruption from the noise.

5.6 ELECTRONIC AMPLIFIERS: RESPONSES AND NOISES

5.6.1 Introduction

The electronic amplifier, as a pre-amplification stage of an optical receiver, plays a major role in the detection of optical signals so that the optimum SNR and then the OSNR can be derived based on the photodetector responsivity. Under coherent detection, the amplifier noise must be much less than the quantum shot noise contributed by the high power level of the LO, which is normally about 10 dB above the average signal power.

Thus, this section gives an introduction to electronic amplifiers for wideband signals applicable to ultra-high-speed, high-gain, and low-noise transimpedance amplifiers (TIA). We concentrate on differential input TIA but address the detailed design of a single-input, single-output amplifier with noise suppression techniques in Section 5.8, including the design strategy for achieving stability in the feedback amplifier, as well as low noise and wide bandwidth. By electronic noise of the pre-amplifier stage, we refer to the total equivalent input noise spectral density, where all noise sources (current and voltage sources) of all elements of the amplifier are referred to the input port of the amplifier. From this, an equivalent current source is found, and the current density is derived. Once this current density is found, the total equivalent at the input can be determined when the overall bandwidth of the receiver is established. When this current is known, and with the average signal power, we can easily obtain the SNR at the input stage of the optical receiver, and hence the OSNR. On the other hand, if the OSNR is required for a specific modulation format, then with the assumed optical power of the signal available at the front of the optical receiver and the responsivity of the photodetector, we can determine the maximum electronic noise spectral density allowable by the pre-amplification stage and thus the design of the amplifier electronic circuit.

The principal function of an opto-electronic receiver is to convert the received optical signals into electronic equivalent signals, then amplify and sample them, and process them to recover the properties of the original shapes and sequence. First, the optical domain signals must be converted to electronic current in the photodetection device, which is either a p-i-n or avalanche photodiode (APD). In this device, the optical power is absorbed in the active region, and both electrons and holes generated are attracted to the positively and negatively biased electrodes, respectively. Thus, the generated current is proportional to the power of the optical signals, hence the name "square law" detection. The p-i-n detector is structured with p+ and n+ doped regions sandwiching the intrinsic layer, where the absorption of the optical signal occurs. A high electric field is established in this region by reverse biasing the diode, and thus electrons and holes are attracted to either side of the diode, generating current. Similarly, an APD works with a reverse biasing level close to the reverse breakdown level of the pn junction (with no intrinsic layer), so that electronic carriers can be multiplied in the avalanche flow when the optical signals are absorbed.

This photo-generated current is then fed into an electronic amplifier whose transimpedance must be sufficiently high and low noise so that a sufficient voltage signal can be obtained and then further amplified by a main amplifier, which is a voltage gain type. For high-speed and wideband signals, transimpedance amplification type is preferred as it offers wideband, much wider than the high impedance type, though the noise level might be higher. With TIA, there are two types: the single input-single output port and two differential inputs with a single output. The output ports can be differential with a complementary port. The differential input TIA offers much higher transimpedance gain (Z_T) and wider bandwidth as well. This is due to the use of a long-tail pair at the input, which provides a reasonably high input impedance that would ease the feedback stability.[25,26,27]

In Section 5.9, a case study of a coherent optical receiver is described from the design to implementation, including feedback control and noise reduction. Although the corner frequency is only a few hundred MHz, the bandwidth is remarkable with the limited transition frequency of the transistors. The design is scalable to ultra-wideband reception sub-systems.

5.6.2 Wideband Transimpedance Amplifiers (TIA)

Two type of transimpedance amplifiers are described. They are distinguished by whether they use a single input or a differential input, which depends on the use of a differential pair or long-tail pair, or a single transistor stage at the input of the TIA.

5.6.2.1 Single Input Single Output

We prefer to treat this section as a design example and experimental demonstration of a wideband and low-noise amplifier in a separate Section 5.8 located at the end of this chapter, which can be considered an appendix. In the next section, the differential input TIA is discussed with large transimpedance and reasonably low noise.

5.6.2.2 Differential Inputs, Single/Differential Output

An example circuit of the differential input transimpedance amplifier is shown in Figure 5.20 in which a long tail pair or differential pair is employed at the input stage. Two matched transistors are used to ensure minimum common mode rejection and maximum differential mode operation. This pair has very high input impedance, and thus the feedback from the output stage can be stable. Thus, the feedback resistance can be increased up to the limit of the stability locus of the network pole. This offers the high transimpedance Z_T and wide bandwidth. Typical Z_T of 3,000 to 6,000 Ω can be achieved with 30 GHz 3 dB bandwidth (see Figure 5.22) can be obtained, as shown in Figures 5.21 and 5.22. Also the chip image of the TIA can be seen in Figure 5.21a. Such TIA can be implemented in either InP or SiGe material. The advantage of SiGe is that the circuit can be integrated with a high speed Ge-APD detector, ADC, and DSP. On the other hand, if implemented in InP, then high-speed p-i-n or APD can be integrated and RF interconnected with ADC and DSP. The differential group delay may be serious and must be compensated in the digital processing domain.

5.6.3 Amplifier Noise Referred to Input

There are several noise sources in electronic systems, including thermal noise, shot noise, and quantum shot noise, especially in optoelectronic detection. Thermal noise arises because the

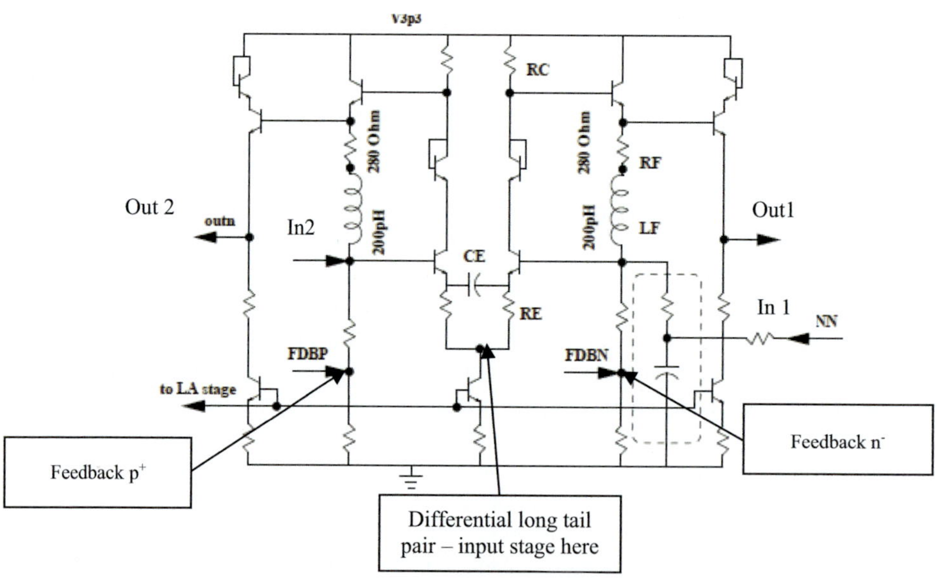

FIGURE 5.20 A typical structure of a differential TIA[50] with differential feedback paths.

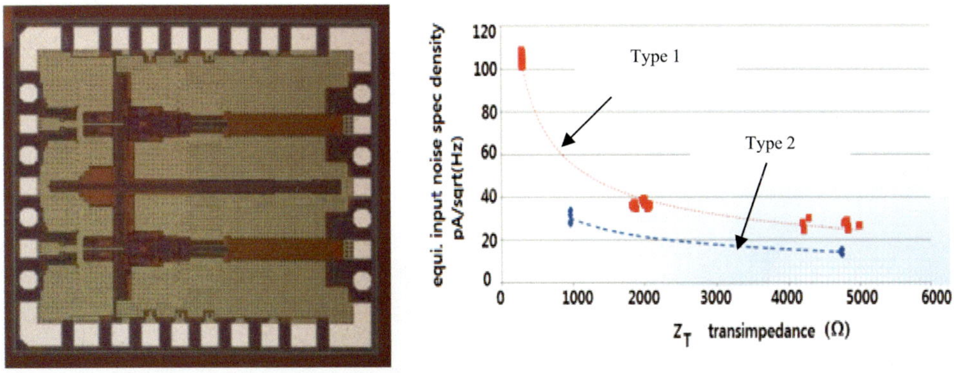

(a) (b)

FIGURE 5.21 Differential amplifiers: (a) chip level image, (b) referred input noise equivalent spectral noise density {Inphi TIA 3205 (type 1) and 2850 (type 2)} (Courtesy of Inphi Inc.[51])

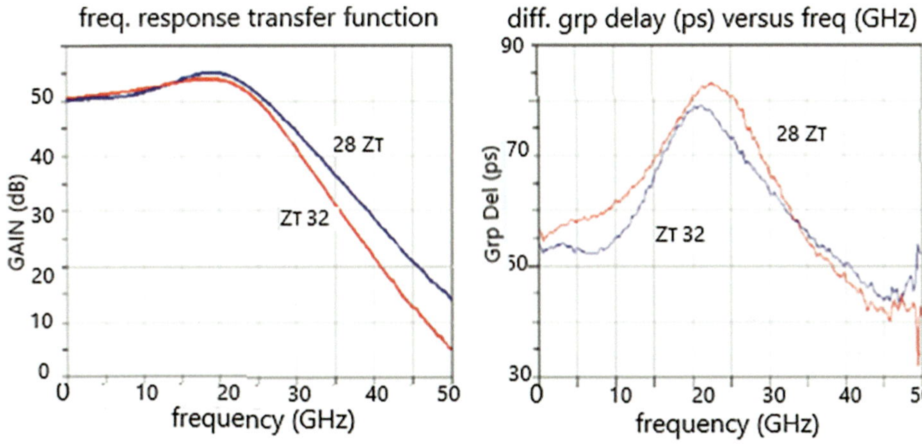

FIGURE 5.22 Differential amplifier: gain frequency response and differential group delay versus frequency with a peaking stage at the output stage.

operating temperature is well above the absolute zero temperature, at which no random movement of electrons occurs, and is related to the resistance of electronic elements. This type of noise is dependent on temperature. Shot noise is caused by the current flowing and random scattering of electrons, so it depends on the strength of the flowing currents, such as the biasing current in electronic devices. Quantum shot noise is generated due to the current emitted from optoelectronic detection processes and is dependent on the intensity of the optical signals or sources imposed on the detectors. This type of noise is dependent on the signals. In the case of coherent detection, the mixing of the local oscillator (LO) laser and signals normally occurs with the strength of the LO being much larger than that of the signal's average power. Thus, the quantum shot noise is dominated by that from the LO.

In practice, an equivalent electronic noise source is the total noise referred to the input of electronic amplifiers, which can be measured by evaluating the total spectral density of the noise distribution over the entire bandwidth of the amplification devices. Therefore, the total

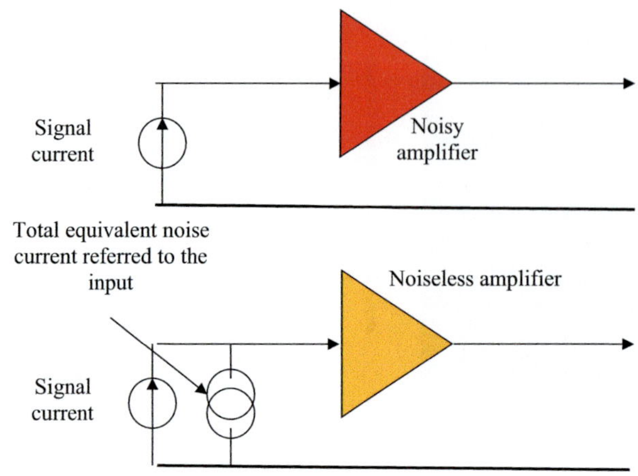

FIGURE 5.23 Equivalent noise spectral density of current sources by referring all noise sources to their equivalent at the input port of the amplifier.

noise spectral density can be evaluated with reference to the input port. For example, if the amplifier is a transimpedance type, the transimpedance of the device is measured first. Then, the measured voltage spectral density at the output port can be referred to the input. In this case, it represents the total equivalent noise spectral density. The common term employed and specified for transimpedance amplifiers is the total equivalent spectral noise density over the midband region of the amplifying device. The midband range of any amplifier is defined as the flat gain region from DC to the corner 3 dB point of the frequency response of the electronic device.

Figure 5.23 illustrates the concept of total equivalent noise sources referred to the input port of a two-port electronic amplifying device. A noisy amplifier with an input excitation current source, typically a signal current generated from the PD after optical-to-electrical conversion, can be represented as a noiseless amplifier in parallel with a noise source whose strength is equal to the total equivalent noise current referred to the input. Thus, the total equivalent current can be found by taking the product of this total equivalent current noise spectral density and the 3 dB bandwidth of the amplifying device. Consequently, one can determine the signal-to-noise ratio at the input of the electronic amplifier given by

$$\text{SNR} = \frac{\text{Square of current generated}}{\text{Square of current generated} + \text{total equivalent noise current power}} \tag{5.47}$$

From this SNR referred to at the input of the electronic front end, one can estimate the eye opening of the voltage signals at the output of the amplifying stage, which is normally required by the ADC for sampling and conversion to digital signals for processing. Thus, one can then estimate the required OSNR at the input of the photodetector and hence the launched power required at the transmitter over several span links with certain attenuation factors.

Detailed analyses of amplifier noises and their equivalent noise sources as referred to input ports are given in the Appendix. It is noted that noises have no flowing direction as they always add and not subtract; thus, the noises are measured as noise power and not as current. Therefore, electrical spectrum analyzers are commonly used to measure the total noise spectral density, or the distribution of noise voltages over the spectral range under consideration, which is defined as the noise power spectral density distribution.

5.7 DIGITAL SIGNAL PROCESSING SYSTEMS AND COHERENT OPTICAL RECEPTION

5.7.1 DSP-Assisted Coherent Detection

Over the years since the introduction of optical coherent communications in the mid-1980s, the invention of optical amplifiers initially left coherent reception behind, until recently, when long-haul transmission suffered from nonlinearity in dispersion compensating fibers (DCF) and SSMF transmission lines due to their small effective area. Furthermore, the advancement of DSP in wireless communication has also contributed to the application of DSP in modern coherent communication systems. Thus, the term "DSP-assisted coherent detection" refers to a process where real-time DSP is incorporated after the opto-electronic conversion of the total field of the local oscillator and the signals. The analog received signals are sampled by a high-speed ADC, and then the digitized signals are processed in a DSP. Currently, real-time DSP processors are intensively researched for practical implementation. The main difference between real-time and offline processing is that the real-time processing algorithm must be effective due to the limited time available for processing.

When polarization division multiplexed (PDM) QAM channels are transmitted and received, polarization and phase diversity receivers are employed. The schematics of such receivers are shown in Figure 5.24a. Further, the structures of such reception systems incorporating DSP with the diversity hybrid coupler in the optical domain are shown in Figure 5.24b, c and d. The polarization diversity section with the polarization beam splitters at the signal and local oscillator inputs facilitates the demultiplexing of polarized modes in the optical waveguides. The phase diversity using 90° optical phase shifter here allows the separation of the in-phase (*I*-) and quadrature (*Q*-) phase components of QAM channels. Using a 2×2 coupler also enables balanced reception using PDP connected back to back, resulting in 3 dB gain in sensitivity. Section 3.5 of Chapter 3 describes the modulation scheme QAM using I-Q modulators for single-polarization or dual-polarization multiplexed channels.

5.7.1.1 DSP-Based Reception Systems

The schematic of a synchronous coherent receiver based on digital signal processing is shown in Figure 5.25. Once the polarization and the *I*- and *Q*-optical components are separated by the hybrid coupler, the positive and negative parts of the *I*- and *Q*-components are coupled into a balanced opto-electronic receiver, as shown in Figure 5.24b. Two PDs are connected back-to-back so that push-pull operation can be achieved, resulting in a 3 dB improvement compared to a single PD detection. The current generated from the back-to-back connected PDs is fed into a transimpedance amplifier so that a voltage signal can be derived at the output. A further voltage-gain amplifier is used to boost these signals to the correct level for the analog-to-digital converter (ADC) so that sampling can be conducted and the analog signals are converted to digital domain. These digitized signals are then fetched into digital signal processors (DSP) and processed in the "soft domain". Thus, a number of processing algorithms can be employed to compensate for linear and nonlinear distortion effects due to optical signal propagation through the optical guided medium, to recover the carrier phase and the clock rate for resampling of the data sequence, etc. Chapter 6 will describe in detail the fundamental aspects of these processing algorithms. Figure 5.25 shows a schematic of possible processing phases in the DSP incorporated in a DSP-based coherent receiver. Besides all the soft processing, optical phase locking, as described here, is necessary to lock the frequencies of the local oscillator and the signal carrier within a certain limit, so that the algorithms for clock recovery can function, for example, within ±2 GHz.

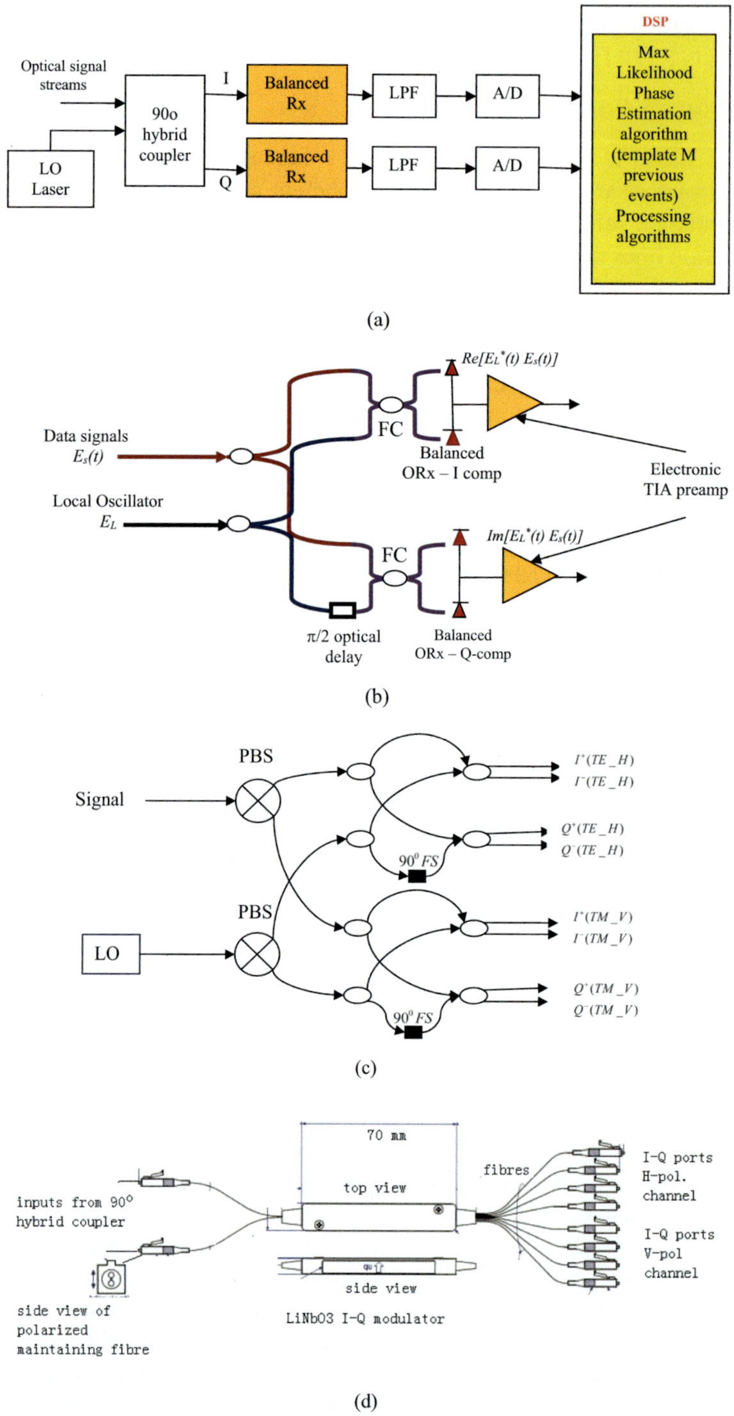

FIGURE 5.24 Scheme of a synchronous coherent receiver using digital signal processing for phase estimation in coherent optical communications. (Adapted from Ref.[52]) (a) Generic scheme, (b) detailed optical receiver using only one polarization phase diversity coupler, (c) Hybrid 90° coupler for polarization and phase diversity, (d) typical view of a hybrid coupler with two input ports and eight output ports of structure in panel (c). TE_V, TE_H=transverse electric mode with vertical (V) or horizontal (H) polarized mode, TM=transverse magnetic mode with polarization orthogonal to that of the TE mode, FS=phase shifter, PBS=polarization beam splitter.

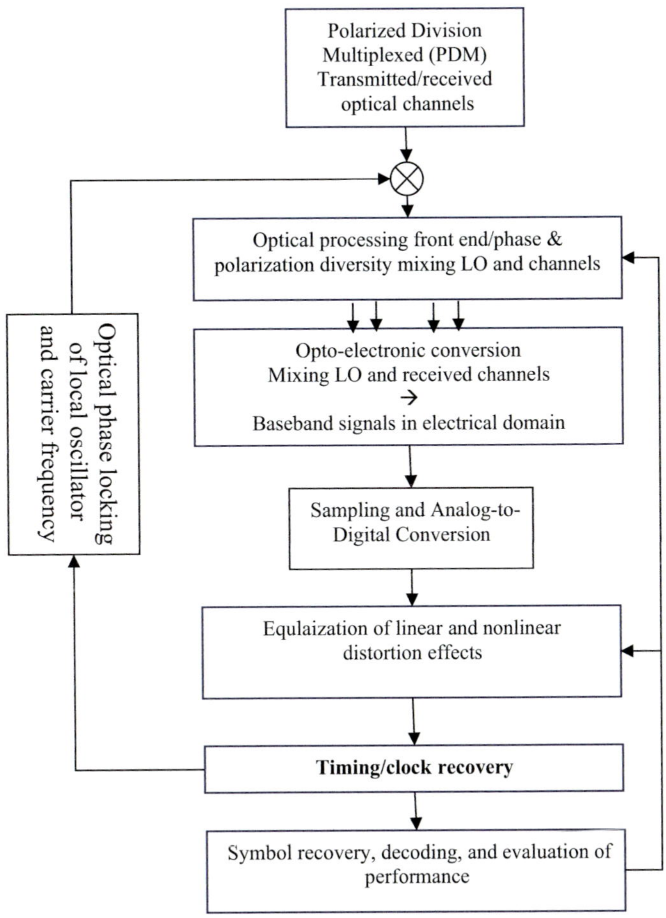

FIGURE 5.25 Flow of functionalities of DSP processing in a QAM coherent optical receiver with possible feedback control.

5.7.2 Coherent Reception Analysis

5.7.2.1 Sensitivity

At ultra-high bit rates, the laser must be externally modulated so that the phase of the lightwave is preserved along the fiber transmission line. Detection can be direct, self-coherent, homodyne, or heterodyne. The sensitivity of the coherent receiver is also crucial for the transmission system, especially for the PSK scheme under both homodyne and heterodyne transmission techniques. This section provides an analysis of the receiver for synchronous coherent optical fiber transmission systems. Consider that the optical fields of the signals and local oscillator are coupled via a fiber coupler with two output ports, 1 and 2. The output fields are then launched into two photodetectors connected back-to-back, and the electronic current is amplified using a transimpedance type and further equalized to extend the bandwidth of the receiver. Our objective is to obtain the receiver penalty and its degradation due to imperfect polarization mixing and unbalancing effects in the balanced receiver. A case study of the design, implementation, and measurements of an optical balanced receiver electronic circuit and noise suppression techniques is provided in Section 5.8 (Figure 5.26).

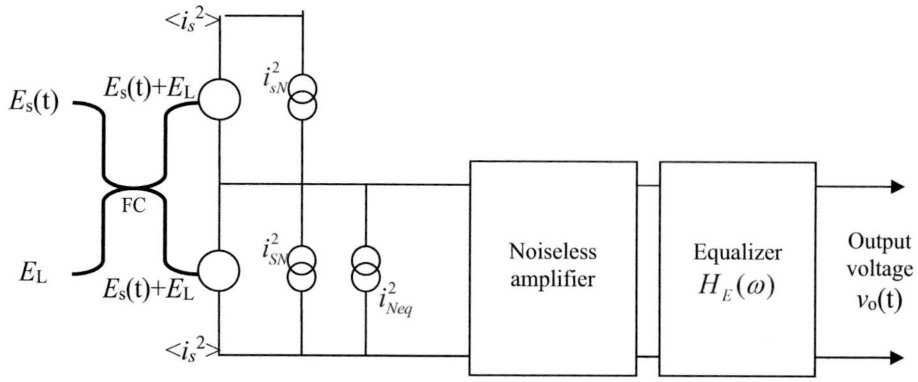

FIGURE 5.26 Equivalent current model at the input of the optical balanced receiver under coherent detection, average signal current, and equivalent noise current of the electronic preamplifier as seen from its input port and equalizer. FC=fiber coupler, Es, E_L=signal and LO electric field, H_E=equalizer frequency transfer function.

The following parameters are commonly used in analysis:

E_S	Amplitude of signal optical field at the receiver
E_L	Amplitude of local oscillator optical field
P_S, P_L	Optical power of signal and local oscillator at the input of the photodetector
$s(t)$	The modulated pulse
$\langle i_{NS}^2(t)\rangle$	Mean square noise current (power) produced by the total optical intensity on the photodetector
$\langle i_s^2(t)\rangle$	Mean square current produced by the photodetector by s(t)
$S_{NS}(t)$	Shot noise spectral density of $\langle i_s^2(t)\rangle$ and local oscillator power
$i_{Neq}^2(t)$	Equivalent noise current of the electronic preamplifier at its input
$Z_T(\omega)$	Transfer impedance of the electronic preamplifier
$H_E(\omega)$	Voltage transfer characteristic of the electronic equalizer followed by the electronic preamplifier

The combined field of the signal and local oscillator via a directional coupler can be written with their separate polarized field components as:

$$\begin{aligned}
E_{sX} &= \sqrt{K_{sX}}E_S\cos\left(\omega_s t-\phi_{m(t)}\right)\\
E_{sY} &= \sqrt{K_{sY}}E_S\cos\left(\omega_s t-\phi_{m(t)}+\delta_s\right)\\
E_{LX} &= \sqrt{K_{LX}}E_L\cos\left(\omega_L t\right)\\
E_{LY} &= \sqrt{K_{LY}}E_L\cos\left(\omega_L t+\delta_L\right)\\
\phi_{m(t)} &= \frac{\pi}{2}K_m s(t)
\end{aligned} \tag{5.48}$$

with $\phi_{m(t)}$ representing the phase modulation and K_m being the modulation depth, K_{sX} K_{sY} K_{LX} K_{LY} are the intensity fraction coefficients in the X and Y direction of the signal and local oscillator (LO) fields, respectively.

Thus, the output fields at ports 1 and 2 of the FC in the X-plane can be obtained using the transfer matrix as

$$\begin{bmatrix} E_{R1X} \\ E_{R2X} \end{bmatrix} = \begin{bmatrix} \sqrt{K_{sX}(1-\alpha)}\cos\left(\omega_s t-\phi_{m(t)}\right) & \sqrt{K_{LX}\alpha}\sin\left(\omega_L t\right) \\ \sqrt{K_{sX}\alpha}\sin\left(\omega_s t-\phi_{m(t)}\right) & \sqrt{K_{LX}(1-\alpha)}\cos\left(\omega_L t\right) \end{bmatrix} \begin{bmatrix} E_s \\ E_L \end{bmatrix} \tag{5.49}$$

$$\begin{bmatrix} E_{R1Y} \\ E_{R2Y} \end{bmatrix} = \begin{bmatrix} \sqrt{K_{sY}(1-\alpha)}\cos\left(\omega_s t-\phi_{m(t)}\right) & \sqrt{K_{LY}\alpha}\sin\left(\omega_L t+\delta_L\right) \\ \sqrt{K_{sY}\alpha}\sin\left(\omega_s t-\phi_{m(t)}\right) & \sqrt{K_{LY}(1-\alpha)}\cos\left(\omega_L t\right)+\delta_L \end{bmatrix} \begin{bmatrix} E_s \\ E_L \end{bmatrix} \tag{5.50}$$

with α defined as the intensity coupling ratio of the coupler. Thus, the field components at ports 1 and 2 can be derived by combining the X and Y components from Equations 5.49 and 5.50, hence, the total power at ports 1 and 2 are given as

$$\begin{aligned} P_{R1} &= P_s(1-\alpha)+P_L\alpha+2\sqrt{P_sP_L\alpha(1-\alpha)K_p}\sin\left(\omega_{\text{IF}}t+\phi_{m(t)}+\phi_p-\phi_c\right) \\ P_{R2} &= P_s\alpha+P_L(1-\alpha)+2\sqrt{P_sP_L\alpha(1-\alpha)K_p}\sin\left(\omega_{\text{IF}}t+\phi_{m(t)}+\phi_p-\phi_e+\pi\right) \\ \text{with}\quad K_p &= K_{sX}K_{LX}+K_{sY}K_{LY}+2\sqrt{K_{sX}K_{LX}K_{sY}K_{LY}}\cos\left(\delta_L-\delta_s\right) \\ \phi_p &= \tan^{-1}\left[\frac{\sqrt{K_{sX}K_{LY}}\sin\left(\delta_L-\delta_s\right)}{\sqrt{K_{sX}K_{LX}}+\sqrt{K_{sY}K_{LY}}\cos\left(\delta_L-\delta_s\right)}\right] \end{aligned} \tag{5.51}$$

ω_{IF} is the intermediate angular frequency and equals the difference between the frequencies of the local oscillator and the carrier of the signals. ϕ_e is the phase offset, $\phi_p-\phi_e$ is the demodulation reference phase error.

In Equation 5.51, the total field of the signal and the local oscillator are added, and then the product of the field vector and its conjugate is taken to obtain the power. Only the term with the frequency falling within the range of the sensitivity of the photodetector would produce the electronic current. Thus, the term with the sum of the frequencies of the signal and the local oscillator would not be detected, and only the product of the two terms would be detected as given.

Now, assuming a Binary PSK (BPSK) modulation scheme, the pulse has a square shape with amplitude +1 or –1, the PD is a p-i-n type, and the PD bandwidth is wider than the signal's 3 dB-bandwidth, followed by an equalized electronic preamplifier. The signal at the output of the electronic equalizer, or the input signal to the decision circuit is given by:

$$\begin{aligned} \hat{v}_D(t) &= 2K_HK_p\sqrt{P_sP_L\alpha(1-\alpha)K_p}\int_{-\infty}^{\infty}H_E(f)\,df\int_{-\infty}^{\infty}(t)\,dt\,\sin\left(\frac{\pi}{2}K_m\right)\cos\left(\phi_p-\phi_e\right) \\ \hat{v}_D(t) &= 2K_HK_p\sqrt{P_sP_L\alpha(1-\alpha)K_p}\sin\left(\frac{\pi}{2}K_m\right)\cos\left(\phi_p-\phi_e\right) \end{aligned} \tag{5.52}$$

$$K_H=1\text{ for homodyne;}\quad K_H=1/\sqrt{2}\text{ for heterodyne}$$

For a perfectly balanced receiver $K_B=2$ and $\alpha=0.5$, otherwise $K_B=1$. The integrals of the first line in Equation 5.52 are given by

$$\int_{-\infty}^{\infty} H_E(f)\,df = \frac{1}{T_B} \qquad \because \; H_E(f) = \operatorname{sinc}(\pi T_B f)$$
$$\int_{-\infty}^{\infty} s\,(t)\,dt = 2T_B \tag{5.53}$$

$V_D(f)$ is the transfer function of the matched filter for equalization, and T_B is the bit period. The total noise voltage as a sum of the quantum shot noise generated by the signal and the local oscillator and the total equivalent noise of the electronic preamplifier at the input of the preamplifier, at the output of the equalizer is given by:

$$\langle v_N^2(t)\rangle = \frac{\left[K_B\alpha S'_{IS} + (2-K_B)S_{IX} + S_{IE}\right]\int_{-\infty}^{\infty}\left|H_4(f)\right|^2 df}{K_{IS}^2}$$

or (5.54)

$$\langle v_N^2(t)\rangle = \frac{\left[K_B\alpha S'_{IS} + (2-K_B)S_{IX} + S_{IE}\right]}{K_{IS}^2 T_B}$$

For homodyne and heterodyne detection, we have

$$\langle v_N^2(t)\rangle = \Re q \frac{P_L}{\lambda T_B}\left[K_B\alpha S'_{IS} + (2-K_B)S_{IX} + S_{IE}\right] \tag{5.55}$$

where the spectral densities S'_{IX}, S'_{IE} are given by

$$S'_{IX} = \frac{S_{IX}}{S'_{IS}}$$
$$S'_{IE} = \frac{S_{IE}}{S'_{IS}} \tag{5.56}$$

Thus the receiver sensitivity for binary PSK and equi-porobable detection and Gaussian density distribution we have

$$P_e = \frac{1}{2}\operatorname{erfc}\left(\frac{\delta}{\sqrt{2}}\right) \tag{5.57}$$

with δ given by

$$P_e = \frac{1}{2}\operatorname{erfc}\left(\frac{\delta}{\sqrt{2}}\right) \quad \text{with}\ldots\delta = \frac{\hat{v}_D}{2\sqrt{\langle v_N^2\rangle}} \tag{5.58}$$

Thus, using Equations 5.55 and 5.58 and 5.52, we have the receiver sensitivity in linear power scale as

$$P_s = \langle P_s(t)\rangle = \frac{\Re q\delta^2}{4\lambda T_B K_H^2}\frac{\left[K_B\alpha S'_{IS} + (2-K_B)S_{IX} + S_{IE}\right]}{\eta K_p(1-\alpha)\alpha K_B^2\sin^2\left(\frac{\pi}{2}K_m\right)\cos^2(\phi_p - \phi_e)} \tag{5.59}$$

5.7.2.2 Shot Noise-Limited Receiver Sensitivity

In the case that the power of the local oscillator dominates the noise of the electronic preamplifier and the equalizer, the receiver sensitivity (in linear scale) is given as

$$P_s = \langle P_{sL} \rangle - \frac{\Re q \delta^2}{4\lambda T_B K_H^2} \tag{5.60}$$

This shot noise limited receiver sensitivity can be plotted as shown in Figure 5.27.

5.7.2.3 Receiver Sensitivity under Non-Ideal Conditions

Under non-ideal condition, the receiver sensitivity departs from the shot noise limited sensitivity and characterized by the receiver sensitivity penalty PD_T as

$$\text{PD}_T = 10 \log \frac{\langle P_s \rangle}{\langle P_{sL} \rangle} \text{ dB} \tag{5.61}$$

$$\text{PD}_T = 10 \log_{10} \left[\frac{K_B \alpha S'_{IS} + (2 - K_B) S_{IX} + S_{IE}}{K_B \alpha} \right] - 10 \log_{10} \left[K_B (1 - \alpha) \right] - 10 \log_{10} \left([\eta] [K_p] \sin^2 \left(\frac{\pi}{2} K_m \right) \cos^2 (\phi_p - \phi_e) \right) \tag{5.62}$$

with η is the local oscillator excess noise factor.

The receiver sensitivity plotted against the ratio f_B/λ for the case of homodyne and heterodyne detection is shown in Figure 5.27a, and the power penalty of the receiver sensitivity against the excess noise factor of the local oscillator is shown in Figure 5.27b. Receiver power penalty can be deduced as a function of the total electronic equivalent noise spectral density, and as a function of the rotation

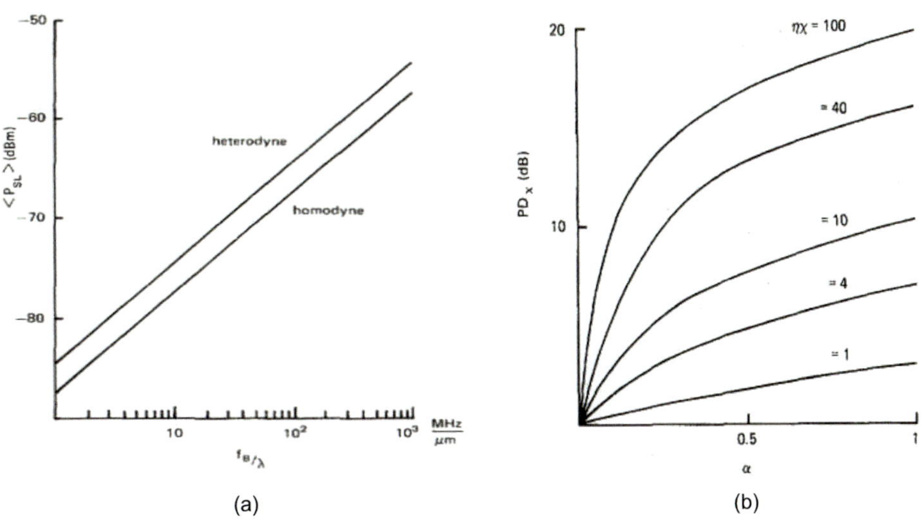

FIGURE 5.27 (a) Receiver sensitivity of coherent homodyne and heterodyne detection, signal power versus bandwidth over the wavelength. (b) Power penalty of the receiver sensitivity from the shot noise limited level as a function of the excess noise of the local oscillator. (Extracted from Ref.[53])

of the polarization of the local oscillator, as found in Ref.[28] Furthermore, in reference[37] the receiver power penalty and the normalized heterodyne center frequency can vary as a function of the modulation parameter, and the optical power ratio at the same polarization angle can also be found.

5.7.3 Digital Processing Systems

A generic structure of the coherent reception and digital signal processing system is shown in Figure 5.28, in which the digital signal processing system is placed after the sampling and conversion from the analog state to digital form. The optical signal fields are, of course, beating with the local oscillator laser, whose frequency is approximately identical to the signal channel carrier. The beating occurs in the square-law photodetectors, where the summation of the two fields is squared, and the product term is decomposed into the difference and summation terms. Thus, only the difference term falls back into the baseband region and is amplified by the electronic pre-amplifier, which is a balanced differential transimpedance type.

If the signals are complex, there are real and imaginary components that form a pair. Another pair comes from the other polarization mode channel. The digitized signals of both the real and imaginary parts are processed either in real time or offline. The processors contain algorithms to mitigate various transmission impairments, such as the imbalance between the in-phase and quadrature components created at the transmitter, the recovery of the clock rate and timing for resampling, carrier phase estimation for the signal phases, and adaptive equalization for compensating propagation dispersion effects using MLSE, etc. These algorithms are built into the hardware processors or memory and loaded into processing subsystems.

The sampling rate must normally be twice that of the signal bandwidth to satisfy the Nyquist criteria. Although this rate is very high for 25 G to 32 GSy/s optical channels, Fujitsu ADC has met this requirement with sampling rates of 56 G to 64 GSa/s, as depicted in Figure 5.40.

The linewidth resolution of the processing for semiconductor device fabrication has progressed tremendously over the years in an exponential trend, as shown in Table 5.1. This progress has been possible due to successes in lithographic techniques using optical carrier at short wavelength such as the UV, the electronic optical beam, and X-ray lithograpohic with appropriate photoresist such as SU-80, which have allowed the line resolution to reach 5 nm in 2020. If we plot the trend on a log-linear scale, as shown in Figure 5.29, a linear line is obtained, meaning that the resolution is reduced exponentially. When the gate width is reduced, the electronic speed increases tremendously; at 5 mm, the sped of electronic CMOS device in SiGe can reach several tens of GHz. Regarding high-speed ADC and DAC, the clock speed is increased by parallel delays and the summation of all the digitized digital lines to form a very high-speed operation. For example, for Fujitsu 64 GSa/s DAC or ADC, the applied clock sinusoidal waveform is only 2 GHz. Figure 5.30 shows the progress in the speed development of Fujitsu ADC and DAC.

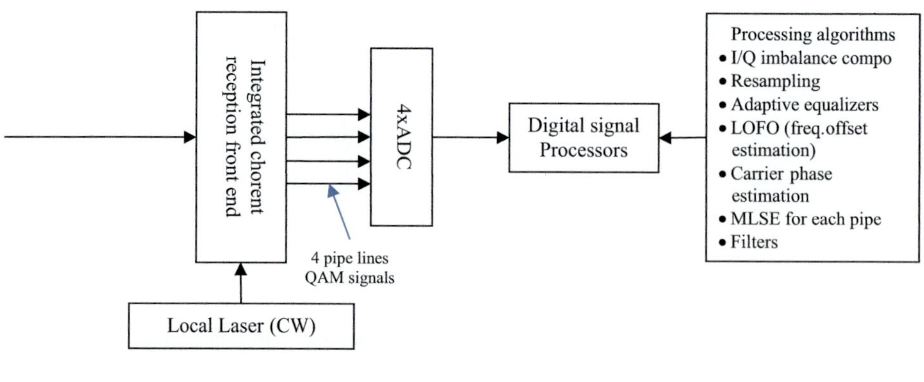

FIGURE 5.28 Coherent reception and digital signal processing system. CW = continuous wave.

TABLE 5.1
Milestones of Progresses of Linewidth Resolution

Semiconductor Manufacturing Processes

- And spatial resolution (gate width)
  - 10 µm – 1971
  - 3 µm – 1975
  - 1.5 µm – 1982
  - 1 µm – 1985
- 800 nm (0.80 µm) – 1989- UV lithography
  - 600 nm (0.60 µm) – 1994
  - 350 nm (0.35 µm) – 1995
  - 250 nm (.25 µm) – 1998
  - 180 nm (0.18 µm) – 1999
  - 130 nm (0.13 µm) – 2000
- 90 nm – 2002: electron lithography
  - 65 nm – 2006
  - 45 nm – 2008
  - 32 nm – 2010
  - 22 nm – 2012
- 14 nm – approx. 2014: X-ray lithography
  - 10 nm approx. 2016: X-ray litho
  - 7 nm – approx. 2018: X-ray litho
  - 5 nm – approx. 2020: X-ray litho

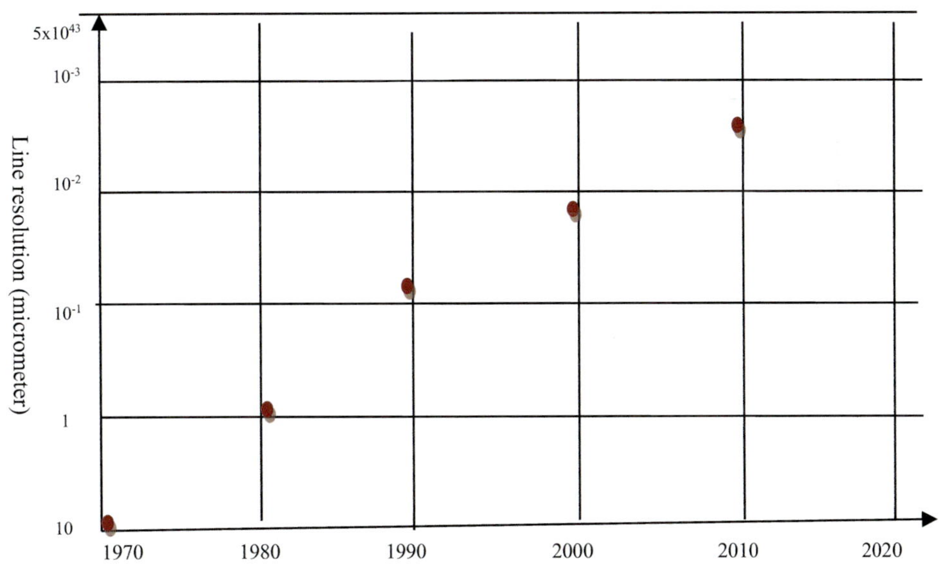

FIGURE 5.29 Semiconductor manufacturing with a resolution of line resolution.

5.7.3.1 Effective Number of Bits

5.7.3.1.1 Definition

Effective number of bits (ENOB) is a measure of the quality of a digitized signal. The resolution of a digital-to-analog or analog-to-digital converter (DAC or ADC) is commonly specified by the number of bits used to represent the analog value, in principle giving 2^N signal levels for an N-bit

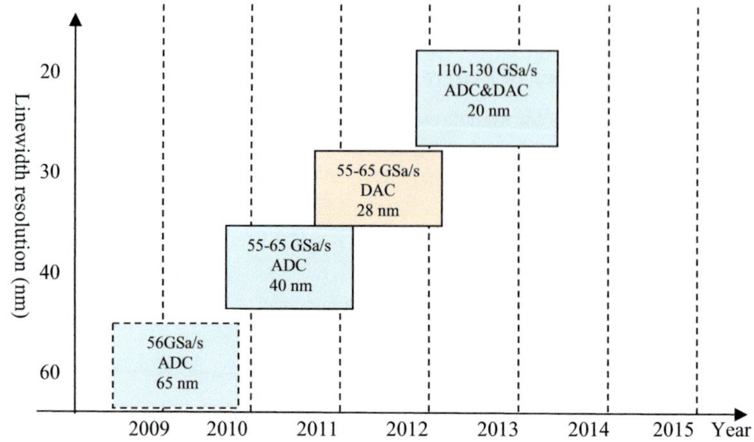

FIGURE 5.30 Evolution of ADC and DAC operating speed with corresponding line width resolution.

signal. However, all real signals contain a certain amount of noise. If the converter is able to represent signal levels below the system noise floor, the lower bits of the digitized signal only represent system noise and do not contain useful information. ENOB specifies the number of bits in the digitized signal above the noise floor. Often, ENOB is used as a quality measure for other blocks such as sample-and-hold amplifiers. This way, analog blocks can also be easily included in signal-chain calculations as the total ENOB of a chain of blocks is usually below the ENOB of the worst block.

Thus we can represent the ENOB of a digitalized system by writing

$$\text{ENOB} = \frac{\text{SINAD} - 1.76}{6.02} \tag{5.63}$$

where all values are given in dB, and signal-to-noise and distortion ratio (SINAD) is the ratio of the total signal, including distortion and noise, to the wanted signal. The 6.02 term in the divisor converts decibels (a $\log_{10}$ representation) to bits (a $\log_2$ representation), while the 1.76 term comes from quantization error in an ideal ADC.[29]

This definition compares the SINAD of an ideal ADC or DAC with a word length of ENOB bits to the SINAD of the ADC or DAC being tested.

Indeed, the SINAD is a measure of the quality of a signal from a communications device, often defined as:

$$\text{SINAD} = \frac{P_{\text{sig}} + P_{\text{noise}} + P_{\text{distorion}}}{P_{\text{noise}} + P_{\text{distorion}}} \tag{5.64}$$

where P is the average power of the signal, noise, and distortion components. SINAD is usually expressed in dB and is quoted alongside the receiver sensitivity, to give a quantitative evaluation of the receiver sensitivity. Note that with this definition, unlike SNR, a SINAD reading can never be <1 (i.e., it is always positive when quoted in dB).

When calculating the distortion, it is common to exclude the DC components. Due to its widespread use, SINAD has collected a few different definitions. SINAD is calculated as one of (i) The ratio of (a) total received power, i.e., the signal to (b) the noise-plus-distortion power. This is modeled by the equation above. (ii) The ratio of (a) the power of the original modulating audio signal, i.e., from a modulated radio frequency carrier to (b) the residual audio power, i.e., noise-plus-distortion powers remaining after the original modulating audio signal is removed. With this definition, it is now possible for SINAD to be less than 1. This definition is used when SINAD is used in the calculation of ENOB for an ADC.

Example: Consider the following measurements of a 3-bit unipolar D/A converter with reference voltage $V_{ref}=8$ V:

Digital Input	000	001	010	011	100	101	110	111
Analog Output (V)	–0.01	1.03	2.02	2.96	3.95	5.02	6.00	7.08

The offset error in this case is –0.01 V or –0.01 LSB as 1 V=1 LSB. The gain error is

$$\frac{7.08+0.01}{1}-\frac{7}{1}=0.09\text{ LSB}$$

LSB stands for least significant bits. Correcting the offset and gain error, we obtain the following list of measurements: (0, 1.03, 2.00, 2.93, 3.91, 4.96, 5.93, 7) LSB. This allows the INL and DNL to be calculated: INL=(0, 0.03, 0, –0.07, –0.09, –0.04, –0.07, 0) LSB, and DNL=(0.03, –0.03, –0.07, –0.02, 0.05, –0.03, 0.07, 0) LSB.

Differential non-linearity (DNL): For an ideal ADC, the output is divided into 2^N uniform steps, each with Δ width, as shown in Figure 5.31. Any deviation from the ideal step width is called differential non-linearity (DNL) and is measured in a number of counts (LSBs). For an ideal ADC, the DNL is 0 LSB. In a practical ADC, DNL error comes from its architecture. For example, in a SAR ADC, a DNL error may be caused near the mid-range due to mismatching of its DAC.

Integral non-linearity (INL) is a measure of how closely the ADC output matches its ideal response. INL can be defined as the deviation in LSB of the actual transfer function of the ADC from the ideal transfer curve. INL can be estimated using DNL at each step by calculating the cumulative sum of DNL errors up to that point. In reality, INL is measured by plotting the ADC transfer characteristics.

INL is popularly measured using either (i) best fit (best straight line) method, or (ii) endpoint method.

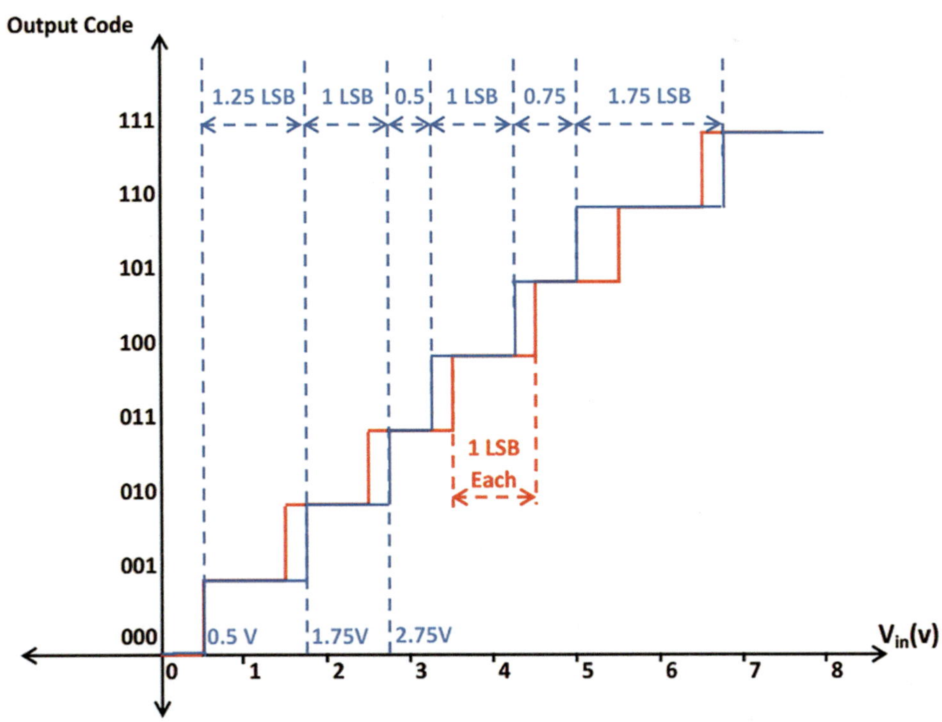

FIGURE 5.31 Representation of DNL in a transfer curve of an ADC.

5.7.3.1.1.1 Best Fit INL The best-fit method of INL measurement considers offset and gain error. One can see in Figure 5.32 that the ideal transfer curve used for calculating the best-fit INL does not go through the origin. The ideal transfer curve here is drawn to depict the nearest first-order approximation to the actual transfer curve of the ADC. The intercept and slope of this ideal curve can provide the values of the offset and gain error of the ADC. Quite intuitively, the best-fit method yields better results for INL. For this reason, this is often the number present on ADC datasheets.

The only real use of the best fit INL number is to predict distortion in time-variant signal applications. This number is equivalent to the maximum deviation for an AC application. However, it is always better to use the distortion numbers rather than INL numbers. To calculate the error budget, end-point INL numbers provide a better estimation. Additionally, this is the specification generally provided in datasheets. Therefore, one should use the end-point INL instead of the best fit INL.

5.7.3.1.1.2 End-Point INL The End-Point method provides the worst-case INL. This measurement passes through the straight line from the origin to the maximum output code of the ADC (refer to Figure 5.6). As this method provides the worst-case INL, it is more useful to use this number compared to the one measured using the best fit for DC applications. This INL number is typically useful for error budget calculation. This parameter must be considered for applications involving precision measurements and control (Figure 5.33).

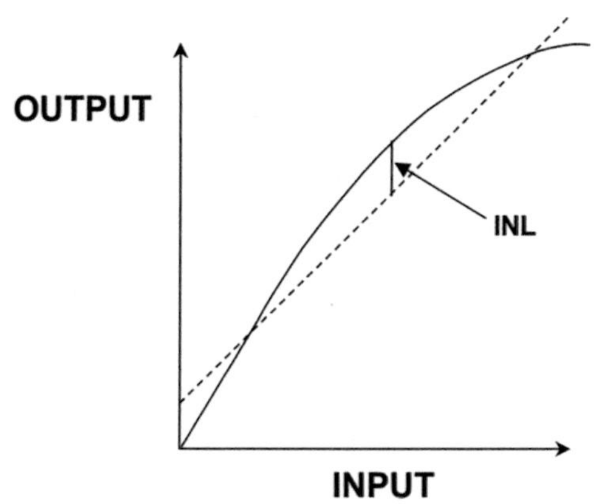

FIGURE 5.32 Best-fit INL.

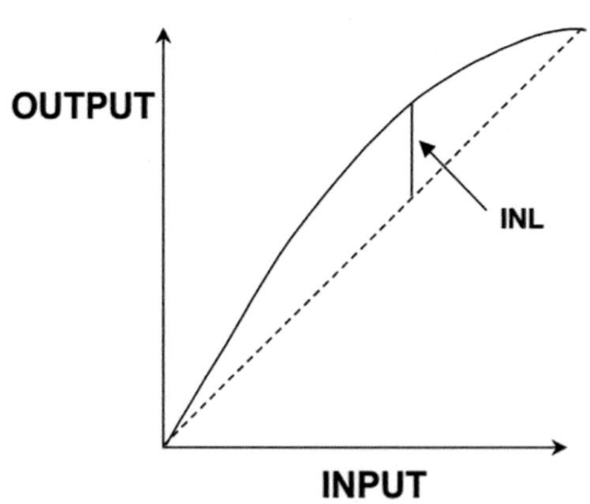

FIGURE 5.33 Endpoint INL.

The absolute and relative accuracy can now be calculated. The ENOB absolute accuracy is calculated using the largest absolute deviation D, in this case, 0.08 V:

$$D = \frac{V_{\text{ref}}}{2^{\text{ENOB}}} \rightarrow \text{ENOB} = 6.64 \text{ bits} \tag{5.65}$$

The effective number of bits relative accuracy is calculated using the largest relative (INL) deviation *d*, in this case, 0.09 V.

$$d = \frac{V_{\text{ref}}}{2^{\text{ENOB}}} \rightarrow \text{ENOB}_{\text{rel}} = 6.47 \text{ bits} \tag{5.66}$$

For this kind of ENOB calculation, note that the effective number of bits can be larger or smaller than the actual number of bits. When the ENOB is smaller than the ANOB, some of the least significant bits of the result are inaccurate. However, one can also argue that the ENOB can never be larger than the ANOB, because you always have to add the quantization error of an ideal converter which is ±0.5 LSB. Different designers may use different definitions of ENOB!

5.7.3.1.2 High Speed ADC and DAC Evaluation Incorporating Statistical Property

The effective number of bits of an ADC is considered the number of bits that an analog signal can be converted into its digital equivalent, accounting for the number of levels represented by the modulo-two levels, which are reduced due to noise contributed by electronic components in such converters. Thus, only an effective number of equivalent bits can be accounted for. Hence, the term effective number of bits (ENOB) is used.

As shown in Figure 5.34b, a real ADC can be modeled as a cascade of two ideal ADCs, additive noise sources, and an automatic gain control (AGC) amplifier.[30] The quantized levels are thus equivalent to a specific ENOB as far as the ADC operates in the linear non-saturated region. If the normalized signal amplitude/power surpasses unity and enters the saturated region, the signals are clipped. The decision level of the quantization in an ADC normally follows a normalized Gaussian probability density function. Thus, we can estimate the RMS noise introduced by the ADC as:

$$\text{RMS noise} = \sqrt{\int_{-\frac{\text{LSB}}{2}}^{\frac{\text{LSB}}{2}} \int_{-\infty}^{\infty} \frac{x^2 \frac{1}{\sqrt{2\pi\sigma^2}} \exp\left(-\frac{(x-y)^2}{2\sigma^2}\right)}{\text{LSB}} dx dy} = \sqrt{\text{LSB}^2/12 + \sigma^2} \tag{5.67}$$

where, σ is the variance; x is the variable related to the integration of decision voltage; similarly, y is for integration inside one LSB. Given the known quantity $\text{LSB}^2/12$ by the introduction of the ideal quantization error, σ^2 can be determined via the Gaussian noise distribution. We can thus deduce the ENOB values corresponding to the levels of Gaussian noise as

$$\text{ENOB} = N - \log_2\left(\frac{\sqrt{\text{LSB}^2/12 + (A\sigma)^2}}{\text{LSB}/\sqrt{12}}\right) \tag{5.68}$$

with A as the RMS amplitude derived from the noise power. According to the ENOB model, the frequency response of ENOB of the DSA is shown in Figure 5.34a with the excitation of the DSA by sinusoidal waves of different frequencies. As observed, the ENOB varies with respect to the excitation frequency, in the range from 5 to 5.5. Having known the frequency response of the sampling device, what is the ENOB of the device when excited with broadband signals? This indicates the different resolutions of the ADC of the receiver of the transmission operating under different noisy and dispersive conditions, thus an equivalent model of ENOB for performance evaluation is

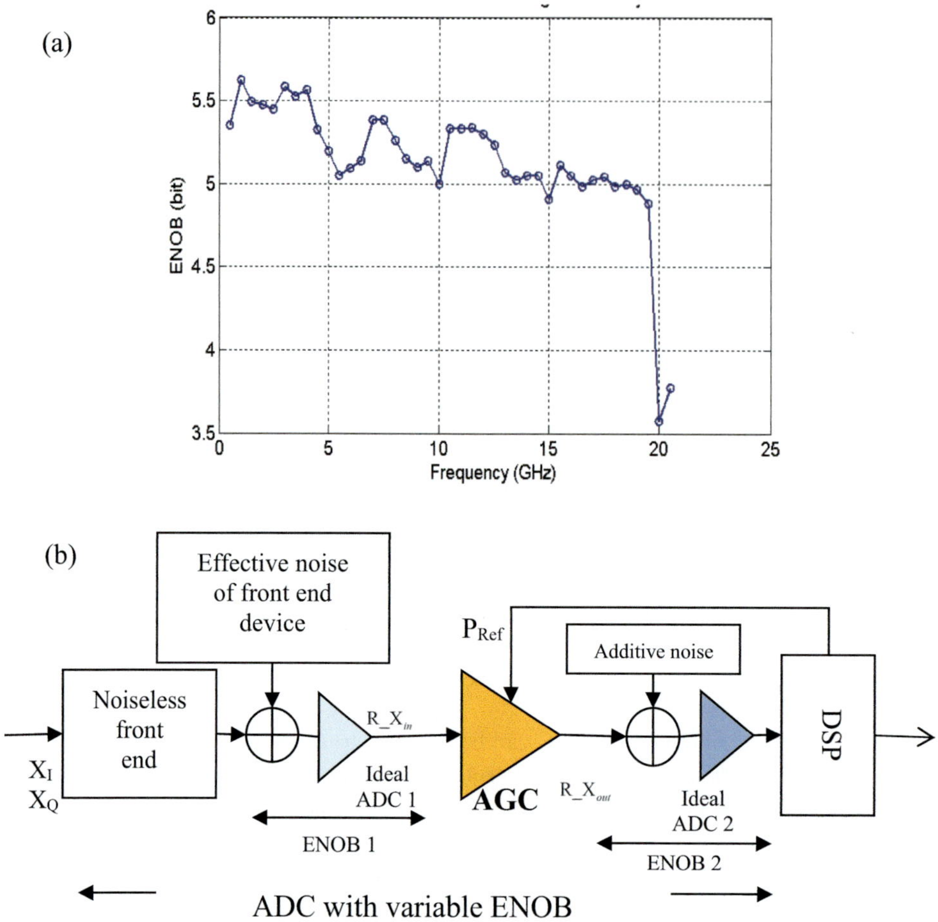

FIGURE 5.34 (a) Measured ENOB frequency response of a commercial real-time DSA of 20 GHz bandwidth and a sampling rate of 50 GSa/s, (b) deduced ADC model of variable ENOB based on the experimental frequency response of panel (a) and spectrum of broadband signals.

essential. We note that the amplitudes of the optical fields arrived at the receiver vary depending on the conditions of the optical transmission line. The AGC has a nonlinear gain characteristic in which the input sampled signal power level is normalized with respect to the saturated (clipping) level. The gain is significantly high in the linear region and saturated in the high level. The received signal R_X_{in} is scaled with a gain coefficient according to $R\_X_{out} = R\_X_{in} / \sqrt{P_{in\_av}/P_{Ref}}$, where the signal-averaged power P_{in_av} is estimated and the gain is scaled relative to the reference power level P_{Ref} of the AGC. A linear scaling factor is then used to obtain the output sampled value R_X_{out}. The gain of the AGC is also adjusted according to the signal energy via the feedback control path from the DSP (see Figure 5.34b). Thus, new values of ENOB can be evaluated with noise distributed across the frequency spectrum of the signals by an averaging process. This signal-dependent ENOB is now denoted as $ENOB_S$.

5.7.3.1.3 Impact of ENOB on Transmission Performance

Shown in Figure 5.35a is the BER variation with respect to the OSNR under back-to-back (B2B) transmission using the simulated samples at the output of the 8-bit ADC with $ENOB_S$ and full ADC resolution as parameters. The difference is due to the noise distribution (Gaussian or uniform).

Figure 5.35b depicts the variation of BER versus OSNR with $ENOB_S$ as the variation parameter in the case of offline data with ENOB of DSA shown in Figure 5.36a. Several more tests were conducted to ensure the effectiveness of our ENOB model. When the sampled signals are presented to the ADC at different amplitudes, controlled and gained nonlinearly adjusted by the AGC, different degrees of clipping effect are introduced. Thus, the clipping effect can be examined for the ADC at different quantization levels but with identical $ENOB_S$, as shown in Figure 5.36a for the B2B experiment. Figure 5.36b, c, and d, and e show, with BER as a parameter, the contour plots of the variation of the adjusted reference power level of the AGC and $ENOB_S$ for the cases of 1500 km long-haul transmission with full CD compensation and non-CD compensation operating in the linear (0 dBm launch power in both links) and nonlinear regimes of the fibers with launch powers of 4 and 5 dBm, respectively. When the link is fully CD compensated, the nonlinear effects further contribute to the ineffectiveness of the ADC resolution and hence moderate AGC freedom in performance is achieved. In the case of non-CD compensation link (Figure 5.36d and e), the dispersive pulse sampled amplitudes are lower with less noise, allowing the resolution of the ADC higher via the nonlinear gain of the AGC. Thus, effective phase estimation and equalization can be achieved. We note that the offline data sets employed prior to the processing using $ENOB_S$ to obtain the contours in Figure 5.36 produce the same BER contours of 2e-3 for all cases, ensuring a fair comparison when the $ENOB_S$ model is used. The opening distance of the BER contours indicates the dynamic range of the $ENOB_S$ model, especially the AGC. It is evident from Figure 5.36a–e that the dynamic range of the model is higher for non-compensating than for full CD compensated transmission, and even for the case of B2B. However, for nonlinear scenarios in both cases, the requirement for $ENOB_S$ is higher for the dispersive channel (Figure 5.36c and e). This may be due to the cross-phase modulation effects of adjacent channels, hence more noise.

5.7.3.2 Digital Processors

The structures of the DAC and ADC are shown in Figures 5.37 and 5.38, respectively. Normally, there would be 4 DACs in an IC, where each DAC section is clocked with a clock sequence derived from a lower-frequency sinusoidal wave injected externally into the DAC. Four units are required for the in-phase and quadrature-phase components of QAM modulated polarized channels, resulting in the notations of I_{DAC} and Q_{DAC} shown in the diagram. Similarly, the optical received signals of PDM-QAM (polarization division multiplexed–quadrature amplitude modulation) would be sampled by a four-phase sampler and then converted to digital form into four groups of I and Q lanes for processing in the DSP sub-system. Due to the interleaving of the sampling clock waveform, the digitized bits appear simultaneously at the end of a clock period that is sufficiently long so that the

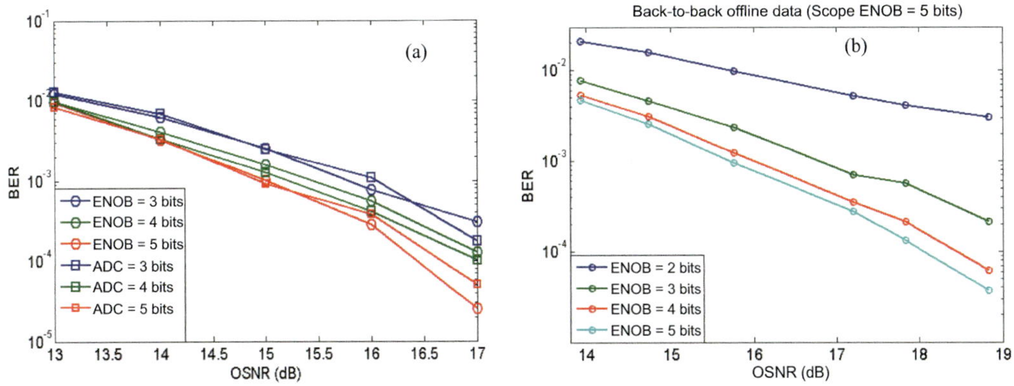

FIGURE 5.35 (a) B2B performance with different $ENOB_S$ values of the ADC model with simulated data (8-bit ADC), (b) OSNR versus BER under different simulated $ENOB_S$ of offline data obtained from experimental digital coherent receiver.[54]

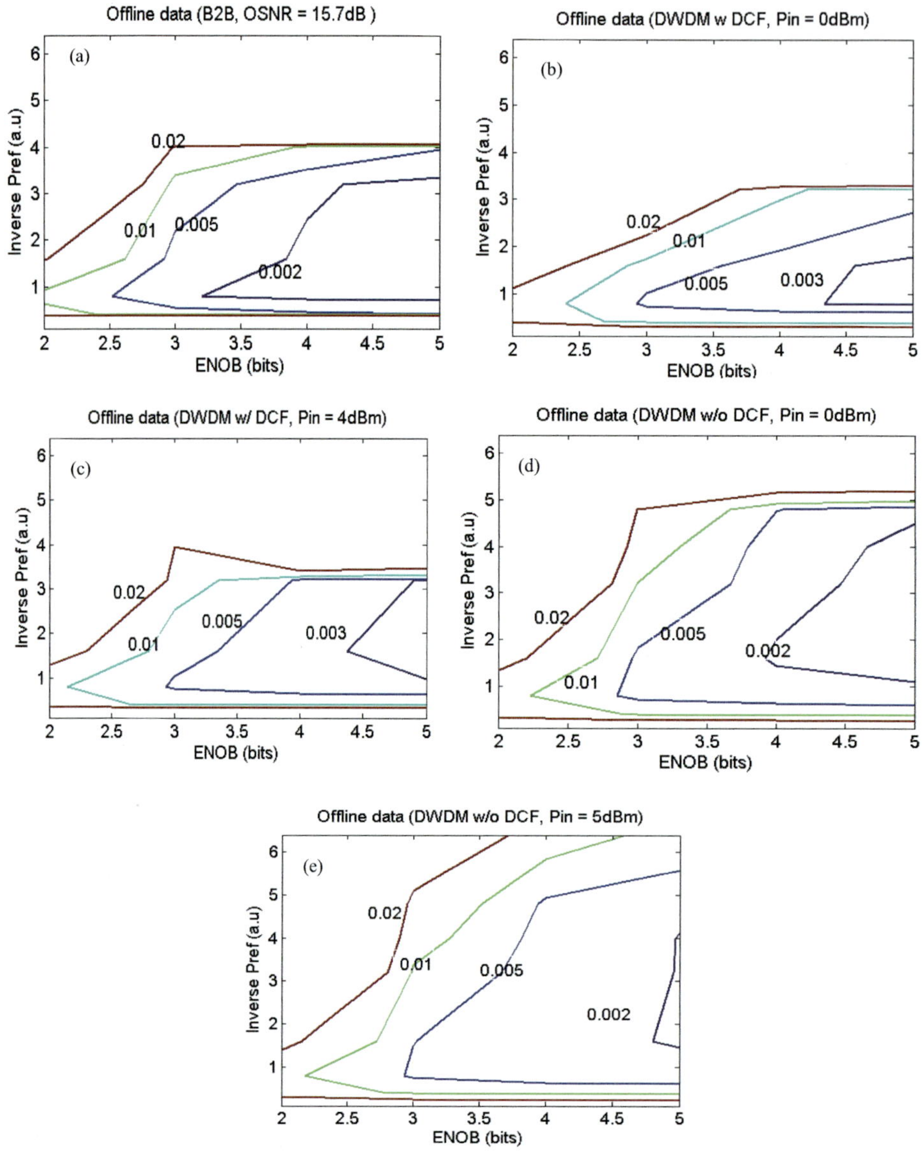

FIGURE 5.36 Comprehensive effects of AGC clipping (inverse of P_{Ref}) and $ENOB_S$ of coherent receiver, experimental transmission under (a) B2B, (b) DWDM linear operation with full chromatic dispersion (CD) compensation, and (b) linear, (c) nonlinear regions with 4 dBm launch power; and (d) non-CD compensation and (d) linear, (e) nonlinear region with launch power of 5 dBm.

sampling number is large enough to achieve all samples. For example, as shown in Figure 5.38, 1024 samples are achieved at a periodicity corresponding to a 500 MHz cycle clock for an 8-bit ADC. Thus the clock has been slowed down by a factor of 128, or alternatively, the sampling interval is 1/(500 MHz·128) = 1/64 GHz s. The sampling is implemented using a CHAIS (CHArged mode Interleaved Sampler) (Figure 5.39).

Figure 5.41 shows a generic diagram of an optical DSP-based transceiver employing both DAC and ADC under QPSK modulated or QAM signals. The current maximum sampling rate of 64 GSa/s is available commercially. An IC image of the ADC chip is shown in Figure 5.40.

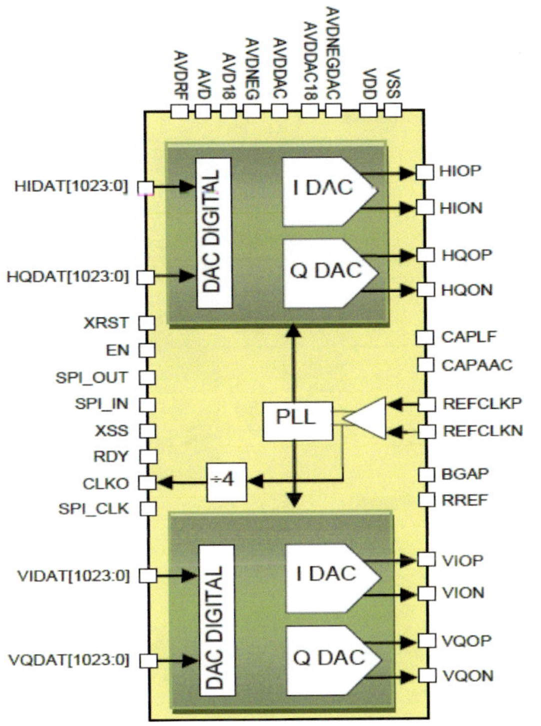

(a)

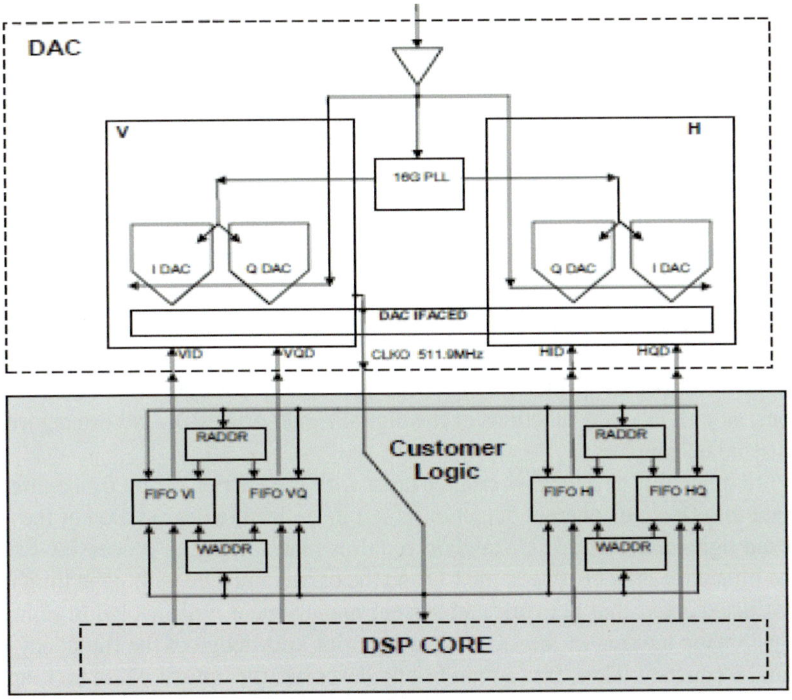

(b)

FIGURE 5.37 Fujitsu DAC structures for four channel PDM_QPSK signals (a) schematic diagram, (b) processing function.

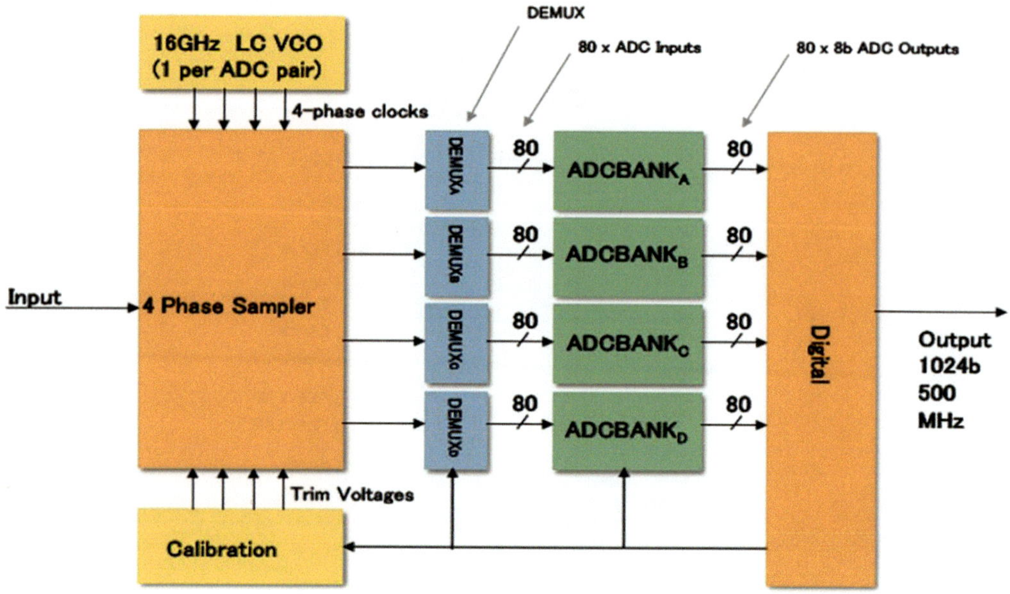

FIGURE 5.38 ADC principles of operations – brand name. CHAIS=CHArge-mode Interleave Sampler.

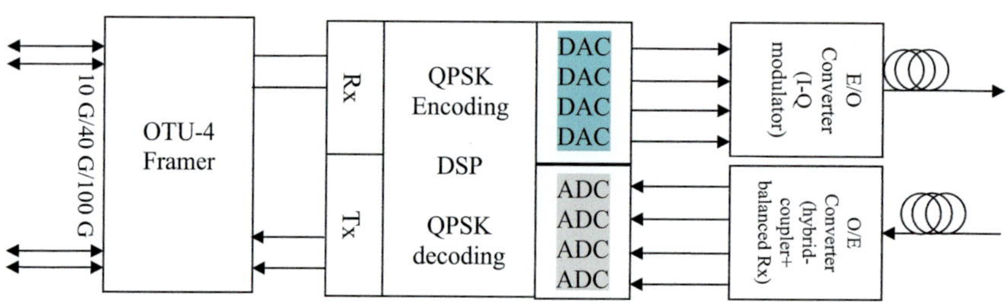

FIGURE 5.39 Schematic of the typical structure of ADC and ADC transceiver sub-systems for PDM-QPSK modulation channels.

5.8 CONCLUDING REMARKS

This chapter has described the principles of coherent reception and associated techniques with noise considerations, as well as main functions of the digital signal processors, whose algorithms will be described in separate chapters.

Furthermore, matching the local oscillator laser with the carrier of the transmitted channel is very important for effective coherent detection, if not it can lead to degradation of the sensitivity of resultant mixed lightwaves. The ITU standard requires that, for digital processing-based coherent receivers, the frequency offset between the LO and the carrier must be within the limit of ±2.5 GHz. In practice, it is expected that network and system management will involve in remote tuning of the LO and automatic locking of the LO with some prior knowledge of the frequency region to set the LO's initial frequency. Thus, this action briefly describes the optical phase locking of the local oscillator source for the intradyne coherent reception subsystem.

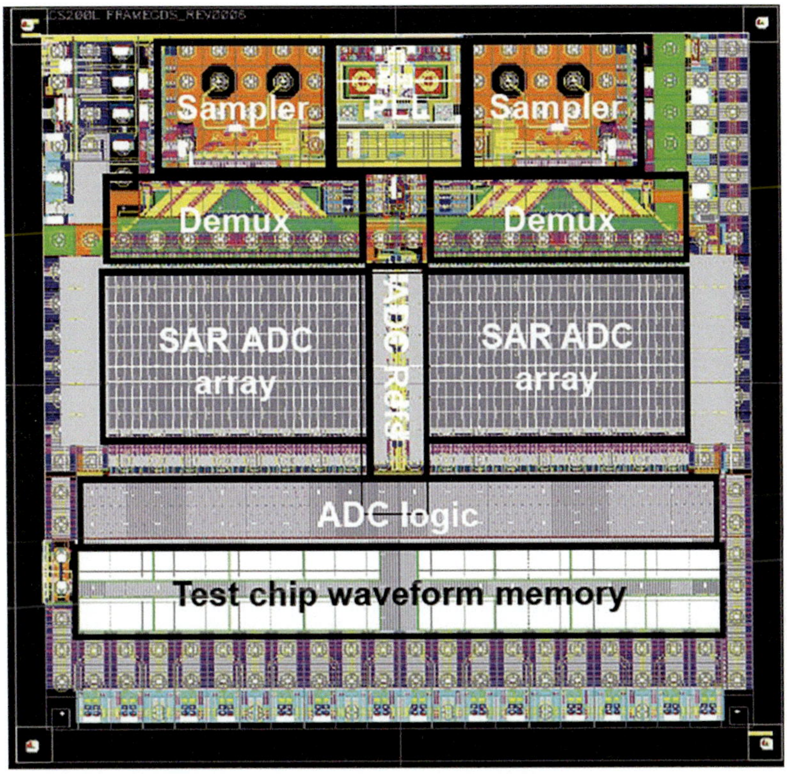

FIGURE 5.40 Fujitsu ADC subsystems with dual converter structure.

5.9 APPENDIX: A COHERENT BALANCED RECEIVER AND METHOD FOR NOISE SUPPRESSION

It has been shown that a balanced optical receiver can suppress the excess noise intensity generated from the local oscillator.[31,32,33] In a balanced receiver, an optical coupler is employed to mix a weak optical signal of with a local oscillator field of average powers of E_S^2 and E_L^2, respectively.

Figure 5.41 shows the generic diagram of a balanced optical receiver under coherent detection scheme in which two photodetectors and amplifiers operate in a push-pull mode.

Usually, the magnitude of the optical filed of the local oscillator E_L is much greater than E_S, the signal field. The fields at the output of a 2×2 coupler can thus be written as:

$$\begin{bmatrix} \tilde{E}_1 \\ \tilde{E}_2 \end{bmatrix} = \begin{bmatrix} \sqrt{1-k} & \sqrt{k}e^{j\pi/2} \\ \sqrt{k}e^{j\pi/2} & \sqrt{1-k} \end{bmatrix} \begin{bmatrix} \tilde{E}_S \\ \tilde{E}_L \end{bmatrix} \tag{5.69}$$

where E_S is the field of the received optical signal, E_L is the field of the local oscillator (possibly a Distributed Feedback Laser), and k is the intensity coupling coefficient of the coupler. For a 3 dB coupler, we have $k = 0.5$. E_S and E_L fields can then be written as:

$$\begin{aligned} E_S &= \sqrt{2}\tilde{E}_S e^{j(\omega_1 t+\phi_1)} \\ E_L &= \sqrt{2}\tilde{E}_L e^{j(\omega_2 t+\phi_2)} \end{aligned} \tag{5.70}$$

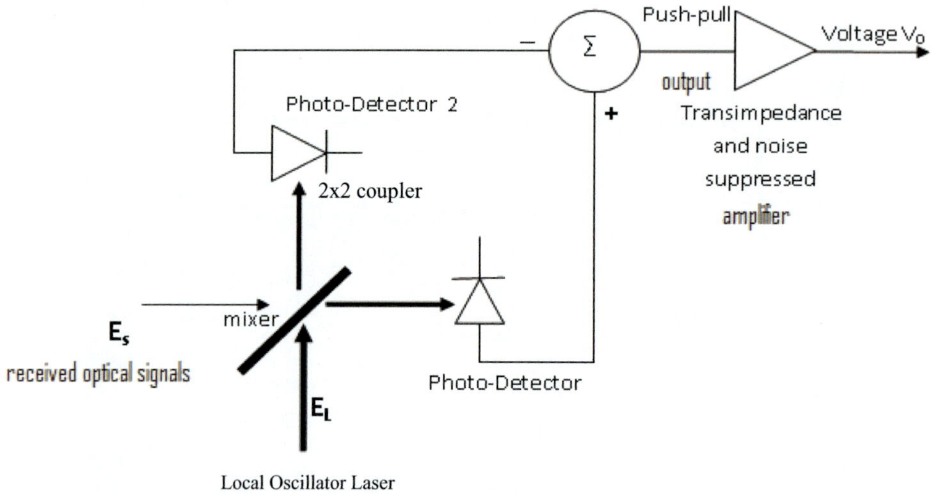

FIGURE 5.41 Generic block diagram of a coherent balanced detector optical receiver.

where $\tilde{E}_S$, $\tilde{E}_L$ are the magnitude of the optical fields of the signal and the local oscillator laser. The electronic current generated from the PDs corresponding to the average optical power and the signals are given by:

$$i_1(t) = \frac{\Re}{2}|E_1|^2 = \frac{\Re_1}{2}\left\{\begin{array}{l}\frac{1}{2}\left(E_S^2 + E_L^2\right) + \\ E_S E_L \cos\left[(\omega_1 + \omega_2)t + \phi_1 + \phi_2 - \frac{\pi}{2}\right]\end{array}\right\} + N_1(t) \tag{5.71}$$

$$i_2(t) = \frac{\Re}{2}|E_2|^2 = \frac{\Re_2}{2}\left\{\begin{array}{l}\frac{1}{2}\left(E_S^2 + E_L^2\right) \\ +E_S E_L \cos\left[\begin{array}{l}(\omega_1 + \omega_2)t \\ +\phi_1 + \phi_2 - \frac{\pi}{2}\end{array}\right]\end{array}\right\} + N_2(t) \tag{5.72}$$

where, ϕ_1, ϕ_2, and ω_1, ω_2 are the phase and frequency of the signal and local oscillator, respectively. $N_1(t)$ and $N_2(t)$ are the noises resulting from the photodetection process, $\Re_1, \Re_2$ are the responsivity of the PDs 1 and 2, respectively. It is assumed that the two photodiodes have the same quantum efficiency, hence their responsivity. Thus, we have the total current of a back-to-back PD pair as referred to the input of the electronic amplified, is

$$i_{\text{Teq}}(t) = \left\{\begin{array}{l}I_{\text{dc1}} - I_{\text{dc2}} \\ +(\Re_1 + \Re_2)E_S E_L \cos\left[\begin{array}{l}(\omega_1 - \omega_2)t \\ +\phi_1 - \phi_2 - \frac{\pi}{2}\end{array}\right]\end{array}\right\} + N_1(t) + N_2(t) \tag{5.73}$$

where the first two terms, I_{dc1} and I_{dc2}, are the detector currents generated by the detector pair due to the reception of the power of the local oscillator (LO), which is a continuous wave source. Thus, these terms appear as the DC constant currents that are normally termed as the shot noise due to the optical power of the local oscillator. The third term is the beating between the local oscillator and the signal carrier, the signal envelope. The noise currents $N_{1eq}(t)$ and $N_{2eq}(t)$ are seen as the equivalence at the input of the electronic amplifier from the noise processes that are generated by the PDs and the electronic amplifiers. Noise components are the shot noise due to bias currents, the quantum shot noise that depends on the strength of the signal and the local oscillator, the thermal noise due to the input resistance of the amplifier, and the equivalent noise current referred to the input of the electronic amplifier contributed by all the noise sources at the output port of the electronic amplifier. We denote $N_{Iep}(t)$ as the quantum shot noise generated due to the average current produced by the PD pair in the reception of the average optical power of the signal sequence.

The difference in the produced electronic currents can be derived using the following techniques: (i) the generated electronic currents of the PDs can be coupled through a 180° microwave coupler and then fed to an electronic amplifier; (ii) the currents are fed to a differential electronic amplifier; or (iii) a balanced PD pair connected back to back and fed to a small signal electronic amplifier.[34] The first two techniques require stringent components and are normally not preferred compared to the high performance of the balanced receiver structure in (iii).

Regarding the electronic amplifier, a transimpedance configuration is selected for the electronic pre-amplifier due to its wide bandwidth and high dynamics. However, it suffers from high noise due to the equivalent input impedance of the shunt feedback impedance, normally around a few hundred ohms. In the next section, the theoretical analysis of noise in the optical balanced receiver is described.

5.9.1 Noises

In the electronic pre-amplifier, the selection of the transistor as the first-stage amplifying device is very critical. Either a FET or BJT could be used; however, the BJT is preferred for wideband applications due to its robustness to noise. The disadvantage of using the BJT compared with the FET is its small base resistance, which leads to high thermal noise. However, for a shunt feedback amplifier, the resistance of the Millers' equivalent resistance referred to as the input of the amplifier is much smaller and thus dominates over the base resistance. The advantage of the BJT over the FET is that its small signal gain follows an exponential trend with respect to the small variation of the driving current derived from the photodetection compared with a parabolic trend for the FET. The FET may offer high input impedance between the gate and source of the input port, but it may not offer much improvement in a feedback amplification configuration in terms of noises. This section focuses on the use and design of BJT multi-stage shunt feedback electronic amplifiers.

Noises of electronic amplifying devices can be represented by superimposing all the noise current generators to the small signal equivalent circuit. These noise generators represent the noises introduced into the circuit by physical sources/processes at different nodes. Each noise generator can be expressed by a noise spectral density current square or power, as shown in Figure 5.42. This figure provides a general model of a small signal equivalent circuit, including noise current sources of any transistor, which can be represented by the transfer $[y]$ matrix parameters. Indeed, the contribution of the noise sources from the first stage of the electronic preamplifier is dominant, along with that of the shunt feedback resistance.

The current generator $di_n'^2/df$ is the total equivalent noise generator contributed by all the noise sources to node 1, including the thermal noises of all resistors connected to the node. Similarly, for the noise source, $di_n''^2/df$ is referred to as the output node.

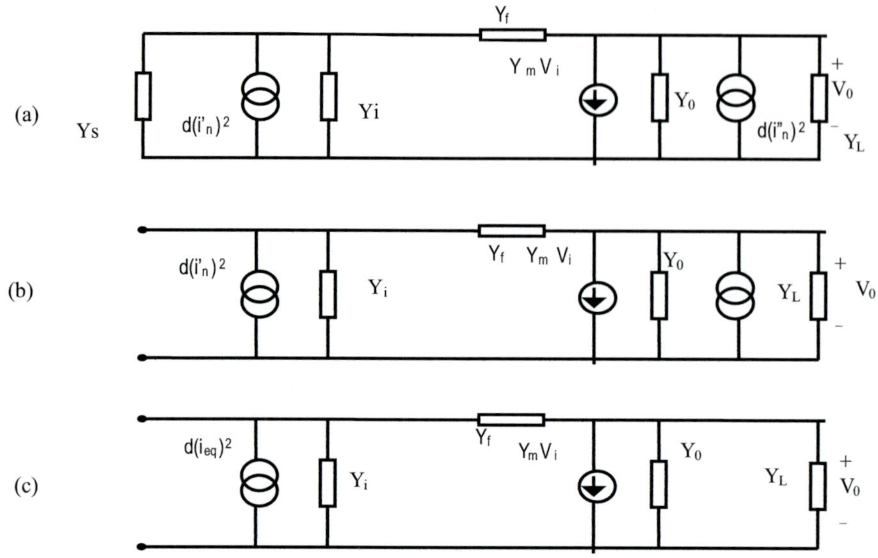

FIGURE 5.42 Small equivalent circuits including noise sources (a) Y-parameter model representing the ideal current model and all current noise sources at the input and output ports, (b) with noise sources at input and output ports, and (c) with a total equivalent noise source at the input.

5.9.2 Noise Generators

Electrical shot noises are generated by the random generation of streams of electrons (current). In optical detection, shot noises are generated by (i) biasing currents in electronic devices and (ii) photocurrents generated by the photodiode.

A biasing current I generates a shot noise power spectral density S_I given by:

$$S_I = \frac{d\left(i_I^2\right)}{df} = 2qI \qquad \text{A}^2/\text{Hz} \tag{5.74}$$

where q is the electronic charge. The quantum shot noise $\left\langle i_s^2 \right\rangle$ generated by the PD by an optical signal with an average optical power P_{in} is given by

$$S_Q = \frac{d\left\langle i_s^2 \right\rangle}{df} = 2q\left\langle i_s^2 \right\rangle \qquad \text{A}^2/\text{Hz} \tag{5.75}$$

If the PD is an APD type, then the noise spectral density is given by

$$S_Q = \frac{d\left\langle i_s^2 \right\rangle}{df} = 2q\left\langle i_s^2 \right\rangle\left\langle G_n^2 \right\rangle \qquad \text{A}^2/\text{Hz} \tag{5.76}$$

It is noted again that the dark currents generated by the PD must be included in the total equivalent noise current at the input after it is evaluated. These currents are generated even in the absence of the optical signal. These dark currents can be eliminated by cooling the PD to at least below the temperature of liquid nitrogen (77°K).

At a certain temperature, the conductivity of a conductor varies randomly. The random movement of electrons generates a fluctuating current even in the absence of an applied voltage. The thermal noise spectral density of a resistor R is given by:

$$S_R = \frac{d\left(i_R^2\right)}{df} = \frac{4k_BT}{R} \quad \text{in A}^2/\text{Hz} \tag{5.77}$$

where k_B is the Boltzmann's constant, T is the absolute temperature (in °K), R is the resistance in ohms, and i_R denotes the noise current due to resistor R.

5.9.3 Total Equivalent Input Noise Current

Our goal is to obtain an analytical expression for the noise spectral density equivalent to a source looking into the electronic amplifier, including the quantum shot noises of the PD. A general method for deriving the equivalent noise current at the input is given by representing the electronic device with a Y-equivalent linear network, as shown in Figure 5.3. The two current noise sources $di_n'^2$ and $di_n''^2$ represent the summation of all noise currents at the input and the output of the Y-network. This can be transformed into a Y-network circuit with the noise current referred to the input as follows:

The output voltages V_0 of Figure 5.3a can be written as:

$$V_0 = \frac{i_N'\left(Y_f - Y_m\right) + i_N''\left(Y_i + Y_f\right)}{Y_f\left(Y_m + Y_i + Y_o + Y_L\right) + Y_i\left(Y_o + Y_L\right)} \tag{5.78}$$

and for Figure 5.3b

$$V_0 = \frac{\left(i_N'\right)_{\text{eq}}\left(Y_f - Y_m\right)}{Y_f\left(Y_m + Y_i + Y_o + Y_L\right) + Y_i\left(Y_o + Y_L\right)} \tag{5.79}$$

Thus, comparing the above two equations, we can deduce the equivalent noise current at the input of the detector as:

$$i_{\text{Neq}} = i_N' + i_N'' \frac{Y_i + Y_f}{Y_f - Y_m} \tag{5.80}$$

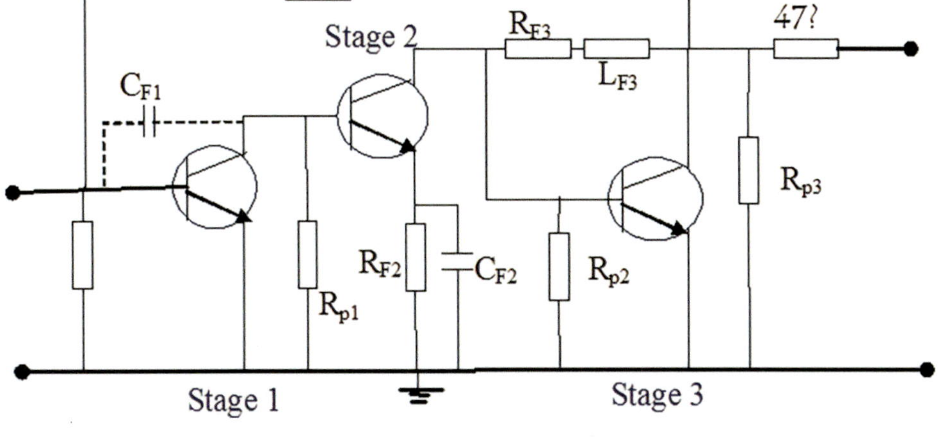

FIGURE 5.43 AC small signal circuit model of a three-stage feedback electronic pre-amplifier.

Then, reverting to mean square generators for a noise source, we have

$$d\left(i_{\text{Neq}}\right)^2 = d\left(i'_N\right)^2 + d\left(i''_N\right)^2 \left|\frac{Y_i + Y_f}{Y_f - Y_m}\right|^2 \tag{5.81}$$

It is therefore expected that if the Y-matrix of the front-end section of the amplifier is known, the equivalent noise at the input of the amplifier can be obtained by using Equation 5.81.

We propose a three-stage electronic pre-amplifier in AC configuration, as shown in Figure 5.43. The design of this amplifier is detailed in the next section. Small-signal and associated noise sources of this amplifier are given in Section 5.9.7. As can be seen, this general configuration is a forward path of shunt, series, and shunt stages, which reduces the interaction between stages due to the impedance levels.[8] Shunt-shunt feedback is placed around the forward path; hence the stable transfer function is the transfer impedance, which is important for transferring the generated electronic photodetected current to the voltage output for further amplification and data recovery.

For a given source, the input noise current power of a BJT front end can be found by

$$i_{\text{Neq}}^2 = \int_0^B d\left(i_{\text{Neq}}\right)^2 = a + \frac{b}{r_E} + c r_E \tag{5.82}$$

where B is the bandwidth of the electronic preamplifier and r_E is the emitter resistance of the front-end transistor of the preamplifier. The parameters $a, b, \& c$ depend on the circuit elements and amplifier bandwidth. Hence, an optimum value of $\mathrm{r_E}$ can be found and an optimum biasing current can be set for the collector current of the BJT, such that i_{Neq}^2 is at its minimum, as

$$r_{\text{Eopt}} = \sqrt{\frac{b}{c}} \ldots\text{hence}\ldots \rightarrow i_{\text{Neq}}^2\Big|_{r_E = r_{\text{Eopt}}} = a + 2\sqrt{bc} \tag{5.83}$$

If two types of BJT, Phillips BFR90A and BFT24, are considered, then a good approximation of the equivalent noise power can be found as:

$$\begin{aligned} a &= \frac{8\pi B^3}{3}\left\{r_B C_s^2 + \left(C_s + C_f + C_{tE}\right)\tau_T\right\}; \\ b &= \frac{B}{\beta_N}; \text{and} \\ c &= \frac{4\pi B^3}{3}\left(C_s + C_f + C_{tE}\right)^2 \end{aligned} \tag{5.84}$$

The theoretical estimation of the parameters of the transistors can be derived from the measured scattering parameters as given by the manufacturer[35] as

For transistor type BFR90A

$$\begin{aligned} &r_{\text{Eopt}} = 59\Omega \quad \text{for} \quad I_{\text{Eopt}} = 0.44\ \text{mA} \\ &I_{\text{eq}}^2 = 7.3\times10^{-16}\ \text{A}^2 \quad \text{hence} \quad \frac{I_{\text{eq}}^2}{B} = 4.9\times10^{-24}\ \text{A}^2/\text{Hz} \\ &I_{\text{eq}} = 27\ \text{nA} \quad \text{thus} \quad \frac{I_{\text{eq}}}{\sqrt{B}} = 2.21\ \text{pA}/\sqrt{\text{Hz}} \end{aligned} \tag{5.85}$$

For transistor type BFT24

$$r_{\text{Eopt}} = 104\,\Omega \qquad \text{for} \quad I_{\text{Eopt}} = 0.24\text{ mA}$$

$$I_{\text{eq}}^2 = 79.2\times10^{-16}\text{ A}^2 \qquad \text{hence} \quad \frac{I_{\text{eq}}^2}{B} = 6.1\times10^{-24}\text{ A}^2/\text{Hz} \tag{5.86}$$

$$I_{\text{eq}} = 30.2\text{ nA} \qquad \text{thus} \quad \frac{I_{\text{eq}}}{\sqrt{B}} = 2.47\text{ pA}/\sqrt{\text{Hz}}$$

Note that the equivalent noise current depends largely on some not well-defined values such as the capacitance, the transit times the base spreading resistance, and the short circuit current gain β_N. The term $\frac{I_{\text{eq}}}{\sqrt{B}}$ is usually specified as the noise spectral density equivalent to the input of the electronic pre-amplifier.

5.9.4 Pole-Zero Pattern and Dynamics

An AC model of a three-stage electronic preamplifier is shown in Figure 5.43, and the design circuit is shown in Figure 5.44. As briefly mentioned above, there are three stages with feedback impedance from the output to the input. The subscripts of the resistors, capacitors, and inductors indicate the order of the stages. The first stage is a special structure of shunt feedback amplification in which the shunt resistance is increased to infinity. The shunt resistance is in order of hundreds of ohms for the required bandwidth, thus, contributing to the noises of the amplifier. This is not acceptable. The shunt resistance increases the pole of this stage and approaches the origin. The magnitude of this pole is reduced by the same amount as that of the forward path gain. Thus, the poles of the close-loop amplifier remain virtually unchanged. As R_F is increased, the pole p_1 decreases, but G_1 is increased. Hence, $\frac{G_1 p_1}{s - p_1}$ remains constant. Thus, the position of the root locus is almost unchanged.

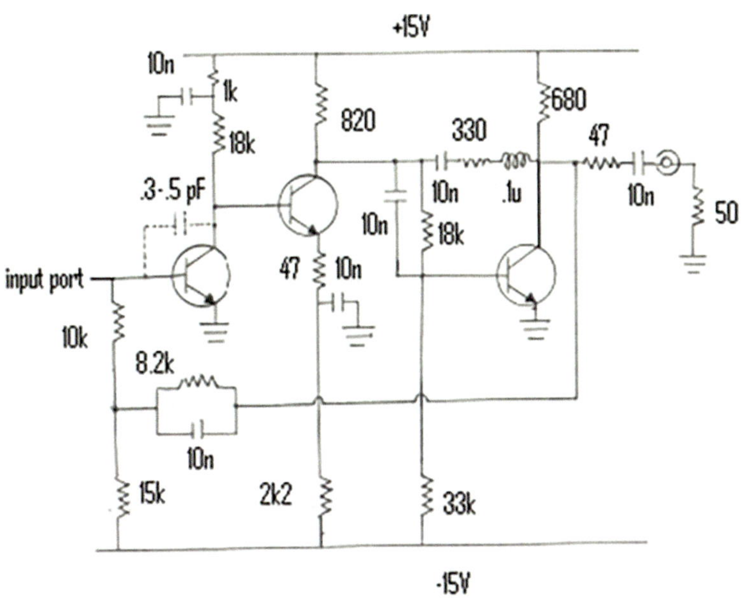

FIGURE 5.44 Design circuit of the electronic pre-amplifier for the balanced optical amplifier.

A compensating technique for reducing the bandwidth of the amplifier is to add capacitance across the base collector of the first stage. This may be necessary if oscillation occurs due to the phase shift becoming unacceptable at $GH = 1$, with G as the open loop gain and H as the feedback transfer function.

The second stage is a series feedback stage with feedback peaking. The capacitance C_{F2} is chosen such that it begins to bypass the feedback resistor at high frequencies. Hence, the feedback admittance partially compensates for the normal high-frequency drop in gain associated with the base and stray capacitances. Also, if the capacitance is chosen such that $R_{F2}C_{F2} = \tau_r$, the transfer admittance and input impedance become single-pole, which would be desirable.[36] The first and second stages are directly coupled (Figure 5.45).

The third stage uses the inductive peaking technique. For a shunt stage with a resistive load, the forward path gain has only one pole, and hence, there is only one real pole in the close loop response. A complex pole pair can be obtained by placing a zero in Z_{F2} and hence a pole in the feedback transfer function $H(\omega)$. Figure 5.46a–c shows the high-frequency singularity pattern of individual stages. Figure 5.46d shows the root locus diagram. It can be calculated that the poles take up the positions in Figure 5.46e when the loop gain GH is 220.

In the low-frequency region, the loop gain has five poles, three zeros on the negative axis and two zeros at the origin. The largest low frequency pole is set at approximately the input coupling capacitor and the input resistance leading to a low frequency cut-off of around tens of kHz.

The singularities of individual stages can be approximated by:

Stage 1: The dominant pole at

$$p = -\frac{1}{\beta_N R_{L1} C_{F1}} \tag{5.87}$$

and the other pole is located at

$$p = -\frac{C_{F1}}{C_{F1} + C_{L1}} \frac{1}{\tau_T + C_{F1}/g_m} \quad \text{and the midband gain} = \beta_N R_{L1} \tag{5.88}$$

with the load of stage 1 of R_{L1}

Stage 2: The dominant pole at

$$p = -\frac{1}{r_B G_{T2} \tau_T}; \quad \text{and the mid-band gain} = G_{T2} = 1/R_{F2} \tag{5.89}$$

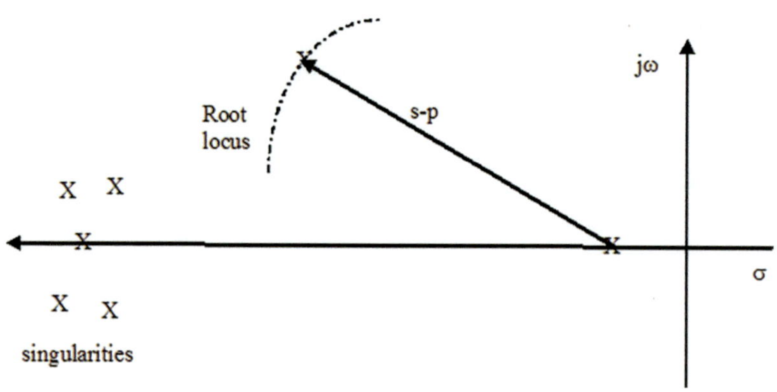

FIGURE 5.45 Effect of first stage pole on root locus of a shunt feedback electronic amplifier. The root locus is given by $1 = (GH)_{\text{mid-band}} \sum \frac{s - z_i}{z_i} \sum \frac{p_i}{s - p_i}$.

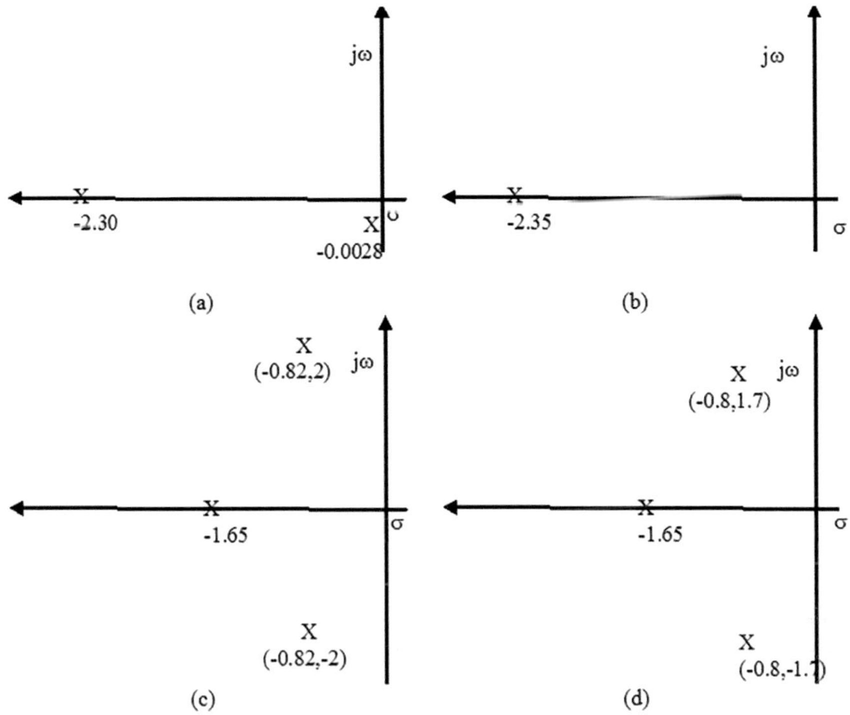

FIGURE 5.46 High-frequency root-locus diagram (a) singularity pattern for first stage, (b) second stage, (c) third stage, (d) root locus for close loop pattern for $GH = 220$.

Stage 3: Complex pole pair at

$$|p|^2 = -\frac{R_L}{\tau_T^2 R_{F3}}; \sigma = -\frac{(R_L + R_{F3})}{2\tau_F R_{F3}}; \tau_F = \frac{L_{F3}}{R_{F3}}$$
$$\text{and a zero at} -\frac{1}{\tau_F}; \text{ with a gain} = -R_{F3} \tag{5.90}$$

The feedback configuration of the circuit of the trans-impedance electronic preamplifier has a 10 or 15 K resistor, whose noise would contribute to the total noise current of the amplifier. The first and second stages are direct-coupled, thus eliminating the level shifter and a significant amount of stray capacitance. The peaking capacitor required for the second stage is on the order of 0.5 pF, hence, no discrete component can be used. This may not require any component, as the stay capacitance may suffice.

The equivalent noise current at the input of the trans-impedance amplifier is approximately proportional to the square of the capacitance at the base of the transistor. Therefore, minimization of the stray capacitance at this point must be conducted by shortening the connection lead as much as possible. As a rough guide, critical points in the circuit where stray capacitance must be minimized are those not at signal ground and have impedance and small capacitance to ground. For instance, at the base of the output stage, the capacitance to ground is more tolerable than at the collector to ground of the second transistor, and say the collector of the last stage, which should ideally have no capacitance to ground.

An acceptable step response can be achieved by manipulating the values of changing C_{F1}, R_{F3}, L_{F3}. Since C_{F1} contributes to the capacitance at the base of the first stage, C_{F1} is minimized. The final value of C_{F1} is 0.5 pF (about half twist of two wires), R_{F1} is 330 ohms and L_{F1} is about 0.1 μF (2 cm of wire). The amplifier is sensitive to parasitic capacitance between the feedback resistor and the first two stages. Thus, a grounded shield is placed between the 10 K resistor and the first and second stages.

The expressions for the singularities for the first two stages are similar as described above. The base-collector capacitance of the third stage affected the position of the poles considerably. The poles of the singularities of this stage are as follows: the mid-band trans-resistance is –RF3, with a zero $@ z = \frac{1}{\tau_F}$, and a complex pole pair given by

$$|p_P|^2 = \frac{1}{L_F C_F + R_F C_F \tau_T + \tau_T \tau_F \dfrac{R_F}{R_L}}; \ \sigma = \frac{1}{2}|p_P|^2 \left(R_F C_F + \tau_T \left(1 + \frac{R_F}{R_L} \right) \right) \tag{5.91}$$

In addition, a large pole given approximately at

$$p = -\frac{1}{L_F C_F \tau_T |p_P|^2} \tag{5.92}$$

Shown in Figure 5.47 are the open loop singularities in (a), the root locus diagram in (b), and the close loop in (c). The two large poles can be ignored without any significant difference. Similarly, for the root-locus diagram, the movement of the large poles is negligible, the pole and zero pair can be ignored, and the movement of the remaining three poles can be calculated by considering just these three poles as the open-loop singularities.

5.9.5 Responses and Noises Measurements

5.9.5.1 Rise-Time and 3 dB Bandwidth

Based on the pole-zero patterns given in the previous section, the step responses can be estimated and contrasted with the measured curve, as shown in Figure 5.48. The experimental setup for the rise time measurement is shown in Figure 5.49 in the electrical domain without using a photodetector. An artificial current source is implemented using a series resistor with minimum stray capacitance. This testing in the electrical domain is preferred over the optical technique, as the rise time of an electrical TDR is sufficiently short to influence the measurement of the rise time. If the optical method is used, the principal problem is that we must be able to modulate the intensity of the lightwave with a very sharp step function. This is impossible when the Mach-Zehnder intensity modulator (MZIM) is used as its transfer characteristics following a square of a cosine function. A very short pulse sequence laser such as mode-locked fiber laser can be used. However, this is not employed in this work, as the bandwidth of our amplifier is not wide enough and the TDR measurement is sufficient to give us a reasonable value of the rise time.

The passband of the transimpedance amplifier is also confirmed with the measurement of the frequency response by using a scattering parameter test set HP 85046A, the scattering parameter S_{21} frequency response is measured and the 3 dB passband is confirmed at 190 MHz.

5.9.5.2 Noise Measurement and Suppression

Two pig-tailed PDs are used and mounted in a back-to-back configuration at the input of the transistor of Stage 1. A spectrum analyzer is used to measure the noise of the amplifier, as shown in Figure 5.50. The background noise of the analyzer was measured to be –88 dBm. The expected noise referred to the input is 9.87×10^{-24} A^2/Hz. When the amplifier is connected to the spectrum analyzer, this noise level increases to –85 dBm, which indicates that the noise at the output of the amplifier is around –88 dBm.

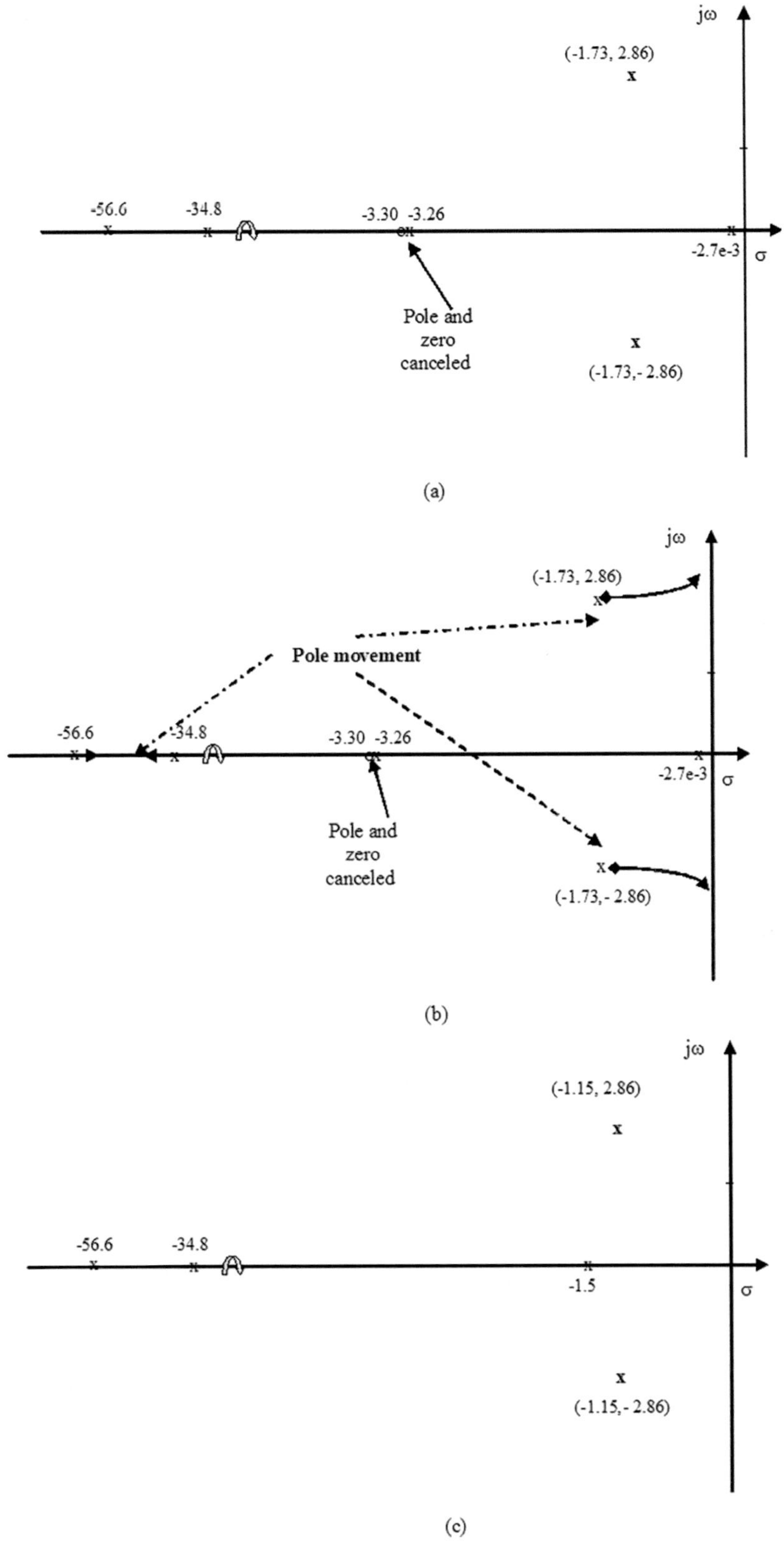

FIGURE 5.47 Pole-zero patterns in the s-plane and their dynamics of Ref. (a) open loop, (b) root locus, and (c) close loop.

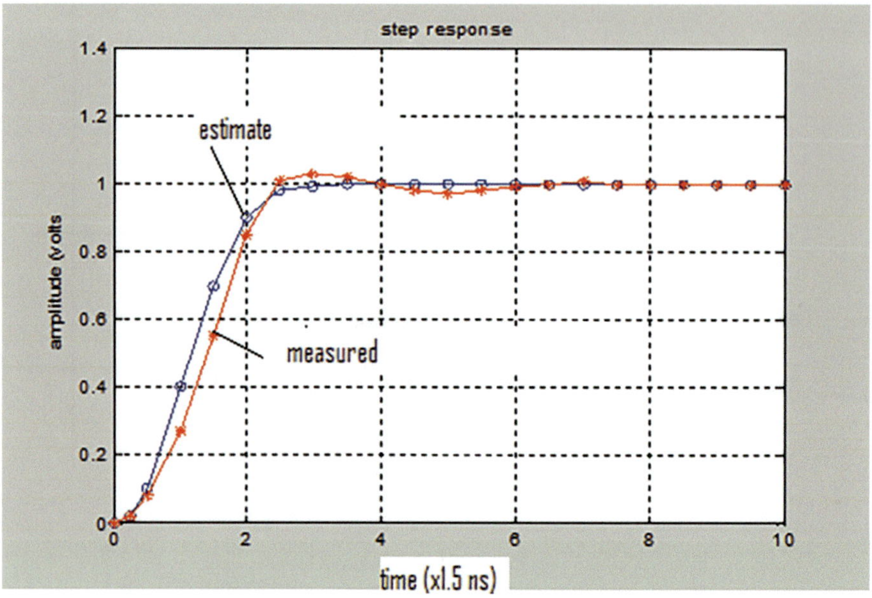

FIGURE 5.48 Step response of the amplifier: o - estimated *- measured.

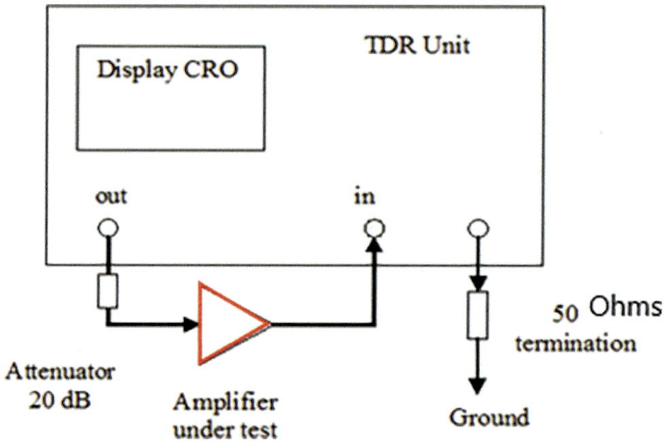

FIGURE 5.49 Experimental setup for measurement of the rise time in the electrical domain using a time domain reflectometer (TDR).

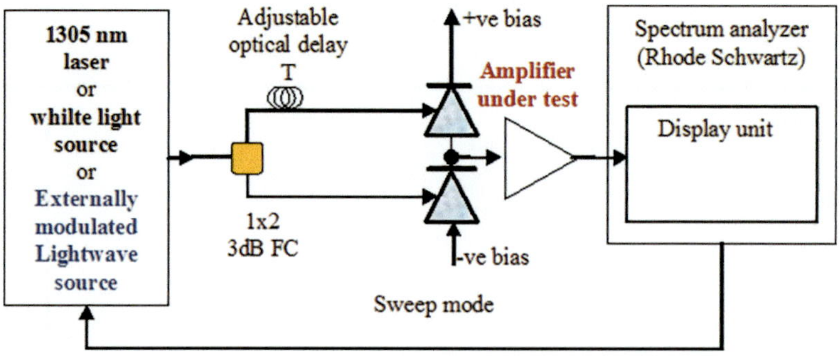

FIGURE 5.50 Experimental setup to measure excess noise impact and cancellation of a balanced receiver. FC=Fiber coupler.

Since this power is measured into 50 Ω, using a spectral bandwidth of 300 kHz, the input current noise can be estimated as:

$$\int di_{\text{Neq}}^2 = \frac{v_N^2}{R_T^2} = \frac{10^{-11}\times 50}{(5k)^2} = 2\times 10^{-17.5}\ \text{A}^2 \rightarrow \frac{di_{\text{Neq}}^2}{df} = \frac{2e^{-17.5}A^2}{3\times 10^5} = 1.06\times 10^{-23}\ \text{A}^2/\text{Hz} \quad (5.93)$$

Thus, the measured noise is very close to the expected value.

5.9.5.3 Requirement for Quantum Limit

The noise required for near quantum limit operation can be estimated. The total shot noise referred to the input is

$$i_{T\text{-shot}}^2 = (\Re_1 + \Re_2)|E_L|^2 \quad (5.94)$$

with $\Re_{1,2}$ as the responsivity of the photo-detectors 1 and 2, respectively. The total excess noise referred to the input is

$$i_{NT\text{-shot}}^2 = (\Re_1 + \Re_2)|E_L|^2; \quad i_{N\text{-excess}}^2 = 2q\gamma(\omega)(\Re_1 + \Re_2)^2|E_L|^4 \quad (5.95)$$

The excess noise from each detector is correlated; so γ is a function of frequency, typically between 10^4 and 10^{10} A^{-1}. A receiver would operate within 1 dB of the quantum limit if the shot noise is about 6 dB above the excess noise and the amplifier noise. The amplifier noise power spectral density is 6.09×10^{-14} A^2/Hz Assuming that complete cancellation of the excess noise can be achieved, the local oscillator required power is given as

$$i_{NT\text{-shot}}^2 > 4i_{\text{NTeq}}^2 \rightarrow |E_L|^2 > 0.24\ \text{mW} \approx -6.2\,\text{dBm} \quad (5.96)$$

Since there is bound to be excess noise due to imperfection of the balancing of the two detectors, the power of the local oscillator is expected to be slightly greater than this value.

The output voltage of the amplifier can be monitored to measure the noises contributed by different processes. When the detectors are not illuminated, we have

$$v_0^2 = Z_T^2 i_{\text{Neq}}^2 \quad (5.97)$$

with Z_T as the transimpedance, and v_0 as the output voltage of the amplifier. It is known that incoherent white light is free from excess intensity noise; so, when the detectors are illuminated with white light sources we have

$$v_0^2 = Z_T^2\left(i_{\text{Neq}}^2 + i_{N\text{-shot}}^2\right) \quad (5.98)$$

When the local oscillator is turned on illuminating the detector pair, then

$$v_0^2 = Z_T^2\left(i_{\text{Neq}}^2 + i_{N\text{-shot}}^2 + i_{N\text{-excess}}^2\right) \quad (5.99)$$

5.9.5.4 Excess Noise Cancellation Technique

The presence of un-canceled excess noise can be shown by observing the output voltage of the electronic pre-amplifier using a spectrum analyzer. First, the detectors are illuminated with incoherent white light, and the output noise is measured. Then, the detectors are illuminated with the laser local oscillator at a power level equal to that of the white light source, so the shot noise would be of the same magnitude. Thus, the increase in the output noise will be due to the un-canceled intensity noise. This, however, does not demonstrate the cancellation of the local oscillator intensity noise.

A well-known method for this cancellation is to bias the local oscillator just above its threshold.[37] This causes the relaxation resonance frequency to move over the passband. Thus, when the outputs of the single or balanced receivers are compared, the resonance peak is observed in the former case but not in the latter case, in which the peak is suppressed. Since the current is only in the μA scale in each detector for a local oscillator operating near the threshold, the dominant noise is that of the electronic amplifier. This method is thus not suitable here. However, the relaxation resonance frequency is typically between 1 and 10 GHz. A better method can be developed by modulating the local oscillator with a sinusoidal source and measuring the difference between single and dual photodetection receivers. The modulation can be implemented at either one of the frequencies to observe the cancellation of the noise peak power or by sweeping the local oscillator over the bandwidth of the amplifier so that the cancellation of the amplifier can be measured. For a fair comparison, the local oscillator power would be twice that of the single detector case, so the received power of both cases can be almost identical, and hence the SNR at the output of the amplifier can be derived and compared.

5.9.5.5 Excess Noise Measurement

An indication of excess noise and its cancellation within the 0–200 MHz frequency range by the experimental setup can be observed, as shown in Figure 5.50. A laser of 1305 nm wavelength is used with an output power of –20 dBm is employed as the local oscillator, so the dominant noise is that of the electronic trans-impedance preamplifier as explained above. The local oscillator is intensity modulated by the sweep generator from the spectrum analyzer with a 3 GHz frequency range (Rhode Schwartz model). A noise suppression pattern of the same trend can be observed with suppression about every 28 MHz interval over the entire 200 MHz range of the trans-impedance amplifier.

Figure 5.51 shows the noise spectral density for the case of single and dual detectors. Note that the optical paths have different lengths. This path difference is used for cancelling the excess noise,

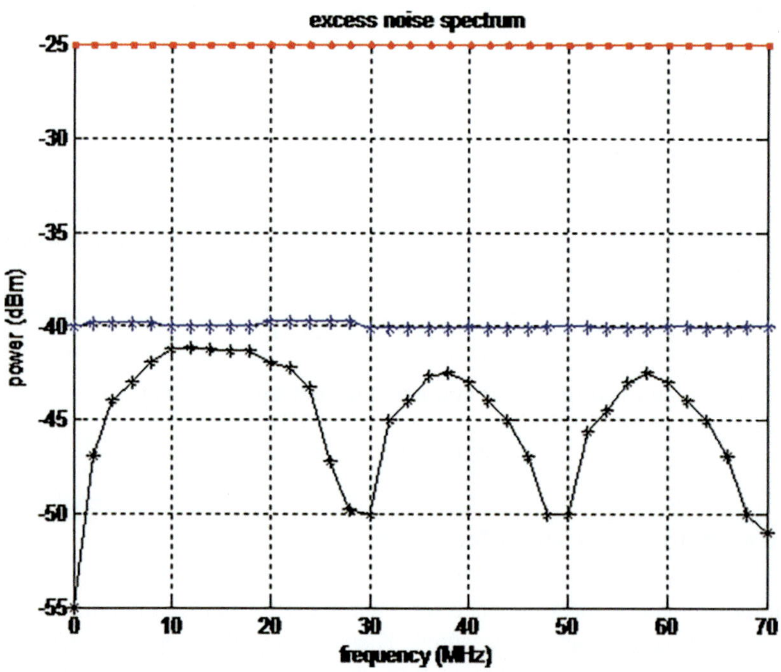

FIGURE 5.51 Excess noise for single detector and dual detectors with noise cancellation using delay and filter at the input port of the preamplifier.

as described in Refs.[38,39] It can be shown that the optical delay line leads to a cancellation of noise following a relationship of $\{1-\cos(2\pi fT)\}$ with T as the delay time of the optical path of one of the fibers from the fiber coupler to the detector. Hence, the maxima of the cancellation occurs at $f=m/T$, where m is an integer. From Figure 5.50, we can deduce the optical path length as $f=30$ MHz or $T=31.7$ ns, thus $d=cT/n_{\text{eff}}=6.8\,\text{m}$ with n_{eff} (~1.51 @1550 nm) as the effective refractive index of the guided mode in the fiber. The total length of the fiber delay is equivalent to about 6.95 m in which the effective refractive index of the propagation mode is estimated to be 1.482 at the measured wavelength. Thus, there is a small discrepancy, which can be accounted for by the uncertainty of the exact value of the effective refractive index of the fiber.

Using this method, Abbas et al.[40] achieved a cancellation of 3.5 dB, while 8 dB was obtained in our setup. This discrepancy can be accounted for by a combination of (i) the photodiode pair is tuned to be identical.

In our initial measurements, we observed some discrepancies due to (i) unmatched properties of the photodetection in the balanced detector pair and (ii) the delay path having three extra optical connections, resulting in more loss. So, even when there is no delay, the excess noise would not be completely canceled. Thus, at some stage, the longer path was implemented with a tunable optical delay path to match the delay path to the null frequency of the RF wave, achieving cancellation of the excess noise. This noise cancellation can lead to an improvement of 16 dB in the OSNR due to the dual-detector configuration and an improvement of nearly 100 km in the length of standard single-mode fiber.

The noise cancellation is periodic following a relationship with respect to the delay time T imposed by the optical path of the interferometer T as $(1-\cos 2\pi fT)$. The optical delay path difference is estimated by adjusting the optical delay line, which is implemented by a fiber path coupled out of a fiber end to open air path and back to another fiber path via a pair of Selfoc lenses.

It is noted that the local oscillator light source can be directly modulated (DD) with a sweep sinusoidal signal derived from the electrical spectrum analyzer (ESA). This light source has also been externally modulated using a MZIM of bandwidth much larger than that of the electronic amplifier (about 25 GHz). The excess noise characteristics obtained using both types of sources are almost identical. Furthermore, with the excess noise reduction of 8 dB and the signal power for the balanced receiver operating in a push-pull mode, the received electrical current would be double than that of a single detector receiver. This means a gain of 3 dB in the SNR by this balanced receiver.

5.9.6 Remarks

In this appendix, we consider the design and implementation of a balanced receiver using a dual detector configuration and a BJT front-end pre-amplifier. In view of recent growing interest in coherent optical transmission systems with electronic digital signal processing,[41,42,43] a noise-suppressed balanced receiver operating in the multi-GHz bandwidth range is essential. Noise cancellation using an optical delay line in one of the detection paths leads to the suppression of excess noise in the receiver.

In this section, we demonstrate through design analysis and implementation, a discrete wideband optical amplifier with a bandwidth of around 190 MHz and a total input noise equivalent spectral density of $10^{-23}\ \text{A}^2/\text{Hz}$. This agrees well with the predicted value using a noise model analysis. It is shown that with sensible construction of the discrete amplifier, minimizing the effects of stray capacitances, and the appropriate application of compensation techniques, a bandwidth of 190 MHz can be easily obtained for transistors with a few GHz transition frequencies. Furthermore, by using an optimum emitter current for the first stage and minimizing the biasing and feedback resistance and capacitance at the input, a low noise amplifier with an equivalent noise current of about $1.0\ \text{pA}/\sqrt{\text{Hz}}$ is achieved. Furthermore, the excess noise cancellation property of the receiver is found to give maximum SNR of 8 dB with a matching of the two photodetectors.

Although the bandwidth of the electronic amplifier reported here is only 190 MHz, which is about 1% of the 100 Gb/s target bit rate for modern optical Ethernet. The bandwidth achieved by our amplifier has reached around the maximum region that can be achieved with discrete transistor stages whose transition frequency is only 5 GHz. The amplifier configuration reported here can be scaled up to a multi-GHz region for a 100 Gb/s receiver using an integrated electronic amplification device without much difficulty. We believe that the design procedures for determining the pole and zero patterns on the s-plane and their dynamics for stability considerations are essential. Readers interested in implementing this using MMIC technology can benefit from these procedures. Should the amplifier be implemented in the multi-GHz range, microwave design techniques employing the noise figure method may be most useful when incorporating the design methodology reported in this section.

5.9.7 Electronic Noise Equations

Refer to the small signal and noise model given in Figures 5.52 and 5.53, the spectral density of noises due to collector bias current I_C, small signal transconductance g_m, and base conductance g_B generated in a BJT can be expressed as:

$$di_1^2 = 4k_BTg_Bdf;\quad di_2^2 = 2qI_cdf + 2k_BTg_mdf$$

$$di_3^2 = 2qI_Bdf + 2k_BT(1-\alpha_N)g_mdf;\ \text{and} \tag{5.100}$$

$$di_4^2 = \sum \text{shot noise of diodes and thermal noise of bias ing}_B\ \text{resistors}$$

From nodal analysis of the small signal equivalent circuit given in Figure 5.53, we can obtain the relationship

$$\begin{bmatrix} Y_s + g_B & -g_B & 0 \\ -g_B & g_B + y_1 + y_f & -y_f \\ 0 & g_m - y_f & y_f + y_2 \end{bmatrix} \begin{bmatrix} V_1 \\ V_2 \\ V_3 \end{bmatrix} = \begin{bmatrix} i_{\text{eq}} - i_{N_1} \\ i_{N_1} + i_{N_3} \\ i_{N_2} \end{bmatrix} \tag{5.101}$$

Hence, V_3, V_2, V_1 can be found by using Euler's rule for the matrix relationship

$$V_3 = \frac{\Delta_{13}}{\Delta}\left(i_{\text{eq}} - i_{N_1}\right) + \frac{\Delta_{23}}{\Delta}\left(i_{N_1} + i_{N_3}\right) + \frac{\Delta_{33}}{\Delta}\left(i_{N_2}\right) \tag{5.102}$$

Hence, the noise currents as referred to the input are

$$d\left(i_1'^2\right) = \left|\frac{Y_s}{g_B}\right|^2 dI_1^2 = \omega_2 C_s^2 r_B 4kTdf \tag{5.103}$$

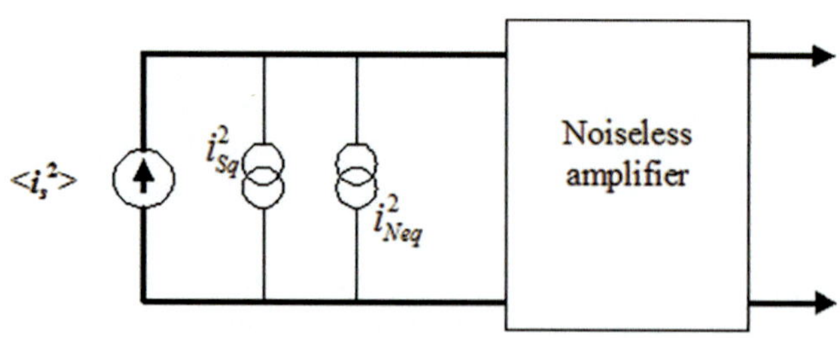

FIGURE 5.52 Equivalent noise current at the input and noiseless amplifier model. i_{sq}^2 is the quantum shot noise, which is signal dependent, i_{Neq}^2 is the total equivalent noise current referred to the input of the electronic amplifier.

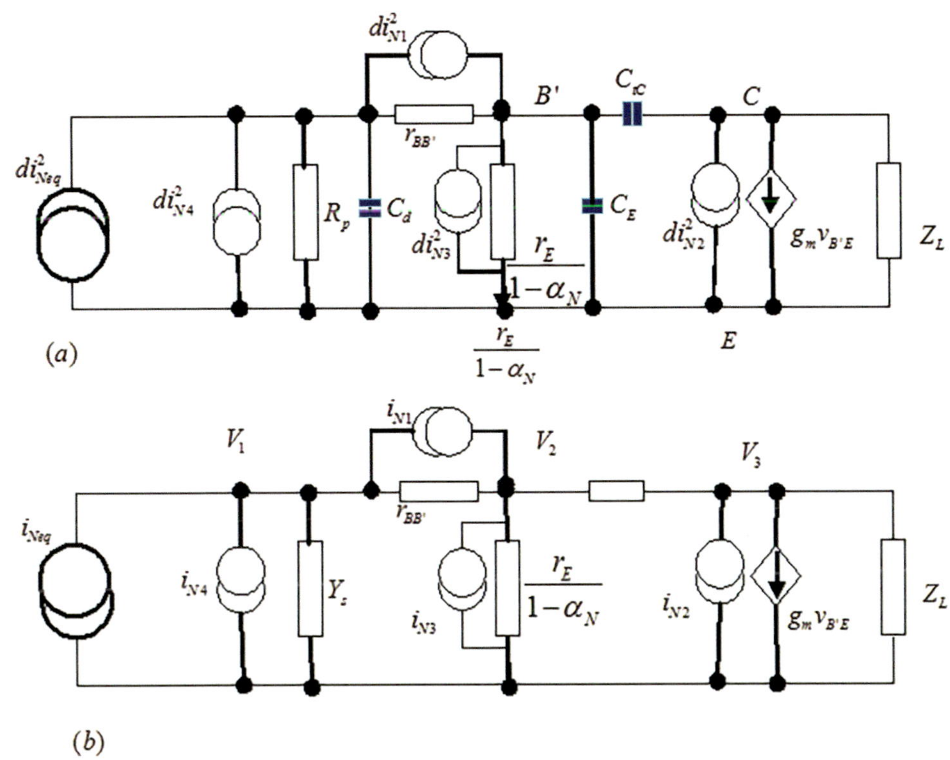

FIGURE 5.53 (a) Approximated noise equivalent and small signal model of a BJT front end, (b) generalized noise and small signal model circuit. Note that $r_B = r_{sd} + r_{BB'}$ and $C_d = C_p + C_i$ with $r_{BB'}$ is the base resistance, r_{sd} is the diode resistance, C_d is the photodiode capacitance, and C_i is the input capacitance.

$$d\left(i_2'^2\right) = \left|\left(Y + y_1 + y_f\right) + \frac{Y_s}{g_B}\left(y_1 + y_f\right)\right|^2 \left|\frac{1}{y_f - g_m}\right|^2 = \left|\frac{1}{y_f - g_m}\right|^2 dI_2^2$$
$$= \left(\frac{1}{\beta_N r_E^2} + \omega^2\left(C_0^2 - \frac{2C_B r_B}{\beta_N r_E}\right) + \omega^4 2C_s r_B C_B^2\right) 2kTr_E df \tag{5.104}$$

$$\text{and} \quad d\left(i_3'^2\right) = \left|\frac{Y_s}{g_B} + 1\right|^2 dI_3^2 = \left(\omega\, C_s^2 r_B^2 + 1\right)\frac{2kT}{\beta r_E} df \tag{5.105}$$

in which we have assumed that

$$\omega C_f \ll g_m$$

$$C_0 = C_s + C_E + C_f + \frac{C_s r_s}{\beta r_E} \approx C_s + C_{tE} + C_f + \frac{\tau_t}{r_E} = C_x + \frac{\tau_t}{r_E} \tag{5.106}$$

$$\text{with} \qquad C_x = C_E + C_f; \quad C_E = C_{tE} + \frac{\tau_t}{r_E}$$

The total noise spectral density refers to the input of the amplifier as shown in Figures 5.52 and 5.53 and is thus given as

$$\frac{di_{\text{Neq}}^2}{df^2}=\frac{di_{\text{1eq}}^2}{df^2}+\frac{di_{\text{2eq}}^2}{df^2}+\frac{di_{\text{3eq}}^2}{df^2}$$
$$=2k_BT\left[\frac{1}{\beta r_E}\frac{\beta+1}{\beta}+\omega^2\left(\frac{2}{r_B}+\frac{1}{\beta r_E}\right)+C_0r_E-\frac{2C_sr_BC_x}{\beta}+\omega^2C_s^2r_B^2C_x^2r_E\right] \tag{5.107}$$

Hence, the total noise power referred to the input is given as:

$$i_{\text{Neq}}^2=\int_0^B di_{\text{Neq}}^2=a+\frac{b}{r_E}+cr_E;\ \text{with}$$

$$a=2k_BT\left[\frac{8\pi^2B^3}{3}\left(C_sr_B+C_x\tau_t+\frac{C_s^2r_B^2}{\beta}\right)+\frac{(2\pi)^4}{5}B^5C_s^2r_B^2 2C_y\tau_t\right]$$
$$b\approx 2k_BT\left[\frac{B}{\beta}+\frac{4\pi^2B^3}{3}\left(\frac{C_sr_B^2}{\beta}+\tau_t^2\right)+\frac{16\pi^4B^5}{5}C_s^2r_B^2\tau_t^2\right] \tag{5.108}$$
$$c\approx 2k_BT\left[\frac{4\pi^2B^3C_x^2}{3}+\frac{16\pi^4B^5}{5}C_s^2r_B^2C_y^2\right]\quad \text{with}\ldots C_y=C_{tE}+\frac{\tau_t}{r_E}$$

where the first term in each coefficient $a, b\ \&\ c$ is dominant for a bandwidth of 190 MHz of the two transistor types Phillip BFR90A and BFT24 under considerations.

If the dependence of C_E on I_E is ignored, Equation (5.108) becomes

$$i_{\text{Neq}}^2=2k_BT\left[\frac{1}{r_E}\frac{B}{\beta}+r_E\left(\frac{4\pi^2B^3}{3}C_0^2\right)+\frac{16\pi^4B^5}{5}C_s^2r_B^2C_x^2+\frac{8\pi^2B^3C_s^2r_B}{3}\right] \tag{5.109}$$

Clearly, the noise power is a cubic dependence on the bandwidth of the amplifier. This is thus very critical for ultra-wide band optical receiver. Therefore, it is very important to suppress the excess noise of the optical receiver due to the quantum shot noise of the local oscillator.

For the transistor BFR90A, the optimum emitter resistance is $r_{\text{Eopt}}=49\,\Omega$, and the total equivalent noise power referred to the input is $4.3\times10^{-16}\ \text{A}^2$, which is moderately high.

NOTES

1 A. H. Gnauck and P. J. Winzer, "Optical phase-shift-keyed transmission," *Journal of Lightwave Technology,* 23, 115–130, 2005.

2 W. A. Stallard, A. R. Beaumont, and R. C. Booth, "Integrated optic devices for coherent transmission," *IEEE Journal of Lightwave Technology*, LT-4(7), 852, July 1986.

3 Synchronous detection is implemented by mixing the signals and a strong local oscillator in association with the phase locking of the local oscillator to that of the carrier.

4 V. Ferrero and S. Camatel, "Optical phase locking techniques: An overview and a novel method based on single sideband sub-carrier modulation," *Optics Express*, 16(2), 818–828, January 21, 2008.

5 I. Garrett and G. Jacobsen, "Theoretical analysis of heterodyne optical receivers for transmission systems using (semiconductor) lasers with nonneglible linewidth," *IEEE Journal of Lightwave Technology*, LT-3/4, 323–334, 1986.

6 G. Nchoson, "Probability of error for optical heterodyne DPSK systems with quantum phase noise," *Electronics Letters*, 20/24, 1005–1007, 1984.

7 S. Shimada, Coherent Lightwave Communications Technology. Chapman and Hall: London, 1995, p. 27.

8 Y. Yamamoto and T. Kimura, "Coherent optical fiber transmission system," *IEEE Journal of Quantum Electronics*, QE-17, 919–934, 1981.

9 K. Iwashita and T. Masumoto, "Modulation and detection characteristics of optical continuous phase FSK transmission system," *IEEE Journal of Lightwave Technology*, LT-5/4, 452–462, 1987.
10 F. Derr, "Coherent optical QPSK intradyne system: Concept and digital receiver realization," *IEEE Journal of Lightwave Technology*, 10(9), 1290–1296, 1992.
11 G. Bosco, I. N. Cano, P. Poggiolini, L. Li, and M. Chen, "MLSE-based DQPSK transmission in 43Gb/s DWDM long-haul dispersionmanaged optical systems," *IEEE Journal of Lightwave Technology*, 28(10), 1573–1581, May15, 2010.
12 D.-S. Ly-Gagnon, S. Tsukamoto, K. Katoh, and K. Kikuchi, "Coherent detection of optical quadrature phase-shift keying signals with carrier phase estimation," *Journal of Lightwave Technology*, 24, 12–21, 2006.
13 E. Ip and J. M. Kahn, "Feedforward carrier recovery for coherent optical communications," *Journal of Lightwave Technology*, 25, 2675–2692, 2007.
14 L. N. Binh, "Dual-ring 16-star QAM direct and coherent detection in 100 Gb/s optically amplified fiber transmission: Simulation," *Optical and Quantum Electronics*, 40, 707–732, August 5, 2008.
15 L. N. Binh, "Generation of multi-level amplitude-differential phase shift keying modulation formats using only one dual-drive Mach-Zehnder interferometric optical modulator," *Optical Engineering*,, 48(4), 2009.
16 M. Nazarathy, X. Liu, L. Christen, Y. K. Lize, and A. Willner, "Self-coherent multisymbol detection of optical differential phase-shift-keying," *Journal of Lightwave Technology*, 26, 1921–1934, 2008.
17 R. Noe, "PLL-free synchronous QPSK polarization multiplex/diversity receiver concept with digital I&Q baseband processing," *IEEE Photonics Technology Letters*, 17, 887–889, 2005.
18 L. G. Kazovsky, G. Kalogerakis, and W.-T. Shaw, "Homodyne phase-shift-keying systems: Past challenges and future opportunities," *Journal of Lightwave Technology*, 24, 4876–4884, 2006.
19 R. Noe, "PLL-free synchronous QPSK polarization multiplex/diversity receiver concept with digital I&Q baseband processing," *IEEE Photonics Technology Letters*, 17, 887–889, 2005.
20 D.-S. Ly-Gagnon, S. Tsukamoto, K. Katoh, and K. Kikuchi, "Coherent detection of optical quadrature phase-shift keying signals with carrier phase estimation," *Journal of Lightwave Technology*, 24, 12–21, 2006.
21 J. P. Gordon and L. F. Mollenauer, "Phase noise in photonic communications systems using linear amplifiers," *Optics Letters*, 15, 1351–1353, 1990.
22 H. Kim and A. H. Gnauck, "Experimental investigation of the performance limitation of DPSK systems due to nonlinear phase noise," *IEEE Photonics Technology Letters*, 15, 320–322, 2003.
23 S. Zhang, P. Y. Kam, J. Chen, and C. Yu, "Receiver sensitivity improvement using decision-aided maximum likelihood phase estimation in coherent optical DQPSK system," In ***Conference** on Lasers and Electro-Optics/Quantum Electronics and Laser Science and Photonic Applications Systems Technologies, Technical Digest (CD) (Optical Society of America, 2008)*, paper CThJJ2.
24 P. Y. Kam, "Maximum-likelihood carrier phase recovery for linear suppressed-carrier digital data modulations," *IEEE Transactions on Communications*, COM-34, 522–527, June 1986.
25 E. M. Cherry and D. A. Hooper, *Amplifying Devices and Amplifiers*, John Wiley: New York, 1965.
26 E. Cherry and D. Hooper, "The design of wide-band transistor feedback amplifiers," Proceedings of the IEEE, 110(2), 375–389, 1963.
27 N. M. S. Costa and A. V. T. Cartaxo, "Optical DQPSK system performance evaluation using equivalent differential phase in presence of receiver imperfections," *IEEE Journal of Lightwave Technology*, 28(12), 1735–1744, June 2010.
28 I. Hodgkinson "Receiver analysis for optical fiber communications systems," *IEEE Journal of Lightwave Technology*, 5(4), 573–587, 1987.
29 http://en.wikipedia.org/wiki/ENOB-cite_note-3, access date, September 2011.
30 N. Stojanovic, "An algorithm for AGC optimization in MLSE dispersion compensation optical receivers," *IEEE Transactions on Circuits and Systems I*, 55, 2841–2847, 2008.
31 G. L. Abbas, V. Chan, and TK Yee, "A dual-detector optical heterodyne receiver for local oscillator noise suppression," *IEEE Journal of Lightwave Technology*, LT_3(5), 1110–1122, 1985.
32 B. Kasper, C. Burns, J. Talman, and K. hall, "Balanced dual detector receiver for optical heterodyne communication at Gb/s rate," *Electronics Letters*, 22(8), 413–414, April 10, 1986.
33 S. Alexander, "Design of wide-band optical heterodyne balanced mixer receivers," *IEEE Journal of Lightwave Technology*, LT-5(4), 523–537, April 1987.
34 L. N. Binh, *Digital Optical Communications*. CRC Press: Boca Raton, FL, Chapter 4, 2008.
35 Phillips AG, *Phillips Handbook of Semiconductor*, S10, 1987.

36 E. M. Cherry and D. E. Hooper, *Amplifying and Low Pass Amplifier Design*. John Wiley: New York, Chapters 4 and 8, 1968.

37 S. Alexander, "Design of wideband optical heterodyne balanced mixer," *IEEE Journal of Lightwave Technology*, LT-5(4), 523–537, 1987.

38 R. Sterlin, R. Battiig, P. Henchoz, and H. Weber, "Excess noise suppression in a fiber optic balanced heterodyne detection system," *Optical and Quantum Electronics*, 18, 445–454, 1986;

39 (a) J. Kahn and E. Ip, "Principles of digital coherent receivers for optical communications," *Proceedings of OFC 2009*, San Diego, CA, paper OTuG5, March 2009; (b) C. Zhang, Y. Mori, K. Igarashi, K. Katoh, and K. Kikuchi, "Demodulation of 1.28-Tbit/s polarization-multiplexed 16-QAM signals on a single carrier with digital coherent receiver," *Proceedings of OFC 2009*, San Diego, CA, Paper OTuG3, March 2009; (c) L. N. Binh, *Digital Optical Communications*. CRC Press, Taylor & Francis Group: Boca Raton, FL, Chapter 11, 2008.

40 G. L. Abbas, V. Chan, and T. K. Yee, "A dual-detector optical heterodyne receiver for local oscillator noise suppression," *IEEE Journal of Lightwave Technology*, LT_3(5), 1110–1122, October 1985.

41 J. Kahn and E. Ip, "Principles of digital coherent receivers for optical communications," *Proceedings of OFC 2009*, San Diego, CA, paper OTuG5, March 2009.

42 C. Zhang, Y. Mori, K. Igarashi, K. Katoh, and K. Kikuchi, "Demodulation of 1.28-Tbit/s polarization-multiplexed 16-QAM signals on a single carrier with digital coherent receiver," *Proceedsings of OFC 2009*, San Diego, CA, Paper OTuG3, March 2009.

43 L. N. Binh, *Digital Optical Communications*. CRC Press, Taylor & Francis Group: Boca Raton, FL, Chapter 11, 2008.

44 R. C. Alferness, "Guided wave devices for optical communication," *IEEE Journal of Quantum Electronics*, QE-17, 946–959, 1981.

45 W. A. Stallard, A. R. Beaumont, and R. C. Booth, "Integrated optic devices for coherent transmission," *IEEE Journal of Lightwave Technology*, LT-4(7), 852–857, July 1986–.

46 G. Nicholson; T.D. Stephens, "Performance analysis of coherent optical phase-diversity receivers with DPSK modulation," *IEEE Journal of Lightwave Technology*, 7(2), 393–399, February 1989.

47 S. Savory and Takis Hadjifotiou, " Laser linewidth requirements for optical DQPSK optical systems," *IEEE Photonics Technology Letters*, 16(3), 930–932, March 2004.

48 K. Iwashita and T. Masumoto, "Modulation and detection characteristics of optical continuous phase FSK transmission system," *IEEE Journal of Lightwave Technology*, LT-5/4, 452–462, 1987 (Fig.1).

49 K. Iwashita and T. Masumoto, "Modulation and detection characteristics of optical continuous phase FSK transmission system," *IEEE Journal of Lightwave Technology*, LT-5/4, 452–462, 1987 (Fig.2&3).

50 H. Tran, F. Pera, D. S. McPherson, D. Viorel, and S. P. Voinigescu, "6-kΩ, 43-Gb/s differential transimpedance-limiting amplifier with auto-zero feedback and high dynamic range," *IEEE Journal of Solid State Circuits*, 39(10), 1680–1689, 2004.

51 Inphi inc., Technical information on 3205 and 2850 TIA device, 2012.

52 (a) S. Zhang, P. Y. Kam, J. Chen, and C. Yu, "A comparison of phase estimation in coherent optical PSK system," *Photonics Global'08*, Singapore, Paper C3-4A-03, December 2008. (b) S. Zhang, P. Y. Kam, J. Chen, and C. Yu, "Adaptive decision-aided maximum likelihood phase estimation in coherent optical DQPSK system," *OptoElectronics and Communications Conference (OECC)'08*, Sydney, Australia, Paper TuA-4, 1–2, July 2008.

53 I. Hodgkinson "Receiver analysis for optical fiber communications systems," *IEEE Journal of Lightwave Technology*, 5(4), 573–587, 1987 (Fig. 1&2).

54 B. N. Mao, et al., "Investigation on the ENOB and clipping effect of real ADC and AGC in coherent QAM transmission system," *Proceedings of ECOC 2011*, Geneva, 2011.

6 DSP-Based Optical Coherent Receiver for SPC

Processing and SPC Systems Performance

CONTENTS

6.1 INTRODUCTION

Digital signal processing (DSP) is a principal functionality of modern optical communication systems, especially in optical coherent communication techniques that are employed to mitigate the frequency mismatch between the carrier frequency and that of the local oscillator laser. DSP systems can be employed at both the coherent optical receivers and the transmitter.

DOI: 10.1201/9781003426677-6

As space photonic communications (SPC) takes advantage of well-developed optical and photonic technology that has matured for fiber optic terrestrial systems and networks, integrated photonic technologies, selected photonic components, and advanced techniques such as modulation, processing, and reception are to be used. Recent advanced techniques are digital signal processing algorithms for both the transmitter and receiver sides, which are so important for various functionalities to mitigate impairments. SPC also faces impairments but with different characteristics. In particular, spatial spreading, air turbulence, and the Doppler frequency shift require special attention in digital signal processing.

At the transmitter, DSP is employed for shaping the pulse to sharpen the passband and, hence, the sampling of the sequence, for compensating for the limited bandwidth of the digital-to-analog converter (DAC) as well as for the transfer characteristics of the electro-optic external modulators, pre-distortion, etc. At the receiver, DSP can perform de-skewing of the electrical path differences between the in-phase and quadrature phase signals, which often occurs in ultra-high-speed systems, mitigation of distortion, compensation for the frequency mismatch between the carrier and the local oscillator laser, mitigation of delay differences of the lightwave rays of different paths due to alignment angles, etc. It is noted that the mixing of the received signals and the local oscillator (LO) laser happens in the photo-detection stage. Details of such coherent reception and complex modulations are described in Chapter 5.

Due to the square law detection process in optoelectronic conversion, this mixing produces several terms: two square terms related to the intensity of the optical signals and the intensity of the local oscillator laser, and two product terms resulting from the beating between the optical signal beam and that of the local oscillator. This process is resultant from the square law detection of the sum of the two lightwaves fields of the local oscillator (LO) laser and the modulated signal field. The square of this summed field leads to three terms, two square terms which are the the DC components; and one crosss product at frequency of a sum of two fields which is very high as compared to the passband of the receiver so not detected and one cross produt term oscillation at the different frequency of the lightwaves which is detected by the receiver. Thus, only the difference product between these two optical frequencies, the signal frequency and that of the local oscillator laser When the optical frequencies of the signals and the oscillator are the same, the detection poess is a homodyne coherent reception. Detailed derivations of the coherent reception can be found in Chapter 5.

Thus, if perfect homodyne mixing is achieved in the photo-detection device, the detected signals, now in the electrical domain, are in baseband superimposed with noise from the detection process. After electronic preamplification and sampling by an analog-to-digital converter (ADC), the signals are in the discrete domain and are processed by digital signal processors, similar to the processing in wireless communication systems. The main difference lies in the physical processes of distortion, dispersion, nonlinear distortion, clock recovery, and the broadband properties of optical modulated signals transmitted over long-haul optical fibers or short-reach scenarios. In SPC, broadband signals are also used, and the carrier power level must be at least 6 dB above the signal band so that the lightwave carrier provides the coherency needed for long-distance transmission. The distance can reach 43,000 km for GEO-GEO or MEO-MEO intersatellite distances. A coherent homodyne reception can offer an additional 20 to 30dB higher than direct intensity detection. Thus, this additional gain can enhance the compensation of the optial transmission loss. However, as we can see at the later stage when considering the gain of the optical preamplifier, the amplifier may be operating in the linear region andsome additional ASE (amplification stimulated emission) noises may be addional.

On the other hand, at the receiver end, DSP follows the ADC with the digitized symbols and processes the noisy and distorted signals using algorithms to compensate for linear distortion impairments such as chromatic dispersion (CD), which is equivalent to beam spatial spreading in the Space Photonic Communication (SPC) medium, polarization mode dispersion (PMD), nonlinear self-phase modulation effects (SPM) generated at the short-length output of a high-power optical booster amplifier (HPA), clock recovery, and local oscillator frequency offset with respect to the relieved channel carrier, etc. A possible flow of sequences in DSP can be illustrated in Figure 6.46 for a modulated channel with in-phase and quadrature components and polarization multiplexing, as recovered in the electronic domain after the optical processing front end, as shown in the initial input boxes. Thus, there are two pairs of electrical signals output from the two polarized channels, with one pair each for the in-phase and quadrature components. Interprocessing of these pairs of signals can be implemented using 2×2 MIMO (multiple input multiple outputs) techniques, which are well known in the digital processing of wireless communications signals.[1]

After the optoelectronic detection of the mixed optical signals, the electronic currents of the PDM channels are amplified using a linear transimpedance amplifier and, if necessary, a further amplification stage to boost the electrical signals to a level where the ADC can digitize them into their equivalent discrete states. Once the digitized signals are obtained, the first task is to ensure that any delay differences between the channels and the in-phase and quadrature components in the electronic and digitization processes are eliminated through de-skewing. Subsequently, all equalization processing can be carried out to compensate for any distortion of the signals due to propagation through the optical transmission line. Furthermore, nonlinear distortion effects on the signals can also be addressed at this stage. Then, clock recovery and carrier phase detection can be implemented, followed by the processes of symbol recovery, decoding, and performance evaluation.

Regarding noise in optical detection, if coherent detection is used, the noise is primarily dominated by shot noise generated by the local oscillator power. However, if direct detection is employed, the noise is mainly quantum shot noise contributed by the signal power, and hence is signal-dependent. These noises will affect the convergence of algorithms. Sections 6.2.2 and 6.3.1 address the methodology.

The Doppler frequency shift can reach 7.5 GHz due to the rotational speed of satellites, such as between LEO and GEO. This frequency shift can be automatically adjusted by the receiver DSP system or pre-shifted at the transmitter side. Note that the frequency shift can occur continuously as the satellites rotate. Therefore, it is preferable to compensate for such shifts at the receiver. A DSP system integrated into the receiver, using a phase recovery algorithm, can handle this shift, allowing the reception to be considered pure homodyne detection. Carrier phase recovery can distinguish the Doppler shift from the original carrier phase of the clock pulses. This process is done continuously, and most DSP-based integrated coherent receivers have such algorithms to ensure that coherent detection meets homodyne reception standards. Once the carrier phase and thus the clock of the pulse sequence are determined, a feedback signal can be used to tune the LO laser or feed into the optical phase-locked loop (OPLL) to adjust the laser. Furthermore, such phase recovery can also tune the LO embedded in the ICPTRx (integrated coherent photonic transceiver – see Chapter 5).

Figure 6.1 shows the schematic structure, including the compact integrated form QSFP (Quad Small Form-factor Pluggable), of the dual-polarized optical transmitter and coherent reception of complex IQ modulation formats. DSP systems are integrated into both the transmitter subsystem and the receiver processing subsystem. The carrier laser and the local oscillator laser can be either integrated or external. these lasers are tunable to set their frequencies within the spectral window. The IQ modulator consists of two polarized optical Mach-Zehnder interferometric modulators (MZIMs) arranged in parallel that are biased and driven by data sequences in the electrical domain. These modulators are implemented using Si-integrated photonics.[2] Similarly, the optical hybrid structure detects and splits the polarized channels and the in-phase and quadrature components before feeding them to the balanced detector pair, and then to the ADC and DSP for processing the polarized signals. Note: the lightwaves of the carrier and LO. The DSP can be either a single unit or two separate units dedicated to the transmitter and the receiver. Algorithms for processing signals are assigned and described in the next sections.

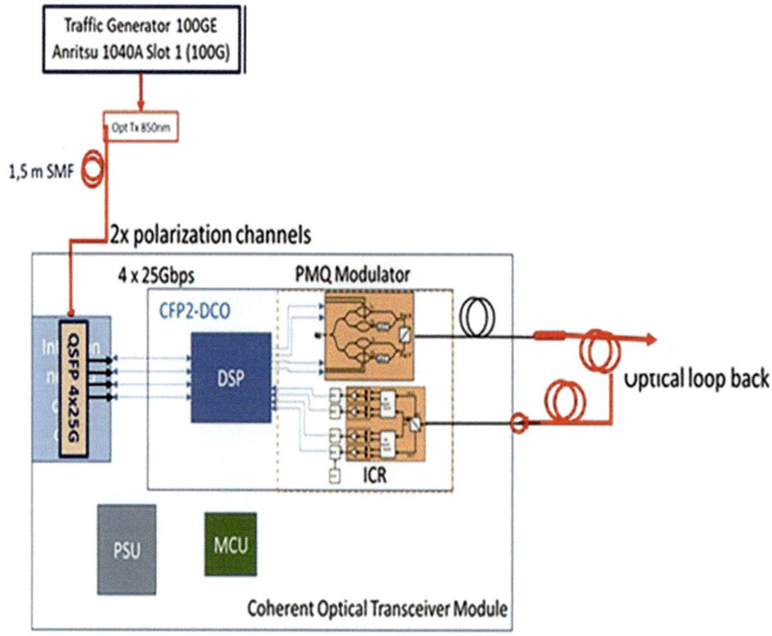

FIGURE 6.1 Schematic of dual-polarized channels transmitting and reception using coherent reception. Insert: QSFP module all integrated optics and DSP module. CFP2=compact form-factor, ICR=integrated coherent receiver, PMQ=polarization quad modulator.

This chapter is thus organized as follows. Section 6.2 provides a basic background on equalization using transversal filtering and zero-forcing, with and without feedback in the linear sense, describing the linear equalizer. When a decision is made in a nonlinear sense and feedback is applied to the input sequence based on certain criteria, the equalization becomes nonlinear, and the equalizer is classified as a nonlinear equalizer or decision-directed equalizer. Tolerance to noise is also discussed, and a simple quality factor for system performance is derived.

Clock recovery is important for determining the timing for sampling the received sequence, and this technique is briefly discussed. Determining the reference phase for coherent reception is also critical for evaluating various quadrature-modulated transmitted signals. Therefore, carrier phase recovery techniques using digital signal processing are described, especially in cases where there is a significant offset between the LO and the lightwaves carrying the signals. This is described in Section 6.5.

When the maximum likelihood sequence estimation (MLSE) is used in the nonlinear equalizer, the equalization can be considered optimal but requires substantial memory. We dedicate Section 6.3 to this algorithm and provide examples to demonstrate the effectiveness of MLSE in minimum shift keying (MSK) self-homodyne coherent reception transmission systems under the influence of linear distortion effects, with the modulation scheme being MSK. This is detailed in Section 6.6.

We can see that DSP algorithms are critical in modern DSP-based optical receivers or transponders. This chapter introduces the fundamental aspects of these algorithms for optical transmission systems over dispersive channels, whether under linear or nonlinear distortion effects.

A flow of the digital processing of received signals is shown in Figure 6.2. First, the polarized channels enter the hardware hybrid coupler, where the polarization is separated into two optical paths, each by a separate polarized optical waveguide. The in-phase and quadrature-phase components are separated by a $\pi/2$ optical phase shifter, which is a thermal shifter embedded in a Si waveguide. This results in four optical outputs from the optical hybrid coupler: two IQ outputs for each polarized channel.

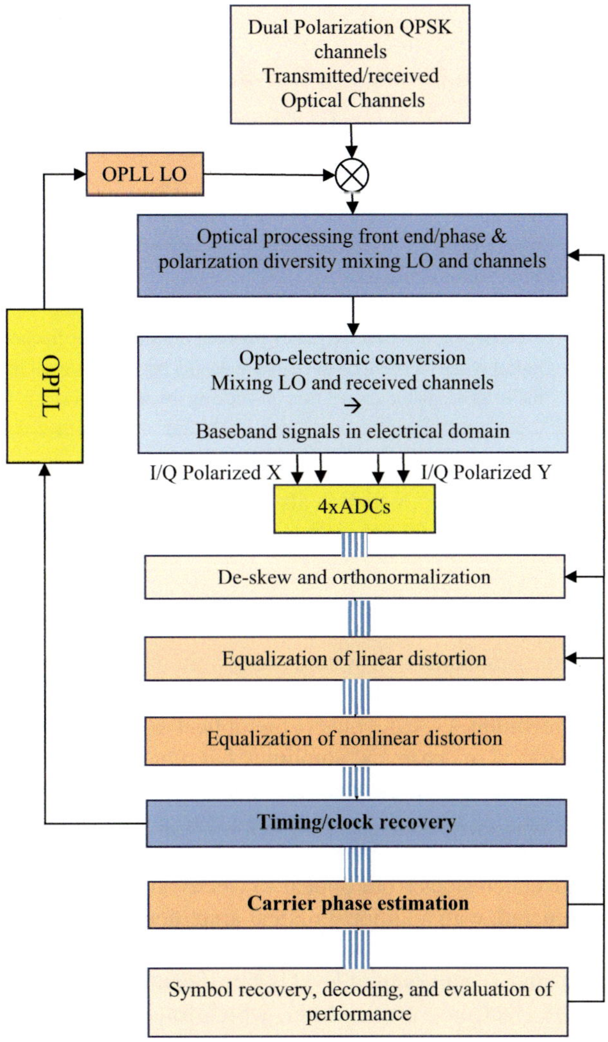

FIGURE 6.2 Flow of functionalities of DSP processing in a QAM coherent optical receiver with possible feedback control. LO = local oscillator, OPLL = optical phase lock loop, ADC = analog-to-digital converter.

The IQ optical modulator outputs of the polarized channels are fed into a balanced photodetector pair for balanced detection, providing an additional 3 dB gain. The differential outputs from this balanced detection are then amplified by a transimpedance amplifier. These signals are sampled via ADC systems and then entered into the SPC system, where the algorithm software executes processing functionalities such as carrier phase recovery for clock recovery, compensation for signal distortion, etc.

The functionalities of each processing block in Figure 6.2 can be categorized as shown in Table 6.1.3

Polarization channels can propagate with different propagation vectors, resulting in different delay times. Thus, the skew times can be serious, and can misalign the sampling points, thereby increasing bit error. To de-skew these different delay times, an algorithm can be used to mitigate or correct the effects of skew.

All algorithms are implemented in the receiver DSP system. The amount of storage required for processing is typically not higher than the maximum length of storage at high sampling speed and

TABLE 6.1
Functionalities of Subsystem Processing in a DSP-Based Coherent Receiver

De-skewing	Alignment of in-phase and quadrature components or polarized channel components temporarily or any different propagation time due to electrical connections
Orthogonalization	Ensuring independence or de-correlation between channels
Normalization	Standardization of amplitude of components to a maximum value of unity
Equalization	Compensation for impairments due to physical effects or imperfections of subsystems
Interpolation	Correction of timing error
Carrier phase estimation	Compensation for phase errors of the carrier
Frequency offset	Correction of the offset frequency between LO and carrier frequencies
OPLL	Optical phase locked loop to ensure matching between channel lightwave carrier and that of the local oscillator carrier. (see the locking mechanism given in Chapter 5)

commonly not higher than 4.0Mbytes, which is also sufficient to evaluate the BER to 10^{-7} at very high sampling rate for the 100 G DP-QPSK modulation format.

6.2 GENERAL ALGORITHMS FOR OPTICAL COMMUNICATIONS SYSTEMS

Indeed, in coherent optical communication systems incorporating DSP, the reception of transmitted signals by mixing them with a local oscillator whose frequency is identical or close to it – using homodyne or intradyne techniques – can convert the optical signals back to the baseband. These signals, which can be distorted in phase and amplitude but remain complex-valued as they were at the input of the transmitter, are now in the electrical domain. Consequently, one can apply several processing algorithms developed for high-speed modems[4] or algorithms used in wireless communication systems.[5]

Naturally, several conversion steps from optical to electrical and vice versa at the receiver and transmitter must be conducted, with additional optical amplifier noise accumulated along the optical transmission line and electronic noise.

In this chapter, we assume that the signal level is significantly higher than the electronic noise level at the output of the electronic preamplifier, which follows the photodetector. Therefore, the processing algorithms are performed in the baseband electrical digital domain after the ADC stage.

6.2.1 Linear Equalization (LE)

Coherent optical communication systems can be considered synchronous serial systems. When transmitting optical signals over uncompensated optical fiber spans in long-haul multi-span links or uncompensated metropolitan networks at very high bit rates, commonly now in the second decade of the 21st century, at 25 GSy/s or 28–32 GSy/s, including forward error coding overload, the received signals are distorted due to linear and nonlinear dispersive effects, leading to ISI (intersymbol interference) in the baseband after coherent homodyne detection at the receiver. In Chapter 5, we have described in detail the processes of coherent detection, especially homodyne reception for optical signals, including noise processes. Over non-dispersion-compensating multi-span fiber links, the most important signal distortions are linear chromatic dispersion and polarization mode dispersion, especially when PDM (polarization division multiplexing) is employed with two orthogonal polarization channels transmitted simultaneously. ISI naturally occurs due to the band-limiting of the signals over the limited bandwidth of the long fiber link, as described in Chapter 2 on the transfer function of optical fibers. Other nonlinear distortion effects have also been included in the transfer function, indicating the unevenness of the passband. These ISI effects degrade the performance of optical communications and need to be addressed with equalization processes, which will be discussed in this chapter.

The optical coherent detection processes and DSP of sampled digital signals can be classified into two separate groups.

The first of these processes involves feeding the received sampled digital signals through an equalizer that corrects the distortion introduced by chromatic dispersion and PMD (polarization mode dispersion) effects, restoring the received signals to a copy of the transmitted signals in the electrical domain. In SPC, the polarization can rotate due to the movement of the transmitting and receiving satellites. The received signals are then detected in the conventional manner applied to any serial digital signals in the absence of ISI. In other words, the equalizer acts as the inverse of the channel, so that together, the equalizer and the channel introduce no signal distortion, and each data symbol is detected as it arrives, independently of each other. The equalizers can be either linear or nonlinear.

For SPC, linear distortion is mainly due to spatial spreading, which results in differential delay between the rays of the beam propagating through space. Additionally, the angle of incidence of the beam on the receiver optics can vary over long distances due to the relative rotation of satellites, as the satellite transmitters and receivers may rotate at different speeds depending on their orbits.

Nonlinear distortion in SPC arises from the fiber connecting the booster amplifier output to the optical head, where the high power is strong enough to cause nonlinear distortion in the short fiber. Crosstalk and distortion can be compensated for during the digital processing stage.

In the second group of detection processes, the decision process is modified to account for the signal distortion introduced by the channel, and no attempt needs to be made to reduce signal distortion prior to the actual decision process. Although no equalizer is required in this case, the decision process may be considerably complex, necessitating complex algorithms to handle this type of decision processing.

In addition to these two decision groups, there are additional issues to address, including the synchronization of sampling at the receivers, clock recovery, and carrier phase estimation. This ensures that a reference clocking instant can be established and that the referenced phase of the carrier can be used to detect the phase states of the received samples. Concurrently, any skew or delay differences between the *I*- and *Q*-components, as well as the polarized multiplexed channels, must be detected and corrected to maintain synchronization in the coherent detection processes. Although these physical processes are controlled by software, their practical implementation is not covered in this chapter.

Furthermore, any delay differences due to varying propagation times of the polarized channels or electrical connection lines require a de-skewing process to eliminate these effects, as indicated in Figure 6.46.

Thus, this section introduces the linear equalization process.

6.2.1.1 Basic Assumptions

In general, a simplified block diagram of the coherent optical communication system can be represented in the baseband, i.e., after optical field mixing and detection by opto-electronic receivers, and after E/O (electrical to optical) conversion at the transmitter, as shown in Figures 6.3 and 6.4. The input to the transmitter is a sequence of regularly spaced impulses, assuming that the pulse width is much shorter than the symbol period. This sequence can be represented by $\sum_i s_i\delta(t-iT)$ with s_i of binary states taking both positive and negative values as the amplitude of the input signal sequence. Thus, we can write

$$s_i = \pm k; \qquad k > 0 \tag{6.1}$$

Thus, the impulse sequence is of the binary polar type. This type of signal representation is advantageous in practice as it can be used to modulate optical modulators, especially the Mach-Zehnder interferometric modulator (MZIM) or I-Q modulator, as described in Chapter 3. This allows the phase of the carrier under the modulated envelope to take positive or negative phase angles, which can be represented on the phasor diagram. In practice, the impulse can be replaced by rectangular, rounded, or Gaussian waveforms by modifying the transmitter filter, which can be a Nyquist raised cosine type to tailor the transmitting pulse shape.

It is further noted that the phase of the optical waves and the amplitude modulated by the *I*- and *Q*-MZIMs in the two paths of the I-Q modulator ensure the negative and positive positions of the *I*- and *Q*-components on the real and imaginary axes of the constellation, as shown in Figure 6.3 (Figure 3.42 of Chapter 3, 4-ASK-1 and 4-ASK-2). Under coherent reception, these negative and positive amplitudes are recovered as they are at the transmitter, though with superimposed noise. It is crucial to synchronize the modulation signals fed to the two MZIMs to avoid degradation of the sampling instant due to skew.

6.2.1.2 Zero-Forcing Linear Equalization (ZF-LE)

Consider a transmission system and receiver subsystem as shown in Figure 6.4 n which the transmitter transfer function can be represented by $H_{Tx}(f)$, the transfer function of the channel in the linear domain by $H_C(f)$, and the transfer function of the equalizer or filter by $H_{Eq}(f)$. These must meet the condition that the overall transfer function of all subsystems equates to unity.

$$H(f) = H_{Tx}(f) H_C(f) H_{Eq}(f) e^{j\omega t_0} \tag{6.2}$$

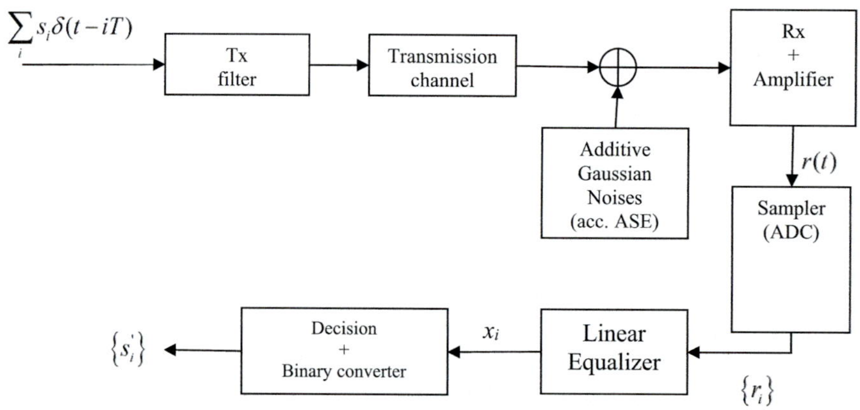

FIGURE 6.3 Simplified schematic of a coherent optical communication system. Note equivalent baseband transmission systems due to baseband recovery of amplitude and phase of optically modulated signals.

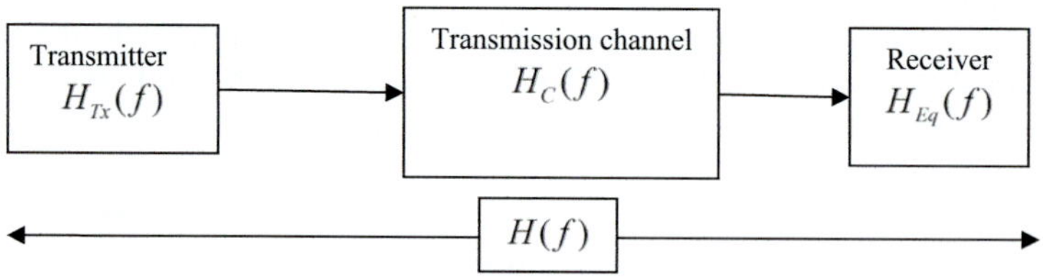

FIGURE 6.4 Schematic diagram of a transmission system, including a zero-forcing equalizer/filter.

Thus, from this equation, if the desired overall transfer response follows that of a Nyquist raised-cosine filter, the equalizer frequency response can be obtained as follows

$$H_{\mathrm{Eq}}(f)=\frac{\mathrm{Traisecos}\left(\frac{f}{1/T},\rho\right)}{H_{\mathrm{Tx}}(f)H_C(f)}e^{j\omega t_0} \tag{6.3}$$

With the raised cosine function defined with β as the roll-off factor, the frequency response of the equalizer is given by

$$H_{\mathrm{Eq}}(f)=\mathrm{Raise\_cos}\left(\frac{f}{1/T},\beta\right)=\begin{cases}1 & 0\le|x|\le\frac{1-\beta}{2}\\ \cos^2\left(\frac{\pi}{2}\frac{|x|-\frac{1-\beta}{2}}{\beta}\right) & \frac{1-\beta}{2}\le|x|\le\frac{1+\beta}{2}\\ 0 & |x|>\frac{1+\beta}{2}\end{cases} \tag{6.4}$$

This implies that the filter is a Nyquist filter of raised cosine shape; ω is the signal baseband frequency, and t_0 is the time at the observable instant. In the case where two channels are observed simultaneously, the right-hand side (RHS) of Equation 6.4 is a unity matrix, and the unitary condition must be satisfied, which ensures the conservation of energy.

From Equation 6.3, the magnitude and phase response of the equalizer can be derived. This condition guarantees the suppression of ISI, thus the equalizer can be referred to as a linear equalizer (LE) with zero-forcing (ZF) filter (LE-ZF).

6.2.1.3 ZF-LE for Fiber as Transmission Channel

The ideal zero-forcing equalizer is simply the inverse of the channel transfer function given in Equation 6.3. Thus, the required equalizer transfer function is the Hermitian of the channel transfer function, such that their product equals unity. Therefore, $H_{\mathrm{Eq}}(f)$ can be written as

$$H_{\mathrm{Eq}}(f)=H_{\mathrm{SSMF}}^{-1}(f)=e^{j\alpha\omega^2} \tag{6.5}$$

With ω as the radial frequency of the signal envelope and α as the parameter dependent on dispersion, wavelength, and the velocity of light in vacuum as described in Chapter 4. However, this transfer function does not maintain conjugate symmetry; that is

$$H_{\mathrm{Eq}}(f)=e^{j\alpha\omega^2}\neq H_{\mathrm{Eq}}^{-1}(f) \tag{6.6}$$

This is due to the square-law dependence of the dispersion parameter of the fiber on frequency.

Thus, the impulse response of the equalizing filter is complex. Consequently, this filter cannot be realized by a baseband equalizer using only one baseband received signal, which explains the limited capability of LEs employed in direct-detection receivers to mitigate chromatic dispersion. On the other hand, this also explains why fractionally spaced LEs, which are used within a coherent receiver, can potentially extend the system's reach to distances limited only by the number of equalizer taps.[6] Under coherent detection, and particularly with QAM formats, both real and imaginary parts are extracted separately and processed digitally. The schematic of a fractionally spaced finite impulse response (FIR) filter is shown in Figure 6.5 where the delay time between taps is only a fraction of the bit period. The weighting coefficients $W_0(n)$ … $W_{N-1}(n)$ are determined and then multiplied with the signal paths $S(n)$ at the outputs of the taps. These outputs are then summed and

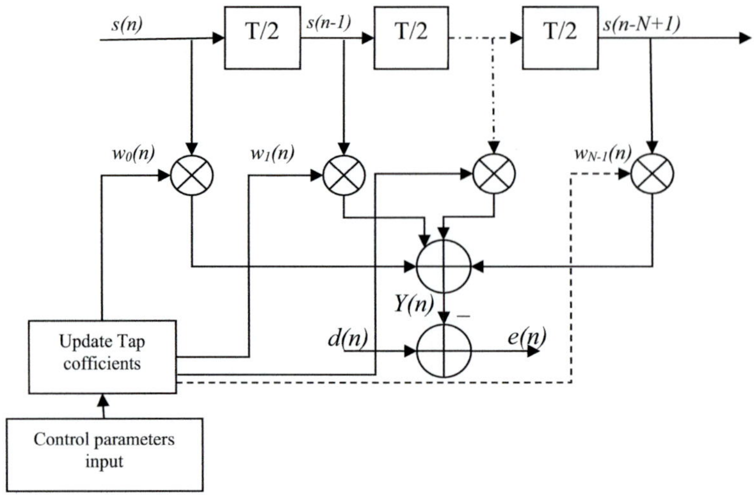

FIGURE 6.5 Schematic of a fractionally spaced FIR (*finite* impulse response) filter.

subtracted from the expected filter coefficients $d(n)$ to obtain the estimation errors. This process is iteratively operated till the final output sequence $Y(n)$ is close to the desired response.

Under ZFE, at the sampling instant, the noise may be increased due to errors associated with the delay of the zero-crossing point of adjacent pulses, leading to ISI effects. As a result, the signal-to-noise ratio (SNR) may not be optimal.

The number of taps must cover the whole length of the dispersive pulses. For example, for a 1000 km SSMF transmission line and the BW of the channel being 25GHz (or ~0.2 at 1550 nm), the total pulse spreading of the pulse is 17 ps/nm/km x 1000km x 0.2 nm=3400 ps. The sampling rate would be 50GSa/s (2 samples per period @ 25 Gb/s assuming modulation results in the BW equal to symbol rate), and then one would have 3400/40 periods or ~85 periods or taps with 170 samples altogether.

Similar fiber-equivalent distortion and equivalent passband of a free-space transmission distance can be derived as briefly given in Chapter 2. It is noted here again that, for vacuum free-space transmission the distortion is mainly contributed by the spatial beam spreading and tus, the ray path different delay time, while for turbulent and vacuum free-space transmission additional multipath distortion contribute to significant signal degradation.

6.2.1.4 Feedback Transversal Filter

Transversal equalizers are considered, so far, as linear feedforward (FFE) transversal filters commonly employed in practice over several decades since the invention of g=digital communications. Equalization may alternatively be achieved by a feedback transversal filter, provided that certain conditions can be met by the channel responses; in this case, the linear transfer function of the fiber transmission line or that of the free-space vacuum or turbulent air transmission path

Let the z-transform of the sampled impulse response of the channel, including fiber field propagation and coherently detected, as

$$Y(z) = y_0 + y_1 z^{-1} + y_2 z^{-2} + \cdots + y_g z^{-g} \tag{6.7}$$

where $y_0 \neq 0$ and let $V(z)$ be the z-transform derived from the scaling of $Y(z)$ so that the first term becomes unity. Thus, we have

$$V(z) = \frac{1}{y_0} Y(z) = 1 + \frac{y_1}{y_0} z^{-1} + \frac{y_2}{y_0} z^{-2} + \cdots + \frac{y_g}{y_0} z^{-g} \tag{6.8}$$

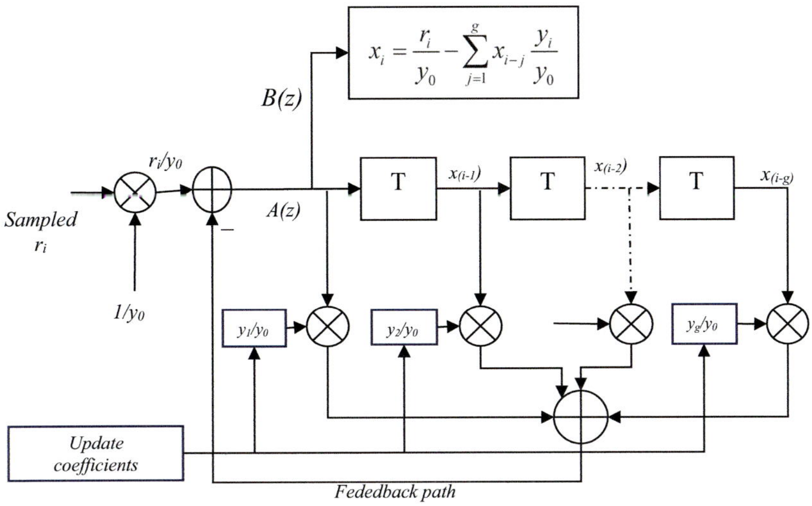

FIGURE 6.6 Schematic structure of a linear feedback transversal equalizer for a channel with an impulse response in the Z-transform of $Y(z)$.

or written in cascade-factorized form

$$V(z) = \left(1+\beta_1 z^{-1}\right)\left(1+\beta_1 z^{-1}\right) \quad \left(1+\beta_g z^{-1}\right) \tag{6.9}$$

where β_i can be real or complex-valued. The feedback transversal filter can then be configured as shown in Figure 6.6, which follows the operation just described. Obviously, the received sequence is multiplied by a factor $1/y_0$ in order to normalize the sequence, then delayed by one period with a multiplication ratio, summed up, and fedback to subtract from the normalized incoming received sequence. The output is tapped from the feedback path, as shown in the diagram, at the point where the transversal operation starts.

6.2.1.5 Tolerance to Additive Gaussian Noises

Consider a linear FFE with $(m+1)$ taps, as shown in Figure 6.7. Assuming that the equalizer has equalized the input signals in the baseband, the output of the equalizer is given as

$$x_{i+h} \simeq s_i + u_{i+h}$$
$$\text{with}\begin{cases} s_i = \pm k; & \text{binary level signal amplitude} \\ u_{i+h} = \text{noise-component} \end{cases} \tag{6.10}$$

Note that noise always adds to the signal level. For positive k, the noise u_{i+h} follows the same sign as the signal, and vice versa for negative values.

The output noise signal of the FFE can thus can be written as

$$u_{i+h} = \sum_{j=0}^{m} w_{i+h-j} c_j \tag{6.11}$$

The noise components $\{w_i\}$ are statistically independent Gaussian random variables with zero mean and variances σ^2 such that the total accumulated noise output, given by Equation 6.11, is also an uncorrelated Gaussian random variable with zero mean and a variance.

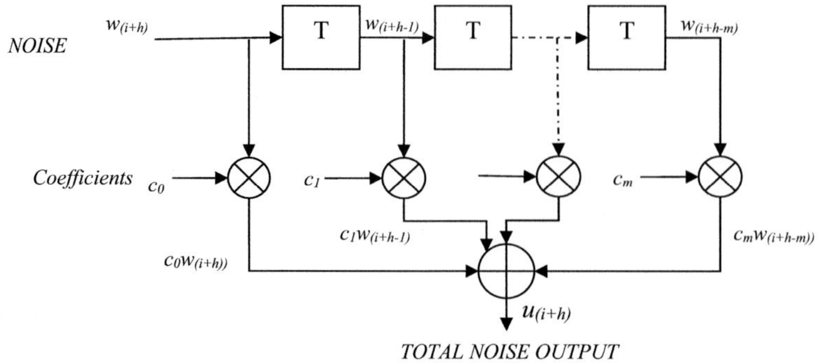

FIGURE 6.7 Noisy signals in an $(m+1)$ tap linear FFE.

$$\eta^2 = \sum_{j=0}^{m} \sigma^2 c_j^2 = \sigma^2 CC^T = \sigma^2 |C|^2 \tag{6.12}$$

where $|C|$ is the Euclidean length of the $(m+1)$ component vector $[C]=\begin{bmatrix} c_0 & c_1 & \dots & c_m \end{bmatrix}$, the sampled channel impulse response. Thus, η^2 is the mean square noise of the process and is also the mean square error of the output sampled values x_{i+h}.

Then, the probability density function of this Gaussian noise is given by

$$p(u) = \frac{1}{\sqrt{2\pi\eta^2}} e^{-\frac{u^2}{2\eta^2}} \tag{6.13}$$

Thus, with the average value of the magnitude of the received signal value of k (equalized without any distortion), the probability of error in the noisy detection process is given by

$$P_e = Q\left(\frac{k}{\eta}\right) = Q\left(\frac{k}{\sigma|C|}\right) \tag{6.14}$$

and the electrical SNR is given by

$$\text{SNR} = 20\log_{10}\left(k/\sigma\right) \text{ dB} \tag{6.15}$$

with $Q(x)$ as the quality function of x, which is the complementary error function as commonly defined. In practice, this SNR can be measured by determining the standard deviation of the constellation points and the geometric distance from the center of the constellation to the central point of the constellation. As will be shown in the next section, the equalized constellation offers a much higher SNR compared to the non-equalized constellation. Thus, under Gaussian random distribution, the noises contributing to the equalization process are also Gaussian, and the probability density function will follow straightforwardly from the Euclidean distance of the channel impulse response.

6.2.1.6 Equalization with Minimizing MSE in Equalized Signals

It has been recognized that a linear transversal filter, which equalizes and minimizes the mean square error in its output signals,[7,8,9] generally provides a more effective degree of equalization than an equalizer that minimizes peak distortion. Thus, an equalizer that minimizes the mean square difference between the actual and ideal sampled sequences at its output for a given number of taps would be advantageous in the presence of noise.

Now revisiting the transmission, the sampled impulse response of the channel, the sampled impulse response of the $(m+1)$ tap linear FFE, and the combined sampled impulse response of the channel and the equalizer, denoted as $[Y]$, $[C]$ and$[E]$ respectively, are given by

$$\begin{aligned}[Y] &= \begin{bmatrix} y_0 & y_1 & \dots & y_g \end{bmatrix} \\ [C] &= \begin{bmatrix} c_0 & c_1 & \dots & c_m \end{bmatrix} \\ [E] &= \begin{bmatrix} e_0 & e_1 & \dots & e_{m+g} \end{bmatrix}\end{aligned} \tag{6.16}$$

Under the condition that the combined impulse response of the channel and the equalizer results in the recovered output sequence being only a pure delay of h periods of the sampling period, the input sampled sequence is given by

$$x_{h+i} \approx s_i + u_{h+i} \tag{6.17}$$

or it is a pure delay superimposed by Gaussian noise from the transmission line. For free space, turbulent air, and scattering similar to fading effects in multipath wireless signals, the probability distribution function for QPSK modulation under multi-scattering delay is given by (see Chapter 2, Equation 2.92)

$$P_{e\text{-QPSK}} = \frac{(\alpha\beta)^{(\alpha+\beta)/2}}{\sqrt{\pi}\Gamma(\alpha)\Gamma(\beta)} \delta^{(\alpha\beta)^{-(\alpha+\beta)/2}} G_{2,3}^{2,2}\left[\frac{\alpha\beta}{\delta}\left| \begin{matrix} 1-(\alpha+\beta)/2,\ 1-(\alpha+\beta)/2 \\ \dfrac{\alpha-\beta}{2}, -\dfrac{\alpha+\beta}{2}, \dfrac{-\alpha+\beta}{2} \end{matrix}\right.\right] \tag{6.18}$$

For vacuum-free space, a Gaussian distribution can be assumed as the noise is mainly contributed by ASE noise. In the case of optical transmission lines, the noise consists of accumulated ASE over many spans of dispersive, optically amplified transmission spans. Thus, the resultant ideal vector at the output of the equalizer can be written as

$$[E] = \begin{bmatrix} \underbrace{0 \quad \dots \quad 0}_{h} & 1 & 0 & 0 & 0 & \dots & 0 \end{bmatrix}; \qquad h = \text{interger} \tag{6.19}$$

In fact, the equalized output sequence would be (as seen from Equation 6.19)

$$x_{h+i} \approx \sum_{j=0}^{m+g} s_{i+h} - je_j + u_{h+i};\ j = 0,1,\dots,(m+g) \tag{6.20}$$

The input sampled values are statistically independent and have an equal probability of taking values of $\pm k$, this indicates that the expected value (denoted by the symbol $\xi[\cdot]$

$$\xi\left[s_{i+h-j}s_{i+h-l}\right] = 0; \qquad \forall j \neq l \qquad \text{and} \qquad \xi\left[s_{i+h-j}^2\right] = k^2 \tag{6.21}$$

The noise superimposed on the sampled values is also statistically independent of the sampled signals and has zero mean. Thus, the sampled signals and noise are orthogonal, so we have the expected value $\xi\left[s_{i+h-j}u_{i+h}\right] = 0$

Thus, a linear equalizer (LE) that minimizes the mean square error in its output signals would also minimize the mean square value of $\left[x_{i+h} - s_i\right]$, which can be written as

$$\xi\left[\left(x_{i+h}-s_i\right)^2\right]=\begin{cases}\xi\left[\left(\sum_{j=0}^{m+g} s_{i+h-j}e_j+u_{i+h}-s_i\right)^2\right]\\ =k^2\left|E-E_h\right|^2+\sigma^2\left|C\right|^2\end{cases}$$

with (6.22)

$$\left|E-E_h\right|^2=\sum_{\substack{j=0\\ j\neq h}}^{m+g} e_j^2+k^2\left(e_h-1\right)^2$$

where $|E-E_h|$ and $|C|$ are the Euclidean distances of the vector $[E-E_h]$ and $C=[C]$, respectively. We can now observe that the first and second terms of Equation 6.22 represent the mean square error in x_{i+h} due to ISI and the mean square error due to Gaussian noise, respectively. Thus, we see that the MSE process minimizes not only the distortion in terms of the mean square error due to ISI but also minimizes that of the superimposed noise.

6.2.1.7 Constant Modulus Algorithm for Blind Equalization and Carrier Phase Recovery

6.2.1.7.1 Constant Modulus Algorithm (CMA)

Suppose that inter-symbol interference, additive noise, and carrier frequency offset are considered. Then a received signal before processing, $x(k)$ can be written as

$$x(k)=\sum_{i=0}^{M-1} h(i)a(k-i)e^{j\phi(k)}+n(k) \tag{6.23}$$

where $h(k)$ is the overall complex baseband equivalent impulse response of the combined transfer function, including the transmitter, unknown channel, and receiver filter. Note that all *E/O* and *O/E* conversion steps in coherent reception are removed, assuming only noise contributions without distortion in both phase and amplitude. $n(k)$ represents the overall noise, which is assumed to be Gaussian, mainly due to ASE noise over cascaded amplifiers with zero mean and a certain standard deviation. The input data sequence $a(k-i)$ is assumed to be independent and identically distributed, $\phi(k)$ is the carrier phase difference between the signal carrier and the LO laser.

Now, let the impulse response of the equalizer be$\mathbf{W}(k)$ as denoted in the schematic given in Figure 6.7. The output of the equalized signals can be obtained by

$$y(k)=\mathbf{X}^T(k)\mathbf{W}(k) \tag{6.24}$$

where $\mathbf{W}(k)=\left[w_0,w_1,\ldots,w_k\right]^T$ the tap weight vector of the equalizer as is described in the above sections of this chapter; and $\mathbf{X}(k)=\left[x(k),x(k-11),\ldots,x(k-N+1)\right]^T$ is the vector input to the equalizer with N as the length of the weight vector.

In fiber transmission, there is pure phase distortion due to chromatic dispersion effects in the linear region, causing the constellation to rotate in the phase plane. Additionally, the carrier phase introduces further spinning rotation of the constellation, as observed in Equation 6.24. To equalize the linear phase rotation effects, it is necessary to cancel the spinning effects caused by the carrier phase difference created by the mixing between the LO and the signal carrier. Therefore, the constant modulus algorithm must be combined with carrier phase recovery so that an additional phase rotation in the reverse direction can be superimposed on the phase rotation caused by channel phase distortion, fully equalizing the constellation.

The constant modulus (CM) criterion can be expressed by the non-negative cost function $J_{\text{cma},p,q}$ parameterized by positive integers p and q such that

$$J_{\text{cma},p,q} = \frac{1}{pq} E\left\{ \left| |y_n|^p - \gamma \right|^q \right\}; \qquad \text{with} \qquad \gamma = \text{constant} \tag{6.25}$$

where $E(\cdot)$ denotes the expected statistical value. The CM criterion is typically implemented as CMA 2-2, where p, $q=2$. Using this cost function, the weight vectors can be updated according to

$$\begin{aligned} \mathbf{W}(k+1) &= \mathbf{W}(k) - \mu \cdot \nabla \mathbf{J}(k) \\ &= \mathbf{W}(k) - \mu \cdot \left(y(k)|y(k)|^p |y(k)|^{p-2} \left(|y(k)|^p - R_p \right) \right) \cdot \mathbf{X}^*(k) \end{aligned} \tag{6.26}$$

with R_p being a constant that depends on the type of constellation. Since the final output of the equalizer system will converge to the original input states, Equation 6.26 can be rewritten as

$$\left(a(k)|y(k)|^p |a(k)|^{p-2} \left(|a(k)|^p - R_p \right) \right) = 0 \tag{6.27}$$

Assuming that the in-phase and quadrature-phase components of $a(k)$ are decorrelated from each other, and using the convolution of the input sequence $a(k)$ with the impulse response of the channel $h(k)$, we have $x(k) = h(k) * a(k)$, the constant R_p can be determined as

$$R_p = \frac{E\left[|a(k)|^{2p} \right]}{E\left[|a(k)|^p \right]} \tag{6.28}$$

The flow of the update procedure for the weight coefficients of the filter in the equalizer is shown in Figure 6.7.

6.2.1.7.2 Modified CMA: Carrier Phase Recovery Plus CMA

As shown in Figure 6.8, the cost function used in the CMA is phase-blind, allowing the CMA to converge even in the presence of phase error. Although this is an advantage of the CMA, at convergence, the equalizer output will exhibit a constant phase rotation. This phase rotation is commonly due to the difference in frequency between the LO and the carrier, or LOFO (LO frequency offset). Additionally, the constellation will spin at the carrier frequency offset rate (considered as phasor rotation) due to a lack of carrier frequency locking. While the phase-blind nature of the CMA is not a significant issue for constant phase rotation, it can severely degrade performance with parameters such as random polarization mode dispersion and fluctuations in the phase of the carrier, e.g., a 100 MHz oscillation of the feedback mirror in the external cavity of the LO. Therefore, carrier phase recovery is essential to determine the LOFO, the offset between the LO and the signal carrier. The block diagram of combined CMA blind equalization and carrier tracking and locking is illustrated in Figure 6.8. The carrier recovery loop uses the error between the output of the equalizer and the corresponding decision. The phase updating rule is given by $\phi(k+1) = \phi(k) - \mu_\phi I\left[z(k)e^*(k)\right]$, where μ_ϕ is the step-size parameter, $e(k)$ is the error signal given by $e(k) = z(k) - \hat{a}(k)$ with $z(k) = y(k)e^{-j\phi(k)}$ as the equalized output with phase error correction and $\hat{a}(k)$ as the estimation of $z(k)$ by a decision device. The carrier tracking loop, as described above, provides an estimate of the phase error, as shown in Figure 6.8b. A differential phase compensation (DPC) can then be implemented.

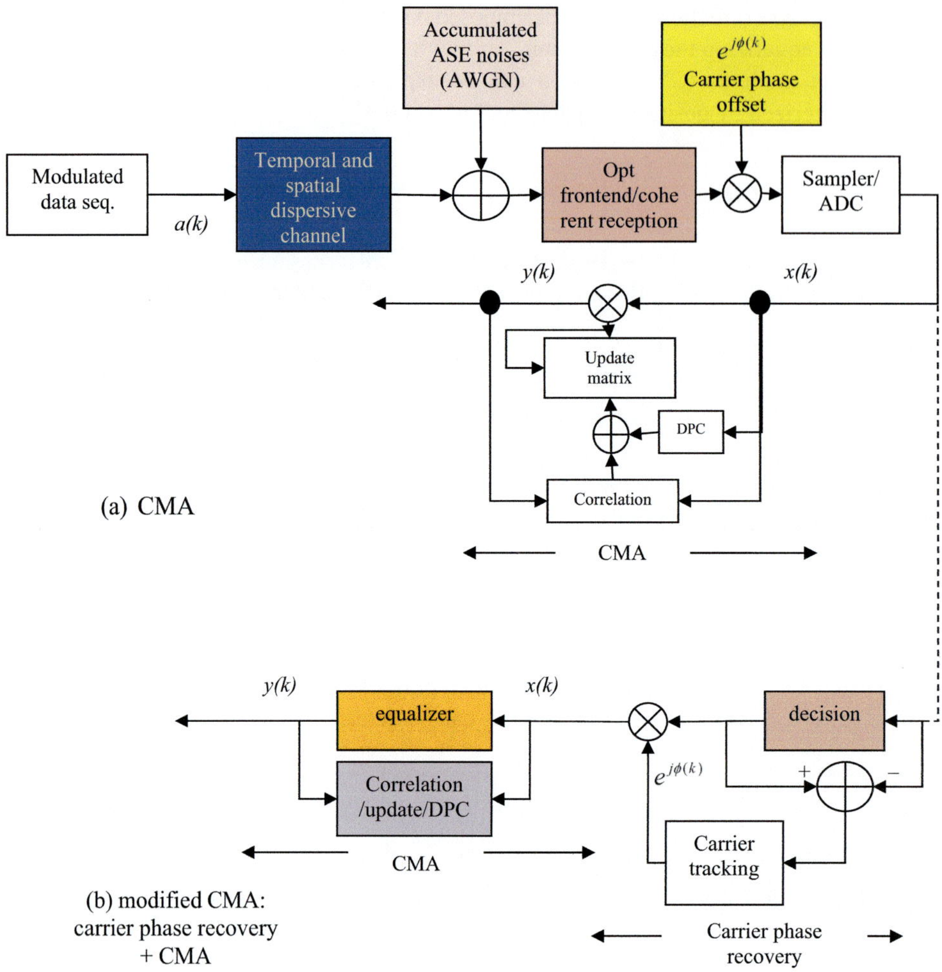

FIGURE 6.8 Equivalent model for baseband equalization (a) blind CMA and (b) modified CMA by cascading CMA with carrier phase recovery. Signal inputs come from the digitalized received samples after coherent reception. DPC = differential phase compensation.

The modified CMA is implemented by first cascading the carrier phase recovery and then the standard CMA. Typical CMA operations with carrier phase recovery are shown in Figure 6.9 for a 16-QAM constellation with optical coherent transmission and reception, as described in Chapter 4, where a 100 MHz oscillation of the LO laser is observed. We can see that the constellation of the non-equalized processing is noisy and cannot be used to determine the bit-error rate of the transmission system. However, in Figure 6.10b if only the CMA is employed, the phase rotation due to the phase offset between the LO and the signal channel carrier remains. When carrier phase recovery is applied, the constellation can be recovered, as shown in Figure 6.10c and d, with and without the decision-directed (DD) processing stage, both offering similar constellations. The DD processing will be described in the section dealing with NLEs or decision feedback equalizers (DFE) in this chapter.

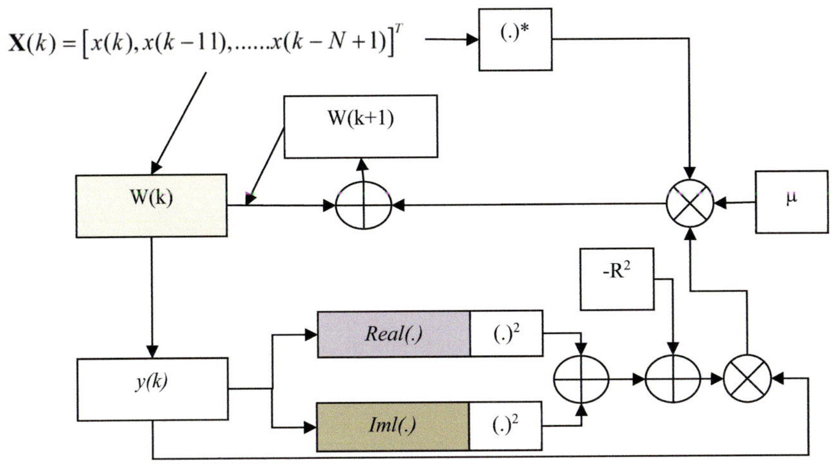

FIGURE 6.9 Constellations of 16-QAM modulated signals with a carrier frequency offset with respect to that of the local oscillator laser. (a) Unequalized output, (b) output equalized by CMA, (c) output by joint CMA and carrier phase decision directed (DD) recovery, and (d) output by modified CMA, carrier phase recovery, and CMA in cascade.

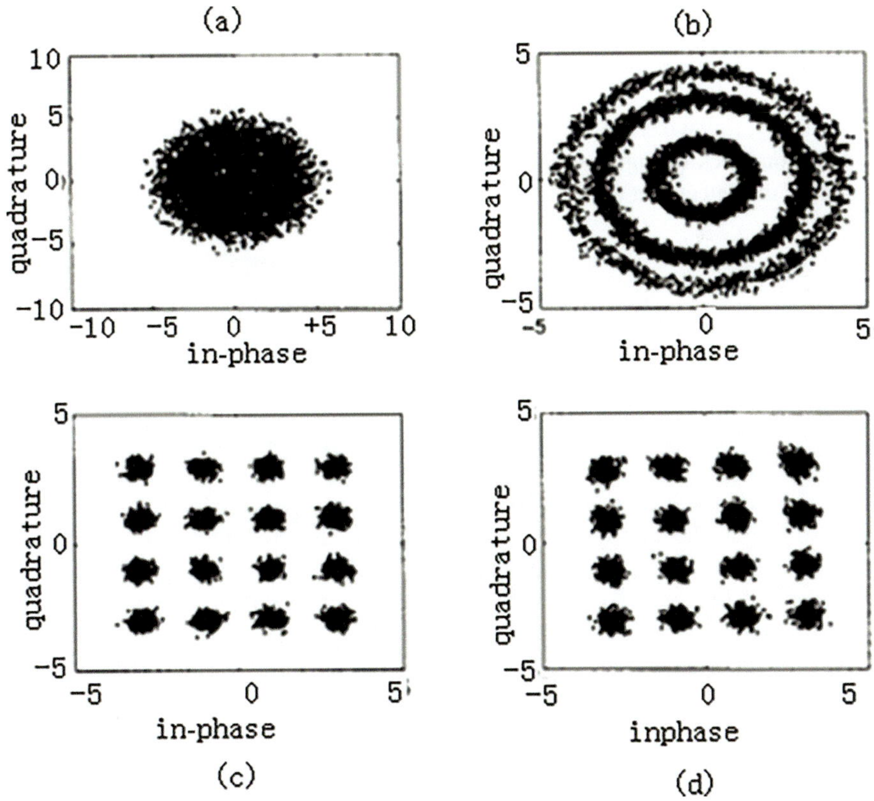

FIGURE 6.10 Block diagram of tap update with CMA cost function.

Reference[10] has reported simulations of coherent receivers using polarization-multiplexed QPSK employing CMA algorithms in both standard and modified forms, as well as original and modified decision-directed algorithms, under the effects of the finite linewidth of the laser (carrier source). The results show a certain penalty of about 0.5 dB on receiver sensitivity under coherent detection and DSP-based processing in the electronic domain, with a 1e-3 FEC level for a linewidth-to-symbol-rate product ranging from 1e-4 to 1e-3.

6.2.2 Nonlinear Equalizer (NLE) or Decision Feedback Equalizers (DFE)

6.2.2.1 Decision Directed Cancellation of ISI

It has been demonstrated in both wireless and optical communication systems over recent years that nonlinear (decision feedback) equalizers achieve better performance than their linear counterparts. The method of operation of these decision-directed equalizers for the cancellation of ISI is as follows.

Consider the transmission systems shown in Figure 6.11a, which is similar to Figure 6.11b, except that the LE is replaced with a nonlinear one. The linear feedforward (FF) equalizer or transversal filter in this NLE is shown in Figure 6.11c. The arrow directing from the decision block across the FF transversal filter indicates the updating of the filter coefficients.

Similar to the LE case, the input sampled data sequence also follows the notations given in Section 6.2 of this chapter. Multi-level data-symbol can be treated the same way without unduly affecting any of the important results in this analysis.

The pure NLE uses the detected data symbols $\{s_i'\}$ to synthesize the ISI component in a received signal $\{r_i\}$ and then removes the ISI by subtraction. This is the process of decision-directed cancellation of ISI.

Mathematically, following the notation used from Equation 6.10 for the section just before the decision/detection block of Figure 6.11a or the sampled sequence at the input of Figure 6.11b, the signals entering the detector at the instant $t=iT$ is given as

$$x_i = \frac{r_i}{y_0} - \sum_{j=1}^{g} s_{i-j}' v_j = s_i + \sum_{j=1}^{g} s_{i-j}' v_j + \frac{w_i}{y_0} - \sum_{j=1}^{g} s_{i-j}' v_j \tag{6.29}$$

So with corrected detection of each sampled value s_{i-j} such that

$$s_{i-j}' = s_{i-j} \qquad \text{for} \qquad j = 1,2,\ldots,g \quad \rightarrow x_i = s_i + \frac{w_i}{y_0} \tag{6.30}$$

So that with correct detection and equalization, we recover the sampled sequence $\{x_i\}$. Thus, it is clear that so long as the data symbols $\{s_i\}$ are correctly detected, their ISI is removed from the equalized signals by the NLE, and the channel continues to be accurately equalized.

The tolerance of this pure NLE, operating with a channel sampled impulse response whose z-transform is given by $\{Y(z)\}$, as defined in Equation 6.5. The equalized sampled signals at the detector input are given by:

$$x_i = s_i + \frac{w_i}{y_0} \tag{6.31}$$

$$s_i = \pm k; \text{ binary amplitude of input seq.}$$

It is assumed that the sampled sequence is detected as +k or −k depending whether x_i taking negative or positive values. An error would occur if the noise levels corrupt the magnitude, which implies

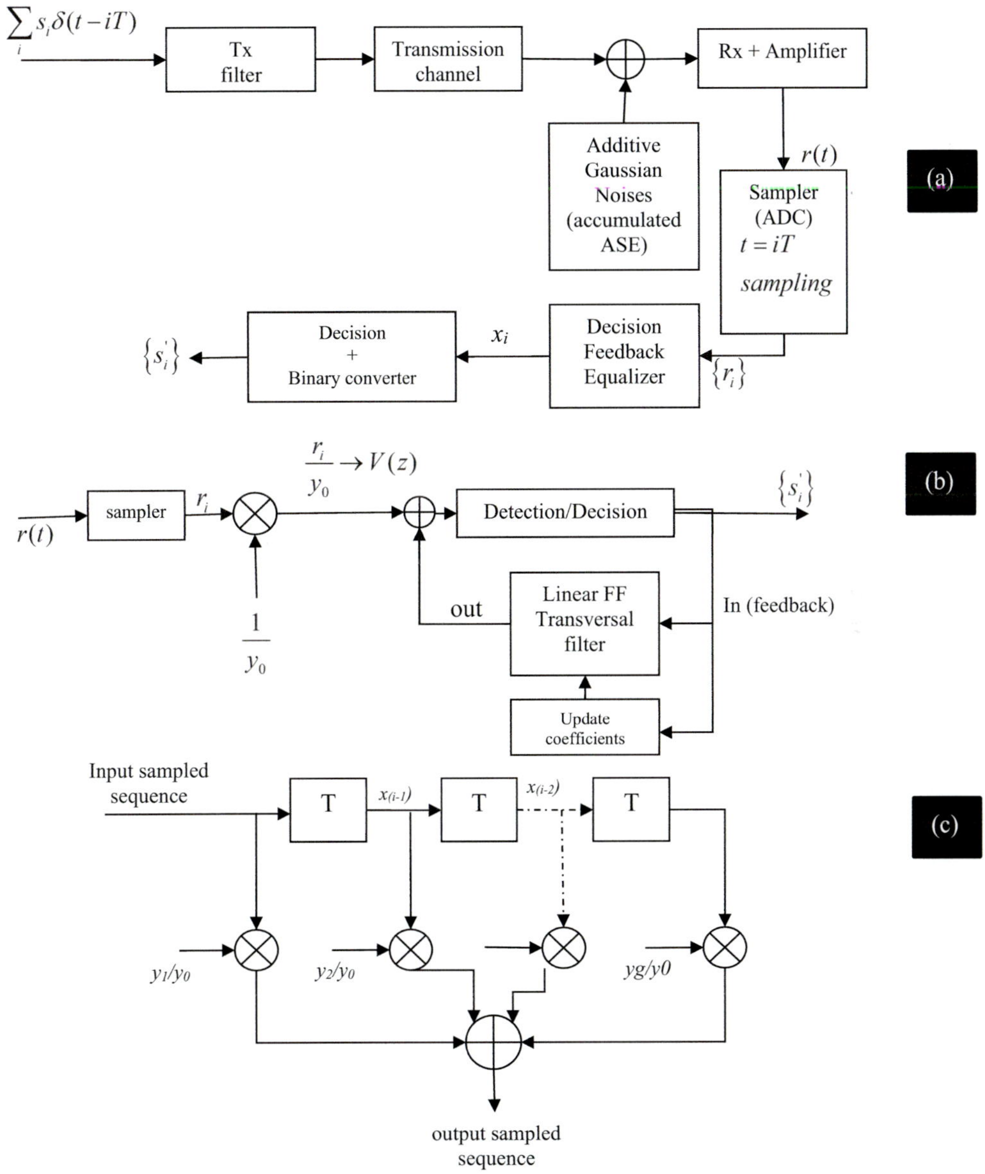

FIGURE 6.11 (a) Simplified schematic of a coherent optical communication system incorporating an NLE. Note equivalent baseband transmission systems due to baseband recovery of amplitude and phase of optically modulated signals. (b) Structure of the decision feedback equalizer employing linear FFE and a decision device. (c) Structure of the FF transversal filter. ASE=amplification stimulated emission.

that the noise magnitude is greater than k. As in the previous section related to the Gaussian noise (Section 6.2.1.5), we can obtain the probability density function of the random variable noise, w, as

$$p(w) = \frac{1}{\sqrt{2\pi y_0^{-2}\sigma^2}} e^{-\frac{w^2}{2y_0^{-2}\sigma^2}} \tag{6.32}$$

The probability of error can be found as

$$P_e = \int_k^{\infty} p(w)\,dw = \int_k^{\infty} \frac{1}{\sqrt{2\pi y_0^{-2}\sigma^2}} e^{-\frac{w^2}{2y_0^{-2}\sigma^2}}\,dw = \int_{k\frac{|y_0|}{\sigma}}^{\infty} \frac{1}{\sqrt{2\pi}} e^{-\frac{w^2}{2}}\,dw = Q\left(k\frac{|y_0|}{\sigma}\right) \tag{6.33}$$

The most important difference between the NLEs and LEs is that the NLEs can handle the equalization process when there are poles on the unit circle of the Z-plane of the Z-transform impulse response of the transmission channel.

A relative comparison between the LE and NLE is that (i) when all zeros of the channel transfer function in the Z-plane lie inside the unit circle in the z-plane, the NLE, with samples including the initial instant, gains an advantage in tolerance to the Gaussian additive noise over the LE. The NLE would now be preferred in contrast to LE as the number of taps of the NLE would be smaller than that required for the LE. (ii) However, when the transmission is purely phase distortion, meaning that the poles are outside the unit circle of the z-plane and the zeros are the reciprocal conjugates of the poles (inside the unit circles), the LE is the matched filter and offers better performance than the NLE, approaching that of a maximum likelihood detector. The complexity of the NLE and LE are nearly the same in terms of taps and noise tolerance. (iii) Sometimes, there may be about a 2–3 dB better performance between NLE and LE processing.

6.2.2.2 Zero-Forcing Nonlinear Equalization (ZF-NLE)

When an NLE is used in place of the LE of the ZE-FE, the equalizer is called ZF-NLE, as illustrated in the transmission system shown in Figure 6.12. This system consists of a linear feed-forward transversal filter, a nonlinear equalizer with a linear feedback transversal filter, and a decision/detector operation.

The decision feedback equalizer performs equalization by zero forcing, which is similar to the ZF-LE with zero forcing described in Section 6.2.1.3. The operational principles for this type of NLE are as follows:

The linear filter in Equation 6.8 partially equalizes the channel by setting to zero all components of the channel's sampled impulse response preceding the one with the largest magnitude, without affecting the relative values of the remaining components.

Similar to the notations assigned for the input sequence and the channel impulse response at g-sampling instants in the z-transform $Y(z)$, as given in Equations 6.10–6.12, we can set that the required equalizer impulse response in the z-domain, $C(z)$ of m-tap would be

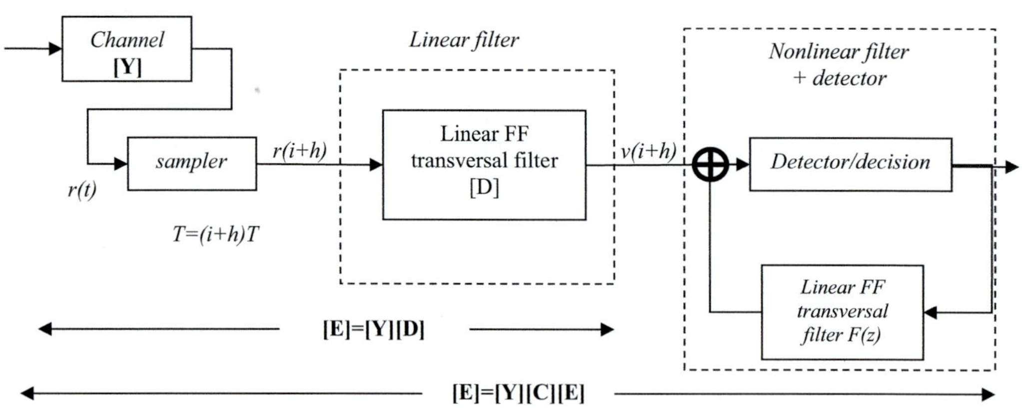

FIGURE 6.12 Decision feedback equalizer (DFE) incorporating a linear and nonlinear filter with overall transfer impulse responses.

$$Y(z)C(z) \simeq z^{-h} \qquad \text{with} \qquad 0 < h < m+g \tag{6.34}$$

where h is an integer in the range of 0 to $(m+g)$ with

$$[Y] = \begin{bmatrix} y_0 & y_1 & \ldots\ldots & y_g \end{bmatrix}; \text{ Channel impulse reponse}$$

$$[C] = \begin{bmatrix} c_0 & c_1 & \ldots\ldots & c_m \end{bmatrix}: LE \text{ impulse response}$$

so that

$$Y(z) = y_0 + \quad y_1 z^{-1} + \quad \ldots \quad + y_g \quad z^{-g}; \text{and} \qquad C(z) = c_0 + \quad c_1 z^{-1} + \quad \ldots \quad + c_m z^{-m} \tag{6.35}$$

Let the LE have $[D]$ as an $(n+1)$-tap gain filter, with the impulse response $D(z)$ of $(n+1)$-taps expressed as

$$\begin{gathered}[D] = \begin{bmatrix} d_0 & d_1 & \ldots & d_n \end{bmatrix}; \text{ impulse response linear transversal filter} \\ D(z) = c_0 + c_1 z^{-1} + \cdots + c_m z^{-m}\end{gathered} \tag{6.36}$$

Thus, the overall transfer function of the channel and the equalizer can be written when setting $[D]=[C][E]$, as

$$Y(z)C(z)E(z) \simeq z^{-h}E(z) \qquad \text{with} \qquad n = m+g-l \tag{6.37}$$

with $[E]$ as the required sampled impulse response, so that the NLE can equalize the input pulse sequence, satisfying the condition that the product of the two transfer functions must be unitary, or

$$\begin{gathered}[E] = \begin{bmatrix} 1 & \dfrac{y_{l+1}}{y_l} & \dfrac{y_{l+2}}{y_l} & \dfrac{y_g}{y_l} \end{bmatrix} \rightarrow z-\text{transform impulse response} \\ E(z) = 1 + \frac{y_{l+1}}{y_l} z^{-1} + \frac{y_{l+2}}{y_l} z^{-2} + \cdots + \frac{y_g}{y_l} z^{-g+l}\end{gathered} \tag{6.38}$$

With y_l as the largest component of the channel sampled impulse response, when this condition is satisfied, then the nonlinear filter ZF-NLE now acts the same as ZF-LE but with the tap gain order reduced to $(g-1)$ taps, whose coefficients are $\frac{y_{l+1}}{y_l}, \frac{y_{l+2}}{y_l}, \ldots, \frac{y_g}{y_l}$ instead of g-taps with gains $v_1, v_2, \ldots, v_g$. The probability of error can be evaluated in a similar manner as described above for additive noises with a Gaussian distribution of the probability density function with zero mean and a standard deviation σ, to give

$$P_e = Q\left(\frac{k}{\sigma|D|}\right) \quad ; \quad |D| = \text{Euclidean length of impulse response}[D] \tag{6.39}$$

6.2.2.3 Linear and Nonlinear Equalization of Factorized Channel Response

In normal case, the z-domain transfer function of the channel $Y(z)$ can be factorized into a cascade of two sub-channels given as

$$Y(z) = Y_1(z)Y_2(z) \tag{6.40}$$

so that we can employ the first LE to equalize the sub-channel $Y_1(z)$ and the NLE to tackle the second sub-channel $Y_2(z)$. The procedure is a combined cascading of the two processing stages, which have been described above.

6.2.2.4 Equalization with Minimizing MSE in Equalized Signals

As we have seen, the equalization using nonlinear equalizers described in the previous section enhances the SNR performance of the system. However, under what criteria can the NLE can make the decision block so that optimum performance can be achieved?, The mean square error can be employed as described above to minimize errors in the decision stage of the NLE.

6.3 MAXIMUM LIKELIHOOD SEQUENCE DETECTION (MLSD) AND VITERBI

Consider the detection of the first data symbol s_0 of the sequence $\{s_i\}$; $i = 0,\ldots,m$ by arranging the delays of the h-sampling intervals increased to $m+g+1$ such that all signals $\{v_i\}$ at the output of the filter D, which are dependent on any of the m received data symbols $\{s_0, s_1,\ldots,s_{m-1}\}$, are involved in the detection process at once. Since s_0 is the first detected value of the sequence of data symbols, no decision directed cancellation of the ISI can be carried out, and the detection process operates on the $m+g$ signals $\{v_0, v_1,\ldots,v_{m+g-1}\}$. The detected value s_0' of the initial data symbol I_s therefore taken as its possible value for which

$$\sum_{i=0}^{m+g=1}\left(v_i - \sum_{j=0}^{g} s'_{i-j}e_j\right)^2 \rightarrow \text{minimum for } \forall \text{ combinations} \tag{6.41}$$

of all possible values of $\{s_0', s_1',\ldots,s_{m-1}'\}$ with the note that $s_i' = s_i = 0$; for $-i,0$ and $i > m-1$. This detection process is the maximum likelihood detection (MLSD) or MLSE (estimation) when the estimation is used in place of detection, of the data symbols $\{s_0, s_1,\ldots,s_{m-1}\}$ from the corresponding received sequence of signals $\{v_0, v_1,\ldots,v_{m+g-1}\}$ at the output of the filter D.

Under the assumed conditions, the detection process minimizes the probability of error in detecting the complete sequence of data symbols. Thus, all the data symbols are detected simultaneously in a single, optimum detection process. Unfortunately, when m is large, especially in DSP operating at ultra-high speed, the processing speed may not allow such operation. Thus, successive processing of the data symbols is necessary and hence requirement of memory storage so that the detection process can be considered equivalent to a simultaneous process. This sequential detection can be implemented using the Viterbi algorithm detector.[11]

The Viterbi algorithm operates as follows. Consider the data transmission as shown in Figure 6.12, in which the detector (by thresholding) and the transversal filter F are replaced by a more complex subsystem. Assuming that the l-level data symbols $\{s_i\}$ are transmitted, after the optical-to-electronic detection and preamplification, the received signals are sampled, and the stored components of $l^g\ h+1$ component vectors of sequences $\{Q_{i-1}\}$, where

$$Q_{i-1} = \begin{bmatrix} x_{i-h-1} & x_{i-h} & & x_{i-1} \end{bmatrix} \tag{6.42}$$

The symbol x_j for any j in the range of $\{0,\ldots,m-1\}$ may take on any possible values of s_j and $h > g$. Each vector Q_{i-1} is formed by the last $(h+1)$ components of the corresponding i-component vector $[X_{i-1} = x_0, x_1, \ldots, x_{i-1}]$, which represents a possible received sequence of data-symbols $\{s_j\}$. Associated with each stored vector $\{Q_{i-1}\}$ is its cost

$$c_{i-1} = \sum_{i=0}^{i-1} (v_j - z_j)^2; \qquad \text{with} \qquad v_j = z_j + u'_j \qquad z_j = \sum_{k=0}^{g} x_{j-k} e_k \tag{6.43}$$

$$x_j = 0 \qquad \text{for} \qquad j < 0 \,\&\, j > m-1$$

The symbol z_j takes the value of v_j, which is received in the absence of noises. The cost c_{i-1} is the Euclidean distance between vectors $V_{i-1} = [v_0, v_1, \ldots, v_{i-1}]$ and $Z_{i-1} = [z_0, z_1, \ldots, z_{i-1}]$. Since Z_{i-1} is the sequence up to the time instant $t = (i-1)T$, resulting from the transmitted sequence X_{i-1} in the absence of noise, the cost c_{i-1} is the measure of the probability that this sequence is correct. Thus, the smaller the value of this cost, the more likely the corresponding sequence X_{i-1} to be correct.

Thus, in summary, in an MLSE decision is performed by calculating the transmitted bit stream with the highest probability. This means that the probability of all bit combinations for the whole sequence is considered, and the combination of highest probability is assumed as the transmitted bit sequence.

6.3.1 Nonlinear MLSE

Maximum likelihood sequence estimation (MLSE) is a well-known technique in communications for equalization and detection of the transmitted digital signals. The concept of MLSE is discussed in brief. An MLSE receiver determines a sequence **b** as the most likely transmitted sequence when the conditional probability $Pr(\mathbf{y} \mid \mathbf{b})$ is maximized, where **y** is the received output sequence. If the received signal **y** is corrupted by a noise vector **n**, which is modeled as a Gaussian source (i.e. $\mathbf{y} = \mathbf{b} + \mathbf{n}$), it is shown that the above maximization operation can be equivalent to the process of minimization of the Euclidean distance d[12]

$$d = \sum_k |\mathbf{y}_k - \mathbf{b}_k|^2 \tag{6.44}$$

MLSE can be carried out effectively with the implementation of the Viterbi algorithm, based on a state trellis structure.

6.3.2 Trellis Structure and Viterbi Algorithm

6.3.2.1 Trellis Structure

Information signal **b** is mapped to a state trellis structure by a Finite State Machine (FSM), giving the mapping output signal **c** as shown in Figure 6.13. The FSM can be a convolutional encoder, a trellis-based detector, or, as presented in more detail later, the fiber medium for optical communications.

A state trellis structure created by the FSM is illustrated in Figure 6.14. At epoch *n*th, the current state *B* has two possible output branches connecting to states *E* and *F*. In this case, these two branches correspond to the two possible transmitted symbols, "0" or "1", respectively. This predefined state trellis applies to 1-bit-per-symbol modulation formats such as binary ASK or DPSK formats. In cases of multi-bit modulation per symbol, the phase trellis has to be modified. For example, in the case of QPSK, four possible branches are leaving the current state B and connecting to the next states, corresponding to the possible input symbols of "00", "01", "10", and "11".

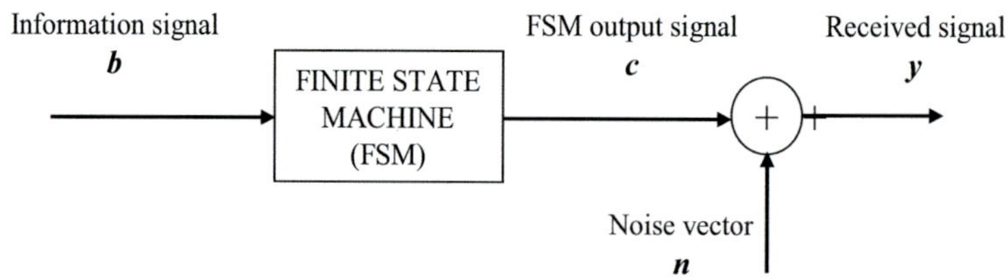

FIGURE 6.13 Schematic of the MLSE equalizer as a finite state machine.

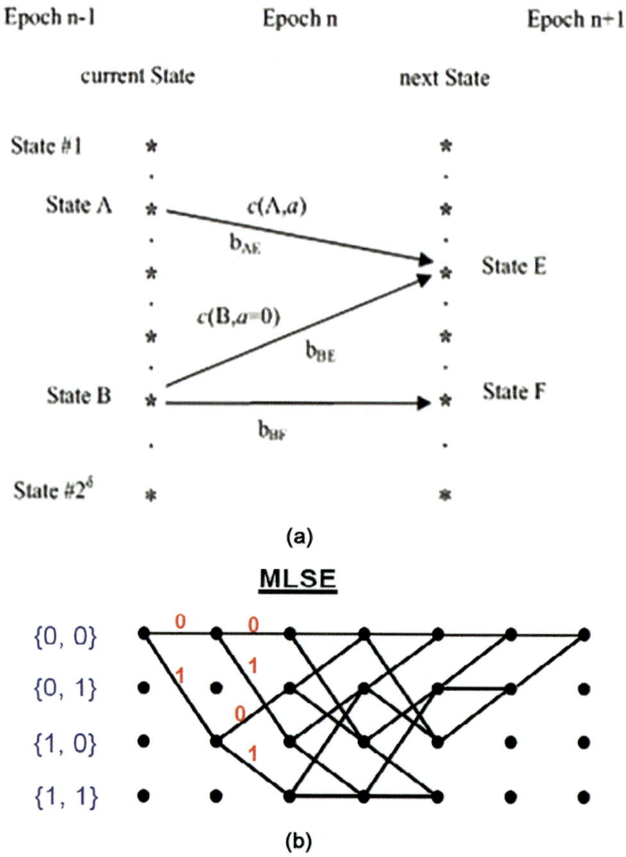

FIGURE 6.14 State trellis of a finite state machine (a) Trellis path, (b) tracing of possible maximum likelihood estimation with assigned bit pattern.

Furthermore, for a simplified explanation, several assumptions are made. These assumptions and notations can be described as follows: (i) Current state B is the starting point for only two branches, which connect to states E and F, denoted as b_{BE} and b_{BF}. In general, b_{B*} represents all the possible branches in the trellis, which starts from state B. In this case, these two branches correspond to the possible binary transmitted information symbols, "0" or "1", respectively. The notation $c(B,a=0)$ represents the encoded symbol (at the FSM output) for the branch BE, b_{BE}, which starts from state B and corresponds to a transmitted symbol a=0. Similar conventions are applied for branch BF, b_{BF}, that is $c(B,a=1)$, and for all the branches in the trellis. (ii) There are only two possible output branches ending at state E (from state A and state B), denoted as b_{AE} and b_{BF}. In general, $b_{*\mathrm{E}}$ represents all the possible branches in the trellis, which end at state E. The number of states 2^{δ} in the trellis structure is determined by the number of memory bits, δ, also known as the constraint length. Figure 6.47b also shows a possible trace of the trellis with the assigned states of the bit pattern.

6.3.2.2 Viterbi Algorithm

In principles, the Viterbi algorithm follows two main phases:

6.3.2.2.1 Phase 1: Calculation of State and Branch Metrics

At a sampling index n, epoch n, a branch metric is calculated for each of the branches in the trellis. For example, considering the connecting branch between the current state B and the next state E, which corresponds to the case when the "0" symbol is transmitted, the branch metric is calculated as

$$\text{BM}_{BE}(n) = \left| y_{(n)} - c_{BE} \right|^2 \tag{6.45}$$

The state metric $\text{SM}_E(n)$of the state E is calculated as the sum between the metric of the previous state, that is state B, and the branch metric obtained above. The calculation of state metrics $\text{SM}_E(n)$ of state E is given by

$$\text{SM}_E(n) = \min_{\forall B \to E} \left\{ \text{SM}_B(n-1) + \text{BM}_{BE}(n) \right\} \tag{6.46}$$

The branch giving the minimum state metric $\text{SM}_E(n)$ for state E is called the preferred path.

6.3.2.2.2 Phase 2: Trace-Back Process

The process of calculating state and branch metrics continues along the state trellis before terminating at epoch $N_{\text{trace},}$ which is usually referred to as the trace-back length. A rule of thumb, as referred to in Refs.[13,14], is that the value of $N_{\text{trace}}$is normally taken to be five times the sequence length. This value comes from the results showing that the solution for the Viterbi algorithm converges, giving a unique path from epoch 1 to epoch N_{trace} for the MLSE detection. (i) At N_{trace}, the terminating state with the minimum state metric and its connecting preferred path is identified. (ii) The previous state is then identified, as well as its previous preferred path. (iii) The trace-back process continues until reaching epoch 1.

6.3.3 Optical Space Medium as a Finite State Machine

A structure of the Finite State Machine shown in Figure 6.47 is now replaced with the optical channel based on the optical space propagation medium (FSM). It is significant to understand how the state trellis structure is formed for an optical transmission system. In this case, the spatial spreading and turbulent effects involve all the ISI sources causing the waveform distortion of the optical signals, which includes CD, PMD, and filtering effects. Thus, the optical FSM excludes the signal corruption by noise as well as the random-process nonlinearities. The ISI caused by dispersion effects and nonlinear phase, ASE, and nonlinear effects[15] in the short fiber output of the HPA, that is under ultra-high power launched into such short length.

6.3.4 Construction of State Trellis Structure

At epoch n, it is assumed that the effect of ISI on an output symbol of the FSM c_n is caused by both executive δ pre-cursor and δ post-cursor symbols on each side. In this case, c_n is the middle symbol of a sequence, which can be represented as

$$c = \left(c_{n-\delta}, \ldots, c_n, \ldots, c_{n+\delta}\right) \;\; b \in \{0,1\};\; n > 2\delta + 1 \tag{6.47}$$

Unlike the conventional FSM used in wireless communications, the output symbol c_n of the optical fiber FSM is selected as the middle symbol of the sequence c. This is because the middle symbol gives the most reliable picture of the effects of ISI in optical fiber channels.

A state trellis is defined as the total number of $2^{2\delta}$ possible data sequences. At epoch n, the current state B is determined by the data sequence

$$\left(b_{n-2\delta+1},\ldots,b_{n-\delta},\ldots,b_n\right) \quad b\in\{0,1\};\; n\geq 2\delta+1 \tag{6.48}$$

where b_n is the current input symbol into the transmission medium FSM. Therefore, the state trellis structure can be constructed from the optical fiber FSM and ready to be integrated with the Viterbi algorithm, especially for the calculation of a branch metric.

The MLSE nonlinear equalizer can effectively combat all the above ISI effects, and it is expected that the tolerance to both CD and PMD of optical MSK systems with OFDR receiver can be improved significantly.

6.4 MAXIMUM A POSTERIORI (MAP) TECHNIQUE FOR PHASE ESTIMATION

6.4.1 Method

Assuming that we want to estimate an unobserved population parameter θ on the basis of observation variable x. Let f be the sampling distribution of x, so that $f\left(x|\theta\right)$ is the probability of x when *the underlying population parameter is* θ. Then, the function that transform $\theta\rightarrow f\left(x|\theta\right)$ is known as the likelihood function, and the estimate $\hat{\theta}_{\mathrm{ML}}(x)=\max_{\theta} f\left(x|\theta\right)$ is the maximum likelihood estimate of θ.

Now, assume that a prior distribution g over θ exists. This allows us to treat θ as a random variable as in Bayesian statistics. Then, the posterior distribution of θcan be determined as

$$\theta\rightarrow f\left(\theta|x\right)=\frac{f\left(x|\theta\right)g(\theta)}{\int_{v\in\Theta} f\left(x|v\right)g(v)\,dv} \tag{6.49}$$

where g is density function of θ, Θ is the domain of g. This is a straightforward application of *Bayes' theorem.* The method of maximum a posteriori estimation then estimates θ as the *mode* of the posterior distribution of this random variable.

$$\hat{\theta}_{\mathrm{MAP}}=\arg\max\frac{f\left(x|\theta\right)g(\theta)}{\int_{v} f\left(x|v\right)g(v)\,dv}=\arg_{\theta}\max f\left(x|\theta\right)g(\theta) \tag{6.50}$$

The denominator of the posterior distribution (so-called partition function) does not depend on θ and therefore plays no role in the optimization. Observe that the MAP estimate of θ coincides with the maximum likelihood estimate when the prior g is uniform (that is, a constant function). The MAP estimate is a limit of Bayes estimators under a sequence of 0–1 loss functions, but generally not a Bayes estimator per se, unless θ is discrete.

6.4.2 Estimates

MAP estimates can be computed in several ways:

- Analytically, the mode(s) of the posterior distribution can be obtained in closed form. This is the case when conjugate priors are used; then

- Either, via numerical optimization such as the conjugate gradient method or Newton's method, which usually require first or second derivatives, are to be evaluated analytically or numerically; or
- Via a modification of an expectation-maximization algorithm. This does not require derivatives of the posterior density.

One of the possible processing algorithms is the maximum likelihood (ML) phase estimation for coherent optical communications.[16] As an example, the differential QPSK (DQPSK) optical system can be considered. In this system, as described in Chapter 5, the phase-diversity coherent reception technique with a mixing of an LO, can be employed to recover the in-phase (I) and quadrature-phase (Q) components of the signals. Such a receiver is composed of a 90° optical hybrid coupler to mix the incoming signal with the four quadruple states associated with the LO inputs in the complex field space. A $\pi/2$ phase shifter can be used to extract the quadrature components in the optical domain. The optical hybrid can then provide four lightwave signals to two pairs of balanced photodetectors to reconstruct I and Q information of the transmitted signal. In case the LO is the signal itself, then the reception is equivalent to a self-homodyne detection. For this type of reception subsystem, the balanced receiver would integrate a one-bit delay optical interferometer so as to compare the optical phases of two consecutive bits. We can assume a complete matching of the polarization between the signal and the local oscillator fields so that only the impact of phase noise be considered. Indeed, the maximum likelihood algorithm is the modern MLSE technique.[17]

The output signal reconstructed from the photocurrents can be represented by

$$r(k) = E_0 \exp\left[j\left(\theta_s(k) + \theta_n(k)\right)\right] + \tilde{n}(k) \tag{6.51}$$

where k denotes kth sample over time interval $\left[kT, (k+1)T\right]$ (T is the symbol period), $E_0 = \Re\sqrt{P_{\text{LO}} p_s}$, $\Re$ is the photodiode responsivity, P_{LO} and p_s are the power of LO and received signal, respectively; $\theta_s(k) \in \{0, \pi/2, \pi, -\pi/2\}$ is the modulated phase, the phase difference carrying the data information; $\theta_n(k)$ is the phase noise during the transmission. $\{\tilde{n}(k)\}$ are complex white Gaussian random variables with $E\left[\tilde{n}(k)\right] = 0$ and $E\left[|\tilde{n}(k)|^2\right] = N_0$.

To retrieve information from the phase modulation $\theta_s(k)$ at time $t=kT$, the carrier phase noise $\theta_n(k)$ is estimated based on the received signal over the immediate past L symbol intervals, that is, based on $\left\{r(l),\ k-L \le l \le k-1\right\}$. In the decision-feedback strategy, a complex phase reference vector $\mathbf{v}(k)$ is computed by the correlation of L received signal samples $r(l)$ and decisions on L symbols $\left\{\hat{m}(l),\ k-L \le l \le k-1\right\}$, where $\hat{m}(l)$ is the receiver's decision on $\exp\left(j\theta_s(l)\right)$

$$\mathbf{v}(k) = \sum_{l=k-L}^{k-1} r(l)\hat{m}^*(l) \tag{6.52}$$

Here, * denotes the complex conjugation. An initial L-symbol long-known data sequence is sent to initiate the processor/receiver. Alternatively, the maximum likelihood decision processor can be trained before the transmission of the data sequence. Importantly, not only can the $\pi/4$-radian phase ambiguity be resolved, but the decision-aided method is now totally linear and efficient to implement, based on Equation 6.51. To some extent, the reference $\mathbf{v}(k)$ assists the receiver to acquire the channel characteristics. With the assumption that $\theta_n(k)$ varies slowly compared to the symbol rate, the computed complex reference $\mathbf{v}(k)$ from the past L symbols forms a good approximation to the phase noise phasor $\exp\left(j\theta_n(k)\right)$ at time $t=kT$.

Using the phase reference $\mathbf{v}(k)$ from Equation 6.52, the decision statistic of the ML receiving processor is given as

$$q_i(k) = \mathbf{r}(k)C_i^* \cdot \mathbf{v}(k), \qquad i = 0,1,2,3 \tag{6.53}$$

where $C_i \in \{\pm 1, \pm j\}$ is the DQPSK signal constellation, $\cdot$ denotes the inner product of two vectors. The detector declares the decision $\hat{m}(k) = C_k$ if $q_k = \max q_i$. This receiver/processor has been shown to achieve coherent detection performance if the carrier phase is constant, and the reference length L is sufficiently long.

The performance of ML the receiving processor can be evaluated through simulation using Monte Carlo simulations in two cases: a linear phase noise system and a nonlinear phase noise system. For comparison, the *M*th power phase estimator and differential demodulation can also be employed.

Nonlinearity in an optical fiber can be ignored when the launch optical power is below the threshold, allowing the fiber optic to be modeled as a linear channel. The phase noise difference in two adjacent symbol intervals, i.e., $\theta_n(k) - \theta_n(k-1)$, obeys a Gaussian distribution with the variance σ^2, determined by the linewidth of the transmitter laser and the LO, given by the Lorentzian linewidth formula $\sigma^2 = 2\pi\Delta v T$, where Δv is the total linewidth of the transmitter laser and the LO.[18] In the simulations, σ can be set to 0.05, corresponding to the overall linewidth $\Delta v = 8$ MHz when the baud rate is 40 symbols/s.

As observed in Figure 6.15, the ML processor outperforms the phase estimator and differential demodulation by about 0.25 dB and 1 dB, respectively. Although the performance gap between the ML processor and *M*th power phase estimation is not very large in the case of an MPSK system, *M*th power phase estimator requires more nonlinear computations, such as an arctan (·), which incurs a large latency in the system and leads to phase ambiguity in estimating the block phase noise. On the contrary, the ML processor is a linear and efficient algorithm, and there is no need to deal with the π/4-radian phase ambiguity, making it more feasible for on-line processing for real systems. We also extend our ML phase estimation technique to the QAM system. The conventional *M*th-power phase estimation scheme suffers performance degradation in QAM systems since only a subgroup of symbols with phase modulation $\pi/4 + n \cdot \pi/2$ $(n = 0,1,2,3)$ can be used to estimate the phase reference. The maximum tolerance of linewidth per laser in Square 16 QAM can be improved ten times compared to the *M*th-power phase estimator scheme.

Let's now consider the case of nonlinear phase noise. In a multi-span optical communication system with erbium-doped fiber amplifiers (EDFAs), the performance of optical DQPSK is severely limited by nonlinear phase noise converted by amplified ASE noise through fiber nonlinearity.[20,21]

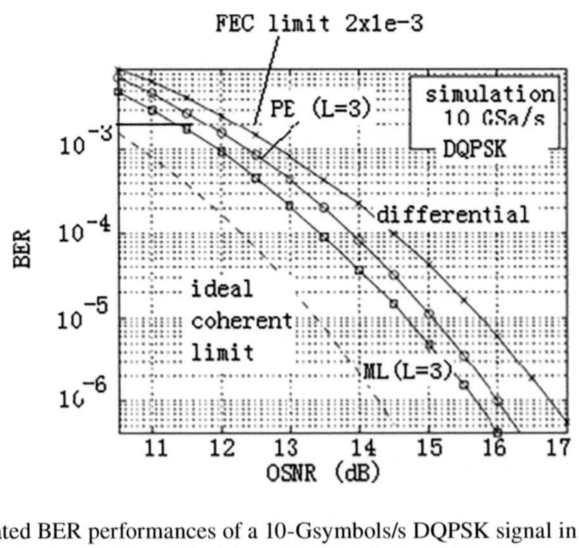

FIGURE 6.15 Simulated BER performances of a 10-Gsymbols/s DQPSK signal in a linear optical channel with laser linewidth 2 MHz ($\sigma = 0.05$) and decision-aided length L=3. (Modified from Ref.[19]) The FEC limit can be set at 2× 1e-3 to determine the required OSNR.

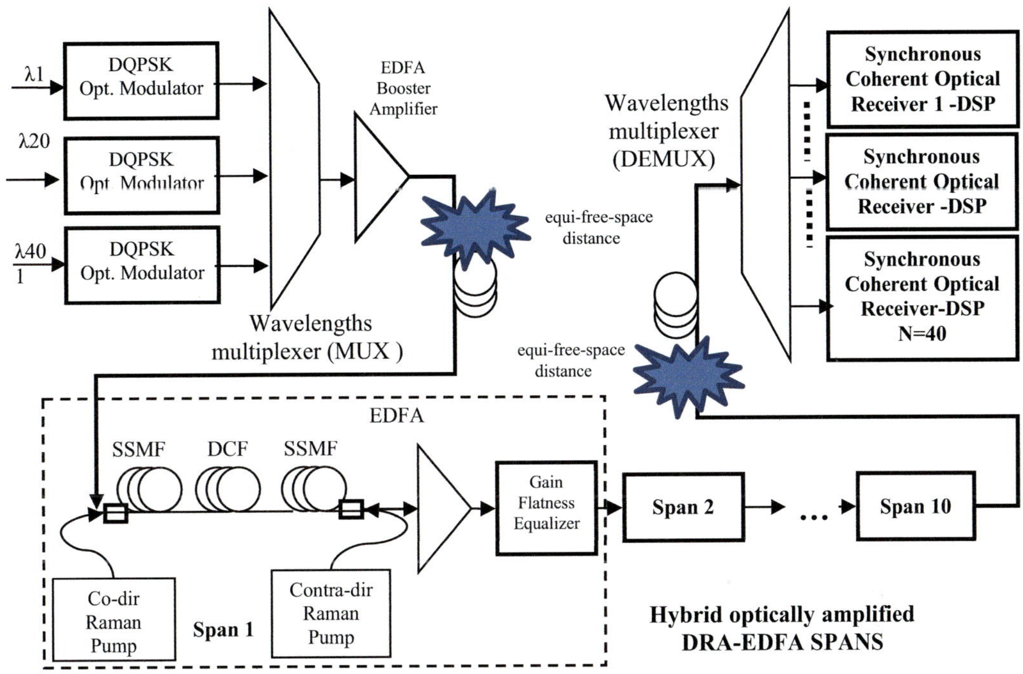

FIGURE 6.16 DWDM Optical DQPSK system with an N-span dispersion-compensated optic fiber transmission link using a synchronous self-homodyne coherent receiver to compensate nonlinear phase noise under both Raman amplification and EDF amplification.

The laser linewidth is excluded to only consider the impact of the nonlinear phase noise on the receiver. The transmission system can comprise a single-channel DQPSK modulated optical signal transmitted at 10-40 G-Symbols/s over N 100-km equally-spaced amplified spans, which are fully compensated for chromatic dispersion and gain equalized along the fiber, as shown in Figure 6.16. The following parameters are assumed for the transmission system: $N=20$, fiber nonlinear coefficient $\gamma=2$ W/km, fiber loss $\alpha=0.2$ dB/km, amplifier gain $G=20$ dB NF of 6 dB, optical wavelength $\lambda=1553$ nm, and bandwidths of the optical and electrical filters of $\Delta\lambda=0.1$ nm and 7 GHz, respectively. The OSNR is defined at the location just in front of the optical receiver as the ratio between the signal power and the noise power in two polarization states contained within a 0.1-nm spectral width region. As shown in Figure 6.17a, the receiver sensitivity of the ML processor improved by about 0.5 and 3 dB over phase estimation and differential demodulation, respectively, at a BER level of 10^{-4}. Moreover, it is noteworthy that differential demodulation exhibited an error floor due to severe nonlinear phase noise with high launch optical power at the transmitter.

To obtain the impact of nonlinear phase noise on the demodulation methods and the noise loading effects, the number of amplifiers can be increased to $N=30$. The Q-factor is used to show a numerical improvement in the ML processor receiver. As shown in Figure 6.17b, with the increase in launch power, the ML receiver/processor approximates the optimum performance because of the reduced variance of the phase noise. At the optimum point, the ML receiver/processor outperforms the Mth power phase estimator by about 0.8 dB, while having about a 1.8 dB receiver sensitivity improvement compared to differential detection. It is also found that the optimum performance occurs at an OSNR of 16 dB when the nonlinear phase shift is almost 1 rad. With launch power exceeding the optimum level, nonlinear phase noise becomes severe again. Only the ML phase estimator can offer a BER beyond 10^{-4}, while both the Mth power phase estimator and differential detection exhibit an error floor before reaching a BER 10^{-4}. From the numerical results, the ML processor carrier phase estimation shows significantly better receiver sensitivity improvement than the

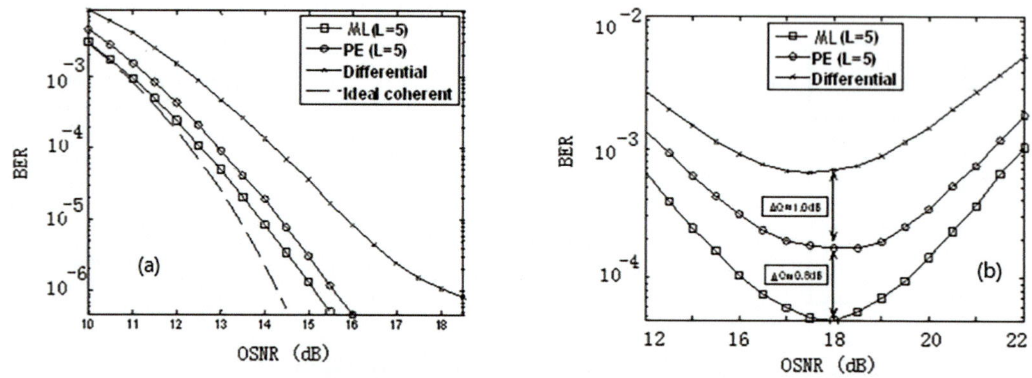

FIGURE 6.17 Monte Carlo simulated BER of a 10-Gsymbol/s DQPSK signal in an N-span optical fiber system with template length $L=5$: (a) $N=20$; (b) $N=30$ under different detection schemes of ML estimator, phase estimator, differential demodulation and ideal coherent detection. ML=maximum likelihood, PE=phase estimation. (Extracted from Ref.[22])

*M*th power phase estimator and conventional differential demodulation for optical DQPSK signals in systems dominated by nonlinear phase noise. Again, we want to emphasize that ML estimator is a linear and efficient algorithm, and there is no need to deal with the $\pi/4$-radian phase ambiguity, making it feasible for on-line processing for real systems.

In summary, the performance of synchronous coherent detection with ML processor carrier phase estimation may offer a separate linear phase noise and nonlinear optical noise system. The ML-DSP phase estimation can replace the optical PLL, and the receiver sensitivity is improved compared to conventional differential detection and *M*th power phase estimator. The receiver sensitivity is improved by approximately 1 and 3 dB in these two channels (20 spans for the nonlinear channel) respectively, compared to differential detection. For the 30-span nonlinear channel, only the ML processor can offer BER beyond 10^{-4} (using Monte Carlo simulation), while both *M*th power phase estimator and differential detection exhibit an error floor before BER reaches 10^{-4}. At the optimum point of the power, the Q-factor of the ML receiver/processor outperforms the *M*th power phase estimation by about 0.8 dB. In addition, an important feature is the linear and efficient computation of the ML phase estimation algorithm, which enables real-time on-line DSP processing.

6.5 CARRIER PHASE ESTIMATION

6.5.1 Introductory Remarks

Under homodyne or intra-dyne reception, matching the local oscillator laser with the signal carrier is critical to minimizing the deviations in reception performance, hence enhancing the receiver sensitivity and to maximizing the transmission system performance. This difficulty remains a considerable obstacle for the first generation of coherent systems developed in the mid-1980s, which employed analog or complete hardware circuitry.

The first few parts of this chapter deal with the recovery of the carrier phase. The later part involves some advanced processing algorithms to address carrier phase recovery under the scenario where the frequency of the LO laser is oscillating, an operational condition to get the stabilization of an external cavity laser, which is essential for homodyne reception with high sensitivity.

In DSP-based optical reception sub-systems, this hurdle can be overcome by processing algorithms installed in a real-time memory processing system, such as an ASIC (application-specific integrated circuit), which may consist of an analog-to-digital converter, a digital signal processor, and high-speed fetch memory.

The frequency offset between the LO and the signal lightwave carrier is commonly due to the oscillation of the local oscillator. An external cavity incorporating an external grating is commonly used, which oscillates by a low frequency of about 300 MHz to stabilize the feedback of the reflected specific frequency line to the laser cavity. However, this oscillation degrades the sensitivity of homodyne reception systems. Thus, the later part of this section illustrates the application of DSP algorithms described in the above sections of this chapter to demonstrate their effectiveness in a real-time experimental setup.

It is noted that the carrier phase recovery must be implemented before any equalization process can be implemented as explained in Sections 6.5.3 and 6.5.4.

It is noted here that the optical phase locking (OPL) technique presented in Chapter 5 of Ref.[23] is another way to reduce the frequency offset between the LO and the signal carrier, operating in the optical domain. In contrast, the technique described in this section is applied to the receiver output in the digital processing domain.

6.5.2 Frequency Offset Oscillations and Q-Penalties

Figure 6.18 shows the typical digital signal processing of medium complexity for a coherent PDM (polarization division multiplexing) optical transmission system. Chromatic dispersion and polarization effects are compensated in the CD and MIMO blocks, respectively. The DV FE is realized in the μc. Several thousands of data are periodically loaded into the microcontroller, which calculates FO (frequency offset). A small bandwidth of the estimation loop delivers only an averaged FO value to the CMOS ASIC part. However, long-time experimental tests with commercially available lasers, for example, the EMCORE laser, show that laser frequency oscillations are sinusoidal with the frequency of 888 Hz to stabilize its central frequency using feedback control, and an amplitude of more than 40 MHz (see Figure 6.19 for measured oscillation of the EMCORE external cavity laser as LO). This oscillation is created by the vibration of the external reflector – a grating surface so that the cavity can be stabilized. The mean offset is close to 300 MHz. So, the DiffV enables the averaging of the FO of 300 MHz, while the residual offset of ±50 MHz is compensated by the V-V algorithm.

In SPC, the frequency offset arises from the Doppler frequency shift caused by the relative rotation speeds of satellites in different orbits, for example, GEO, LEO, or MEO. This shift is continuous due to the continuous rotation of the satellites around the Earth. Thus, the main FO compensation algorithm is important for S systems. Chapter 10 presents tests of the ICPTRC (integrated coherent photonic transceiver) in which the DSP algorithm is embedded to compensate for the FO effect.

Penalties caused by a residual offset of 50 MHz have been investigated for the case of V-V CPE. Figure 6.20 shows the simulation values of the Q penalties versus the V-V averaging window

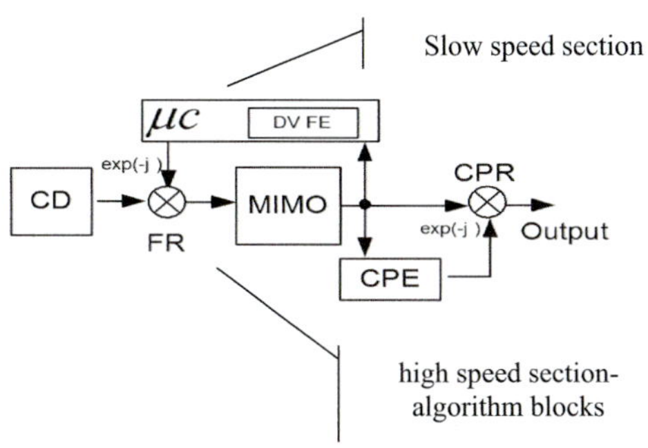

FIGURE 6.18 Typical digital signal processing structure in PDM coherent receivers, including a low-speed processor and a high-speed digital signal processing system. CPE = carrier phase estimation, MIMO = multiple input multiple output, CPR = carrier phase recovery, FR = frequency recovery, CD = chromatic dispersion, or spatial broadening of space beam.

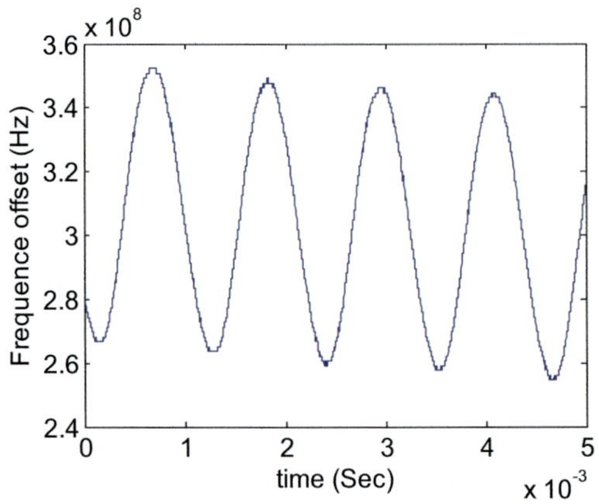

FIGURE 6.19 Laser frequency oscillation of an external cavity laser as measured from an external cavity laser.

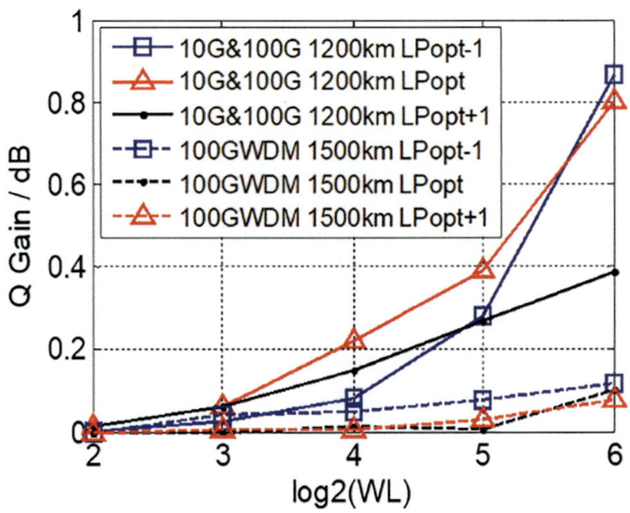

FIGURE 6.20 Penalties of V&V CPR with residual FO of 50 MHz.

length (WL) for two scenarios under offline data processing. In the first case, we multiplexed a 112 G PDM-QPSK channel with twelve 10 G on-off keying (OOK) neighbors (10 and 100 G over a 1200 km link), while the second scenario includes the transmission of eight 112 G PDM-QPSK channels (100 G WDM – 1500 km link). Channel spacing was 50 GHz (ITU-grid). The FO was partly compensated with the residual offset of 50 MHz to check the V&V CPR performance. Three values of launch power have been checked: optimum launch power (LP_{opt}), LP_{opt}-1 dB (slightly linear regime), and LP_{opt}+1 dB (slightly nonlinear regime). Maximum Q penalties of 0.9 dB can be observed, strongly depending on the averaging window length parameter WL.

6.5.3 Correction of Phase Noise and Nonlinear Effects

Kikuchi's group[24] uses electronic signal processing based on the *M*th power phase estimation to estimate the carrier phase. However, DSP circuits for *M*th power phase estimator need nonlinear computations, thus impeding the potential possibility of real-time processing in the future. Furthermore, the *M*th power phase estimation method requires addressing the π/4-radian phase ambiguity when estimating the phase noise in adjacent symbol blocks. In contrast, the electronic DSP based on the ML processing for carrier phase estimation approximates the ideal synchronous coherent detection in optical phase modulation systems and requires only linear computations. This eliminates the need for optical PLL and is more feasible for online processing in real systems. Initial simulation results show that the ML receiver/processor outperforms the *M*th power phase estimator, especially when nonlinear phase noise is dominant, thus significantly improving receiver sensitivity and tolerance to nonlinear phase noise.

Liu's group[25,26] uses optical delay differential detection with DSP to detect differential BPSK (DPSK) and DQPSK signals. Since direct detection only detects the intensity of the light, the improvement of DSP is limited after direct detection. However, after the synchronous coherent detection of our technique, DSP methods such as electronic equalization of chromatic dispersion and PMD offer better performance since the phase information is retrieved. As the level of phase modulation increases, it becomes increasingly difficult to apply optical delay differential detection because of the complex implementation of the receiver and the degraded SNR of the demodulated signal. In contrast, the ML estimator/processor can still demodulate other advanced modulation formats, such as 8-PSK, 16-PSK, 16-QAM, and so on.

6.5.4 Forward Phase Estimation QPSK Optical Coherent Receivers

Recent progress in DSP,[27,28] with the availability of ultra-high sampling rates allow the possibility of DSP-based phase estimation and polarization management techniques that make the coherent robust and practical. This section is dedicated to emerging technologies that will significantly influence the optical transmission and detection of optical signals at ultra-high speeds.

Recalling the schematic of a coherent receiver in Figures 6.21 and 6.22, which show a coherent detection scheme for QPSK optical signals mixed with the local oscillator field. In the case of modulation formats such as QPSK or DQPSK, an $\pi/2$ optical phase shifter is needed to extract the quadrature component. Thus, the real and imaginary parts of phase-shift keyed signals can be deduced at the output of a balanced receiver from a pair of identical photodiodes connected back-to-back. Note that for a balanced receiver the quantum shot noise contributed by the photodetectors is doubled, as noise is always represented by the noise power and no current direction must be applied.

When the received signal is raised to the quadratic power, the phase of the signal disappears because $e^{j4\theta_s(t)} = 1$ for DQPSK whose states are $0, \pi/2, \pi, 3\pi/2, 2\pi$. The data phase is excluded, and the carrier phase can be recovered with its fluctuation with time. This estimation is computed using a DSP algorithm. The estimated carrier phase is then subtracted from the detected phase of the

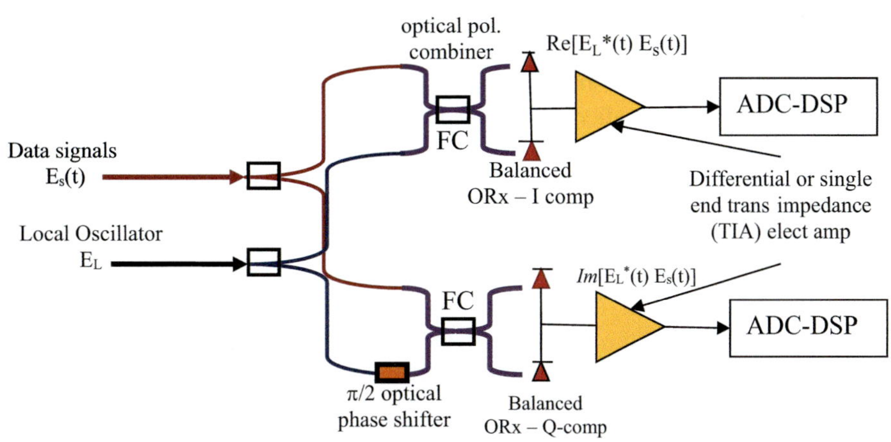

FIGURE 6.21 Schematic of a coherent receiver using a balanced detection technique for *I-Q* component phase and amplitude recovery, incorporating DSP and ADC. Re=real part, Im=imaginary part.

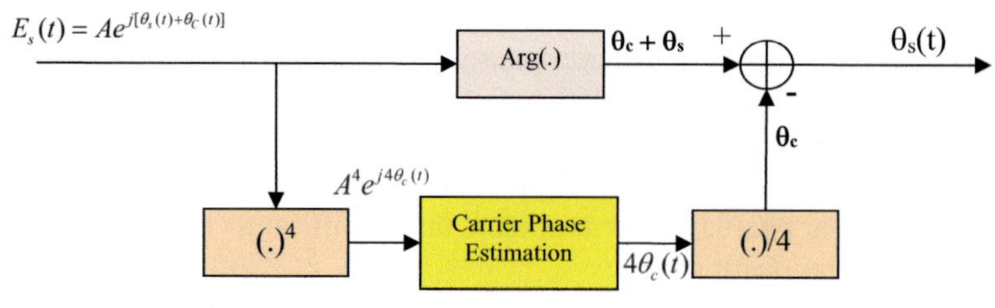

FIGURE 6.22 Schematic of a QPSK coherent receiver.

signal, leaving only the phase states of the signals, as indicated in the diagram. This is a feedforward (FF) phase estimation, which is most suitable for DSP implementation, either offline or in real-time.

6.5.5 Carrier Recovery in Polarization Division Multiplexed Receivers: A Case Study

Advanced modulation formats in combination with coherent receivers incorporating digital signal processing subsystems enable high capacity and spectral efficiency.[29,30] Polarization division multiplexing (PDM), quadrature amplitude modulation (QAM), and coherent detection dominate the next generation high-capacity optical networks since they allow information encoding in all the available degrees of freedom[31] with the same requirement for the OSNR. The coherent technique is homodyne or intradyne mixing the signal carrier with LO. However, as just mentioned, a major problem in homodyne coherent receivers is the carrier recovery (CR), which ensures that matching between the channel carrier and the LO to maximize the signal amplitude and phase strength, hence improving the SNR in both the amplitude and phase planes. This section illustrates an experimental demonstration of CR in a PDM-quadrature phase shift keying (QPSK), transmission scheme carrying 100 Gbits/s per wavelength channel, including two polarized modes [*H* (horizontal) and *V* (vertical) polarized modes] of the linear polarized mode in standard single-mode optical fiber (SSMF), that is 28 GSy/s×2 (2 bits/symbol)×2 (polarized channels) using some advanced DSP algorithms, especially the Vitebi-Viterbi (V-V) algorithm with MLSE nonlinear decision feedback procedures.[32]

Feed-forward carrier phase estimation (FFCPE)[33] has been commonly considered the solution to this problem. The fundamental operation principle is to assume a time-invariant of the phase offset between a carrier and local laser during N ($N>1$) consecutive symbol periods.

In coherent optical systems using tunable lasers, the maximum absolute frequency offset can vary by 5 GHz. To accommodate such a large frequency offset (FO), a coarse digital frequency estimation (FDE) and recovery (FDR) techniques[34,35] can be employed to limit the frequency offset to within the allowable range so that phase recovery can be implemented/managed using fine feedforward carrier phase estimation (FFCPE) and recovery (FFCPR) techniques.[36] The algorithm presented in Ref.,[24] so-called "Differential Viterbi" (DiffV), estimates the phase difference between two successive complex symbols. This method enables the estimation of large frequency offsets of Multiple Phase Shift Keying (M-PSK) modulated signals up to $\pm f_s/(2M)$, where f_s stands for the symbol rate frequency. A larger frequency offset is possible to estimate using the technique presented in Ref.,[6] which generates insignificant estimation error that can be handled using the feedforward FFCPE and FFCPR. However, the feedforward FFCPE, or "Viterbi and Viterbi" (V-V), would not perform well when the absolute residual frequency offset at the input of the estimation circuit is larger than $\pm f_s/(2MN)$. This problem becomes more serious when the laser frequency oscillates, and the coarse feedforward equalization (FFE) can be realized on a slow-speed processor to decrease the complexity of integrated digital electronic circuits, thereby releasing the valuable processing time for processing the signals at an ultra-high sampling rate, which is a very important factor in the cost and power consumption in ultra-high-speed coherent DSP-based optical receivers.

An algorithm for coupling carrier phase estimations in PDM systems is presented in Ref.[37] The algorithm requires two loop filters, and the coupling factor should be carefully selected. In this section, an enhanced carrier recovery concept covers large frequency offset and enables almost zero residual frequency offset for the V-V CPE. Also, a novel polarization coupling algorithm with reduced complexity is extracted from the work in Ref.[22]

6.5.6 Algorithm for Carrier Phase Recovery

A robust FFCPE and FFCPR with a polarization coupling circuit (MIMO) is shown in Figure 6.24. The FFCPE and FFCPR are developed using the resources of the V-V, and by simple resource reallocation, an enhanced CR is obtained. The proposed method is designed to maximize CR performance with moderate realization complexity. The method consists of four main parts: coarse carrier frequency estimation (CCFE) by processing in slow-speed DSP subsystems (indicated by μc) and recovery (CCFE-R), feedback carrier frequency estimation and recovery (FBCFE-R, dark-light line in Figure 6.25), XY phase offset estimation and compensation (H-V (or XY)-POE-C, -light line in Figure 6.24), and joint H-V feedforward carrier phase estimation and recovery (JFF-CPE-R).

The M-F block conducts the *M*th power operation and averaging in the case of the M-PSK modulation format.[6] For the K-QAM signal, the M constellation points on the ring with a specific radius may be selected and used in the *M*th power block. More rings can be selected to improve the estimation procedure (e.g., inner and outer rings in 16-QAM can be used). So, the use of the method is not limited to PSK schemes.

Large FO $\left(\text{FO} > \pm f_s/(2\,\text{MN})\right)$ can be estimated using the method presented in Ref.[38] (denoted by FE1 in Figure 6.23) that can estimate FO within ±10 GHz FO (linear curve in the case of 112 G PDM-QPSK – polarization division multiplexed QPSK). A short data sequence is transferred into the μc for calculating the coarse range of the FO. Depending on the FO estimation sign, one can superimpose certain FO (e.g. 4 GHz) of the opposite sign. Using linear interpolation (FE2), the current FO can be estimated to be an error within a few hundred MHz. Similarly, depending on the available electrical bandwidth and the receiver structure, compensation for such a large FO can be conducted either in CMOS ASIC or by controlling the frequency of the local oscillator, commonly an external cavity laser.

The residual FO after CCFE and CCFER may have values of several hundred MHz, plus 50 MHz generated by low processing speed and laser frequency oscillations. The FBCFE estimates the

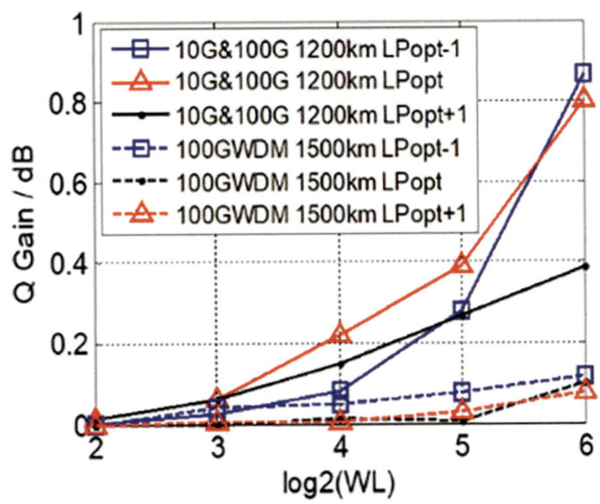

FIGURE 6.23 Penalties of the Viterbi & Viterbi CPR with residual FO of 50 MHz.

residual frequency, which is compensated by the CFR feedback loop. The frequency estimation range of FBCFE&R is $\pm f_s/(2\,\mathrm{MN})$, due to its implementation based on the Differential Viterbi (DiffV) algorithm.

The FBCFE&R compensates for almost the total frequency offset. The residual FO and the influence of finite laser line width can be compensated by the JFF-CPE&R. This operation is supported by ideally coupling the carrier estimations from both polarizations. Before coupling the estimations, both X and Y constellations are aligned using the XY-POE&C. The phase difference between adjacent outputs of the lower unwrapping block can then be used for alignment. It is sufficient to rotate the Y polarization for the constellation angle mismatch.

The algorithm employing MIMO and V-V CPE is compared with the CR, as presented in Figure 6.23, in which the uncompensated FO after the MIMO block is set to 50 MHz to simulate a realistic scenario. The measured FO after the MIMO is around 300 MHz. All data processing algorithms are done block-wise with parallelization of 64 symbols per block and a realistic processing delay is added.[22]

Figure 6.24 shows the gain in the quality factor (Q) for the two measurement cases. Since the FO is recovered by the feedback CR, the penalties shown in Figure 6.25 are completely compensated, with additional gain resulting from estimations of coupling.

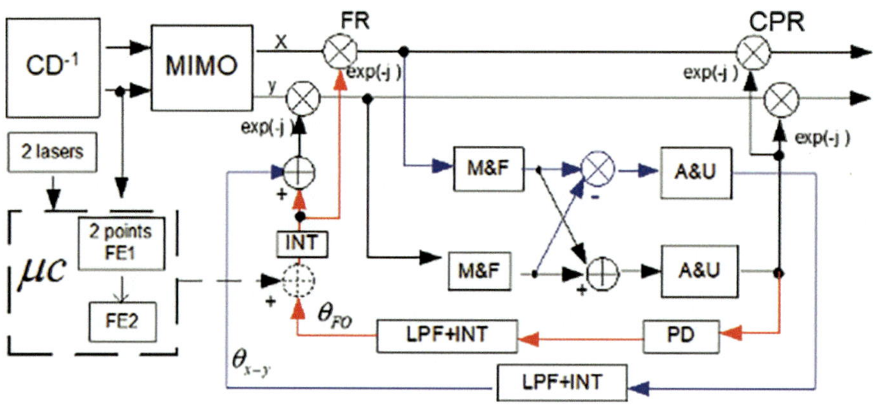

FIGURE 6.24 CR Q-gain versus the natural log of WL.

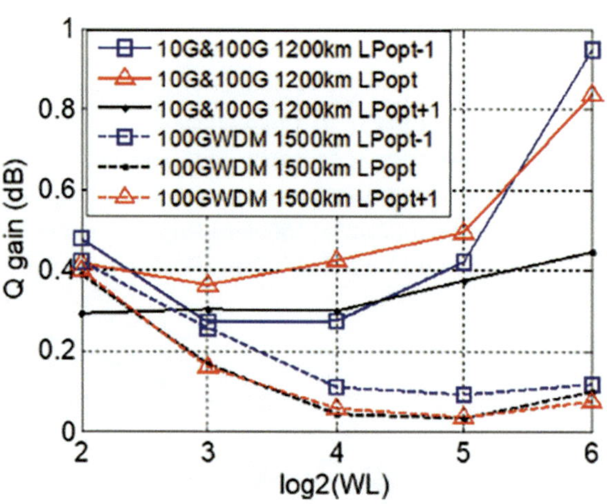

FIGURE 6.25 Modified structure of algorithms for CR. M&F=$M^{\text{-th}}$ power operation and averaging filter, A&U=arctan and unwrapping, PD=phase difference calculation, LPF=low-pass filter, INT=integrator. (Extracted from Ref.[22]).

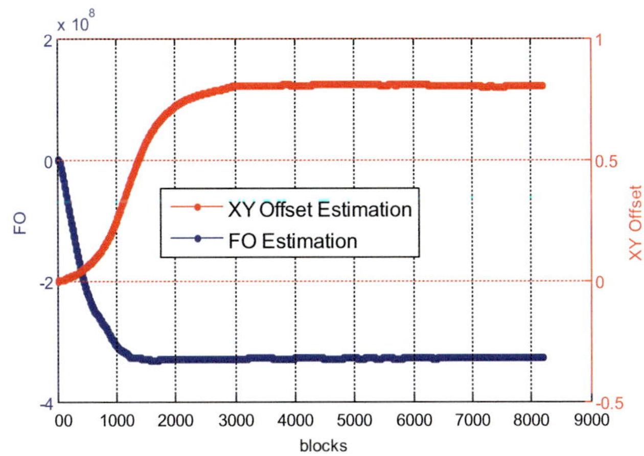

FIGURE 6.26 The 10 G hybrid off-line experiments LOFO and X-Y phase offset estimation by FO variation with respect to the number of blocks. FO = frequency offset – vertical axis.

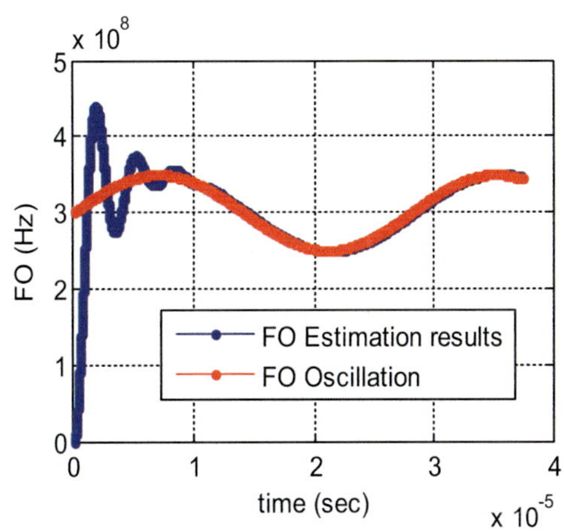

FIGURE 6.27 The 36 kHz oscillation variation of the laser frequency.

The 10 and 100 G hybrid offline optical transmission with homodyne coherent reception experiments are used to demonstrate the convergence of the feedback loops in the modified algorithm shown in Figure 6.26. As shown in Figure 6.27, the frequency offset of 327 MHz can be acquired after 3.4 μs, or 1500 blocks of processing,[22] reaching the steady-state plateau region. The XY constellation offset is acquired after 3000 blocks due to the demand for correct FO compensation. The estimated XY phase offset is 0.8 rad. The CR can cope with a laser frequency oscillation of 36 kHz. Acquisition and tracking times are shown in Figure 6.27, with the CR being able to track the laser frequency changes after 100 ms. Note that the feedback loop can cause a delay between real oscillations and estimation (delay between the two lines in Figure 6.27), resulting in some penalties.

High-precision carrier recovery is achieved through the use of V-V circuits with appropriate allocation and connections. By employing both feedback and feedforward CR, significant improvements can be achieved. Laser frequency offset can be compensated through complex circuit design and appropriate parameter selection. The ideal coupling of estimations from two polarizations further enhances CR performance while reducing less complexity.

6.6 MLSE EQUALIZER-MSK SPACE PHOTONIC TRANSMISSION SYSTEMS

6.6.1 MLSE Equalizer for Optical MSK Systems

6.6.1.1 Configuration of MLSE Equalizer in Optical Frequency Discrimination Receiver (OFDR)

Figure 6.28 shows the block diagram of a narrowband filter receiver integrated with non-linear equalizers for the detection of 40 Gb/s optical MSK signals. The MSK (minimum shift keying) transmitter was described in Chapter 3 and in the Appendices of Chapter 9. In optical modulation techniques, it uses 2 bits/symbol modulation as in the case of QPSK, except that the phase change from one state to another is continuous rather than a discrete step change. Continuous phase change implies frequency step change. So, when the frequency difference between states equals a quarter of the bit rate, the minimum distance between them is such that the channel states can be recovered with minimal ISI. Thus, the two frequency states are orthogonal, and the modulation is termed minimum shift keying (MSK).

In the reception of MSK channels, two narrowband filters can be used to discriminate between the upper sideband (USB) and the lower sideband (LSB) channel frequencies, which correspond to logic ‘1’ and ‘0’ transmitted symbols, respectively. A constant optical delay line, which is easily implemented in integrated optics, is introduced on one branch to compensate for any differential group delay $t_d = 2\pi f_d \beta_2 L$ between f_1 and f_2, where$f_d = f_1 - f_2 = R/2$, β_2 represents group velocity delay (GVD) parameter of the fiber, and L is the fiber length, as described in Chapter 2. Alternatively, de-skewing can be implemented in the digital domain to compensate for this delay difference. If the differential group delay is fully compensated, the optical lightwaves in two paths arrive at the photodiodes simultaneously. The outputs of the filters are then converted to electrical domains through the photodiodes. These two separately detected electrical signals are sampled before being fed as inputs to the non-linear equalizer.

It is noted that the MSK reception system described here can be a self-homodyne detection or a coherent system with the signal channel employed as the LO. Alternatively, the reception subsystem can be a coherent one with the LO oscillator having the power level equal to the average power of the signal.

6.6.1.2 MLSE Equalizer with Viterbi Algorithm

At epoch k, it is assumed that the effect of ISI on an output symbol of the FSM c_k is caused by both executive δ pre-cursor and δ post-cursor symbols on each side. First, a state trellis is constructed with 2δ states for both detection branches of the OFDR. A lookup table per branch corresponding

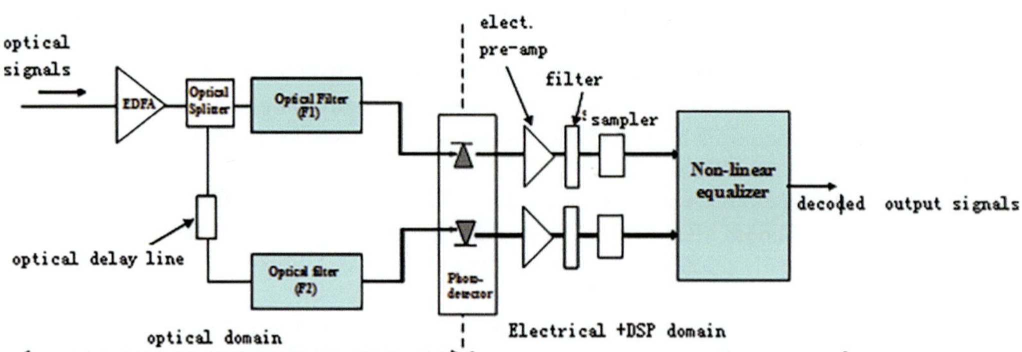

FIGURE 6.28 Block diagram of narrowband filter receiver integrated with non-linear equalizers for the detection of 40 Gb/s optical MSK signals.

to symbols "0" and "1" transmitted and containing all the possible $2^{2\delta}$ states of all 11 symbol-length possible sequences is constructed by sending all the training sequences incremented from 1 to $2^{2\delta}$.

The output sequence $c(k) = f(b_k, b_{k-1}, \ldots, b_{k-2\delta-1}) = (c_1, c_2, \ldots, c_{2\delta})$ is the non-linear function representing the ISI caused by the δ adjacent pre-cursor and post-cursor symbols of the optical fiber FSM. This sequence is obtained by selecting the middle symbols c_k of 2δ possible sequences with a length of $2\delta + 1$ symbol intervals.

The samples of the two filter outputs at epoch k y_k^i, $i = 1, 2$ can be represented as $\mathbf{y}^i(k) = \mathbf{c}^i(k) + \mathbf{n}_{\mathrm{ASE}}^i(k) + \mathbf{n}_{\mathrm{Elec}}^i(k)$, $i = 1, 2$. Here, n_{ASE}^i and n_{Elec}^i represent the ASE noise and the electrical noise, respectively.

In linear transmission of an optical system, the received sequence $\mathbf{y}_n$ is corrupted by the ASE noise of the optical amplifiers, $\mathbf{n}_{\mathrm{ASE}}$ and the electronic noise of the receiver, $\mathbf{n}_E$. It has been proven that the calculation of branch metric and hence state metric is optimum when the noise distribution follows the normal/Gaussian distribution, meaning the ASE accumulated noise and the electronic noise of the reception sub-system are collectively modelled as samples from Gaussian distributions. If noise distribution departs from the Gaussian distribution, the minimization process is sub-optimal.

The Viterbi algorithm subsystem is implemented on each detection branch of the OFDR. However, the MLSE with the Viterbi algorithm may be too computationally complex to be implemented at 40 Gb/s with the current integration technology. However, commercial products have been available for 10 Gb/s optical systems. Thus, a second MLSE equalizer using the technique of reduced-state template matching is presented in the next section.

In an optical MSK transmission system, narrowband optical filtering plays the main role in shaping the noise distribution back to the Gaussian profile. The Gaussian-profile noise distribution is verified. Thus, branch metric calculation in the Viterbi algorithm based on minimum Euclidean distance over the trellis can achieve the optimum performance. Additionally, the computational effort is less complex than ASK or DPSK systems due to the issue of non-Gaussian noise distribution.

6.6.1.3 MLSE Equalizer with Reduced-State Template Matching

The modified MLSE is a single-shot template matching algorithm. First, a table of $2^{2\delta+1}$ templates, $\mathbf{g}^k, k = 1, 2, \ldots, 2^{2\delta+1}$ corresponding to the $2^{2\delta+1}$ possible information sequences, I^k of length $2\delta + 1$, is constructed. Each template is also a vector of size $2\delta + 1$, obtained by transmitting the corresponding information sequence through the optical channel and obtaining the $2\delta + 1$ consecutive received samples. At each symbol period n, the sequence $\hat{I}_n$ with the minimum metric is selected as $\mathbf{c}(k) = \arg\min_{b(k)} \{m(\mathbf{b}(k), \mathbf{y}(k))\}$. The middle element of the selected information sequence is then output as the nth decision, $\hat{I}_n$ i.e. $\hat{I}_n = \hat{I}_n(\delta+1)$. The minimization is performed over the information sequences, $\mathbf{I}^k$, which satisfies the condition that the $\delta - 2$ elements are equal to the previously decoded symbols $\hat{I}_{n-\delta}, \hat{I}_{n-(\delta-1)}, \ldots, \hat{I}_{n-2}$, and the metric $m(\mathbf{I}^k, \mathbf{r}_n)$ is given by $m(\mathbf{I}^k, \mathbf{r}_n) = \{\mathbf{w} \cdot (\mathbf{g}^k - \mathbf{r}_n)\}^{\mathrm{T}} \{\mathbf{w} \cdot (\mathbf{g}^k - \mathbf{r}_n)\}$, where $\mathbf{w}$ is a weighting vector chosen carefully to improve the reliability of the metric. The weighting vector is selected so that when the template is compared with the received samples, less weight is given to the samples further away from the middle sample. For example, numerical results show that a weighting vector with elements $\mathbf{w}(i) = 2^{-|i-(\delta+1)|}$, $i = 1, 2, \ldots, 2\delta + 1$ gives good results. Here, $\cdot$ represents Hadamard multiplication of two vectors, $(\cdot)^T$ represents transpose of a vector, and $|\cdot|$ is the modulus operation.

6.6.2 MLSE Scheme Performance

6.6.2.1 Performance of MLSE Schemes in 40 Gb/s Transmission Systems

Figure 6.28 shows the simulation system configuration used to investigate the performance of both schemes when used with the narrowband optical Gaussian filter receiver for the detection of

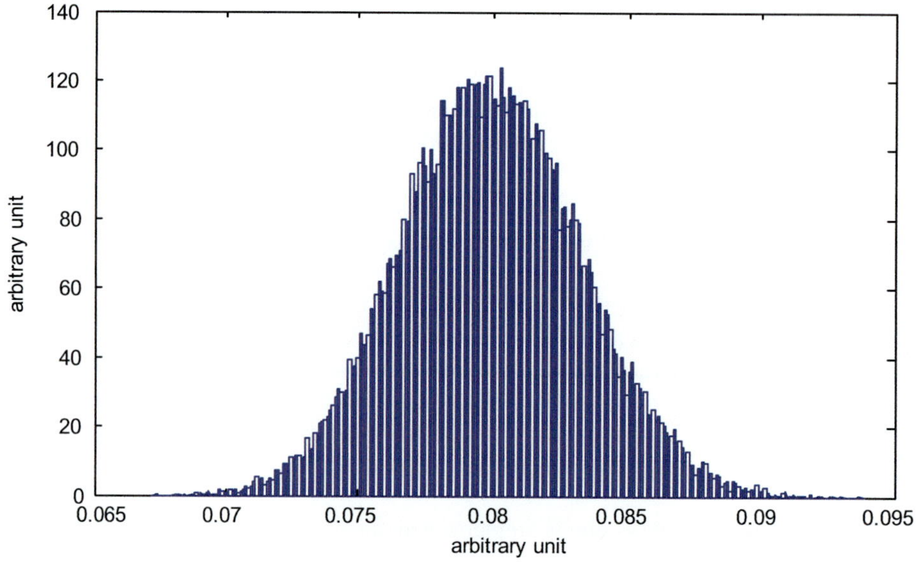

FIGURE 6.29 Noise distribution following a Gaussian shape due to narrowband optical filtering.

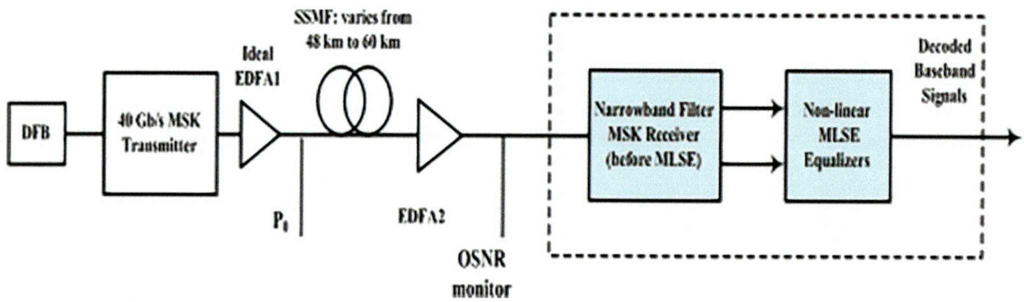

FIGURE 6.30 Simulation setup for performance evaluation of MLSE and modified MLSE schemes for detection of 40 Gb/s optical MSK systems. DFB laser=distributed feedback laser, MSK=minimum shift keying, EDFA=Erbium-doped fiber amplifier, OSNR=optical signal-to-noise ratio (power term – unit dB dimensionless).

non-coherent 40 Gb/s optical MSK systems. The input power into the fiber (P_0) is –3 dBm, which is much lower than the non-linear threshold power. The EDFA2 provides 23 dB gain to maintain the receiver sensitivity of –23.2 dBm at a BER = 10^{-9} (Figure 6.29). As shown in Figure 6.30, the optical received power (P_{Rx}) is measured at the input of the narrowband MSK receiver, and the OSNR is monitored to obtain the BER curves for different fiber lengths. The length of the SSMF is varied from 48 to 60 km in steps of 4 km to investigate the performance of the equalizers to the degradation caused by fiber cumulative dispersion. The narrowband Gaussian filter with a time-bandwidth product of 0.13 is used for the detection filters.

The electronic noise of the receiver can be modeled with the equivalent noise current density of an electrical amplifier of $20\,\text{PA}/\sqrt{\text{Hz}}$ and dark current of 10 nA. The key parameters of the transmission system are given in Table 6.2.

The Viterbi algorithm used with the MLSE equalizer has a constraint length of 6 (that is 2^5 states) and a trace-back length of 30. Figure 6.31 shows the BER performance of both non-linear

TABLE 6.2
Key Simulation Parameters Used in the MLSE MSK Modulation Format

Input power: $P_0=-3$ dBm	Narrowband Gaussian filter: $B=5.2$ GHz or BT$=0.13$
Operating wavelength: $\lambda=1550$ nm	Constant delay: $t_d=\lvert 2\pi f_D \beta_2 L\rvert$
Bit rate: $R=40$ Gb/s	Pre-amp EDFA of the OFDR: $G=15$ dB and NF$=5$ dB
SSMF fiber: $\lvert\beta_2\rvert=2.68\text{e-}26$ or $\lvert D\rvert=17$ ps/nm/km	$i_d=10$ nA
Attenuation: $\alpha=0.2$ dB/km	$N_{eq}=20$ pA/(Hz)$^{1/2}$

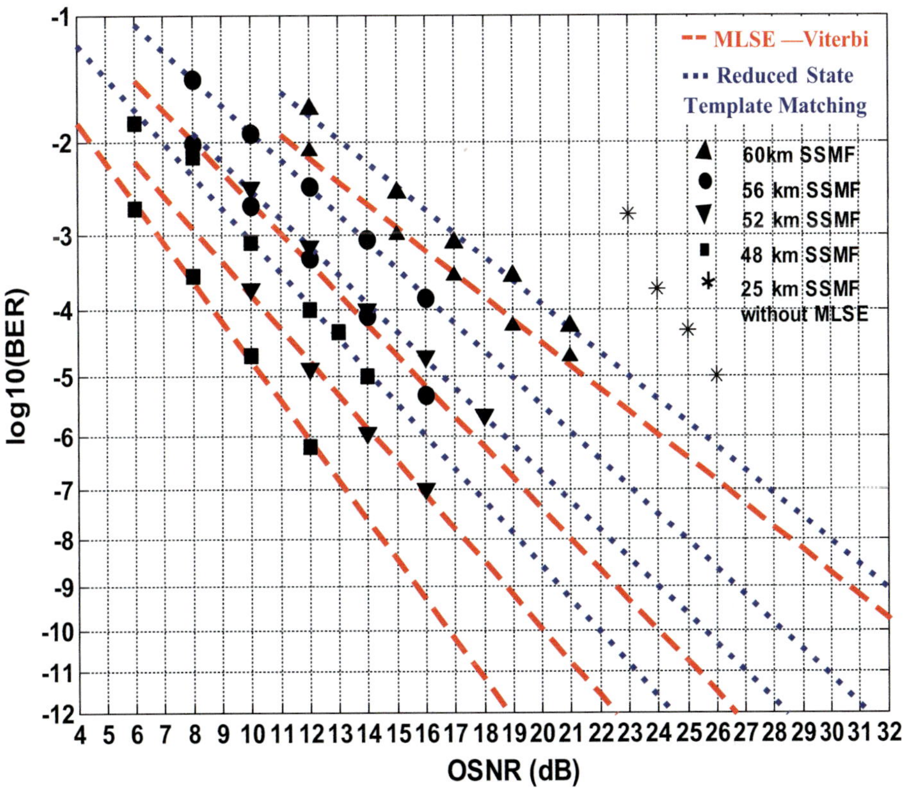

FIGURE 6.31 Performance of modified Viterbi-MLSE and template matching schemes, BER versus OSNR.

equalizers plotted against the required OSNR. The BER performance of the optical MSK receiver without any equalizers for 25 km SSMF transmission is also shown in Figure 6.32 for quantitative comparison. The numerical results are obtained via Monte Carlo simulation (triangular markers as shown in Figure 6.31), with the low BER tail of the curve linearly extrapolated. The OSNR penalty (at BER$=10^{-9}$) versus residual dispersion corresponding to 48, 52, 56, and 60 km SSMF are presented in Figure 6.32. The MLSE scheme outperforms the modified MLSE schemes, especially at low OSNR. In the case of 60 km SSMF, the improvement at BER$=10^{-9}$ is approximately 1 dB, and in the case of 48 km SSMF, the improvement is about 5 dB. In the case of transmission of 52 and 56 km SSMF, MLSE with the Viterbi algorithm has 4 dB gain in OSNR compared to the modified scheme. With residual dispersion at 816, 884, 952, and 1020 ps/nm, at BER of 10^{-9}, the MLSE scheme requires 12, 16, 19, and 26 dB OSNR penalties, respectively, while the modified scheme requires 17, 20, 23, and 27 dB OSNR, respectively.

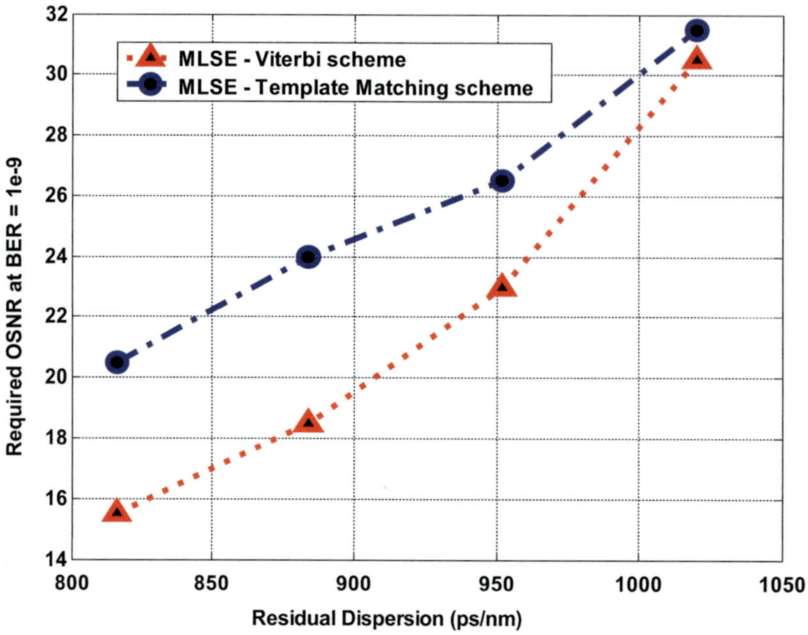

FIGURE 6.32 Required OSNR (at BER = 1e-9) versus residual dispersion in Viterbi-MLSE and template-matching MLSE schemes.

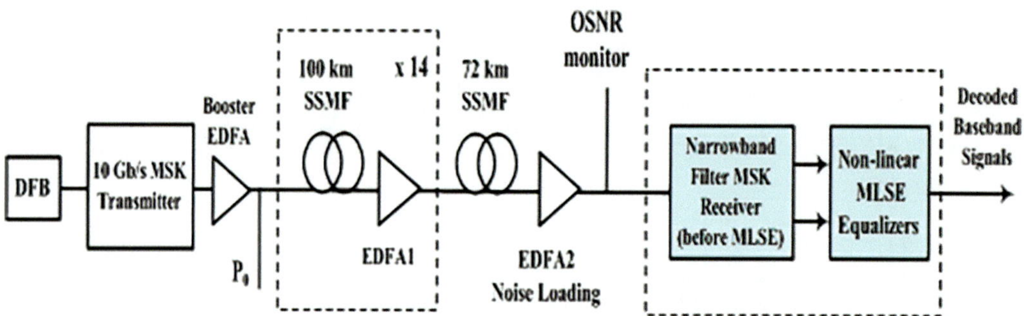

FIGURE 6.33 Simulation setup for transmission of 10 Gb/s optical MSK signals over 1472 km SSMF with MLSE-Viterbi equalizer integrated with the narrowband optical filter receiver.

6.6.2.2 Transmission of 10 Gb/s Optical MSK Signals over Long Haul SSMF Uncompensated Optical Link

Figure 6.33 shows the simulation setup for 10 Gb/s transmission of optical MSK signals over 1472 km SSMF. The receiver employs an optical narrowband frequency discrimination receiver integrated with a 1024-state Viterbi-MLSE post-equalizer. The input power into the fiber (P_0) is –3 dBm, which is lower than the fiber non-linear threshold power. The optical amplifier EDFA1 provides an optical gain to completely compensate for the attenuation of each span. EDFA2 is used as a noise-loading source to vary the required OSNR. The receiver's electronic noise is modeled with an equivalent noise current density of the electrical amplifier of 20 $20\,\text{pA}/\sqrt{\text{Hz}}$ and dark current of 10 nA for each photodiode. A narrowband optical Gaussian filter with a two-sided bandwidth of 2.6 GHz (one-sided BT = 0.13) is optimized for detection. A back-to-back OSNR = 8 dB is required

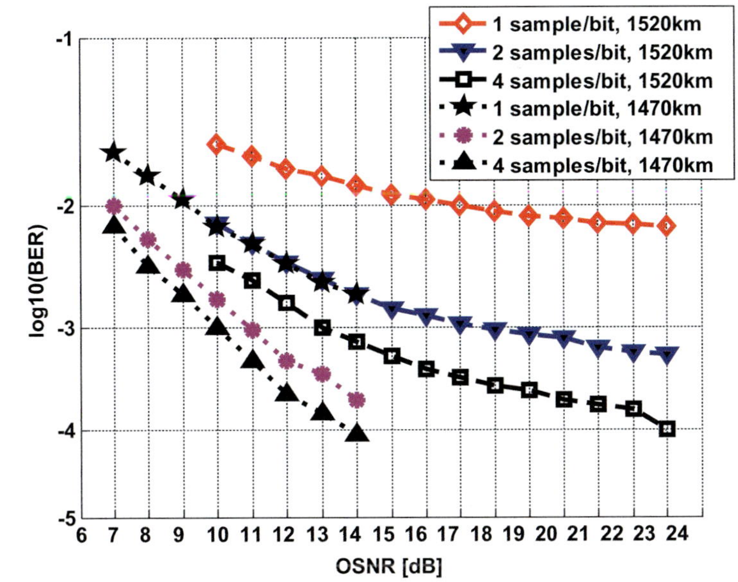

FIGURE 6.34 BER vs required OSNR for transmission of 10 Gb/s optical MSK signals over 1472 and 1520 km SSMF uncompensated optical links.

for a BER at 10^{-3} for each branch. The correspondent received power is –25 dBm. This low OSNR is possible due to the large suppression of noise after being filtered by narrowband optical filters. A trace-back length of 70 is used in the Viterbi algorithm. Figure 6.33 shows the simulation results of BER versus the required OSNR for 10 Gb/s optical MSK transmission over 1472 and 1520 km SSMF uncompensated optical links with 1, 2, and 4 samples per bit, respectively. The 1520 km SSMF with one sample per bit is seen as the limit for the 1024-state Viterbi algorithm due to the slow roll-off and the error floor. However, with 2 and 4 samples per bit, error values lower than the FEC limit of 10^{-3} can be obtained.

Thus, 1520 km SSMF transmission of 10 Gb/s optical MSK signals can reach error-free detection with the use of a high-performance FEC. In the case of 1472 km SSFM transmission, the error events follow a linear trend without signs of an error floor, and therefore, error-free detection can be comfortably accomplished.

Figure 6.34 also shows significant improvement in OSNR of 2 and 4 samples per bit over 1 sample per bit counterpart with values of approximately 5 and 6 dB, respectively. In terms of OSNR penalty at BER of 1e-3 from back-to-back setup, 4 samples per bit for 1472 and 1520 km transmission distance suffer 2 and 5 dB penalties, respectively.

6.6.2.3 Performance Limits of Viterbi-MLSE Equalizers

The performance limits of the Viterbi-MLSE equalizer to combat ISI (inter-symbol interference) effects are investigated against various SSMF lengths of the optical link. The number of states used in the equalizer is incremented accordingly to this increase and varies from 2^6 to 2^{10}. This range was chosen to reflect the current feasibility and the future advance of electronic technologies that can support high-speed processing of the Viterbi algorithm in the MLSE equalizer. In addition, these numbers of states also provide feasible simulation time.

In addition, one possible solution to ease the requirement for improving the performance of the equalizer without increasing the complexity is by multi-sampling within the one-bit period. This technique can be performed by interleaving the samplers at different times. Although a greater number of electronic samplers are required, they only need to operate at the same bit rate as the received

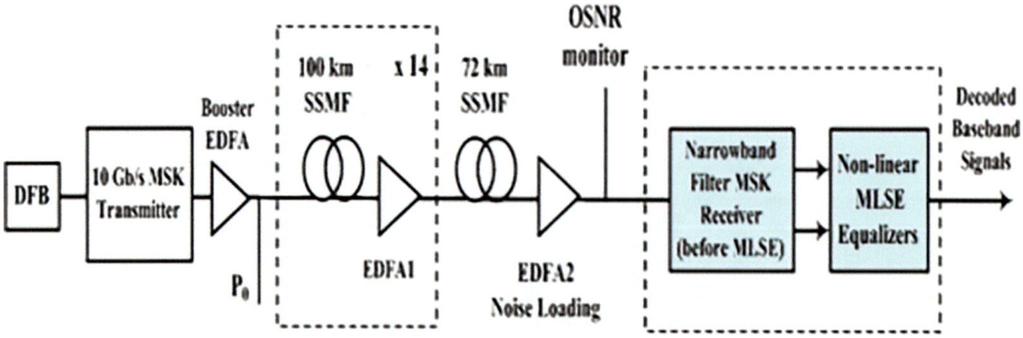

FIGURE 6.35 Simulation setup of OFDR-based 10 Gb/s MSK optical transmission for study of performance limits of MLSE-Viterbi equalizer.

MSK electrical signals. Moreover, it will be shown later that there is no noticeable improvement with more than 2 samples per bit period. Hence, the complexity of the MLSE equalizer can be affordable while improving the performance significantly.

Figure 6.35 shows the simulation setup for 10 Gb/s optical MSK transmission systems with lengths of uncompensated optical links varying up to 1472 km SSMF. In this setup, the input power into fiber (P_0) is set to be –3 dBm, thus avoiding the effects of fiber nonlinearities. The optical amplifier EDFA1 provides an optical gain to compensate for the attenuation of each span completely. The EDFA2 is used as a noise-loading source to vary the required OSNR values. Moreover, a Gaussian filter with a two-sided 3-dB bandwidth of 9 GHz (one-sided BT=0.45) is utilized as the optical discrimination filter because this BT product gives the maximized detection of eye openings. The receiver electronic noise is modeled with the equivalent noise current density of the electrical amplifier of 20 PA/Hz and a dark current of 10 nA for each photodiode. A back-to-back OSNR of 15 dB is required for a BER of 10^{-4} on each branch, and the correspondent receiver sensitivity is –25 dBm. A trace-back length of 70 is used for the Viterbi algorithm in the MLSE equalizer. Figures 6.35–6.37 show the BER performance curves of 10 Gb/s OFDR-based MSK optical transmission systems over 928, 960, 1280, and 1470 km SSMF uncompensated optical links for different numbers of states used in the Viterbi-MLSE equalizer. In these figures, the performance of Viterbi-MLSE equalizers is given by the plot of the BER versus the required OSNR for several detection configurations: balanced receiver (without incorporating the equalizer), the conventional single-sample per bit sampling technique, and the multi-sample per bit sampling techniques (2 and 4 samples per bit-slot). The simulation results are obtained by the Monte Carlo method.

The significance of multi-samples per bit-slot in improving the performance of MLSE equalizer in cases of uncompensated long distances is shown. It is found that the tolerance limits to the ISI effects induced by the residual CD of a 10 Gb/s MLSE equalizer using $2^6, 2^8, 2^{10}$ states are approximately equivalent to lengths of 928, 1280, and 1440 km SSMF, respectively. The equivalent numerical figures in the case of 40 Gb/s transmission correspond to lengths of 62, 80, and 90 km SSMF, respectively.

In the case of 64 states over 960 km SSMF uncompensated optical link, the BER curve encounters an error floor that cannot be overcome even by using high-performance FEC schemes. However, at 928 km, the linear BER curve indicates the possibility of recovering the transmitted data with the use of high-performance FEC. Thus, for 10 Gb/s OFDR-based MSK optical systems, a length of 928 km SSMF can be considered the transmission limit for the 64-state Viterbi-MLSE equalizer. Results shown in Figure 6.36 suggest that the length of a 1280 km SSMF uncompensated optical link is the transmission limit for the 256-state Viterbi-MLSE equalizer when incorporated with an OFDR optical front-end. It should be noted that this is achieved when using multi-sample per bit sampling schemes.

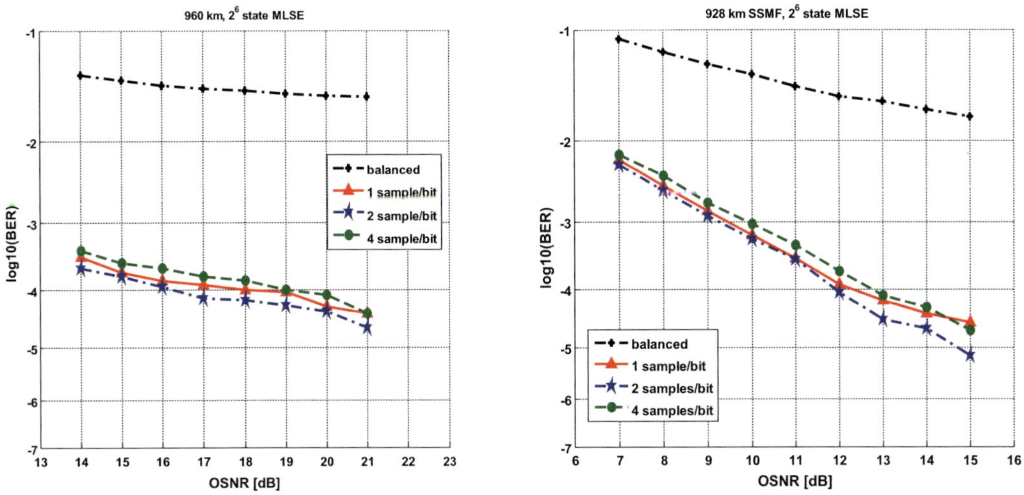

(a) (b)

FIGURE 6.36 Performance of 64-state Viterbi-MLSE equalizers for 10 Gb/s OFDR-based MSK optical systems over (a) 960 km and (b) 928 km SSMF uncompensated links.

It is observed from Figure 6.38 that the transmission length of 1520 km SSMF with one sample per bit is seen as the limit for the 1024-state Viterbi algorithm due to the slow roll-off and the error floor. However, 2 and 4 samples per bit can obtain error values lower than the FEC limit of 10^{-3}. Thus, 1520 km SSMF transmission of 10 Gb/s optical MSK signals can reach error-free detection with the use of a high-performance FEC. In the case of 1472 km SSFM transmission, the error events follow a linear trend without sign of error floor, and therefore, error-free detection can be achieved. Figure 6.37 also shows the significant OSNR improvement of the sampling techniques with two and four samples per bit compared to the single-sample per-bit counterpart. In terms of OSNR penalty at BER of 1e-3 from back-to-back setup, 4 samples per bit for 1472 and 1520 km transmission distance suffer 2 and 5 dB penalties, respectively.

It is found that for incoherent detection of optical MSK signal based on the OFDR, MLSE equalizer using 2^4 states do not offer better performance than the balanced configuration of the OFDR itself, i.e., the uncompensated distance is not over 35 km SSMF for 40 Gb/s or 560 km SSMF for 10 Gb/s transmission systems, respectively. It is most likely that in this case, the severe ISI effect caused by the optical fiber channel has spread beyond the time window of 5-bit slots (2 pre-cursor and 2 post-cursor bits), the window which 16-state MLSE equalizer can handle. Thus, the number of taps or states of MLSE must cover the full length of the dispersive pulse.

The trace-back length used in the investigation is chosen to be 70, which guarantees the convergence of the Viterbi algorithm. The longer the trace-back length, the larger memory is required. With state-of-the-art technology for high storage capacity nowadays, memory is no longer a big issue. Very fast processing speed at 40 Gb/s operations hinders the implementation of 40 Gb/s Viterbi-MLSE equalizers in the meantime. Multi-sample sampling schemes offer an exciting solution for implementing a fast signal-processing process. This challenge may also be overcome shortly together with the advance of the semiconductor industry. At present, the realization of Viterbi-MLSE equalizers operating at 10 Gb/s has been commercially demonstrated.[39]

Fiber-equivalent free-space transmission istance can be derived and this distance must meet the attenuation budget for link error-free performance. Thence, the BER and channel powr penalty for the transmission performance can be obtained as described here.

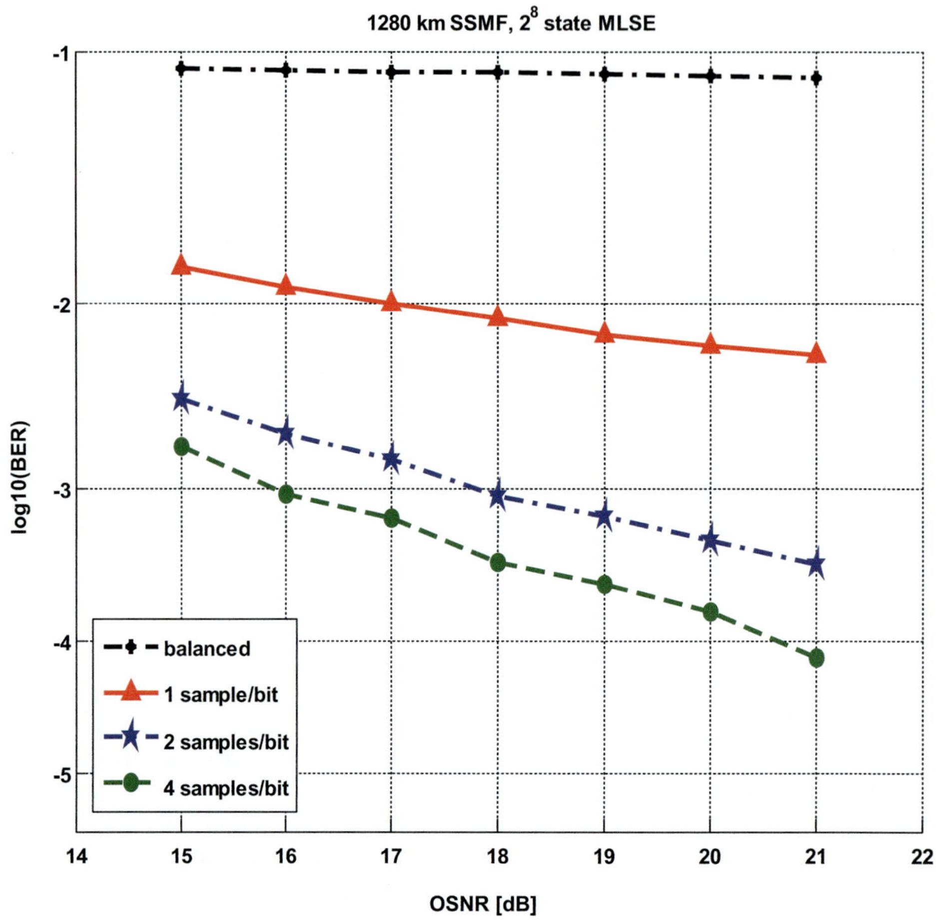

FIGURE 6.37 Performance of 256-state Viterbi-MLSE equalizer for 10 Gb/s OFDR-based MSK optical systems over a 1280 km SSMF uncompensated link.

6.6.2.4 Viterbi-MLSE Equalizers for PMD Mitigation

Figure 6.39 illustrates the simulation test-bed for the investigation of MLSE equalization of the PMD effect. The transmission link consists of several spans comprised of 100 km SSMF (with $D = +17$ ps/nm km, $\alpha = 0.2$ dB/km) and 10 km DCF (with $D = -170$ ps/nm km, $\alpha = 0.9$ dB/km). Launched power into each span (P_0) is –3 dBm. TheEDFA$_1$ has a gain of 19 dB, hence providing input power into the DCF to be –4 dBm, which is lower than the non-linear threshold of the DCF. The 10 dB gain of EDFA$_2$ guarantees the input power into the next span unchanged with an average power of –3 dBm. An OSNR of 10 dB is required for receiver sensitivity at BER of 10^{-4} in case of back-to-back configuration. Considering the practical aspect and complexity of a Viterbi-MLSE equalizer for PMD equalization, a small number of 4-state bits, or effectively 16 states, were chosen for the Viterbi algorithm in the simulation study.

The performance of MLSE against the PMD dynamic of optical fiber is investigated. BER versus required OSNR for different values of normalized average differential group delay (DGD) (Δt) are shown in Figure 6.40. The mean DGD factor is normalized over a one-bit period of 100 and 25 ps for 10 and 40 Gb/s bit rate, respectively. The numerical studies are conducted for a range of values from 0 to 1 of normalized DGD (Δt), which equivalently corresponds to the instant delays of up to 25 or 100 ps in terms of 40 and 10 Gb/s transmission bit rate, respectively.

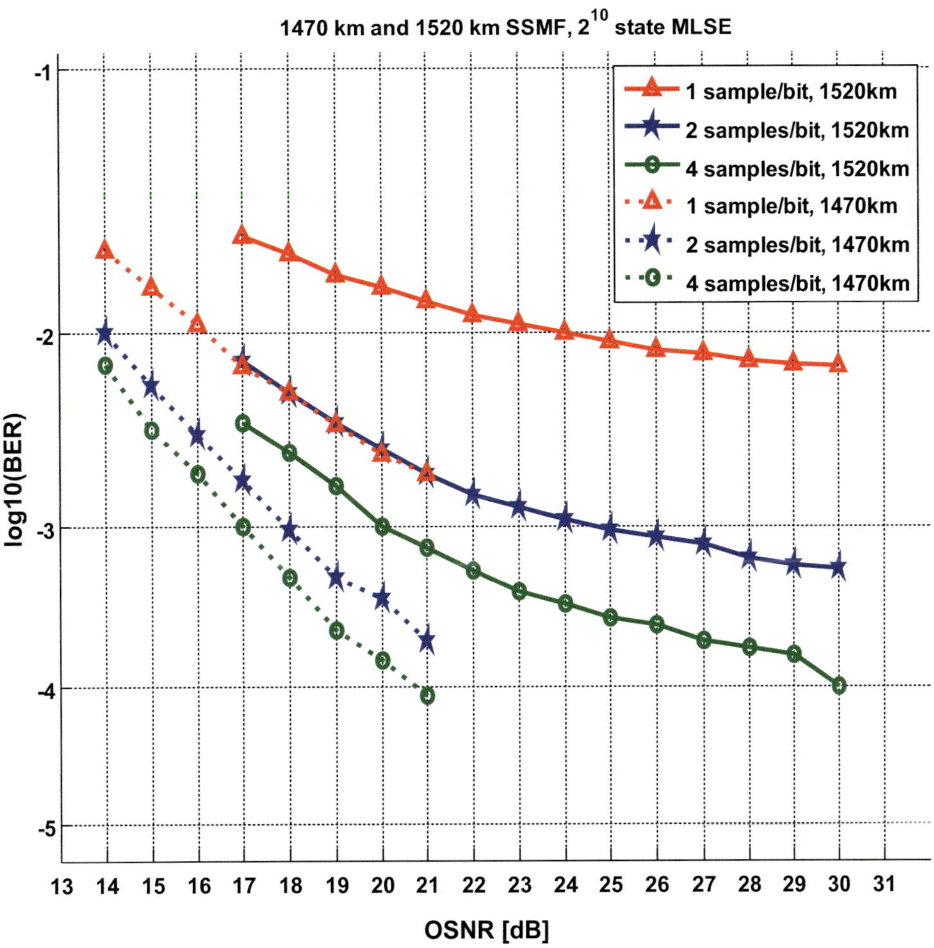

FIGURE 6.38 Performance of 1024-state MLSE equalizer for 10 Gb/s optical MSK signals vs required OSNR over 1472 and 1520 km SSMF uncompensated optical links.

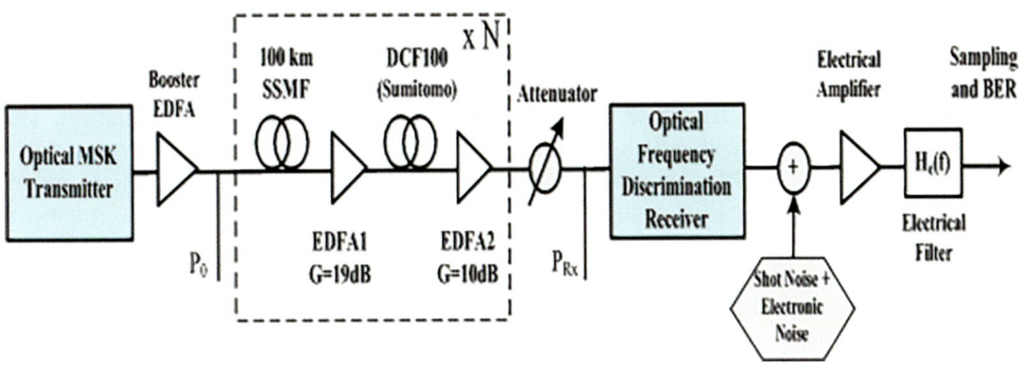

FIGURE 6.39 Simulation test-bed for investigation of the effectiveness of the MLSE equalizer to PMD.

The advantages of using multiple samples per bit over the conventional single sample per bit in MLSE equalizer are also numerically studied, and the Monte Carlo results for normalized DGD values of 0.38 and 0.56 are shown in Figures 6.41 and 6.42. However, the increase from two to four

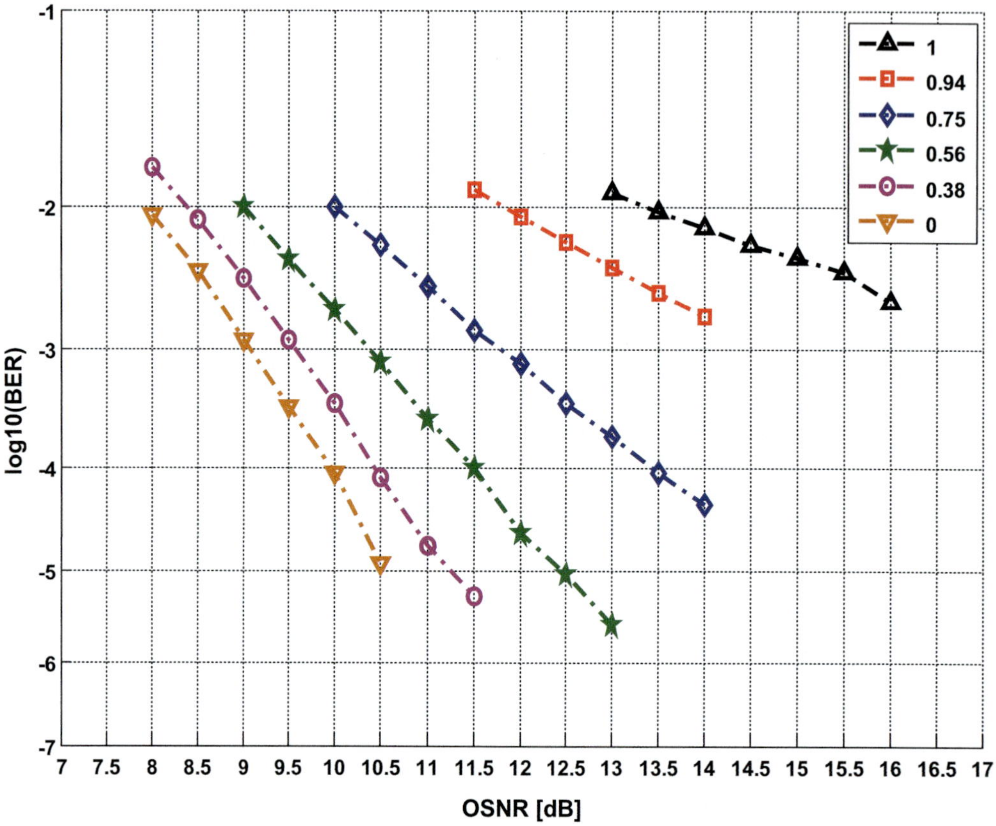

FIGURE 6.40 Performance of MLSE equalizer versus normalized ($\Delta\tau$) values for one sample per bit

samples per bit does not offer any gain in the performance of the Viterbi-MLSE equalizer. Thus, two samples per symbol are preferred to reduce the complexity of the equalizer. Here and for the rest of this chapter, multi-samples per bit implies the implementation of two samples per symbol. The performance of the MLSE equalizer for different normalized DGD values with two samples per bit is shown in Figure 6.43. In Figure 6.44, the important remark is that a 16-state MLSE equalizer implementing 2 samples per symbol enables the optical MSK transmission systems to achieve a PMD tolerance of up to one-bit period at BER=10^{-4} with a required OSNR of 8 dB. This delay value starts introducing a BER floor indicating the limit of the 16-state MLSE equalizer. This problem can be overcome with a high-performance FEC. However, the DGD mean value of 0.94 can be considered the limit for acceptable performance of the proposed MLSE equalizer without the aid of high-performance FEC. Figure 6.45 shows the required OSNR for MLSE performance at 10^{-4} versus normalized (Δt) values in configurations of a balanced receiver, 1 sampler per bit, and 2 samples per bit.

A balanced receiver with no MLSE equalizer incorporated requires an OSNR=5 dB to obtain a BER = 10^{-4} when there are no effects of the PMD compared to the required OSNR of 3 and 1 dB in cases of 1 sample per bit and 2 samples per bit, respectively. The OSNR penalty for various normalized (Δt) values of the above three transmission system configurations are shown in Figure 6.45.

It can be observed that the OSNR penalties (back-to-back) of approximately 3 and 1 dB apply to the cases of the balanced receiver and 1 sample per bit, respectively with reference to the two samples per bit configuration. Another important remark is that the 16-state Viterbi-MLSE equalizer, which implements two samples per bit, enables optical MSK transmission systems to achieve

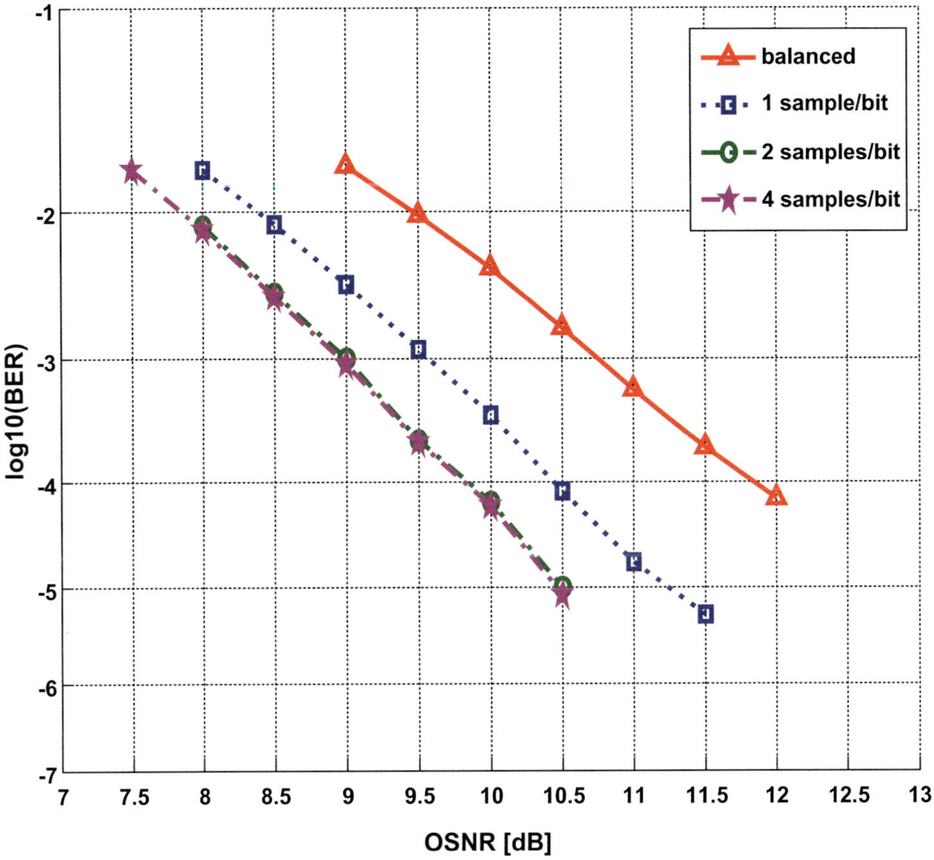

FIGURE 6.41 Comparison of MLSE performance for configurations of 1, 2, and 4 samples per bit with normalized ($\Delta\tau$) value of 0.38.

a PMD tolerance of up to one bit period at a BER of 10^{-4} with a power penalty of about 6 dB. Moreover, the best 2-dB penalty occurs at 0.75 for the value of normalized (Δt). This result shows that the combination of OFDR-based MSK optical systems and Viterbi-MLSE equalizers, particularly with the use of multi-sample sampling schemes, was found to be highly effective in combating fiber PMD dynamic impairment and is better than recently reported PMD performance for OOK and DPSK modulation formats.[40]

6.6.2.5 On the Uncertainty and Transmission Limitation of Equalization Process

The fundamental limitations of quadratic phase media on transmission speed are formulated in signal space for significant applications in signal equalization. These limitations are quantitatively derived for both coherent and incoherent optical systems. This section investigates the dispersive effect of optical fiber on the transmission speed of the media. The main effect of dispersion is pulse broadening (and frequency chirping) as the signal pulses propagate through the quadratic phase media. This causes ISI, thereby limiting the speed of transmission. There are various ways to mitigate the harmful effect of ISI. One way is to use optical fiber with anti-dispersive properties to equalize the dispersion. In practice, this optical scheme reduces signal level and thus requires amplification. The amplifier introduces noise, which in turn limits the channel speed (Shannon's information capacity theorem). Another way of combating ISI is through digital electronic equalization means such as MLSE, which is applicable for both coherent and incoherent

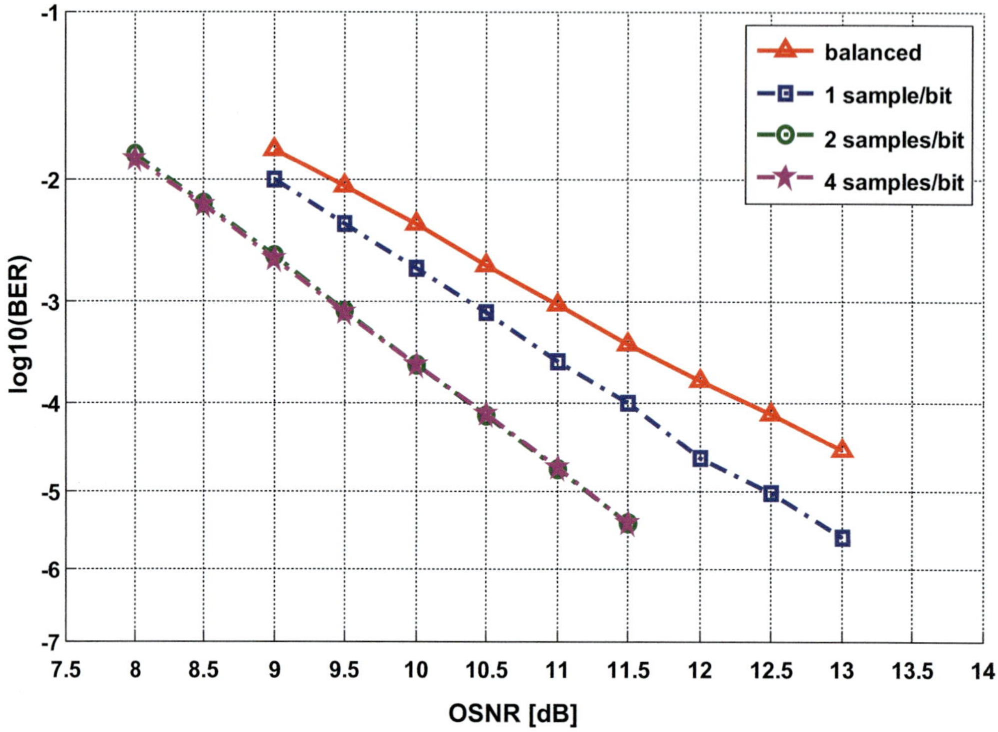

FIGURE 6.42 Performance of Viterbi-MLSE equalizer for configurations of 1, 2, and 4 samples per bit with a normalized (Δt) value of 0.56.

detection schemes.[41,42,43,44,45] These reported results clearly show that there is a fundamental limit on the achievable transmission speed for a given fiber length.

The section studies the limitations of the quadratic phase channel, the single-mode optical fiber, in terms of signal space. The signal space is chosen so that it consists of binary signals in an 8-bit block code. As the 8-bit symbols propagate down the quadratic phase channel, naturally the waveform patterns of the symbols become less and less distinctive, making it more and more difficult to discriminate between symbols. How do we quantify the detrimental effect of the quadratic phase channel? Although it is obvious that the more information we have at hand, the more accurately one can obtain the BER. This system performance is approached entirely from a digital communication perspective, and the results derived are the fundamental limits imposed by the quadratic phase channel (pure phase distortion but quadratic or square dependence on frequency due to the linear dispersion). It is virtually independent of the detection scheme used. Two mechanisms that limit the transmission speed are explained. One is brought about by the finite time window available for detection. In all practical schemes, the decoder must decode each symbol within a finite time. This requires the algorithm the least complex as possible. The other is brought about by not using the phase information for detection. Depending on the complexity of the detection scheme chosen, it is quite often that the phase information may be lost inadvertently when the optical signal is converted into an electrical signal. Another issue is the power consumption of the ASIC (application-specific integrated circuit) for implementing the digital algorithm for real-time applications. The more complex the algorithms are, the higher the number of digital circuits and hence the more power consumption. It is expected that all ASICs must consume <70 W in real-time processing.

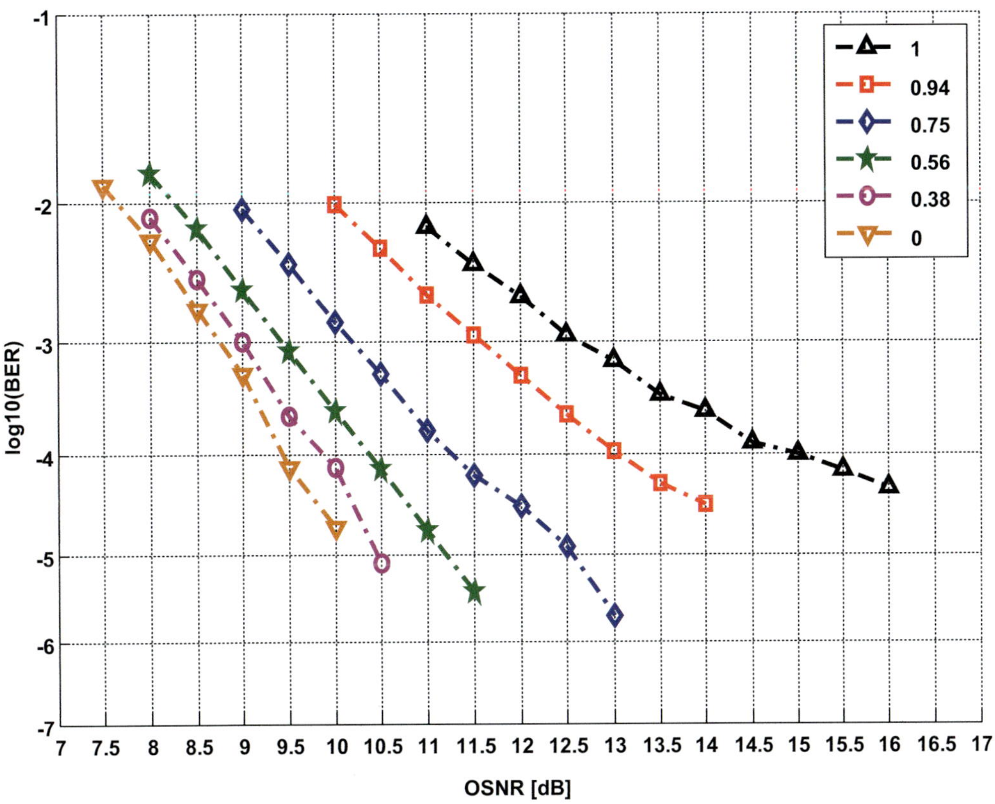

FIGURE 6.43 Performance of Viterbi-MLSE equalizer for different normalized (Δt) values with 2 and 4 samples per bit.

We thus can see that DSP algorithms are most critical in modern DSP-based optical receivers or transponders. This chapter introduces the fundamental aspects of these algorithms for optical transmission systems over dispersive channels whether under linear or nonlinear distortion physical effects.

6.6.2.6 Equalization with Minimizing MSE in Equalized Signals

As we have seen, equalization using nonlinear equalizers was described in the previous section. The decision or detection process will enhance the SNR performance of the system. However, under what criteria that the NLE can make the decision block so that an optimum performance can be achieved. The mean square error can be employed as described above, to minimize the errors in the decision stage of the NLE.

6.6.2.6.1 Optical Fiber as a Finite State Machine

A structure of the Finite State Machine is shown in Figure 6.46 and is now replaced with the optical channel based on optical fiber. It is significant to understand how the state trellis structure is formed for an optical transmission system. In this case, the optical fiber involves all the ISI sources causing waveform distortion of the optical signals, including CD, PMD, and filtering effects. Thus, the optical fiber FSM excludes signal corruption by noise as well as random-process nonlinearities. The ISI is caused by dispersion effects, nonlinear phase, ASE, and nonlinear effects[46]

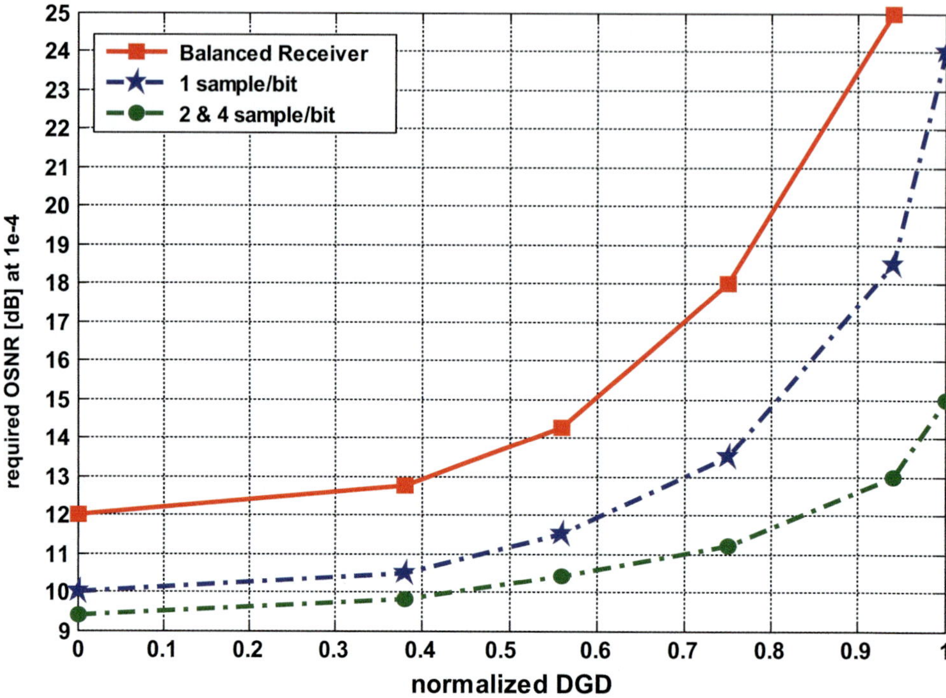

FIGURE 6.44 Required OSNR for MLSE performance at 10^{-4} versus normalized (Δt) values in configurations of balanced receiver and 1 and 2 samples per bit.

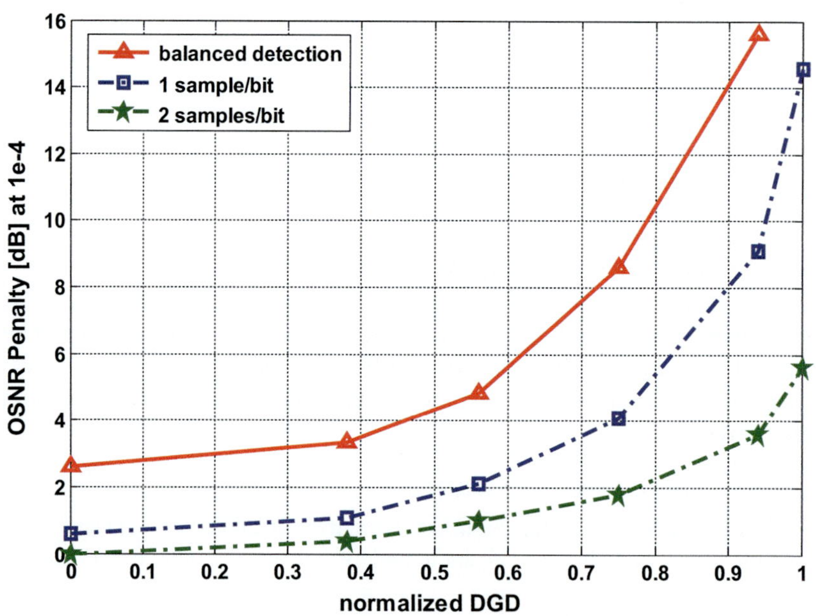

FIGURE 6.45 OSNR penalties of MLSE performance at 10^{-4} for various normalized (Δt) values in configurations of balanced receiver, single-sample, and 2 samples per bit sampling schemes.

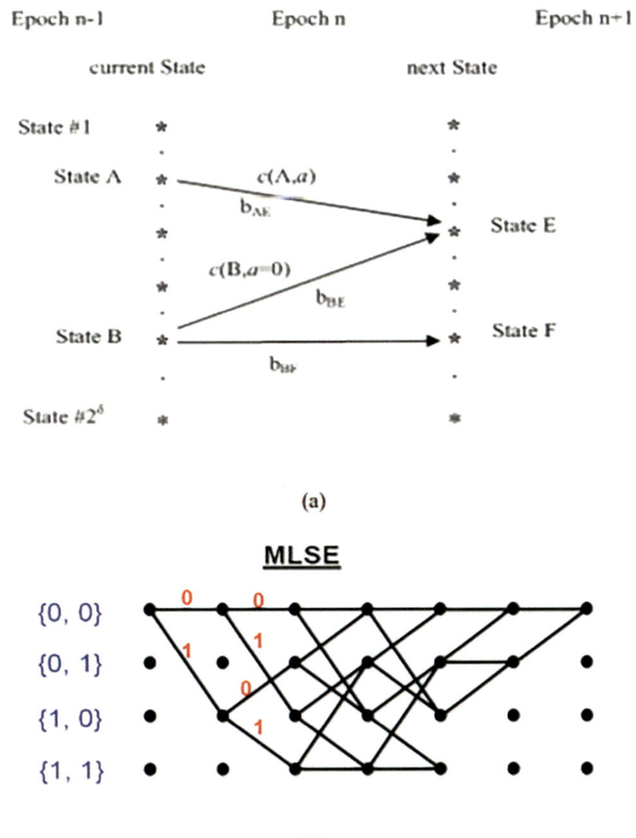

FIGURE 6.46 State trellis of a finite state machine (a) Trellis path, (b) tracing of possible maximum likelihood estimation with assigned bit pattern.

6.7 SHARED EQUALIZATION BETWEEN TRANSMITTER AND RECEIVERS

It is commonly known in optical communications that DSP is employed to compensate for and mitigate the propagation distortion of advanced modulation formats, especially in fiber optic transmission lines. In a space photonic transmission system, impairments rise from random air disturbances due to Mie scattering when the particles are bigger than the carrier wavelength and the scattering due to molecules. Second, the significant distortion occurs due to the nonlinearity of the fiber at the output of HPA at the transmitter before launching into space via the OH.

Therefore, it is advantageous to use the equalization process at both the transmitter and the receiver. Obviously, the compensation at the transmitter for SPC is at the optical modulator placed before the HPA to mitigate any nonlinear phase distortion by rotating the phase of the optical waves in such a way to oppose the distortion and rotation of the phase of the lightwaves caused by nonlinear phase distortion. In addition, the mitigation of the beam spatial spreading due to the finite linewidth of the modulated laser beam is necessary. Hence, at the receiver, the compensation of the propagation distortion, including the spatial spreading and turbulence due to the turbulent layer before reaching the vacuum region, which is highly random and fading due to multipath propagation through the random turbulence, is required. The fading effects, similar to those in wireless propagation in high-rise city, building, are analyzed in Chapter 2, and the expected variance is known with a probability of error.

It is noted here that the spatial spreading in SPC is very similar to the CD/spatial distortion in fiber optic terrestrial or SPC transmission, respectively. The spreading of the beam due to finite line width and possibly from the signal modulation bands, and the fading due to multi-ray propagation and multiple arrival times.

Figure 6.47a and b depict the arrangement of equalizers at the transmitter only and combined at both the transmitter and receiver sides, respectively. There are advantages when equalization can be conducted at both sides of the transmitter and receiver in the digital domain. At the transmitter, the equalizer performs pre-distortion to compensate for the chromatic dispersion by chirping the phase. However, the dispersion can never be determined exactly, so it is always necessary to have an equalizer in the reception subsystem to address not only residual CD effects but also others issues such as PMD and nonlinear impairment (Figure 6.47).

6.7.1 Equalizers at the Transmitter

Consider the arrangement where the equalizer acts as a pre-distortion and is placed at the transmitter, as shown in Figure 6.47a and Figure 6.48. An ultra-high-speed DAC would be best and most flexible for generating and pre-emphasizing digital signals to compensate for the linear CD distortion effects. The nonlinear equalizer in this case converts the sequence of data symbols $\{s_i\}$ into a nonlinear channel $\{f_i\}$ at the output of the transmitter, normally an optical modulator driven by the output of the nonlinear electrical equalizer, as shown in Figure 6.47b and re-modeled in Figure 6.49. These symbols are sampled and transmitted as sampled impulses $\{f_i\delta(t-iT)\}$. The SNR in this case is given by $E\left[f_i^2\right]/\sigma^2$ with the energy being the expected average energy of the pre-distorted signal sequence and must be equal to k^2.

A nonlinear equalizer (Figure 6.48) is necessary at the output before the final recovery of the system. This processor performs a modulo-m operation on the received sequence $x[\text{modulo}]-m = x - jm$ with j as an appropriate integer. The z-transform $F(z)$ at the output of the nonlinear distortion equalizer is given by

$$F(z) = M\left[S(z) - F(z)\left(y_0^{-1}Y(z) - 1\right)\right] \tag{6.54}$$

Here, $S(z)$ is the z-transform of the sampled input signals, and $Y(z)$ is the z-transform of the channel transfer function. Then, the z-transform at the output of the nonlinear processor at the receiver can be written as:

$$X(z) = M\left[F(z)Y(z) + W(z)\right] = M\left[F(z)\left(y_0^{-1}Y(z) - y_0^{-1}W(z)\right)\right]$$

$$\text{or } X(z) = M\left[S(z) - y_0^{-1}W(z)\right] \tag{6.55}$$

that is

$$x_i = M\left[y_0^{-1}r_i\right] = M\left[s_i + y_0^{-1}w_i\right]$$

where x_i denotes the signal at the output of the nonlinear processor and all ISI has been eliminated except the noise contribution. The nonlinear processor M operates as a modulo-m as

$$M[q] = \left[(q+2k)\text{modulo}-4k\right] - 2k = q - 4jk$$
$$-2k \le M[q] \le 2k \tag{6.56}$$

Thus, an error would occur when

$$(4j-3)k \le y_0^{-1}w_i \le (4j-1)k \qquad \forall j \tag{6.57}$$

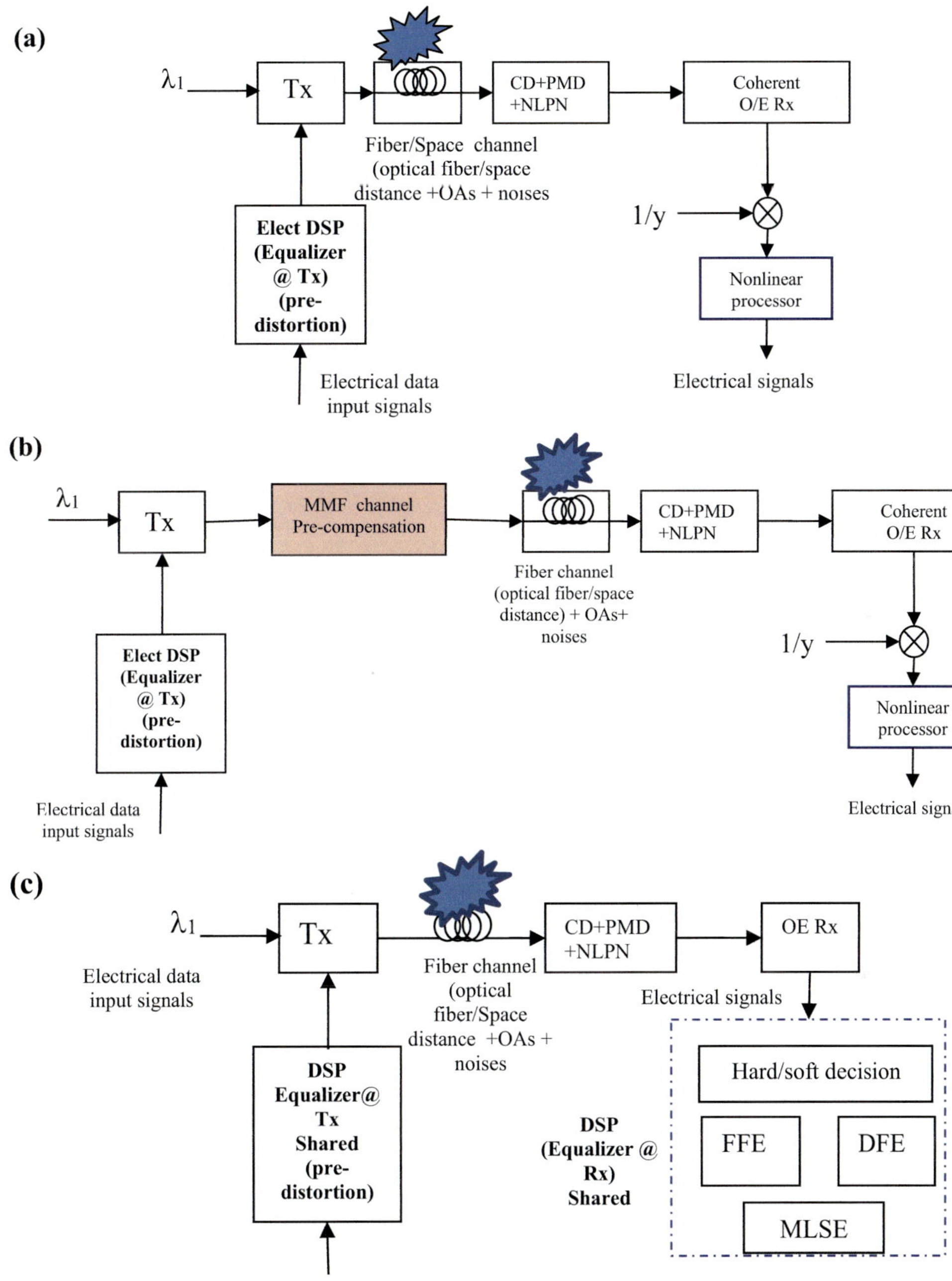

FIGURE 6.47 Arrangement of electronic digital signal processing for equalization in optical transmission systems. Electronic DSP for pre-distortion, Electronic DSP at the Rx is for post-equalization (a) DSP -based pre-compensation at transmitter; (b) MMF (Multi-mode Fiber) as optic pre-compensation; (c) and shared pre-distortion and post-equalization at the Tx and Rx (a) Equalization at the transmitter-side, (b) equalization at the receive-side and (c) equalization shared between Tx and Rx sides, equalizers at Tx can be similar to (a) and (b).

The probability of error is then given as

$$P_e = 2\int_k^{\infty} \frac{1}{\sqrt{2\pi y_0^{-2}\sigma^2}} e^{-\frac{w^2}{2y_0^2\sigma^2}} dw = 2Q\left(\frac{k|y_0|}{\sigma}\right) \tag{6.58}$$

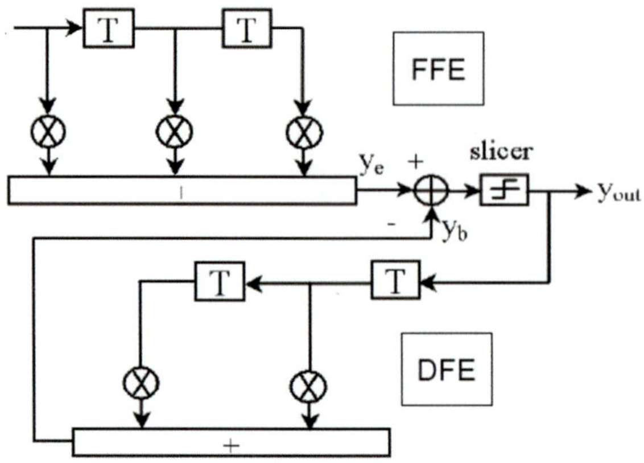

FIGURE 6.48 Schematic of a nonlinear feedback equalizer following an optical receiver combined FFE and DFE equalization-slicer or decision/detector for data decision.

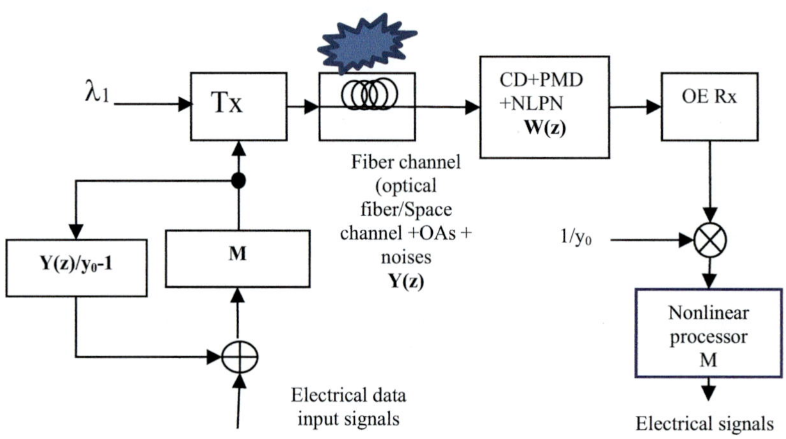

FIGURE 6.49 Schematic of a nonlinear feedback equalizer at the transmitter.

6.7.2 Shared Equalization

Long-haul transmission requires equalization where the transmission is extended as long as possible. Equalization at the receiver can be supplemented with that at the transmitter to further increase the reach.

The equalizer at the transmitter of the above section can be optimized by inserting another filter at the receiver, as shown in Figure 6.47c. Consider the channel transfer function $Y(z)$ that can be written as a cascade of two linear transfer functions $Y_1(z)$ and $Y_2(z)$ as

$$Y(z) = Y_1(z)Y_2(z)$$

$$\text{with } \begin{cases} Y_1(z) = 1 + p_1 z^{-1} + \cdots + p_{g-f} z^{-g+f} \\ Y_2(z) = q_0 + q_1 z^{-1} + \cdots + q_f z^{-f} \end{cases} \tag{6.59}$$

$Y_1(z)$ has all the zeros inside the unit circle and $Y_2(z)$ has all the zeros outside the unit circle. Then, we could find a third system with a transfer function that has all the coefficients in reverse of those of $Y_2(z)$, given as

$$Y_3(z) = q_f + q_{f-1}z^{-1} + \cdots + q_0 z^{-f} \tag{6.60}$$

With the reverse coefficients, the zeroes of the $Y_3(z)$ now lie inside the unit circle.

The linear FFE filter D of the nonlinear filter formed by cascading an FFE and a DFE, as shown above, can minimize the length of the vector of D, and minimizing e_0 can be satisfied provided that

$$D(z) = z^{-h} Y_2^{-1}(z) Y_3(z) \tag{6.61}$$

which represents an orthogonal transformation without attenuation or gain. The linear equalizer with a z-transform $Y_2(z)Y_3^{-1}(z)$ is an all-pass pure phase distortion.

This transfer function of the all-pass pure phase distortion is indeed a cascade of optical interferometers, known as half-band filters.[47,48] This function can also be implemented in the electronic domain. The equalizer can then be followed by an electronic DFE. Hence, the feedback linear filter $F(z)$ in Figure 6.48 can be written as

$$F(z) = Y_1(z) Y_3(z) \tag{6.62}$$

The probability of error can be similarly evaluated and given by

$$P_e = \int_k^{\infty} \frac{1}{\sqrt{2\pi y_0^{-2}\sigma^2}} e^{-\frac{w^2}{2y_0^2\sigma^2}} dw = Q\left(\frac{k|q_f|}{\sigma}\right) \tag{6.63}$$

with q_f taken as the first coefficient of $Y_3(z)$.

6.8 CARRIER PHASE ESTIMATION IN QPSK OPTICAL COHERENT RECEIVERS

Carrier frequency and its bandwidth, or alternatively the linewidth and the wavelength, of the signal, determines the coherency of the transmitted signals. Thus, the recovery of this carrier frequency is very critical, especially for the case of space transmission. The recovery must be accurate to less than 10KHz tolerance with repect to the main carrier central frequency which s approximately 193.51 THz for 1550nm wavelength. The received optical signals are converted into electrical domain by the coherent reception process and amplified to an appropriate level. They are then sapled processed in the DSP system.

Advanced progress in DSP,[49,50] with the availability of ultra-high sampling rates allow the possibility of DSP-based phase estimation and polarization management techniques that make the coherent robust and practical. This section dedicated to emerging technologies that will significantly influence the optical transmission and detection of optical signals at ultra-high speeds. Recalling the schematic of a coherent receiver in Figure 6.50, it shows a coherent detection scheme for QPSK optical signals mixed with the local oscillator field. In the case of modulation formats such as QPSK or DQPSK, a $\pi/2$optical phase shifter is needed to extract the quadrature component. Thus, the real and imaginary parts of a phase shift keyed signals can be deduced at the output of a balanced receiver from a pair of identical photodiodes connected back-to-back. Note that for a balanced receiver, the quantum shot noise contributed by the photodetectors is doubled as noise is always represented by the noise power and no current direction must be applied (Figure 6.51).

When the received signal is raised to the quadratic power, the phase of the signal disappears because $e^{j4\theta_s(t)} = 1$ for DQPSK whose states are $0, \pi/2, \pi, 3\pi/2, 2\pi$. The data phase is excluded and

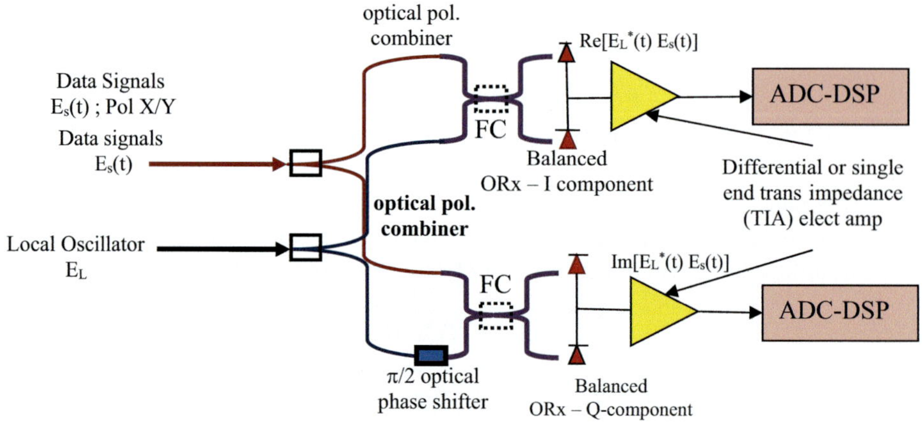

FIGURE 6.50 Schematic of a coherent receiver using a balanced detection technique for *I*-*Q* component phase and amplitude recovery, incorporating DSP and ADC.

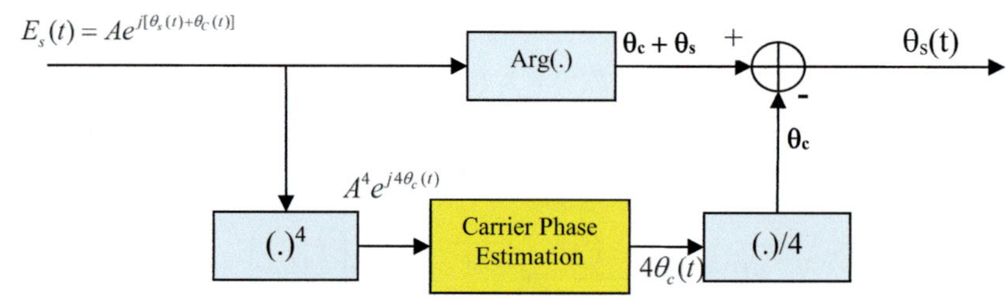

FIGURE 6.51 Schematic of a QPSK coherent receiver.

the carrier phase can be recovered with its fluctuation with time. This estimation is computed using a DSP algorithm. The estimated carrier phase is then subtracted from the detected phase of the signal, leaving only the phase states of the signals as indicated in the diagram. This is a feedforward (FF) phase estimation, which is most suitable for DSP implementation either offline or in real-time. Detailed algorithms for the phasean be found in Chapter of Ref. [51].

6.9 CONCLUDING REMARKS

This chapter has addressed DSP processing in SPC transmission systems to

- Recover the carrier frequency through carrier phase recovery, then use this to match the frequencies of the carrier and the LO laser, hence using true homodyne coherent reception to achieve a 3 dB higher sensitivity.
- Mitigate channel distortion caused by beam spreading due to the finite linewidth of the lightwave carrier, or equivalently, CD distortion in optical fiber lines, spatial spreading, and time group delay.
- Present a Viterbi algorithm to mitigate distortion effects over the transmission channels.
- Devise a strategy for equalization at either the transmitter or receiver, and propose a shared equalization between the Tx (transmitter) and Rx (receiver) to effectively equalize the mitigation of distortions known at either transmitter end or receiving end.
- Compensate for distortion caused by multipath delay, and propose an optical technique using a short length MMF to implement minor pre-compensation of the delay multi-path due to beam spreading in SPC systems.

- Describe MLSE algorithms in detail, which are the main algorithms to mitigate distortion effects and are applicable in SPC for any unknown disturbance. These MLSE algorithms are adapted from wireless techniques to mitigate the multi-path fading distortion. Examples are given to demonstrate the effectiveness of this technique.
- Present frequency shifting techniques to mitigate the Doppler frequency shift so that homodyne detection can be truly implemented. Furthermore, the OPLL technique can also be used as an alternative technique to lock the shifted carrier and lock it into the local oscillator laser.
- Present additional algorithms which can be updated in the memory of the integrated coherent DSP-based optical transceiver.

NOTES

1 *N. Benuveto and G. Cheubini, Algorithms for Communications Systems and Their Applications. John Wiley & Sons: Chichester, England, 2004.*
2 L. Vivien and L. Pavesi Eds., *Handbook of Silicon Photonics*. CRC Press, Taylor & Francis: Boca Raton, FL, 2013.
3 S. Savoy, "Digital coherent optical receivers", *IEEE Journal of Selected Topics in Quantum Electronics*, 16(5), 1164–1178, 2010.
4 A. P. Clark, *Equalizers for High Speed Modem*. Pentech Press: London, 1985.
5 L. Hanzo, S. X. Ng, T. Keller, and W. Webb, *Quadrature Amplitude Modulation: Froom Basics to Adaptive Trellis Coded, Turbo-Equalized and Space-Time Coded OFDM CDMA and MC-CDMA Systems*, 2nd Ed. John Wiley & Sons: Chichester, England, 2004.
6 J. Winters, "Equalization in coherent lightwave systems using a fractionally spaced equalizer," *IEEE Journal of Lightwave Technology*, 8(10), 1487–1491, October 1990.
7 R. W. Lucky, J. Salz, and E. J. Weldon, *Principles of Data Communications*. McGrawhill: New York, pp. 93–165, 1968.
8 A. Papoulis, *Probability, Random Variables and Stochastic Processes*. McGrawhill: New York, 1965.
9 S. Bennetto and E. Biglieri, *Principles of Digital Transmission with Wireless Applications*. Kluver Academic: New York, 1999.
10 R. Noe, T. Pfau, M. El-Darawy, and S. Hoffmann, "Electronic polarization control algorithms for coherent optical transmission", *IEEE Journal of Selected Topics in Quantum Electronics*, 16(5), 1193–1199, September 2010.
11 A. P. Clark, "Pseudo-binary Viterbi detector", *Proceedings of IEE*, Part F, 131, 208–218, 1984.
12 J. G. Proakis and M. Salehi, *Communication Systems Engineering*, 2nd Ed. Prentice Hall, Inc: Hoboken, NJ, 2002.
13 J. G. Proakis and M. Salehi, *Communication Systems Engineering*, 2nd Ed. Prentice Hall, Inc: Hoboken, NJ, 2002.
14 J. G. Proakis, *Digital Communications*, 4th Ed. McGraw-Hill: New York, pp. 185–213, 2001.
15 L. N. Binh, *Digital Optical Communications*. CRC Press: Boca Raton, FL, Chapter 5, 2009.
16 S. Zhang, P. Y. Kam, J. Chen, and C. Yu, "A comparison of phase estimation in coherent optical PSK system," *Photonics Global'08*, Singapore, Paper C3-4A-03, December 2008.
17 A. P. Clark, *Equalizers for Digital Modem*. Pentec Press: London, 1985.
18 G. Nicholson, "Probability of error for optical heterodyne DPSK system with quantum phase noise," *Electronics Letters*, 20(24), 1005–1007, November 1984.
19 S. Zhang, P. Y. Kam, J. Chen, and C. Yu, "A comparison of phase estimation in coherent optical PSK system," *Photonics Global'08*, Singapore, paper C3-4A-03, December 2008.
20 J. P. Gordon and L. F. Mollenauer, "Phase noise in photonic communications systems using linear amplifiers," *Optics Letters*, 15, 1351–1353, 1990.
21 S. Zhang, P. Y. Kam, J. Chen, and C. Yu, "A comparison of phase estimation in coherent optical PSK system," *Photonics Global'08*, Singapore, paper C3-4A-03, December 2008.
22 L. N. Binh, *Digital Processing: Optical Communications and Coherent Receiving Techniques*. CRC Press, Taylor & Francis: Boca Raton, FL, 2014.
23 L. N. Binh, *Digital Processing: Optical Transmission and Coherent Receiving Techniques*. CRC Press, Taylor & Francis, Boca Raton, FL, Chapter 5, 2014.
24 D.-S. Ly-Gagnon, S. Tsukamoto, K. Katoh, and K. Kikuchi, "Coherent detection of optical quadrature phase-shift keying signals with carrier phase estimation," *Journal of Lightwave Technology*, 24, 12–21, 2006.

25 X. Liu, X. Wei, R. E. Slusher, and C. J. McKinstrie, "Improving transmission performance in differential phase-shift-keyed systems by use of lumped nonlinear phase-shift compensation," *Optics Letters*, 27(18), 1351–1353, September 2002.
26 X. Wei, X. Liu, and C. Xu, "Numerical simulation of the SPM penalty in a 10-Gb/s RZ-DPSK system," *IEEE Photonics Technology Letters*, 15(11), 1636–1638, November 2003.
27 E. Ip and J. M. Kahn, "Digital equalization of chromatic dispersion and polarization mode dispersion", *IEEE Journal of Lightwave Technology*, 25, 2033–2043, 2007.
28 R. Noe, "PLL-free synchronous QPSK polarization multiplex/diversity receiver concept with digital I&Q baseband processing", *IEEE Photonics Technology Letters*, 17, 887–889, 2005.
29 L. N. Binh, *Digital Optical Communications*. CRC Press, Taylor & Francis Group: Boca Raton, FL, 2009.
30 J. G. Proakis, *Digital Communications*, 4th Ed. McGraw-Hill: New York, 2001.
31 Y. Han and G. Li, "Coherent optical communication using polarization multiple-input-multiple output", *Optics Express*, 13, 7527–7534, 2005.
32 N. Stojanovic, Y. Zhao, B. Mao, C. Xie, F. N. Hauske, and M. Chen, "Robust carrier recovery in polarization division multiplexed receivers", *OFC 2013*, Annaheim, CA, March 2013.
33 H. Meyr, et al., *Digital Communication Receivers*. John Wiley & Sons: Chichester, England, 1998.
34 A. Leven, et al., "Frequency estimation in intradyne reception", *IEEE Photonics Technology Letters*, 19, 366–368, 2007.
35 Z. Tao, et al., "Simple, robust, and wide-range frequency offset monitor for automatic frequency control in digital coherent receivers", *ECOC 2007*, Berlin, Germany, paper Tu3.5.4, 2007.
36 A. J. Viterbi and A. M. Viterbi, "Nonlinear estimation of PSK-modulated carrier phase with application to burst digital transmission", *IEEE Transactions on Information Theory*, 29, 543–551, 1983.
37 M. Kuschnerov, et al., "DSP for coherent single-carrier receivers", *IEEE Journal of Lightwave Technology*, 27, 3614–3622, 2009.
38 Z. Tao, et al., "Simple, robust, and wide-range frequency offset monitor for automatic frequency control in digital coherent receivers", *ECOC 2007*, Berlin, Germany, paper Tu3.5.4, 2007.
39 http://www.coreoptics.com/product/prod_ic.php, USA March 2010.
40 T. Sivahumaran, T. L. Huynh, K. K. Pang, and L. N. Binh, "Non-linear equalizers in narrowband filter receiver achieving 950 ps/nm residual dispersion tolerance for 40Gb/s Optical MSK transmission systems", In *Proceedings of OFC'07*, paper OThK3, Anaheim, CA, USA, 2007.
41 P. Poggiolini, G. Bosco, M. Visintin, S. J. Savory, Y. Benlachtar, P. Bayvel, and R. I. Killey, "MLSE-EDC versus optical dispersion compensation in a single-channel SPM-limited 800 km link at 10 Gbit/s", *ECOC2007*, paper1.3, Berlin, 2007.
42 N. Alic, G. C. Papen, and Y. Fainman, "Performance of maximum likelihood sequence estimation with different modulation formats", *Proceedings of LEOS'04*, pp. 49–50, San Juan, Puerto Rico, 7–11 November 2004.
43 T. Sivahumaran, T. L. Huynh, K. K. Pang, and L. N. Binh, "Non-linear equalizers in narrowband filter receiver achieving 950 ps/nm residual dispersion tolerance for 40Gb/s optical MSK transmission systems", *Proceedings of OFC'07*, paper OThK3, Anaheim, CA, USA2007.
44 V. Curri, R. Gaudino, A. Napoli, and P. Poggiolini, "Electronic equalization for advanced Modulation formats in dispersion-limited systems", *IEEE Photonics Technology Letters*, 16(11), 2004.
45 G. Katz, D. Sadot, and J. Tabrikian, "Electrical dispersion compensation equalizers in optical long-haul coherent-detection system", In *Proceedings of ICTON'05*, paper We.C1.5, Barcelona, Spain, 2005.
46 L. N. Binh, *Digital Optical Communications*. CRC Press: Boca Raton, FL, Chapter 5, 2009.
47 K. Jinguji and M. Oguma "Optical half-band filters", *IEEE Journal of Lightwave Technology*, 18(2), 252–259, February 2000.
48 L. Binh and V. A. T. Tran, "Design of photonic half-band filters using multirate DSP technique", Technical report MECSE-27-2004, Master thesis 2004, Department of Electrical and Compter System Engineering. (formerly located at) http://www.ds.eng.monash.edu.au/techrep/reports/2004/MECSE-27-2004.pdf.
49 E. Ip and J. M. Kahn, "Digital equalization of chromatic dispersion and polarization mode dispersion", *IEEE Journal of Lightwave Technology*, 25, 2033–2043, 2007.
50 R. Noe, "PLL-free synchronous QPSK polarization multiplex/diversity receiver concept with digital I&Q baseband processing", *IEEE Photonics Technology Letters*, 17, 887–889, 2005.
51 L. N. Binh, *Digital Optical Communications*. CRC Press: Boca Raton, FL, Chapter 5, 2009.

7 Optical Amplifiers and Routers for Space Photonic Transmission

CONTENTS

DOI: 10.1201/9781003426677-7

Optical fiber amplifiers dominate the loss compensation in fiber optic communications systems, especially the Erbium-doped types for communication spectra based on silica fibers. Several models have been developed for such photonic amplification devices. However, there are no comprehensive models developed on the MATLAB™ Simulink platform. This paper describes in detail the development of a Simulink model for Erbium-doped fiber amplifiers (EDFAs) operating under steady-state and dynamic conditions.

The equations that govern pump and signal propagation in the doped fiber are derived and tabulated with the corresponding amplifier parameters. The model blocks are then described in the SIMULINK environment. Simulation results are compared with experimental values, and their close agreement indicates the validity of the Simulink EDFA models. These simulated results and techniques are given in the appendix of this chapter.

We also present practical aspects of amplifiers for long-reach space transmission systems, including the optical amplification for photonic space transmission, power booster amplifiers, and highly sensitive optical preamplifiers. Thus, the main amplifiers for space photonic communications, the high-power amplifier (HPA), and low-noise amplifier (LNA) are described to explain their uses and applications.

In summary, this chapter addresses optical amplification techniques. One focuses on the HPA for channel signal boosting, increasing the average power from a few milliwatts to a few watts or at least 30–40 dB with high output power. The other focuses on the low-noise and low input level to –65 dBm to receive ultra-low average power level to achieve channel signal average power sensitivity that is acceptable with error-free reception, typically at –40 dBm for 100 Gbps DP-QPSK modulation formats.

The two amplifiers HPA (as a booster amplifier) or LNA (as a preamplifier), must consist of two or three stages of amplification, either unidirectional or bidirectional. The doping of rare earth impurities must be done with appropriate metals such as erbium for 1550 nm, Nd (Neodymium) for 1060 nm or Er-Yb (thus EYDFA) for higher bands of the C-band. The pump lasers are at 980 or 1480 nm for C-band amplification. The doping concentration is high enough so that the threshold pumping energy is high enough to produce high due to high absorption so that high power can be produced

On the other hand, for LNA, the gain must be in the saturation or linear range with two pumping stages. An optical tunable amplifier must be situated between the two amplification stages to filtering the noises of the stages. The roll-off band of –40 dB for the tunable filter (TOF) must be sharp to ensure a low-level signal operating level. Thus, for dense wavelength division multiplexing (DWDM) operation, a bank of TOF must be used, and WSS devices can offer such filtering.

This chapter also presents optical switching and routing with independence in optical wavelength, optical direction, and optical free contention.

7.1 INTRODUCTORY REMARKS

Since 1989 rare-earth metal doped fibers have been exploited as the optical amplification medium and laser sources, both in experimental and theoretical aspects. It is no doubt that EDFAs are the amplifying devices employed in commercial optical transmission systems. Several models have also been reported and developed for the design and system simulation.

The developments of optical amplifiers have been intensive since its invention in 1988 by Poole et al. Of South Hampton University. Since then, tremendous progress has been made in designing and demonstrating optical amplifiers in several ultra-high-speed and ultra-long transmission systems.

The mathematical laboratory MATLAB has also been progressively developed and improved over the last two decades. It has now been considered the standard mathematical toolbox in teaching and research and development laboratories around the world. Recently, the SIMULINK toolboxes have been available. This gives an additional and powerful tool for the simulation of physical systems. Simulation and experimental implementations are important for test and prototype demonstration of the amplifications and evaluation of transmission performance.

Why should Simulink be developed when several commercial simulation packages such as VPI Systems, etc. are available. Simulink is a modular toolbox approach attached to MATLAB, which is now used commonly as a universal computing tool in global university systems and engineering laboratories. Its communications block set contains many toolboxes for digital communications in the baseband as well as the passband. Thus, these toolboxes can be used as a modular set for the generation and detection of lightwave signals. As the frequency of the lightwaves is much higher than the baseband signals, it is not likely that the carrier is contained under the envelope of the signals because an extremely high sampling rate would be required and is currently not possible in computing systems. Thus, complex envelope signals are used to represent the phase of the carrier. Therefore, the simulation models for digital optical communications can be modeled accurately and with ease due to the modularity of Simulink. The Simulink techniques and samples of models may be made available as a toolbox in MATLAB in the future so that global laboratories can equip and facilitate teaching and research activities. Thus, the Simulink models described in this book would be easily integrated into standard MATLAB software packages in university, research, and engineering laboratories, as well as in the field, especially in the toolset. The Digital Optical Communications MATLAB Simulink models can thus be the most universally and globally affordable for researchers and students of advanced levels of applied physics and telecommunications engineering.

Data centers over terrestrial global internet are about to enter 1.6 Tbps internet networking with 224 GB and multi-level complex modulation formats. In a similar trend, the inter-satellite internet networks are reaching the capability to transmit multi-tera-bits/second with baud rate at 25 or 112 or 224 GB. Without optical amplifiers, such as booster and low noise pre-amplifiers, it is not possible to boost and cascade the transmission of channels over many inter-satellite or multi-optically amplifier spans.

This section describes the comprehensive and successful implementation of the EDFA using Simulink MATLAB 7 R14. Furthermore, we present the practical characteristics of the two main optical amplifiers: the power booster HPA and the low-input, low-noise optical preamplifier.

The chapter is organized as follows:

The fundamental and theoretical expressions for representing optical amplification, the rate of transitions of erbium ions, and lightwave propagation over the amplifying medium are described in Section 7.2. Both forward and backward pumping configurations are given.

Section 7.3 provides a brief account of the Simulink models for the amplifiers in steady-state and dynamic environments.

Section 7.4 gives an account of HPA and LNA. Section Appendix 6 of Chapter 12 illustrates samples from the Simulink simulator; and details the simulation results to validate the developed models, comparing them with experimental data obtained from EDFAs constructed in the laboratory as well as those specified by commercial manufacturers.

Section 7.5 presents a design of both HPA and LNA in the same structure.

Section 7.6 introduces the use of optical add-drop mux for multi-port optical amplifications, sharing the same structures as the amplifiers. Tunable optical filters (TOF) with a sharp roll-off spectrum, used as add/drop multiplexers, are described for integration into low-noise optical amplifiers.

The proposed MATLAB Simulink models are proven to be user-friendly and easy to use. They would be a valuable tool for researchers and advanced levels graduate students in the fields of photonics and optical communications.

Sections 7.8 and 7.9 present the structure of OADM and CDC-ROADM, the colorless, contentionless, and directionless reconfigurable optical add-drop multiplexer for SPC interconstellation networking.

There is a table that tabulates the duality of space and time for space nd integrated guided wave opical component and operations. Table 1 is included as the table of the book on this duality and placed at near the end of this book. Optical amplifiers are also included in this table.

7.2 FUNDAMENTAL AND THEORETICAL ISSUES OF EDFAS

In this section, basic theoretical analyses of the EDFA are described for the development of EDFA simulation modeling. Several technical terms, such as pumping scheme, population inversion factor, ASE noise, and others, are defined.

7.2.1 EDFA Configuration

The EDFA is mainly constructed from a short length of optical fiber whose core has been doped with less than around 0.1% erbium, an optically active rare earth element with unique intrinsic properties for optical amplification.[1] The optical fiber used is called Erbium-Doped Fiber. The optical amplifier engineer will choose the appropriate erbium-doped fiber during the EDFA development. Different doped fibers have different characteristics of absorption and emission factors, which will be identified and discussed in a later section.

An EDFA consists of an erbium-doped silica fiber and other photonic components such as the laser pump source, the optical isolators, the WDM coupler, etc. as shown in Figure 7.1a. An enhanced EDFA module consists of an erbium-doped fiber, a wavelength-dependent coupler (WSC), and a polarization-insensitive optical isolator (OI). An optical filter may be required to improve EDFA performance[2] depending on specific applications. The optical filter is used to reduce the amplified spontaneous emission (ASE) noise and protect the amplifier from saturation caused by ASE accumulated in an in-line-amplification system. The OI is used for achieving a stable amplified operation to prevent spurious oscillations and reflect back into the amplification section to avoid oscillation. The pump source is used for pumping the erbium ions to the upper energy level; several proper pumping wavelength laser diodes exist at the present. Both 1480 and 980 nm pumping laser diodes can provide a high pump gain; the investigation of those two pumping wavelengths will be described in a later section. The WSC/WDM is used to combine the pump sources and the incoming signal. The pump light waves are guided and propagated along the EDF length, causing the depleted erbium ions to be excited by the absorption of energy. As the signal propagates through the EDF, it stimulates emission of light from the excited ions, thus amplifying the signal. The tap coupler is used to separate and tap the pumping power in order to protect the receiver, and it is normally used in pre-amplifier.

Two pumping configurations can be used in EDFAs,[3,4] the backward pumping and dual backward and forward pumping, as shown in Figure 7.1b and c.

The only difference among the three configurations is the pumping propagation direction and whether there is only one or dual bidirectional pumping lasers.[5] As the name of the configuration implies, in forward pumping, the signal light co-propagates with the pump light, whereas in backward pumping, it counter-propagates relative to the pump light. Dual dumping would involve bi-directional pumping. Considering noise performance, forward pumping can produce optical amplification with low noise but may offer lower amplification gain due to fluctuations in the pump power and signal power, which is much lower than in the dual bidirectional case, where much more power for absorption can be supplied. On the other hand, backward pumping may be effective for high power output but suffers from higher ASE noise. Thus, there is an enhanced model in which bi-directional pumping or dual pumping can be combined.

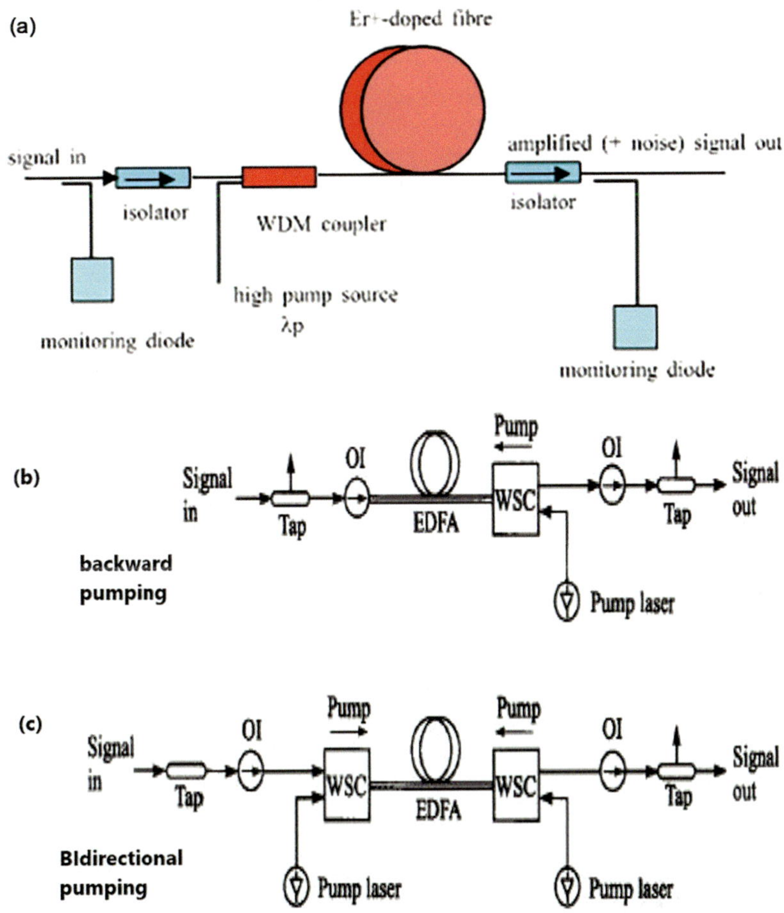

FIGURE 7.1 Optical amplifiers using (a) standard pumping with monitoring detector for control; (b) backwards/contra-pumping; and (c) dual (or bi-directional) pumping. WSC = wavelength selective coupler or WDM coupler, Tap = coupler with small coupling ratio tapping for monitoring, OI = optical isolator, EDFA = erbium-doped fiber amplifier.

7.2.2 EDFA Operational Principles

The basic operating principles of EDFA is that Erbium ions can exist in several energy states. When they are in a high energy state, photons could be stimulated to give up some of their energy (as light) and return to a lower-energy (more stable state), resulting in photonic amplification.[6]

7.2.2.1 Pump Wavelength and Absorption Spectrum

The pump wavelength can be either 980 or 1480 nm due to the availability of these laser sources. The 980 nm lasers can supply high power, up to a few hundred of milliwatts (mW), and are much more commercially available. A WDM optical coupler can couple 980 nm into the cross port where the signal is propagating, enabling amplification via the rare-earth-doped region. Why do most of EDFAs use these two wavelengths for pumping and not others? To the answer this, we should refer to the energy levels of the triply ionized erbium ion, Er^{3+}, the absorption band of Er^{3+}, and the pump efficiency (Figures 7.2 and 7.3).

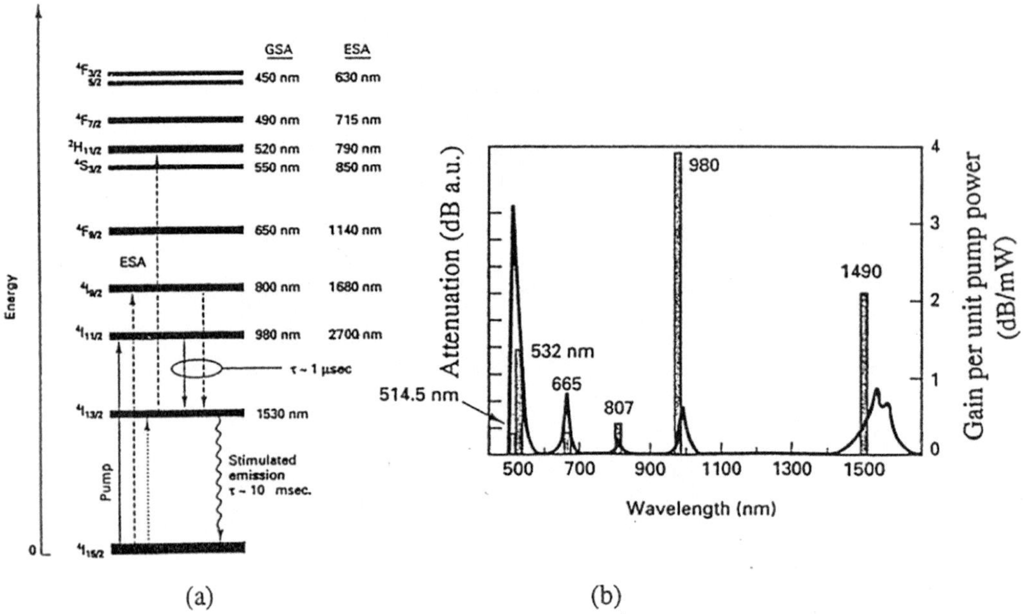

FIGURE 7.2 (a) Energy levels of Er^{3+} ions and (b) gain and attenuation/absorption spectra.

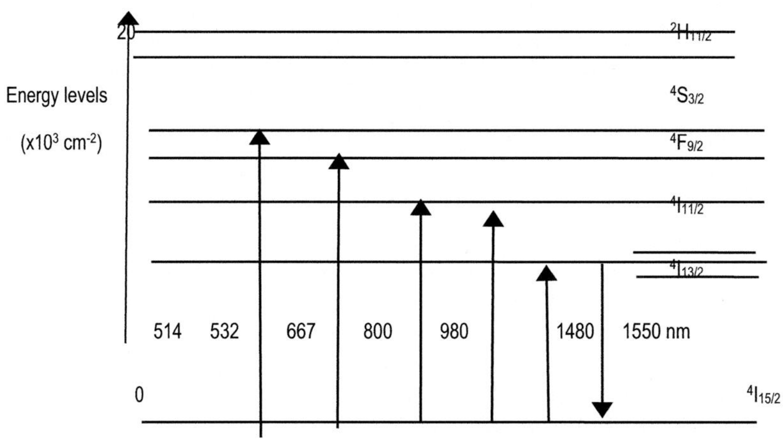

FIGURE 7.3 Erbium ion energy levels and pump spectra in simplified model and the energy level for 1550 nm spectral window jump. Up arrows are reserved for pump lasers of different wavelengths.

There are several states to which the erbium ions could be pumped using sources of different wavelengths such as those operating at 1490, 980, and 800 nm. However, in real lightwave systems, the pump wavelength must provide a high power to achieve a high gain per pump. The commonly available laser diodes can operate at 800, 980, and 1480 nm, but the pump efficiency can exceed 1 dB/mW with low attenuation depending on the pump frequency. The only pump wavelength laser sources that can give a high pumping efficiency with lower attenuation are those operating at 980 or 1490 nm. In practice, the 980 nm pumping source is commonly used due to its high gain coefficient (4 dB/mW). The model given in this report employs both 980 and 1480 nm pumping sources. The different effect of those two wavelength sources is mainly caused by the absorption and emission factors.

7.2.2.2 Pump Mechanism

The amplification mechanism is the key to developing the EDFA simulation model. The optical amplification gain is supplied by the excited erbium ions (Er^{3+}) when the amplifier is pumped to achieve population inversion. Depending on the energy states of the dopant, pumping schemes can be classified as two or three schemes. The main difference between the two-level and three-level pumping schemes is related to the energy state occupied by the stimulated-emission event. In the case of a three-level scheme, the lower level is ground, whereas it is an excited state with a fast relaxation time in the two-level scheme.

As shown in Figure 7.4, Erbium ions are pumped to an upper energy level by the absorption of lightwave energy from the pump source at either 980 or 1480 nm to regain the equilibrium distribution[7] (Figures 7.5–7.8).

Stage 1:

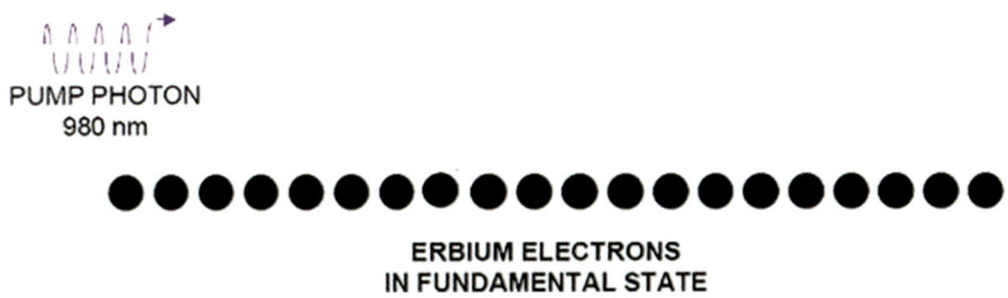

FIGURE 7.4 The pump photons start propagating through the erbium fiber, all the erbium electrons now are on the fundamental state (ground state). (Adopted from Ref.[8])

Stage 2:

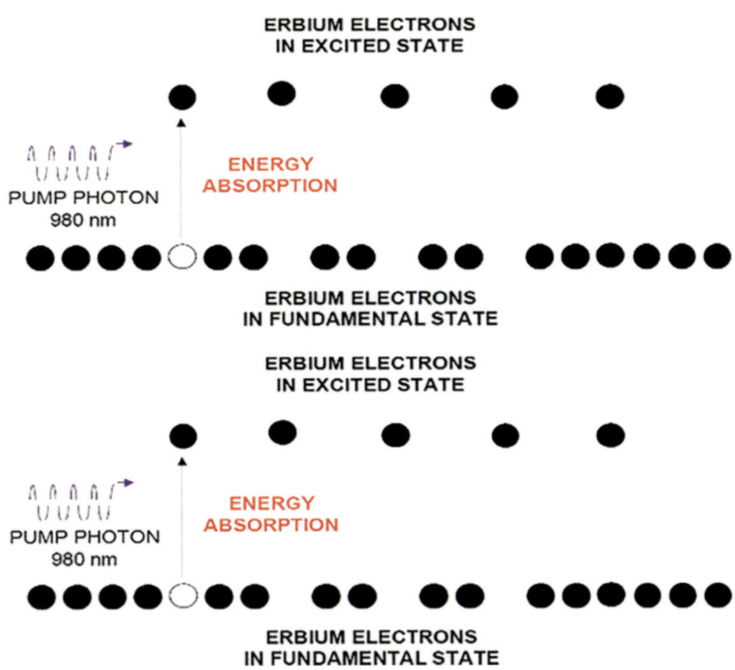

FIGURE 7.5 The erbium electrons are pumped to higher-level energy states. (Adopted from Ref.10)

Stage 3:

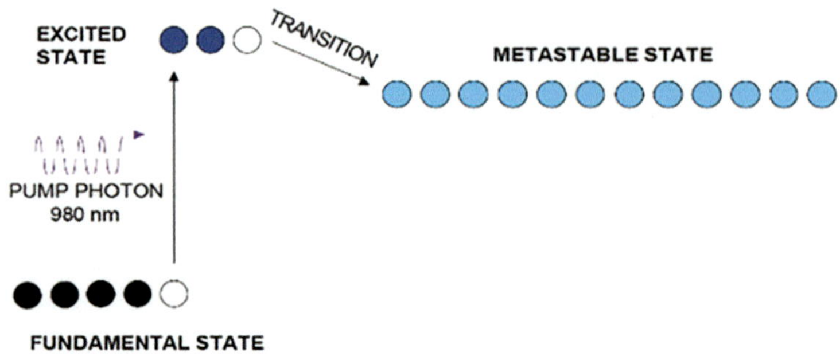

FIGURE 7.6 The erbium electrons revert to lower states (metastable state) spontaneously after some mean lifetimes $t - 1$ ms (spontaneous emission). (Adopted from Ref.10)

Stage 4:

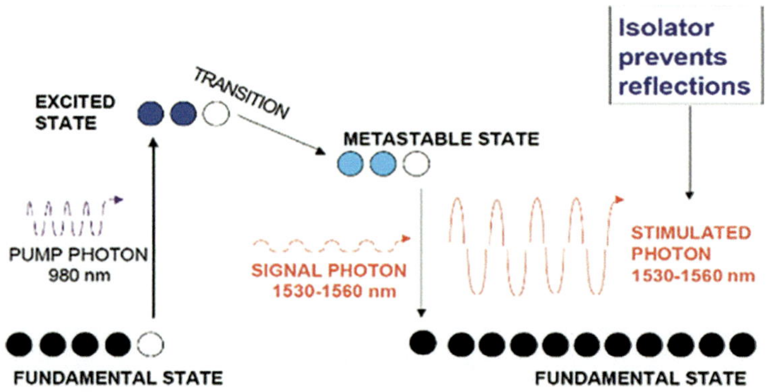

FIGURE 7.7 Erbium electron reverts to the fundamental state spontaneously (spontaneous emission) and is stimulated to do so by some incident photons (signal) of exactly the energy corresponding to the drop to lower state (stimulated emission). (Adopted from Ref.[9])

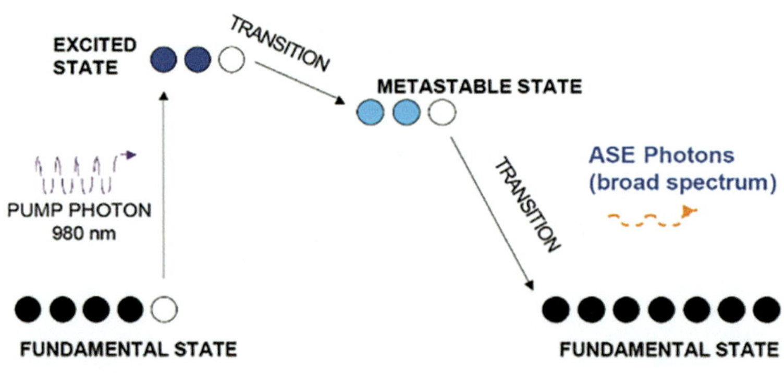

FIGURE 7.8 ASE photons demonstration. (Adopted from Ref.[12])

The coherence of pump sources and signals means the incident photon and the stimulated one have the same frequency, phase, propagation direction, and polarization.[10] The optical gain is achieved when the population inversion reaches sufficiency high level at which the absorption from the ground state is considerably reduced. This can be readily achieved with the available pump lasers because the meta-stable state exists in Er^{3+}, thus rendering the population inversion possible. In addition, the photon lifetime from the meta-stable state to the fundamental state must be longer than that from the excited state to the metastable state, normally around 10–10.5 ms, the lifetime from an excited state to meta-stable state is normally about 1 ms. Therefore, the ions are accumulated in the metastable; hence the ion population in the excited state is negligibly small. This fact is critical in the design of the simulator.

7.2.2.3 Amplifier Noises

During the signal amplification process, noises are generated. The dominant amplification noise source is the ASE noise (amplifier-stimulated emission noise). When ions are in transition from the excited state to the fundamental state, some excited erbium ions decay to the fundamental state with spontaneous emission before amplifying the injecting signal photons. Thus, photons are emitted with random phase and direction.[11]

The ASE spectral range is very wide and covers almost the C- and L- bands, encompassing the gain spectrum of the amplifier, as shown in Figure 7.9.

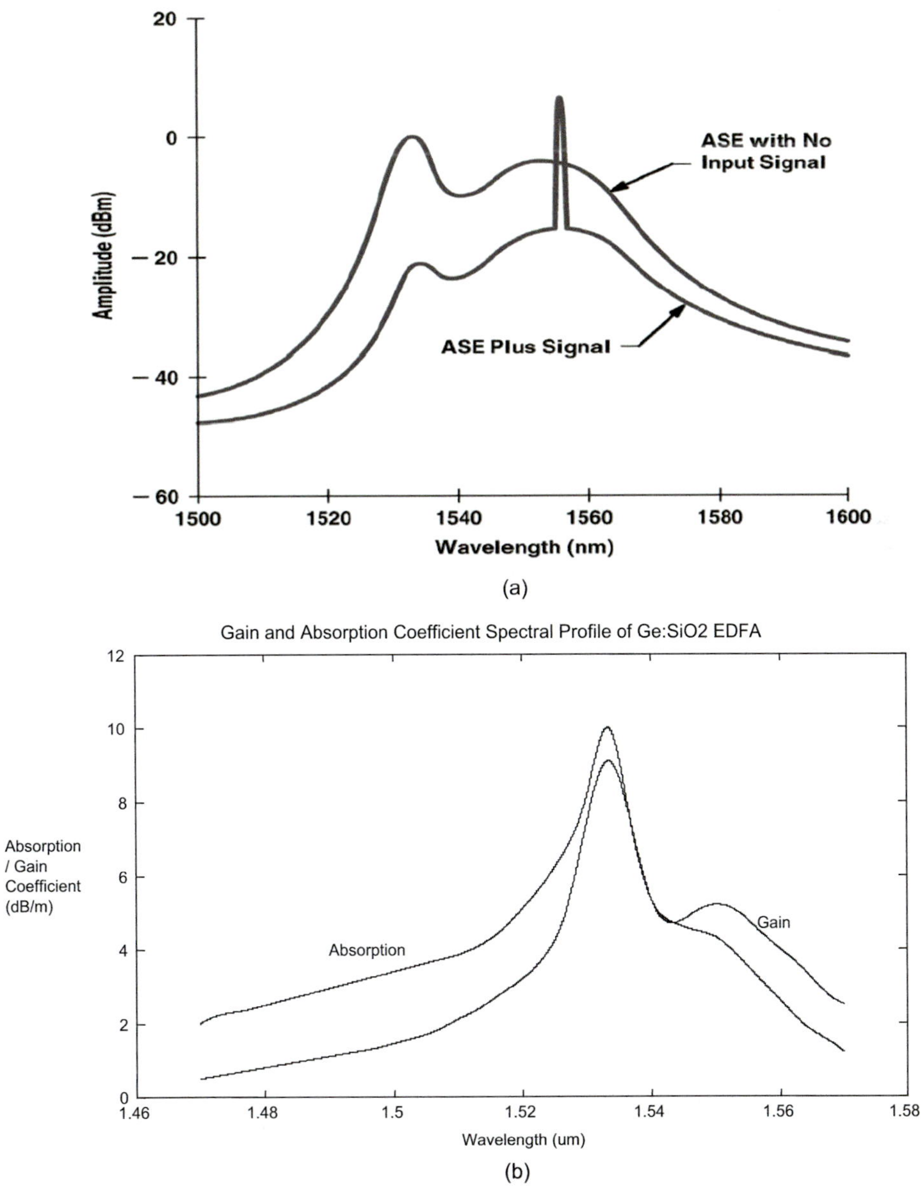

FIGURE 7.9 (a) ASE noise spectrum. (Adopted from Ref.[13]) (b) Absorption and gain spectra.

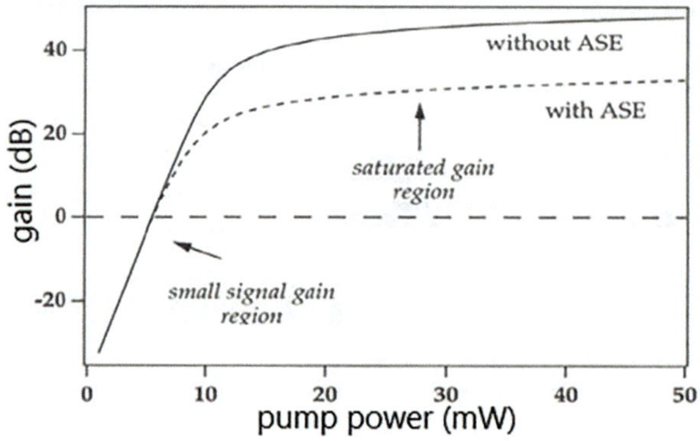

FIGURE 7.10 Optical gain reduction by ASE with and without contribution (ideal).

It will reduce the available gain for the signal field. Figure 7.9 shows the saturated gain has been reduced by the ASE. This is due to the incoherence of the excited erbium ions with the signal photons. The ASE noises cause degradation to the signal and noise ratio (SNR) (Figure 7.10).

EDFAs are employed in long-haul transmission systems, the ASE level grows after a cascade of amplifiers and begins to saturate the optical amplifiers, hence reducing the signal gain. This is because total power is almost equal to the pump power in addition to the ASE power.

7.2.3 Amplifier Gain Modulation

Gain modulation of EDFAs can provide an attractive approach for implementing status and control communication in modern DWDM communication networks. The desired control signal is imposed as amplitude-over-modulation on the payload communication signals propagating through the system via modulation of the EDFA pump power.[14,15,16,17]In this document, we report the impacts of pump modulation and signal modulation on EDFA gain.

7.2.4 EDFAs in Long-Haul Transmission Systems

Optical amplifiers can be employed in three different types for optical transmission systems: as a booster amplifier, an in-line amplifier regenerator, and as an optical pre-amplifier at the receiver or transceiver, with amplifiers in unidirectional or bidirectional systems in terrestrial or space mediums, as shown in Figure 7.11a–c. Single directional transmission is simple and well-known in terrestrial fiber optic systems. In space transmission (Figure 7.11b and c), the transmission occurs over space channel with unidirectional and bidirectional transmission. Bidirectional transmission is normally selected due to the advantages of using only one unit for alignment in either direction, reducing the complexity, weight, and power consumption of the alignment unit. For bidirectional transmission and reception, couplers and circulators are used, and its schematic is shown in Figure 7.12, in which a 4-port circulator is employed for both launching the signal to the HPA and into space, and accepting the signal to the LNA and in circulating it (see Figure 7.12b) to the optical receiver.

The booster amplifier increases and provides a high optical power signal before transmission. In long-haul optical communication systems, the use of a booster amplifier can increase the link power budget and reduce the number of regenerators required. A three-port high-power optical circulator is shown in Figure 7.12b for bidirectional transmission and reception.

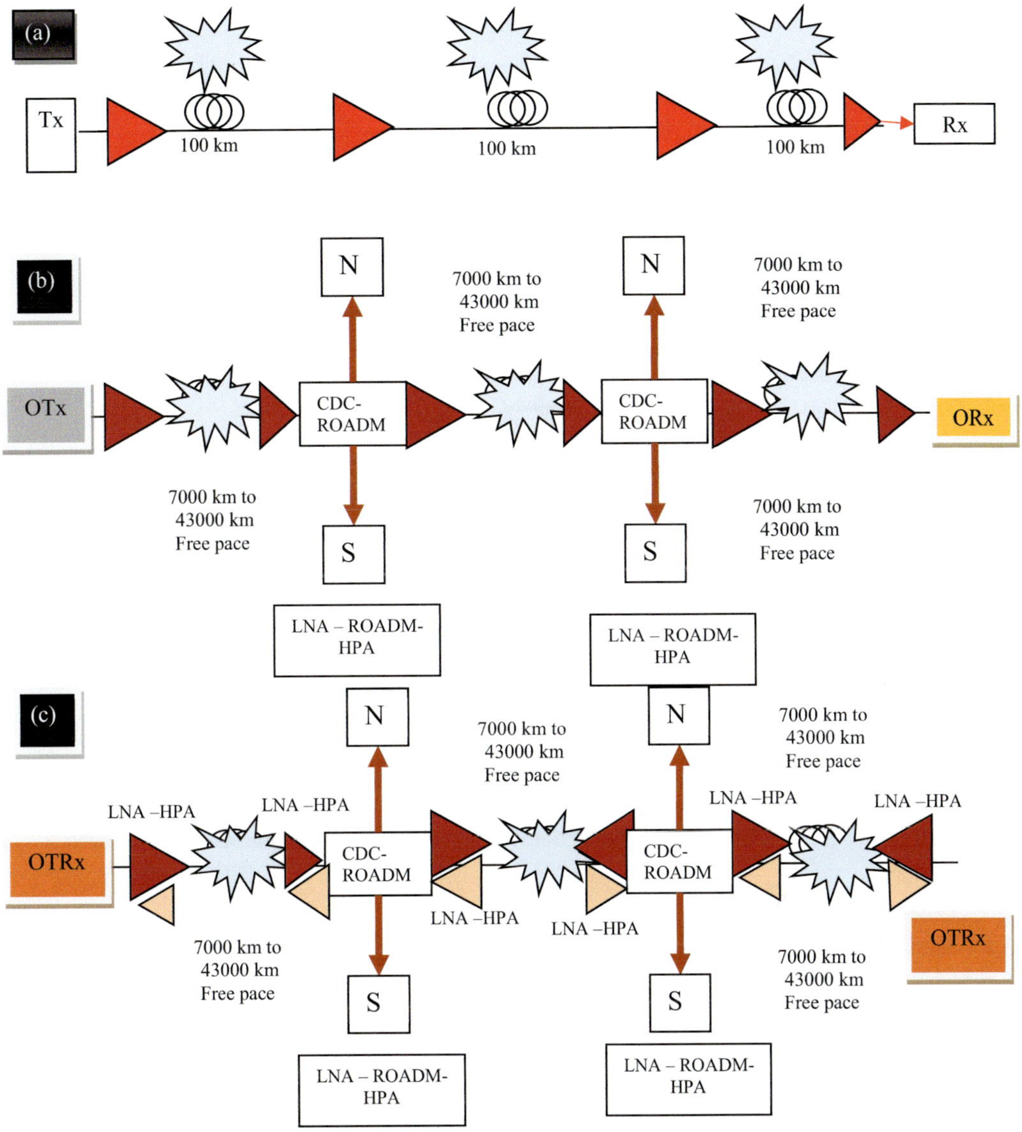

FIGURE 7.11 Integration of three types of amplifiers in a multi-span fiber spans (a) SPC space hops single direction (b); and (c) bidirectional transmission using two HPA (large red triangles) and two LNA (light dark triangle) per hop for amplification in both directions so a pair of LNA – HPA at the receiving and launching side to and from space hop. N=north, S=south, ORX=optical receiver, OTx=optical transmitter; OTRX=optical transceiver.

In-line optical amplifiers: The main function of an in-line amplifier is to compensate for the fiber loss in the fiber span or in the dispersion-compensating fibers, thereby eliminating the need for O/E regeneration.

Optical preamplifier: The main function of the optical preamplifier is to increase the power level of a signal before detection, thus improving the receiver sensitivity.

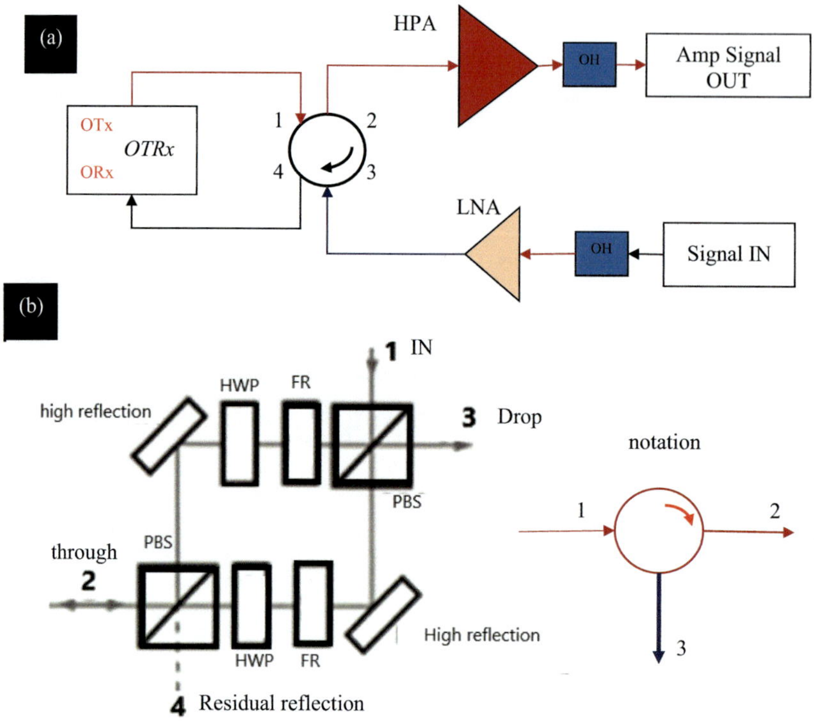

FIGURE 7.12 (a) Schematic of bidirectional amplifier unit using circulator, LNA, and HPA; (b) schematic of a bulk optical circulator. PBS=polarization beam spitter, FR=Faraday rotator, HR=high reflection mirror, HWP=half-wave plate, 1=input port, 2=bidirectional input/output, 3=output port, OH=optical head.

7.3 SIMULATION MODEL

In this section, the mathematical expressions for developing the EDFA simulation model are described. In general, the simulation models can be classified into steady-state and dynamic models and/or perturbation modes.[18] In this report, only the first two models are used for EDFA simulation. The models of single EDFA are investigated; a cascade of EDFAs can be implemented without difficulty and left for users to construct.

7.3.1 Amplifier Parameters

Several parameters can clarify the principles of the simulation model, as shown in Tables 7.1 and 7.2.

A dynamic model is developed based on the one-dimensional ordinary non-linear differential equation for the time-dependent population of excited state erbium ions in EDFAs.[19] With the three-level states, the amplification in erbium-doped fiber is based on the stimulated erbium ions; thus, the state population would have discrepancies. Simplifying the schemes to be a three-level pumping, the three populations of erbium atoms are ground state (the fundamental state with population density n_1), level state 2 (the excited state population density, also called meta-stable state with signal frequency n_2), and level state 3 (the excited state population with pump frequency n_3). The rate equation for these three states can be defined and given as[6,13,20]

TABLE 7.1
Erbium-Doped Fiber Parameters

Physical Meaning	Expression	Symbol & Unit
Erbium-Doped Fiber Parameters (Part I)		
Core radius of erbium-doped fiber		r (μm)
Core cross-sectional area of EDF		A (m^2)
Overlap factor of EDFA at wavelength (λ)		Γ_λ
Ion density of erbium-doped fiber		ρ (ions/m^3)
Florescence lift time of EDFA		τ (ms)
Absorption cross-section at wavelength (λ)		σ_λ^α (m^2)
Emission cross-section at wavelength (λ)		σ_λ^α (m^2)
Absorption coefficient at wavelength (λ)	$\rho\Gamma_\lambda\sigma_\lambda^\alpha$	σ_λ (m^{-1})
Absorption over length L at wavelength (λ)	$\alpha_\lambda \times L$	
Plank constant	6.626e-34	h (J s)
Speed of light		c (m/s)

TABLE 7.2
Glossary of EDFA Parameters

Physical Meaning	Expression	Symbol & Unit
Erbium-Doped Fiber Parameters (Part II)		
Length of the erbium fiber		L (m)
Population density in fundamental state		n_1 (ions/m^3)
Population density in excited state (assume 980 nm pump)		n_3 (ions/m^3)
Population density in meta-stable state		n_2 (ions/m^3)
Signal power		P_{in} (mW)
Pump power		P_{out} (mW)
Signal power in simulation	$N_s = \frac{P_{\text{signal}}^{\text{in}} C}{h\lambda_s}$	N_s (photons/s)
Pump power in simulation	$N_p = \frac{P_{\text{pump}}^{\text{in}} C}{h\lambda_p}$	N_p (photons/s)

Ground state 1 population:

$$\frac{\partial n_1}{\partial t} = -W_{1-3} * n_1 - W_{1-2} * n_1 + W_{2-1} * n_2 + \frac{n_2}{\tau_{2-1}} + W_{3-1} n_3 + \frac{n_3}{\tau_{3-2}} \tag{7.1}$$

State 2 populations:

$$\frac{\partial n_2}{\partial t} = W_{1-2} * n_1 - W_{2-1} * n_2 + W_{3-2} * n_3 - \frac{n_2}{\tau_{2-1}} \tag{7.2}$$

TABLE 7.3
Transition Rates between States of a Stimulated Er^{3+} Ion

Physical Meaning	Symbol	Expression	Comment
Transition rate from state 1 to state 2	W_{1-2}	$W_{1-2} = \frac{\Gamma_s \sigma_{1-2} N_s}{A}$	Transition associated with the signal
Transition rate from state 1 to state 3	W_{1-3}	$W_{1-3} = \frac{\Gamma_p \sigma_{1-3} N_p}{A}$	Transition associates with the pump (980 nm)
Transition rate from state 2 to state 1	W_{2-1}	$W_{2-1} = \frac{\Gamma_s \sigma_{2-1} N_s}{A}$	Transition associated with the signal (C-band) 1525–1565 nm (L-band) 1566–1605 nm
Transition rate from state 3 to state 2	W_{3-2}	$W_{3-2} = \frac{\Gamma_p \sigma_{3-2} N_p}{A}$	Transition associated with the pump (980 nm)
Transition rate from state 3 to state 1	W_{3-1}	$W_{3-1} = \frac{\Gamma_p \sigma_{3-1} N_p}{A}$	Transition associated with the pump (980 nm)

And finally, the state 3 population equation (Table 7.3):

$$\frac{\partial n_3}{\partial t} = W_{1-3} * n_1 - W_{3-2} * n_3 + W_{3-1} * n_3 - \frac{n_3}{\tau_{3-2}} \tag{7.3}$$

The transition from state 3 to 2 (W_{3-2}) is much more influential on the EDFA gain than the transition down to the fundamental state (W_{3-1}). The transition rate W_{3-1} can be neglected. The transition W_{1-3} is associated with the pump source. Furthermore, we can refer to Figure 7.8, the florescence life time, as transition (W_{3-2}) is short enough (1 μs) compared with that of the transition (W_{2-1}) (10 ms), thus the erbium ions can be accumulated in state 2. Hence, the transition rate W_{1-3} should very close to transition W_{3-2} due to the shorter life time. We thus have $W_{3-2} \cong W_{1-3}$. Substituting this into Equation 7.2, we have

$$\frac{\partial n_2}{\partial t} = W_{1-2} * n_1 - W_{2-1} * n_2 + W_{1-3} * n_3 - \frac{n_2}{\tau_{2-1}} \tag{7.4}$$

In fact, n_3, n_2, *and* n_1 are independent. However, as mentioned above, the erbium ions are accumulated in upper state 2; the population density in n_3 thus can be negligible. The total population density ρ of the fiber is a constant value. Therefore, if $n_3 \approx 0$, then n_1 and n_2 are no longer independent with each other, and under the constraint, the sum of the two equals the total erbium ions with the fiber $n_1 + n_2 = \rho$. In particular, the population are functions of both time and linear distance along the erbium-doped fiber.[3,6,13,21]

Furthermore, the relationship between the input power and the population must be determined next. Figure 7.13 illustrates how the input power evolute along the EDF.

Refer to Figure 7.13; the input light is amplified along the distance during the time. Hence, the optical power through the erbium-doped fiber is determined[22] by

$$\frac{\partial P_p}{\partial z} = -\Gamma_p * \sigma_{1-3} n_1 * P_p - \gamma_p * p_p \tag{7.5}$$

$$\frac{\partial P_s}{\partial z} = -\Gamma_s * \sigma_{1-2} n_1 * P_s + \Gamma_s * \sigma_{2-1} n_2 * P - \gamma_s * p_s \tag{7.6}$$

where γ denotes the scattering and other losses of the EDF. As shown in Figure 7.1, X indicates the co-directional photons. O represents the photon of a different direction, in this case contra-directional,

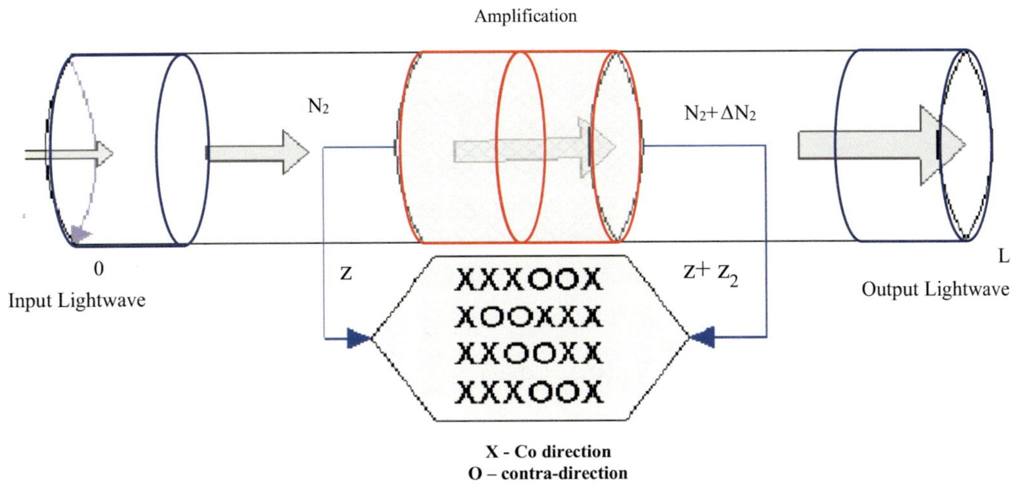

FIGURE 7.13 Power evolution of the signal during the amplification process along EDF.

thus referring to the contra-pump gain photons or generating noise, reflection power, and scattering. It is noted that noise is random, so in both directions, such noise is added to or superimposed on each other.

7.3.2 EDFAs Dynamic Model

In our simulator, the focus is on the dynamic modeling of EDFAs rather than steady-state modeling, as "EDFAs with high output powers for multiple channel optical network are strongly saturate and the resultant effective time constant are reduced to the order of tens of microsecond which is 1000 times shorter than its lift time."[6] The time-dependent problem is especially important in optical networks where optical channels are constantly reconfigured. Dynamic modeling can predict the EDFA gain more accurately during multi-channel amplification. This section describes the essential Bononi-Rusch approach and develops it to accommodate time-varying pump and communication signals. Multi-channel amplification modeling principle also will be considered in this section.

First, we note that Equations 7.1–7.6 can be typically solved by obtaining the population in different states as a function of the transition rates $W_{1-2} = \frac{\Gamma_s \sigma_{1-2} N_s}{A}$, $W_{1-3} = \frac{\Gamma_p \sigma_{1-3} N_p}{A}$, and $W_{2-1} = \frac{\Gamma_s \sigma_{2-1} N_s}{A}$.

Substituting into Equations 7.5 and 7.6, we have

$$\frac{\partial P_p}{\partial z} = -W_{1-3} * n_1 * A - \gamma_p * p_p \tag{7.7}$$

$$\frac{\partial P_s}{\partial z} = -W_{1-2} * n_1 * A + W_{2-1} * n_2 * A - \gamma_p * p_p \tag{7.8}$$

Multiplying Equation 7.4 by the fiber effective core area and substituting Equations 7.7 and 7.8,[11,23] we have

$$\frac{\partial N_2}{\partial t} = P_s(0,t) * \left(1 - e^{G_s}\right) + P_s(0,t) * \left(1 - e^{G_p}\right) - \frac{N_2}{\tau} - \gamma_p * p_p - \gamma_s * p_s \tag{7.9}$$

where G_s and G_p are the amplification gains of the power evolution, respectively. The increment of the power along the EDFA is non-linear and assumed to vary exponentially, we have[24]

$$P_p(L,t) = P_p(0,t)e^{\frac{\Gamma_p\sigma_{1-3}}{A}N_2-\Gamma_p\sigma_{1-3}\rho L-\gamma_p} \tag{7.10}$$

$$P_s(L,t) = P_s(0,t)e^{\frac{\Gamma_s(\sigma_{1-2}+\sigma_{2-1})}{A}N_2-\Gamma_s\sigma_{1-2}\rho L-\gamma_s} \tag{7.11}$$

Assuming $\gamma = 0$ leading to

$$G_p = \frac{\Gamma_p(\sigma_{1-3}+\sigma_{3-1})}{A}N_2 - \Gamma_p\sigma_{1-3}\rho L \tag{7.12}$$

$$G_s = \frac{\Gamma_s(\sigma_{1-2}+\sigma_{2-1})}{A}N_2 - \Gamma_s\sigma_{1-2}\rho L \tag{7.13}$$

Input channels with different wavelengths have different EDFA gains due to the spectral dependence of the absorption and emission cross-sections and the overlap factor. For the Three-level transition schemes (980 nm pump), σ_{1-2} and σ_{2-1} are the absorption and emission factors, respectively. At the signal wavelength, σ_{1-3} and σ_{3-1} correspond to the absorption and emission, respectively, due to the pump wavelength, and σ_{3-1} is assumed to be negligible over the short transition time. For the two-level scheme, the pump source is normally operating in the spectral region 1480–1490 nm, the rate equations are similar to those obtained for the three-level scheme, but the absorption and emission factors $\left[\sigma_{1-3}\,\sigma_{3-1}\right]$ are to be matched with the corresponding wavelength. Rewrite these matrices to $\left[\sigma_a\,\sigma_e\right]$. The $\left[\sigma_a\,\sigma_e\right]$ for the pump wavelength is $\left[\sigma_{a_p}\,\sigma_{e_p}\right]$, with σ_{e_p} as non-zero and refers to the spectrum of absorption and emission factors. For multi-channel amplification, the modeling equations are integrated and summarized as

$$\frac{\partial N_2}{\partial t} = \sum_{\lambda_o}^{\lambda_n} P_\lambda(0,t)*\left(1-e^{G_\lambda}\right) - \sum_{\lambda_o}^{\lambda_n}\gamma_\lambda * p_\lambda - \frac{N_2}{\tau} \tag{7.14}$$

7.3.3 EDFA Steady-State Modeling Principles

It is well-known that one of the advantages of the EDFA is its slow dynamics due to the long spontaneous life time of around 10 ms. This is also the reason why steady-state models play an important role in the dynamic analysis of EDFA properties for system applications.[25] The EDFA steady-state modeling has been widely used. An EDFA operates in a steady-state condition when $\frac{\partial N_2}{\partial t} = 0$, that is when there are no population increments, and the EDFA gain reaches the saturation mode. Thus, to model the EDFA in steady state, we have

$$0 = \sum_{\lambda_o}^{\lambda_n} P_\lambda^{\text{saturation}}(0,t)*\left(1-e^{G_\lambda^{\text{saturation}}}\right) - \sum_{\lambda_o}^{\lambda_n}\gamma_\lambda * p_\lambda^{\text{saturation}}(0,t) - \frac{N_2^{\text{steady-state}}}{\tau} \tag{7.15}$$

this can be simplified to

$$P_p(L,t) = \sum_{\lambda_0}^{\lambda_n} P_{\lambda_p}^{\text{steady}}(0,t)e^{\frac{\Gamma_p\sigma_{a\lambda p}}{A}N_2^{\text{steady}}-\Gamma_p\sigma_{a_{\lambda}p}\rho L-\gamma_p} \tag{7.16}$$

$$P_s(L,t) = \sum_{\lambda_0}^{\lambda_n} P_{\lambda_s}^{\text{steady}}(0,t) e^{\frac{\Gamma_s(\sigma_{a\lambda_s} + \sigma_{e\lambda_s})}{A} N_2^{\text{steady}} - \Gamma_s \sigma_{a\lambda s} \rho L - \gamma_s} \tag{7.17}$$

This model thus can predict the onset of the steady state of the amplification gains and pump absorption in an optical amplifier containing an arbitrary number of pumps and signals for co- and contra-propagating scenarios.

7.3.4 Population Inversion Factor

Population inversion indicates the number of ions inverted from state 1 to state 2 excited by the pumping energy. To obtain the amplification gain, the number of photons gained stimulated emission from the metastable state to the fundamental state must be greater than those of the absorption from the fundamental state to state 2. This condition implies that the population in a metastable state must be maintained at a greater level than that of a fundamental state. The degree of population inversion can thus be expressed by population inversion factor n_{sp}[13,26] as

$$n_{sp} = \frac{N_2}{N_2 - \dfrac{\sigma_e}{\sigma_a} N_1} \tag{7.18}$$

where n_{sp} is related to the pump and signal power and is the key factor of the amplifying spontaneous emission noise (ASE).

7.3.5 Amplifier Noises

In the previous section, the production of the ASE photons was described. The noise statistics of the amplified lightwaves follow a Gaussian distribution. By considering noise as a random variable generated by various independent sources, it can be assumed that, in general, the probability density function of their sum also approaches Gaussian distribution. The effect of amplifier noise can, therefore, be modeled as a Gaussian random process with a noise variance.

7.3.5.1 ASE Noise Model

The ASE noise as a random noise can be modeled by generating complex spectral components whose real and imaginary random variables are Gaussian with variance P_{ase} corresponding to the average noise. The ASE noise spectral power can be estimated over a bandwidth Δf as

$$P_{ase} = n_{sp}(G-1)hf_s\Delta f = N_0\Delta f \tag{7.19}$$

where it is assumed that a sufficiently large bandwidth on ASE detection is used and sensitive within the spectral region 1480–1605 nm, G is the small signal amplification gain. To model the ASE noise, a Gaussian wave with ASE mean power and variance are used, which are described in the next section.

7.3.5.2 Other Noise Sources

In practice, an optical BPF is usually incorporated in the EDFA to reduce the ASE noise contribution. ASE-induced noise is contributed by two noise sources:

- Spontaneous-spontaneous beat noise

  sp-sp beat noise is caused by the self-interference of the generated spontaneous photons and is given as:

$$P_{sp\text{-}sp} = 2n_{sp}^2(G-1)^2 e^2 \Delta f \tag{7.20}$$

- Signal-spontaneous beat noise

  s-sp beat noise is produced by the beating between the signal photons and those of the spontaneous emission and is given as

$$P_{\text{s-sp}} = 4e^2 \frac{P_{\text{in}}}{hf} n_{\text{sp}}(G-1)G \tag{7.21}$$

The signal $S_{\text{s-sp}}$ term dominates the ASE noise, thus we have

$$N_{\text{s-sp}} = 4e^2 N_{\text{in}} n_{\text{sp}}(G-1)G \tag{7.22}$$

Extensive models for the simulation of EDFA using MATLAB Simulink are given in the section Appendix 6 of Chapter 12.

7.4 OPTICAL AMPLIFICATIONS FOR SPACE PHOTONIC TRANSMISSION

In space photonic transmission systems, telescope optics are used at both the transmitter and receiver due to the high attenuation factor of the link, about 73 dB, including the attenuation and optical gain of the transmitting and receiving optics. This factor cannot be compensated by the gain and power at either the transmitter or receiver alone but requires a combination of optical amplifiers in these subsystems.

At the transmitter, integrated photonic components must inevitably be employed in association with the optical amplifier booster to gain enough power to launch into the space link. A simple structure of an optical amplifier can be shown in Figure 7.14 and satellite installation in Figure 7.15. The optical signal power coming out of an integrated laser-modulator may not reach the right level for the optical power booster; therefore, an optical preamplifier, a typical optical amplifier at a moderate input level, is used to boost the power to an acceptable level. For example, an input level of 1.0 mW (or 0 dBm) is required for a HPA to achieve an output power of 40 dBm (10 W) with a 30–40 dB gain amplifier operating in saturation mode. Commonly, the average power per channel is about –5 dBm with a modulation format of DPM-QPSK at 100 Gbps, thus, a moderate optical amplifier with 20 dB gain is required for ten wavelength channels at 100 Gbps to achieve a total of 1.0 Tbps. Therefore, the total average power input into the HPA of 0 dBm would give a total optical power of the beam in free space of 43 dBm. The output of an optical transmitter is about –11 dBm for a laser power of 0 dBm and an insertion loss of 11 dB from an IQ modulator. Thus, a moderate gain of 20 dB optical power amplifier is required to provide the input power for the saturation operation of the HPA. We normally bias the HPA with a CW laser with an output power of 0–10 dBm so that we can set the HPA into saturation and ensure all channels input into the HPA can reach saturation level, especially when there may be only a few100 Gbps channels.

The main characteristics of an HPA transmitter and receiver in a satellite station can be the following: (i) Doped fiber materials: Erbium-Ytterbium (Er:Yb) doped fiber with core and cladding pump to give booster amplification, EYDFA; (ii) Output power 0.5–10 W; (iii) Compatible with standard DWDM Transceivers; in association with receiving pre-amplifier; (iv) Erbium low noise optical amplifier with ~4 dB noise figures and an input power level of –45 to –65 dBm; (v) Conduction-cooled vacuum operation over 9000 hours; (vi) LEO (low earth orbit) nuclear radiation tested with microprocessor rated for 30 K rad or about a few hundred keV over 10 mm square area, and low power consumption and high wall-plug efficiency of ~26 W power consumption for 5 W output optical power.

The co-doping of Ytterbium allows the higher gain and boosts the saturation level higher and accordingly the higher input level so that output high power can be achieved. Pumping can be done both in the core and cladding pump to provide sufficient gain and power level at the output port. Erbium ions are required so that the energy level jumping allows the gain of the photons at the 1550 nm spectral window. The Yb^{3+} ions co-doping in core and cladding allow higher input signal power required and then higher gain and higher reaching power levels in such amplifiers.

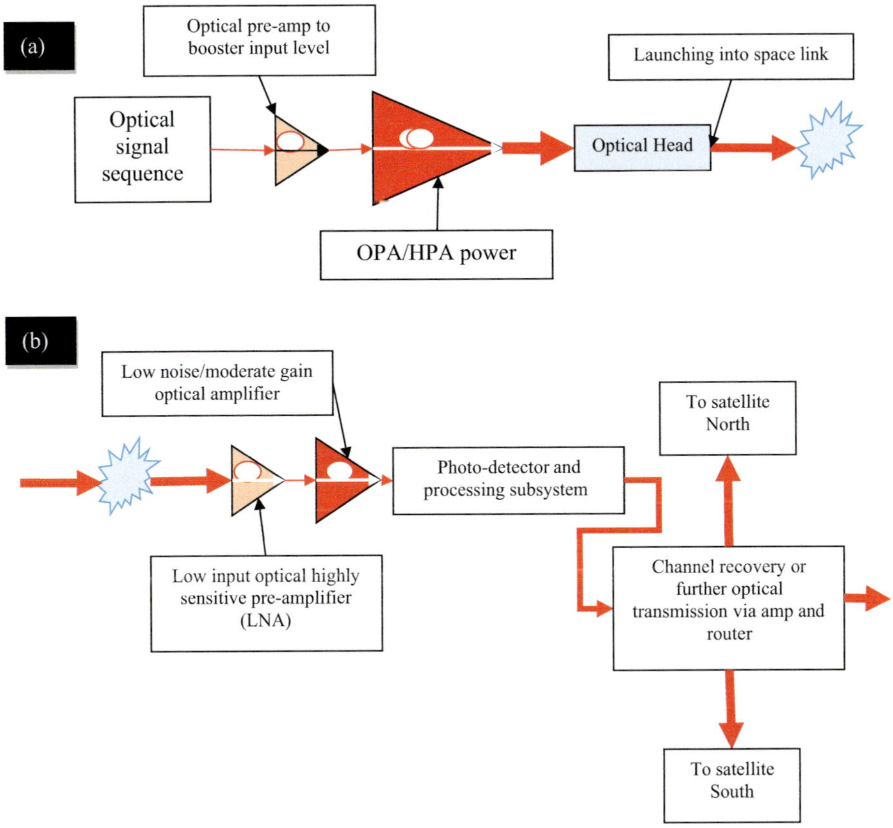

FIGURE 7.14 Schematics of panel (a) optical transmitter booster and (b) optical preamplifier at the receiving subsystem for further transmission via optical routers.

FIGURE 7.15 Optical amplifiers and satellite transmission.

On the side of the receiving end, an optical pre-amplifier must be used to receive a low sensitive channel power level. An example of this connection budget is as follows:

- Output average power of HPA = 10 W or 40 dBm
- Propagation loss and gain of the link 36,000 km from the ground (OGS) to the satellite and 36,000 km between satellites in a GEO constellation = –73 dB, obtained by attenuation of 273 dB (using Friis equation and no finite linewidth additional loss), optical gain by Optical Head = 100 dB per site (Tx and Rx).

- Optical power at the front of the receiver=–33 dBm.
- Margin of operation=–10 dB (to allow for fluctuation of the beam intensity and the laser finite linewidth).
- Power sensitivity of the receiver=–43 dBm per wavelength channel.

LNA has characteristics=low level with low noise –4 dB Noise figure, gain=45 dB, input power from –45 to –65 dBm.

The low noise figure and high gain of the LNA are highly critical to the link, allowing it to act as a pre-amplifier for a pre-amplified photodetector receiver. Such LNAs are available with bidirectional pumping and noise filtering with narrow and precise optical filtering devices.

7.4.1 High Power Optical Amplifications

7.4.1.1 Remarks

Definition: Fiber lasers and fiber amplifiers with high output powers of more than 1.0 W. In SPC, the average power level of 30 dBm per channel at the output of the HPA and –45 to –60 dBm at the input of the Low Noise Amplifier (LNA) is sufficient. Thus, for tera-bps, we need ten channels of 100 Gbps each; therefore, the output of the HPA should be at 40 dBm (10 W), which is a typical amplification level for moderately high output power HPA.

Whereas the first fiber lasers in the 1990s could deliver only a few milliwatts of output power, there have subsequently been rapid developments that have led to high-power fiber lasers and, particularly, amplifiers with output powers of tens or hundreds of watts, sometimes even several kilowatts from a single fiber. This potential arises from a very high surface-to-volume ratio (avoiding excessive heating) and the guiding (waveguide) effect, which avoids thermo-optical problems even under conditions of significant heating. In SPC, we consider the output power to be on the order of 1.0 W per wavelength channel. At this power level, even nonlinear effects are created. Leading to distortion of signals over 1–2 m of single-mode fiber, which is connected from the HPA to the OH before launching into space.

There is no generally agreed minimum power level for the attribute "high power". It may mean more than 100 W of average power, but some lasers are labeled as high-power even for only 10 W.

Fiber laser technology now competes strongly with other high-power laser technologies based on solid-state bulk lasers, such as thin-disk lasers, etc. A comparison of all the different pump launching techniques is complex but simple enough; in fact, there is only the gain medium, reflectors, filters, and pump techniques, such that much absorption can be achieved to supply energy for high power output. Figure 7.16a and b shows the schematic structure of an HPA amplifier or high-power laser, depending on the reflector FBG properties and its spectral window.

The structure of the gain in photon-in, photon-out can be arranged with multiple pumps to supply sufficient power for the absorption and stimulated emission. Crystal fiber can be used with the main core doped with rare earth ions and cladding pumping as well. These aspects are different to obtain maximum transfer efficiency, minimize the emission loss, and ensure ease of manufacturing, as well as flexibility in handling. Additionally, considerations include possible back-reflections, leakage of light from the fiber core back to the pump source to avoid oscillation if amplification is only required, and maintaining polarization if desired.

Typical specifications of HPA are as follows:

- Wavelength Range 1545–1574 nm
- Input Power Level –10 to 10 dBm
- Output Power Up to +40 dBm or 10 W average across all channels
- Input Isolation >30 dB Output Isolation >30 dB
- Gain 40 dB @ –10 dBm optical input fiber type HI1550,
- Noise Figure=7 dB
- High Power Connector and Collimator

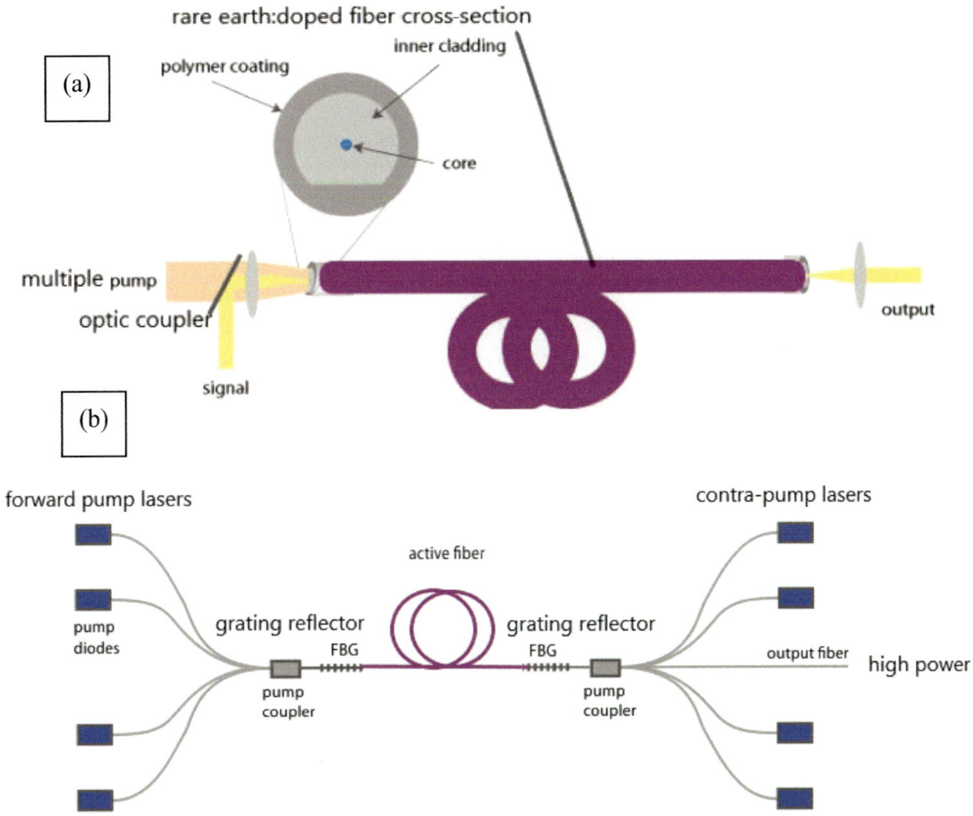

FIGURE 7.16 (a) HPA schematic structure with coupling optics and pumping; (b) HPA output distribution and coupling for safe operation and testing (b) dual directional pumping and high-power output, FBG=fiber Bragg grating; FBG acts as reflector and filter to reduce ASE (amplification stimulatedEmission) noises.

7.4.1.2 Double-Clad Rare-Earth Doped Fibers and Output Beam

High-power fiber lasers and fiber amplifiers are nearly always realized with rare-earth-doped double-clad fibers, which are pumped with fiber-coupled high-power diode bars or other kinds of laser diodes. The pump light is launched into an inner cladding rather than into the (much smaller) fiber core, in which the laser light is generated. The laser light can have very good beam quality – even diffraction-limited beam quality if the fiber has a single-mode core. Therefore, the brightness of the fiber laser output can be orders of magnitude higher than that of the pump light, even though the output power is of course somewhat smaller. Double-clad fiber lasers can thus effectively be used as brightness converters. Typical optical-to-optical power efficiencies are above 50%, sometimes even above 80%; the overall electrical-to-optical power conversion efficiency of a diode-pumped fiber laser can be of the order of 50% for a Yb-based fiber laser pumped with efficient high-power diode lasers (Figure 7.17).

For the highest powers, the core area needs to be fairly large (i.e., → large mode area fibers or LEAF of Corning Inc.) because the optical intensities would otherwise become too high, and often also because a double-clad fiber with a large ratio of cladding to the core area has a weak pump absorption. For core areas up to the order of a few thousand μm^2, it is feasible to have a single-mode core. Larger mode areas with still fairly good output beam quality are possible with a slightly multimode core, where most of the light propagates in the fundamental mode. (The excitation of higher-order fiber modes can be suppressed to some extent; e.g., by coiling the fiber to leak higher-order

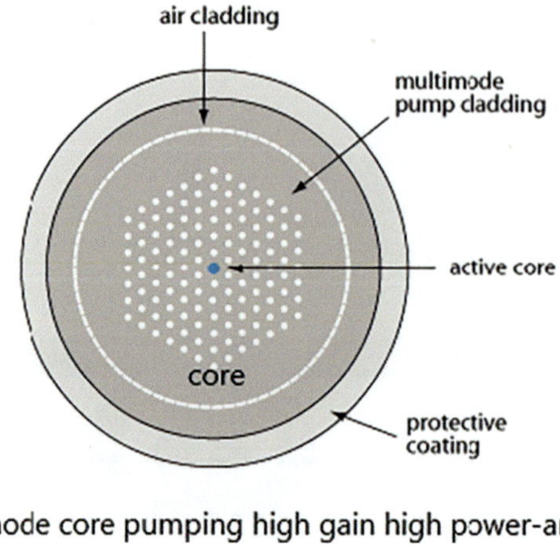

FIGURE 7.17 Schematic structure of air-silica clad photonic crystal fiber and multi-pump cladding crystal fiber for high gain, high power amplifier. Double-clad fiber with an air cladding.

guided modes into the cladding, except if one obtains strong mode coupling at high power levels.[27]) For even larger mode areas, the beam quality can no longer be nearly diffraction-limited, but it can still be fairly good compared with, e.g., rod lasers operating at similar power levels.

The required fiber length is often several tens of meters, i.e., a few tens of rounds of rare-earth-doped fiber, dictated by the need to absorb the injected pump light with sufficiently high efficiency. Although the fiber length can, in principle, be reduced, e.g., by using a fiber core with a higher doping concentration, a larger fiber core, or a smaller pump cladding (for a smaller cladding to core ratio), there are various limits to this approach. These limits arise from factors such as the doping concentration, the beam quality of the pump source, beam quality requirements for the signal, or thermal considerations, as cooling is expected to increase the amplification efficiency.

7.4.1.3 Lasers and Amplifiers

As stated above and given in Figure 7.16, the fundamental structure option for generating high-power laser light consists of a gain power medium that ensures sufficient absorption of power to generate stimulated emission and hence sufficient gain by reflection for further gain. The reflector and filter ensure stimulated emission at a certain wavelength spectrum. However, high-power fiber devices are very often composed of laser-amplifier combinations in coherence, i.e., with a master oscillator fiber amplifier (MOFA) architecture. This concept has several advantages. It is easier to control the emission properties of a lower-power seed laser in terms of linewidth, laser noise, wavelength tunability, pulse generation, etc. Furthermore, the amplifier fiber needs to handle only power approximately equal to the output power, whereas in a laser, the intracavity propagating and reflecting power is higher, with an output coupling of about 70%. Furthermore, it can be advantageous to use a modular design approach, where amplifier stages can be combined as required. A seed laser is normally preferred to initiate the amplification, thus setting the operating point to achieve saturation gain before channels are launched to obtain high power output.

7.4.2 Nanosecond Pulses

7.4.2.1 Basics

Nanosecond pulse lasers using fiber amplifiers under mode locking can be very useful in SPC operations under cloudy conditions. The short pulse and high peak power of several kilowatts can pierce through the clouds, despite substantial loss, and propagate to the satellite. The peak power of the pulse sequence is still sufficient to travel to the satellite constellation. The average power may still be a few hundred milliwatts, which is sufficient to propagate the data sequence to the GEO satellite constellation.

Many high-power lasers for material processing are Q-switched bulk lasers, generating intense nanosecond pulses. The direct application of that technique to fiber lasers can be limited in terms of achievable peak power, basically due to the strong fiber nonlinearities. However, the high gain of fiber devices allows for the realization of very flexible MOPA devices with a gain-switched laser diode, to obtain picosecond diode lasers as the seed laser. This approach, unlike Q-switching of a laser, allows modification of the pulse duration independently of the pulse repetition rate. Still, the possible pulse energies are far lower than those achievable with bulk lasers. For many applications, however, the performance is sufficient.

7.4.2.2 Ultra-Short Pulses

The large gain bandwidth of fiber amplifiers allows for the amplification of ultrashort pulses. This can be exploited in both mode-locked fiber lasers and ultrashort pulse fiber amplifiers.

Particularly, in this pulse duration regime, substantial challenges arise from the strong fiber nonlinearity, as high-energy femtosecond pulses have huge peak powers. In addition to the risk of fiber damage, strong nonlinear distortions can result from fiber nonlinearity, and the high level of chromatic dispersion (including higher-order dispersion) can also be problematic. Particularly, when pulse quality matters, these issues need to be carefully considered.

One possible approach is to make fiber-based chirped-pulse amplification systems, where the pulse duration within the amplifier is strongly increased reducing the effects of nonlinearities accordingly. A less powerful alternative to this approach is the amplification of parabolic pulses, where up-chirped pulses experience the gain and nonlinearity of the amplifier fiber, evolving in a self-similar fashion. Their close-to-linear chirp makes it possible to obtain a high pulse quality with dispersive compression.

An interesting direction is the development of all-fiber ultrashort pulse sources, possibly allowing for a final free-space pulse compressor (e.g., a transmission grating), but eliminating the need to launch pulses from free space into a fiber.

7.4.3 Limiting Factors

Various possible limiting factors for the performance of high-power fiber amplifiers and lasers are briefly discussed in the following sections. The factors that limit the performance of a particular device depend very much on the operation regime.

7.4.3.1 Available Pump Power

Some devices are limited by the amount of pump power launched into the fiber. Note that even when using double-clad fibers, the beam quality of the pump source needs to be sufficiently high. Effectively, that means that only pump sources with sufficiently high radiance (brightness) can be used.

Fortunately, the continuous development of diode bars and broad-area laser diodes results in devices with higher and higher radiance, so that the corresponding performance limits are extended increasingly.

7.4.3.2 Heat Generation

A common problem for high-power lasers is heat generation in the gain medium. Fiber amplifiers and lasers, however, are affected by this only at very high power levels, basically because of (i) their often quite high power conversion efficiency (resulting in a relatively low heat load), (ii) the distribution of the heat load over a relatively long length (e.g., several meters) and (iii) the waveguide function of the fiber core.

The heat load in a fiber may become a problem, particularly in cases where the fiber length has to be minimized, e.g., to mitigate nonlinear effects. This means that using fibers with relatively high doping concentrations allows them to turn over the required power within a shorter length. The temperature rise in the fiber core can then be quite substantial. Besides, the resulting thermal lensing may significantly modify the mode properties, e.g., reducing the effective mode area and increasing the number of guided modes. Cooling is normally used to improve amplification conversion efficiency.

The dissipated power per unit length often reaches values on the order of 100 W/m, which causes significant heating for air-cooled fibers. Water cooling may be used for further power increases, although it cannot remove the thermal lensing problem. However, in SPC, such cooling is preferred using either TEC (thermal electric cooling) or cooling fins to minimize power consumption waste.

If the fiber is not strictly single-mode, which is often the case in high-power devices, particularly in the amplifier's last stage, heating can trigger another problem: thermally-induced transverse mode instability (TMI), generated based on stimulated thermal Rayleigh scattering (STRS). A typical operational characteristic is that the amplifier or laser can suddenly produce a substantially reduced output power because of a significant transfer of power into higher-order fiber-guided modes. The distribution of optical power shows an oscillatory or chaotic behavior, leading to substantial errors. The tendency for mode instability increases in fibers with particularly large fiber cross-sectional areas.

The physical mechanism underlying thermally induced mode instability involves the generation of a long-period grating in the fiber (LFBG), which couples light from the fundamental mode to higher-order modes. At the same time, modal interference occurs in the grating itself, leading to the reflection of modes from many unwanted wavelength channels, and thence a reduction of power in the desired channels. Such multi-modal interference leads to a spatially varying intensity pattern and, via Kramers–Kronig, to a refractive index pattern. An important aspect is the slight difference in optical frequency between the involved modes, which leads to a beat note and the movement of the refractive index grating.

7.4.3.3 Nonlinear Limitations

As described in Chapters 2, 3, and in an Appendix of Chapter 10, various high-power fiber devices are essentially limited in performance by consequences of fiber nonlinearities of different types.

In particular, single-frequency devices suffer from stimulated Brillouin scattering (SBS), four-wave mixing (FWM), Stimulated Raman Scattering (SRS), etc. The problem is that the intense signal light in an amplifier, for example, leads to a high Brillouin gain for counterpropagating light with a slightly lower optical frequency. Once that nonlinear gain reaches approximately 90 dB, quantum noise is amplified to a substantial power level. The resulting counterpropagating wave then extracts power from the signal, which in turn reduces the nonlinear gain; that interaction usually leads to chaotic power fluctuations. There are various mitigation techniques. For example, it is common to apply a first-phase modulation to the input signal, which can be removed after the amplifier and reduces the Brillouin gain by increasing the signal linewidth beyond the Brillouin gain bandwidth. One may also optimize the fiber design, particularly concerning the effective mode area and used length, and exploit a temperature gradient along the fiber.

Similarly, a nonlinear gain arises from stimulated Raman scattering (SRS), where various parameters are different from those of Brillouin scattering: the gain coefficient is much lower, the nonlinear gain bandwidth is much higher, and the nonlinear gain occurs for co-propagating light. SRS is often the dominant limiting factor for the amplification of ultrashort pulses, where SBS is not relevant due to the broader signal bandwidth and the limited spatial overlap region for counter-propagating short pulses.

A rather hard limit for the possible peak power in fiber is set by nonlinear self-focusing. Above a certain critical power, this effect leads to a catastrophic collapse of the beam profile, usually followed by instant destruction of the fiber material. Interestingly, the critical power is not increased by using a fiber with a larger effective mode area, even though the optical intensity for the undisturbed mode is reduced. For silica fibers, which have a relatively low nonlinear index, the critical power is relatively high (several megawatts), but it still presents a serious limitation in some of the cases with ultrashort pulse amplification. Even when applying chirped-pulse amplification, the pulse energy is limited to the order of 10 mJ because there are practical limits to the usable chirped pulse duration.

There can also be the problem of laser-induced damage at fiber ends. That problem can be substantially mitigated by using coreless end caps, where the beam can expand to a substantially larger area before reaching the more sensitive glass-air interface.

Note that the optimization of high-power file devices often involves a trade-off between different kinds of limitations; the best results are usually obtained by finding an optimal compromise.

7.4.4 Remarks

This chapter demonstrates the successful implementation of the modeling of the EDFA operating under dynamic and steady states using MATLAB Simulink. The model has been used to investigate the characteristics of the EDFAs over the C- and L-band spectral regions. Predicted results, obtained by varying the parameter settings, show good agreement with experimental data. The MATLAB Simulink toolboxes for simulating dynamic systems offer significant advantages of being well-known and available in teaching and research laboratories worldwide. Multiple wavelength signal propagation and ASE noise simulations have been considered in our EDFA SIMULATOR. Although we demonstrate only the Simulink [EDFA Module] using one wavelength for the pump and one for the signal, it is straightforward to modify the simulation to achieve multiple-channel propagation as an EDFA simulator.[28] The only limitations are the computer memory and the CPU runtime. Cascaded EDFA systems can also be connected, and the effects of propagation loss and noise in such systems can be modeled with ease. The gain modulation of the EDFA has not been discussed and will be reported in a future article.

Primarily a powerful simulator of modern optical amplifier modules and their various components has been established. Future work will aim to incorporate a number of optimization components into the EDFA simulator, thereby further enhancing its utility. Multiple-channel propagation simulation in the EDFA Module has also been incorporated. The developed EDFA SIMULATOR and the EDFA Modules will provide the framework for the future development of Raman amplifier simulation using MATLAB Simulink.

High-power amplifiers and lasers using multiple pumps and multi-core fibers are described, including issues of manufacturing and operation of such high-power amplifier devices, as well as potential nonlinear effects in such high-output power amplifiers.

7.4.5 Low Level Low Noise Preamplifier (LNA)

Figure 7.18 depicts the schematic of an optical amplifier with a single-stage Er-doped fiber, including reflectors and a wavelength-dependent coupler at the input or codirectional pump laser. A tunable optical filter (TOF) is not included, resulting in lower suppression of ASE noises. The gain spectrum and the NF as a function of wavelength are depicted in Figures 7.19 and 7.20 for erbium-doped fiber. As observed in Figure 7.19, the gain spectrum is not flat in the region <1545 nm. Therefore, to extend the flat gain in this region, one has to use a gain equalizer to flatten, as described in Chapter 9.

If only one channel is to be amplified with low noise and hence a highly sensitive input signal level, such as −60 dBm average power of the channel, then a single narrow band TOF, such as the one whose schematic is given in Figure 7.21b would suffice. However, in tera-bps with 100 G DPM-QPSK, there is a need for ten channels spaced 50–100 GHz, so a parallel TOF must be used. The cascade of a number of TOFs, such as the one given in Figure 7.21, will suffer many losses, which can create unequal power across all channels. This would limit the number of hops in space for SPC. Therefore, a parallel TOF, such as the AWG with a high dynamic range and high roll-off of

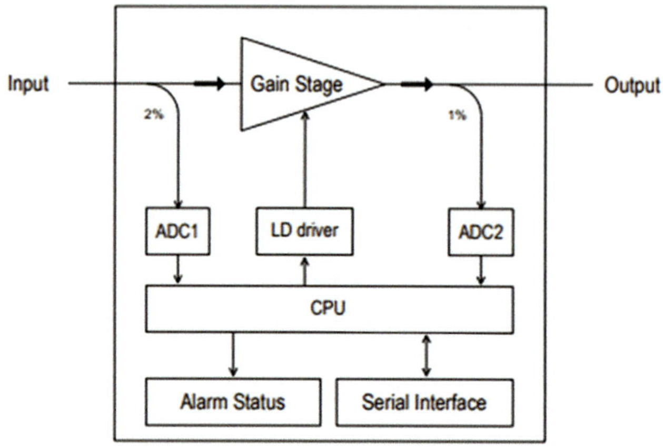

FIGURE 7.18 Structure of an LNA with pump and control feedback for stabilization. This structure does not include the TOF (tunable optical filter to suppress the ASE noises. ADC=analog-to-digital converter, LD=laser diode, CPU=central processing unit, 1% or 2% power=coupling factor.

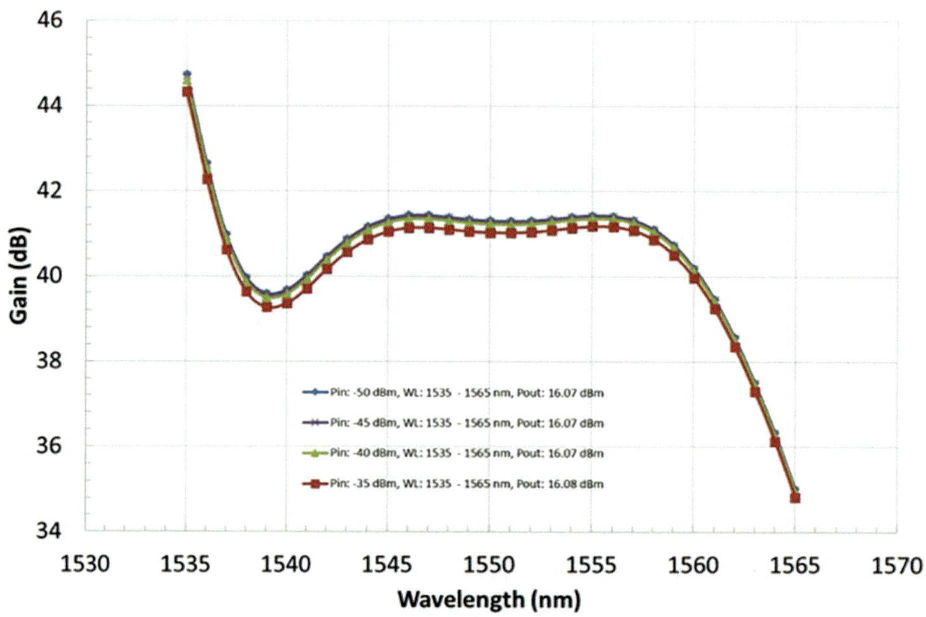

FIGURE 7.19 LNA gain versus wavelength with input power as a parameter. Gain at different input powers (–35 to –50 dBm) and constant LD pump power; gain=~ 43 dB at the central C-band 1545–1560 nm.

the order of seven to nine filters over 0.1 nm, would be preferred, and this is possible in Si_3N_4 or SOI-integrated photonic technologies. These parallel TOF devices are very critical for ultra-high bit rate SPC systems, especially when routing is done by CDC-ROADM (contention-less, direction-less, colorless reconfigurable ADD/DROP multiplexer). The CDC-ROADM can be referred to in Chapters 3 and 6.

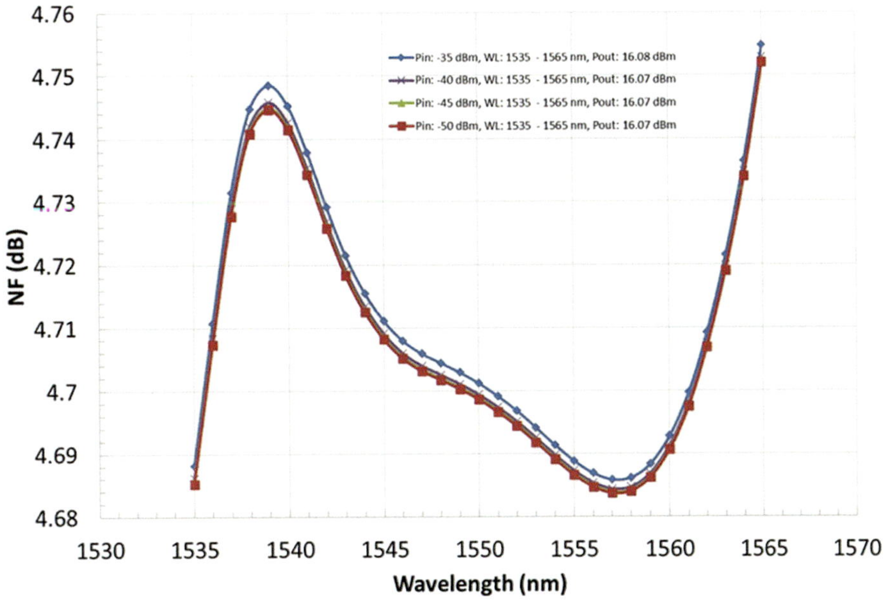

FIGURE 7.20 NF at different input powers and constant LD (laser diode) pump power plotted versus wavelength, that is noise figure spectrum. Saturated output power ~ 16 dBm at saturation. Minimum average signal power at –50 dBm.

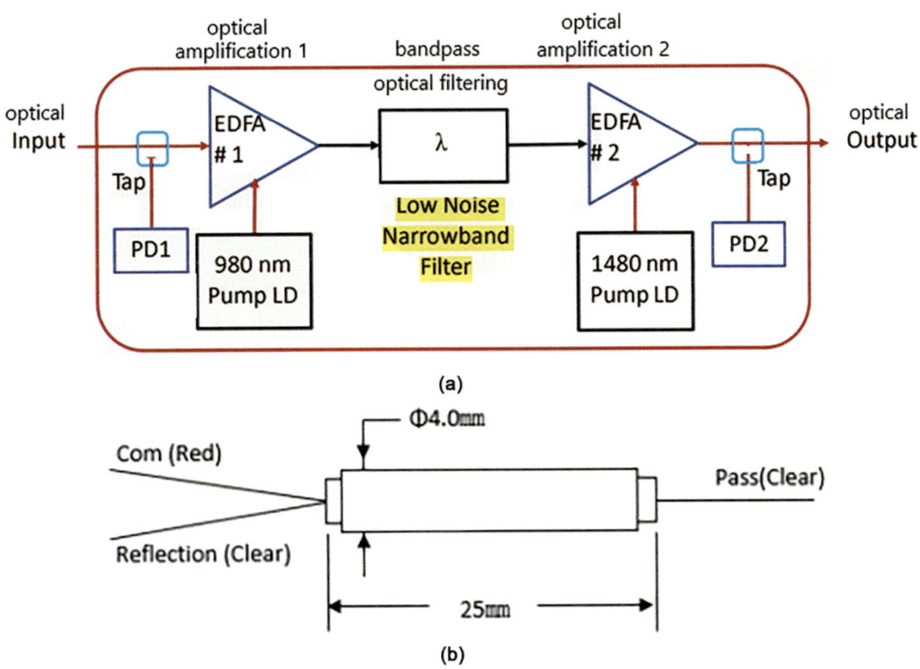

FIGURE 7.21 Structure of a low noise amplifier with two-stage pumps Er: doped fiber and an optical filter (a) Dual EDFA and optical filtering for low noise amplification. (b) Optical filter structure as ADD/DROP multiplexer. (a) Dual cascade stage EDFA and optical filtering for LNA. TOF (tunable optical filter) is inserted between two optical amplification stages to suppress ASE noise below the noises to the minimum ASE level (due to spontaneous emisson) of about 3 dB or a bit higher to 4 dB – the sharp roll-off to obtain required signal power level can be accepted to –65 dBm. (b) Typical optical reflection filter as an optical ADD/DROP multiplexer and usable as tunable optical filter (TOF). It is normally for two to three wavelength dropped or added.

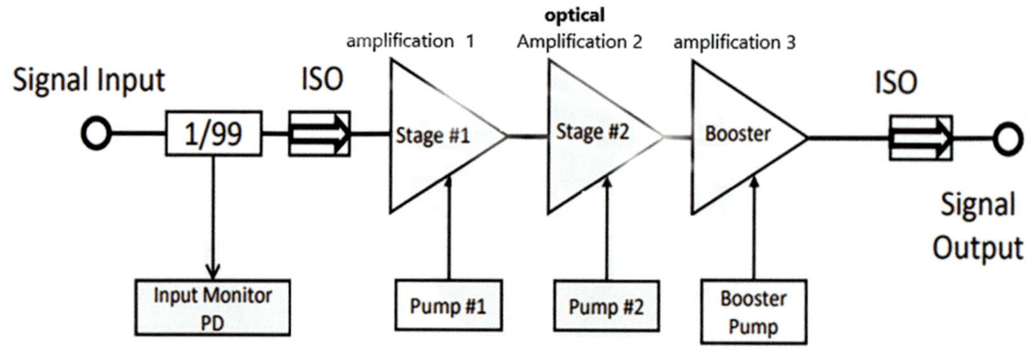

FIGURE 7.22 High power booster amplifier (HPA) with three-stage pumping of EYDFA sections for high power amplifications. Note *n* optical filters.

7.5 EDFA MODERATE-LEVEL HIGH-POWER BOOSTER AMPLIFIER AND LOW-NOISE PRE-AMPLIFIER

7.5.1 HPA Structure and Circuit

A schematic of a three-stage pumping HPA (high-power amplifier) is shown in Figure 7.22. The first two stages are used to boost the optical channels to a level at which the booster amplification section can gain enough power to output an average optical power level on the order of thousands of watts, normally 1.0–20 kW. The laser pump sources #1 and #2 are used to pump the first two stages from –20 to 0–10 dBm of optical channel power. The third stage of amplification with a gain of 30–40 dB can boost the channel powers to an average of 30–40 dBm in total. This means the power of each channel is reduced by the decimal logarithm of the number of multiplexed channels. For example, if an HPA with 40 dBm output power is used with 10 DWDM channels whose total average power is 0 dBm, i.e., –10 dBm average power per channel, then the output power per channel is 30 dBm with 40 dB optical gain. Therefore, the required input power per channel is at least –10 dBm. If this average is higher than –10 dBm then the output power remains at this saturation value of 30 dBm per channel.

If the average power of a channel is less than –10 dBm, then the channel is operating with a linear gain of amplification. This linear gain region cause fluctuations in the signal level; thus, the receiving amplifier must have automatic gain control to ensure the electronic signal level is at an acceptable input level to the ADC, and thence, the DSP sub-system.

We will illustrate the operation of such an HPA under a seedling pump channel whose wavelength is in the saturated gain region of the booster amplifier. Note that the amplifiers here can output moderate to high power on the order of 30–43 dBm or 1–20.0 W, which is sufficient for an SPC system. At this level, the nonlinear distortion can be acceptable with 1–2 dB penalty in sensitivity. The NF (noise figure) of this amplifier is about 6.5 dB.

In a space photonic communication system, it is much more reliable for the optical DWDM channels to operate at the saturation level. A seed laser is normally used so that the number of DWDM channels can be switched on or off, and they can all operate at the saturation level set by the seed laser (Figure 7.23).

7.5.2 Structure and Mechanical Engineering and Optical Spectra

The HPA shown in Figure 7.20 can be packaged as shown in Figure 7.21. Figure 7.24a and b shows the plane view of the mechanical packaging such that the heat dissipation of the HPA can be absorbed and cooled by such aluminum block.

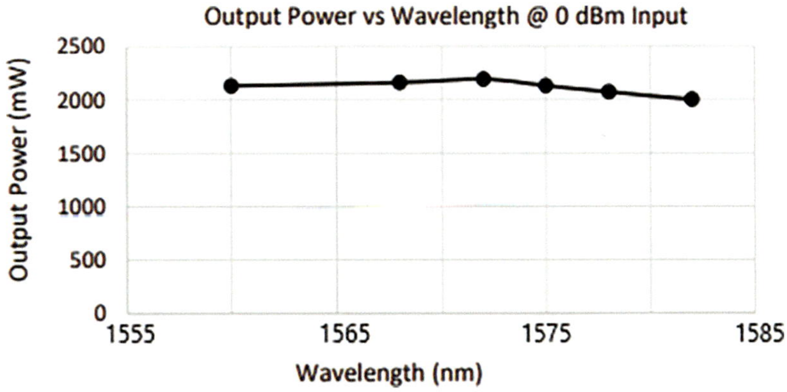

FIGURE 7.23 The optical gain and output power spectrum of an HPA is shown in Figure 7.22. For 1565 nm spectral window of 30 nm.

Figure 7.24b and c shows the input and output fibers for launching signal channels and extracting amplified boosting channel outputs. The extension of the fiber output is necessary for mounting into the optical head to launch the beam into space. This extended fiber generates nonlinear crosstalks. Within an extended fiber length of about 2.0 m, we have already observed nonlinear self-phase modulation effects and four-wave mixing. Other nonlinear effects, such as Stimulated Brillouin scattering (SBS), are also observed with pulse sequence jittering. The high-band SBS of around 9 GHz from the channel carrier is also observed but its amplitude is much lower than that of the channel spectra, so it is ignored.

This shows that the nonlinear effects are quite serious, and mitigation of such nonlinear distortion must be deployed at both the transmitter and receiver. Such compensation and mitigation at both the transmitter and receiver, or only at the receiver using the Viterbi-MLSE algorithm, have been extensively described in Chapter 6.

Figure 7.25 shows the optical and electro-control circuits for both HPAs and LNAs. The Tx control is used to control the pump power by adjusting the driving current to the laser diode, normally about a few hundred milli-amps. Two or three pump lasers are used to boost the output power of HPA to 1–20 W level.

In operational aspects for stable output saturated average power of the channels, we use a seed laser whose power is sufficient to obtain the saturated gain. For example, for an output saturated power of 30 dBm and a gain of 40 dB, the minimum power of the seed laser must be at least –10 dBm. This seed laser is turned on at all times or operates in the CW mode.

The rare-earth-doped fiber sections can get very hot, and the housing must be arranged in such a way as to cool these fibers without using TEC (thermo-electric Peltier cooling), as shown in Figure 7.26. The Er-doped fiber and then add/drop optical filter of the LNA can be housed in the same winding of the HPA in this structure.[30]

7.5.3 Expected Spectra

The spectrum of the optical filtering multiplexer, as an add/drop multiplexer for optical filtering to obtain low noise amplification, is shown in Figure 7.27a and b. Channel #21 is at 1560.61 nm and channel #22 is at 1559.79 nm. The channel grids can be examined in Appendix 7 of Chapter 12.

7.5.3.1 Optical Pre-Amplifier Section MPB – SDA- 2023-0013 et al.

The expected optical spectrum of the output lightwaves at ports P5 and P6, as well as the reflection port of the optical band pass filter (OBPF - AC-Photonics) at P6, is shown in Figure 7.27c.

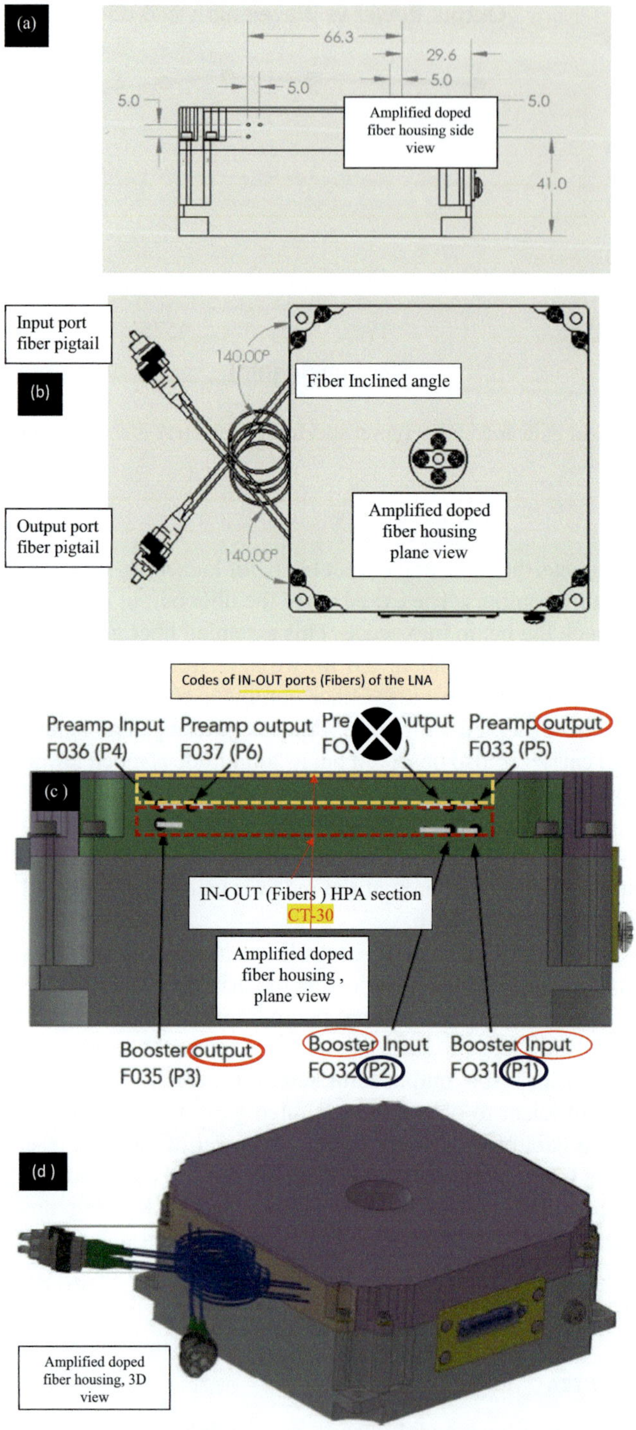

FIGURE 7.24 Optical interface OA_MBP_EDFA_HPA_LNA. (a) Mechanical drawings of SPA-CT30R40-MY-C2 (or SPA-CT30R40-MY-C3) unit showing the positions of the optical interfaces (side view); (b) mechanical drawings of SPA-CT30R40-MY-C2 unit showing the positions of the optical interfaces (top view); (c) 3D side view of optical fiber inputs/outputs. P1 = Booster in P1; **For HPA**: P2 = Booster IN 2; P3 = Booster OUT P#; **For LNA**: P4 = LNA IN; P5 (far right) = LNA IN; P6 = Output 1 LNA; P7 = Output 2 LNA. (d) 3D model of enclosure with cover (125 mm wide square base by 53 mm high) showing one electrical connector, one ground connection, and six fiber-optic interfaces.[29]

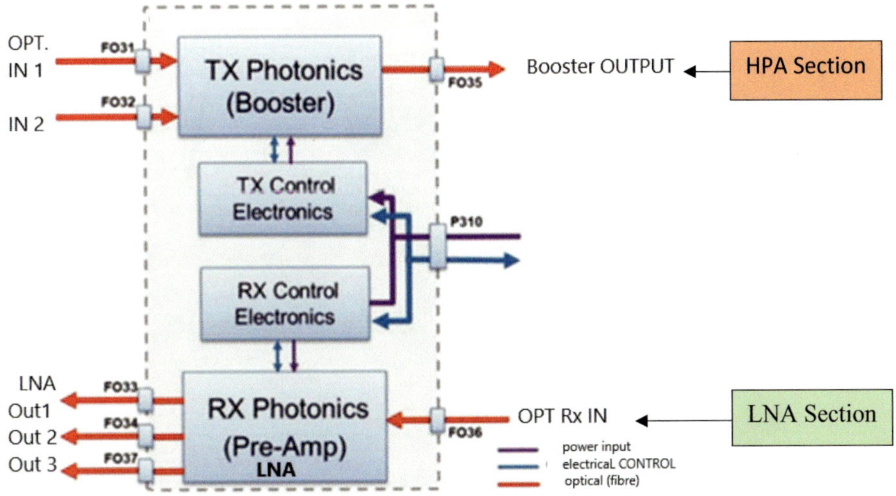

FIGURE 7.25 Power drive and control of a combined EDFA HPA-LNA optical booster amplifier and low noise preamplifier (LNA) with one pre-amplifier and three channel outputs, with only two channel inputs for high power booster outputs.

Splicing of the fiber of the reflection port (clear fiber) and circle biding and open and extend for splicing.

The general characteristics of the whole C-band, in which the channel central carriers can be shown, are depicted in Figure 7.27. The add and drop channels can be expected in the spectrum. This depends on the channels located in the spectral grid as given in Appendix 7 of Chapter 12, Section 12.5. It is also dependent on the grid spacing of either 100 or 50 GHz. The 100 GHz spacing is standard and numbered accordingly, while the 50 GHz spacing is interleaved between two 100 GHz channels.

7.5.4 Optical Circuit and Mechanical Aspects of the Low Noise Amplifier (LNA)

An Optical Pre-Amplifier LNA circuit is shown in Figure 7.28. It is a dual pump amplifier with two Er:Yb (Erbium Ytterbium) co-doped fiber sections. There are two pumps with wavelengths of 980 nm. The pump lasers are cooled by TEC (thermo-electric cooler or Peltier p-n junction cooling driven by a 2 A current driver). A control circuit is used for setting the driving conditions of the amplifiers. The unit contains both the HPA and LNA, and the control platform can be seen in Figure 7.29.

Three optical isolators (OIs) are used to ensure that no reflection is sent back to the cavity, which consists of optical amplification, to avoid oscillations (Figures 7.30).

Appendix 6 of Chapter 12 tabulates the optical frequencies and wavelengths of the channel central carriers and the spacing of 100 and 50 GHz. The channel order number can be assigned and the optical filter ADD/DROP can be allocated and matched accordingly to obtain low noise input pre-amplification.

Obviously, there are losses in the throughput channels after passing through the TOF (tunable optical filter or the add/drop multiplexer). The arrangement of the two output ports in the optical pre-amplifier LNA shown in Figure 7.28 is acceptable, but dropping more ports can only be possible with a DWDM demultiplexer, such as the array waveguide grating multi-demultiplexer to avoid the accumulated loss in serial add/drop mux devices, such as the one shown in Figures 7.32, 7.33, and 7.34. The operations and applications of these TOFs are given in Section 7.6.

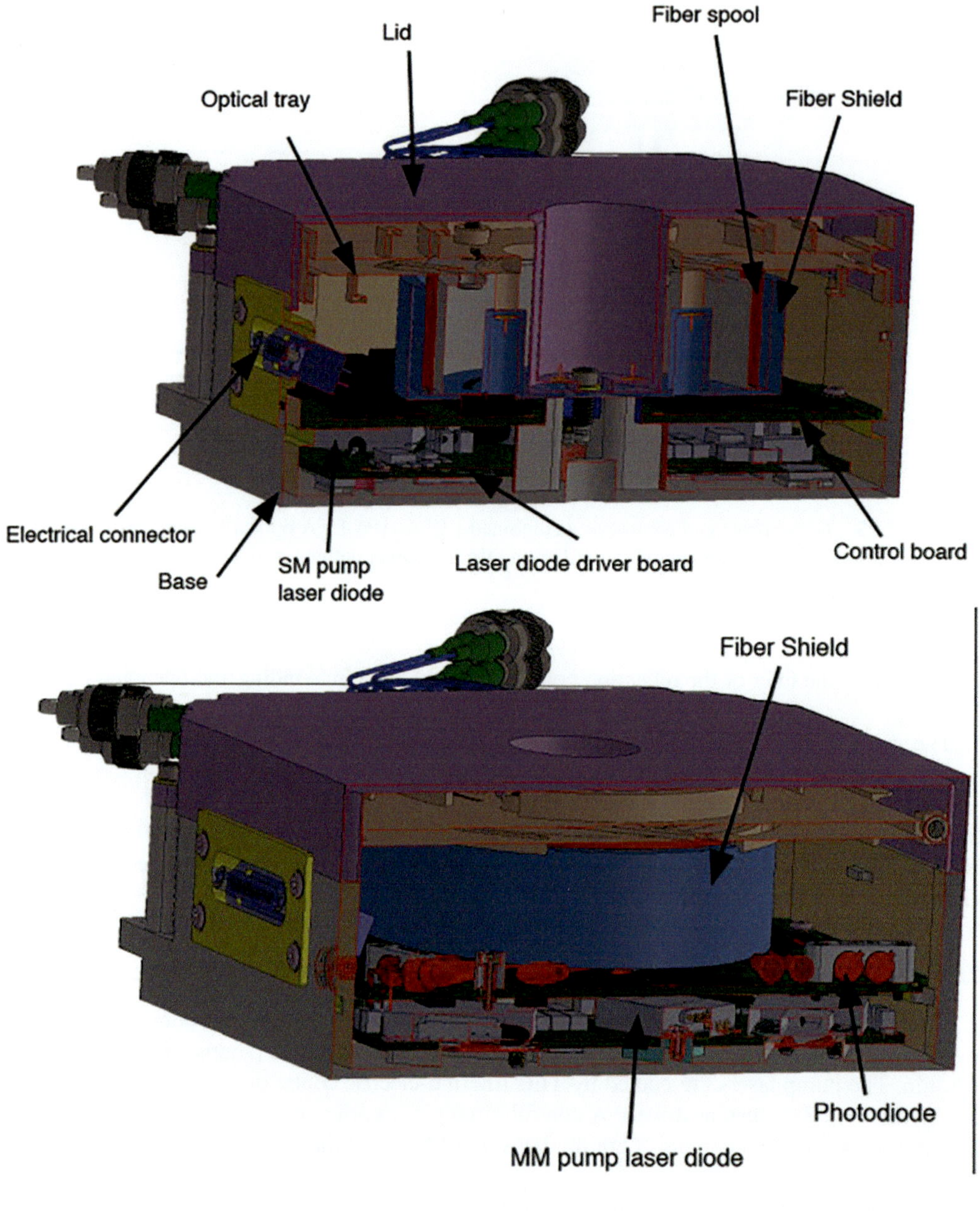

FIGURE 7.26 Internal housing arrangements of module parts of MBP EDFA module HPA-LNA[31] for cooling and monitoring to minimize the heat problem and ensure efficient power transfer from pump absorption to stimulated emission. MM = multi-mode.

7.5.5 Tunable Optical Filter

The package of optical filtering ADD/DROP multiplexer OF, whose specification is shown in Figure 7.31. This package similar to the one for a tunable optical filter (TOF). The reflection port is CLEAR, where the wavelength channel is dropped. The other port is the through port, which passes all the other wavelength channels except the dropped wavelength.

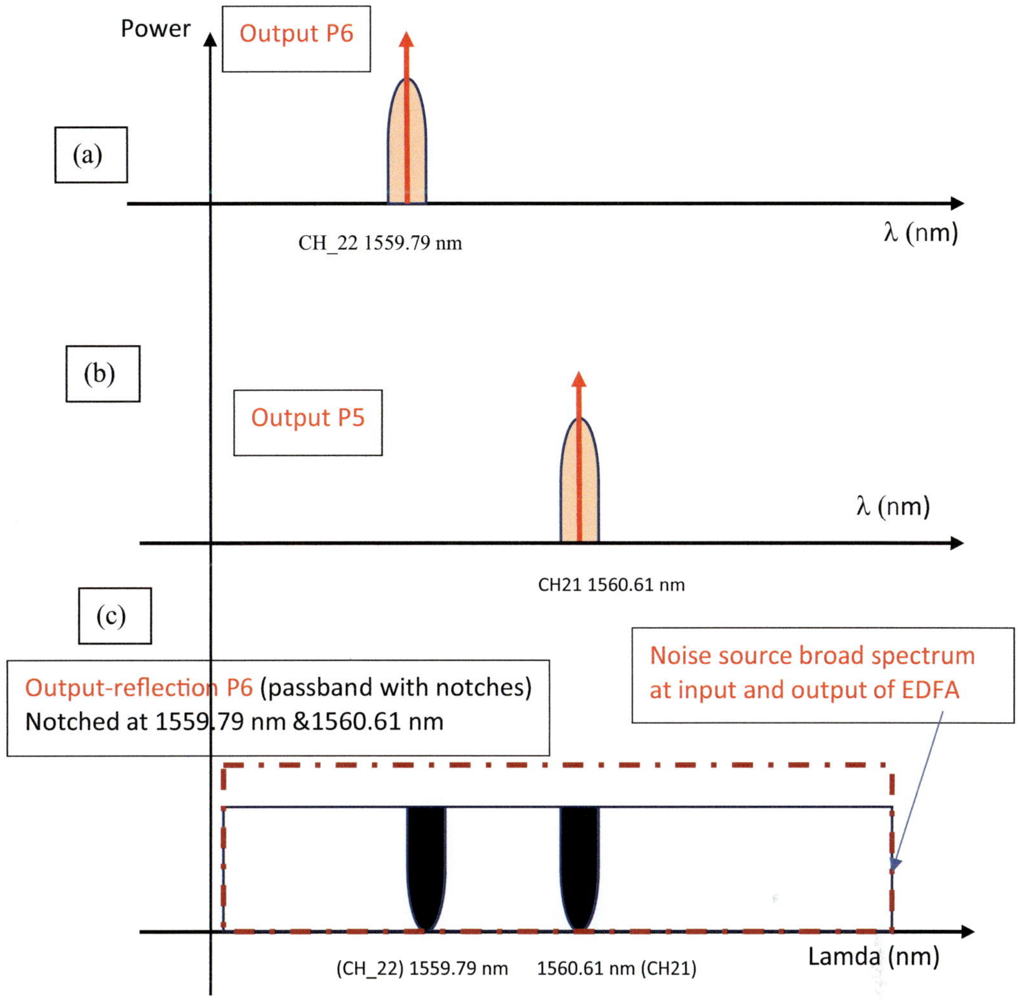

FIGURE 7.27 Spectrum of P5 (PORT 3 of Figure 7.28) output (a), P6 (PORT P2 of Figure 7.28) output, and reflection port (no P3 for HPA & Pre-amp) of AC photonic filter port (b and c).

The tunable OF is used to eliminate the ASE noise out of band of the filter, which reduces the total accumulated noise and thus provides the low noise and low-level signal average power input to the amplifier. Such a low input level can go down to –65 dBm, which is the level of signal arriving at the input of the amplifier after the optical head at the reception site.

The OF must meet the requirements for use in a DWDM transmission system with 50 GHz spacing between channels or interleaved 100 GHz spacing. That is, the passband of the optical filter must be <50 GHz to prevent crosstalk to other optical channels.

Another optical ADD/DROP filter is shown in Figure 7.32, which has different ports labeled in and out. The functionality is the same as that given in Figure 7.31. This illustrates how one can use such discrete optical integrated components to construct a multi-channel ROADM in a space module, as current technology is not developed for space optical communication but terrestrial optical network. SPC systems can employ these components, but in most cases, discrete integration of photonic components are catered for terrestrial optical systems.

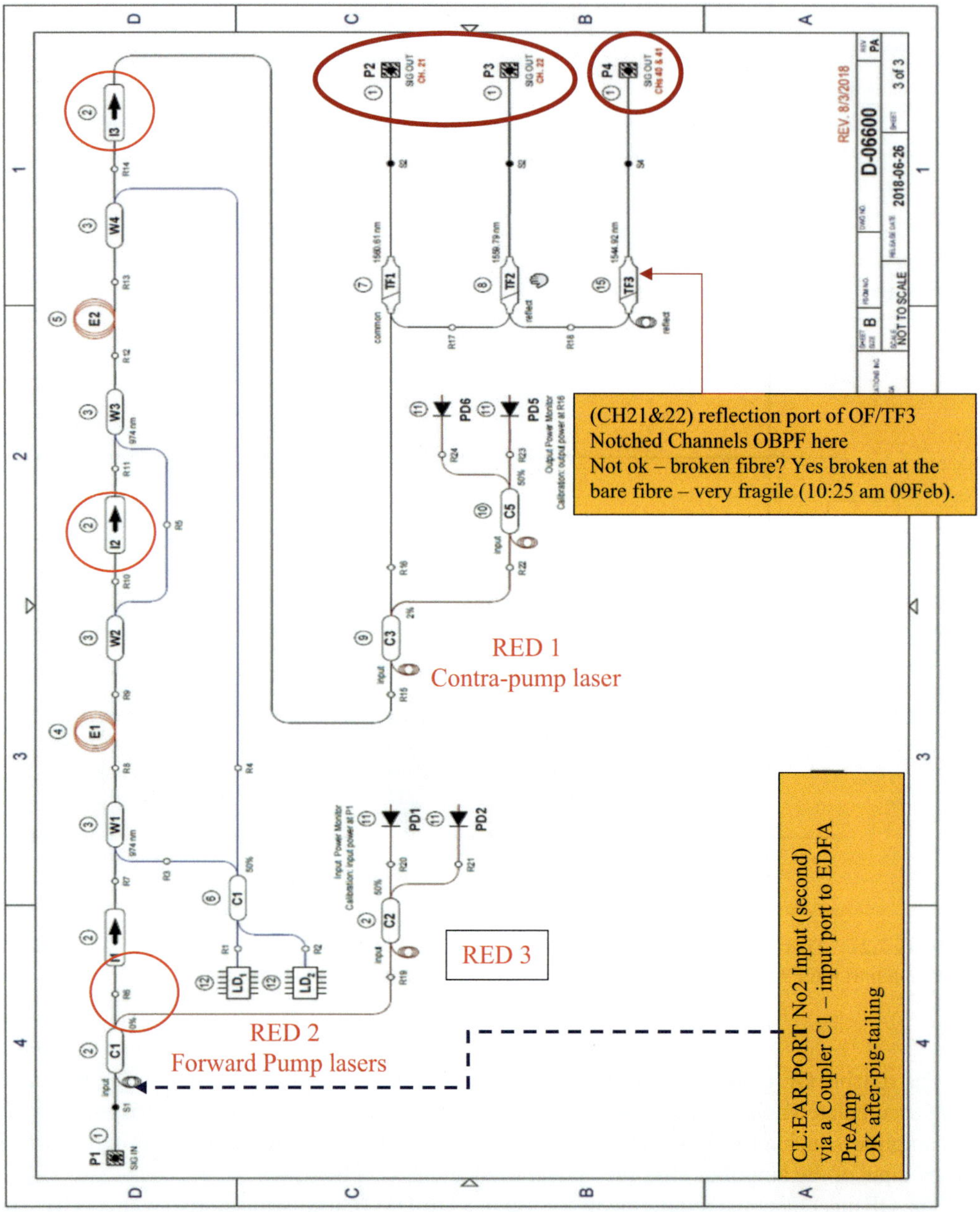

FIGURE 7.28 Pre-amplifier optical circuit: notes on photonic components of the LNA optical pre-amplifier. The components are: Three isolators (I1, I2, I3); six couplers (C1, C1, C2, C3, C5, W1, W2, W3, W4); two/three (1×2) three-port optical filters (TF1 (CH21), TF2 (CH22), TF3 (CH 40 & C41) – reflection port is a passband with two-three notches at CH21, CH22, and CH40 & CH41, using the 50 GHz ITU-grid. W=WDM coupler, C=common coupler (fiber); TF=tunable filter.

FIGURE 7.29 Open Optical Power Amplifier (OPA-HPA) with pig-tail single-mode fiber of a few meters length and Pre-Amp LNA. Note: This packaging was for both HPA and Optical Pre-Amplifier sections. The OPA(HPA) has been exposed to radiation conditions in space installation.

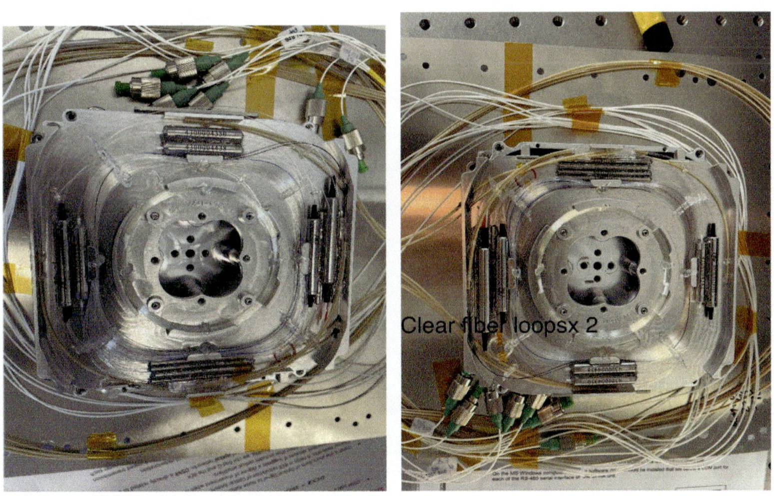

(a)

(b)

FIGURE 7.30 OPEN radiated OA, which was radiated with a nuclear shower in space situation, with 1545 nm, spectral location of channel no. 22 and channels 40 and 41 located at (a) 1560 nm and (b) 1559.7 nm, respectively.

An Optical Add/Drop multiplexer - tunable optical fulters

50GHz DWDM Device Spec

Customer PN	Product PN	Product Name	Description
6729SP	DWD-7C221020-01	1×2 DWDM Device	50GHz, C22, Glass tube, bare fiber, none connector.
6730SP	DWD-7C211020-01	1×2 DWDM Device	50GHz, C21, Glass tube, bare fiber, none connector.
6731SP	DWD-7C411020-01	1×2 DWDM Device	50GHz, C41, Glass tube, bare fiber, none connector.
6732SP	DWD-7C401020-01	1×2 DWDM Device	50GHz, C40, Glass tube, bare fiber, none connector.

Performance Specifications

Parameters		Specifications	Unit
Operating Wavelength		C-Band	nm
Center Wavelength (λc) @-3dB		ITU C22\C21\C41\C40	nm
Channel Spaceing		50	GHz
Channel Passband (@-0.5dB Bandwidth)		λc±0.05	nm
Pass Channel Insertion Loss		≤1.0	dB
Reflection Channel Insertion Loss		≤0.35	dB
Isolation	Adjacent	≥30	dB
	Non-adjacent	≥40	dB
Insertion Loss Temperature Sensitivity		≤0.5	dB
Center Wavelength Accuracy		±0.02	nm
Wavelength Temperature Shifting		≤0.002	nm/°C
Polarization Dependent Loss		≤0.2	dB
Polarization Mode Dispersion		≤0.2	ps
Ripple		≤0.5	dB
Directivity		≥50	dB
Return Loss		≥45	dB
Maximum Power Handing		300	mW
Operating Temperature		-40~+85	°C
Storage Temperature		-40~+85	°C
Fiber Type		G657A1	
Fiber Length		1.5±0.05	m
Connector		None	
Package size		Φ4.0*25	mm

Mechanical Dimensions

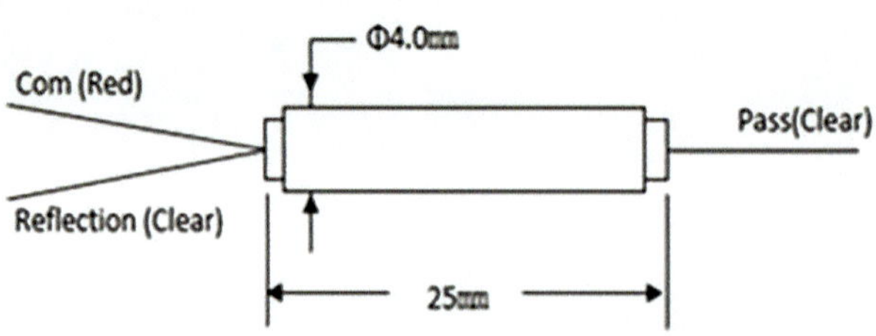

FIGURE 7.31 An optical ADD/Drop filter and its operational characteristics.

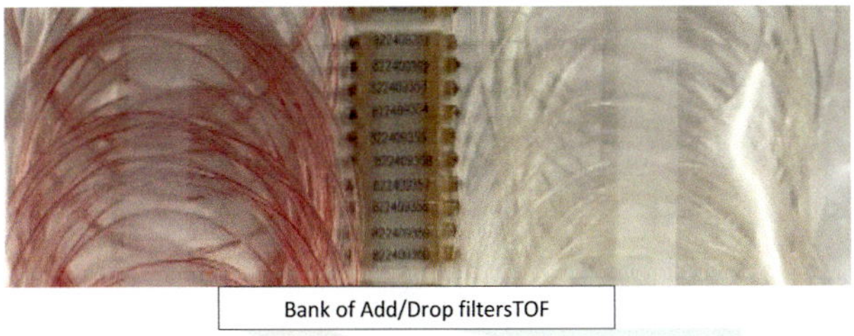

DWDM 1x2 (50GHz)

ACP's 50GHz Dense Wavelength Division Multiplexer (DWDM) utilizes thin film coating technology and proprietary design of non-flux metal bonding micro optics packaging to achieve optical add and drop at the ITU wavelength. It provides ITU channel center wavelength, low insertion loss, high channel isolation, wide pass band, low temperature sensitivity and epoxy free optical path . It can be used for wavelength add/drop in telecommunication network system.

(a)

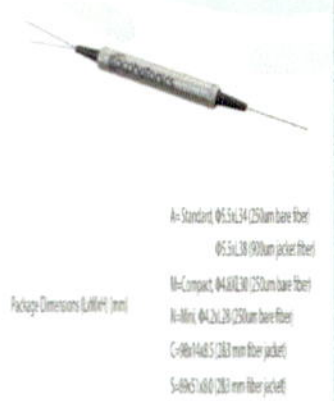

Parameter	Mux/Demux
Operating Wavelength	1530.33 ~ 1560.61 nm (21 ~ 59 ITU grid)
Minimum Channel Spacing	50GHz
Center Wavelength (CWL) Accuracy	±0.04nm
Channel Passband (@-0.5dB bandwidth)	≥ 0.14nm
Insertion Loss* (Add / Drop Ch.)	≤ 1.0dB
Insertion Loss* (Express Ch.)	≤ 0.60dB
Express Channel Isolation	≥ 20dB (Adjacent)
	≥ 35dB (Non-adjacent)
Add/Drop Channel Isolation	12dB(Typ.) ≥ 10dB(Min.)
Add / Drop Channel Ripple	≤ 0.6dB
Insertion Loss Temperature Sensitivity	≤ 0.003dB/°C
Wavelength Temperature Shifting	≤ 0.001nm/°C
Polarization Dependent Loss	≤ 0.10dB
Polarization Mode Dispersion	≤ 0.10ps
Directivity	≥ 45dB
Return Loss	≥ 45dB
Optical Power	≤ 300mW
Operating Temperature	0 to +70°C
Storage Temperature	- 40 to +85°C
	A= Standard, Φ5.5xL34 (250um bare fiber)

(b)

MECHANICAL DIMENSIONS

A package:

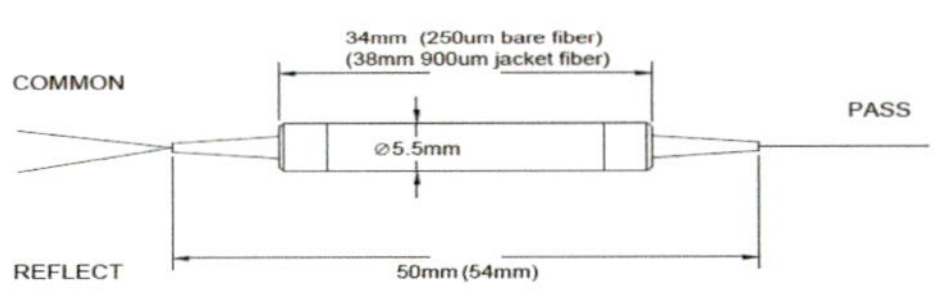

M package:

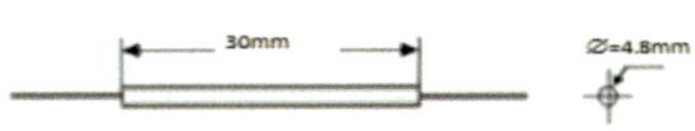

N package:

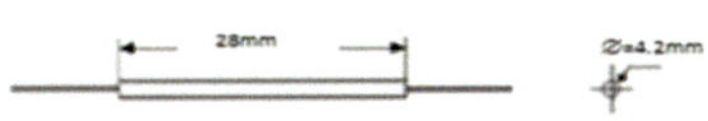

(c)

FIGURE 7.32 (a) A photonics ADD/DROP filter general features; (b) add/drop filter characteristics; (c) mechanical dimensions for installation.

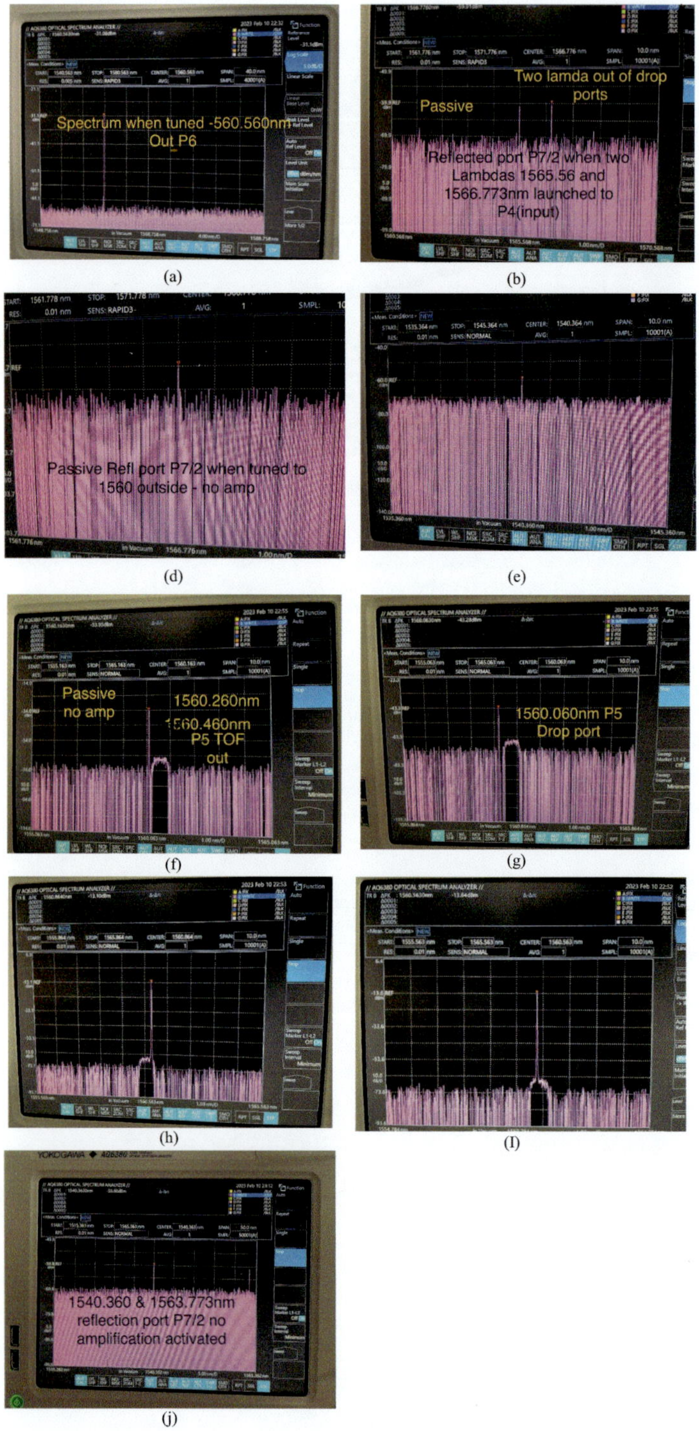

FIGURE 7.33 Sequence of initial test and spectra with tunable laser sources by sweeping a twin source across the expected spectral region from out of band to edge of band and the across the band and out of band or cut-off region of the filter. (a) Single lightwave carrier specrum at input port; (b) Two carriers at drop port; (c) single carrier at drop port; (d) drop port tuned to different region outside the carrier; (e) single carrier and optical drop port filter band nder no amplification (f) Filter passband and carrier at drop port (carrier and filter passband uner amplification (h) carrier and filter passband as monitored by optical spectrum analyser (j) no amplification of carrier, carrier and optical filter spectrum.

7.5.6 Spectra to Identify the Channel Wavelength Carrier and Signal Band

The tests of the LNA can be conducted as shown in Figure 7.28. Tapped port/reflection port of TOF at the P7 optical filter, the ADD&DROP filter "OADM", and the splice CLEAR PORT of TOF 3 in Figure 7.28 are used. By monitoring the spectra of Ports P2, P3, and P4 (Channels 22, 40, and 41), the spectra can be obtained. A broadband ASE source and a twin laser source, tunable across the C-band with a power of 9.5 dBm per source line, are used to test the optical characteristics of the amplifiers. The OSA setting is at a 10 pm resolution, with a minimum resolution of 5.0 pm. The centered wavelength and spectral span are set to obtain maximum accuracy (Figure 7.33).

We note that at the high power output of the HPA, high-power-resistant fiber must be spliced to prevent the high power from overheating the fiber, such as the SMF-28 from Corning Inc.; especially no pig-tail ferrule is to be used, as the high-power beam could damage and result in dangerous operation. Safety goggles must be used wherever required when handling high-power laser beam in the invisible region.

7.6 OADM (Optical Add-Drop Multiplexer) and Optical Amplifications

The enormous growth in the demand for bandwidth is pushing the utilization of optical cable and free space infrastructures to their limits. Similarly, the SPC for satellite constellations and inter-constellation systems is designed for flexible delivery to any location on Earth at ultra-high capacity reaching multi-era-bits/s capacity. To fulfill this requirement, optical add-drop multiplexers (OADMs) have been introduced in metro/access fiber networks, serving as the simplest elements to introduce wavelength management capabilities by enabling the selective add and drop of optical channels. Since the OADM is based on low-loss, low-cost passive devices and does not need any power supply, a reliable, cost-effective, and scalable network can be achieved with its help. There is no doubt that SPC networking would be advantageous to adopt and adapt such ROADM technologies in networking in space environment.

As a key component for DWDM and UW-WDM (ultra-wide wavelength division multiplexing) optical networks, OADM is used for selectively dropping and inserting optical signals into a transparent DWDM network. We know that the main function of an optical multiplexer is to couple two or more wavelengths into the same fiber. If a demultiplexer is placed and properly aligned back-to-back with a multiplexer, it is clear that in the area between them, two individual wavelengths exist. This makes it possible for OADM to remove or insert individual wavelengths.

Similarly, the employment of optical add and drop in low-noise amplifiers is crucial. As we can see in the design of LNA, a tunable optical amplifier is essential to reduce the noise figure of the previous amplification stage so that the noise figure can be reduced in the next stage. The tunable optical filter has a similar structure to that of an OADM. This section illustrates the use of OADM and its dynamics in low-noise amplification structures.

7.6.1 Configurations and Functions of the OADM

An OADM generally consists of three parts: an optical demultiplexer, an optical multiplexer, and a method of reconfiguring the paths between the optical demultiplexer and the optical multiplexer, as well as a set of ports for adding and dropping signals. The multiplexer is used to couple two or more wavelengths into the same fiber. Reconfiguration can be achieved by a fiber patch panel or by optical switches that direct the wavelengths to the optical multiplexer or to drop ports. The demultiplexer separates the multiple wavelengths in a fiber and directs them to many fibers. The channel numbers should match the ITU-grid of 50 or 100 GHz spacing (see Appendix 7, Section 12.5 of Chapter 12) for ease of available optical transceivers, which are presently available and developed for terrestrial fiber optic transmission systems (Figure 7.34).

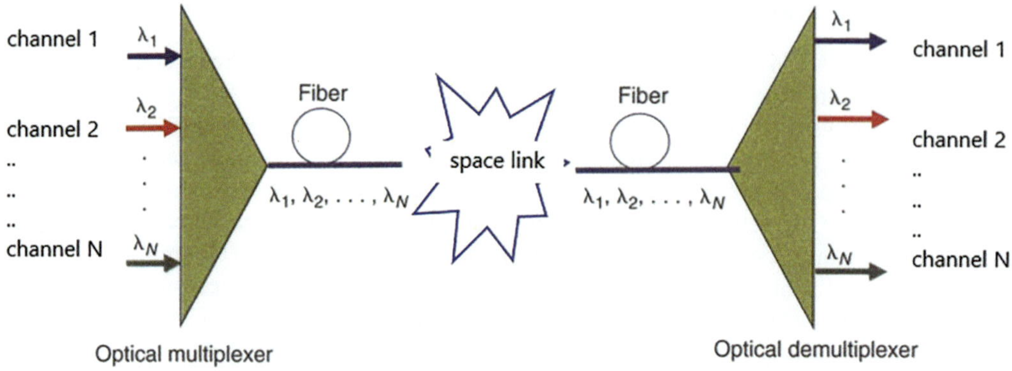

FIGURE 7.34 Components of a traditional OADM embedded in a space photonic link using DWDM multiplexer and demultiplexer (WSS structure as MUX/DEMUX).

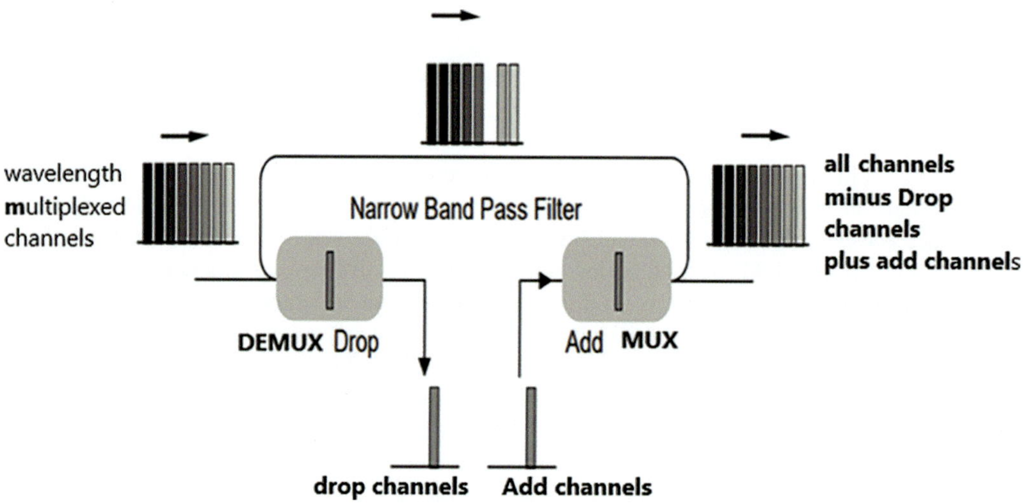

FIGURE 7.35 Configuration of OADM with TFF. Arrows indicate the directions of lightwaves flowing.

7.6.2 Configurations of OADM

There are two basic configurations of an OADM: using dielectric thin-film filters (TFF) and fiber Bragg gratings (FBG). For an OADM configuration with TFF as TOF, an arbitrary signal wavelength is branched/dropped from wavelength-multiplexed signals via a narrow band-pass filter (BPF), whereby only the desired signal wavelength is transmitted while others are reflected. Meanwhile, an arbitrary signal wavelength can be inserted/added into wavelength-multiplexed signals via a narrow BPF, whereby the desired signal wavelength is combined with the reflected signal wavelengths. To implement the add/drop function, a number of TOF can be used and arranged as shown in Figure 7.35.

While configuring an OADM with FBG, the wavelength-multiplexed signals enter an FBG through a circulator, where only one arbitrary signal wavelength is reflected while others are transmitted, as shown in Figure 7.36. The reflected signal wavelength is branched/dropped into a port other than the one where the wavelength-multiplexed signals enter. In the case of wavelength multiplexing an arbitrary signal wavelength, the signal wavelength incident on the circulator is reflected by the FBG and is inserted/added into the wavelength-multiplexed signals that are transmitted via

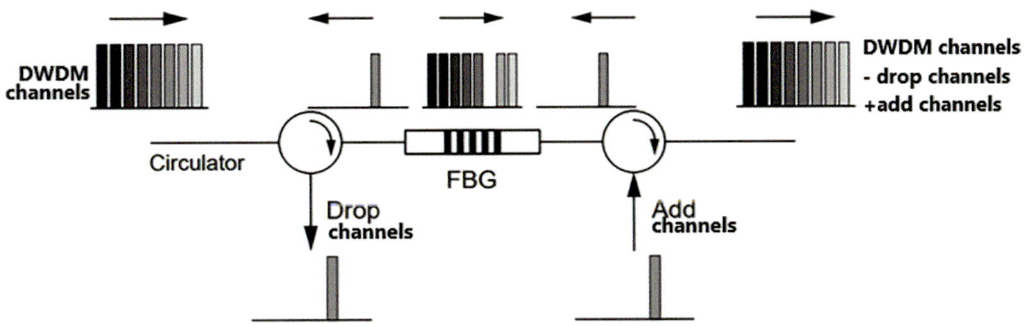

FIGURE 7.36 Configuration of OADM with FBG (fiber Bragg gratings) employed as reflector and bandpass filter.

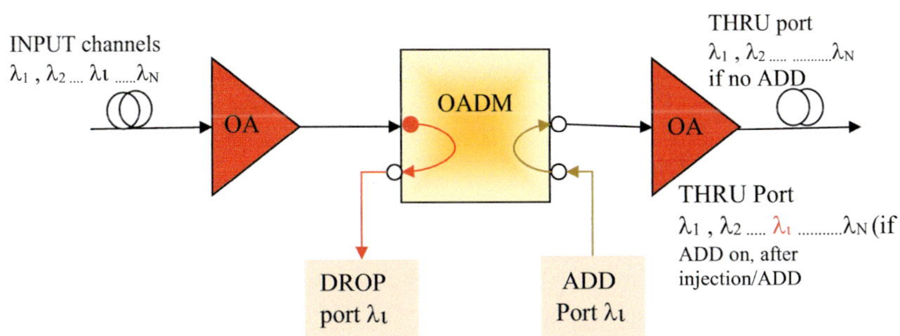

FIGURE 7.37 The function of an OADM with DROP and ADD wavelength λ_i.

the optical circulator. The FBG must be cooled and controlled so that there is no drifting of the centered filtering wavelength. The sharpness of the FBG filter is not as sharp compared with multi-layer thin-film narrow-bandwidth (Figure 7.35), such as the add/drop multiplexer TFF. An acceptable sharp frequency roll-off to the 3 dB pass band would be about 40 dB per 0.1 nm spectral bandwidth.

7.6.3 Main Functions of OADM

As the name indicates, an OADM can add one or more new wavelength channels to an existing multi-wavelength WDM signal, or drop/remove one or more channels, passing those signals to another network path. The OADM selectively removes (drops) a wavelength from a multiplicity of wavelengths in a fiber, thus separating the signals from the particular channel. It then adds the same wavelength in the same direction of data flow, but with different data content. The OADM is commonly used in WDM ring systems and in long-haul networks for more flexible configuration and higher capacity (Figure 7.37).

7.6.4 Different Types of OADM

OADMs are classified as FOADM (Fixed Optical Add-Drop Multiplexer) and ROADM (Reconfigurable Optical Add-Drop Multiplexer). In a fixed-wavelength OADM, the wavelengths are selected and remain the same until human intervention changes them. In a reconfigurable wavelength OADM, the wavelengths between the optical demultiplexer/multiplexer can be dynamically directed from the outputs of the demultiplexer to any of the inputs of the multiplexer.

7.6.4.1 Fixed Optical Add-Drop Multiplexers

FOADMs were originally developed to improve the delivery of "express" traffic through networks, without requiring expensive OEO regeneration. FOADMs use fixed filters that add/drop a selected wavelength "band" and pass the rest of the wavelengths through the node. Static wavelength-filtering technology eliminates the cost and attenuation associated with demultiplexing all DWDM signals in a signal path. The solution is called FOADM because the wavelength(s) added and dropped are fixed at the time of filter installation on the optical path through a node. No additional filters can be added without interrupting the express wavelengths traveling through the node.

7.6.4.2 Reconfigurable Optical Add-Drop Multiplexers

ROADMs were developed to provide flexibility in rerouting optical streams, bypassing faulty connections, allowing minimal service disruption, and adapting or upgrading the optical network to different WDM technologies. It uses a Wavelength Selective Switch (WSS), which has an eight-dimensional cross-connect and provides quick service start-up, remote cross-connect, and WDM mesh networking. The ROADM scheme also allows inputting or outputting a single wavelength or wavelength group via the fixed port. In ROADM systems, we do not need to convert optical signals to electrical signals and route those signals using conventional electronic switches before converting them back to optical signals just like FOADM does. ROADM can be configured as required without affecting traffic.

7.6.4.3 Application of OADM

In conventional long-haul photonic transmission systems, emphasis has been placed on the capacity and the distance of transmission. In metro/access networks, however, low cost and system flexibility are more strongly required. Similarly to long-haul cases, space photonic transmission places a strong emphasis on capacity, reliability, and distance. OADM can verify both demands. The main application areas for OADM are in MAN (metropolitan area network) and satellite constellation communications, such as LEO, MEO, or GEO, featuring high flexibility, easy upgrade, and amplification, as well as reconfiguration.

As an ideal multi-service transport platform in SPC applications, OADM also allows different wavelength multiplexed signals at different locations. Another application for OADM is in optical cross-connection (OXC). The proposed equipment allows different networks to connect dynamic, on-demand wavelength resources, and a wider range of network interconnection. OADM and OXC only need to download the information in the nodes, to send to the person who handles the equipment, including ATM switchboard, SDH switchboard, IP (Internet Protocol) router, etc., thereby greatly improving the efficiency of the node to process information. Narrower passband can be implemented[32] using interferometric and micro-ring integrated structures.

Figure 7.38 shows the structure of a CDC-ROADM for the SPC node in satellite for routing and add/drop to North, East West, and South. Optical amplifiers required to compensate for losses in such reconfigurable routers are not shown. Before launching into space to the next satellite of the same constellation or other constellation, booster amplifiers are required.

Figure 7.39 incorporates the CDC-ROADM from Figure 7.38 into a number of input channels with different wavelength carriers and dropping to different channels as well as adding channels into the space node. The grid for wavelength carriers can be flexible, that is the spacing can be more compact or more spread out in terms of frequency, depending on the baud rate and, thus, the aggregate bit rate of the channels.

7.6.4.4 Summary

To reduce the cost of large-capacity transmission, whereas conventionally most signal processing has been done after optical-to-electrical conversion, it is now required to process signals in optical form. OADM is one of the key devices to implement such optical signal processing. The use of OADM makes it possible to freely add or drop signals with arbitrary wavelengths over multiplexed optical signals by assigning a wavelength to each destination. Moreover, it also helps simplify the component configuration of optical amplifiers through reduced optical attenuation for the express

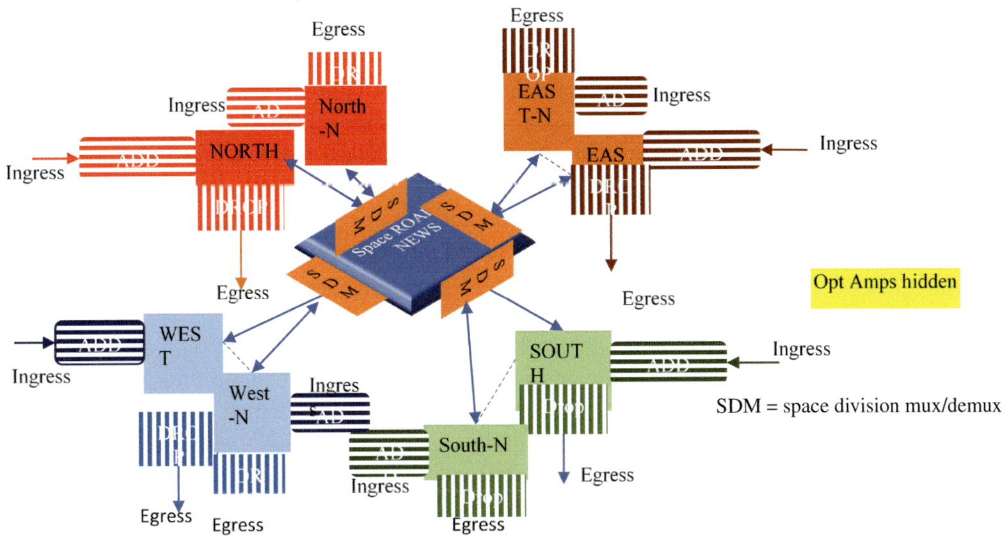

FIGURE 7.38 Schematic diagram: North-East-West-South-ROADM (CDC-Flex). Opt Amps = optical amplifiers, including both HPA at the transmitter site and LNA types for booster amplifiers and optical pre-amplifiers at the receiver site. SDM = space division multiplexer/demultiplexer. See also Chapter 12 Appendix 5 Section 12.3.

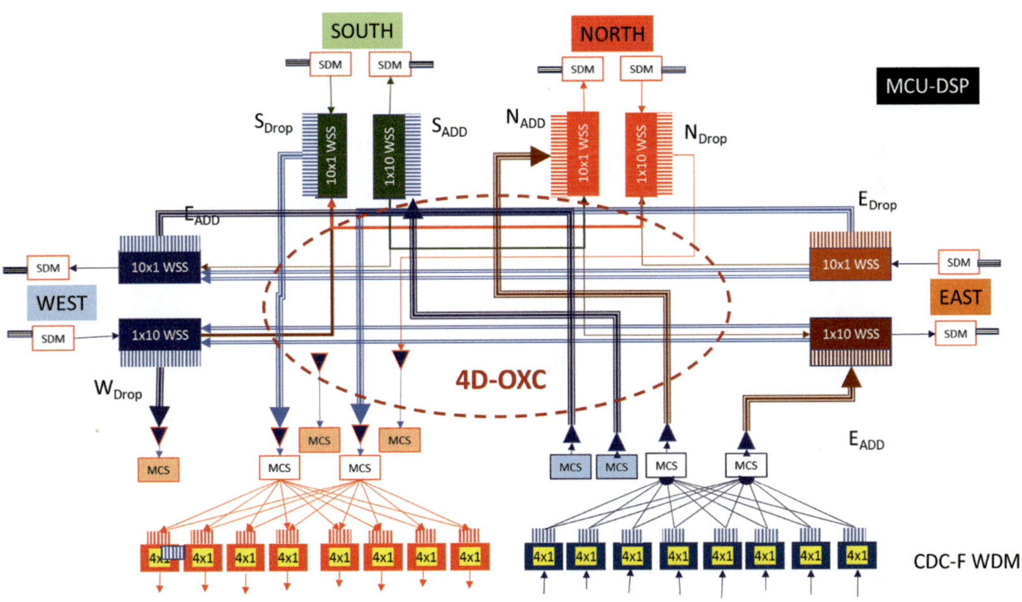

FIGURE 7.39 NEWS space optical routing add/drop mux node. SDM = spatial division mux; WSS = wavelength selective switch; OXC = optical cross-connects; CDC-F = contentionless, directionless, colorless flexible grid; MCU = microcontroller unit; DSP = digital signal processor. See also in Chapter 10.

channels – optical channels that neither add nor drop at nodes – in OADMs, thereby decreasing the total cost of networks. OADM is still evolving. Although these components are relatively small, they will play a key role in producing compact, monolithic, and cost-effective devices.

7.7 NON-RECONFIGURABLE AND RECONFIGURABLE OADM (OPTICAL ADD-DROP MULTIPLEXER) AND ROADM

The enormous growth in the demand for bandwidth for space photonic communications networking, is pushing the utilization of fiber and space communications infrastructures to their limits. To fulfill this requirement, OADMs were introduced in metro/access networks and internetworking in space photonic communication, serving as the simplest elements to introduce wavelength management capabilities by enabling the selective add and drop of optical channels between satellites and optical ground stations. Since the OADM is based on low-loss, low-cost passive devices, and does not need any power supply, a reliable, cost-effective, and scalable network can be achieved with its help.

OADM and ROADM (reconfigurable OADM) are the key component for DWDM and UW-WDM optical networks, for selectively dropping and inserting optical signals into a transparent DWDM network. The main function of an optical multiplexer is to couple two or more wavelengths into the same fiber. If a demultiplexer is placed and properly aligned back-to-back with a multiplexer, it is clear that in the area between them, two individual wavelengths exist. This makes it possible for OADM to remove or insert individual wavelengths. There are non-reconfigurable and reconfigurable OADMs which are associated with tunable or fixed optical filters and optical cross connect (OXC). The flexibility of these multiplexers and the color dependence or non-dependence determine their uses in the optical networks, especially in space optical networking.

The insertion losses of the multiplexers require additional optical amplifiers at moderate gain, thus the optical signal to noise ratio of the channels must meet the required OSNR for further transmission. These aspects are briefly addressed in this section.

7.7.1 Configurations and Functions of OADM

An OADM generally consists of three parts: an optical demultiplexer, an optical multiplexer, a method of reconfiguring the paths between the optical demultiplexer and the optical multiplexer, as well as a set of ports for adding and dropping signals. The multiplexer is used to couple two or more wavelengths into the same fiber. Then the reconfiguration can be achieved by a fiber patch panel or by optical switches that direct the wavelengths to the optical multiplexer or to drop ports. The demultiplexer separates the multiple of wavelengths in a fiber and directs them to many fibers (Figure 7.40).

7.7.2 Optical Add/Drop Filter

7.7.2.1 Optical Add/Drop Filter as a Multiplexer for OADM

A three-port tunable optical filter using a micro-ring resonator or multilayer thin-film filter with angled input generates an add/drop multiplexer, which can be shown as a device in Figure 7.41. Such an optical add/drop can use multi-layer thin-film technology to offer high isolation and a sharp roll-off band to ensure that residual crosstalk is minimum.

A typical spectrum of the three-port OADM filter can be illustrated in Figure 7.42 showing just two wavelength channels are dropped to two ports and the transmission port spectra. Such an add/drop multiplexer is commonly employed in OADM, incorporating an optical pre-amplifier as shown in Figure 7.44. Three optical filtered ports drop three channels CH22, CH40, and CH41, with the rest transmitting through. For multi-channel operations, such as C-, super C-band, and L-band add/drop spectra, as shown in Figure 7.43.

7.7.2.2 Fixed and Partly Reconfigurable OADMs

The three-port filters can be employed in an OADM, as shown in Figure 7.32a and b. These OADMs are non-reconfigurable. There are two basic configurations of an OADM: using a TFF (Figure 7.32a)

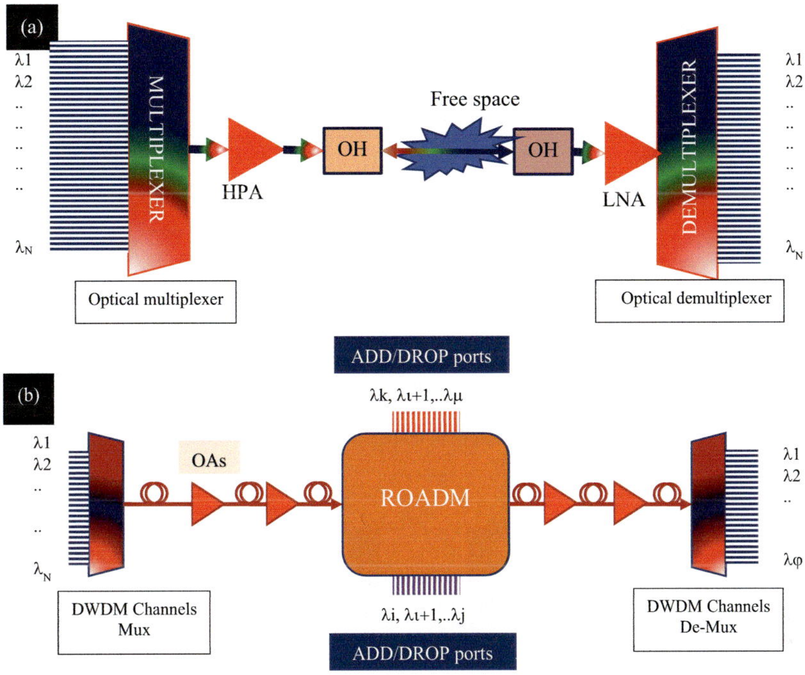

FIGURE 7.40 Components of (a) a traditional Mux and Demux for multi-wavelength (DWDM) transmission, and (b) optical transmission with CDC-ROADM. OA = optical amplifiers, circles = fibers.

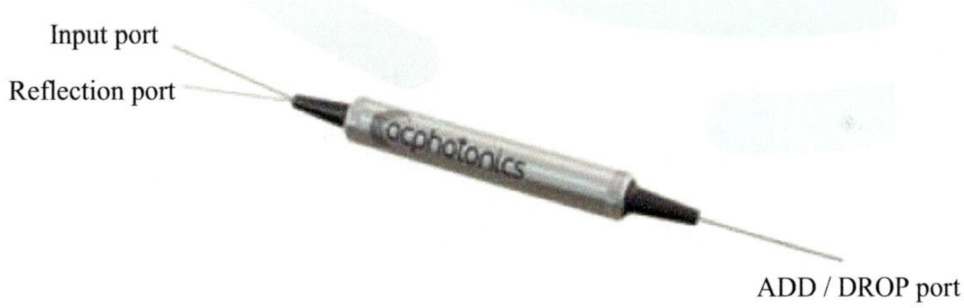

FIGURE 7.41 An OADM filter with a drop/add port, an input port, and a reflected port as transmission port (minus the drop channel).

and a FBG. For the OADM configuration with a TFF, an arbitrary signal wavelength is branched and dropped from wavelength-multiplexed (DWDM) signals via a narrow band-pass filter (BPF), whereby only the desired signal wavelength is transmitted while others are reflected. An arbitrary signal wavelength can be inserted or added into wavelength-multiplexed signals via a narrow BPF, where the desired signal wavelength is transmitted along with the reflected signal wavelengths. When configuring an OADM with an FBG, as referred to Figure 7.32b, the wavelength-multiplexed signals enter an FBG through a circulator, where only one arbitrary signal wavelength is reflected while others are transmitted. The reflected signal wavelength is branched/dropped into a port other than where the wavelength-multiplexed signals enter. In the case of wavelength multiplexing an arbitrary signal wavelength, the signal wavelength incident on the circulator is reflected by the FBG and is inserted/added into the wavelength-multiplexed signals that are transmitted via the circulator. The disadvantage of FBG is that the group delay is linear, which leads to dispersion of the channel. At a later stage, WSS offer massive DWDM channels using free space and micro-resonators.

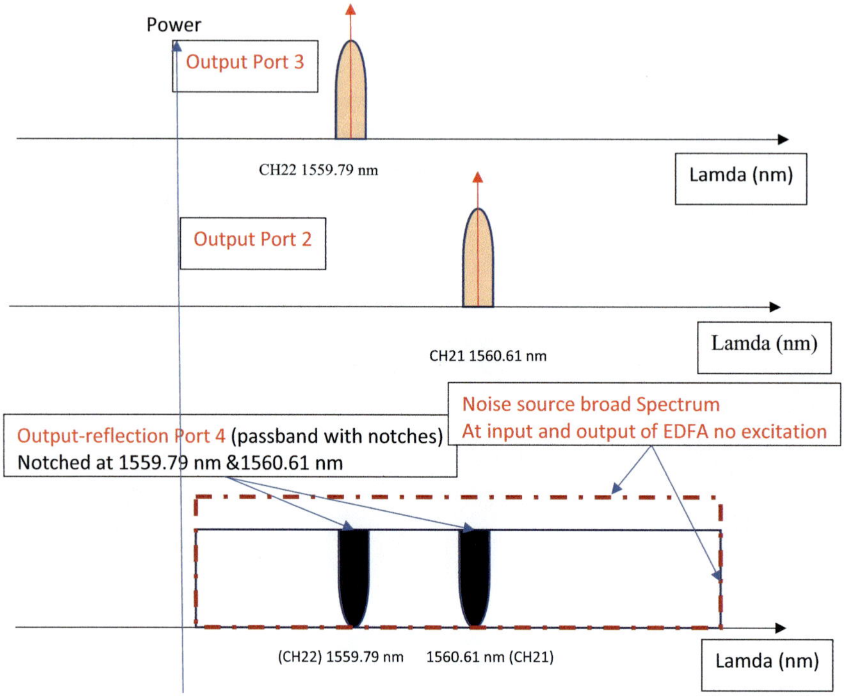

FIGURE 7.42 Spectrum of PORT 3 output, P2 output and reflection port of "AC Photonic" filter port.

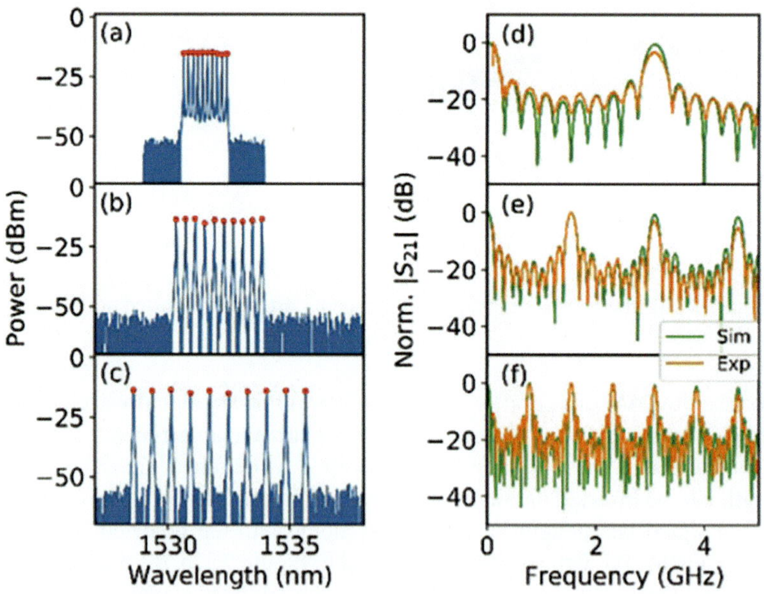

FIGURE 7.43 Multi-channel ADD/DROP in a DWDM network.spectrum of (a) 10 multiplexed channels at input of the multiplexer; (b) 10 multiplexed channels at output of drop port under low loss ; (c) separation of multiplexed channels at drop port whose passbands are sufficient wide ; (d) close view of the spectrum of a multiplexed channel; € close view of the demutiplexed channels at drop ports (f) close f=view of spectrum of 6 channels at drop port.

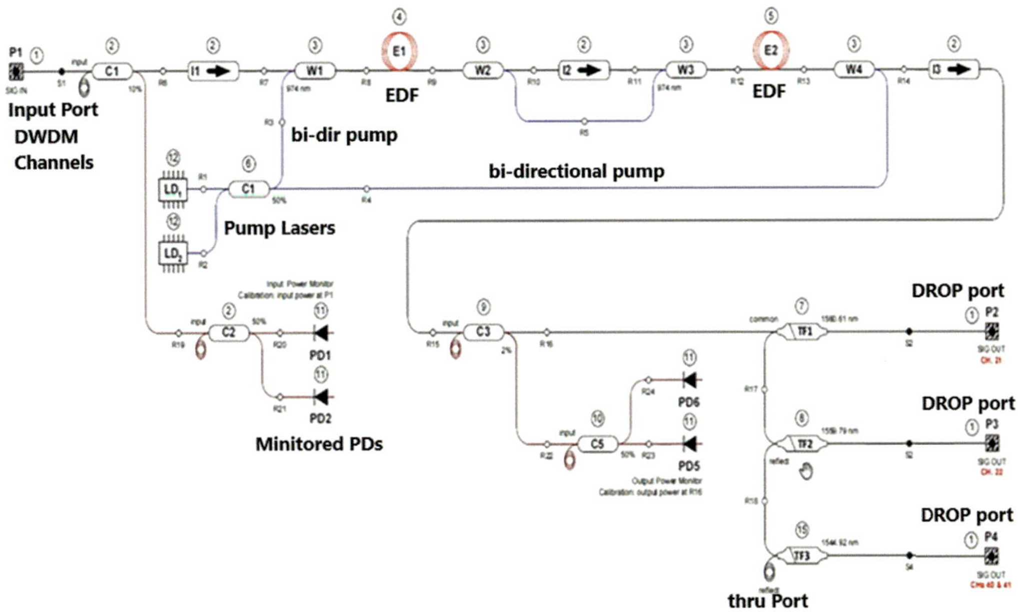

FIGURE 7.44 Structural schematic of an optical amplifier with OADM filter to drop three wavelength channels and a throughput port. EDF=Erbium doped fiber for amplification E1/E2, PD=photodetectors,

7.7.2.3 Main Functions of OADM

As the name indicates, an OADM can add one or more new wavelength channels to an existing multi-wavelength WDM signal or drop/remove one or more channels, passing those signals to another network path. The OADM selectively removes (drops) a wavelength from a multiplicity of wavelengths in a fiber, thereby separating the signals from the particular channel. It then adds the same wavelength in the same direction of data flow, but with different data content. OADMs are commonly used in WDM ring systems and long-haul networks for more flexible configuration and higher capacity. Figure 7.45 illustrates the functionalities of such an OADM, in which OAs are incorporated to compensate for any insertion losses of the devices.

7.7.3 Different Types of OADM

OADMs are classified as FOADM (Fixed Optical Add-Drop Multiplexer) and ROADM (Reconfigurable Optical Add-Drop Multiplexer). In a fixed-wavelength OADM, the wavelength has been selected and remains the same until it is manually changed through human intervention. In a reconfigurable wavelength OADM, the wavelengths between the optical demultiplexer and multiplexer can be dynamically directed from the outputs of the demultiplexer to any of the inputs of the multiplexer.

7.7.3.1 Fixed Optical Add-Drop Multiplexers

FOADMs were originally developed to improve the delivery of "express" traffic through networks, without requiring expensive OEO regeneration. FOADMs use fixed filters that add/drop a selected wavelength "band" and pass the rest of the wavelengths through the node. Static wavelength-filtering technology eliminates the cost and attenuation of demultiplexing all DWDM signals in a signal path. The solution is the FOADM because the added and dropped wavelength(s) channels are fixed at the time of add/drop filter installation on the optical path through a node. No additional filters can be added without interrupting express wavelengths traveling through the node.

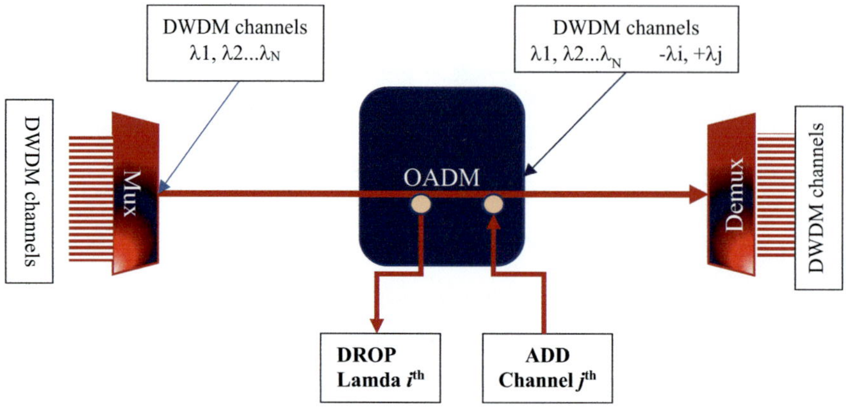

FIGURE 7.45 The schematic function of OADM. OA = optical amplifiers (pre- and post-), lamdas = wavelength channels, OADM = optical ADD/DROP multiplexer.

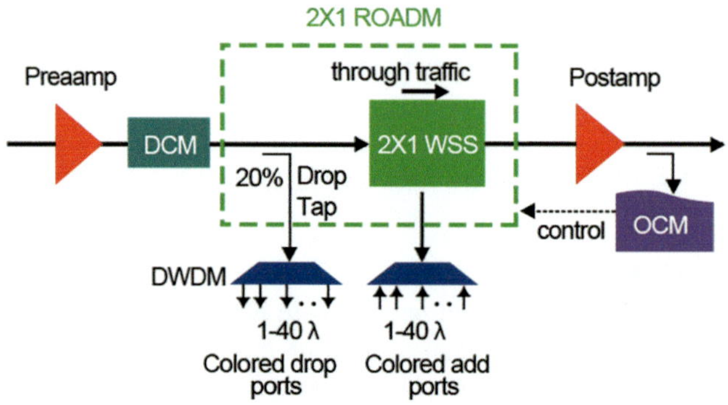

FIGURE 7.46 The principle of reconfigurable OADM. OCM = optical channel monitoring. DCM = dispersion compensator module. WSS = wavelength selective switch. Post-amp, pre-amp = optical amplifiers.

7.7.3.2 Reconfigurable Optical Add-Drop Multiplexers

ROADMs have been developed to provide flexibility in rerouting optical streams, bypassing faulty connections, allowing minimal service disruption, and adapting or upgrading the optical network to different WDM technologies. They use a WSS. The WSS has an eight-dimensional cross-connect and provides quick service start-up, remote cross-connect, and WDM mesh networking. The ROADM scheme also allows inputting or outputting a single wavelength or wavelength group via the fixed port. In ROADM systems, we don't need to convert optical signals to electrical signals and route those signals using conventional electronic switches, then convert them back to optical signals just like FOADM does. ROADM can be configured as required without affecting traffic, as shown in Figure 7.46.

7.7.3.3 Application of OADM

In conventional long-haul transmission systems, emphasis has been placed on how much capacity and how far the system can transmit. In metro/access networks, however, low cost and system flexibility are strongly required. OADM can verify both demands. The main battlefield of the OADM application is in MAN (metropolitan area network), featuring high flexibility, easy upgrade, and amplification. As an ideal multi-services transport platform in MAN (metropolitan area networks) applications, OADM also allows different wavelength multiplexing signals at different locations. Another application for OADM is in OXC. The proposed equipment enables different networks to connect dynamic, on-demand wavelength resources and a wider range of network

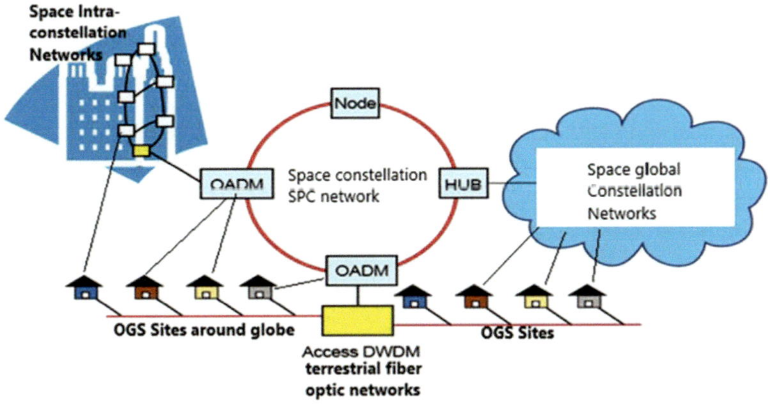

FIGURE 7.47 Application of OADM in space global networks of intra- and inter-constellation orbital optical networks and OGS (optical ground station.

interconnection. OADM and OXC only need to download the information in the nodes, to send to the person that handles the equipment, including ATM (Asynchronous Transfer Mode) switchboard, SDH (Synchronous digital hierarchy) switchboard, IP router, etc., or fast switches and routers in data centers to connect servers or racks of servers, which greatly improve the efficiency of the data center node to process information (Figure 7.47).

7.7.4 Remarks

To reduce the cost in large-capacity transmission, whereas conventionally most signal processing has been done after optical-to-electrical conversion, it is required to process signals in optical form. And OADM is one of the key devices to implement such optical signal processing. Use of OADM makes it possible to freely add or drop signals with arbitrary wavelengths over multiplexed optical signals by assigning a wavelength to each destination. Moreover, it also helps simplify the component configuration of optical amplifiers through reduced optical attenuation for the express channels - optical channels neither add nor drop at nodes - in OADMs, thereby decreasing the total cost of networks or data centers or nodes of inter-communications between nodes in data centers. OADM is still fast evolving, although these components are relatively small, it will play many key roles in producing compact, monolithic, and cost-effective devices in DCI (data center intercommunications).

7.8 SPACE NEWS-ROADM PHOTONIC ROUTING NODE

7.8.1 Demands and Configurations

Multi-dimensional ROADM (MD-ROADM) networkings have recently been implemented and revolutionized the management and design of the structures of optical networks. MD-ROADMs are powerful tools increase the capacity, flexibility, and scalability of all optical networks. An MD-ROADM is a type of optical multiplexer capable of switching and routing light signals across *multiple* optical paths, allowing for a greater degree of flexibility in the design of an optical network and more efficient use of available optical resources. Furthermore, MD-ROADMs optimize the performance of an optical network by allowing for more efficient utilization of bandwidth and better management of network resources.

The most common type of MD-ROADM is the WSS. A WSS is an optical switch that can switch light signals across *multiple* wavelengths. This allows for more efficient use of optical resources, as the WSS can divide the optical signal into multiple paths or channels, each carrying different wavelengths of lightwaves, thus allowing greater flexibility in the design of an optical network, as each wavelength can carry different types of traffic.

The advanced type of MD-ROADM is the colorless, directionless, and contentionless CDC-ROADM. This type of ROADM can switch light signals across multiple optical paths without the need for wavelength-specific components.

CDC-ROADMs are reconfigurable optical-networking devices used to route optical signals across different dimensions within a network. They are designed to reduce contention and maximize network capacity by allowing the transmission of multiple signals on the same wavelength. CDC-ROADMs also enable wavelength-division multiplexing (WDM) to increase the capacity of the network by combining multiple signals into a single wavelength. They provide multiplexing, switching, and routing of optical signals/channels. Additionally, they can reconfigure optical paths in an optical network, making them ideal for dynamic and agile networks.

7.8.1.1 Geometrical Structures of Multi-Directional ROADM

ROADMs have been widely deployed in the long-haul, metro WDM, and internal and inter-space data center networks (space DCNs) in recent years. They enable flexibility in the injection or adding and dropping of any or all multiplexed channels at the wavelength level without manual configuration or O-E-O (optical to electrical and then converted back to optical) switches. In a mesh or meshed ring optical networks, multi-dimensional ROADMs (MD-ROADMs) can also provide cross-connection functions for multiplexed signals among different paths, and they can also be termed wavelength optical cross-connects (WOXCs). Present MD-ROADMs can be constructed by interconnecting individual single-degree ROADMs. However, such a node cannot provide full flexibility for add/drop and switching. This section investigates the requirements for the next generation of MD-ROADM nodes and their modes of operation.

A multi-dimensional ROADM is an optical switch used for wavelength routing in optical networks. It is a key component of optical networks, allowing the efficient management of the optical spectrum and allows for the addition and removal of wavelengths from a particular fiber. The main advantage of multi-dimensional ROADMs is their ability to create larger, more complex networks. They can add or drop multiple wavelengths from a single fiber, increasing the number of connections within a network. Additionally, multi-dimensional ROADMs can create networks with multiple nodes, allowing for more efficient data transmission.

Multi-dimensional ROADMs are also more cost-effective than traditional ROADMs, as they require fewer components and are more easily installed and maintained. Furthermore, they are more reliable due to their advanced technology, which makes them less susceptible to failure. Multi-dimensional ROADMs are used in various applications, including telecommunications, data centers, and research high-performance computing research networks. In telecommunications, they are used to create high-speed networks with fewer components and more reliable connections. In data centers, they enable high-capacity networks with greater flexibility. In research networks, they create networks with greater scalability and flexibility.

For SPC, inter-satellite constellation networks are quite different, as there is no way that data centers can be installed in a satellite or multiple satellites but in distributed OGS. Thus, the SPC network looks more like a data center at the edges. The technological developments of routers and OADM, as well as ROADMs, can assist the SPC operations and structuring into sky-wide global data communications networking.

Colorless means that add/drop ports are not wavelength specific, and directionless feature enables any transponder to be connected to any degree. In this section we review and analyze different multi-degree ROADM node architectures that offer full colorless and directionless features.

Figure 7.48 shows the structure of ROADM conventional node of degree 1, 2, …, N. The node consists of WSSes in each input set of ports and output set of ports. Under WSS operations, there are through ports and routing ports. The through ports can lead wavelength channels through the switch and reserve the channel wavelength as well as their bandwidth for other devices. On the other hand, the routed channels can be directed to predetermined wavelength connections as indicated by the cross-connection path in Figure 7.48.

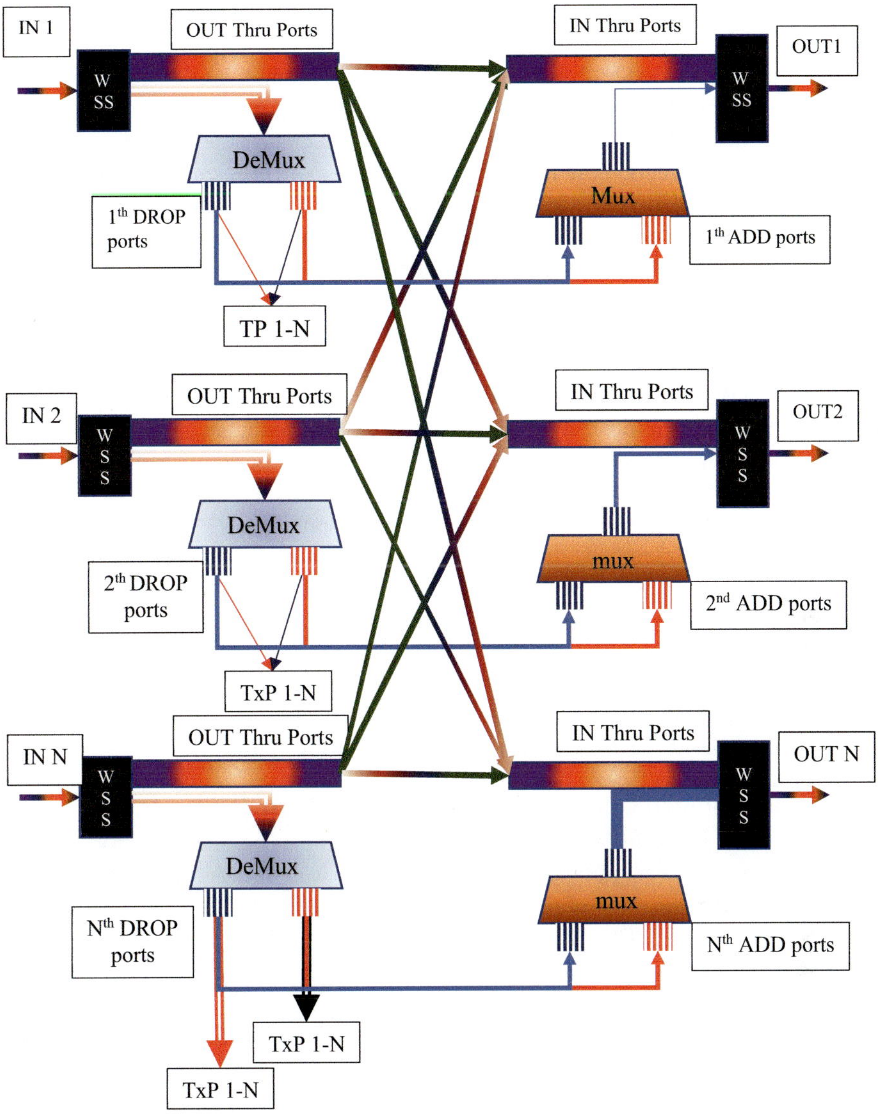

FIGURE 7.48 Structures of a conventional ROADM node. TxP=transponder, S=South, WSS=wavelength selective switch.

The current generation MD-ROADM nodes have two major limitations due to (i) the channels being configured and switched are wavelength dependent. This means that each transponder corresponds to a fixed wavelength because the wavelength corresponding to each demultiplexer output port is predetermined at the manufacturing stage, even though the receiver of the transponder can be active to any spectral window within the transmission band. Furthermore, even when the transponder Tx subsystem can be tunable across the entire operational band, all transponders must be predetermined with a specific wavelength, hence not colorless. (ii) The second limitation is that the add/drop operation is not direction-less but strongly dependent on the channel flow direction. This prevents the routing of ADD channels, leading to the non-flexibility of the nodes to be reconfigured to integrate with other ROADMs but requiring manual reconfiguration. For an MD-ROADM to offer a directionless feature, it must have the properties that "for any channel dropped at the local ROADM node, the corresponding ADD channel can transport to any output port, regardless of the position of coming input port". The muxes (multiplexers) and demuxes (demultiplexers) are employed for multiplexing and demultiplexing channels of different wavelengths (assigned in spectral grids) and for ADD or DROP to predetermined ports. The mux and demuxes in Ports N are set to drop to or add from transponders (TxP). It is clear that the wavelength channel in this case must be pre-determined and not flexible in selecting the wavelength or assignment of such spectral groups to be routed. The directionless feature permits much more efficient sharing of transponders in a node among different paths and significantly improves the flexibility of protection schemes.

Then, Figure 7.49a Conventional N-degree ROADM node shows the structure of CDC-ROADM nodes of similar degrees 1 to N. The main aim is to enable pre-filtering and software-defined service provisioning such that the ROADM node is not fixed to predetermined wavelengths, thus eliminating the need for manual operation. The TxPes are aggregated into two multiplexers and demultiplexers, and all routing for ADD and DROP are combined. The WSSes can switch the aggregated channels to the DWDM muxers and demuxers. The wavelength channels are aggregated together via a massive wavelength multiplexer or wavelength selective switches, and groups of selective wavelengths can be assigned to channels, which can be dropped or added to the desired destinations. Using WSSes at the ROADMs, one can select desired channels of specific wavelengths to route in or out of the CDC-ROADM and direct them to transponders.

Figure 7.48 shows the interconnections of CDC-ROADM Nodes to an optical cross-connect (OXC) device to generate an MD-ROADM node. A number of input and output ports of the OXC are reserved for shared TxPs so that exchanged and switched channel data can be performed via this OXC. Therefore, the MD-ROADM can offer a much higher degree of ADD/DROP with extended wavelength operations, hence the term multi-dimension. The structure of the MD-ROADM can be extended to many configurations, and the use of WSSes, ROADMs, and transponders can be reconfigured to evolve the MD-ROADM into generic systems (Figure 7.50).

7.8.1.2 Mathematical Applications in Tensor or Indexing Problems

Due to the complexity of the MD-ROADM, tensor analyses can be employed for indexing, thus allowing computerized indexing of input and output ports as well as programming the OXCs ports. The transmittances and attenuation of the through ports and switched ports can also be integrated, and automatic control via an MCU (Micro-controller unit) can be used to perform these functions.

The next section addresses these automatic expansions and the interconnection of MD-ROADMs for large-scale systems, especially for applications in data centers. Under the high demands of ultra-dense data transfer and online servicing online such massive switching and routing systems are necessary. Furthermore, optical interconnection and switching in the optical ports will reduce power consumptions, hence lowering the demands for power/energy supplies.

The mathematical representation of ROADM must be able to address the indexing of the following features: flexible add/drop of wavelengths, N-degree ROADM, flex-grid ready, rates of support 10/100/200 and 400 G wavelengths, up to 110 super C-band add/drop wavelengths, WDM spacing of 50 or 100 GHz, and A-to-Z provisioning of wavelengths and protection.

7.8.2 Scalable ROADM Nodes Using WSSes + OXC&OXC$_S$

When it comes to network infrastructure, scalability is key. This is especially true for ROADMs. ROADMs provide flexibility and scalability to optical networks, allowing for the addition and removal of wavelengths without disrupting existing traffic. To ensure scalability, ROADMs must be designed to expand and contract as needed. This means that the ROADM should handle an increased number of wavelengths without needing to be completely replaced. It should also adjust its configuration to accommodate new wavelengths or changes in existing wavelengths.

Scalable ROADMs are essential for creating an efficient and reliable network infrastructure. They provide the flexibility needed to quickly add or remove wavelengths, while still maintaining the existing traffic, allowing networks to quickly respond to changes in demand or traffic patterns.

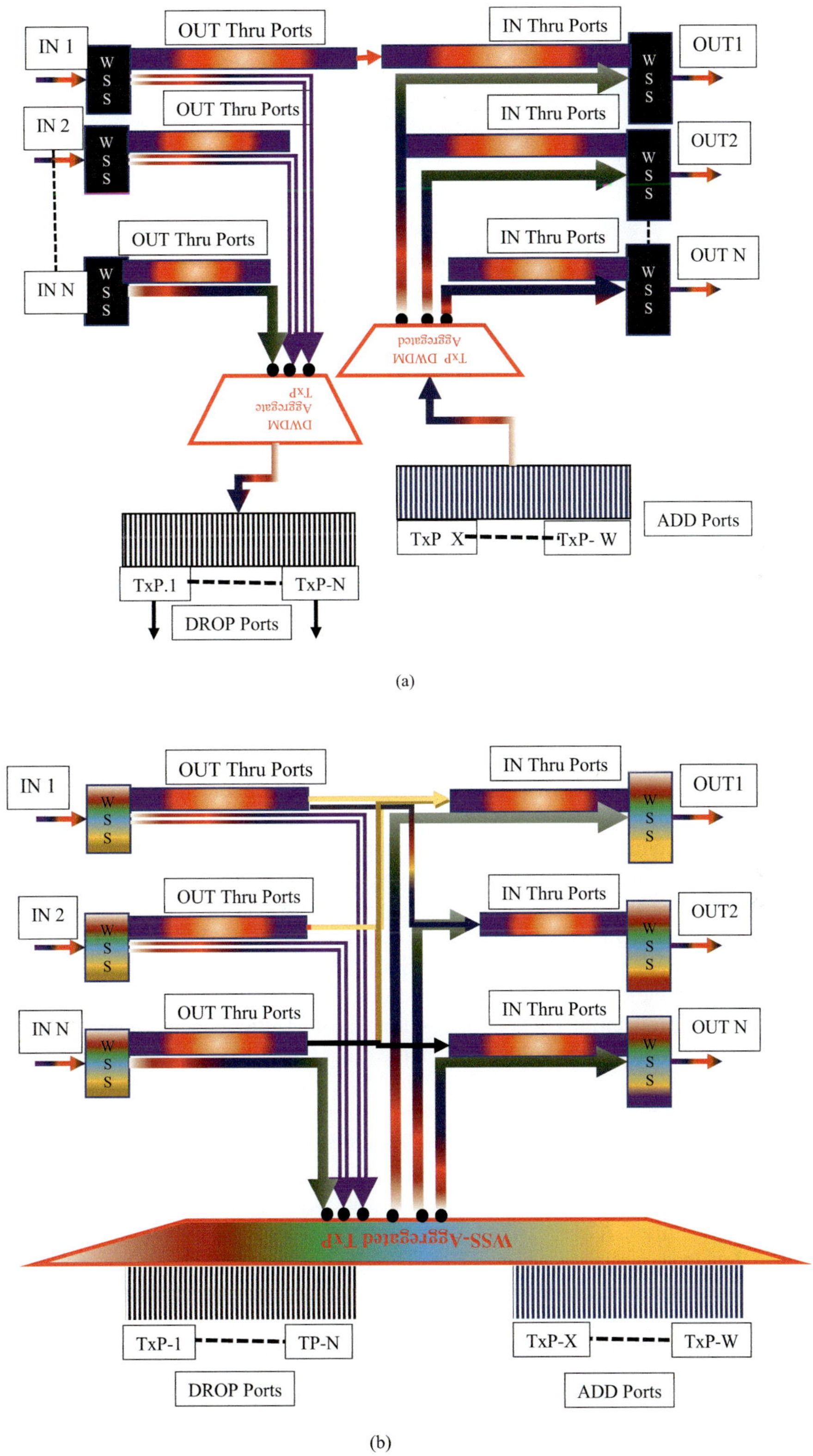

FIGURE 7.49 (a) Conventional N-degree ROADM node; (b) Schematic structure of a hybrid conventional ROADM and CDC-ROADM node. WSS = wavelength selective switch, TxP/TP = transponder.

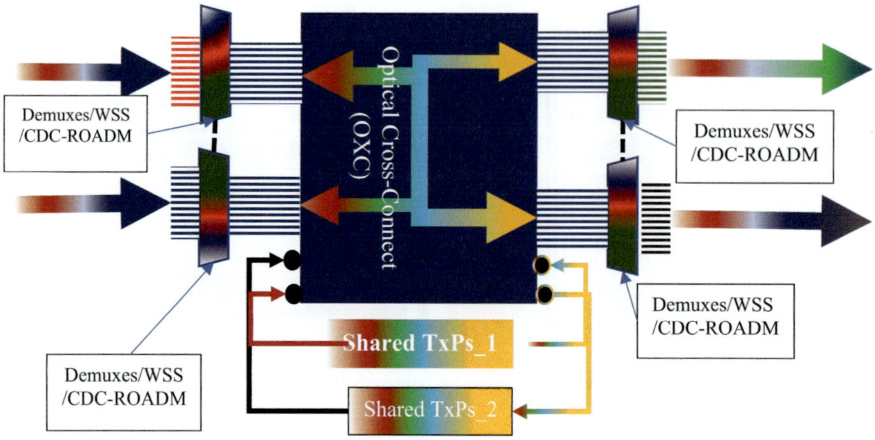

FIGURE 7.50 CDC-ROADM model with shared TxP and OXC for interconnection and switching between ROADMs, muxes, demuxes, and WSSes. Several MD-ROADMs can evolve from this generic structure. OXC=POC (photonic/optical cross-connect).

Scalable ROADMs also provide cost savings by reducing the need for expensive upgrades or replacements. By allowing for the easy addition and removal of wavelengths, ROADMs help keep networks running smoothly and efficiently while providing the flexibility needed to respond to changes in demand. Finally, scalable ROADMs provide greater security and reliability. By allowing for the addition and removal of wavelengths, ROADMs can help protect against malicious attacks or service outages. This helps ensure that networks remain secure and reliable even in the face of unexpected changes or disruptions.

This section describes and provides structures for a number of scalable ROADM nodes. First, the ROADM node integrates WSSes and Tx muxes and demuxes to Rx with optical interconnect. Second, optical cross-connect (OXC) acts as the central part to exchange and connect wavelength channels with great flexibility to input lightwave channels and route to desired output ports with flexible wavelength assignment. Third, non-blocking $(m \times n)$ switches form a non-blocking central part and accept muxed or demuxed channels to the desired output with the selection of wavelengths.

Regarding multiplexers and demultiplexers, WSSes are becoming increasingly popular for their ability to serve as both multiplexers and demultiplexers. A multiplexer is a device that enables multiple input signals to be transmitted over a single output channel, allowing multiple signals to be sent and received simultaneously. A demultiplexer, on the other hand, allows a single input signal to be divided into multiple output signals. WSSes are ideal for multiplexing and demultiplexing because they can handle multiple simultaneous connections. This means that a single WSS can be used to send and receive multiple streams of data at once. This is especially useful when dealing with high-traffic applications, such as video streaming, as it enables the server to handle multiple requests at once. WSSes also offer low latency and high throughput, making them ideal for applications that require fast response times. Additionally, they are secure, as they use the latest encryption protocols to protect data from malicious actors. Overall, WSSes are an ideal choice for applications that require multiplexing and demultiplexing, as they can handle multiple simultaneous connections, offer low latency and high throughput, and are secure. If you're looking for a reliable and efficient way to send and receive multiple streams of data at once, then a WSS is worth considering. The integration of such WSSes and optical cross-connect can form ROADM with a high degree of flexibility and scalability.

7.8.2.1 ROADM Using WSSes

WSS is a wavelength selective switch that can be configured to act as an optical multiplexer or demultiplexer or as an optical add or drop multiplexer by programming the direction of flow of the input or output ports. The WSSes can be integrated with optical splitters or optical space switches, as shown in Figure 7.51a to form reconfigurable OAM (ROADM). The $N \times M$ spatial switching and shuffling of the wavelength channels then feed into the WSS to arrange the channel groups for further transmission and demultiplexing to reach the assigned destination. The WSS can be reconfigured to match the assigned destination of the channels. The OXC can be used to the shuffling and cross-connecting instead of the multiple number of MxN space switches as shown in Figure 7.84b. The structure of an OXC can be constructed as shown in Figure 7.51c.

7.8.2.2 Scalable ROADM

A scalable ROADM node is shown in Figure 7.52a, in which two types of OXCs are employed, one is the master OXC and the other facilitates communication through the ROADMs to the slave-OXCs, which then switches and interconnects to selected wavelength channels. The scale of such ROADM can be scalable by increasing either the number of planar ports of the master OXC or slave OXCs. There are 4 ROADMS located as slaves in NEWS directions through which any drop wavelength channels from a ROADM can be then added to the next and then switched to predetermined output ports of the master OXC.

Therefore, any wavelength channel can be routed to any desired port from there to TxP. It is noted that the ROADM performs best if it is either CDC-ROADM or MD-ROADM, as described in the previous section, Section 7.8.1.1.

Figure 7.52b shows the structure of a planar OXC using integrated photonic technology, in which the guided waves are switched via a switching-guided optical coupler and possible wavelength filtering, depending on the coupling length and strength of the coupling region on the wavelength. Further, Figure 7.52c illustrates a 3D MEMS switch in which two planes of MEMS can be controlled and switched to an appropriate port via a free-space region. This free-space coupling is achieved via the GRIN lens collimator, and the insertion loss is <0.1 dB per coupling. It is also noted that MMI (multimode planar coupler) can be used in Figure 7.52b to distribute the channels and then switched or filtered by the interferences through the MMI (multi-mode interferometer) section.

An extensive scaled NEWS-ROADM is shown in Figure 7.53 when additional space division multiplexers or demultiplexers are integrated with scalable NEWS-ROADM. Figure 7.54 shows further extension of the NEWS-ROADM with flexible CDC-ROADM, which generates a massive optical router and switches for ultra-high capacity systems in internal data center communications and services of ultra-fast Internet channels. N×M WSSes means that N input ports are switched and filtered to M output ports. Selections of wavelengths and filter bandwidth can also be possible. The interconnection buses are communicated directly between WSSes. The 4×1 or 1×4 CDC-F WDM muxes and demuxes are used to distribute or combine the wavelength channels for routing and dropping to appropriate destinations. The central integrated ROAA is a NEWS ROADM type, and it uses the SDM (Spatial Division Multi/De-multiplexers) to split the channels in ingress and egress ports. The MCU is responsible for the control of the setting of switching and routing. The speed is not fast but stable for setting the ports in and out for routing. Therefore, a medium speed MCU is enough to conduct or manage these functions.

Figure 7.55a depicts the inter-satellite constellations, in which each satellite would use an optical routing system, the CDC-ROADM to distribute and add different channels from different satellite orbits. Figure 7.55b shows a more detailed structure of a NEWS-ROADM with WSSes as the ROADM and switches to distribute wireless channels for terrestrial as well as space RF distribution of 5G and 6G channels to users or satellites and optical ground stations. Optical amplifiers are

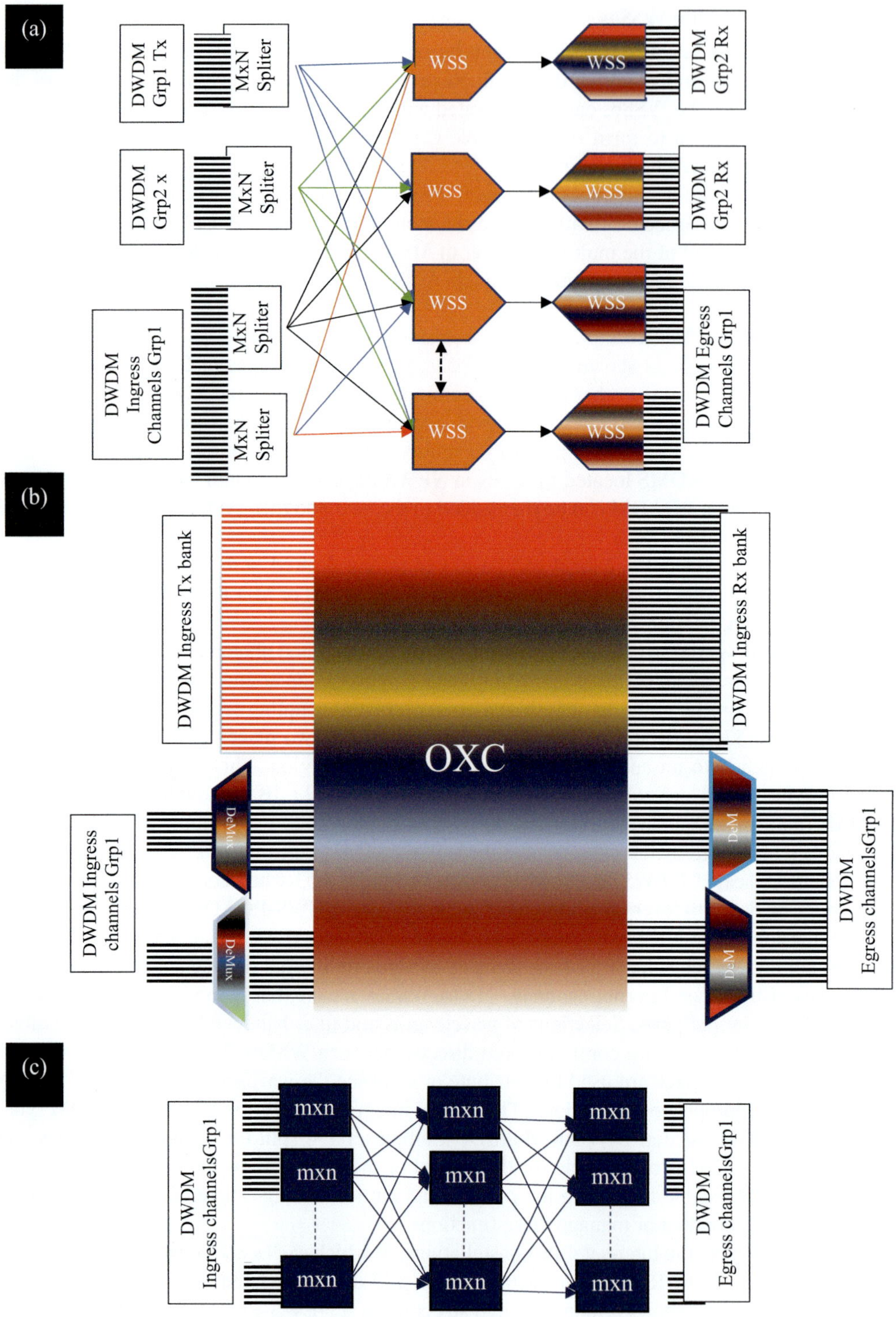

FIGURE 7.51 Schematics of (a) WSS-based ROADM; (b) 3D OXC switch-based ROADM; (c) non-blocking switching mesh network-based ROADM Nx(mxn) N devices with m by n cross-switch and connect. Grp = Group, DWDM = Dense wavelength division multiplexing.

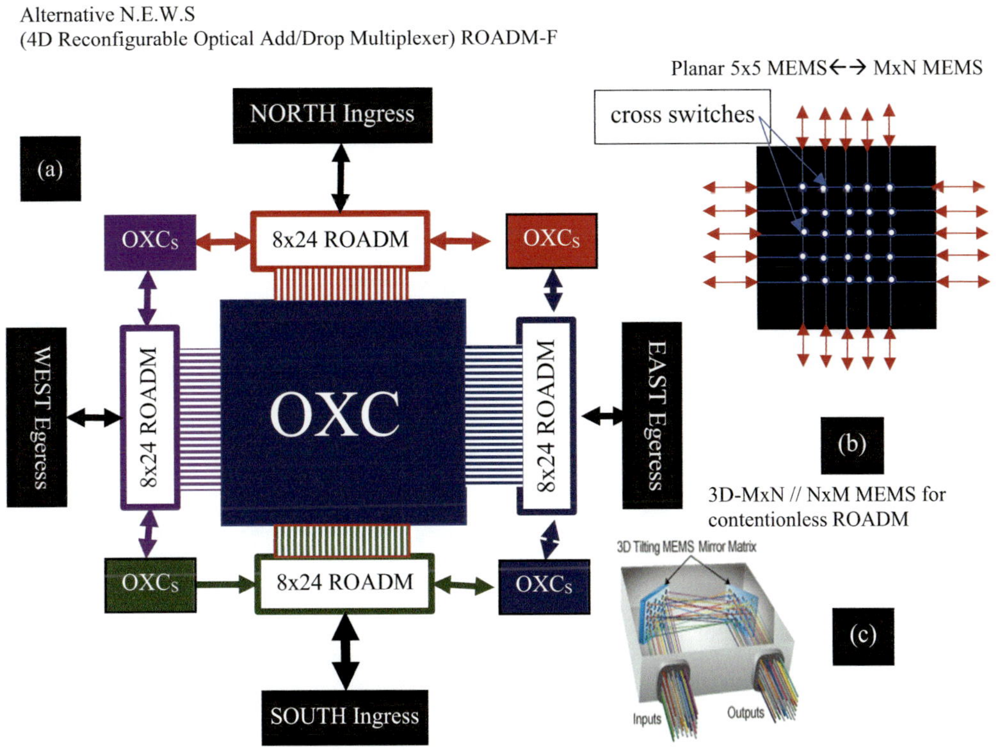

FIGURE 7.52 Scalable space ROADM by WSSes+OXC (a) Structure of optical routing and add/drop of wavelength channels; (b) planar 5×5 MEMS interconnection/switching for optical cross-connect; (c) 3D MEMS routing and switching.

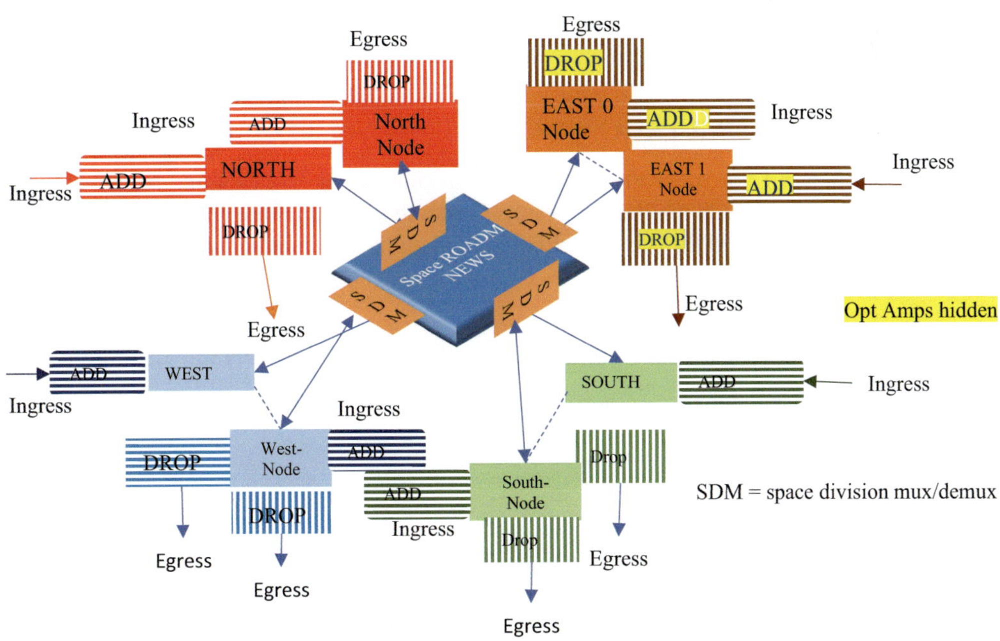

FIGURE 7.53 Schematic diagram: North-East-West-South (NEWS)-ROADM (CDC-Flex). SDM=space division multiplexer or demultiplexer, Egress=DROPPED to output destinations, Ingress=ADD to ports for switching and routing, N=Node.

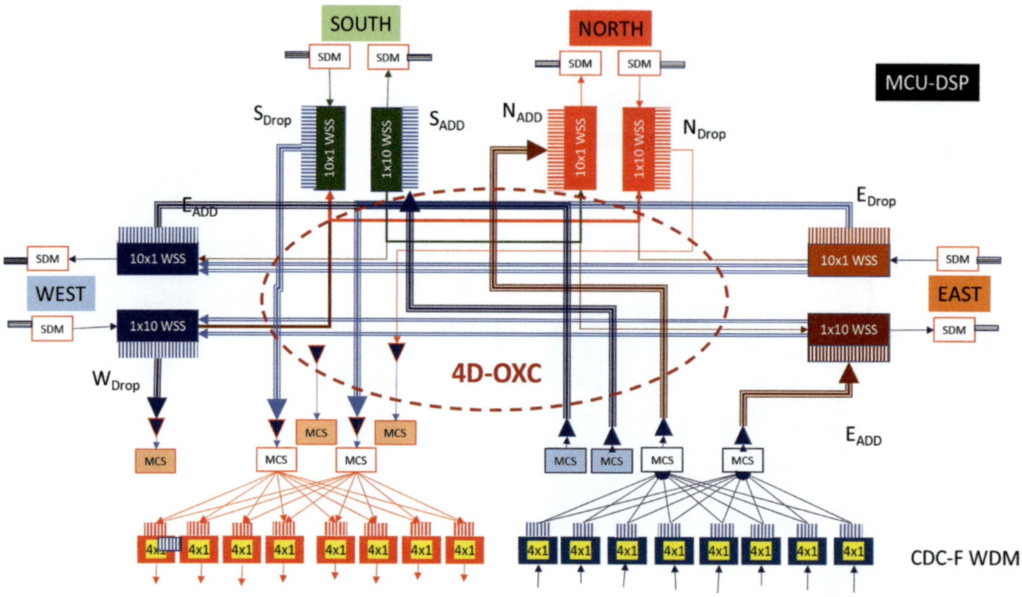

FIGURE 7.54 NEWS space optical routing add/drop mux node. SDM=spatial division mux/demux, WSS=wavelength selective switch, OXC=optical cross-connect, CDC-F=contentionless, directionless, colorless, flexible grid.

essential to compensate for the attenuation of channels as they pass through many stages of optical routing and switching. Optical power amplifiers are also used to boost the average power of the channels to other satellites or ground stations. This forms an all-optical routing node in a satellite deep space communication network.

APPENDIX A.1: EDFA SIMULATION MODEL

Appendix 6 of Chapter 12 provides the evaluation of the numerical data resulting from our EDFA simulator and EDFA module. Experimental results are compared with those generated by our models. The principal advantage of the Simulink EDFA simulator is its versatility and ease of use. These results are by no means exhaustive.

APPENDIX A.2: ITU GRID FOR CHANNEL CARRIER LOCATIONS

The 100 GHz DWDM wavelength channels ITU Grid and the 50 GHz DWDM wavelength channels ITU Grid are provided in Appendix 7 of Chapter 12.

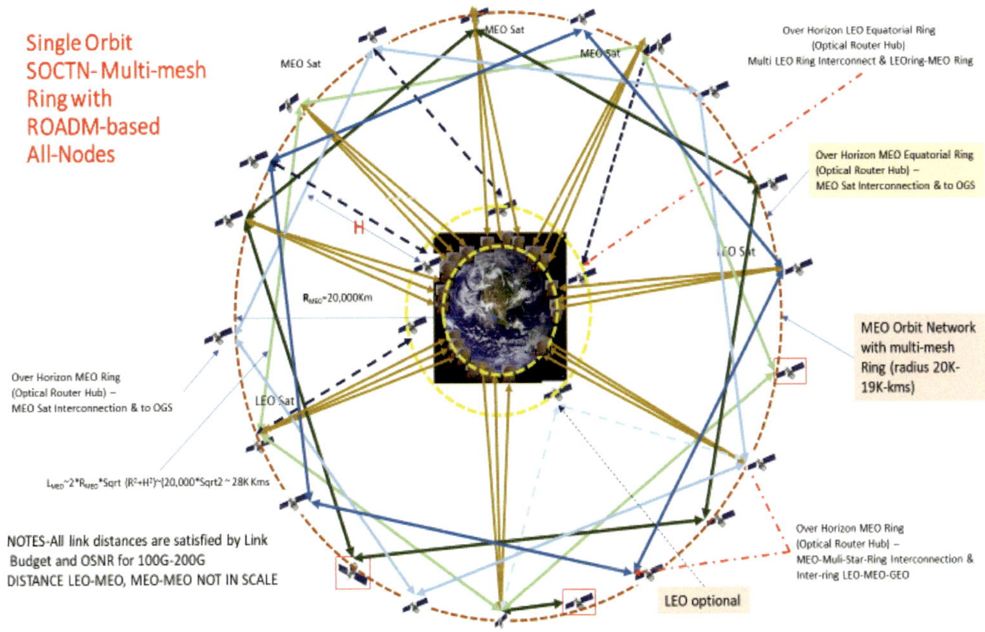

(a) Inter-orbital constellations and optical transmission paths for SOCS networking. Shown are GEO, LEO and MEO satellite constellations.

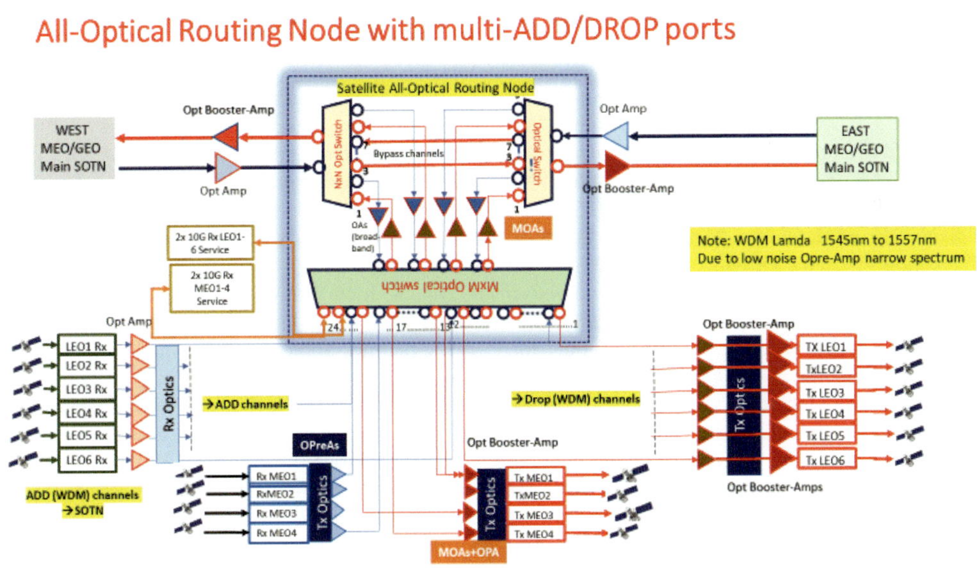

(b) All optical routing Space Node

FIGURE 7.55 (a) Schematic of an inter-satellite constellation in a space communication network; inter-orbital constellations and optical transmission paths for SOCS networking. Shown are GEO, LEO, and MEO satellite constellations; (b) optical routing in an optical node using CDC-ROADM in inter-satellite transport networking. TX=transmitter, Rx=coherent receiver, MEO=medium Earth orbit, GEO=geostationary Earth orbit (satellite), Amp=optical amplifier; SOCS=space optical communication systems.

NOTES

1 A. A. Saleh, R. M. Jopson, J. D. Evankow, and J. Aspell, "Modelling of gain erbium-doped fibre amplifier," *IEEE Photonic Technology Letters,* 2, 714, October 1990.
2 W. J. Miniscalco, "Erbium-doped glass for fibre amplifiers at 1500nm", *IEEE Journal of Lightwave Technology,* 9, 234, 1991.
3 V. Mehta, "Amplified spontaneous emission tutorial", ECE1654: Optical Networks, University of Toronto, 2003.
4 L. N. Binh "Lecture notes on optical repeater, regenerator and amplifier", ECE 4405, Monash University, 2004.
5 V. Mehta, "Amplified spontaneous emission tutorial", ECE1654: Optical Networks, University of Toronto, 2003.
6 G. P. Agrawal, *Fibre-Optic Communication Systems,* 2nd Ed. John Wiley & Sons: Sydney, 1997.
7 W. J. Miniscalco, "Erbium-oped glass for fibre amplifiers at 1500nm", *IEEE Journal of Lightwave Technology,* 9, 234, 1991.
8 V. Mehta, "Amplified Spontaneous Emission tutorial", ECE1654: Optical Networks, University of Toronto, 2003.
9 V. Mehta, "Amplified Spontaneous Emission tutorial", ECE1654: Optical Networks, University of Toronto, 2003.
10 E. Kreyszig, *Advanced Engineering Mathematics,* 7th Ed. John Wiley & Sons: New York, 1993.
11 W. J. Miniscalco, "Erbium-doped glass for fibre amplifiers at 1500nm", *IEEE Journal of Lightwave Technology,* 9, 234, 1991.
12 Celion Network Inc, "Overview of optical fibre transmission using DWDM technology", white paper, 2003.
13 V. Mehta, "Amplified spontaneous emission tutorial", ECE1654: Optical Networks, University of Toronto, 2003.
14 S. Novak and R. Gieske, "Simulink model for EDFA dynamics applied to gain modulation", IEEE Journal of Lightwave Technology, 20(6), 986–992, June 2002.
15 S. Novak and R. Gieske, "Analytic model for gain modulation in EDFAs", *IEEE Journal of Lightwave Technology,* 20(6), 986–992, June 2002.
16 S. Yamashita and M. Nishihara, "L-band erbium-doped fibre amplifier incorporating an inline fibre grating laser", *IEEE Journal of Quantum Electronics,* 7(1), 44–48, January/February 2001.
17 Y. Sum, J. L. Zyskind, and A. K. Srivastava, "Average inversion level, modelling, and physics of erbium-doped fibre amplifiers," *IEEE Journal of Selected Topics in Quantum Electronics,* 3(4), 717, August 1997.
18 Y. Sum, J. L. Zyskind, and A. K. Srivastava, "Average inversion level, modelling, and physics of erbium-doped fibre amplifiers," *IEEE Journal of Selected Topics in Quantum Electronics,* 3(4), 717, August 1997.
19 S. Novak and R. Gieske, "Analytic model for gain modulation in EDFAs", *IEEE Journal of Lightwave Technology,* 20(6), 975–985, June 2002.
20 G. P. Agrawal, *Fibre-Optic Communication Systems,* 2nd Ed. John Wiley & Sons: Sydney, 1997.
21 G. P. Agrawal, *Fibre-Optic Communication Systems,* 2nd Ed. John Wiley & Sons: Sydney, 1997.
22 S. Novak and R. Gieske, "Simulink model for EDFA dynamics applied to gain modulation", *IEEE Journal of Lightwave Technology,* 20(6), 986–992, June 2002.
23 Celion Network Inc, "Overview of optical fibre transmission using DWDM technology", white paper, 2003.
24 I. Nusinsky and A. A. Hardy, "Analysis of the effect of up conversion on signal amplification in erbium-dope fibre amplifier", *IEEE Journal of Quantum Electronics,* 39(4), 548–554, April 2003.
25 Y. Sum, J. L. Zyskind, and A. K. Srivastava, "Average inversion level, modelling, and physics of erbium-doped fibre amplifiers," *IEEE Journal of Selected Topics in Quantum Electronics,* 3(4), 717, August 1997.
26 G. P. Agrawal, *Fibre-Optic Communication Systems,* 2nd Ed. John Wiley & Sons: Sydney, 1997.
27 ADVA EDFA specification data sheet, by optical service & Solution (ADVA).
28 Mathworks Inc, https://au.mathworks.com/help/simulink/gs/model-based-design.html, MATLAB Simulink Dynamics Model Design Handbook; Cambridge, England
29 MPB Inc., "Interface control document transmitter and receiver erbium fibre optical amplifiers for Mynaric's flight demonstrator program", MPB Comm Inc., p. 9, 2023, access February 2024.
30 L. N. Binh, "Interface control document transmitter and receiver erbium fibre optical amplifiers", Technical report, Mynaric Laser Comm GmbH, Muenchen, Germany 2023.
31 MPB Inc., "Interface control document transmitter and receiver erbium fibre optical amplifiers for Mynaric's flight demonstrator program", MPB Comm Inc., p. 31, 2023, access February 2024.
32 B. Corcoran, et al., "A wavelength selective switch for optical add/drop multiplexing of sb-bands within Nyquist WDM super-channels", *Proceedings of European Conference on Optical Communications,* ECOC 2015, paper ID 0328, Valencia, Spain, 2015.

8 Nonlinear Effects in SPC for DWDM Channel Transmission

CONTENTS

DOI: 10.1201/9781003426677-8

Nonlinearity along the transmission path of a photonic communication system is highly undesirable, as several distortions of the channel information can occur, increasing the system's error rate. In space photonic communications (SPC), it is thought that such vacuum media would have no nonlinear effect. Correct, but due to the long-reach free space transmission path, the loss is high, and thus, to compensate for such high loss, optical amplification at both the transmitter and receiver sites is required. A booster amplifier is used at the transmitter site, and at the receiver, low-noise, low-level input amplifiers are used. The high-power booster amplifiers are used at the transmitter, and their output is connected to the optical head by a single-mode fiber whose diameter is small enough that the intensity across the cross-sectional area is extremely high at an average input power of a few watts. This is the main source of nonlinearity distortion effects in SPC.

We also illustrate the experimental observation of nonlinear products generated by high-power lightwaves propagating through a short length of single-mode fiber connected to the HPA output.

This chapter addresses these distortions and the system penalties for the SPC system.

8.1 INTRODUCTORY REMARKS

8.1.1 Nonlinearity in SPCS: Where Does It Come From?

It is noted that when launching to the space channel, the OPA output may need to be extended for connection to the optical head (OH).

High-power lightwaves propagating through a fiber area of 65 μm^2, resulting in ultra-high intensity, lead to nonlinear effects such as nonlinear phase shift, SBS (stimulated Brillouin scattering),

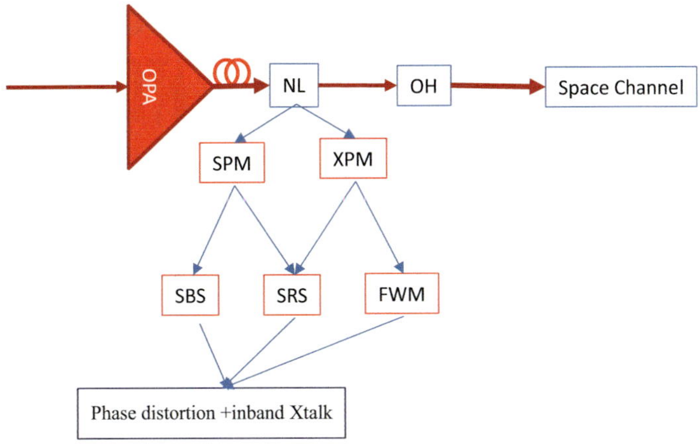

FIGURE 8.1 Nonlinear effects over a short-length fiber (single mode) connecting the HPA output to the optical head for launch into space. SBS=stimulated Brillouin scattering, NL=nonlinear, SPM=self-phase modulation, XPM=cross-phase modulation, SRS=stimulated Raman scattering, FWM=four-wave mixing.

SRS (stimulated Raman scattering), SPM (self-phase modulation), XPM (cross-phase modulation), and FWM (four-wave mixing). These nonlinear distortion effects are shown in Figure 8.1.

The following are the rules of thumb for estimating nonlinear effects and their impact on system performance.

- For low-power fiber <10 dBm modulated sequence or 3 dBm continuous waves, the effective length for nonlinear effects (SPM) is 22 km. So, → if the fiber is launched with lightwaves 1,000 times higher, then the effective length=22 m. Five meters would take effects about 6 dB below the penalty. Fibers are Germanica (Ge_2O) doped silica (SMF-28 or 28e) leading to significant NL distortion effects.
- For SBS (Brillouin's), distance is 1–10 km fiber → HPA SBS ~1–10 m (SMF28 or SMF-28e). Nonlinear coefficients are proportional to lightwave intensity, which is the power/fiber effective area.
- Therefore, to connect the fiber output of OPA to OH, such nonlinear effects must be avoided. Alternatively, split the OPA stages into two, with the last stage (high power) placed close to the OH.

8.1.2 WDM Channels and Nonlinear Effects (SBS, FWM, XPM)

8.1.2.1 Wavelength Multiplexing for Tbps SPC Transmission

As shown in Figure 8.2, the wavelength carrier can be assigned to a location on the spectral grid in the C- and L-bands with a spacing of 50 or 100 GHz grid as the IEEE standard given in an appendix of Chapter 5. The channel spacing can be 12.5, 25, 50, or 75 GHz, depending on the baud rate, and hence the bandwidth of the system. For 64 GBaud, 200 Gbps can be achieved using 25 GB 16- QAM or 2 or 400 Gbps with QPSK and dual polarization. The spacing can occupy 100 GHz channel spacing, and so on 1.6 Gbps can be transported using two carriers each with 800 Gbps, as shown in the last row of Figure 8.2.

The work presented in this book concentrates on 100 Gbps DPM_QAM (4-QAM QPSK or 16 QAM).

Thus, to make up a total capacity of SPC transport, we need at least ten carriers and 100 Gbps DPM-QPSK.

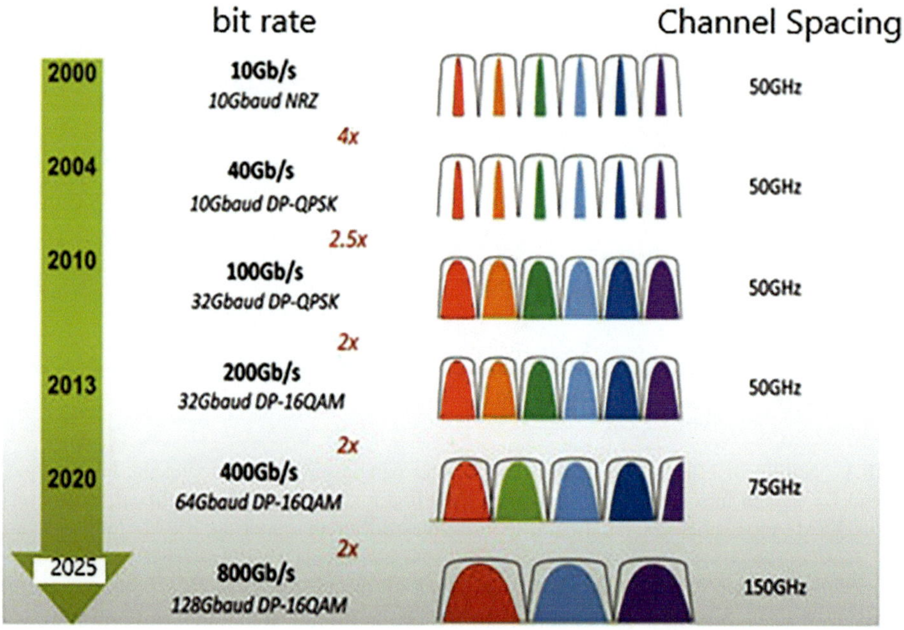

FIGURE 8.2 Yearly evolution of single-carrier data rate, modulation format and channel bandwidth. In multiplexing with channel spacing.

The summation of the total average power of all DWDM channels can lead to nonlinearity after the booster amplifier. Such nonlinear distortion is described in Section 8.2.

8.1.2.2 Absorption and Turbulence

As we have described, the absorption properties of the air layer when transmitting from an optical ground station to a satellite optical node are quite transparent over the C-L band of silica fiber (see Chapter 3). The low-attenuation spectral window can range from 1530 to 1650 nm, with only minor attenuation near the edges of this window. Thus, it is feasible to use transceiver modules developed for terrestrial and data center transmission systems. The advantages of using these transceivers are that they are robust and quite compact, with low energy consumption of <8 W per module or the electrical wall plus is <80 W per module when installed in satellite optical nodes.

The OTRx for long-haul at 100 Gbps using dual polarization-QPSK complex modulation format can offer significant sensitivity in coherent reception. The main question is whether, given that the bit rate can reach 1.0 Tbps per wavelength channel and is now ready for deployment in data center interconnections and intra-data center transmission, this would be possible for SPC, given that optical amplifications are available as booster amplifiers and low-noise, low-level sensitive optical pre-amplification at the reception. In inter-satellite connection SPC links, the vacuum link has no dispersion and is only limited by the ASE noise of the two amplifiers. We will provide our observations and comments on the employment of such a Tbps module per wavelength carrier for SPC in the concluding chapter.

8.2 BASICS OF NONLINEAR EFFECTS IN GUIDED MEDIUM

Six WDM channels at a total of 10 W are to be transported in fiber after the high-power amplifier, with a length ranging from 3 to 5 m for installation in the satellite housing area. The effective fiber length is ~ 5 m to about 22 m under 30 or 40 dBm total average power of guided lightwaves in the short length of single-mode fiber. The total power of all WDM channels propagating in the fiber is equal to the total average power at the output of the HPA. SMF-28, with an effective area ~65 μm^2 (area of the mode spot size – see Appendix 1 of Chapter 11), or larger effective core fibers (LEAF) of 80 μm^2, are used as fiber connecting the output of HPA booster amplifier to the optic head.

Nonlinear effects: SBS, FWM, SPM, XPM, and parametric amplification (in far-out spectral regions of interest). Inter-channel crosstalk, and thus in-band crosstalk, are primary concerns in WDM. In space transmission, high-power beams are required to overcome high propagation loss and the optical heads. The short fiber, under intense power, creates induced noise in both in-band and inter-channel regions.

This section outlines the nonlinear effects and induced noise and their impact on system performance, especially the total amount of in-band crosstalk so that no more than 1 dB is allowable due to the induced noise. The channels and the optical head, incorporating a short fiber, are shown in Figure 8.3 with multiplexed channels.

Figure 8.4 illustrates the physical mechanism of the SBS phenomenon in SMF under strong pumping, which results from the high-power modulated lightwave carriers of Tbps capacity.

- SBS nonlinear scattering in optical fibers is due to the inelastic scattering of a photon to a lower energy photon (conservation of momentum). The energy difference is absorbed by molecular vibrations or phonons in the medium (acoustic waves).
- The other wave is known as the Stokes wave. The signal is considered the pump wave, as its average power is on the order of 1000 mW. High-energy photons at the anti-Stokes

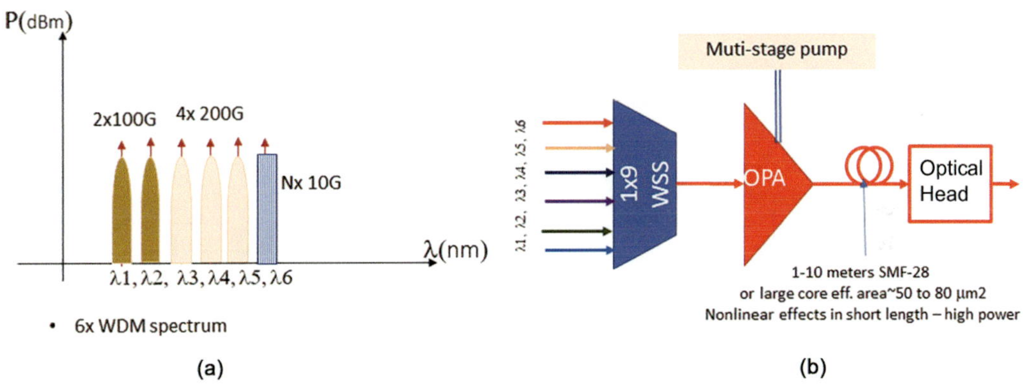

FIGURE 8.3 (a) Six multiplexed channels under design and (b) illustration of nonlinear effects expected to occur over the short SMF length before launching into OH and then space channel. WSS=wavelength-selective switch (arranged as multiplexer or demultiplexer) SBS.

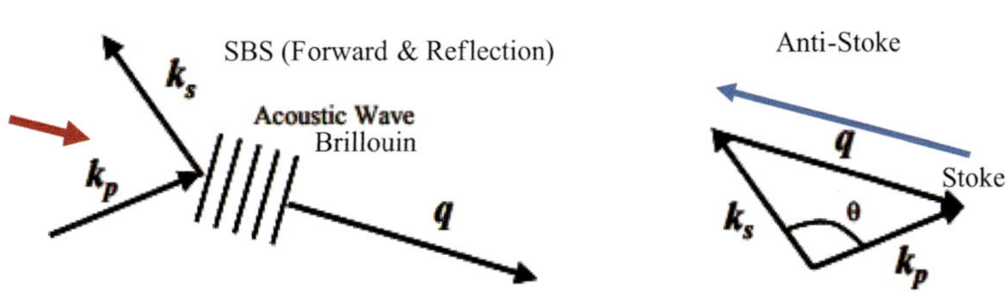

FIGURE 8.4 Physical mechanism of SBS with acoustic wave grating and Bragg phase matching to generate Stoke and anti-Stoke waves whose frequencies are shifted in both upper and lower sideband by about 9–12 GHz, these are falling into the in-band of the 100 and 200 G line rate channels.

frequency are also created if phonons of the right intensity and amplitude level meet the conservation of momentum.

- Furthermore, SBS results from the scattering of the incident lightwave by acoustic phonons (or acoustic waves). These are the shock waves created by the ultra-high intensity pulse of the modulated data sequence, propagating along the fiber core, with acoustic photons at 40 MHz, (thus the name acoustic waves). These acoustic phonons do not have an infinite lifetime and will be themselves scattered by impurities, thence, decay of the acoustic wave. The acoustic wave decay follows a Lorentzian distribution, with linewidth being the inverse of the decay time. No work has been published on such linewidth broadening of the laser under SBS, to the best of the author's knowledge.
- The phonon spectrum leads to the Brillouin gain spectrum. The phonon spectrum, leading to scattered lightwave shape and linewidth, affect the spatial spreading of the laser beam in SPC transmission. In the linear regime, the scattered lightwave intensity is the convolution of the Brillouin gain and the incident lightwave spectrum, during a pulse period, and so on.
- The SBS shift (momentum conservation due to phase matching via acoustic photon waves) is 11 GHz @1550 nm (cross-section area 50 µm^2 SMF28). This shift is enough to occupy half of the 100 Gbps DP-QPSK channel spectrum, thus seriously interfering with the adjacent channels.
- The phonon bandwidth is ~40 MHz for SBS scattering-gain, with an SBS frequency shift of ~9–11 GHz. This generates crosstalks in DWDM channels. A measurement test setup is shown in Figure 8.5a and b. These resonant acoustic frequencies create jittering in the channels and in-band crosstalks among the multiplexed channels. The main mechanism of this phenomenon is the modulation of the lightwaves by these acoustic waves.

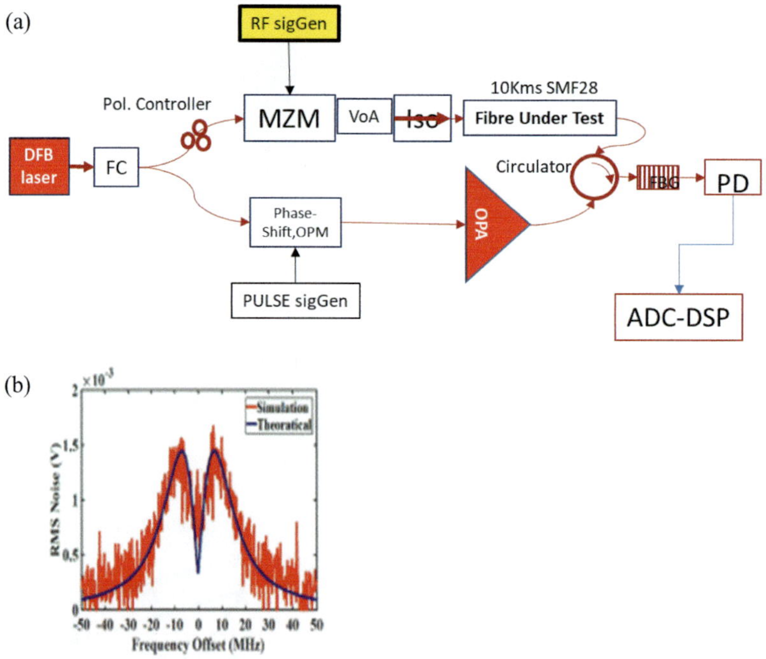

FIGURE 8.5 (a) Measurement setup to assess SBS noise spectrum; (b) noise spectrum generated by SBS, especially resonant peaking at a few tens of MHz.

We note the following related to the experimental investigations:

- Theoretical and SBS-induced noise in gain-based Brillouin spectral characteristics, low-frequency RF domain PD +Amp. → RF specific waves; and loss-based Brillouin optical time-domain analyzers.
- The gain/loss-based noise (blue solid curve) consists of phase-to-intensity (P-I) noise and spontaneous Brillouin scattering (SpBS)-p, that is under pumping at the onset of a threshold of SBS.
- SBS generates pulse sequence jitter, leading to sampling errors at the optical Rx. Thus, a high-pass filter may be required if care is not taken to eliminate such nonlinear pumping in the short-length SMF.

8.2.1 Self-Phase Modulation

High-power splitwaves propagating in a guided medium create a change in the refractive index superimposed on the refractive index of the waveguide, consequently, affect its own phase, hence the name self-phase modulation. Modulation indicates the envelope of the lightwaves by a modulation format.

We can estimate the SPM (self-phase modulation) effects over the short-length SMF as follows:

- L_{eff}=[1− exp(−αL)]/α; with α=0.15 dB/km=2.3e-7 cm^{-1}, thus the nonlinear coefficient affecting the propagation constant of guided lightwave, thus → $\left(\gamma = \left(2\pi n_2^2\right)/\lambda A_{\text{eff}}\right)$; n_2=2.7×10^{-20}m^2/W (nonlinear index)

The *guided wave propagation constant* under nonlinear effects can be written as the summation of the linear part and its nonlinear part as

$$\beta_{\text{NL}} = \beta + \gamma P$$

where β, $\beta_{\text{NL},}$ are the propagation constants of the guided mode in the fiber under linear and nonlinear conditions, respectively. P_s is the total average lightwave power of all channels guided in the fiber.

So, for $P_{s=}$10,000 mW; n_2=2.16e-11 mm^2/mW (nonlinear coefficient); 1=1550 mm; A$_{eff}$=65 μm^2; then g=0.06736e-11 mm/mW.

The nonlinear phase change over (5 m, P=10,000 mW), then DSFPM=γ P L_{eff}=0.06736e-11×10,000×500e3=0.336e-2 rad or 3.36 mrad. DSFPPM=dispersion shifted fiber polarization maintaining.

For the QPSK constellation, this results in rotating an angle of 33.6 mrad or ~19° angle of arc at each point of the constellation. The frequency of the lightwave carrier is also chirp by a frequency of 1.7 GHz for 10 W and 5 m of fiber (SSMF-standard single mode fiber – Corning SMF-28 or 28e types). Thus, the LO frequency of the receiver must allow this shift to match that of the shifted carriers. DSP algorithms can be used to automatically compensate for this phase rotation. The propagation constant of the light-guided wave in fiber is also chirped with a propagation constant change of 1.374e-2/m. This chirp, in turn, can be considered as continuous frequency offset, which is within the range of 10 GHz or (±5 GHz) for Commercial coherent receivers, but reduces the LOFO range of the TRx.

We intend to conduct the experimental testing by connecting the output fiber of the HPA to an optical spectrum analyzer (OSA). Then, by using

- Two unmodulated carriers launched to the input of the HPA, observe and measure the second harmonic product XPM as a second-order modulation product.

- One wavelength laser input and HPA under pump for saturation to generate phase distortion or chirp of the single lightwave carrier; the OSA must have high-resolution features.

Three light-wave carriers are input to the HPA under maximum pumping to reach saturation to measure the increased power of the other carriers and compare it with their level under no pumping. FWM distortion is observed via the third-order intermodulation product.

Then, obtain IIP3 (intersecting point of second harmonic and third harmonics) or the intersection point between the second-order modulation (obtained above) and the third-order term. Thus, determine the spectral region in which the distortion has taken effect.

8.2.2 Four-Wave-Mixing (FWM) and Cross Phase Mixing (XPM)

Figure 8.6 illustrates the case of FWM with six multiplexed channels and notes the following:

- Four-wave mixing (FWM) is an inter-modulation phenomenon, where interactions between two or three wavelengths produce two or one new wavelengths. It is like the third-order intercept point (IIP3) in electrical systems. It is a parametric nonlinear process, in which the energy of the incoming photons is conserved. FWM is a phase-sensitive process. The conversion efficiency of FWM is strongly affected by the phase-matching conditions of the two incoming waves and the generated other two waves.
- FWM in fibers is related to self-phase modulation (SPM intensity nonlinear effects – Kerr) and cross-phase modulation (XPM – also Kerr's). These phenomena originated from

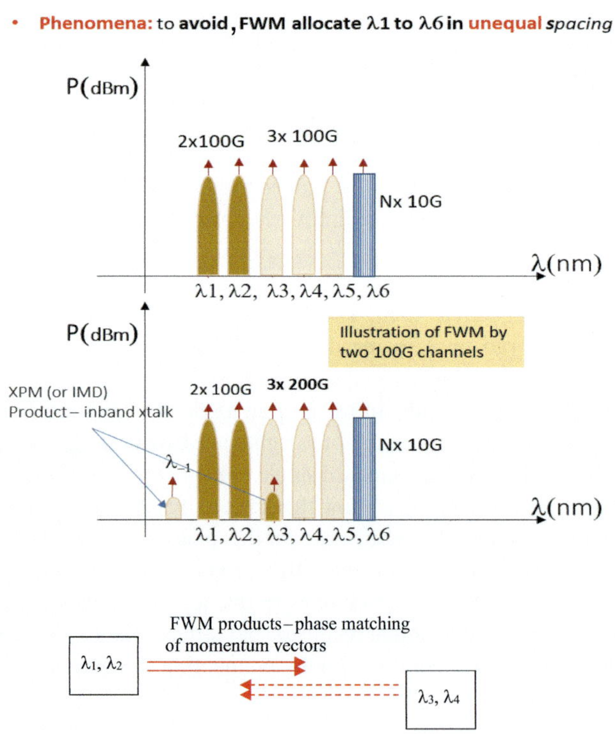

FIGURE 8.6 Illustration of four-wave mixing by six multiplexed channels under high power launching into a short fiber length from the output of HPA. Insert below is the phase matching of the four waves equally spaced.

(Kerr) nonlinearity and are different only in terms of the degeneracy of the waves involved. Modulational instability (MI) can be interpreted as an FWM Effect.

- FWM is serious when the lightwave carriers are equally spaced, as shown in the RHS, where the wavelengths are assigned to the 100, 200, and 10 G channels.
- FWM of two carriers λ_1 and λ_2 (power about 3–6 dB above the QPSK power spectra), FWM products are located in the carrier λ_3 and λ_{-1}
- Similarly, for the spectral components of the 100 G, interference or crosstalks can be seen in the channel 13. For example, the phase-matching cases (freq. ν) y applying the conservation of momentum: $\nu_3=\nu_1-(\nu_2-\nu_1)=2\nu_1-\nu_2$ and $\nu_4=\nu_2+(\nu_2-\nu_1)=2\nu_2-\nu_1$,
- Alternatively, three waves can be combined with one wave in traveling in the opposite direction (e.g., a reflected wave) to generate FWM product.

Thus, it is expected that many FWM products will be generated from the five light waves spaced equally at 200 GHz.

How long does the fiber take, and how high does the power need to be to reach the level at which FWM occurs?

Our estimations are: n_2=2.16e-11 µm²/mW. So, for an effective core area of 65 µm² (SSMF[1] Fiber core eff. area) → 1.40e-9 /mW. So, at 40 dBm or 10,000 mW, this leads to a nonlinear refractive index change (n_2 * power/A_{eff})=1.4e-5, where IMD=intermodulation distortion.

8.2.3 Cross-Phase Modulation (XPM)

XPM of the six multiplexed wave channels is illustrated in Figure 8.7 and notes the following:

- Cross-phase modulation (XPM): XPM is a nonlinear optical effect where one wavelength of light can affect the phase of another wavelength of light through the optical Kerr effect (SPM).
- In WDM with six channels in a short length of fiber (say, 5 m) for space Tbps applications with amplitude-phase modulation and coherent reception, the effect of XPM is a two-step process: (i) the signal is phase modulated by the co-propagating second signal. (ii) Nonlinear dispersion leads to a transformation of the phase to amplitude modulation distortion of other WDM carriers. (iii) Furthermore, dispersion walk-off between the channels and polarized components → reduces the effect of XPM.
- XPM → intermodulation effects on the other WDM channels, or like IIP2 (Intersection point 2). The second-order intermodulation products → fall into other channels that are two-grid spacings away, resulting in in-band crosstalk.

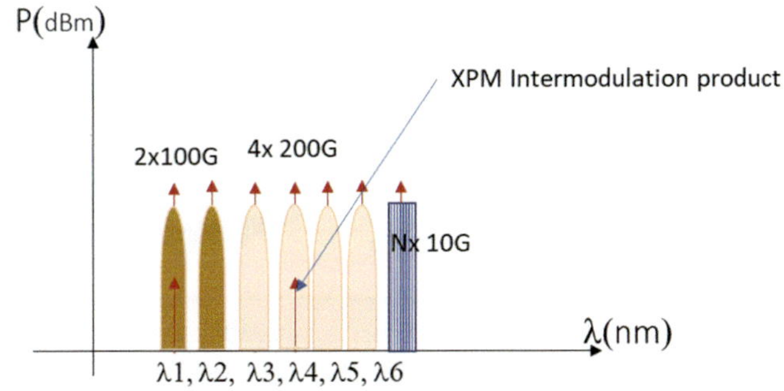

FIGURE 8.7 XPM illustration.

8.2.4 SRS (Stimulated Raman Scattering)

SRS is different from SBS due to the generation of the Stokes forward propagation – transferring of energy of the original waves to another Stokes shift of around 11 GHz in such a short distance of high-power pumping from the high-power output of the high-power HPA.

This Raman shift is 13 THz away (in Ge_2O-doped silica fiber of 50 μm² sectional area), resulting in no interference or crosstalk in the channels under consideration, which are about 3 nm (1553–1556 nm) away from the C-band, and thus less problematic. However, this scattering SRS would generate a power penalty on the optical channels before launching them into the space channel.

8.2.5 SBS in Short Length and High-Power Launch

SBS is the interaction between the phononic waves, i.e., the vibrational waves caused by the high-power pumping of the multi-wavelength channels. The Brillouin waves then interact and phase-match with the incoming optical waves to generate other waves far away at the channel frequency of about 9 GHz, and other low frequency waves due to the dislocation of the atomic molecules of the fiber's silica atomic structure. These stray waves generate or modulate the envelope of the optical waves. Therefore, amplitude envelops modulation of the optical channels. This is the noise component that we can observe in Figures 8.8 and 8.9. The interaction can be observed in both the forward and reflected phononic waves as shown in Figure 8.10.

These noises or amplitude modulation in randomness lead to the jittering of the optical channel timing, resulting in an error in sampling points and affecting the bit error rate of the channels. In SPC systems, the SBS noises are mainly generated in the short-length fiber of single mode under the launching of high average power of the amplified channels. Again, the average optical power is about 30–40 dBm over a cross-sectional area of only 65 μm², resulting in a very high intensity to the silica molecules.

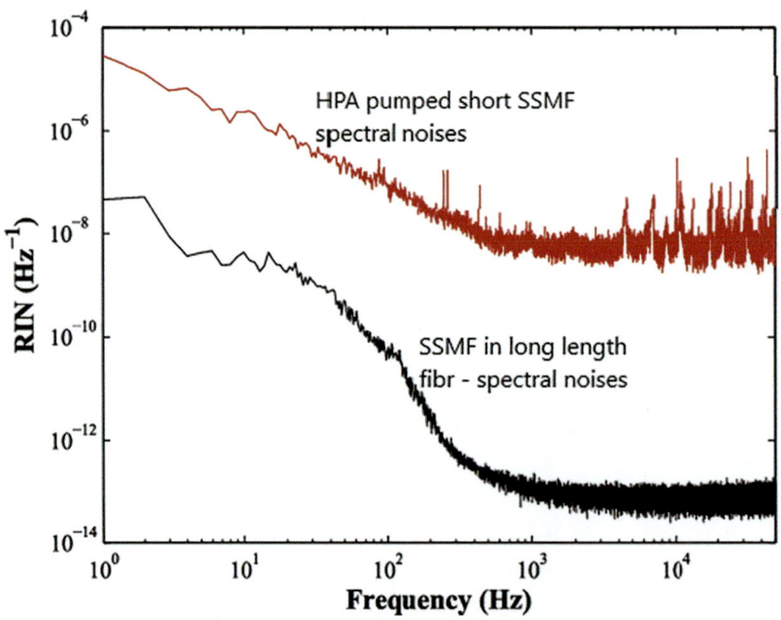

FIGURE 8.8 Relative intensity noises (RIN) produced with high power channels launched into short-length SSMF after HPA, before launching into free space. Note the high RIN for short-length, high-power compared to long length and moderate power.

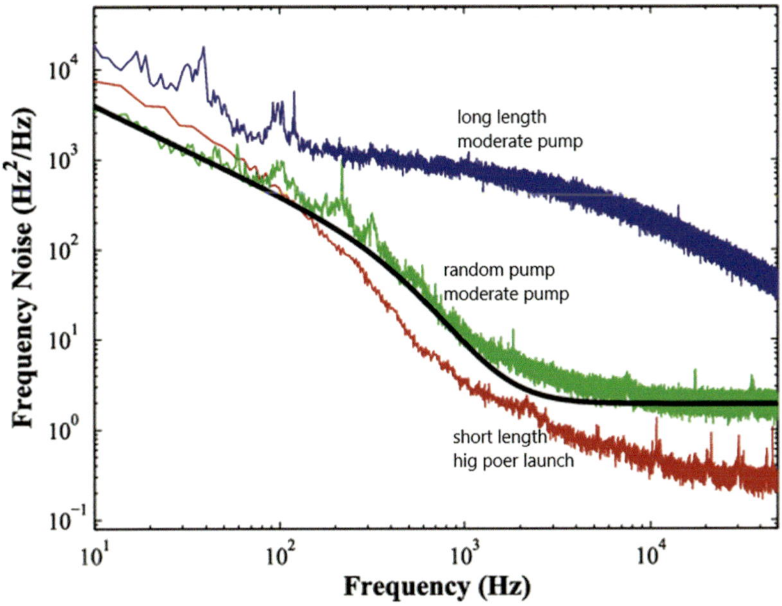

FIGURE 8.9 Spectral noise of SBS interaction over short-length high power pumping.

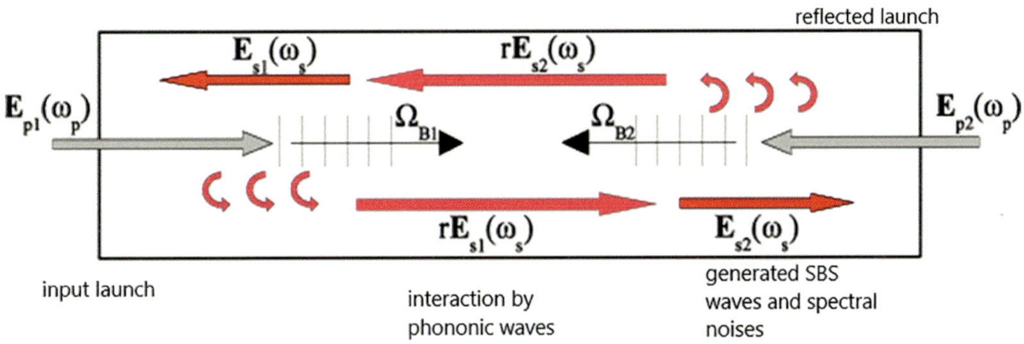

FIGURE 8.10 Phononic interaction and noise generation by forward and reflected waves in short-length high-power pumping by high-power aggregate channels in short-length fiber after HPA, leading to SBS noises via phononic interaction, i.e., phononic dynamic gratings and the optical waves of the DWDM channels.

Usually, the amplitude modulation is about 3 kHz on the envelope of the optical channels. To mitigate this amplitude modulation, the optical channels can be detected and electrically filtered using a high-pass filter, which can be implemented in the DSP system of the receiver.

8.2.6 FWM in High Power Launch and Short Fiber Length

As discussed so far, the high power launched into a short-length fiber of minute cross-sectional area leads to nonlinear interactions such as self-phase modulation. FWM is one of the main nonlinear effects in such medium, used in space photonic systems. The short-length fiber connected to the output of the HPA is necessary to connect the HPA to the optical head and launch the beam into space medium. This short length is influenced by the four-wave-mixing (FWM) process, where two optical channels are phase-matched by their bidirectional waves to generate a fourth wave whose frequency is equal to the frequency spacing between them. Therefore, these generated FWM noise components can fall into the passband of another channel, where frequency spacing equals to the spectral difference of the two interacting channel carrier frequency.

In the following, we represent our experimental setup and evaluation of the FWM in such fiber connected HPPA.

8.3 NONLINEAR EFFECTS IN SPC UNDER WAVELENGTH MULTIPLEXED CHANNELS

8.3.1 Short Fiber from OPA to OH and Nonlinear Penalty

It is noted that when launching into the space channel, the OPA output may have to be extended for connection to the OH.

High power in optical waves can lead to nonlinear effects such as nonlinear phase shift, SBS, SRS, SPM, XPM, and FWM. These distortion effects are shown in Figure 8.1.

The rule of thumb for estimating nonlinear effects and impact on system performance.

For low-power fiber <10 dBm modulated sequence or 3 dBm continuous waves, the effective length of nonlinear effects (SPM) is 22 km. So, → if fiber is launched with lightwaves of 1000 times higher than the effective length, the efffective length = 22 m. A Five-meter fiber would take effects about 6 dB below the penalty threshold. Fibers are Germanica (Ge_2O)-doped silica (SMF-28 or 28e), leading to significant NL distortion effects.

For SBS (Brillouin's) distance 1–10 km fiber → HPA SBS ~1–10 m (SMF28 or SMF-28e). Nonlinear coefficients is proportional with lightwave intensity, which is power/fiber effective area.

Hence, to connect the fiber output of the OPA to the OH and avoid such nonlinear effects, one option is to split the OPA stages into two, with the last stage (high power) placed close to the OH (Figure 8.11).

8.3.2 WDM Channels and Nonlinear Effects (SBS, FWM, XPM)

Five WDM channels at 10 W each are to be transported in fiber after the high power amplifier, with the fiber length ranging from 3 to 5 m for installation in the satellite housing area. The effective fiber length is ~ 5 m to ~ 22 m under 30–43 dBm of total average power of guided lightwaves in the short length of single-mode fiber. The total power of all WDM channels propagating in the fiber is equal to the total average power at the output of the HPA.

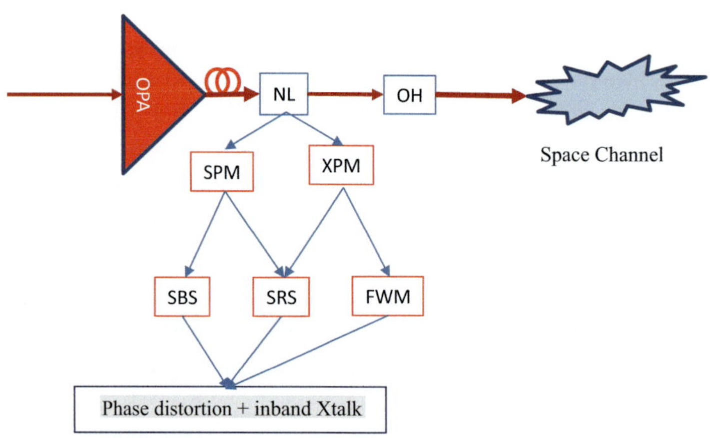

FIGURE 8.11 Nonlinear effects over a short length fibre (single mode) connecting the HPA output to optical head for launch into space. SBS = stimulated Brillouin scattering, NL = nonlinear, SPM = self-phase modulation, XPM = cross-phase modulation, SRS = stimulated Raman scattering, FWM = four-wave mixing.

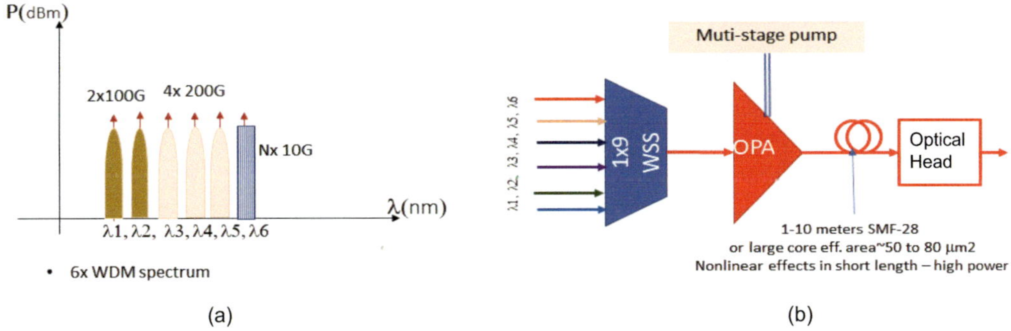

FIGURE 8.12 (a) Six multiplexed channels under design and (b) illustration of nonlinear effects expected to occur over the short SMF length before launching into OH and then space channel.

SMF-28 (effective area ~65 μm^2) or larger effective core fibers have an effective core area of ~80 μm^2. See Appendix 2 of Chapter 11 for fiber details.

Nonlinear effects include SBS, FWM, SPM, XPM, and parametric amplification (to far outside the spectral region of interest). Inter-channel crosstalks and thus in-band crosstalks are primary concerns in WDM. In space transmission, high power beams are required to overcome high propagation loss and the OH. The short fiber under intense power creates induced noise in both in-band and inter-channel signals.

This section outlines the nonlinear effects described above, along with the induced noises and their impact on system performance, especially the total amount of in-band crosstalks. No higher than 1 dB is allowable due to the induced noises. The channels and the OH, incorporating a short fiber, are shown in Figure 8.3 with multiplexed channels (Figure 8.12).

8.3.3 Wavelength Assignments

In the case that FWM, XPM, or SPM distortion effects occur, the optical frequencies of the multiplexed channels should be allocated such that their spacings are unequal. This prevents the crosstalk from fallen into one of the bands of the channels.

There is a significant difference in the wavelength assignment of channels due to the absorption characteristics of the turbulence layer in the atmospheric region. Furthermore, the attenuation and beam spreading caused by the refraction index of the turbulence are expected. As a result, the channels from the OGS to the satellite are received, converted back to electrical signals, and then modulated again using an integrated optical modulator and transmitted again to other satellite optical node. This is the optical repeater or transponder at the first satellite node, providing access to the ground information channels.

Thus, the channel wavelength carriers can be assigned to new wavelengths based on the low absorption of the medium on that orbit.

In vacuum media, the nuclear radiation that may damage the transceiver and router components is much more serious, especially at the coupling points between fiber and guided channel waveguide. This damage may create misalignment of the optical fields of the two waveguides, the circular fiber and the non-symmetrical channel waveguides of the modulator.Furthermore, the doped impurity metals in the laser cavity would alter the resonant frequency, thus the carrier may be shifted to the other location of the wavelength grid of the low loss optical spectrum of the transmission air medium . This is one of the main reasons that 50 GHz grid is selected for 100 Gbps, and a 100 GHz for 56 GB 200 Gbps with ten wavelength carriers, is used. If we want 1.0 Tbps with 128 GB per wavelength carrier, then the spacing should be 150 or 200 GHz.

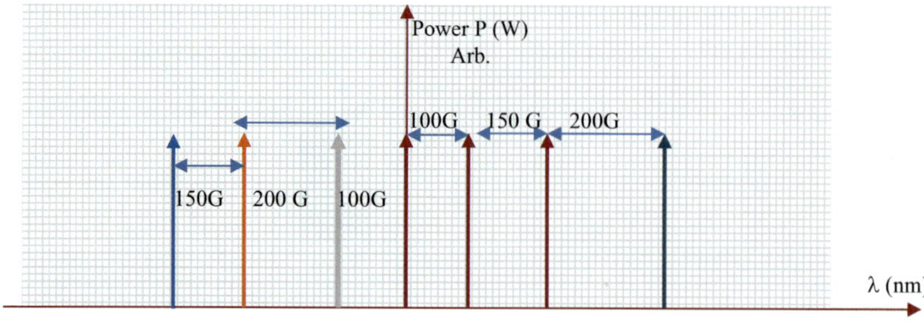

FIGURE 8.13 A possible wavelength carrier allocation to avoid nonlinear interaction, with unequal frequency spacings. Note: At 1550 nm 100 GHz spacing is approximately 0.8 nm spacing.

The nonlinear FWM creates unwanted crosstalks of channel frequencies that are equally spaced, so there are a wavelength scheme to ensure minimum nonlinear crosstalks, as shown in Figure 8.13.

8.3.4 Penalty on OSNR Due to In-Band Crosstalk by NL Scattering

The in-band crosstalk can be illustrated in Penalty on OSNR due to In-band Crosstalk by NL Scattering, Figure 8.14a, under the envelope of the signal band. These crosstalks are caused by different NL distortion effects generated as discussed above. The PDF (probability distribution function) of the summation of all nonlinear effects is shown in Figure 8.31c, which allows the estimation of the signal penalty as the power of the crosstalks to that of the signal power contained under the signal power spectrum $X_c(\omega)$.

We note that if the crosstalk is <15 dB, the penalty is about 1 dB, so this is the limit that we must consider when designing the length of the SMF fiber in the high-power booster amplifier, to avoid high power penalties between channels.

8.3.5 Remarks

This section reaffirms the design parameters of SPCS over a multi-hop structure, whose laboratory equivalent structures are designed and specified for transmission capacity >1.0 Tbps. We show also that a maximum capacity of >5.0 Tbps can be possible under[2] SPC systems.

The multi-hop transmission power budget is outlined with a method of estimation. The receiver sensitivity, under the impact of impairments such as the Doppler shift due to the rotational speed of the satellite relative to the transmitter/receiver, is considered. It has been proven that the commercial receiver can mitigate this effect up to 5.0 GHz and tune the receiver LO frequency to match the tracking of the Doppler shift. This is critical, as the Doppler shift may change continuously twice a day in the case of an LEO-GEO link.

Furthermore, we have conducted several preliminary tests of commercial optical transceivers operating at 100 G DP-QPSK. Nonlinear impairments caused by the short length of fiber under the pumping of high-power laser before launching into space channels were identified, and their penalty on the OSNR was estimated, setting a 1-dB limit for such nonlinear distortion. The methods for mitigating and avoiding such nonlinear penalties are provided.

Having confirmed the preliminary design of multi-hop SOCS (Space Optical Communications Systems), we set up a laboratory-based SOCS to demonstrate its operation and the nonlinear effects on the transmission crosstalk of the DWDM channels in photonic communications systems. The transmission penalty is given in Section 8.3.4.

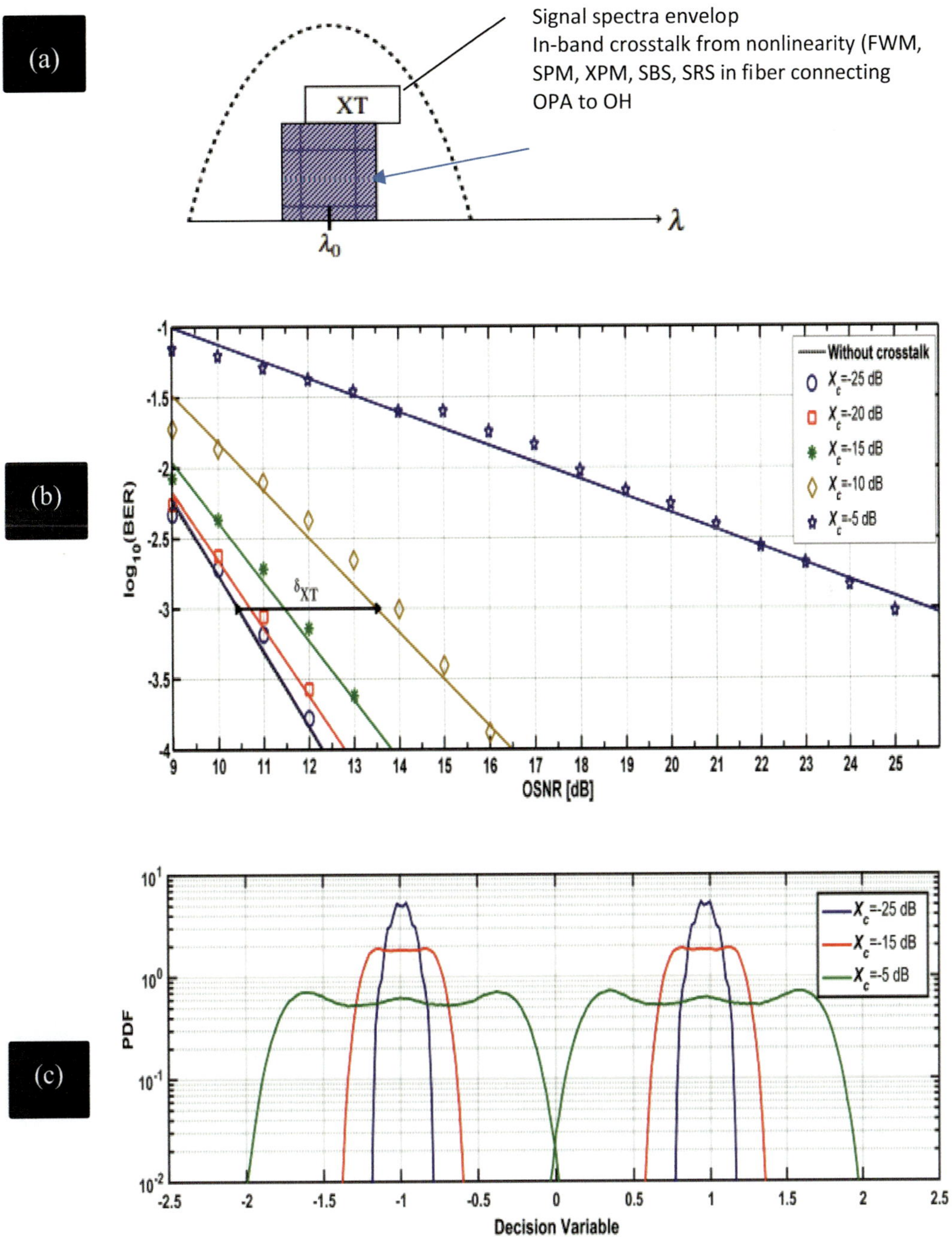

FIGURE 8.14 Signal spectrum and the in-band crosstalk (blue stars) noises falling into signal band, the in-band crosstalk.

This section reaffirms the design parameters of SPC over a multi-hop structure whose laboratory equivalent structures are designed and specified for >1.0 Tbps transmission capacity. We also show that a maximum capacity of >5.0 Tbps can be possible under our SPC[3] space optics systems.

The multi-hop transmission power budget was given in Chapter 2. It outlines a method of estimation for operating margin including the penalty of powr degradation due nonlinear effect crosstalk. The receiver sensitivity, under the impact of impairments such as the Doppler shift due to the rotational speed of satellite relative to the transmitter/receiver is considered and proven the mitigation capability of commercial receiver to 5.0 GHz and the ability to tune the receiver LO frequency to match to the tracking of the Doppler shift. This is critical as the Doppler shift may change continuously twice a day for the case of LEO-GEO link.

Further we have conducted in later Sections 8.3.4 and 8.8, several preliminary tests of commercial optical transceivers operating at 100 G DP-QPSK. Nonlinear impairments due to short length of fiber under pumping of high-power laser before launching into space channels are identified, their penalty on the OSNR is estimated and for setting a 1-dB limit of such nonlinear distortion. The methods for mitigation and avoidance of such nonlinear penalties are given.

8.4 THE 100 G DUAL POLARIZATION QPSK SIGNAL AND TRANSMISSION PERFORMANCE

8.4.1 System Structure

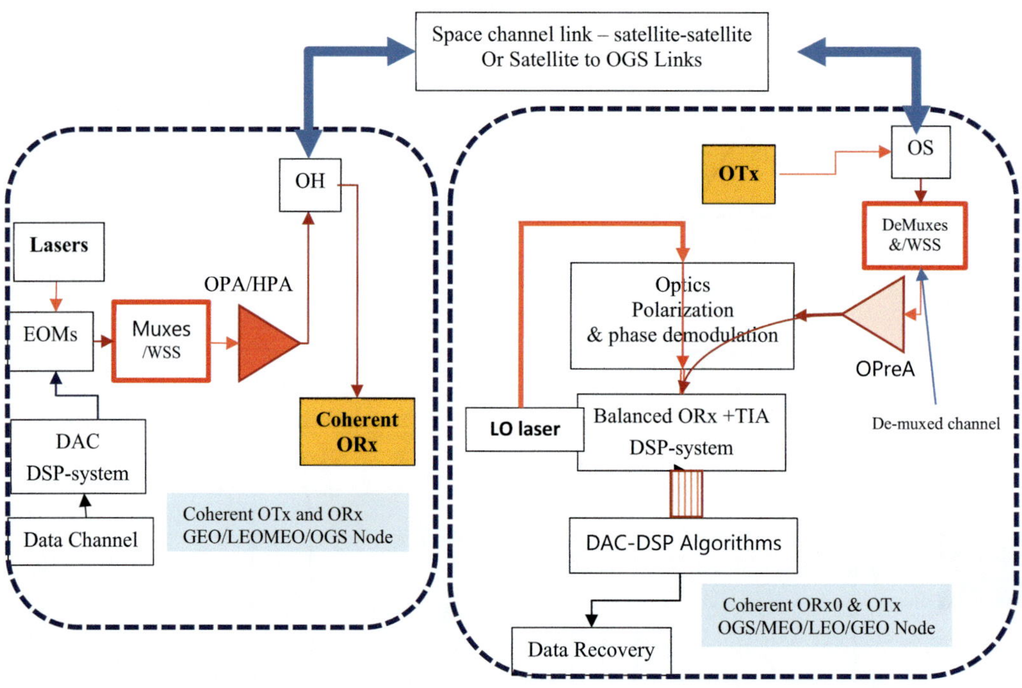

FIGURE 8.15 Space photonic transmission system. OPA/HPA = optical power amplifier, OPreA = low-input low noise optical pre-amplifier, TIA = transimpedance amplifier (differential input, differential or single output, LO = local oscillator, ORx = optical receiver, OTx = coherent optical transmitter, OH = optic head (bi-directional), EOM = electro-optic modulator, PIC = photonic integrated circuit, OMux/ODemux = optical wavelength multiplexer/demultiplexer, WSS = wavelength selective switch. Note: Optical path in ‘ light-dark lines’, electrical path = ‘dark lines’.

8.4.2 Wavelength Assignment

The wavelengths of the multiplexed channels are assigned on the 100 GHz grid, as illustrated in Table 8.1, assuming no nonlinear effects so equal spacing is set between the adjacent channels (Figure 8.15). Figure 8.16 illustrates the channels and modulated signals, as well as the optical

TABLE 8.1
Assigned Wavelengths and Channel Spacing for Space Photonic Communication Transport Networks (SPCTN)

ITU C-Band, 50, 100 and 200 GHz DWDM Grid and Assignment for Space OTN

C-Band Channel ITU-No	100 GHz Grid Wavelength (nm)	Frequency (THz)	SOCTN Wavelength Assignment 200 GHz-Grid (nm)	Remarks
1	1528.77	196.1		
...14	1538.98	194.8	1538.98	
15	1539.77	194.7		Outside opt amp band for optical amplifiers for SPC system
16	1540.56	194.6	1540.56	Provisional assignment
17	1541.35	194.5		Provisional assignment
18	1542.14	194.4	1542.14	Provisional assignment
19	1542.94	194.3		Provisional assignment
20	1543.73	194.2	1543.73	Provisional assignment
21	1544.53	194.1		Provisional assignment
22	1545.32	194	1545.32	Provisional assignment
23	1546.12	193.9		Provisional assignment
24	1546.92	193.8	1546.92	Provisional assignment
25	1547.72	193.7		Provisional assignment
26	1548.51	193.6	1548.51	Provisional assignment
27	1549.32	193.5		Provisional assignment
28	1550.12	193.4	1550.12	Provisional assignment
29	1550.92	193.3		Provisional assignment
30	1551.72	193.2	1551.72	Provisional assignment
31	1552.52	193.1		Provisional assignment
32	1553.33	193	1553.33	Provisional assignment
33	1554.13	192.9		Provisional assignment
34	1554.94	192.8	1554.94	Provisional assignment
35	1555.75	192.7		Assigned
36	1556.55	192.6	1556.55	Assigned
37	1557.36	192.5		Assigned
38	1558.17	192.4	1558.17	Assigned
39	1558.98	192.3		
40	1559.79	192.2	1559.79	Provisional assignment
41	1560.61	192.1		
42	1561.42	192	1561.42	Provisional assignment
43	1562.23	191.9		
44	1563.05	191.8	1563.05	Reserved
45	1563.86	191.7		Provisional assignment
46	1564.68	191.6	1564.68	Provisional assignment
47	1565.5	191.5		Provisional assignment
48	1566.31	191.4	1566.31	Provisional assignment
49	1567.13	191.3		Provisional assignment
50	1567.95	191.2	1567.95	Provisional assignment

gain passband of the amplifiers, the booster, and the pre-amplifier of the SPC link. Ideally, 50 GHz spacing would be used if 100 G DP-QPSK modulation format is applied to all channels, as the band width is only 18–25 GHz. Thus, the double sideband can fit into such spacing. In the case of serious nonlinear crosstalk, unequal spacing is used. This is described later after the detailed investigation of the nonlinear crosstalk in Section 8.8. In SOC, transmission with a high-power amplifier, the channel spacing must be arranged to minimize or avoid nonlinear crosstalk.

8.4.3 System Structure and Set-Up for Evolution of 100 G DP-QPSK over the Optically Amplified Hop

The test system which is equivalent to an optically amplified SPC hop is set up as shown in Figure 8.17. The system consists of the following:

- 100 G DP-QPSK, i.e., dual polarization quadrature phase shift keying, signals are generated from the integrated module DP04QFSP to generate and transmit complex modulation (IQ modulation of 4-point constellation signals). This means that two polarized channels are generated, each channel modulated by integrated optical IQ modulator.
- The transmitter is formed by two lasers, one high power of a few mW coupled to an integrated IQ modulator consisting of four integrated MZM modulators, as provided in appendix on optical modulators, and $\pi/2$ phase shifter so that I and Q-components in optical signals are generated.
- The optical receiver, which is also integrated into the transceiver, is a DSP-based integrated photonics and electronic structure. Both the receiver and transmitter are integrated into an optical transceiver.
- Two lasers, one acts as the optical carrier for modulation of the QPSK format, and the other acts as the local oscillator for coherent reception.
- Optical amplifiers: one HPA with an average output power of 2.0 W or 33 dBm, and one optical preamplifier LNA (low noise amplifier).
- One variable optical attenuator (VOA) is used to attenuate signals, representing the optical transmission loss of the free space medium.
- It is necessary to use optical couplers to tap signals for the measurement and monitoring of the transmitting optical channels.
- We can use the VOA and the control of the *attenuation factor* to control the power into the pre-amplifier and optimize the noises and the overall gain, and hence the average signal into the OSA.
- The optical booster amplifier HPA operates under a laser bias coupled at the input. This bias laser input has an optical frequency whose location is far from the carrier frequency. The power of this laser is sufficient to pass the HPA operating in the saturation level. This ensures that the channel gains are sufficient and equal across the gain spectrum of the HPA. The average power per channel is equal to the average saturated power of the HPA minus the logarithmic number of the channel number. This is true if all channels are distributed in the flat gain band of the HPA. We can assign the wavelengths of the six channels in the 50 GHz spacing and locate them in the gain flat band of the HPA. These spectral windows are also in the transparent region of the vacuum transmission medium.

The wavelength of 1536.61 nm is set for the seed laser, which provides the initial activation of the booster amplifier to operate the HPA at saturation so that signals can pass through the power amplifier and then enter the pre-amplifier. This wavelength also falls into one of the wavelength slots at

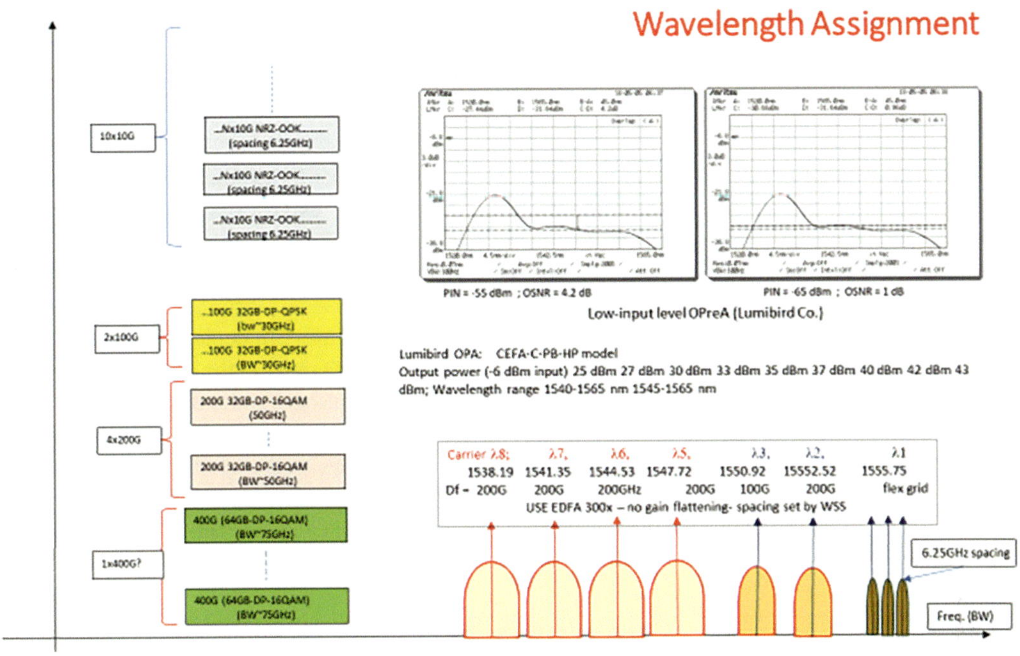

FIGURE 8.16 Illustration of wavelength assignment and amplification spectral window for Tbps SOTN. OPreA=optical pre-amplifier, OPA=optical power amplifier, WSS=wavelength selective switch, EDFA=erbium doped fibre ampifier, DP-QPSK=dual polarization- quandrature phase shit keying, QAM=quadrature amplitude modulation.

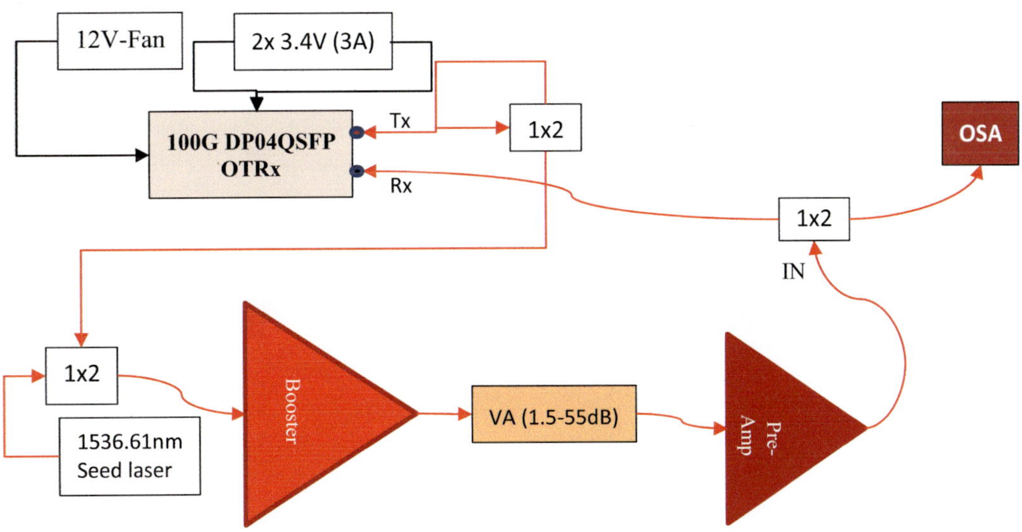

FIGURE 8.17 Optical and signaling setup for 100 G transmission over OP MEO_MEO with booster and preamp of MPB model and part number SDA-1116-006. Operating by GUI (1–3) with monitored power, gain and driving current of LD1/LD2 and LD3 drving current for operations of amplifiers. OTRx=optical transceiver

the output of the pre-amplification subsystem, allowing monitoring to indicate that the HPA is operating at saturation. Therefore, such a seeding laser used to set the HPA at saturation can be called the biasing laser. Optical amplifiers are high-power boosters and low-noise, low-level pre-amplifier.

The OSA (optical spectrum analyzer) displays all channels as well as the nonlinear products of the nonlinearity, as expected in our analysis above.

8.4.4 Optically Amplified Free-Space Hop Test

One free-space consists of two optical head whose gain is in the order of about 110 dB each and a total propagation loss of 172 dB for 8000 km space distance (by using Friis estimation with an assumption of single frequency source or no linewidth of the laser). Thus, a total insertion loss from the output of the HPA to the front face of the receiver detector is about 45–65 dB. We assume a loss of 45 dB.

Now, setting VOA = 45 dB, or the attenuation factor at 45 dB, Figure 8.18 depicts the OSA optical spectrum of the channel. The optical spectrum of the transmitted channel at the output port, which is filtered by an OADM (optical add/drop multiplexer) in demultiplexer mode, is shown. The 3 dB bandwidth of the OADM filter is about 65 GHz, and the channel bandwidth at 100 Gbps QPSK is about 18 GHz, which is higher than half of the baud rate. This is due to the additional bits for FEC (forward error correction). If 30% FEC is used then the channel bandwidth is about half of the 65 GHz bandwidth of the filter. If soft FEC is used, then 432% overhead of the bits is applied, which leads to nearly half of the filter bandwidth (Figures 8.19–8.21).

8.4.5 Evolution of Spectra of 100 G DP-QPSK over Hop Stages

At the output of the Tx of 100 G DP-QPSK module, it is coupled via a 1×2 FC (3.4 dB) (Figure 8.22).

At output of Booster amplifier –20 dB FC port (Figures 8.23–8.25).

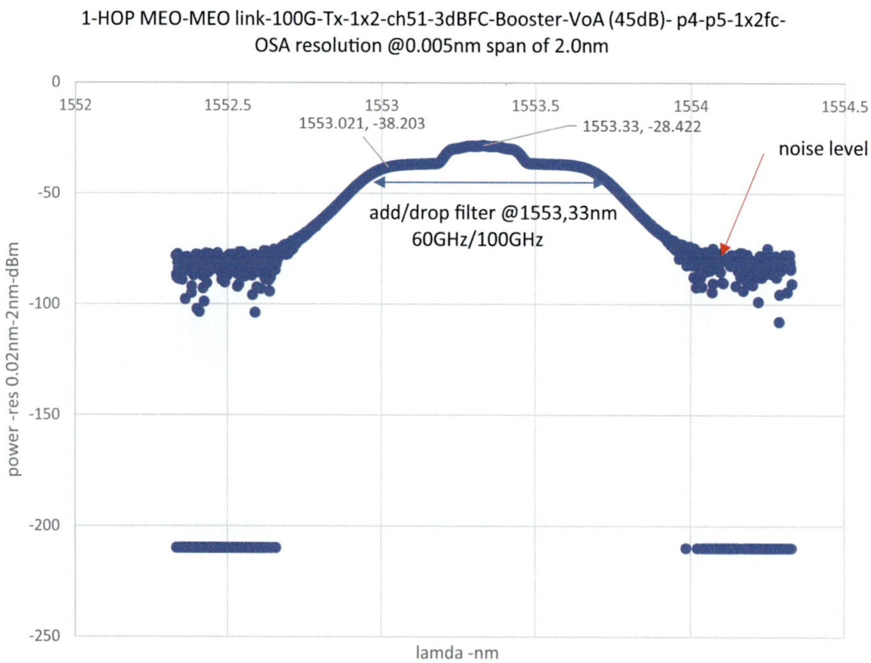

FIGURE 8.18 VOA = 45 dB optical spectrum of the output of the transmitted channel at output of port filtered by an OADM (optical add/drop multiplexer) or demultiplexer mode.

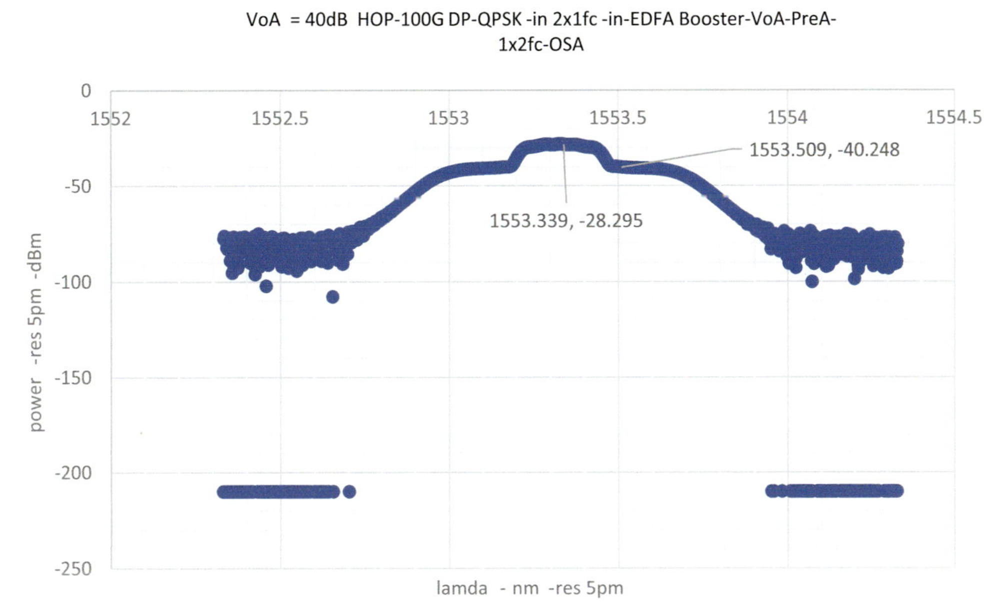

FIGURE 8.19 VOA = 40 dB.

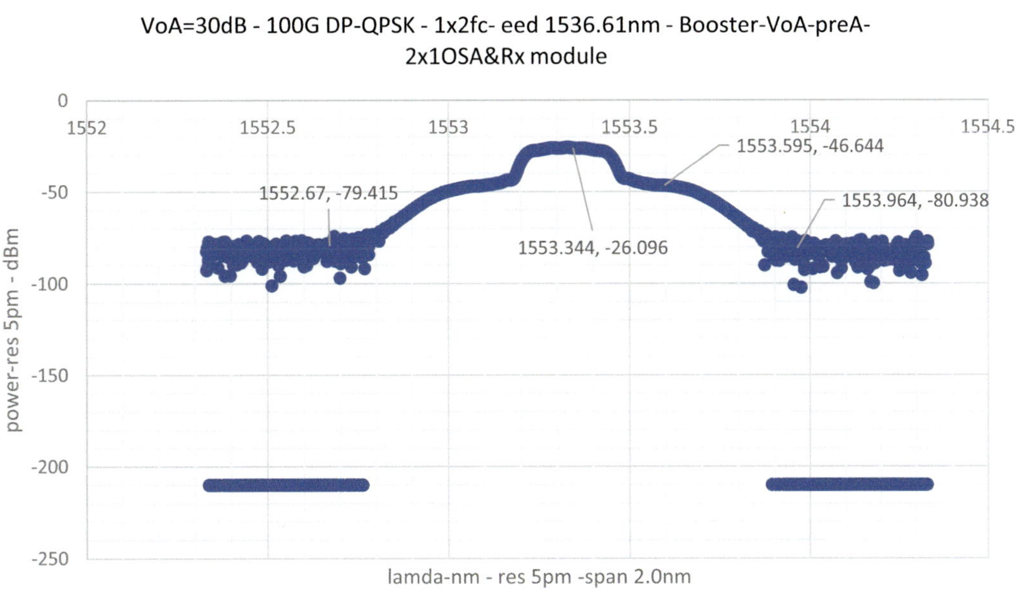

FIGURE 8.20 VOA = 30 dB HOP link, broader spectrum is output of optical filter of OADM and narrower spectrum is channel signal.

8.4.6 Hop Testing Using Carrier at CH51 (1536.61 nm)

Carriers at 1553.33 nm (CH30) and 1536.61nm (CH51) are used to test and understand the operational characteristics of the amplified hop. The setup is as shown in Figure 8.26, in which (i) two CW lasers at 1553.33 and 1536.61 nm are generated and coupled via a 2 × 1 SMF coupler, with each power reduced by 3.6 dB. Thus, the total power per carrier is 5.6 dBm when the output power of the lasers is set at 9.5 dBm (not less) (Figures 8.27–8.33).

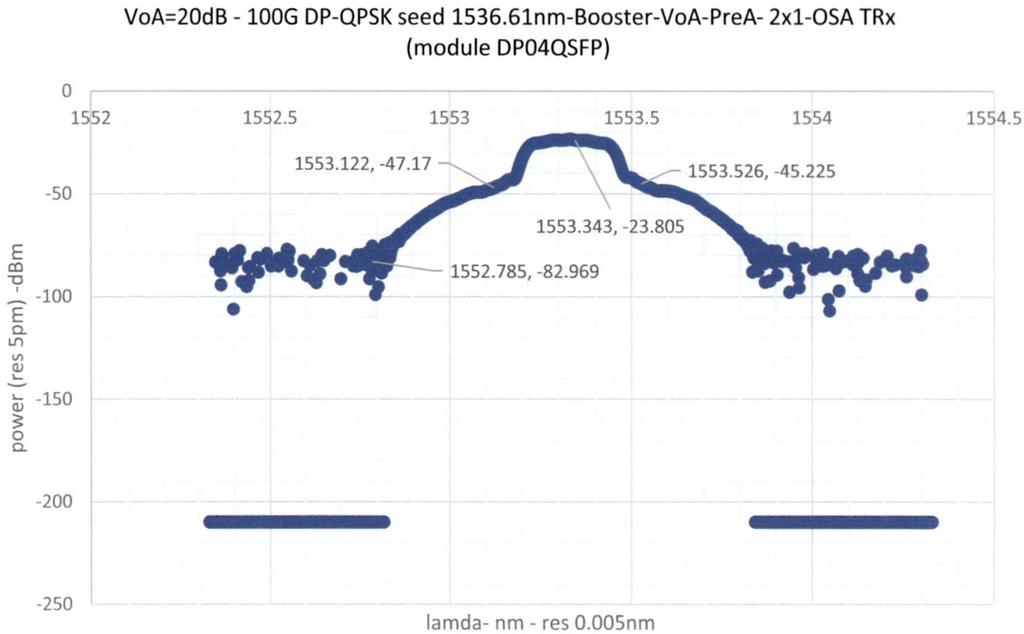

FIGURE 8.21 Setting VOA = 20 dB, HOP link using DPßQPSK 100G D'DP04QSFP.

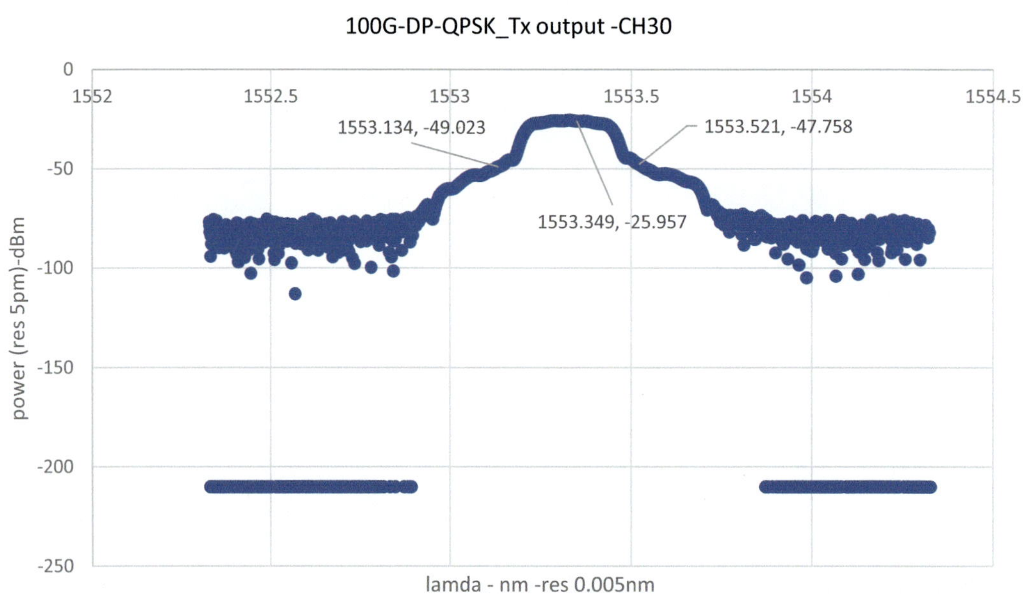

FIGURE 8.22 DP-QPSK 100 G output signal spectrum at Tx output.

8.5 THE 100 Gbps DP-QPSK INTEGRATED OPTICAL TRANSCEIVER MODULE

This sub-title provides an account of testing the module of an integrated coherent transceiver (ICTR) that offers 100 Gbps dual polarization channels, each operating with complex modulation QPSK (2 bits per symbol), and a baud rate of 32 GB, including the 25 GB plus overhead from soft-FEC.

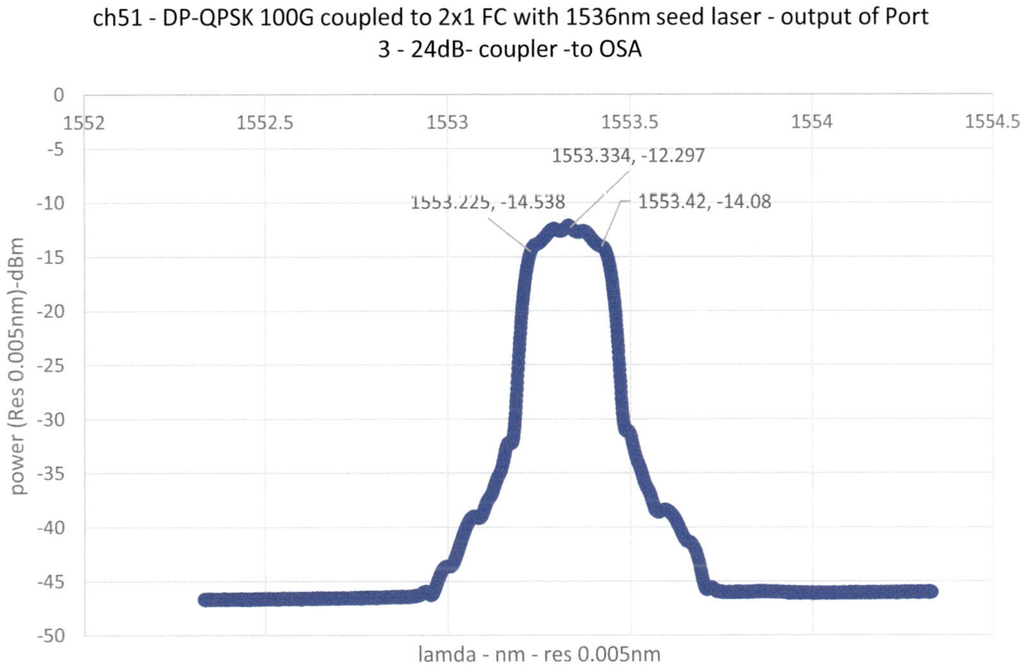

FIGURE 8.23 The 100 G DP-QPSK amplified through booster – via a 2×1 fiber coupler with seed laser at 1636.61 nm, booster output port P3 → then to 20 dB coupler → OSA.

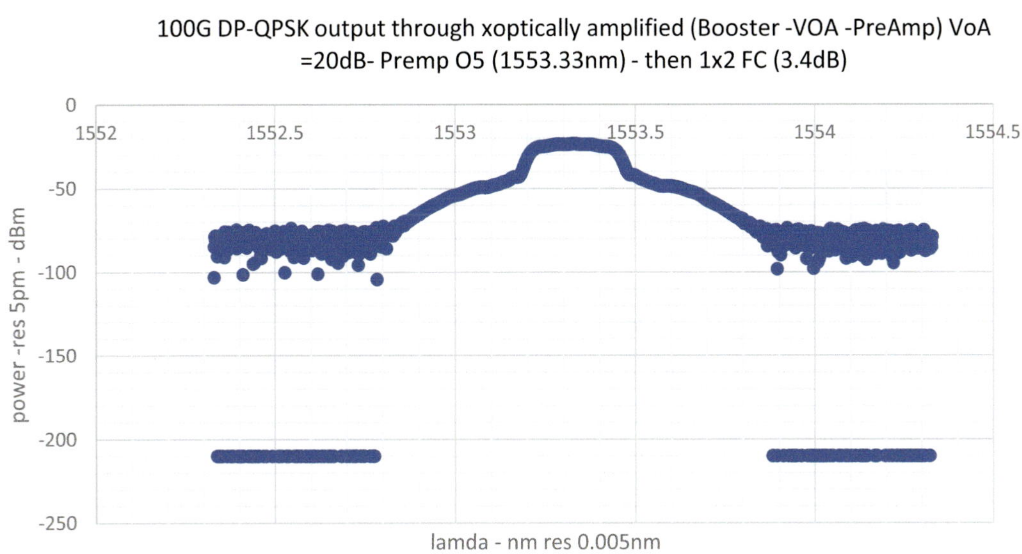

FIGURE 8.24 OSA spectrum of 100 G-DP-QPSK at 1553.33 nm monitored at P5 of preamplifier via 1×2 coupler, Tx output 100 G DP-QPSK to 2×1 FC, then fed to 10 dB attenuator (FA) then to booster of gain 24.34 dB, output high power to integrating sphere (IS) and PM monitored. The 24 dB output of FC is then coupled to a variable attenuator (VoA).

The lasers for both the carrier and the local oscillator for coherent beating have a narrow linewidth of 10 kHz, which is quite sufficient for coherent reception.

The receiver sensitivity of the module is obtained versus the BER (bit error rate), and hence the OSNR (optical signal-to-noise ratio). Two important tests related to operating the ICTR are the local

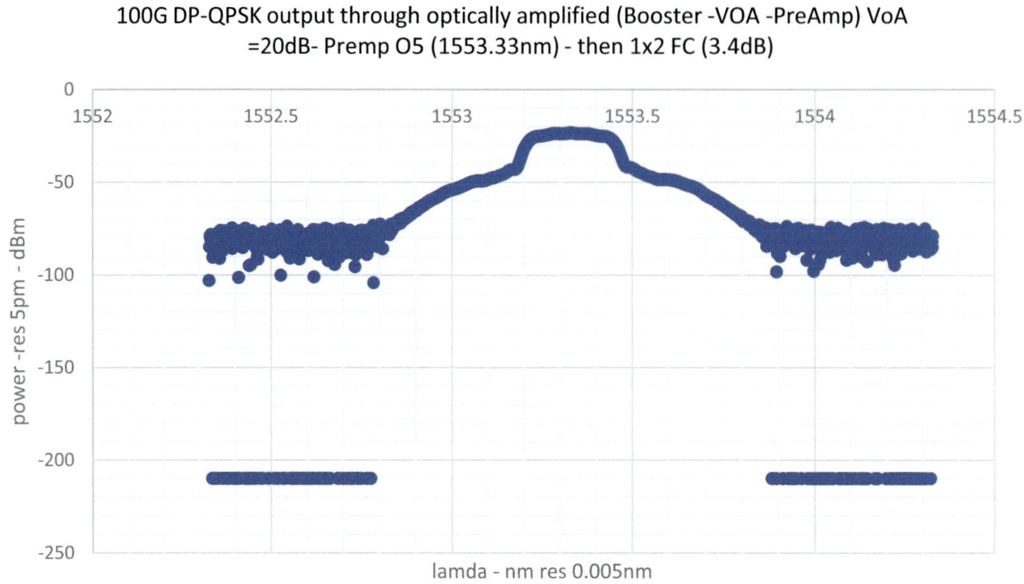

FIGURE 8.25 Spectrum of 100 G DP-QPSK output at P5 of optical preamplifier after passing through a 2×1 FC (3.4 dB), then 10 dB FA and then to VOA of 20 dB then through the preamplifier → 1×2 FC splitter, with one port to Rx of DP04QSFP Rx and another to OSA,as displayed in this figure.

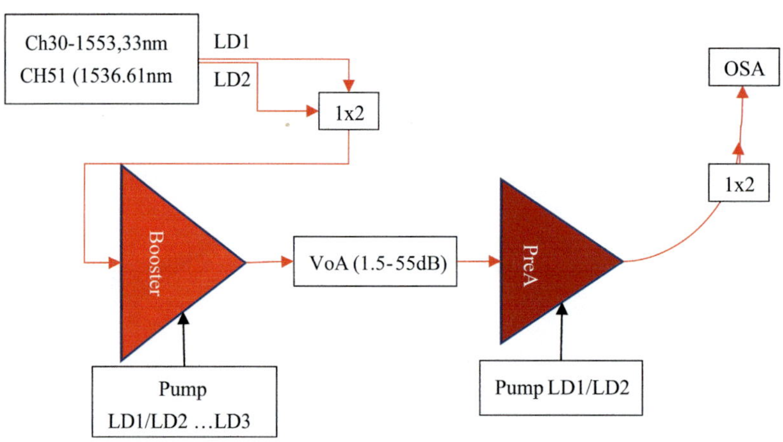

FIGURE 8.26 Setup to test the optically amplified free-space hop, channel No. 51.

oscillator frequency offset, which evaluates the module's sensitivity under frequency shifts due to the Doppler effect from different rotational speeds of the module and the transmitted satellite or the OGS (optical ground station). An LOFO (local oscillator frequency offset) of up to 4.5 GHz can be recovered by the ICTR module.

Radiation test of the module is also given to show the resilience of the module when operating in space transmission networks.

Furthermore, the optical routing modules and node architectures are provided to construct Tbps space photonic networking over satellite constellations.

The performance of the OTRx under no Doppler shift and with Doppler shift is given in Sections 8.5.3 and 8.5.4. More detailed performance, including nonlinear effects and crosstalk, is discussed in Chapter 10.

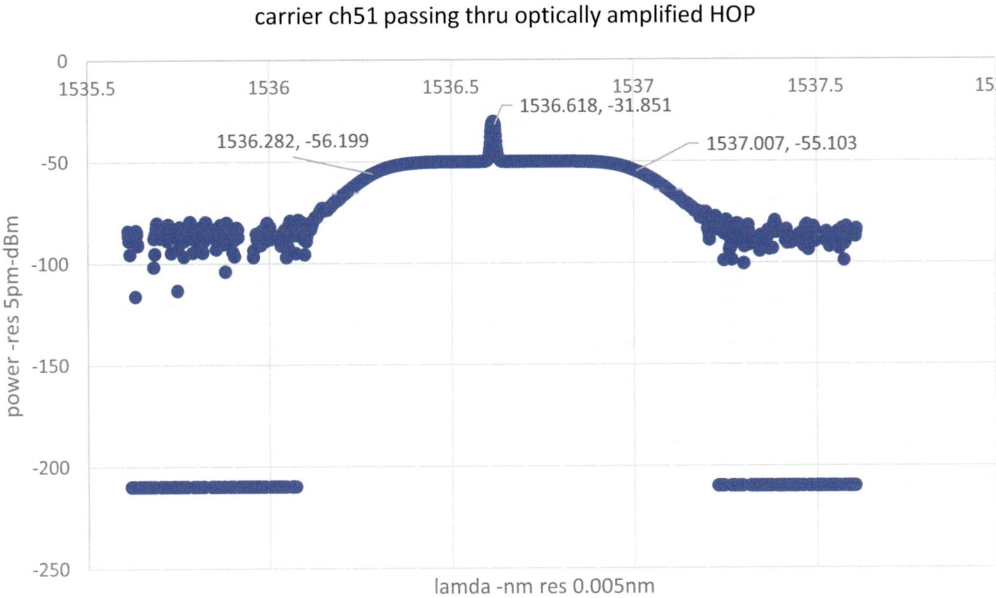

FIGURE 8.27 HOP1 and data_OptAmplified_HOP_CH51 channel 51 (at 1536.618 nm) through P5 of PreAmp – output filtered 100 GHz of preamp- CH30 1553.33 nm. VOA = 20 dB. The 3 dB badndwidth of of indicated at 1536.282 and 1537.007 nm.

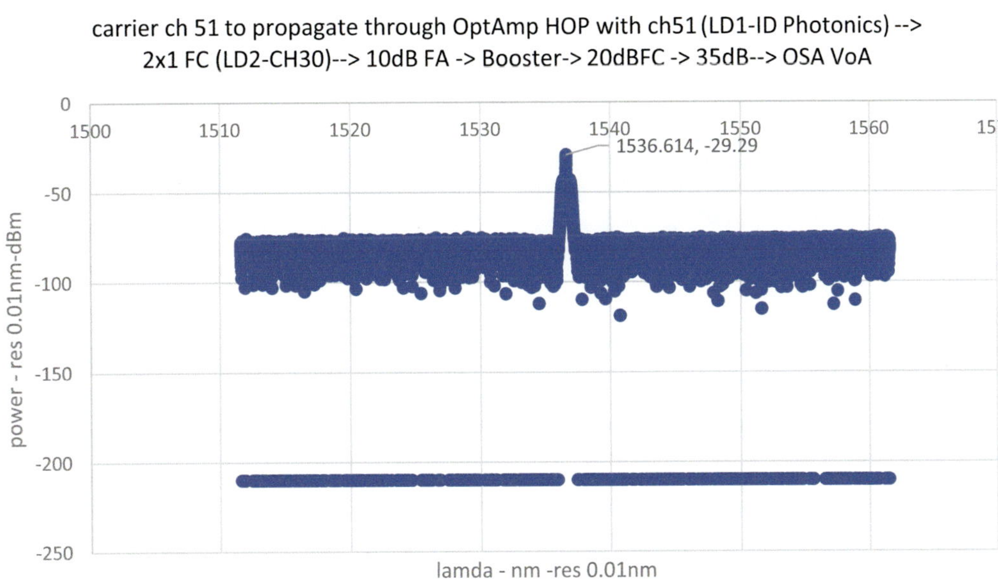

FIGURE 8.28 Output spectrum of single carrier (10 kHz linewidth-ID photonics dual lasers unit) propagating through 2×1 FC – 10 dB – booster, then 20 dB coupling port – VoA (30 + 5 dB)-OptPreA → P5 OSA.

8.5.1 Frequency Shift in SPCS

Figure 8.34 shows the Doppler frequency shift between the satellites of GEO and LEO at heights of 43,200 and 7000 km, respectively. The frequency shift can range from –7.5 to +7.5 GHz in 32 minutes. This frequency shift can cause coherent reception to go off track if no optical phase locking is

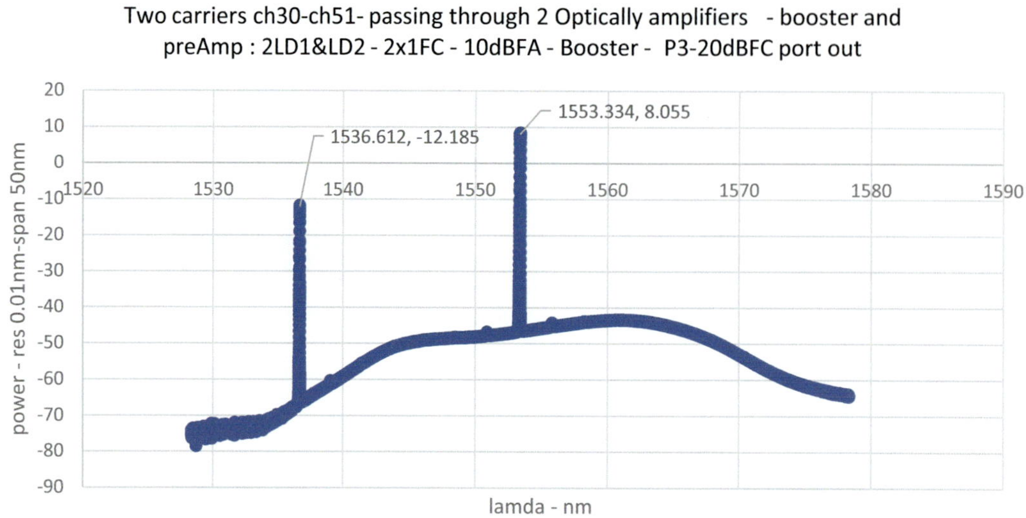

FIGURE 8.29 Spectrum covering two carriers CH30 and CH51 going through 2×1 FC – 10 dB FA, then passing through booster and OSA spectrum. Span = 50 nm – res. 0.01 nm.

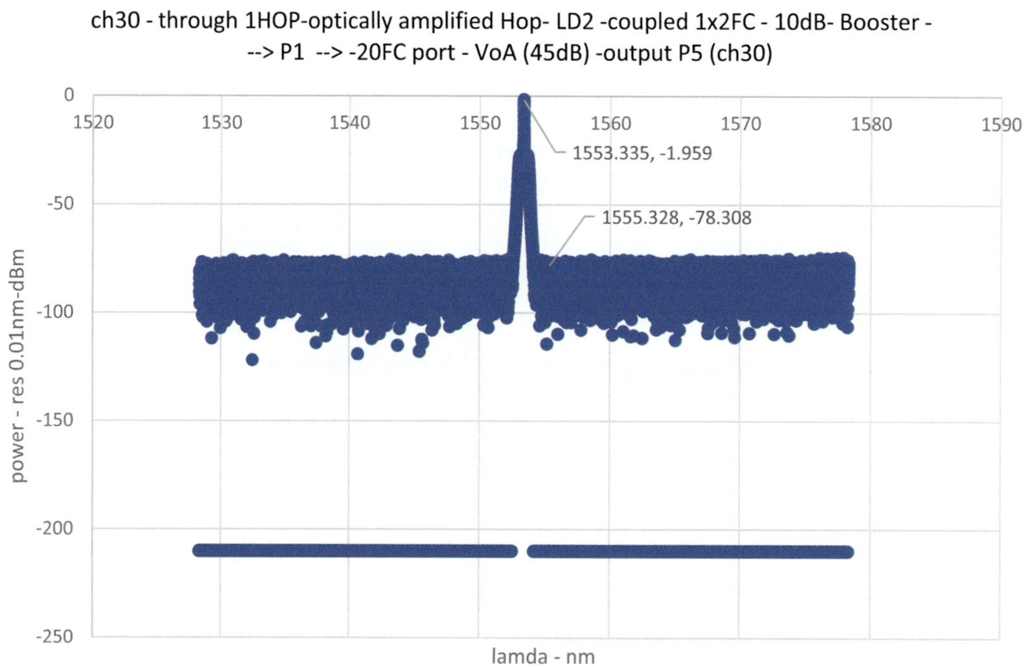

FIGURE 8.30 Channel 30 – best amplified channel in the 1550 nm C-band to be used in the Hop transmission. Spectrum after the HOP link via 2×1 FC – 10 dB – Booster – 20 dB FC port → VoA (45 dB) – OSA monitoring spectrum.

used to adjust the local oscillator to the correct frequency shift, so that a pure homodyne reception can be obtained. Thus, at least 3 dB power gain is achieved compared to the heterodyne case if no LOFO compensation or phase recovery algorithm by DSP.

Thus, the Doppler shift is highly critical for the receiver to operate properly and that the homodyne receiver can be obtained, that is 6 dB gain compared to the heterodyne detection. This

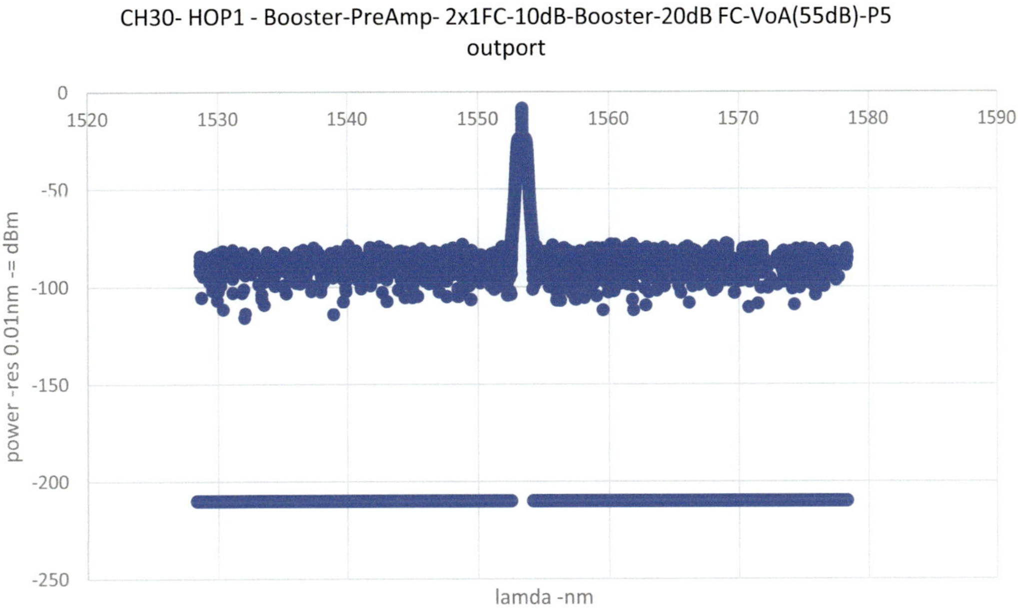

FIGURE 8.31 Channel 30 output spectrum after passing through HOP optically amplified, 2×1 FC (LD1 & LD2) → 10 dB → Booster – VOA (55 dB) P4 output to OSA.

subsection provides an account of how to overcome Doppler shift in PC involving satellites and Earth station between satellites in a number of constellations in which laser communications are installed. Section 8.5.4 presents the test performance of an integrated module of optical transceiver to evaluate its potential uses in SPC multi-hop transmission.

We ignore the Doppler effect on the polarized channels since we are using polarization multiplexing and demultiplexing. Details of the investigation of the Doppler effect on polarized channels can be found in Ref.[4]

Conventional frequency pre-shift technology cannot distinguish between the large Doppler shift and the frequency offset caused by crystal oscillator, which significantly hinders system performance improvement in high mobility environments when the frequency offset is present. Ref.[4] develops a joint frequency offset and Doppler shift estimation algorithm for high mobility environments based on orthogonal angle domain subspace projection. This algorithm can track the fast time-varying Doppler shift and separate the Doppler shift from the frequency offset. The orthogonal angle domain subspace projection algorithm projects the Doppler spread caused by high mobility into equivalent Doppler shifts in each orthogonal subspace. It then utilizes the equivalent Doppler shifts of all orthogonal subspaces to jointly estimate the maximum Doppler shift and frequency offset. Finally, these estimates are used to compensate for the frequency error in the received signal. Moreover, the maximum Doppler shift estimation, along with the projection vectors and the received power in each orthogonal subspace, is fed back and exploited in the subsequent transmitted signal. Reference simulation results show that the proposed method can estimate the two parameters accurately and eliminate almost all errors in the rapidly changing Doppler area. When SNR is up to 15 dB, compared with the conventional receiver with automatic frequency control, the receiver using our proposed method achieves better system performance even in the rapidly changing Doppler area. This is quite significant, and polarization locking of the optical clock should be implemented for inter-satellite communication.

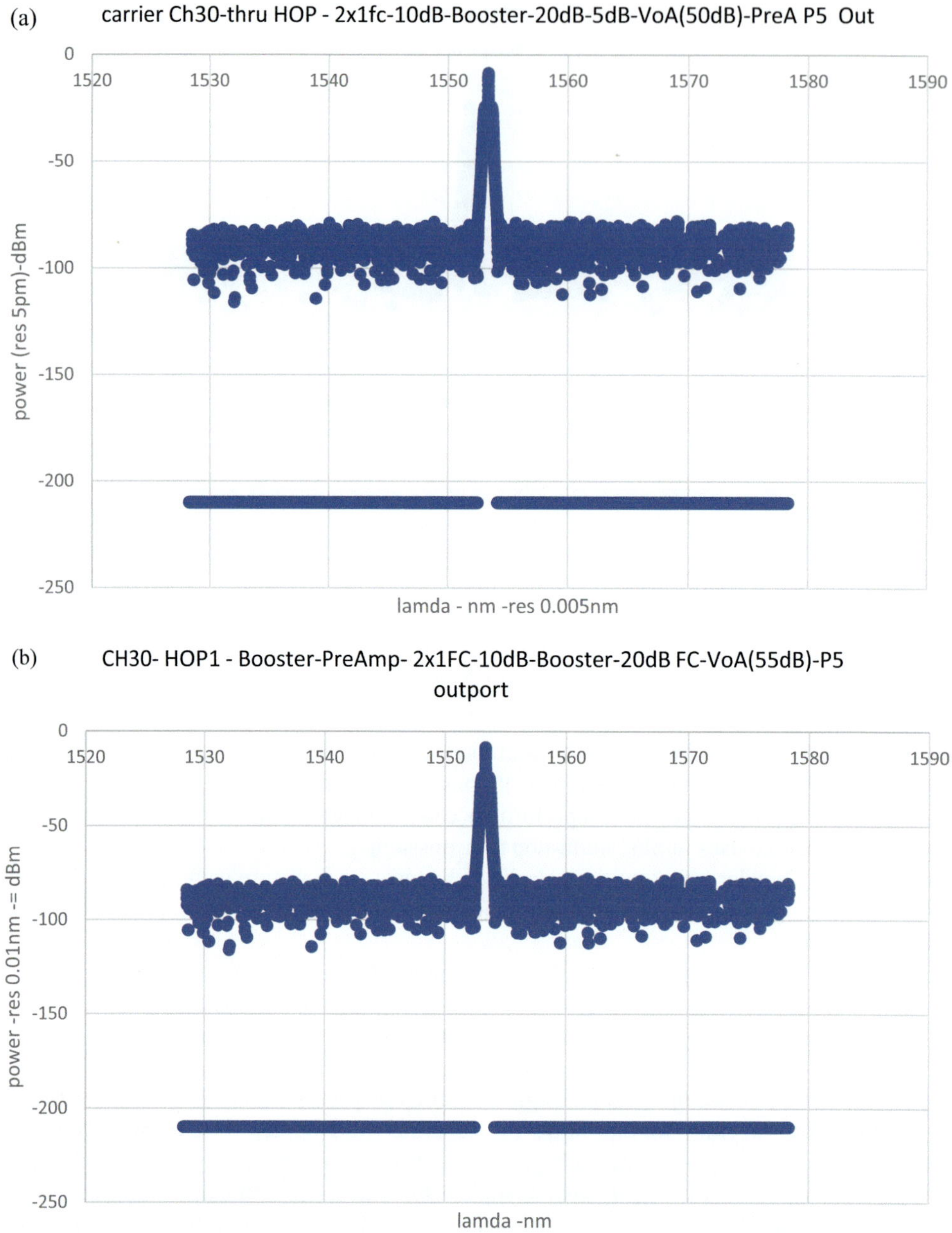

FIGURE 8.32 VOA = 55 dB (45 + 10 dB), (a) channel 30 at output of optical pre-amplifier LNA with attenuation factor of 50 dB and (b) 55 dB.

8.5.2 Doppler Shift for Receiver Node

8.5.2.1 Estimation

Question: If I know how fast a satellite is moving through space, how can I calculate the Doppler shift for the pass?

The Doppler shift Δf is a function of the relative velocity Δv (a scalar quantity) of the two moving objects and the emitted frequency f_0. If we know the relative velocity (either at one instant in time or as a function of time), it's just a matter of using the formula.

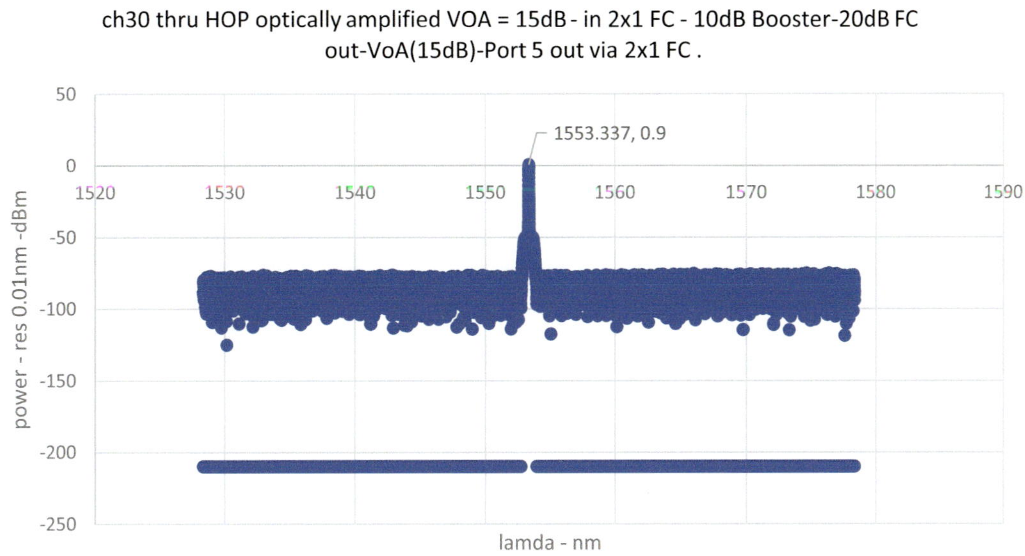

FIGURE 8.33 Spectrum of channel 30 at 1553.35 nm at output of LNA optical preamplifier with low-noise and high-sensitivity input power.

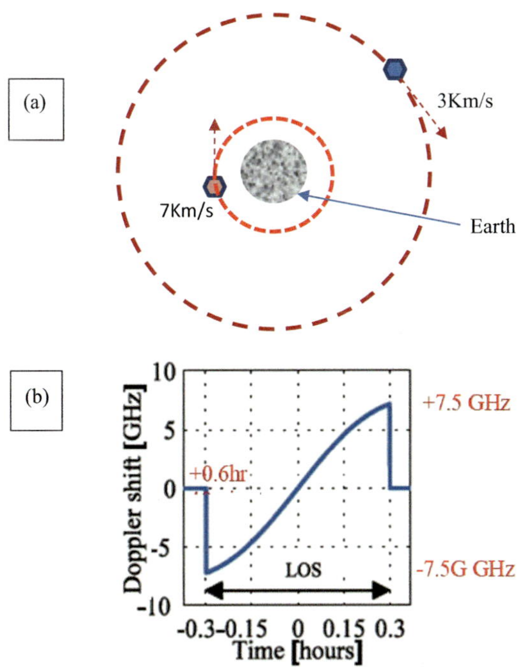

FIGURE 8.34 (a) Rotation of satellites in orbit GEO and LEO with speeds of doubling each other; (b) Doppler shift in GHz during the rotation relative between the satellites. Duration is 0.6 hours with frequency shift of 15 GHz when the LEO (at 7000 km above earth surface) and GEO satellites rotate at twice the earth rotational speed.

If we do not know the relative velocity, then we need to calculate it.

We need much more than the speed to find this. The orbital elements of a satellite, the object's location, and the time and date must be plugged into a satellite tracking program to perform the calculation. Doing it by hand can also be done, but it isn't trivial.

Note that usually we measure Δf with a radio, an SDR, and an accurate clock, compute Δv and then try to figure out the satellite's orbit by guessing which one fits best the observed Δv.

The velocity v is a vector quantity; the scalar quantity is the magnitude of v, which is the speed. Also note that the equation is commonly simplified to:

$$\Delta f = \Delta v c f_{\text{carrier}} = \Delta v \lambda = \|\Delta v\| \lambda \cos\theta = f_m \cos\theta \tag{8.1}$$

where θ is the angle of arrival at the transmitter, f_m is the maximum Doppler shift (commonly called the Doppler spread), and $\|v\|$ is the scalar speed at which the receiver is traveling relative to a stationary transmitter (or vice versa).

The equation shows that Δf can vary from $-f_m$ to $+f_m$ as θ varies, causing the received frequency to vary (this is called frequency dispersion) and contributing to fast fading. If the only thing you know is the speed at which the satellite is traveling, then all you can calculate is f_m, which represents the worst-case scenario for Δf. So, if your system is resistant to frequency shifts of up to f_m, then the system will operate well.

To obtain the Doppler shift, we need to know the actual emitted frequency and the range rate (the first derivative of range, which is the distance between you and the satellite).

To compute the range rate, we need two 3D vector quantities: the satellite's position P and velocity v relative to the stationary or moving position. The range rate is then given by the dot product of P and V divided by the magnitude of P. The vector P divided by its magnitude is the unit position vector, so what you're really computing is the component of the spacecraft's velocity along the direction to you; how fast it moves perpendicular to that direction doesn't really matter.

Although most satellite programs seem to compute range rate by finite differencing range, it can be computed analytically for more precise results. Standard orbit models like SGP produce velocity as well as position as functions of time, and it is also fairly straightforward to compute the position and velocity of a ground station on the rotating Earth in inertial space.

The Doppler shift causes coherent detection to become heterodyne. We investigate these effects on the optical transceiver sensitivity and bit error rate of the 100 Gbps DP-QPSK transceiver.

Note: This discussion ignores relativistic effects. At relativistic speeds, there will be an additional red shift (downward Doppler shift) due to time dilation making the spacecraft's oscillator appear to run more slowly. For details, see Appendix Section 8.13 of this chapter.

8.5.2.2 Doppler Shift and Reception Techniques

Doppler spectral shift is critical in coherent optical communication transceivers because it creates a heterodyne detection instead of homodyne reception. Doppler shift results from the relative velocity between the satellite and the IGS or the speed difference with respect to other satellites. Thus, when mixing the carrier and the LO laser, the signal baseband is shifted by Doppler shift. The Doppler frequency shift can seriously affect the mode of coherent reception, including homodyne or intradyne and even homodyne reception.[5] If the bandwidth of the receiver is not wide enough to accommodate this shift, then the signal cannot be detected or will suffer distortion.

As mentioned in previous chapters, optical phase lock loop (OPLL) may offer some advantages for manufacturing universal satellite optical receivers that can operate in any orbital constellation, including MEO-LEO, OGS-MEO/LEO/GEO, and MEO-GEO.[6] This is especially relevant in unmanned space vehicles traveling to and from other orbital satellites.

8.5.2.3 Optical Phase Lock Loop to Mitigate Doppler Shift

An analog-based OPLL is shown in Figure 8.35. The locking process is as follows: (i) The received digital signal of PDM-QPSK channel is received and extracted from its polarized states and the I- or Q-components. Assuming that the received signals are in the optical domain, that is, they are either at pre- or post-stage of the optical hybrid coupler. (ii) This optical signal is then detected by a photodetector or a pair of balanced detectors, if available. An integrator or RC circuit is used to extract

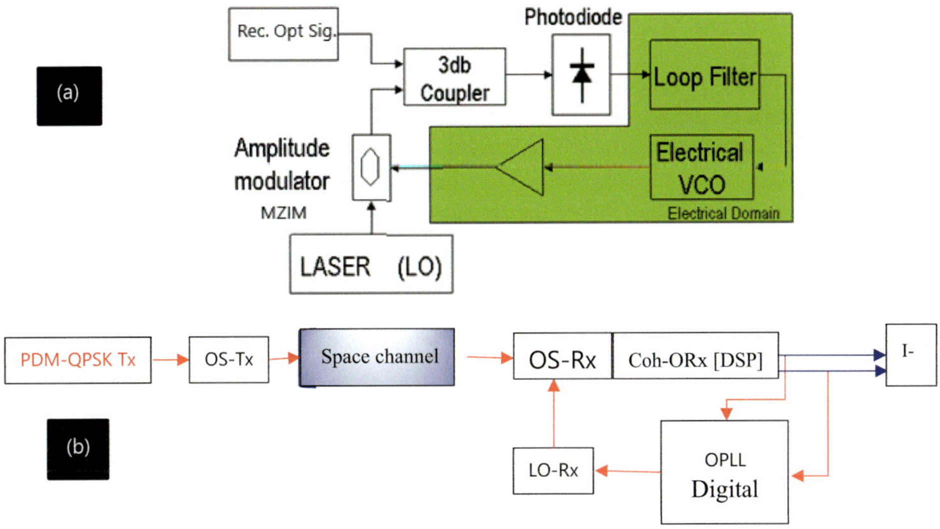

FIGURE 8.35 (a) Schematic of an optical phase locking loop (OPLL) in analog electronic hardware form; (b) OPLL integrated in a PDM-QPSK DSP-based transceiver reception subsystem, in digital processing form by using FFT and LOFO phase recovery algorithm and feeding to LO laser of the transceiver (also in digital form). LO=local oscillator, MZIM=Mach-Zehnder interferometric modulator, VCO=voltage control oscillator, rec. opt. sig=received optical signals.

the level of the intensity of the optical signal. This is then fed into an electrical voltage control oscillator, whose output is used to drive an optical modulator, such as a MZI, with a bandwidth that should be higher than the maximum Doppler shift. This is then used as the optical local oscillator, feeding back to the hybrid coupler of the transceiver.

Similarly, in the case of digital signal processing, the received signals are in digital form after passing through the optical hybrid coupler and the digital conversion processes. An inverse Fast Fourier Transform is performed to extract the Doppler shift frequency via the phase recovery algorithm as described in Ref.[7] This Doppler shift is then extracted via a two-tap finite impulse response (FIR) filter in the DSP-based TRx (transceiver). The shift is amplified and fed into an external or integrated optical modulator in the ORx section, whose input is the Rx LO laser. This effectively tunes or shifts the LO frequency to match the Doppler-shifted optical carrier of the received channel. Alternatively, once the Doppler shift is detected after the FIR filtering, the LO can be tuned to the required optical frequency via the use of a look-up table for tuning the laser LO frequency. This is the simplest way to lock the two laser sources.

In summary, an OPLL in an ORX-DSP sub-system can operate by: (i) digitally detecting and filtering to extract the Doppler frequency shift; and (ii) using this information and a look-up table to tune the LO laser, due to the LO frequency to match the channel carrier frequency exactly. This means that for this real-time Doppler shift tuning of the LO laser, an extra step of digital signal processing is required, along with a look-up table for frequency tuning of the LO laser.

The main difficulty with this OPLL processing is that users of Acacia DSP-based transceivers cannot access the embedded software algorithms for data processing.

8.5.2.4 Pros and Cons of OPLL in Optical Coherent Transceiver

The following are the pros and cons of OPLL:

Doppler shift estimation with LEO to GEO, whose rotation speeds are 7 and 3 km/s, respectively, and the position of the LEO is perpendicular to the GEO direction relative to the

Earth center. This position gives the maximum angle of the LOS. This angle changes accordingly due to the position of the LEO satellite. The Doppler varies between left shift and right shift, with a total of 2×7 GHz. If OPLL is required, then the VCO should be automatically tuned to the correct frequency, according to the voltage detection level and fed to the tunable LO laser.

Further note that the Doppler shift will be known, as it is determined by the orbital dynamics of satellites, so it can be pre-determined for the set-point. This may not require OPLL but rather a look-up table in the DSP of the transceiver, tuning accordingly depending on the orbits around which the satellites are moving.

In the case of a Doppler shift of 7 GHz and no locking to tune the LO laser, the DSP-ORx can be out of the DSP intradyne detection range, which is specified at a maximum of 5 GHz (Acacia 100/200 G module). If there is no locking between the carrier frequency and that of the LO laser with LOFO still within the intradyne coherent reception range (<5 GHz for Acacia DP-QPSK TRx), then the receiver sensitivity suffers a loss about 3–6 dB compared to homodyne reception using OPLL.[8] The Doppler shift varies depending on the location of the LEO satellite with respect to a GEO satellite. So, one can set the LO frequency accordingly, either as a fixed value at the manufacturing stage or by using a look-up table during operation. However, if we want to be exact in homodyne to gain extra 3 dB, then OPLL should be employed. This offers real-time locking.

OPLL can be difficult to operate in orbital SPC and would not come "out of the box" with the current transceivers unless it is integrated into the DSP-ORx module.

Doppler pre-compensation can be the way, by pre-shifting the optical carrier. However, the Doppler is not fixed but changes according to the time of day, especially in the case of satellite SPC from LEO to MEO or GEO-LEO, where the relative velocity depends on the rotational speed and the tangential component. However, this is to be tuned to every module installed in the satellite's optical port. While OPLL can be integrated in the optical coherent transceiver so that the frequency of the local oscillator can be matched to that of the carrier.

Orbits are well-defined and known, so we can measure and track the Doppler or relative velocities in the classical channel and feed this back for lambda Doppler compensation. This is done particularly for known orbit satellites. However, this section gives a comparison of ORx with OPLL and without OPLL as a variation of technological options for optical transceivers in SPCS.

Recommendations for using OPLL in SPC systems:

The optimum solution is that if OPLL is used, then such a receiver can be versatile for any orbital satellite, especially in unmanned space vehicles. The LO is tuned accordingly via the OPLL. More complex but simpler in manufacturing, however, is tuning the laser. Alternatively, a look-up table can be loaded into the DSP receivers for different satellite orbits. In the case of an unmanned vehicles in space, Doppler frequency shift , due to different rotational velocities of satellites, may not be known, and such optical phase locking would be necessary for homodyne coherent detection .

Intradyne only requires 'close enough' wavelength/frequency tracking/control. As the name implies, an optical phase-locked loop also requires phase locking in addition to frequency locking. This is non-trivial and , to my understanding, would require a completely new transceiver. However, an advanced DSP-based transceiver can be used to extract Doppler shift via an FFT operation on the received data sequence, which is then fed to the voltage-controlled oscillator (integrated in the analog section) of the processing integrated circuit.

So, depending on the application and manufacturing cases of the optical transceiver for specific satellites or universal satellites, OPLL may or may not be employed to improve the receiver sensitivity.

8.5.3 TRx Back-to-Back Set-Up 100 Gbps PDM-QPSK Integrated Module Transmission

Case scenario: CONSTELLATION – LOFO TEST of Acacia 100 G 28 GSy/s and BER with various parameters recorded.

Test OTRx Module: ACACIA 100 G: TRx module No.: 075-0199-002_Rev_0.1_DP04CFP2-Mxx_ProductSpecification by CISCO Acacia Inc

The integration of the coherent OTRx module in a satellite optical node, including both the transmitter and receiver modules and routers is shown in Figure 8.36. As a transceiver, both the transmitter and receiver are in one module.

The DSP block can be made up of two separate DSP ASIC (application-specific integrated circuit) chips, one dedicated to the Tx side and one for the Rx DSP. The input into the DSP is in digital form from the ADC (analog-to-digital converter), which accepts channel complex signals of the in-phase and quadrature-phase components of the X- and Y-polarized channels which are the outputs of the transimpedance amplifiers. The transimpedance amplifiers accept the electronic signals coming out of the balanced photodetector pair (BPD), which are identical PDS and connected back-to-back. The optical signals are derived from the channel carriers and modulated signal band mixing with the local oscillator laser, also integrated on the same OTRx chip.

The received optical transmitted signals are fed into a hybrid π/2 dual polarization optical coupler. The *I* and *Q* components are separated by the π/2 optical phase shifter through pyroelectric effects. These orthogonal optical components are then coupled with their polarization counterparts from the local oscillator laser. These optical waves are mixed *in* the PD. They beat each other coherently to give the product of their amplitudes and the co-sinusoidal amplitude variation. This means that the optical fields of the local oscillator and that of the channel signal are added vectorially. The total optical field is squared as detected by the photodetector. Such squaring results into two square terms (DC components) of the optical waves amplitude and a product of these two oscillation frequencies that results in a sum and subtraction of the frequency terms. The sum of the frequencies is very high and located far from the response of the detector. Thus, only the subtraction term of the two frequencies cn be detected by the detector to produce detected currents and amplified voltage signals. These electronic signals are then sampled and digitized for input to the DSP for processing using embedded algorithms.

The test set-up is shown in Figure 8.37. A signal test generator is required to provide an optical data sequence input to the OTRx (or ICOTRx[9]). Such an optical sequence can be internally generated. The programmable DSP system of the OTRx section can modulate the lightwaves fed into the integrated IQ modulator of two polarizations. The four DACs provide the analog signals equivalent to the digital codes, modulating the MZM of the IQ Modulator. The optical polarized channels are first split into their polarized optical waveguides, which guide them into the integrated optical hybrid coupler. The *I* and *Q* components are separated with the π/2 phase shift and then mixed with the local oscillators. These two parallelly mixed lightwaves beat in the PDs to generate coherent electronic signals for further digitalization and processing in the DSP, including the FEC. Thus, there is an overhead sequence of bits to maximize the BER versus the optical signal-to-noise ratio.

In SPC, the test is required to evaluate the OSNR required under high attenuation factors of the space link and the Doppler frequency shift, which can reach up to 7.5 GHz, is quite substantial. This section and the next report on the performance of the integrated OTRx. All fibers connecting photonic components are SSMF (standard single-mode fiber). No optical amplifiers are used in this

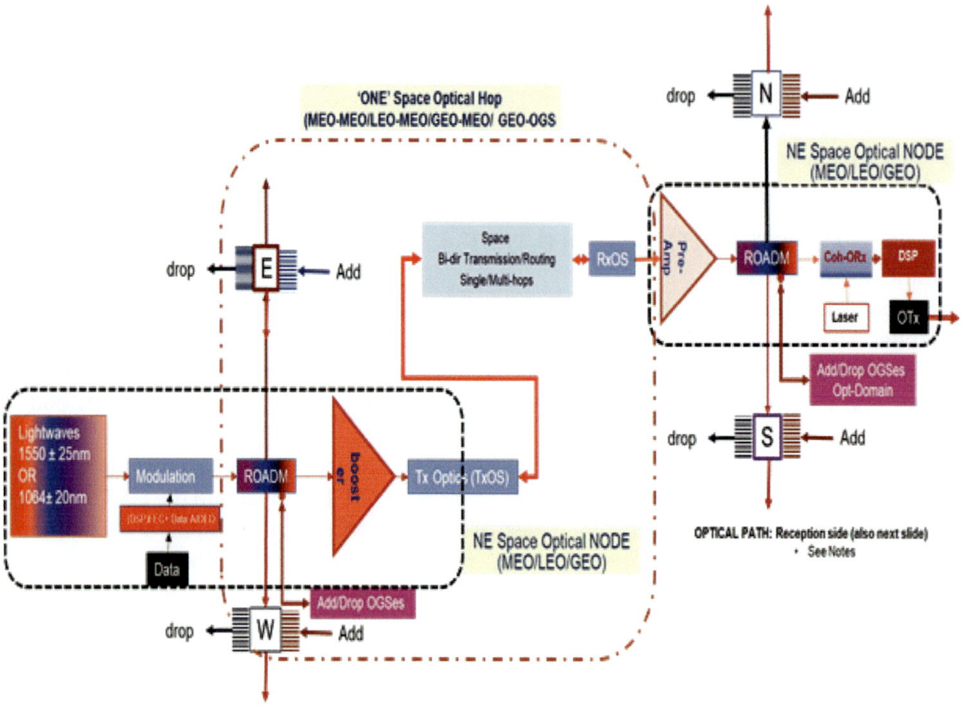

(a)

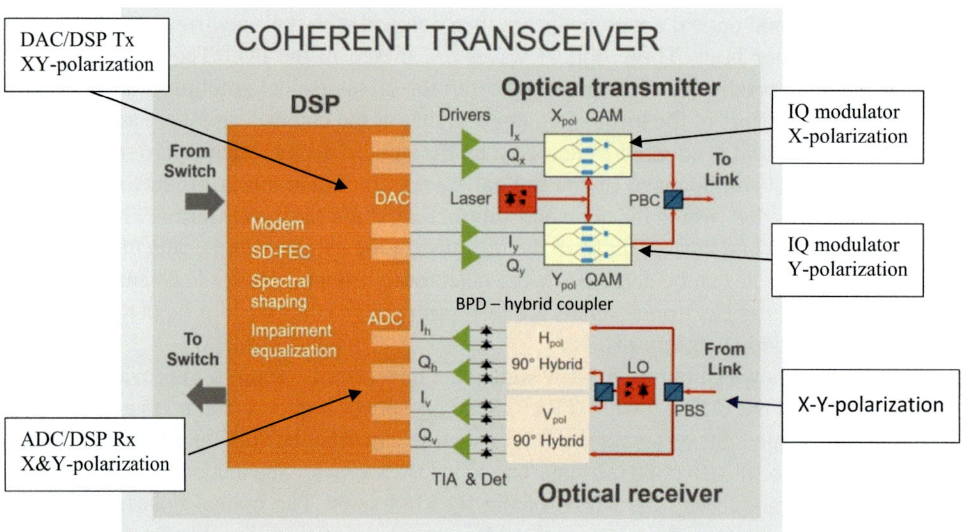

(b)

FIGURE 8.36 (a) Generic schematic of multi-hop SOCS (Space Optical Communication System) in which OTRx are integrated into both the receiving end, i.e., optical lightwaves entry and at the output transmitting to launch into next SPC link. (b) The 100/200 G coherent transceiver with integrated DSP, including TIA and balanced photodetector pair. Legends: RxOS=optic system in front of optical reception subsystem, ROADM=reconfigurable optical add/drop multiplexer, IQ modulator using SOI and ABC. Independent lasers for Tx and LO ORx. SD-FEC, spectral shaping and impairment equalization. ICR=integrated coherent receiver (and transmitter) with four lanes (IQ (X) and IQ (Y)).

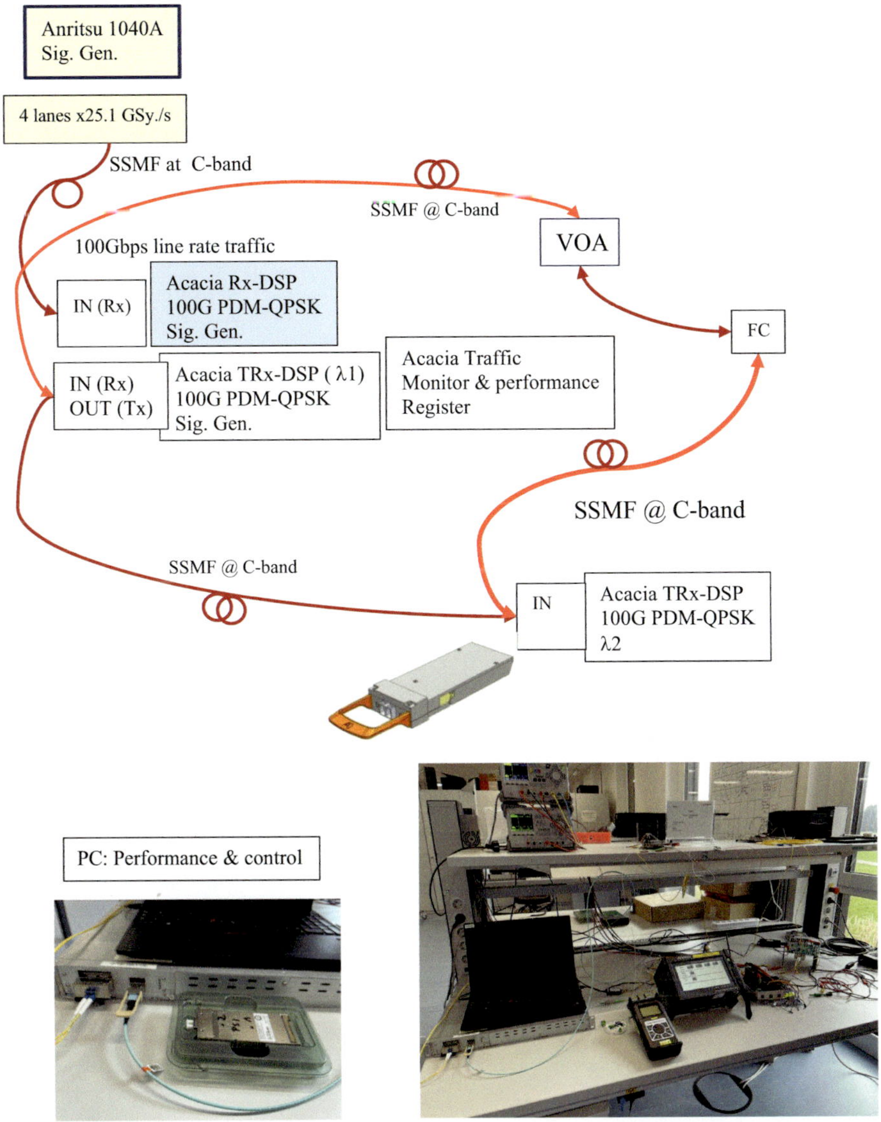

FIGURE 8.37 Schematic and photo of the 100 G optical transceivers (Acacia 100 G PDM-QPS module) with two wavelength and bidirectional transmission in a B2B (back-to-back) incorporating a VOA (variable optical attenuator). SSMF=standard single mode fiber or corning SMF-28. Insert is the QFSP packet size of OTRx module.

test. Later sections report on the link with optical amplifiers and low-noise amplifiers with high attenuation factor VOA.

The test of receiver sensitivity (in dBm) plotted against the $\log_{10}$ (BER) is shown in Figures 8.38 and 8.39, with and without using OPLL or phase recovery to determine the clocking under homodyne or intradyne reception. The aggregate bit rate is 112 Gbps using two polarized channels, 28 GBaud, and QPSK format, or 2 bits/symbol. At −42 dBm, the equivalent sensitivity is 5 photons per bit.[10]

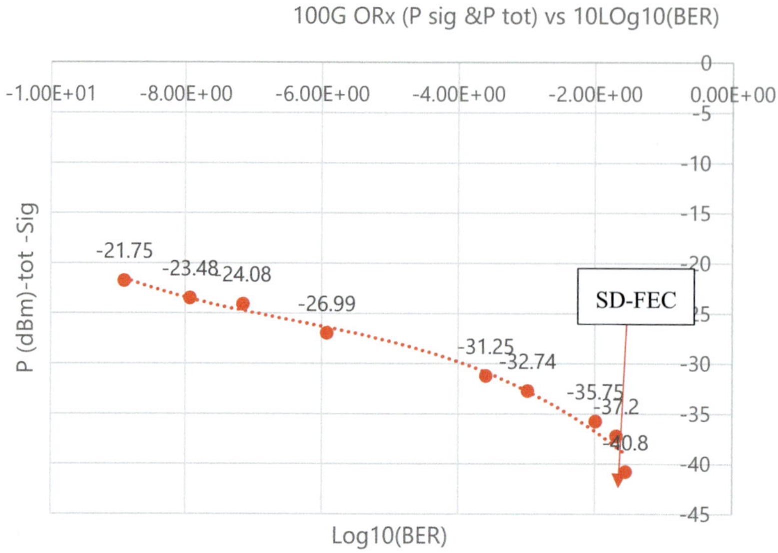

FIGURE 8.38 Plots of BER versus signal average power (dBm) without any optical amplifier as a pre-amp coherent optical reception subsystem. Note: SD-FEC 2e-1 achieved at –42 dBm for 100 Gbps (PDM-QPSK @25 GSy/s) or about 5 photons/bits as expected from SoA & CA WP1100 as reported. P_{tot}=total optical received power as registered by TRx including noise, P_{sig}=extracted signal power. Note: No optical pre-amplifier is used. The 10 $\log_{10}$(BER)=–2 at –37.2 dBm.

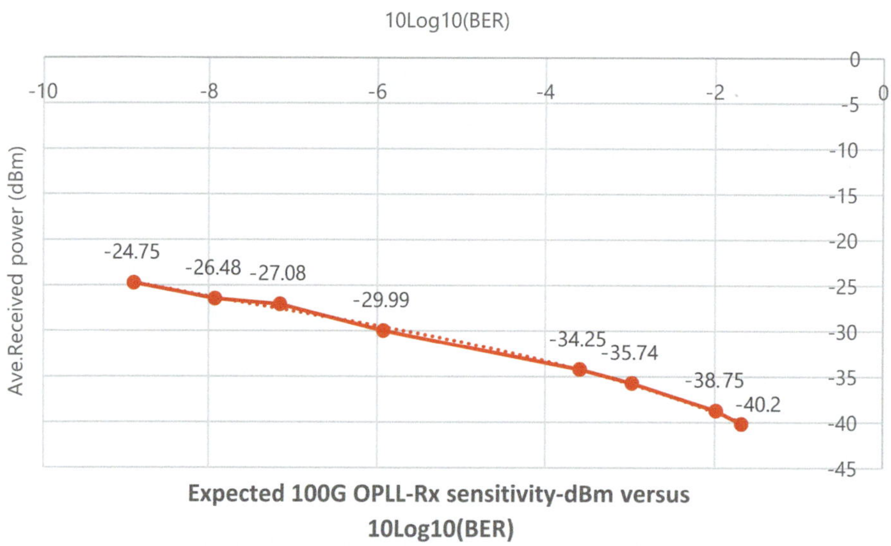

FIGURE 8.39 Plots of BER versus signal average power (dBm) without any optical amplifier as a pre-amp optical reception sub-system, but incorporating OPLL in DSP processing. Note: Expected BER based on optical transceiver 100 G of Acacia PDM-QPSK transceiver 25.2 GSy/s 100 G PDM-QPSK transceiver. The 10 $\log_{10}$(BER)=–2 dB at –38.75 dBm receiver sensitivity.

8.5.4 Performance of 100 Gbps DP-QPSK Under Doppler Shift LOFO at High Receiver Sensitivity

Settings: Rx sensitivity=–26 dBm; Tx power –5.1 dBm; LOFO=1000 to 4.5K MHz; 6000 MHz (NaA) (Figures 8.40–8.51, Tables 8.2 and 8.3).

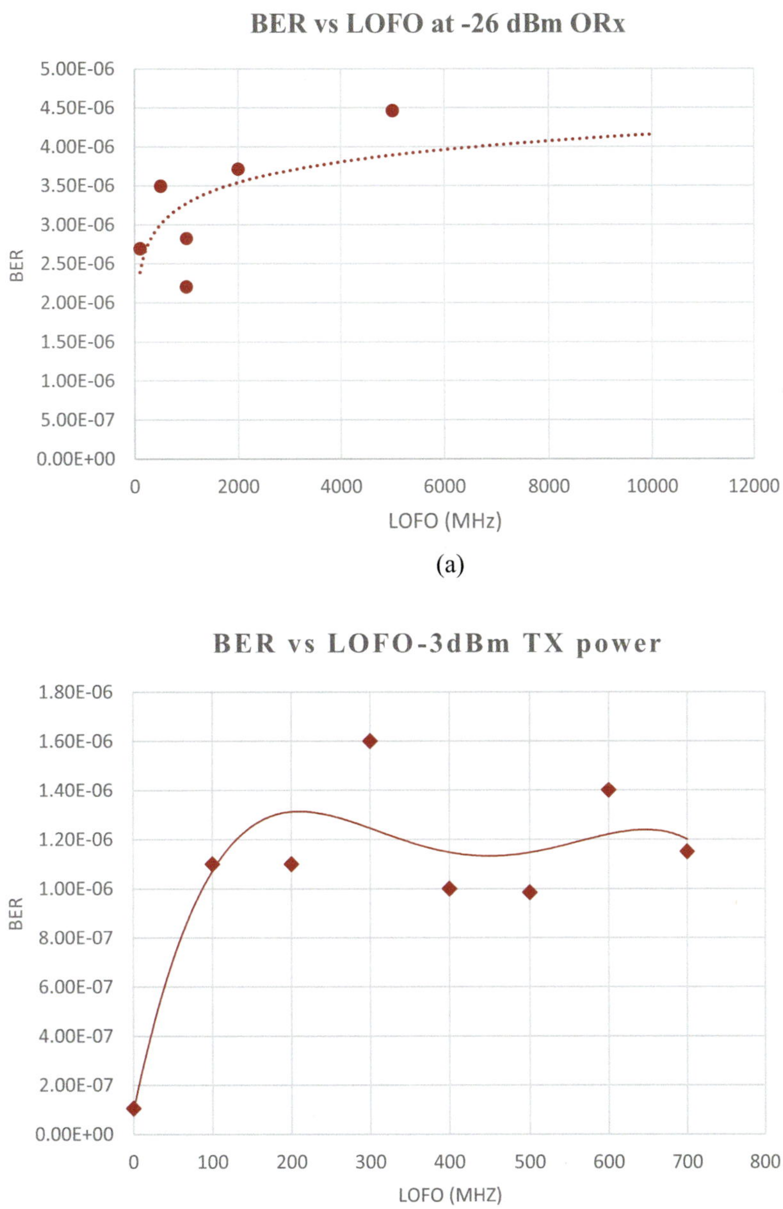

(a)

(b)

FIGURE 8.40 BER vs LOFO for –26 dBm total P ORx and (a) Tx P=5 dBm. (b) –3 dBm.

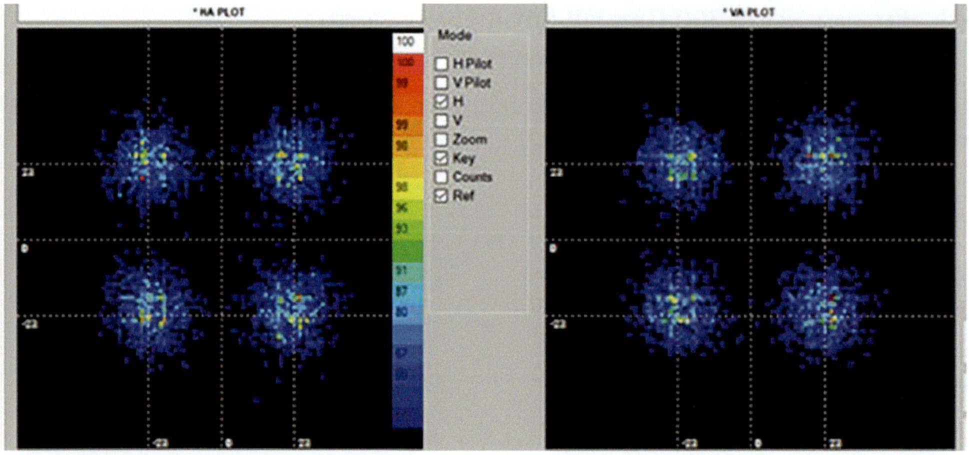

FIGURE 8.41 QPSK eye constellations of module Acacia 100 G, QPSK constellation under Doppler shift LOFO = 1000 GHz frequency offset, Tx Power = –3.3 dBm, BER = 1.6e-6, PRx = –26 dBm, 28 GSy/s. Use constellation.

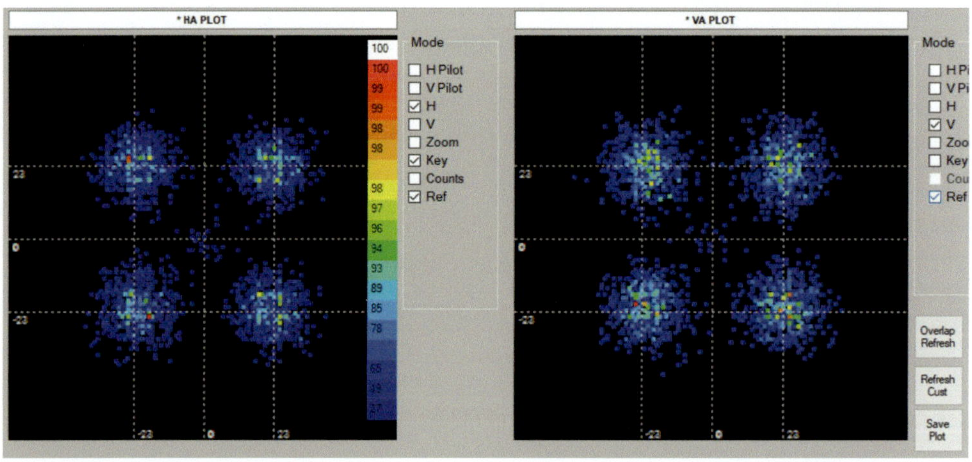

FIGURE 8.42 QPSK constellation under LOFO = 5000 MHz, other conditions are same as operating given in Figure 8.41.

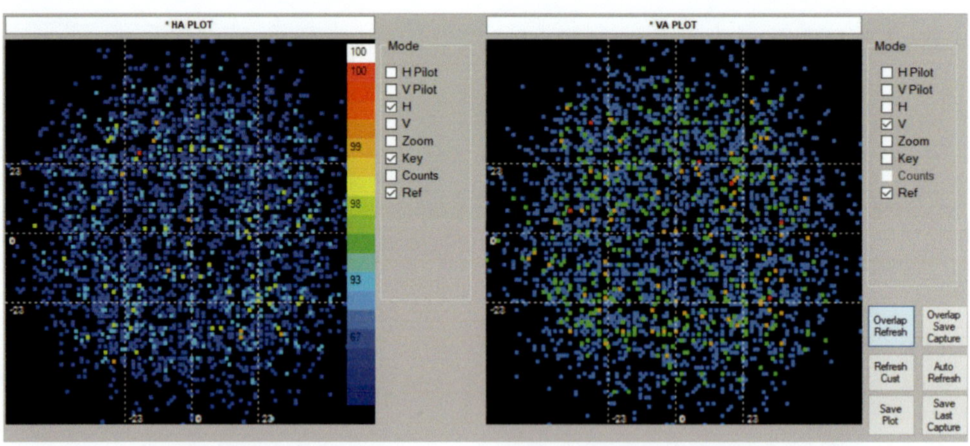

FIGURE 8.43 LOFO = 7000 MHz, same condition as Figure 8.41. 'NaA' = impossible for SD-FEC.

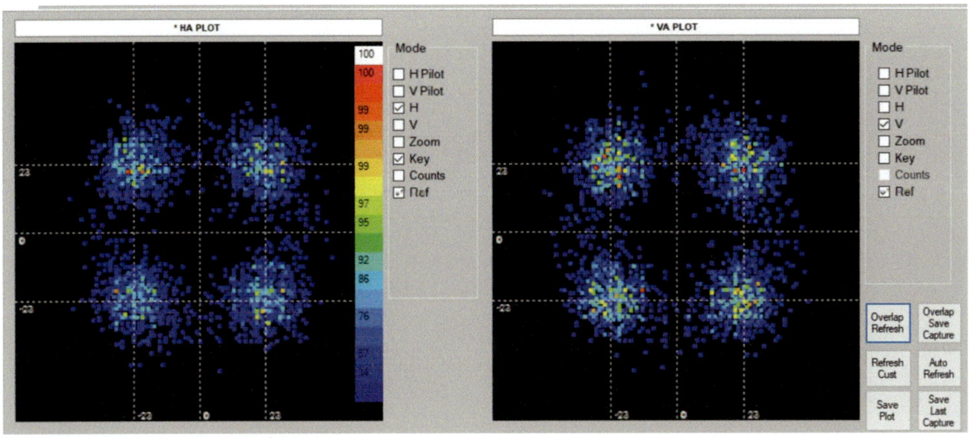

FIGURE 8.44 LOFO=5500 MHz, same condition as Figure 8.41. BER=not NaA. Note the rotation of constellation indicates the unmatching phase of the constellation and its original signal carrier phase.

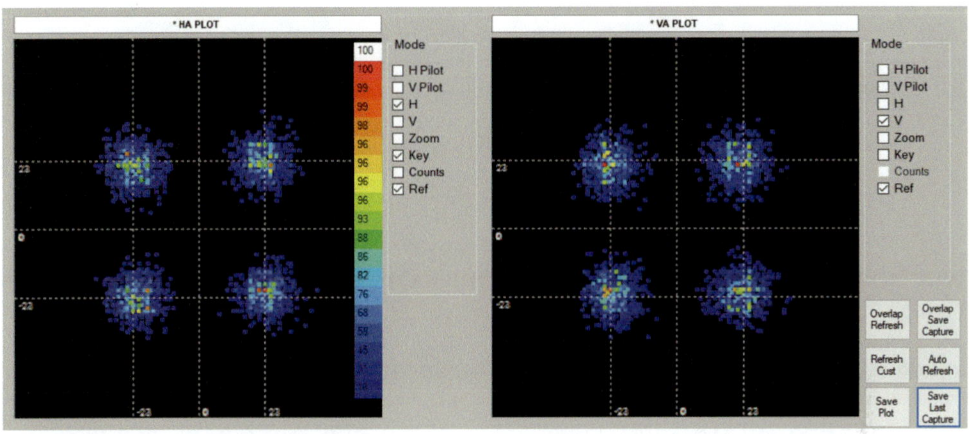

FIGURE 8.45 LOFO=2000 MHz, same condition as Figure 8.41. BER=not NaA.

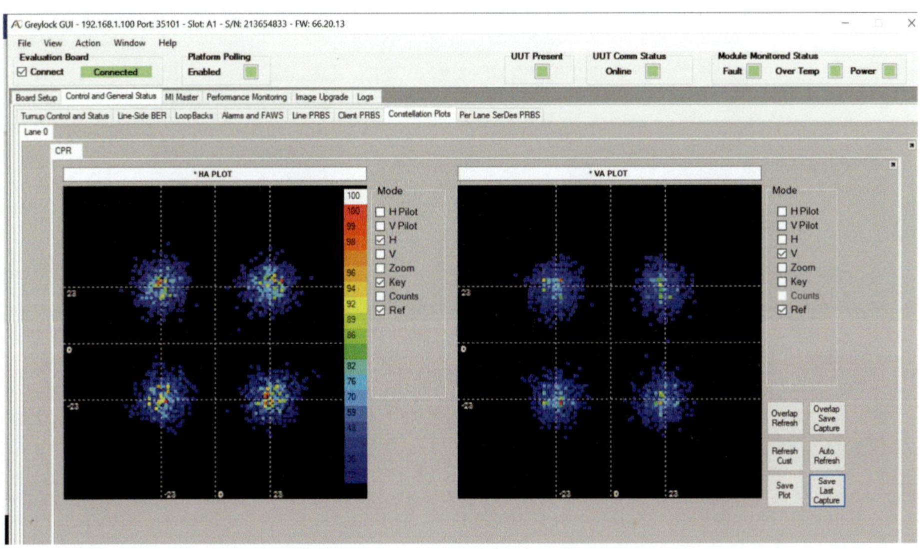

FIGURE 8.46 LOFO=4000 MHz, same condition as Figure 8.41. BER=not NaA at LOFO 7000 MHz.

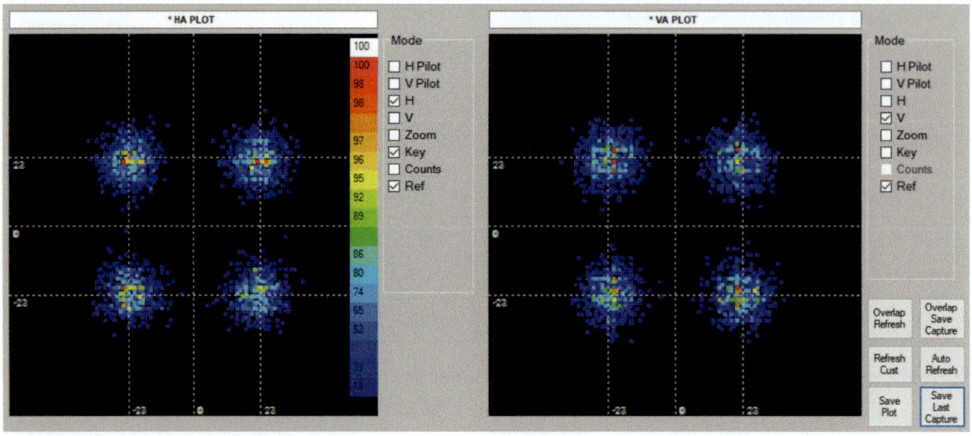

FIGURE 8.47 LOFO = 4500 MHz, same condition as Figure 8.41. BER = 3.8e-6.

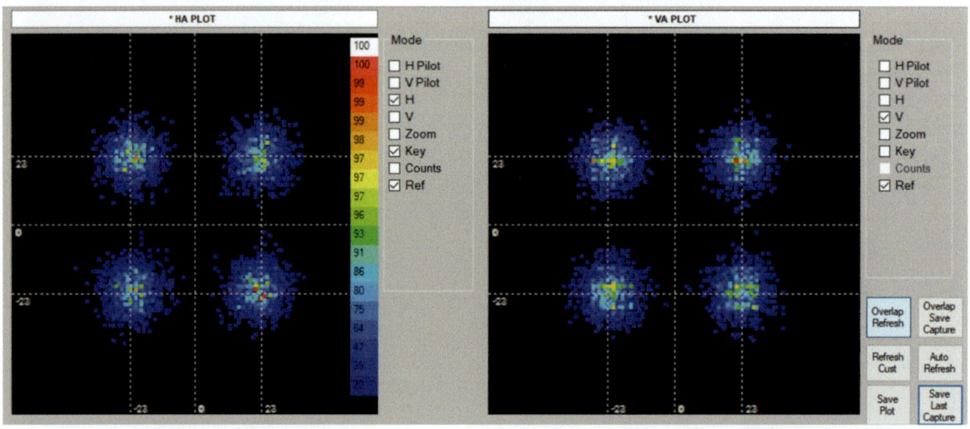

FIGURE 8.48 LOFO = 5500 MHz, same condition as Figure 8.41. BER = 3.9e-6. Rx P = –26.9 dBm.

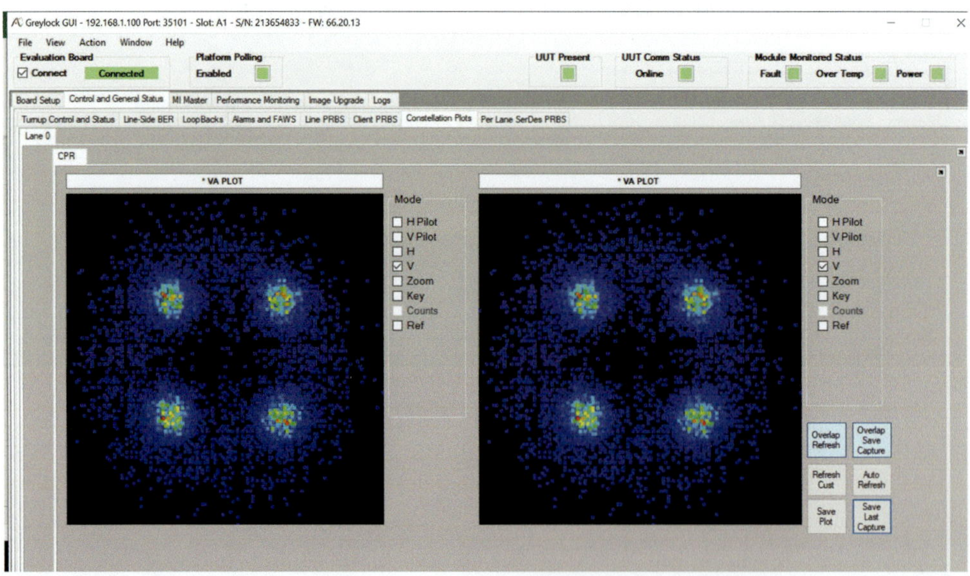

FIGURE 8.49 Constellation at LOFO = 4500 MHz; BER = 2.42e-2, Rx signal power = –25.8 dBm; OSNR = 32.3 dB. Note on phase rotation of constellation.

8.5.4.1 Performance LOFO vs BER at Minimum and Maximum Receiver Sensitivity

setting	Tx -5 dBm	Opt Freq (MHz)		
LOFO (MHz)	BER	194,505	Rx SigP -40dBm	Total P Rx (dBm)
LOFO		LOFO set		
5000	2,42E-02	5000		-38,12
4500	2.43e-2	4500		-38,17
4000	2.44e-2	4000		-38,2
3500	2.44e-2	3500		-38,1
3000	2.42e-2	3000		-38,2
2500	2.44e-2	2500		-38,28
2000	2.43e-2	2000		-38,07
1500	2.42e-2	1500		-38,3
1000	2.43e-2	1000		-38,14
0	2.41e-2	0		-38,04

(a)

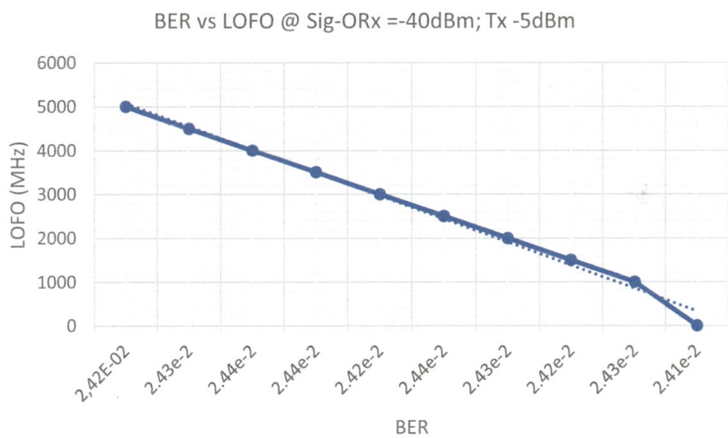

(b)

FIGURE 8.50 (a) Data table, (b) plots of LOFO frequency (in MHz) against BER at minimum receiver power (signal). (b) LOFO=2500 MHz but 2500+180 MHz recorded. Approximately 180 MHz offset differs from setting and actual recording.

(a)

(b)

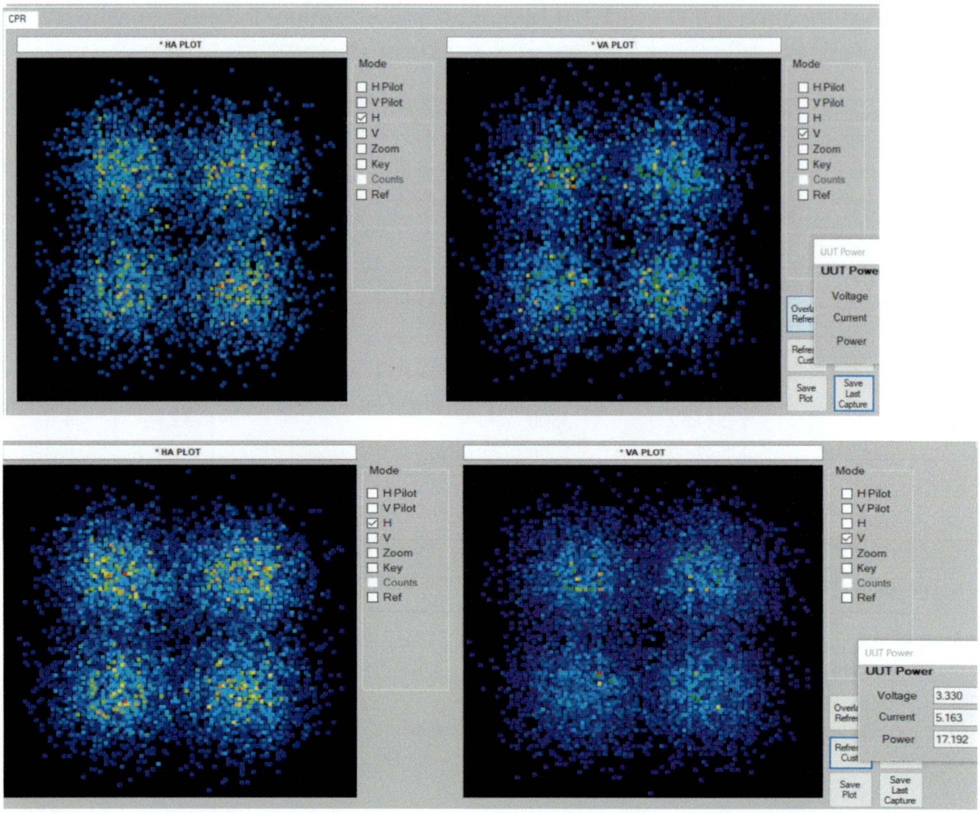

FIGURE 8.51 Rx Power at –38.2 dBm VOA at maximum. Constellation of QPSK at (a) LOFO = 5000 MHz and (b) LOFO = 2500 MHz but 2500 + 180 MHz recorded. Approximately 180 MHz offset diff from setting and actual recorded. Rx signal power = –40 dBm, OSNR = 32.3 dB.

TABLE 8.2
Data Plots as Data Recorded from Acacia 100 G TRx

LOFO (MHz)	BER	–5.1 dBm Tx	LOFO (MHz)	Orx (dBm)
1000	9.20e–06		1000	–27.85
100	2.69e–06		100	–26
500	3.49e–06		500	–26.9
1000	2.82e–06		1000	–26.5
2000	3.71e–06		2000	–27
5000	4.46e–06		5000	–26.9
10000			10000	–27
500	9.84e–07		500	–25.7
400	1.00e–06		400	–25.78
300	1.60e–06		300	–25.7
200	1.10e–06		200	–25.8
100	1.10e–06		100	–25.76
0	1.06e–07		0	–25.75
600	1.40e–06		600	–25.78
700	1.15e–06		700	–25.8

TABLE 8.3
Receiver Sensitivity and BER as Function of LOFO with the Following Settings: SigP (dBm) ORxn=–40 dBm; Total P received ORx (dBm)

Setting LOFO (MHz)	Tx –5 dBm BER	Opt Freq (MHz) 194,505 LOFO Set	Rx P_{sig} –40 dBm	Total *P* Rx (dBm)
5000	2.42e–02	5000		–38.12
4500	2.43e–2	4500		–38.17
4000	2.44e–2	4000		–38.2
3500	2.44e–2	3500		–38.1
3000	2.42e–2	3000		–38.2
2500	2.44e–2	2500		–38.28
2000	2.43e–2	2000		–38.07
1500	2.42e–2	1500		–38.3
1000	2.43e–2	1000		–38.14
0	2.41e–2	0		–38.04

8.5.4.2 Discussions on LOFO

Issues/Questions: Given that space optical transmission systems connect satellites or between ground stations and satellites. The moving speeds (centrifugal and tangential) of the satellites are relatively appreciable with respect to that of the speed of lightwaves. Thus, a frequency shift due to Doppler effects occurs. This Doppler frequency shift can vary during the day depending on the rotational angular speed of the satellite, ranging from –7 to +7 GHz. This range is outside the processing range of Acacia TRx or DSP processing for intradyne reception.

Question: Are there monitoring procedures and compensating feedback mechanisms to lock the carrier and the LO of the receiver?

Answer: Yes, the enhanced software package for enhanced processing is available in which the frequency shift is monitored and compensated by feedback control of the LO (integrated in the module) frequency shift to lock the two frequencies of the signal carrier and the LO to be identical . This software base locking of the LO can ease the difficulties of homodyne coherent reception by software tuning and processing. Such carrier phase algorithms are now readily available in integrated coherent receiver and transmitter.

Question: Once such an enhancement processing package is delivered, will an additional processing program be developed to implement this compensation?

Concluding Remarks: If the Doppler frequency shift (DFS) on the optical carrier can be compensated and locked close to homodyne reception, it is expected that:

A 3 dB improvement in dynamic margin can be obtained.

The reception subsystem can be setup once for all satellites, with pre-settings using a lookup table and online correction or compensation for the DFS.

Refer to Figure 8.35b for the procedure to compensate for the offset frequency between the LO laser and the DSF of the carrier.

8.5.5 DSP Systems for SPC High Capacity and LOFO

In such SPC networking, the DSP-based transceivers are important to mitigate the impairments in the transmission path as well as nonlinear distortion effects.

Figure 8.52 shows the arrangement of algorithms embedded in the transceiver to perform several signal processing tasks in the digital domain. The optical channels entering the reception

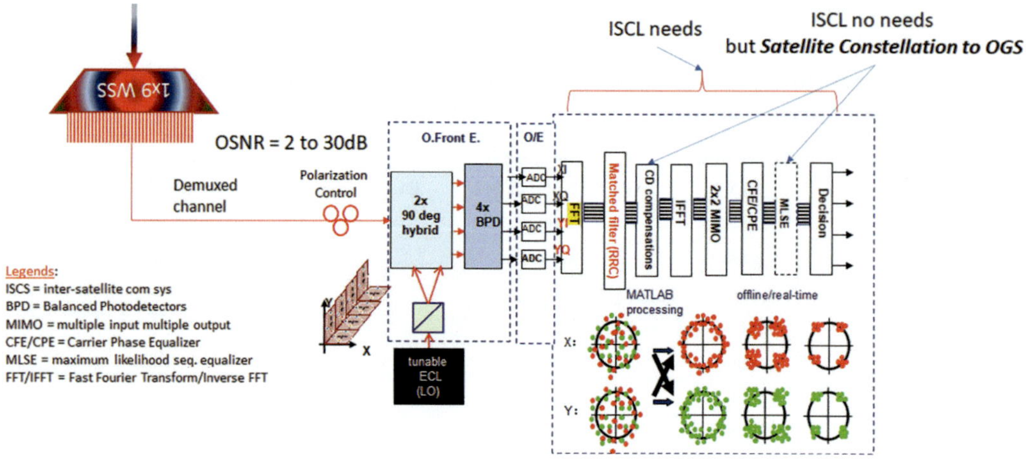

FIGURE 8.52 Embedded processing algorithms in the digital processor of a DSP-based ORx. WSS = wavelength selective switch equivalent to an optical wavelength demultiplexer, OSNR = optical signal to noise ratio, O-Front-E = optical-front-end to split polarized channels and the in-phase and quadrature components, LO = local oscillator laser, OSNR = optical signal to noise ratio, MLSE = maximum likelihood sequence equalizer, CPE = carrier phase equalizer.

section of the SPC link and constellation are amplified by an LNA as the optical preamplifier, then detected by a coherent receiver. In this receiver, a balanced pair of detectors is connected back-to-back and beat with a local oscillator laser, allowing the product of the LO and the channel sequence to be reproduced in an intra-spectral electronic region of the electronic passband amplifier. AS shown in the Figure 8.52 shows that the π/2 optical hybrid couplers are assigned to each polarized channel. They couple the local oscillator beam and the optical data channel, phase shift them, and split the in-phase and quadrature components of the complex signals. These are then received by the balanced photodetectors and beating together to produce sinusoidal signals along with noise from coherent residual power and electronic noise. The signal sequences are sampled by the ADC and then processed by the DSP with long sequence memory, which is long enough for determining the bit error rate. The software algorithms are listed in Figure 8.52 and show a typical sequence of processing depending on the detection and recovery of either the ultra-high frequency or power-limited case, and the quantum detection case. First, the sequence is sampled by the ADC, then a serial-to-parallel conversion transfers it to parallel bit patterns. An FFT is used to obtain the frequency domain of the signals. Processing is mostly preferred in spectral domain. The RRC matched filtering algorithm is used to ensure the signals can be accommodated in the DSP systems and to equalize any band limited subsystems. The MIMO algorithm can be used for autocorrelation and cross-correlation of the polarized channels to mitigate skew effects in propagation through the space channels, especially under turbulence or angle entering the optical heads due to the rotational speed of the satellite. Furthermore, mitigation of Doppler effects helps obtain pure homodyne coherent detection through carrier phase recovery.

The DSP embedded in the OTRx is dedicated to each channel, so each channel must be received by a specific DSP-based OTRx, and the local oscillator laser is tuned to the correct frequency for that specific wavelength channel.

Figure 8.53 shows a typical number of sampled constellations in the complex plane of QPSK signals under various cases: high fidelity Figure 8.53a, Doppler shift effects on the QPSK constellation or rotation of the constellation Figure 8.53c, and signals embedded in noise.

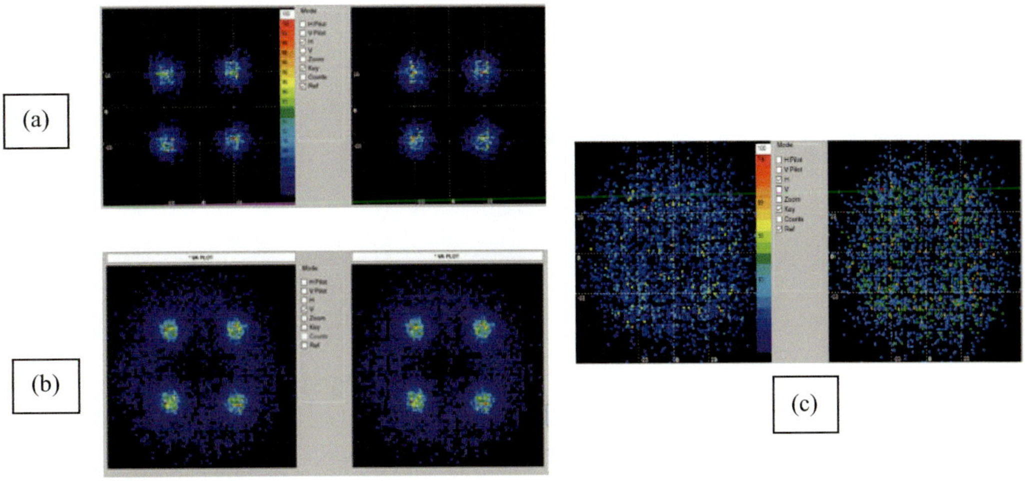

FIGURE 8.53 PDM-QPSK Constellations at (a) LOFO=2 GHz, (b) 4.5 GHz, and (c) 6 GHz; at BER=2.42e-2, Rx signal power=–25.8 dBm, OSNR=32.3 dB. Note on phase rotation of constellations of (b), i.e., phase mismatched by several 2π (cycles). PDM=polarization division multiplexing, QPSK=quadrature phase shift keying, LOFO=local oscillator frequency offset.

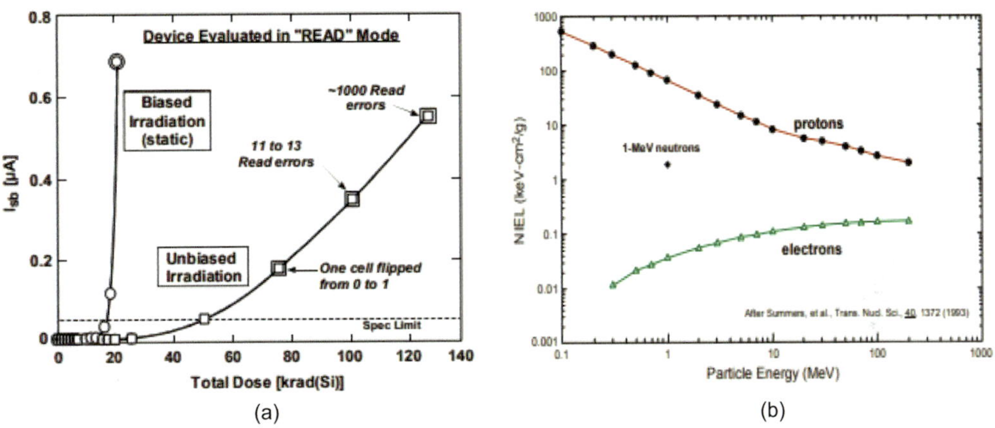

FIGURE 8.54 (a) Failure of a CMOS device due to field oxide inversion; (b) dependence of non-ionizing energy loss on proton energy in silicon.

8.6 RADIATION DAMAGE

8.6.1 The 100 G TRx Structure and Exposure

Radiation damage is caused by proton radiation in space bombarding the tera-bits/s transceiver (Figure 8.54). With a pulse duration of 200 Gbps line rate, the pulse period is only about 20 ps. The modulators and modulation formats for the Tbps transmitter are shown in Figure 8.55. The coherent receiver structure and packaged block diagram are presented in Figure 8.56a and b, respectively. Radiation damage can enforce errors in four main sections of the transceiver, as shown in Figure 8.56a. The following outlines the expected influences on changes in the transceiver functions:

Two lasers are independently tuned to generate different laser lightwaves for the optical PDM-IQM and coherent reception sub-system. The lasers are fabricated with compound semiconductors, and changes are expected in the driving current, thereby affecting the optical power.[11,12]

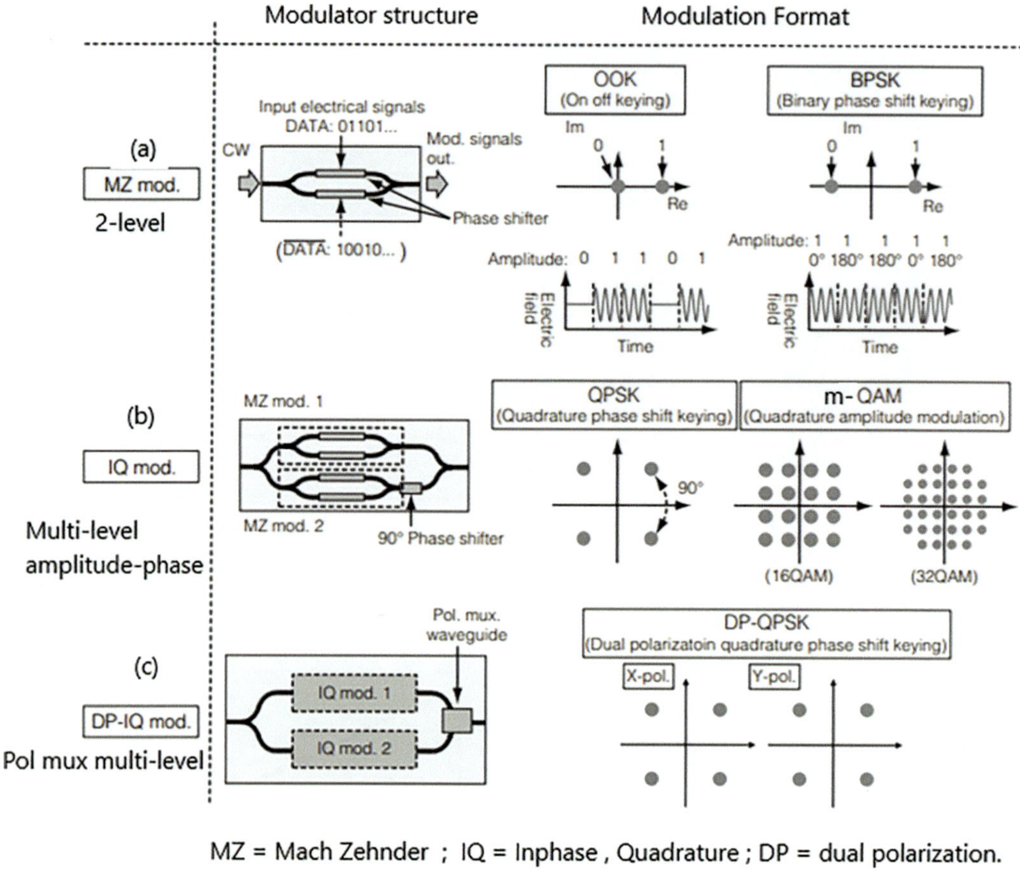

FIGURE 8.55 Modulation formats based on complex modulation, amplitude and phase QAM, polarization multiplexing and high order levels to reach Tera-bps. Left hand sides are complex modulator structures for OOK, BPSK, QPSK, mQAM, and PDM-QPSK. (a) single MZ (Mach-Zehnder) modulator under OOK (or NRZ) and QPSK modulation with constellation and modulated carriers; (b) two MZ modulators plus an optical phase shifter integrated in one MZ path for complex modulation schemes QPSK and m-QAM; and (c) Structure of DP-QPSK (dual-polarization QPSK) modulator by combining two QPSK modulators , one each for the two orthogonal polarizations of the lightwaves.

Mechanisms for displacement damage in optoelectronics are essentially the same as those in conventional semiconductor devices. For modern photonic devices, especially those in integrated optic structures, it is necessary to examine material types such as silicon in SOI Si photonics, as well as compound semiconductors, including GaAs, AlGaAs, and InGaP, for the spectral window of 1525–1565 nm C-band.

Dislocation of doped impurity in gain sections leads to an alteration of stimulated emission, hence higher driving current would be required for effectively shifting the lasing curve power output versus driving current, especially the laser threshold.

Broadening of the laser linewidth occurs due to alterations in the DBR (Distributed Bragg Reflectors) at the two ends of the gain cavity, especially the 100% reflection mirror.

Alterations in the tuning cavity, which is formed by etching a cavity within the gain medium whose thickness determines the resonant lasing frequency of the laser. The radiation of the Gama ray would alter by exciting the electron pairs, thus depositing charges on the lasing electrodes, and thereby affecting the electro-optic coefficients. See Figure 8.54 for damage on electronic devices.

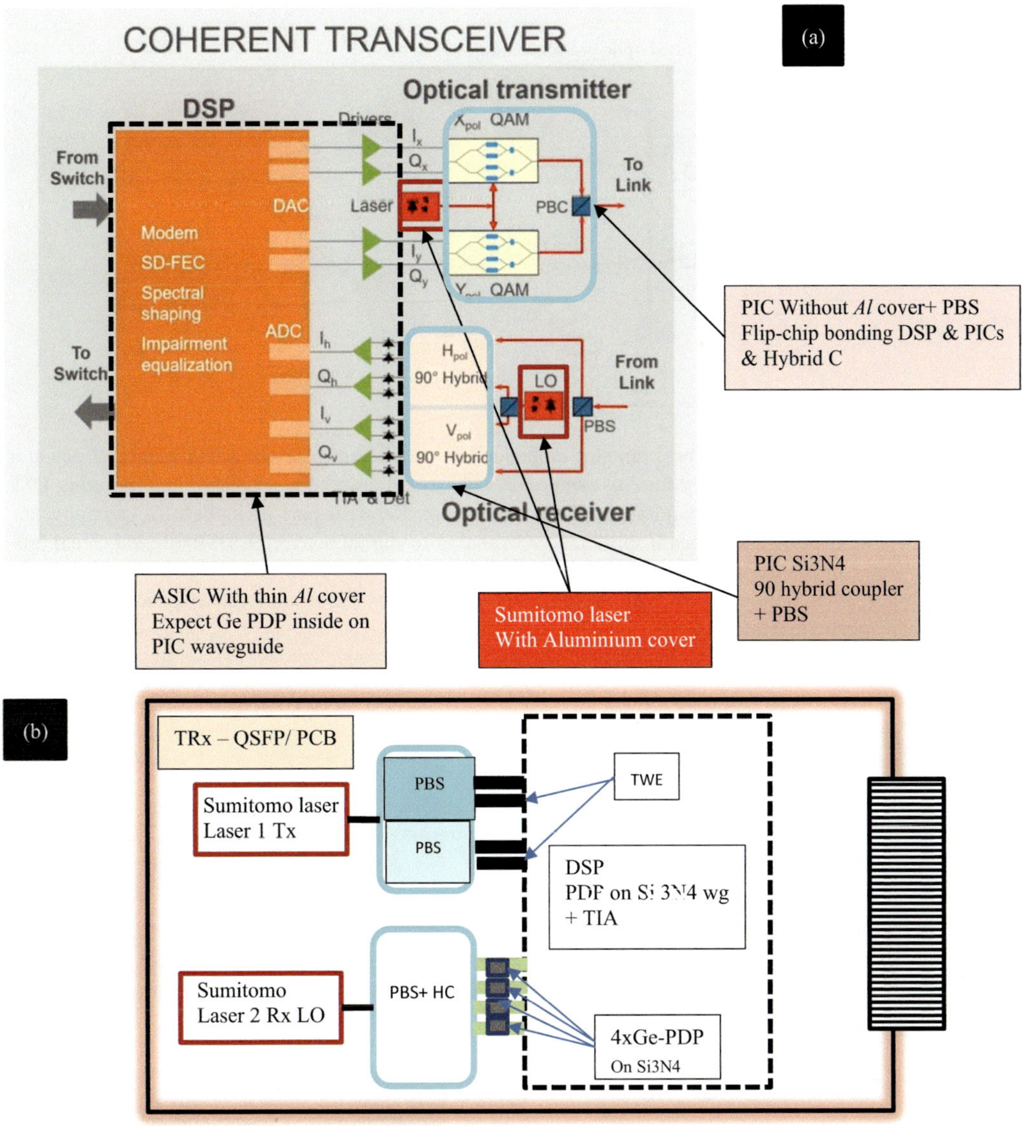

FIGURE 8.56 Integrated photonic TRx structure and exposure. PDM-QPSK 100 G module. (a) Schematics with layout (b) Layout with actual positions of PICs, HC, DSP, PDP, all in flip-chip bonding. Electronics on PCB not shown.

8.6.2 Integrated Photonic Transceiver (IPTRx): Analog and Digital Processing

This section comprises (i) a digital processor, the DSP, to generate the data sequence and the algorithm for processing digitized data received; (ii) ADC and DAC to provide signals to the DSP and to power amplifiers to drive the optical modulators; (iii) an analog electronic sub-section composed of transimpedance amplifiers to boost the received coherent mixed signals between LO laser and modulated carriers. The DSP is based on CMOS digital logic, and hence, effects due to electron-hole pair generation (Figures 8.57 and 8.58).

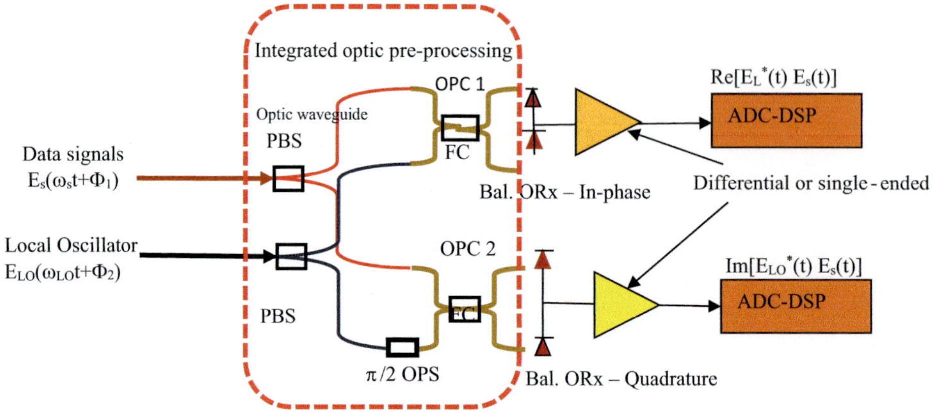

FIGURE 8.57 Schematic of a hybrid coupler embedded in a coherent receiver using balanced detector and an integrated optic pre-processor for *I-Q* components phase and amplitude recovery, incorporating DSP and ADC. Im=imaginary part or quadrature phase component, Re=real part or inphase (*I*-) components, PBS=integrated optic polarization beam splitter, ADC=analog-to-digital converter, DSP=digital signal processor, TIA=transimpedance amplifier, ω_s, ω_{LO}=Signal carrier and local oscillator frequencies, OPS=optical phase shifter, OPC=optical polarization combiner (same pol), TE=transverse electric (horizontal), TM=transverse magnetic (vertical).

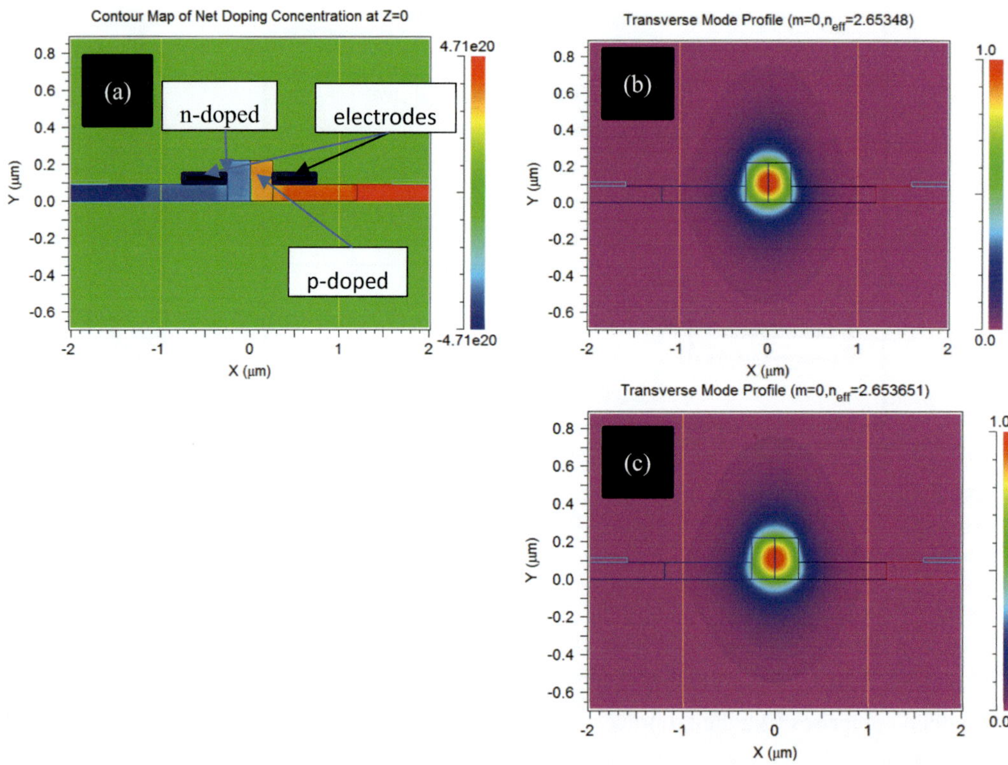

FIGURE 8.58 SOI optical waveguide and guiding for IQ modulation (a) cross-section of Si on insulator waveguides with n- and p-doped regions; (b) intensity distribution of TE guided mode, and (c) TM mode. Electrodes are for RF modulation signals and DC bias via a bias-T.

8.6.3 IPTRx: Optical Modulation and RF Signaling

IQ modulators are based on SOI integrated photonic technology. Various IQMs based on GaAs MQW (Multi-quantum well) and SOI technology are shown in Figure 8.59, Figure 8.60. IQMs based on a Lithium Niobate substrate are also efficient modulators but due to their size and non-matching material with respect to Si and GaAs, they are not commonly used in integrated TRx QSFP modules.

For Silicon based IQM is formed using Si-waveguide structure on box SiO_2, which is an oxidized layer on Si wafer, as shown in Figure 8.60a. The guided TE and TM modes are illustrated in Figure 8.58. As observed the mode intensity distribution is tight due to the difference in refractive indices between the waveguide and the substrate. The waveguide becomes a pure dielectric when it is reverse biased.

Under radiation, it is expected that charges would be deposited on the electrodes on top of the Si waveguide. Hence, electro-optic effects change the biasing conditions, pushing the linear region of the transfer transmittance curve of the MZM toward near nonlinear regions, depending on the strength of radiation dose. This would lead to higher-order harmonics, interfering with spectral components of the same signal band, resulting in distortions and errors (Figure 8.61).

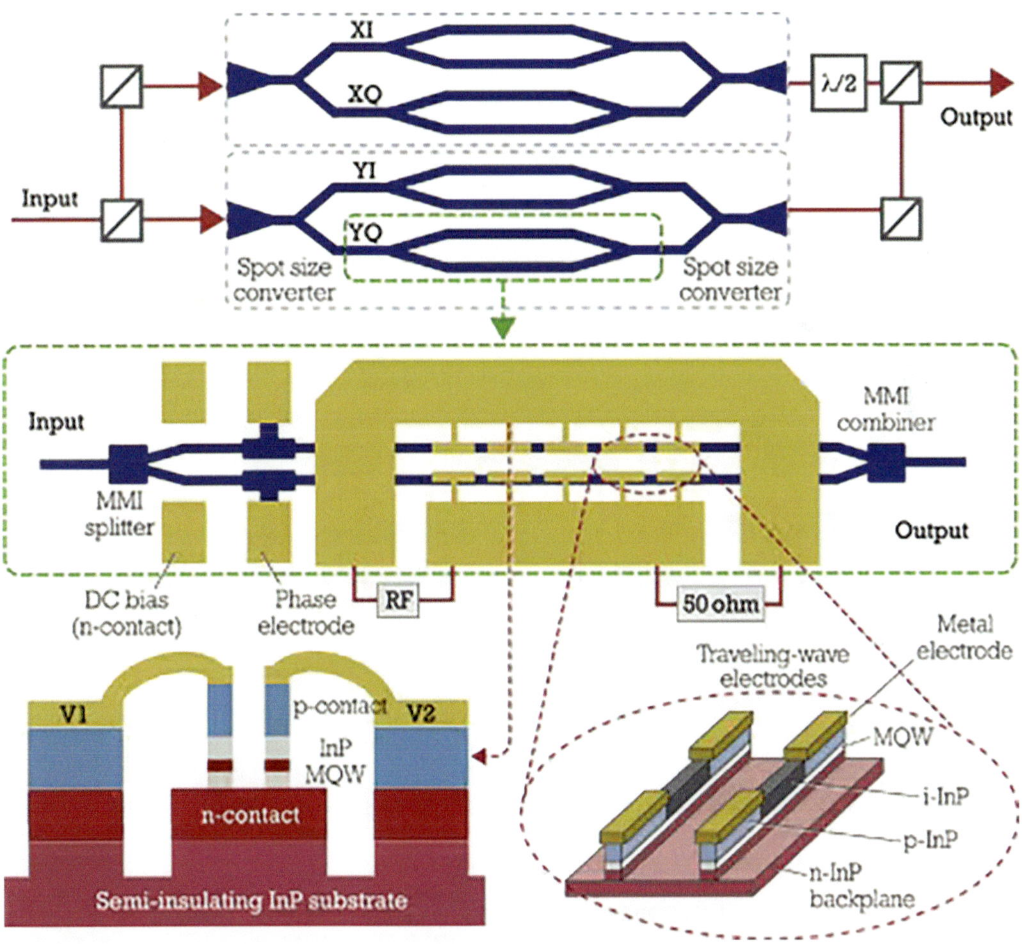

FIGURE 8.59 IQM based on MQW GaAs structure – Bockham UK (now iXBlue). MQW = multi-quantum well.

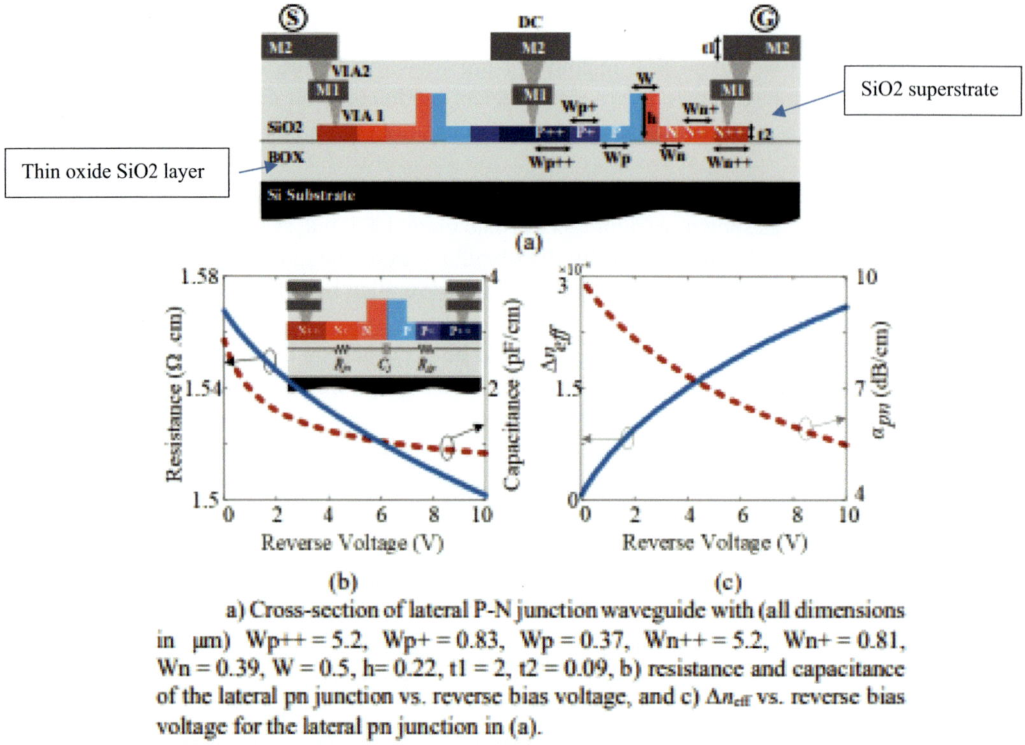

FIGURE 8.60 IQM based on SOI structure (Acacia).[13] T=thickness, W=width, W_p=width of p-doped region, W_n=width of n-doped region.

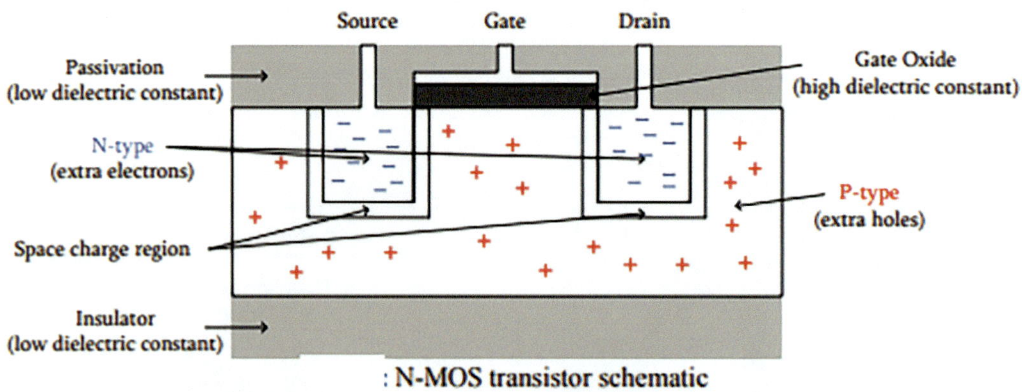

FIGURE 8.61 n-CMOS under radiation effects: distribution of charges across the gate channel. Pinch-off channel changes accordingly depending on the radiation dose.

The free-moving charges, holes, and electrons in the SiO_2 cover regions on top of the electrode and waveguides would deposit on the electrodes of the MZMs in the IQM, thus changing the biasing and the 3 dB operating bandwidth of the device.

8.6.4 IPTRx: Hybrid Coupling Pre-Opto-Electronic Processing

The π/2 hybrid coupler is used to separate the polarized components of the channel signals and that of the LO laser so they can be mixed accordingly. The phase of the *I*- and *Q*-components is separated via thermal effects, the pyroelectric effects, by applying voltage to the electrodes.

Commonly, this section is fabricated by using Si_3N_4 on SiO_2 substrate so that larger spot sizes of the waveguides can be used to minimize the coupling loss and propagation loss, as shown in Figure 8.57. When the receiver/transmitter module is under nuclear radiation, the charges would be deposited on the electrodes which can be biased at π/2 phase shift of the lightwaves of the signals and the LO lasers. This phase change would shift the difference in amplitudes of the *I*- and *Q*-components and thus a diamond-type constellation of the QPSK would be expected as shown in the figure, Figure 8.57. Thus, errors would occur.

8.6.5 Radiation Damage Test of 100 G TransceiverTesting radiation damage strategy for PDM-QPSK transceivers.

Step 1: Test Sections 8.6.1–8.6.4 separately by isolating radiation exposure.

Focus radiation beam on Section 8.6.1, observe and record the frequency shifting of the LO and carrier. Keep the monitoring GUI active and record the constellation and the frequency LOFO. Determine the frequency shift's impact on BER as compared to that obtained with no radiation dose.

For Section 8.6.2, we expect some changes to the TIA due to damage to the biasing condition of the main amplifiers and the shunt feedback resistance. It is expected that a differential TIA is used, and thus significant changes to the differential pair, particularly the long-tail pair in such TIA.

For Section 8.6.3, IQ modulators are severely affected under radiation as the biasing voltages change the distinction ratio of the SI waveguide MZM. Furthermore, distortion would occur, and monitoring the BER at different doses of radiation would give some features of the damages.

For Section 8.6.4: The hybrid couplers would be affected along with the phase shifter electrodes, thus distorting the QPSK constellation. The main axis of the diamond would provide some indication of the damage, revealing whether the *I*- or *Q* was affected most.

Notes on the structure of PDM-QPSK 100 G module by Acacia and tests of effects by radiation damages:

- DSP (spectral shaping (RRC) factor alpha=0.2 to 0.8 (400–200 G then 100 G): Unit covered with a thin Al plate.
- PDP with expected Ge-detectors is integrated on Si_3N_4 waveguides of the hybrid couplers, then flip-chip bondedwith DSP and TIAs.
- Hybrid coupler with thermal electrode for π/2 phase shifter to extract *I*/*Q* components.
- PBS integrated into the hybrid coupler and IQM (PDM).
- PICs using a Si-on-Insulator (SiO_2) substrate. IQ modulators (one pair for H- and V- polarized modes) are formed using Si-waveguides and MZM structures with a π/2 phase shift, as shown in Figure 8.56a and blocks of Figure 8.56b. Traveling wave electrodes for modulation of IQM are flip-chip bonded to DSP and analog circuits chip.
- Lasers (Tx and ORx LO) are supplied by Sumitomo and flip-chip bonded to couple to the hybrid coupler (PBBS) and to the IQM Si PIC.
- PICS and hybrid couplers are exposed, with no Al cover.

8.6.5 Test Procedures

For 100 G PDM-QPSK transceiver OTRx to be tested, its sections of TRx will be tested one-by-one as follows:

- Stage 1: Sumitomo lasers for Tx and LO coherent Rx – how to determine the damage caused by Tx laser or LO laser?
- Stage 2: PDM_IQM modulators (4 x PDM-IQM – for two polarizations and *I*- and *Q*- driving electrodes with phase shifter on the Q-MZM section.
- Stage 3: Hybrid coupler: PBS, phase shifter to extract *I* and *Q* sections of *H*- and *V*- polarizations. Furthermore, Ge-PDP as balanced detectors.
- Stage 4 : : DSP for both Tx and Rx parts. Effects on balanced detection, DSP on Tx parts, and DSP on reception parts.

8.7 DEMONSTRATION PLATFORM

WSSes (2×9) are used for, wavelength multiplexing and demultiplexing with space spans consisting of OPA, VOA and OPreA. The number of spans should be arranged so that total gain and losses are equalized to minimize the amplifier noise. This minimizes noise accumulation on channels. Tx sections of TRx are used to generate line-rate signals to the WSS, as shown in Figure 8.62.

The outputs of the WSS are demultiplexed into WDM channels and then coupled to the next cascade: Space span 1, then 2, and so on up to span *N* via the OPA as the power booster, followed by a VOA representing the total space propagation loss, and an OPreA for optical pre-amplification.

The 200 GHz spacing is assumed, and the channel line rate can be 1×200 G 52GSy/s QPSK; 2×100 G 28 G-PDM-QPSK, and Nx10D IM/DD. Thus, the total capacity is >1 Tbps.

Demultiplexing is implemented using WSS (2×9), feeding into the coherent Rx section of the transceiver TRx.

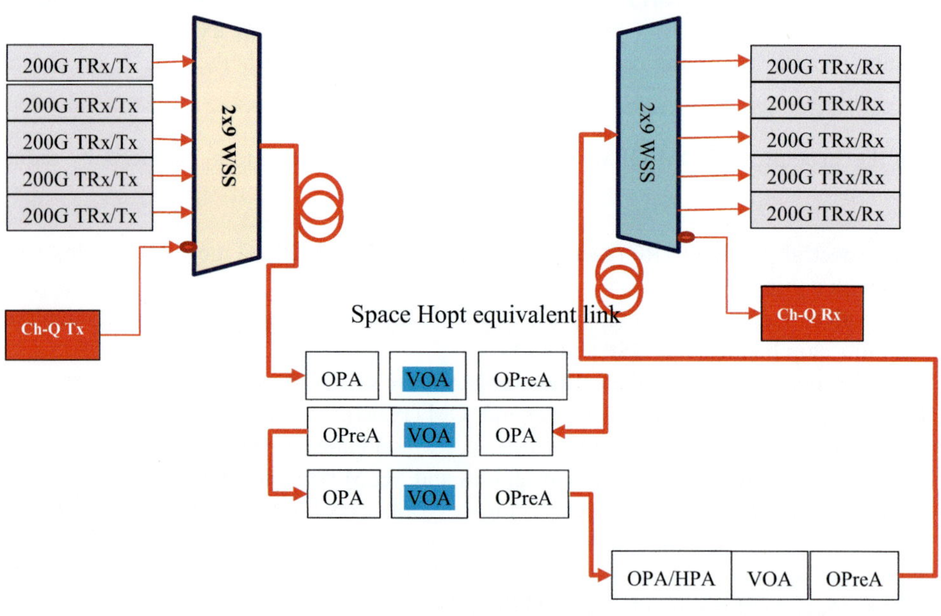

FIGURE 8.62 WSS mux/demux and space spans of a demonstration platform. OPA/HPA = optical power booster amplifier, OPreA = optical preamplifier or LNA -low noise amplifier, WSS = wavelength selective switch, TRx = transceiver, VOA = variable optical Attenuator, λ_Q = wavelength reserved for quantum channel.

8.8 NONLINEAR EFFECTS AND SIGNAL DEGRADATION IN ONE HOP FIBRE-EQUIVALENT FREE-SPACE PLATFORM

This section presents a brief study of the nonlinear effects on the propagation of guided optical mode in standard single-mode fiber (Corning SMF-28 or G.652 ITU-T), which is amplified by a booster amplifier whose output can reach 2.0 W or 33 dBm. With the effective area of the guided mode being only 50 μm^2, the nonlinear refractive index change of the fiber can be approximately n_2I with I as the guided beam intensity; so, n_2I is about 5.5*10^{-10}/W /m. Therefore, integrating over 2 m of SSMF with a power of 2 W, the refractive index change is about 2.2e-10. This small change in refractive index would then create a conversion efficiency of about 0.1% or about 30 dB down from the input wave.

The nonlinear conversion includes SPm and XPM. SPM generates the phase distortion in lightwaves or modulated lightwaves, causing dispersion of ultra-high-speed signals. SBS, with acoustic waves, in turns, generates a low-frequency envelope on the high-speed signals and shifts the main carrier and sidebands by about 9.5 GHz from the signal baseband center. The envelope modulation generates the jitter in the detection of the high-speed signal.

Furthermore, the XPM generates phase matching between two waves in case such multi-carrier channels exist. This, in turn, generates two-wave mixing to generate a third wave and thence a fourth wave or the four-wave mixing.

We report here the nonlinear effects in such SMF under the pumping of high-power signals and carriers. Such a high-power beam is required for a free-space propagation of laser beams in optical inter-satellite communication and free-space optical networks.

Section 8.2 gives a summary of the nonlinear effects in guided waveguides, as well as a basic understanding of nonlinear effects in fibers.

8.8.1 Case 1: Two Carriers Spaced by 0.4 nm from Channel #30 (1536.61 nm)

Two carriers are set as Carrierl at 156.21 nm and carrier 2 at 1538.81 nm, with a 15 dBm seed laser input to Port 2 of the booster amplifier so that the amplifier is saturated and the gain is obtained at saturation level. This is to ensure that no fluctuation of the signal occurs. The spacing between the two carriers is equal to 0.4 nm = 50 GHz.

The setup schematic is shown in Figure 8.63. The interaction of the two carriers due to nonlinear effects can be observed in Figures 8.64–8.66.

Data channel is 100 Gbps DP-QPSK (Dual polarized QPSK format) signal, and carrier 1536.21 nm is coupled with a 2×1 FC, then fed into P1 of the MPB booster. There is no 10 dB carrier at 1536.4 nm; VOA = 20 dB.

8.8.2 Case 2: Two Carriers Spaced by 0.4 nm from Channel 30 (1536.61 nm)

Similar to the setup in Case 1 shown in Figure 8.63, we set the following for Channel 30. A 100 Gbps DP-QPSK data channel is placed and centered at 1536.61 nm. Two carriers are set at Carrier 1 at 156.21 nm and carrier 2 at 1538.81 nm. To saturate the booster amplifier, a seeding laser is set at 15 dBm to Port 2 of MPB booster amplifier. Two carriers are spaced by 0.4 nm or 50 GHz in the frequency domain

The 100 Gbps DP-QPSK signal and carrier 1536.21 nm coupled 2×1 FC then feeding into P1 of MPB booster. VOA = 20 dB. The signal is under saturation gain as well as the two carriers.

Note: LNA output is filtered by an add/drop filter of passband = 100 GH but 60 GHz roll off.

The two carriers and the FWM product can be observed in Figures 8.67–8.69. The crosstalk product can be embedded in the data channel, as observed. Higher power from the two carriers generates an FWM crosstalk product, which can also be observed and measured. We also observe the FWM product on the left side, above the noise level, at abut 30 dB below the carrier power.

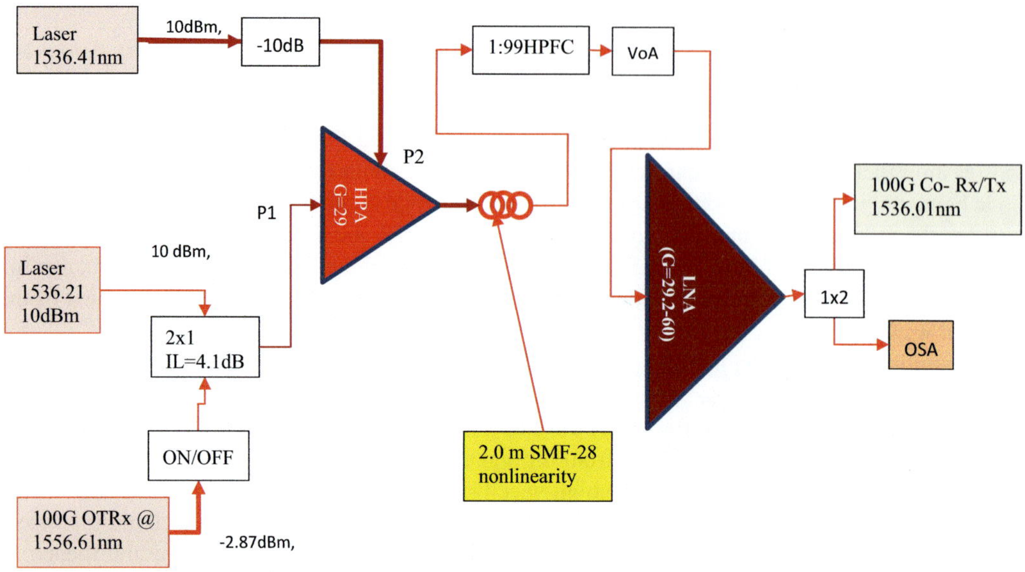

FIGURE 8.63 Set-up to test and observe NL FWM products with carriers and 100 G channel. The 10dB at 1537.01 nm. Under high gain booster amplifier with 2 W output pumping into 2.0m of SMF-28, high nonlinear phase changes lead to SBS, SPM, and XPM. IL=insertion loss.

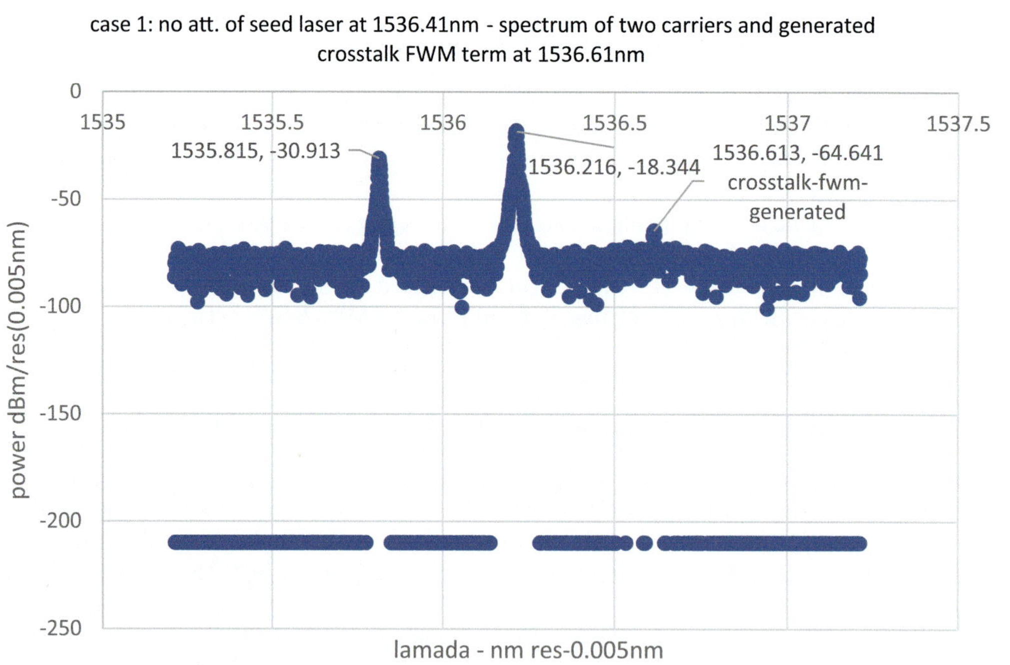

FIGURE 8.64 CASE1 spectrum of two carriers and FWM-generated term at 1536.61 nm – showing crosstalk if whole signal channels are multiplexed.

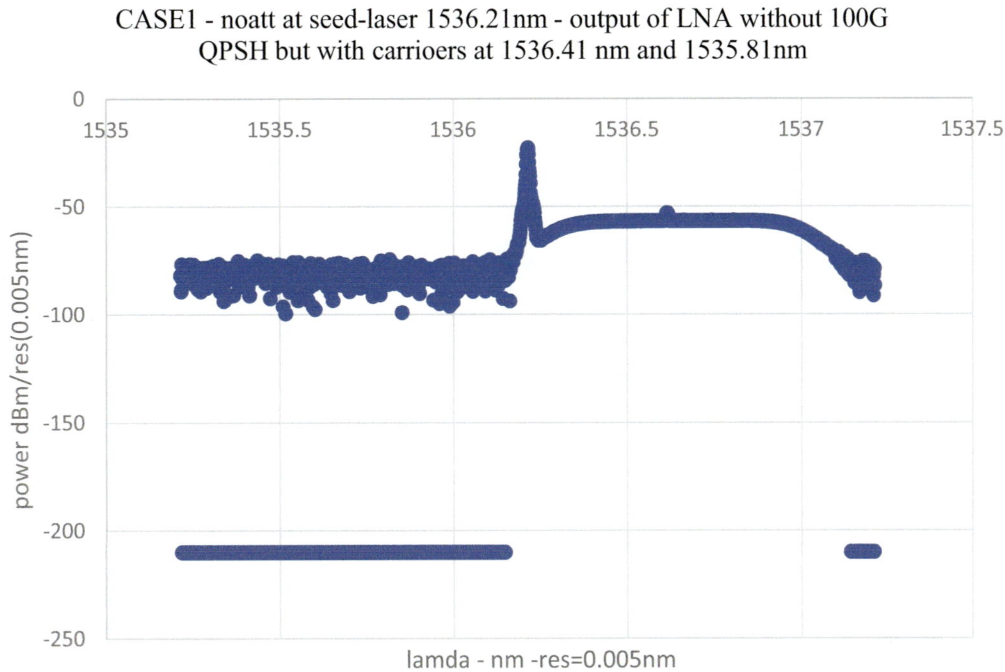

FIGURE 8.65 Spectrum at output of LNA showing carrier at 1536.21 nm and 100 GHz passband (OBPF-centered at 1536.1 nm) optical filter (optical ADD/DROP filter) – no 100 G QPSK signal. Carrier crosstalk is observable about 8 dB above noise band.

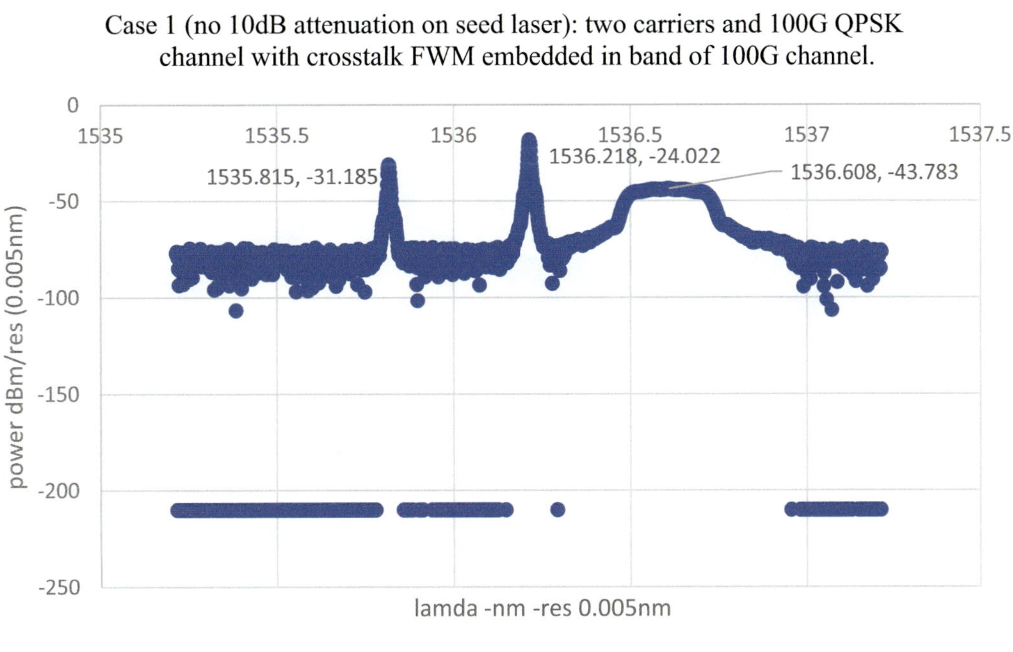

FIGURE 8.66 Case 1 (no 10 dB attenuator on the seed laser into Port 1 of MPB booster amp) spectrum of two carriers and 100 G QPSK before LNA preamp.

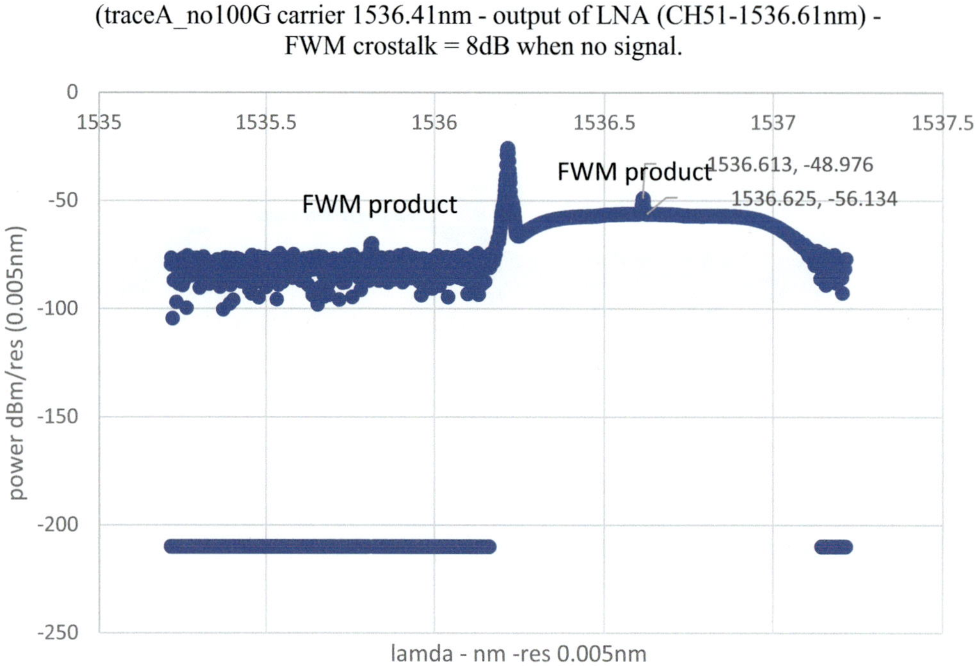

FIGURE 8.67 Spectrum at output of LNA showing carrier 1536.21 nm and 100 GHz passband (OBPF-centered at 1536.1 nm) optical filter (optical ADD/DROP filter) – no 100 G QPSK signal. Carrier cross talk observable about 8 dB above noise band. Seed-laser with 10 dB attenuator.

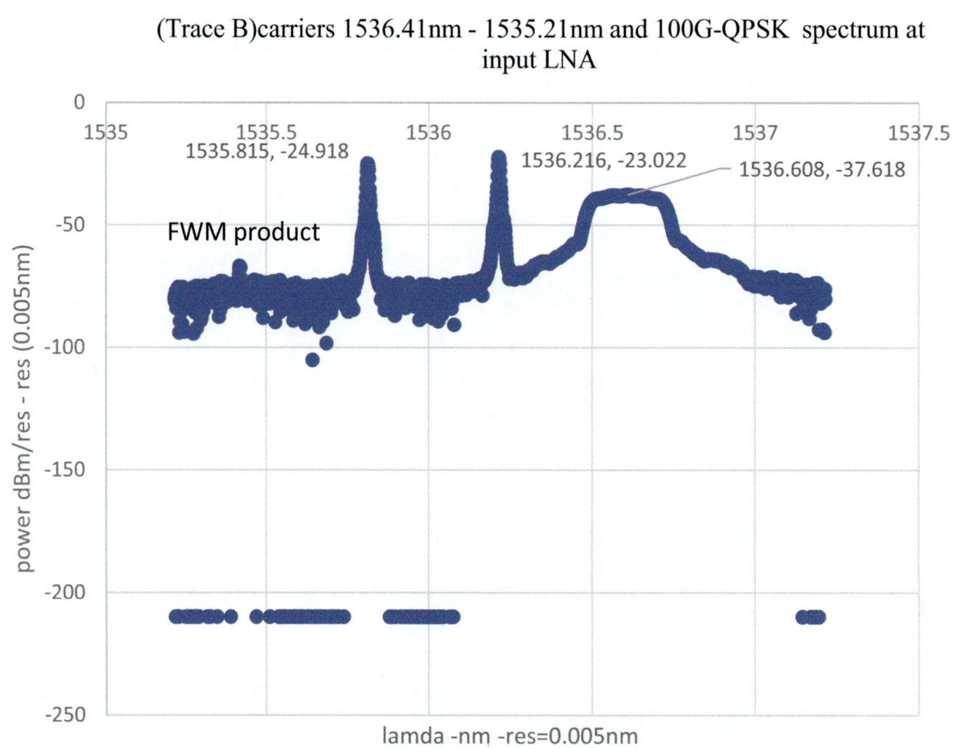

FIGURE 8.68 Carriers (1536.21 and 1535.81 nm) and 100 G QPSK signal (centered at 1536.61 nm), spectrum at input of LNA. Seed-laser –10 dB into port 2 of HPA MPB booster amp, saturating the booster.

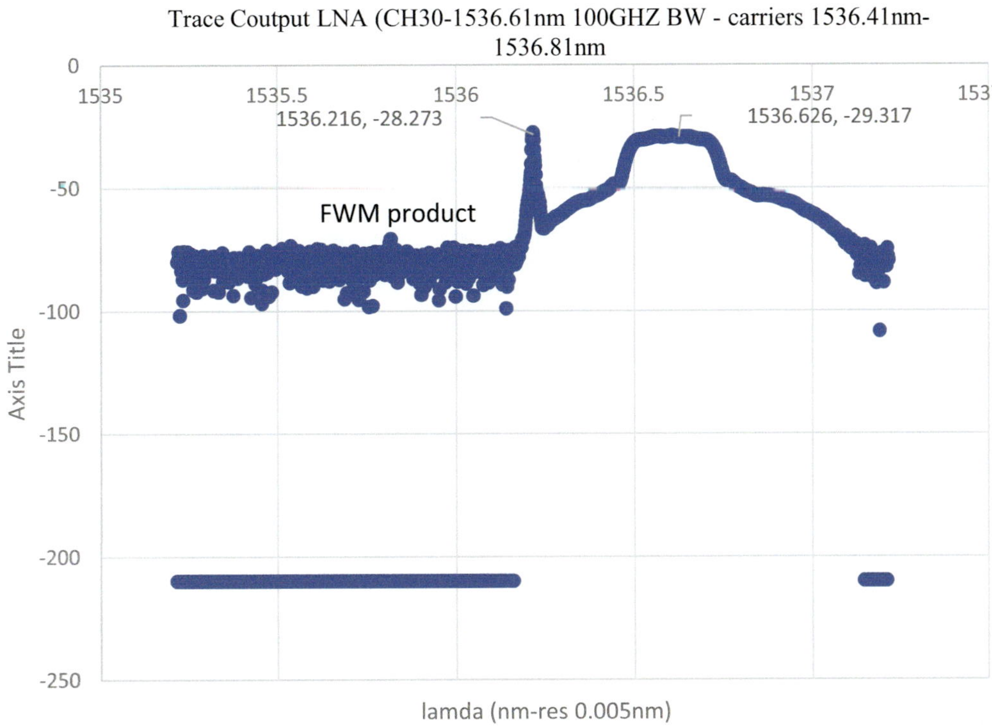

FIGURE 8.69 Signal and carrier at output of CH30 gf LNA output with input from Figure 8.68.

We conclude that the conversion efficiency of the FWM is about 0.1%. Although this is small, when contributing to weak signals, it can create appreciable errors in the BER measurement.

The setup for the interaction of two carriers and their FWM product allows us to gauge the expected crosstalk in the data channel, which we can later use to measure the error rate contribution due to this nonlinear effect.

8.8.3 Case 3: Two Carriers at 0.2 nm Spacing Interactions

Settings are as follows: (i) Channel 30 is centered at 1536.61 nm. (ii) Two carriers are set: Carrier 1 at 1536.21 nm, attenuated by 10 dB before mixing with data channel 100 G QPSK, and Carrier 2 at 1538.81 nm. A 15 dBm seedling laser is connected to Port 2 of the MPB booster amplifier to operate at saturation at all times. Two carriers are spaced at 0.4 nm or 50 GHz in frequency domain. (iii) Data channel of 100 G DP-QPSK signal on top of the carrier at 1536.21 nm coupled with 2×1 FC and fed into P1 of MPB booster.

The output of the booster amplifier is connected to a length of SMF 28 single-mode fiber of 2.0 m, and the output is coupled with a high-power fiber coupler, 1% output power port is fed to the OSA to observe the FWM products of the data channel as well as that of the two carriers, as shown in Figure 8.70.

Several variations of the attenuation factor of the VOA are used to vary the power of the carrier's transmission through the high-power fiber so that the FWM mixing product can be observed and measured. The intensity of the lines of the spectrafrom light to dark, are superimposed in Figure 8.70 in which the FWM products of the carriers can be observed.

Comments:

- Middle light intensity curve: 100 G QPSK and two carriers after HPA. Two 4 FM products at equal distances up and down of the carriers at 1536.41 and 1536.31 nm (or 1.25 GHz spacing).

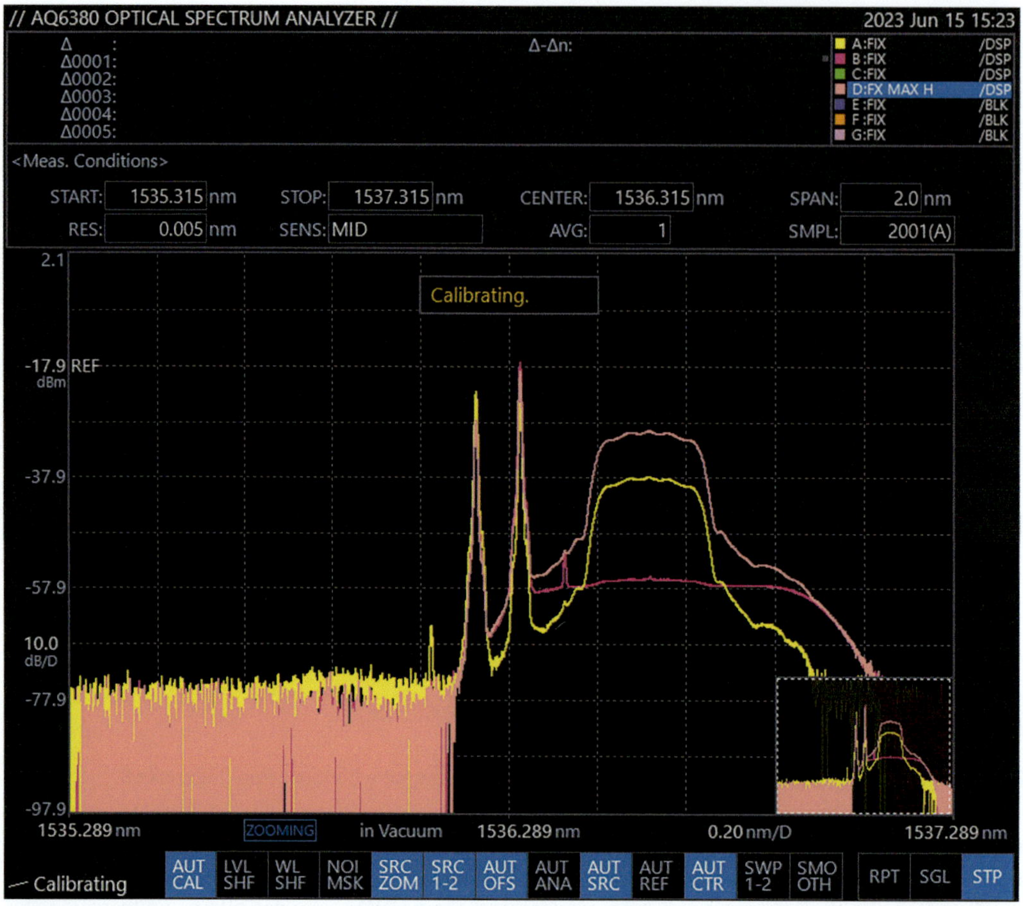

FIGURE 8.70 Spectra of nonlinear effects, FWM in full hop MPB HPA-amp and LNA-preamp. (i) Yellow – carriers and 100 G in front of LNA (i.e., post-booster HPA); (ii) dark-light lines no 100 G signal, two carriers at 1536.41 and 1536.31 nm – before passing through HPA booster with input VOA (30 dB) (MPB); (iii) light-line curve- signals and carriers after passing through preamp LNA at port 5 (channel 30 and 100 G bandwidth). The 8 dB FWM product carrier superimposes on filtered band. No VOA at input of LNA (i.e., after HPA booster).

- Dark intensity curve: Output of LNA (Channel 30 -Port 5); see passband region and the generated FWM product of about 8 dB, which is much higher than the signal band.
- Light intensity curve: LNA-amplified 100 G and carriers. FWM product of 8 dB embedded in the signal band (Figures 8.71 and 8.72).

8.8.4 Case 4: High and Low Attenuation (VOA) at Input of LNA

8.8.4.1 Case 4.1: Carrier Unequal via VoA- 1537.01 nm

Two carriers are set at two sides of 1536.61 nm and spaced by 0.4 nm from 1536.61 nm, above and below the attenuated carrier at 1537.01 nm, to generate unequal nonlinear interactions. The setup is shown in Figure 8.73.

The output of the booster amplifier is connected to a length of SMF-28 single-mode fiber of 2.0 m, and the output is coupled to a high-power fiber coupler. The 1% output power port is fed to the OSA to observe the FWM products of the data channel as well as those of the two carriers, as shown in Figure 8.74.

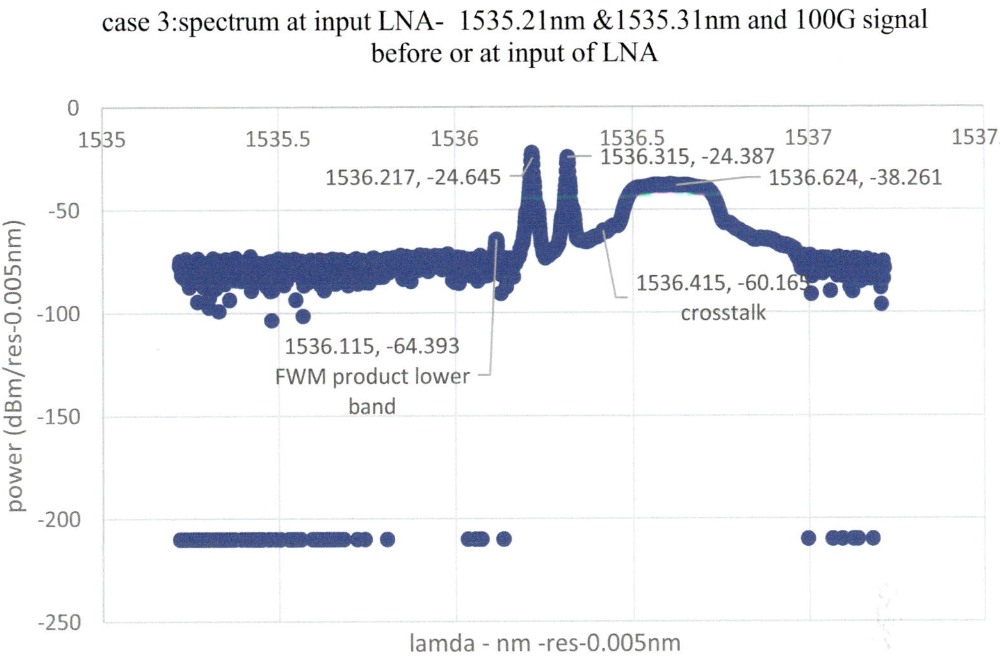

FIGURE 8.71 Signal at 100 G and two carriers at input of LNA before VOA, that is, before passing through VOA and then LNA.

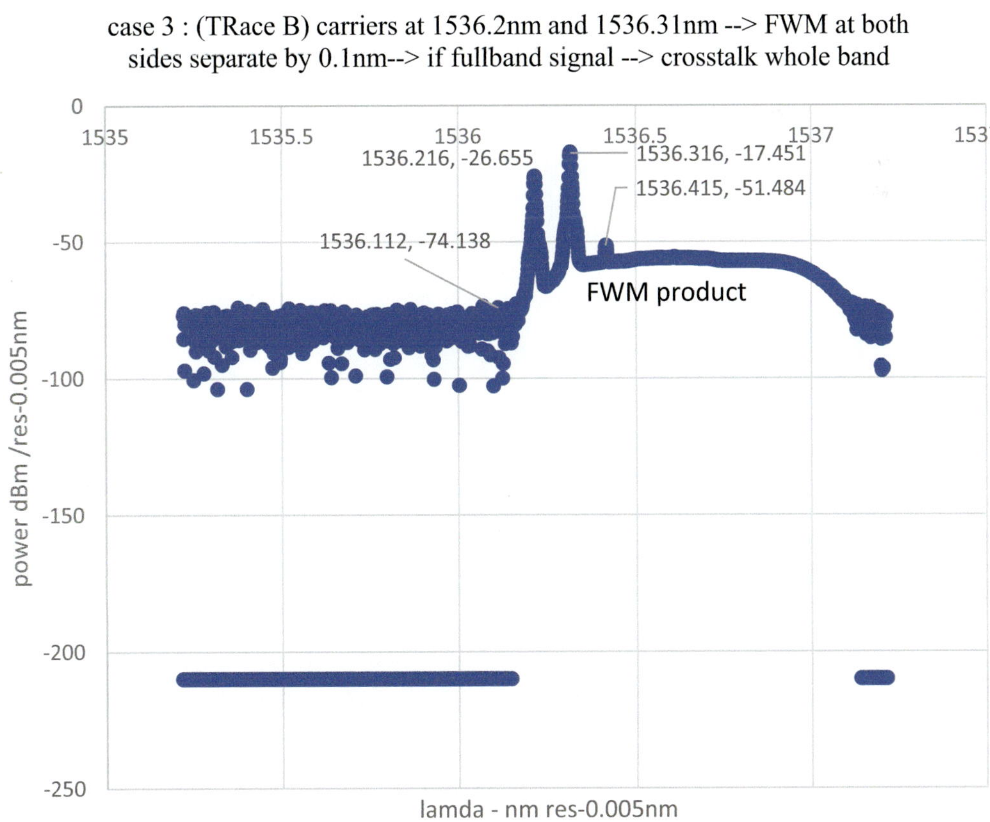

FIGURE 8.72 Signals at 100 G and two carrier's spectra at input of LNA, that is, at the output of HPA and VOA.

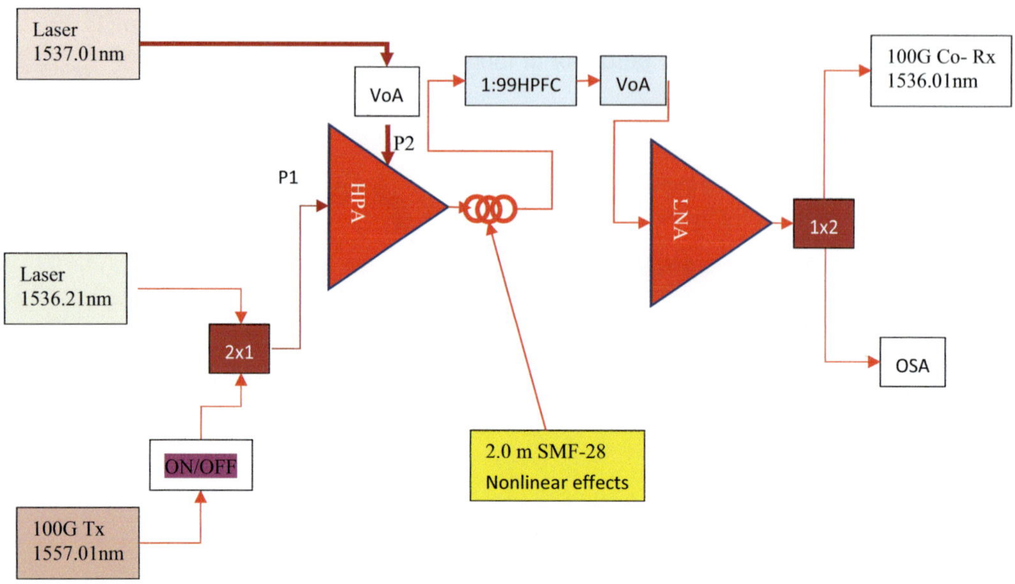

FIGURE 8.73 Setup to test and observe NL FWM products with carriers and 100 G channel. Spacing different, either upper or lower of 1536.61 nm.

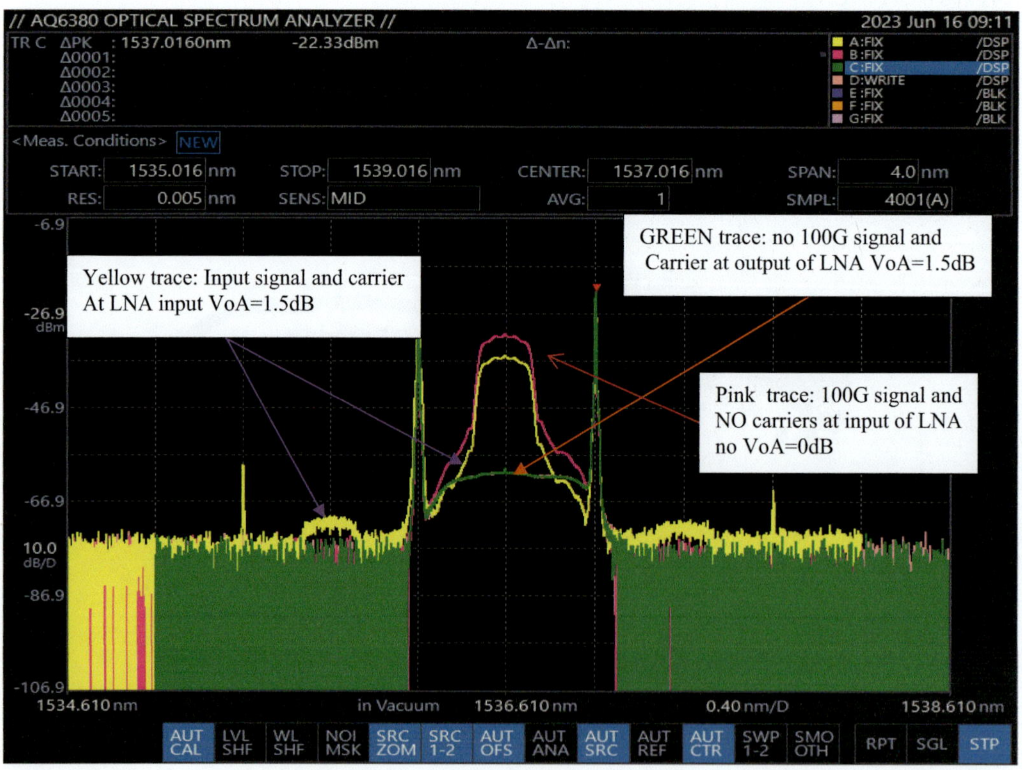

FIGURE 8.74 (i) Yellow trace: two carriers at 1536.21 and 1537.01 nm and signal at LNA output (yellow), (ii) green trace=two carriers at LNA output with 100 G-QPSK channel removed, (iii) red purple: trace of 100 G channel at LNA input (direct connection). VoA=1.5 dB.

A number of variations in the attenuation factor of the VOA are used to tune the power of the carrier's transmission through the high-power fiber so that the FWM mixing product can be observed and measured. The colors of the spectra are superimposed in Figure 8.74, in which the FWM products of the carriers as well as the data channel can be observed.

Comments: The yellow trace shows the FWM product at 1535.81 nm, which is the conversion of FWM from the carrier at 1536.21 nm and the channel centered at 1566.61 nm modulated by 100 G QPSK. Note also the FWM product at 1536.61 nm – 3×0.4 nm. This FWM product is about 5 dB above the FWM product at 1536.61 nm+3×0.4 nm.

We also observe the channel noise from the FWM process at 1536.61 nm+2×0.4 nm as the upper band FWM product.

Further to the graphics obtained for the cases of two carriers at 0.4 nm from the channel centered at 1536.61 nm, OSA spectra of the channels at variable value of VOA of 0 and 38 dB are shown. This is depicted in Figure 8.75.

8.8.4.2 Case 4.2: VOA 1536.21 nm

Two carriers at 1536.21 nm (via a 3 dB coupler with a 100 G channel) and 1535.81 nm (15 dBm into P2 of the booster), VOA = 1.5 dB and VOA = 39 dB – input and output of LNA are recorded.

FIGURE 8.75 The 0.4 nm spacing from 100 G central carrier and 0.8 nm between dual carriers (A) green: input 100 G and dual carriers with VOA = 0 dB (no VOA) at input of LNA preamplifier, that is after HPA – FWM product at 0.8 nm of the two carriers and 100 G band FWM double side of the 100 G with spacing of 0.4 nm – minor product beside the 100 G FWM product is the ~10 GHz. SBS = Stimulated Brillouin Scattering product. (B) Purple-red: VOA = 38 dB and spectrum of 100 G, dual carrier after LNA amplification – noises level increased. (C) Pink: 100 G and dual carriers at LNA output with VOA = 0 dB (no VOA). (D) Yellow: 100 G signal and dual carriers at the output of LNA with no HPA and passing through LNA.

8.8.5 Nonlinear Interaction with Variation of Carrier via VOA

The NL interaction and crosstalk can be evaluated by examining the optical spectrum of the 100 Gbps channels and two test carriers, as the following:

- The system setup is shown in Figure 8.76, in which one of the carriers is attenuated by a VOA with value varied from 1.5 to 20 dB in steps of 10 dB.
- Figure 8.77 shows the received spectrum at the output of the booster for the case that VoA = 1.5, 10 and 20 dB. The crosstalk of the 100 G signal level can be estimated and tabulated in Chapter 2.

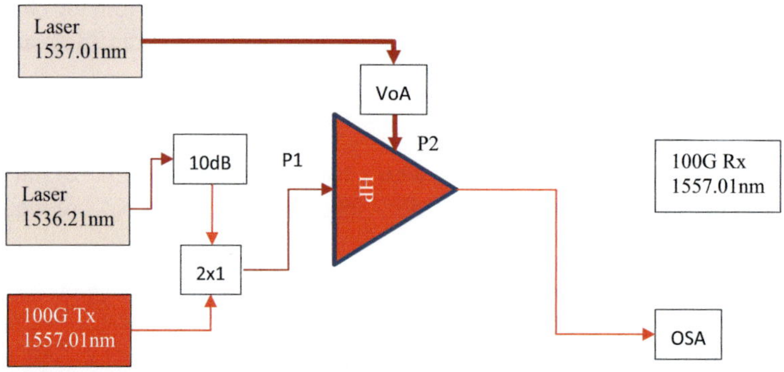

FIGURE 8.76 Set-up to test and observe NL FWM products with carriers and 100 G channel. Observation at input port P4 of LNA preamp. VoA at 1537.01 nm and 100 G coupled with 1536.21 nm (with 10 dB).

FIGURE 8.77 The 100 G DP-QPSK (1536.61 nm) signal and dual carriers (1536.21 and 1537.01 nm) at the front of LNA or at the “output of HPA booster and VOA (dB)” with step of 10 dB, i.e., 1.5 dB (yellow), 10 dB (red), 20 dB (green), 30 dB (unchanged from 20 dB). VOA > 20 dB traces are overlapped.

Note on Figure 8.77:

- Carriers at 1536.21 and 1537.01 nm VOA is placed at 1537.01 nm carrier. The carrier 1536.21 nm is coupled with 100 G signal via a 2×1 FC (4.1 dB-IL).
- The FWM product by the two carriers is located 0.8 nm from the carrier, as the spacing between the carriers is 0.8 nm.
- The FWM product is located at 0.4 nm from (1536.21 to 0.4) nm and (1537.01+0.4 nm) with a full 100 G band at a baud rate of 25 GBaud with a bandwidth of 25 GHz plus 30% FEC or 32.5 GHz (Double sideband) (Figure 8.78–8.88).

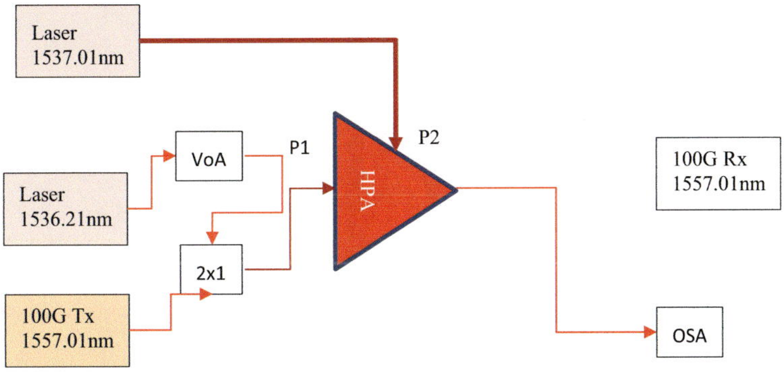

FIGURE 8.78 Set-up to test and observe NL FWM products with carriers and 100 G channel. Observation at input port P4 of LNA preamp. VOA placed after 1536.21 nm and combined with 100 G.

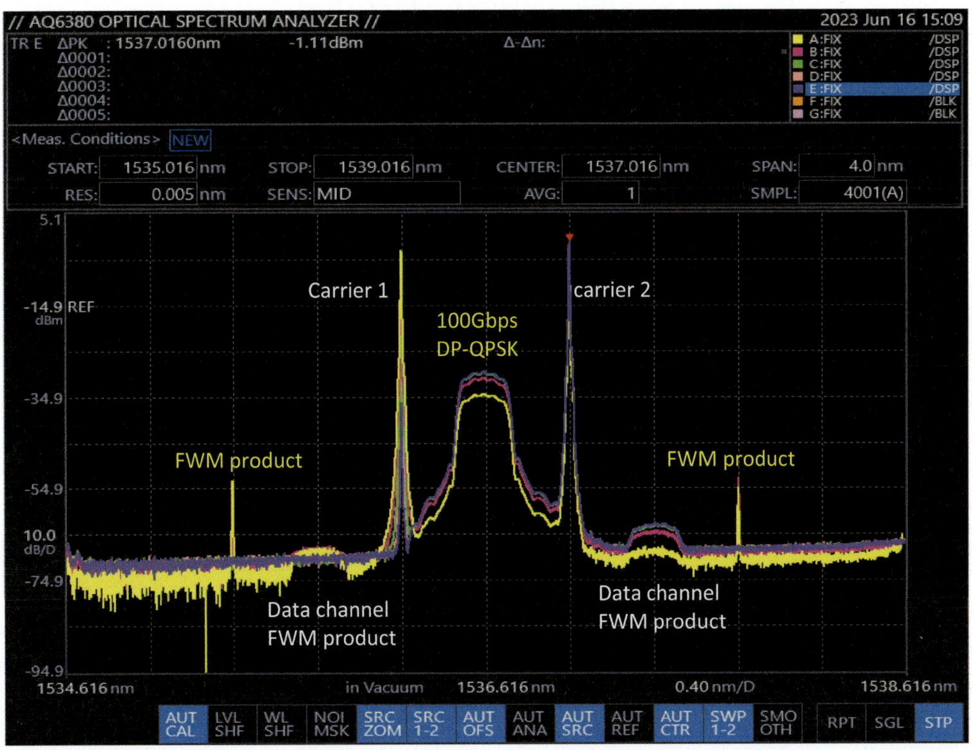

FIGURE 8.79 Signal and carriers (1536.21 nm -VoA and 1537.01 nm) at the front of LNA or at the output of HPA booster and VOA (dB) with steps of 10 dB for VoA, i.e., 1.5 dB (light), 10 dB (dark line), 20 dB (dark-light line), 30 dB (unchanged from 20 dB) dB. VoA > 20 dB traces are overlapped.

8.8.5.1 Case NL1: VOA = 1.5 dB on Carrier

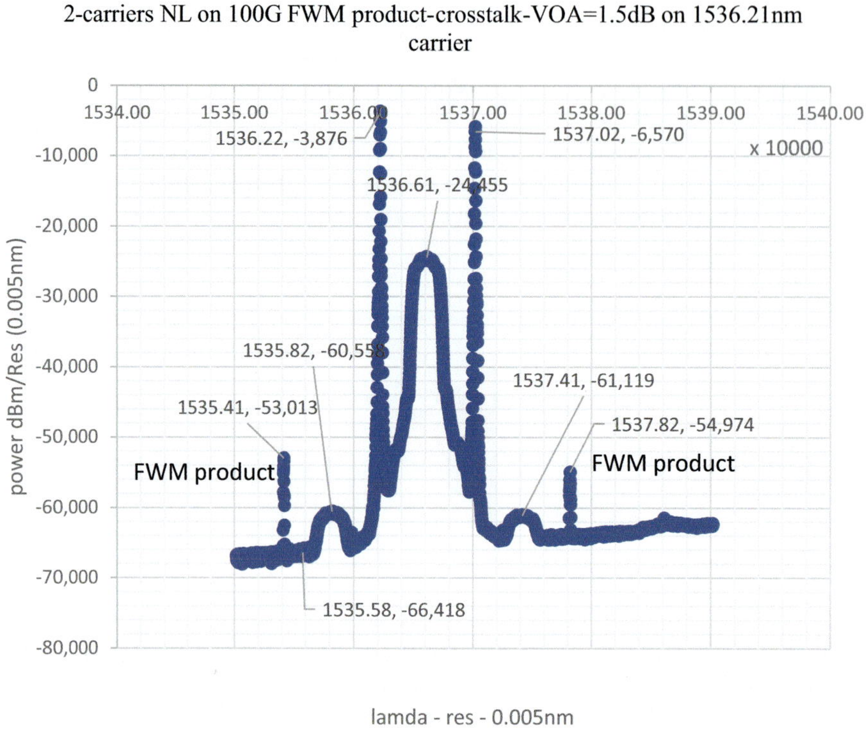

FIGURE 8.80 The 1.5 dB VOA on carrier and NL FWM products between carriers and 100 G channel and FWM product of two carriers. Only 1.5 dB difference in the crosstalk and carrier crosstalk when the difference in carrier power is 1.5 dB.

8.8.5.2 Case NL2: VOA = 10 dB on Carrier 1536.21 nm

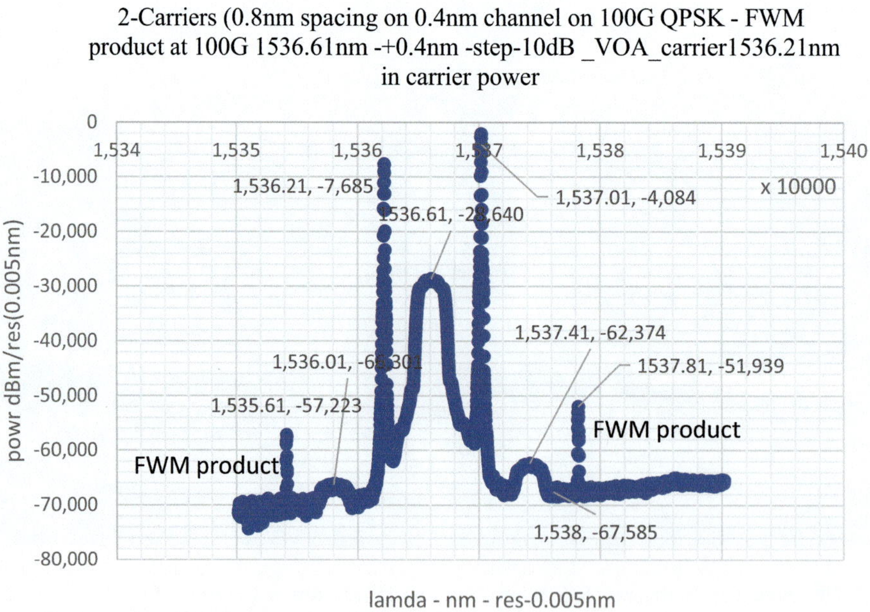

FIGURE 8.81 The 10 dB VOA on carrier at 1536.21 nm, interaction with 100 G and carrier at 1537.01 nm.

Comments: The FWM product of 100 G at 1537.21 or 0.4 nm from 1536.61 is an interaction of the 100 G at 1536.61 nm and the 10 dB down carrier at 1537.01 nm. In addition, there is a nonlinear FWM product of the two carriers to generate FWM carrier at 1537.81 and 1535.61 nm. The power level of the 100 G crosstalk bands differs by about 5.0 dB (−62,374 + 57,223 dB). The crosstalk level is about 5 dB above the noise floor.

Similar to the case where VOA = 1.5 dB on the carrier, a 10 dB difference between the carriers is only 5 dB in 100 G crosstalk and also the carrier FWM products.

The crosstalk level about 30 dB below the 100 G power level. Thus, the conversion efficiency is about – OSNR signal = 23 dB, crosstalk noise of 5 dB or the penalty on OSNR = 4 dB from 10 dB to 20 dB carrier difference.

8.8.5.3 Case NL3: VOA = 20 dB and Higher on Carrier

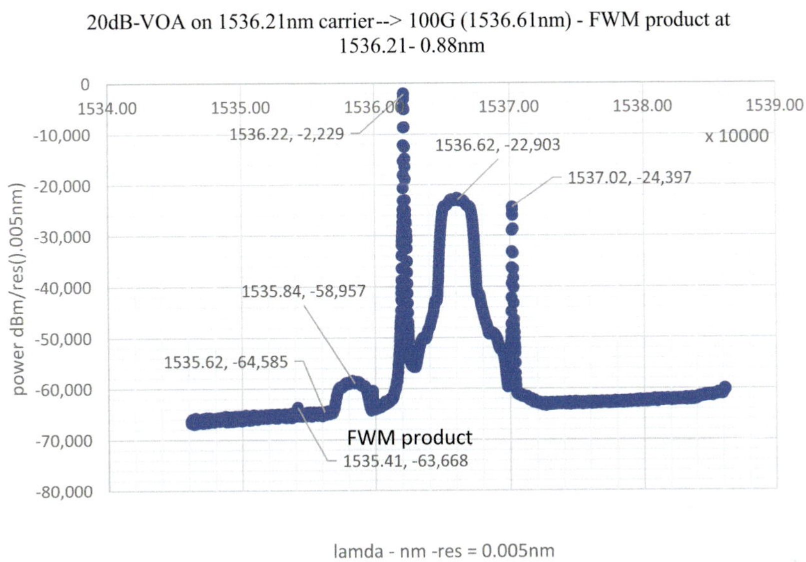

FIGURE 8.82 The 20 dB VOA on carrier at 1537.02 nm and nonlinear 100 G crosstalk band, as well as carrier FWM product.

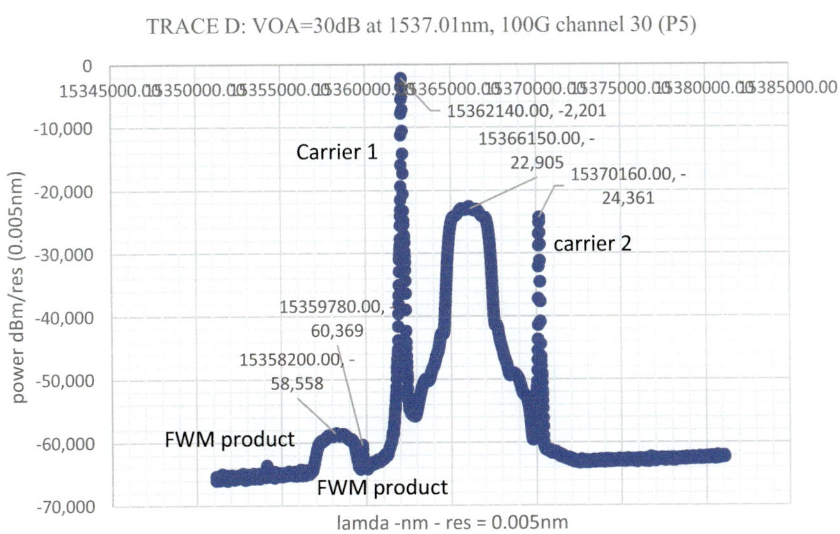

FIGURE 8.83 The 30 dB VOA on carrier at 1537.01 nm.

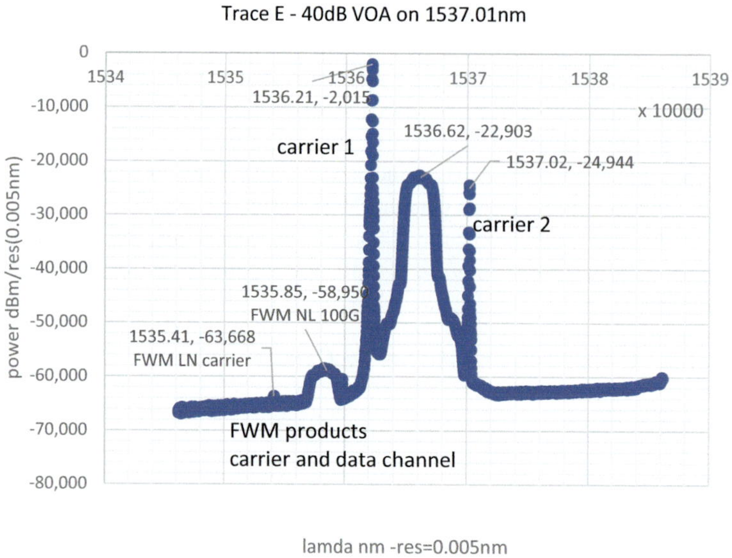

FIGURE 8.84 TRACE E (Figure 8.79): VOA = 40 dB, NL FWM product of 100 G channel and FWM of one band carrier.

8.8.5.4 Case NL4: One Carrier Plus 100 G Channel @1537.01 Setting VOA = 1.5–30 dB

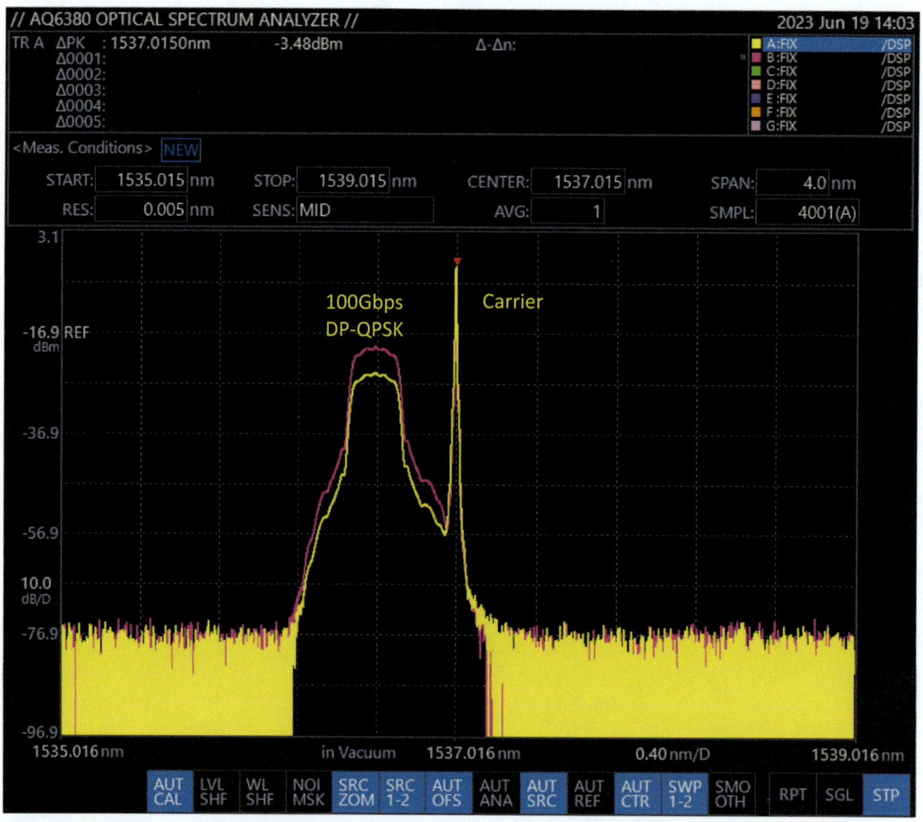

FIGURE 8.85 Spectrum of two cases VOA = 1.5 dB (light-line) and VoA = 10 dB (pink) at output of LNA. All 100 Gbps DP-QPSK data and FWM carriers are filtered by ADD/DROP OBPF of LNA output P5 (100 GHz), so no carriers here.

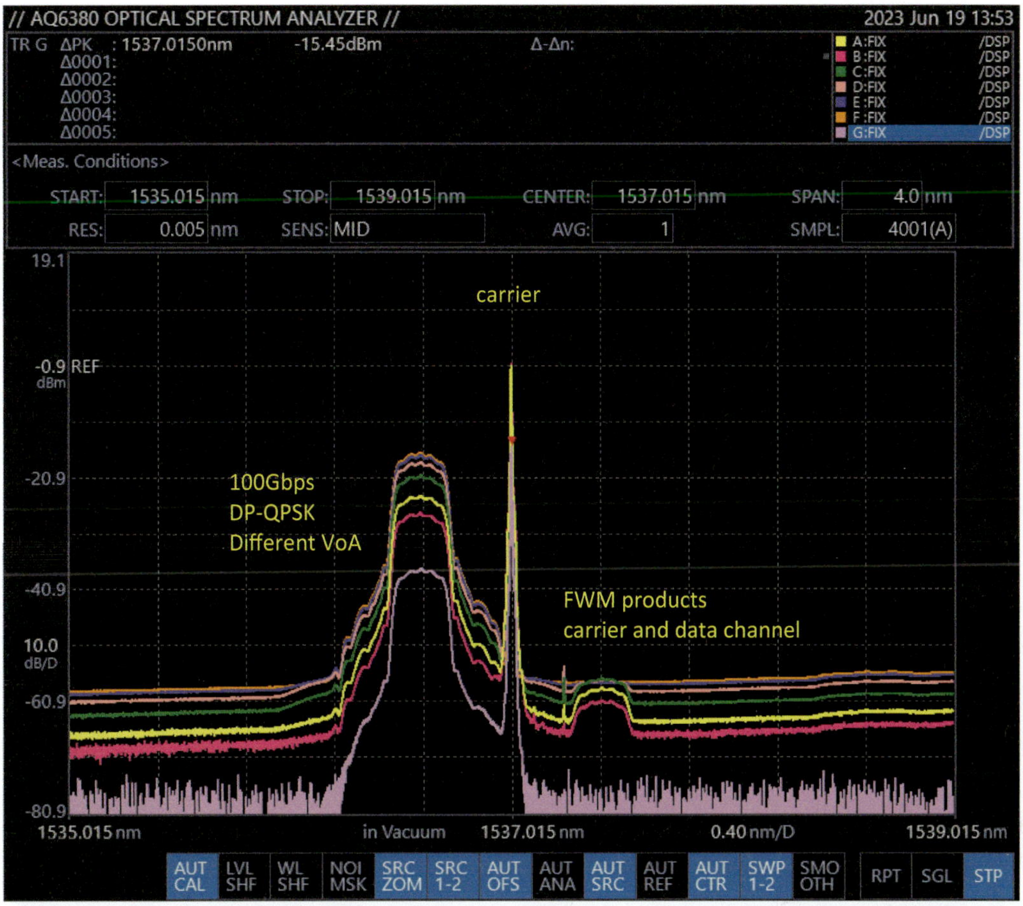

FIGURE 8.86 Trace G: output of LNA at VOA = 30 dB – all filtered. Input of LNA – i.e., output of HPA for (see also notation on OSA screen), (A) 1.5 dB; (B) 5 dB; (C) 10 dB; (D) 15 dB; (E) 20 dB; (F) 25 dB; (G) 30 dB – HPA off with no power seeding. Seeding at 1537.01 nm. FWM product 100 G (1.5 dB B = 8 dB above noise level at –72 dBm); (B = 5 dB – FWM 100 G = 8 dB above noises at –66 dBm); (C) VOA = 10 dB – 100 G FWM = 6 dB above noise level at –61 dBm); (D) VOA = 15 dB, 100 G FWM = 4 dB above noise level at –56 dBm); (E) no FWM 100 G, noise at –52 dBm.

8.8.5.5 Case NL5: Two Carriers Plus 100 Gbps IN/Out of LNA

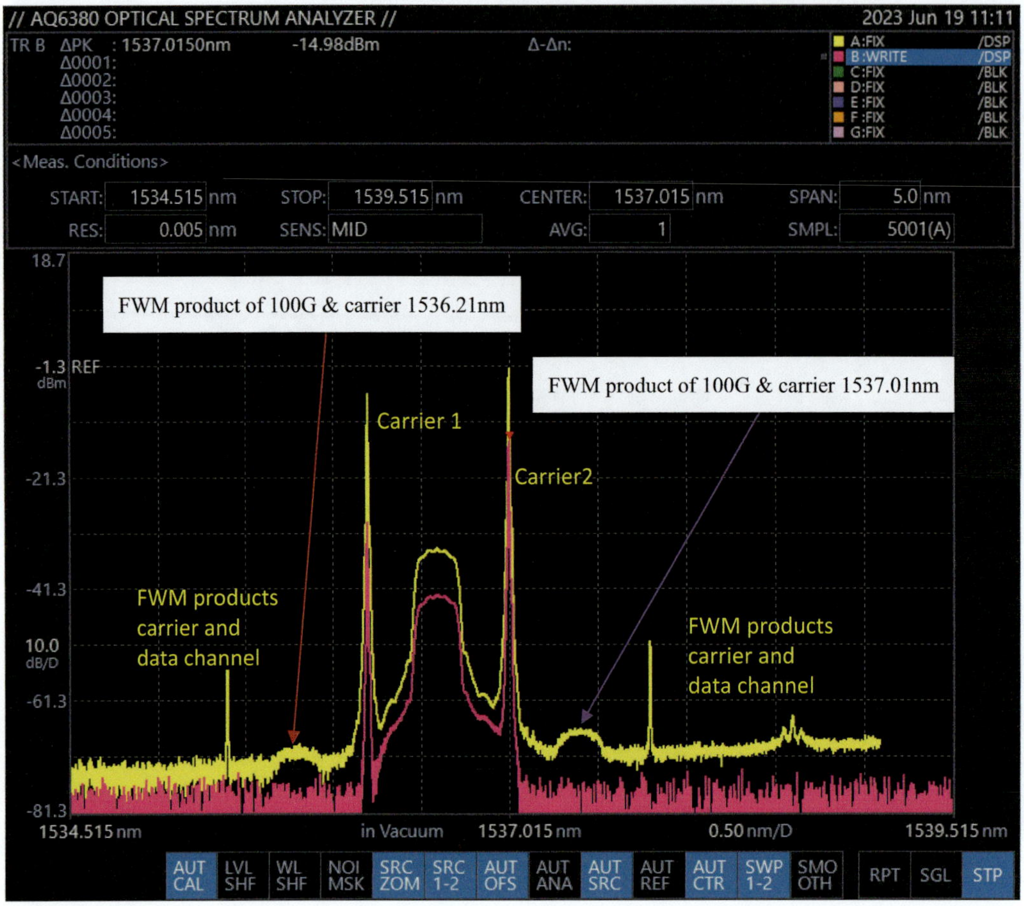

FIGURE 8.87 Two carriers at 1536.21 nm (coupled with 100 G) and 1537.01 nm (seed-laser) with power 10 dBm, coupled with 100 G via a 2×1 coupler. (A) 100 G+two carriers and FWM products of two carriers and carriers and 100 G – upper band and lower sideband. (B) Output of LNB (signal filtered- however BER reduced significantly. Gain (HPA)=23.4; G(LNA)=29.2. Carrier in band of OBPF of LNA port 5. The 100 G power –3 to 4.1 dB (fiber coupler) – carrier at 10 dBm -FC (4.1 dB) → 5.9 dBm -both are at saturation power level of HPA output. OSA of LNA output is obtained via a 1×2 coupler (4.1 dB loss).

8.8.6 BER and OSNR under FWM Effects

8.8.6.1 Set-Up

We monitored the BER and OSNR of the 100 G module when the two carriers were spaced at 1536.81 and 1537.01 nm, so that the FWM product was about –30 dB below the power level of the two carriers. This FWM product falls exactly at the center of the 1536.61 nm of the 100 G channel. The link was set up with one hop, and the Rx OSNR and the BER were monitored.

8.8.6.2 OSNR and BER vs Rx Sensitivity (dBm)

The BER and OSNR are plotted with respect to the input power injected into 100 G Rx as given in Figures 8.89 and 8.90.

8.8.6.3 Discussion

As we can observe:

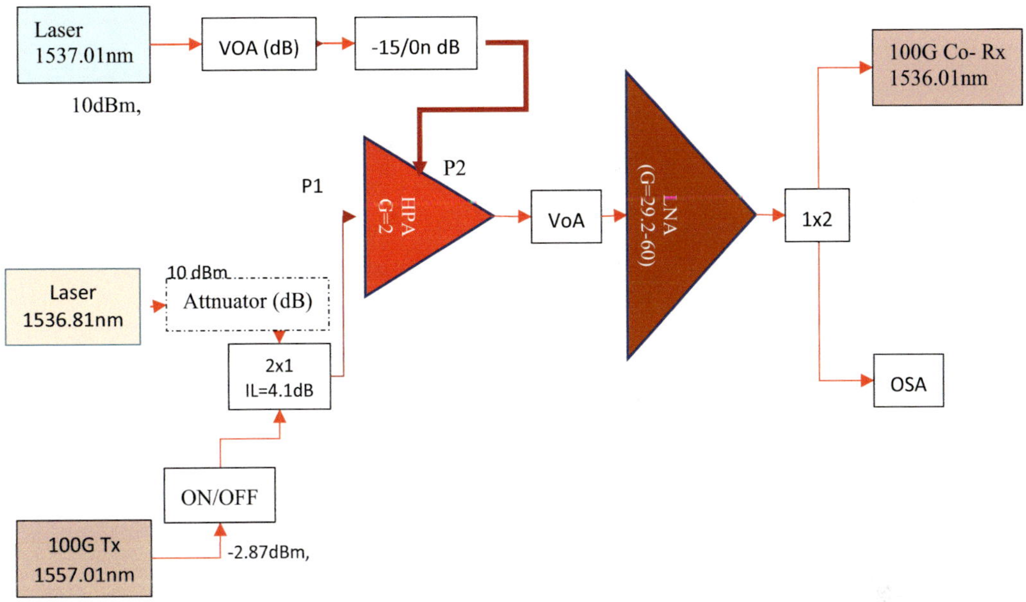

FIGURE 8.88 Set-up of one-hop transmission to evaluate the nonlinear distortion in 100 G QPSK transmission with two carriers at 1536.81 and 1537.01 nm to generate FWM carrier product at 1536.61 nm. VOA can be exchanged at two outputs of the carrier lasers.

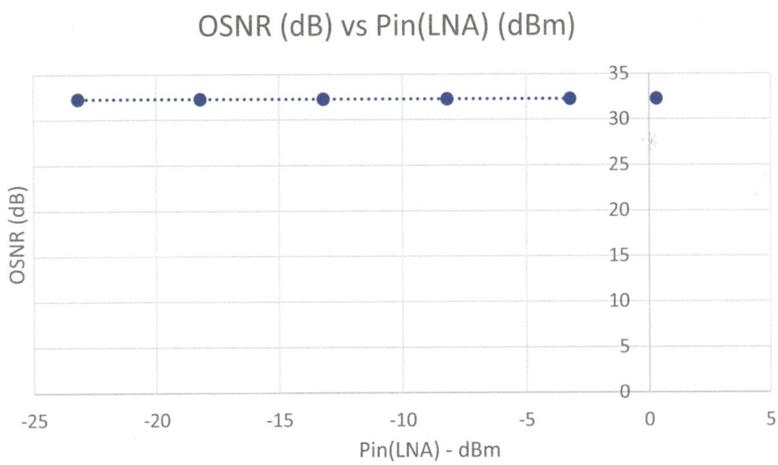

FIGURE 8.89 OSNR of Rx 100 G plotted against input power into Rx.

1. The OSNR does not change with respect to the received power, which is at the optimum value of 32.3 dB OSNR. The OSNR would change to 16.7 dB if the seedling laser power is lowered with –35 to 40 dB attenuation. However, at this level, the HPA cannot operate stably due to the power being lower than the required power for amplification process.
2. The BER changes accordingly when the received power is greater than –10 dBm. This is due to the FWM products creating distortion in the received 100 G signal.
3. TheOSNR is predetermined by processing the algorithm, and additional noise due to the FWM cannot be distinguished. Hence, it remains constant at 32.3 dB OSNR.

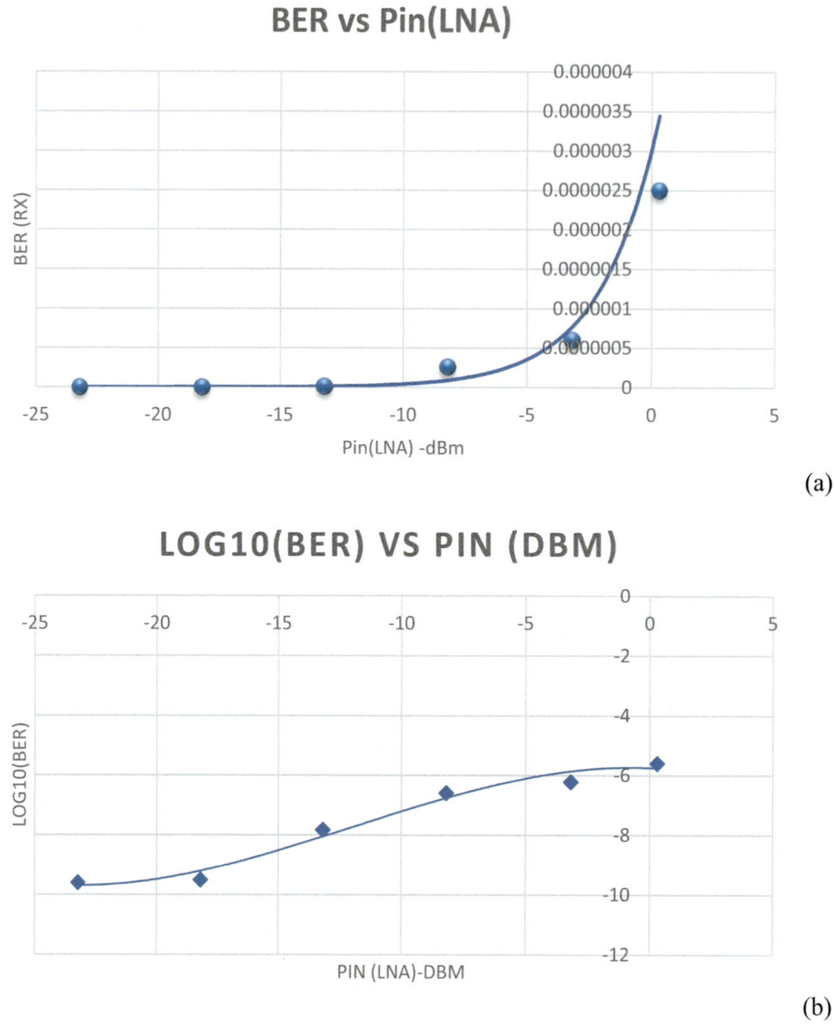

FIGURE 8.90 BER of 100 G Rx as a function of received power (dBm) after 1×2 fiber coupler into both Rx and OSA. (a) BER plotted against received power (in dBm) (b) $\log_{10}$ (BER) versus input power into LNA.

8.8.6.3.1 Penalty of 100 G Due to Nonlinear FWM Effects

As discussed previously, the nonlinear coefficient of the high-power pumped fiber is $n_{2^*}P/A_{eff}$, where P is the total power of the lightwaves propagating in the fiber and the effective area of the fiber mode (50 μm square for SSMF – standard SMF). In the case of four-wave mixing, the power is a multiplied product of the amplitude of the 100 G and the pump power of the carriers or that of the two carriers to generate the FWM product at the 1536.61 nm channel 100 G. We observe that the FWM product is about 30 dB below the power of the two carriers. Thus, a 0.1% efficiency of the FWM effect is obtained.

As shown in Figure 8.87, the carrier peak power is –3 dBM, and the 100 G is –33 dBm/res (0.005 nm). Thereby, the total power 100 G is –15.6 dBm into the Rx detector. So, if we lower the carrier to an amount of 20 dB, we would operate at the level of another channel of 100 G. The crosstalk amplitude power at this level is about 5–8 dB (at –2 dBm carrier power), the amplitude is reduced by ten times due to the amplitude product of the two channel and the carrier. The crosstalk can then be estimated to be of an amount of about 10% of the crosstalk amplitude or in power about 2 dB. So, we can estimate that the penalty could be about 3 dB of the OSNR.

Observing the BER, we can see that a change of one unit of $\log_{10}$ (BER) can result in the reduction of the input power of about 15 dB or 1e-3 in BER. Thus, a reduction in OSNR of about 2–3 dB is estimated. This amount can be higher when the FWM nonlinear effect originates from the whole 100 or 200 G of the next two channels adjacent to the considered channel.

8.8.6.4 Two Waves Mixing

Two-wave mixing in the nonlinear fiber operation can be observed in Figure 8.91.

As shown in Figure 8.91, the mixing of the 1537.01 nm carrier and 100 G channel whose carrier is located at 1536.61 nm. The noise level has increased or decreased accordingly depending on the average power of the 100 G channel when the variable attenuator is varied. The 100 G channel spectrum is shown in Traces A-F. Trace G shows the spectrum of the 100 G channel at the output of the LNA.

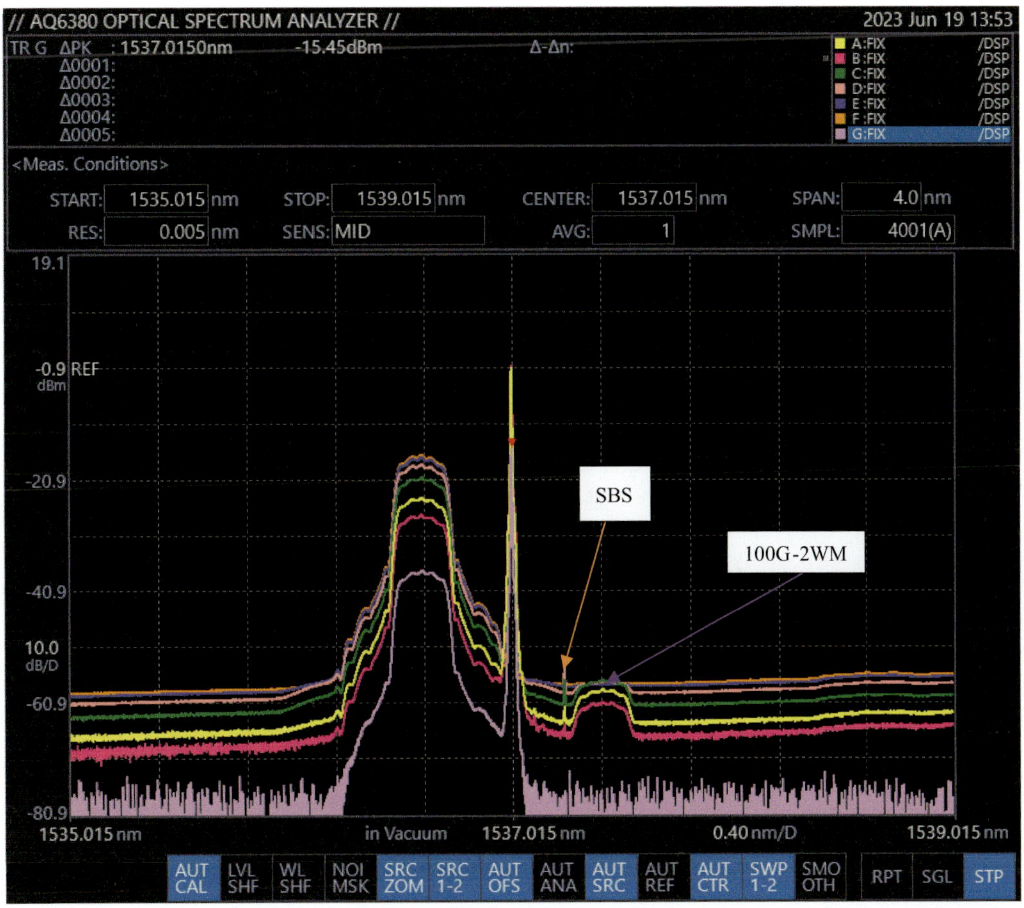

FIGURE 8.91 Two-wave mixing nonlinear product between carrier and 100 G channel. Carrier power varied in steps of 5 dB from 1.5 dB minimum of VoA of Figure 8.88, i.e., 1.5, 5, 10, … dB noted by traces ABCDE. Trace G is the output of LNA pre-amp. Carrier at 1537.01 nm or 50 GHz from 100 G central wavelength at lowest power level of carrier. 2WM = Two wave mixing.

8.8.7 FWM Product as Function of Mixing of Two Carrier

We conduct the mixing of two carriers such that the FWM product falls into the 100 G signal band. The arrangement of the two carriers is: f_{c1} = 1537.41 nm and f_{c2} =1537.01 nm (seed laser carrier).

The signal carrier is at 1536.61 nm. The FWM mixing product as a function of the carrier power is given in Table 8.4.

The VOA is placed at the position just after the carrier fc1 so that the variation of this carrier interacts with that of the seed laser f_{c2} to create the FWM product, which falls exactly at the center of the 100 G channel (1536.61 nm). The plot of the f_{c1} power versus the FWM product is given in Figure 8.92. The power of the FWM product at 1536.61 nm (ch# 51) is appreciable and varies from –58 to –45 dBm for the carrier f_{c1} of power ranges from –24 to 5.8 dBm.

TABLE 8.4
Power of the carriers and the FWM product

	Initial Power (dBm)	HPA Out (dBm)	FA (dB)	VoA (dB)	fc2 (seed)	Power (dBm)-fc1	fc3(4 WM) @1536.61 nm
fc1 (1537.41)	5.9	5.8		30	–9.2	–24.2	–58.29
fc2 (1537.01-seed)	10	5.8	–15	25	–9.2	–19.2	–53.66
				20	–9.2	–14.2	–49.86
fc3-FWM) (1536.61nm)				15	–9.2	–9.2	–45.71
				10	–9.2	–4.2	–43.71
				5	–9.2	0.8	–45.48

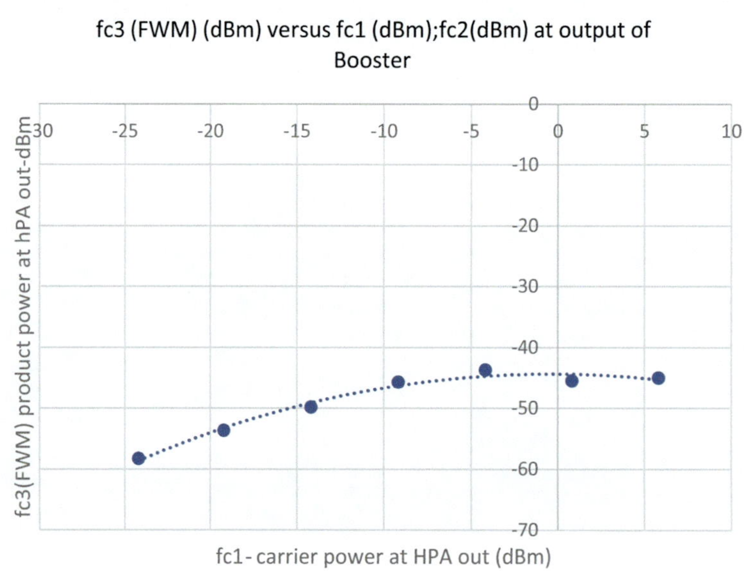

FIGURE 8.92 FWM product power at output of HPA versus the power of the carrier f_{c1} located at 100 GHz spaced from the 100 G channel to interact with a seed laser at 1537.01 nm.

The channel signal level of the 100 G channel would be about –35 to –42 dBm when reaching the preamp located at the end of the optical ISOL. Thus, we expect the FWM product to be in the range of –60 to –65 dBm, which is also within the lowest input range of the amplifiers. Therefore, the output power of the FWM will be as discussed in the following. With an input channel 100 G power of 5.8 dBm (saturated output power of HPA) and using the same HPA so the power of the FWM product is expected to be about 25–30 dB below the signal power and varied based on results displayed in Figure 8.92. This generates a penalty of about –8 dB below the 100 G band, which is integrating over the DSB of the 100 G channel.

8.8.8 Penalty on OSNR Due to In-Band Crosstalk by NL Scattering

In Penalty on OSNR due to In-band Crosstalk by NL Scattering Figure 8.14a, the in-band crosstalk is illustrated under the signal band envelope, resulting from various NL distortion effects as discussed earlier. The probability distribution function (PDF) of the summation of all nonlinear effects can be seen in Figure 8.31c, allowing for the estimation of the signal penalty due to the power of the cross-talks in relation to that of the signal power contained under the signal power spectrum Xc. It should be noted that if in-band noises remain below 15 dB, the penalty will be around 1 dB. This is a critical consideration for designing the length of the SMF fiber at the output of the booster high-power amplifier (HPA or OPA) to prevent in-band crosstalk on the channels. In instances where FWM, XPM, or SPM distortion effects occur, assigning the optical frequencies of the multiplexed channels with unequal spacings should prevent crosstalk from falling within any of the bands of the channels (Figure 8.93).

8.9 MULTI-TERA BITS/S TRANSMISSION

8.9.1 Modulation and Multiplexing

Reaching a capacity of Tbps can be achieved by polarization multiplexing and wavelength multiplexing in the optical domain. Further increases in the aggregate bit rate can be achieved by employing multi-level modulation in the complex plane, as shown in Figure 8.94. For example, with 256 QAM or 256 constellation points in the complex plane, the number of bits per symbol is 8. Thus, for five wavelength channels and dual polarization multiplexing, we can obtain an 80×symbol rate. So, for a 64 GSy/s rate, the aggregate capacity is 5.12 tera-bits/s.

Naturally, the increased number of bits per symbol must satisfy the required SNR so that error-free transmission can be achieved. Under modern FEC techniques, a BER of 2.e-2 can be achieved with an SNR of only 3.5 dB for the QPSK modulation format, as shown in Figure 8.95.

We also show the region in which the QKD system can operate; this is the encircled area that can only be inspected in detail in Figure 8.95, when quantum noise fluctuations and the Heisenberg uncertainty principle are applied. Further improvement in receiver sensitivity can be achieved by statistical modulation. Such a modulation scheme and the effective bits per symbol for 16 QAM modulation are shown in Figure 8.95. In statistical modulation, the constellation points are generated by modulation in such a way that their geometrical Euclidean distance between points is minimized so that the overlapping of the probability distribution is minimum , thus reducing the BER. Under such probability feature, some constellation points may be missing and thus the effective number of bits per symbol can be reduced, thus a minor reduction of the effective number of bits per symbol, thence the reduction of the resultant aggregate bit rate. Figure 8.9 shows the schematic structure of a high-speed transceiver, which consists of two sections, the optical transmitter and the coherent reception and electronic digital signal processing (DSP). The DSP serves as both the transmitter and receiver sections (Figure 8.96) to generate pre-distortion, hence to compensate for

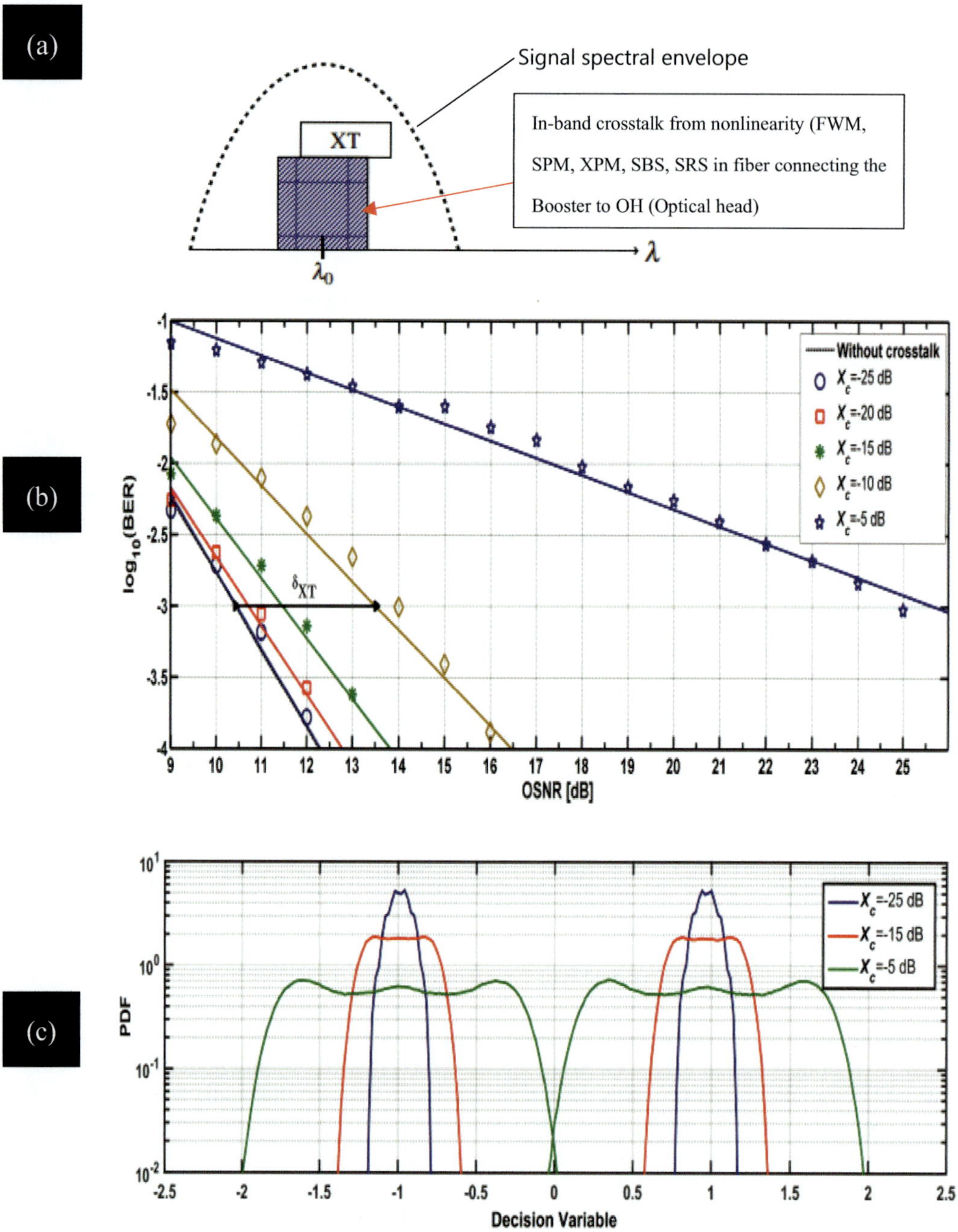

FIGURE 8.93 (a) Signal spectrum and the in-band crosstalk (blue block) noise falling into signal band, the in-band crosstalk; (b) Log10(BER) plotted against OSNR of the receiver; (c) Penalty of performance (POF) versus the decision level on the probability of error between constellation points.

any distortion of the transmission media or post-compensation for and distortion of the transmission. Besides these sections, there is an optical hybrid sub-section for processing optical signals, polarization demultiplexing, quadrature phase shift, and local oscillator mixing with signals prior to the electronic photo-detection. Therefore, the QAM modulator for both polarizations uses integrated optic Si-photonics.

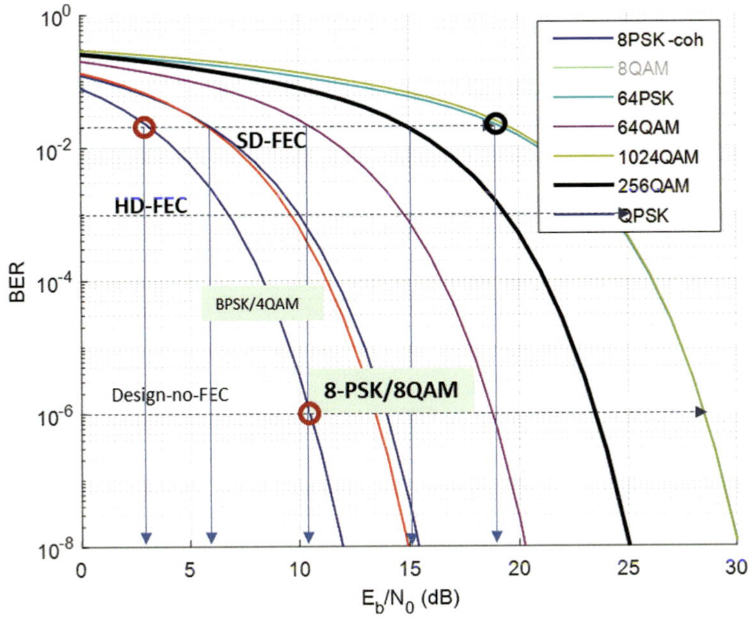

FIGURE 8.94 Bit Error Rate (BER) versus SNR (E_b/N_0) for different modulation formats. HD-FEC = hard decision Forward Error Coding, SD-FEC = soft decision FEC.

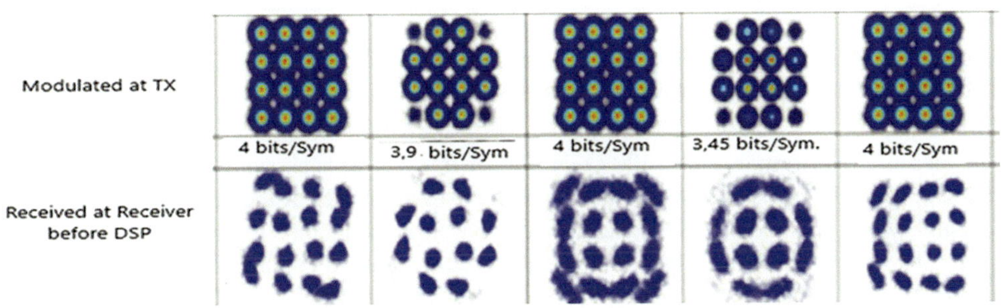

FIGURE 8.95 Constellations and effective bits/symbol for 16 QAM under statistical modulation.

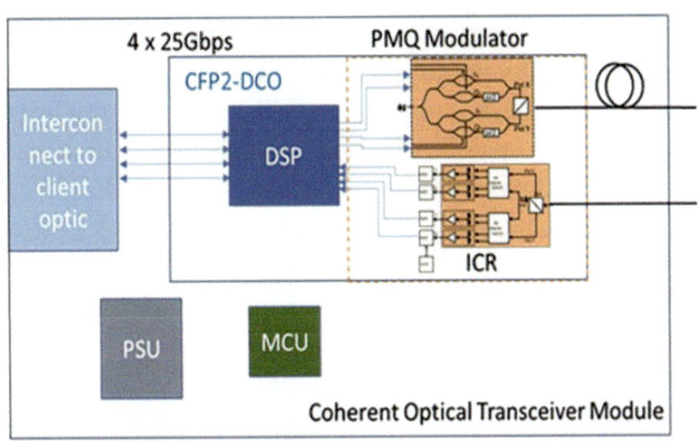

FIGURE 8.96 Typical structure of an integrated DP-QPSK optical transceiver to obtain 100 G, 200 G or 400 Gbps. DSP = digital signal processor, ICR = integrated coherent receiver, CFP2-DCO = Compact Form Factor 2 digital coherent optics, MCU = micro-control unit, PSU = power saving unit, PMQ = Polarization multiplexed quadrature modulator.

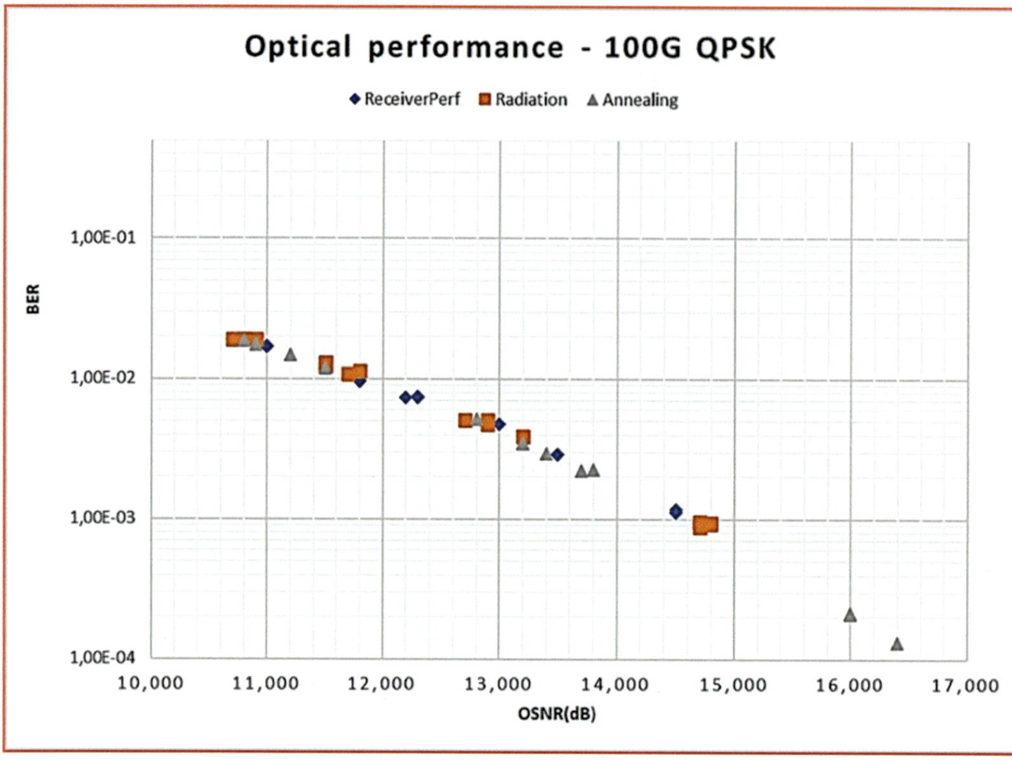

FIGURE 8.97 Measure receiver sensitivity as a function of optical SNR of DP-QPSK receiver. Quantum level of 5 photons/bit at –42 dBm for 100 Gbits/s module.

The measured performance of a 100 G DP-QPSK transceiver is shown in Figure 8.97, in which the BER is plotted against the OSNR with specific average power at the receiver. The integrated optical and electronic module is evaluated for space transmission and employs DP-QPSK modulation format with integrated DDSPs and lasers for both the optical carrier and local oscillator. A 5 photons/bit can be achieved at –42 dBm for 100 Gbits/s with a symbol rate of 25 GSy/s, that is reaching the quantum limit of the coherent receiver. The local oscillator power is injected at 5 dBm per polarization to boost sensitivity. The integrated optical transceiver module is shown in Figure 8.96 in which the upper section shows the IQ optical modulator and the lower section shows the optical hybrid, which splits the polarized beams and performs orthogonal phase shifting of the in-phase and quadrature channels, followed by photo-detection, amplification, and digital processing.

Statistical modulation can be used to improve the BER and sensitivity by ensuring the maximum distance between consecutive bits or points on the constellation. The constellation diagrams are shown in Figure 8.42. While the number of bits per symbol can be reduced, as stated in Figure 8.42, the gains in probability of error reduction and the sensitivity are much greater.

8.9.2 Optical Amplifiers

Due to the long transmission distance, the total loss of the signals can reach 80–90 dB after accounting for the gains of the optical systems due to telescope optics at the transmitter and reception systems. Such a high attenuation factor must be compensated by using a booster optical amplifier at the transmitter to launch the laser beam into space and an optical preamplifier in front of the coherent receiver. The optical power at the booster amplifier can reach 10–20 W (40–43 dBm) and the gain of the optical pre-amplifier can reach 50 dB with a low sensitive input level down to –5 dBm. In addition, the coherent reception can boost the signal gain by 20–30 dB with a moderate average power of the local oscillator laser.

8.9.3 Impairments and Mitigation over Space and Long-Haul Transmission Channels

SOCS and long-haul installed fiber systems can face a number of impairments. Distortion of the optical wave-fronts occurs due to optical systems required to launch optical beam through the free-space channel from the transmitter and reception at the receiver. The WFE can be directly related to the error vector magnitude in the digitally processed data at the receiver. A digital pre-distortion at the transmitter can be used to mitigate this distortion effect.

The frequency shift of the optical carriers occurs due to the DS effects under different rotation speeds of the satellite constellation of different orbits. For example, the DS can be –7.5 to +7.5 GHz twice a day between the satellites in GEO and LEO rotating at speed 7 and 3 km/s, respectively. These DSs can be mitigated by tuning the frequency of the LO (local oscillator) laser so that homodyne detection can be achieved without the DS effect.

Crosstalk and in-band crosstalk occur due to nonlinear effects in fibers, which follow the optical booster amplifiers, even with a short length of a few meters. The self-phase modulation or cross-phase modulation effects play a critical role in this crosstalk and intermodulation effects. Such crosstalk should be limited to <1 dB by using a large effective area (spot size) fiber.

Polarization rotation via propagation of free-space beam can be mitigated using an adaptive optics system and control signals from the DSP system.

Dispersion and random polarization mode dispersion of lightwaves passing through fibers interconnection modules of integrated transmitters and receivers, ROADM etc can cause signal distortion. These signal degradation can be compensated by digital processing algorithms which are stored in the memory section of the DSP integrated coherent modules. These compensation algorithms can be found in Chapter 6.

8.10 CHANNEL SPACING AND PACKING CAPACITY FOR MULTI-TBPS TRANSMISSION OVER FREE SPACE

8.10.1 Pulse Shaping and Packing and Sensitivity

In free-space optical systems, for bidirectional transmission and reception, a dichroic beam Tx/Rx separator can be designed with different spectral windows for each direction of channels for Tx & Rx to allow a dual-polarization optical transmission system per optical carrier. This means that each polarization of the lightwaves can be modulated by different information sequence. Thus, two data sequences can be carrieried by one lightwave beam, thus doubling the total transmission capacity. This enables the use of coherent COTS/terrestrial modules. However, it requires a minimum "dead band" between Tx/Rx of >10 nm (on axis) and 20 nm (±2° internal range). Note that the transmittance and reflectance – OLPF and OBPF (optical low/band pass filter) – with a gap for bidirectional Tx and Rx possibly by multi-layer thin-film filters. The best performance of the optical amplifiers for space applications, developed under current technology, can operate in the window of 1545–1565 nm. This spectral window allows >12 × 100 G or >8 × 200 G for either range (see Chapter 2). Channel numbering can set under the ITU-T grid with 100 GHz spacing, but this is not necessarily compulsorily. Channels can be specified by frequency in GHz and channel wavelength in nm. For example, channels No. 14–22 and channels 34–42 are used for transporting/transmitting of 100 or 200 G data channels. Service channels at 10 G bps are allocated to channels 13 and 43 using intensity modulation and direct detection. Narrower spacing, such as 50 GHz, can be used, with no restriction in space laser communications systems. Indeed, unequal spacing should be used to avoid nonlinear crosstalk in fiber at output of booster high power amplifiers.

Under pulse shaping, 50 GHz spacing can be used for 200 G (16 QAM), allowing for 2×600 GHz to offer 24 channels of 200 G to obtain 4.8 Tbps or 24×400 G (16 QAM50[14]) for 9.6 Tbps. Furthermore, such a "red/green" system provides improved optical isolation between Rx/Tx paths, exceeding 100 dB even without pilot/tracking tone electrical signal filtering. This tremendously reduces false light/straylight constraints as optics is capable of handling (very) high-power levels (with high TRL). This will not constrain link scenarios (East/West in green/red) and supports two optical ground stations (red and green) at each location (also for throughput).

The obtained receiver sensitivity is –42 to –26 dBm at 100 Gbps with QPSK25 or 5 photons/bit at the minimum level. For 200 G QPSK50, it ranges from –35 to –21 dBm with a LOFO to 5 GHz with a minimum penalty of the receiver sensitivity.

Figure 8.98a1 and a2 depicts the experimental packing of 10 DWDM channels with spacing of 100 GHz and an aggregate bit rate of 100 G bps using dual polarization and QPSK format. Furthermore, pulse shaping with a raised-cosine factor of 0.7 is used to obtain sharp roll-off and distinguished sampling points. Figure 8.98b1and b2 depicts the packed DWDM channels without pulse shaping to demonstrate the effectiveness of pulse shaping.

Pulse shaping can be provided and implemented using the DSP embedded on the transmitter side. The electrical pulse sequence is shaped with a raised-cosine factor and then applied to the optical modulation. This translates to the optical intensity, and after the coherent detection, the pulse shaping can still be preserved, resulting in a sharp optical spectrum and the sampling to recover the data sequence.

8.10.2 Nonlinear Crosstalks and Channel Packing

Nonlinearity over a short length of SMF connecting the output of the booster amplifier and the optical head telescope can be serious due to SBS, SRS , Kerr effects, and cross-phase modulation. These intermodulation effects are caused by the nonlinear coefficients n_2 of the GeO_2 doped silica. These effects are serious at a few mW average power over a long length of the SMF (1–22 km), but for 1000 mW effective power level, the nonlinear distortion and crosstalks can exist to a length of a few meters and even shorter. To avoid such nonlinearity, the fiber length should be much shorter than a meter or the multi-stage optical amplifiers should be redesigned.

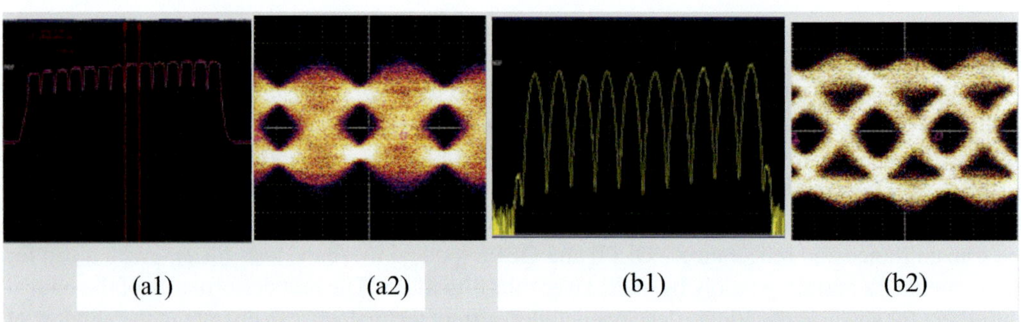

FIGURE 8.98 Packing of 100 G channels with 50 GHz spacing between channels to obtain >1.0 Tbps (a1) spectrum of 12 channels under pulse shaping with root raised cosine (RRC) and (b1) ten channel spectrum when no pulse shaping is used. (a2) Eye diagram of a channel under RRC pulse shaping and (b2) eye diagram of channel sequence under no pulse shaping.

8.11 POYNTING AND TRACKING

Figure 8.99 depicts a simplified block diagram of a digital control system for tracking the stabilization of the pointing system of the modulated laser beam. The main objectives of such a system are to reduce the residual error with less micro-vibrations from the optical platform, improve the performance of the control loop, lower noise from free-space motion and other actuators, and improve tracking detector performance. The frequency responses and noise spectral density of the pointing angle are shown in Figure 8.100, where a 3 dB passband can be found. By integrating over the passband, we can find out the total signal penalty as follows: Pointing noise spectral density ($mrad^2$/Hz) = 5e-3 at 80 Hz, and integration over the spectral area (dotted box in Figure 8.100) gives ~ 0.22 µrad. Thus, WFE can be estimated as 0.22/7 µrad, or (3.1%) ~ 0.5 dB penalty. This WFE can be directly related to error vector magnitude (EVM) in the signal processing of complex signal as illustrated in Figures 8.100 and 8.101 (see also DSP processing of Chapter 6. This dynamic penalty can be included in the operating margin of link budget design for digital coherent communication system.[15] This dynamic penalty can be offset by setting operating safety margin allowance of the system to about 2 dB. The overall system dynamic and allowable operating margin can be set at –6 to –10 dB within the receiver sensitivity of a transceiver operating at 100 Gbps with the modulation format QPSK25 such as 25 GBaud DP-QPSK of 15 dB from –26 to –42 dBm and less for QPSK50 with a bit rate of 200 Gbps.

The word "pointing" is used instead of pointing to indicate the beam as a coherent electromagnetic field, as we require the lightwave beam to be coherent so that spatial spreading is minimized, and coherent reception is used to gain further 30 dB. Thus, the pointing vector transmission over long-haul space distances.

The frequency passband of the pointing occurs at around 28–180 Hz. This creates jitter in the high data rate sequence, thus causing errors in sampling. A one-tap filter in the DSP at the receiver can mitigate this effect. The DSP processing developed for coherent reception has been enhanced and permits true coherent optical communications, as it can mitigate the frequency offset between the lightwave channel carrier and that of the LO laser, distortion due to dispersion, jitter, skew effects of the two polarized channels, etc.[16] The optical heads and the beam pointing can generate errors in the QAM constellation, which, in turn, degrades the bit error rate and causes a penalty on the OSNR, as shown in Figures 8.101 and 8.102. Indeed, WFE can be directly related to the EVM, which is commonly used in the determination of BER for complex signals using either real-time sampling or a sampling oscilloscope, the digital signal analyzer (DSA).[17]

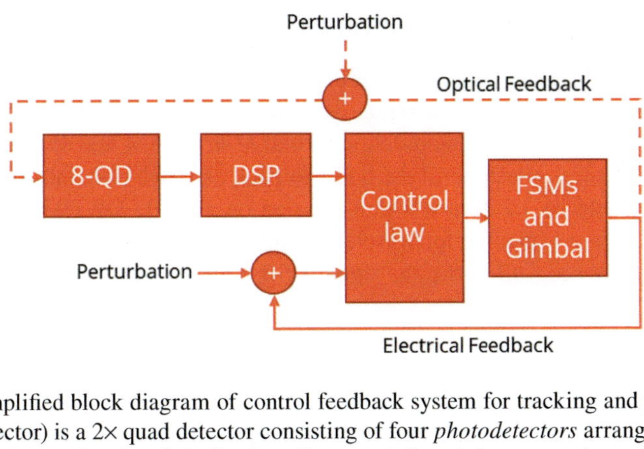

FIGURE 8.99 Simplified block diagram of control feedback system for tracking and pointing system. The 8-QD (quadrant detector) is a 2× quad detector consisting of four *photodetectors* arranged in two differential pairs crossing each other to give signals indicating alignment left or right or up or down. DSP is a digital signal processor.

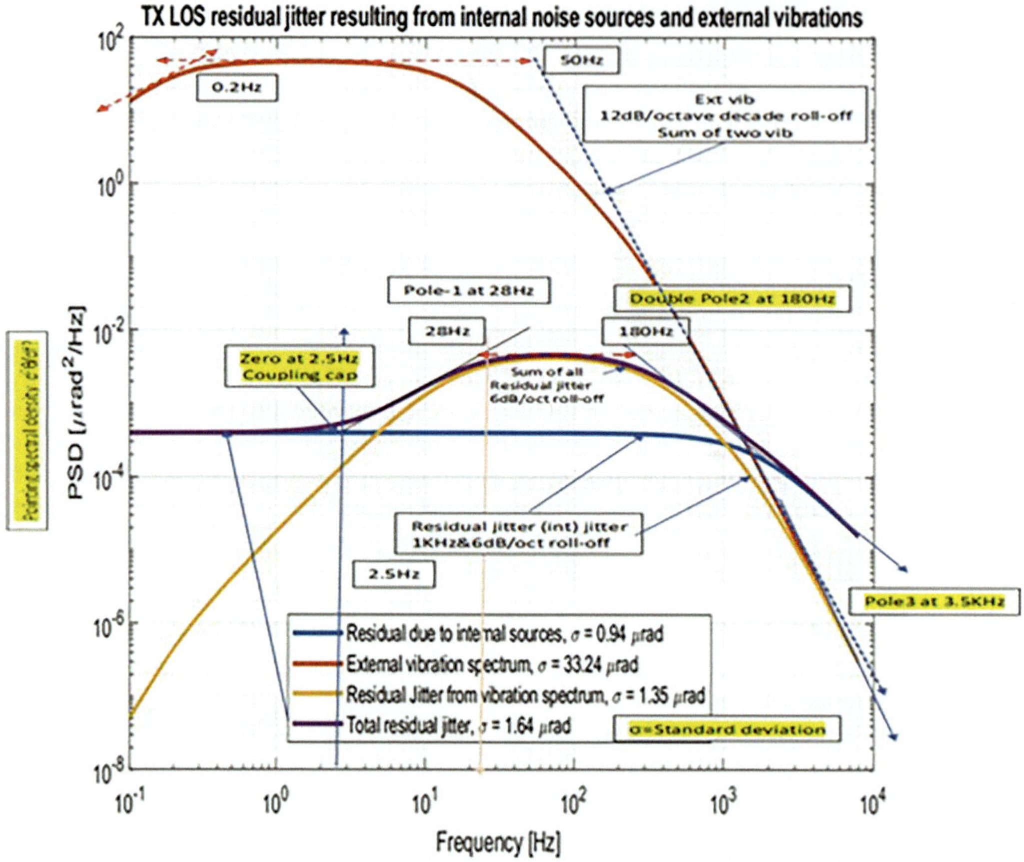

FIGURE 8.100 A typical pointing angle noise spectral density (PSD) of the controlled feedback system of Figure 8.99. Poles and zeroes of the feedback control system for pointing stability are identified. A complex pole pair is nullified by complex zeroes to avoid oscillation.

8.12 CAN 1.0 TERA-bps PER WAVELENGTH CHANNEL FOR SPC SYSTEMS AND NETWORKS?

Transmission over fiber distance, especially for data centers and some medium distances in terrestrial fiber optic networks, the transmission rate can reach 1.2 Tera-bps per wavelength carrier, and even 1.6 Tbps when a higher order of modulation format M-ary QAM and dual polarization is used.

The next medium future milestone in the quest to pack as much data into space channel transmission as possible would be 1.2 Terabit/s transmission on a single DWDM wavelength. The 100 and 200 Gbps modules are evaluated for wavelength spacing of 100 GHz and proven possible with the carrier peak power about 6 dB above the signal DSB with baud rate of 32 and 64 GB with 32% FEC overhead.

The main constraint to SPC is the pseudo-coherence of the modulated beam and propagation loss over space vacuum medium distances, as well as turbulence effects if passing through the atmospheric region. The losses can be compensated by the HPA booster amplifier and the low-noise low-input LNA preamplifier. Therefore, the ASE noise builds up over one hop and then consecutive SPC hops, which would limit the transmission bit rate of the SPC system under the Nyquist theorem on capacity limit.

However, the current generation of 1.2 T OTRx (optical transceivers) utilizes two separate 75 GHz-spacing DWDM wavelengths each running at 600 Gbps and occupying a bandwidth of

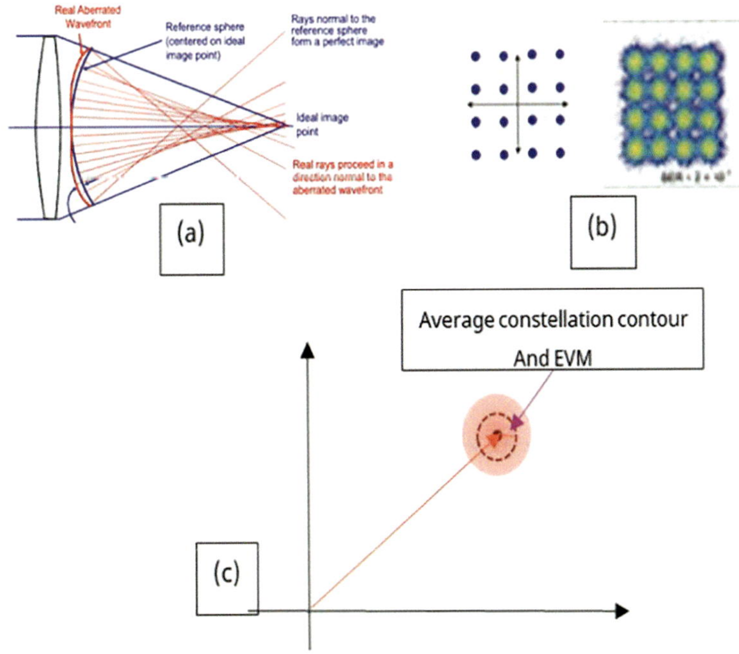

FIGURE 8.101 Illustration of EVM (error vector magnitude) and WFE similarity for inter-relationship to signal penalty. (a) Wave-front error (WFE), (b) digital complexed signal in ideal and distorted constellations, and (c) direct relation between WFE and EVM.

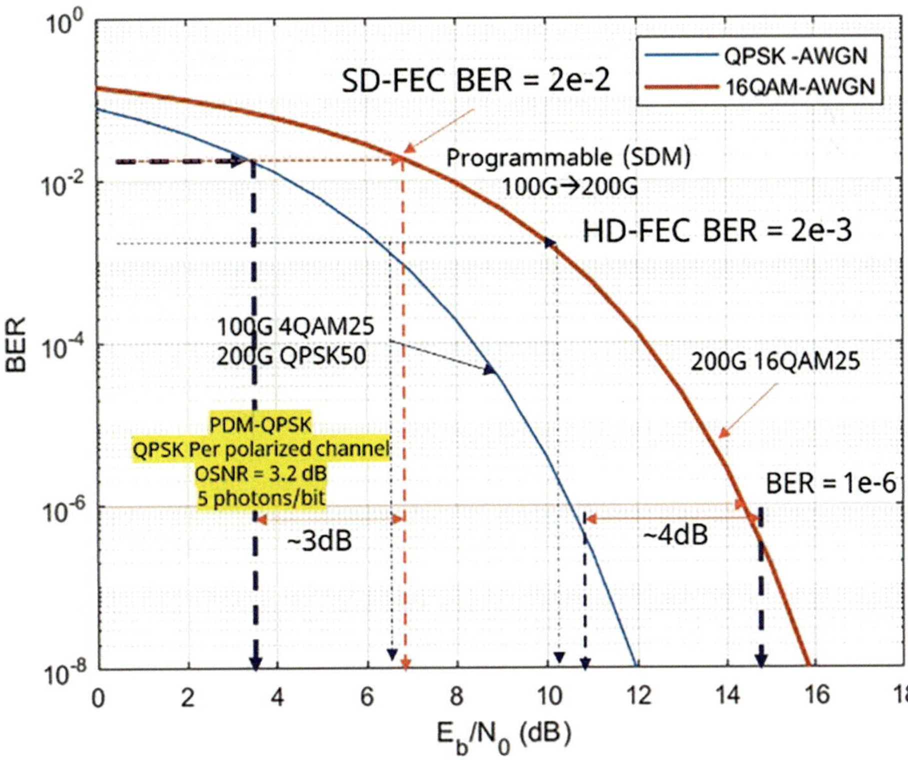

FIGURE 8.102 BER versus energy/noise per bit of QPSK and 16 QAM under FEC of soft decision and hard-decision, as well as designed rate of 1e-6.

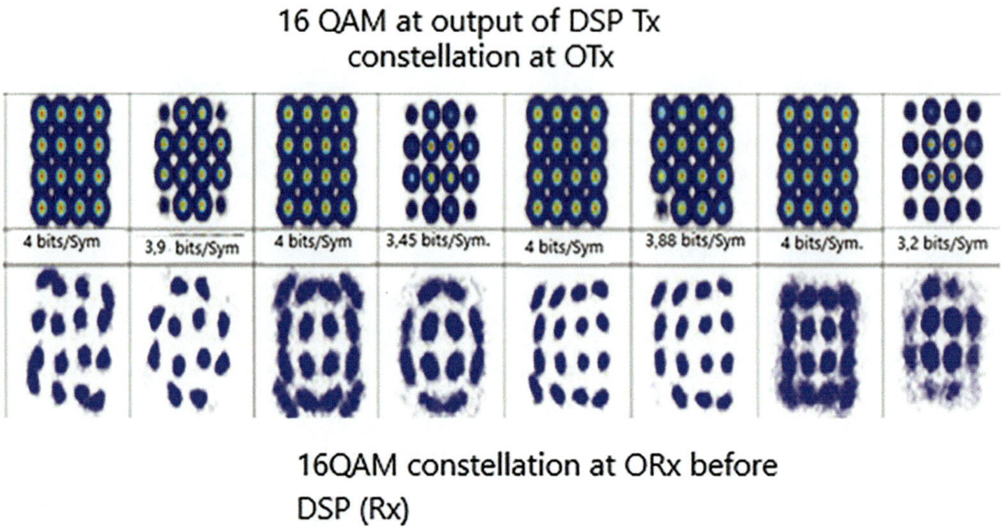

FIGURE 8.103 Constellation of 16 QAM by statistical modulation or the consecutive constellation points are such that maximum Euclidean distance is obtained.

75 GHz with sharp roll-off by pulse shaping. The baud rate is 128 GB PDM-QPSK for 64 GB if 16 QAM is used, to achieving a total capacity of 1.2 Terabits/s. The baud rate of 128 GB can offer 6 with the assistance of statistical modulation of QPSK, 16 QAM, etc., as shown in Figure 8.103 in which the Euclidean distance is maxima between the two consecutive points. Thence, the bit rate is reduced slightly.

This means that while one space transponder can support the full 1.2 T, it uses two wavelength channels in the fiber, meaning two lasers, two modulators, and two receivers. So, from a cost perspective, it would be better to transmit 1.2 T on a single wavelength rather than two carriers due to the need for coherence in the signal band and the power of the carriers we assume to be pseudo-coherent over that distance between satellites within constellation and inter-constellation.

Since coherent transmission was introduced a decade ago,[18] two factors have allowed the continuing increase in the data rate of a single wavelength.

First, coherent reception allows the use of higher order modulation, the complex modulation, to increase the number of bits encoded in a single "on-off" cycle or symbol rate, also called the baud rate. Furthermore, the use of dual polarizations of the lightwaves can double the capacity and the employment of digital processing DSP algorithms[19] to enforce the homodyne detection by phase recovery algorithm. For example, non-coherent on-off keying systems have one bit per baud, while initial coherent systems had 4 bits per baud (including the use of two polarizations). Another example is the 64 QAM, which gives 6 bits per symbol.

Second, to increase the data rate by increasing the baud rate itself, a wider bandwidth of electronic and photonic components is required. Initial coherent systems utilized 32 GBaud, but the current 600 G/lambda operates at 64 GBaud, with the band width now reaching 40–45 GHz.

Third, if we combine the benefits of both high baud rate and high order modulation, such as 64 Gbaud and 64 QAM, in polarization division multiplexing, then we note that 64 GBaud times 6 bits per symbol times 2 polarizations gives 768 Gbps, which is well above 600 Gbps. The baud rate can reach even 128 GBaud but is limited by the bandwidth of optical modulators with stable operation. Coherent systems require some amount of overhead for FEC, so we do not get the full amount of bandwidth for usable data, that is 28%–32% overhead.

So, how can we achieve 1 Tbps per wavelength?

Either the baud rate or the constellation can be increased. There are advantages and disadvantages to each, in particular in the SPC system over long-haul space hops. Increasing the constellation tends to decrease the distance that can be transmitted, because the Euclidean distance between the symbols becomes smaller. The advantage of increasing constellation is to have a higher spectral efficiency. For example, increasing to 256-QAM while maintaining the baud rate at 64 GBaud would yield 1.2 Tbps of total bandwidth, but the net data rate would be reduced due to the overhead required for FEC.

Alternatively, the other approach is to increase the baud rate itself. Increasing from 64 to 100 GBaud would give 1.2 Tbps of total capacity, along with reduction of the overhead by FEC. Depending on the actual system requirements, some combination of the two is most likely to be employed. For example, 90 Gbaud and 128-QAM give 1.26 Tbps total capacity, which can support 1 Tbps per wavelength with 20% FEC overhead.

Or, increasing either of these two parameters requires more computation by the DSP to extract the data, and hence leads to higher power consumption. However, increasing the baud rate also requires increasing the speed of the ADC front end as well as the speed of the DSP itself. These higher-speed electronics can now be delivered by AIC semiconductor chips. As the semiconductor industry moves to more advanced process nodes with smaller features, the resultant chips become faster and lower in power. The first generation of coherent systems used DSPs based on a 35 nm process, which supported 100 Gbps based on 32 GBaud and QPSK. The current 600 G per wavelength systems are based on 16 nm process nodes and utilize 64 GBaud and 64 QAM. Now DSPs based on the 7-nm resolution integrated electronic circuit are becoming available. These ultra-high speed digital processors have attracted the most attention due to their low power, which enables compact pluggable modules like OSFP and QSFP-DD. It is believed that yet another generation of 7 nm DSP will support 100+ GBaud and high-order modulation for metro and long distances. These ASIC technological developments would allow the SPC system transmission at such Tbps over one carrier wavelength.

With the inevitability of the DSP, although not without extensive effort, the only question about achieving 1.0 Tbps per wavelength carrier is the availability of the optical components that can work with the DSP. Faster modulators and receivers will be required to modulate and demodulate the analog signals from and for the DSP to convert back to digital form. Fortunately, good progress has been made on the optics as well. However, the limit to 60 GHz is only on $LiNbO_3$ integrated modulators. The integrated SOI photonic technology is limited to 45 GHz. Lithium niobate modulators can only be integrated with others through another intermediate platform, Si_3N_4. These integrated platforms are not yet ready for commercialization, especially due to photorefractive effects in such integrated substrates.

Initial systems were shipped with integrated modulators and receivers on the SOI integrated optic platform that supported 32 GBaud. These were termed Class 20 due to the fact that the 3 dB analog bandwidth was 20 GHz. In general, the possible digital transmission rate is higher than the analog bandwidth because the roll-off is gradual and some response exists at the higher frequencies.

The current 600 Gbps per wavelength systems use Class 40 receivers and modulators, which support 64 GBaud.

Now, Class 50 receivers and modulators are being introduced, which have 3 dB bandwidth of around 50 GHz and can support 90–100 GBaud.

One other element is critical to achieving higher than 1.0 Tbps per wavelength.

As mentioned above, when higher-order modulation is used, the symbol constellation points come closer and closer together. If they overlap, the DSP cannot distinguish one from the other.

One of the reasons for the spots smearing is phase noise in the lasers used and the ASE noise accumulated by the HPA and LNA of SPC hops, as described above. Therefore, it is critical to use lasers with very low phase noise, or linewidth, and ASE noise when using 64 QAM or 128 QAM. Fortunately, these narrow linewidth lasers also exist and are in mass production. However, optimization of HPA and LNA is not yet available for 1.0 Tbps per wavelength. The typical linewidth of an external cavity laser is well below 10 kHz and can minimize the OSNR penalty with higher QAM. But the ASE are still quite substantial for 1.0 Tbps per wavelength carriers for SPC systems.

8.13 APPENDIX: DOPPLER SHIFT ESTIMATIONS

The Doppler shift Δf is a function of the relative velocity Δv (a scalar quantity) between the two moving satellites or rotating satellites and optical ground station (OGS), with respect to the emitted frequency f_0 of the referenced transmitter which can be situated in either a satellite or an OGS

If we know the relative velocity (either at one instant in time or as a function of time), it's just a matter of using the formula:

$$\Delta f = \frac{\Delta v}{c} f_0 \tag{8.2}$$

The orbital elements of a satellite and the time and date are required to estimate the tracking and hence the Doppler shift.

The velocity v can be a vector quantity, where both the centrifugal and tangential components of the rotational velocity are considered. The scalar quantity is the magnitude of v, which is the speed of the referenced transmitter. We also note that the equation is commonly simplified to:

$$\Delta f = \Delta vc\, f_{\text{carrier}} = \Delta v\, \lambda = \|\Delta v\| \lambda \cos\theta = f_m \cos\theta \tag{8.3}$$

where θ is the angle of arrival at the transmitter, f_m is the maximum doppler shift, commonly known as the Doppler spread, and $\|\Delta v\|$ is the scalar speed of the receiver, which travels relative to a stationary transmitter (or vice versa).

The equation shows that Δf will vary from $-f_m$ to $+f_m$ as θ varies, making the received frequency variation; this is called frequency dispersion and contributes to fast fading. If the only thing we know is the speed at which the satellite is traveling, then all we can calculate is f_m, which represents the worst-case scenario for Δf. So, if our system is resistant to frequency shifts of up to f_m, then it can operate well.

To obtain the Doppler shift, we need to know the actual emitted frequency and the *range rate* (the first derivative of range, or the distance between the referenced transmitter and the satellite). In our example, the Doppler shift is assumed for two satellites in LEO and GEO with a relative velocity of doubling each other.

To compute the range, we need two 3D vector quantities: the satellite's position P and velocity V relative to you. The range rate is given by the dot product of P and V divided by the magnitude of P. The vector P divided by its magnitude is the *unit position vector*, so what you're really computing is that component of the spacecraft's velocity along the direction to you; how fast it moves perpendicular to that direction is not as important.

Although most satellite programs seem to compute range rate by finite differencing range, it can be computed analytically for more precise results. Standard orbit models can produce velocity as well as position as functions of time, and it is also fairly straightforward to compute the position and velocity of a ground station on the rotating Earth in inertial space.

So far, we have ignored the centrifugal velocity component, which is orthogonal to the vectoral velocity, in the Doppler shift (Figures 8.104–115).

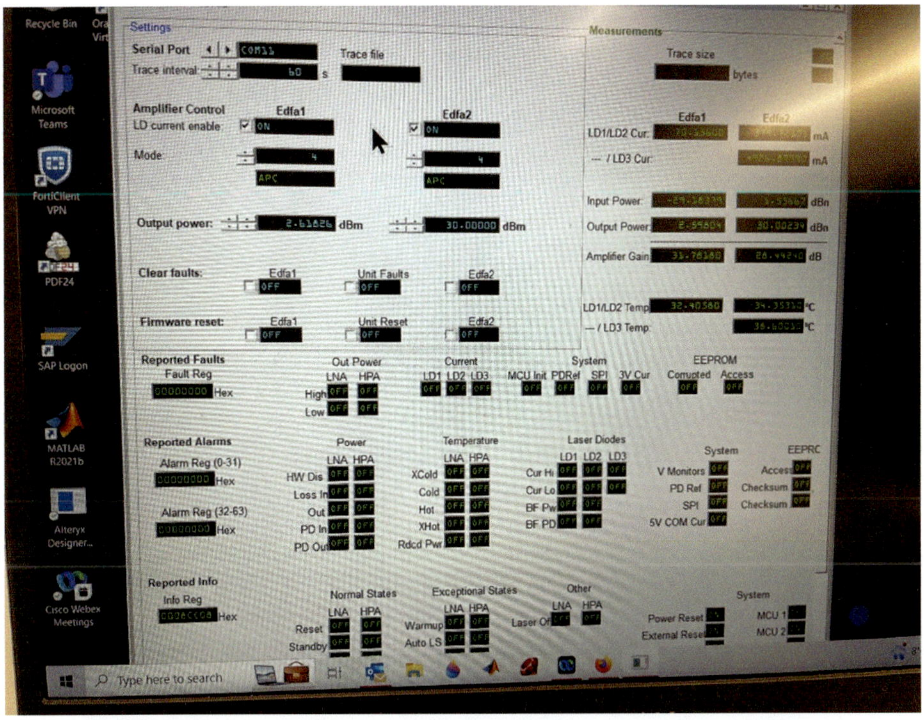

FIGURE 8.104 GUI screen shot of the optical amplifiers – both booster and preamplifier. Note the drive currents for preamplifier EDFA 1 and power amplifier EDFA@ (lD1/L@ and LD3 (~4.9 A).

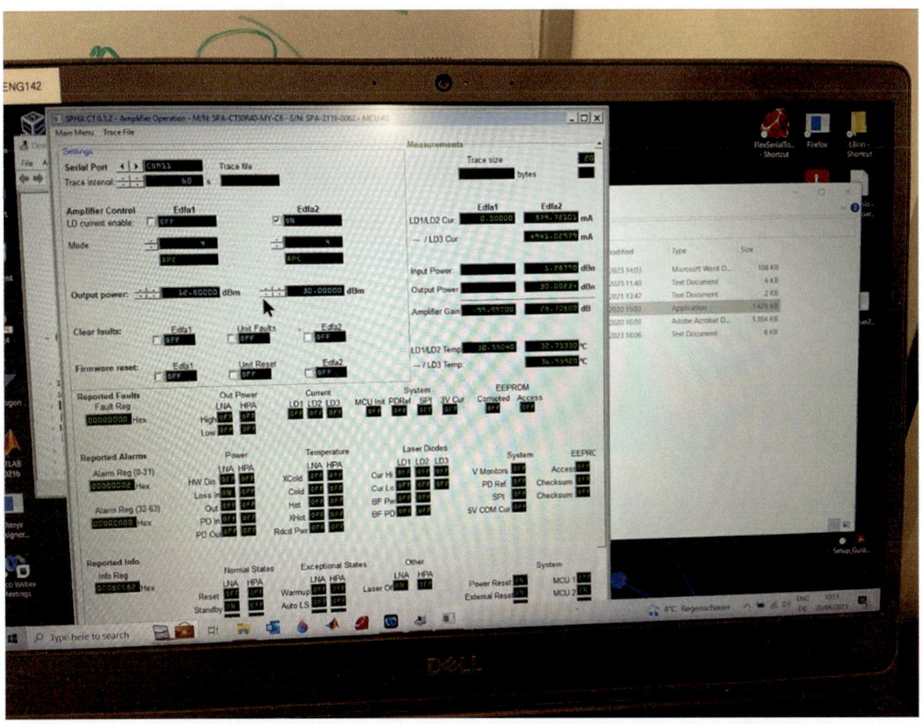

FIGURE 8.105 GUI of amplifier when preamplifier is not functioning.

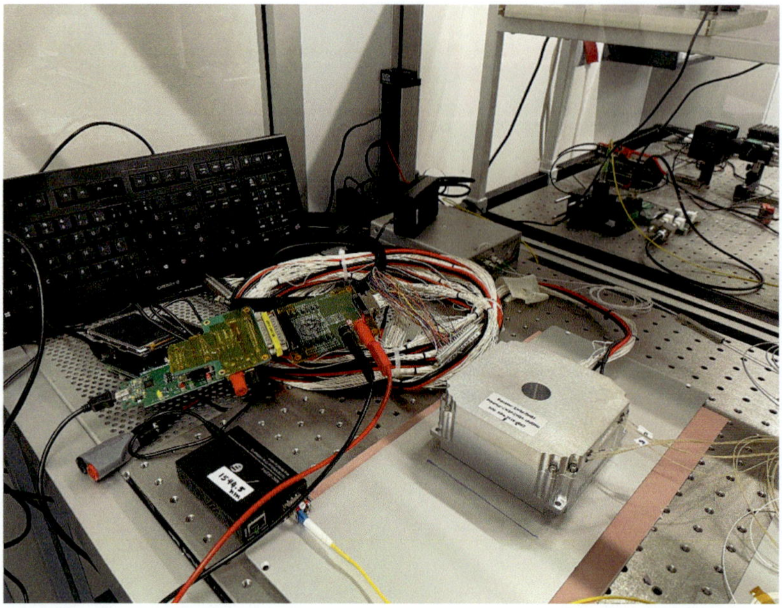

FIGURE 8.106 MPB optical amplifiers, both booster and preamplifier S/N SPA-2119-006.

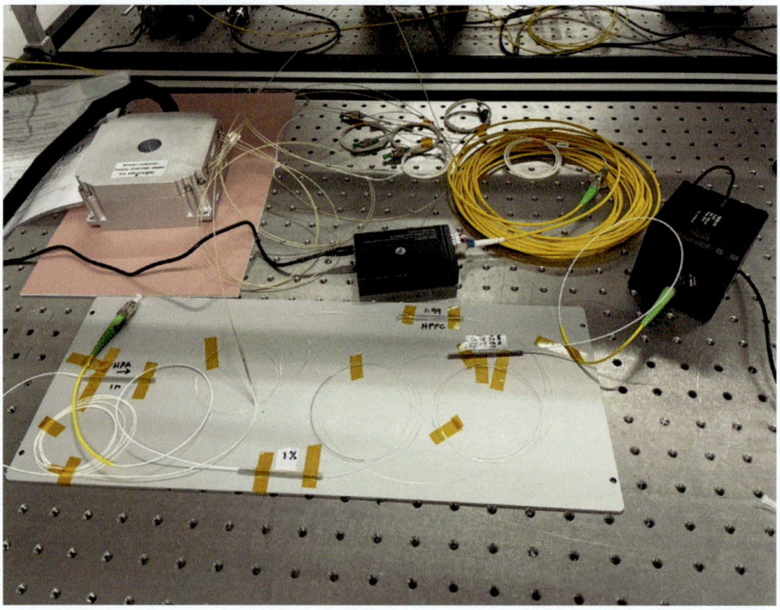

FIGURE 8.107 Amplifiers and layout with –20 dB HPFC and beam dump integrating sphere.

8.14 APPENDIX: SET-UP OF TEST SYSTEM

8.14.1 Carrier Test Set-Up

Figures 8.104 to 8.108 illustrate the experimental platform set-up for one equivalent SPC optical hop. The figure captions are self explaining with the GUI for controlling the optical amplifications with laser pump power tuning for optimization, the physical arrangement of both high power

FIGURE 8.108 Power and interface to drive the amplifier.

FIGURE 8.109 Optically amplified equivalent fiber free-space hop.

amplifier and optical pre-amplifier, the fiber equivalent optical hop in which the optical attenuator can be tuned to the total loss of the transmission distance, and the feedback control to set the transmitter laser . The set up illustrated here is for the test of the optical carriers for both the transmitter and the local oscillator laser for coherent receiver.

FIGURE 8.110 Interface and driver for driving Booster and optical preamp.

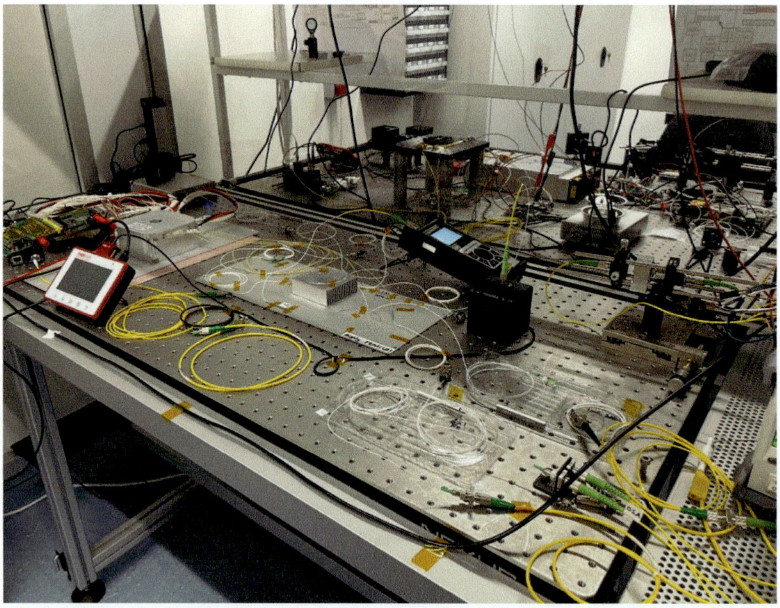

FIGURE 8.111 Optical section of HOP with associated fiber optics and optical amplifiers, booster and preamp.

8.14.2 Equivalent Hop Set Up Test

Figures 8.109 to 8.110 illustrate the testing set up of a fiber-equivalent transmission hop in space. Variable attenuator is used to set the equivalent optical transmission loss of the free-space vacuum hop. Modulated lightwaves are split with a coupling ratio of a high-power optical coupler so that a low power eye-safe level can be obtained for further transmission in the link.

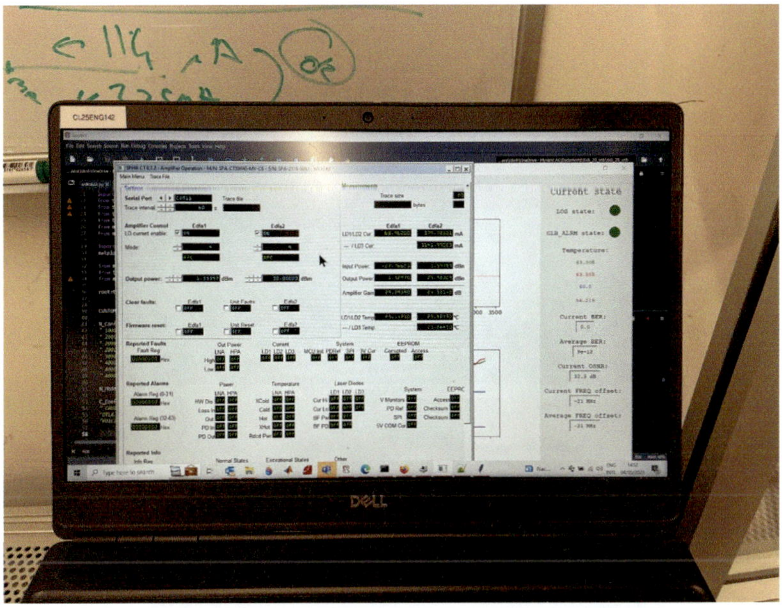

FIGURE 8.112 GUI of high-power booster amplifier MPB amplifier operations and drivers LD1, LD2, LD3.

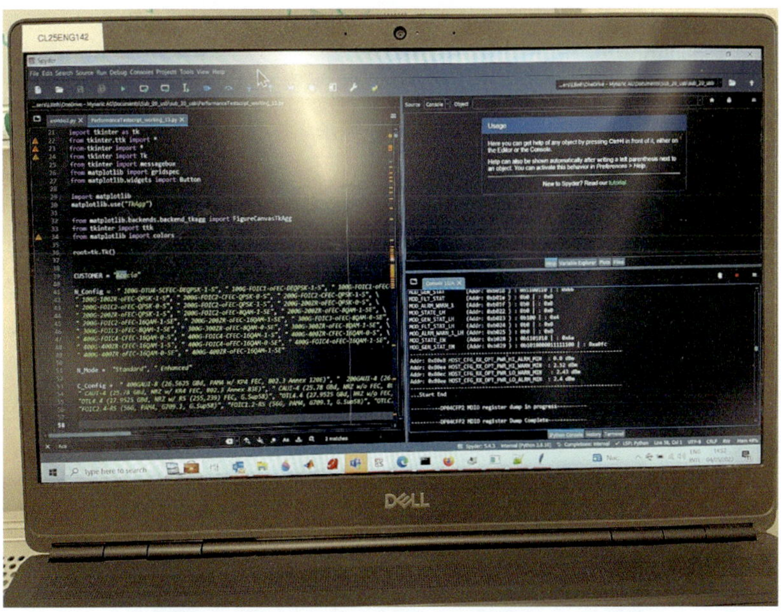

FIGURE 8.113 Platform for executing Python to activate 100 G module DP04QSFP.

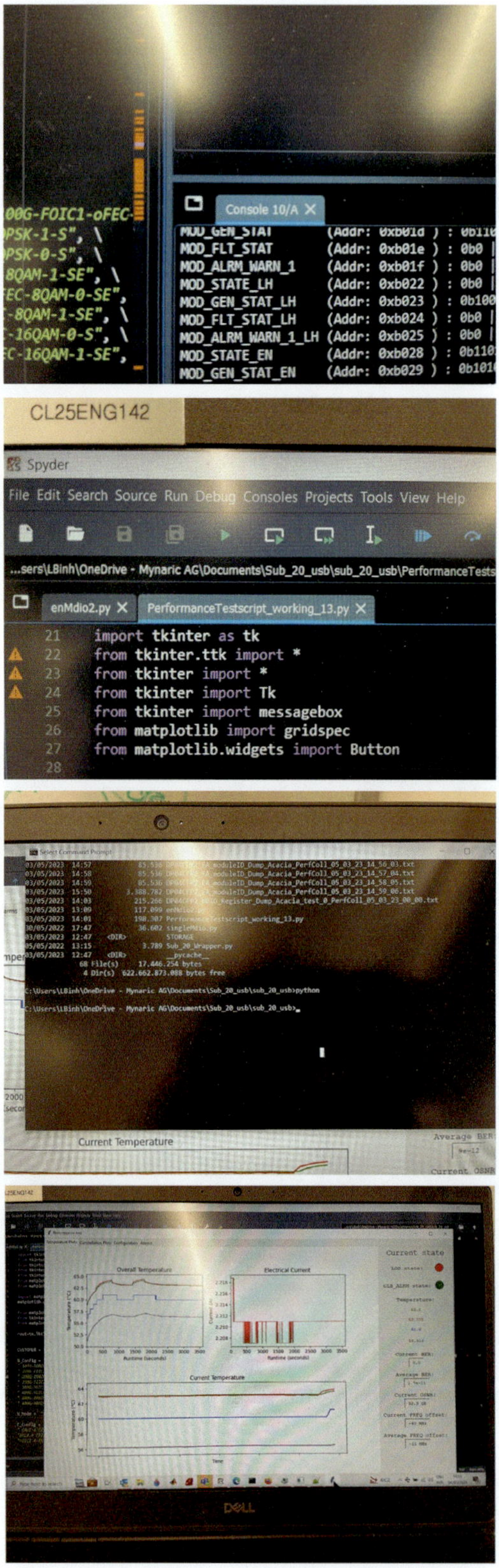

FIGURE 8.114 (a) To run and activate 100 G QPSK transceiver module; (c) folder to execute 100 G-DP-QPSK module; (d) performance screenshot; (b) the program for execution of driver in Python and performance monitoring.

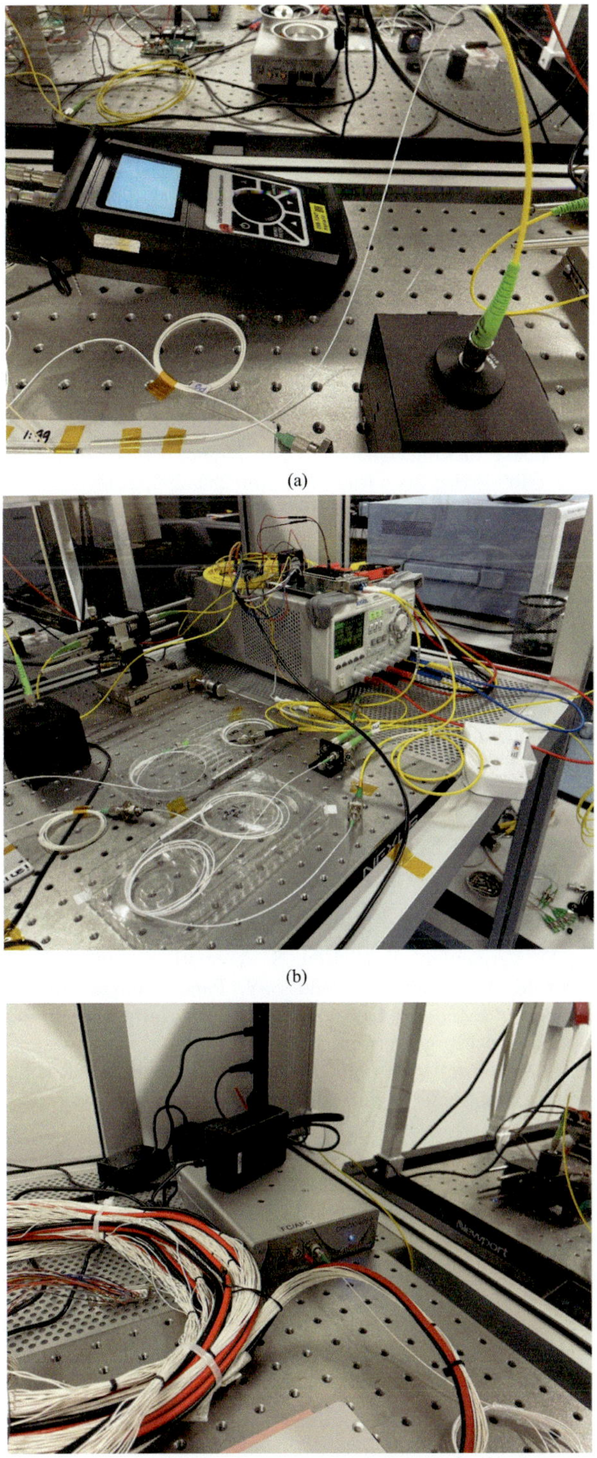

(a)

(b)

(c)

FIGURE 8.115 Miscellaneous optical components of a single SPC HopP platform. (a) VoA and integrating sphere to dump high-power beam, (b) splitter in from TRX 100 G and splitter out to OSA, (c) cabling for driving MPB amplifiers (booster and preamp).

NOTES

1 SSMF = standard single mode fiber. Effective area is the cross-sectional area of the guided mode with a gaussian intensity distribution profile which is usually larger the core geometrical radius due to the fact that there is only 70% of the intensity contained in the geometrical core and extensive intensity distributed in the cladding region.
2 Mynaric Laser Communications GmbH, Neu-Aubing, Bayern, Deutchland.
3 Mynaric Laser Communications GmbH, Neu-Aubing, Bayern, Deutchland.
4 R. Zeng, et al., "Joint estimation of frequency offset and doppler shift in high mobility environments based on orthogonal angle domain subspace projection", *IEEE Transactions on Vehicular Technology*, 67(3), 2254–2266, March 2018.
5 D. A. de Arruda Mello and F. A. Barbosa, *Digital Coherent Optical Systems: Architecture and Algorithms*. Springer International Publishing: Cham, Chapter 8, 2021. Book Series: Optical Networks.
6 GEO = geostationary orbit, LEO = low earth orbit, MEO = medium earth orbit.
7 L. N. Binh, *Digital Processing in Coherent Fiber Optic Transmission and Coherent Reception Techniques*. CRC Press, Taylor & Francis: Boca Raton, FL, 2014.
8 L. N. Binh, "Optical phase locking, Chapter 5", In *Digital Processing: Optical Transmission and Reception Techniques*. CRC Press: Boca Raton, FL, pp. 255–295, 2014.
9 ICOTRx = integrated coherent optical transceiver, this includes the integrated optical modulators of dual polarized lightwaves, whose input signals are DSP generated electrical sequences for modulating polarized channels, DSP based electronic processing of analog signals received from the detectors and electronic amplifiers and analog to digital converter. Lastly an optical hybrid coupler that receive the optical channels and the local oscillator laser, then $\pi/2$ phase shifter so split the in-phase and quadrature components of complex modulated signals. Two lasers are also integrated one for transmitter section and one for laser local oscillator for coherent receiving section.
10 This is equivalent to –42 dBm for 100 Gbits/s. The calculation is conducted by obtain the energy of one photon at 1550 nm (the operating wave length). Then multiplying this photon energy over the one-bit period of 100 Gbps.Thence the number of photons per bit.
11 B. Todd and S. Uznanski, "Radiation risks and mitigation in electronic systems", *Proceedings of the CAS-CERN Accelerator School: Power Converters*, Baden, Switzerland, 7–14 May 2014, edited by R. Bailey, CERN-2015-003 (CERN, Geneva, 2015).
12 A. H. Johnston, "Radiation damage of electronic and optoelectronic devices in space", *4th International Workshop on Radiation Effects on Semiconductor Devices for Space Application*, Tsukuba, Japan, October 11–13, 2000.
13 C. Doerr, L. Chen, D. Vermeulen, T. Nielsen, S. Azemati, S. Stulz, G. McBrien, X.-M. Xu, B. Mikkelsen, M. Givehchi, C. Rasmussen, and S.-Y. Park, "Single-Chip Silicon Photonics 100-Gbs Coherent Transceiver", *Acacia*, December 22, 2016, https://acacia-inc.com/acacia-resources/single-chip-silicon-photonics-100-gbs-coherent-transceiver/9.
14 16QAM50 → Number of constellation points = 16 or 4 bits/symbol; 50 = baud rate or in Giga-Sy/sec.; QAM = quadrature amplitude modulation, i.e., complex modulation and reception by coherent technique.
15 Modulation accuracy- EVM, Keysight Inc.
16 Mynaric Lasercomm, private communications.
17 Keysight Inc, EVM and Digital Modulation, https://rfmw.em.keysight.com/wireless/helpfiles/89600b/webhelp/subsystems/digdemod/Content/digdemod_symtblerrdata_evm.htm, access date July 2022.
18 L. N. Binh, *Advanced Digital Optical Communications*, 2nd Ed. CRC Press, Taylor & Francis: Boca Raton, FL, 2015.
19 L. N. Binh, *Digital Processing: Optical Transmission and Coherent Receiving Techniques*. CRC Press, Taylor & Francis: Boca Raton, FL, USA, 2014, Chapters 4 and 6

9 SPC Power-Limited Quantum Transmission

CONTENTS

Communication technologies have been developed and are commonly categorized by two main techniques: the maximum capacity of the transmission system given the limited bandwidth of the whole system, and when power is limited, or the available average power of the information channel is well below the noise level. This is the quantum continuous variable mode for secure transmission. Naturally, the quantum transmitter and receiver employ continuous variable quantum key distribution, in which the bit is considered as a whole and the number of samples is as high as possible to recover the bit state statistically. Thus, we employ Shannon's theorems to consider this case to prove quantum transmission and receiving with absolute security.

We also present the case of bandwidth-limited capacity for tera-bits/sec, which is to obtain the maximum capacity under the limited bandwidth available with current technology, in contrast to the power-limited case, as seen in quantum communications.

This chapter mainly investigates the quantum transmission case while also contrasting the bandwidth-limited case, allowing us to see the main differences between these two extreme transmission techniques.

9.1 INTRODUCTORY REMARKS

Given that optical and photonic technologies have progressed tremendously in fiber optic communication terrestrial systems over the last six decades, reaching more than 50 Tbps over the C- and L-bands, advanced modulation and multiplexing techniques incorporating digital signal processing with ultra-high sampling rates have allowed symbol rates reaching >200 GSy/s, By combining higher-order modulation formats and coherent reception techniques, the bit rate using a single

DOI: 10.1201/9781003426677-9

wavelength can reach over 1.0 Tbps in inter-satellite constellation networks, taking advantage of these developments in terrestrial fiber optic networks and systems. As we have seen in the above sections, high-capacity transmission using 100 G with dual polarization and complex modulation (QPSK format) represents a case where the bandwidth of the subsystems can be bandwidth-limited, and we strive to achieve the highest possible transmission.

However, only recently has the transmission of such ultra-high capacity over terrestrial and space channels, especially over multi-stage optically amplified hops or inter-orbital satellite constellations, been considered.[1,2] Simultaneously, the secure transmission of secret keys at the quantum level of optical bits is critical for ensuring coding and decoding with absolute security over the same channels using optical technology.

Thus, the transmission of these secret keys requires the signal's average power level to be at the noise level and to contain only a fraction of a single photon's energy or power. This is the power-limited case of SPC transmission.

This section 9.1, therefore, presents a summary of recent works on coherent optical transmission technology for transporting both ultra-high-capacity channels and moderate bit rates at low power, at the quantum level, for secret key transmission, key rate, and quantum key distribution (QKD) over space inter-orbital satellite constellations with multiple hops, all-optical routing, and switching at optical satellite nodes. We show that with signals at the quantum level or power-limited signaling, a sampling theorem combining both Shannon's channel capacity and the Heisenberg uncertainty principle under quantum noise fluctuation enables recovery of a signal-to-noise ratio (SNR) of 10^{-N} is possible, with $N>2$ for an appropriate sampling factor.

In the next section, we provide an overview of channel capacity under bandwidth-limited scenarios and power-limited cases, and their relationship with the systems' sampling rate. Quantum noise fluctuations and the sampling for channel capacity to reach a signal-to-noise ratio of –30 dB can be provided. Sections 9.2 Section 9.30 review the technological development of Tbps transmission capacity over SPCS (Space Photonic Communication Systems), with initial results obtained for modules providing bit rates of 200 Gbps per wavelength under digital coherent optics for optical transceivers. In contrast to the bandwidth limited communication systems, quantum communications are the cases in which the average power of the signals is very lower to the quantum level of a single photon. Section 9.4 0 gives insights into the design of quantum optical transceivers and QKD protocols to ensure absolute security.

9.1.1 Shannon "Channel" and Bit Rate

Figure 9.1 shows a generic optical transmission link over a space channel. The channel bandwidth is considered to include the bandwidth of the sampler. The sampling rate can be much higher than the bit rate of the information data sequence, which demands an increase in the bandwidth of the transport link. The link can be between satellites of one constellation to other satellites of the same constellation, inter-constellation, or OGS to satellites in the constellation and vice versa.

The input DWDM channels can be assigned in the transparent spectral window at 1550 nm so that the total number of bits can reach Tbps with 100 or 200G dual-polarization QPSK or 16- QAM. They are multiplexed by an optical multiplexer or a wavelength-selective switch (WSS), whose passband is tunable to a bandwidth that can accommodate the channel bandwidth without creating any distortion. The characteristics of WSS are given in the tests demonstrated in Chapter 4. WSSes can also be employed in CDC-ROADM as described in Chapter 4. Such ROADM is also included in this SPC transmission node, where the routing of a selected channel or a group of channels of different wavelengths can be directed to one of the four directions: North, South, East, and West.

It is noted that the channels are amplified to a high-power level in order to achieve a few watts average power and are detected at the end of a space hop by a low noise amplifier (LNA) whose sensitivity can reach as low as –65 dBm at 100 Gbps DPM-QPSK. Sensitivity could be higher at lower bit rates or at the secret key rate.

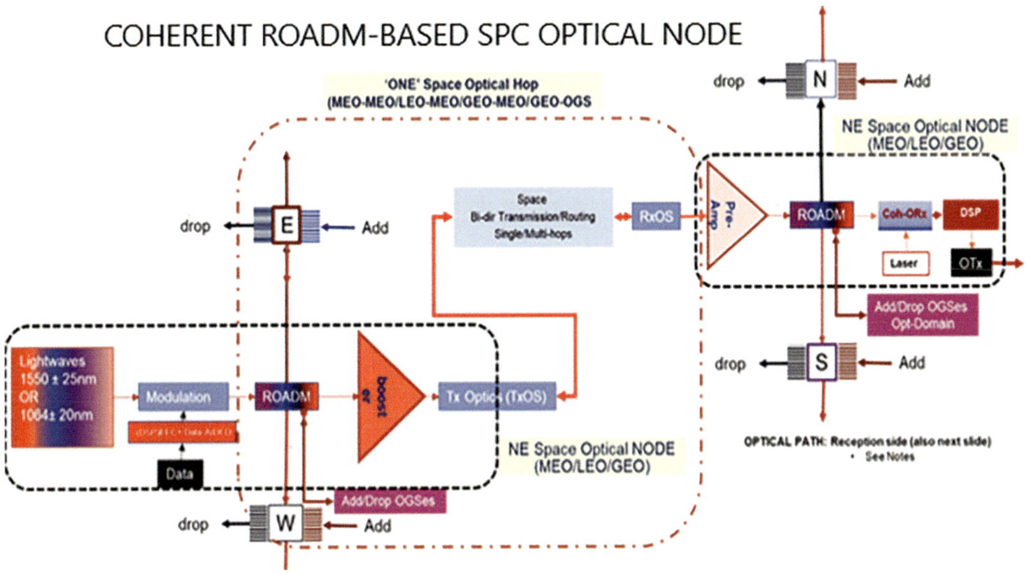

FIGURE 9.1 A generic schematic of SPC system link with transport transceiver and optical router to four directions North, South, East, and West. Tx=transmitter, BPF=band-pass filter, DSP=digital signal processor, B_w=bandwidth, ROADM=reconfigurable optical add/drop multiplexer, S=South, N=North, E=East, W=West, RxOS=optical system at receiving end, TxOS=optical system at transmitting end, LEO/MEO/GEO=low/Medium/Geo orbits, Coh-ORX=coherent optical reception, DSP=digital signal processing.

This high power and low sensitivity in detection put limits on transmission without leaking secret keys in the high-power radiation and at the detection level of the quantum channel at reception. This places a further limit on the reception of quantum channels.

Now taking to the limit of the $\frac{E_b}{N_0}$ → quantum limit or ~1 → 0 or ≪1; by using (l'Hospital's rule) i.e.,

$$\frac{e^{\eta}-1}{\eta}\underset{\eta\to 0}{\to} 2\ln(2) \to 0.6931 \left(\text{or} \quad -1.592\,\text{dB}\right) \tag{9.1}$$

thus by plotting the SNR (E_b/N_0) versus η. The channel capacity given by Equation 9.2 can be plotted as shown in Figure 9.2, in particular the region in dotted circles.

The channel capacity given by Shannon,[3] denoted as C, is the capacity or the system bit rate, B_w is the bandwidth of the transport system, E_b is the energy per bit, N_0 is the total noise energy per bit, and n is the ratio between the transport bit rate and the system bandwidth.

$$\text{Then} \to \frac{C}{B_w} = \eta = \log_2\left[1+\frac{E_b}{N_0}\eta\right] \to \frac{E_b}{N_0} = \frac{e^{\eta}-1}{\eta} \tag{9.2}$$

which the signal-to-noise ratio is very low, even at a level where the signal energy per bit is well below the noise level. The region marked with a dotted circle is where the signals are limited by the average power, which can be well below the noise level, or the OSNR can be less than zero, 0.

However, in the above Equation 9.2, the noises are considered to be Gaussian, which is not quite correct when considering the noise fallen into the quantum level. So, in the limit of B_w/C (bandwidth over capacity- bit rate) reaching large ratio (for example, 10 → 1000 E_b/N_0 → 0.691… or –1.599 dB,

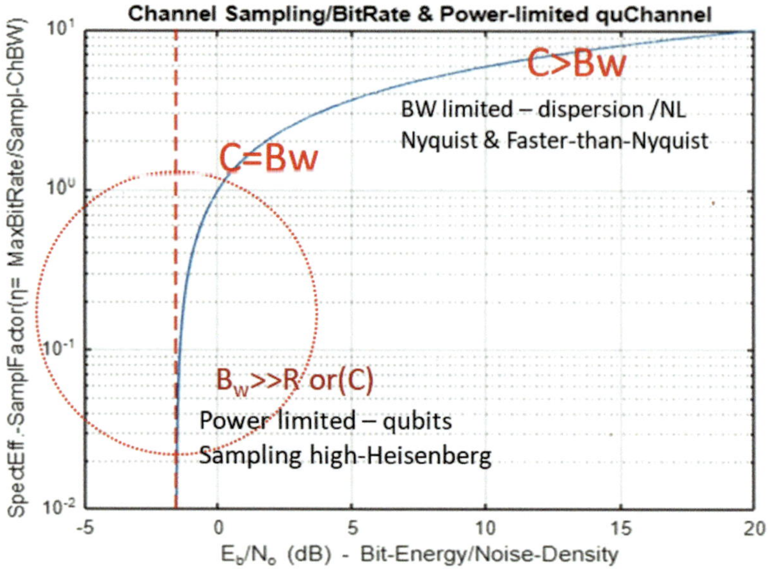

FIGURE 9.2 Channel capacity versus signal energy per bit over noise per bit.[3] R=bit rate, B_w=bandwidth, C=capacity or bit rate under transmission.

that is the noise power>signal average power in the bit period. E_b/N_0=power efficiency=energy density (per bit)/noise density. The energy/noise per bit is used here in order to be independent of the channel bandwidth that can be scaled appropriately for specific transmission systems. Therefore, the sampling rate can be much greater than transmission bit rate, thence highly possible to recover the signal in theory. This also satisfies Nyquist sampling theorem.

The bandwidth/bit rate is the inverse of the current popular term, number of bits per hertz as the transmission efficiency of capacity. That is, the maximum number of bits that can be transmitted over the channel per 1 Hz of bandwidth. This indicates the efficient transmission such that any impairments of the transmission link can be overcome and the detection of signals in the presence of noise.

In the case of power-limited scenarios, the signal average power is lower than that of the noise, resulting in secret keys that both parties can confirm. Thus, the final secret key rate is lower than the bit rate transmitted between the two parties. The SNR per bit period is on the negative axis. Therefore, the quantum techniques and signal processing presented here are required to recover the bit sequence and the correct signals.

In the scenario at the quantum level, where the sampling rate is increased and the time window becomes very narrow, the Heisenberg uncertainty principle must be applied. This is given by

$$\Delta E \Delta t \geq h/2\pi \tag{9.3}$$

Where ΔE is the energy contained in the time window Δt and h is the Plank constant. So, how do we set the noise and quantum signal to achieve absolutely secure transmission? This is considered under noise modeling in Section 9.4.

9.1.2 Bandwidth-Limited System

By inspecting the regions of Figure 9.2 in which the OSNR is much >1 dB, we see that the limit of the bit rate or channel capacity under a Nyquist sampling rate requires sufficient power to achieve the maximum bit rate that can be transmitted over the channel. This is the region in which Tbps can be achieved by first obtaining the maximum symbol rate under the available bandwidth of the SPCS sub-systems. Such multiplexing and higher-order modulation formats are shown in Figure 9.3, the generic schematic of the transmission system. The modulation is coherent and uses complex M-ary QAM (quadrature amplitude modulation) with M-constellation points. Furthermore, polarization multiplexing diversity is also employed. Thus, the advanced modulation technique is PDM-M-QAM, or M-ary QAM with pol-mux schemes.

Thus, by multiplexing the optical carriers and using polarization diversity, one can increase the aggregate capacity to over several Tbps. The section "Bandwidth-limited Multi-Terabits/s Transmission" will expand on these aspects, demonstrating how, under the transmission characteristics of inter-satellite links, the maximum Tbps can be achieved (Figure 9.4).

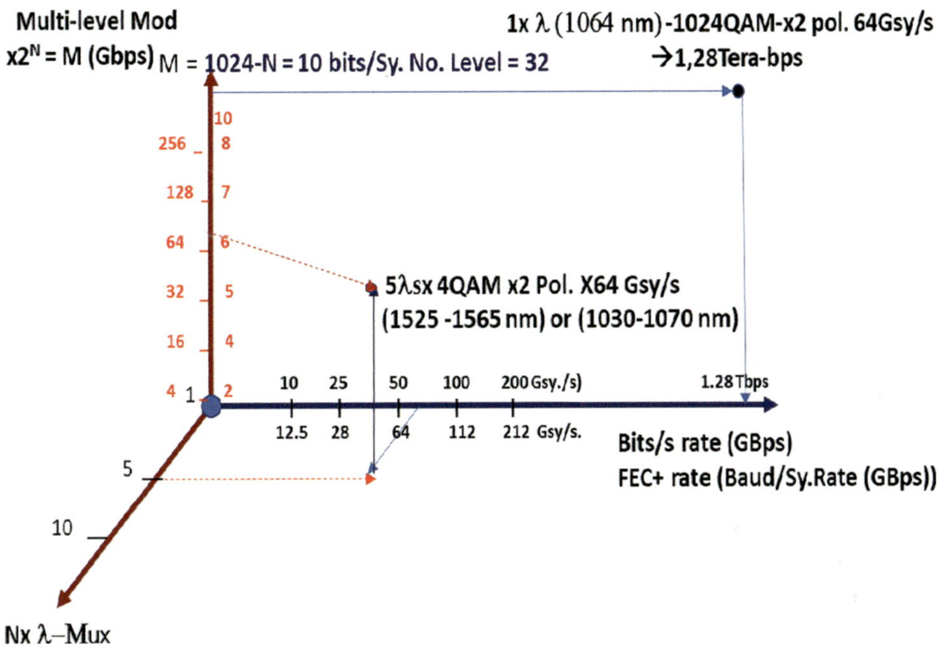

FIGURE 9.3 Optical multiplexing and higher-order modulation formats to reach aggregate-capacities of multiple Tbps.

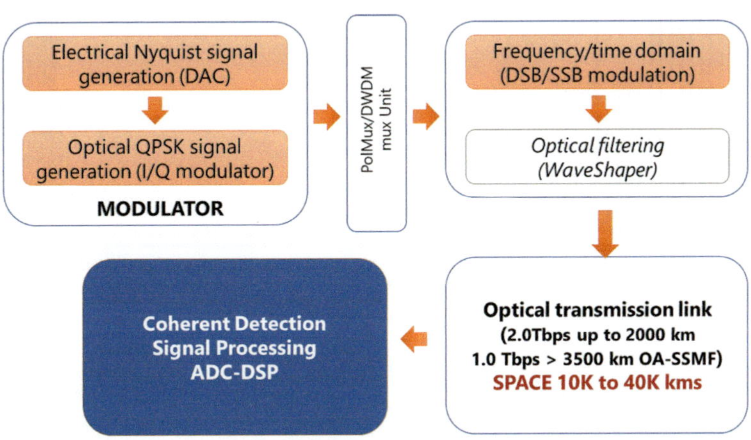

FIGURE 9.4 Generic schematics of an optically multiplex tera-bits/s transmission system.

Figure 9.5 illustrates the multiplexing of wavelength channels and corresponding bandwidth with bit rates under dual polarization M-ary QAM modulation formats. To reach an aggregate Tbps, for example, we need to multiplex five wavelength carriers, each with 200 Gbps. We note that as the modulation level increases, the required OSNR also increases, as shown in Figure 9.7. The transmittance of the inter-satellite link is shown in Figure 9.5b), which helps determine the spectral window over which the carriers are transmitted so that terabits/s capacity can be achieved.

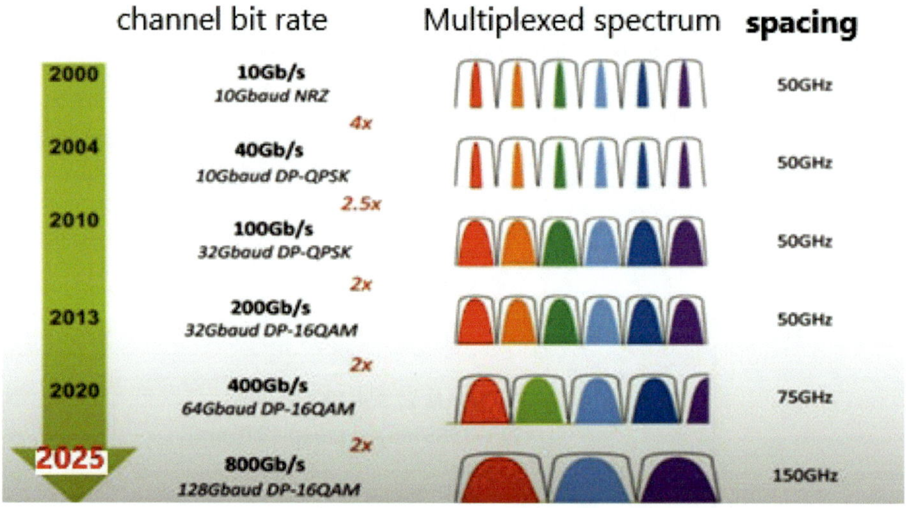

(a)

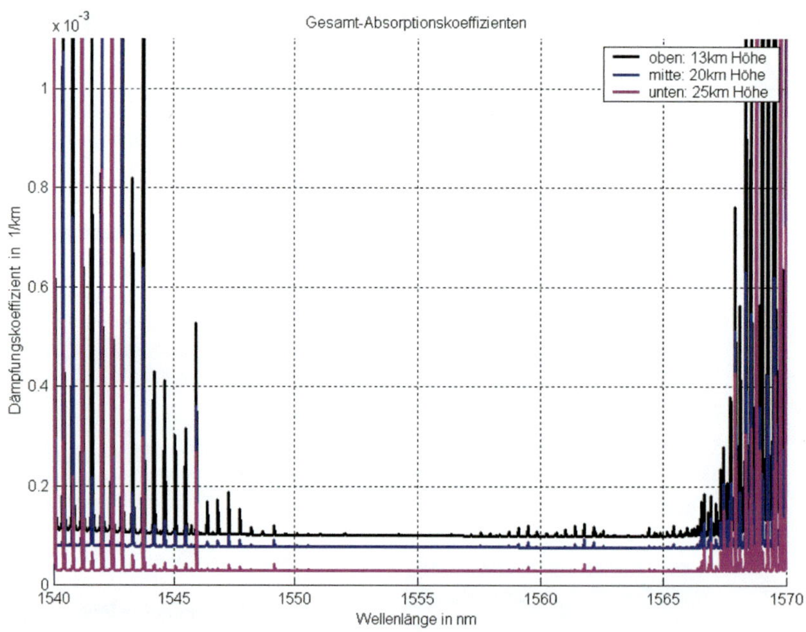

(b)

FIGURE 9.5 (a) Wavelength division multiplexing and required double sideband bandwidth in optical spectral window, (b) transmittance property of atmospheric medium as function of optical wavelength in the 1550 nm window. A wavelength channel can be selected to transport quantum bits sequence or secret keys.

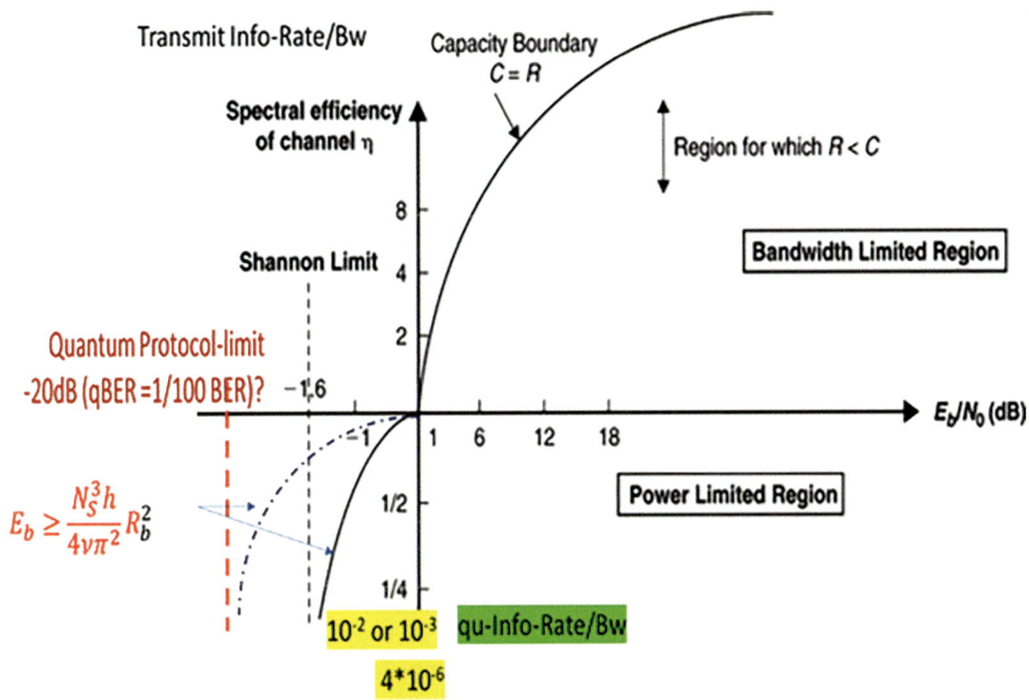

FIGURE 9.6 Plots of modified channel capacity (bit rate/bandwidth) against SNR per bit (in power terms). E_b=energy per bit, R_b= bit rate, Ns=sampling rate scaling factor.

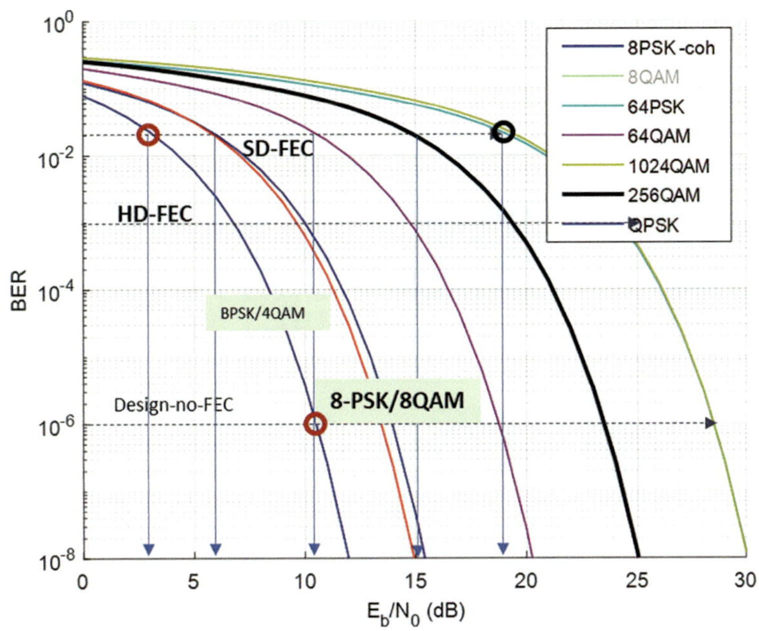

FIGURE 9.7 Bit Error Rate (BER) versus SNR (E_b/N_0) for different modulation formats. HD-FEC=hard decision-forward error coding, SD-FEC=soft decision FEC.

9.1.3 Power-Limited Transmission System

This is the case when the signal average power is very low, at the quantum level. This is the region in which the SNR ≪ 0 dB, as indicated in Figure 9.6. By expanding the power-limited region with an increasing sampling rate and applying the Heisenberg uncertainty principle along with quantum noise fluctuations, we can obtain the modified channel capacity for this case, as shown in Figure 9.6. It is noted that as the sampling rate increases, the channel bandwidth also increases, and the signal average power that can be detected relative to the noise power level decreases by a cubic factor.

9.2 BANDWIDTH-LIMITED MULTI-TERA BITS/S TRANSMISSION

9.2.1 Modulation and Multiplexing

Reaching a capacity of Tbps can be achieved through polarization multiplexing and wavelength multiplexing in the optical domain. Further increases in the aggregate bit rate can be achieved by employing multi-level modulation in the complex plane, as shown in Figure 9.4. For example, with 256 QAM or 256 constellation points in the complex plane, the number of bits per symbol is 8. Thus, with five wavelength channels and dual polarization multiplexing, we can achieve 80 times the symbol rate. For a 64 GSy/s rate, the aggregate capacity is 5.12 terabits/s.

Naturally, an increased number of bits per symbol must satisfy the required SNR so that error-free transmission can be achieved. Under current FEC (forward error correction) techniques, a BER of 2e-2 can be achieved by soft-decision FEC, with only the SNR of only 3.5 dB for the QPSK modulation format and some overhead bits, as shown in Figure 9.7.

We also show the region in which the QKD system can operate, but this is the encircled area that can only be inspected in detail in Figure 9.7, where quantum noise fluctuation and the Heisenberg uncertainty principle are applied. Further improvement in receiver sensitivity can be achieved by statistical modulation. Such a modulation scheme and the effective bits per symbol for 16 QAM modulation are shown in Figure 9.8. In statistical modulation, the constellation points are arranged in such a way that the Euclidean distance between them is maximized to minimize overlapping, thus reducing the system BER. Under this probability feature, some constellation points may be missing, which reduces the effective number of bits per symbol and, consequently, the aggregate bit rate.

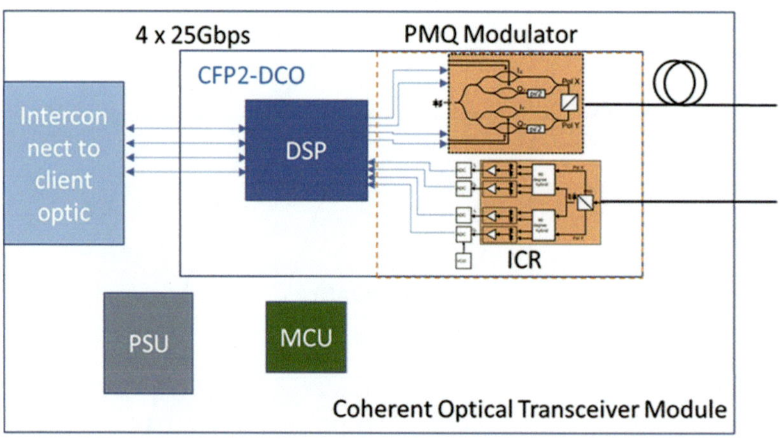

FIGURE 9.8 Typical structure of an integrated DPM-QPSK optical transceiver to obtain 100 G, 200 G or 400 Gbps. DSP = digital signal processor, ICR = integrated coherent receiver, CFP2-DCO = compact form factor 2 digital coherent optics, MCU = micro-control unit, PSU = power saving unit, PMQ = polarization multiplexed quadrature modulator, DPM = polarization division multiplexing.

this statistical modulation can offer higher sensitivity with a gain of about 2 to 3 dB in average detetion average power. Although this small gain but the processing does not create any loss in the transmitting average power. Furthermore, the DSP integrated at the transmitter side can generate the statistical sequence such that the Euclidean distance between adjacent constellation points can be maximized. Figure 9.9 shows the schematic structure of a high-speed transceiver, which consists of two sections: the optical transmitter and the coherent receiver, along with electronic digital signal processing (DSP). Figure 9.10 shows the sensitivity of the receiver and the bit-error-rate (BER) of an integrated transceiver. The DSP serves both the transmitter and receiver sections (Figure 9.11) to generate pre-distortion, compensating for any distortion in the transmission media or providing post-compensation for distortion. In addition to these sections, there is an optical hybrid sub-section for processing optical signals, polarization demultiplexing, quadrature phase shift, and local oscillator mixing with signals prior to electronic photo-detection. The QAM modulator for both polarizations uses integrated Si-photonics.

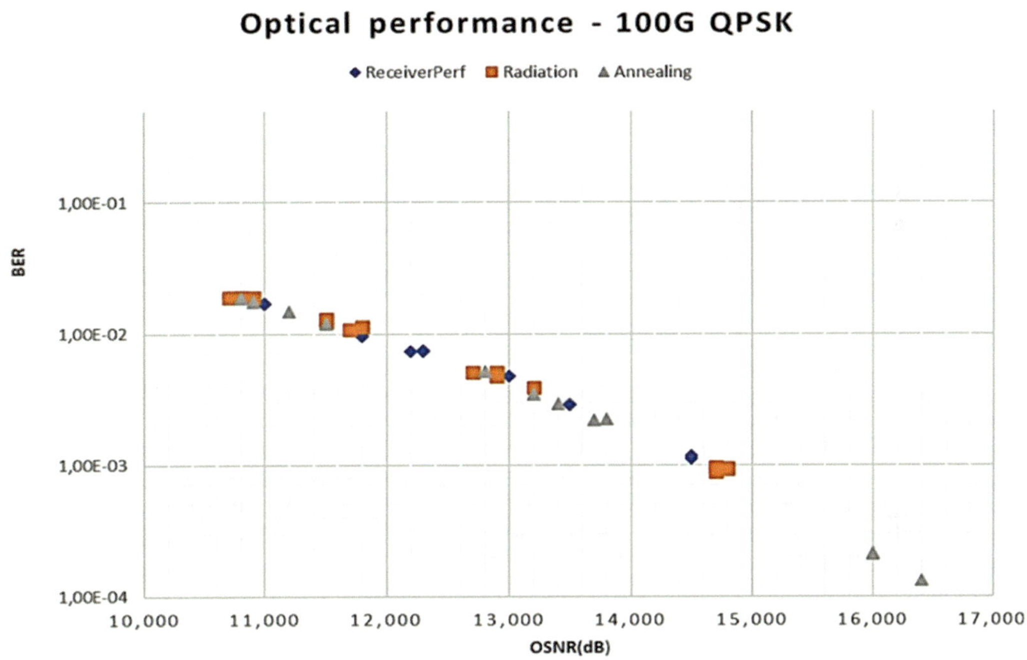

FIGURE 9.9 Measured receiver sensitivity as a function of optical SNR for a DP-QPSK receiver. Quantum level of five photons/bit at –42 dBm for 100 Gbits/s.

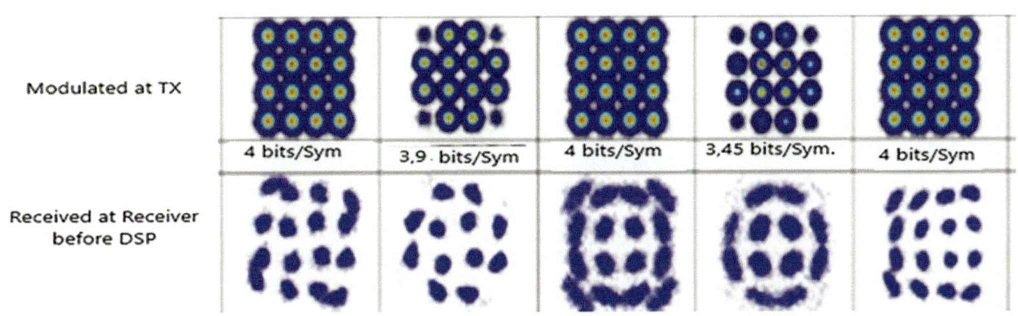

FIGURE 9.10 Constellations and effective bits/symbol for 16-QAM under statistical modulation.

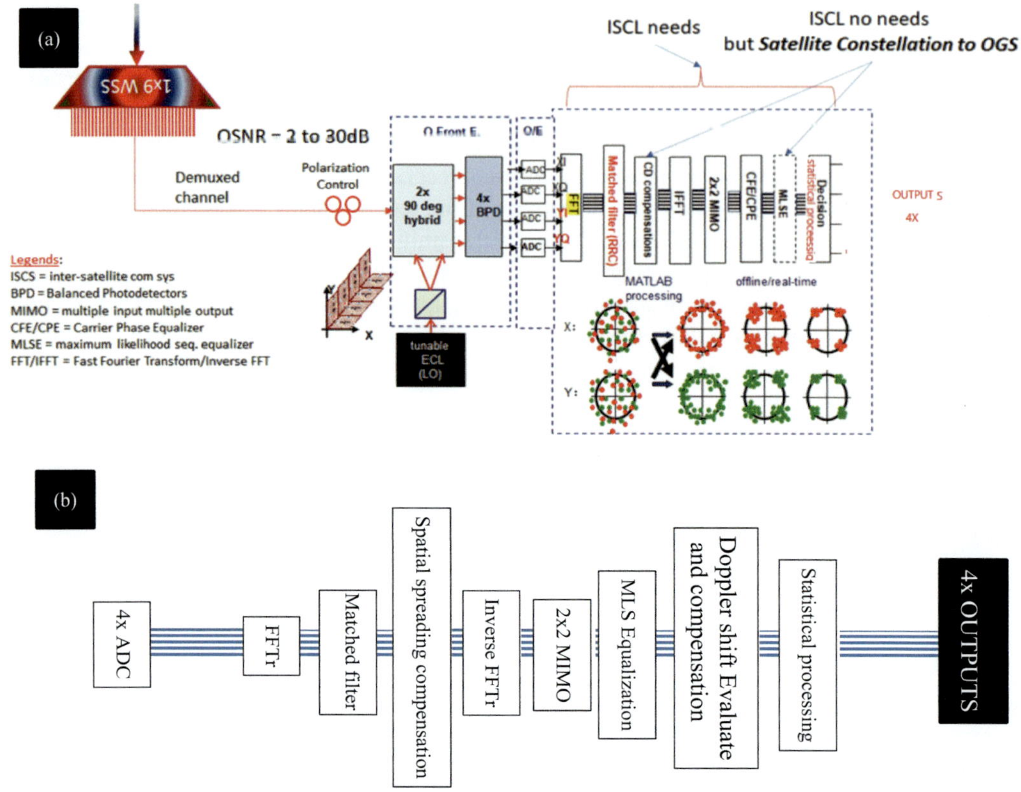

FIGURE 9.11 Embedded processing algorithms in the digital processor of a DSP-based ORx. Statistical processing algorithm is included for the SPC quantum channel. (a) Schematic system and generic processing, (b) detailed algorithms for processing. WSS=wavelength selective switch (equivalent to an optical wavelength demultiplexer), OSNR=optical signal to noise ratio, O-Front-E=optical-front-end to split polarized channels and the in-phase and quadrature components, LO=local oscillator laser, , MLSE=maximum likelihood sequence equalizer, CPE=carrier phase equalizer, MIMO=multiple input multiple output (2×2 in this case for two polarized channels).

The measured performance of a 100 G DP-QPSK transceiver is shown in Figure 9.9, where the BER is plotted against the OSNR (optical signal-to-noise ratio) with specific average power at the receiver. The integrated optical and electronic module is evaluated for space transmission, employing the DP-QPSK modulation format with integrated DDSPs and lasers for both the optical carrier and local oscillator. A 5 photons/bit can be achieved at –42 dBm for 100 Gbits/s with a symbol rate of 25 GSy/s, reaching the quantum limit of the coherent receiver. The local oscillator power is injected at 5 dBm per polarization to boost sensitivity. The integrated optical transceiver module is shown in Figure 9.8, where the upper section displays the IQ optical modulator and the lower section shows the optical hybrid, which splits the polarized beams and performs orthogonal phase shifting of the in-phase and quadrature channels, followed by photo-detection, amplification, and digital processing.

Statistical modulation can be used to improve the BER and sensitivity by ensuring the maximum distance between consecutive bits or points on the constellation. The constellation diagrams are shown in Figure 9.10. Although the number of bits per symbol can be reduced, as stated in Figure 9.10, the gain in probability of error and sensitivity can be higher.

9.2.2 Optical Amplifiers

Due to the long transmission distance, the total loss of the signals can reach 80–90 dB after accounting for the gains of the optical systems, including the telescope optics at both the transmitter and reception systems. Such a high attenuation factor must be compensated for by using a booster optical amplifier at the transmitter to launch the laser beam into space and an optical preamplifier in front of the coherent receiver. The optical power at the booster amplifier can reach 10–20 W (40–43 dBm), and the gain of the optical preamplifier can reach 50 dB with a low sensitivity input level down to –5 dBm. In addition, the coherent reception can boost the signal gain to 20–30 dB with a moderate average power of the local oscillator laser.

The high-power amplifier should be operated at all times and should not be operating in the linear region, as this would vary the amplitude of the channels, creating uneven gain and power across the channels. Consequently, a multi-hop SPC link is limited by the minimum power and saturating receiver for maximum power channels, leading to severe BER. To avoid operating in this linear gain region, we use a seedling laser at a level that will saturate the booster high-power amplifier (HPA). This way, the input channel power at a reasonable level is boosted to the saturation level of the seedling laser.

Furthermore, the amplifier at the receiving optical head is required to amplify the low-level average power channel, which ranges from –45 to –65 dBm after propagation over a long distance in a vacuum. Thus, the LNA must be capable of receiving low-level signal power and minimizing optical noise through optical filtering with sharp roll-off and cutoff frequency.

See Chapter 7 for more details on optical amplifiers.

9.2.3 Impairments and Mitigation over Space and Long-Haul Transmission Channels

SPCS and long-haul installed fiber systems can face a number of impairments as follows:

Distortion of the optical wave-fronts can occur due to optical systems required to launch the optical beam through the free-space channel from the transmitter to the receiver. These wave front errors (WFE) can be directly related to the error vector magnitude in the digitally processed data at the receiver. Digital pre-distortion at the transmitter can be used to mitigate this distortion effect.

Frequency shift of the optical carriers can occur due to Doppler shift (DS) effects caused by different rotation speeds of the satellite constellations in different orbits. For example, the DS can range from –7.5 to +7.5 GHz twice a day between the satellites in GEO and LEO, rotating at speed of 7 and 3 km/s, respectively. These DS effects can be mitigated by tuning the frequency of the local oscillator (LO) laser, allowing for homodyne detection to be achieved without the DS effect.

Crosstalk and in-band crosstalk can occur due to nonlinear effects in fibers that follow optical booster amplifiers, even over short lengths of a few meters. Self-phase modulation and cross-phase modulation effects play a critical role in these crosstalk and intermodulation effects. Such crosstalk should be limited to <1 dB by using large effective area (spot size) fiber.

Polarization rotation during the propagation of free-space beam can be mitigated using an adaptive optics system and control signals from the DSP system.

Dispersion and random polarization dispersion and compensation techniques can be implemented with ease using the advanced DSP processing power of the integrateddigital signal processor n both the transmitter and receiver sections.

9.3 DSP SYSTEMS FOR SPC HIGH CAPACITY AND POWER-LIMITED CHANNELS

In such SPC networking, DSP-based transceivers are important for mitigating impairments in the transmission path as well as nonlinear distortion effects. Figure 9.11 shows the arrangement of algorithms embedded in the transceiver to perform a number of signal processing tasks in the digital domain. The optical channels entering the reception section of the SPC link and constellation are amplified by an LNA as the optical preamplifier, then detected by a coherent receiver in which a balanced pair of detectors is connected back to back and beats with a local oscillator laser so that the product of the LO and the channel sequence can be reproduced in an intra-spectral electronic region of the electronic passband amplifier. As shown in Figure 9.11, the $\pi/2$ optical hybrid couplers are assigned to each polarized channel, then couple the local oscillator beam and the optical data channel and phase shift to split the in-phase and quadrature components of the complex signals. They are then received by the balanced photodetectors and beat together to produce sinusoidal signals plus noise from the coherent residual power and electronic noise. These signal sequences are then sampled by the ADC (analog-to-digital converter) and processed by the DSP with long sequence memory, which is long enough for determining the bit error rate. The software algorithms listed in Figure 9.11 show a typical sequence of processing depending on the detection and recovery of either the ultra-high frequency or power-limited case, the quantum detection case. But first, the sequence is sampled by the ADC, then a serial-to-parallel conversion transfers it to parallel bit patterns, followed by the use of an FFT algorithm.

To obtain the frequency domain of signals, processing is mostly preferred in the spectral domain. The RRC matched filtering algorithm is used to ensure the signals can be accommodated in the DSP systems and for the equalization of any band-limited sub-systems. The MIMO (multiple input multiple output) algorithm can be used to perform autocorrelation and cross-correlation of the polarized channels to mitigate skew effects in propagation through the space channels, especially under turbulence or angle changes due to the rotational speed of the satellite. Furthermore, the mitigation of Doppler effects is important to achieve pure homodyne coherent detection through carrier phase recovery.

The DSP embedded in OTRX is dedicated to each channel, so each channel must be received by a specific DSP-based OTRX, and the local oscillator laser is tuned to the right frequency for that specific wavelength channel. There is a special algorithm for SPC quantum transmission, involving statistical processing to distinguish overlapping constellations in cases when the OSNR <0 dB or the noise-to-signal ratio is positive, as shown in Figure 9.12e. This processing block is included in Figure 9.11. The normal positive OSNR constellations of QPSK are shown in Figure 9.12a–d.

Figure 9.12 shows a typical number of sampled constellations in a complex plane of QPSK signals under the cases of high fidelity (a), Doppler shift effects on QPSK constellation or the rotation of the constellation, then (c) shows the signals embedded in noises.

9.4 QUANTUM KEY DISTRIBUTION

9.4.1 QKD System over Space Channels and Quantum Signals

Consider the following parameters of a QKD transmission system over a space channel by SPC:

- Quantum Key Rate = 1–100 Mbits/s → random key generator <10 Mbits/s. Note: at 1550 nm, over 100 Mbps rate one-bit period contains several thousands of optical wave periods;
- then we have C = Max Key Rate = $(B_w/2)$. $1/N_s$; (Equation 9.1)
- here N = sampling ratio = sampling rate/key rate; B_w = sampler bandwidth. (e.g., 100 GSa/s implies a bandwidth of B_w = 50 GHz);

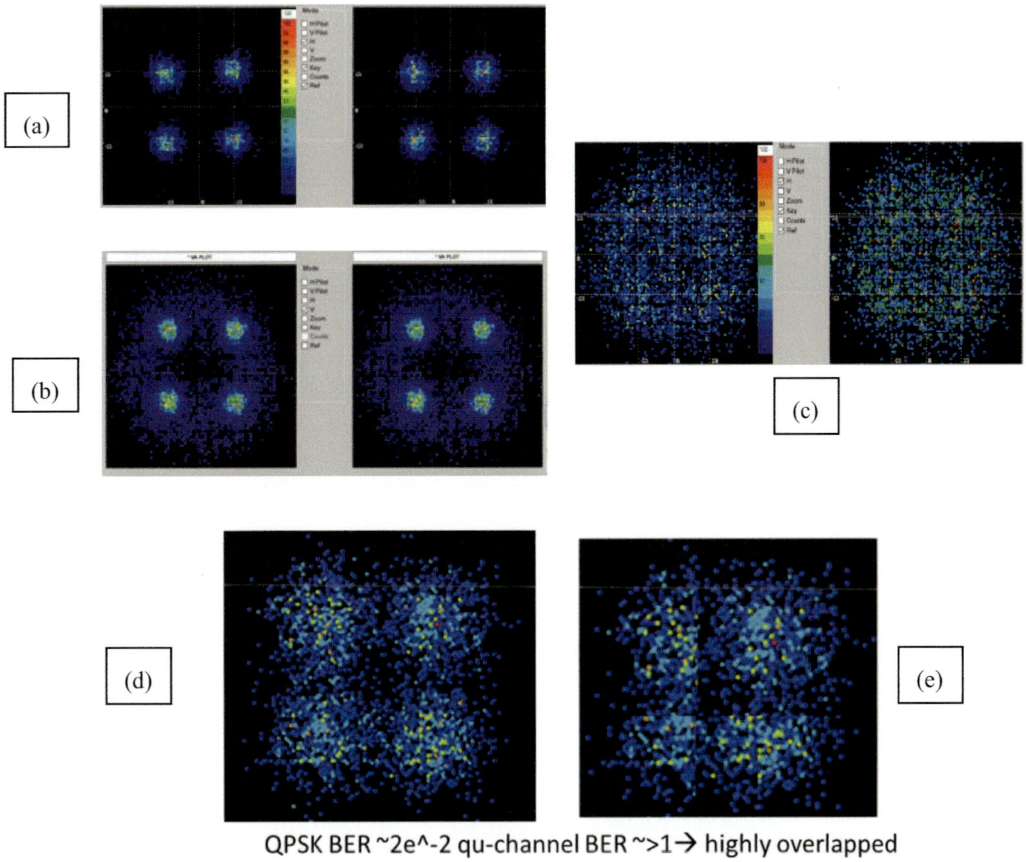

FIGURE 9.12 PDM-QPSK constellations at (a) LOFO=2 GHz, (b) 4.5 GHz, (c) 6 GHz; at BER=2.42e-2, Rx signal power=–25.8 dBm, OSNR=32.3 dB. (d) QPSK under five photons/bit received average power; (e) quantum receiver at signal power much lower than noise average power. Note that phase rotation of constellations in panel (b), i.e., phase mismatched by several 2π (cycles). PDM=polarization division multiplexing, QPSK=quadrature phase shift keying, LOFO=local oscillator frequency offset.

- so, if the key rate=1 Mbps, then the number of samples is N_s=100e9/1e6=100,000; if C=100 Mbps, thence the number of samples becomes N_s=1000; still well above and reaching the Shannon limit.
- The sampling window Δt must satisfy the Heisenberg uncertainty principle and quantum state for secure transmission with minimal eavesdropping – or else the BER will increase significantly when the photonic signal energy is reduced due to eavesdropping. An illustration of the time window and energy per temporal window is shown in Figure 9.13.
- Thence, applying the Heisenberg uncertainty principle, the energy contained in one sampled window is given as
- $\Delta E = E_b/N_s \geq (h/2p)(T_0/N_s)$ (Equation 9.2)
- This leads to $E_b/(N_s^2 \cdot R_b) \rightarrow E_b \geq (h/2\pi)(R_b \cdot N_s^2)$ (equation (9.3) the averaged value of which we can assume.
- However, under noise fluctuation, the expected value of finding a quantum state, thence the expected value <E> gives a fluctuation

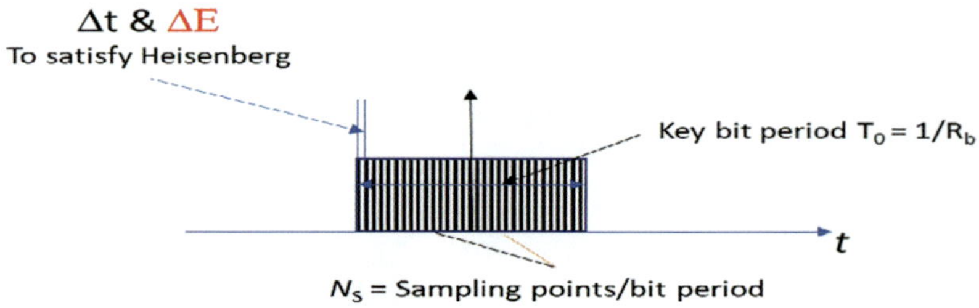

FIGURE 9.13 Illustration of the narrow sampling temporal window and the energy per sampled slit.

$$|.>= \text{Coherent states}: \ |\alpha\rangle = e^{-\frac{|\alpha|^2}{2}} \sum_{n=0}^{\infty} \frac{\alpha^n}{\sqrt{n!}} |\alpha\rangle \qquad (9.)\ (9.1)$$

where α = fluctuation; n=photon number; $\bar{n} = |\alpha|^2$=average photon number.

- Hence, using the Heisenberg uncertainty principle gives energy fluctuation in one sample window

$$\sigma_E = \text{var}(\Delta E) = h/2\pi\omega\, \text{sqrt}(< n >) = h/2\pi\ \omega\, \text{sqrt}$$
$$\left(E_b/\left(N_s h\omega/2\pi\right)\right) \geq \left(h/2\pi\right)/\left(T_0/N_s\right), \qquad (9.5)$$

therefore,

$$\sigma_E = \text{var}(\Delta E) = h/2\pi\omega\, \text{sqrt}(< n >) = h/2\pi\omega\, \text{sqrt}$$
$$\left(E_b/\left(N_s h\omega/2\pi\right)\right) \geq \left(h/2\pi\right)/\left(T_0/N_s\right), \qquad (9.6)$$
$$\text{Thence,}\quad \omega R_b\, \text{sqrt}\left(E_b/\left(h/2\pi\ \omega\right)\right)/N_s^{(3/2)} \geq 1;$$

with n = number of photons in one sampled window:
thus we have

$$E_b \geq \frac{N_s^3 h}{4\nu\pi^2} R_b^2; \qquad (9.7)$$

where N_s=sampling number in a bit period; R_b=information/key rate; ν=optical frequency; h=Plank's constant.

The QKD method described in this section employs the continuous variable mode, meaning the bit over the whole bit period is used and not the photon counting one by one sample. The amplitudes of the samples over one-bit period are statistically computed, and based on this, the decision of the '1' or '0' is decided. Thus, the QKD described in this section is a CV-QKD (continuous variable quantum key distribution) type, which is much more practical compared to discrete variable QKD (DV_QKD).

We can illustrate the estimation with two numerical examples as follows:

Example 9.1: R_b=10 Mbps and taking **1 photon per symbol** → Ns=250,000 (or 250 K×R_b) → fulfilling the uncertainty relation thus by the Heisenberg limit, we obtain. E_b=1.2 * 10^{-19}J or one photon=1.27766 * 10^{-19}J or about one photon in a bit @250 GSa/s!

Numerical 9.2: R_b=10 Mbps, then energy per bit E_b over 1 s of 10 Mbps key rate and sampling factor N_s=250 or sampling rate=2.5 GSa/s, of 1550 nm → E_b= 1.25e-28 J or about 1e-9 of one photon → If Ns=2500 or 25 GSa/s then energy per bit=1e-6 of one photon at $R_{b=}$10 Mbps**.**

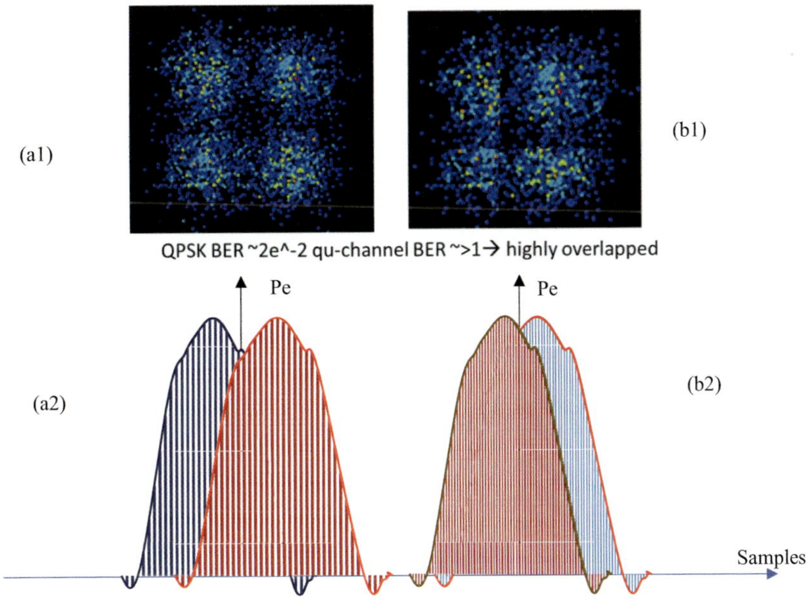

FIGURE 9.14 Constellation of (a1) QPSK under five photons/bit received average power; (b1) quantum receiver at signal power much lower than noise average power; (a2) and (b2) the probability of error of the detected QPSK constellation bits for ultra-high capacity under minimum sensitivity of five photons per bit and CV-QKD under high sampling rate. Pe=probability of error.

One photon energy$=h\nu=$1.053e-34$\times$3e8/1.55e-6=2.03e-19 J $\rightarrow$ Number of photons per bit$\geq$0.94 photons to satisfy Heisenberg's for 10 MKey/s key rate 1550 nm.

Under coherent reception, the QPSK modulation format channel is accepted by the transceiver, whose structure is given in Figure 9.9. A receiver sensitivity of −42 dBm, for 100 Gbps, or 5 photons per bit period, can be the minimum at the lowest noise contribution. The total noise is a summation of receiver noise, optical noise from lightwave emission of the transmitter, and nonlinear noise crosstalks.

In quantum receivers, the total noise spectral density must be minimized so that they are at the quantum level, allowing the quantum signal to breach minimum level or be embedded at or below the quantum level. The received average power is much less than 5 photons per bit.

Figure 9.14a and b show the constellations received under these two transmission conditions. The probability of error can be 0.02 and 0.1, respectively, for the minimum BER before soft FEC and quantum channels. In the quantum constellation, the four-point constellations are almost overlapped and completely cover each other. However, the statistical distribution, as shown in Figure 9.14b, allows us to distinguish the four states when the number of samples is much higher than the Nyquist sampling rate. Thus, the sampling rate, DSP algorithm, protocol and public confirmation allow us to transmit secret keys with absolute security.

9.4.2 Transmitter, Receiver Sensitivity, Sampling and Protocol

9.4.3 QKD Transmission System

A QKD transmission using coherent reception techniques are similar to the Tbps transmission system, as shown in Figure 9.15. The inter-satellite or long-haul distance can reach 40,000–50,000 km over which a total attenuation factor can be 80–90 dB.

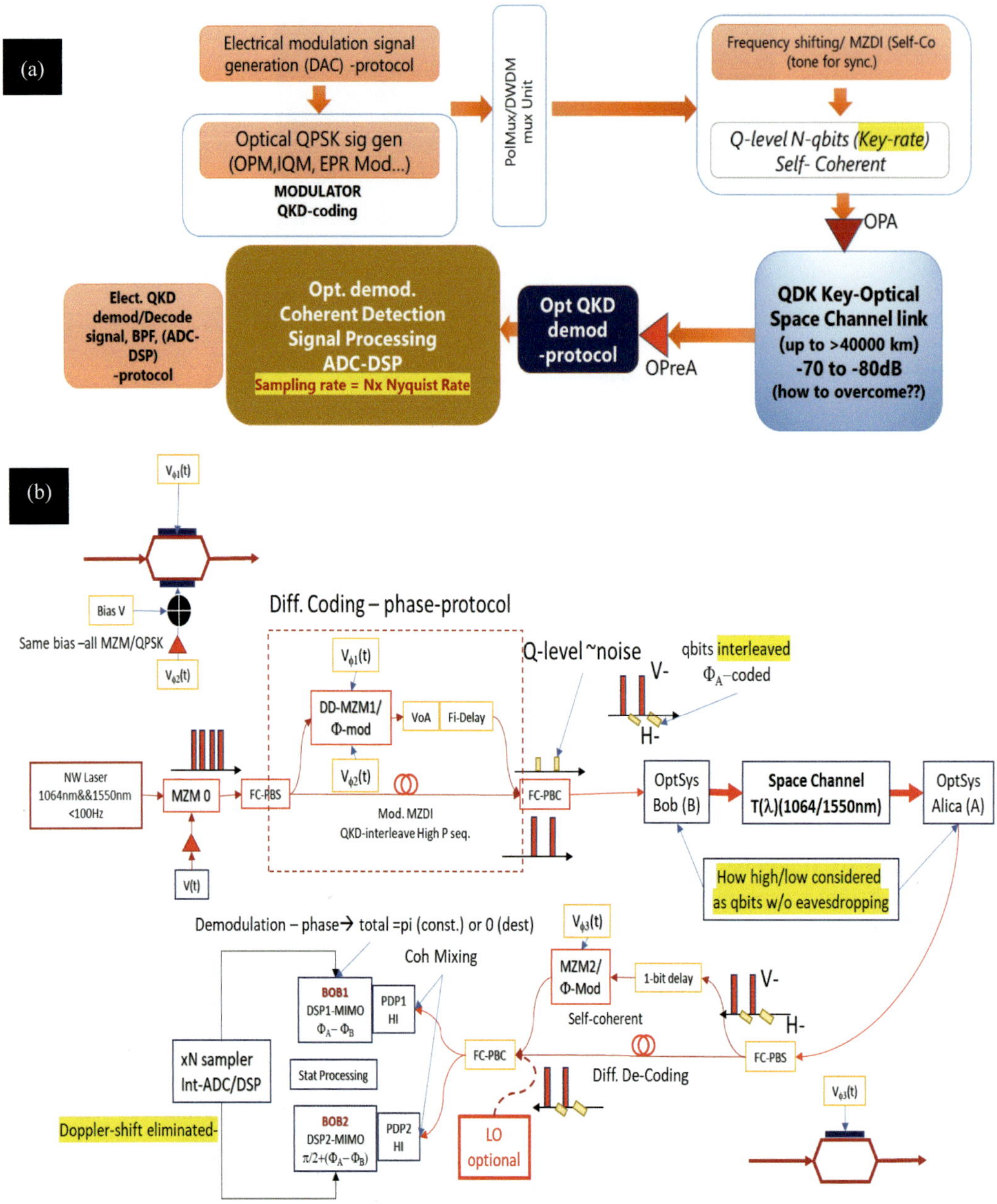

FIGURE 9.15 (a) Generic optical transceiver for complex quantum signaling for QKD; (b) implemented CV-QKD optical system. Receiver section is using self-coherence detection with option of coherent reception via the use of an external LO laser. Insert: Photonic MZM modulator. (b) a typical optical implementation of the QKD transmission in.

If the QKD signal is at quantum level or its average power is around –80 dBm at 10 Mbits/s, then no receiver can detect it at (–80 dBm – 90 dB = –170 dBm), such quantum signals at the receiving end of the space channel. This is a major challenge for QKD over long-haul space channels, especially when there are multi-hop satellite links, as shown in Figure 9.15a. The main sub-section of the QKD systems is shown in Figure 9.15b.

DSP systems can be embedded in both transmitter and receiver to generate the bit or key bit sequences, implementing pre-distortion or pre-compensation.

Several modulation formats can be designed for appropriate quantum signaling, such as complex M-ary-QAM, differential M-ary QAM, MSK (minimum shift keying or continuous phase modulation format with sawtooth or triangular amplitude distribution over the bit period), etc. For a self-coherent transmitter, there needs to be two consecutive bits whose phases can be coded in such a way that they can be decoded to recover with higher sensitivity.

Optical amplification at the transmitter and receiver is not fully shown.

Modulated DWDM channels are multiplexed and then amplified to a high power level to be launched into space via an optical head which is controlled by a beam alignment feedback control system.

The optically pre-amplified sequence is detected and fed to the digitally processed subsystem to meet the protocol as described. The confirmed secret keys are obtained. A more detailed description of the optical and quantum reception can be provided next.

- Polarization of the two channels is split after the polarization fiber coupler. Each polarization is delayed with respect to the other by one bit period for differential modulation and demodulation. Now, polarization switches can operate up to 40 GHz, so the bit period can be very short. However, the key rate is only a few hundred of MHz, leading to a secret secure key rate of only a few mega-keys/sec.
- A narrow-linewidth laser is used and coupled to the optical modulator for transporting the quantum bits, which are modulated by the electrical bits of 'Alice' that are not known by 'Bob' at the receiving end.
- Bobl and Bob2 are used to generate signals that modulate two OPMs with a phase shift to detect the phase-modulated signals transmitted by 'Alice', using either constructive or destructive interference with the coherent reception with a local oscillator laser.
- Photonic transmitter: consisting of a narrow-linewidth laser coupled to two parallel optical modulators (MZM), an optical one-bit delay line is inserted into one path so that differential coding can be implemented.
- Optical reception and demodulation: The optically pre-amplified sequence is detected self-coherently or coherently if an LO is injected to mix with the optical signals. The optical sequence is then split into two parallel paths again and each delayed from the other by one bit period.
- An optical modulator is also employed in one path so that the bias and RF wave can be used to find the state of phase of the path. Thus, the constructive or destructive interference lightwaves occurs at the output of the interferometer of the differential structure, allowing for the detection of the differential phase of the encoded bits.

If coherent detection is preferred, the optical demodulated optical signals are mixed with the LO laser in the photodetectors, electronically amplified, then sampled and digitally processed in the DSP. The transmission and reception signals by 'Alice' and 'Bob' are then decoded via a protocol provided in the section Protocols.

Figure 9.15 depicts the transmission system when either polarization or phase of the modulated lightwaves are used. Thus, a sequence of secure bits is decided for further coding of the tera-bits channels and inserted or interleaved quantum bit sequence. The quantum bit sequence must be hidden in such a way that the optical amplification in long-haul space optical communication link remains at quantum states and not possible to be detected by the high power radation of the optically amplified beam. This optically amplified optcal link was described n details in Chapter 2, Chapter 5 and Chapter 10.

The advantages of self-coherent detection are that both consecutive bits suffered the same disturbances and delays, so transmission delay due to rotation of satellite and misalignment in inclination can be eliminated. However, additional losses are added due to beam splitting and the insertion losses of a high device count.

DSP subsystems embedded in the receiver would assist in the recovery and demodulation of the differential codes and protocol processing to obtain secure key bits.

9.4.4 Protocols

9.4.4.1 Polarization Coding/Modulation and Polarization De-coding or De-modulation

Alice generates a random bit string and a random binary sequence of coding bases. Assuming now that the quantum information is transmitted using either vertical (|)-horizontal (rectilinear ---) or ±45° diagonal (/) polarization of single photons.

A '0' = '|..|' rectilinear basis, and a '1' = '/' = diagonal basis. Further, a '0' in the bit string would be represented by horizontal or –45° polarization, depending on the basis, and a '1' would be represented by vertical | or +45° polarization.

Alice codes a stream of photons randomly as dictated by the sequences she generates and the previous definitions. Alice then sends the stream of photons to Bob. Alice retains both of her random sequences.

As Bob receives the photons, he randomly decides the basis in which he will measure their polarization. Essentially, he also generates a random binary sequence of coding bases. It is crucial that Bob's sequence is entirely independent of Alice's. Bob then measures the polarization of the photons he receives according to his basis.

In the case where Bob selects the same basis as Alice, he will correctly measure the bit that Alice (Tx party) has generated. When Bob selects a different basis from Alice, he may or may not obtain the same bit, as the polarization he measures will be purely random.

A ±45° diagonal polarization has equal components of horizontal and vertical polarization and hence has an equal probability of being detected as a horizontally or vertically polarized photon in a rectilinear basis. An optical polarization switch can produce this functionality.

Similarly, for a rectilinearly polarized photon measured in a diagonal basis, when Bob (as the receiving party) has selected the opposite basis to Alice, he obtains no meaningful data. The maximum meaningful data Bob can obtain is half of the random sequence that Alice transmits – those photons for which he and Alice (the transmitting party) used the same basis – although in reality, this will be somewhat lower due to losses incurred during transmission and detector inefficiencies.

Figure 9.16 outlines a scheme to complement that of Figure 9.15b, in which an additional Bob 2 is used to strengthen the decoding of the secure bits. By modulating the two optical modulators (OPMs), one can obtain the confirmation of the lightwave phases of the two differential bits. This helps eliminate any disturbances on the low-intensity bit.

9.4.4.2 BB84 Protocol

The BB84 protocol is illustrated in Figure 9.18 and tabulated in Table 9.1, showing step by step the coding, transmission, and public confirmation to reach an identical set of key sequences for one-time encryption. The rest of the BB84 protocol is implemented over a public communication channel, with any eavesdropper, Eve, free to listen, but not alter transmissions between Alice and Bob. Therefore, the classical communication channel is used to determine which photons are received and which are detected with the correct basis. Bob tells Alice which bits he detected and the basis he used to measure them, but not the result he obtained from the measurement. Alice responds by telling Bob which bits he used as the correct basis for decision making. Both Alice and Bob discard all other bits. Assuming the transmissions are not disturbed, Alice and Bob should now share a random sequence of bits that are identical. To test for Eve's presence on the quantum transmission link, Alice and Bob share a random subset of this sequence publicly to determine any anomalies. If excessive anomalies exist, the sequence is discarded; otherwise, the sequence is retained and used as the key for the one-time pad encryption technique. Table 9.1 tabulates the above protocol steps.

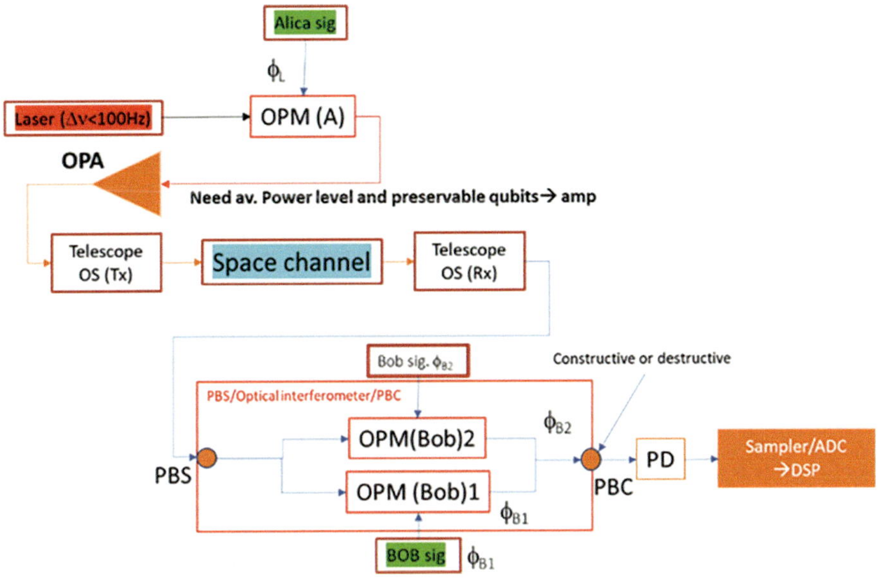

FIGURE 9.16 Schematic of optical transmission over space for QKD. OPM=optical phase modulator, PBS=polarization beam splitter, PD=photodetector, OS=optical system or optical head to launch or accepting lightwaves, OPA=optical power amplifier or HPA (high power booster amplifier).

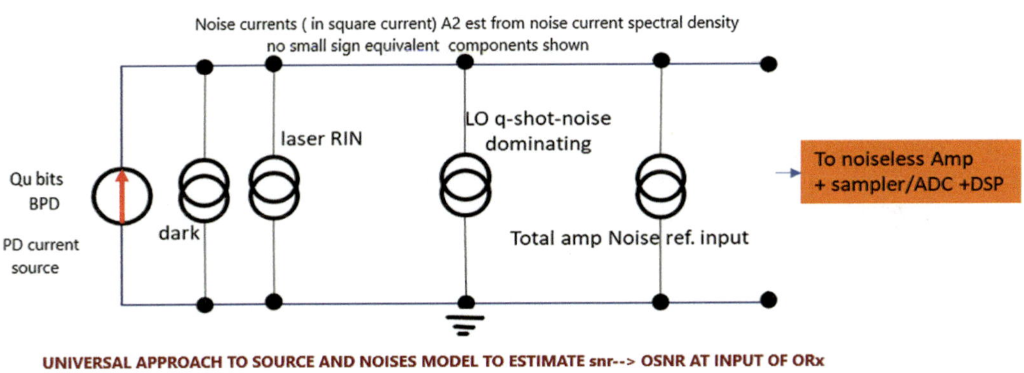

FIGURE 9.17 Noise model for CV-QKD estimation of noise and the quantum power level to be employed.

- Bob phase via OPM(Bob) +Alice Phase (OPM(A) = pi destructive interference – output =0 → no signal
- Otherwise if <<= 0 or 2pi the constructive interference → '1'
- Based on this construct BB84 → apply Bob1 and Bob 2 signalling to construct or deplete the output intensity of interferometer and thus protocol.

FIGURE 9.18 QK protocol in general , BB84 protocol can be found in reference material [6, 7].

TABLE 9.1
BB92 Protocol for Quantum Transmission and Public Confirmation of Key Bits

Quantum Transmission	Bit Sequence			
Alica random bits	1 0 0 1	1 1 0 0	1 0 0 1	1 0 0
Bob's random bits	1 0 1 1	0 1 1 0	1 0 1 1	0 1 0
Detected bits	. 0 . 1	. 1 . 0	. . . 1	. . 0
Open Public Discussions	**Public Confirmation of Received Sequence**			
Bob confirmation of bits (agreed)	. . Y Y	. Y Y	. . . Y	. . Y
Raw keys (IF no eavesdropping)	. 0. 1	. 1 . 1	. . 1	. . 1

$LiNBO_3$ OPM/IQM = ER → 30 dB, OPM = Optical Phase modulation, ER = extinction ratio, IQM = in-phase quadrature modulator.

9.4.5 Routing in ADD/DROP Multiplexer

The routing of quantum keys is also as important as their transmission. However, the losses over the routing modules are reasonably high, and optical amplification must be performed. We have proposed using dark quantum states for transmission over high-loss space hops.[4]

9.5 NOISES AND SENSITIVITY

In order for the transmitted bits to reach the absolute level, the receiver's noise must be at the quantum level so that any eavesdropping can be detected by 'Bob', leading to the discarding of the bits. Thus, the 'Bob' receiver must reach the absolute minimum level so that QKD system can be assured of the absolute secure level. In addition, the sampling system must be at the maximum rate so that the DSP can recover the QKD key bits, whose levels can be several times below the noise level, as described in Sections 9.2 and 9.4.

Referring to Figure 9.17 that represents the small model of all noise sources in noise spectral densities, the followings sources are included: (i) the dark current noise of the photodetector if it is a balanced photodetector pair, this noise source is doubled, as noises are random and added, not subtracted, even if the two PD of the PDP are connected back to back; (ii) the laser relative intensity noise; (iii) the LO quantum shot noise spectral density; (iv) the total equivalent noise current spectral density looking into the two input ports of the electronic amplifier; and (v) the current source as the detected current of the quantum signals sequence.

The total noise spectral density as referred to the input port of the electronic amplifier, including the LNA noises, can be written as a summation of all electronic currents derived from the conversion of optical energy to electronic currents. The total electronic current is equivalent as referred to the input port of the electronic preamplifier. The differential noise current spectral densities are then integrated over the 3 dB bandwidth of the system or the electronic amplifier to obtain the total equivalent noise current as referred to the input port of the electronic amplifier. Hence, the OSNR required at the input port of the electronic receiver can be obtained, and thus the OSNR required can be a converted from electronic current to optical power using the E/O conversion sensitivity factor.

$$\frac{di_{nT}^2}{df} = \frac{di_d^2}{df} + \frac{di_{\text{RIN}}^2}{df} + \frac{di_{\text{LO}}^2}{df} + \frac{di_{T,\text{eq}}^2}{df} + \frac{di_{\text{ASE}}^2}{df} \tag{9.8}$$

where

i_{nT}^2 = the total noise current power
i_d^2 = diode dark current

i_{RIN}^2 = laser relative intensity noise, noises contributed by the spectral linewidth of the laser or the phase noise due to fluctuation of the frequency of the lightwaves

i_{LO}^2 = local oscillator residual noises

i_{ASE}^2 = amplification stimulated emission noises

$\frac{d\left(\ldots i_{nT}^2\right)}{df}$ = the differential term $\frac{d}{df}$ represent the spectral density

All noise currents are expressed in the square of current as they are random variables, and so they must be added regardless of the directions in which they flow. So, a square of the noise quantities ensures this property. The noise term in Equation 9.8 is noise squared current spectral densities. These can then be integrated over the bandwidth of the system, and the square root can be taken to get the noise current. This current is composed of a number of currents, as shown in Figure 9.17. The model is a current equivalent small signal model, or the system is operating in the linear region. The current model is employed here because our signals are detected and converted from optical power to photo-current, as shown at the input of the model in Figure 9.17. Thus, once the noise currents are found, the signal-to-noise ratio can be found by the ratio of the signal current to the total noise current referred to the input port, as given in Equation 9.8. Any nonlinearity can be approximated, and the penalty on the performance, that is, a reduction of the SNR and thus, equivalently the OSNR.

The ASE noise current spectral density can also be lumped into the current source produced by the signal current so that we do not have to include it in the above expression. Detailed calculations using this technique can be found in Ref.[5] The noise current source is expressed in squared terms, as the noises are random variables and they always add up. Therefore, the square terms are used to ensure that they are always positive and add in either direction of flow.

9.6 REMARKS

This chapter has addressed

- The transmission technology for ultra-high capacity channels and testing of modules to transport 100 Gbps employing dual-polarized channels and the QPSK modulation format, as well as the bandwidth limited scenario
- The transmission and reception of power-limited quantum bits over long-distance inter-orbital constellations of satellites.

The channel capacity and optical SNR are analyzed for both bandwidth-limited and power-limited cases, including the role of sampling systems, so that under quantum signal reception, the signal average power can be much lower than the noise level. Illustrations of Tbps and QKD transmission optical systems are provided, and test results for reception systems are illustrated. Techniques for estimating the OSNR based on the information capacity of Nyquist and beyond the Nyquist limit for conventional calculations of channel capacity are outlined, especially when the power-limited case or OSNR $\ll 1$ is the operational limit.

Implementation and analysis of the noise of the reception are crucial for selecting and designing low-noise electronic amplifiers and assessing the system's performance under noise spectral densities.

QKD protocols and effective secret keys for coding of data sequence are also important.

Integrated coherent transceivers have been extensively developed for terrestrial transmission. The common modulation format is DP-QPSK. If one does not have the facilities and technologies to develop integrated coherent photonic circuits for QKD transmission systems, then the QPSK format should be used with the modulation of amplitude ASK on both real and imaginary axes. The optical booster amplifier may contribute to noise levels that can be much higher than the quantum signal itself. Therefore, an appropriate modulation technique must be able to deal with extensive high power in SPC. We consider using the low state of the pulse as the quantum signal, which can be embedded in this slot and preserve the quantum states such that no one can detect under high power amplification. We call this dark quantum states.[8]

NOTES

1 HydRON optical communication for broadband in space, https://www.esa.int/ESA_Multimedia/Images/2021/07/HydRON_optical_communication_for_broadband_in_space.
2 L. N. Binh, C. Fechtmann, M. Koegler, A. Hof, and J. Horwath, "Transmission of multi-Tera-bits/sec to quantum bits over space intra-and inter-orbital links", *Proceedings of ICSO*, Croatia, September 2022.
3 C. E. Shannon, "A mathematical theory of communication", *The Bell System Technical Journal*, 27, 379–423, 623–656, July, October, 1948.
4 L. N. Binh, Dark quantum states for quantum key distribution in long haul space laser communications, Patent pending.
5 L. N. Binh, *Noises in Optical Communications and Photonic Systems*. CRC Press, Taylor & Francis: Boca Raton, FL, 2016.
6 C. H. Bennett and G. Brassard. "Quantum cryptography: Public key distribution and coin tossing". In *Proceedings of IEEE International Conference on Computers, Systems and Signal Processing*, volume 175, p. 8. New York, 1984
7 Charles H. Bennett and Gilles Brassard. "Quantum cryptography: Public key distribution and coin tossing", *Theoretical Computer Science. Theoretical Aspects of Quantum Cryptography – celebrating 30 years of BB84*, 560, Part 1: 7–11. arXiv:2003.06557. doi:10.1016/j.tcs.2014.05.025
8 L. N. Binh, "Dark quantum states for QKD over long haul quantum transmission", Patent pending.

10 SPC-Equivalent System Transmission Performance

CONTENTS

DOI: 10.1201/9781003426677-10

The demonstration by the long-haul space photonic transmission experiment is very important to show the realizability of such a system for inter-satellite photonic transmission and routing. This chapter outlines the experimental SPC (Space Photonic Communication) transmission system of multi-channel and dual polarization QPSK (quadrature Phase Shift Keying) modulation format with 100 Gbps per wavelength carrier, so 10 optical carriers can aggregate to a total capacity of 1.0 Tbps.

We show how to do the testing when constructing a one-hop inter-satellite equivalent photonic link. This includes the back-to-back (B2B) system transmission and components testing, such as the wavelength selective switches (WSS) connected at both the transmitter and receiver to evaluate the system performance. Then equivalent SPC link is constructed using the HPA (high-power amplifiers) and low-noise optical amplifiers (LNA) plus variable optical attenuators (VoA). The OSNR (optical signal to noise ratio) and BER (Bit-Error-Rate) versus signal average optical power are obtained for various cases and demonstrate the validity of the SPC system link. Furthermore, B2B tests and WSS used as multiplexers and demultiplexers are measured and described for Tbps transmission systems.

10.1 SYSTEM TEST STRUCTURE

The equivalent SPC system is arranged using the following components: (i) an optical integrated module transceiver in which the optical modulator is a dual-polarization IQ modulator, as given in Appendix 1 of Chapter 11; (ii) the lightwaves are generated by an external laser or integrated laser embedded in the OTRx module, the laser linewidth is about 10.0 kHz and tunable over the entire C-band; (iii) fiber optic coupler to split the generated optical data channel for spectrum monitoring using the OSA (the vacuum free space medium can be equivalently set by inserting a VoA); (iv) an optical preamplifier is placed in front of the receiver input of the integrated OTRx module; (v) the output of the optical preamplifier is fed into the transceiver module which consists of a balanced photo-detector (PDP) pair placed at the output of an n optical hybrid π/2 coupler. A tunable optical laser of narrow linewidth is used as a local oscillator. It is also fed to the input of the hybrid coupler so that both the polarized data channels and dual polarization LO (local oscillator)laser can be co-propagating with channel signals. They are beating with each other in the PDP to produce RF signals whose frequency may be shifted by an amount representing the frequency difference between the lightwave carrier and that of the LO laser. The functioning of the interconnected components of the system is further explained in Figure 10.1.

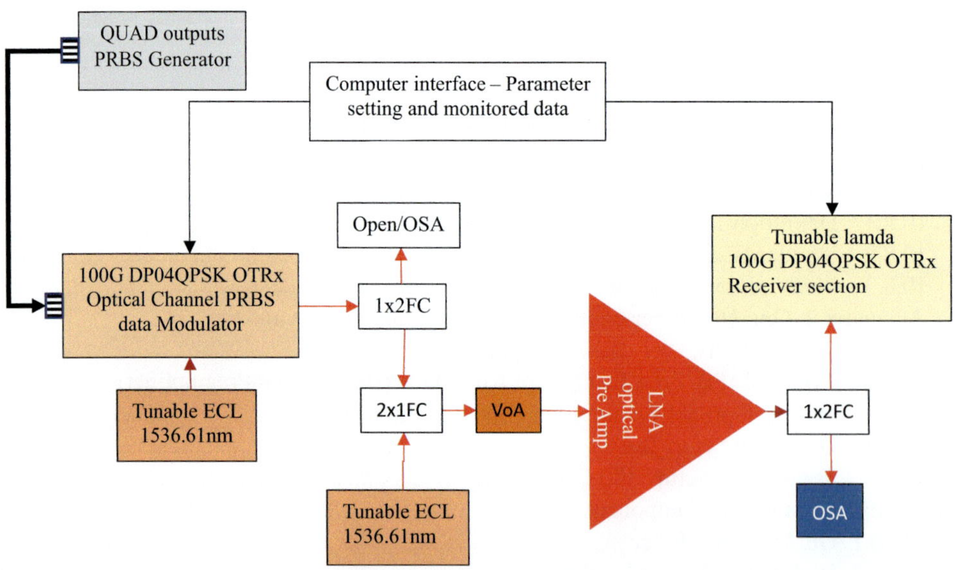

FIGURE 10.1 Schematic of test system using integrated optical transceiver with low noise optical preamplifier whose filtered passband (3 dB BW) is 5.9 nm. OSA is with a minimum resolution of 5 pm.

The operating conditions of the optical and electronic components of the test system are:

Signal data sequence PRBS (pseudo-random binary sequence) to modulate QPSK optical modulator (see Appendix 1 of Chapter 11), dual polarization of X (horizontal)- and Y (vertical)-directions to generate 100 Gbps DP-QPSK[1]– commercial OTRx module DP04QSFP with channel setting in (100 GHz grid ITU-T) at 1553.33 nm (Ch #30 ITU-T[2]) and 50 GHz grid (or approx. 0.4 nm) per number unity step, that is if channel # 20, then the wavelength is 1553.33 – with a spacing of 0.8 nm or 100GHz in the frequency domain, 1552.53 nm and so on[3]....In the C-band, approximately the wavelength step of 0.8 nm is equivalent to 100 GHz in frequency scale.

Four PRBS data electrical channels are required for dual-polarization IQ signaling, two each for each IQ modulator, that is, the in-phase and quadrature inputs to drive the MZM modulator, see Chapter 4 on optical modulation for details. Note that the orthogonality of the I and Q channels is implemented by using the $\pi/2$ phase shift built into the IQ modulator; that is, a phase of $\pi/2$ shifts the branch of the MZM by applying an electrical voltage via the pyroelectric effect to change the refractive index of the optical waveguide. See the detailed structure of such IQ modulator in Appendix 1 of Chapter 11.

Transmission is used via splitting the optical channel at the output of the high-power amplifier using a high-power fiber coupler (FC) of 2×1 FC with an insertion loss of 4.01 dB. This is to ensure that the fiber is resilient to the high power of a few thousand mW at the output of the booster amplifier HPSA. The gain of the HPA can be read out from the amplifier-monitored power meter. Thus, from the input average power of the optical channels to the output of the amplifier and then the variable attenuator and the 1×2 fiber coupler to split the signal is estimated as (25.4 dB) - VoA (dB) – [1×2] FC (4.1 dB).

Optical Amplifiers = An HPA is used as booster amplification, and LNA as the pre-amplifier with a noise figure of 4 dB and low level acceptable at –45 to –65 dBm over the whole C-band (1545–1565 nm).

The wavelength of a channel is set at 1553.33 nm or assigned as Channel #30 on the ITU grid (see Appendix 6 of Chapter 12). The optical filter passband of the low-noise optical preamplifier also determines this channel wavelength. See the structure of the low-noise preamplifier given in Chapter 9 and Chapter 12 and the optical add/drop mux of a single channel.

For Channel #30, the wavelength of the generating lightwaves is at 1553.33 nm.

Parameters that can be monitored by the OTRx module for optical receiving signals are: (i) OSNR by Rx monitoring parameter at the output interfaces shown in Section 10.9; (ii) OSNR by optical spectrum techniques, OSNR1, OSNR2, OSNR3, OSNR4, and OSNR5, which are described in the next section (these values are estimated by different methods to ensure the best approximated experimental values for OSNR); (iii) BER (bit-error-rate) is also obtained as well as the QPSK constellations that can be obtained as displayed in Chapters 5 and 6. This BER can be estimated using the EVM technique as described in Chapter 5 and Figure 10.7.

The OSNR and BER must be obtained simultaneously to confirm the system's performance. We can see in the later section that when the system is affected by nonlinearity due to the high-power lightwaves propagating in the short fiber, the booster's output is connected to the optical head. The OSNR remains high, but the BER is decreasing. Thus, the nonlinear cross-talk degenerates the signal quality but not the OSNR, as the optical signal average power cannot distinguish the cross-talk components.

See Chapters 5 and 6 on the DSP system's receiver hardware and soft processing, which follows the coherent reception signals.

As indicated, changing the data in the storage file and the number of channels of the communications interface can set different channels in the ITU-T grid. (See Appendix 7 -Section 12.5 of Chapter 12 for tables of ITU-T grids)

10.2 OSNR MEASURING TECHNIQUES

This section evaluates the system OSNR by integrating the signal and noise spectra as obtained by using the OSA, which is outside the module OTRx. This is to confirm with the BER monitored by internal processing of the transceiver module. E named this spectrum integration technique (SIT);

see Section 10.4.4.1. The results presented here, which eliminate the booster amplifier HPA but only the LNA, demonstrate the validity of our SIT method. The OSA produces an Excel file in which the sampled power is monitored with the resolution. The lowest resolution of OSA nowadays can be as low as 5.0 pm. Thus, the integration of a channel's spectrum can be easily determined.

We note that the OSA displays the optical spectral density on its screen; that is, the display of the power is measured in the close setting of the resolution. So, the reading on the vertical scale of the OSA is the power spectral density. Thus, the display must be read as power/resolution. Thence, the power is obtained by (power/resolution)×(resolution) which is set as a variable in the OSA before sweeping.

It is noted further that users are to determine the spectrum under noisy and less noisy conditions. This is critical to obtaining accurate results. Using the rule of thumb, we consider the spectrum section if the total penalty is more than 1 dB; if not, we can ignore it.

In the following, we examine and investigate using five techniques for measurements of the OSNR.

OSNR1: By integrating the signal and noise spectral densities over the signal band, which is defined as the "20 dB" point of the signal spectrum as noted by the OSNR 1 mark in Figure 10.2. This is the main source of the signal energy. This is valid well if the noise accumulated reaches this level. However, when the signal-to-noise ratio is large, there is a large difference between the receiver sensitivity monitoring level and this signal integration and noise power using this technique.

OSNR 2, OSNR 3: Similar to OSNR 1, OSNR 2 and OSNR 3 are obtained by integrating the power spectrum of signal noise power at the level as indicated in Figure 10.2. Even at the normalization of the noise over 0.1 nm bandwidth, there are noticeable differences between OSNR 2 and OSNR3 with respect to the receiver's measurements.

OSNR 4: Obtained by integrating the signal and noise power spectra over the signal bandwidth. In the case of 100 G DP-QPSK, the baud rate, including 30% FEC (Forward Error Correction)overhead bits, is 32.5 Gbps, and thus, its DSB (double sideband) spectral window is ~0.26 nm in the C-band. Integrating but no normalization of the noise bandwidth, there are differences between OSNR 4 and OSNR (Rx) obtained by the transceiver digital signal processing of the receiver section.

OSNR5: This method is then derived by (i) taking the power OSNR from the noise spectral density provided by the OSA. Since both quantities are obtained over the same resolution, then the ratio in dB is indeed the difference between the levels of the signal band and that of the noise band. In the case of a low-noise optical amplifier, the filtering band of the optical filter indicates the noise band as distinct from that of the signal band as indicated in Figure 10.2.

This technique, OSNR5, can be illustrated in Figure 10.2 (also in Section 10.4.4.1) for the test system case depicted in Figures 10.1 and 10.2. Once the signal-to-noise ratio is obtained from the power spectral density, a normalization of the signal bandwidth (0.26 nm) and noise band (0.1 nm) by 10*log10 (0.26/0.1) to reach the final OSNR5 (0.1 nm). This technique is quite compatible with the digital technique of error vector magnitude (EVM),[4] as described in Section 10.3 and Figure 10.7.

As illustrated, EVN is the linear ratio between the original (noiseless) amplitude and the RMS of the noisy sampled points. These points include the signal and the superimposed noise amplitude, indicating the amplified noises.

So, in the interpretation of the power spectrum, we can determine the noise power spectral density level at 3 dB of the noise band, which can be distinctly found in the case of the optical amplified signal of low noise amplification type, i.e., including an optical filter of order higher than 5 so that sharp roll-off can suppress the noises outside the passband and not interfering adjacent channels. As shown in Figure 10.2, the 3 dB point below the noise band can be found by inspecting the noise

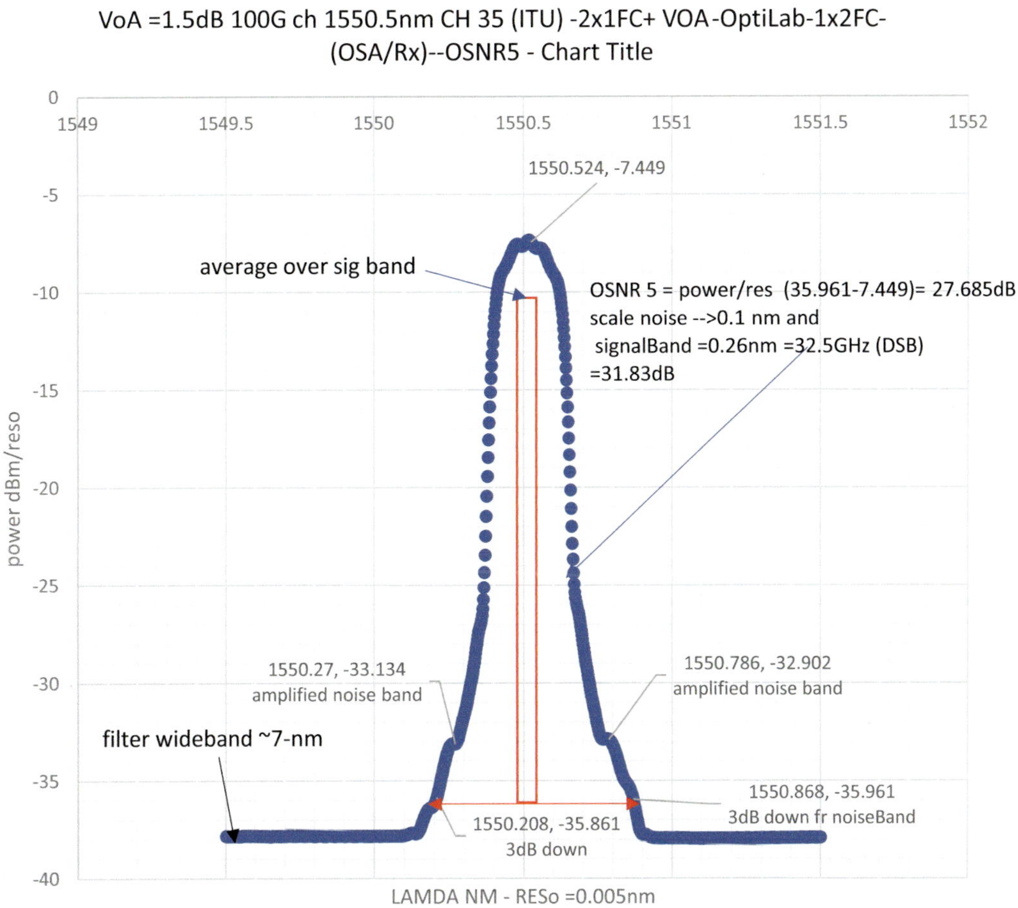

FIGURE 10.2 Illustration of measurement technique for OSNR5.

TABLE 10.1
Average Recever Power and OSNR 1–3, 5 with Variation of Attenuation Factor

VoA (dB)	OSNR1 (20 dB) - dB	OSNR2 (OF Band) - dB	OSNR3 (Out-of-Band) - dB	Ave. OTRx OSNR (Rx - 0.1 nm) - dB	OSNR5 (0.1 nm)
1.5	28.3	30.1	33	31.2	31.83
5					
10	21.5	23.5	24.5	23.4	23.3
15					
20	18.8	19.4	20.1	19.2	17.523
25	13.2	14.5	14.5	10.1	10.7

band. The 3 dB point below the noise band is taken as any noise level below 3 dB can be ignored, the same as EVM can be evaluated by averaging.

We can observe that the amplified noise level is about 8 dB above the filtered noise band.

Table 10.1 tabulates all OSNRs and the receiver average power by OSA and Rx as obtained.

Plots of OSNR5 (0.1 nm[5]) versus VoA with Rx and OSNR5 are given, as shown in Figure 10.3. Both sets of OSNR can see a very close approximation over standard 0.1 nm noise bandwidth.

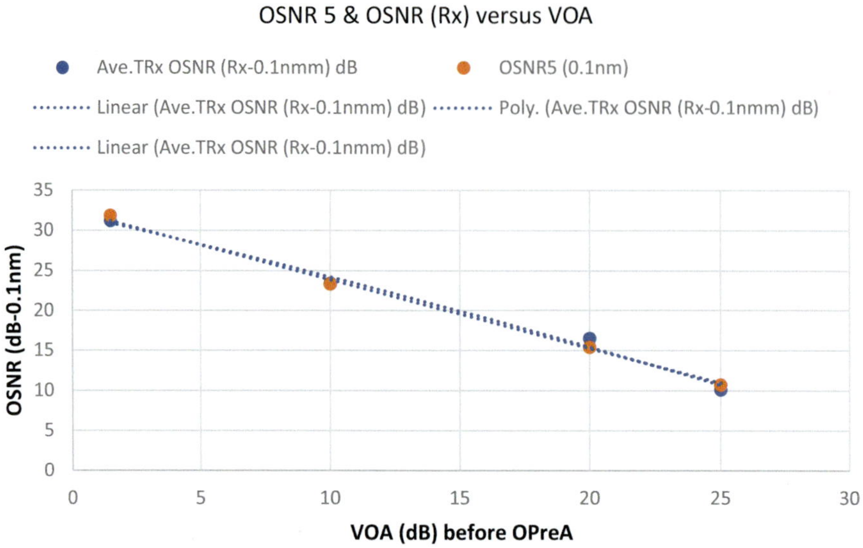

FIGURE 10.3 OSNR (0.1 nm) signal-to-noise ratio (OSNR5) of Rx sensitivity (OSNR-Rx) and OSA platform measured with a central strip of 0.1 nm or scaled Power spectral density with noise level at 3 dB below the noise level of the amplifier. The test system is shown in Figure 10.1 preamplifier=OptiLab -type 7 nm passband.

10.2.1 OSNR Measurement and Compatible to EVM Technique

This section illustrates the measurement of the spectra in different cases of VoA. Both in the cases of high signal-to-noise ratio and low signal-to-noise ratio, that is, when the noise level reaches the signal level where noises overcome –20 dB. At –20 dB amplitude (10 dB in power), the signal reaches the lowest level in the frequency domain. The resolution of the OSA is not small enough to see this level, but approximately, we can determine the level. The estimation of the OSNR is derived here for laboratories that do not have extensive equipment and only have access to the integrated, coherent photonic transceiver (ICPTRx), which provides the final estimations of BER, constellation and received power. In this experiment, we access only such facility of the ICPTRx. Thus, the spectrum evaluation technique is described here to evaluate various effects on the SPC transmission systems, especially the accumulated Hop noises contribution on each multi-hop transmission link cascade stage. This is important when such a link is established in real space to save the weight of the equipment carrying the load on satellite and simplify monitoring the channel signals for updating the link networks (Figure 10.4).

10.2.2 Performance OSNR by Integration Technique without Noise and Signal Bands Scaling

Section 10.2.1 shows the OSNR1,2,3 calculated by integrating the signal and noise spectra without normalization to the standard 0.1 nm noise bandwidth, as shown in Figure 10.5. This shows a significant difference of about 10 dB when the input power into the amplifier (preamplifier) is low. That is when the noises corrupt the measurement integration.

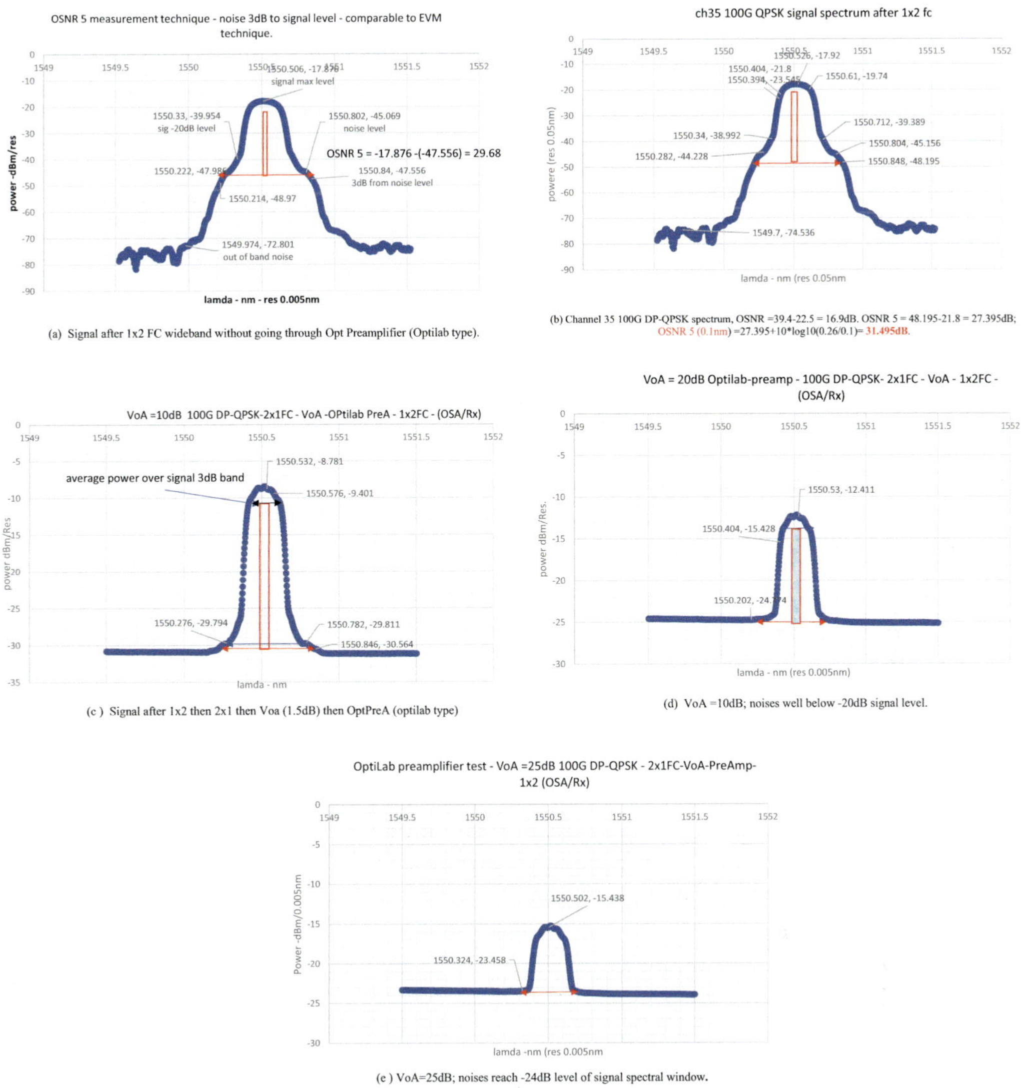

(a) Signal after 1x2 FC wideband without going through Opt Preamplifier (Optilab type).

(b) Channel 35 100G DP-QPSK spectrum, OSNR =39.4-22.5 = 16.9dB. OSNR 5 = 48.195-21.8 = 27.395dB; OSNR 5 (0.1nm) =27.395+10*log10(0.26/0.1)= 31.495dB.

(c) Signal after 1x2 then 2x1 then Voa (1.5dB) then OptPreA (optilab type)

(d) VoA =10dB; noises well below -20dB signal level.

(e) VoA=25dB; noises reach -24dB level of signal spectral window.

FIGURE 10.4 (a) Signal after 1×2 FC wideband without going through Opt Preamplifier (Optilab type). (b) Channel 35 100 G DP-QPSK spectrum, OSNR = 39.4–22.5 = 16.9 dB. OSNR 5 = 48.195 − 21.8 = 27.395 dB; OSNR 5 (0.1 nm) = 27.395 + 10 * $\log_{10}$ (0.26/0.1) = **31.495 dB**. (c) Signal after 1×2, then 2×1, then VoA (1.5 dB), then OptPreA (optilab type). (d) VoA = 10 dB; noises well below –20 dB signal level. (e) VoA = 25 dB; noises reach –24 dB level of the signal spectral window.

10.2.3 Optical Amplified Noises in Receiving Signals

10.2.3.1 Remarks

Regarding the in-band optically amplified noises, as reported in the PDR report, amplified noise exists in an optically amplified hop link. These noises are unpolarized. When the s100 G and in-band noises enter the integrated optic receiver of the transceiver module, we can see, as depicted in Figure 10.6, that at the reception front end, the local oscillator laser and the signals are coupled into a rectangular cross-section waveguide, which is polarized into TE and TM modes (transverse electric and transverse magnetic field). Therefore, the unpolarized optical noises are filtered at this point.

(A)

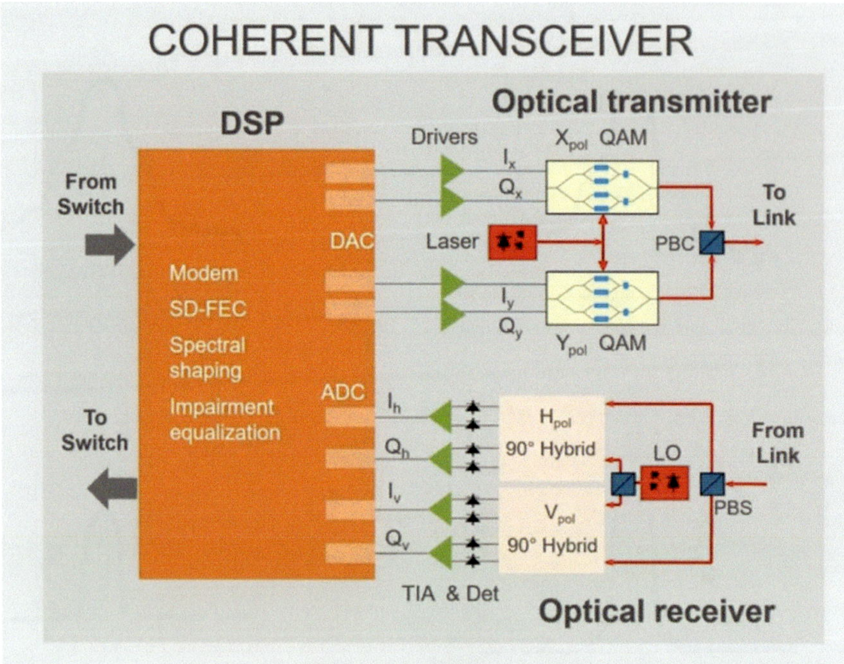

(B)

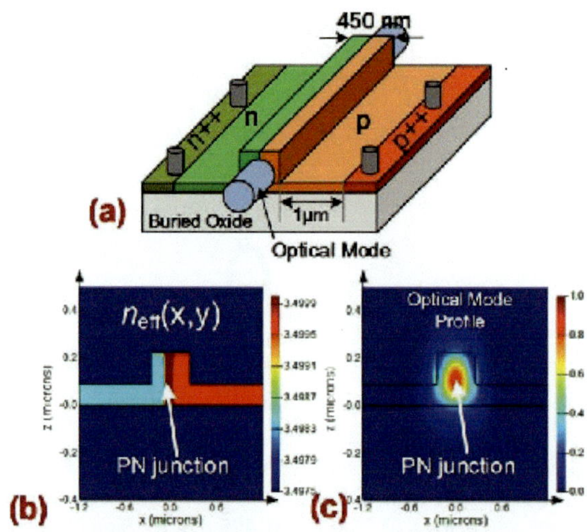

FIGURE 10.5 OSNR versus VoA using integration of signal and noise spectra without scaling to noise 0.1 nm and the values obtained from module processed OSNR (average) by EVM technique. The determination of OSNR5 is compatible with the EVM technique, as described in Section 10.7 of Case VoA = 1.5 dB, channel at wavelength of 1550.50 nm.

The TE/TM are mixed with the polarized channels of the DP-QPSK channels in the pairs of the balanced photodetectors whose outputs are amplified by TIA (transimpedance amplifier) and then sampled and held in ADC (analog to digital converters) and fed into digital signal processors in which the processing algorithms are executed and the outputs derived with carrier phase detection to determine the LO frequency offset with respect to that of the carrier. Compensation for this offset is done to achieve true homodyne detection. Compensation for any distortion, such as dispersion in free space, does not exist. The delay difference between the two polarized channels can be evaluated, and both channels are aligned so that the in-phase and quadrature-phase components can be synchronized.

(a)

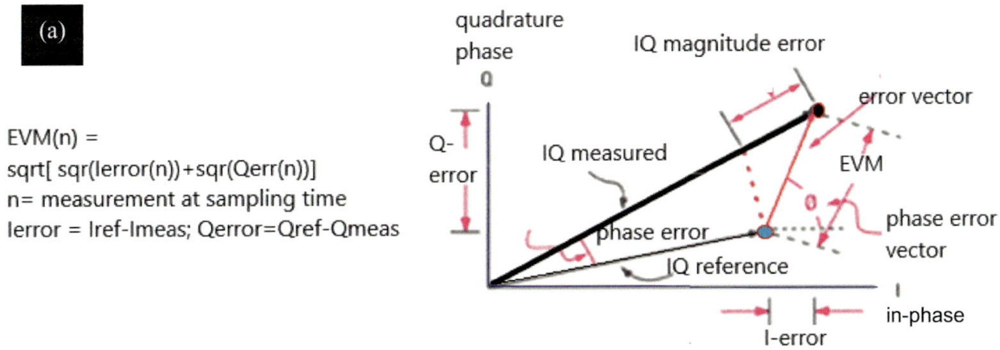

(b)

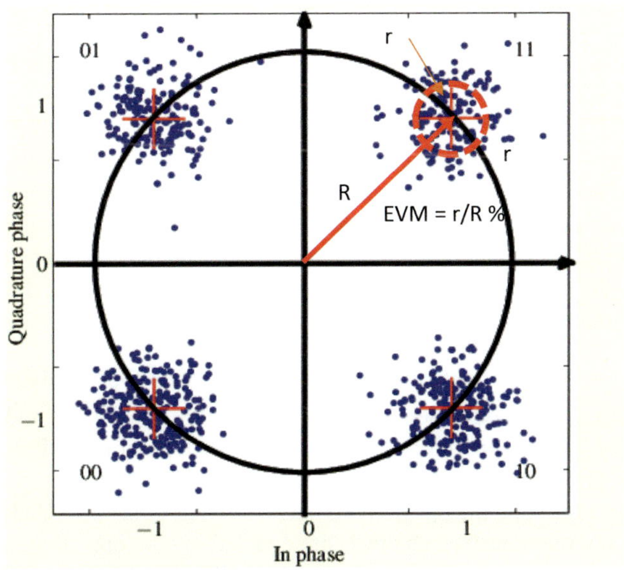

FIGURE 10.6 (A) The 100/200 G integrated, coherent photonic transceiver (ICPTRx) with integrated DSP, a hybrid coupler with phase shift, polarization splitter, and TIA and balanced photodetector pair. IQ modulator using SOI and ABC. Independent lasers for Tx and LO ORx. SD_FEC, spectral shaping and impairment equalization. Cross section of the hybrid coupler shown in (B); (B) SOI (Si over insulator) optical waveguides for hybrid coupler, the pre-photonic structure of Rx of coherent TRx (a) device structure and fiber–waveguide coupling; (b) physical silicon p-n junction cross-section of waveguide (c) fundamental guided mode including both TE and TM mode, when under reverse bias. DAC=digital to analogue converter, ADC=Analogue to digital converter, LO=local oscillator laser, QAM=quadrature amplitude modulation, I_x/I_y=in-phase optical components in X/Y polarized channels (Tx side), Q_x/Q_y=quadrature optical components, I_h/I_v=in phase components received optical components in electrical domain after coherent reception via hybrid coupler and mixing in PDP (Phodetector pair), *h*,*v* = horizontal and vertical polarization; Q_h/Q_v = quadrature components for polarization vertical and horizontal. *I*=in-phase, Q=quadrature components, *n*, *p*=*n*- and *p*-doped Si region, *n*++ and *p*++=highly doped p region – conductive. (Panel (B) Extracted from Ref.[6])

Digital signals with sampled points are evaluated using the error vector magnitude (EVM) or the percentage of the EVM over the magnitude of the vector that represents the QPSK constellation point. The EVM (see Figure 10.7) is then interpreted into the SNR and, hence, the OSNR with the known quantum efficiency or the responsivity of the balanced detectors.

Therefore, we can see that the OSNR provided by the transceiver represents the signal-to-noise ratio without or with a minimum contribution of such in-band optical noises due to the polarization property of the hybrid coupler.

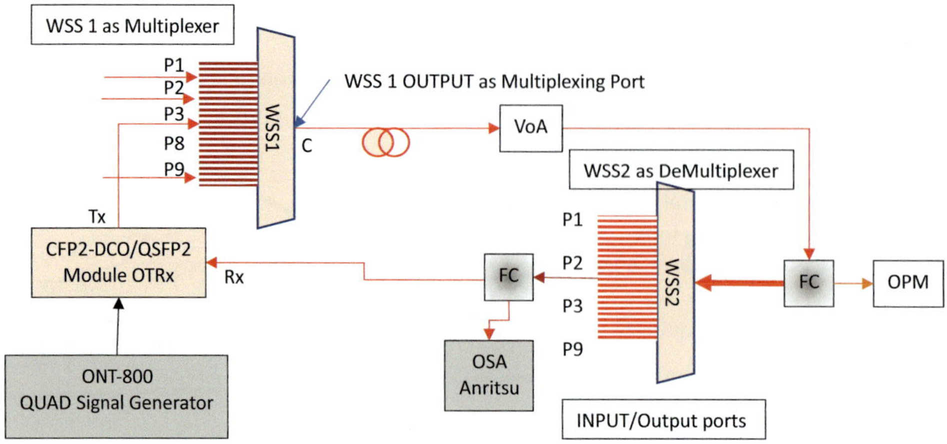

FIGURE 10.7 (a) Illustration of EVM in phase modulated transmission signals (original) and error; (b) QPSK constellation and EVM in %.

So, while we obtain the optical spectrum of the optically amplified signals over the OSA, the ratio OSNR can include both the signal and the optically amplified noises hidden under the signal envelope. To compare the OSNR from the OSA signals and that of the transceiver, we must determine the correct noise level that is compatible with that of the receiver. The selection is the noise level of 3 dB below the filtered band of the optical preamplifier and that of 3 dB of the signal band. These OSNR measurements are illustrated in the above two sub-sections.

10.2.3.2 Non-Polarized Optical Amplification Noises and OSNR Measurements

Optical amplification noises embedded in optical signals are important, and they must be determined, especially for polarization multiplexed channels such as in DP_QPSK modulation formats in 100 G and 200 G. IN polarized lightwaves transmission under optically amplified links such as in free space fiber equivalent link, the optical ASE noises are non-polarized. Under coherent detection, the local oscillator and the polarized channels are coupled into an optical hybrid coupler, normally a PLC (planar lightwaves circuit) such as the SOI-based PLC, as shown in Figure 10.6. This PLC hybrid coupler is commonly formed and structured to be highly polarization dependent, and all non-polarized lightwaves are filtered except the TE or TM polarized fields. Thus, the optical amplification ASE noises are filtered in this case, and the DSP-based optical receiver of the DP04QSFP module processes the channels and interprets the EVM into the OSNR (0.1 nm) with 0.1 nm noise bandwidth. This noise also includes the optical equivalent electronic noises, which are normally much lower than the quantum shot noises of the mixing product of the two waves of LO and signals. The EVM and QPSK constellations are illustrated in Figure 10.7.

In the Pegasus platform, the OSA monitors the optical channels to be estimated to match the OSNR given by the module. Under the OSNR quantity described above, the OSNR calculated from the signal spectrum can be used without much detailed knowledge of the polarized amplification noises. Hence, the techniques used in previous sections of this report. This applies to all OSNR of cascaded stages.

10.3 BACK-TO-BACK (B2B) TRANSMISSION OF DWDM 100 G DP-QPSK MODULATED CHANNELS

The SPC system is developed here, and this chapter demonstrates the steps for the testing stage by stage of the one-hop optical transmission using fiber equivalent structure. To ensure the validity of the SPC link, we will conduct the following steps: (i) B2B using only the mux and demux, which are the WSS. This is to ensure the working principles of the optical multiplexer and demultiplexer and the setting of the filter bandwidth of the WSS to meet the demand of multiplexing wideband signals so that the aggregate t rate can teach Tbps.

This section reports a B2B transmission system to ensure that the WSSes, WSS1 and WSS2, structured as WDM multiplexers and demultiplexers of a space optical communication system (SOCS), operate at the Tbps total capacity. The system carries 6 carriers modulated with DP-QPSK at 25 or 50 GSy/s to give 100 G and 200 Gbps per wavelength channel. This report briefly describes the B2B transmission setup and the performances of these channels. Part 1 provided test results of the WSS transfer characteristics, the transmittance versus wavelength, to demonstrate the filtering property of the multiplexer and demultiplexer whose passband can be tuned to pass the signal band and filter the noise band. Due to the difference in the signal band and noise band in the measurement of noises under an optically amplified and filtering transmission system, the OSNR is strongly dependent on the determination of the in-band noise power.

The next section then gives a measurement technique for coherent polarization diversity transmission of channels, the DP-QPSK, under such amplification and filtering component contribution.

10.3.1 Transmission Setup

The schematic diagram of the transmission B2B is shown in Figures 10.8 and 10.9.

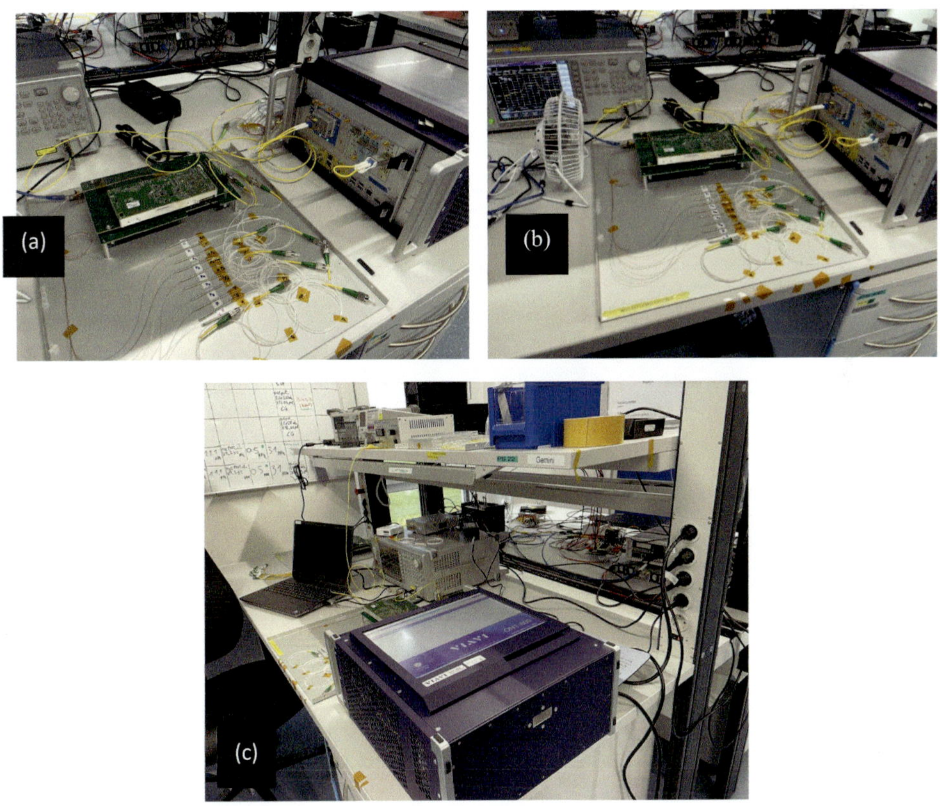

FIGURE 10.8 B2B transmission setup schematic. WSS1/WSS2=wavelength elective switches configured as multiplexer and demultiplexer; Rx=receiver, OSA=optical spectrum analyzer, VoA=variable optical attenuator, FC=fiber coupler, port C=common port of WSS, ONT=optical network transporter, C=common, CFP2-DCO=digital coherent optics Optical Transceiver, P1…P7=port 1–9 as input ports to be multiplexed by WSS1, P1…P9=demultiplexed ports by WSS2, DCO=digital coherent optical module, ONT-800=model name as optical signal generator and receiver module, OPM=optical power meter.

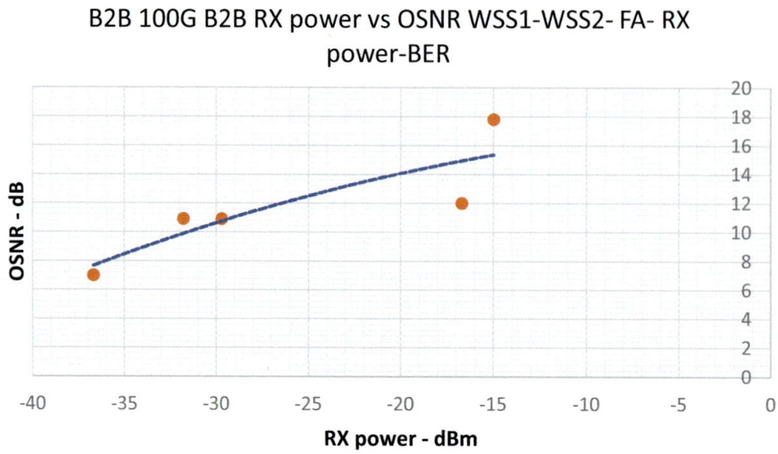

FIGURE 10.9 Photos of the setup of transmission B2B using WSS1 (multiplexer) and WSS@ (demultiplexer), signal source=Viavi (a) w/o OSA; (b) with OSA to monitor signal spectrum. Both cases can access BER under HD-FEC; (c) two system structures as in Figure 10.8.

NOTES: File in Excel obtained from OSA in excel csv format – to convert to column, go to "DATA" then select row to columns and set delimiter to "dot" or "columns" (Figure 10.10).

10.3.2 B2B Performance with Modulated Data Sequence

10.3.2.1 The 100 G PDM-QPSK

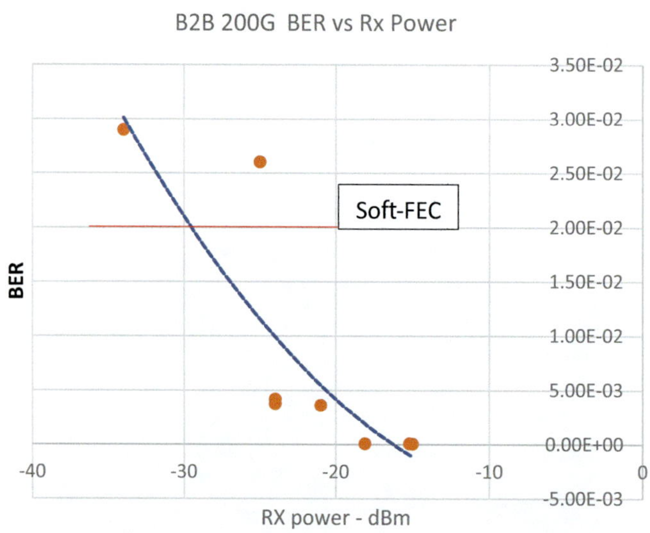

FIGURE 10.10 B2B testing 100 G PDM-QPSK WSS1-WSS2 OTRx (Rx). OSNR obtained in Acacia Rx of Acacia OTRx CFP2-DCO. Input power into the Rx.

10.3.2.2 The 200 G 16 QAM-PDM

The transceiver test can be done at a 56 GB baud rate and the signal generator can operate at 28 GB and/or 56 GB. The system setup is shown in Figure 10.11. The modulation format is QPSK and dual-polarized channel with a baud rate of 56 GB. The BER was monitored or measured in the DSP-based OTRx. EVM is the evaluation method of sampled signal average power that is potted against the averaged received power in dBm. The received power can be varied by varying the attenuation factor of the VoA.

10.3.2.3 The 400 G 16 QAM-PDM (56 GB DP-16 QAM)

The transceiver test can be done at a 56 GB baud rate. The signal generator can operate at 28 GB and/or 56 GB. The system setup is shown in Figure 10.8. The modulation format is 16 QAM and dual-polarized channels with a baud rate of 56 GB. The BER was monitored or measured in the DSP-based OTRx. EVM is the evaluation method of sampled signal average power that is potted against the averaged received power in dBm. The received power can be varied by varying the attenuation factor of the VoA (Figures 10.12 and 10.13).

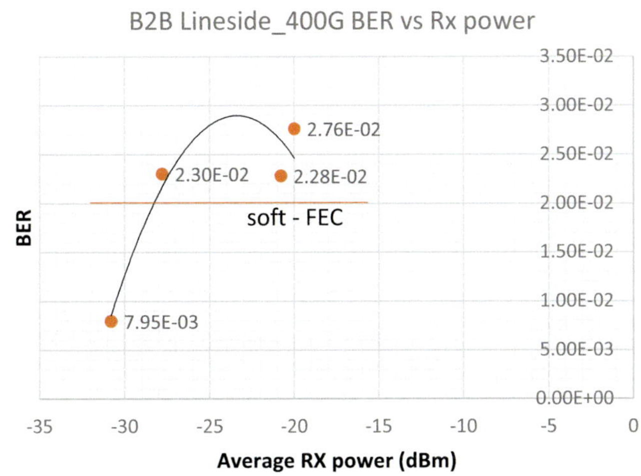

FIGURE 10.11 B2B testing 200 G WSS1-WSS2 OTRx (Rx) – 200 G 50 GSym/s (56 GB with FEC overhead) DP-QPSK (dual polarization quadrature amplitude modulation four constellation points and two bits/symbol).

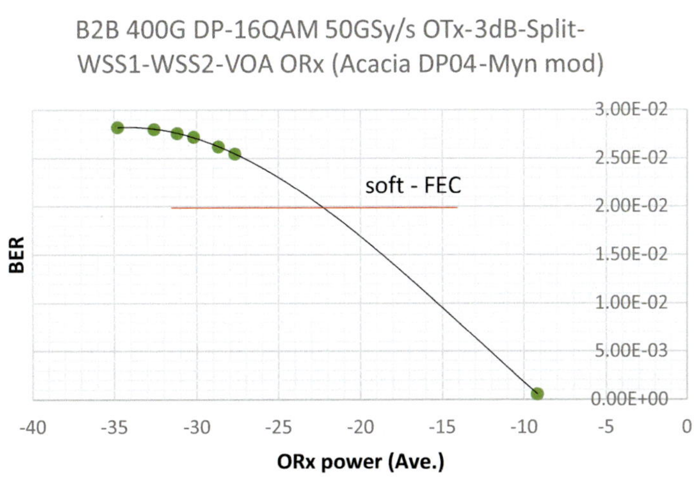

FIGURE 10.12 BER versus Rx power (dBm) B2B line side 400 G Viavi -Acacia-module modulating QSFP DP04 QSFP Acacia unit. Line side (average)- dBm -Rec of the pattern generator and receiving processing.

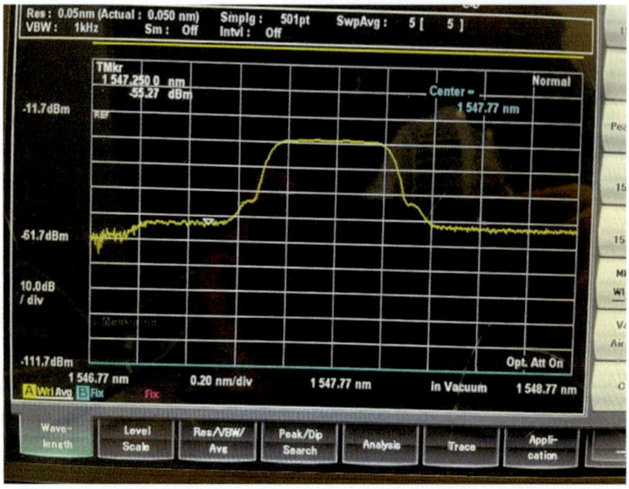

FIGURE 10.13 BER versus receiver average power – B2B with 400 G from ONT-800 Viavi – measured on 400 G PDM-16 QAM (Acacia module by Mynaric). VOA is set to a minimum power level. DP=dual polarization.

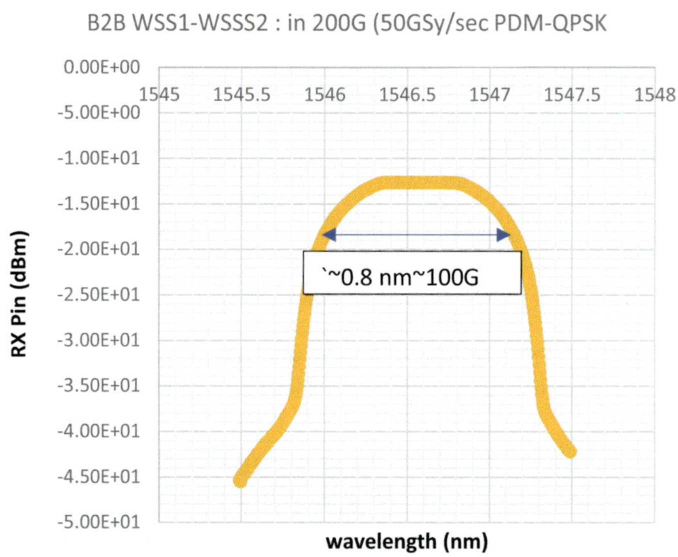

FIGURE 10.14 OSA spectrum display of channel at 1547.250 nm center and 2.1×0.2 nm/div=0.42 nm or ~50 GHz 3 dB bandwidth. Signals without passing through the B2B system.

10.3.3 Spectra of Transmitted Channel

The modulated channels are generated from the modules DP04QPSK and CFP2-DCO AC1200[7]. The symbol rates can be set to 25 or 50 GSym/s Internet format plus FEC extra bits for DP-QPSK to generate 100 or 200 Gbps and 400 Gbps with DP-16QAM. Figure 10.14 shows the photo of the signal spectrum when not passing through the B2B WSS1-WSS2 and attenuator (Figures 10.15–10.17).

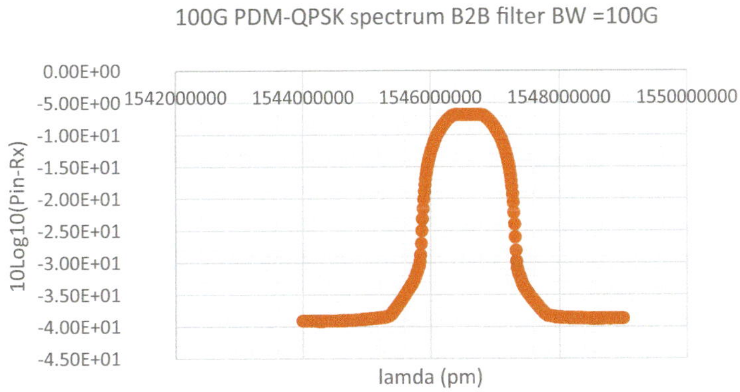

FIGURE 10.15 The transmitted spectrum of 100 G PDM-QPSK setting at 100 GHz bandwidth WSS and the central frequency settings of WSS1 and WSS2 are identical. WSS is setting central freq. or lambda=center wavelength of 1546.49 nm, division 0.5 nm/div.

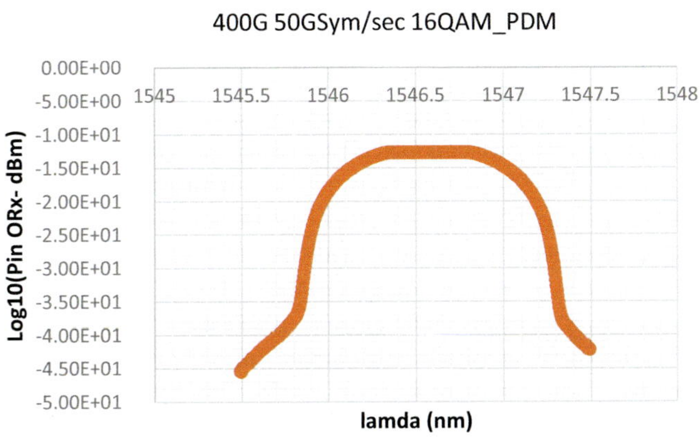

FIGURE 10.16 The 100 G PDM-QPSK B2B WSS1-WSS2 at 100 GHz passband: received power (dBm) versus lambda (pm) OSA=Anritsu 1.0 nm per division.

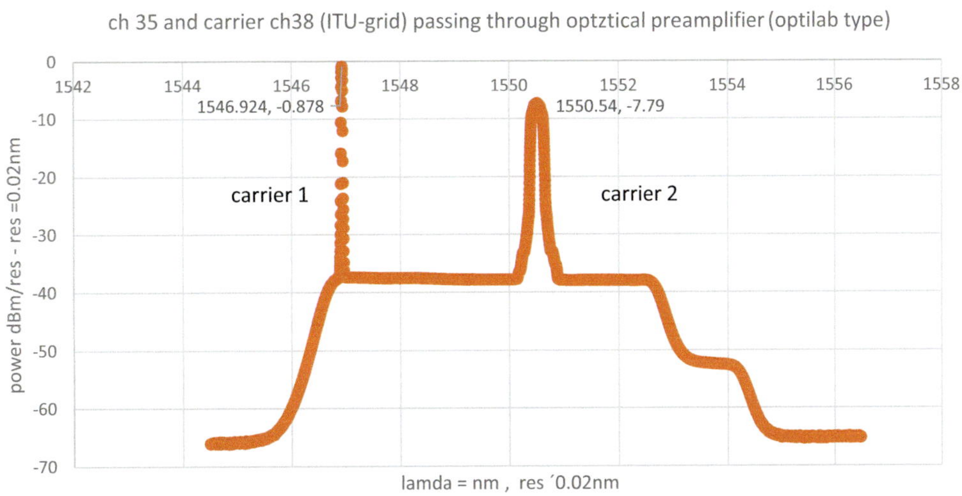

FIGURE 10.17 The 100 G PDM -QPSK B2B WSS1-WSS2 at 50 GSym/s PDM-16 QAM 400 GHz passband: received power (dBm) versus lambda (nm) OSA=Anritsu-type (30 pm min resolution). WSS filter bandwidth=100 GHz. No spectrum distortion was observed.

10.4 ONE-HOP LINK WITH OPTILAB[8] OPTICAL PREAMPLIFIER AND MPB BOOSTER (FOR BUDGET ESTIMATION ONLY)

10.4.1 REMARKS

This section presents the experimental results of a fiber-equivalent free-space hop (MEO-MEO) using an optical preamplifier type and booster amplifier. However, the experiment was not fully implemented due to some technical issues. Thus, the MPB[9] characteristics (experimentally measured) are used to estimate the power budget only. Experiments will be conducted at a later stage, and results will be supplemented.

Optilab preamplifier is a low noise of a wide passband of about 5.9 nm, which is used to filter the ASE (amplification stimulated emission) noise of the EDFA amplifier to reduce the total accumulated noise. The noise level of this preamplifier reaches −66 dBm, about 10 dB higher than that of the MPB optical preamplifier. The passband roll-off is sufficiently high of order >5 is sufficient to ensure that the signals of 100 or 200 G DP-QPSK can be multiplexed to form an aggregate of 1.0 Tbps without suffering any distortion. The gain of the amplifier is not controlled to obtain the optimized operating condition; that is, the gain is not increased by higher pumping power into the gain medium when the input signals reach below the saturation level of the EDFA. Unlike the preamplifier by Optilab type, the MPB low noise amplifier (LNA) section of MPB EDFA amplifiers can automatically set the pump laser power to achieve higher gain. The Full One-Hop link using booster and MPB type preamplifier is reported in Section 10.5.

Figure 10.18 depicts the optical passband spectrum of the Optilab preamplifier (LNA) in which there is an optical carrier at 1546.924 nm and a modulated lightwave channel centered at 1550.54 nm. The noise level of this preamplifier is at about −66 dBm and the 3 dB passband is about 5.9 nm or approximately 737.5 GHz. Thus, for a spectral grid size of 75 GHz for either 100 or 200 G QPSK, the number of channels that can be accommodated is about 10 in this band.

In this section, we present the performance of three 100 G channels: one located at 1553.33 nm, which falls into the roll edge of the amplifier; one channel located at 1552.5 nm, which is at the edge of the passband, but no distortion can occur and one at 1550.52 nm, which is located at the mid-band of the optical amplifier, hence no distortion.

The photographs of the platform setup employing the Optilab low-noise optical preamplifier are given in Section 10.3.1.

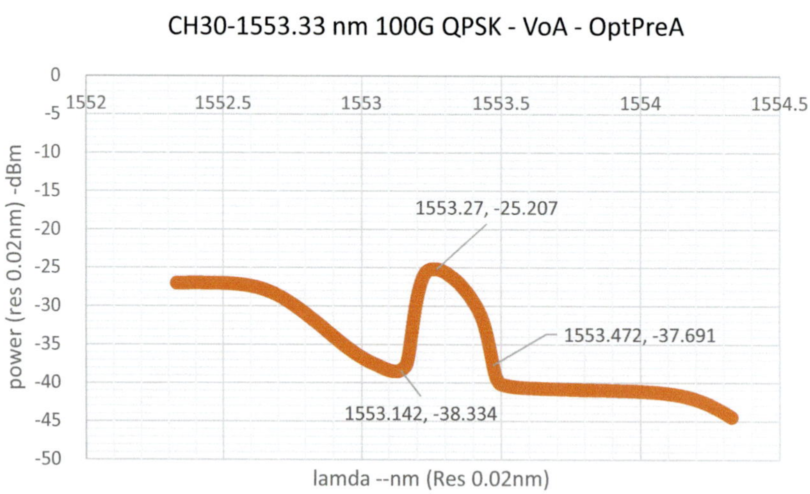

FIGURE 10.18 Optilab optical preamplifier spectrum with two inputs channel #35 (ch35) ITU-T 100 GHz spacing grid (see Appendix 6 of Chapter 12) and 100 G DP-QPSK plus carrier at ch38 of 1546.924 nm.

10.4.2 Channel 1553.33 nm or Channel # 30 ITU-T 50 GHz-Grid under Distortion

10.4.2.1 Spectrum

IN SPC, the different rotation velocity between two satellites or between a satellite and the earth's ground creates the Doppler effect. The Doppler frequency can be as high as 7.5 GHz. This makes the channel signal band move towards the roll-off band of the WSS passband of a port. This section investigates the effects of such frequency shift of the signal band when transmitting through the WSS>

Although the channel's wavelength at central is 1553.33 nm, it does not fall into the upper filter band, as indicated in Figure 10.27. The Optilab LLC. optical preamplifier, the configuration of which is shown in Figure 10.1, is tested. The signal spectra at different VoA values are shown in Figures 10.19–10.21 for attenuation of 1.5, 10 and 20 dB; the receiver cannot detect the channels after 20 dB of VoA.

The OptPreA (LNA) has a rolling-off at the edge of this spectral window.

- VoA = 1.5 dB

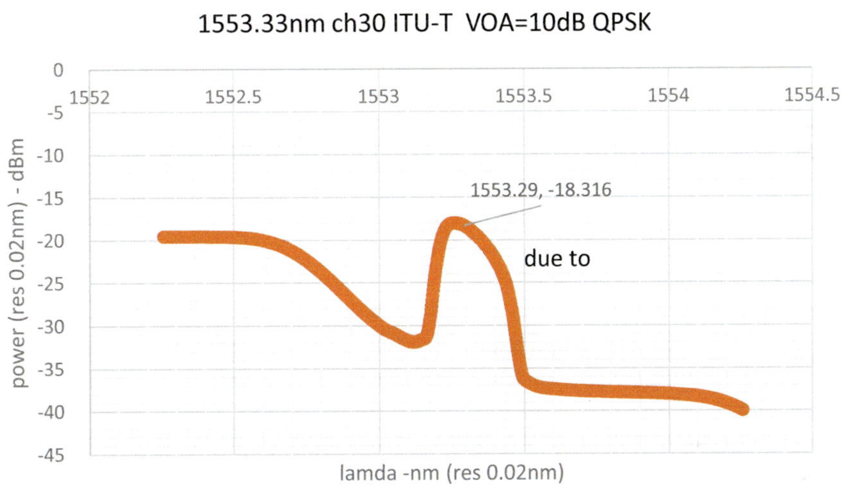

FIGURE 10.19 CH#30 1553.33 nm into VoA via 2×1 FC then VoA → OptPreA (OPtilab (biased at pump laser min current gain = 30 dB) VoA = 1.5 dB.

- VoA = 10 dB

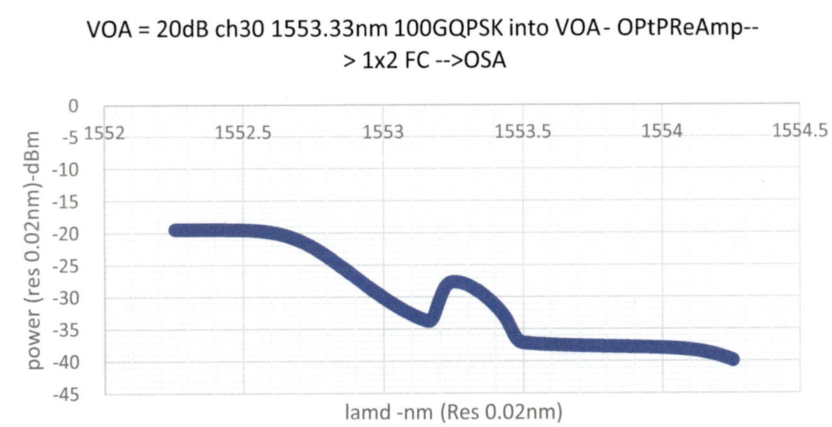

FIGURE 10.20 OptilabPreA-1553.33 nm Sig 100 G_In_10 dB VoA, VoA = !0 dB, CH#30 1553.33 nm to VoA and then OptPreA → OSA.

- VoA=20 dB 1553.33 nm 100 G

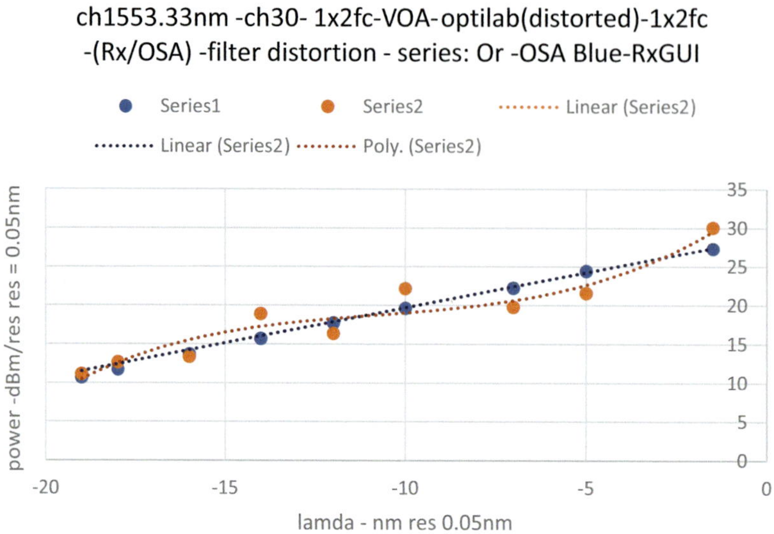

FIGURE 10.21 Filename=20230511-optilabPreA-1553.33 nm Sig 100 G_In_10 dB VoA, VoA=20 dB, CH#30 1553.33 nm to VoA and then OptPreA → OSA.

10.4.2.2 Channel 1553.33 nm Performance

Note: system and test arrangement: 100 G DP-QPSK tuned to Ch30 1553.33 nm transmitting through Optilab LLC. Optical Preamplifier.

The OSNR at the Rx versus Rx input average power is obtained. However, the signal 100 G DP-QPSK has fallen into a roll-off edge of the optical preamplifier LNA's optical filter. The link is tested to gauge the performance of the optical preamplifier. The OSNR is measured from the spectra of the output signal and recorded from the GUI monitor of the transceiver DP04QSFP, as depicted in Figure 10.22. As observed, there are several discrepancies between the OSNRs of the GUI and those from OSA processing. The distortion is only approximated. AS sampled spectrum of the output signal 100 G at VoA=10 dB can be seen in Figure 10.23. Note that the total received power is the power spectral density multiplied by the resolution of the OSA to obtain the average channel power. Thus, there is some penalty due to the shift towards the roll-off band. This penalty can be mitigated by setting the WSS passband wider and 15 GHz allowance for the Doppler shift and setting a penalty of 3 dB due to this shift.

Similarly channel # 30 whose wavelength is at 1553.33 nm is also measured and the spectrum is shown in Figure 10.23: When an optical preamplifier is monitored by using a 1×2 FC. The spectrum with two inputs CH#35 ITU-T grid and 100 G DP-QPSK plus an unmodulated carrier of ch#38 of wavelength 1546.924 nm can be observed in this figure. The two carriers passing through the WSSes are clearly shown in the spectrum.

The whole one hop is under the setting of the WSSes passband, and the carrier is designed so that distortion can occur. The receiver average input power is potted versus log10(BER), which is obtained in the processor of the OTRx, receiver section, for 100 G DP-QPSK, which is tuned to 1553.33 nm (Channel 30) and passing through the optical preamplifier. The passband of the amplifier and the WSS can be observed in Figure 10.24, and we can observe the spectrum distortion as the WSS is adjusted so that the roll-off band is in proximity to the signal band (Figure 10.25).

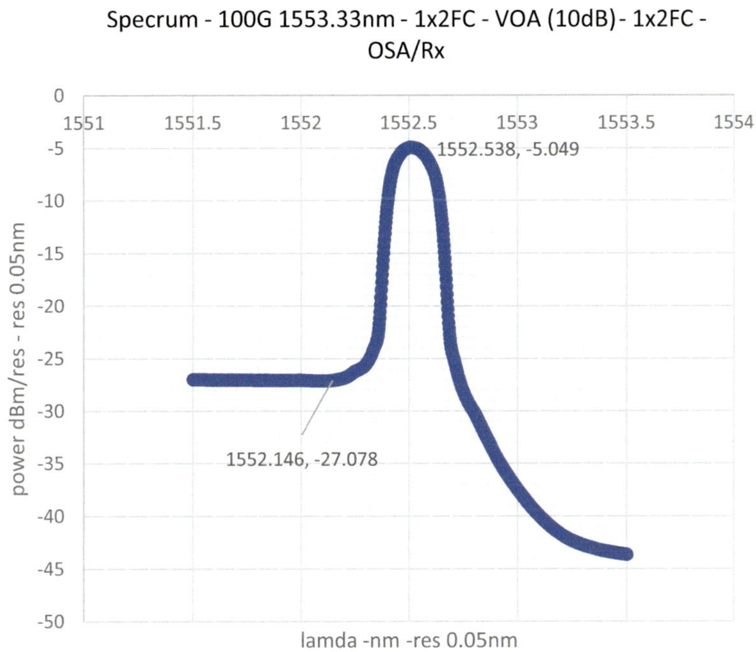

FIGURE 10.22 Receiver sensitivity and OSNR versus Rx input power) dBm=of 100 G CH30 1553.33 nm to 1×2 FC - VoA - 1×23 FC - Rx-OSA.

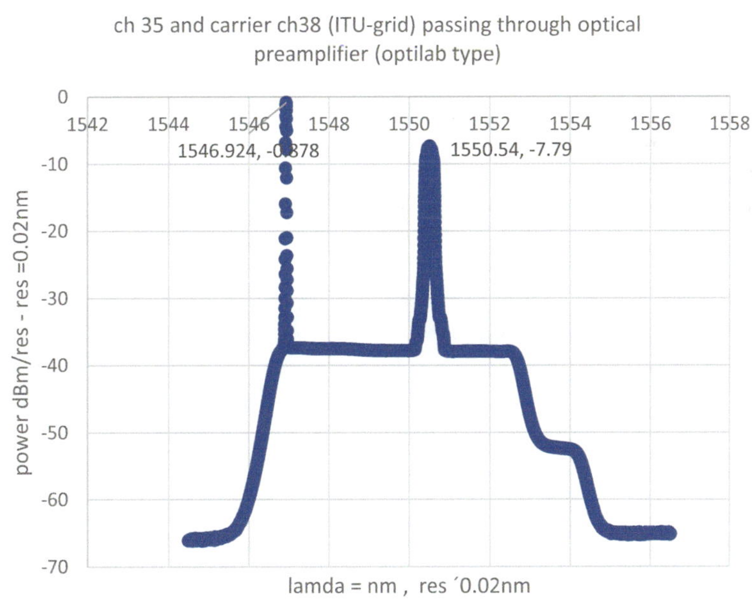

FIGURE 10.23 Sampled spectrum of 1553.33 nm CH30 (generated from DP04QSFP module whose wavelength tuned to 1553.33 nm -ch. no.18) after transmitting through optical preamplifier of optilab and 1×2 FC – VoA (10 dB) -optical preA –1×2 FC then to Rx (GUI module monitor) and OSA (processed). Distortion due to the filter roll-off (see Figure 10.24).

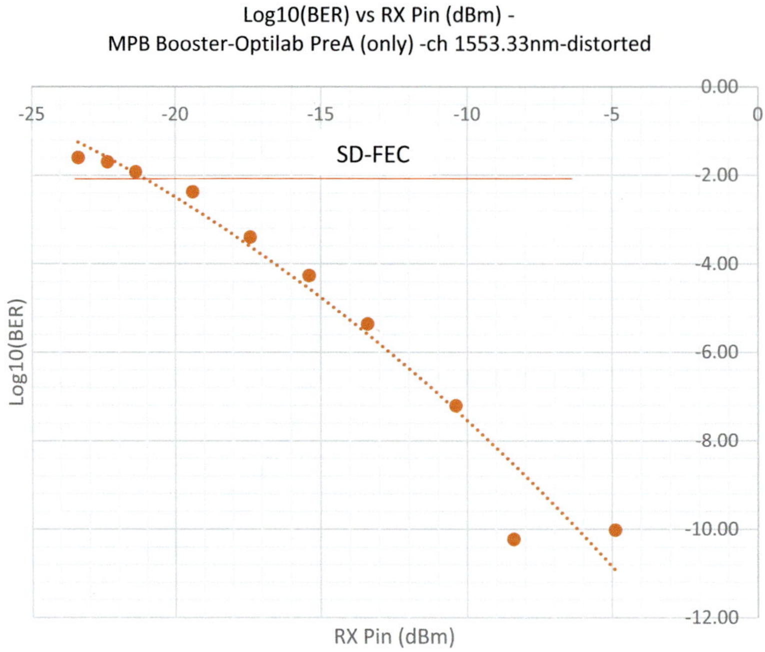

FIGURE 10.24 Optilab preamplifier spectrum with two inputs CH35 ITU-T grid and 100 G DP-QPSK plus carrier at CH38 1546.924 nm.

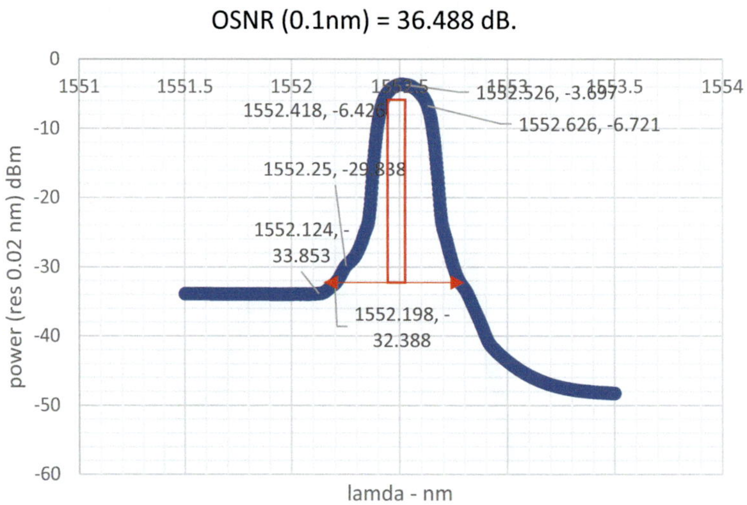

FIGURE 10.25 Receiver input power (ave.) versus $\log_{10}$ (BER) (GUI recorded) for 100 G DP-QPSK tuned to 1553.33 nm (Channel 30) and passing through the Optilab optical preamplifier whose passband can be observed from Figure 10.24.

10.4.3 Channel 1552.50 nm or Channel 20(MYN) -Channel 31 ITU-T 50 GHz-Grid

10.4.3.1 Received Spectrum

- VoA = 1.5 dB (Figure 10.26)

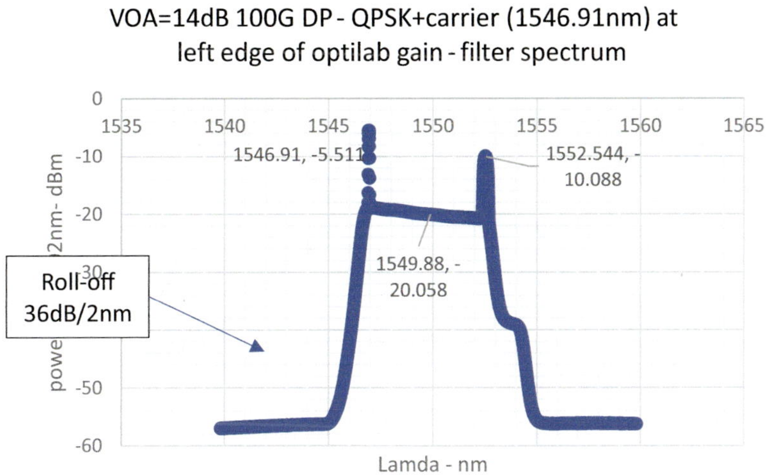

(a)

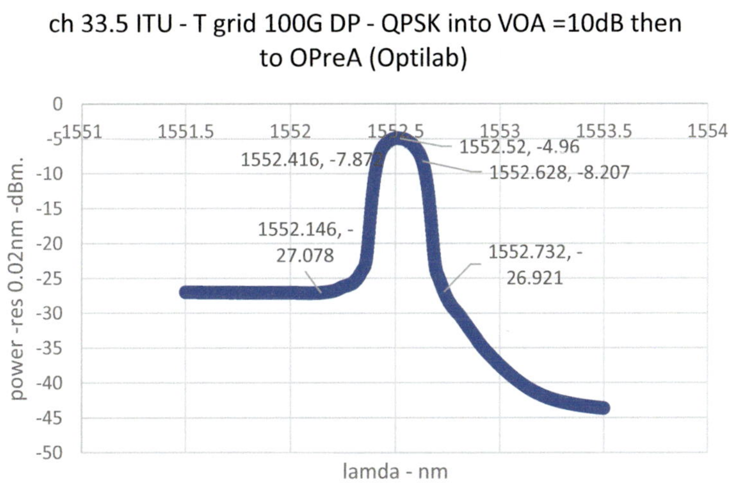

(b)

FIGURE 10.26 VoA = 1.5 nm Channel 20 (Myn) channel 30.5 (ITU-T 50 G grid). OSNR 5 (dB) = 32.388 − 6.72.

- VoA = 10 dB attenuation and optical amplifiers of Optilab OptPreA with Signal bit rate of 100 Gbps DP_QPSK and lightwaves of wavelength 1552.56 nm. Signals are fed into a 10 dB VoA (Figure 10.27)

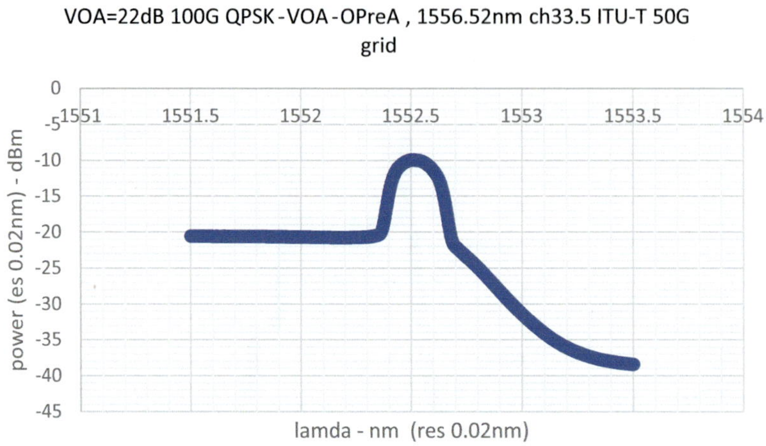

FIGURE 10.27 VoA = 10 dB, CH#33.5 ITU-T 1552.52 nm 100 G DP-QPSK signals into VOA (10 dB) then to OSA (via 1 × 2 FC). The bandwidth of the filter is 1552.544 – 1546.91 nm = 5634 nm or approximately 7,040,425 GHz, thus with 52.5 GHz per channel (100 G DP-QPSK) (including FEC overhead), the number of channels can be 13 channels or for 200 GBaud or 65 GHz including soft FEC overhead, the bandwidth per channel can be set at 75 GHz so the channel number can be nearly ten channels. Tbps with 5 × 200 Gbps/channel can be easilyfitted into this band. (a) Overall spectrum of carrier and 100G QPSK signals. Noise filtering is about 10 dB higher than LNA (low noise amplifier). (b) 100 G with VoA =10dB. Spectrum OSA after OPreA (LNA).

- VoA = 22 dB (Figure 10.28)

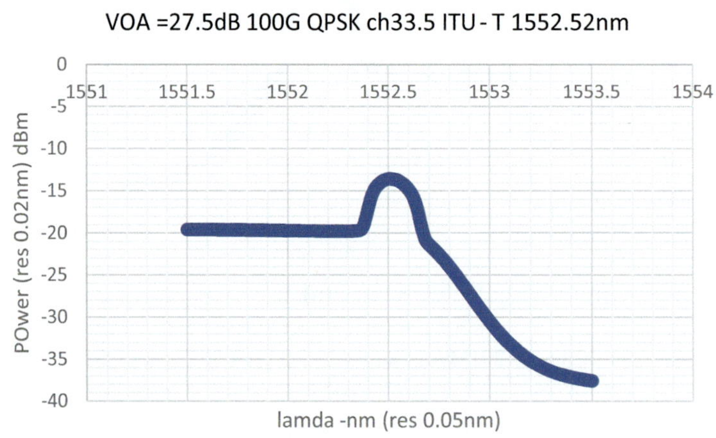

FIGURE 10.28 VoA = 22 dB, CH#33.5 1552.52 nm 100 G-QPSK into VoA – OpreA → 1 × 2 FC (3.4 dB) → OSA, spectrum.

- VoA = 27.5 dBm (Figures 10.29–10.31)

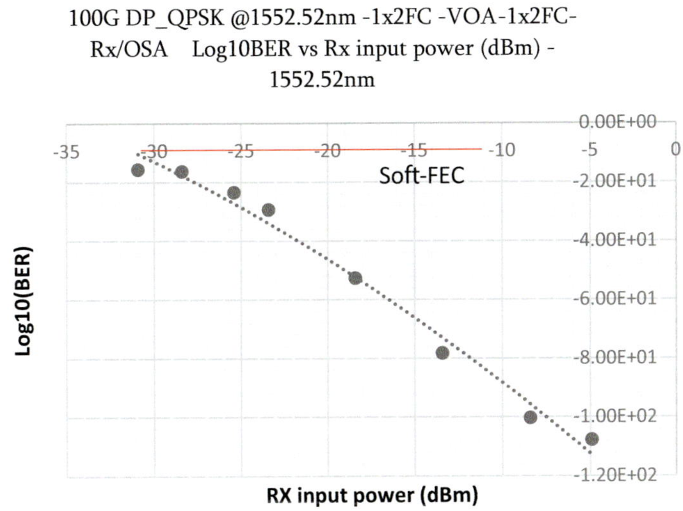

FIGURE 10.29 VoA = 27.5 dB.

10.4.3.2 Performance

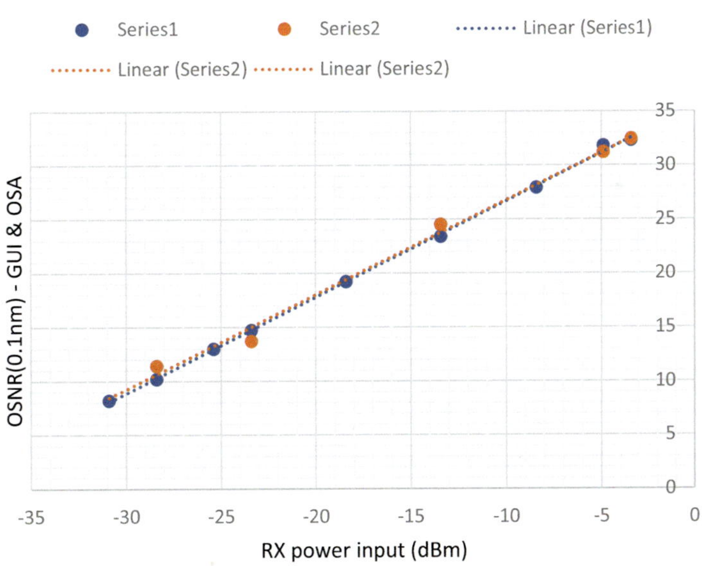

FIGURE 10.30 The $\log_{10}$ (BER) versus Input power to receiver (derived from the function of VoA attenuation factor-dB, the power budget).

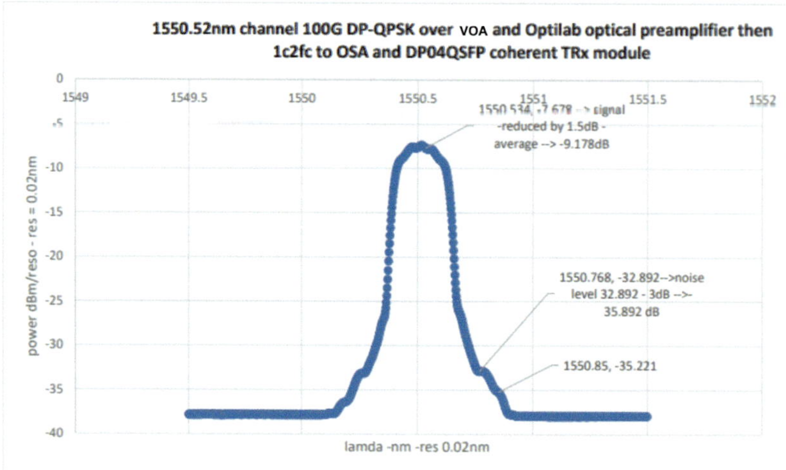

(a)

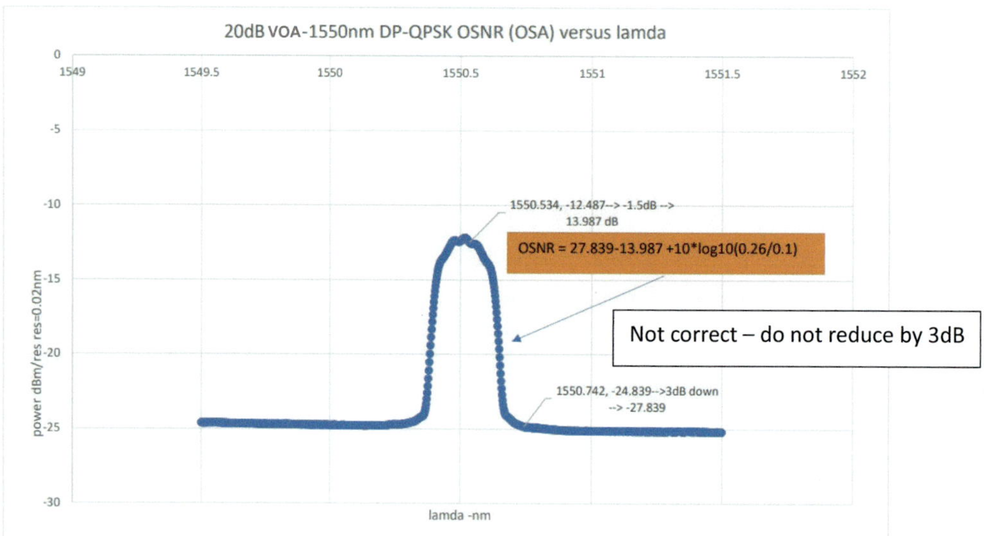

(b)

FIGURE 10.31 The 1552.52 nm OSNR (GUI) and OSNR (OSA) versus input power into Rx module. Seriel: OSNR by module GUI, Serie2=OSNR determined by spectrum obtained from OSA.

10.4.4 The 100 G-DP-QPSK Tuned to Channel at Wavelength of 1550.52 nm

10.4.4.1 OSNR Determination Method from Spectra of Signals at Rx

Calculation method to determine OSNR of reception by DP04QSFP: OSA for OSNR: (peak - 1.5 dB) -(noise - 3 dB) → scaling to 0.1 nm noise BW (bandwidth), thence scaling to noise bandwidth of 0.1 nm using +log10(0.26/0.1)=4.1 dB. Section 10.2 gives details of the OSNR determination method. Note that if the noise is at the noise minimum level, then there is no deduction of 3 dB. The following illustrates two signal spectra in Figure 10.32.

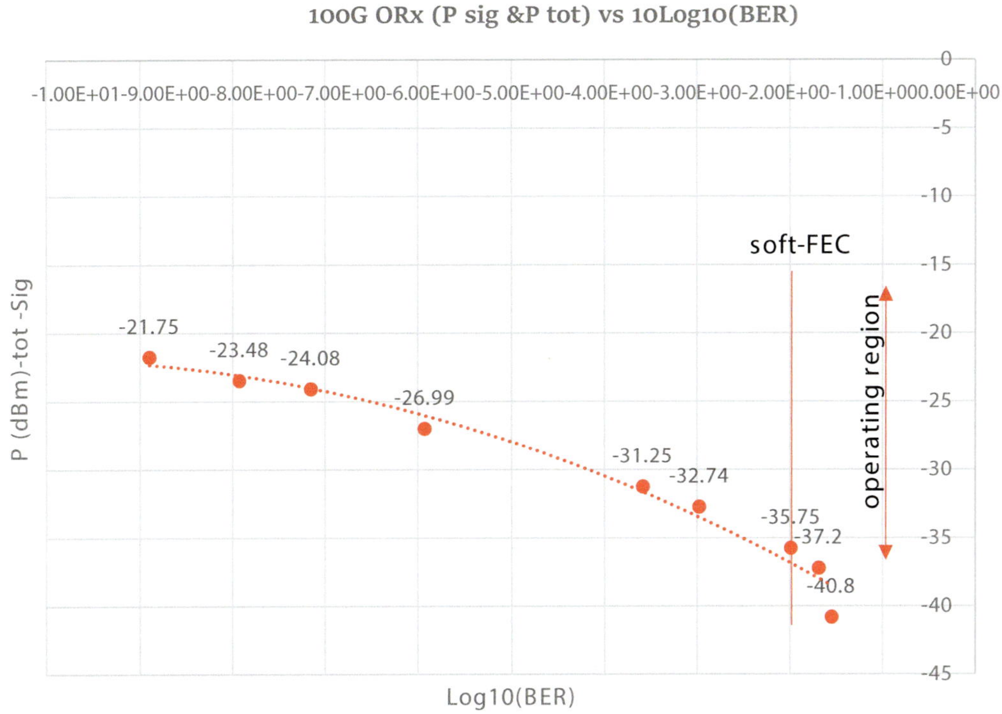

FIGURE 10.32 Spectra of the 100 G DP_QPSK after transmitting through the optical preamplifier with VoA of 20 dB. Evaluation of OSNR (OSA) is shown in the inserts. (a) Case of distinction of noise lowest floor and noise level of signal. When the signal noise level above the lowest level of noise floor, so take a minus 3dB of the noise floor level in the estimation of the OSNR, thus there are some grey area (uncertainty) of processing here; (b) case when noise is at the lowest noise floor level. Note: noise level at minimum noise level, so no 3 dB deduction (b). (a) The noise level above the lowest noise floor level is reduced by 3 dB of noise level. So, there is some grey area. When the reduction of 3 dB is below the lowest noise level floor, then take the lowest noise floor level. VoA = 1.5dB only. (b) When noise is at the lowest level, there is no reduction of 3 dB.

10.4.4.2 Performance

In Figure 10.33, the sensitivity of the Rx of the TRx Acacia DP04QSFP (dual polarized QPK with 4 channel outputs and 4 inputs transceiver 28 GBaud). This B2B, with WSS1 and WSS2 inserted, is measured with the received average power versus the BER as plotted. Thus, the power input (dBm) or received average power versus $\log_{10}$(BER) is depicted in Figure 10.34. This shows the optical filtering effect in the WSSes does not affect the receiver sensitivity. This is critical as one has to obtain both the OSNR and the BER with variation of the average power of the channel to confirm that the system is operating in the linear region, not the nonlinear crosstalk (Figures 10.35 and 10.36).

10.4.4.3 BER, OSNR

10.4.5 Link Power Budget-One Hop Link -with LNA Preamplifier (Experiment)-MPB HPA Booster (for Budget Estimate Only)

Although the test is interrupted by the problem of grounding the LNA optical preamplifier, we can interpret the total hop link budget as follows:

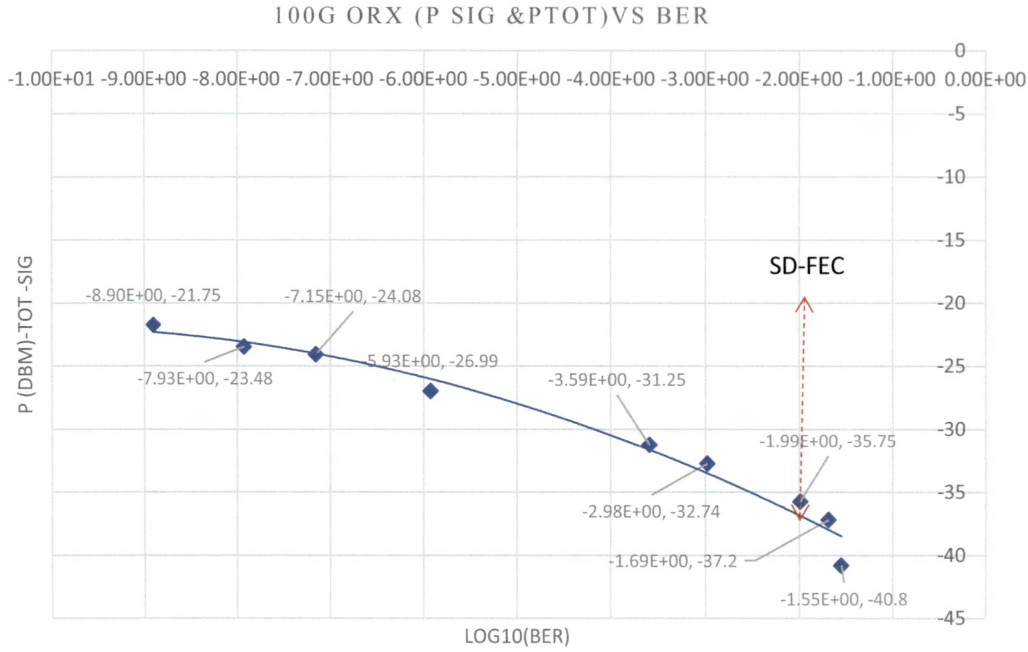

FIGURE 10.33 Plots of BER versus signal average power (dBm) without optical amplification as a pre-amp optical reception sub-system. Note: SD-FEC 2e-1 achieved at –42 dBm for 100 Gbps (PDM-QPSK @25 GSy/s) or about 5 photons/bits as expected from SoA & CA WP1100 as reported. P_{tot} – total optical received power registered by TRx, including noises, P_{sig}=extracted signal power. Note: No optical pre-amplifier is used. 10 $\log_{10}$ (BER)=–2 at –37.2 dBm.

- Optical power amplification (booster): The output of the booster amplifier is saturated with 1800 mW output at the high-power port of the 1:99 HPFC (high-power fiber coupler). The output at the 1% is saturated at 5.8 dBm. The coupling coefficient is 25.1 dB from the input power. The input power into the booster is set at 3 dBm to 4.01 dB (1×2 FC) and 4.1 dB of 2×1 coupler to reach –4.8 dBm. AT this input value to the booster, the saturated output power is 5.8 dBm with a gain of 24.4 dB of the booster and a pump power of 360 mA (Stage 1&2) and 4.8 A pump of the stage 3.
- Using this value at 5.0 dBm, we can assume and set input into the LNA preamplifier. We obtain the power budget for the transmission of 100 G DP-QPSK polarized channels, as shown in Table 10.2.

10.5 ONE-HOP LINK WITH OPTICAL PREAMPLIFIER AND BOOSTER

Remarks: This section gives the test performances of a complete one-hop fiber equivalent free space transmission in which both booster amplifier and optical preamplifier are MPB type. The 100 G QPSK signals and single carrier (non-modulated) are used in the test wherever appropriate.

Section 10.5.2 provides optical spectra of 100 G QPSK signals at the front of each receiver subsystem.

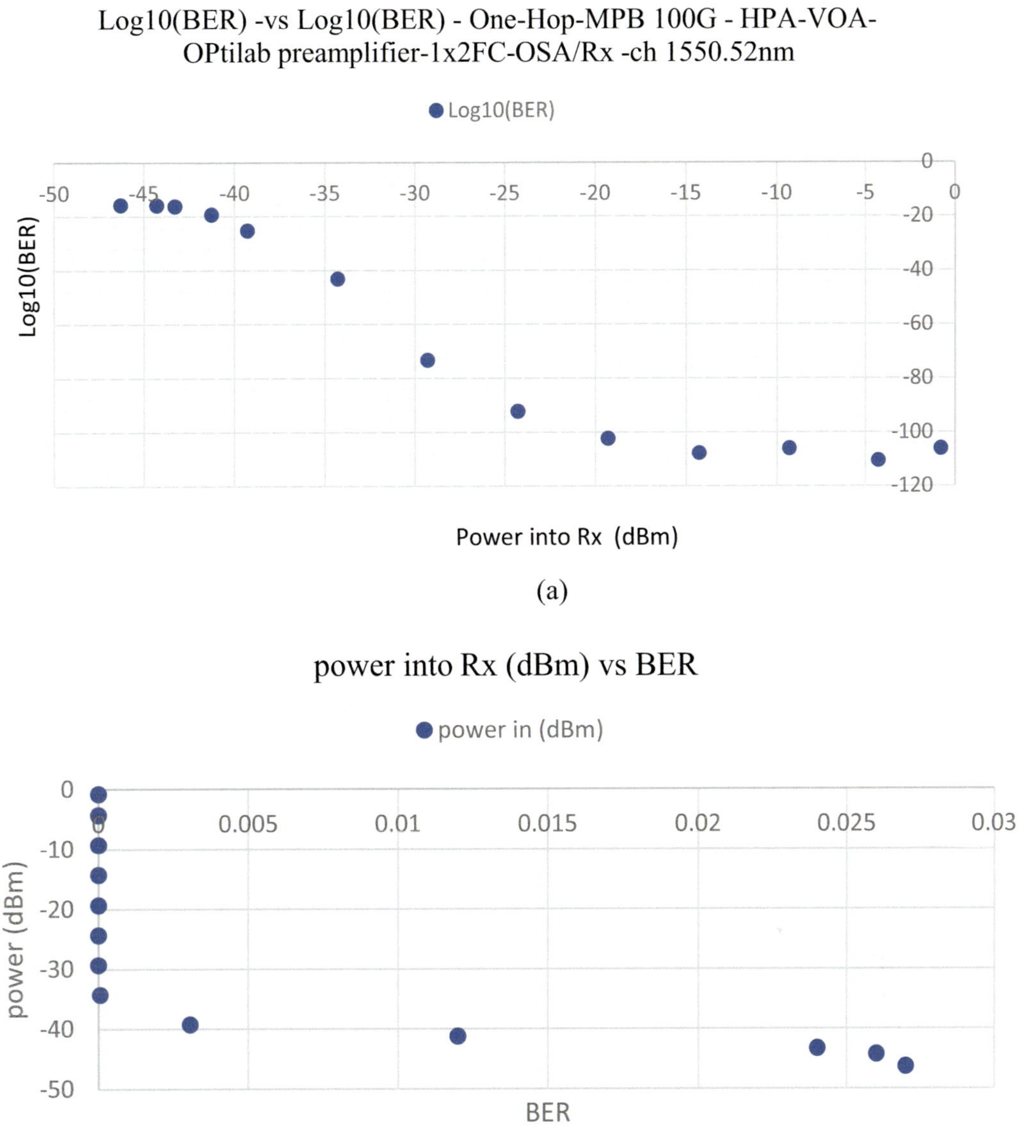

FIGURE 10.34 100G transceiver (B2B with WSS1 and WSS2) receiver sensitivity test. Plots of BER versus signal average power (dBm) without any optical amplifier as a pre-amp optical reception sub-system. Note: SoftDefined-FEC 2e-1 achieved at –42 dBm for 100 Gbps (PDM-QPSK @25 GSy/s) or about 5 photons/bits as expected from SoA & CA WP1100 [10] as reported. P_{tot} – total optical power received as registered by TRx, including noises. $P_{tot\text{-}sig}$=extracted signal total average power. Note: No optical pre-amplifier is used. 10 $\log_{10}$ (BER)=–2 at –37.2 dBm.

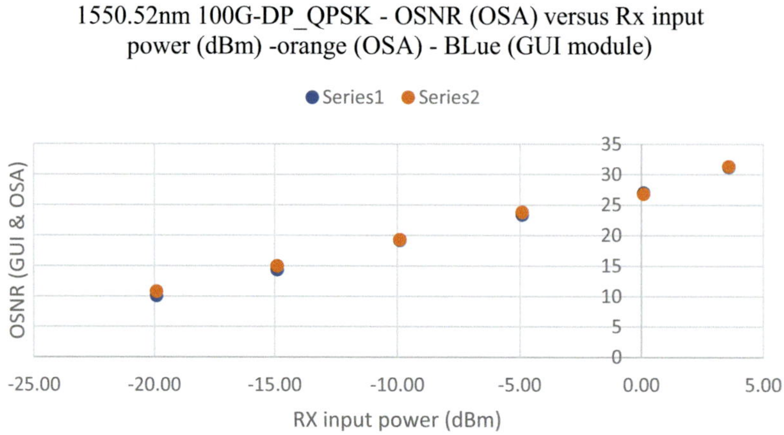

(a)

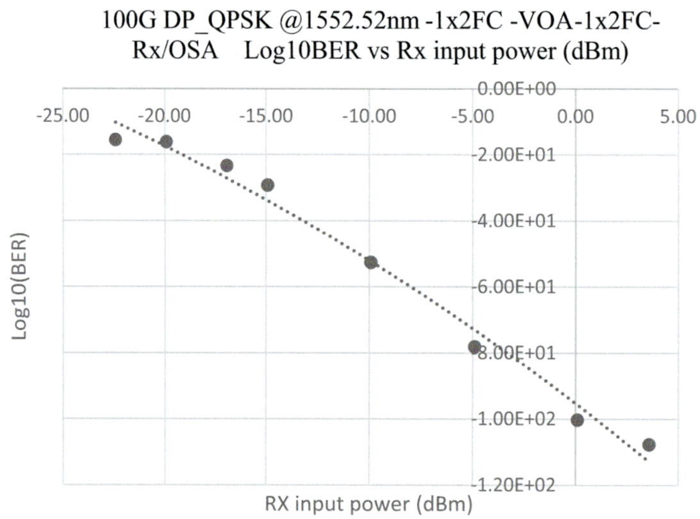

(b)

FIGURE 10.35 Transceiver receiver sensitivity with Tx tuned to specific wavelength (1550.52 nm) and transmitting through Optical lab preamplifier. Input power follows that of the output of the 1:99 fiber coupler. **System Structure**: One Hop MPB amps with 10 G DP04QSFP Rx with system schematics of Figure 10.1, that is with amplification noises (a) power in versus $\log_{10}$ (BER); (b) Power In (dBm- derived from setup) versus BER (signal channel@ 1550.52 nm 100G DP-QPSK and OPreA LNA).

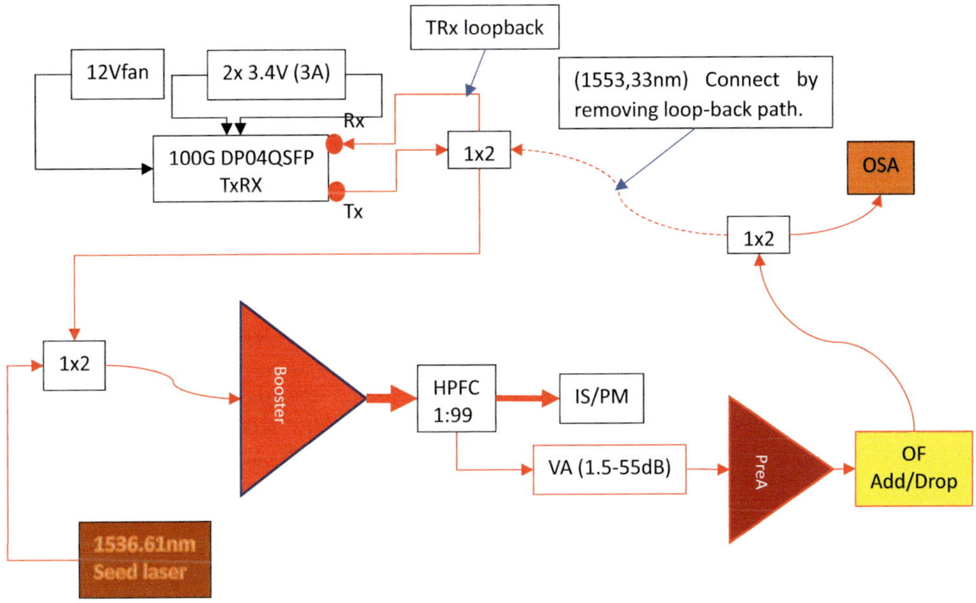

(a)

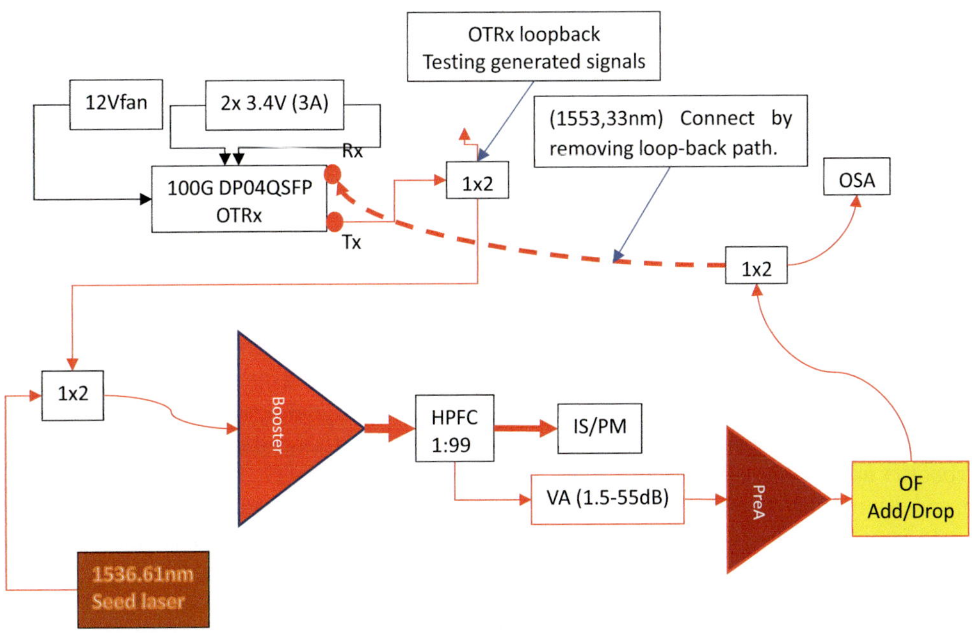

(b)

FIGURE 10.36 The wavelength 1550.52 nm is selected to transport information channel with 100 Gbps using dal polarization and modulation QPSK, that is DP-QPSK, over VoA and optical preamplifier (LNA OptPreA), OSNR (montored by GUI to the transceiver) and OSNR (monitored by the spectrum analyser OSA) versus power received by TRx (Rx section) coherent reception. Blue (mid-light lines)=OSNR readout from module evaluation using EVM method – light continuous line=OSNR evaluated by spectral properties of signal and noise obtained from OSA of the system (see Figure 10.1). (a) OSNR (GUI) and OSNR (OSA) versus input power into RX module and OSA. (b) $\log_{10}$ (BER) versus input power into Rx module – channel 1550.52nm no distortion of the signal as it is in the passband of the filter of the optical preamplifier as shown in Figure 10.23.

TABLE 10.2
System (100 G DP-QPSK and VoA, FA, Fiber couplers (FC) and Optical Preamplifiers

Parameters Setting	Equivalent		Units	Notes
DP04QSFP integrated coherent module			C-band tunable	OTRx module
Channel #30		1553.33	nm	ITU-grid 100 GHz
channel #30.5		1553.52	nm	
channel #51		1536.61	nm	
Symbol rate/BW-DSB		32.5	Gsy/s	25 G+FEC 30%
DSB BW	0.26	32.5	GHz	32.5×800 nm/100=260 GHz equivalent in nm
ASE noise BW (nm)	0.1		nm	Standard BW
OSNR (Tx Sig)		32.3	dB	GUI & OSA measured
3 dB-FC (IL)		–4.1	dB	3 dB split/combiner
Booster (MPB) G		24.4	dB	Amp gain (min) - LD1/LD2 &LD3
Highpower FC (1:99)		–25	dB	C 1:99 coupling factor (dB)
Booster saturated power @HPFC –20 dB port		5.8	dBm	Saturated power at the output of HPFC 20 dB port out (HPFC)
Booster output *P* at 99% out-port HPFC		30.8	dBm	
HPFC out				
Fixed Attenuator (FA)		–10	dB	
VoA			dB	Var atten 1.5–55 dB
OpreA Gain		29.9	dB	Not optimized

10.5.1 The 100 G QPSK Signal and Transmission Performance

10.5.1.1 System Structure and Setup for Evolution of 100 G DP-QPSK over the Optically Amplified Hop

The test system is set up as shown in Figure 10.37. The system consists of the following: (i) 100 G DP-QPSK signals are generated from the module DP04QFSP. (ii) The transceiver is a DSP-based integrated photonics and electronic structure. (iii) Uses of the VoA and the control of the attenuation factor. Control the power into the preamp and hence the average signal power measured by the OSA.

Figure 10.37 depicts an optical and signaling setup for 100 G transmission over OP MEO_MEO with booster and preamp of MPB S/N SDA -1116-006[11]. Operating by GUI[12] with monitored power, gain, and driving current of LD1/LD2 and LD3 driving current for operations of amplifiers. 1536.61 nm seed laser is used to activate the booster amplifier so that the level signals can be operated through the power amplifier and then enter the preamplifier. Optical amplifiers and TOF (tunable optical filter) as Add/Drop=Add/Drop tunable optical filter built into the Optical Preamplifier LNA (linear/low noise amplifier of OptilabInc. type) - Channel #30 at wavelength of 1553.33 nm[13] is the selected channel for the TOF whose passband is 60 GH (as measured) over the effective slot of 100 GHz. TOF Add/Drop filter is built into the optical preamplifier section.

Note the loopback path of the OTRx. That is, the Tx signal output of the DP-QPSK DP04QSFB is 3 dB split into two paths. One path is used for further transmission into the Hop and one is looped back into the Rx to ensure the functionality of the OTRx. The transmission path is "built" and proven.

The procedures for the optically amplified Hop to be set up stage-by-stage are : (i) The booster or high-power amplifier of MPB dual HPA and LNA is tested first. The path between the Booster and the OptPreA (LNA) is broken, and OSA is connected to a high-power fiber coupler (HPFC) output. The coupling coefficient is 27 dB, although it is measured at low power with a 1:99 intensity

TABLE 10.3
Power Budget of One-Hop Fiber-Equivalent Link (Pegasus Exp Platform)

	Optilab PreAmp used only, Booster HPA – MPC type							
	DP_QPSK Tx (dBm)	**1x2 FC (dB) &2x1FC**	**2x1 FC (dB)**	**VoA (dB)**	**OPreA (Optilab) Gain**	**Rx (dBm into 0.005nm)**	**Module BER**	**budget**
Power /reso	−14						1.20E-02	
								PreA (OptiLab)
Peak power (dBm)	−14.00	−8.2	−4.1	−42	29.9	−38.40		−79.30
Average power (dBm) -out HPFC (OSA-int)	5.06							
	Expected after booster			max	Current setting – min.			
Signal BW (30% overhead FEC) 32.5GHz/25GB	0.26				29.9			
Reso (nm)	0.02				51			
Signal average power (dBm)	3.21 mw	5.06 dBm						
HPFC	−25 dB							
One Hop link – 100G DP-QPSK- 1x2FC -2x1FC (+carrier) - MPB Booster VoA- Optilab-Preamp-1x2FC (Rx/OSA)								
From previous test of Ch#30 of MPB HPA and Optical PreAmp								
Signal 100G P(Ave) (dBm) -after booster	**1x2FC**	**2x1FC**	**Booster (dB)**	**1:99 HPFC**	**VoA**	**sig P@Rx**	**total loss budget(dB)**	**sig @Rx (dBm)**
5.8	−8.2	−4.1	24.4	−25.1	−42	−10.4	−79.4	−10.4
					to go to line 6 and continue		better if optimize the booster HPA and Optilab preamp	
NL penalty	Estimation from TWM and FWM prouct							To be determined
Resultant						−15.438	0.213567999	11.10553595

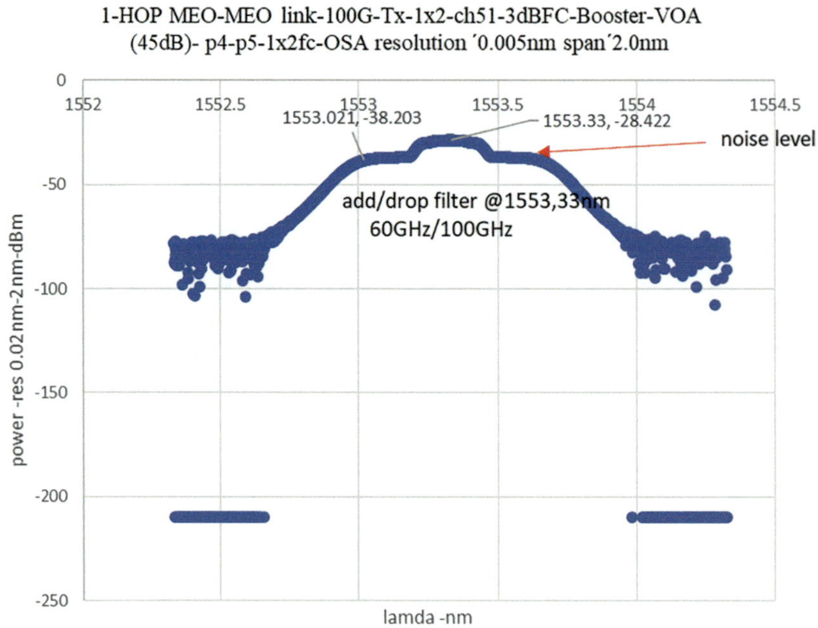

FIGURE 10.37 (a) Before injecting transmitted 100 G signal – initial circuit to check the functionality of the OneHop link. OF=optical filter as ADD/drop multiplexer installed in LNA preamplifier. (b) After ensuring the functionality of the OneHop link, 100 G DP-QPSK signals are injected into the receiving section of the transceiver 100 G DP04QSFP. IS=isolator, PM=power meter, VA=variable optical attenuator, HPFC=high power fiber coupler, DP=dual polarization, QSFP=quad special form package, 04=quad input/output ports, OF=optical filtering for add/drop multiplexer, OTRx=optical transceiver.

coupling ratio; (ii) The optical spectrum of the 100 G channel is monitored at this point to obtain the spectra modulated channels. These spectra are shown in Section 10.5.2 and figures therein. (iii) The high-power amplified signals are fed into the variable optical attenuator (VoA) and then to OSA.

Similarly, the spectra are obtained for the channel (1553.33 nm) 100 G after a complete transmission over the optically amplified Hop. These spectra are depicted in Section 10.5.1.2.

The TOF is built into the OptPreA, whose technical specifications are given in Appendix 2 Section 10.9.

10.5.1.2 Optically Amplified Free-Space Hop Test

Remarks: The One-Hop performance Excel file includes the OptPreAmp and MPB amplifiers cases (rows 70 and higher number). It links to Excel files in the storage cloud. The link attenuation can be used to set the equivalent transmission loss. The output spectrum is monitored and an estimation of the sensitivity is given in the following:

1. Case VoA=45 dB (Figure 10.38)

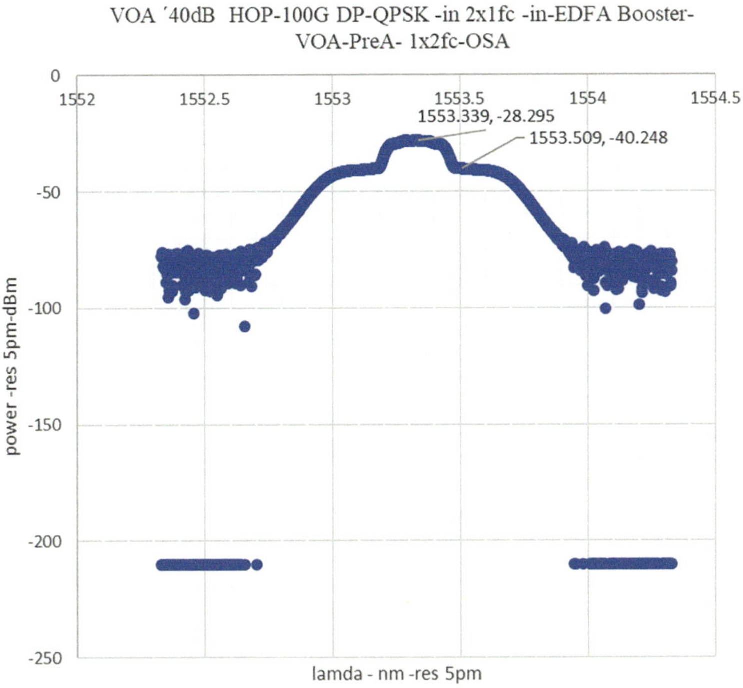

FIGURE 10.38 VoA=4 5 dB.

2. Case VoA=40 dB (Figure 10.39)

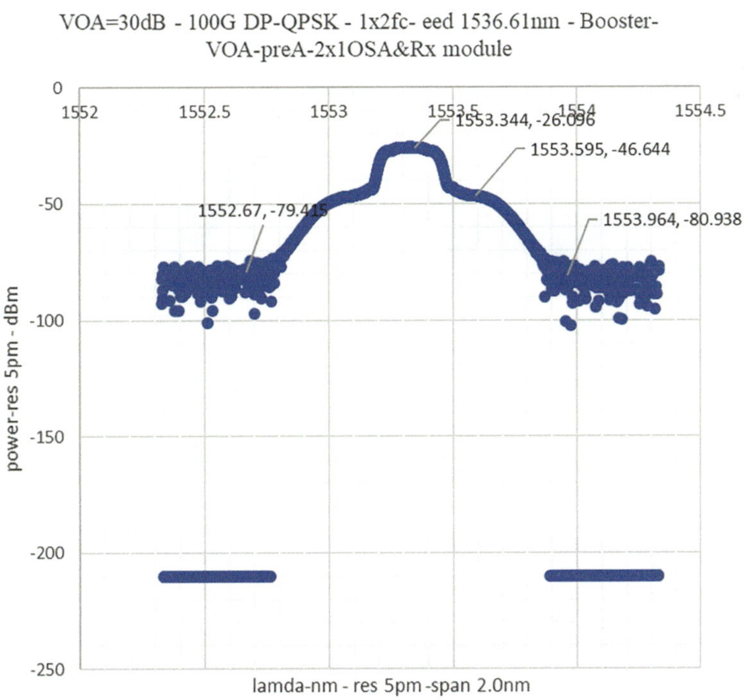

FIGURE 10.39 VoA=40 dB.

3. Case VoA = 30 dB (Figures 10.40 and 10.41)

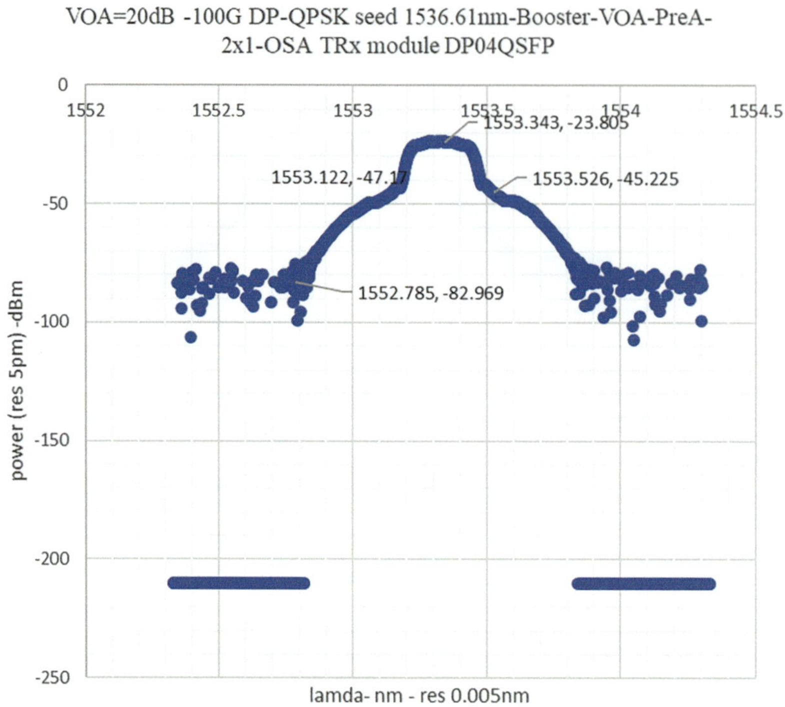

FIGURE 10.40 VoA = 30 dB HOP link. See Chapter 7 for VoA.

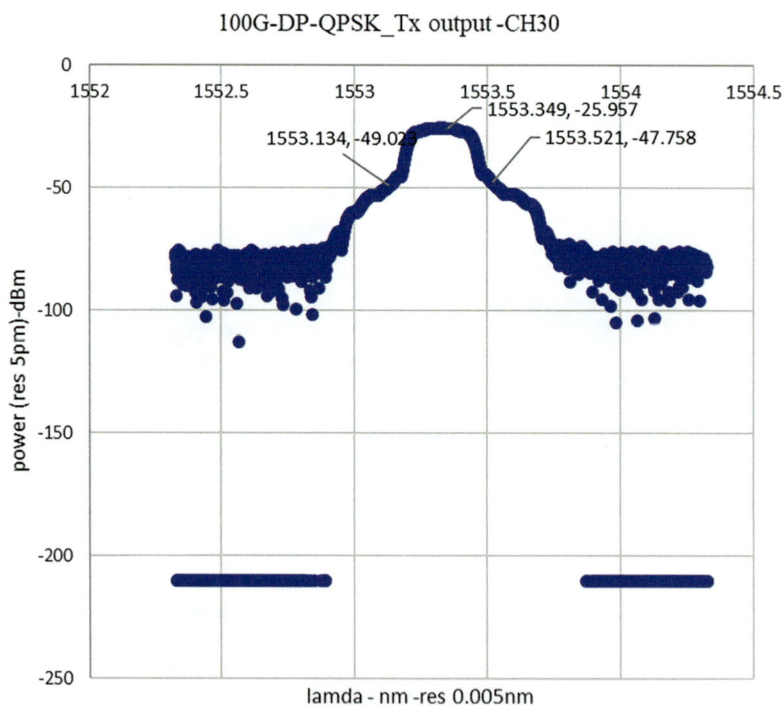

FIGURE 10.41 VoA = 20 dB, HOP link using DPßQPSK 100 G D'DP04QSFP9.

10.5.1.3 Evolution of Spectra of 100 G DP-QPSK over Hop Stages

At the output of Tx of 100 G DP-QPSK module coupled via 1×2 FC (3.4 dB). One path is fed to the SA and the other to the transceiver's Rx input to confirm the consistency in receiver sensitivity. Furthermore, the BER can be obtained. The average power as the sensitivity and OSNR and BER must be obtained to confirm the receiver's performance of the selected channel. If there are any nonlinear effects, the BER is reduced, and the OSNR can remain unchanged or even higher. Thus, monitoring these parameters in SPC is essential to ensure that the nonlinearity can be observed and considered.

a. Monitored spectrum at the output of Tx of OTRx (Figure 10.42)

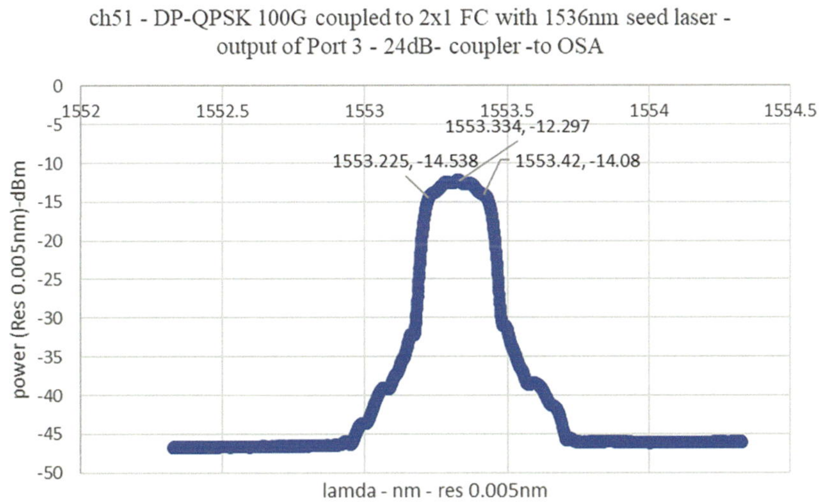

FIGURE 10.42 DP-QPSK 100 G output signal spectrum at Tx output.

b. Monitored spectrum at output Port P3 and Port P5 of Booster amplifier –20 dB FC port (Figures 10.43 and 10.44)

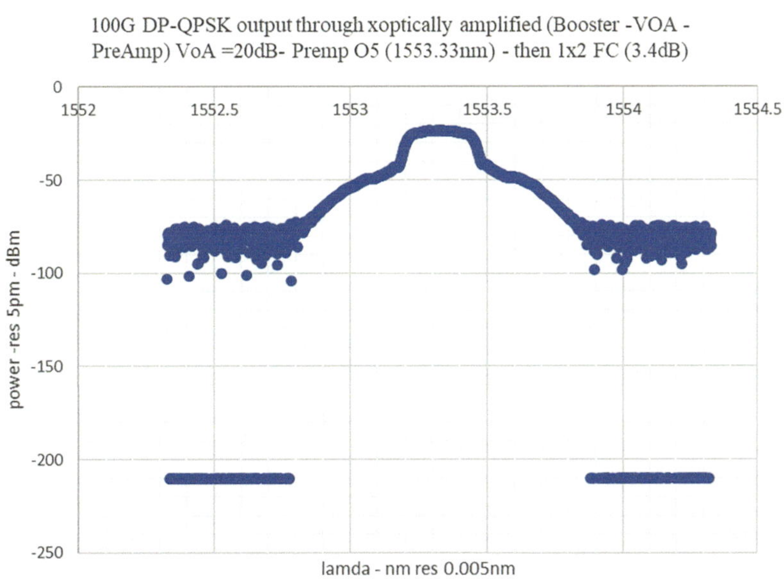

FIGURE 10.43 The 100 G DP-QPSK amplified through Booster – via a 2×1 fiber coupler with seed laser 1536.61 nm, booster output Port P3 → then to 20 dB coupler → OSA.

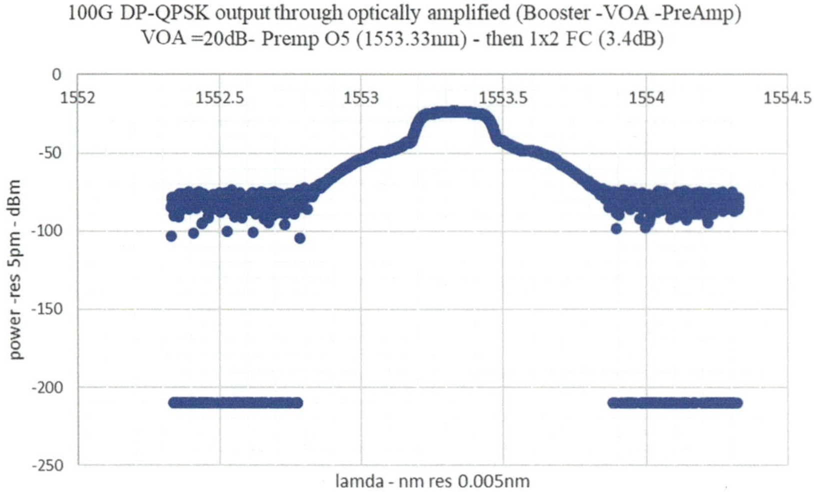

FIGURE 10.44 OSA spectrum of 100 G-DP-QPSK at central wavelength of 1553.33 nm monitored at P5 of Preamplifier via 1×2 coupler, Tx output 100 G DP-QPSK to 2×1 FC then feed to 10 dB attenuator (FA) then to booster of gain 24.34 dB, output high power to integrating sphere (IS) and PM monitored. The 24 dB output of FC is then coupled to a variable attenuator (VoA).

c. Monitored spectrum at the input of Rx of OTRx from Port P5 (Figure 10.45)

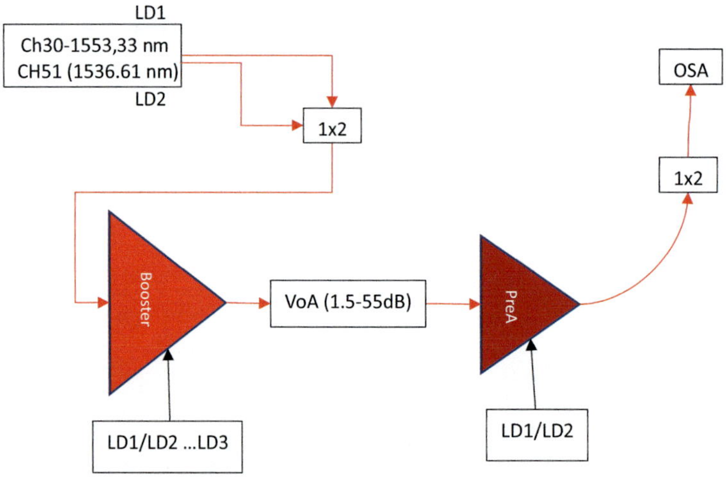

FIGURE 10.45 Spectrum of 100 G DP-QPSK output P5 of Optical preamplifier after passing through a 2×1 FC (3.4 dB), then 10 dB FA, and then to VoA of 20 dB, then through the preamplifier → 1×2 FC splitter, one port to Rx of DP04QSFP Rx and one to OSAas displayed in this figure.

10.5.1.4 Hop Testing Using Carrier at Channel 51 (1536.61 nm)

Optical carriers only under no modulation are set at wavelength of 1553.33 nm (CH#30) and 1536.61 nm (CH#51) are used to test and understand the operational characteristics of the amplified hop. The setup is shown in Figure 10.46, in which (i) two CW lasers are tuned to wavelengths 1553.33 and 1536.61 nm carrier are generated and coupled via a 2×1 SMF coupler[14] whose power is lower by 3.6 dB each. Thus, the total power per carrier is 5.6 dBm when the lasers' output power is set at 9.5 dBm (not less) (Figures 10.47 and 10.48).

Remark: Channel #51 is located at wavelength of 1536.61 nm is dropped from the booster amplified main gain spectrum, as shown in Figure 10.49. It is about 20 dB below Channel #30 located at wavelength of 1553.33 nm (Figures 10.50–10.53).

10.5.1.4.1 Channel 51 of P5 after Transmission through the Complete One-Hop Link

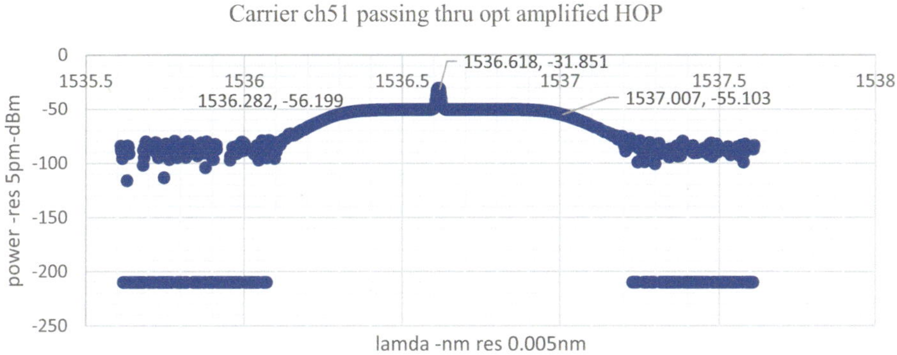

FIGURE 10.46 Set up to test the optically amplified FreeSpace Hop.

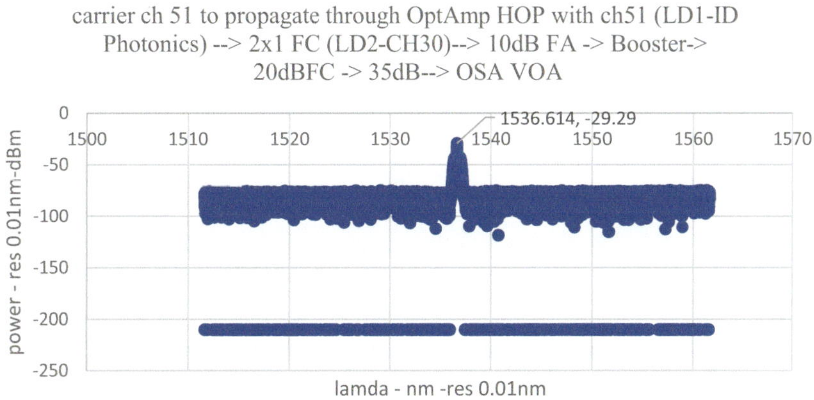

FIGURE 10.47 HOP1 and channel 51 through P5 of PreAmp – output filtered 100 GHz of preamp- CH30 1553.33 nm. VoA = 20 dB.

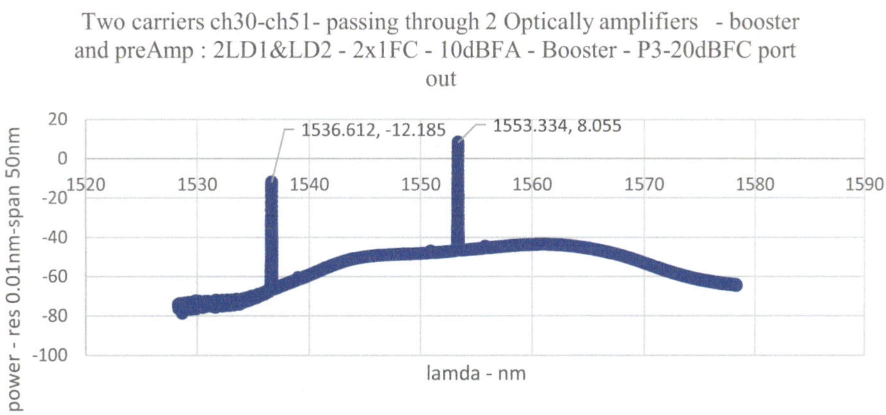

FIGURE 10.48 Output spectrum of single carrier (10 kHz linewidth-ID photonics dual lasers unit) propagating through 2×1 FC – 10 dB – Booster, then 20 dB coupling port -VoA (30+5 dB)-OptPreA → P5 OSA.

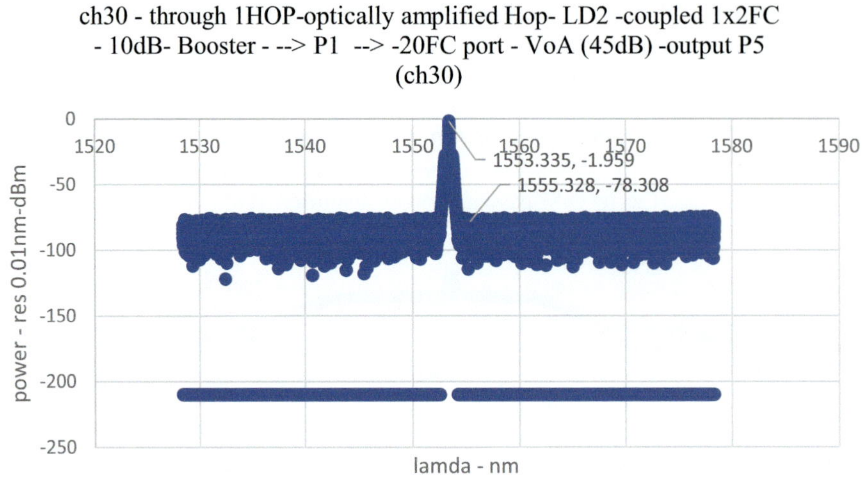

FIGURE 10.49 Spectrum covering two carriers, CH30 and CH51, going through 2×1 FC – 10 dB FA, then passing through Booster, then OSA spectrum. Span=50 nm – res. 0.01 nm.

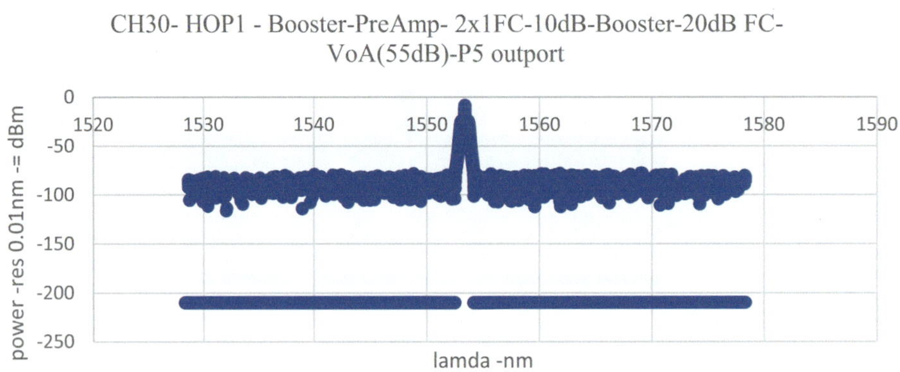

FIGURE 10.50 Channel 30 – best-amplified channel in the 1550 nm C-band to be used in the Hop transmission. Spectrum after the HOP link via 2×1 FC – 10 dB – Booster – 20 dB FC port → VoA (45 dB) – OSA monitoring spectrum.

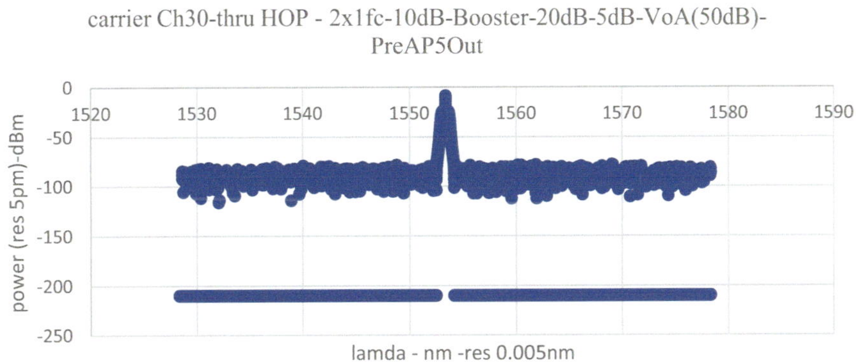

(a)

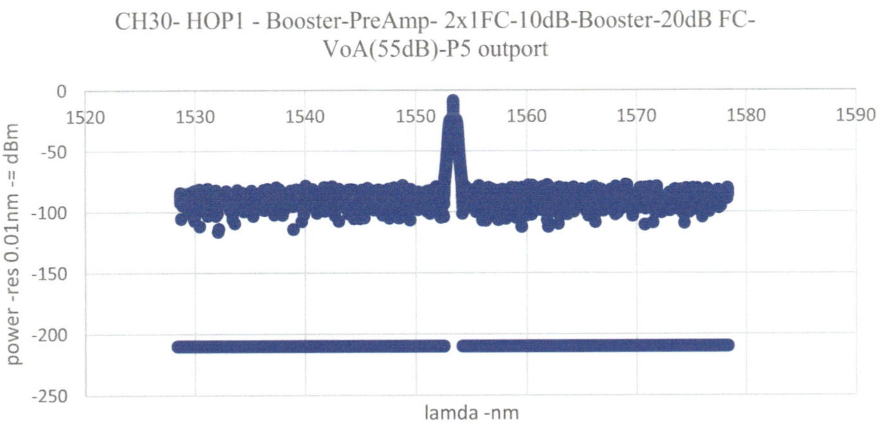

(b)

FIGURE 10.51 Channel #30 output spectrum after passing through HOP optically amplified, 2×1 FC (LD1&LD2) → 10 dB → Booster – VoA (=55 dB) P4 output to OSA.

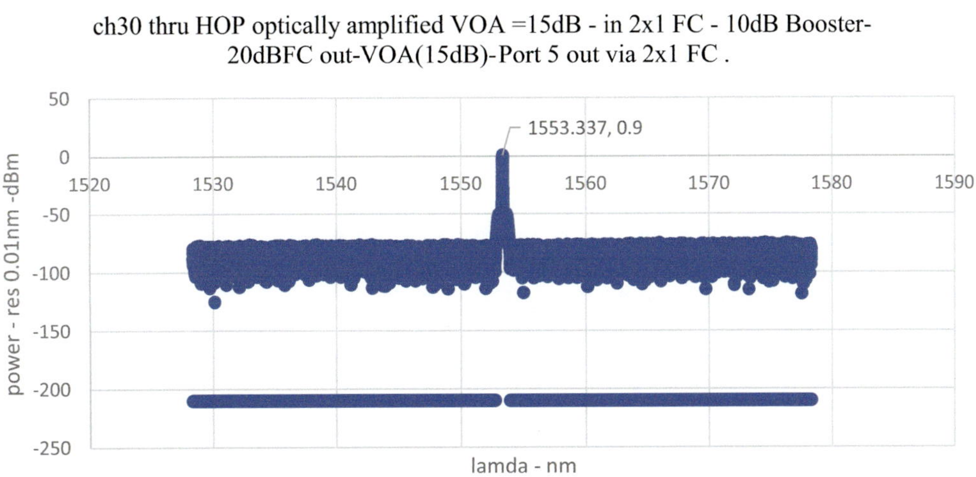

FIGURE 10.52 VOA=55 dB (45+10) (a) & (b) in minute different time. Spectra of channel No. 30, booster and attenuation VoA of (a) 50 dB, (b) 55 dB.

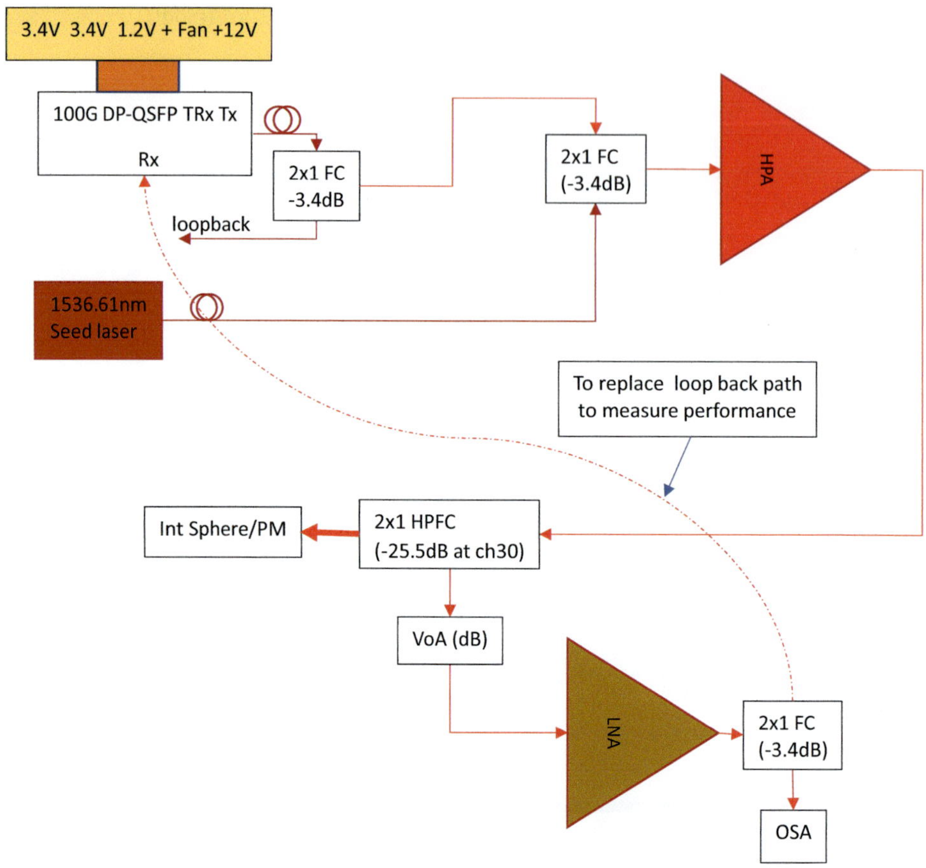

FIGURE 10.53 One-Hop transmission, monitored Channel #51 after passing pre-amplifier with the carrier in CW mode passing through the hop. OneHop_Link_Testing_Set-up Optical High Power Amplifier (HPA)_HOP_CH51-channel #30-HoP1-LD1&2 (10 dB) Port1-Port3-15 dB_Port4.

10.5.2 Test of Signals at Different Stages of the Optically Amplified One-HOP

10.5.2.1 Tests Conducted on 05-05-2023: Signals at Front and after Optical PreAmp (LNA)

Test conducted with structure: please refer to Figure 10.46: Set up to test the optically amplified FreeSpace Hop.

The point to obtain/probe the signal 100 G spectrum is at the front of the OptPreAmp (LNA) or after the VoA unit, whose attenuation can be tuned from 1.5 to 50 dB. The following figures depict the spectra of the 100 G-DP-QPSK at different attenuations.

The system configuration is shown below in Figure 10.54.

Channel at (1536.61 nm) (1536.61 nm)

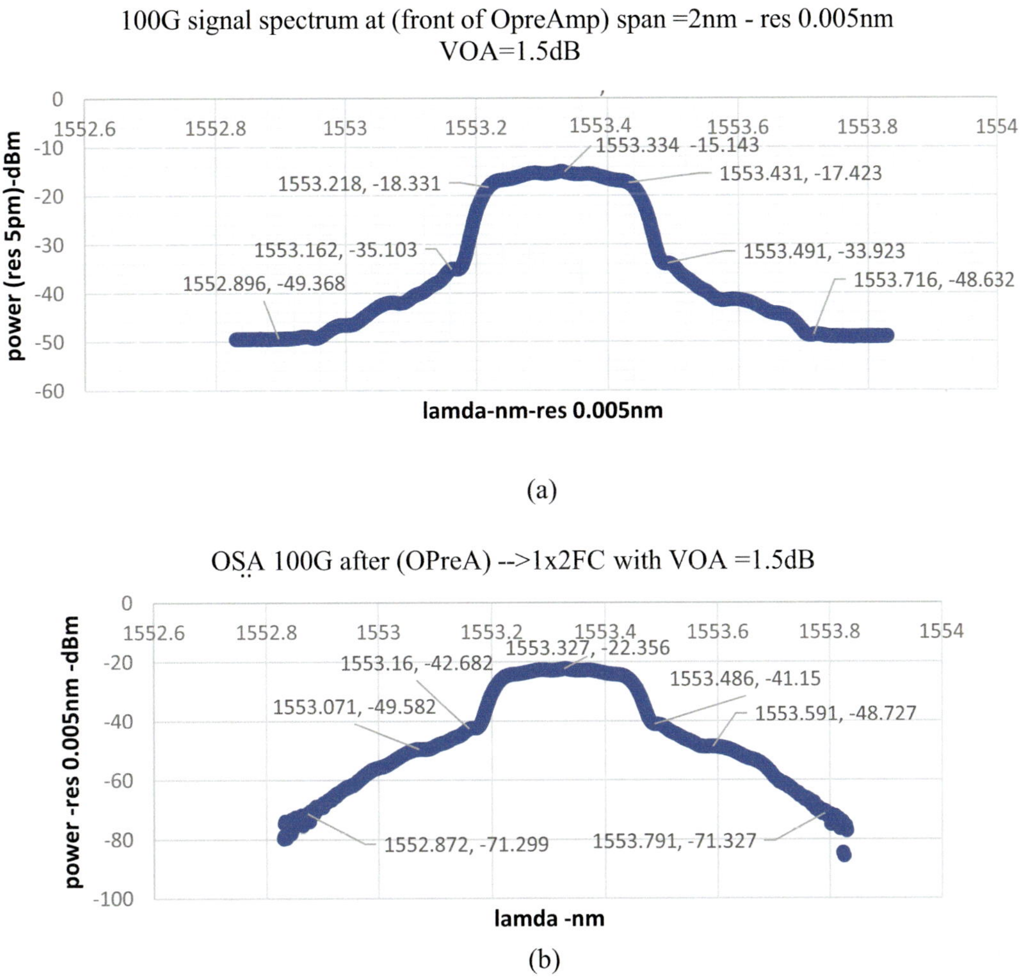

FIGURE 10.54 Setup schematic of one-Hop fiber-equi free space Hop (MEO-MEO) and 100 G DP-QPSK transmission performance.

a. Case 1: VoA = 1.5 dB (Figure 10.55)

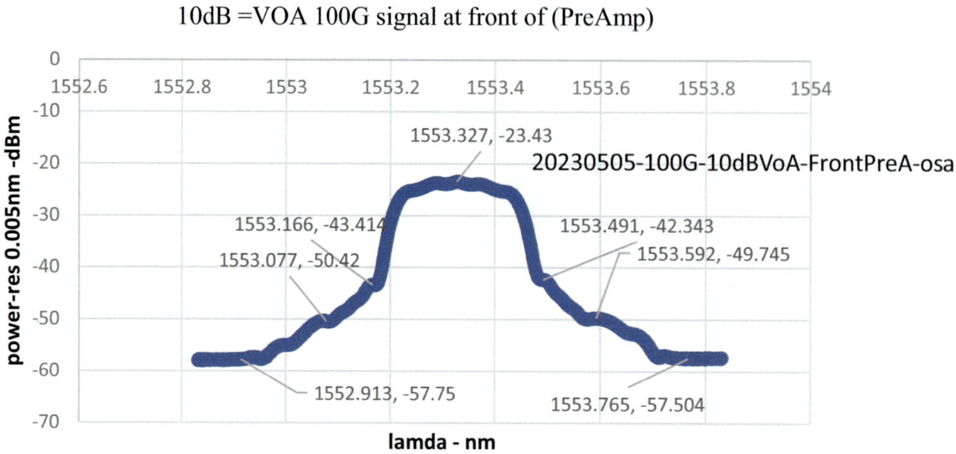

(a)

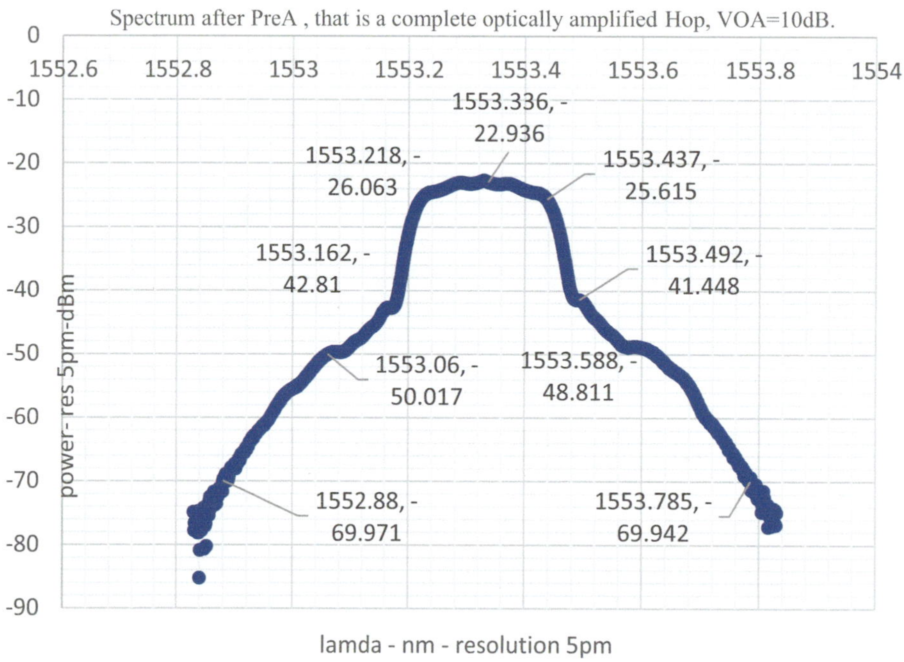

(b)

FIGURE 10.55 The 100 G DP-QPSK signal spectrum after complete optically amplified HOP with VoA = 5 dB (a) at the front of PreAmp; (b) after preamp into 1×2 FC splitter then to OSA, note channel 1553.33 nm = CH30 with tunable optical filter of 100 GHz but effective 60 GHz roll off. (a) At the front of OPt PreAm after VoA (1.5 dB). (b) Spectrum after PreA, complete HOP (with HPA and PreA).

TABLE 10.4
Calculation of the Spectrum Signal Spectrum over Noise Spectrum Power Spectral Density

Resolution (nm)	Spectral Window (nm)	Spectral Window (sig) -nm	Total Sig Power (mW)	Power (dBm)	Noise Spec Density (mW/nm)	Noise Power (in band)	Noise Power (dBm)	OSNR (dB)	OSNR (DP04QSFP) dB
0.005	0.31	3.10e-01	1.62e+01	1.21e+01	1.50e-04	9.30e-03	−2.03e+01	3.24e+01	32.3

The signal spectrum is evaluated at the central spectral area where the noise takes effect. The in-band noise is evaluated at the signal noise level, and the power is the total power spectral density of the spectrum. The OSNR is estimated this way and compared to the value obtained from the signal monitor of the module DP04QSFP, as shown in Table 10.4, which is very close. This is true for the case of high OSNR, or the signal power is far higher than the noise level (Table 10.5).

b. Case 2 VoA = 10 dB (Figure 10.56)

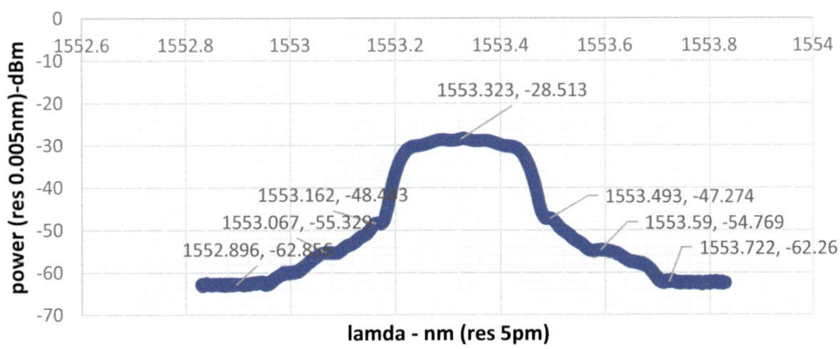

(a)

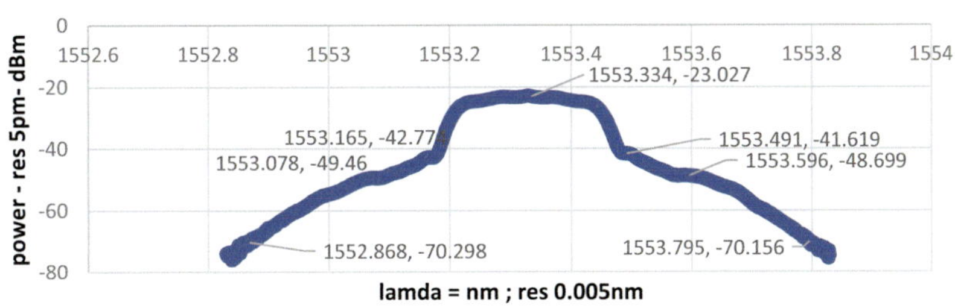

(b)

FIGURE 10.56 VoA = 10dB; 100 G DP-QPSK signal spectrum after completing transmission through the optically amplified HOP with VoA = 10 dB. (a) Case VoA = 10 dB; spectrum at the front of OPreA (MPB) 20230505-100 G-10 dBVoA-FrontPreA-osa. (b) VoA = 10 dB, spectrum after optical preamp and at the front of the receiver of optically amp HOP via 1×2 FC. 20230505-100 G-10 dB VoA-HOP-osa-2calOSNR_TOF. The spectrum after PreA is complete HOP (with HPA and PreA)S.

c. Case 3 VoA = 15 dB (Figure 10.57)

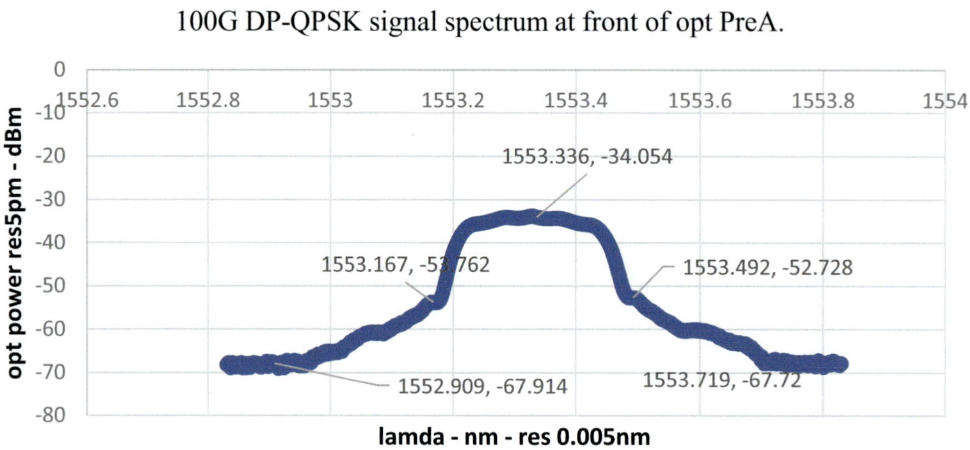

(a)

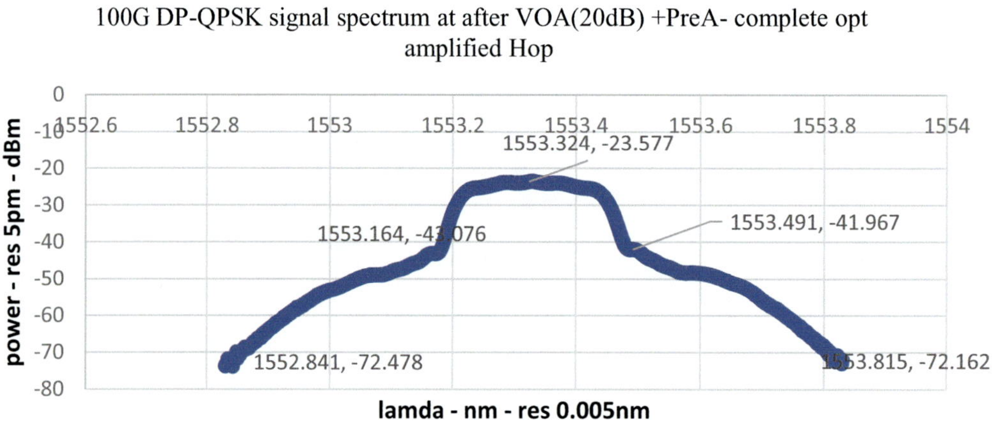

(b)

FIGURE 10.57 VoA = 15 dB; 100 G DP-QPSK signal spectrum after completing transmission through the optically amplified Hop with VoA = 15 dB, (a) spectrum at the front of VoA (15 dB) and OPreA (MPB) -100 G-10 dBVoA-FrontPreA-OSA; (b) signal/noise spectrum at the front of the receiver of optically amplified Hop via 1×2 FC: 100 G-10 dB VoA-Hop-OSA-OSNR_TOF (WSS filter). Spectrum after PreA that is, complete HOP (with HPA and PreA).

d. Case 4 VoA = 20 dB (Figure 10.58)

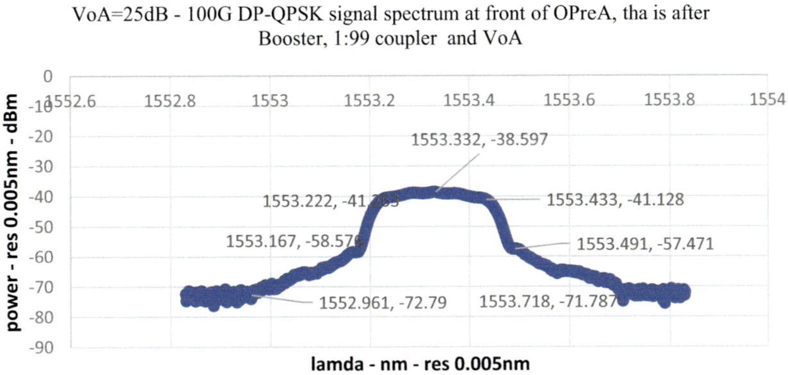

FIGURE 10.58 VoA = 20 dB; 100 G DP-QPSK signal spectrum at the front of PreA and after completing transmission through the optically amplified HOP with VoA = 20 dB. (a) Spectrum at the front of VoA (20 dB) and PreAmplifier (LNA); (b) spectrum at a point after VoA cascade with Opt PreA (LNA). (a) Spectrum at the front of VoA (20 dB) and PreAmp. (b) Spectrum at a point after VoA + OPt PreA.

e. Case 5 VoA = 25 dB (Figure 10.59)

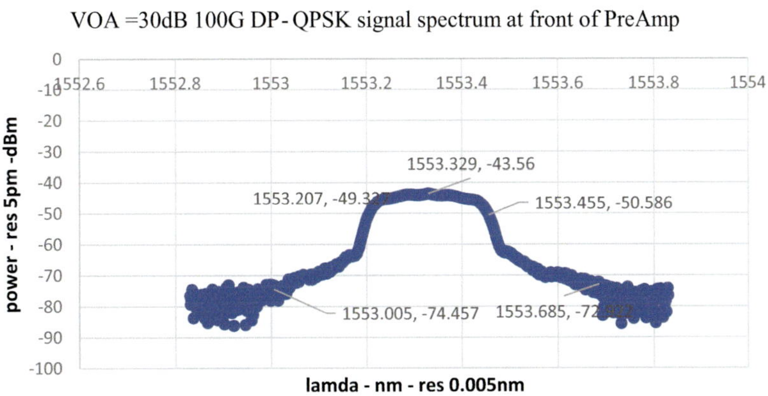

(a)

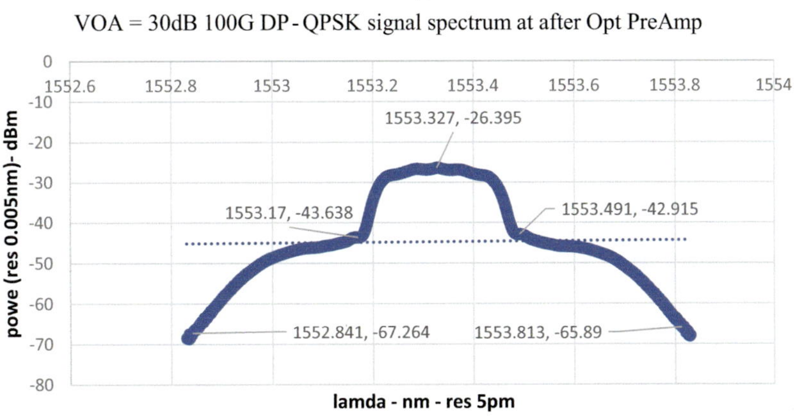

(b)

FIGURE 10.59 VoA = 25 dB; 100 G DP-QPSK signal spectra at front of preamp and after completing transmission through the optically amplified Hop with VoA = 25 dB. Spectrum after OptPreA or complete optically amplified Hop transmission.

f. Case 6: VoA = 30 dB (Figure 10.60)

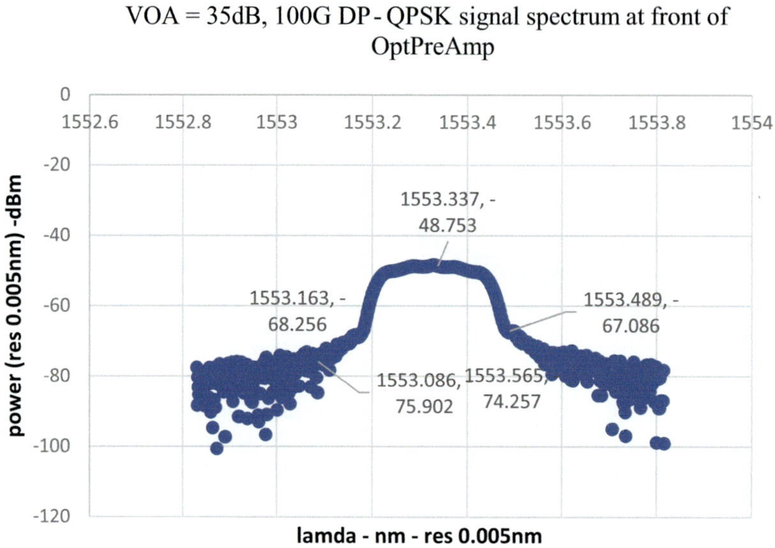

(a)

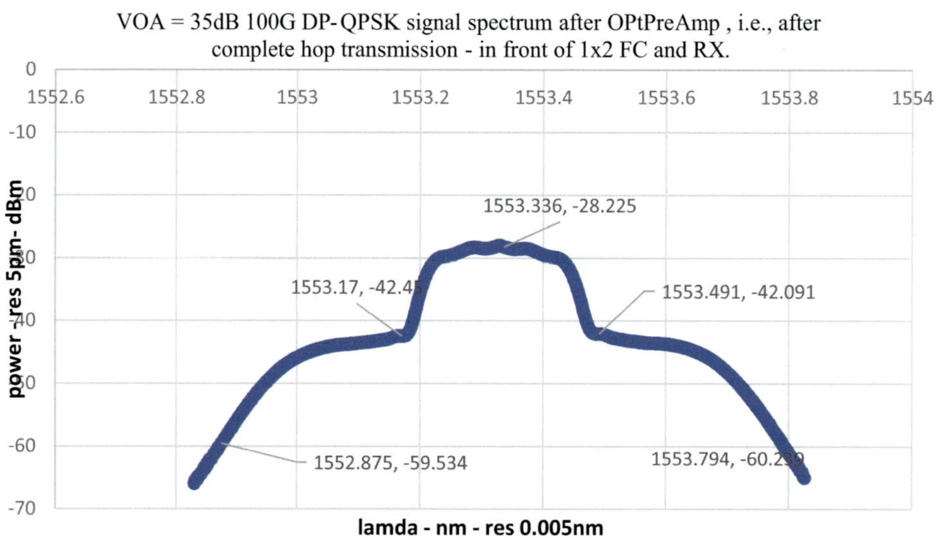

(b)

FIGURE 10.60 VoA = 30 dB; 100 G DP-QPSK signal spectra (a) at the front of the optical preamplifier and (b) after opt preamplifier, that is after completing transmission through the optically amplified HOP with VoA = 30 dB. (a) at the front of OptPreAmp (LNA), i.e., after Booster+ VoA. (b) Spectrum after preamp, i.e. after optically amplified Hop.

g. Case 7 VoA=35 dB (Figure 10.61)

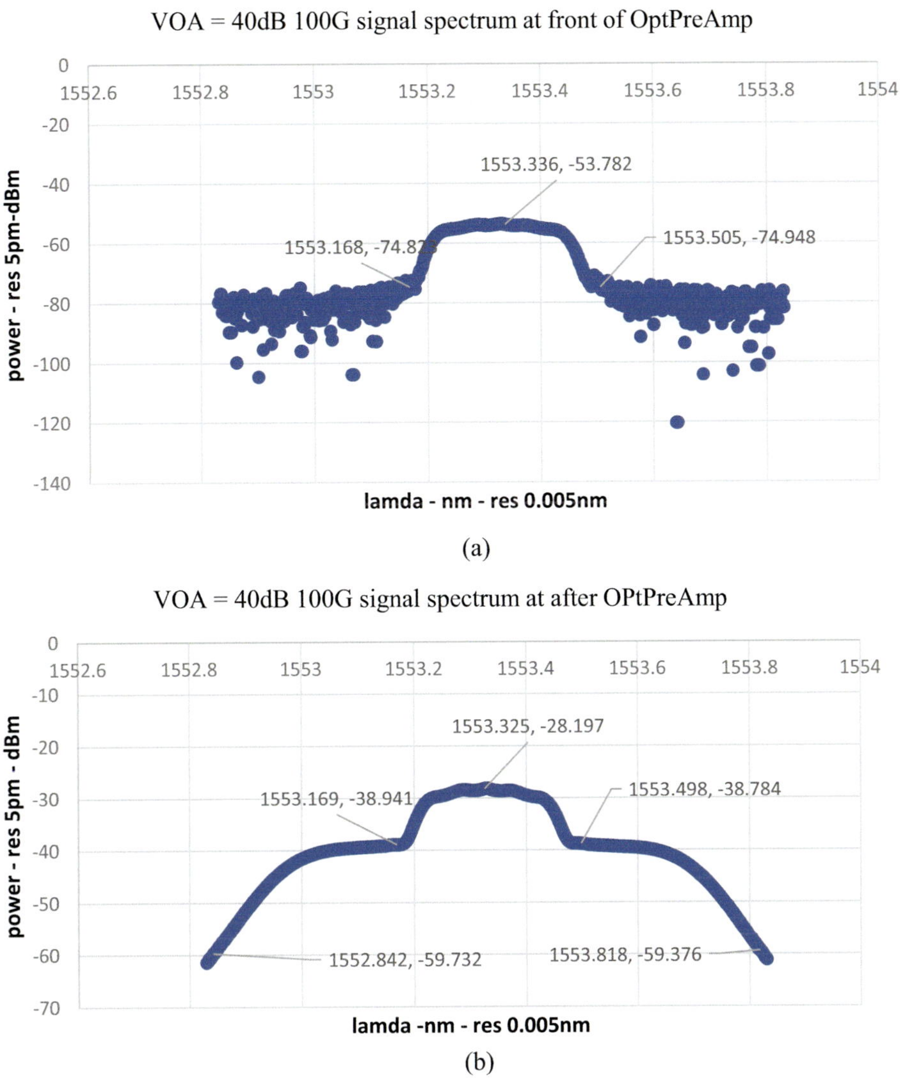

FIGURE 10.61 VoA=35 dB; 100 G DP-QPSK signal spectra (a) at the front of preamp and (b) after preamplifier, after completing transmission through the optically amplified HOP with VoA=35 dB. (a) Signal spectrum at the front of OptA, i.e. after VOA (35 dB) and booster. (b) VoA=35 dB, 100 G signal spectrum after OPtPreAmp.

h. Case 8 VOA=40 dB (Figure 10.62)

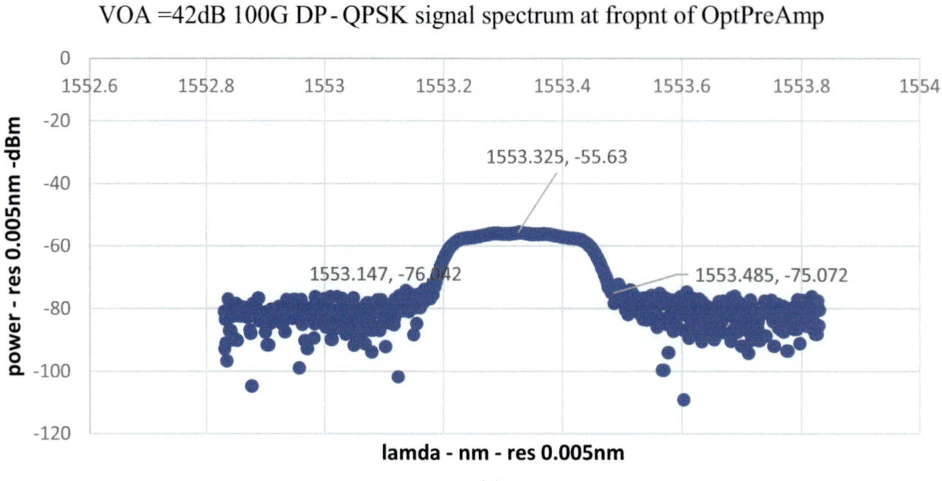

(a)

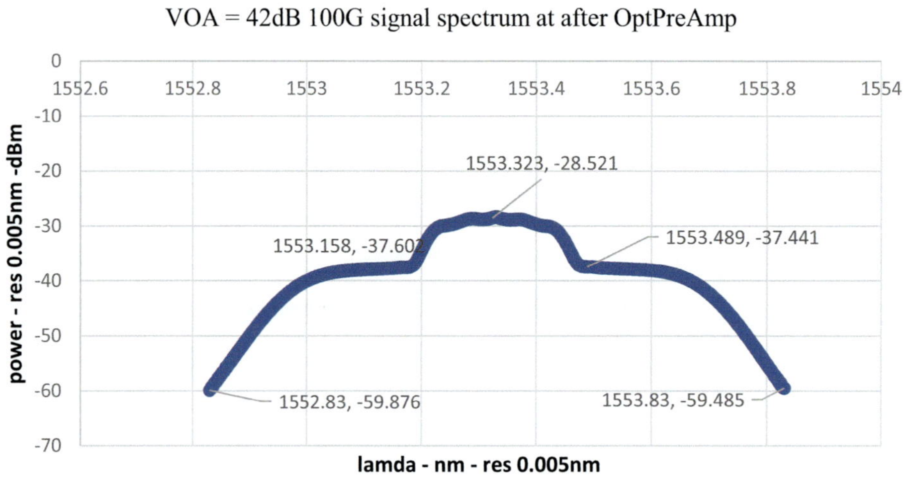

(b)

FIGURE 10.62 VoA=40 dB; 100 G DP-QPSK signal spectra (a) at the front of preamp and (b) after preamplifier, after completing transmission through the optically amplified HOP with VoA=40 dB. (a) 100 G DP-QPSK, signal spectrum at the front of Opt PreAmp (LNA). (b) Signal spectrum 100 G after Opt PreAmp (LNA), VoA=40 dB.

TABLE 10.5
OSR Estimations Using ODA Spectrum Signal Spectrum at Front of Optical Preamplifier LNA (a) At Front of Optical Preamplifier, (b) After Optical Preamplifier

(a) At Front of Optical Preamplifier

VoA (dB)	Resolution (nm)	Spectral Window (nm)	Spectral Window (Sig) - nm	Total Sig Power (mW)	Power (dBm)	Noise Spec Density (mW/nm)	Noise Power (in band)	Noise Power (dBm)	OSNR (dB)
40	0.005	0.167		2.88e-03	−2.54e+01	1.16645e-07	3.89593e-06	−5.40939e+01	2.87e+01

(b) After Optical Preamplifier

Reso (nm)	Spectrum Window Sig (nm)	Noise Spec Window	Signal Power (mW)	Noise Power (mW)	OSNR (dB)	OSNR (Module) - dB	Power Spec Difference@ Central Lamda OSNR (dB)
0.005	0.307	0.307	7.34e-02	6.34e-03	1.06e+01	15	10.56

i. Case 9 VoA=42 dB (Figure 10.63)

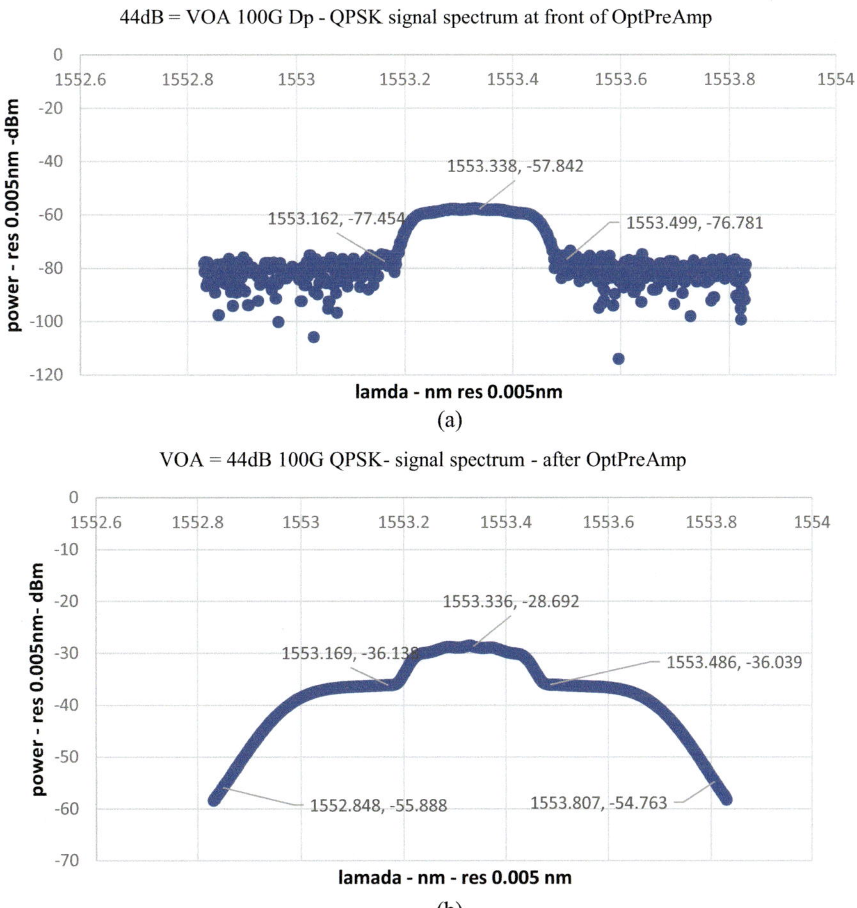

FIGURE 10.63 VoA=42 dB; 100 G DP-QPSK signal spectra (a) at the front of preamp and (b) after preamplifier, after completing transmission through the optically amplified HOP with VoA=42 dB. (a) In front of OptPreA. (b) 100 G after OptPreA with VoA=42 dB.

j. Case 10 VoA = 44 dB (Figure 10.64)

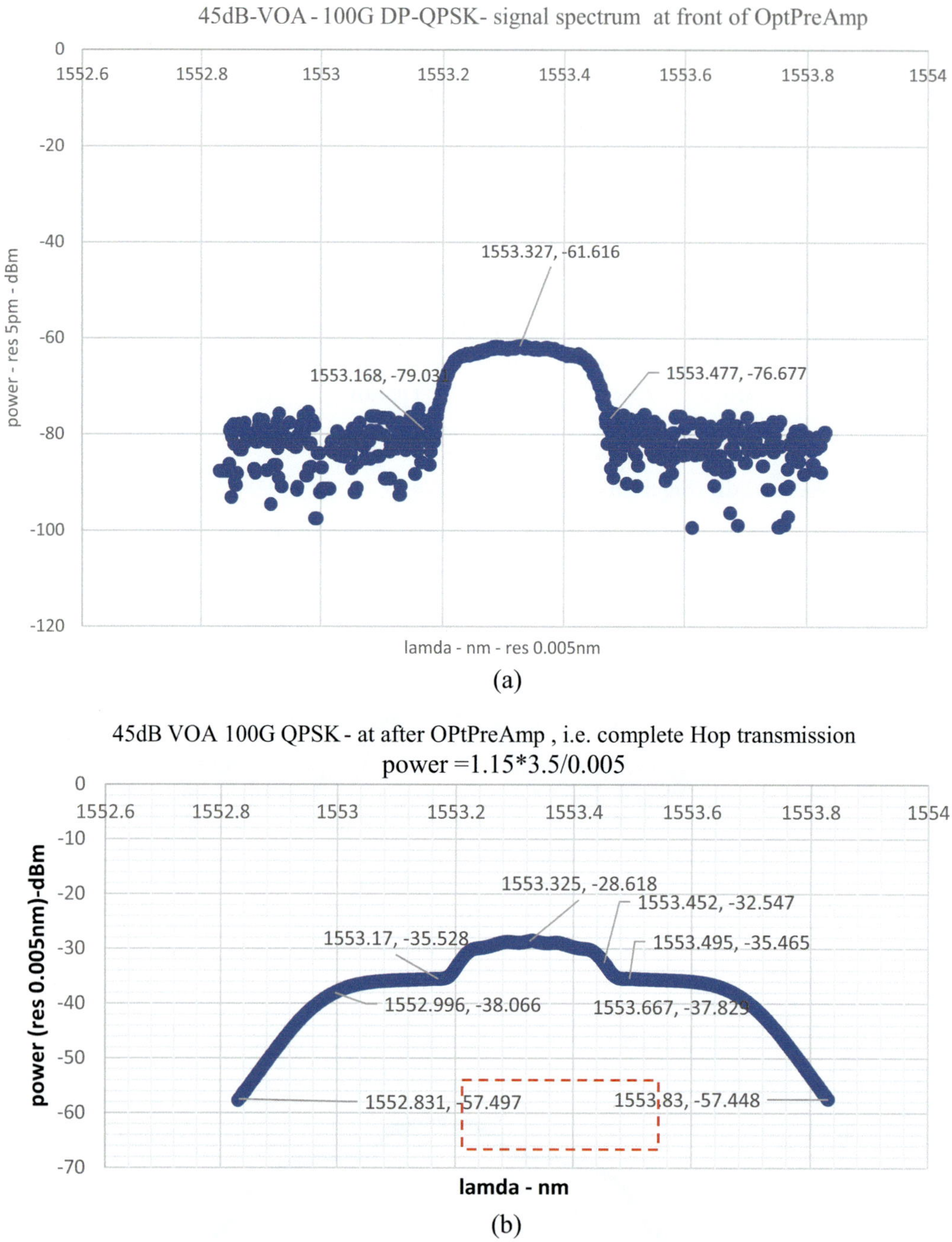

FIGURE 10.64 VoA = 44 dB; 100 G DP-QPSK signal spectra (a) at the front of preamp and (b) after preamplifier, after completing transmission through the optically amplified HOP with VoA = 44 dB. (a) In front of OPtPreAmp. (b) 100 G signal spectrum after OptPreAmp.

k. Case 11 VoA=45 dB (Figure 10.64)

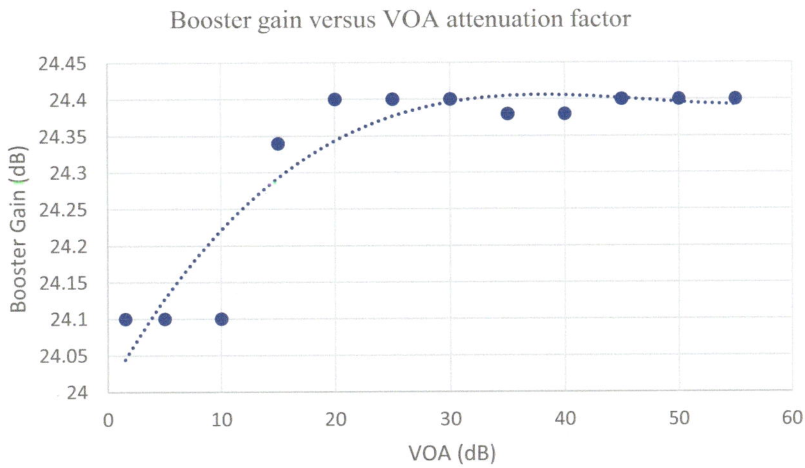

FIGURE 10.65 VoA=45 dB; 100 G DP-QPSK signal spectra (a) at the front of preamp and (b) after preamplifier, after completing transmission through the optically amplified HOP with VoA=45 dB. (a) 100 G signal spectrum in front of optical PreAmp. (b) VoA=45 dB after OptPreAmp – optical pre-amplification in front of the receiver.

After this VoA=45 dB, the transmission system can no longer recover the signal. Even at this level, the BER=2e-2 and OSNR=8.2 dB, well below the 10.5 dB threshold for 2e-2 BER.

10.5.2.2 Tests using Signals Preamp after 1 × 2 FC (fiber coupler)

Already integrated in the above section, the 100 G spectra under various attenuation factors of the VoA attenuator (Figures 10.66–10.70).

10.5.2.3 Numerical Data as Recorded from GUI of DP04-QSFP and GUI MPNB Amplifiers

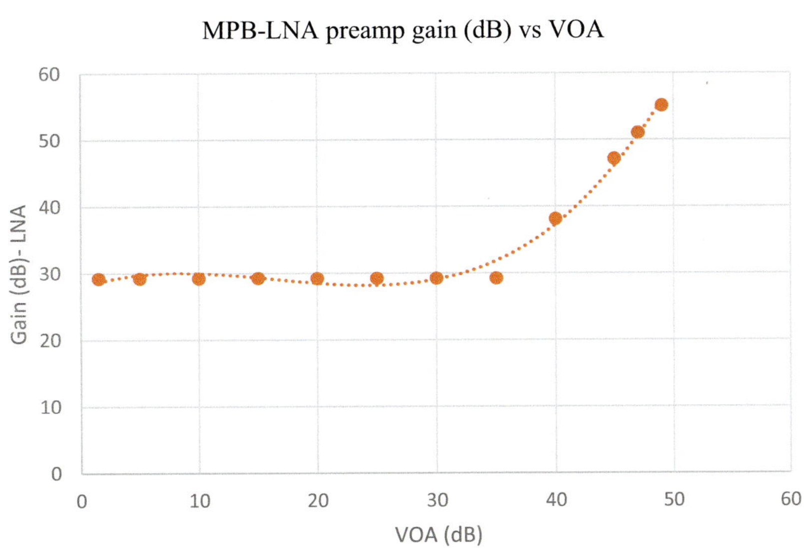

FIGURE 10.66 One-Hop optical amplification booster-preamp. Boosters gain versus VoA (no effects).

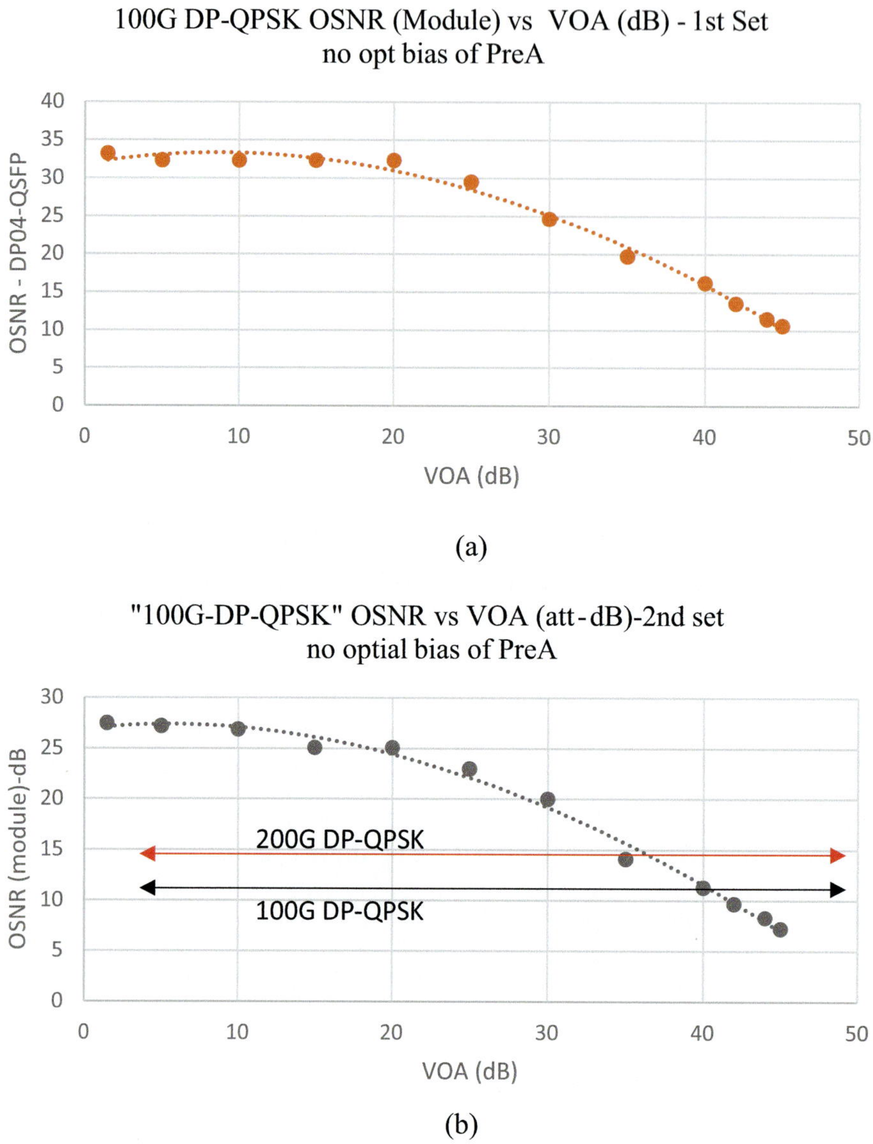

FIGURE 10.67 One-Hop optical amplification booster and optical preamplifier (LNA), preamplifier gain versus VoA (no effects).

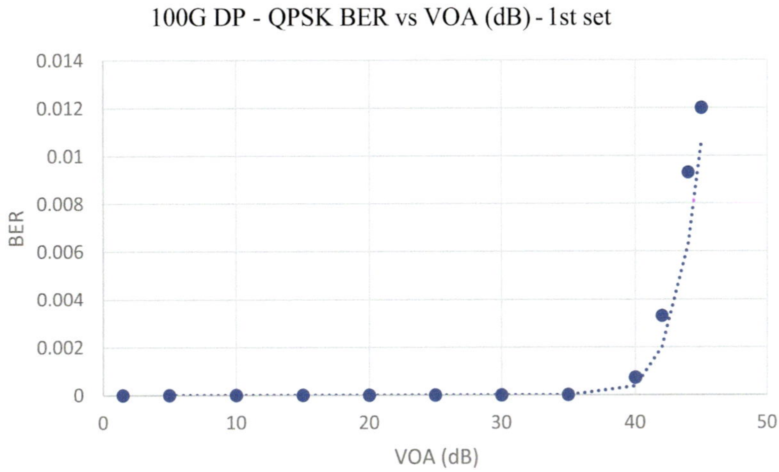

(a)

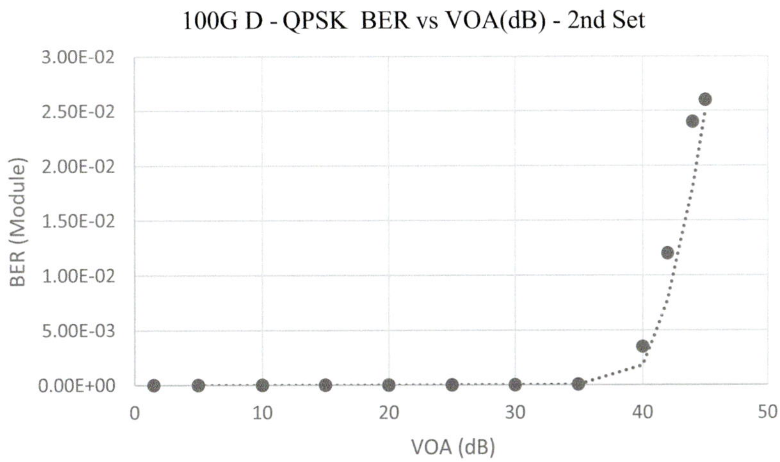

(b)

FIGURE 10.68 OSNR versus VoA as recorded from DP04QSFP. (a) OSNR (from DP04QSFP) versus VoA attenuation factor without optical biasing. (b) BER (from DP04QSFP module) versus VoA Attenuation factor, with optical biasing.

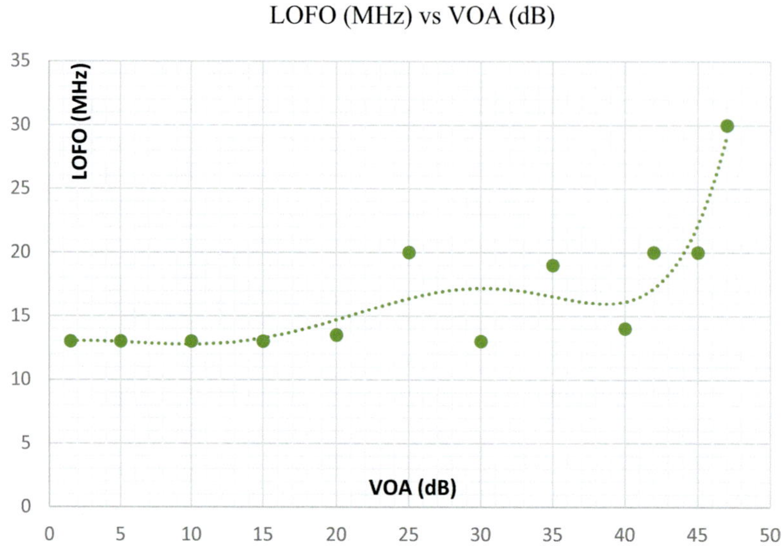

FIGURE 10.69 BER under soft FEC versus VoA attenuation factor as recorded from DP04QSFP. (a) BER (from DP04QSFP) versus VoA attenuation factor as monitored by module. (b) BER (by module monitor) versus VoA Attenuation factor repeating after one day to prove module stability.

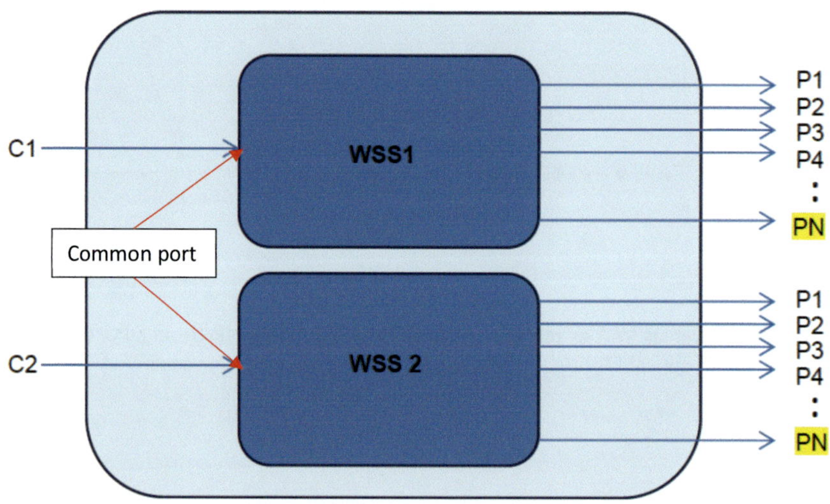

FIGURE 10.70 DP04QSFP LOFO versus VoA.

10.6 BACK-TO-BACK WSS1 (MUX)-WSS2 (DEMUX) TRANSMISSION

10.6.1 WSS Structure and Setup

A 1×9 wavelength selective switch (WSS)[15], if arranged as a multiplexer or demultiplexer, consists of (i) A common port or C-port which acts as a combiner (as multiplexer) or a splitter (demultiplexer); (ii) 9 ports as inputs (as ports to be multiplexed) or output (demultiplexed outputs) ports.

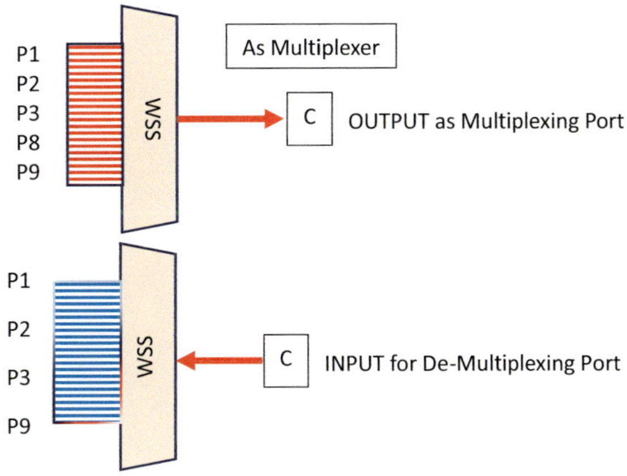

FIGURE 10.71 Schematic of a twin WSS, dual WSS. C=Common port, i.e. multiplexed signals output or to be demuxed signal INPUT. P1, P2, …, P9 are input ports or output ports. Module # must be specified to identify which module is to be driven.

Ports P1-P9 are assigned as ADD/DROP ports in bidirectional mode, as shown in Figure 10.71.

The symbols representing the WSS as a multiplexer (Mux) or demultiplexer (Demux) are shown in Figure 10.72, and the photos of the arrangement of the WSS ports are shown in Figure 10.73a and b, including the evaluation/control board. The common port can act as either input multiplexed optical channels or the WSS, which is arranged as a demultiplexer. Similarly, if all optical channels are into Ports 1 to 9, the C port acts as the multiplexed port. The communication from the I+USB PC port to the UART 40-pin interface is shown in Figure 10.73c via the Tx and Rx communication points.

The WSS can be bidirectional; thus, it can act as an optical filter and add-or-drop switching device. This section demonstrates WSS as a multiplexing and demultiplexing device and optical filtering for the SPC system. The WSS as an add/drop multiplexer or demultiplexer is integrated into MEMS, and CDC-ROADM is presented in Chapters 3 and 12.

Figure 10.72 depicts the structures and port assignment of the 1×9 WSS with one common (C) port and 9 inputs or output ports (P1, P2, …, P9). These ports can be used as multiplexed or demultiplexed, respectively. They can be commonly assigned as ADD/DROP ports. Thus, if a bidirectional circulator is integrated, we can obtain these ports as ADD ports, that is, to be inserted or DROP ports to be demultiplexed ports. From the input common, i.e., reverse flow direction. Note the bidirectional transmissivity of the WSS.

10.6.2 Notes on Setting WSS[16]

The Twin WSS comprises two independent $1 \times N$ port switches in a single module. The module is composed of an optical block and a block of control electronics. The module dynamically attenuates and/or switches an individual wavelength channel or sets of channels independently. The devices in the Micro-Twin family operate over the full range of one frequency band (C, L, or extended frequency band). Channels can only be configured to and from ports associated with their common input/output port. Channels cannot be configured between WSS modules.

The channel edge frequencies $f1$ and $f2$ identify the channels. The channel bandwidth can be any multiple of slot size. The center frequency must be defined so that the edges of the channel defined sit on a 12.5 GHz grid, as illustrated in Figure 10.74.

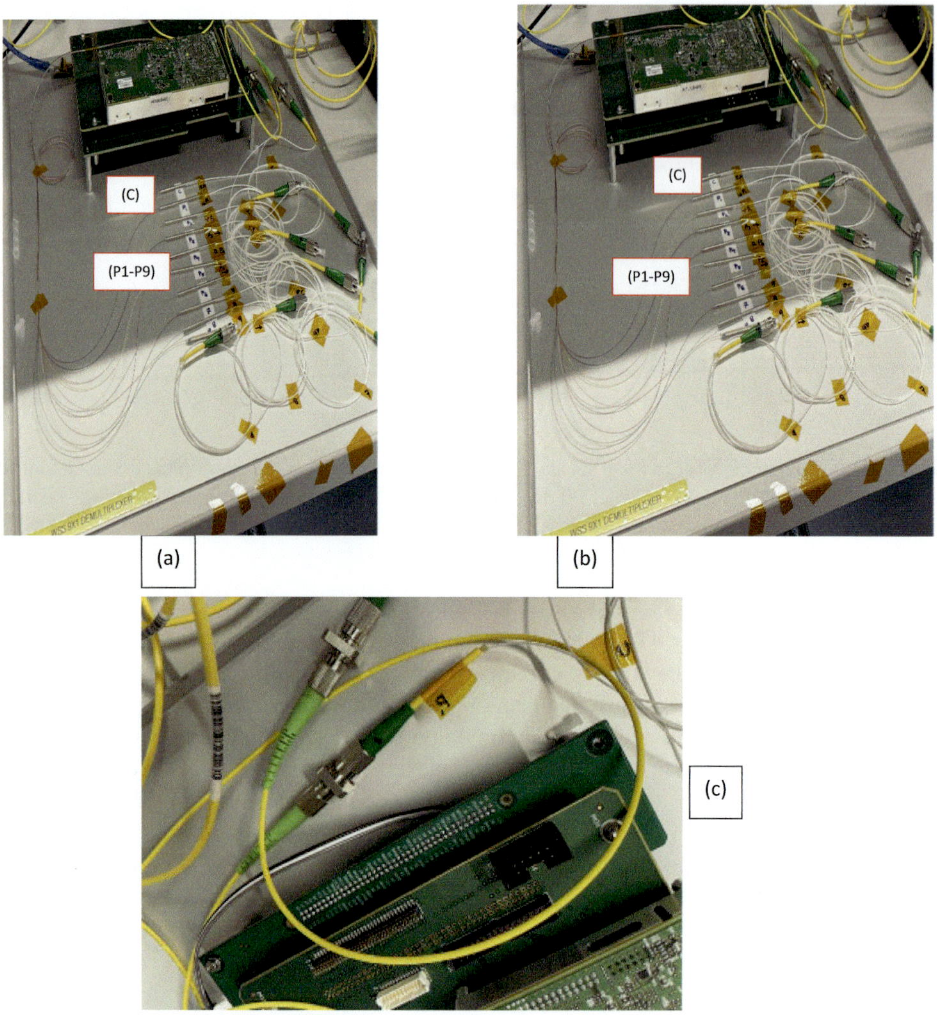

FIGURE 10.72 Structures and port assignment of WSS (1×9) or one common (C) port and nine inputs or output ports (P1, P2, …, P9).

In Flex mode, to increase the width of an existing channel, it must be re-provisioned with its new center frequency and bandwidth. In Fixed Grid mode, a slot can be added or removed from a channel on its edges by at least one slot size width.

Once a slotted channel has been created, the individual slots can change their attenuation independently, allowing a shape to be applied to the channel. The slots are accessed by addressing commands to CH.M.N.S. Within a channel, the slots are numbered from 1 to P (where P=channel bandwidth/slot Size), where 1 corresponds to the slot at the lowest frequency.

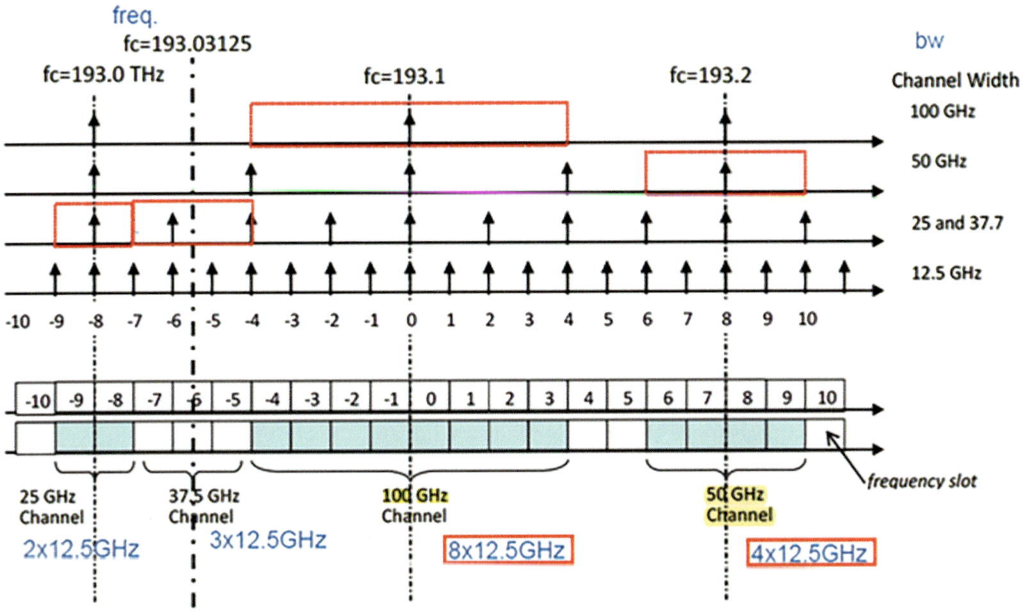

FIGURE 10.73 Lumentum 1×9 WSS (a) multiplexer with port assignments (c)=common, P1,..., P9 input (mux) or output (demux) ports (b) demultiplexer WSS, with evaluation board; and (c) Communication from USB (PC) to UART port (Tx and Rx).

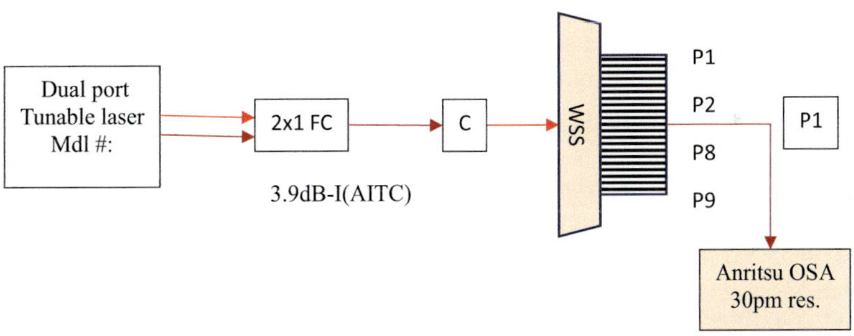

FIGURE 10.74 Fixed grid channel provisioning.

10.6.3 Commands

A set of commands must be designed to control and configure the WSS as a multiplexer or demultiplexer, an add/drop multiplexer with different functionalities of the common and input/output ports. This section gives a brief overview of such commands for WSS.

10.6.3.1 Types

The following commands are supported:

- ADD: creates an object in the system identified by an instance identifier
- DELETE: removes an object from the system identified by an instance identifier
- SET: change an attribute value of an object (for example, attenuation or port)
- GET: get an object or an attribute of an object
- ACTION: perform an activity for which an immediate response is not expected (restart, etc)

10.6.3.2 Command Examples

Note that the GET command supports a wildcard character *, but the SET command does not support a wildcard character.

Also, note that the wildcard will return all attributes and cannot be used to return a subset of the attributes for all channels (once the module reads a wildcard, the command processing ceases).

So GET: CH.1.* and GET:CH.1.*:ADP will return the same response that includes all attributes.

If a SET:CH command is issued, but no attenuation value is specified, the default will be set to the BLOCK state. If no optional attributes are given, the module will return all attributes.

The module will respond with OK after each command is successfully completed (except for queries such as GET:IDN, which will return specific values).

ADD:Ch.1.3:ADP=3: central frequency at fc=191000.0 GHz:BW=50.0 GHz: att=6 dB:→ Creates a 50 GHz channel for Twin module 1 configured to Port 3 at 6 dB attenuation with a channel ID of 3.

SET:CH.1.3:ADP=8 → Re-configures channel 1.3 from port 3 to port 8 in the BLOCK state (since no attn value is specified).

SET: CH.1.3:att=10 → Sets the attenuation of channel 3–10 dB attenuation

GET:CH.1.3 → Returns all of the attributes associated with channel 3

GET:CH.1.1:F1:F2 → Returns the start and stop frequencies of the first channel on module 1.

GET: CH.1.* → Returns the attributes of all of the channels configured on WSS # 1

DELETE: CH.1.3 → command to delete channel of central frequency of 191000.0 GHz from WSS#1

SET:Module.1:slotSize=TF → True-Flex. Sets the mode of module 1 to

GET:Temp → Returns the case temperature of the module with six-digit precision (response example: TEMP=55.532703 degree Fahrenheit)

Concatenated Message Format: All channel message commands above can be concatenated with a semicolon character ";". If compound messages are used, they must have the same verb. These commands shall be processed together. **Note** that concatenated GET commands are NOT supported.

10.6.4 Testing

10.6.4.1 Test 1: CW Laser Determination and Setting of the ADD/DROP Port of WSS

Figure 10.75 depicts the setup for testing the ADD/DROP port passband using a dual-tuneable carrier laser source and OSA of 30 pm resolution to observe the pass band without attenuation, thus no distortion of the WSS ports as assigned. The dual laser outputs two laser output ports, which can be tuned independently and sweeping of one source gives the output of the passband of a desired WSS port. The WSS operates bidirectionally, so we can send the carriers of the dual laser into the common port and monitor the output ports as the central passband can be set appropriately to obtain the carriers at the appropriate port.

Figure 10.76 shows the full setup with the source, a modulated channel using DP-QPSK format and two polarized optical channels. The test setup employs a commercial VIAVI ONT with a 400 G optical channel; the format can be either 56 GB and DPQPSK Or P16QAM with 28 GB. The two polarized channels are combined and split using an optical PBS unit where the *V* (vertical) - component

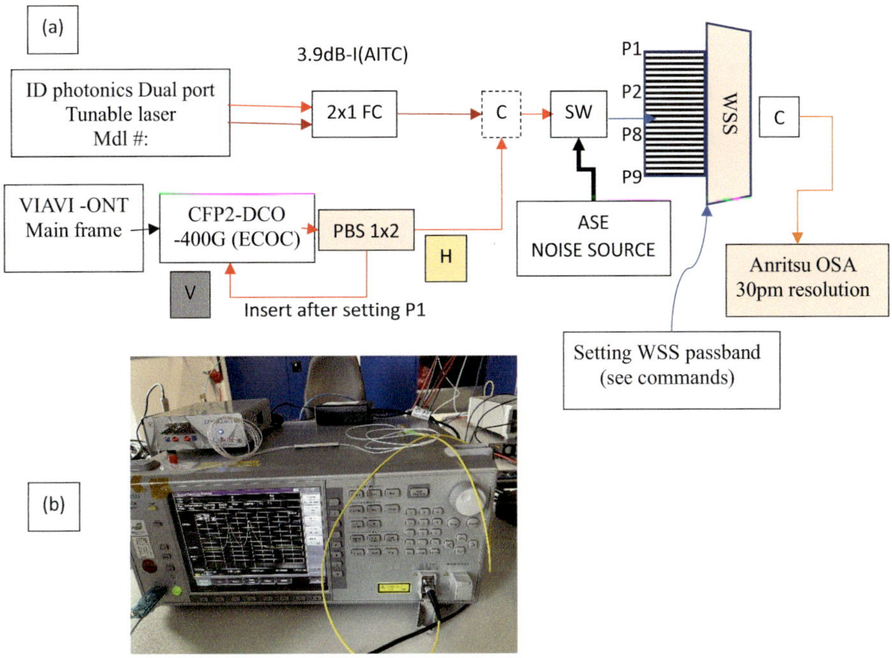

FIGURE 10.75 Setting ADD/DROP port passband using a dual tunable carrier laser (ID photonics) and OSA of 30 pm resolution to observe the pass band without attenuation, thus no distortion of the WSS ports as assigned.

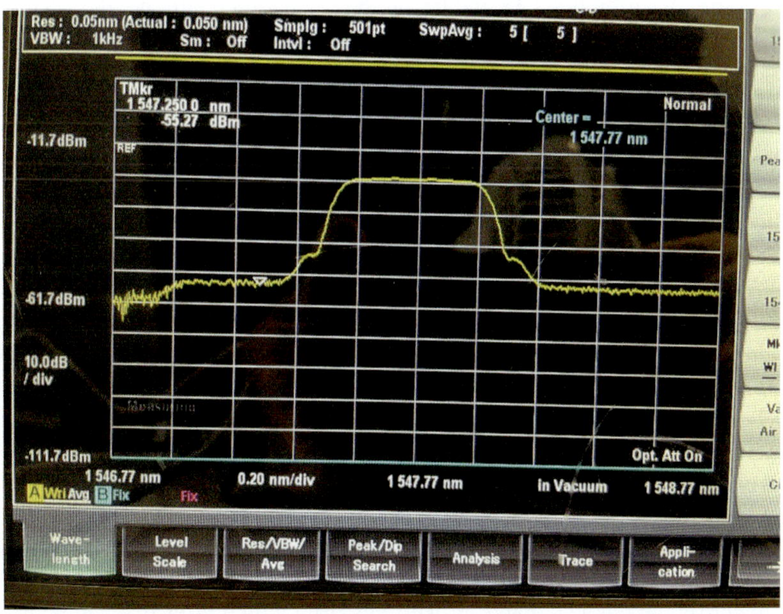

FIGURE 10.76 Test using VIAVI ONT with 400 G channel (optical). This is via an optical PBS unit where the *V* (vertical) - component is feedback to the optical receiver. The *H* (horizontal)-component is fed into the common port to drop to the WSS port 1 and then the OSA to observe the spectrum. The reverse is also true. (a) Set up block schematic; (b) photo of setup for monitoring dual laser generated spectrum before launching into WSS. C=the common port of WSS, where *C* is the output to direct any input port signal to *C* with a passband to be specified. If *C* is used as input, one can be directed to any specific port assigned to the passband bandwidth and central wavelength.

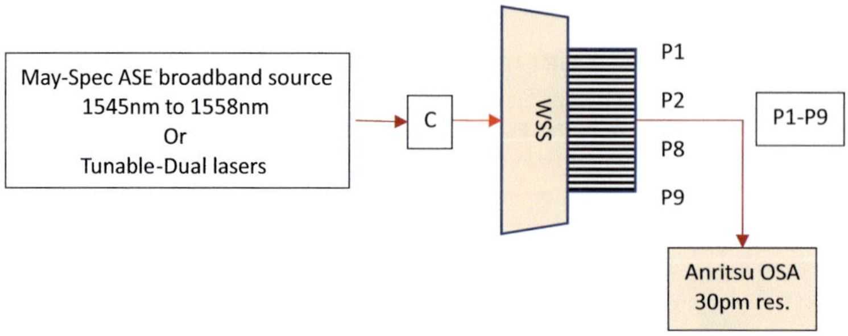

FIGURE 10.77 Obtained 400 G 16 QAM 50 Gsy./s[17] spectrum at channel at wavelength of 1547.77 nm (see ITU grid given n Section 12.5 – Appendix 7 of Chapter 12) on OSA Anritsu.

is feedback to the optical receiver, and the *H* (horizontal)-component is fed into the Common port to drop to the WSS port 1 and vice versa. Then, the OSA is used to observe the spectrum. The reverse is also true. (i) Set up block schematic and (ii) photo of setup for monitoring dual laser generated spectrum before launching into WSS.

Injecting the broadband noise source can obtain additional noise levels, which allows us to set the channel's OSNR and determine the receiver sensitivity and the BEER of the transmission system.

Figure 10.77 shows the 400 G 16 QAM 50 Gsy/s spectrum and the channel wavelength of 1547.77 nm (ITU grid) on an OSA Anritsu.

It is noted on the operation of the WSS in the following

- WSS turn on via the power supply to the WSS evaluation board with the green light LED on
- Note the communication ports Tx and Rx on the evaluation board, not the RS232 port.
- Commands from GUI with the following commands: GET, SET, ADD, etc.
- Communication port (inspected from the Devices manager) to the WSS board. Be sure that the correct COM port is assigned to communicate to WSS.

Once the power is turned off from the WSS, then no data can be stored, and all settings are to be done next time → an initiation file with all settings is to be loaded to the WSS board.

10.6.4.2 WSS Setting and Testing of Filtering Passbands

It is required to pass signals of channels operating with 100 G QPSK 25-28 GSy/s., this modulated lightwave channel exhibits a bandwidth of 18–22 GHz. If 200 Gbps with modulation format using dual polarization DP - QPSK at a baud rate of 50–56 GSy./s (or GBaud)[18], then its bandwidth is about 38 to 50 GHz. If pulse shaping is used as mentioned in Chapter 6 the signal bandwidth can be sharper and narrower. Therefore, the passband of the WSS ports must be set accordingly to pass through these channels without affecting the integrity of the signals, that is, no distortion due to the amplitude and phase of the WSS port channels. Note that we assume a linear phase optical filter of the pots of the WSS input or output ports.

This section reports the measured passband characteristics of the WSS. The set is implemented as shown in Figure 10.78. The lightwave sources are either BB ASE sources whose power level is about –22 dBm across the whole C-band or two lines from a dual laser that can be tunable across the whole C-band. The sources are fed to the common port C, which acts as the multiplexer output or input to be demultiplexed by the WSS. The WSS operates as a bidirectional optical device.

Figure 10.79a shows the output of the dual lightwave source whose wavelengths are placed within the passband of a channel bandwidth of the OBPF or the WSS. Figure 10.79b and c depict the output spectrum at port P1 (set as an add port) with the two carriers placed symmetrically about the central passband and shifted asymmetrically so that one of the carriers is at the fallen edge of

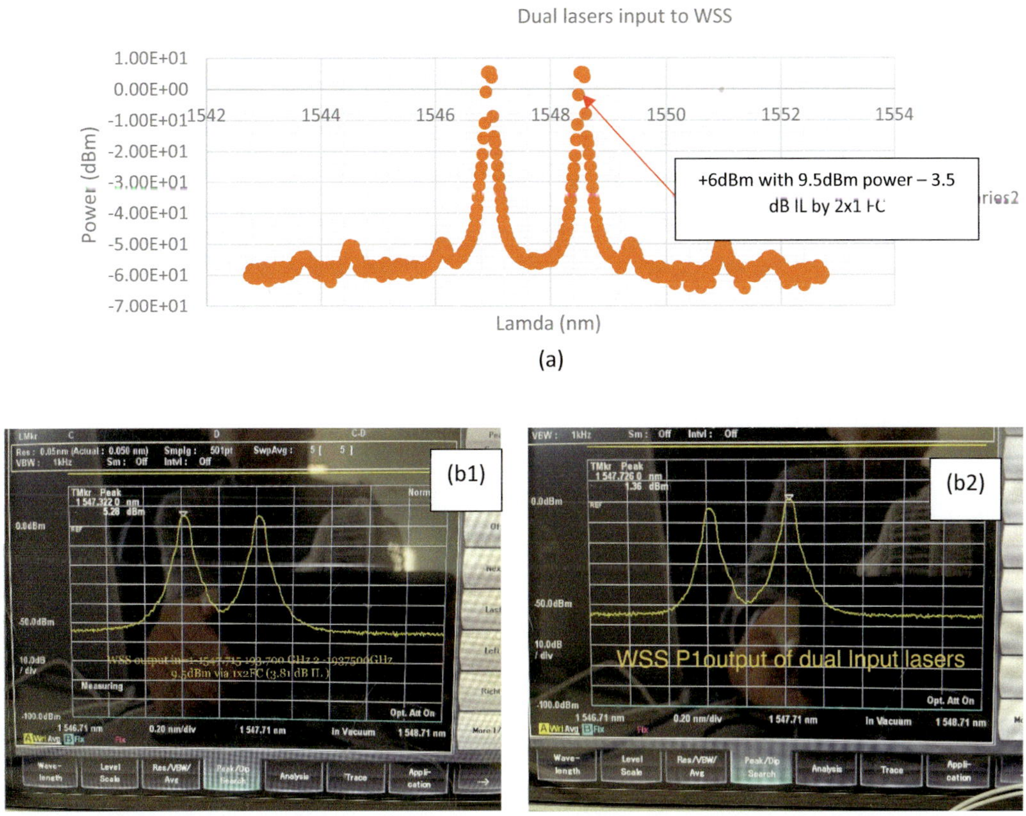

FIGURE 10.78 shows the schematic diagram of the setup to characterize the WSS filter characteristics. Characterization setup for WSS ports with passband filters. Figure 10.78 shows the sampled plot obtained from the OSA when the input into the common port C and the dual laser lines are injected into WSS. The optical frequency is 193,700 GHz (or in wavelength located at 1547.72 nm) and the BW is 200 GHz.

the passband so that we can determine the 3 dB corner frequency and thence the roll-off band of the WSS passband of each port. This ensures no distortion of the channel even when the Doppler shift occurs. This frequency shift happens quite often from a few hundred kHz to 7 GHz within one hour and twice a day when the link is set between the GEO and LEO satellites. Thus, the setting of the passband of the WSS ports is critical so that such Doppler shift can be accommodated without suffering from distortion due to the roll-off band, whose phase delay is highly critical.

Figure 10.80 sets the bandwidth t maximum range of 200 GHz, accommodating 3 to 4 100 G DP-QPSK channels. Then, the characterization of such wideband of the passband of the WSS port is made.

Similarly, Figure 10.79 shows the spectrum of the output of port P1 Central carrier under the injection of the carrier of FC = 193,700 GHz (1547.72 nm) and a bandwidth of the WSS of 200 GHz. The frequencies of the dual-laser input lines are tuned to 93,700 GHz and 193,750 GHz at an equal power level of 9.5 dBm via 1×2 FC (fiber coupler) with an IL (insertion loss) = 3.81 dB.

10.6.4.3 ASE Broadband Source Spectrum

The monitoring and measurement of the WSS port filter passband can be confirmed by using a broadband optical source feeding into the common port, as shown in Figure 10.81, which shows the spectrum of the broadband source. Each output port, P1 to P9, can be measured using an OSA connected to each port to obtain the spectrum described in the next section.

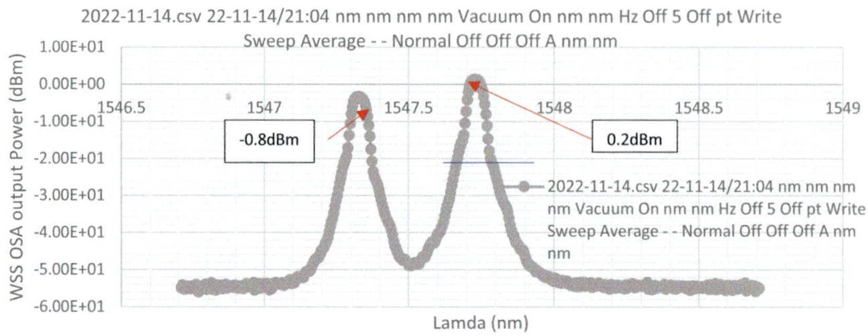

FIGURE 10.79 (a) Digital plot (.CSV file from Anritsu, of Dual laser lines before injecting into WSS. @ fc = 193,700 GHz (1547.72 nm) and BW = 200 GHz; (b) foto (b1) input dual laser spectrum (b2) WSS output spectrum with input (b1) with port P1 ADD.

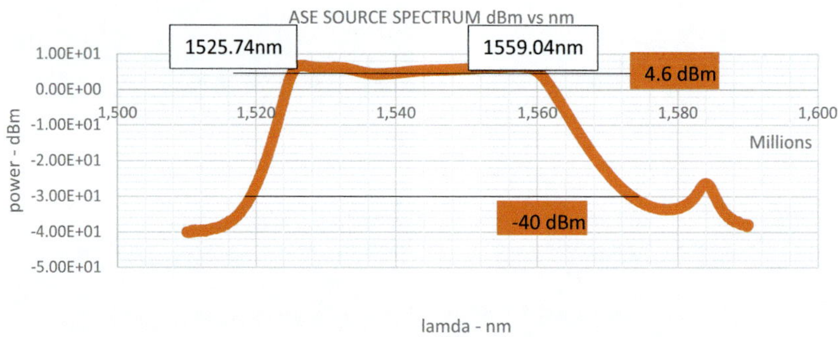

FIGURE 10.80 Output of P1 central carrier@ fc = 193,700 GHz (1547.72 nm and BW = 200 GHz. Dual laser input = 193,700 GHz and 193,750 GHz at 9.5 dBm via 1×2 FC with IL (insertion loss) = 3.81 dB.

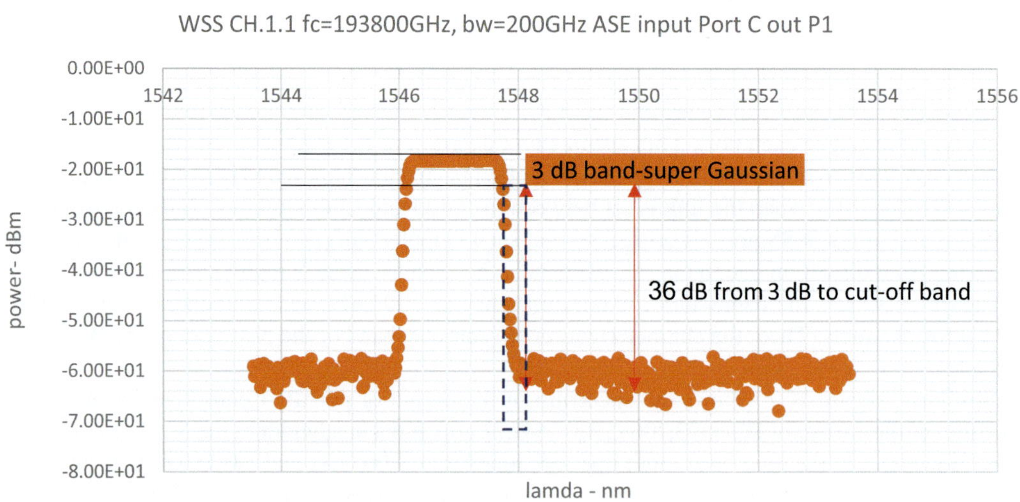

FIGURE 10.81 Optical spectrum of NOISE SOURCE (ASE – BaySpec Co. of San Francisco USA) source for WSS MUX and DEMUX. The setting of the ASE source at the optical power reading of 14.07 mW (maximum total power).

10.6.4.4 Channel P1 WSS Set at Different Bandwidth (200–25 GHz)

The test of the passband of the WSS can further be implemented by using a broadband ASE noise source whose spectrum compensates for that of an EDFA when no signal input is applied. The broadband source is shown in Figure 10.81. The source characteristics spread across the C-band, where most DWDM channels are located.

Initially, the passband of the filter of a channel of the WSS is set to 200 GHz, which is very large enough to observe the passband flatness of the filter embedded in the device, as shown in Figure 10.82. The fluctuation of the passband is less than 0.1 dB across the whole band. This is to ensure that for DWDM channels, we can insert dual channels so that the CDC- ROADM can be simplified if the number of multiplexed channels is higher than 9. It is noted that the baud rate of the Tbps transmission system can be 25, 50, 75 and 100 GB without the overhead of FEC. So, the bandwidth can be nearly half of the baud rate. So, 200 GHz is sufficient for testing and determination of the flatness of a channel of the WSS.

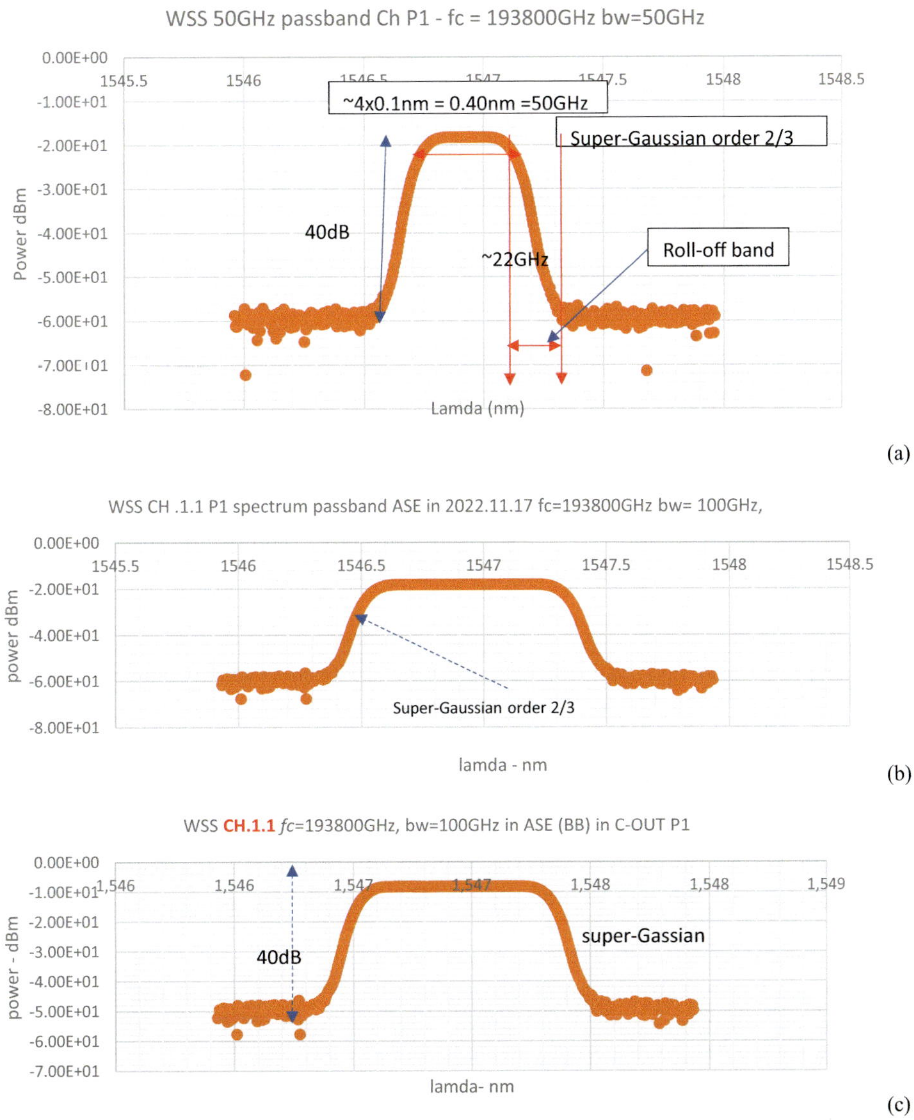

FIGURE 10.82 WSS passband characteristics – f_c = 193,800 GHz input ASE broadband source. WSS P1 setting $f_{c=}$193,800 GHz BW=200 GHz. 1.5 unit×0.4 nm=0.6 nm or 75 GHz. Res=0.2 nm?

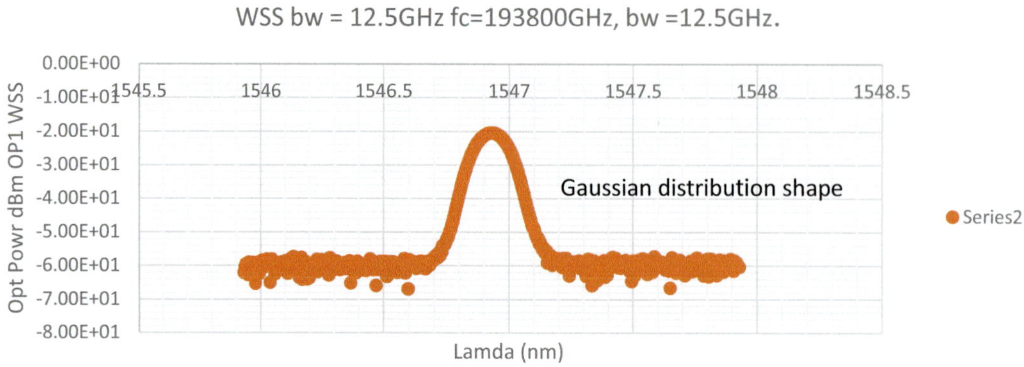

FIGURE 10.83 WSS passband of bandwidth of 100 GHz with input ASE broadband source. WSS P1 setting fc = 193,800 GHz BW = 100 GHz, att = 0 (a) 50 GHz; (b) 100 GHz of channel 1 module 1 of a twin; (c) 100 GHz channel port 1 WSS module 1 of a twin, note setting pass band at a different point.

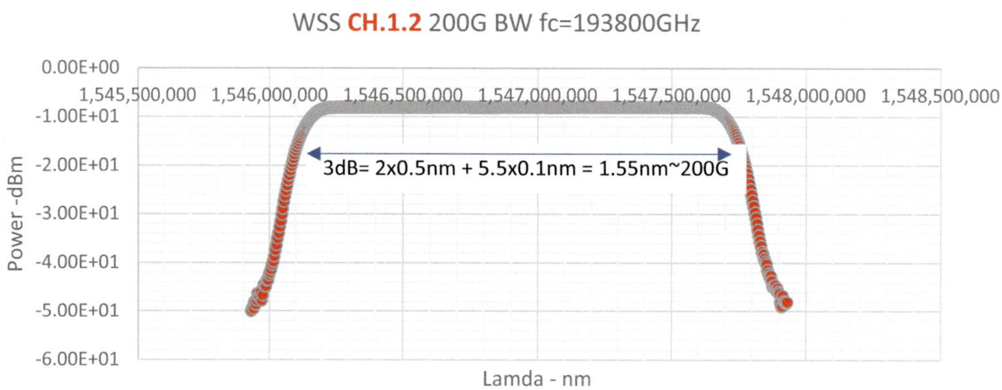

FIGURE 10.84 WSS passband 12.5 GHz with ASE broadband input source – channel P1 BW = 12.5 GHz; input WSE wideband source at –20 dBm power flat level. OSA setting of Span = 3 nm or 3 × 125 GHz full span.

The passband is reduced, and the passband characteristics are observed, as shown in Figure 10.83 and Figure 10.84 of 12.5 GHz. The passband is narrow and can create distortion of 100 G DP-QPSK format. However, this passband can be useful for lower rates.

10.6.4.5 Channel P2 Set WSS at Different BW 100, 50 and 25 GHz

Figures 10.85–10.87 depict the output of port P2 of the WSS with the tuning of the passband of the filter of the port with 200, 100 and 50 GHz, respectively. The Gaussian and super-Gaussian shape of the output spectra show the high and sharp roll of the cot-off band.

10.6.4.6 Channel P3, P4 …, P9 Set WSS at Different BW 100, 50, 25 GHz

Similarly to P1 and P2, we have measured the output spectra of the ports P3 and P4 to P9 and we can observe consistent spectra as shown in Figures 10.88–10.90 or Port P3. The spectra of the ports P4 to P9 are not repeated here but are very much the same as P3.

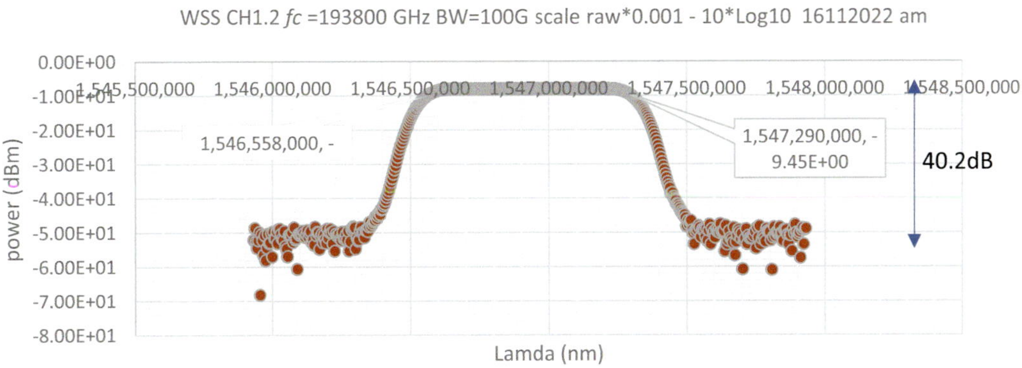

FIGURE 10.85 WSS setting **CH.1.2** (module 1 port output P2) port BW=200 GHz at fc=193,800 GHz. BW=200 GHz. Output spectrum of selected port 1.2 (1=WSS 1 of a pair of WSSes–demultiplexer and 2=WSS2.

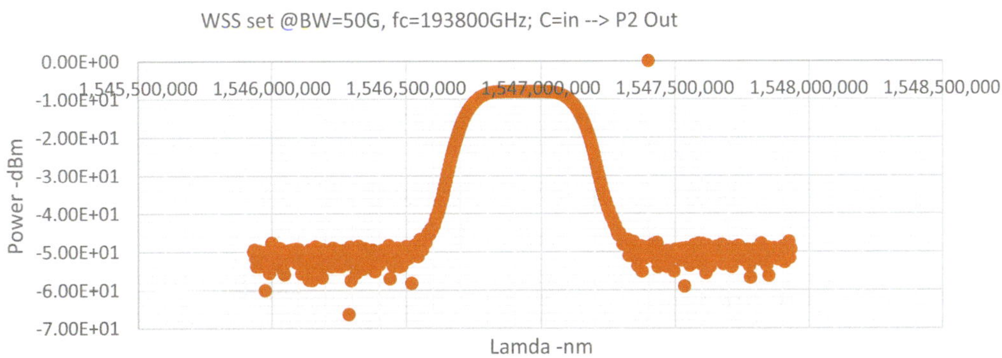

FIGURE 10.86 WSS channel P2 ADD/DROP module 1 input source ASE BaySpec. WSS setting **CH.1.2** (P2) port BW=100 GHz at central frequency of fc=193,800 GHz. Measured at 3 dB=91.5 GHz from the two indicated points for 3 dB from the mid-center point.

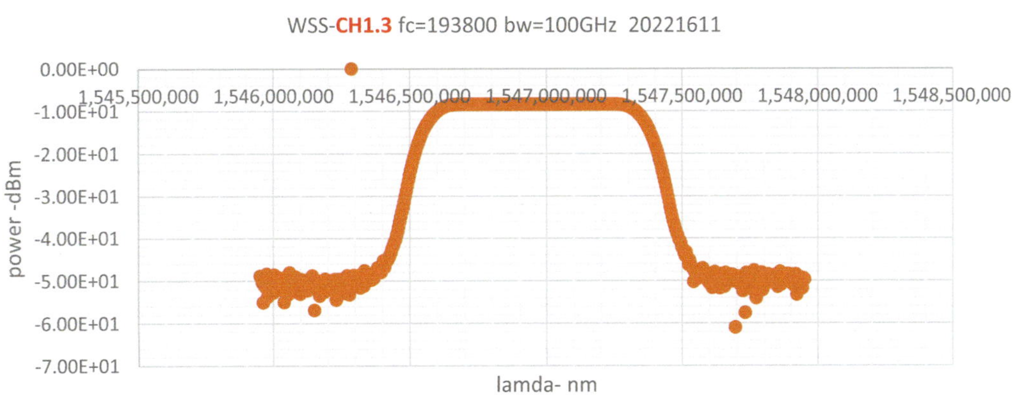

FIGURE 10.87 WSS channel P2 ADD/DROP module 1 input source ASE BaySpec. WSS setting CH.1.2 (P2) port BW=50 GHz at central frequency of fc=193,800 GHz, f_1=193,775 G f_2=193,825 GHz.

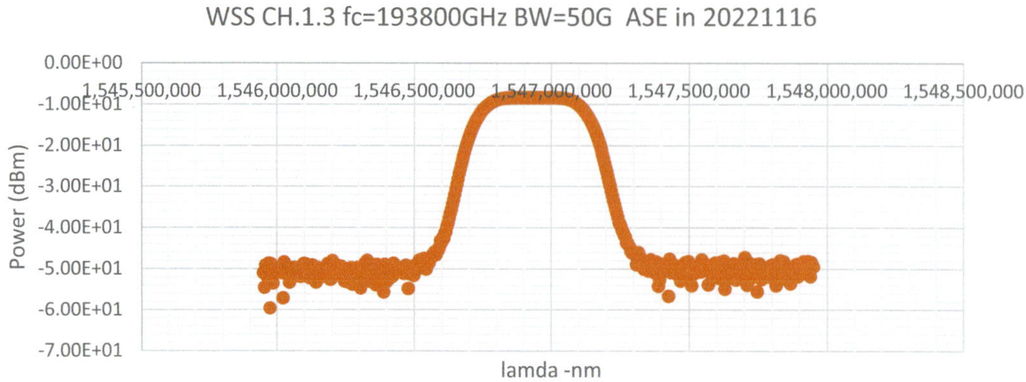

FIGURE 10.88 WSS channel P3 ADD/DROP module 1 input source ASE BaySpec. WSS setting CH.1.3 (P3) port BW = 100 GHz at central frequency of fc = 193,800 GHz, f_1 = 193,750 G f_2 = 193,850 GHz.

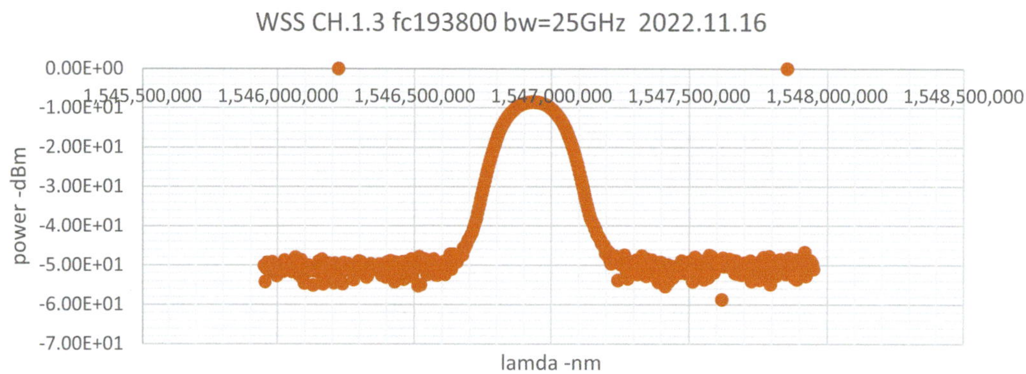

FIGURE 10.89 WSS channel P3 ADD/DROP module 1 input source ASE BaySpec. WSS setting CH.1.3 (P3) port BW = 50 GHz at central frequency of fc = 193,800 GHz, f_1 = 193,725 G f_2 = 193,825 GHz.

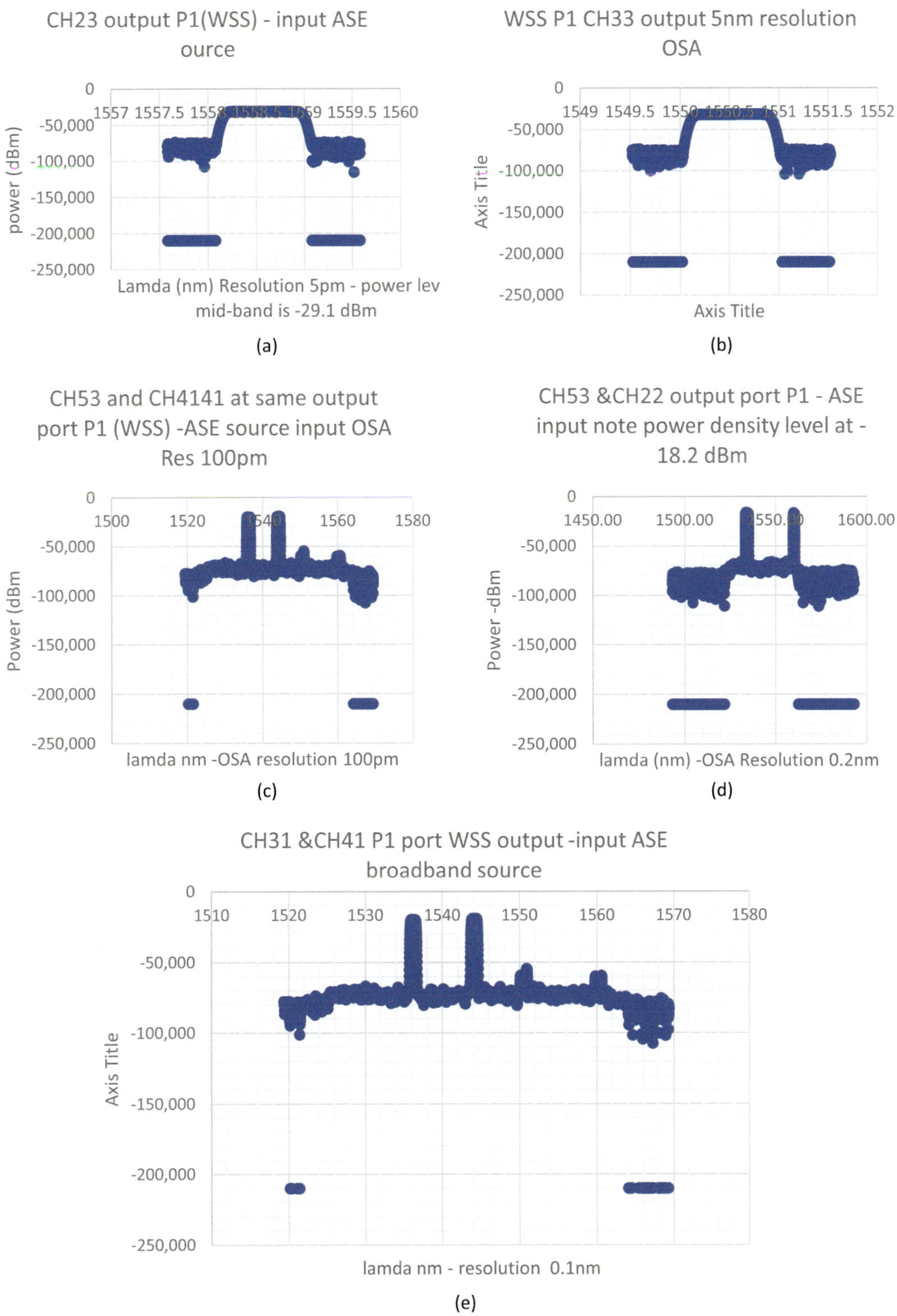

FIGURE 10.90 WSS channel P3 ADD/DROP module 1 input source ASE BB Source. WSS setting CH.1.3 (P3) port BW=25 GHz at central frequency of fc=193,800 GHz, f_1=193712.5 G f_2=193812.5 GHz.

10.6.5 Testing of WSS Behavior over the Whole C-Band

Further tests on the performance of the WSS over the whole <c-band are reported in this section. Due to numerous results of the filtering and the output ports, the obtained spectra of outputs of port 1, when different wavelength carriers are launched into Port 1 of the WSS, are given. For other ports, we provide a table of whether it is okay or not to indicate the agreeable outputs as specified by Lumentum LLC[19], the manufacturer of the WSS.

Similarly, we randomly set the ports of WSS and the channels to be launched and monitored. We display only a number of typical channels here, as shown in Table 10.6.

The above results, derived from Figure 10.91 and tabulated in Table 10.6, indicate the consistent behavior of the WSS Module 1 (Muxer or multiplexer) or Module 2 (Demuxer or demultiplexer). They are used in the free-space transmission link in the budget design of multi-hop systems.

10.7 REMARKS

This chapter presents a detailed experimental platform of one hope free-space inter-satellite optical constellation. An equivalent hardware model is established with the dual polarization QPSK and 16 QAM modulation format, generated from the integrated, coherent optic modules integrated with the optical coherent receiver and DSP generated and modulation of complex signals of dual-polarized channels.

WSSes are used as multiplexers and demultiplexers, with pulse shaping to sharpen channels that are multiplexers to give an aggregate bit rate of Tbps.

Tests of WSSes port filters' passband and roll-off band to prove that they are appropriate for multiplexing channels very effectively. Dual laser tunable wavelengths can be used to evaluate the detailed passband of the WSS filter. A broadband lightwave source is also used to obtain the passband of all channels under the WSS demultiplexer or as a combined multiplexer to obtain the multiplexing functionality of the WSS.

The performance in terms of OSNR and BER, as well as the receiver sensitivity in terms of average received power with respect to the variation attenuation of the space hop.

Hardware and software platforms are given to illustrate the setup with an integrated optical transceiver and the booster and pre-amplifiers for an optical SPC system of one hop. These data on signals and noises are used to obtain the power budget for multi-hop transmission.

10.8 APPENDIX 1: ERROR VECTOR MAGNITUDE

10.8.1 Definition

EVM is the root mean square (RMS) of the error vectors computed and expressed as a percentage of the square root of the mean power of the ideal signal. The error vector is the magnitude of the vector at the detected symbol location. This connects the *I/Q* reference-signal phasor to the *I/Q* measured-signal phasor and is computed as follows. Refer to the phasor diagram shown in Figure 10.92.

10.8.2 Standards

The following standards are referred to (i) IEEE Standard 802.11b-1999[20] 16 September 1999 Paragraph 18.4.7.8 Transmit modulation accuracy. (ii) IEEE Standard 802.11a-1999 16 September 1999 Paragraph 17.3.9.7 Transmit modulation accuracy test. (iii) ANSI/IEEE Standard. 802.11 First Edition 1999-00-00 Paragraph 15.4.7.9 Transmit modulation accuracy.

TABLE 10.6

Table of Output Port with Random Testing of Output Channels and Consistency of WSS Ports and Spectra

WSS Test Notes on the Passband Output Using ASE Input Source - 1*N*=module 1 port *N*	P1	P1.2	P3	P1.4	P1.5	P1.6	P1.7	P1.8	P1.9	Notes: OK - Identical Spectrum
Check status over whole range CH#53-54 to CH10-11	OK REC									
CH#53-54	Recorded	OK as P1	OK	OK	OK	OK	OK	OK	OK	Power level −29.03 dBm ASE input level ~ −21dBm center 1534.250 nm BW 100 GHz or 0.8 nm - ok consistent all ports

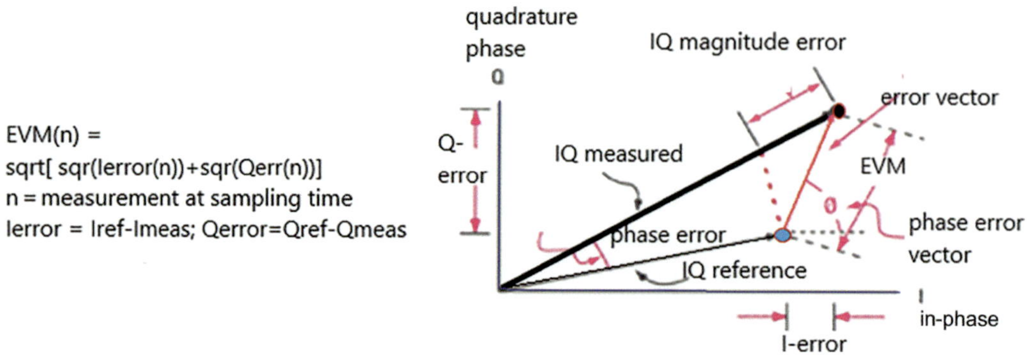

FIGURE 10.91 Optical spectrum of WSS port 1 output with input ASE broadband noise source (1525–1559 nm). The 5 pm resolution of the OSA gives a low power spectral density of –29.1 dBm. To obtain the total channel, we must integrate this noise spectral density over the 3 dB bandwidth of the spectrum of the channel. (a) CH23, (b) Ch33, (c) two channels, CH41 and CH53 in the same WSS slot port P1, (d) CH53 and CH22 at P1 output with OSA resolution of 0.2 nm and (e) resolution of 0.1 nm.

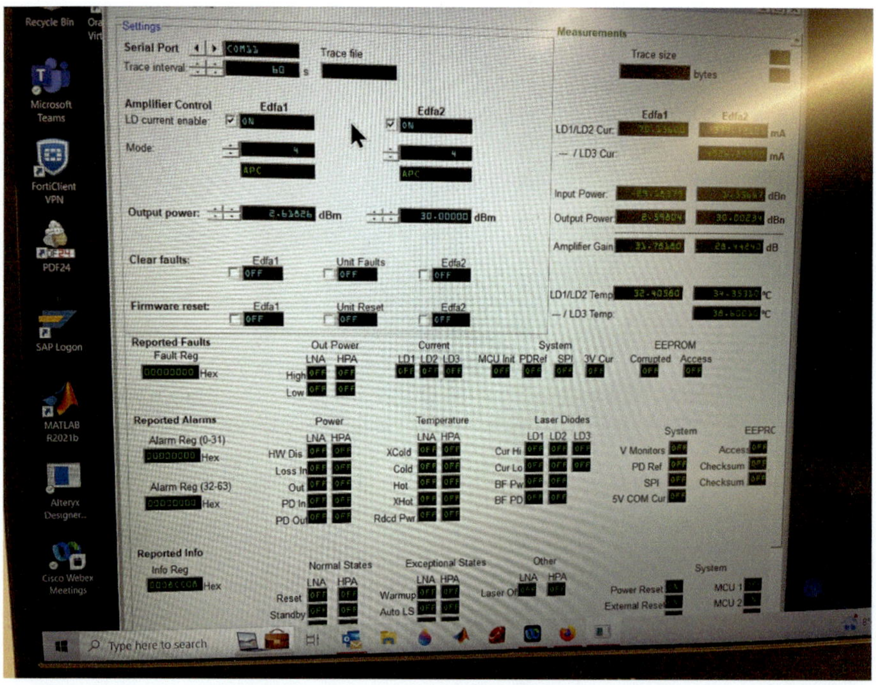

FIGURE 10.92 Illustration of EVM in complex modulated transmission signals (original as a referenced point) and error.

10.8.3 Measurement Subtleties

There are measurement trade-offs when using the digitizing oscilloscopes instead of the Keysight Model 89,600 (or a combination of PSA or ESA and Keysight Model 89,600). It is highly recommended that you read the application note Keysight Infinium Oscilloscope Performance Guide

Using Keysight Model 89601A[21] Vector Signal Analysis with processing software platform. This application note explains the benefits and trade-offs of using the oscilloscope as a front end for the 89601A software. In the Keysight 89,600 model, EVM is calculated only at the symbol times (the instant in time when symbols are detected). The computation does not include points between symbols. Therefore, the points/symbol setting does not affect the value. Related measurements can be the eye diagram and signal constellation.

10.9 APPENDIX 2: ONE-HOP FIBER-EQUIVALENT TRANSMISSION SYSTEM AND THE SETUP OF TEST SYSTEM

10.9.1 Type 1 (MPB-Type Test Setup) Amplifiers (Booster and Preamplifier)

Figures 10.93 and 10.94 depict the interface settings for the booster amplifier and the low-noise amplifier. Settings can be made to the power input and output levels and the gain of each amplifier. For LNA, the input can be varied, and the gain can be in the saturation mode or linear mode. Hence, the ASE noises are also changed accordingly. Figure 10.95 shows the setup for the driving booster amplifier and the amplifier itself with cooling and mounting of the amplifier's EDF. Figure 10.96 depicts the electronic driver of the pump diode and the electronic monitoring ports. Figure 10.97 is the interface with a high current amplifier to drive the booster.

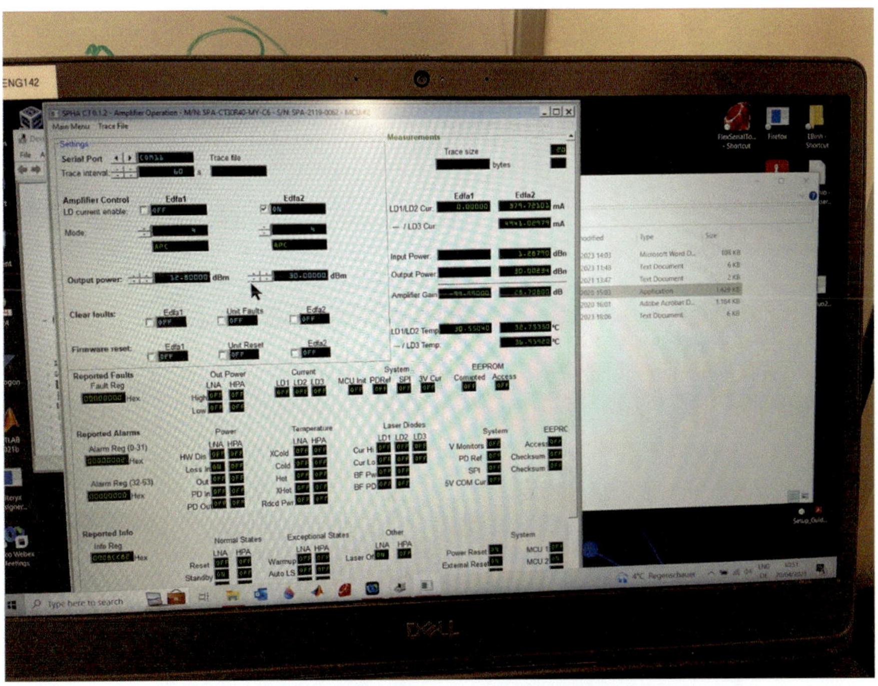

FIGURE 10.93 GUI screenshot of the optical amplifiers – both booster and preamplifier. Note the drive currents for preamplifier EDFA 1 and power amplifier EDFA with pump lasers LD1 and LD3 (~4.9 A driving current).

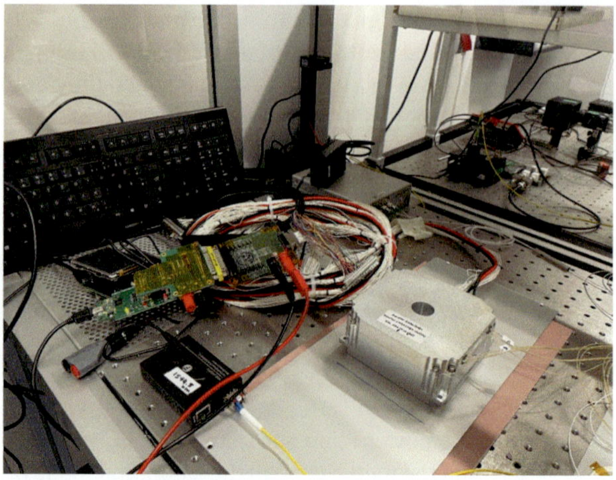

FIGURE 10.94 GUI of booster amplifier when preamplifier not excited.

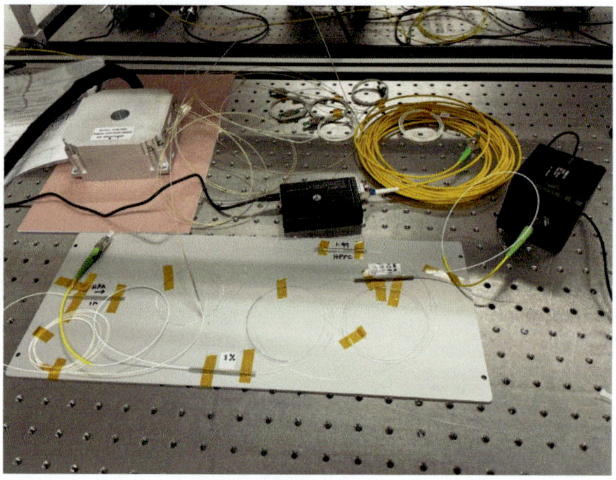

FIGURE 10.95 MPB optical amplifiers (HPA), both booster (HPA) and preamplifier (LNA).

FIGURE 10.96 Amplifiers and layout with –20 dB HPFC and beam dump (high power) integrating sphere.

(a)

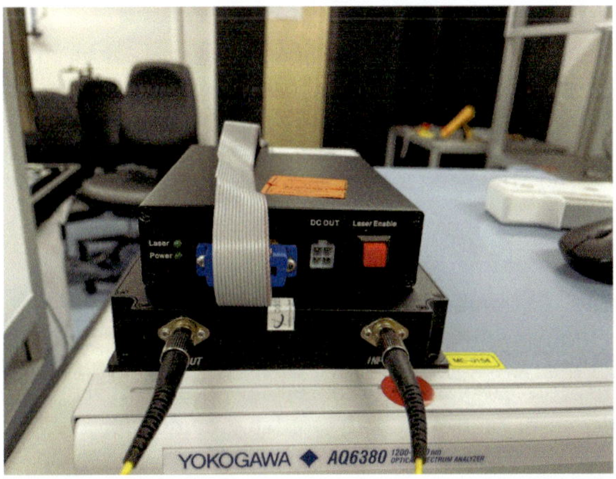

(b)

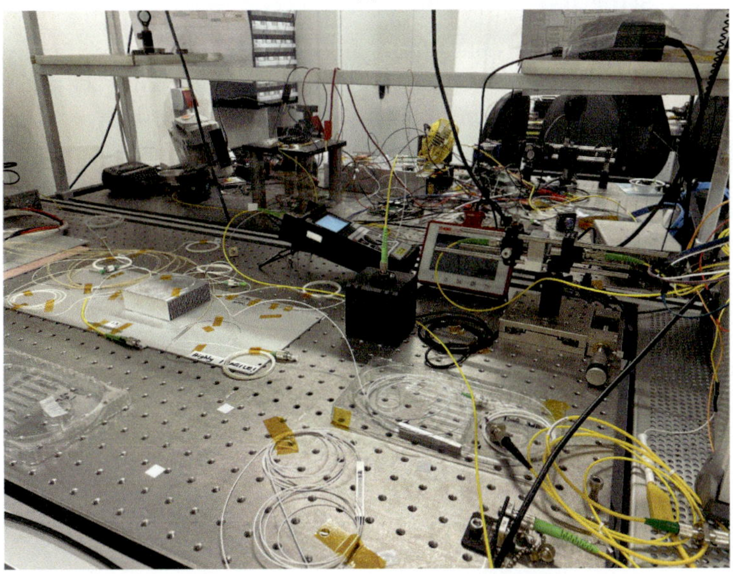

(c)

FIGURE 10.97 Power and interface to drive the amplifiers, HPA and LNA.

FIGURE 10.98 Optilab optical preamplifier (on top of Yokogawa optical spectrum analyser, in connection to the transmission test platform. (a) Overview of the test setup; (b) Optilab preamplifier; (c) MPB booster amplifier associated with both Optilab preamplifier and MPB low noise amplifier. (a) Optilab optical preamplifier (on top of Yokogawa Optical Spectrum Analyzer, in connection to the transmission test platform. (b) Optical preamplifier (Type 2), (c) MPB booster amplifier (Type 1) – ready for one-hop connection – but not yet tested in the complete One-Hop link.

10.9.2 Type 2 (Optilab) Preamplifier (LNA) in Platform

Figure 10.98 shows the optical preamplifier (on top of the Yokogawa Optical Spectrum Analyzer) in connection to the transmission test platform. (a) overview of the test setup; (b) Optical preamplifier; (c) MPB booster amplifier associated with both Optilab preamplifier and MPB low noise amplifier.

10.9.3 One-HOP Setup Test

Figure 10.99 is the photograph of an optically amplified equivalent fiber Free-Space hop setup. All the hardware of the hop has amplifiers of the booster and low noise amplifier, and the attenuators have the coupler, power absorber, and meter. Figure 10.100 shows the interface and driver for the Booster and optical preamp.

Figure 10.101 shows the optical section of the space equivalent HOP with associated fiber optics, optical amplifiers, boosters, and preamps.

Figure 10.102 is the graphic interface GUI MPB amplifier operations and drivers LD1,2,3. EDFA1 = LNA control and EDFA2 = Booster control GUI. The booster has bidirectional pumps of two pump lasers for high power output.

Figure 10.103 is the 'Snyder' software platform for executing 100 G module DP04QSFP, which produces the dual channel with QPSK Modulation format plus FEC to produce aggregate 100 Gbps.

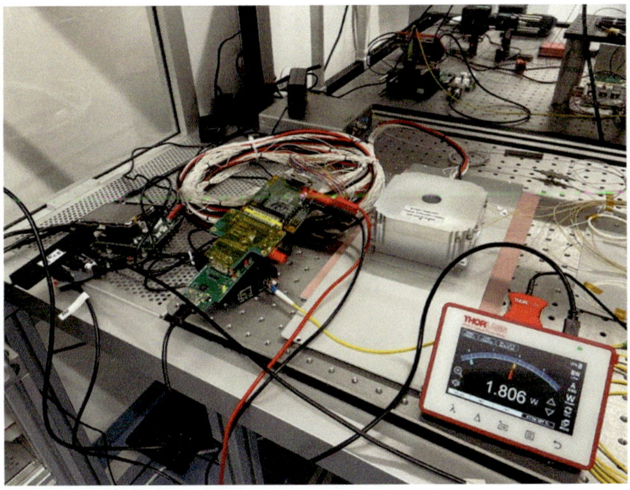

FIGURE 10.99 Optically amplified HOP equivalent fiber free-space hop.

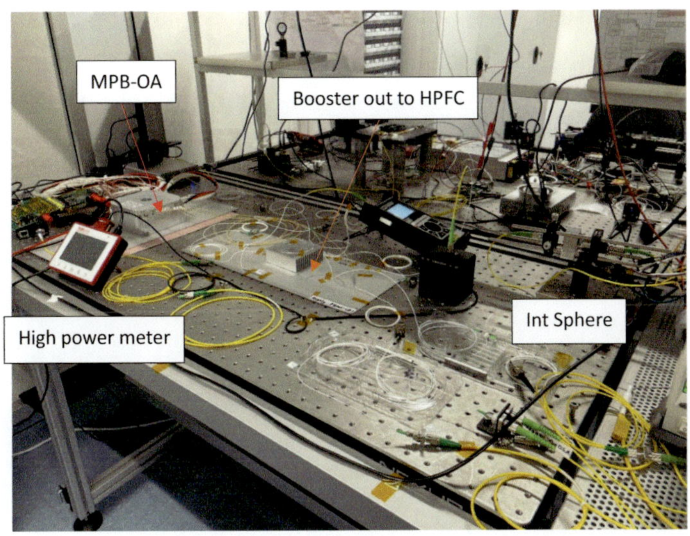

FIGURE 10.100 Interface and driver for driving Booster and optical preamp.

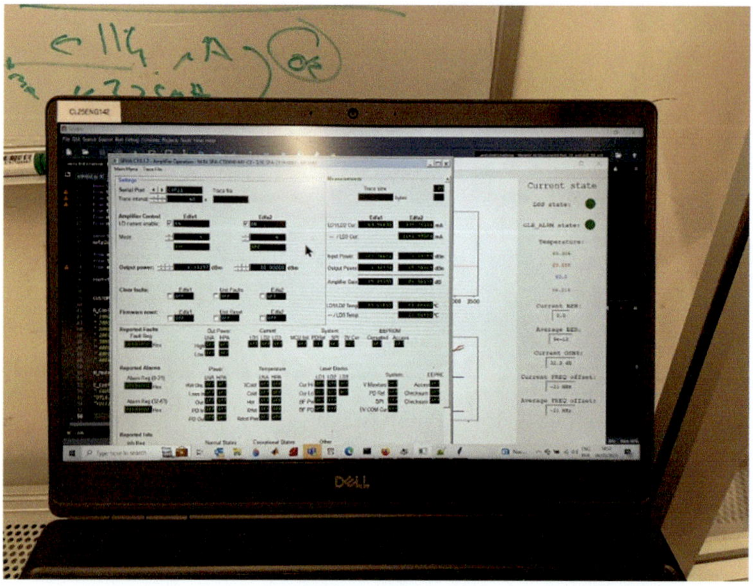

FIGURE 10.101 Optical section of HOP with associate fiber optics and optical amplifiers, booster and preamp.

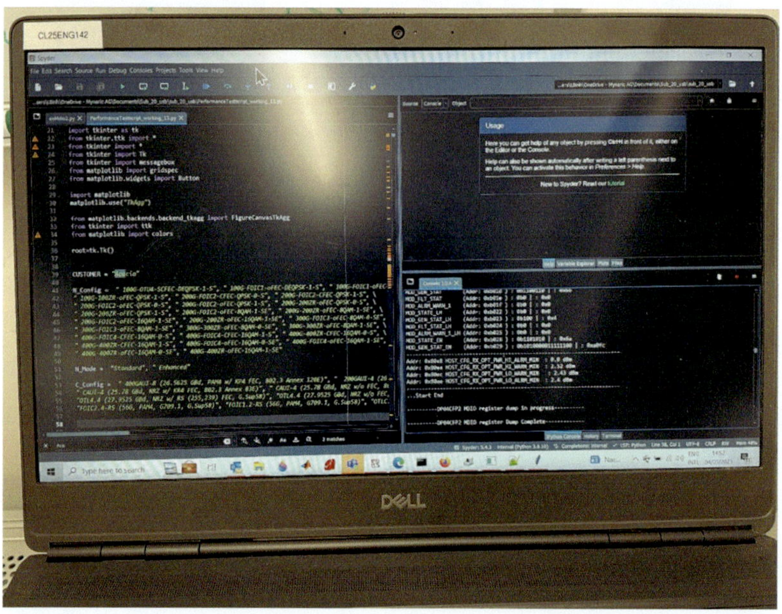

FIGURE 10.102 GUI MPB amplifier operations and drivers LD1,2,3. EDFA1=LNA control, EDFA2=Booster control GUI.

(a)

(b)

(c)

(d)

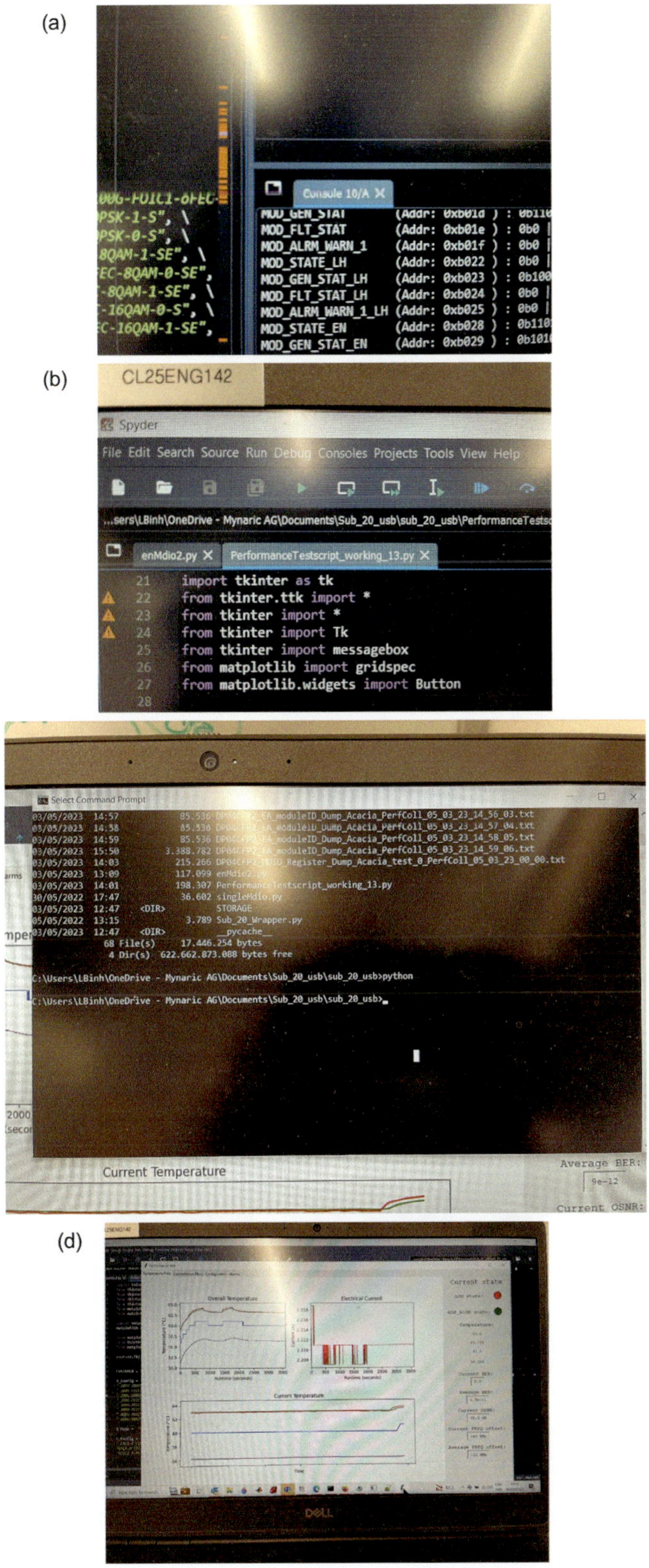

FIGURE 10.103 Software platform for Python execution for 100 G module DP04QSFP.

10.9.4 Transceiver Setup: 100 G-DPQPSK

Figure 10.104 shows details of the platform to run and activate the 100 G QPSK transceiver module (a) Python interface to activate the lasers of transceiver sections; (b) the program for execution of driver in Python and performance monitoring; (c) Folder to execute 100 G-DP-QPSK transceiver module; (d) performance screenshot. Figure 10.105 shows various optical components of the HOP platform.

Python platform to activate the OTRx and monitor performance parameters, OSNR (0.1 nm), Rx power, etc.

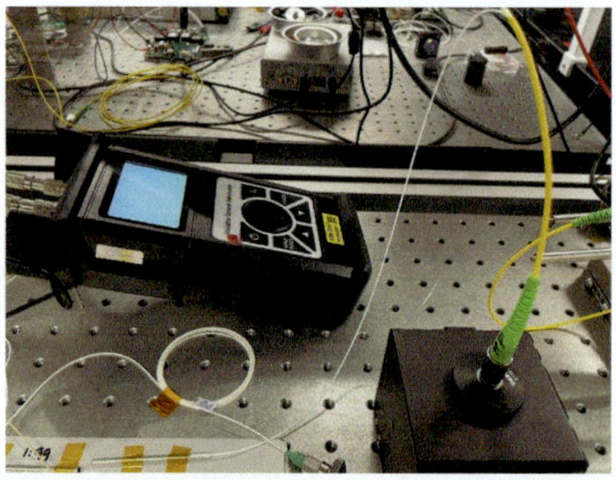

(a)

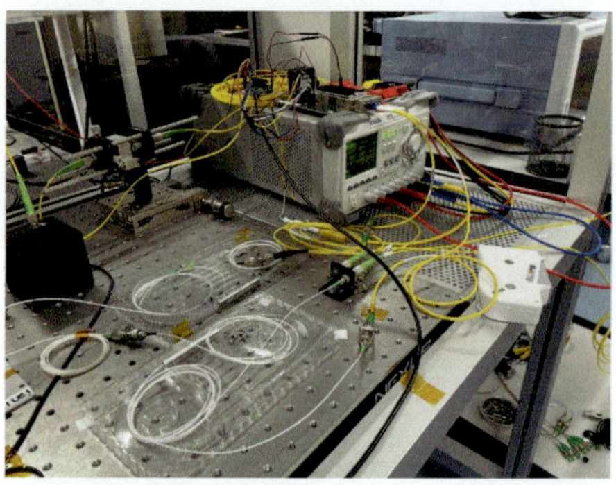

(b)

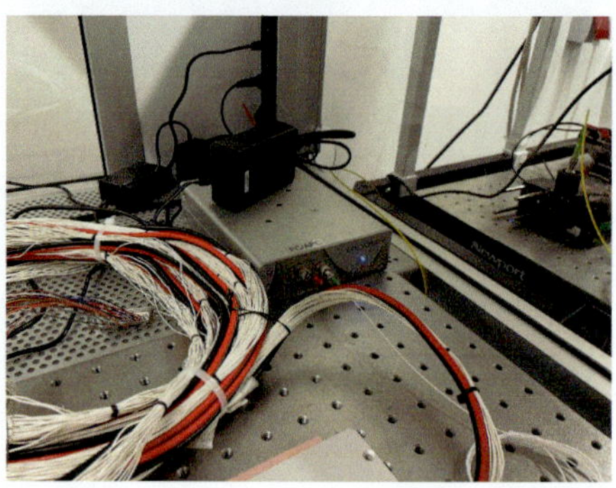

(c)

FIGURE 10.104 To run and activate 100 G QPSK transceiver module (c) folder to execute 100 G-DP-QPSK module; (d) performance screenshot; (b) the program for execution of driver in Python and performance monitoring.

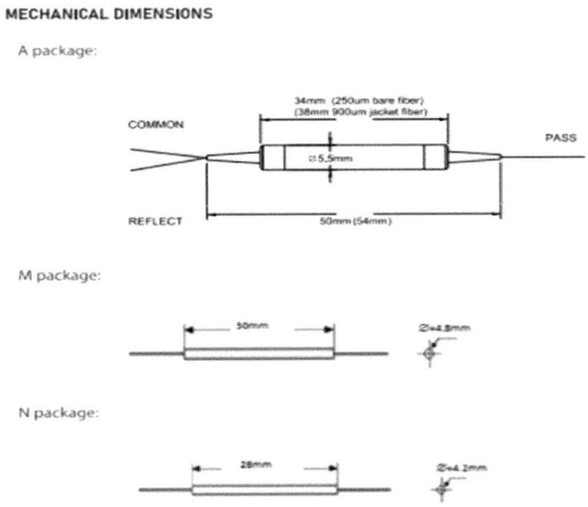

(a)

Parameter	Mux/Demux
Operating Wavelength	1530.33 ~ 1560.61 nm (21 ~ 59 ITU grid)
Minimum Channel Spacing	50GHz
Center Wavelength (CWL) Accuracy	±0.04nm
Channel Passband (@-0.5dB bandwidth)	≥ 0.14nm
Insertion Loss* (Add / Drop Ch.)	≤ 1.0dB
Insertion Loss* (Express Ch.)	≤ 0.60dB
Express Channel Isolation	≥ 20dB (Adjacent) ≥ 35dB (Non-adjacent)
Add/Drop Channel Isolation	12dB(Typ.) ≥ 10dB(Min.)
Add / Drop Channel Ripple	≤ 0.6dB
Insertion Loss Temperature Sensitivity	≤ 0.003dB/°C
Wavelength Temperature Shifting	≤ 0.001nm/°C
Polarization Dependent Loss	≤ 0.10dB
Polarization Mode Dispersion	≤ 0.10ps
Directivity	≥ 45dB
Return Loss	≥ 45dB
Optical Power	≤ 300mW
Operating Temperature	0 to +70°C
Storage Temperature	- 40 to +85°C
Package Dimensions (LxWxH) (mm)	A= Standard, Φ5.5xL34 (250um bare fiber) Φ5.5xL38 (900um jacket fiber) M=Compact, Φ4.8XL30 (250um bare fiber) N=Mini, Φ4.2xL28 (250um bare fiber) C=98x14x8.5 (2&3 mm fiber jacket) S=89x51x8.0 (2&3 mm fiber jacket)

(b)

FIGURE 10.105 Miscellaneous optical components of the HOP platform. (a) VoA (numerically control) and integrating sphere to dump the high-power beam. (b) Splitter in from TRx 100 G and splitter out to OSA. (c) Cabling for driving MPB amplifiers (booster and preamp) using GUI. The tunable laser is used as a seedling laser to saturate the booster amplifier, and thence, the optical biasing is used to saturate the booster amplifier to ensure stable operation.

10.10 APPENDIX 3: OPTICAL COMPONENTS

10.10.1 OF Add/Drop Filter

Note the 100 GHz bandwidth – the cut-off band is 100 GHz, the passband is 5–60 GHz. Figure 10.106 shows the technical specifications including mechanical dimension of the optical add/drop multiplexer.

10.10.2 High Power Fiber Coupler

Figure 10.107 shows the high-power fiber optic coupler, which handles the booster's high-power output. The fiber contributes to the nonlinearity effects on the channel signals.

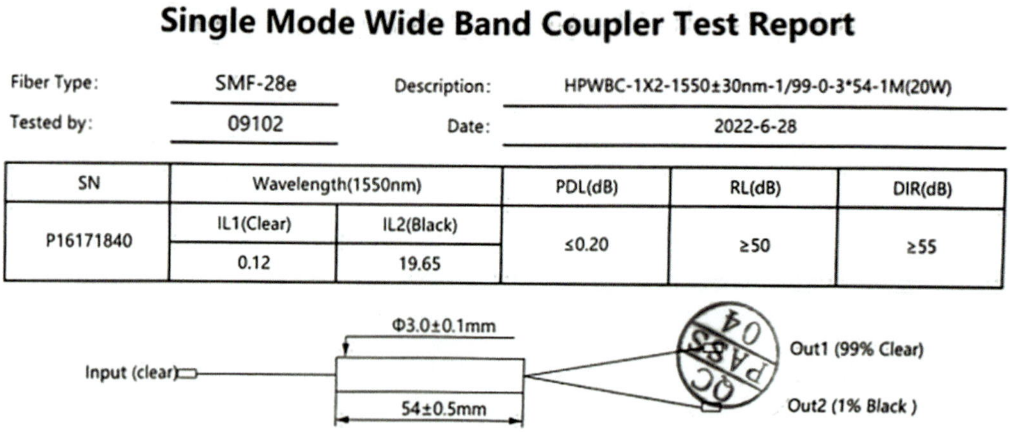

Single Mode Wide Band Coupler Test Report

Fiber Type: SMF-28e Description: HPWBC-1X2-1550±30nm-1/99-0-3*54-1M(20W)

Tested by: 09102 Date: 2022-6-28

SN	Wavelength(1550nm)		PDL(dB)	RL(dB)	DIR(dB)
P16171840	IL1(Clear)	IL2(Black)	≤0.20	≥50	≥55
	0.12	19.65			

FIGURE 10.106 Parameters of the optical add/drop multiplexer can be used as an optical band pass filter of sharp roll-off. (a) Mechanical dimension, (b) technical parameters.

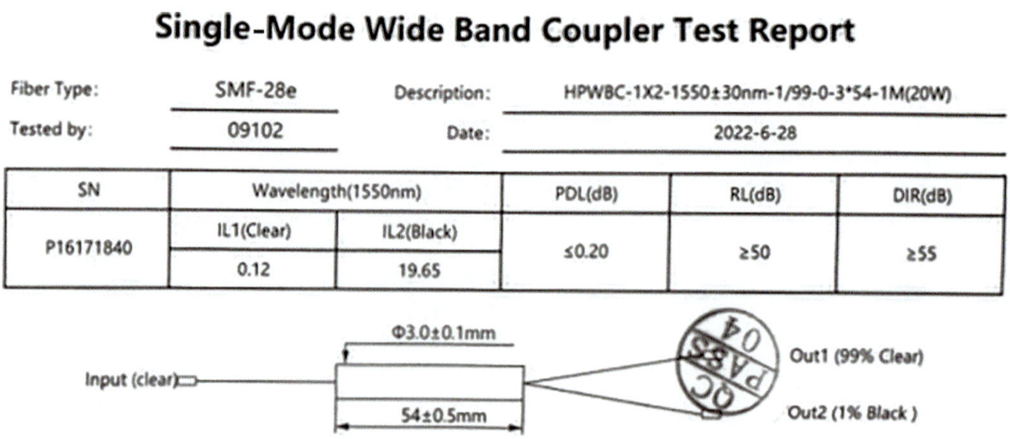

Single-Mode Wide Band Coupler Test Report

Fiber Type: SMF-28e Description: HPWBC-1X2-1550±30nm-1/99-0-3*54-1M(20W)

Tested by: 09102 Date: 2022-6-28

SN	Wavelength(1550nm)		PDL(dB)	RL(dB)	DIR(dB)
P16171840	IL1(Clear)	IL2(Black)	≤0.20	≥50	≥55
	0.12	19.65			

FIGURE 10.107 Structure and coupling coefficient of a high-power fiber coupler.

NOTES

1 100 Gbps = 25GBaud × 2 polarizations × 2 bits per symbol. So there are two polarized channels, each modulated with complex modulation QPSK or 2 bits per baud, 25 GBaud to give a total 100 Gbps. If forward error coding is used then the baud rate would be 28 GB with 7% overhead.
2 Refer to Section 12.5 of Chapter 12.
3 See Appendix 7 of Chapter 12 for grid spacing and channel number under ITU-T standard.
4 Keysight "modulation accuracy – EVM", https://www.keysight.com/us/en/lib/resources/user-manuals/modulation-accuracyevm-332561.html, access date May15, 2023.
5 Standard noise bandwidth is defined by 0.1 nm or 12.5 GHz at 1550 nm.
6 M. J. Shawon, R. Wang, and V. Saxena, "Design and modeling of silicon photonic ring-based linearized RF-to-optical modulator", In *Proceedings of IEEE 61st International Midwest Symposium on Circuits and Systems (MWSCAS),* Windsor, Ontario, Canada, 2018.
7 Model number of integrated transceiver module of dual polarization (DP) and modulation quadrature phase shift keying (QPSK) with 4 lanes of electrical lines. CFP2 is a Compact Factor Package type 2.
8 Optilab is a commercial name.
9 MPB is a commercial name.
10 A company product name
11 Commercial model number of high power optical amplifier
12 See Figure 10.9
13 See Section 12.5 of Chapter 12 , Appendix 7 for ITU-T grid and wavelength assignments
14 Note: coupler is wideband and transparent to the spectral region of interests for space photonic communications.
15 a one common port and 9 output ports (1 x 9) or reverse. If the WSS common port are inputted with multiplexed channels the output ports carry 9 separate channels if the central pass frequency of each port is matched to the central frequency of the channel. On the other hand if the WSS is operating at reverse direction, that is the 9 channels are inpu into the 9 ports then the common port C output a lightwave channel consisting of 9 multiplexed channels whose wavelengths are matched with the central wavelength set for the ports.
16 Lumentum Inc., "TrueFlex MicroTwin 1 × N WSS Interface Operation Manual (includes both 26 and 80 pin modules).
17 GSy./s = Giga-symbols per second
18 GBaud = GSy./s = Giga Symbols per second
19 https://www.lumentum.com/en
20 https://standards.ieee.org/ieee/802.11b/1166/
21 https://www.keysight.com/au/en/product/89601A/vector-signal-analysis-software-node-locked-license.html

11 Appendices 1 to 2 *Optical Modulators, Integrated Lasers*

CONTENTS

The chapter provides two appendices, 1 and 2, which describe the devices, fabrication, and operational principles of optical modulation. Appendix 2 gives a detailed account of lasers and integrated lasers, especially the autonomous locking of emitted lasing spectral lines using the external cavity of an MRR-DBR structure.

Optical modulation using integrated guided optical waveguides and electro-optic effects is critical for forming low RF-driven signals for high bit rate operation up to 100 Gbps with complex modulation.

A laser with a narrow linewidth is essential as an optical carrier to transport data over long-haul transmission in SPC inter-satellite constellation networks; the coherency of the polarized channels is particularly critical.

Appendices 1 and 2 present the technological developments of these devices over the last 50 years for fiber optic networks. They are well-suited as sources and modulators for SPC transmission and networking.

In Chapter 12, Appendices 3–7 are provided, focusing on fibers, integrated optical waveguides, and the nonlinear effects in these optical waveguiding structures. The chapter also covers optical routing technologies using MEMS and optical cross-connects, as well as thin-film resonators for multi-wavelength routing and the addition or dropping of channels in all-optical networking in space. ITU grids with 50 and 100 GHz channel spacing are discussed. Optical amplification devices using simulation are also included.

11.1 APPENDIX 1: OPTICAL MODULATORS

For SPC systems, preserving the laser's narrow linewidth is crucial to ensure minimal spatial broadening of the beam. Therefore, direct modulation of the laser is not preferred because modulating the current injected into the laser cavity can cause minute changes in the resonant cavity, leading to unstable frequency and increased laser phase noise.

DOI: 10.1201/9781003426677-11

To avoid phase noise and broadening of the laser linewidth, external modulators are required. This means the laser is biased at a certain constant current with constant output power and is run in continuous mode. The output lightwaves of the laser are then coupled to an optical modulator. The phase or intensity of the lightwaves can then be changed via electro-optic, magneto-optic, acousto-optic interaction with RF voltages, acoustic waves, or magnetic fields. Thus, modulation sequences of binary or multilevel amplitudes and phases can be applied to produce lightwave digital sequences for transmission.

This appendix provides some insights into electro-optic modulation for high-speed transmission. Such modulators have been proven to be the most effective for high-speed modulation of lightwaves. However, we first briefly review the bulk modulator type and then the integrated optic modulator, followed by the integration of several optical modulators on one chip to provide complex modulation and polarization multiplexing to increase transmission capacity.

11.1.1 Bulk Modulators

At the early stage of optical communications using free space as the transmission medium, optical modulation was implemented via bulk crystal structures and optical-electrical interactions such as the electro-optic (EO), acousto-optic (AO), and magneto-optic (MO) effects.[1]

The modulation of the phase and magnitude of lightwaves is contributed to by (i) the optical fields and electrical, acoustic, or magnetic fields, which change the refractive index of the crystal media. This, in turn, affects the phase of the lightwaves and can even switch or convert the polarized direction of the lightwaves.[2,3]

Crystals that exhibit EO, AO, or MO properties can be used as optical modulators as shown in Figure 11.1. The applied signal must be arranged so that there is either an electric field, acoustic waves, or a magnetic field, enabling changes in the refractive index in the region through which the light waves travel. This allows the optical phases of the light waves to be modulated by the index difference.

Once the laser can be integrated into optical structures, modulation of the optical light waves can be classified into two main techniques: direct modulation of the driving current of the laser by superimposing the RF signal on top of the biasing current of the laser, and external modulation by keeping the laser operating in continuous waves (CW) while coupling the light waves to an external modulator. In this method, the light waves are modulated, and the RF signals are applied to manipulate them. These modulation techniques for laser communication can be classified into three types, as shown in Figure 11.2.

Further to the integration of lasers and modulators, we can observe many integrated optical structures, such as multi-carrier optical generators, complex multi-level optical modulators integrated

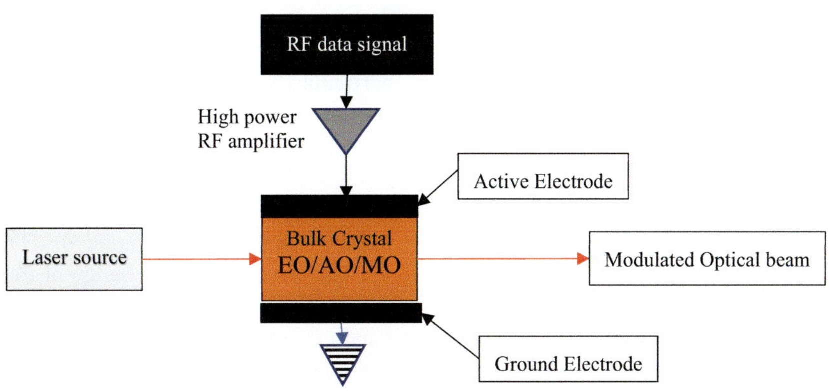

FIGURE 11.1 Bulk crystal optical modulator.

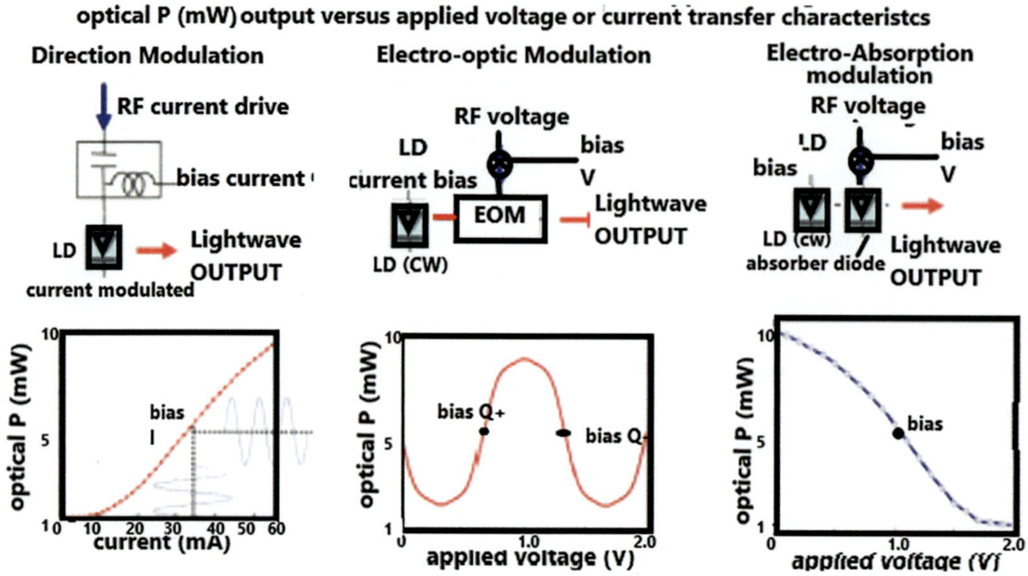

FIGURE 11.2 Optical modulation techniques (a) direct modulation, (b) electro-optic modulation – external, (c) electro-absorption (external) modulation. Structures on top and transfer curve and signaling on bottom.

with lasers, and lasers used for local oscillation to enable coherent reception. There are also multi-RF carrier generation systems for 6G and 7G optical networking. The transfer curve of the modulation RF signal versus the optical output shows that direct modulation suffers from high nonlinearity due to the exponential characteristics of power versus driving current. Electro-optic and electro-absorption transfer curves are more linear with minor nonlinearity. The operating frequency of integrated optical structures can be much higher than that of bulk modulators.

11.1.2 Integrated Modulator

As we can see in Figure 11.1, the bulk crystal needs to be thick to ensure stability, which allows for a larger electrode distance and hence a higher required signal strength. Therefore, a high-power RF amplifier is required at high frequencies. This puts a limit on the operating frequency or the pass-band of the amplifiers, and consequently, the RF band. This leads to the development of integrated optical modulators with structures based on planar designs, where the optical waveguide width is in the micrometer range. Thus, the RF voltage amplitude for EO modulation can be as small as a few volts or even less than one volt, enabling high-frequency operations. This section provides a brief description of such integrated optical modulators, particularly lithium niobate and silicon-based integrated modulator types.

11.1.2.1 $LiNbO_3$ Modulator

Lithium niobate materials are uniaxial crystals that along the c-axis of the crystal, the refractive indices as seen by an observer are two: the ordinary and the extraordinary components. This crystal thus has a significant electro-optic (EO) coefficient for the effective index change of the medium under an applied voltage, and hence an electric field. This field-induced EO index change can be sufficient for modulating the phase of the lightwave path enough to achieve a π phase shift with respect to the idle light path. Thus, the light waves at the combined output of an interferometer can be either destructive or constructive, corresponding to the ‘0’ or ‘1’ state of the light waves.

The schematic of an optical MZIM is shown in Figure 11.3a and b shows the hardware module of such an integrated modulator. The optical waveguides are created using micro-electronic

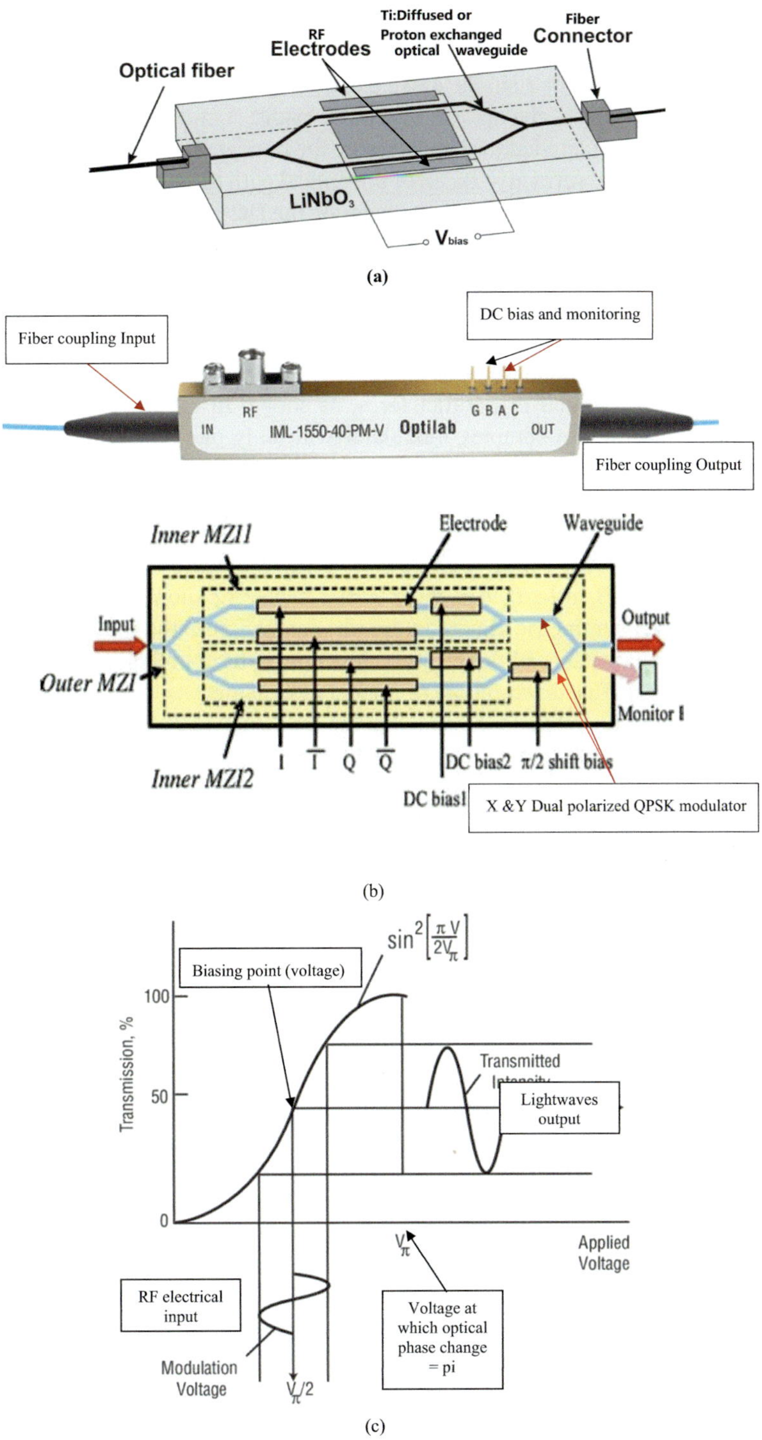

FIGURE 11.3 Optical $LiNbO_3$ modulator with optical fiber pigtail, RF input, and output ports with hidden T-coupler integrated in the module. (a) Schematic structure of an MZIM, (b) module of integrated single-polarized and complex MZIM with input RF port and optical ports (above), (below) is an integrated IQ complex optical modulator with polarization multiplexing. (c) Combined four MZIM for complex modulation with in-phase (*I*-) and quadrature (*Q*-) components; with dual-polarized light waves channels, total 4 MZIM structures.

fabrication steps on a $LiNbO_3$ substrate. The waveguides are formed by either Ti: diffusion or proton exchange processes. The input waveguide is split into two optical paths, and the electrodes are formed by gold deposition and plating, which are thick for both RF and DC voltage applied to the electrodes. Either a single drive or a single active electrode can be used. Similarly, dual electrodes can be formed on the two sides of the optical waveguide paths.

Thus, once a DC voltage is applied to the electrode, the optical phase shift is created via the EO interaction, which changes the refractive index of the optical path through which the light waves are passing. By superimposing RF traveling wave voltages on the DC bias, oscillating or data pulses can switch the light waves on and off accordingly, with the amplitudes varying based on the strength of the RF waves. This allows us to set the operating point of the light waves and achieve high-speed operation. Consequently, the modulation formats can be adjusted as desired.

Various MZIM types are used in this book, including single or double drives and different biasing points, to generate desired modulation formats.

The MZIMs can be integrated in parallel form, as shown in Figure 11.3c to create parallel *I*- and *Q*-components, forming a complex modulator with only one polarization. By placing this I-Q modulator in parallel with another modulator having an orthogonal polarization, we can form a two-polarization multiplexed IQ modulator. The insertion loss, including coupling from fiber to a planar waveguide, propagation loss, and splitting as well as coupling from planar waveguide to fiber output, is generally higher, about 11 dB for this type of modulator. For a single MZIM, the insertion loss is about 4.3 dB in total. Appendix 3 of Chapter 12 shows that the fiber mode spot size distribution follows a symmetric Gaussian distribution, whereas the mode field distribution is asymmetric. This is the main cause of the insertion loss due to the coupling between a circular fiber waveguide and the rib or diffused waveguide of optical modulators.

The RF electrodes must be long and thick so that the frequency response can provide a passband of about 45 GHz. Typically, one can find a 3 dB bandwidth of 35 GHz and a loss of about 4.3 dB.

The total length of the interferometric section of the MZIM is about 11.0 mm long. The thicknesses must be sufficient so that RF traveling waves can propagate and interact with lightwaves to offer wideband characteristics.

The splitting angle of the optical waveguides is about 0.7° of arc, due to the refractive index difference between the waveguide and that of the substrate being small, approximately 0.2. The phase evolution at this splitting or combining junction is also critical for such an interferometric structure. Thus, the total length of the MZIM is about 18–20 mm because there needs to be a small section of straight line for coupling the fiber to the diffused waveguide. The splitting angle is small because the refractive index change between the guided region and the substrate is about 0.2. Note that the refractive index difference between the fiber cladding and its core is only 0.3% for low loss and low dispersion factor.[4] The π phase shift in optical phase ensures that the bias quadrature point *Q*-component is $\pi/2$ phase shift with respect to the real in phase *I*-component.

The difference and mismatch in the mode spot sizes of the coupling fiber and the integrated waveguide lead to higher insertion loss of the modulator. Typical insertion loss for a single MZM is about 4–5 dB, and 8.0 dB for a dual-polarized modulator for complex modulation formats such as QPSK. If dual-polarized channels are desired, then the total insertion loss is further increased by another 3–4 dB. Therefore, commonly an optical amplifier such as an EDFA is used to compensate for this loss and boost the signal average power sufficiently for transmission.

The operation of the MZIM can be described as follows: (i) biasing the modulator by applying voltage to the electrode or electrodes for DD-MZIM so that the voltage difference equals to $V_{\pi,}$ which is the voltage at which the phase difference between the light waves in the two interferometric paths equals to π. The typical V_π of MZIM is about 3.5 V. The transfer curve of the MZIM follows a $\sin^2(x)$ to x with $x=(\pi/2)*(V/V_\pi)$. Thus, the transfer function is softly nonlinear (via the cosine transfer function of the electric field and the output optical field). Therefore, care must be taken into account when estimating the amplitude of the applied RF waves in order to avoid such nonlinear distortion.

11.1.2.2 External Integrated Modulator

The modulation of light waves via an external optical modulator can be classified into three types, depending on the specific effects that alter the light wave's properties, especially the intensity or phase of the light wave carrier. In an external modulator the intensity is normally manipulated by alerting the phase of the carrier light waves guided in one path of an interferometer. The Mach-Zehnder interferometric structure is the most common type, particularly the lithium niobate type.[5,6,7,8]

An EA modulator employs the Franz-Keldysh effect[9], which is observed as a lengthening of the wavelength at the absorption edge of a semiconductor medium under the influence of an electric field.[10,11,12] In quantum structures, such as multi-quantum well structures, this effect is called the Stark effect, or the electro-absorption (EA) effect. The EA modulator can be integrated with a laser structure on the same integrated circuit chip. For a $LiNbO_3$ modulator, the device is externally connected to a laser source via an optical fiber.

The total insertion loss of a semiconductor intensity modulator is about 8–10 dB including fiber-waveguide coupling loss which is rather high. However, this loss can be compensated by a semiconductor optical amplifier (SOA) that can be integrated on the same circuit. In comparison, the total insertion loss of a LiNbO3 modulator is about 3–4 dB which is more manageable, especially since Erfiber amplifiers (EDFAs) are now readily available.

The driving voltage for an EA modulator is usually lower than that required for $LiNbO_3$. However, the extinction ratio is not as high as that of the $LiNbO_3$ type which is about 25 dB compared to 10 dB for the EA modulator. This feature contrasts the operating characteristics of the $LiNbO_3$ and EA modulators. Although the driving voltage for an EA modulator is about 2–3 V and 5–7 V for $LiNbO_3$, the former type would be preferred for intensity or phase modulation formats due to its extinction ratio that offers a much lower "zero" noise level and hence a high-quality factor.

11.1.2.2.1 Optical Phase Modulators

The phase modulator is a device that manipulates the "phase" of optical carrier signals under the influence of an electric field created by an applied voltage. When no voltage is applied to the RF electrode, the number of periods of the light waves, n, exists in a certain path length. When voltage is applied to the RF electrode, one or a fraction of one period of the wave is added, which now means $(n + 1)$ waves exist in the same length. In this case, the phase has been changed by 2π and the half voltage of this is called the driving voltage. In the case of long-distance optical transmission, the waveform is susceptible to degradation due to non-linear effects such as self-phase modulation etc. A phase modulator can be used to alter the phase of the carrier to compensate for this degradation. The magnitude of the change in phase depends on the change in the refractive index created via the electro-optic effect that in turn depends on the orientation of the crystal axis with respect to the direction of the electric field established by the applied signal voltage.

An integrated optic phase modulator operates in a similar manner except that the lightwave carrier is guided only through an optical waveguide, which consists of diffused or ion-exchanged confined regions for $LiNbO_3$ and rib-waveguide structures for the semiconductor type. Two electrodes are deposited so that an electric field can be established across the waveguiding cross-section in a direction that maximizes the refractive index change through electro-optic interaction or Electro-Absorption effect, as shown in Figure 11.4. For ultra-fast operation, one of the electrodes is a traveling wave type or hot electrode, and the other is a ground electrode. The traveling wave electrode must be terminated with a matching impedance at the end to avoid RF wave reflection. Usually, a quarter wavelength impedance is used to match the impedance of the traveling wave electrode to that of the 50 Ω transmission line.

A phasor representation of a phase-modulated lightwave can be described by circular rotation at a radial speed of ω_c. Thus, the vector with an angle ϕ represents the magnitude and phase of the lightwave.

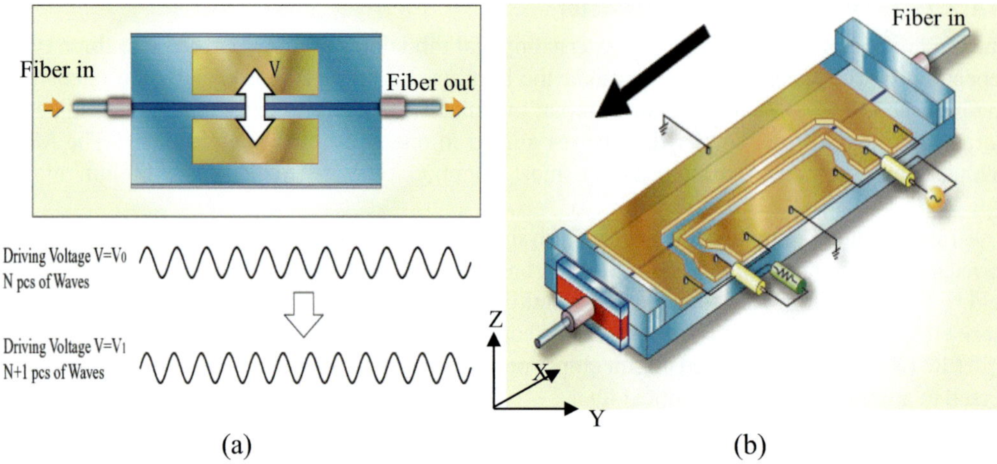

FIGURE 11.4 Electro-optic phase modulation in an integrated modulator using $LiNbO_3$. Electrode impedance matching is not shown (a) schematic diagram, (b) integrated optic structure.

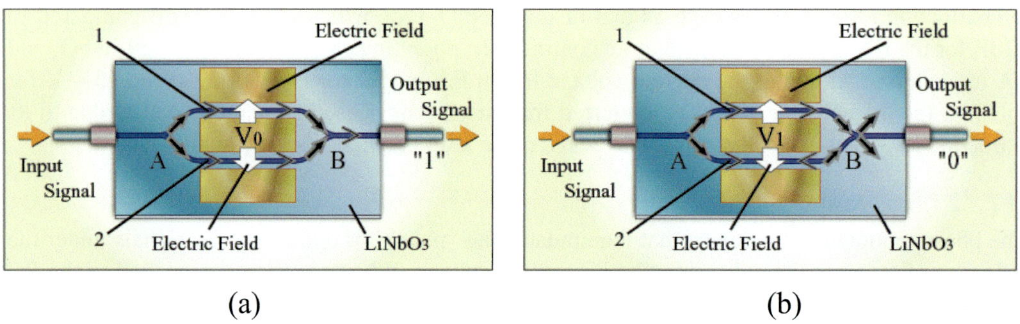

FIGURE 11.5 Intensity modulation using interferometric phenomena in guide wave structures in $LiNbO_3$ (a) ON – constructive interference mode, (b) destructive interference mode – OFF. Optical guided wave paths 1 and 2. An electric field is established across the optical waveguide.

11.1.2.2.2 Intensity Modulators

The modulation of the intensity of lightwaves is based on turning the lightwaves on and off. This is done through interferometric action or by using an interferometric structure with two optical paths. One of the paths is modulated by an RF wave signal to generate a phase change in the lightwave. If this phase change equals π radians, the combined output is destructive, resulting in the "OFF" state. On the other hand, if the phase change is zero '0' or 2π radians, constructive interference occurs, resulting in the "ON" state of the combined output lightwaves (Figure 11.5).

The basic structured LN interferometric modulator comprises (i) two waveguides, (ii) two Y-junctions, (iii) DC (bias) and RF traveling wave electrodes, (iv) impedance matching for RF waves, (v) optical tapping for feedback control of the stability of the biasing or operating point, and finally a thin film deposited on top of the waveguide area for placing the electrode so that the traveling velocity of the lightwaves and RF waves can be matched, thus achieving higher bandwidth.

Optical signals coming from the lightwave source and/or its pigtail are launched into the LN modulator through a polarization-maintaining fiber. They are then equally split into two branches at the first Y-junction on the substrate. When no voltage is applied to the RF electrodes, the two signals are re-combined constructively at the second Y-junction and coupled into a single output. In this case, the output signals from the LN modulator are recognized as "ONE". When voltage is applied to the RF electrodes, due to the electro-optic effects of the Lithium Niobate ($LiNbO_3$) crystal substrate, the waveguide refractive index changes due to the stress and strain of different molecular structures when titanium atoms are diffused into the lithium niobate crystal structure. Consequently, the carrier phase in one arm is advanced and similarly retarded in the other arm. Therefore, when the two signals are re-combined destructively at the second Y-junction, the lightwaves are combined destructively and transformed into higher-order modes, which are radiated at the output of the junction. If the phase retardation is an odd multiple of π, the two signals are completely out of phase, the combined signals are radiated into the substrate, and the output signal from the LN modulator is recognized as a "ZERO". The voltage difference that induces this "ZERO" and "ONE" is called the driving voltage of the modulator and is one of the important parameters in determining the modulator's performance.

11.1.2.2.3 Phasor Representation and Transfer Characteristics

Consider an interferometric intensity modulator that consists of an input waveguide, which is then split into two branches and recombined into a single output waveguide. If the two electrodes are initially biased with voltages V_{b1} and V_{b2} then the initial phases exerted on the lightwaves would be $\phi_1 = \pi V b_1 / V_\pi = \phi_2$ which are indicated by the bias vectors shown in Figure 11.6b. From these positions, the phasors swing according to the magnitude and sign of the pulse voltages applied to the electrodes. They can be switched to two positions that can be either constructive or destructive. The output field of the lightwave carrier can be represented by:

$$E_0 = \frac{1}{2} E_{\text{iRMS}} e^{j\omega_c t} \left(e^{j\phi_1(t)} + e^{j\phi_2(t)} \right) \tag{11.1}$$

where ω_c is the carrier radial frequency, E_{iRMS} is the root mean square (RMS) value of the magnitude of the carrier and $\phi_1(t)$ and $\phi_2(t)$ are the temporal phases generated by the two time-dependent pulse sequences applied to the two electrodes. As the voltage levels vary according to the magnitude of the pulse sequence, one can obtain the transfer curve as shown in Figure 11.6a. This phasor representation can be used to determine the exact biasing conditions and magnitude of the RF or digital signals required to drive the optical modulators to achieve 50%, 33% or 67% of the bit period, pulse sequence, and shapes. The shapes would follow a $(\cos)^2$ profile due to the transfer function, which is the optical power output versus the applied voltage, see Figure 11.6a.

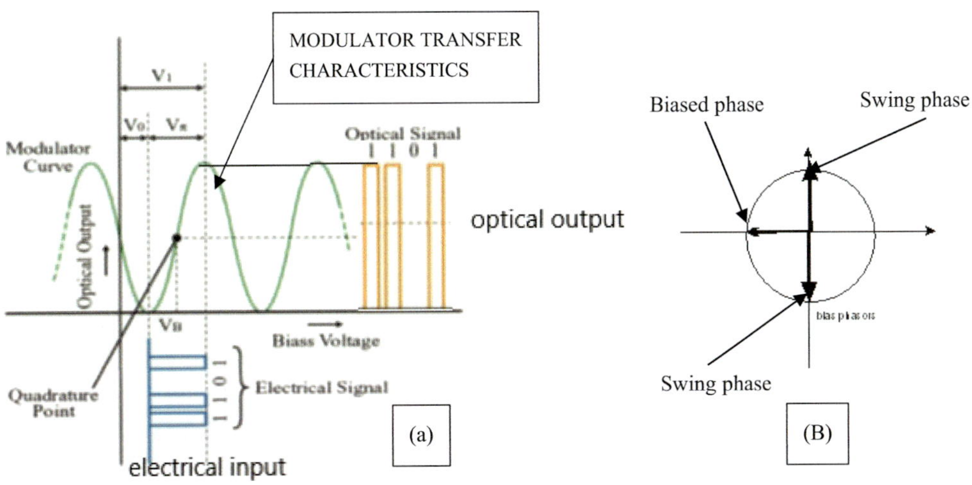

FIGURE 11.6 Electrical to optical transfer curve of an interferometric intensity modulator (a) optical power versus applied voltage–transfer characteristics, (b) phasor representations of biased state and the swinging phase states due to signal amplitude applied to the electrode to create phase shifts.

The power transfer function of a Mach-Zehnder modulator can be expressed as[13]

$$P_0(t) = \alpha P_i \cos^2 \frac{\pi V(t)}{V_\pi} \tag{11.2}$$

where $P_0(t)$ is the transmitted output power, α is the modulator's total insertion loss in the linear scale, P_i is the input power (usually from the laser diode coupled via a fiber), $V(t)$ is the time-dependent signal applied voltage, and V_π is the driving/bias voltage required to create and exert a π phase shift on the lightwave carrier, thus affecting the output lightwave.

It is necessary to set the static bias or operating point on the transmission curve through the bias electrode, via a T-bias. It is common practice to set the bias point at the 50% transmission point "a" here or a $\pi/2$ phase difference between the two optical waveguide branches, known as the quadrature bias point. As shown in Figure 11.6, electrical digital signals are transformed into optical digital signals by switching the voltage at both ends of the quadrature points, both positive and negative. Alternatively, one can use the minimum transmission point (or the π phase bias point) or the maximum transmission point (the zero-phase bias difference point). Figure 11.6a shows the quadrature bias point and the swinging of the signal sequence.

11.1.2.2.4 Bias Control

One factor that affects the modulator's performance is the drift of the bias voltage. For the Mach-Zehnder Interferometric Modulator (MZIM) type, this drift is particularly critical when the modulator is biased at the operating point, whether at the quadrature point or at the minimum or maximum points on the transfer curve.

The DC bias drift can occur in a modulator due to the deposition of static charges on the substrate surface. This drift phenomenon often occurs in $LiNbO_3$ due to the buildup of charges on the electrode surface, leading to a large field drop across the waveguide and changes in the DC charge affecting the refractive index of the waveguide. This results in a phase change, causing the interference of the light waves from the two branches to occur at different applied voltage levels. Under this drift, the transmission curve gradually shifts over the long term.[14,15] In the case of the $LiNbO_3$ modulator, controlling the bias point is vital as it will shift over time. To compensate for the drift, it is necessary to monitor the output signals and feed them back into the bias control circuits to adjust the DC voltage, so that the operating points remain consistent, as shown in Figure 11.7, for example, the quadrature point. It is the manufacturer's responsibility to minimize DC drift so that the DC voltage remains within limits throughout the lifetime of the device.

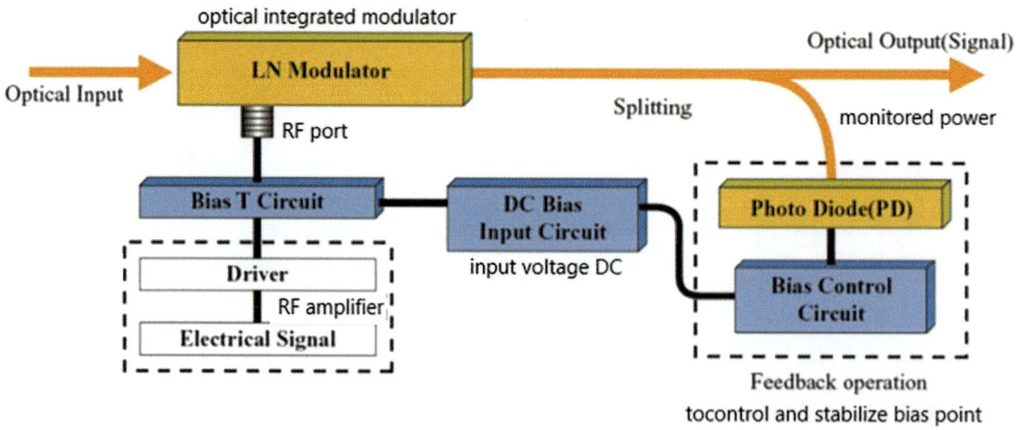

FIGURE 11.7 Arrangement of bias control of integrated optical modulators.

11.1.2.2.5 Chirp-Free Optical Modulators

Due to the symmetry of the crystal refractive index in the uniaxial anisotropy of the class m of $LiNbO_3$, the crystal cut and the propagation direction of the electric field affect both modulator efficiency, denoted as driving voltage and modulator chirp. The uniaxial property of $LiNbO_3$ is shown in Figure 11.8. As shown in Figure 11.9, in the case of the Z-cut structure, the 'hot' electrode

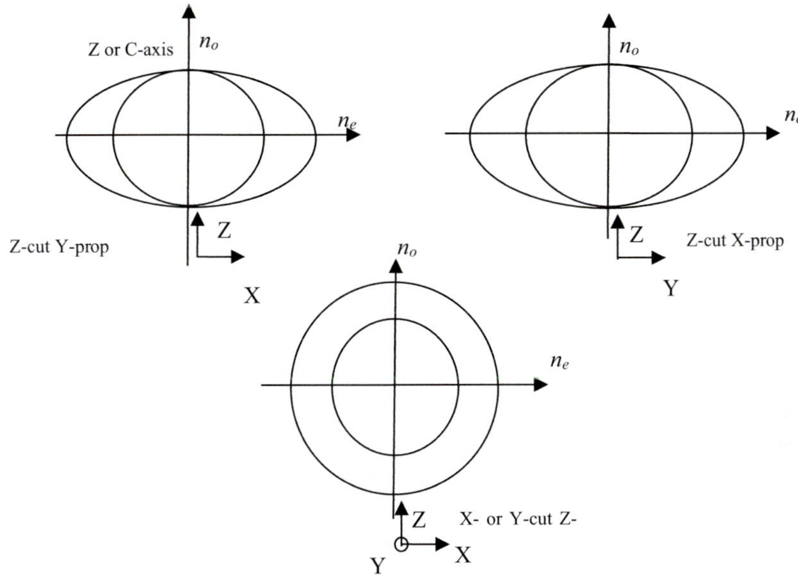

FIGURE 11.8 Refractive index contours of $LiNbO_3$ uniaxial crystal with Z- or C- denoting as the principal axis (a) lightwaves propagation in the Z-axis polarized along Y-cut $LiNbO_3$, (b) prop- direction Z-axis and X-cut crystal, (c) Y-prop and Y-cut crystal. Orientation of the C-axis to optimize the use of electro-optic coefficients of the crystal.

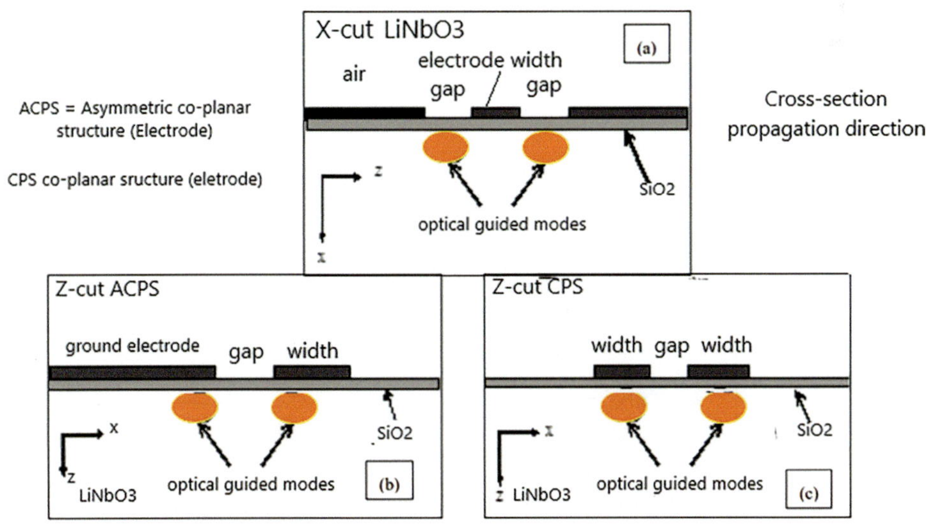

FIGURE 11.9 Commonly used electrode structure and crystal orientation for interferometric modulation to maximize the use of the overlap integral between the optical guided mode and the electric field distribution for largest electro-optic coefficients (a) X-cut crystal orientation; (b) Z-cut $LiNbO_3$ crystal structure and electric field in vertical; and (c) Z-cut and electric field in the vertical direction. w=RF waveguide width, g=gap, CPW=co-planar waveguide for RF traveling wave.

is placed on top of the waveguide. The RF field flux is more concentrated when passing through the waveguide cross-section with the electric field pointing downwards and the optical field parallel to the waveguide's x-direction. This results in maximizing the overlap between RF and optical fields. However, the overlap between RF in the ground electrode and the waveguide is reduced in the Z-cut structure, so the overall improvement of driving voltage for the Z-cut structure, compared to the X-cut, is approximately 20% higher.

The different overlapping area for the Z-cut structure results in a chirp parameter of 0.7, whereas in X-cut and Z-propagation, the chirp is almost zero due to its symmetric structure. Chirping refer to the variation in the frequency or phase of the light waves with the signal voltage level, leading to dispersion of the modulated light waves and, consequently, distortion of optical output signals. A number of commonly arranged electrodes and waveguide structures are shown in Figure 11.9 to maximize the interaction between the traveling electric field and the optical-guided waves. Furthermore, a buffer layer, normally SiO_2, is used to match the velocities between these waves to optimize the optical modulation bandwidth.

11.1.2.2.6 Structures of Photonic Modulators

Figure 11.10a and b show the 3D (three-dimensional) structure of an MZ intensity modulator using single and dual electrode configurations, respectively. The structure with dual electrodes indicates

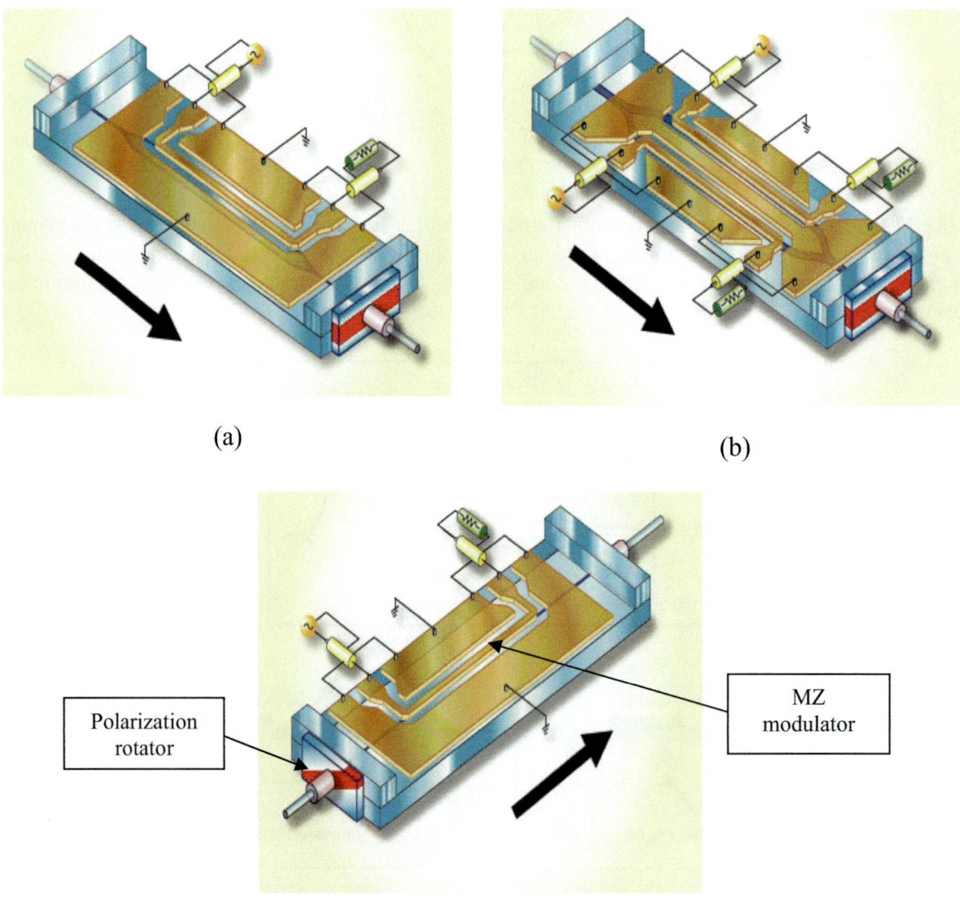

FIGURE 11.10 Intensity modulators using $LiNbO_3$ (a) single drive electrode, (b) dual electrode structure, (c) electro-optic polarization scrambler using $LiNbO_3$.

TABLE 11.1
Typical Operational Parameters of Optical Intensity Modulators

Parameters	Typical Values	Definition/Comments
Modulation speed	10 Gb/s	Capability to transmit digital signals
Insertion loss	Max 5 dB	Defined as the optical power loss within the modulator
Driving voltage	Max 4 V	The RF voltage required to have a full modulation
Optical bandwidth	Min 8 GHz	3 dB roll-off in efficiency at the highest frequency in the modulated signal spectrum
ON/OFF extinction ratio	Min 20 dB	The ratio of maximum optical power (ON) and minimum optical power (OFF)
Polarization extinction ratio	Min 20 dB	The ratio of two polarization states (TM and TE guided modes) at the output

that there are two active electrodes and one common ground electrode. The thin line electrode is called the "hot" electrode or traveling wave electrode. RF connectors are required for launching the RF data signals to establish the electric field necessary for electro-optic effects. Impedance termination is also required. Optical fiber pigtails are attached to the end faces of the diffused waveguide. The mode spot size of the diffused waveguide is not symmetric; hence, some diffusion parameters are controlled to maximize the coupling between the fiber and the diffused or rib waveguide. Due to this mismatch between the mode spot sizes of the circular and diffused optical waveguides, coupling loss occurs. Furthermore, the difference between the refractive indices of the fiber and $LiNbO_3$ is quite substantial, leading to Fresnel reflection loss as well.

Figure 11.10c shows the structure of a polarization modulator, which is essential for multiplexing two polarized data sequences to double the transmission capacity, e.g., from 40 to 80 Gbps. Furthermore, this type of polarization modulator can be used as a polarization rotator in a polarization dispersion compensating subsystem.[16,17] Currently, these types of modulators with IQ modulation have been reported.[18]

It is noted that such operating speed can only be implemented in integrated planar structures and not in bulk modulator structures, due to the separation distance between the electrodes (Table 11.1).

11.1.2.2.7 Typical Operational Parameters of External Modulators

11.1.3 Si-Based Optical Modulator

Optical modulators formed in silicon are the keystone to many low-cost optical applications. Increasing the data rate of the modulator benefits the efficiency of channel usage and decreases power consumption per bit of data, especially in data center environments. SOI-based optical modulators are considered the most important and economical devices for future optical networks, especially in access networking. The main driving force is the CMOS-compatible process used in the production of such modulators in integration with electronic integrated circuits.

11.1.3.1 SOI Optical Phase Modulator

A typical phase modulator based on SOI (silicon on oxide) integrated photonic technology is shown in Figure 11.11. The device is formed in rib waveguides with a height of 220 nm and a width of 400 nm. A 100 nm slab height was used as it provides lower access resistance to the junction and lower sidewall roughness interaction compared to a 50 nm slab height. The rib section of the device and the slab on one side are p-type doped, while the slab on the other side of the rib is n-type doped. The target doping concentration of the *p*-type region (3×10^{17} cm^{-3}) is lower than that of the *n*-type region (1.5×10^{18} cm^{-3}) so that the depletion extends mainly into the rib region during reverse bias conditions. It is noted that the width of the rib is about one-third of the guided lightwave wavelength.

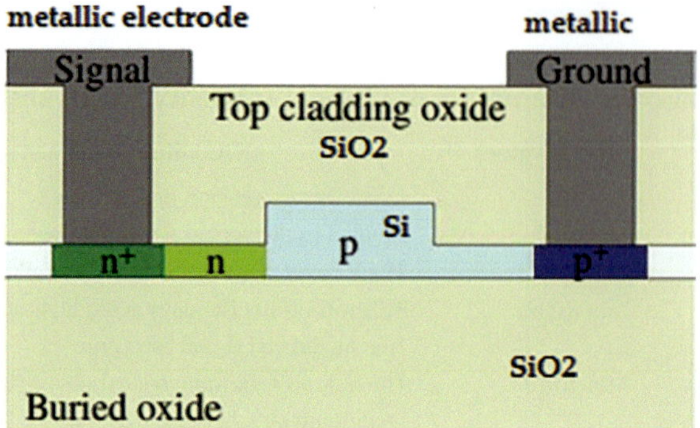

FIGURE 11.11 Cross-sectional view of a Si-On-Insulator (SOI) optical phase modulator.

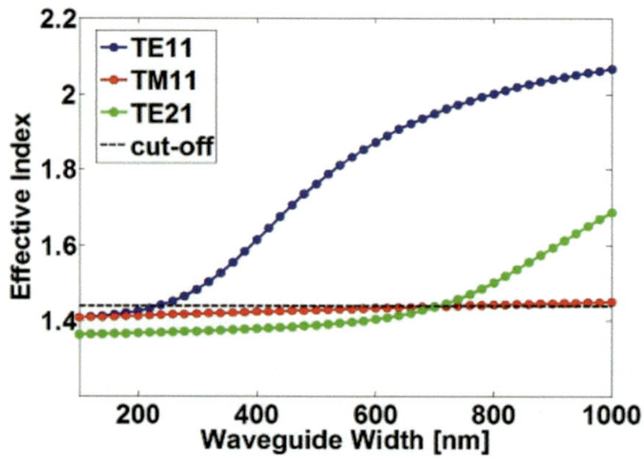

FIGURE 11.12 The effective index of the guided modes TE (TE_{11} and TE_{21}) and TM (TM_{11}) in the SOI ridge waveguide as a function of the waveguide width. The effective refractive index n_{eff} plotted against the waveguide width of various waveguide modes at 1550 nm wavelength, guided modes TE_{11}, TM_{11}, and TE_{21}, by simulation. TE_{21} cut off at about 700 nm or single mode for width 120–700 nm width.

This is reasonable since the wavelength is equal to the free-space wavelength divided by the effective refractive index of the guided wave in silicon, which is about 4.4.

This configuration also allows for reduced optical loss. The *p*-type and *n*-type regions contact highly doped *p*-type and *n*-type regions, respectively, which in turn connect to coplanar waveguide (CPW) electrodes used to drive the device at high speed. The phase modulators are incorporated into asymmetric MZI structures with an arm length mismatch of 180 μm to convert between phase and intensity modulation. A key advantage of the device design is the simplicity of the fabrication process required for its formation.

The effective index of the Si on SOI ridge waveguide for TE and TM modes can be estimated as shown in Figure 11.12. Figure 11.1319 shows the guided mode field distribution with the height of the waveguide as a parameter. This indicates that the mode still holds the single-mode conditions but with different group velocities, and hence the dispersion of the lightwave channel would vary as shown in Figure 11.14. Dispersion is important if one wants to use the waveguide as a wideband phase matching for nonlinear interactions in four-wave mixing for phase conjugation.

Figure 11.12 plots the effective index of the guided mode against the waveguide width. This dispersion curve can be used to design and fabricate the waveguide structure so that the desired group delay and dispersion can be optimized, including the fabrication tolerance. TE_{11} and TM_{11} modes are desired, and the thickness and width of the waveguides must be above the cut-off point so that the second-order mode TE_{21} does not appear, or the waveguide width must be <700 nm.

The effective index is the index observed along the propagation direction of the guided lightwaves, derived from the eigenvalue of the wave equation.

Due to the high refractive index difference between the Si waveguide and the substrate SiO_2, the guided mode is well confined, as shown in Figure 11.13. Figure 11.14 then shows the theoretical and experimental relationship between the group index (and hence the dispersion factor) and the operating wavelength. By differentiating the group index with respect to the wavelength, one can find the dispersion factor, which is critical for high-speed operation in the tera-bps rate.

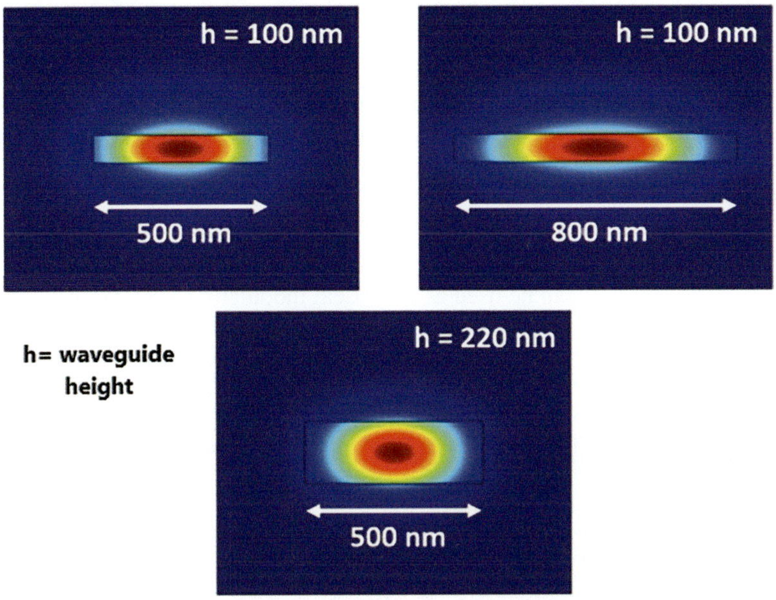

FIGURE 11.13 Structure and near field intensity image of guided modes Different thickness *h* of waveguide 100–220 nm. Note the nanometer scale.

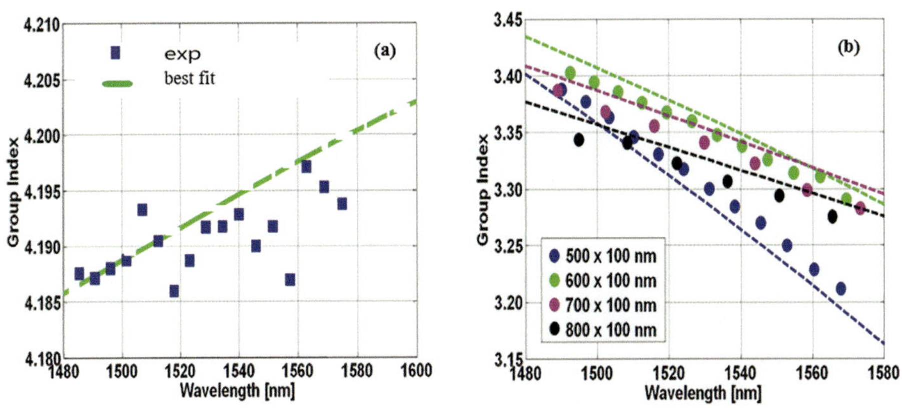

FIGURE 11.14 (a) Experimentally derived (light squares) group index values for standard $500 \times 220\,nm^2$ Si waveguides, and values predicted from simulations (light continuous lines). (b) Experimental (dots) group index values for 100 nm height and 500–800 nm width waveguides, and values predicted from simulations (dashed lines). Note that the y-axis scale is ten times smaller in (a) with respect to (b). Note that high refractive index leads to good confinement of guided modes.

11.1.3.2 Si MZI Modulator

Figure 11.15 shows the plane-view structure of an MZI modulator in which the incoming lightwaves are split into two branches via a multi-mode interference (MMI) coupler. The MMI region ensures that the lightwaves can be focused into the waveguide over which the electric field can interact. The phases of the guided lightwaves are modulated by applying signal voltages to the CPW electrodes. A lateral PN junction is embedded into a 450 nm-wide waveguide of an MZI modulator using the standard 180 nm CMOS process. The P-N junction interface is accurately offset by 100±20 nm from the waveguide center by improving the locating accuracy of the doping masks. Both the P- and N-doped regions have a concentration of $2\times10^{17}\,cm^{-3}$. The fabricated MZI modulator with the ground-signal-ground (GSG) traveling electrodes can be observed in Figure 11.15a. The optical transmission power of the MZI is shown in Figure 11.15b under different biasing voltages. The 56 Gbit/s NRZ-PRBS sequence is applied to the traveling wave electrode of the MZI modulator; the observed eye diagram is shown in Figure 11.16 with a BER (bit-error rate) of 1e-7. Note the noise level of this eye diagram. After transmission through 2 km, the BER is about 1e-3 that can be error-free under FEC.

11.1.3.3 Micro-Ring Modulator (MRM)

A micro-ring resonator (MRR) can be used as an optical modulator (MRRM). The MRRM, in an integrated waveguide structure, offers much lower modulation voltage levels and very short-length electrodes, as shown in Figure 11.17, due to the operating region when the ring is at the resonant

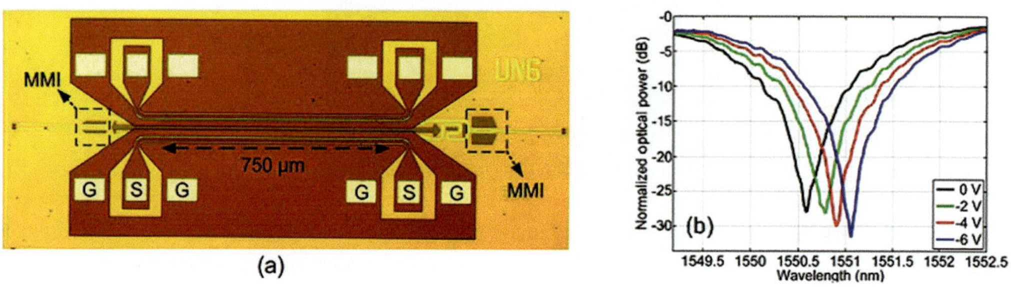

FIGURE 11.15 MZI modulator (a) Top-view microscope figure of the 750 μm-long MZI modulator with the GSG (ground-signal-ground) electrodes. (b) Optical transmission of the asymmetric MZI modulator under different reversed bias voltages. MMI = multi-mode interference, G = Ground, S = signal, white pads for signal gold wire bonding.

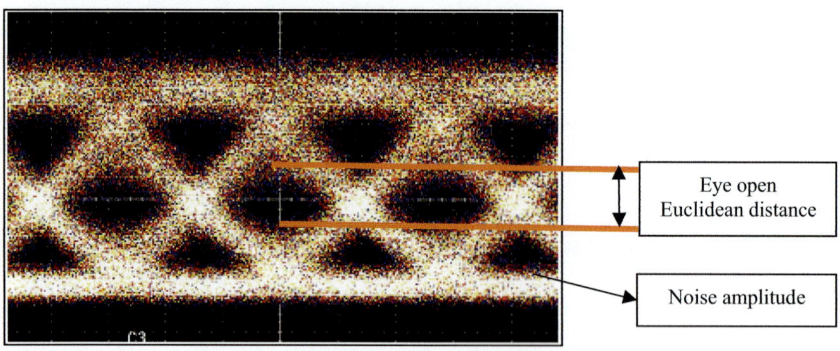

FIGURE 11.16 NRZ eye diagram of the 56 Gbps modulation bit sequence. Eye open distance and noise amplitude.

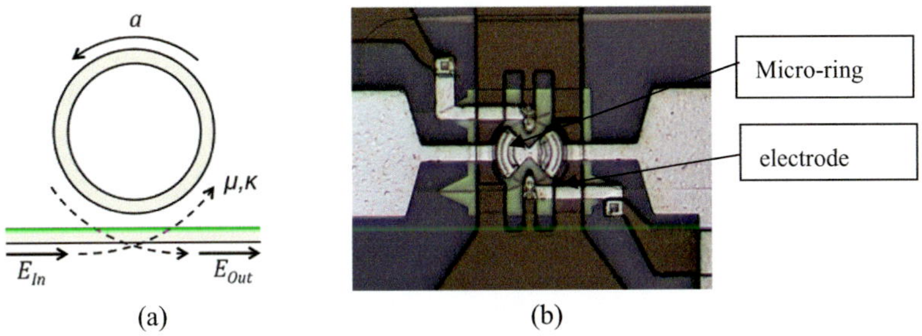

FIGURE 11.17 Micro-ring modulator (MRM) (a) sketch of the MRM with field propagation and coupling directions; (b) plane view of the fabricated RRM.

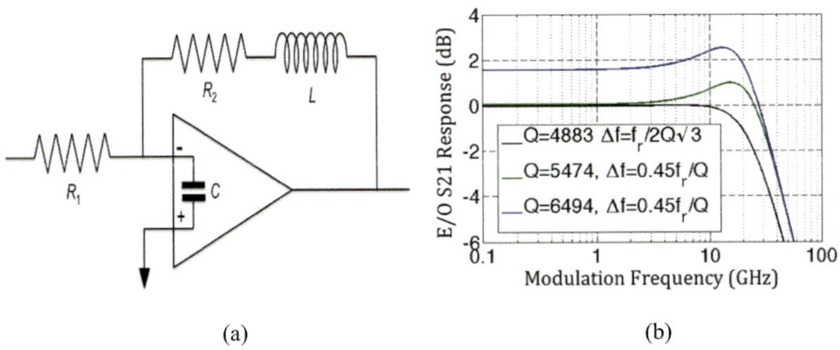

FIGURE 11.18 RRM peaking technique for equalization of optical transfer responses (a) circuit diagram of an idealized peaking amplifier RLC with an extended transfer function, that is a complex pole pair in the frequency transfer function, (b) ideal transfer function of equalized modulator by peaking to a 3 dB bandwidth to 40 GHz. Note: peaking of the transfer function.

wavelength. The resonant condition is such that the total phase distributed around the ring equals a multiple of half wavelengths. Thus, any applied voltage that departs from this resonant condition reduces the intensity circulating in the ring, hence modulating the amplitude. The MRRM frequency passband can be very wide, up to a few tens of GHz (Figure 11.18).

The investigated MRM has a radius of 10.3 mm and is composed of a 400 nm wide ridge waveguide partially etched into the 220 nm thick silicon device layer of a silicon-on-insulator wafer, with a remaining silicon slab height of 90 nm around the waveguide on an insulating SiO_2 thin layer. A reverse-biased *p-i-n* (*p*-type intrinsic *n*-type) junction can be fabricated using a nominal implant concentration of *p*-type at 3×10^{17} cm^{-3} and *n*-type of 1e^{17} cm^{-3}, overlapping by 50 nm, serving as a phase shifter. Electrical connectivity is achieved via highly doped regions with a doping concentration about one thousand times higher than that of the lightly-doped regions. The total junction capacitance is 17 femto-farads (*fF*) at 21 V bias, and the series resistance is 215 Ω, resulting in an electrically limited RC-bandwidth of 43 GHz.[20] It is noted that the electrode for the MRM is not a CPW type but more or less a lumped type. The limited bandwidth of this type of electrode can be equalized by using a peaking technique, which is commonly used in electrical circuit theory.[21] This extends the 3 dB bandwidth to about 40 GHz, permitting the modulator to operate at 56 Gbps or 64 Gbps without difficulty. The main issue with the MRM is the accurate positioning of the wavelength of the optical carrier due to resonant shifting when applying electrical signals, as shown in Figure 11.19a and b. Thus, a biasing voltage is required in association with the RF signals for tuning the resonant wavelength and aligning the applied signals to ensure linearity in phase and amplitude.

The resonance of the ring modulator ensures the exact frequency of the optical waves. The bandwidth of the resonator must be wide enough to accommodate the signal band to avoid distortion. Therefore, accurate control of the resonance frequency is critical; otherwise, there would be distortion in the channel if the resonance peak is shifted.

The frequency shift of the resonant peak can be expressed as

$$\Delta\omega = \frac{n_{\text{eff}}^3 r\omega^2}{2\sqrt{3}\pi c_0} V_{\text{pp}} \tag{11.3}$$

Where ω is the angular frequency of the optical waves, n_{eff} is the effective refractive index as seen by the guided optical waves, c_0 is the velocity of light in vacuum, r is the electro-optic coefficient, and V_{pp} is the peak-to-peak signal voltage applied to the ring. The driving voltage level can be further reduced by constructing a dual-electrode structure, allowing for a differential push-pull mode of operation, which reduces the complexity of the RF signaling.

The resonant characteristics of the micro-ring can be modified by combining it with other rings or interferometers, such as the one shown in Figure 11.20. Additionally, the free spectral range of the RRM must be considered to ensure that no crosstalk occurs in multiple wavelength channel systems.[22]

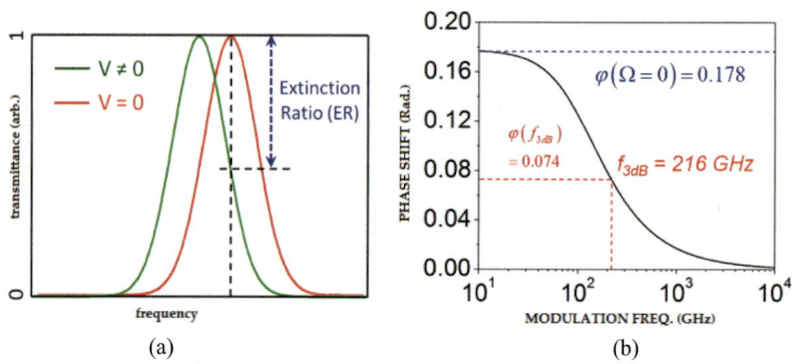

FIGURE 11.19 The transfer function of RRM and its variation with applied voltage across the electrode along the ring perimeter. (a) Intensity amplitude response; (b) phase response.

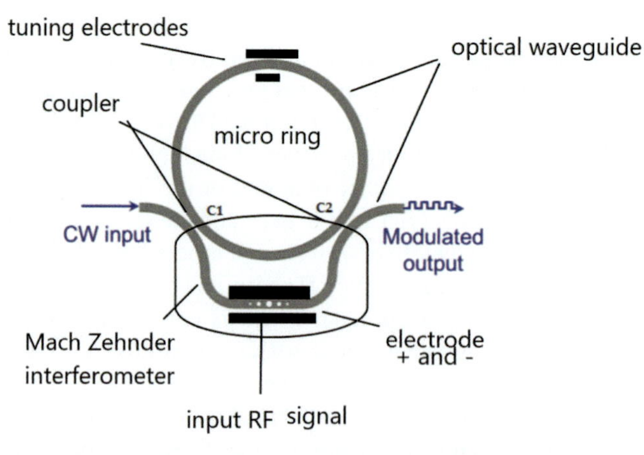

FIGURE 11.20 Schematic of the dual photonic crystal/micro-ring cavity modulator (top view); C1 and C2: integrated couplers for coupling lightwaves into and out of the integrated ring.

11.1.3.4 Si CMOS Modulator

Silicon has long been the optimal material for electronics, but it is only relatively recently that it has been considered as a material option for photonics due to its propagation loss compared to other crystallized electro-optic materials such as lithium niobate. Furthermore, one of the key limitations of using silicon as a photonic material has been the relatively low speed of silicon optical modulators compared to those fabricated from III to V semiconductor compounds. To date, the fastest silicon-waveguide-based optical modulator demonstrated experimentally has a modulation frequency of only 20MHz although it has been theoretically predicted that a 1-GHz modulation frequency might be achievable in some device structures.

This section describes an approach based on a metal–oxide–semiconductor (MOS) capacitor structure shown in Figure 11.21, embedded in a silicon waveguide that can produce high-speed optical phase modulation. It is demonstrated that such an all-silicon optical modulator has a modulation bandwidth exceeding 1GHz. As this technology is compatible with conventional complementary MOS (CMOS) processing, monolithic integration of the silicon modulator with advanced electronics on a single silicon substrate becomes possible. Modulation of the refractive index in (Si) silicon can be achieved via the free carrier plasma dispersion effect. In contrast to the conventional way of inducing a free carrier density change in silicon (through injection or depletion of electrons and holes in the intrinsic region of a forward-biased silicon p-i-n diode or a three-terminal device), we demonstrate the use of an MOS capacitor phase shifter to realize the charge density modulation. Under accumulation conditions, the majority carriers in the silicon waveguide modify the refractive index so that a phase shift is induced in the optical mode. The advantage of using an MOS capacitor phase shifter is the high achievable modulation speed, as there are no slow carrier generation and/or recombination processes involved in the accumulation operation. The silicon-waveguide-based MOS capacitor phase shifter comprises an n-type doped crystalline silicon slab (the silicon layer of the 'silicon-on-insulator' wafer) and a p-type doped polysilicon ridge with a gate oxide sandwiched between them. The polysilicon is formed by first depositing amorphous silicon using low-pressure CVD (chemical vapor deposition) at 508°C. This amorphous layer is then crystallized into polysilicon by high-temperature annealing. The annealing process is designed to minimize polysilicon grain boundaries, which can cause optical loss and limit the electric activation of dopants.

The frequency response of such a typical modulator is shown in Figure 11.22, which demonstrates that up to a 1.0GHz 3 dB-bandwidth can be achieved. The modulator is structured in an MZM with the phase change due to an applied voltage given by:

$$\Delta\Phi = \frac{2\pi}{\lambda}\Delta n_{\text{eff}} L \tag{11.4}$$

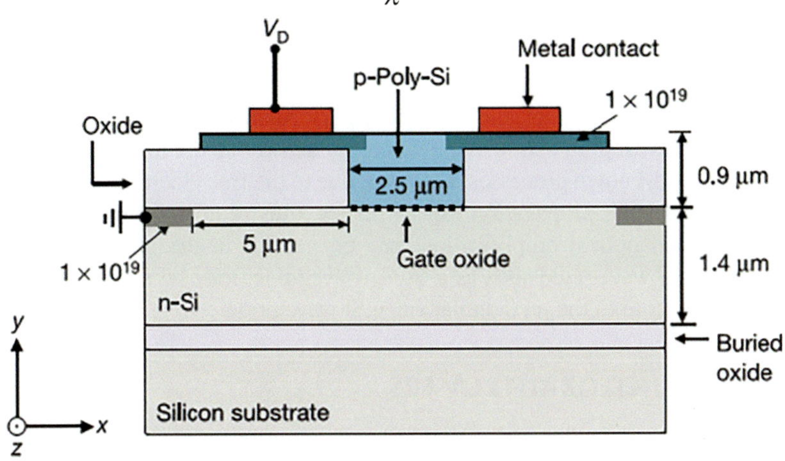

FIGURE 11.21 Cross-sectional schematic of a CMOS modulator.

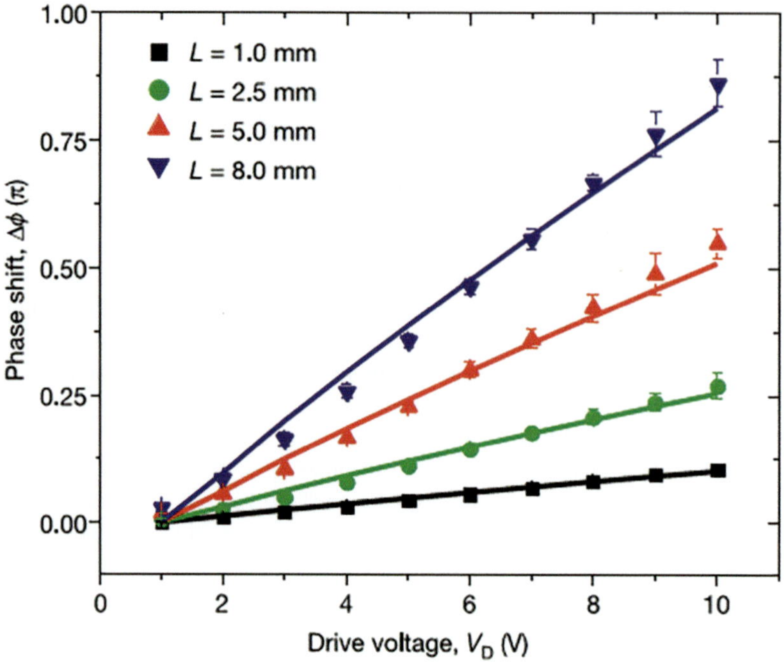

FIGURE 11.22 Frequency response of the CMOS modulator (b) of the MZIM employing the phase modulation in one of the branches of the interferometer (a).

Where the change in the effective refractive index due to the applied voltage to the CMOS is given by the change in the impurity of the gate carrier density. This change in phase is illustrated in Figure 11.23. The voltage-induced charge density change ΔN_e (for electrons) and ΔN_h (for holes) is related to the drive voltage applied to the drain and the gate forward bias and is given by:

$$\Delta N_e = \Delta N_h = \frac{\varepsilon_0 \varepsilon_r}{q t_{ox} t}\left[V_D - V_{FB}\right] \tag{11.5}$$

$$\begin{aligned} \Delta n_e &= -8.8 \times 10^{-22}\, \Delta N_e \\ \Delta n_h &= -8.5 \times 10^{-18} \left(\Delta N_h\right)^{0.8} \end{aligned} \tag{11.6}$$

In summary, the Si-based modulator can offer high performance in modulation voltage and frequency with low bias and amplitude of driving voltage at a lower cost. Furthermore, it can be fabricated using standard well-known processes. However, due to the free electrons in the semiconductor substrate and waveguide, the propagation loss is higher, thus requiring higher lightwave power. Consequently, integrated optical amplification may be needed in the integrated device system. Additionally, a coupling waveguide such as a Si_3N_4 waveguide, may be necessary for transitional coupling from the fiber core to the rectangular ridge Si waveguide.

11.2 APPENDIX 2: INTEGRATED LASERS

Lasers are lightwave generators that are crucial for carrying information data through transmission media and maintaining coherence to minimize spatial beam spreading. This helps in reducing the receiving power needed for a fixed optical reception area in space photonic communication systems.

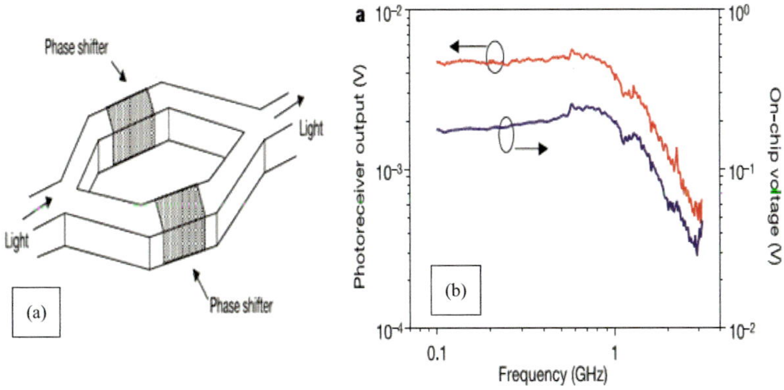

FIGURE 11.23 Phase shift ΔΦ versus drive voltage V_D of the MOS capacitor of the phase shifter (PS) at a wavelength of 1550 nm for different PS lengths as a parameter. The symbols represent the measured phase shifts, and the solid lines are the simulated phase.

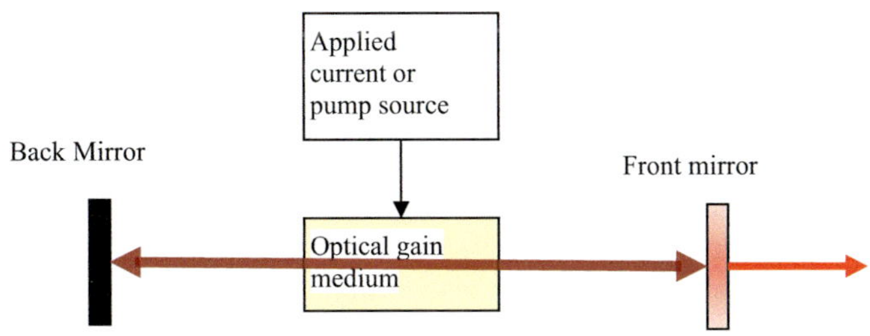

FIGURE 11.24 Basic structure of a generic laser.

This section provides a brief overview of lasers, particularly their optical spectra and the sharpness of the linewidth, which are important for coupling with optical modulators for data modulation of the lightwave source. External modulators are discussed in Section 11.1.

A laser has a basic structure consisting of an optical gain medium and two mirrors, with reflection coefficients of 100% and about 30% for the back mirror and front mirror, respectively, as shown in Figure 11.24. The gain medium is an optical gain material that excites electrons from the ground state to higher energy states. When the excited electrons gain sufficient energy, for example, from a seed photon source, they drop to lower energy states and release energy in the form of photonic lightwaves. This can be the spontaneous emission, which has a very broad spectrum, spanning several tens of nanometers. The spontaneous photons move back and forth through the gain medium, further stimulating the generation of photons. Once the photons have sufficient energy to pass through the front mirror, lightwaves are generated.

The optical gain medium can be a semiconductor gain material with applied current, such as In GaAsP for the 1550 nm spectral region, or Nd: doped or Er:doped fiber pumped with 980 or 1480 nm laser sources for the generation of 1060 or 1550 nm, respectively.

The mirrors can be designed as gratings, such as gratings, integrated gratings of Bragg feedback, and photorefractive gratings in fiber, etc. The grating can reflect only the photonic lightwaves whose wavelength can pass through the passband of the grating. Thus, only the selected wavelength photonic lightwaves can be further amplified and output to the front mirror.

If the mirrors are not grating filtering devices then a broadband mirror can be used. This type of laser cavity is the Fabry-Perot laser, whose spectrum is wide and generates several longitudinal modes.

This section provides a brief overview of the laser, particularly its optical spectra and the sharpness of the linewidth, which are important for coupling to the optical modulator for data modulation of the lightwave source. External modulators are discussed in Section 11.2.5.

11.2.1 Fabry-Perot and DFB Semiconductor Lasers

A laser consists of a gain medium placed between two mirrors, which reflect the lightwaves back and forth until the lightwaves gain sufficient energy to be emitted out of the front mirror, which has a reflection coefficient of about 30%. Initially, the gain medium is biased, and a driving current (for semiconductor lasers) supplies current to excite electrons to the upper sideband (USB) so that stimulated emission can occur after the initial spontaneous emission. The spontaneous photons bounce back and forth in the cavity, and further photon gain is achieved. The photons then gain sufficient power until their energy is high enough to pass through the front mirror and generate the lasing beam output.

A structure of an FP laser is shown in Figure 11.25a, in which an external modulator, EAM (Electro/absorption Modulator), is also integrated to form a modulator package such as TOSA (Transmitter Optical Sub-Assembly). The gain medium is integrated into the active guided layer of MQW (Multi-Quantum Well) through which the lightwaves are guided and reflected back and forth through the front mirror and back mirror. The lightwaves bouncing back and forth in the resonant cavity are shown red, until resonance is achieved, emitting out of the 30% front mirror or grating. In this structure, the mirror has no grating, making it a broadband type. The EAM is also integrated into the cavity, and the applied voltage of the information data can be used to turn the laser on and off by varying the absorption coefficient of the guided region. An emitted spectrum of the FB laser integrated in line with an EAM is shown in Figure 11.25b. There are several discrete lightwave modes, whose equal separation frequency can be estimated by the resonant mode of the cavity length and the average propagation phase velocity in such a guided medium. The dynamics of the generation of the resonant longitudinal modes of an FP laser are shown in Figure 11.26. First, under the gain and the resonant cavity, the oscillation modes of the guided medium are shown at the top left of the Figure 11.26. The gain spectrum has a Gaussian distribution, which depends on the material and doping properties of the semiconductor or the doping impurity of the rare earth element in fiber. This gain spectrum dictates the photonic lightwaves. Multiplying two such curves gives the laser spectrum, assuming that the mirror passband is wider than that of the gain spectrum.

To reduce the linewidth of the laser spectrum to only one longitudinal mode, we have to use reflectors in the form of Bragg reflection gratings, as shown in the structure given in Figure 11.27a, with its cross section in panel (b) and the emitting lasing spectrum in panel (c). The Single frequency of a longitudinal mode is determined by the Bragg frequency of the DFB structure, with a high rejection band to ensure that only one mode is emitted. The bandwidth of the Bragg grating determines the linewidth of the laser, which is about 1–2 MHz. Higher precision in fabrication, including more accurate gratings and longer cavity lengths, can reduce the linewidth to about 10–100 kHz. These linewidths can resolve into pseudo/coherent states of modulated laser for launching and transmission into a space photonic link. Narrower linewidths of lightwave sources, in the range of a few Hertz, can be achieved if a fiber laser integrated with Fiber Bragg gratings is used.

Table 11.2 lists the parameters and contrasts of typical optical modulators, including $LiNbO_3$, Si-based, and EML (Electro-absorption Modulated Laser) with DML (Direct Modulation Laser), where the laser is directly modulated by varying the injection current according to the data sequence (Figure 11.28). The biasing of this DML is not constant but changes alternately with the data sequence. The threshold current is the minimum current required to excite spontaneous emission, followed by a higher bias current to generate the desired power. On top of this bias current, the data signal sequence is applied. This is a direct modulation technique used with a semiconductor laser. The current varies between on and off states. This variation causes broadening of the laser linewidth due to the cavity's expansion and contraction, thus exciting the cavity's

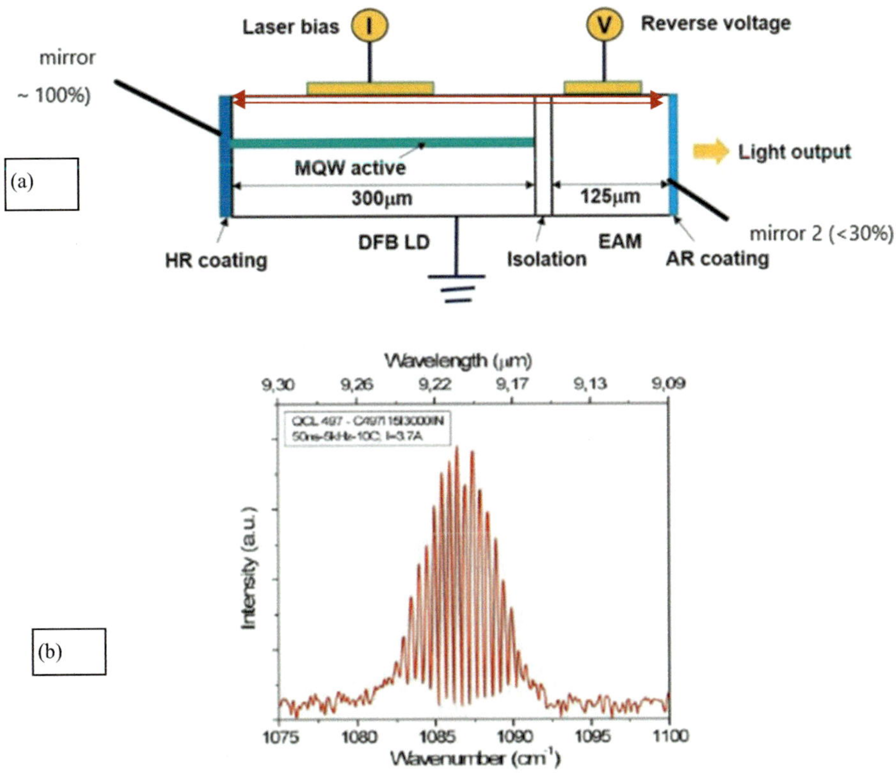

FIGURE 11.25 (a) Structure of an FP laser. (b) Spectrum of emitted lightwaves. AR=anti-reflection, DFB=distributed Bragg reflector, LD=laser diode, EAM=electro/absorption modulator, HR=high reflection.

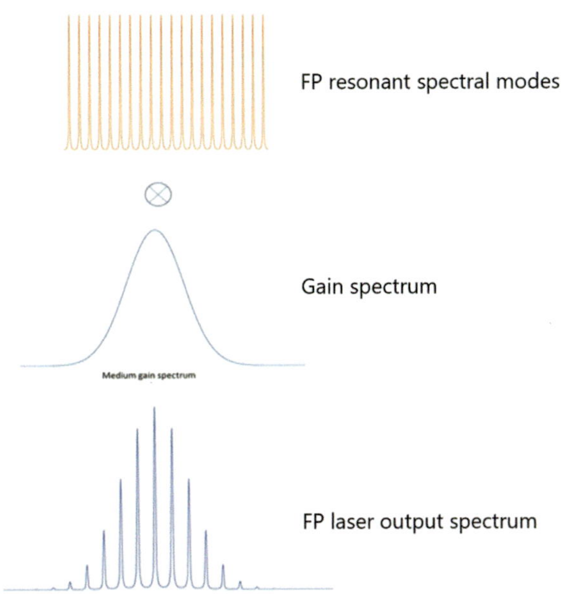

FIGURE 11.26 Dynamics of the generation of longitudinal resonant modes in an FP laser cavity. Multiplication of the gain spectrum and mode spectral distribution (light line) gives the output lasing spectrum (darker line)

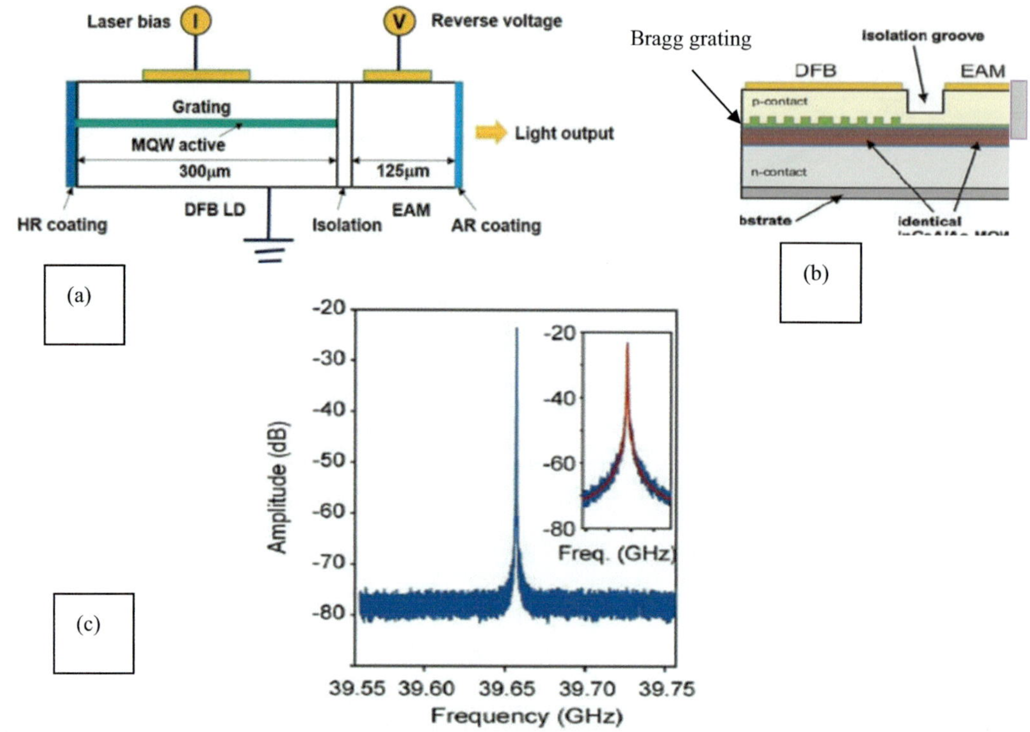

FIGURE 11.27 Structure and emitting spectrum of a DFB laser – EAM transmitter. (a) DFP and EAM in integration; (b) cross-section of a DFB-EAM device; (c) emitted spectrum of single frequency, frequency scale indicates the modulated frequency centered at the lasing frequency *f* 193.51 THz.

TABLE 11.2

Parameters and Contrast of Typical Optical Modulators, $LiNbO_3$, Si-Based, and EML (Semiconductor) with DML (Direct Modulation Laser)

Parameters	$LiNbO_3$ MZM	SOI MZM	EML	DML
ER (Ext R-dB)	20–30	→20	<10	10–20
BW	→60 GHz	→45 G	<40 G	<8 GHz
V_π (V)	3.5	~3	2–3	NRZ (40–50 mA)
Current (mA)	Low	Low	Low	40–50
Linearity	Best	Good	Good but limited	Worst
Laser & spectral width	External cavity laser - <1.0 kHz	External – no integrated (<1 kHz)	Integrated-laser inline 10–20 MHz 40 pm (5G) (EML Eudyna-type)	Laser modulated (1–2 nm)
Technology	External cavity laser (δ_f< 10 kHz)	SOI MZM	FB-EML (30–40 pm ~5 GHz) DFB EML (<1.0 MHz)	DFB modulated
IL (dB)	<4.5	<2	0	NA
Technology	Edge coupling loss	Coupling and laser-integrated Si_3N_4	Inline – advantage	Current above *I*th (laser threshold current)

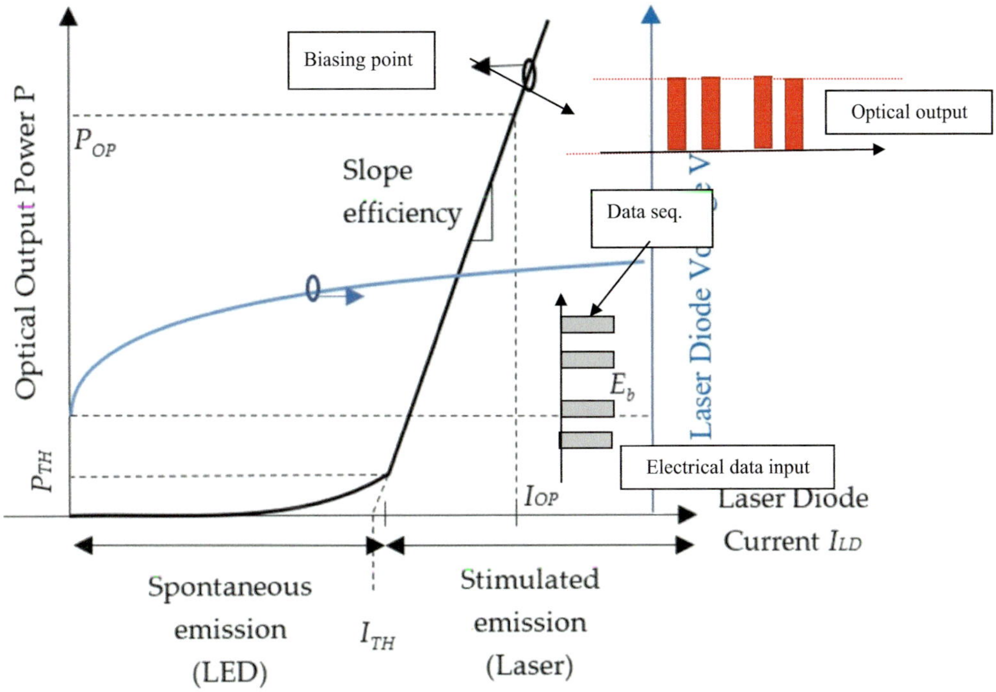

FIGURE 11.28 Laser driving characteristics with biasing point and superimposed signal sequence. In small signal linear level.

resonant mode. This is the primary reason why direct modulation cannot be employed in space photonic communications for ultra-high bit rates.

External modulation indicates that the laser is operating at the bias point to generate continuous waves (CW), with the linewidth remaining unchanged. The laser is then coupled to the modulator mode and modulated via electro-optic effects through the applied signal sequence superimposed on the bias point of the modulator. The modulated signals must be adjusted to ensure that their minimum and maximum levels remain in the linear region to avoid distortion. Operating in the nonlinear region can sometimes offer advantages, such as in an MZIM biased at the minimum transmission point, where the swinging of the data sequence would then regenerate a suppression of the carrier and frequency doubling of the optical wave output. Thus, one can generate high-frequency optical clock pulses at half the frequency of the clock (Figure 11.6).

11.2.2 Fiber Lasers

A fiber laser is a laser in which the active gain medium is an optical fiber doped with rare-earth elements such as erbium (for the 1550nm spectral window), ytterbium (1300 to–1600nm and possibly co-doped with erbium), neodymium (1060nm), dysprosium, praseodymium, thulium, and holmium. They are related to doped fiber amplifiers, which provide light amplification by boosting power or low-power input light waves, or serve as highly sensitive, low-noise amplifiers in space photonic communications systems. The fiber core for optical amplifiers can be doubly doped to increase the output power from a few watts to tens of watts.

Figure 11.29 shows the structure of a fiber laser with a bidirectional pump. The rare-earth-doped fiber section can be either a single-cladding or double-cladding doped fiber. A single-pump fiber laser set-up on an optical table is shown in Figure 11.29a. Two WFCs (wavelength fiber couplers) are

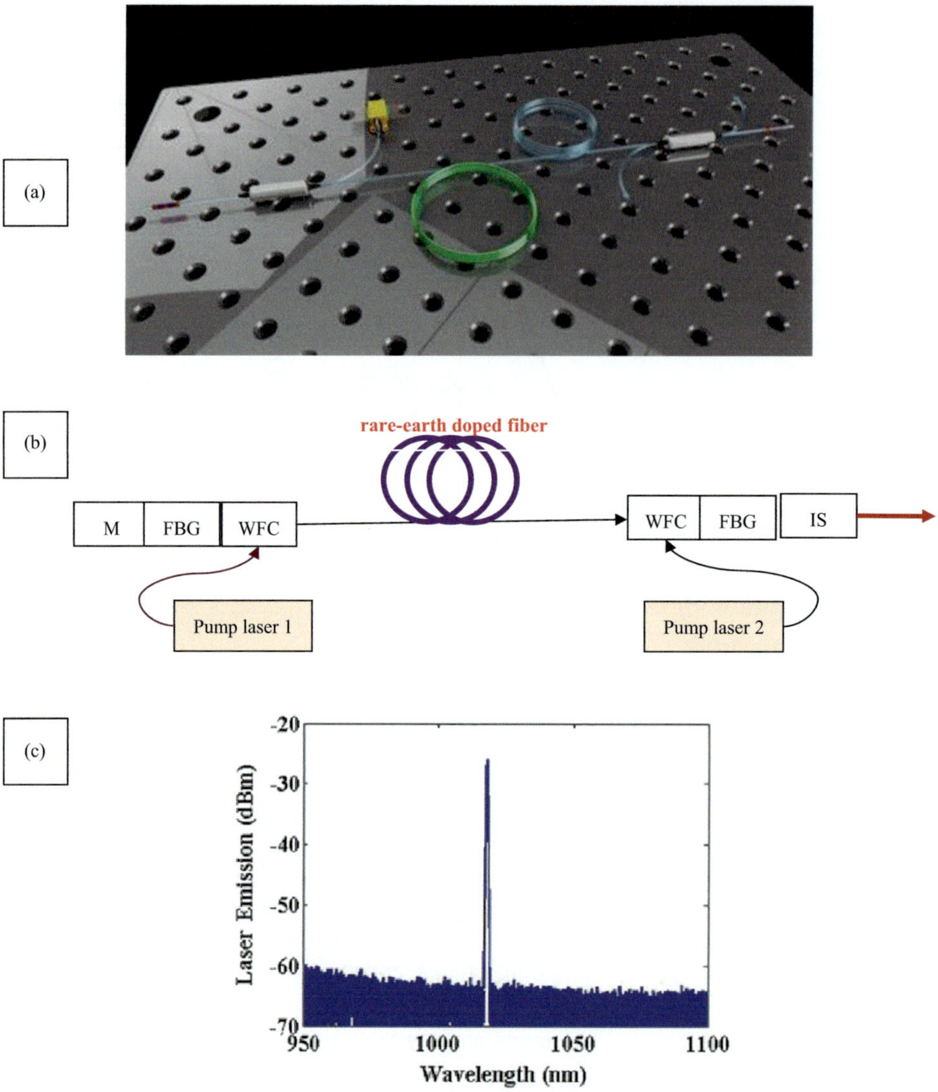

FIGURE 11.29 (a) Hard ware structure of a single pump fiber laser; (b) structure of a double pump fiber laser, and (c) typical single mode laser at 1018.5 nm of Nd:doped fiber laser. FBG = Fiber Bragg Grating, WFC = wavelength depend on fiber coupler, IS = isolator, M = mirror.

designed with a coupling coefficient such that it is very small for one wavelength but much higher at the pump laser wavelength. A bidirectional pump is typically used for low-input lightwaves in highly sensitive optical amplification.

Lasing occurs when the fiber cavity is formed by two fiber Bragg gratings (FBG) at the two ends of the doped fiber section. The emitting wavelength depends on the doped rare-earth metal elements. For 1550 nm, the doped materials should be erbium and or erbium- ytterbium. The energy levels of such atomic structures can be multiple so the excitation can be broad and then narrowed by the optical passband of the FBG.

The linewidth of a fiber laser can be controlled by tuning and feedback of the FBG to a few Hz. This is critical and useful for the coherency of space photonic communications. Such a narrow linewidth laser can minimize beam spreading, thereby improving photo-receiving power for detection. A typical spectrum of an Nd: fiber laser is shown in Figure 11.29c.

A section is dedicated to addressing the special operational characteristics and principles of optical amplifiers for boosting power and pre-amplification of the receiver or for further power boosting. For space photonic communications, the amplifiers are somewhat different, and the section will clearly highlight the distinct properties, especially the double-clad doped fibers for operations and applications. Fiber lasers are important as well, as they can provide output power of a few tens of watts and even a few hundred watts, which can be sufficient for compensating power loss, along with optical pre-amplification for inter-GEO/MEO/LEO-satellite links.

This section outlines (i) Optical modulators, especially the $LiNbO_3$ uniaxial type, for effective modulation via its electro-optic effects, and (ii) Lasers for space photonic communications long-haul systems. Only the basic fundamentals of these devices are briefly covered, focusing on their suitability for space long-haul communications. Although these devices were developed for fiber optic communications, they are most efficient for space communications; otherwise, there are no bulk devices for this advanced space photonics.

11.2.3 Laser Intensity and Phase Noises

Lasers can exhibit various kinds of "noises", with manifold influences on applications. This section discusses where such noise can come from, how it can be quantified, and how to mitigate or minimize its effects in optical transmission systems, both in guided and free-space media, especially in DSP-based subsystems employing coherent optical reception.

11.2.3.1 Remarks

"Noise" in lasers[23] is a term for random fluctuations of various output parameters. This is a frequently encountered phenomenon that has a profound impact on many applications in photonics, particularly in the area of precision measurements. For example, interferometric position measurements can be directly affected by fluctuations in the optical phase, or spectroscopic measurements of transmission, where intensity fluctuations limit the possible sensitivity. Similarly, the data rate and transmission distance for fiber-optic links are at least partly limited by noise issues.

Scientific communities, especially those in optical communications, feel more or less uneasy about laser noise. One reason is that the causes are often hard to evaluate and eliminate. Further technical and mathematical difficulties are related to the measurement and quantification of noise; specifications found in several commercial data sheets reveal less about the laser than desired. Furthermore, it is desirable but not always easy to estimate the influence of different kinds of noise in some applications. For such reasons, trial-and-error approaches without a thorough understanding are often used, for example, to minimize noise influences, but these approaches often turn out to be ineffective or inefficient. This appendix explains some basics of laser noise, focusing on intensity and phase noise; many aspects, of course, can be applied to other types of noise.

11.2.3.2 Intensity Noise

Intensity noise is usually understood as fluctuations in the laser output power (not the optical intensity itself), and in most cases, it is normalized to the average power. The measurement is based on recording the temporally varying output power, using a photodiode, for example. Normalization is the simplest aspect; other aspects, to be discussed below, are subtler. Frequently encountered specifications like "±1%" may appear simple but are quite misleading. They suggest that the power always stays within 1% of its average value, while in reality, there is usually a smooth probability distribution without sharp edges. A good way to avoid this problem is to specify the RMS (root mean square) values, which is the square root of the average of the squared power fluctuations, or mathematically

$$\delta P_{\mathrm{RMS}} = \sqrt{\left\langle \left(P(t) - P_{\mathrm{av}}\right)^2 \right\rangle} \tag{11.7}$$

where P_{av} is the average power. This is usually applied to *relative* power fluctuations, thus specifying relative intensity noise (RIN). The RMS RIN is, of course, dimensionless.

Another problem is that the recorded fluctuations can strongly depend on the measurement bandwidth. For example, a laser may exhibit fast fluctuations in the output power, which can be detected by a fast photodetector, while being averaged out and thus not detected by a slower detector. Furthermore, a limited measurement time may not be sufficient to detect slow fluctuations (drifts); essentially, one does not know how far the average power in the chosen time interval deviates from the average over longer times. For these reasons, a simple number without specifying the measurement bandwidth (lower and upper noise frequency) is actually meaningless and should never appear in a data sheet. The measurement bandwidth may extend from a very small frequency, limited by the inverse measurement time, to some maximum frequency, determined by the speed of the detector. It is also noted that any noise specification is actually a statistical measure, which is only estimated by one data trace for a given time interval or (preferably) with an average over several traces. Additionally, the laser noise may depend on the ambient conditions. Therefore, one should know whether a specification applies to constant room temperature, after a long warm-up time, and in a vibration-free environment.

If it is of interest to determine the extent to which different noise frequencies contribute to the overall noise, a power spectral density (PSD) $S_I(f)$ is most useful. Conceptually, one can imagine that many noise measurements are conducted for different (small) noise frequency intervals. This is how many electronic spectrum analyzers record noise spectra: the detector is subsequently tuned to different noise frequencies. However, there are also techniques based on Fourier transforms that allow estimation of the entire noise spectrum from a single (or a few) data traces recorded in the time domain. For example, one may use a sampling card to record a signal proportional to the laser power over a certain time with a certain temporal resolution. Before performing a Fourier transform, a so-called "window function" must normally be applied to avoid certain artifacts. The lowest noise frequency in such a measurement is the inverse of the total duration, while the maximum frequency is at most half the sampling rate. The latter limit results from the Nyquist theorem.

Once we have the power spectral density $S_I(f)$ of the relative intensity, we can calculate the RMS relative intensity noise according to

$$\frac{\delta P}{\bar{P}} = \sqrt{\int_{f_1}^{f_2} S_I(f)\,df} \tag{11.8}$$

where f_1 and f_2 are the lower and upper noise frequencies, respectively. (This assumes a one-sided PSD; physicists often use two-sided PSDs, where the integration must also include negative frequencies.) The units of $S_I(f)$ are 1/Hz, indicating that $S_I(f)$ represents the noise contribution per frequency interval. It is also common to specify $10\log_{10}(S_I(f))$, arriving at units of dBc/Hz, meaning dB relative to the carrier in a 1-Hz noise bandwidth.

Figure 11.30 shows the simulated relative intensity noise spectrum (i.e., $S_I(f)$ versus frequency) of a diode-pumped single-frequency Nd:YAG solid-state laser at 1080 nm, using logarithmic scales for both axes. The RIN exhibits a characteristic peak around 140 kHz, related to so-called relaxation oscillations, which result from the dynamic interaction of the intracavity power and the laser gain. Obviously, a detector with a 20 kHz bandwidth would not be able to detect such fast oscillations and would thus record much weaker noise than a 200 kHz detector. Laser diodes exhibit relaxation oscillations with much higher frequencies (multiple GHz) and stronger damping due to their short carrier lifetime and short resonator. Generally, different laser types can exhibit very different noise properties, as characteristic parameters may be totally different.

Around the relaxation oscillation frequency f_{ro}, a laser is particularly sensitive to external noise influences, such as from the pump diode, and even to quantum noise. Well above f_{ro}, noise of any kind has little impact. Well below f_{ro}, pump noise may directly affect the output. For the dark light

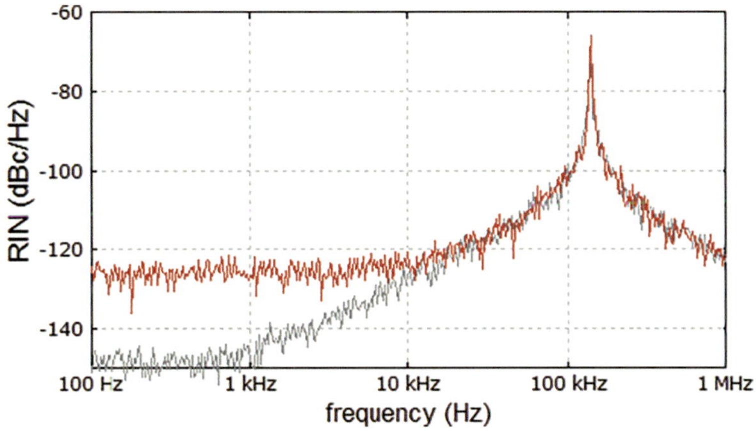

FIGURE 11.30 Simulated intensity noise spectrum of a diode-pumped miniature Nd:YAG laser. Light line curve: with quantum-limited pump source. Dark line curve: with 30 dB excess noise of the pump source.

line curve in Figure 11.30, some frequency-independent excess noise from the pump diode has been assumed; in reality, there can be an increase toward lower frequencies, often determined by the quality of the diode driver. Even for a perfect shot-noise-limited pump source (light line curve), noise around f_{ro} is relatively strong. The impact of quantum noise depends on various parameters; in particular, it is stronger in cases with low intra-cavity power, high resonator losses, and a short round-trip time of the resonator. Additional contributions to intensity noise in a laser can result from acoustic influences. For example, a cooling system may cause vibrations of the laser resonator, which translate into intensity and phase noise. In many cases, such noise contributions appear as sharp peaks in the spectrum.

An entirely different phenomenon is noise from mode beating in lasers where multiple resonator modes are oscillating. The beat frequencies that occur are differences between resonator mode frequencies, which may be quite high (multiple GHz) for laser diodes and very low (a few kHz) for long fiber lasers. The pattern of beat frequencies can reveal whether or not the laser operates on longitudinal modes only.

Minimization of laser noise can be achieved through entirely different approaches:

- A first approach is to minimize external noise influences, such as by using single-frequency pump diodes operated with a carefully stabilized current source and ensuring a mechanically stable laser setup.
- A second possibility is to optimize laser parameters so that the impact of quantum noise and/or external noise influences is minimized. For example, one can minimize quantum noise influences by using a long low-loss laser resonator or by moving the relaxation oscillation frequency to a region where noise is less disruptive to the application.
- Finally, laser noise can be reduced with a feedback system that automatically adjusts, for example, the pump power based on measured output power fluctuations. The characteristics of the feedback loop need to be optimized based on the knowledge of laser parameters.

Obviously, the effectiveness of these approaches depends greatly on the circumstances, which should be analyzed carefully beforehand. Without a proper understanding of the sources and types of noise, such efforts are likely to fail or at least be inefficient.

The intensity noise of nearly all light sources is limited by shot noise. In an intuitive (and somewhat simplified) picture, this can be understood as the random occurrence of photons (packets of light energy). Even if the probability of detecting a photon within a short time interval is constant, the recorded photon absorption events exhibit some randomness if there are no correlations between photons. This leads to a PSD of the relative intensity noise of $2h\nu/P_{av}$, where $h\nu$ is the photon energy. The relative intensity noise of a laser is often well above the shot noise level, but the latter rises if the output is increasingly attenuated (for example, with a linear absorber). With sufficiently strong power attenuation, the intensity noise will reach the shot noise limit. Squeezed states of light can exhibit intensity noise below the shot noise limit but require special methods to be generated and tend to lose that special property, for example, when the light is attenuated.

11.2.3.3 Phase Noise and Linewidth

Phase noise is related to fluctuations in the optical phase of the output. As simple as this sounds, the optical phase may not even be defined for a laser oscillating on multiple resonator modes. We thus assume we are dealing with a single-frequency laser, where essentially all power is in a single resonator mode. (For multimode lasers, phase noise may be considered for different modes separately).

Due to various influences, even a single-frequency laser will not exhibit a perfect sinusoidal oscillation of the electric field at its output. There are fluctuations in both the power (see above) and the optical phase φ. The latter can be quantified with a phase noise PSD $S_\varphi(f)$, which has units of rad^2/Hz or Hz^{-1} as radians are dimensionless. Initially, we may consider only quantum noise, which, in a simplified picture, is described as the random phase of photons added by spontaneous emission. This leads to a "random walk" of the optical phase, which is related to a phase noise PSD $S_\varphi(f) \sim f^{-2}$. The divergence at $f=0$ is related to the unbounded drift of the phase, which, unlike the output power, lacks a "restoring force". It also causes a finite value of the linewidth, which is the width of the main peak in the power spectral density of the optical field. A famous paper from 1958 by Schawlow and Townes presented a simple formula to calculate that linewidth and showed that the linewidth decreases with increasing intra-cavity power, decreasing resonator losses, and increasing resonator length. However, the quantum-limited linewidth is often difficult to reach, as other noises may dominate. In many cases, the linewidth, being a single value, is of primary interest for applications. However, the complete phase noise spectrum may be required in other cases and may reveal important noise contributions even if these are not relevant for the laser linewidth, which is mainly determined by low-frequency noise.

The measurement of phase noise or linewidth is substantially more difficult than that of intensity noise, partly because the phase evolution must be compared with a reference. A conceptually simple but often impractical method involves measuring a beat note of the laser with a second laser that has a similar optical frequency (keeping the beat frequency low enough) and much lower phase noise. Alternatively, two similar lasers may be used, and the relative phase noise provides an estimate of the noise of a single laser.

An often convenient variant is the delayed self-heterodyne method, using a setup as shown in Figure 11.31. Here, one shifts the frequency of the lightwave beam with an acousto-optic modulator and derives a reference from the laser itself by introducing a substantial time delay using an optical fiber. For a very long delay (say, about 1.0 km, depending on the expected linewidth), the noise of the delayed reference would be uncorrelated with the laser noise, allowing for a simple analysis. Noise measurements are still possible with shorter delays but require a more sophisticated mathematical analysis. This should also consider the possibility that the phase noise PSD does not exhibit an f^{-2} law.

In principle, the phase noise spectrum can also be obtained from the optical spectrum, as recorded, for example, with a scanning high-finesse Fabry-Pérot interferometer. Here, the optical resonator provides a phase reference. This method can work for a semiconductor laser, but it is challenging to achieve a resonator linewidth below that of a typical solid-state laser.

11.2.3.4 Phase Noise Measurement of a Narrow Linewidth CW Laser

Two different laser phase noise measurement techniques are compared. One of these techniques is based on a conventional and low-cost delay line system, which is typically used for the linewidth measurement of semiconductor lasers. The results obtained with both techniques

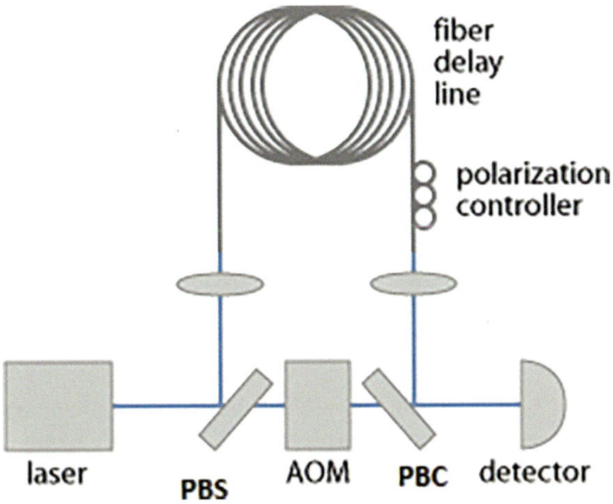

FIGURE 11.31 Setup for delayed self-heterodyne measurements. The mirrors next to the acousto-optic modulator (AOM) have a reflectivity of e.g. 50%. The detector records the beat note between the frequency-shifted part and the delayed part of the laser light. This contains information on phase noise and linewidth. PBS = Polarization beam splitter, OPC = Polarization beam combiner.

on a high-spectral-purity laser agree well and confirm the usefulness of the low-cost technique. Additionally, an extraction of the laser linewidth using computer-aided design tools is performed.

The delay line technique is a low-cost and efficient approach for measuring the frequency fluctuations of frequency sources. Long optical delay lines (a few kilometers) are generally used in optics to achieve complete signal de-correlation and measure the laser linewidth by simply mixing this signal with the original laser signal on a photodiode.[24] In the microwave range, the same long optical delay lines are used in frequency discriminators, thus featuring ultrahigh sensitivity.[25]

When a high-spectral-purity laser is measured using this type of setup, the measurement conditions are similar to those observed in microwave frequency discriminators: the coherence length is no longer shorter than the delay line length, and the data measured at the system output represent the frequency fluctuation spectrum of the laser.[26,27,28] However, compared to the frequency discriminators used for microwave source measurement, an optical high-sensitivity frequency discriminator faces different challenges related to the frequency stability of the laser during the measurement period.

In this Letter, two techniques for measuring laser short-term frequency fluctuations using an optical delay line are compared. One technique is based on a homodyne delay line method and uses a fast Fourier transform (FFT) signal analyzer, while the other technique is based on a self-heterodyne approach and involves spectrum analysis close to an intermediate frequency of 80 MHz using a costly piece of equipment: an RF phase noise measurement bench. The measurement bench is depicted in Figure 11.32. The signal is delayed on one path using a 2 km fiber spool, followed by an optional polarization controller. In the homodyne approach, this signal is directly recombined with the reference path, and the output of the coupler feeds a photodiode, which performs the quadratic detection. When the laser under test is a high-quality laser, such as a fiber laser or an externally stabilized semiconductor laser, which typically features a linewidth smaller than 10 kHz, the signal on both arms remains correlated, and the output is proportional to the frequency fluctuations of the laser at low frequency offsets (frequency discriminator).

11.2.3.4.1 Analytical Derivations

A classical approach to describe this system is to replace the frequency noise with a deterministic sinusoidal modulation.[29] In this case, the signal in both arms before the coupler can be written as

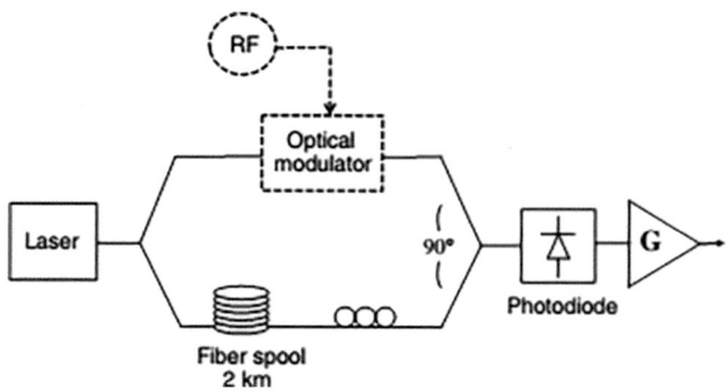

FIGURE 11.32 Measurement bench for both approaches; the acousto-optic modulator is used in the self-heterodyne technique and removed for the homodyne technique. G=amplifier gain.

$$\begin{aligned} A_1(t) &= \frac{A}{2}\cos\left(2\pi f_0(t-\tau)+\frac{\Delta f}{f_m}\cos\left(2\pi f_m(t-\tau)\right)\right) \\ A_2(t) &= \frac{A}{2}\cos\left(2\pi f_0(t-\tau)+\frac{\Delta f}{f_m}\cos\left(2\pi f_m t\right)\right) \end{aligned} \tag{11.9}$$

where τ is the delay between the two arms, f_0 is the optical frequency, f_m is the noise frequency, and Δf is the modulation amplitude (frequency noise). The quadratic detection leads to (mixing term only)

$$I(t) = S\frac{A^2}{4}\cos\left(-2\pi f_0\tau + \frac{2\Delta f}{f_m}\sin\left(2\pi f_m(t-\tau/2)\right)\sin\left(\pi f_m\tau\right)\right) \tag{11.10}$$

with S being the photodiode sensitivity. Equation 11.30 may be written as

$$I(t) = S\frac{A^2}{4}\cos(\varphi + x) \tag{11.11}$$

with φ being a constant phase and x a small amplitude component (noise). The derivative of this function with respect to x.

Near $x=0$, we can linearize the output current and determine the ratio of the output current amplitude and Δf, which is the sensitivity factor of the frequency discriminator, K_m

$$K_m = S\frac{P}{4}\sin(\varphi)2\pi\tau\frac{\sin\left(\pi f_m t\right)}{\left(\pi f_m t\right)} \tag{11.12}$$

with P being the optical power received by the photodiode. While varying the phase φ, the output current changes from its minimum, which should be zero in a perfect interferometer, to its maximum SP. In practice, the minimum is never equal to 0, and SP is generally approximated by the peak-to-peak excursion of the output current. Thus, $SP=2$ represents the peak excursion of this current, which we denote as ΔI peak.[22]

Ideally, the phase φ between the two arms should be adjusted to $\pi=2$ in order to maximize frequency fluctuation detection. This is what is classically performed in a microwave frequency discriminator. However, in an optical experiment, such precise phase control after a 2 km delay is impossible due to temperature variations. The phase φ rotations caused by the slow temperature drift of the fiber spool can be observed at the system output. In our case, 360° rotations were reached every 200 to 300 ms, meaning that, over a 15-minute acquisition time, complete phase rotations will occur approximately 4000 times. Due to such a large number of phase rotations, and using a large number of averaged spectra (about 200 spectra), we expect the phase states to be evenly distributed across 360° during the measurement. This approach is very similar to the setup described in Ref.,[25] in which a phase modulator was introduced into the system. However, with kilometer-long delay lines, the phase modulator is no longer necessary; a natural phase shift occurs due to the slow temperature drift of the fiber spool. When a large set of phase states is considered, the effective K_m can be calculated from Equation 11.12 by integrating the quadratic value of KmK_mKm over a complete phase rotation (as a power spectrum is being detected):

$$K_{m-\text{eff}} = \frac{K_m}{\sqrt{2}} = \frac{\Delta I_{\text{peak}}}{\sqrt{2}} 2\pi\tau \frac{\sin\left(\pi f_m t\right)}{\left(\pi f_m t\right)} \tag{11.13}$$

In our experiment, the photodiode is loaded with a resistance R, and the voltage signal is measured using a 100 kHz FFT spectrum analyzer (Advantest R9211B). The system calibration is performed by measuring ΔI peak using the oscilloscope mode of the FFT analyzer, and then $K_{m-\text{eff}}$ is computed for a set of frequencies fm. In our case, the delay resulting from the optical line is 9:5 μs. The laser frequency fluctuation spectrum is then calculated,

$$S_{\Delta f} = \frac{S_{V_{\text{output}}}}{R^2 K_{m-\text{eff}}^2} \tag{11.14}$$

$$K_{m-\text{eff}} = \frac{K_m}{\sqrt{2}} = \frac{\Delta I_{\text{peak}}}{\sqrt{2}} 2\pi\tau \frac{\sin\left(\pi f_m t\right)}{\left(\pi f_m t\right)} \tag{11.15}$$

The system sensitivity is determined by $K_{m-\text{eff}}$ and the output noise due to any source of noise different from $S_{\Delta f}$

$$L\left(f_m\right)_{\text{dBc/Hz}} = 10\log_{10}\left(\frac{S_{\Delta f}\left(f_m\right)}{\left(2f_m^2\right)}\right) \tag{11.16}$$

with $S_{\Delta f}$ laser amplitude modulation (AM) noise, photodiode noise, Brillouin scattering in the fiber, etc. Generally, laser AM noise and photodiode noise are negligible in a long-line interferometer. However, with a delay in the kilometer range, the Brillouin threshold is lowered, and this noise may become the system's noise floor. A compromise must therefore be found between long delay (high $K_{m-\text{eff}}$), high optical power, and the onset of scattering phenomena in the fiber.

The self-heterodyne technique applied to high-quality lasers was presented in Ref.[30] and is briefly described here. The measurement setup schematic is depicted in Figure 11.32, incorporating an optical AO (acoustooptic) modulator to shift the output signal around the 80 MHz RF (TeO_2 AO crystal). The signal is then analyzed using a phase noise measurement bench at 80 MHz. Using the same approach developed for the homodyne case, it can be demonstrated[28] that the phase fluctuations of this 80 MHz RF signal are proportional to the laser frequency fluctuations $S_{\Delta f}$ at low-frequency offsets. More precisely,

$$L_{\text{RF}}\left(f_m\right) = 20\log_{10}\left(\sqrt{2}\pi\tau\frac{\sin\left(\pi f_m\tau\right)}{\left(\pi f_m\tau\right)}\right) + 10\log_{10}\left(S_{\Delta f}\right) \tag{11.17}$$

From Equations 11.14 and 11.15, we can calculate the laser phase noise from the measured RF signal phase noise as

$$L_{\text{laser}}\left(f_m\right) = L_{\text{RF}}\left(f_m\right) - 20\log_{10}\left(2\sin\left(\pi f_m\tau\right)\right) \tag{11.18}$$

The calibration in this case is performed by replacing the 80 MHz output signal with a frequency-modulated source of the same frequency and amplitude as the source under test, which is automatically handled by modern phase noise measurement platforms. Compared to the first approach, this method is quite straightforward but requires much more sophisticated and costly equipment, particularly an RF phase noise measurement bench. To compare the two approaches, the phase noise characterization of a commercial 1:55 μm fiber laser is performed. Using the homodyne approach, 200 spectra are averaged over three different and overlapping frequency bandwidths (600 spectra for the entire measurement).

The long acquisition time (10–15 minutes) allows for random phase distribution, and the spectrum becomes remarkably well-defined and stable after about 50 averages. The FFT analyzer is a low-cost processor, and, moreover, no RF synthesizer or optical modulator is required. The self-heterodyne approach can be performed using a 200 mW power source at 80 MHz sinusoidal waves to drive the AO modulator, and the signal is measured at the output by a signal source analyzer, which is a high-performance phase noise test set incorporating self-calibration. The results of both measurements are shown in Figure 11.33. A good agreement is observed, validating our approach for the homodyne technique.

From the frequency noise or phase noise spectrum, it is possible to evaluate the laser linewidth. Although it is common to specify a laser by its linewidth, for very high spectral purity sources, this parameter cannot be measured directly and therefore has limited practical significance. What is effectively measured is the frequency noise or phase noise spectrum, and the linewidth evaluation requires a computation process based on these data. The relationship between phase noise and power spectrum is a complex problem that has been addressed in many papers, and[31,32,33,34,35,36] represent only a small selection of them. The only case that can be computed analytically is for white frequency noise8 which results in a Lorentzian line shape. However, for the type of laser discussed in this Letter, almost all the laser power is dominated by the 1 =*f* part of the spectrum, rendering the Lorentzian formula ineffective. The computation of the linewidth in the case of pure 1 =*f* noise is more complex[30–33] and only a computer-based approach can solve the problem for an arbitrary phase noise spectrum with a combination of different slopes[32]

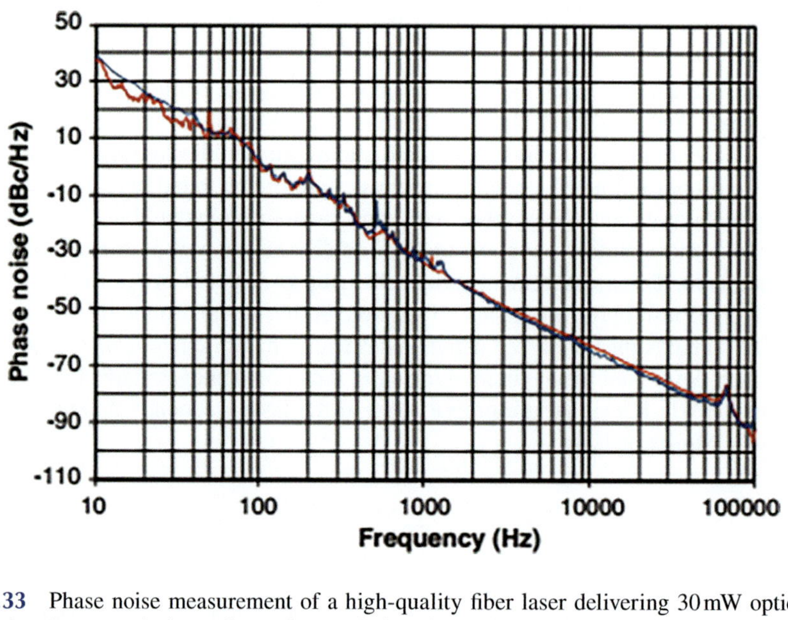

FIGURE 11.33 Phase noise measurement of a high-quality fiber laser delivering 30 mW optical power at $\lambda = 1{:}55$ μm using the two techniques (homodyne technique in medium dark line and self-heterodyne technique in darker line).

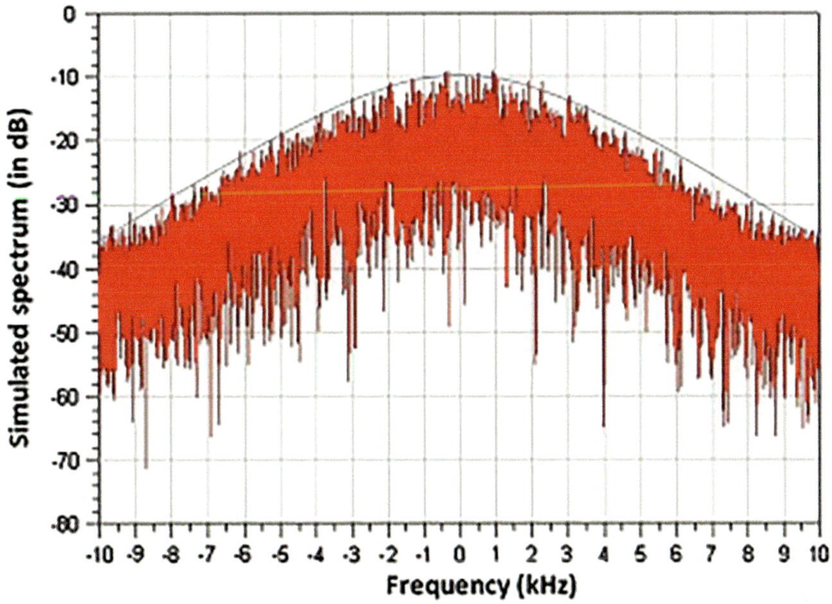

FIGURE 11.34 Laser power spectrum versus the offset from the carrier frequency simulated from the phase noise data using Agilent Advanced Design System and the envelope simulation technique.

To estimate the laser linewidth, we used an original approach based on the simulation of the power spectrum using commercially available microwave computer-aided design software, Agilent Advanced Design System (ADS). This software allows the description of a noisy frequency source through its phase noise spectrum. Additionally, it includes a simulation module based on the envelope technique[34] which enables the computation of slow perturbations of the carrier, such as modulations by deterministic or noisy signals. The result of the simulated spectrum for our laser is shown in Figure 11.34. As the simulated spectrum changes with the window used for both the phase noise description and the baseband temporal analysis window, we considered phase noise data between 100 Hz and 100 kHz for the simulation and set a maximum computation time of 0:5 s. The simulated result gives a 3 dB full linewidth of about 5.0 kHz. This result was compared to the simplified approach in Ref.[33] The frequency noise spectral density of our laser can be roughly described by the following equation:

$$S_{\Delta f} = \frac{k}{f}; k = 10^6 \text{ Hz} \tag{11.19}$$

taking into account an integration time T_0 ¼ 0:5 s, a linewidth of 6 kHz is computed for this laser, which is in relatively good agreement with the result of the envelope simulation approach.

11.2.3.5 Integrated Laser and Modulator

Integration of photonic components is important to reduce noise and system complexity, enhancing compactness. Recently, the demand for compact modules has led to the availability of TOSA and ROSA (Transmitter and Receiver Optical Sub-Assembly) through the integration of laser structures and modulators using electro-absorption as the modulation medium. These sub-assembly modules offer highly compact system integration for data centers and fiber-to-the-X (FTTX) applications (Figure 11.35).

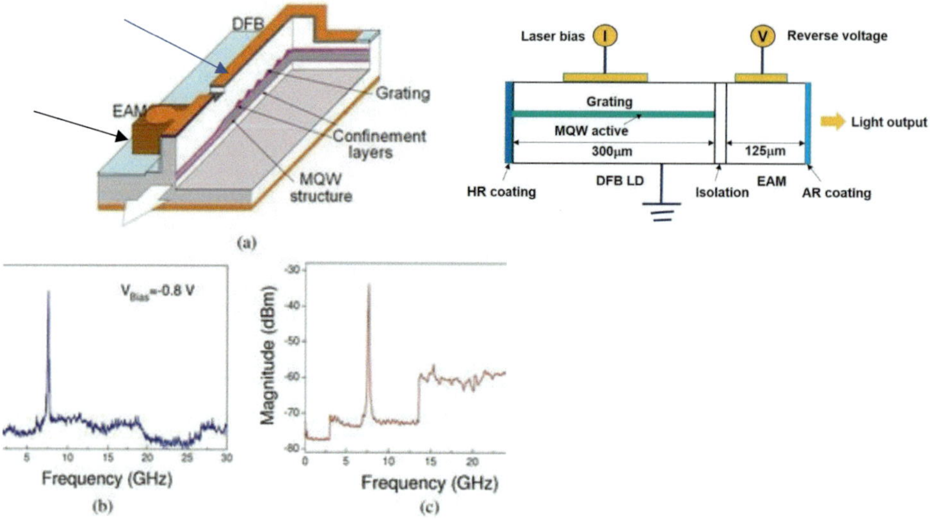

FIGURE 11.35 Schematic of TOSA and ROSA using II-VI compound semiconductor sub-assembly modules. (d) Schematic structure; (b) integrated DFB and EA modulator with multi-quantum well as an active medium waveguide for laser and (c) emission spectrum of laser, (a) sectional view of the TOSA.

11.2.3.6 Remarks

Lasers exhibit intensity and phase noise resulting from various internal and external influences, with the impact strongly influenced by the internal dynamics. Noise can be minimized in various ways, and the best approach depends on the specific circumstances. Intensity noise is generally easy to measure, but proper specifications require a good understanding of certain mathematical aspects, particularly the influence of measurement time and bandwidth, as well as power spectral densities. Phase noise measurements involve additional difficulties, partly due to the need for a phase reference and the mathematical complexity of a diverging phase noise PSD. The laser linewidth is directly related to phase noise but contains much less information than the entire phase noise spectrum. Quantum noise alone in a laser leads to the Schawlow–Townes linewidth, which, however, is often not reached due to additional noise influences. An experimental setup to determine these parameters of a single-mode semiconductor laser is also provided.

11.2.4 Self-Locking Frequency Integrated Laser

11.2.4.1 Motivation

The well-known challenges in low-cost networking for 5G terrestrial mobile and space photonic communications networks include the need for lasers to be uncooled and locked into a specific frequency range. This ensures that the laser carrier remains within the allowable tolerable frequency range for both coherent and non-coherent reception in optical transmission systems. The operational temperature range is –40°C to 85°C. Self-locking offers a significant advantage as it eliminates the need for manual adjustment or remote control of the laser's central wavelength and linewidth.

This section refers to the following Refs.[37,38,39,40,41,42]

Common traditional lasers (DFB, FP by Quantum Well) offer gain mediums that are not very stable with respect to temperature. Quantum Dot (QD) lasers, however, provide a gain spectrum with a consistent gain spectral window across the temperature range. This property allows for the incorporation of optical control feedback and external cavities to tune and stabilize the laser to a degree that is acceptable for 5G front-haul and SPC networking.

Uncooled, stable, and tunable lasers offer significant advantages and economic benefits for access, 5G front-haul, metro, and space inter- and intra-constellation orbital networking in the current market explosion.

One existing solution is:

a. Tunable hybrid laser with a silicon-photonic switch: Patent and references[43,44,45]
b. Demonstration of 12.2% wall plug efficiency in an uncooled single-mode external-cavity tunable
c. External cavity laser with automatic tuning: UTQDECL (ultra-tunable quantum dot external cavity laser) uses a QD (quantum dot) gain chip, offering a highly stable and flat gain spectrum with a stable spectral window when the temperature varies from −40°C to 85°C. This common spectral window gain is utilized in a feedback external cavity with monitoring and simple DSP control for a tunable external cavity laser under uncooled operation.
d. Micro-ring resonator pairs plus planar etched grating: This setup aims to obtain a narrow linewidth by providing feedback to a reflective optical gain, forming an uncooled lasing cavity of high quality, exceeding the requirements for metro/access optical networking.

The claims in this section are supported by simulation, design, and experimental demonstration of a novel structure with low-loss coupling via a spot-size converter and a pair of micro-ring resonators.

The unique gain spectrum of the QD gain chip is used in reflection mode.

The external cavity structure consists of two MRR plus a common planar spatial grating, which further narrows down the selected optical frequency for tuning and lasing. This configuration enables stable temperature operation over the range of −40°C to +85°C for the self-locking of the integrated laser's frequency.

11.2.4.2 Structure

The schematic of the self-frequency locking structure is shown in Figure 11.45. The design has been made to keep the final device as simple as possible to minimize the impact of temperature variations in the vacuum of the SPC transmission medium. It should be noted that any building block in the device will be sensitive to temperature variation; therefore, the challenge is to minimize the number of required blocks. The design also simplifies the upgrade for multi-wavelength operation. Basic steps of the locking mechanism:

- An initial adjustment is required to tune the PS1 and PS2 sections to achieve the target wavelength at room temperature.
- The thermo-optic response of the PS1 and PS2 sections must also be initially characterized at the control structures.
- The control ring is used to obtain the monitored power but will also contribute to improving the side mode suppression ratio (SMSR).
- The variation in the monitored power, which ideally should be zero, will be used to simultaneously adjust the PS1 and PS2 sections to counteract cooling or heating effects.

Thus, the laser design requirements are:

- The optical bandwidth of the DBR reflector must be broad enough to be insensitive to temperature variations in the chosen range (as in the QD-Gain chip).
- A 2π phase shift is required at the PS2 section to counteract both cooling and heating effects.
- For wider wavelength tuning ranges, a Vernier ring structure could be used. Such a structure will also be designed during the project's design stage. However, the principle of operation of the locking mechanism will not change.

The operational principles of a number of options are described in the following sub-sections.

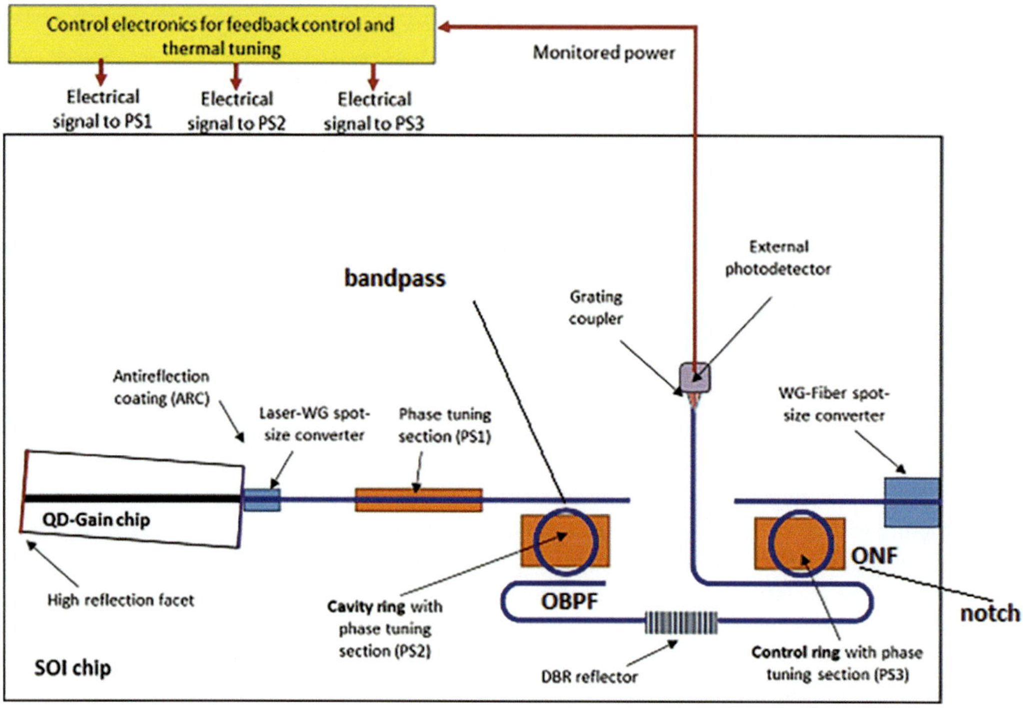

FIGURE 11.36 Idea 1- using two (2×2 μR) in cascade including a DBR. A notch port or drop port is used for the determination of the lasing control feedback. The signal dip is shown in Figures 11.37 and 11.38.

11.2.5 Structure Option 1

Two 2×2 micro-ring resonators (MRR) connected with thru (bandpass) and drop (notch) ports in cascade (Figure 11.36).

11.2.5.1 Principles- Mechanism 1

- The gain section is coated with a highly reflective coating with a coefficient of ~100%. If the reflection is only 99%, a PD can be installed at the back to monitor the optical power output. The gain section has a very wide spectrum, which can be on the order of 60–80 nm in the O-band (supplied by Innolume GmbH – Dortmund, Germany). The output power from this gain section can reach up to 120 mW. The useful spectral window is on the order of 40 nm, within which there is no gain variation due to temperature changes from –10°C to +85°C.
- All structures are fabricated on a Si-on-insulator substrate. Thus, Si waveguides can be either wide or narrow. In this case, with no high-speed modulation, a 4 μm wide waveguide can be used to minimize propagation loss.
- The structure to control the frequency of the emitting/lasing lightwaves and its stability consists of two sections:
  - Section 1: A 2×2 micro-ring (2×2 μR) resonator is used, where the pass through port is a bandpass filtering port connected to a DBR (distributed Bragg Reflector) as section 2 to ensure that the lightwave predetermined by the DBR is reflected to the cavity, including the gain section. A second (2×2 μR) is then connected in cascade with the DBR and the first (2×2 μR), but the signals are extracted from the drop port. The combined spectrum of the two MRRs can be seen in Figure 11.37. The fall and rise edges of the notch band are symmetrical, provided the coupling of the two waveguide couplers is identical.

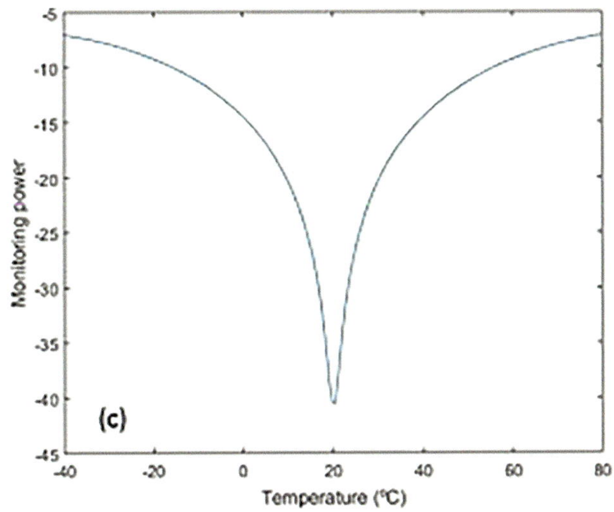

FIGURE 11.37 Transmission characteristics for feedback control.

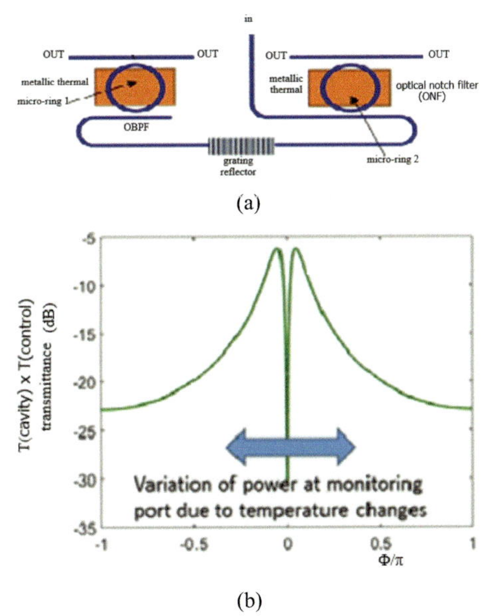

FIGURE 11.38 Optical transfer characteristic curves to be employed by the electronic control feedback of the hybrid laser–cascading. OBPF=Optical bandpass filter, ONF=optical notch filtering (a) integrated planar structure of the micro-ring resonance for locking of desired lightwaves at specific wavelength; (b) Transmittance characteristics of the micro ring structure with a pass band and a notch band centered at the central passband.

- When cascading the OBPF and ONF (see Figure 11.38a), the resulting transmittance from the input port to the output port is obtained, as shown in Figure 11.38b. As observed, the dip of the notch filtering point is located symmetrically at the peak of the resonance peak of the OBPF.

11.2.6 Structure Option 2 for SPC Systems in Space: Autonomous Tuning of Wavelength Using Integrated [2 × 2] and [3 × 3] (MRR)

The operational principles of this Option 2 by referring to Figure 11.39 can be described in the following steps:

The quantum dot gain chip is biased and driven such that it is free-running; effectively, it is a gain medium with reflection at the back so that 100% of lightwaves can be reflected. An angled mirror is formed to ensure that no other scattering is reflected into this QD cavity.

A phase tuning section is integrated to tune the phase of the lightwaves so that the micro-ring resonator can be affected by the random phase of the lightwaves output of the QD gain section.

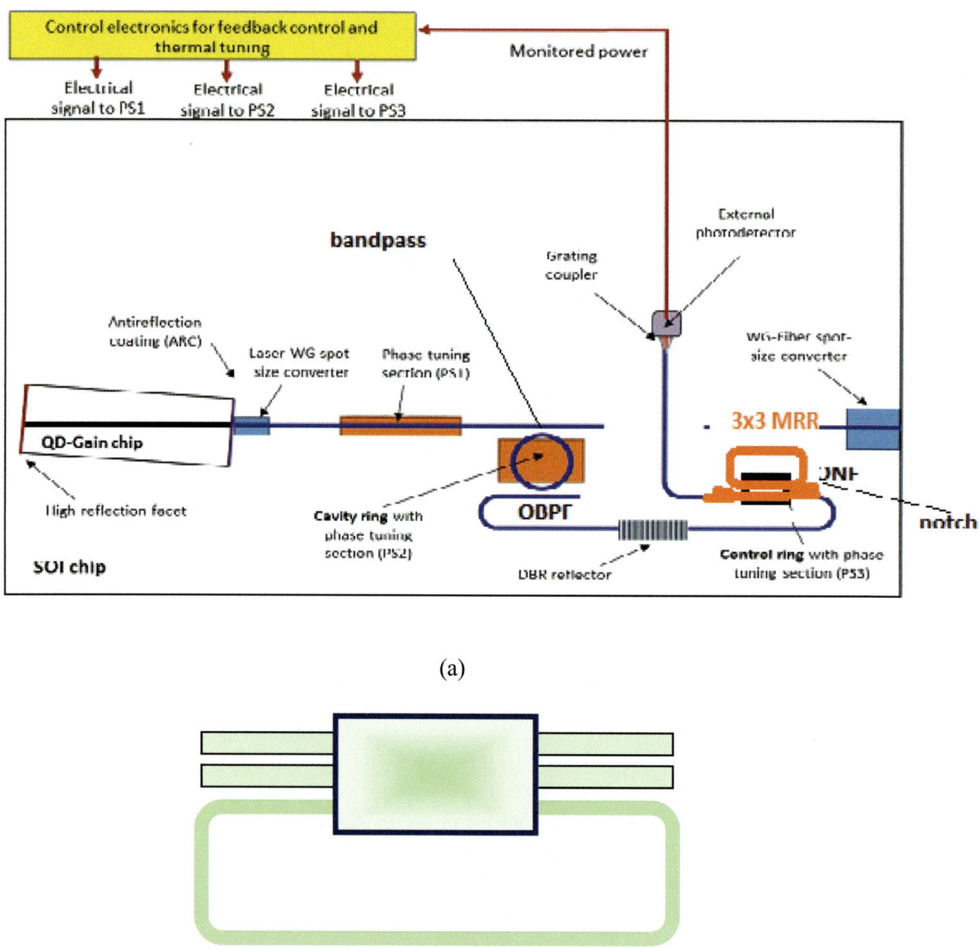

(a)

(b)

FIGURE 11.39 (a) Cascade structure of one (2×2 MPR) and a (3×3 μR FB) for sharp notch to determine fine wavelength lasing (see notch filtering transmission characteristics given in Figure 11.41); (b) detailed structure of (3×3 μR FB) using SOI (silicon on insulator). QD=quantum dot, OBPF=optical bandpass filter.

The cavity reflection from the other side is set up by using a grating section and one optical bandpass filter using an MRR, which is also tuned by an applied voltage that can be derived from feedback electrical signal monitoring.

Another micro-ring resonator is connected and operates as a notch filter so that the frequency of this notch determines the sharpness of the emitted laser via a 3×3 coupler. This ensures that the resonant port of Bandpass filtering can be matched to the micro-ring OBPF. The phase of 3×3 is π/3 (see Figure 11.39b), and this must also be considered and controlled by the electrical feedback signals.

Any fluctuation of the resonant frequency can be autonomously controlled; thus, this laser is very useful in SPC mounted on satellites and would face no problem with temperature variations or any radiation effects on the gain medium.

The structure can be fabricated using silicon on insulator (SOI) in which the refractive index of the waveguide is much higher than that of the substrate insulator (SiO_2). Thus, the bending radius can be much smaller, and hence the device size is quite small in some of tens of micrometers.

11.2.6.1 Principles and Derivation of the 3×3 Transmittance

The transfer matrix of the 3×3 MMI coupler without optical feedback from port b_3 back to port a_3 is given as

$$\begin{pmatrix} b_1 \\ b_2 \\ b_3 \end{pmatrix} = \begin{bmatrix} m_{11} & m_{12} & m_{13} \\ m_{21} & m_{22} & m_{23} \\ m_{31} & m_{32} & m_{33} \end{bmatrix} \begin{pmatrix} a_1 \\ a_2 \\ a_3 \end{pmatrix} = \left[M_{3\times3}\right] \begin{pmatrix} a_1 \\ a_2 \\ a_3 \end{pmatrix}$$

$$\text{where}: \left[M_{3\times3}\right] = \begin{bmatrix} -e^{-j2\pi/3} & e^{-j2\pi/3} & -1 \\ e^{-j2\pi/3} & -1 & e^{-j2\pi/3} \\ -1 & e^{-j2\pi/3} & -e^{-j2\pi/3} \end{bmatrix} \tag{11.20}$$

$$b_1, b_2, b_3 = \text{output port } 1, 2 \,\&\, 3;$$

$$a_1, a_2, a_3 = \text{input port } 1, 2 \,\&\, 3 \text{ of MMI } 3\times3 \text{ coupler}$$

Note that a 120° phase shift occurs between coupling port instead of π/2 for 2×2 coupler.

Now, with optical feedback, the analysis can be given as follows:

The phase passing through the MMI planar waveguide section is given by

$$\phi_0 = -\beta_0 l_{MMI} + \frac{9\pi}{24}$$

$$a_3 = \alpha e^{j\theta} b_3;\; \alpha = e^{-\alpha_0 L};\, l_{FB} = \text{feedback length} = 2\pi R + l_{MMI};$$

$$R = \text{radius of feedback circle; in unit dB/cm} \tag{11.21}$$

$$\theta = \beta_0 l_{FB} = \text{phase ccumulated over feedback path (optical Si waveguide)}$$

$$\beta_0 = \frac{2\pi n_{eff}}{\lambda};\, n_{eff} = \text{effective index of the guide mode; } \lambda = \text{operating wavelength}$$

Thus, the complex amplitudes of the optical fields of the output port b_1 and b_2 of the 3×3 coupler with optical feedback from port output 3 back to port input a_3, resulting in a resonator, can be written as

$$\begin{pmatrix} b_1 \\ b_2 \end{pmatrix} = \begin{bmatrix} M_{11} & M_{12} \\ M_{21} & M_{22} \end{bmatrix} \begin{pmatrix} a_1 \\ a_2 \end{pmatrix}$$

$$M_{11} = m_{11} + \frac{m_{13}m_{31}\alpha e^{j\theta}}{1 - m_{33}\alpha e^{j\theta}}; M_{12} = m_{12} + \frac{m_{13}m_{32}\alpha e^{j\theta}}{1 - m_{33}\alpha e^{j\theta}}$$

$$M_{22} = m_{11} + \frac{m_{23}m_{32}\alpha e^{j\theta}}{1 - m_{33}\alpha e^{j\theta}}; M_{21} = m_{21} + \frac{m_{23}m_{32}\alpha e^{j\theta}}{1 - m_{33}\alpha e^{j\theta}} \tag{11.22}$$

$$\rightarrow$$

$$T_{b_1/a_1} = \left|\frac{b_1}{a_1}\right|^2 = \text{power transfer port } a_1 \text{ to port } b_1 \text{ under optical feedback}$$

where the coefficients $m_{11} \ldots m_{33}$ are given by the matrix $[M_{3\times3}]$, that is the coupling coefficients of the 3×3 matrix without the optical feedback.

11.2.6.2 The 2×2 Transmittance with Optical Feedback → Micro-Ring Resonator

The coupling matrix under the case with optical feedback, as shown in Figure 11.40, is given as

$$\begin{pmatrix} b_1 \\ b_2 \end{pmatrix} = \begin{bmatrix} \tau & \kappa \\ -\kappa^* & \tau^* \end{bmatrix} \begin{pmatrix} a_1 \\ a_2 \end{pmatrix}$$

$$* = \text{conjugation}; \tau = \text{thru coupling coefficient}; \kappa = \text{cross coupling coefficient}$$

$$\tau^2 + \kappa^2 = 1 \tag{11.23}$$

$$a_2 = e^{j\varphi} b_2;$$

$$\varphi = \frac{2\pi n_{\text{eff}}}{\lambda} L;$$

$$L = \text{length of the optical feedback path} = \text{ring resonator length}$$

Thus, the transmittance transfer function is given as:

$$T_{11} = \left|\frac{b_1}{a_1}\right| = \frac{\alpha - 2\alpha\tau\cos\varphi + \tau^2}{1 - 2\alpha\tau\cos\varphi + \alpha^2\tau^2} \tag{11.24}$$

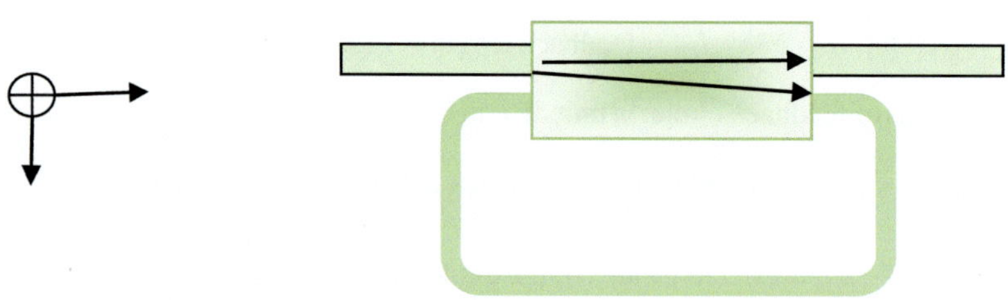

FIGURE 11.40 Detailed structure of (2×2 μR FB) using SOI. MMI=multi-mode interference coupler with planar waveguide section with interference coupling.

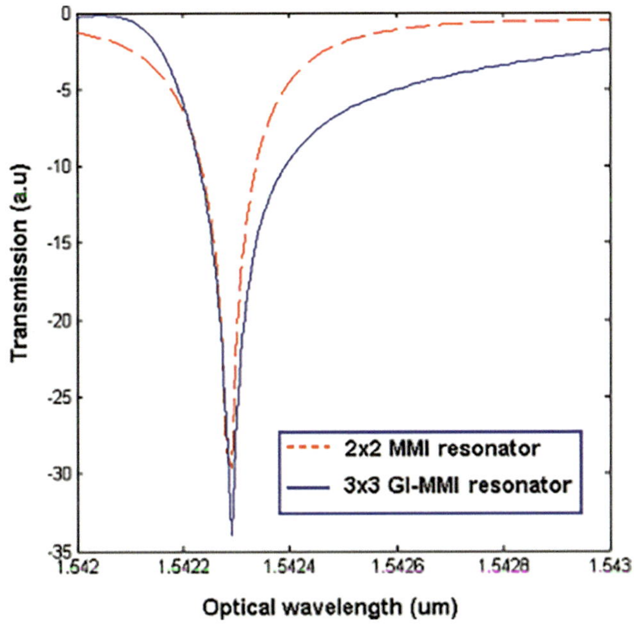

FIGURE 11.41 Notch wavelength determined by the transfer transmittance of the (3×3 micro-ring FB) structure.

11.2.6.3 Responses of 2×2 and 3×3 FB MMI Micro-Ring Structures

The transmittances of the types of ring resonators based on 2×2 and 3×3 with optical feedback are shown in Figure 11.41 using Equations 11.22 and 11.24 given in the previous section.

11.2.6.4 Signal Flow Graph and Mason Rule Derivation[46]

11.2.6.4.1 Signal Flow Graph Representation

The transfer function or the transmittance in terms of the field of the optical lightwaves propagating, coupling, reflecting, etc., can be represented by the signal flow graph, as described in Ref.[44,47] A 2×2 MRR can be re-drawn, as shown in Figure 11.42. The optical couplers (2×2) are assigned with port numbering, complex coupling coefficients, and the propagation loss of the optical paths. The cross-coupling intensity coefficient is κ, and the cross-coupling coefficient is $\tau^{1/2} = \sqrt{1-\kappa}$. The transmittance over one round trip of the resonance path of the ring can be written as

$$\Upsilon = e^{-\alpha L} z^{-1} \tag{11.25}$$

where the Z-transform or sampling variable is defined with a unit propagation of the lightwave of radical frequency ω over one round trip length L, the resonance optical feedback length. Thus, it can be written as

$$\begin{aligned} z &= e^{j\omega t} = e^{-j\beta L} \\ \beta &= \frac{2\pi n_{\text{eff}}}{\lambda} \end{aligned} \tag{11.26}$$

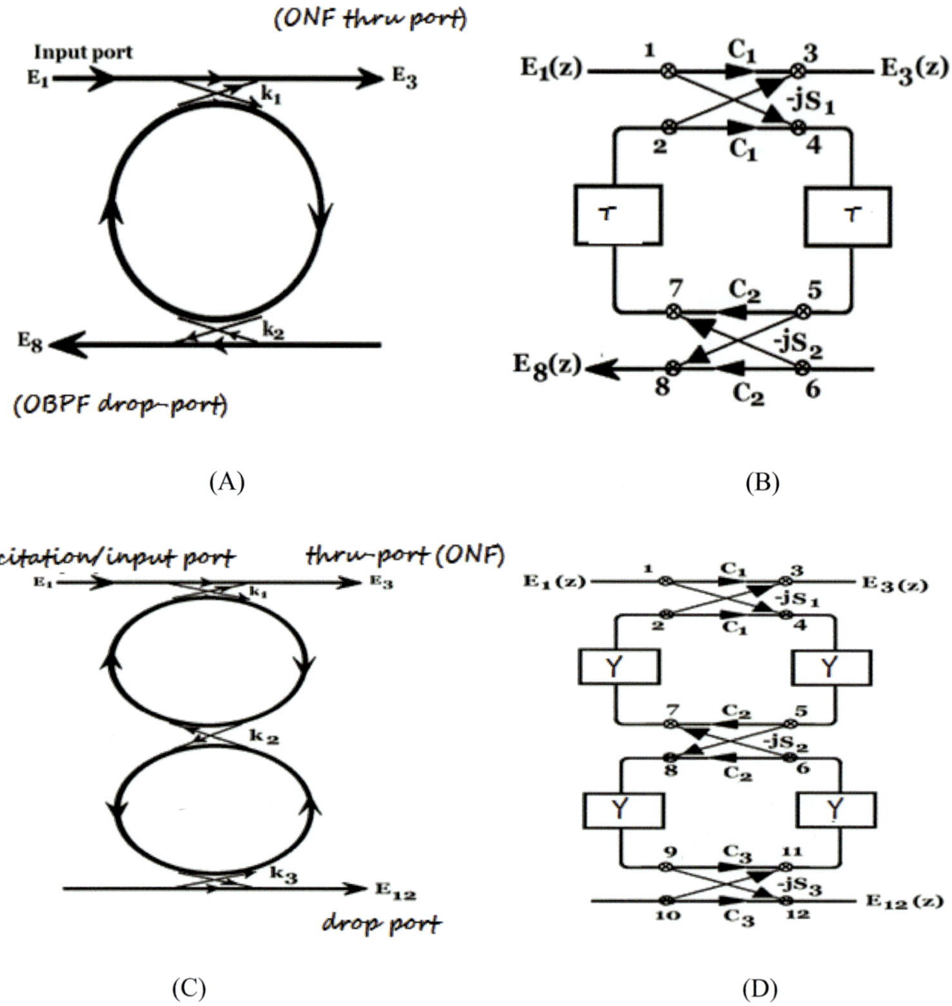

FIGURE 11.42 Signal flow graph representation of the micro-ring resonator using a single ring (a) and (b) and a double ring (c) and (d). $\Upsilon = e^{-\alpha L/4} z^{-1/2}$ Transmittance over half of the resonance loop. The loop length is assumed the same for all loops.

11.2.6.4.2 Mason Gain Formula and Transfer Function

11.2.6.4.2.1 Basic Gain and Transfer Function Establishment Rules According to the Mason gain formula, which is generally applicable to any type of network but was developed for electrical networks in the 1960s, the transfer function from node 1 to node N, H_{1N}, can be derived by

$$H_{1N} = \frac{1}{\Delta}\sum_{i=1}^{N} T_i \Delta_i \tag{11.27}$$

where T_i is the transmission gain of the ith forward path from the input port to the output port, Δ is the determinant of the network, or effectively, the natural response of the network regardless of the input and output port and Δ_i is the determinant Δ evaluated on the network with the loops that touch the path T_i eliminated.

All transfer functions between nodes of the networks would have this same common denominator determinant, which is given by

$$\Delta = 1 - \sum_i T_i + \sum_{i,j} T_i T_j - \sum_{i,j,k} T_i T_j T_k + ... \tag{11.28}$$

The transmission gains T_i of the *i*th ***loop.*** In the summation packet only, non-touching optical loops are included and no other touched loops, that is, loops do not have common nodes. The minus sign is for a sum of products of an odd number of loop gains.

11.2.6.4.2.2 Signal Flow Graph (SFG), Mason Rule of Single Loop MRR Referring to the signal flow graph shown in Figure 11.42, we can tabulate the loop gain and transmission path transmission for the MRR of a single loop (Figure 11.42a and b) (Table 11.3).

Thus, using Equation 11.28, we can obtain the transfer function, the transmittance between nodes 1 and 3 (the thru port or the notch filtering port)

$$\frac{E_3(z)}{E_1(z)} = H_1^1 = \frac{\tau_1 - \tau_2 e^{-\alpha L/2} z^{-1}}{1 - \tau_1 \tau_2 e^{-\alpha L/2} z^{-1}}; \quad \frac{E_8(z)}{E_1(z)} = H_{\text{drop}}^1 = \frac{\kappa_1 \kappa_2 e^{-\alpha L/4} z^{-1/2}}{1 - \tau_1 \tau_2 e^{-\alpha L/2} z^{-1}} \tag{11.29}$$

Similarly, the transfer function/transmittance in terms of the optical field from Port 1 to Port 8 (drop port – OBPF), $\frac{E_8(z)}{E_1(z)}$ can be obtained, as (via the use of the matrix components) given in Table 11.4.

TABLE 11.3
Loop Gain and Transmission Path for Input Port to Thru Port (OBPF) for MRR of Figure 11.42

Parameters	Expressions
Loop gain (only one loop)	$L_1^1 = \tau_1 \tau_2 e^{-\alpha L/2} z^{-1}$
Forward path gain/transmittance and related determinant	$T_{1t}^1 = -\tau_1 \kappa_1^2 e^{-\alpha L/2} z^{-1}; \Delta_1 = 1$
	$T_{2t}^1 = -\tau_1; \Delta_2 = 1 - \tau_1 \tau_2 e^{-\alpha L/2} z^{-1}$
Overall determinant	$\Delta_2 = 1 - \tau_1 \tau_2 e^{-\alpha L/2} z^{-1}$

TABLE 11.4
Loop Gain and Transmission Path for Input Port to Drop Port (OBPF) for MRR of Figure 11.42

Parameters	Expressions
Loop gain (only one loop)	None
Forward path gain/transmittance and related determinant	$T_{1d}^1 = -\kappa_1 \kappa_2 e^{-\alpha L/4} z^{-1/2}$
Overall photonic network determinant	$\Delta_2 = 1 - \tau_1 \tau_2 e^{-\alpha L/2} z^{-1}$

11.2.6.4.2.3 Signal Flow Graph (SFG), Mason Rule in Double Loop MRR The loop gains and the transmittance of the forward optical paths can be listed in Table 11.5.

Thus, the transfer function/transmittance between Ports 1 and 3 (thru-port-ONF) and Ports 1 and 12 (drop port -OBPF) can be written as

$$\frac{E_3(z)}{E_1(z)} = H_1^2 = \frac{\tau_1 - \tau_2 e^{-\alpha L/2} z^{-1} - \tau_1\tau_2\tau_3 e^{-\alpha L/2} z^{-1} + \tau_2 e^{-\alpha L} z^{-2}}{1 - \tau_1\tau_2 e^{-\alpha L/2} z^{-1} - \tau_2\tau_3 e^{-\alpha L/2} z^{-1} + \tau_1\tau_3 e^{-\alpha L/2} z^{-1}} \tag{11.30}$$

$$\frac{E_{12}(z)}{E_1(z)} = H_{\text{drop}}^1 = \frac{j\kappa_1\kappa_2\kappa_3 e^{-\alpha L/2} z^{-1}}{1 - \tau_1\tau_2 e^{-\alpha L/2} z^{-1} - \tau_2\tau_3 e^{-\alpha L/2} z^{-1} + \tau_1\tau_3 e^{-\alpha L/2} z^{-1}} \tag{11.31}$$

The filter response can be plotted with the coupling coefficient as a parameter, as shown in Tables 11.5 and 11.6. Refer also to Figures 11.43 and 11.44.

Note the different uses of MRR bandpass filters in cascade in Figure 11.45 and the use of notch and bandpass filters in the next Option 2 structure.

TABLE 11.5

Loop Gain and Transmission Path for Input Port to Drop Port (OBPF) for Double Ring MRR of Figure 11.42c and d

Parameters	Expressions
Loop gain	$L_1^2 = \tau_1\tau_2 e^{-\alpha L/2} z^{-1}$
	$L_2^2 = \tau_2\tau_3 e^{-\alpha L/2} z^{-1}$
	$L_3^2 = \tau_1\tau_3\kappa_1^2 e^{-\alpha L} z^{-2}$
Forward path gain/transmittance and corresponding determinant	$T_{1t}^2 = -\kappa_1\tau_2 e^{-\alpha L/2} z^{-1}$
	$\Delta_1 = 1 - L_2^2 = 1 - \tau_2\kappa_1^2 e^{-\alpha L/2} z^{-1}$
	$T_{2t}^2 = -\tau_3\kappa_2\kappa_1 e^{-\alpha L} z^{-2}$
	$\Delta_2 = 1$
	$T_{3t}^2 = \tau_1$
	$\Delta_3 = 1 - \tau_1\tau_2 e^{-\alpha L/2} z^{-1} - \tau_2\tau_3 e^{-\alpha L/2} z^{-1} - \tau_1\tau_3\kappa_3 e^{-\alpha L/2} z^{-1} + \tau_1\tau_2^2\kappa_3 e^{-\alpha L} z^{-2}$
Non-touching loop	$L_{12}^2 = \tau_1\tau_2^2\tau_3\kappa_1^2 e^{-\alpha L} z^{-2}$
Overall photonic network determinant	$\Delta = 1 - \tau_1\tau_2 e^{-\alpha L/2} z^{-1} - \tau_2\tau_3 e^{-\alpha L} z^{-2} + \tau_1\tau_3 e^{-\alpha L} z^{-2}$

TABLE 11.6

Loop Gain and Transmission Path for Input Port to Drop Port 12 (OBPF) for the Double MRR of Figure 11.42c and d for the Double Ring MRR given in Figure 11.42c and D

Parameters	Expressions
Loop gain	$L_1^2 = \tau_1\tau_2 e^{-\alpha L/2} z^{-1}$
	$L_2^2 = \tau_2\tau_3 e^{-\alpha L/2} z^{-1}$
	$L_3^2 = \tau_1\tau_3\kappa_1^2 e^{-\alpha L} z^{-2}$

(*Continued*)

TABLE 11.6 (*Continued*)
Loop Gain and Transmission Path for Input Port to Drop Port 12 (OBPF) for the Double MRR of Figure 11.42c and d for the Double Ring MRR given in Figure 11.42c and D

Parameters	Expressions
Forward path gain/transmittance and corresponding determinant	$T_{1t}^2 = -\kappa_1\tau_2 e^{-\alpha L/2} z^{-1}$
	$\Delta_1 = 1 - L_2^2 = 1 - \tau_2\kappa_1^2 e^{-\alpha L/2} z^{-1}$
	$T_{2t}^2 = -\tau_3\kappa_2\kappa_1 e^{-\alpha L} z^{-2}$
	$\Delta_2 = 1$
	$T_{3t}^2 = \tau_1$
	$\Delta_3 = 1 - \tau_1\tau_2 e^{-\alpha L/2} z^{-1} - \tau_2\tau_3 e^{-\alpha L/2} z^{-1} - \tau_1\tau_3\kappa_3 e^{-\alpha L/2} z^{-1} + \tau_1\tau_2^2\kappa_3 e^{-\alpha L/2} z^{-1}$
Non-touching loops	$L_{12}^2 = \tau_1\tau_2^2\tau_3\kappa_1^2 e^{-\alpha L} z^{-2}$
Overall photonic network determinant of SFG (signal flow graph)	$\Delta = 1 - \tau_1\tau_2 e^{-\alpha L/2} z^{-1} - \tau_2\tau_3 e^{-\alpha L/2} z^{-1} + \tau_1\tau_3 e^{-\alpha L/2} z^{-1}$

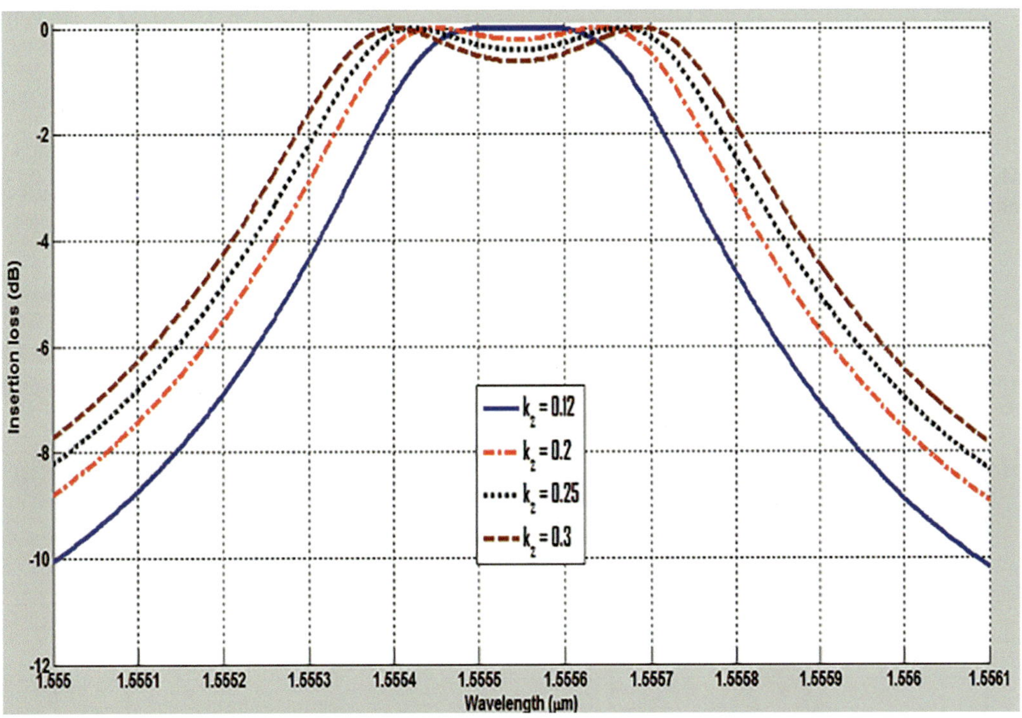

FIGURE 11.43 $k = \kappa$=coupling coefficients. Plots of the drop port from the excitation input put port No. 1 with $\kappa_1 = \kappa_3 = 50\%$; $\kappa_2 = 0.12, 0.2$, and 0.25 as a variable.

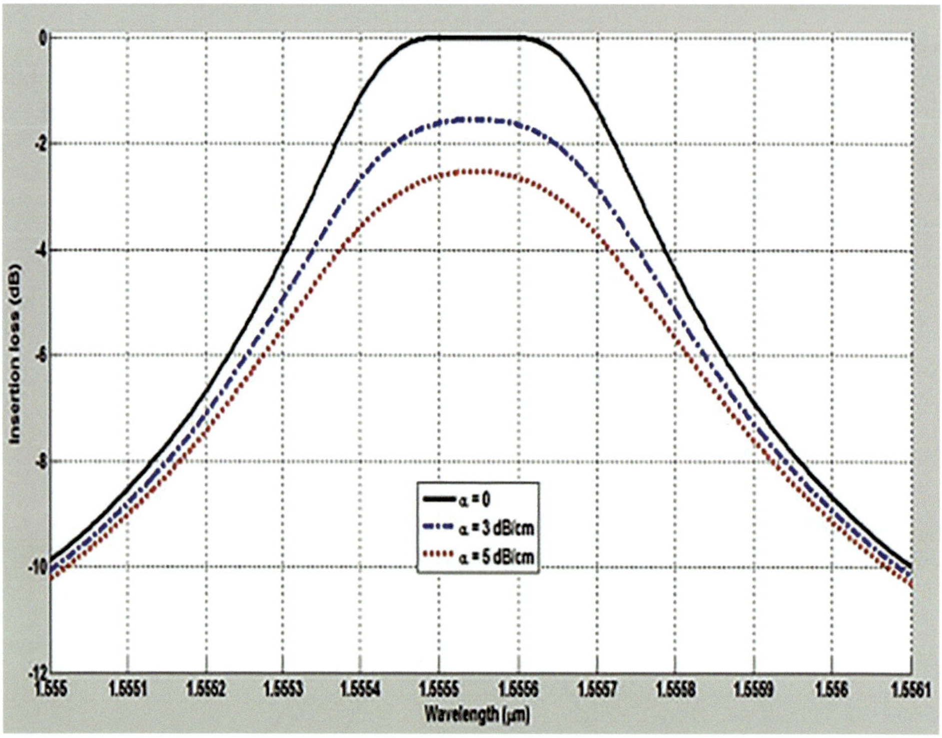

FIGURE 11.44 $k = \kappa$=coupling coefficients. The impact of attenuation coefficients filter shape and amplitude of the drop of the double ring MRR with $\kappa_1 = \kappa_3 = \kappa = 0.3$ and $\kappa_2 = 0.13$. Attenuation coefficient of 0, 3 and 5 dB/cm.

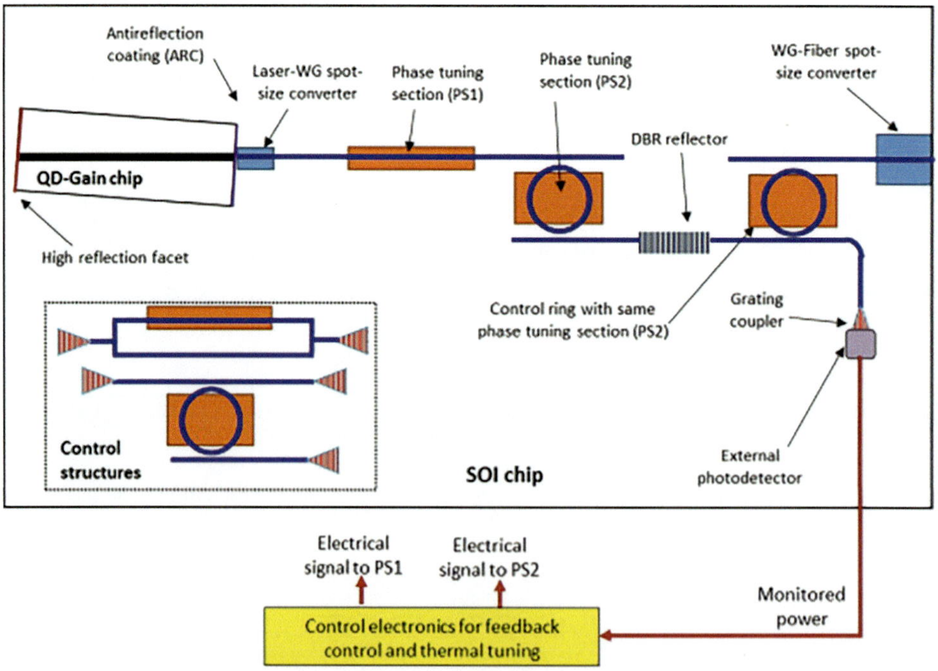

FIGURE 11.45 Schematic of self-locked frequency laser using micro-ring tunable and control feedback for locking to pre-set frequency.

11.2.6.5 MRR-DBR for Temperature Control via Optical Feedback

Figure 11.46 shows the structure of an MRR incorporating a DBR (distributed Bragg reflector), which is a Bragg grating in reflection mode but operates in transmission mode within the reflection band. The reflected lightwaves determine the lightwaves of specific wavelength (frequency). The bandpass of the MRR ensures that any fluctuation in temperature will allow such generated lightwaves to pass through. The resonant passband of this structure is shown in Figure 11.47.

The MRR-DBR structure is integrated in an external cavity of the laser. The spectra of the generated lightwaves can be seen in Figure 11.48. This shows both the bandpass and the resonant peaks of the generated lightwaves.

A fabricated structure of the number of sections of the MRR-DBR-QD gain is shown in Figures 11.49 and 11.50.

Figure 11.51a–c shows the modeling of the external cavity using MRR-DBR under threshold operating conditions caused by the bias of the gain medium, as well as the laser cavity model and the behavior of the integrated cavity with the MRR-DBR. This figure also shows both the resonant spectra and the micro-structure of the laser lines of the dynamics of the fabricated laser.

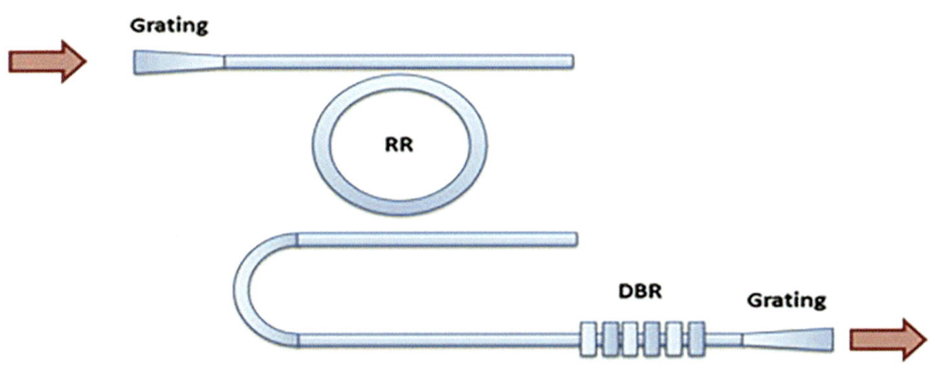

FIGURE 11.46 Micro-ring Resonator (MRR) and Distributed Bragg Reflector (DBR) for tuning automatic feedback locking of frequency stability.

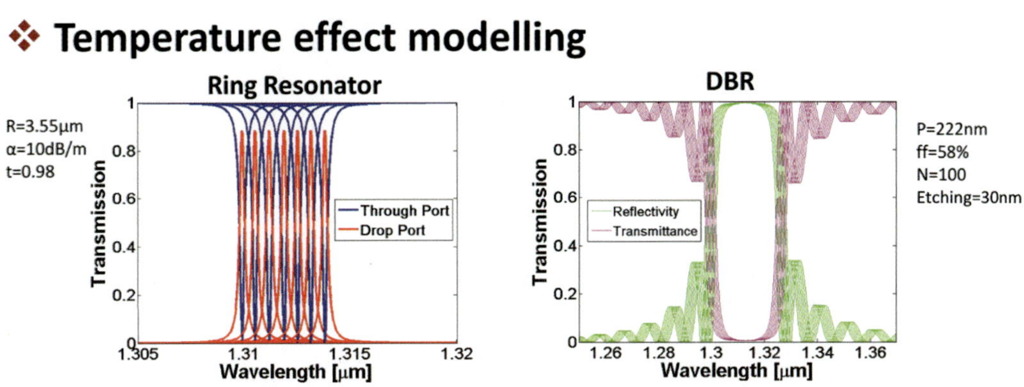

FIGURE 11.47 Resonant peak passband and notched band of the two ports of the MRR-DBR structure.

❖ External cavity modelling

- To achieve single mode lasing ring filters with a large free spectral range (FSR) are required such that there is only one resonance within the 1dB reflection bandwidth of the DBR grating
- Large SMSR will be attributed to the narrow bandwidth of the ring resonator structure.

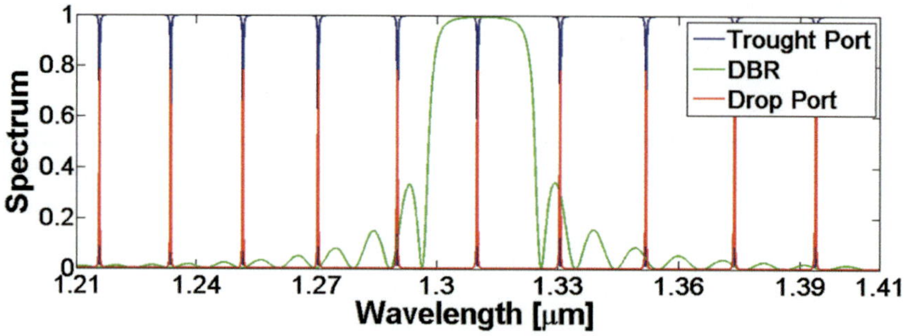

Ring Resonator:

- Radius=3.55μm →FSR=20.7nm
- t=0.99 → FWHM=0.146nm

DBR:

- Etching=30nm →FWHM=26.2nm

FIGURE 11.48 Optical spectra of lightwaves generated from the external cavity MRR-DBR autonomous self-tuning.

❖ External cavity modelling

- The FSR of that external cavity (filter) should be preferably larger o nearly than the gain bandwidth of the active section (QD-chip), to provide a predictable laser wavelength.

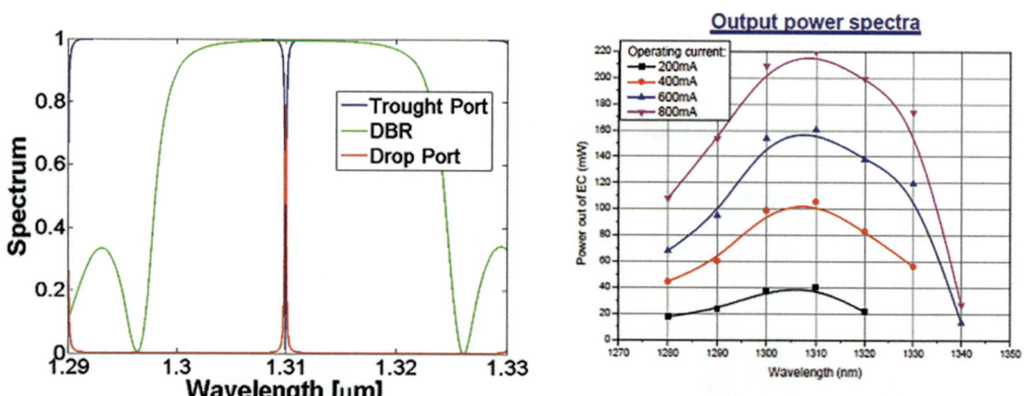

Innolume GC-1310-60-TO-200-A (Straight stripe gain chip for 1280-1340nm tuning range)

FIGURE 11.49 External cavity MRR-DBR plus a QD gain integrated medium for tuning within the cavity to ensure wavelength matching and prevent unlocking by temperature. Optical range of locking.

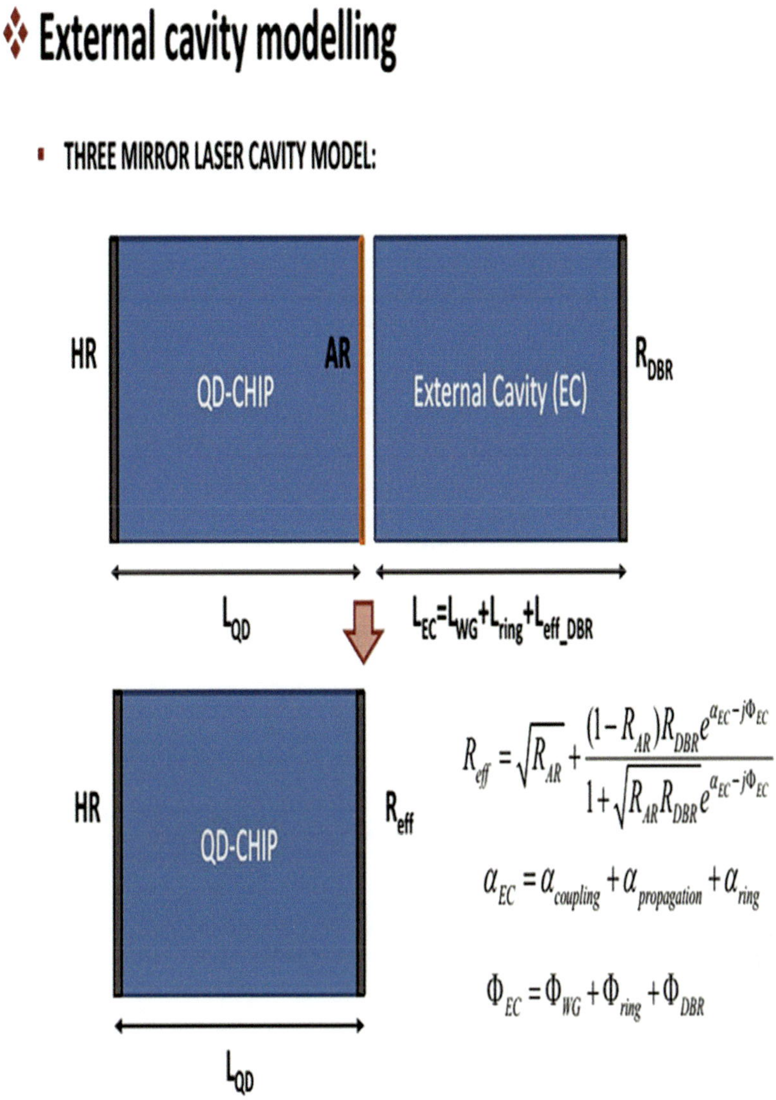

FIGURE 11.50 Fabricated integrated optical circuit structure and parameters. HR=high reflection, QD=quantum dot, DBR=distributed Bragg reflector.

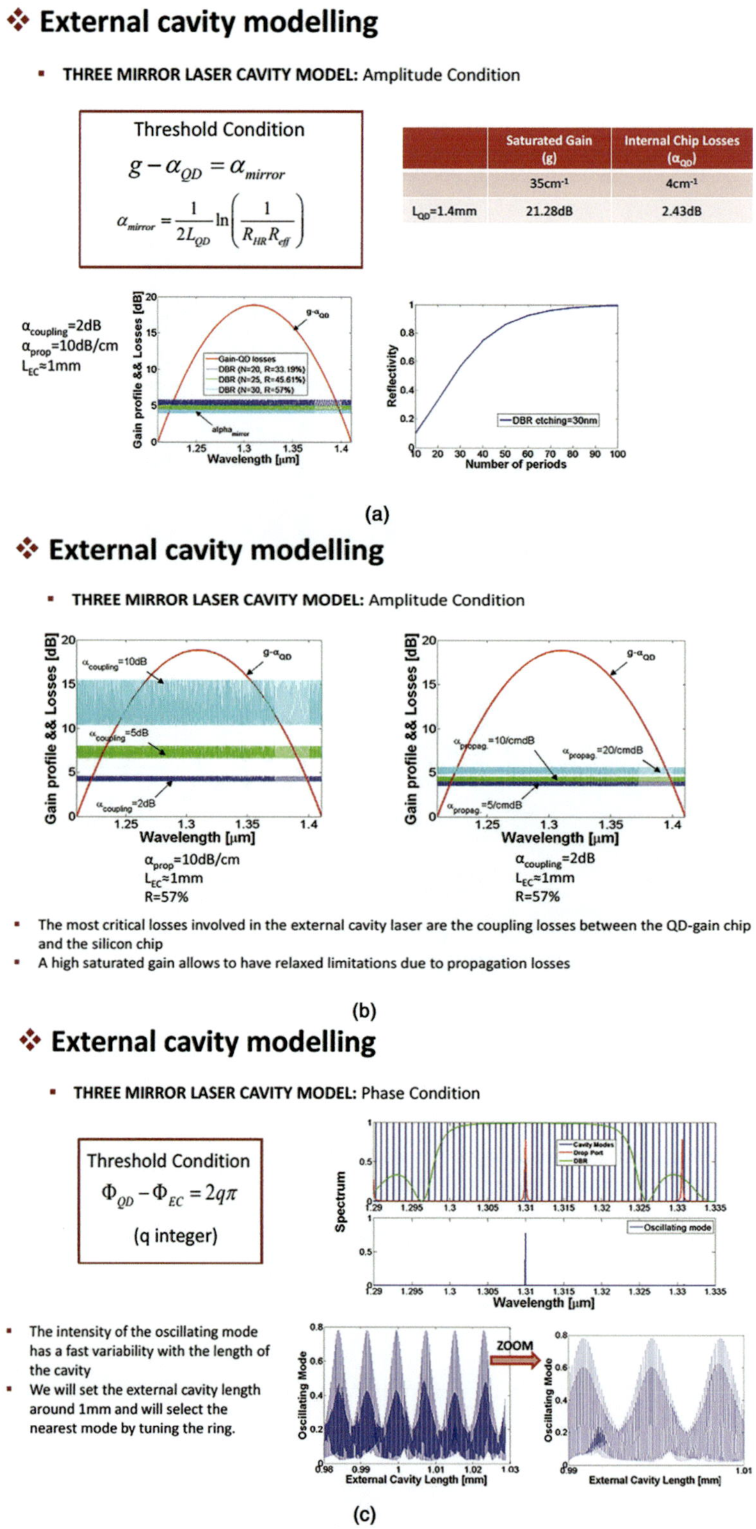

FIGURE 11.51 External cavity for feedback control for automatic locking (a) external cavity modelling for Passband and notch band and geometrical dimensions of manufactured micro-ring ; (b) close up view of the gain profile versus the lightwave wavelength.

11.2.6.6 Control Implementation

Figures 11.51 and 11.52 show the implementation of the final structure of the MRR-DBR cavity laser with the electronic control board for the operation of autonomous self-tuning laser to generate specific narrow linewidth laser for SPC systems operating in remote satellite networks.

11.2.6.7 Locking Algorithm by Transmission/Dropping Optical Feedback

Similar to emitter feedback in biasing electronics, the algorithm for controlling the operation of the external MRR-DBR laser by the micro central unit allows for electronic control of the laser in autonomous mode. These algorithms are shown in Figures 11.53 and 11.54. The feedback control is very similar to the classical emitter feedback for bias stability in an electronic small-signal amplifier.[19] The locking mechanism can be followed by the steps outlined in these figures. These electronic signals are obtained via photodetectors integrated and monitored at the optical output ports of the integrated optical laser.

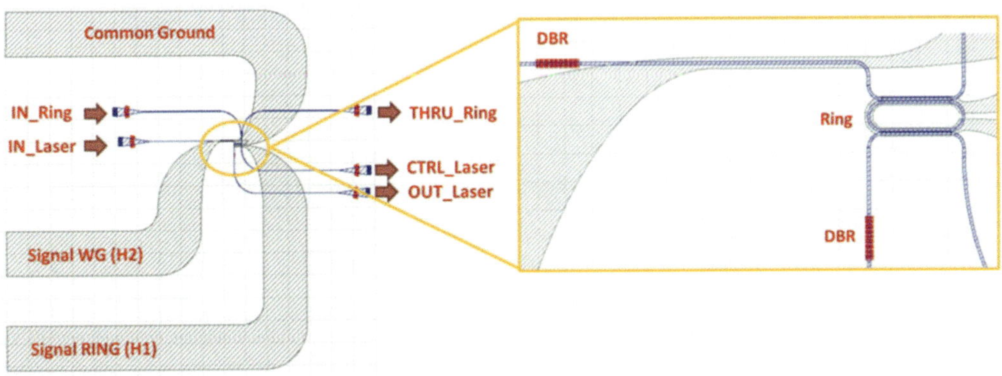

(a)

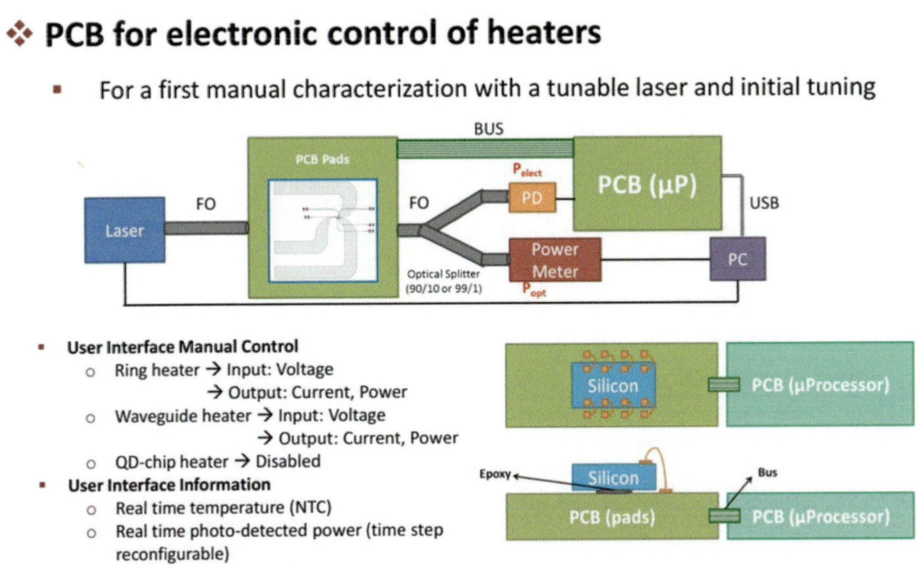

(b)

FIGURE 11.52 Optical integrated circuit (a) layout and (b) electrodes for autonomous control and tuning.

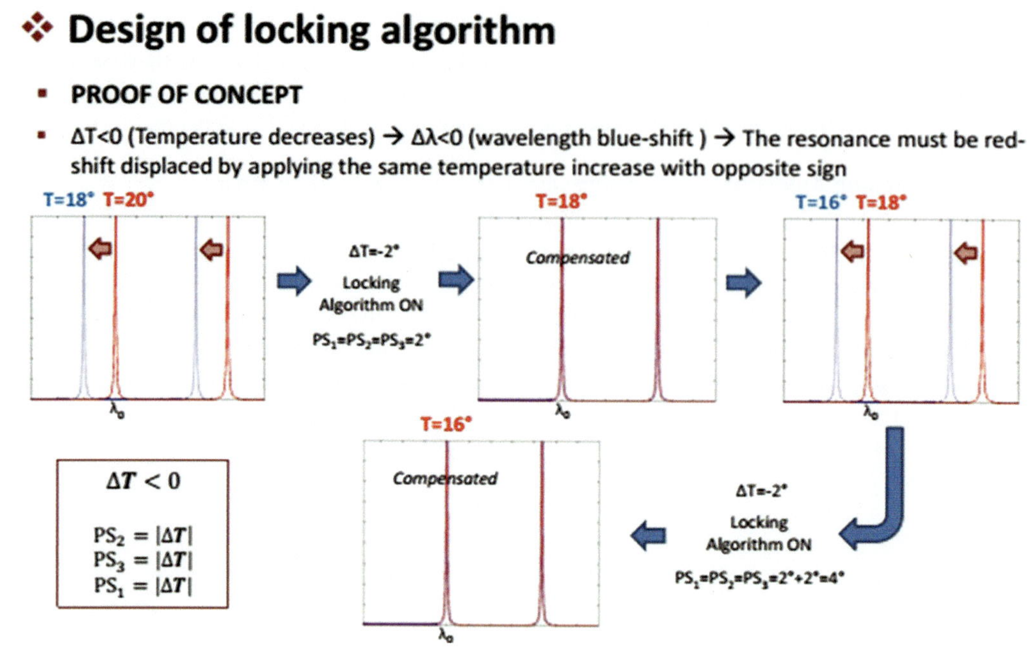

FIGURE 11.53 Similar to emitter feedback in biasing electronics, schematic of locking mechanism.

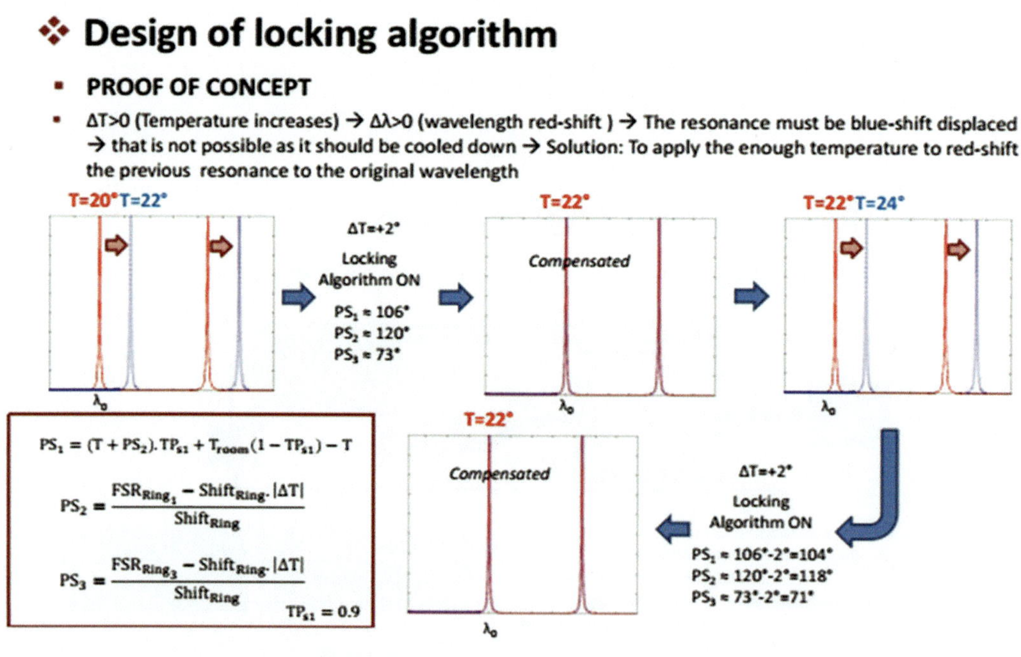

FIGURE 11.54 Design of locking mechanism.

Any increase or decrease in temperature will shift the resonant frequency of the lasing peak to left or right, and the control pulls this back by thermal heating of the electrode embedded in the micro ring. This is called feedback biasing stability of the external cavity. The emitting wavelength is set by a look-up table, and it can be set to any desired spectral line, hence controlling that exact wavelength.

The steps for fabrication of the laser can be seen in Figure 11.55. Figures 11.56 and 11.57 show the fabricated laser and its mounting on a jig for testing, along with the experimentally obtained signals to prove the reality of the laser.

❖Fabrication of samples for QD-flip-chip process optimization

- Previous steps (8-10 in table) to the QD flip-chip have been successfully carried out
- Status: Waiting for the receiving of the preforms (step 12 in table)
- Dummy chips are being processed with the same pads than QD-chips to optimize the flip-chip process (alignment and electrical contact)

Nº	Main steps	Area
1	SOI preparation	FABRICATION
2	Alignment marks	
3	Waveguides	
4	DBR	
5	Gratings	
6	Upper-cladding reduction	
7	SSCs windows & Pool	
8	Pool & QD Alignment marks	
9	SiO_2 thin-film	
10	Pads & Heaters	
11	Sample dicing	PACKAGING
12	Preforms & Solder bumps	
13	PD flip-chip	
14	QD flip-chip	
15	Wire bonding	
16	Fiber pigtailing	
17	PCB assembly & wire bonding	

FIGURE 11.55 Fabrication steps of MRR-DBR integrated circuit based on SOI integrated photonic technology.

❖ **EXPERIMENTAL RESULTS**

$$P_{out} / P_{in}[dB] \approx 2\alpha_{SCC}[dB] + \alpha[dB/cm]L_{wg}[cm]$$

(L_{wg}=0.5cm)

α≈4±2 dB/cm → α_{SSC}≈3.5±0.5dB

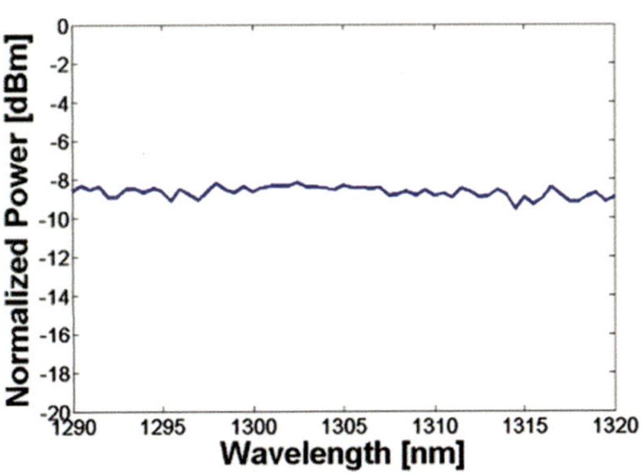

FIGURE 11.56 Experimental testing of fabricated external cavity for autonomous frequency locking of the external MRR-DBR cavity laser Left to right: emitting spectra of QD gain cavity and attenuation factor, mechanical mounting jig for coupling and testing.

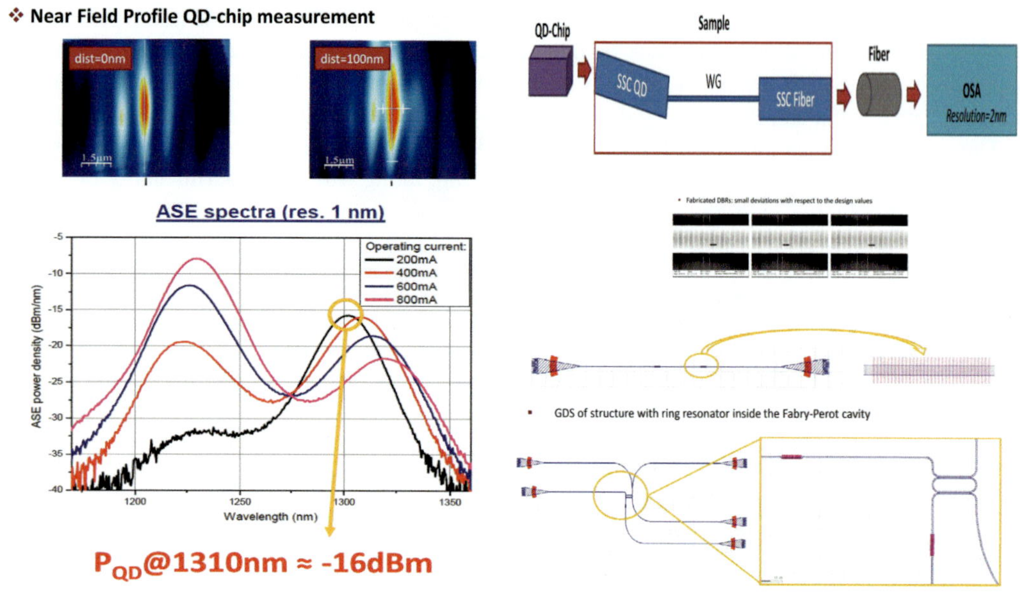

FIGURE 11.57 Measurement, testing, and integration of the external cavity MRR-DFB laser. Left to right: above left emitting intensity of the laser, Left bottom – spectra emitted of integrated laser with QD gain in 1310 nm region, Right above the structure of the QD gain medium and coupled to an external cavity, right bottom micro integrated structure of MRR-DBR external cavity laser.

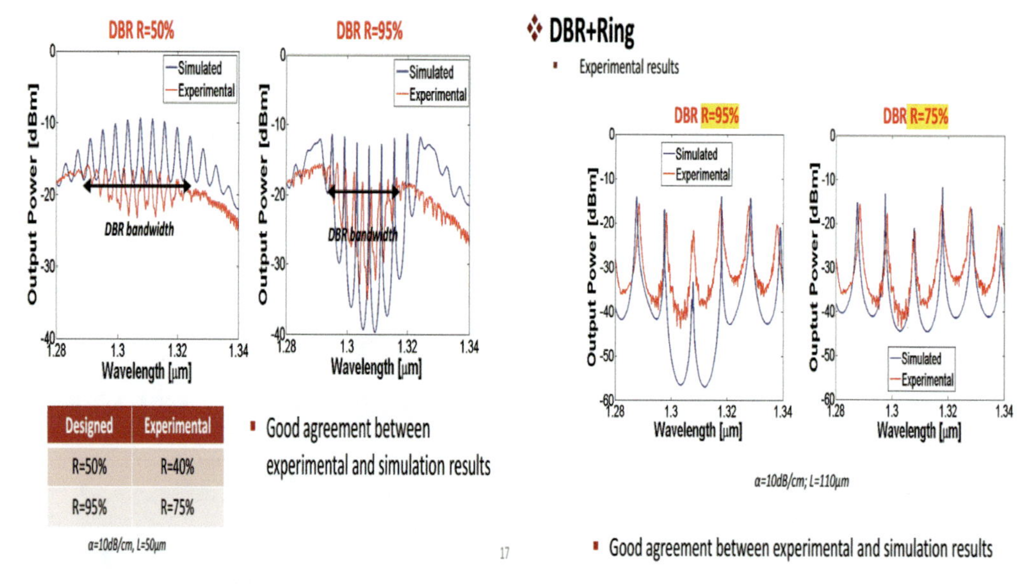

FIGURE 11.58 MRR and DBR performance confirmation.

Figure 11.58 shows the performance spectra of the MRR and the DBR structures, which confirm the simulated results obtained in Figures 11.48 and 11.49.

11.2.6.8 Fabrication of Autonomous ECL Self-Frequency Locking Laser

Figure 11.59 shows the fabricated array of MRR-DBR external cavity QD-lasing chips, allowing each structure to be tested and demonstrating the yield of the design and fabrication process.

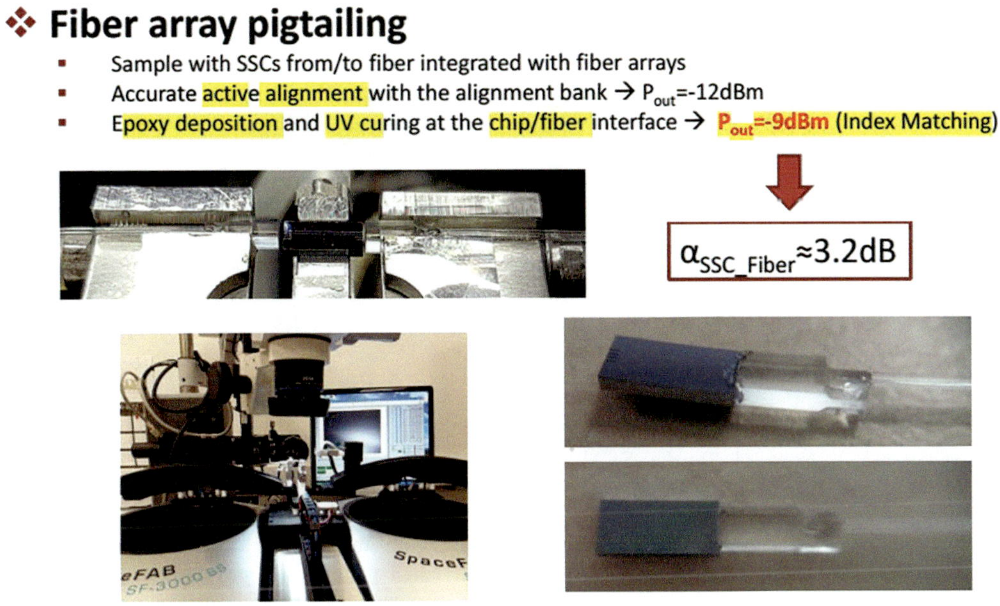

FIGURE 11.59 8x Array of QD-UTECL.

11.2.6.9 Concluding Remarks

Novel QD-UTECL offers the following:

- Low loss coupling between silicon waveguides and optical fiber or other dielectric waveguides. The only loss now is the Fresnel loss due to different refractive indices between the two media.
- Structure of the external cavity incorporating double resonant tuning filters for optical feedback
- Under any temperature variation or fluctuation, the QD-UTECL (unique technology external cavity laser) is locked to a desired optical frequency with high precision.
- Simple fabrication and low cost with silicon and SiGe CMOS technology.

The novelties of this structure MRR_DBR external cavity QD laser with push pull-feedback control are:

- Tunability and lightwave emission over a large range in the C-band, O-band, or any band that can be designed to match.
- Novel optical feedback and optical monitoring provide electronic signals for autonomous control feedback.
- A complete and functional autonomous uncooled tunable external cavity laser (ECL) is demonstrated, and the yield is sufficiently high.

NOTES

1 M. Bertolotti, "Integrated optics: theory and technology", *Journal of Modern Optics*, 34 (1), 3, 1987. https://doi.org/10.1080/09500348714550041
2 Brimrose Ltd., Free Space Acousto-Optic Modulators, https://www.brimrose.com/free-space-ao/acousto-optic-modulators?gad_source=5&gclid=EAIaIQobChMIhPe8u5rfgwMVPBWDAx1KVA0nEAAYAiAAEgKed_D_BwE, access date January 15, 2024.
3 L. N. Binh, et al., "A wide-band acousto-optic TE-TM mode converter using a doubly confined structure", *IEEE Journal of Quantum Electronics*, 16, 964–971, 1980.
4 L. N Binh, *Guided Wave Photonics*. CRC Press, Taylor & Francis: Boca Ratton, FL, 2014.

5 R. C. Alferness, "Optical guided-wave devices", Science, 234(4778), 825–829, November 14, 1986.

6 M. Rizzi and B. Castagnolo, "Electro-optic intensity modulator for broadband optical communications", *Fiber and Integrated Optics*, 21, 243–251, 2002.

7 (a) H. Takara, "High-speed optical time-division-multiplexed signal generation", *Optical and Quantum Electronics*, 33(7–10), 795–810, July 10, 2001. (b) E. L. Wooten, K. M. Kissa, A. Yi-Yan, E. J. Murphy, D. A. Lafaw, P. F. Hallemeier, D. Maack, D. V. Attanasio, D. J. Fritz, G. J. McBrien, and D. E. Bossi, "A review of lithium niobate modulators for fiber-optic communications systems", *IEEE Journal of Selected Topics in Quantum Electronics*, 6(1), 69–80, January/February 2000.

8 K. Noguchi, O. Mitomi, H. Miyazawa, and S. Seki, "A broadband Ti:$LiNb0_3$ optical modulator with a ridge structure", *Journal of Lightwave Technology*, 13(6), 1164–1168, June 1995.

9 Yun-Da Hsieh et al., "Electro-absorption modulation in GeSn alloys for wide-spectrum mid-infrared applications", *Communications Materials*, 2, Article number: 40 (2021), 09 April 2021. Open Access at https://www.nature.com/articles/s43246-021-00144-z#citeas

10 Canadian patent Number CA2031134A1, by authors Erwan Bigan, Michel Allovon, Paul Voisin

11 Fred Heismann et al., "Lithium niobate integrated optics: selected contemporary devices and system applications", in *Optical Fiber Telecommunications* (3rd ed.), Volume B, 1997.

12 M. Suzuki, Y. Noda, H. Tanaka, S. Akiba, Y. Kuahiro, and H. Isshiki, "Monolithic integration of InGaAsP/InP distributed feedback laser and electro-absorption modulator by vapor phase epitaxy", *IEEE Journal of Lightwave Technology*, LT-5(9), 127, September 1987.

13 Note this equation is representing for single drive MZIM – it is the same for dual drive MZIM provided that the bias voltages applied to the two electrodes are equal and opposite in signs. The transfer curve of the field representation would have half the periodic frequency of the transmission curve shown in Figure 11.6.

14 H. Nagata, Y. Li, W. R. Bosenberg, and G. L. Reiff, "DC drift of X-cut $LiNbO_3$ modulators", *IEEE Photonics Technology Letters*, 16(10), 2233–2335, October 2004.

15 H. Nagata, "DC drift failure rate estimation on 10 Gb/s X-cut lithium niobate modulators", *IEEE Photonics Technology Letters*, 12(11), 1477–1479, November 2000.

16 R. Krahenbuhl, J. H. Cole, R. P. Moeller, and M. M. Howerton, "High-speed optical modulator in $LiNbO_3$ with cascaded resonant-type electrodes", *IEEE Journal of Lightwave Technology*, 24(5), 2184–2189, 2006.

17 A. Sano, T. Kobayashi, A. Matsuura, S. Yamamoto, S. Yamanaka, Z. Yoshida, Y. Miyamoto, M. Matsui, M. Mizoguchi, and T. Mizuno, "100 x 120-Gb/s PDM 64-QAM transmission over 160 km using line-width-tolerant pilotless digital coherent detection", *Proceedings of OFC 2012*, New Orleans, LA, 2010.

18 P. Dong, C. Xie, L. Chen, L. L. Buhl, and Y.-K. Chen, "12-Gb/s monolithic PDM-QPSK modulator in silicon", *Optics Express*, 20(26), B624–B629, 2012.

19 R. Marchetti, et al., "Group-velocity dispersion in SOI-based channel waveguides with reduced-height", *Optics Express*, 25(9), 9761, May 1, 2017.

20 J. Muller, F. Merget, S. S. Azadeh, J. Hauck, S. R. Garcıa, B. Shen & J. Witzens, "Optical peaking enhancement in high-speed ring modulators", *Scientific Reports*, 4, 1–9, 2014.

21 E. M. Cherry and D. Hooper, *Amplifying Devices and Low Pass Amplifier Design*. John Wiley: New York, 1968.

22 I. S. Amiri, A. Afroozeh, H. Ahmad, *Integrated Micro-Ring Photonics: Principles and Applications as Slow Light Devices, Soliton Generation and Optical Transmission*. CRC Press: Boca Raton, FL, 2017. ISBN 9781138027831.

23 R. Paschotta, H. R. Telle, and U. Keller, "Encyclopedia of laser physics and technology: Articles on laser noise, intensity noise, phase noise and others in noise of solid state lasers", In A. Sennaroglu (ed.), *Solid-State Lasers and Applications*. CRC Press: Boca Raton, FL, Chapter 12, pp. 473–510, 2007.

24 D. Derickson, *Fiber Optic Test and Measurement*. Prentice-Hall: Hoboken, NJ, 1998.

25 B. Onillon, S. Constant, and O. Llopis, *Proceedings of the 2005 IEEE International Frequency Control Symposium and Exposition* (IEEE, 2005), pp. 545–550, 2005.

26 H. Ludvigsen, M. Tossavainen, and M. Kaivola, *Optics Communications*, 155, 180–186, 1998.

27 H. Ludvigsen and E. Bodtker, *Optics Communications*, 110, 595–598, 1994.

28 W. V. Sorin, K. W. Chang, G. A. Conrad, and P. R. Hernday, *Journal of Lightwave Technology*, 10, 787–793, 1992.

29 H. Packard, "Phase noise characterization of microwave oscillators," product note 11729C-2, Hewlett Packard, 1985.

30 S. Camatel and V. Ferrero, "Narrow linewidth CW laser phase noise characterization methods for coherent transmission system applications", *IEEE Journal of Lightwave Technology*, 26, 3048–3055, 2008.

31 P. Gallion and G. Debarge, "Analysis of the phase-amplitude coupling factor and spectral linewidth of distributed feedback and composite-cavity semiconductor lasers", *IEEE Journal of Quantum Electronics*, 20, 343, 1984.
32 L. B. Mercer, "1/f frequency noise effects on self-heterodyne linewidth measurements", *IEEE Journal of Lightwave Technology*, 9, 485, 1991.
33 O. Llopis, P. H. Merrer, H. Brahimi, K. Saleh, and P. Lacroix "Phase noise measurement of a narrow linewidth CW laser using delay line approaches", *Optics Letters*, 36(14), 2713, July 15, 2011.
34 A. Godone, S. Micalizio, and F. Levi, "Metrological characterization of the pulsed Rb clock with optical detection", *Metrologia*, 49(4), 313, 2012.
35 G. Di Domenico, S. Schilt, and P. Thomann, "Simple approach to the relation between laser frequency noise and laser line shape", *Applied Optics*, 49(25), 4801–4807, September 1, 2010.
36 E. Ngoya and R. Larcheveque, "Envelop transient analysis: A new method for the transient and steady state analysis of microwave communication circuits and systems", *Proceedings of IEEE MTT-S International Microwave Symposium*, IEEE, p. 1365, San Francisco, CA, USA, 1996
37 S. J. McNab, N. Moll, and Y. A. Vlasov, "Ultra-low loss photonic integrated circuit with membrane-type photonic crystal waveguides," *Optics Express*, 11, 2927–2939, 2003.
38 Y. Shani, C. H. Henry, R. C. Kistler, K. J. Orlowsky, and D. A. Ackerman, "Efficient coupling of a semiconductor laser to an optical fiber by means of a tapered waveguide on silicon", *Applied Physics Letters*, 55, 2389, 1989.
39 M.-J. Picard, C. Latrasse, C. Larouche, Y. Painchaud, M. Poulin, F. Pelletier, and M. Guy, "CMOS-compatible spot-size converter for optical fiber to sub μm silicon waveguide coupling with low-loss low-wavelength dependence and high tolerance to misalignment", *Proceedings of SPIE9752, Silicon Photonics XI*, 97520W, San Francisco CA., USA, , 2016.
40 T. Shoji, T. Tsuchizawa, T. Watanabe, K. Yamada, and H. Morita, "Spot-size converter for low loss coupling between 0.3 μm square Si wire wave guide sand single mode fibers", *15th Annual Meeting of the IEEE Lasers and Electro-Optics Society*, vol. 1, pp. 289–290, 2002.
41 J. Cardenas, C. B. Poitras, K. Luke, L. W. Luo, P. A. Mortonand, and M. Lipson, "High coupling efficiency etched facet tapers in silicon waveguides," *IEEE Photonics Technology Letters*, 26(23), 2380–2382, 2014.
42 W. R. Almeida, R. R. Panepucci, and M. Lipson, "Nano-taper for compact mode conversion," *Optics Letters*, 28, 1302–1304, 2003.
43 Patents: (a) Xuezhe Zheng, Jin Yao, Ying Luo, Ashok V. Krishnamoorthy, "Tunable hybrid laser with a silicon-photonic-switch", Patent: US 9780524 B1 (10/2017) and (b) III-V/Si hybrid laser stabilization using micro-ring feedback control, *IEEE Photonic Journal*, 8(5), October 2016.
44 Xuezhe Zheng, Jin Yao, Ying Luo, Ashok V. Krishnamoorthy, "Fast tunable hybrid laser with a silicon-photonic switch", Patent US 9780524 B1, October 2017.
45 Jin-Hyoung Lee, Daniel Y. Lee, Ivan Shubin, Jock Bovington, Stevan S. Djordjevic, Shiyun Lin, Ying Luo, Jin Yao, John E. Cunningham, "III-V/Si hybrid laser stabilization using micro-ring feedback control," *IEEE Photonic Journal*, 8(5), October 2016. DOI:10.1109/JPHOT.2016.2601447
46 L. N. Binh, *Photonic Signal Processing: Techniques and Applications*. CRC Press, Taylor & Francis: Boca Raton, FL, p. 237, 2007.
47 S. J. Mason, "Feedback theory-further properties of signal flow graphs," *Proceedings of IRE*, 44(7), 920–926, 1956.

12 Appendices 3–7
Fiber and Integrated Waveguides, EDFA Simulation, and ITU Grids

CONTENTS

12.1 APPENDIX 3: GUIDED OPTICAL WAVEGUIDES

12.1.1 Introduction

The term “integrated optics” was first coined by Miller in 1969,[2] as an analogy to the electronic integrated circuits for lightwave technology. Indeed, the term currently used in the industry is planar lightwave circuits (PLC). Since that proposed term, there have been tremendous signs of progress. Various integrated optical devices have been researched, developed, and deployed in practical optical transmission systems and networks. Extensive surveys have been conducted over the years[3,4] and even defined and described on the Internet.[5,6] Both linear and nonlinear integrated optics[7] have been explored.

Recent experimental demonstrations have pushed the information transmission bit rate per channel to 100 Gb/s[8,9,10] with the multiplexing of several wavelength channels reaching tens of

DOI: 10.1201/9781003426677-12

tera-bits/sec.[11,12,13] At such speed, the need for optical modulation, switching, optical pre-processing necessitates the use of the nonlinearity of integrated optical devices, and the compensation of dispersion of single-mode optical fibers using fiber or integrated optical components, and equalization of the losses of fibers through optical amplification. Before the availability of Erbium-doped fiber amplifiers (EDFA), attempts were made to increase the span distance between repeaters of amplitude-modulated single fiber transmission systems by employing coherent techniques,[14,15] in which a narrow linewidth and high power are used to mix the received signals to improve the receiver sensitivity. This linewidth requirement on the local oscillator laser limits the deployment of such coherent transmission in real practice. Until recently, coherent optical communications have attracted much interest once again as a possible technique to increase the transmission spans further.[16,17] Naturally, modulation formats such as amplitude shift keying, phase shift keying, and frequency shift keying are employed to reduce the effective signal bandwidth to minimize the dispersion effects in single mode optical fibers. The detection of ultra-high-speed optical signals can thus be in direct or coherent detection. This attracts the employment of integrated optical devices in the optical transmitters, receivers, and online components.

The basic component of these integrated photonic devices is the optical waveguide formed by a thin film or diffused waveguiding layer structure on some substrate. From the point of view of mode propagation, design optimization requires accurate estimation of the propagation constant, dispersion characteristics, mode size, and group velocities, depending on the types of applications. These requirements led us to develop simple, accurate, and efficient methods of analysis or single-mode waveguides. This is the main motivation of this first section.

Most waveguides found in integrated optics, especially lithium niobate types, are fabricated by a diffusion process, commonly formed by the diffusion of impurities into various substrate ferroelectric materials, such as lithium niobate, lithium tantalite, etc. The complementary error function represents the distribution of impurities from the surface of the substrate into its depth[18] and the time of diffusion of metallic impurities. A Gaussian profile is expected when the diffusion time is sufficiently long,[19] which is used to fit the experimental values of Se in the CdS crystal. In addition, various profiles can be used to form optical waveguides using molecular beam epitaxy (MBE) and metallic-organic chemical vapor deposition (MOCVD) techniques, as in the waveguiding structures for laser diodes of separate confinement of the heterojunction.[20]

Exact analytical solutions are available for the step,[21] exponential,[22] hyperbolic secant,[23] clad linear,[24] clad-parabolic,[25] and Fermi[26] profiles. In general, approximate analytical or exact numerical methods are required to analyze general classes of profiles. For some practical profiles, universal charts describing the mechanism of waveguiding have already been presented by several authors. These have been obtained by variational analysis,[27] WKB,[28] multilayer staircase.[29] However, these curves are only accurate for multimode waveguides, except for the last two references. Single-mode planar optical waveguides are very important for integrated optic circuits for applications in advanced ultra-high-speed optical communications, for example, the pioneering works by Karotsky.[30] However, analysis methods are confined predominantly to those originally used in multimode waveguides. In the single-mode regime, the variational and WKB methods are expected to perform poorly. In the former, the solutions are strongly affected by the choice of trial fields. In the latter, a more accurate prediction of the phase changes at the turning point is required. In the Runge–Kutta method for outward integration, instability is caused by the solution and error increasing at large “x”.[31] This problem is resolved by approximating the fields at sufficient depth in the waveguide with an evanescent field.[32] However, this requires knowledge of the location of solution matching.

Thus, this section recognizes that general analytical techniques must be numerical in nature due to the stringent accuracy requirements for single-mode waveguides. In the case of the single-mode waveguide, it is used as a nonlinear interaction medium and the phase matching is very critical. Thus, accurate estimation of the dispersion curves plays a very important part in the conversion efficiency.[33,34]

This section is organized as follows: Section 12.2 addresses the formation of the problems for solving single-mode optical waveguides in which all parameters are expressed in terms of normalized quantities. A novel relationship between a newly defined mode spot size and the normalized waveguide parameter is described. This is very much close to the definition of the mode spot size as defined for single-mode optical fibers commonly specified. Section 12.3 describes the simplest method for calculating the propagation constant and the model field. The stepping function recurrence method of integration originally developed by Killingbeck[35,36] is extensively used. This method is then applied to asymmetrical optical waveguide structures.

For diffused waveguides or graded index profile distributions of the refractive index from the surface of the waveguide to the deepest position, two widely used methods for waveguide modal analysis are given. The variational method with a simple Hermite-Gaussian field was first introduced by Korotsky[29] to estimate the mode spot size in a diffused channel waveguide. In Section 12.4, we show the estimation method to obtain the modal characteristics of all diffused waveguide profiles and show that it is inaccurate and computationally intensive for the calculation of the dispersion characteristics, but very close to the value of the mode spot size.

The spot size of an optical waveguide is the geometrical measurement of the optical field at the point at which the amplitude is e^{-1} of the peak value. The mode distribution profile is normally defined and assumed to follow a Gaussian shape profile. It is important for estimating the effective overlap integral between the electric field and the optical confined field across the waveguide cross-section, as well as the modulation index of the applied signal strength. The guided wave phase fronts are also critical in the design of the power split such that there are no phase distortion and mismatching as well as the polarized modes. The design details of such waveguides can be found in Ref.[37]

12.1.2 Planar Single Mode Waveguide

A planar dielectric waveguide with the geometry shown in Figure 12.1 can support modes with two orthogonal polarizations: the transverse electric (TE) and transverse magnetic (TM) field components of the guided modes. In practice, either polarized mode can be excited. Provided certain boundary conditions are met, these modes are bounded and propagating along the z-axis, each with a unique effective phase velocity. If we allow for uniformity in the refractive index and geometrical dimensions in the propagation direction, the phase velocity is only a function of the transverse index profile. In this section, we consider the simplest configuration with index variation only in the –x direction. Note that the notation of the coordinate system follows the right-hand rule (RHS).

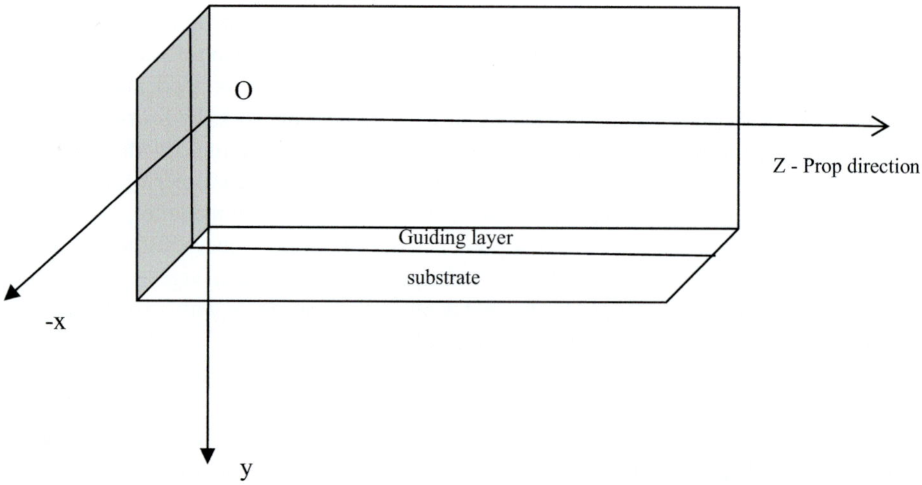

FIGURE 12.1 Schematic structure of a planar dielectric waveguide.

12.1.2.1 TE/TM Wave Equation

The wave equation for the guided modes stems from the Maxwell equations given in the appendix, together with the associated constituent relations. We consider the steady-state solutions in the dielectric medium free from any sources and losses. By omitting the common factor $e^{j(\omega t-\beta z)}$ from the equations we can write, for a medium characterized by the refractive index $n(x)$, the well-known wave equation:

$$\frac{d^2E_y}{dx^2}+\left[k_0^2n^2(x)-\beta^2\right]E_y=0$$

and

$$\frac{d^2H_y}{dx^2}+\left[k_0^2n^2(x)-\beta^2\right]H_y=\frac{1}{n^2(x)}\frac{d^2n^2(x)}{dx^2}\frac{dH_y}{dx} \tag{12.1}$$

where k_0=wave number in free space; β=propagation constant of the wave along the z-axis; $E_y(x)$=transverse electric field; H_y=transverse magnetic field. The wave equation can be obtained by applying Maxwell equations subject to boundary conditions for either the E- or H- field, that is the continuity at the interfaces of the guiding layer and the substrate, as shown in Figure 12.1.

12.1.2.2 Continuity Requirements and Boundary Conditions

It is common to expect that the dielectric continuity is met at the interfaces between the two media, the substrate and the waveguide, or/and the superstrate region and the waveguide. At the interface of the discontinuity, the wave equations are not valid. However, the identity of the modes must be matched; that is, preservation can be achieved by matching the optical fields and their derivatives on either side of the dielectric media discontinuity.

For the TE modes, these boundary conditions impose the continuity of E_y and $\frac{dE_y}{dy}$ across the interface. For the TM modes, we require continuity H_y and $\frac{1}{n^2(x)}\frac{dH_y}{dy}$. In addition, the bound modes must satisfy the conditions that E_y, H_y vanishes at $x=\infty$, i.e., no oscillations of such fields or infinite energy contained in the guided mode to satisfy the energy conservation. Combining these conditions gives rise to the eigenvalue equations from which the propagation constant can be estimated as a function of the wavelength, the thickness of the waveguide region, and the refractive indices of the three media.

Note that the principal objective of the eigenvalue equation is to estimate the maximum value of the propagation constant so that the dependence of this propagation parameter on the optical frequency/wavelength is minimum, resulting in minimum dispersion of the waves at a different wavelengths normally found for other spectral components of a modulated lightwave channel in optical communication system. A maximum value of the propagation constant along the z-direction means that the direction of the wave vector is close to the propagation axis.

12.1.2.3 Index Profile Construction

For the purpose of computation and analysis, it is customary to write the refractive index profile in the general form as:

$$n^2(x)=\begin{cases} n_s^2\left[1+2\Delta S\left(\frac{x}{d}\right)\right] & x\geq 0 \\ n_c^2 & x<0 \end{cases}$$

where

$$\Delta=\frac{n_0^2-n_s^2}{2n_s^2} \tag{12.2}$$

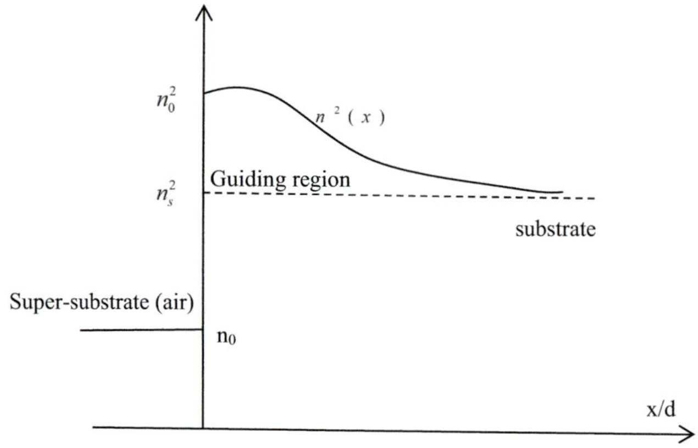

FIGURE 12.2 Distribution of refractive index profile of the square of the refractive index of asymmetrical waveguide.

Δ is the profile height or relative index difference with respect to that of the substrate, and n_c, n_0, and n_s are the refractive indices of the superstrate cover, the guiding layer, and the substrate, respectively. $S(x)$ is the profile shape function, and d is the diffusion depth of a graded index distribution. Figure 12.2 shows a typical representation of the graded index profile. It turns out that further normalization of the shape profile can be represented as

$$S\left(X = \frac{x}{d}\right) = \frac{n^2(x) - n_s^2}{n_0^2 - n_s^2} \tag{12.3}$$

This definition of the profile shape is unaffected by the symmetry of the waveguide. In a symmetric waveguide structure with an axis of symmetry at $x=0$; these equations are equally valid.

12.1.2.4 Normalization and Simplification

The presence of the non-zero term in the RHS of Equation 12.3 complicates the analysis. It is identically zero for a step index profile. Exact solutions to this equation are available for the exponential, hyperbolic secant, and inverted-x profiles.[38] However, for smooth profiles normally encountered in practice, several authors[39] have found through perturbation analysis that the RHS of the equation can be neglected. However, the fundamental mode of an infinite parabolic profile faces a 44% error in the group velocity of the guided mode.

Following Kolgenik and Ramaswamy,[40] we can introduce the normalized parameter for the waveguide as follows:

$$\begin{aligned} V &= \frac{dk_0}{n_0^2 - n_s^2} \\ A &= \frac{n_s^2 - n_c^2}{n_0^2 - n_s^2}\text{; and} \\ b &= \frac{n_e^2 - n_s^2}{n_0^2 - n_s^2} \end{aligned} \tag{12.4}$$

where A is defined as the asymmetry factor, β=normalized propagation constant, $n_e=\beta/k_0$ is the *effective* refractive index of the guided mode as seen along the propagation axis, and V is the normalized frequency.

For a guided mode it requires

$$n_s < n_e < n_0 \tag{12.5}$$

Thus, the normalized propagation constant must satisfy $b < 1$. The real advantages of normalization lead to design optimization. Substituting the normalized parameters into the wave Equation 12.6, we obtain:

$$\frac{d^2\varphi}{dx^2} + V^2\left[S(X) - b\right]\varphi = 0$$

where (12.6)

$$\varphi(X) \equiv E_y; H_y$$

for the TE and TM modes, respectively. The propagation for the TM modes can be accurately represented by the TE modes, except at cut-off or for waveguides of large symmetry.[41] However, if extreme accuracy or mode splitting is required, then Equation 12.25 can be modified to the changes involving only a slight modification of the boundary condition.

12.1.3 Modal Parameters

The solution of the normalized wave Equation 12.6, along with the boundary conditions, enables the determination of various optical parameters. The following are the most commonly used for the design of single optical guided-wave devices:

12.1.3.1 Mode Size

The mode size Γ_a for an asymmetrical field is defined as the full-width half maximum (FWHM) power intensity. For a full description of the field the peak position of the intensity I_p and the field asymmetrical factor Γ_1/Γ_2, defined with respect to I_p is required as shown in Figure 12.3. The knowledge of the mode size is critical to match with that of the single mode optical fiber for inline integration with fiber transmission systems. These parameters are defined in terms of the optical power of the field due to practical reasons because the mode size is normally monitored using a

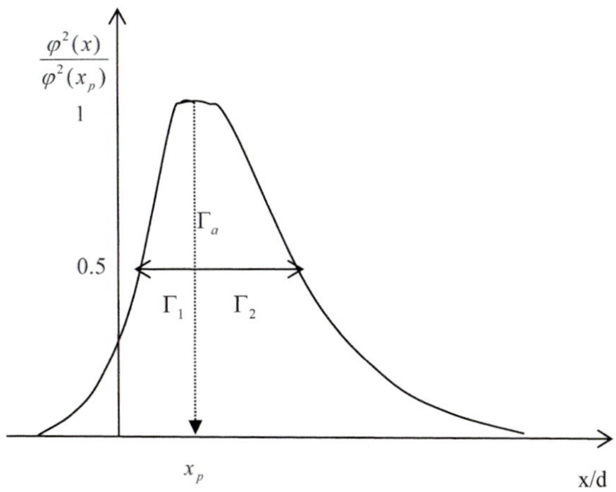

FIGURE 12.3 The minimum set of parameters required for characterization of fundamental mode field in an asymmetrical planar waveguide.

CCD (charge couple device) camera through which the intensity of the mode field is converted into the charge current and displayed or digitized for data processing.

12.1.3.2 Propagation Constant and Effective Refractive Index

The variation of the normalized propagation constant b as a function of the normalized frequency parameter V, is normally required for the design and characterization of optical waveguide. For example, this relationship specifies the diffusion depth required once the mode index at a specific operating wavelength is obtained or given. The purpose of this section is to present the relationship between the modal field and the propagation constant.

If we integrate Equation 12.6 with respect to X, then we have

$$\int_{-\infty}^{+\infty}\left\{\frac{d^2\varphi}{dx^2}+V^2\left[S(X)-b\right]\varphi\right\}dX=0$$

or (12.7)

$$\int_{-\infty}^{+\infty}\varphi''\,dX+V^2\int_{-\infty}^{+\infty}\left\{\left[S(X)-b\right]\varphi(X)\right\}dX$$

where the dashes denote the derivative with respect to X. If we further impose the condition that $\varphi(-\infty)=\varphi(+\infty)=0$, the integral in the left becomes zero and thus we obtain

$$b=\frac{\int_{-\infty}^{+\infty}S(X)\varphi(X)\,dX}{\int_{-\infty}^{+\infty}\varphi(X)\,dX} \tag{12.8}$$

This equation indicates that the dispersion characteristics for an arbitrary index profile must be a smooth curve. The rule of refractive index used by Tien[42] to explain a host of new wave phenomena in integrated optical waveguides is that light tends to propagate in the region where the refractive index is the largest. In the context of planar optical waveguides with arbitrary index profiles, this rule suggests that for a given profile at a given frequency, the mode field adjusts itself so that the maximum value of b is achieved. This corresponds to the minimum phase velocity allowed for the mode. This rule is indeed a direct statement of Fermat's law in ray optics and is a special case of a generalized rule in quantum mechanics formulated in the form of the well-known Feynman path integral, of which Maxwell's equations are also satisfied.[43]

12.1.3.3 Waveguide Dispersion and Spot Size

A second and potentially useful relation between b and the modal field can be established from the stationary expression for the normalized propagation constant b. This can be obtained from Equation 12.25 after multiplying it by φ and integrating with respect to X from negative to positive infinity, that is, very far from the interface. After integrating the results by parts and imposing the boundary conditions$\varphi(-\infty)\varphi'(-\infty)=\varphi(+\infty)\varphi'(\infty)=0$ we obtain the well-known stationary relation:

$$bV^2=\frac{V^2\int_{-\infty}^{+\infty}S(X)\varphi^2(X)\,dX-\int_{-\infty}^{+\infty}\varphi^2(X)\,dX}{\int_{-\infty}^{+\infty}\varphi^2(X)\,dX} \tag{12.9}$$

This is the basic equation for the variational analysis. It possesses the unique property that, for any trial fields which satisfy the boundary conditions above, the quotient remains stationary provided the mismatch between the trial and actual fields is small.[44]

Thus, we can write

$$\frac{d\left(bV^2\right)}{dV^2} = \frac{\int_{-\infty}^{+\infty} S(X)\varphi^2(X)dX}{\int_{-\infty}^{+\infty} \varphi^2(X)dX} \tag{12.10}$$

$$\frac{1}{2}\left(bV + V\frac{d(bV)}{dV}\right) = V\frac{\int_{-\infty}^{+\infty} S(X)\varphi^2(X)dX}{\int_{-\infty}^{+\infty} \varphi^2(X)dX} \tag{12.11}$$

Taking the derivative second time and using Equations 12.9 and 12.11, we obtain

$$\frac{1}{2}\left(V\frac{d^2(bV)}{dV^2} + 2\frac{d(bV)}{dV}\right) = V\frac{d}{dV}\left(b + \frac{2}{V^2 W_m}\right) + \frac{1}{2}\left(b + \frac{d(bV)}{dV}\right) \tag{12.12}$$

in which we define a new spot size parameter as

$$W_m = \frac{2\int_{-\infty}^{+\infty} \varphi^2(X)dX}{\int_{-\infty}^{+\infty} \left[\varphi'(X)\right]^2 dX} \tag{12.13}$$

Further algebraic manipulations lead to the simple relationship

$$V\frac{d^2(bV)}{dV^2} = 4\frac{d}{dV}\left(\frac{2}{VW_m^2}\right) \tag{12.14}$$

This relation is analogous to the relationship between Petermann's spot size and the waveguide dispersion in single-mode optical fibers.[45] The preceding analysis was first performed by Sansonetti [46], which inspired Petermann to define a new spot size for the characterization of single-mode optical fibers based on spot size measurement.

The relation in Equation 12.33 can be found for an optical waveguide with a profile following the Hermite Gaussian variational field of

$$\varphi(X) = \begin{cases} A_0\alpha_0^{1/2} e^{-\alpha_0 X^2/2}; & X \geq 0 \\ 0; & X < 0 \end{cases} \tag{12.15}$$

$$W_m^2 = \frac{4}{\alpha_0^2} = \Gamma_a^2 \tag{12.16}$$

where α_0 is the variational spot size parameter, and A_0 is a scaling constant. The RHS of Equation 12.34 corresponds to Γ_a^2 as defined by Korotsky [47] as given above. Although α_0 is an approximated mode spot size, several experimental measurements show excellent agreement between the theoretical and experimental values W_m and Γ_a.

12.2 APPENDIX 4: FIBERS-GEOMETRICAL AND LIGHTWAVES GUIDING

This section describes the fundamental understanding of guiding lightwaves in planar and circular optical waveguides. A brief fundamental derivation of the wave equation from Maxwell's equations is given for symmetric planar waveguides leading to the phenomenon of guiding in optical fibers, and circular optical waveguides.

Only geometrical parameters of optical fibers are presented here. The attenuation and dispersion effects of modulated lightwave signals can be referred to in the first section (Section 12.1.5). However, the nonlinear effects in such single fibers are discussed here so that the distortion impacts on signals can be treated in the next section. Although the length of the fiber is short, about a few meters, in optical amplifiers for space transmission, the high power launched out of the amplifiers leads to nonlinear phase distortion, and Brillouin acoustic modulation of the channel leads to jittering effects.

12.2.1 Introductory Remarks

Optical fiber communications have, for a long time, no longer been the term used in research centers or university research and teaching laboratories but have been extensively deployed throughout the Earth, both terrestrial and undersea. It has now been more than 35 years since optical fiber communications began progressing, and the technology has brought about a significant revolution in the information-technologically oriented society beyond the 21st century. Indeed, it was a simple idea proposed in an article in the IEE Proceedings by Kao and Hockham.[48] The idea was to guide optical waves in a wave-like light guide, with the term dielectric optical waveguide being used for the first time. It was considered the most fundamental component of optical communications technology, guiding lightwaves whose frequency is extremely high, thus allowing ultra-wideband regions to accommodate information channels.

This section is dedicated to describing and giving a fundamental understanding of optical waveguiding and the structures, as well as the conditions under which the waveguide can be designed to guide lightwaves over extremely long distance. The basic components for optical fiber communications networks are the optical guiding media for the transmission of optical signals in the form of lightwaves, called optical waveguides or optical fibers. These lightwave transmission media must satisfy the following conditions:

It must be able to guide lightwaves over a long distance without losing significantly the optical energy. Secondly, signals transmitted through them must not be distorted, and they must be structured in such a way that they can support several optical electromagnetic field modes. Furthermore, it must preserve the coherency of the guided mode.

The main objectives of this section are: (i) first, to give an introduction and concepts of guided modes in optical waveguides in which lightwaves are confined to one dimension only, the planar optical waveguides, (ii) second the optical waveguides are extended to a circular geometrical structure, the optical fibers. The properties of the step-index optical fiber, which supports only one mode, the single-mode optical fiber, are described and used as the fundamental elements for studying non-step-index fiber types and are installed throughout the global networks. The non-circular or non-step optical profile optical fibers can be transformed into an equivalent circular optical fiber by using an equivalent step-index profile transform technique described later in this section. Hence, normal estimation of the modal characteristics can be obtained without much difficulty.

The advanced design of optical fibers for dispersion shifted, dispersion flattening, and dispersion compensated fibers are also briefly given. These fibers have gained much attention lately due to their potential applications in optical soliton communication systems, dispersion-equalized optical fiber systems, etc. However, these topical issues may not be relevant to space optical communications, except for the nonlinear effects over a short length of fibers under ultra-high-power boosting conditions.

We must note here that this section presents only the geometrical structures and index distribution of circular optical waveguides and guiding conditions, as well as some optical properties of the guided modes. In the next section, we will describe the mechanisms of distortion and attenuation of optical signals when they are transmitted over long lengths of optical fibers.

This section does not aim to give a full treatment of the theoretical development of optical waveguides but rather the essentials for students of physics and engineering fields, allowing them to apply to some applications such as the transmission of lightwave-modulated signals and possibly fiber sensors.

This section is organized as follows: Section 12.2 describes the fundamental aspects of planar optical waveguides with a derivation of the wave equation from the first principle via the formation of Maxwell's equations. Symmetric planar optical waveguides are given as a simple example of optical waveguiding to guide readers through the guided modes of both planar and circular optical waveguides. The *V*-parameter is identical for both types of waveguides, making it easier for readers to appreciate the importance of this parameter in optical waveguides.

Section 12.3 describes optical fibers or circular optical waveguides, including their geometrical structures and conditions for supporting one and only one mode. An equivalent step index profile method is also presented to design different index profile fibers. Gaussian approximation of the guided single-mode optical waveguide is provided to derive the operation parameters of optical fibers without resorting to mathematical complexity through the use of the wave equation. This is derived from practice when measuring the mode field distribution across a single mode optical fiber, one can find that its radial distribution is very close to that of a Gaussian function. Thus, all one can do is substitute this function to the wave equation to obtain the propagation constant and, hence, other guided parameters.

Section 12.4 provides a brief introduction to the manufacturing of optical fibers, allowing readers to appreciate the technological development of optical fibers.

12.2.2 Dielectric Slab Optical Waveguides

The dielectric slab waveguide is the simplest optical waveguiding structure, first investigated in the early 1960s. It is from this structure that the proposal for an optical fiber for optical communications originated.

The optical slab waveguide is now widely used as the basic structure for optical integrated circuits such as semiconductor lasers and optical modulators. This section covers the fundamental aspects of optical waveguiding with a symmetrical structure, where the optical guiding layer is sandwiched between two cladding regions, the substrate, and superstrate, of smaller dielectric constant or refractive index than the core region.

12.2.3 Structure

A dielectric slab waveguide consists of a slab of dielectric material of refractive index n_1, embedded in a material of index n_2, as shown in Figure 12.4.

Assuming that the structure is extended to infinite in the y and z directions, and a guiding thickness of $2a$ and that the materials are isotopic and lossless (i.e. the permittivity is real and scalar), this leads to a one-dimensional wave equation for the electric field of the lightwaves resulting in a simplified mathematical solution. The main reason for this approximation is that a circular optical fiber can be easily analyzed by confining the optical waveguide in the y-direction into a circular structure. The wave equation for this type of structure involves the cylindrical coordinates and is more complicated to visualize analytically.

12.2.4 Numerical Aperture

If we assume that total internal reflections at the boundaries are required for guiding, what is the acceptance angle such that lightwaves can be launched? The ray path entering the optical fiber core for total internal reflection is shown in Figure 12.5.

$$n_0 \sin\theta_0 \le n_1 \cos\theta_c \tag{12.17}$$

By applying Snell's law at the air-core and core-cladding boundaries of the dielectric waveguide, the total internal reflection can take place only if

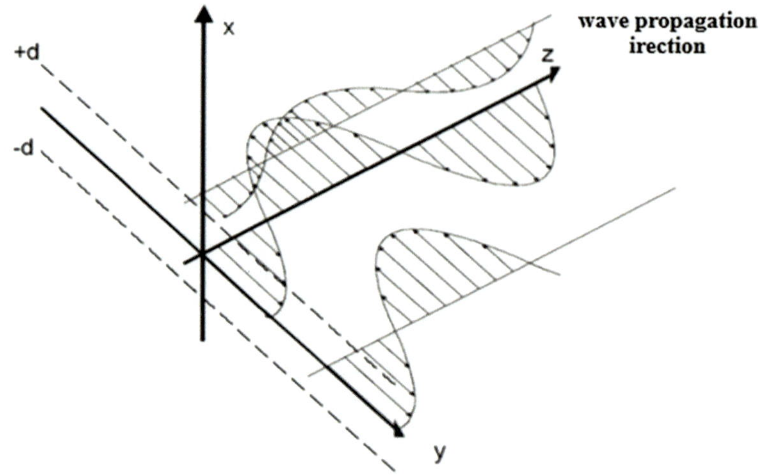

FIGURE 12.4 Evolution of the real part of the E-field in a slab waveguide (t=0).

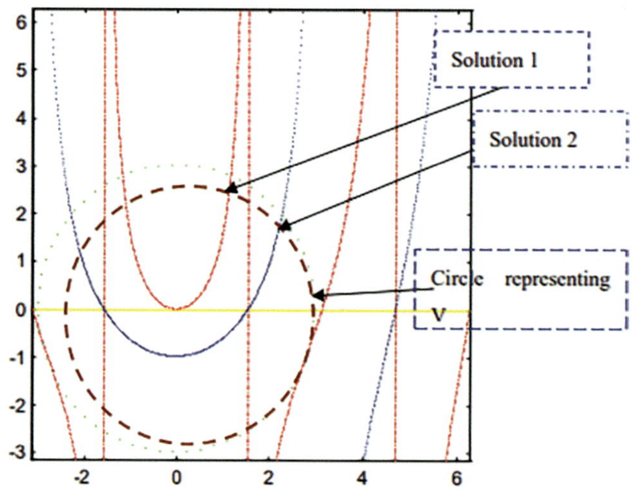

FIGURE 12.5 Graphics and solution of Equations 12.9, 12.10 and 12.12. The curve indicated by ________ is expressing the expression $v = u\tan u$, ___ _ __ $v = u/\tan u$ and __ ___ _ $V^2 = u^2 + u^2$.

$$n_1 \sin\theta_c = n_2 \sin 90° = n_2 \tag{12.18}$$

where θ_c is the critical angle such that

$$\text{NA} = \left(\sin\theta_0\right)_{\max} = n_1 \cos\theta_c \tag{12.19}$$

Thus, the numerical aperture (NA), which is defined as the maximum value of sin θ_0 is given by Equation 12.17.

$$\text{NA} = \left(n_1^2 - n_2^2\right)^{1/2} \tag{12.20}$$

then by substituting (12.18), we have

$$\text{NA} = \left(n_1^2 - n_2^2\right)^{1/2} \tag{12.21}$$

when the refractive indices n_1 and n_2 are different by a small index difference, then the NA is approximated by

$$\mathrm{NA} \simeq n_2(2\Delta)^{1/2} \tag{12.22}$$

where Δ is the refractive index difference between the refractive indices between the core and the cladding regions? Thus, the refractive index difference determines the magnitude of the NA and hence the acceptance angle that an optical fiber can accept incident lightwaves. In practice, designers want to increase the numerical aperture to maximize optical power that can be coupled to an optical fiber. However, it can be seen later that the larger the NA, the larger the V-parameter and optical fibers can become multimode, thus suffering larger delay. So, in designing a single mode optical fiber, we have to consider several parameters, which affect its performance from a communication systems point of view. Thus, we have to resort to the wave equation to understand the behavior of lightwaves in a single mode optical fiber to optimize its transmission characteristics.

12.2.5 Modes of Symmetric Dielectric Slab Waveguides

The number of optical guided modes of lightwaves transmitted in an optical waveguide is crucial because it indicates the concentration of energy in the lightwaves. This is similar to elastic waves propagating in a string that is vibrated with either open or closed ends. To determine the number of optical guided modes, one has to find the conditions under which the lightwaves would conform, this condition is the eigenvalue equation derived from the wave equation, which satisfies the boundary conditions.

Each solution of the optical wave equation corresponds to a propagation wave number for each guided wave, which are derived from the eigenvalues of the equation. Thus, the propagation constants take discrete values.

12.2.5.1 The Wave Equations

Assuming a monochromatic, single frequency or wavelength, (meaning a single, and only one, frequency component of lightwaves) wave propagates in the z-direction, its electric field component is given by:

$$E(x,y,z) = E(x)e^{(\omega t-\beta z)} \tag{12.23}$$

i.e., field dependent on x, and uniform along y direction, β is the propagation constant along z propagating direction. Then, in the absence of charges and currents, Maxwell's equation reduces to:

$$j\beta E_y = -j\omega\mu H_x \tag{12.24a}$$

$$0 = -j\omega\mu H_y \tag{12.24b}$$

$$\frac{dE_x}{dx} = -j\omega\mu H_z \tag{12.24c}$$

$$-j\beta H_x - \frac{dH_x}{dx} = j\omega\varepsilon E_y \tag{12.24d}$$

A visualization of the E-field guided in the slab waveguide is shown in Figure 12.4. By substituting Hx from Equation 12.24a and Hz from Equations 12.24c and 12.24d, we get

$$\frac{d^2E_y}{dx^2} + \left(\beta^2 - \omega^2\mu\varepsilon\right)E_y = 0 \tag{12.25}$$

where μ and ε are the permeability and permittivity of medium n_1 or n_2. Similarly, a wave equation involved Hy is given by

$$\frac{d^2H_y}{dx^2} = +\left(\beta^2 - \omega^2\mu\varepsilon\right)H_y = 0 \tag{12.26}$$

Equations 12.25 and 12.26 can be rewritten using $k=\omega/c$ and c= $(\mu_0\varepsilon_0)^{1/2}$, c is the light velocity in a vacuum and k is the wave number in a vacuum, as

$$\frac{d^2E_y}{dx^2} = -\left(k^2n_j^2 - \beta^2\right)E_y \tag{12.27}$$

$$\frac{d^2H_y}{dx^2} = -\left(k^2n_j^2 - \beta^2\right)H_y \tag{12.28}$$

where $n_j = n_1$ or n_2 $(j=1, 2)$

From Equations 12.27 and 12.28, we observe that the field variation along 0x will always exhibit:

- sinusoidal behavior when$k^2n_j^2 > \beta^2$ oscillating: lightwaves are guided.
- Exponential (decay) behavior when

$k^2n_j^2 < \beta^2$ evanescent lightwaves radiate into the cladding region.
In another word, the EM field oscillates in regions where the longitudinal propagation constant is smaller than the plane-wave propagation constant in this region and evanescent with an exponential-like behavior elsewhere.

12.2.5.2 Optical Guided-Modes

Optical waves are guided along the waveguide when their EM fields are oscillatory in the slab waveguide region and exponentially decay in the cladding region, that is

$$kn_2 < b < kn_1 \tag{12.29}$$

We need now to define a *transverse* propagation constant u/a and *transverse* decay constant v/a as

$$\frac{u^2}{a^2} = k^2n_1^2 - \beta^2 \tag{12.30}$$

$$\frac{v^2}{a^2} = -k^2n_2^2 + \beta^2 \tag{12.31}$$

Thus, we can observe from Equation 12.7 that

$$\frac{u^2}{a^2} + \frac{v^2}{a^2} = k^2\left(n_1^2 - n_2^2\right) \tag{12.32}$$

or alternatively

$$V^2 = u^2 + v^2 = k^2a^2\left(n_1^2 - n_2^2\right) \tag{12.33}$$

in which the *V*-parameter is defined as the normalized frequency (*V*), dependent only on the guide's refractive indices, the radius, and the operating light frequency, that is the wavelength. The *V* parameter and its expression are identical for both planar and optical fiber waveguides, and this will be discussed in the next section. The refractive index difference term is usually much less than one for fiber, that is, the difference is in the order of a few 0.01%. The *V*-parameter can be normally approximated by a simpler expression.

Equations 12.27 and 12.33 show that the field *Ey* for TE modes and *Hy* for TM modes are linear combinations of cos(ux/a) and sin(ux/a) inside the core layer ($|x| \leq a$) and take an exponentially decaying form outside ($|x| > a$) with exp($-vx/a$) for $x > a$ and exp(vx/a) for $x < -a$.

U is the vector representation of the propagation vector constant along the vertical direction, which is the imaginary part of the propagation constant. On the other hand, *v* is the horizontal direction vector of the propagation constant.

We therefore have two distinct cases: the even TE modes and odd TE modes. Their solutions are given as follows:

(a) *Fiber guided even TE modes*:

For $|x| \leq a$: inside core region, oscillating waves, thus the field solution is

$$E_y \text{ or } H_x = A\cos\frac{ux}{a} \tag{12.34}$$

$$H_z = \frac{j}{\omega\mu_o}\frac{\delta E}{\delta x} = -\frac{j}{\omega\mu_o}\frac{u}{a}A\sin\frac{ux}{a} \tag{12.35}$$

For $|x| > a$: outside core region, exponential decaying waves

$$E_y = Ce^{-\frac{V(x-a)}{a}} \tag{12.36}$$

Other H_x and H_z can also be expressed in similar expressions. Applying the boundary condition (continuity of the *E*- field) at the interface between the cladding and core, i.e., at $x = a^+$ and $x = a^-$. Using Equations 12.34–12.36, we have

$$\begin{aligned} -j\frac{u}{\omega\mu}\sin u &= -j\frac{v}{\omega\mu}\cos u \\ \rightarrow v &= u\tan(u) \end{aligned} \tag{12.37}$$

This is usually called the eigenvalue equation of the wave equation, which dictates the number of guided (discrete) modes and the values of the propagation constants for these guided even TE modes.

(b) Odd TE modes.

Similarly, the solutions for guided TE odd modes can be found as follows:

$$x > |a|$$

$$E_y \text{ or } H_x = A\sin\frac{ux}{a} \tag{12.38a}$$

$$H_z = \frac{j}{\omega\mu_o}\frac{\delta E}{\delta x} = -\frac{j}{\omega\mu_o}\frac{u}{a}A\cos\frac{ux}{a} \tag{12.38b}$$

$$|x| \leq a$$

$$E_y = Ce^{-\frac{V(x-a)}{a}} \tag{12.38c}$$

and the eigenvalue equation can be found by applying the boundary conditions:

$$v = \frac{u}{\tan u} \tag{12.38d}$$

Equation 12.38 is obtained by using the continuity of the normal H field component at the boundaries. This is the eigenvalue equation for guided TE *odd* modes of a symmetric planar optical waveguide.

12.2.5.3 Graphical Solutions for Guided TE Modes (Even and Odd)

Combining Equations 12.33, 12.34, and 12.38, we observe that the waveguides can support only discrete modes and the propagation constant β related to u and v parameters can be found by solving graphically the intersection between circles of V and curves representing Equations 12.37, 12.38, and 12.33. These solutions are illustrated in Figure 12.5. The field distribution of the even and odd TE mode can be found in Figure 12.6.

12.2.5.4 Cut-Off Properties

From Figure 12.5, we observe that:

For $V=0$: i.e., zero optical frequency or $\lambda=0$, it is impossible

For $\lambda>0$: we observe that we always have at least one guided mode, TE0.

$V<\pi/2$: one guided mode

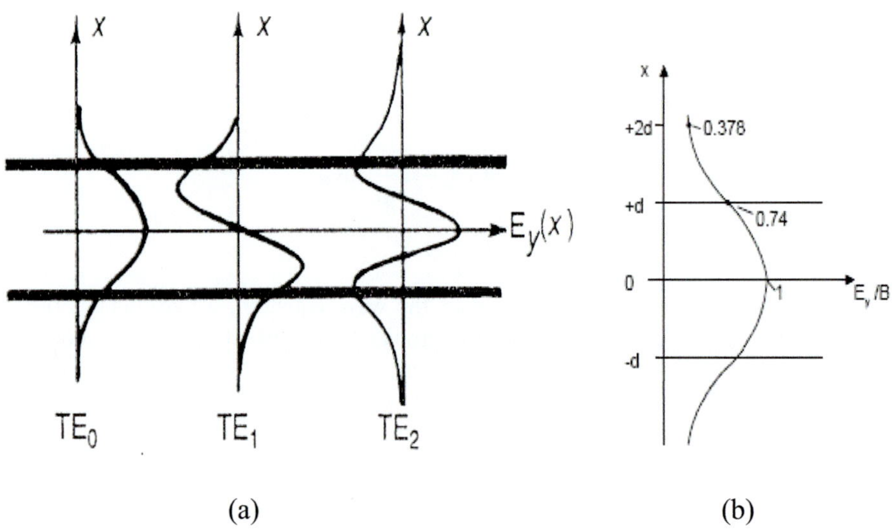

FIGURE 12.6 (a) Field distribution of TE-modes of order 0, 2 (even modes), and 1(odd mode), (b) field distribution of the fundamental mode.

$V \to \pi/2$: odd TE1 mode emerged (second mode)

$V=\pi$: we have third mode (TE_2) emerged, and so on….

That is, each time V reaches a multiple integer of $\pi/2$, a new mode reaches its cut-off which corresponds to $v=0$ and $\beta=kn_2$. We note that for TE guided modes, the optical waveguide can support at least one. However, for TM-guided modes, the waveguide can support its thickness or the optical wavelength, and for the refractive index difference, the v (lowercase v) must be at least $\pi/2$.

This is true for planar optical waveguides only. For circular optical waveguide such as in the case of circular optical fiber, the cut-off condition is different. We will address this type of optical waveguide in the next section.

It is noted that theoretical waveguide (planar type) always supports an optical guided mode. The cut-off is for cutting off the higher order mode to make the waveguide support only one mode.

12.2.6 Optical Fiber: General Properties

12.2.6.1 Geometrical Structures and Index Profile

An optical fiber consists of two concentric dielectric cylinders. The inner cylinder, or core, has a refractive index of $n(r)$ and a radius a. The outer cylinder, or cladding, has an index n_2 with $n(r)>n_2$, and a larger outer radius. The diameter of the core is about 4 9 μm and that of the cladding diameter is 125 μm, which are typical values for silica-based single-mode optical fiber. A schematic diagram of an optical fiber is shown in Figure 12.7.

The refractive index n of an optical waveguide is usually changed with radius r from the fiber axis ($r=0$) and is expressed by

$$n^2(r) = n_2^2 + \mathrm{NA}^2 s\left(\frac{r}{a}\right) \tag{12.39}$$

where NA is the numerical aperture at the core axis, while $s(r/a)$ is the profile function that characterizes any profile shape ($s=1$ at maximum) with a scaling parameter (usually the core radius).

12.2.6.1.1 Step-Index Profile

In a step-index profile, the refractive index remains constant in the core region, thus

$$s\left(\frac{r}{a}\right) = 1 \quad \text{for} \quad r \le a \tag{12.40}$$

$$s\left(\frac{r}{a}\right) = 0 \quad \text{for} \quad r > a \tag{12.41}$$

So, we have a step-index profile

$$n^2(r) = n_1^2 \quad \text{for} \quad r < a \tag{12.42}$$

$$\text{and} \quad n^2(r) = n_2^2 \quad \text{for} \quad r > a \tag{12.43}$$

12.2.6.1.2 Graded – Index Profile

We consider hereunder the two most common types of graded-index profiles: power-law-index and the Gaussian profile.

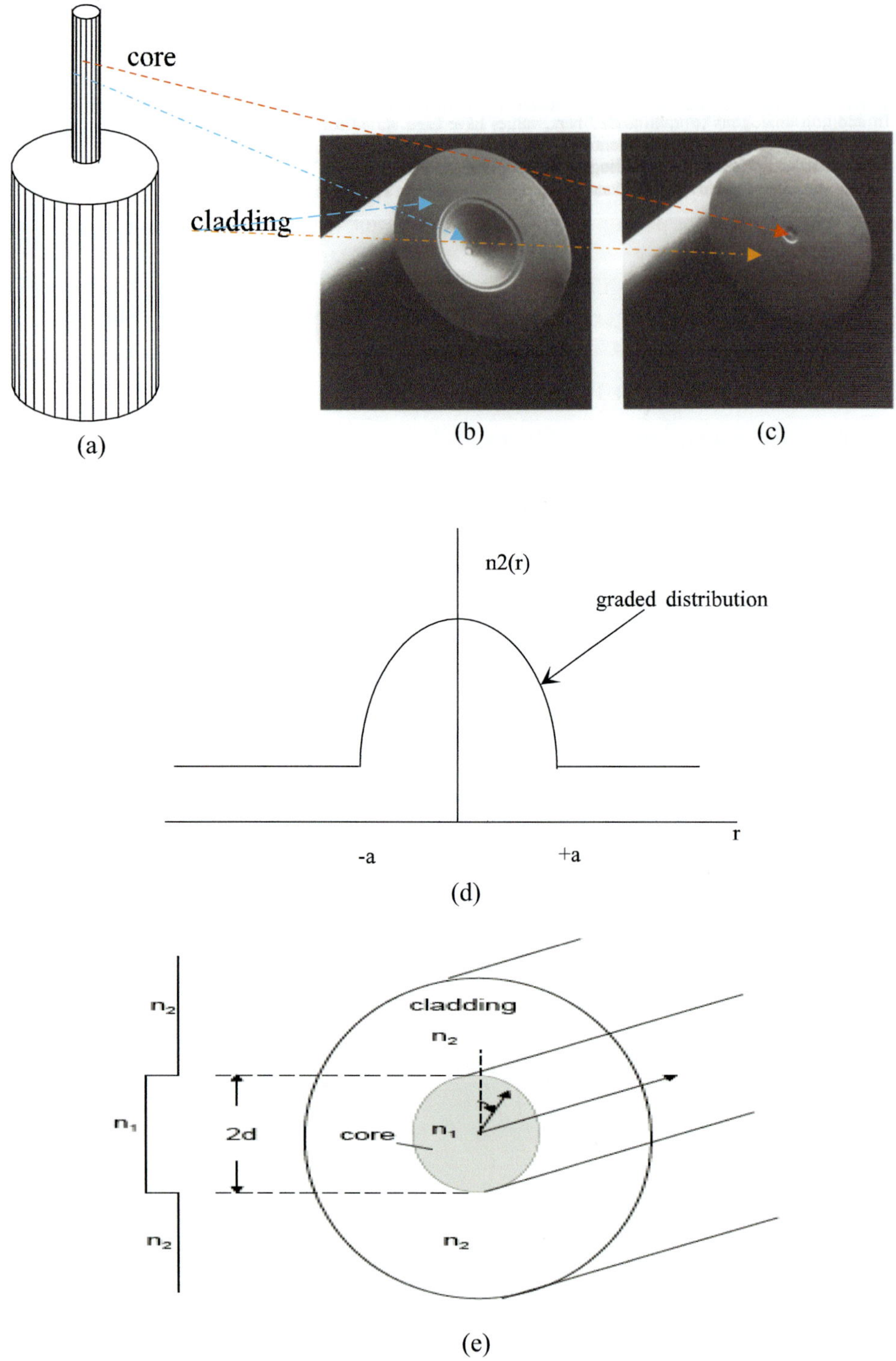

FIGURE 12.7 (a) Schematic diagram of the step-index fiber: coordinate system, structure. The refractive index of the core is uniform and slightly larger than that of the cladding. For silica glass the refractive index of the core is about 1.478 and that of the cladding about 1.47 at 1550 nm wavelength region. (b) Cross-section of an etched fiber – multimode type – 50 µm diameter. (c) Single-mode optical fiber etched cross-section. (d) Graded index profile. (e) Fiber cross-section and step-index profile $d=a=$radius of the fiber.

12.2.6.1.3 Power-Law-Index Profile

The core refractive index of optical fiber usually follows a graded profile. In this case, the refractive index rises gradually from the value n_2 of the cladding glass to the value n_1 at the fiber axis. Therefore, *s*(*r*/*a*) can be expressed as

$$s\left(\frac{r}{a}\right)=\left\{1-\left(\frac{r}{a}\right)^{\alpha}\right. \quad \text{for} \quad r \leq a \quad \text{and} \quad =0 \quad \text{for} \quad r<a \tag{12.44}$$

where α=power exponent. Thus, the index profile distribution *n*(*r*) can be expressed in the usual way by using Equations 12.44 and 12.40, then by substituting $\mathrm{NA}^2 = n_1^2 - n_2^2$, we obtain

$$n^2(r)=\begin{cases} n_1^2\left[1-2\Delta\left(\frac{r}{a}\right)^{\alpha}\right] \text{....}for\text{...}r \leq a \\ n_2^2 \text{..........}for\text{.........}r>a \end{cases} \tag{12.45}$$

where $\Delta = \mathrm{NA}^2/n_1^2$ is the relative refractive index difference between the core and cladding regions.

Observing the equation describing the profile shape in Equation 12.45, there are three special cases:

- α=1: the profile function *s*(*r*/*a*) is linear, and the profile is called a triangular profile.
- α=2: the profile is a quadratic function with respect to the radial distance, and the profile is called the parabolic profile.
- $\alpha = \infty$: then, the profile is a step shape.

12.2.6.1.4 Gaussian - Index Profile

In the Gaussian--index profile, the refractive index changes gradually from the core center to a distance very far from it and *s*(*r*) can be expressed as

$$s\left(\frac{r}{a}\right) = e^{-\left(\frac{r}{a}\right)^2} \tag{12.46}$$

12.2.6.2 The Fundamental Mode of Weakly Guiding Fibers

The electric and magnetic fields *E*(*r*, ϕ, *z*) and *H*(*r*, ϕ, *z*) of the optical fibers in cylindrical coordinates can be found by solving Maxwell's equations. However, only lower order modes of ideal step-index fibers are important to the present optical fiber communication systems. The fact is that $\Delta < 1\%$, thus optical waves are weakly guided, and *E* and *H* are then approximate solutions of the scalar wave equation.

$$\left[\frac{\delta^2}{\delta r^2}+\frac{1}{r}\frac{\delta}{\delta r}+k^2 n_j^2\right]\varphi(r)=\beta^2\varphi(r) \tag{12.47}$$

where n_j=n_1, n_2, and $\psi(r)$ is the spatial field distribution of the nearly transverse EM waves

$$E_x = \psi(r)e^{-i\beta z}$$
$$H_y = \left(\frac{\varepsilon}{\mu}\right)^{1/2} E_x = \frac{n_2}{Z_0}E_x \tag{12.48}$$

with Ey, Ez, Hx, Hz negligible, $\varepsilon = n_2^3\varepsilon_0$ and $Z_0=(\varepsilon\mu(^{1/2}$ is the vacuum impedance. That implies that the waves can be seen as a plane wave traveling down along the fiber tube. These plane waves are reflected between the dielectric interfaces, in other words, it is trapped and guided in and along the core of the optical fiber.

12.2.6.2.1 Solutions of the Wave Equation for Step-Index Fiber

The field spatial function $\psi(r)$ would have the form of Bessel functions (from Equation 12.47) as

$$\varphi(r) = A\frac{J_0(ur/a)}{J_0(u)} \quad \text{for } 0 < r < a \tag{12.49}$$

$$\varphi(r) = A\frac{K_0(vr/a)}{K_0(v)} \quad \text{for } r > a \tag{12.50}$$

where J_0 and K_0 are Bessel functions of the first kind and modified of the second kind, respectively, and u and v are defined similarly as given in Equations 12.30 and 12.31. Thus, following Maxwell's equations relation, we can find that E_z can take two possible solutions, which are orthogonal as

$$E_z = -\frac{A}{kan_2}\begin{pmatrix} \sin\phi \\ \cos\phi \end{pmatrix}\begin{cases} \dfrac{uJ_1\left(u\dfrac{r}{a}\right)}{J_0(u)} & \text{for } 0 \le r < a \\ \dfrac{vK_1\left(\dfrac{vr}{a}\right)}{K_0(v)} & \text{for } r > a \end{cases} \tag{12.51}$$

The terms u and v must satisfy simultaneously two equations.

First, the definitions of:

$$u^2 + v^2 = V^2 = ka\left(n_1^2 - n_2^2\right)^{1/2} = kan_2(2\Delta)^{1/2} \tag{12.52}$$

Second, the boundary condition to be met is the eigenvalue equation:

$$u\frac{J_1(u)}{J_0(u)} = v\frac{K_1(v)}{K_0(v)} \tag{12.53}$$

where Equation 12.53 is obtained by applying the boundary conditions at the interface $r=a$ (Ez is the tangential component and must be continuous at this dielectric interface). Equation 12.53 is usually called the eigenvalue equation. The solution of this equation would give the values of β, which would take discrete values, and are the propagation constants of the guided lightwaves.

Equation 12.37 shows that the longitudinal field is in the order of u/kan_2 with respect to the transverse component. In practice, $\Delta \ll 1$; by using Equation 12.52, we observe that this longitudinal component is negligible compared with the transverse component. We thus consider the mode to be of *transverse polarization.* The fundamental mode is then usually denominated and known as LP_{01} mode (LP=Linearly Polarized), for which the field power distribution is shown in Figure 12.8. The graphical representation of the eigenvalue Equation 12.53, calculated as the variation of $\beta/k=b$ as the normalized propagation constant and the V-parameter, is shown in Figure 12.9c.

12.2.6.2.2 Gaussian Approximation

We note again that the E and H are approximate solutions of the scalar wave equation, and the main properties of the fundamental mode of weakly guiding fibers can be observed as follows:

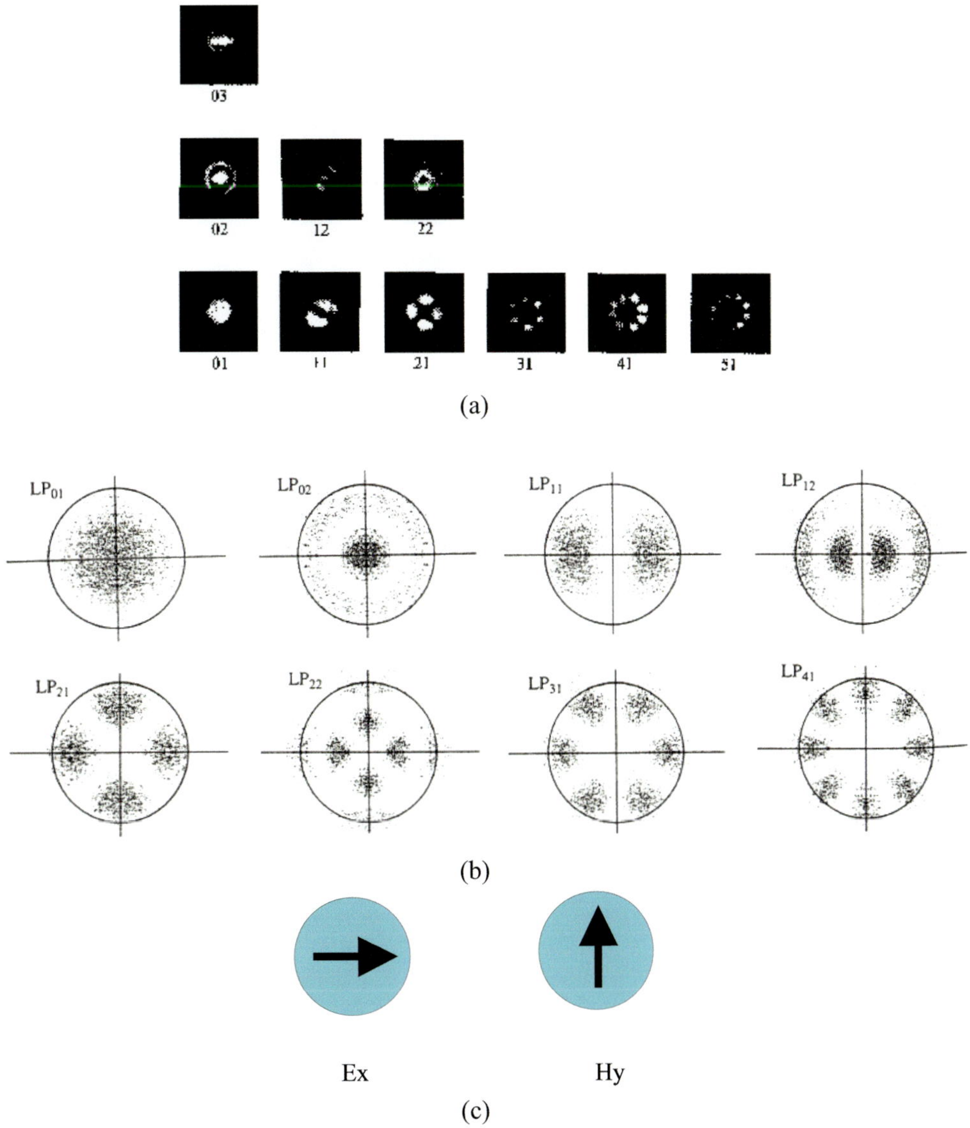

FIGURE 12.8 (a) Spectrum of guided modes in a multi-mode fiber. (b) Calculated intensity distribution of LP guided modes in step-index optical fibers with $V=7$. (c) Electric and magnetic field distribution of an LP_{01} mode polarized along horizontal and vertical directions of the fundamental mode of a multimode fiber.

- the propagation constant β (in the z-direction) of the fundamental mode must lie between the core and cladding wave numbers. This means that the effective refractive index of the guided mode lies within the range of the cladding and core refractive indices.
- accordingly, the fundamental mode must be nearly a transverse electromagnetic wave as described by Equation 12.48.

$$\frac{2\pi n_2}{\lambda} < \beta < \frac{2\pi n_1}{\lambda} \tag{12.54}$$

- the spatial dependence $\psi(r)$ is a solution of the scalar wave Equation 12.47.

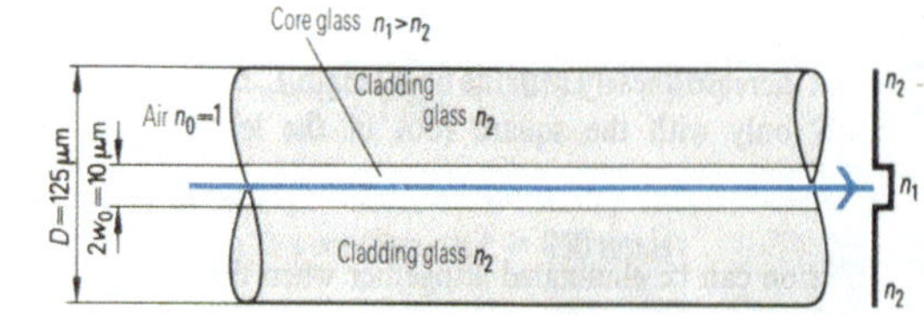

(b)

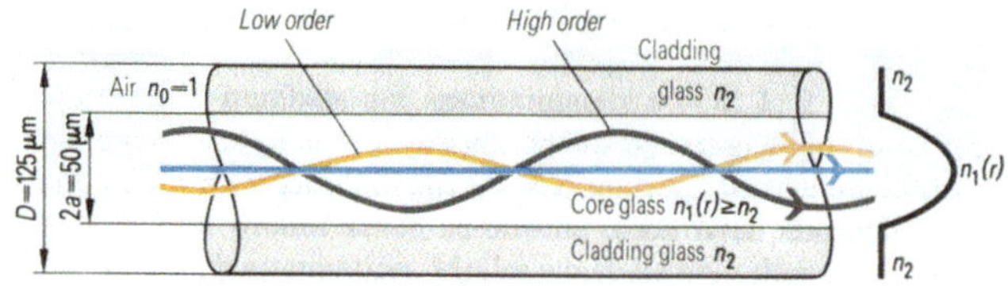

(c)

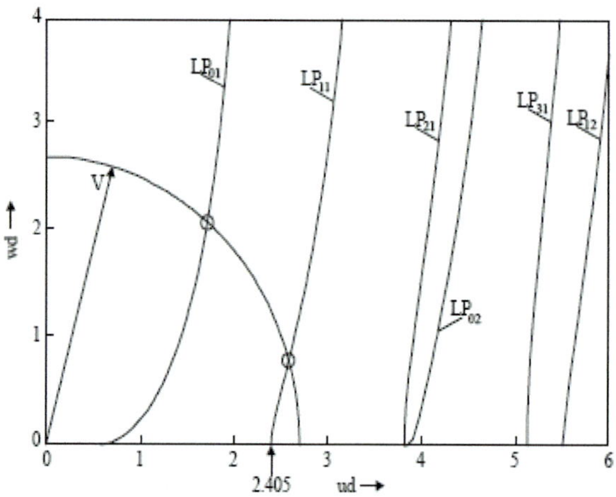

(d)

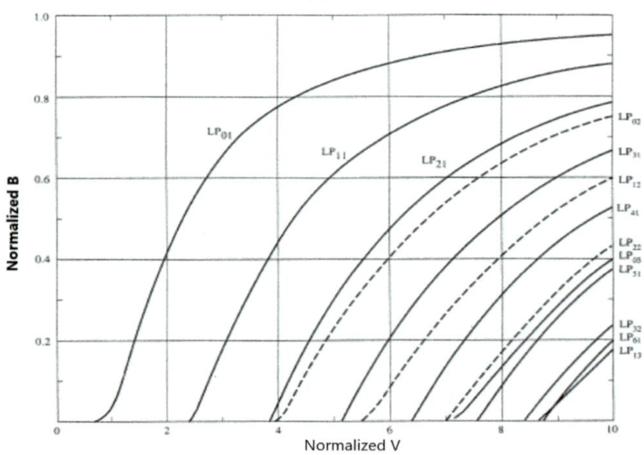

(e)

FIGURE 12.9 (a) Guided modes as seen in the transverse plane of a circular optical fiber. (b) "Ray" model of lightwave propagating in single mode fiber. (c) Ray model for multi-mode graded-index fiber. (d) Graphical illustration of solution for eigenvalues (propagation constant - wavenumber of optical fibers). (e) *b*-*V* characteristics of guided fibers.

The main objective is to find a good approximation for the field $\psi(r)$ and the propagation constant β. These can be found through the eigenvalue equation and Bessel's solutions as shown in the previous section. It is desirable if we can approximate the field to a good accuracy to obtain simple expressions and have a clearer understanding of light transmission on single-mode optical fiber without going through graphical or numerical methods. Furthermore, experimental measurements and numerical solutions for step and power-law profiles show that $\psi(r)$ is approximately Gaussian in appearance. We thus approximate the field of the fundamental mode as

$$\varphi(r) \cong Ae^{-\frac{1}{2}\left(\frac{r}{r_0}\right)^2} \tag{12.55}$$

where r_0 is defined as the spot size, i.e., at which the intensity equals to e^{-1} of the maximum. Thus, if the wave Equation 12.47 is multiplied by $r\varphi(r)$ and using the identity

$$r\varphi\frac{\delta^2\varphi}{\delta r^2}+\varphi\frac{\delta\varphi}{\delta r}=\frac{\delta}{\delta r}\left(r\varphi\frac{\delta\varphi}{\delta r}\right)-r\left(\frac{\delta\varphi}{\delta r}\right)^2 \tag{12.56}$$

then, by integrating from 0 to infinitive and using $\left[r\Psi\frac{d\varphi}{dr}\right]_0^\infty=0$, we have

$$\beta^2=\frac{\int_0^\infty\left[-\left(\frac{\delta\varphi}{\delta r}\right)^2+k^2n^2(r)\varphi^2\right]r\delta r}{\int_0^\infty r\varphi^2\,\delta r} \tag{12.57}$$

The procedure to find the spot size is then followed by substituting $\psi(r)$ (Gaussian) in Equation 12.55 into 12.57, then differentiating and setting $\frac{\delta^2\beta}{\delta r}$ evaluated at r_0 to zero, which means the propagation constant β of the fundamental mode must give the largest value of r_0.

The mode spot size is important for the estimation of nonlinear effects as it contains the majority of the energy of the guided mode. This mode spot size is normally larger than the fiber core radius and contains about 70% of the total energy. There are fibers with spot size larger such as the LEAF (large effective area fiber). The spot size of LEAF is about 30.5% larger than that of standard single mode fiber (SSMF). Thus, its resilience to nonlinearity is increased accordingly.

Knowing r_0 and β, the fields Ex and Hy (Equation 12.48) can be fully specified.

12.2.6.2.2.1 Case 1: Step-Index Fiber Substituting the step-index profile given by Equation 12.48 and $\psi(r)$ in Equation 12.55 into 12.57 leads to an expression for β in term of r_0 given by

$$V = ka\,\mathrm{NA} \tag{12.58}$$

The spot size is thus evaluated by setting

$$\frac{\delta^2\beta}{\delta r_0}=0 \tag{12.59}$$

and r_0 is then given by

$$r_0^2=\frac{a^2}{\ln V^2} \tag{12.60}$$

Then, by substituting (12.60) with (12.58), we obtain

$$(a\beta)^2 = (akn_1)^2 - \ln V^2 - 1 \tag{12.61}$$

This expression is physically meaningful only when $V > 1$ (r_0 is positive).

12.2.6.2.2.2 Case 2: Gaussian Index Profile Fiber Similarly, for the case of a Gaussian index profile, by following the procedures for step-index profile fiber, we can obtain

$$(a\beta)^2 = (an_1k)^2 - \left(\frac{a}{r_0}\right)^2 + \frac{V^2}{\left(\frac{a}{r_0}+1\right)} \tag{12.62}$$

and

$$r_0^2 = \frac{a^2}{V-1} \text{ by applying and setting } \frac{\delta^2\beta}{\delta r_0} = 0 \tag{12.63}$$

That is maximizing the propagation constant of the guided waves. The propagation constant is thus at maximum when the "light ray" is very close to the horizontal direction. Substituting (12.63) into (12.62), we have

$$(a\beta)^2 = (akn_1)^2 - 2V + 1 \tag{12.64}$$

thus Equations 12.63 and 12.64 are physically meaningful only when $V > 1(r_0 > 0)$.

12.2.6.2.3 Cut-Off Properties

Similarly, to the case of planar dielectric waveguides, from Figure 12.9, we observe that when we have $V<2.405$, only the fundamental LP_{01} exists. Thus, we have (Figure 12.10).

It is noted that for single-mode operation, the V parameter must be less than or equal to 2.405; in practice, $V<3$ can also be acceptable. Indeed, the value 2.405 is the first zero of the Bessel function $J_0(u)$. In practice, one cannot really distinguish between V values between 2.3 and 3.0. Experimental observations show that optical fiber can still support only one mode. Thus, designers usually take the value of V as 3.0 or less for obtaining single-mode operation at a specific wavelength.

The V parameter is inversely proportional to the optical wavelength. Thus, if an optical fiber is launched with lightwaves whose optical wavelength is smaller than the operating wavelength at which the optical fiber is single mode, then the optical fiber supports more than one mode. The optical fiber is said to be operating in a multimode region.

Thus, one can define the cut-off wavelength for optical fibers as follows: the wavelength (λ_c), above which only the fundamental mode is guided in the fiber, is called the cut-off wavelength λ_c. This cut-off wavelength can be found by using the V-parameter as $Vc = V$ (at cut-off)$=2.405$, thus

$$\lambda_c = \frac{2\pi a\,\text{NA}}{V_c} \tag{12.65}$$

It is obvious from Equation 12.65 that the spot size of the optical fiber with a V-parameter of 1 is extremely large. This is a very important point that the design of the optical fiber at a near unit value of the V-parameter must be avoided. In practice, we observe that the spot size is large but finite (observable). In fact, if V is smaller than 1.5, the spot size becomes large.

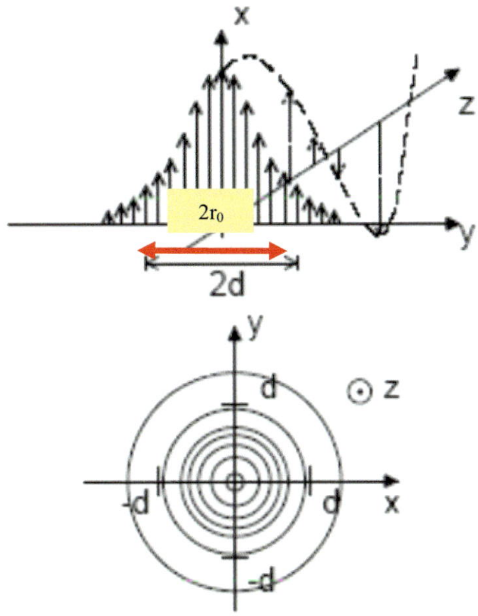

(a)

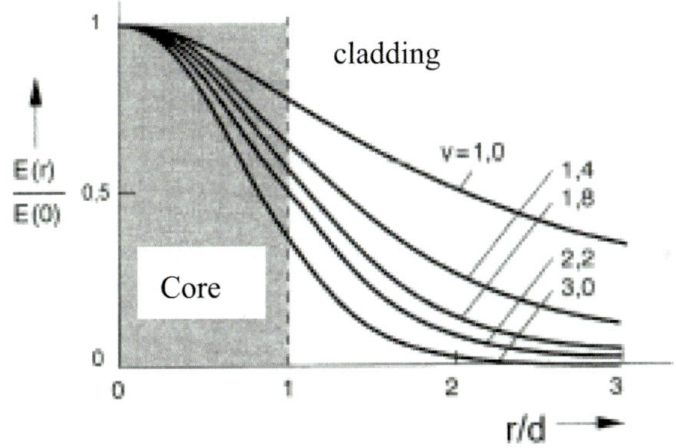

(b)

FIGURE 12.10 (a) Intensity distribution of the LP_{01} mode. (b) Variation of the spot size – field distribution with radial distance r with V as a parameter.

Exercise 12.1

An optical fiber has the following parameters: a core refractive index of 1.46, a relative refractive index difference of 0.3%, a cladding diameter of 125 µm, and a core diameter of 8.0 µm. (i) Find the fiber NA and hence the fiber acceptance angle. (ii) What is the cut-off wavelength of this fiber? (iii) What is the number of optical guided modes that can be supported if the optical fiber is excited with lightwaves of a wavelength of 810 nm? (iv) If the cladding diameter is reduced to 50 and 20 µm, comment on the field distribution of the guided single-mode.

In practice, the fibers tend to be effectively single mode for larger values of V say $V<3$ for the step profile, because the higher order modes suffer radiation losses due to fiber imperfections. Thus, if $V=3$, from Equation 12.52, we have $a<\frac{3}{2}\lambda\,\mathrm{NA}$. In this case, $\lambda=1$ µm, and the numerical aperture NA must be very small ($\ll$1) for radius a to have some reasonable dimension. Usually, Δ is about 1% or less for standard single-mode optical fibers in telecommunications systems.

12.2.6.2.4 Power Distribution

The axial power density or intensity profile $S(r)$, the z-component of the Poynting vector is given by

$$S(r) = \frac{1}{2} E_x H_y^* \tag{12.66}$$

Substituting (12.48) then to (12.66), we have

$$S(r) = \frac{1}{2}\left(\frac{\varepsilon}{\mu}\right)^{1/2} e^{-\left(\frac{r}{r_0}\right)^2} \tag{12.67}$$

The total power is then given by

$$P = 2\pi \int_0^\infty rS(r)dr = \frac{1}{2}\left(\frac{\varepsilon}{\mu}\right)^{1/2} r_0^2 \tag{12.68}$$

and hence the fraction of power $\eta(r)$ within 0 to r across the fiber cross-section is given by

$$\eta(r) = \frac{\int_0^r rS(r)dr}{\int_0^\infty rS(r)dr} = 1 - e^{-\left(\frac{r^2}{r_0^2}\right)} \tag{12.69}$$

Table 12.1 provides the expressions for P and $\eta(r)$ of step-index and Gaussian profile fibers (by substituting the appropriate values of r_0 into Equations 12.68 and 12.69).

As a rule of thumb, it is experimentally confirmed that an optical fiber is most effective for guided mode when the optical power contained in the core is about 80% of the total power.

TABLE 12.1
Analytical Expressions for Total Optical Guided Power and Its Fractional Power Confined Inside the Core Region for Step-Index and Gaussian Index Profiles

	Step-Index	Gaussian
$S(r/a)$ for $V > 1$	$\frac{1}{2}\left(\frac{\varepsilon}{\mu}\right)^{1/2} e^{-\left(\frac{r}{a}\right)^2 \ln V^2}$	$\frac{1}{2}\left(\frac{\varepsilon}{\mu}\right)^{1/2} e^{-\left(\frac{r}{a}\right)(V-1)}$
Power P for $V > 1$	$\frac{1}{2}\left(\frac{\varepsilon}{\mu}\right)^{1/2} \frac{a^2}{\ln V^2}$	$\frac{1}{2}\left(\frac{\varepsilon}{\mu}\right)^{1/2} \frac{a^2}{V-1}$
	$\frac{1}{2}\left(\frac{\varepsilon}{\mu}\right)^{1/2} \frac{a^2}{\ln V^2}$	$\frac{1}{2}\left(\frac{\varepsilon}{\mu}\right)^{1/2} \frac{a^2}{V-1}$
$\eta(r)$ for $V > 1$ $\eta(r($ for $V > 1$	$1 - e^{-\left(\frac{r}{a}\right)^2 \ln V^2}$	$1 - e^{-\left(\frac{r}{a}\right)^2 (V-1)}$
	$= 1 - \frac{1}{V^2}$ for $r = a$	

Exercise 12.2

Using Gaussian approximation for the intensity distribution of the fundamental mode of the single-mode optical fiber with $V=2$, find the fraction of power in the core region with $a=4\ \mu m$.

Exercise 12.3

Find the radius a for maximum confinement of light power, i.e. maximum r_0, for step-index and parabolic profile optical fibers.

12.2.6.2.5 *Approximation of Spot-Size r_o of a Step-Index Fiber*

As stated above, spot-size r_0 would play a major role in determining the performance of single-mode fiber. It is useful if we can approximate the spot size as long as the fiber is operating over a certain wavelength. When a single-mode fiber is operating above its cut-off wavelength, a good approximation (greater than 96% accuracy) for r_0 is given by

$$\frac{r_0}{a} = 0.65 + 1.619V^{-3/2} + 2.879V^{-6} = 0.65 + 0.434\left(\frac{\lambda}{\lambda_c}\right)^{+3/2} + 0.0419\left(\frac{\lambda}{\lambda_c}\right)^{+6}$$

$$\text{for} \quad 0.8 \le \frac{\lambda}{\lambda_c} \le 2.0 \qquad \text{step-index-fiber} \tag{12.70}$$

Exercise 12.4

What is the equivalent range for the V-parameter of Equation 12.70? Inspect the b-V and $V^2\dfrac{d^2(Vb)}{dV^2}$ versus V and b, if possible, do a curve fitting to obtain the approximate relationship for r/a and V (MATLAB procedure is recommended).

Exercise 12.5

Refer to the technical specification of Corning SMF-28 and LEAF (large effective area fiber)

(i) State the core diameter of the fibers, the spot size or mode field diameters of the fibers, and then, estimate the effective areas of these fibers. (ii) What are the ratio of the effective area and the physical area of the cores of the fibers?

12.2.6.2.6 *Equivalent-Step-Index (ESI) Description*

As we have seen in Section 12.2.6.2.1, there are two possible orthogonally polarized modes (Ex, Hy) and (Ey, Hx) that can be propagating simultaneously. These modes are usually approximated by a single linearly polarized (LP) mode. These modes' properties are well-known and well-understood for step-index optical fibers and analytical solutions are also readily available.

Unfortunately, practical SM optical fibers never have a perfect step-index profile due to the dopant diffusion and polarization variation. Under some special conditions, these non-step index fibers can be approximated by *an equivalent-step-index (ESI)* profile technique.

Several index profiles of modern single-mode fibers, e.g., non-zero dispersion shifted fibers, are shown in Figure 12.11. The ESI profile is determined by approximating the fundamental mode electric field spatial distribution $\psi(r)$ by a Gaussian function, as described in Section 12.2.1.6.

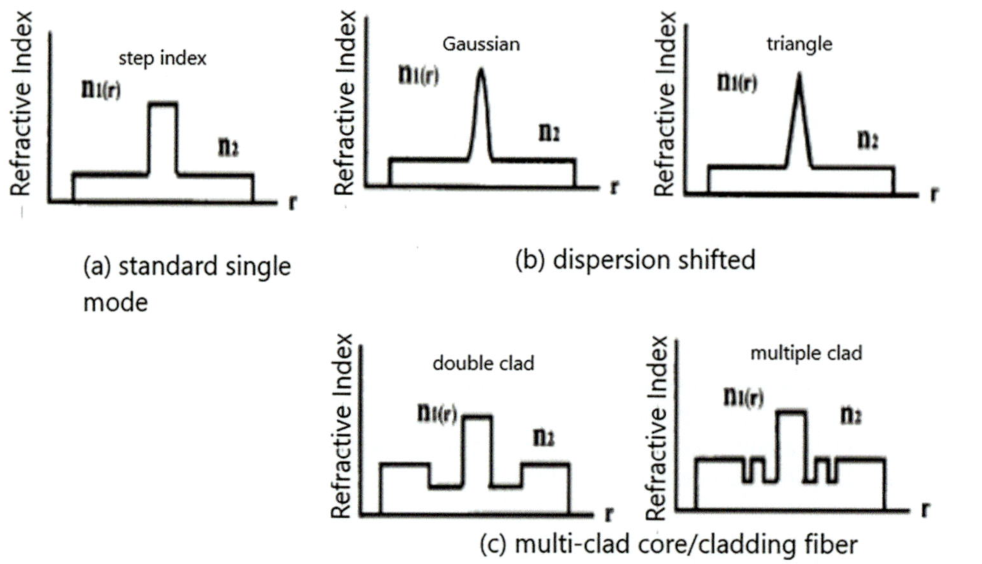

FIGURE 12.11 Index profiles of several modern fibers, e.g., dispersion shifted single mode fibers. (a) Standard single mode, (b) dispersion-shifted, and (c) multi-clad core/cladding fiber.

The electric field can thus be specified by the e^{-1} width of this function or *mode spot size* (r_0). Alternatively, the term mode field diameter (MFD) is also used and is equivalent to twice the size of the mode spot size r_0.

12.2.6.2.7 Definitions of ESI Parameters

The ESI description can be used to design SM fiber with graded index, *W*- or segmented core profiles (under some limitations). These non-step index profiles can be described by ESI parameters denoted as follows:

- V_e = effective or equivalent *V*-parameter
- a_e = ESI core radius
- λ_{ec} = ESI cut-off wavelength
- Δ_e = equivalent relative index difference

These equivalent index ESI parameters are important and specified by the manufacturers of fiber to simplify and ease of understanding of the parameters. Thus, they are discussed here for readers, who are not familiar with fiber parameters and how they can read the parameters of the manufacturer's specifications.

These ESI parameters are related to two moments M_0 and M_1 defined as

$$M_n = \int_0^\infty \left[n^2(r) - n^2(a)\right] r^n \, dr \tag{12.71}$$

For $n = 1, 2$. The effective V_e parameter and effective core radius r_e are given by

$$V_e^2 = 2k^2 \int_0^\infty \left[n^2(r) - n^2(a)\right] r \; dr \tag{12.72}$$

$$V_e^2 = 2k^2 M_1 \quad \text{and} \quad a_e = 2\frac{M_1}{M_0} \tag{12.73}$$

It is followed by Equations 12.72 and 12.73, the parameters λ_{ec} and Δ_e by setting

$$V_e^2 = 2k^2 a_e^2 n_1^2 \Delta_e \tag{12.74}$$

and $V_e = 2.405$ (cut-off condition for step-index). Therefore, the cut-off wavelength for an ESI profile fiber is:

$$\lambda_{ec} = \frac{2\pi\sqrt{2M_1}}{2.405} \tag{12.75}$$

It is to be noted that V_e as given in Equation 12.74 is equivalent to the mode volume. Physically, the significance of V_e can be compared to the average density of a disk with a local density equal to $[n^2(r) - n^2(a)]$.

12.2.6.2.8 Accuracy and Limits

The ESI approximation is generally accurate to within 2% at least over the wavelength range $0.8 < \lambda / \lambda_c < 1.5$. For most practical purpose this range is the operating wavelength to minimize the dispersion property of SM optical fibers.

12.2.6.2.9 Examples on ESI Techniques

12.2.6.2.9.1 Graded-Index Fibers These index profiles of graded fibers are given by Equation 12.45. We thus have

$$n^2(r) - n^2(a) = s(r/a) = 1 - \left(\frac{r}{a}\right)^\alpha \tag{12.76}$$

substituting (12.76) into (12.72) gives

$$\frac{V_e}{V} = \left(\frac{\alpha}{\alpha + 2}\right)^{1/2} \tag{12.77}$$

where $V^2 = k^2 a^2$. NA is the V parameter of a step-index fiber with the core index at the fiber axis of n_1. Hence, we have

$$\lambda_{ec} = \frac{V}{2.405}\left(\frac{\alpha}{\alpha + 2}\right)^{1/2} \tag{12.78}$$

Exercise 12.6

For a single-mode optical fiber with a triangular profile index distribution whose equivalent V-parameter is equal to 2 at 1550 nm wavelength, what is the V-parameter value at the center of the core of the fiber? If the diameter of the core of the two fibers is kept identical, then what is the ratio of the refractive indices at the core center of the fibers? Repeat for parabolic profile.

12.2.6.2.9.2 Graded-Index Fiber with a Central Dip The fiber index profile with a central dip and grade gradually increases to the outer cladding is shown in Figure 12.12.

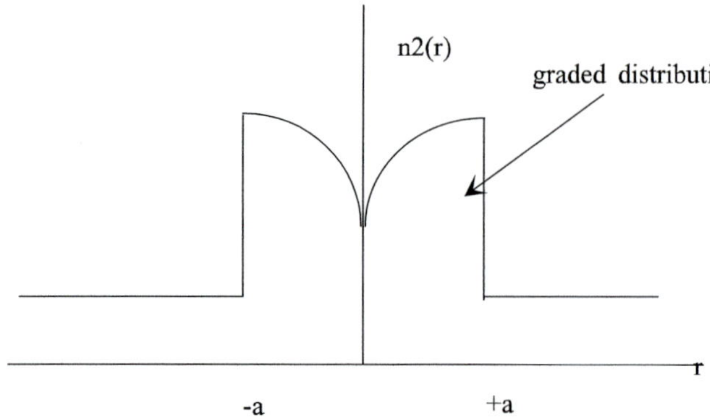

FIGURE 12.12 Refractive index profile of a graded index fiber with a central dip. This is typical profile of manufactured fiber if a good collapsing of the preform is not achieved.

Similar to Equation 12.44 for graded index fiber with maximum index at the core axis, we have

$$S(r/a) = 1 - \gamma(1-x)^{\alpha} \quad \text{for} \quad 0 < r < a \tag{12.79}$$

where γ is the depth, and $0<\gamma<1$. When $r=0$, we have a step-index profile, and when $r=1$, the central axis refractive index is equal to the cladding index.

Using Equations 12.71 and 12.72, V_e can be easily found and given by:

$$\frac{V_e^2}{V^2} = 1 - \frac{2\gamma}{(\alpha+1)(\alpha=2)} \tag{12.80}$$

The general technique to find the ESI parameters for optical fibers can be started by rising the stationary expression in Equation 12.57 for expressing $\beta(\lambda)$ of the actual fiber as compared to its equivalent propagation constant β_e as:

$$\beta^2 = \beta_e^2 + k^2 \frac{\int_0^{\infty}\left[n^2(r) - n_e^2(r)\right] r\psi^2(r)\,dr}{\int_0^{\infty} r\psi^2(r)\,dr} \tag{12.81}$$

where $n_e^2(r)$ is the equivalent counterpart of $n(r)$ when the fiber is expressed in its equivalent step form. The field expression $\psi(r)$ is assumed (in fact to be obtained) to be similar for the actual fiber and its step equivalence. Once the field $\psi(r)$ can be replaced by the approximate exact field shape, we can find V_e and a_e that minimize $\beta^2 - \beta_e^2$ in Equation 12.81. Generally, these parameters are functions of both V and a. Thus, it is not possible to get one ESI technique applicable to a wide range of wavelengths, and complicated numerical calculations are required.

12.2.6.3 Optical Fiber Manufacturing and Cabling

This section provides a brief description of the manufacturing of optical fibers and the cabling of several fibers for optical communication systems. The manufacturing techniques and cabling process affect the transmission and physical properties of the fibers. We focus on the aspects necessary for a general understanding of optical transmission systems.

As described in previous sections, the standard single-mode optical fiber (SSMF) structure is a cylindrical core with a refractive index slightly higher than that of the cladding region. For optical communications operating in the 1300 and 1700 nm wavelength regions, the silica material is the base material. A "pure" silica tube is the starting structure, and a combination of silica, Germanium

Oxide GeO_2, and P_2O_5 are then deposited inside the tube. The refractive index changes of the deposit can be different for alternative doped core materials. For example, the fiber with zero-dispersion shifted wavelength is for low dispersion distortion. Other dopants, such as B_2O_3 and fluoride, can also be used to reduce the refractive index of some small regions of the core. Once the deposition of the impurities is done (see Figure 12.15), the tube is collapsed to produce a fiber preform as shown in Figure 12.13a and b (Figure 12.14). Also illustrated in Figures 12.13a and 12.15a, is an image of the fiber drawing machine and fiber drawing tower. The refractive index of the perform is also shown, as noted in its caption and detailed in Figure 12.15.

Furthermore, the splicing or optical soldering of two fibers can be done by high intensity spark that melts the two glass fibers and thus splicing them. Image processing is used to align the fiber cores and even the polarization of the guided lightwaves between the two fibers. Once the splicing is complete, the measurement *f* scattering loss of the splice can be obtained (Figure 12.16).

This section has introduced the fundamental concepts of optical slab waveguides and the circular optical waveguides or optical fibers. The basic properties of the fiber structures, their profile, the spot size, the cut-off wavelength, and the Gaussian approximation are described. The Gaussian approximation simplifies the understanding of the optical guided mode. It also allows us to obtain

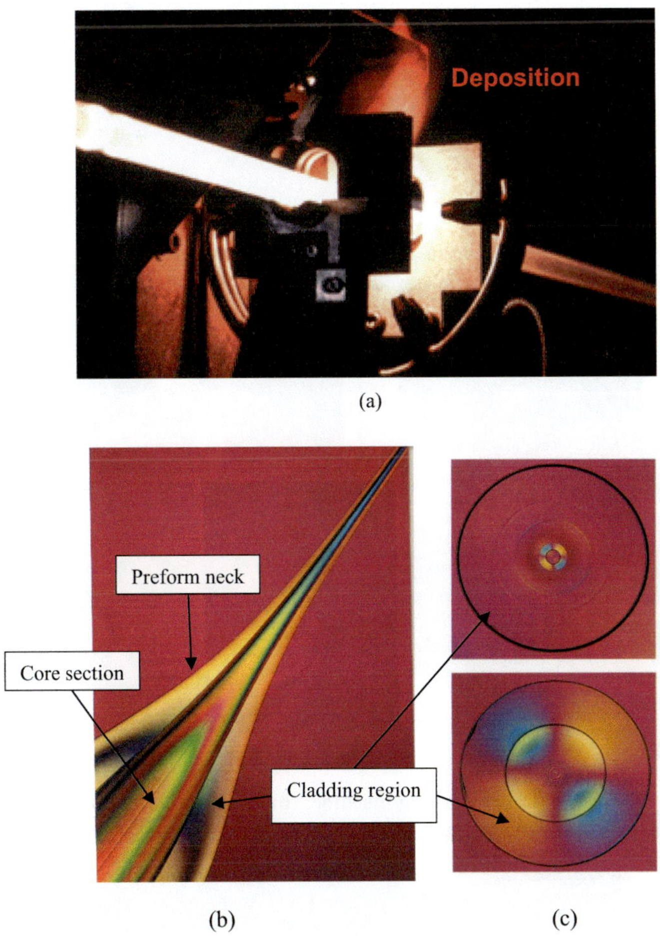

FIGURE 12.13 Images of (a) a fiber deposition and fabrication of fiber preform, deposition of core material and collapsing, (b) fiber perform, before drawing into fiber stands, (c) cross-section of fiber perform with refractive index profile exactly the same as the fiber index profile of single mode (upper) and multimode (lower) types.

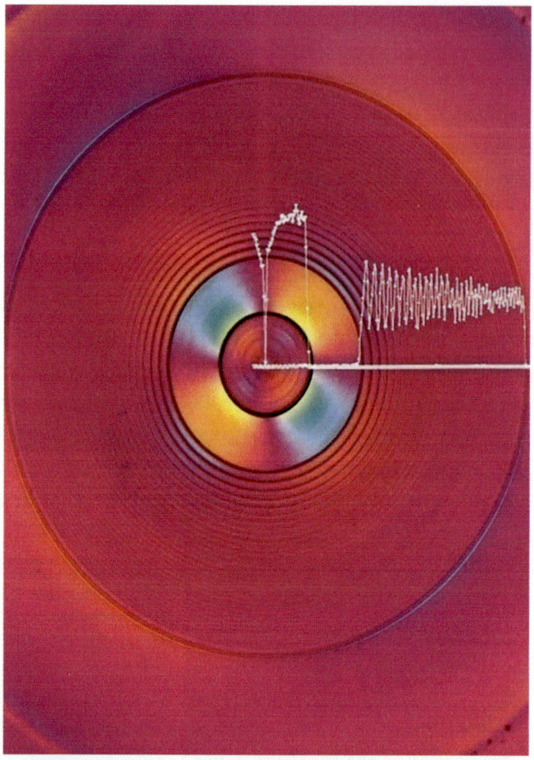

FIGURE 12.14 Real index profile across an SM fiber perform. Note: It is non-step-like profile, so why modeled as a step-index structure? The preform is used to draw tens of kilometers of fiber.

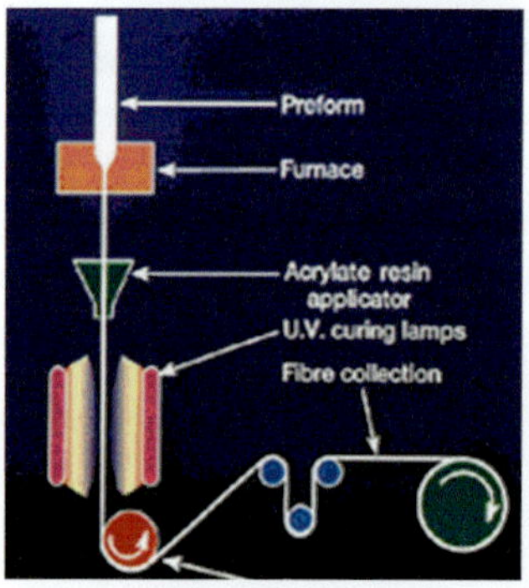

(a)

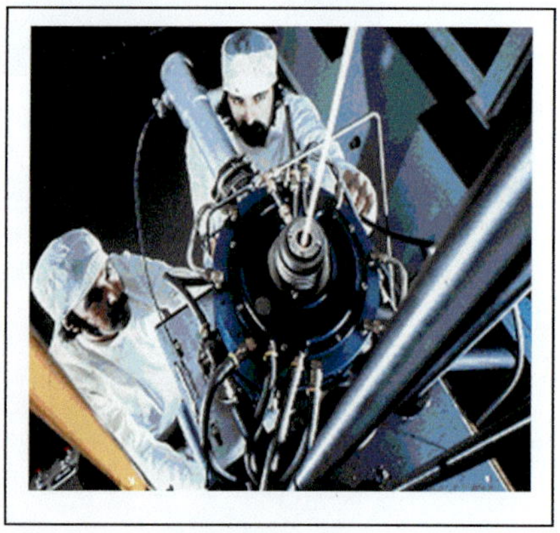

(b)

FIGURE 12.15 (a) Schematic of fiber drawing machine. (b) Foto of fiber microwave furnace and diameter monitoring and the feedback control system.

FIGURE 12.16 Splicing of two bare optical fibers.

directly the optical mode distribution and thus several other approximations required to obtain the simplest form of important parameters of single-mode optical fibers.

Once the basic properties of a single-mode optical fiber are found, they form the basic set of parameters so that optical fibers whose effective index profiles are non-step can be found based on the ESI technique that would converts the parameters to an equivalent step-like profile and hence other optical properties.

Only structural and wave properties of lightwave signals traveling in optical fibers are presented here. As an optical communications system engineer, we need to understand and develop techniques for analyzing and identifying the transmission of digital and analog signals through optical fibers, including the attenuation and broadening of optical signals after transmission through the medium, namely attenuation and broadening via dispersion of lightwaves pulses.

12.2.6.4 Technical Data of Single Mode Optical Fibers

12.2.6.4.1 Remarks

Optical fibers are the main transport medium for ultra-high-speed and ultra-high-capacity data interchange data centers, serving massive data demands. Initially, multimode fibers (MMF) with a step-index were employed, but now graded index types, such as OM4, are used to reduce intermodal delay broadening, thus widening the bandwidth over short distances. However, most transport is done over single-mode fibers (SMF). Over the years, there have been quite a number of types of SMF, and eventually, a standard SMF (SSMF) was established: the G.652 fiber. Corning Inc. has produced this SSMF as SMF-28 and SMF-28e, as well as large effective area fibers (LEAF). The difference between SMF-28e and SMF-28 is that the OH-absorption (water) peak at 1400 nm is depressed, as shown in Figure 12.18, allowing channel carriers in the 1300 –1500 nm spectral window to be used. At 1300 nm, the chromatic dispersion factor is almost zero, enabling the transmission of high-speed data sequences with many carriers in this spectral window, the O-band. The fiber attenuation in this O-band is higher than in the C-band (1525–1565 nm) and L-band (1565–1620 nm).

Emerging service networks, such as 5G, cloud computing, video retrieval, and ultra-HD video, continuously drive increasing service traffic, imposing higher requirements on the transmission rate and performance of transmission networks. The demand for improved single-wavelength rate and performance requires more fiber spectrum resources. Due to the slow deployment of fiber, the most effective way to cope with the ever-increasing network traffic is to improve single-fiber capacity by expanding the fiber spectrum.

Widening the spectrum over which attenuation is lowest is necessary for a longer transmission distance. Presently, the extended C+-band or super-C band is spectrally widened from the traditional C-band of 80 channels with 50 GHz spacing to 120 channels with 75 GHz spacing, as shown in Figure 12.17. This extension of the C-band, combined with appropriate gain from optical

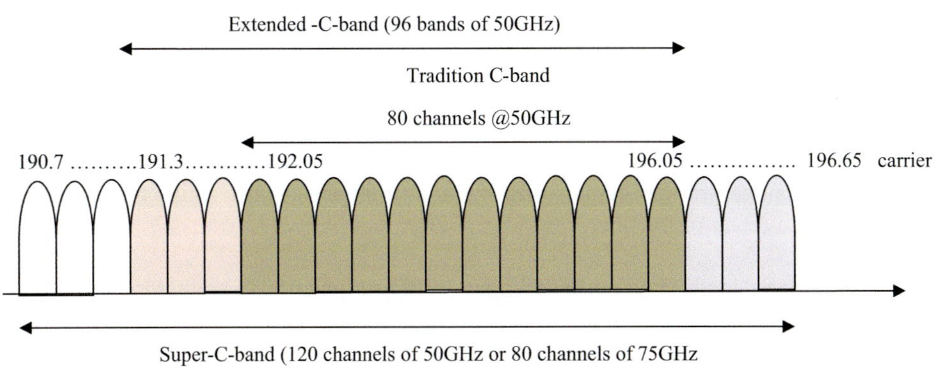

FIGURE 12.17 Traditional C-band, extended C-band, and super-C band over the 1550 nm spectral window.

amplifiers such as EYDFA (Erbium-Ytterbium-doped Fiber Amplifiers) in combination with Raman amplifier types, allows the full super-C band to be fully employed, offering a unique medium for transporting ultra-high-capacity information systems, especially in DCN (Data Center Networking) and DCI (Data Center Interconnect).

12.2.6.4.2 Single Mode Fibers and Standard

This appendix gives the main features of the fibers, including the standard single-mode fiber (SSMF), G.652 standard type. The parameters cover geometrical dimensions, dispersion factors, attenuation as a function of wavelength, and polarization mode dispersion. These data sets give users a picture of how much the signals transmitted through these fibers can suffer from distortion and degradation of their total average power. Figure 12.18 shows the main technical parameters of SSMF, including the refractive index profile, the geometrical parameters of the core, the cladding, and the outside coating for fiber strengthening. The attenuation as a function of wavelength is also shown to indicate the absorption peak at 1400 nm due to water OH contamination during the collapsing of the preform. Figure 12.19 shows the technical parameters of the enhanced SSMF, where the water absorption peak is suppressed to allow low-loss transmission over the O-band and C-band. The dispersion properties of the two SSMF types are quite similar. Specifically, the dispersion is near zero at 1310 nm, and the dispersion factor is about 17 ps/(nm· km), which corresponds to a pulse broadening of 17 ps for a 1.0 nm bandwidth (or 125 GHz) signal propagating over one kilometer of fiber. The dispersion due to group velocity differences at the wavelength carriers over the signal bandwidth can be scaled appropriately for signals of different bit rates or symbol rates, and hence bandwidths, and propagation distances.

Furthermore, the LEAF fiber, or large effective area fibers, are also provided with their technical data. This large effective area fiber allows higher power in all multiplexed channels to be supported with minimal distortion due to nonlinear effects, especially self-phase modulation.

All parameters here are extracted from a number of fiber manufacturers, especially Corning Inc.

Standard Single Mode Optical Fibers (SSMF): The main parameters for standard SSMF are:

(I) Cut-off wavelength: $\lambda c \leq 1260$ nm

(II) Mode Field Diameter (MFD) and mode spot size (r_0)

Mode Field and Spot Size (μm)	
MFD(@1310 nm)	9.2±0.5 μm
MFD (@1550 nm)	10.5±0.5 μm
Dispersion (units ps/nm/km)	
D (1550 nm)	≤18.0
D (1310 nm)	≤22.0
Zero dispersion wavelength $<\lambda_0$	$1304<\lambda_0<1324$ nm

MFD is defined as the diameter of the guided optical field distribution over the fiber cross-section, which is normally a Gaussian distribution, at the $1/e$ intensity with respect to the maximum intensity of the guided mode at the center.

Similarly, the mode spot size is defined based on the radius of the guided mode intensity distribution. It depends on the first- or second-order method of moment to determine the $1/e$ value with respect to that at the center of the fiber.

a. Cut-off wavelength, mode field diameter (MFD), geometrical parameters
b. Attenuation and dispersion factor with wavelength dependence relationship
c. Dispersion and group delay, scattering
d. Index profile and attenuation spectrum

Cable Cutoff Wavelength (λ_{cc})

$\lambda_{cc} \leq 1260$ nm

Mode-Field Diameter

Wavelength (nm)	MFD (μm)
1310	9.2 ± 0.5
1550	10.5 ± 0.5

Dispersion

Wavelength (nm)	Dispersion Value [ps/(nm•km)]
1550	≤ 18.0
1625	≤ 22.0

Zero Dispersion Wavelength (λ_0): 1304 nm $\leq \lambda_0 \leq$ 1324 nm
Zero Dispersion Slope (S_0): ≤ 0.092 ps/(nm²•km)

Polarization Mode Dispersion (PMD)

	Value (ps/√km)
PMD Link Design Value	≤ 0.04*
Maximum Individual Fiber PMD	≤ 0.1

*Complies with IEC 60794-3: 2001, Section 5.5, Method 1, (m = 20, Q = 0.01%), September 2001.

Performance Characterization Specification

Characterized parameters are typical values.

Core Diameter	8.2 μm
Numerical Aperture	0.14 NA is measured at the one percent power level of a one-dimensional far-field scan at 1310 nm.
Effective Group Index of Refraction (N_{eff})	1310 nm: 1.4606 1550 nm: 1.4620
Fatigue Resistance Parameter (N_d)	20
Coating Strip Force	Dry: 0.6 lbs. (3N) Wet, 14-day room temperature: 0.6 lbs. (3N)
Rayleigh Backscatter Coefficient (for 1 ns Pulse Width)	1310 nm: -77 dB 1550 nm: -82 dB

(a) Cut-off wavelength, mode field diameter, geometrical parameters

FIGURE 12.18 SMF-28 Standard SMF, fiber refractive index profile, and spectral attenuation distribution. Note the absorption peaks at 1410 nm due to OH water contamination. (Adapted from public open technical specification of fiber manufacturer.)

(Continued)

Optical Specifications

Attenuation

Standard Attenuation Cells

Wavelength (nm)	Attenuation Cells (dB/km) Premium *	Standard
1310	≤0.35	≤0.40
1550	≤0.25	≤0.30

* Lower attenuation available in limited quantities.

Point Discontinuity

No point discontinuity greater than 0.10 dB at either 1310 nm or 1550 nm.

Attenuation at the Water Peak

The attenuation at 1383 ± 3 nm shall not exceed 2.1 dB/km.

Attenuation vs. Wavelength

Range (nm)	Ref. λ (nm)	Max. α Difference (dB/km)
1285 - 1330	1310	0.05
1525 - 1575	1550	0.05

The attenuation in a given wavelength range does not exceed the attenuation of the reference wavelength (λ) by more than the value α.

Attenuation with Bending

Mandrel Diameter (mm)	Number of Turns	Wavelength (nm)	Induced Attenuation* (dB)
32	1	1550	≤0.50
50	100	1310	≤0.05
50	100	1550	≤0.10

*The induced attenuation due to fiber wrapped around a mandrel of a specified diameter.

Cable Cutoff Wavelength (λ_{ccf})

$\lambda_{ccf} \leq 1260$ nm

Mode-Field Diameter

9.2 ± 0.4 μm at 1310 nm

10.4 ± 0.8 μm at 1550 nm

Dispersion

Zero Dispersion Wavelength (λ_0):

1302 nm $\leq \lambda_0 \leq$ 1322 nm

Zero Dispersion Slope (S_0):

≤ 0.092 ps/(nm²·km)

$$\text{Dispersion} = D(\lambda) :\approx \frac{S_0}{4}\left[\lambda - \frac{\lambda_0^4}{\lambda^3}\right] \text{ps/(nm·km)},$$

for 1200 nm ≤ λ ≤ 1600 nm

λ = Operating Wavelength

Polarization Mode Dispersion

Fiber Polarization Mode Dispersion (PMD)

	Value (ps/√km)
PMD Link Value	≤ 0.1*
Maximum Individual Fiber	≤ 0.2

* Complies with IEC SC 86A/WG1, Method 1, September 1997.

The PMD link value is a term used to describe the PMD of concatenated lengths of fiber (also known as the link quadrature average). This value is used to determine a statistical upper limit for system PMD performance.

Individual PMD values may change when cabled. Corning's fiber specification supports network design requirements for a 0.5 ps/√km maximum PMD.

(b) Attenuation and dispersion factor with wavelength dependence relationship

FIGURE 12.18 (*Continued*) SMF-28 Standard SMF, fiber refractive index profile, and spectral attenuation distribution. Note the absorption peaks at 1410 nm due to OH water contamination. (Adapted from public open technical specification of fiber manufacturer.)

Performance Characterizations

Characterized parameters are typical values.

Core Diameter: 8.2 µm

Numerical Aperture: 0.14

NA is measured at the one percent power level of a one-dimensional far-field scan at 1310 nm.

Zero Dispersion Wavelength (λ_0): 1313 nm

Zero Dispersion Slope (S_0): 0.086 ps /(nm²·km)

Refractive Index Difference: 0.36%

Effective Group Index of Refraction, **(N_{eff} @ nominal MFD):**

1.4677 at 1310 nm
1.4682 at 1550 nm

Fatigue Resistance Parameter **(n_d):** 20

Coating Strip Force:

Dry: 0.6 lbs. (3N)
Wet, 14-day room temperature: 0.6 lbs. (3N)

Rayleigh Backscatter Coefficient **(for 1 ns pulse width):**

1310 nm: -77 dB
1550 nm: -82 dB

(c) Dispersion and group delay, scattering

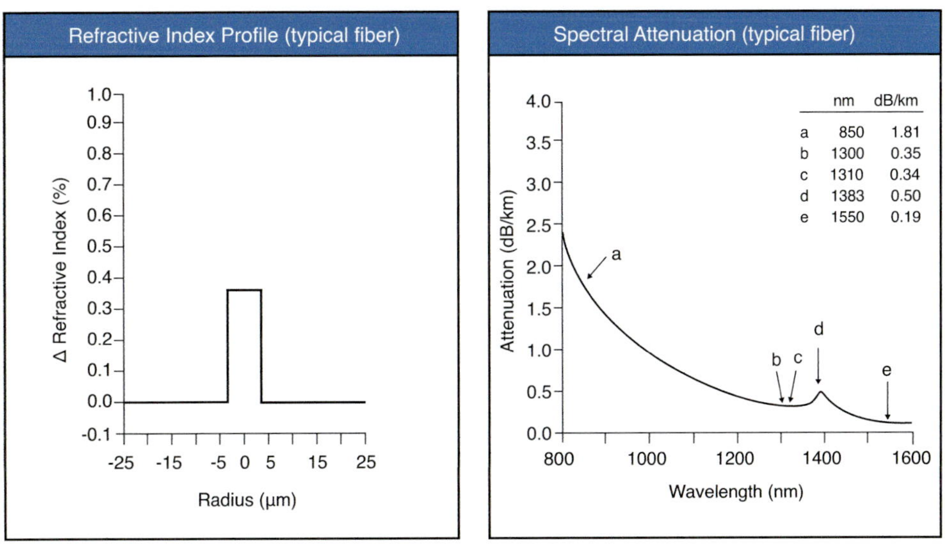

(d) Index profile and attenuation spectrum.

FIGURE 12.18 (*Continued*) SMF-28 Standard SMF, fiber refractive index profile, and spectral attenuation distribution. Note the absorption peaks at 1410 nm due to OH water contamination. (Adapted from public open technical specification of fiber manufacturer.)

12.2.6.4.3 Enhanced Standard Single Mode Optical Fibers

Optical Specifications

Premium Attenuation

Wavelength (nm)	Attenuation (dB/km)
1310	≤ 0.35
1383	≤ 0.31*
1550	≤ 0.25

*Attenuation increases due to hydrogen aging at this wavelength will be ≤0.01 dB/km and evaluated in accordance with the IEC 60793-2 test procedure.

Point Discontinuity

No point discontinuity greater than 0.10 dB at either 1310 nm or 1550 nm.

Attenuation vs. Wavelength

Range (nm)	Ref. λ (nm)	Max. α Difference (dB/km)
1285 - 1330	1310	0.05
1525 - 1575	1550	0.05

The attenuation in a given wavelength range does not exceed the attenuation of the reference wavelength (λ) by more than the value α.

Attenuation with Bending

Mandrel Diameter (mm)	Number of Turns	Wavelength (nm)	Induced Attenuation* (dB)
32	1	1550	≤0.50
50	100	1310	≤0.05
50	100	1550	≤0.10
75	100	1625	≤0.50

*The induced attenuation due to fiber wrapped around a mandrel of a specified diameter.

Cable Cutoff Wavelength (λ_{ccf})

$\lambda_{ccf} \leq 1260$ nm

Mode-Field Diameter

9.2 ± 0.4 μm at 1310 nm
10.4 ± 0.8 μm at 1550 nm

Dispersion

Zero Dispersion Wavelength (λ_0):
1302 nm $\leq \lambda_0 \leq$ 1322 nm

Zero Dispersion Slope (S_0):
$\leq$ 0.092 ps/(nm²•km)

$$\text{Dispersion} = D(\lambda) :\approx \frac{S_0}{4}\left[\lambda - \frac{\lambda_0^4}{\lambda^3}\right] \text{ps/(nm•km)},$$

for 1200 nm $\leq \lambda \leq$ 1625 nm

λ = Operating Wavelength

Polarization Mode Dispersion

Fiber Polarization Mode Dispersion (PMD)

	Value (ps/√km)
PMD Link Value	≤0.1*
Maximum Individual Fiber	≤0.2

* Complies with IEC SC 86A/WGI, Method 1, September 1997.

The PMD link value is a term used to describe the PMD of concatenated lengths of fiber (also known as the link quadrature average). This value is used to determine a statistical upper limit for system PMD performance.

Individual PMD values may change when cabled. Corning's fiber specification supports network design requirements for a 0.5 ps/√km maximum PMD.

Performance Characterizations

Characterized parameters are typical values.

Core Diameter: 8.2 μm

Numerical Aperture: 0.14

NA is measured at the one percent power lev of a one-dimensional far-field scan at 1310 nn

Zero Dispersion Wavelength (λ_0)**:** 1313 nm

Zero Dispersion Slope (S_0)**:** 0.086 ps/(nm²•km)

Refractive Index Difference: 0.36%

Effective Group Index of Refraction (N_{eff})**:**

1.4677 at 1310 nm
1.4682 at 1550 nm

Fatigue Resistance Parameter (n_d)**:** 20

Coating Strip Force:

Dry: 0.6 lbs. (3N)
Wet, 14-day room temperature: 0.6 lbs. (3N)

FIGURE 12.19 SMF-28e fiber refractive index profile and spectral attenuation distribution. Note: no absorption peak at 1410 nm – suppression of water absorption peaks. The refractive index profile and attenuation spectral property are given in the inserted figures at the end. (Adapted from public technical specification of fiber manufacturer.)

(*Continued*)

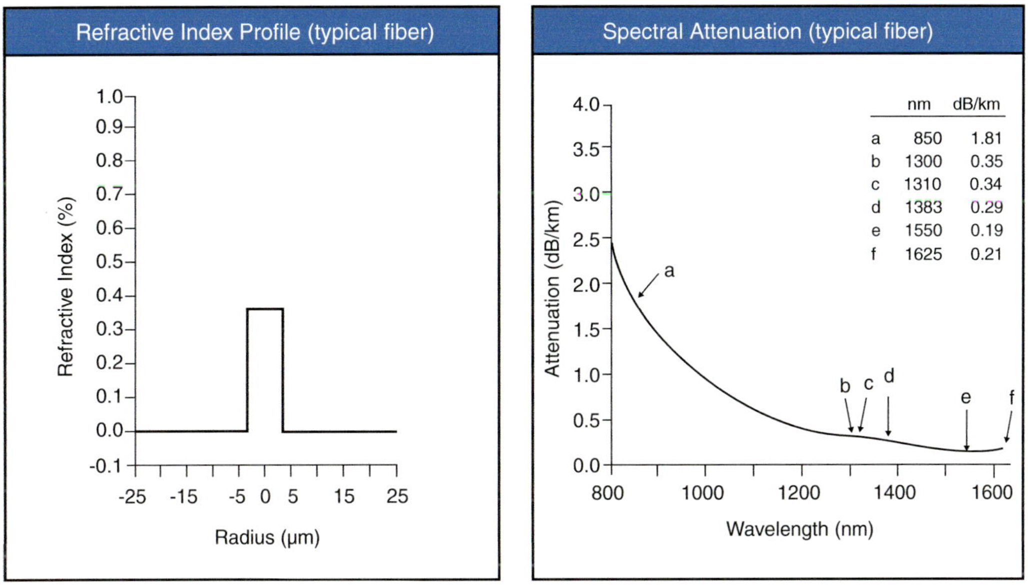

FIGURE 12.19 (*Continued*) SMF-28e fiber refractive index profile and spectral attenuation distribution. Note: no absorption peak at 1410 nm – suppression of water absorption peaks. The refractive index profile and attenuation spectral property are given in the inserted figures at the end. (Adapted from public technical specification of fiber manufacturer.)

12.2.6.4.4 Large Effective Area Fiber

LEAF fibers have a large effective area, which is the area over which the fundamental mode field diameter (or twice the mode spot size) of a near Gaussian distribution is covered. This large effective mode area helps reduce nonlinear effects, such as phase distortion, when many information channels with different carriers are propagating in parallel using the dense wavelength division multiplexing (DWDM) technique. The main properties of LEAF are shown in Figure 12.20.

12.2.6.5 Large Effective Area Fibers for Ultra-High Capacity

Optical transmission technology has, in recent years, advanced rapidly with the combination of multilevel modulation formats, coherent detection, and digital signal processing (DSP).[49] 100 Gb/s optical transport systems have been widely deployed, and 400 Gb/s and 1 Tb/s super-channels have been demonstrated in laboratories and field trials.[50] While transmission technology is necessary to transmit large capacities at higher data rates over longer distances, optical networking is now gradually evolving toward elastic/flexible optical transport networks (flex OTN) or flex-grid optical network architecture[51,52] to make better use of optical network resources to accommodate the ever-increasing traffic demand. In flex OTN, the required minimum spectral resources are allocated adaptively based on traffic demand and network conditions, and the flex OTN may also be able to trade-off between spectral efficiency (SE) and the all-optical reach distance. Currently, the Flex OTN can be achieved, by relying on the flex OTN transponder techniques that are permissible to vary symbol rate, code rate, or modulation formats. One example of such elastic transponders is to support several modulation formats, such as polarization-division-multiplexed quadrature phase shift keying (PD-QPSK) and PD- 16 quadrature amplitude modulation (PD-16 QAM).[53] A combination of transmission and networking technologies may help to make effective use of available optical networking resources. However, the most fundamental improvement to

Attenuation With Bending

Mandrel Diameter (mm)	Number of Turns	Wavelength (nm)	Induced Attenuation (dB)
32	1	1550 & 1625	≤0.50
75	100	1550 & 1625	≤0.05

The induced attenuation due to fiber wrapped around a mandrel of a specified diameter.

Mode-Field Diameter

9.20 to 10.00 μm at 1550 nm

Dispersion

Total Dispersion: 2.0 to 6.0 psec/(nm•km) over the range 1530 to 1565 nm

4.5 to 11.2 psec/(nm•km) over the range of 1565 to 1625 nm

Fiber Polarization Mode Dispersion (PMD)

	Value (ps/√km)
PMD Link Design Value	≤0.04*
Maximum Individual Fiber	≤0.1

**Complies with IEC SC 86A/WG1, Method 1, September 1997 (n=24, Q=0.1%)*

The PMD link design value is a term used to describe the PMD of concatenated lengths of fiber (also known as PMDQ). This value represents a statistical upper limit for total link PMD.

PMD values may change when fiber is cabled. Corning's fiber specification supports emerging network design requirements for high-data-rate systems operating at 10 Gbps (TDM) rates and higher.

Environmental Specifications

Environmental Test Condition	Induced Attenuation (dB/km) 1550 nm
Temperature Dependence -60°C to +85°C*	≤ 0.05
Temperature – Humidity Cycling -10°C to +85°C* and up to 98% RH	≤ 0.05
Water Immersion, 23°C	≤ 0.05
Heat Aging, 85°C*	≤ 0.05

Operating Temperature Range: -60°C to +85°C

**Reference Temperature = +23°C*

Dimensional Specifications

*Standard Length (km/reel) 4.4 - 25.2**

**Longer spliced lengths available at a premium.*

Glass Geometry

Fiber Curl: ≥ 4.0 m radius of curvature
Cladding Diameter: 125.0 ± 1.0 μm
Core/Clad Concentricity: ≤ 0.5 μm
Cladding Non-Circularity: ≤ 1.0%

Defined as:

$$\left[1 - \frac{\text{Min. Cladding Diameter}}{\text{Max. Cladding Diameter}}\right] \times 100$$

Coating Geometry

Coating Diameter: 245 ± 5 μm
Coating/Cladding Concentricity: < 12 μm

Mechanical Specifications

Proof Test

The entire length of fiber is subjected to a tensile proof stress ≥ 100 kpsi (0.7 GN/m²)*

**Higher proof test available at a premium.*

Performance Characterizations

Characterized parameters are typical values.

Effective Area (A_{eff})
72 μm²

Effective Group Index of Refraction (N_{eff})
1.469 at 1550 nm

Fatigue Resistance Parameter (n_d)
20

Coating Strip Force
Dry, 2.8 N (0.6 lbs)
Wet, 14 days room temperature:
2.7 N (0.6 lbs)

Consistency with Global Standards

The values in this product information sheet demonstrate Corning® LEAF® fiber's conformity with ITU-T Recommendation G.655, IEC 60793-2 for B4 class fibers and Belcore/Telcordia GR-20-CORE.

(a)

Dispersion Calculation

$$\text{Dispersion} = D(\lambda) = \left(\frac{D(1565\text{ nm}) - D(1530\text{ nm})}{35} \cdot (\lambda - 1565)\right) + D(1565\text{ nm})$$

λ = Operating wavelength up to 1565

$$\text{Dispersion} = D(\lambda) = \left(\frac{D(1625\text{ nm}) - D(1565\text{ nm})}{60} \cdot (\lambda - 1625)\right) + D(1625\text{ nm})$$

λ = Operating wavelength from 1565–1625

(b)

FIGURE 12.20 Technical specification of corning LEAF fiber -distributed to clients as product (a) Operational characteristics (b) calculation of dispersion as a function of operating wavelength (Adapted from public technical specification of fiber manufacturer.)

the efficiency in the utilization of optical network resources is to optimize optical fiber cabling infrastructures to provide better capacity upgrade pathways and ensure sufficient transmission margins.[54] Another effective way is to improve amplification technologies that are widely utilized in an optical network.[55,56]

Historically, there is no doubt that transmission fibers have been critical to the enormous success of optical communication technology. Optical fibers, as one of the enabling technologies used for optical networks, have evolved over several decades, and their optical properties have been engineered for low system cost and high transmission performance. Recently, major optical fiber cable companies have focused on research and development (R&D) on a new class of transmission fibers[3–7,57,58,59,60,61] that have large effective areas (A_{eff}) and ultra-low attenuation for 400 G and beyond PDM coherent transport systems. In this section, we briefly summarize the recent development of new large-area, low-loss fibers and advanced amplifier technologies for next-generation high-capacity terrestrial long-haul optical networks.

This section describes the key optical fiber properties of the new class of fibers and briefly discusses their impacts on the transmission performance for 400 Gb/s multilevel modulated coherent transport systems. Practical considerations for large-area fibers, such as splicing and cabling for terrestrial optical networks, are briefly addressed. Additionally, experimental results on the transmission performance of new large-area low-loss fibers in several advanced high-order modulation format systems will also be presented.

Next-generation optical networks are being planned, with 400 Gb/s and beyond to be deployed for transport systems, which employ coherent digital detection and PDM high-level modulation formats such as 16 QAM. In such systems, chromatic dispersion and polarization-mode dispersion (PMD) can be digitally compensated in the electrical domain. However, with wavelength division multiplexing (WDM), cross-nonlinearities cause neighboring channels to interact not only based on their power but also on the state of polarization (SOP) of the signals. This is particularly problematic in polarization-multiplexed coherent systems, which are more sensitive to cross-polarization modulation (XPolM),8 associated with nonlinear polarization rotation in fiber transmission links. XPolM does not affect NRZ non-coherent transmission systems but impacts coherent systems because polarization tracking in digital receivers cannot keep up with fast (symbol-to-symbol) polarization changes. As a result, polarization-multiplexed coherent systems, especially those using high-order QAM, are more susceptible to fiber nonlinear effects. Early research indicated at least 2–3 dB more nonlinear power penalties in a QPSK coherent system compared with direct detection systems. This XPolM nonlinear effect is stochastic from fiber transmission links, impairing the effectiveness of digital compensation. Additionally, high-level modulation formats require a much higher optical signal-to-noise ratio (OSNR). For instance, to achieve the same performance at the same transmission distance, upgrading from QPSK-based 100 Gb/s to 16 QAM-based 200 and 400 Gb/s requires OSNR improvements of about 6.5 dB and 10 dB, respectively. Therefore, it is desirable to use new fiber types with low nonlinear coefficients to maintain long-haul transmission capability while scaling up spectral efficiency (SE) and per-channel data rate. Recently, a new class of transmission fibers with ultra-large effective areas (A_{eff}) in the range of 120–155 μm^2 and an attenuation loss of 0.1460 dB/km has been reported for terrestrial and submarine systems.[57–61] To reduce the fiber nonlinearity, which is proportional to n_2/A_{eff}, where n_2 is the nonlinear refractive index, it is necessary to increase the A_{eff} of the fiber while not jeopardizing the bending performance. Step-index profiles with small index differences and large core diameters can be used to design large A_{eff} fibers. However, the light confinement in the core becomes weaker, increasing the cut-off wavelength and degrading both macro- and micro-bending performance. Placing a depressed-index region in the cladding, or placing a trench slightly apart from the core can improve the bending performance and reduce the cut-off wavelength.[58] A low Young's modulus primary coating[60] can also be used to improve the micro- and macro-bending performance and permit a larger fiber A_{eff}.

A simple figure of merit (FOM) was proposed to compare the new class fibers with standard single-mode fibr (SSMF), which can be expressed as follows:

$$\text{Figure of merit} = 10\log_{10}\frac{A_{\text{eff}}}{A_{\text{eff}}^{\text{SSMF}}} - \left(\alpha - \alpha^{\text{SSMF}}\right)L - 10\log_{10}\frac{L_{\text{eff}}}{L_{\text{eff}}^{\text{SSMF}}} \tag{12.82}$$

where the FOM (in dB) is the OSNR improvement of an optical fiber transmission system for new types of fibers compared with SMF, the first term considers the improvement from the effective area (A_{eff}), and the second term is from the fiber attenuation (α in dB/km); here, L and L_{eff} are the span and the fiber effective length, respectively. From Equation 12.82, it is seen that increasing the fiber A_{eff} is the most fundamental improvement for coherent transport, and reducing fiber attenuation can improve span loss; however, the reduction of fiber attenuation can increase the fiber's effective nonlinear length, hence increasing the accumulated nonlinear impairments. Figure 12.21 shows an example of contour maps of relative FOM normalized to SSMF (α dB/km=0.19 dB/km, A_{eff}=80 μm², span length L=80 km) as a function of α dB/km and A_{eff}; clearly, the OSNR performance in large-area low-loss fiber can be improved substantially. It should also be noted that the relative OSNR improvement in FOM depends on the span length of the transmission systems. In addition, it is also important to have high chromatic dispersion in new fibers to reduce nonlinear effects in the PDM coherent systems where chromatic dispersion can be digitally compensated. We have recently developed large A_{eff} TeraWave fiber for 100 G and beyond terrestrial long-haul optical networks. The TeraWave fiber is a single-mode fiber with a germanium-doped core and a depressed-index inner cladding region and is fabricated similarly to an All-Wave zero-water-peak fibr. Considering the features of terrestrial cabling such as craft splicing, closures, and macro- and micro-bending, the A_{eff} of TeraWave fiber is optimized to be 125 μm². The average fiber loss, dispersion, and dispersion slope at 1550 nm are 0.185 dB/km, 20.0 ps/nm/km, and 0.06 ps/nm²/km, respectively. TeraWave fiber is ITU-T G.654.B11 compliant; it uses the DLUX Ultra coating for excellent micro-bending performance, and it meets all macro-bending requirements in G.652.D and G.654.B. Volume splicing study using a commercially available splicer (for example, FiTel S178A) with a standard splice recipe shows that the averaged splice loss between TeraWave to TeraWave is 0.04 dB/splice and TeraWave to SMF is about 0.15 dB/splice. With the optimization of splicing programs using commercially available splicers, the splicing loss between TeraWave and SSMF is below 0.10 dB/splice.

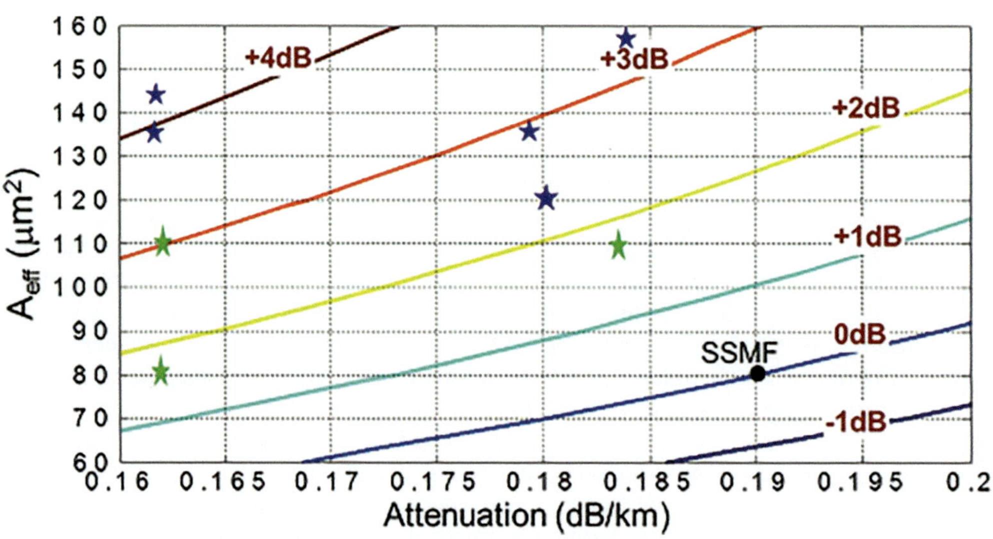

FIGURE 12.21 Relative figure of merit of large-area low-loss fiber normalized to SSMF for 80 km span transmission. (After Ref.[62] with permission.)

Recently, TeraWave ultra-large area single-mode optical fibers were deployed, featuring a core effective area of 125 μm². These fibers offer ultra-low-loss performance, with an average fiber loss of 0.168 dB/km and a dispersion of 20.0 ps/nm/km at 1550 nm. The fiber is optimized for long-haul transmission in the C+L-bands (1530–1625 nm) at 100 Gb/s, 400 Gb/s, and beyond, for terrestrial optical networks. It supports greater distances between regeneration and amplification sites, helping to lower the overall cost of deploying coherent systems. We systematically and experimentally investigated the system performance of TeraWave fiber in a 485 Gb/s coherent optical orthogonal frequency-division multiplexed (CO-OFDM) super-channel long-haul transmission system[57] and compared it with the SSMF. The CO-OFDM super-channel coherent system has advantages, including high spectral efficiency (SE), reduced guard band, and lower modulation baud rate, making it a potential candidate for future high-capacity Tb/s-per-channel optical networks. However, CO-OFDM super-channel systems are more susceptible to fiber nonlinearity compared with other schemes. This is due to the multi-subcarrier configuration with a small guard band in CO-OFDM transmission, which leads to impairments from inter-subcarrier nonlinear interference. In the experiment, large-area fibers were employed, and the 485 Gb/s CO-OFDM signal was generated using PDM-16 QAM with five-subcarrier modulation. The experiment was conducted with a dispersion uncompensated link, consisting of 80 km fiber spans, using two amplification schemes: erbium-doped fiber amplifiers (EDFAs) and hybrid EDFA/Raman amplifiers. The comparison showed that the optimum signal launch power into fiber spans was about 2 dB higher in TeraWave fiber than in SSMF for transmitting 485Gb/s under the modulation scheme CO-OFDM (coherent optical - orthogonal frequency division multiplexing) over 1600 km which consists of 20 spans of 80 km SSMF length that incorporates both EDFA-only and hybrid Raman/EDFA amplification schemes. With EDFA-only amplification, TeraWave fiber offers a ~2 dB higher optimum Q^2 factor than SSMF after 1600 km of transmission. This advantage is about 1 dB higher with the hybrid amplification scheme (Figure 12.22). It was also found that TeraWave fiber allows transmission over distances more than 60% longer than SSMF at a similar Q^2 factor performance for 400 Gb/s CO-OFDM systems.

We further experimentally investigated the transmission performance of TeraWave fiber in a typical 256 Gb/s PDM-16 QAM DWDM system with 37.5 and 50 GHz channel spacing. The 256

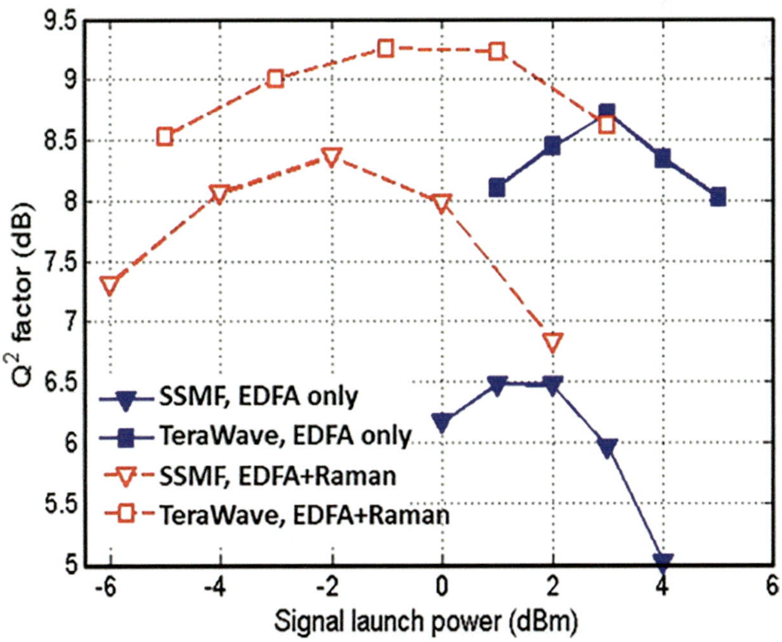

FIGURE 12.22 Measured Q^2-factor versus launch powers of 485 Gb/s CO-OFDM signal transmission over 1600 km of ULAF and SSMF.[57]

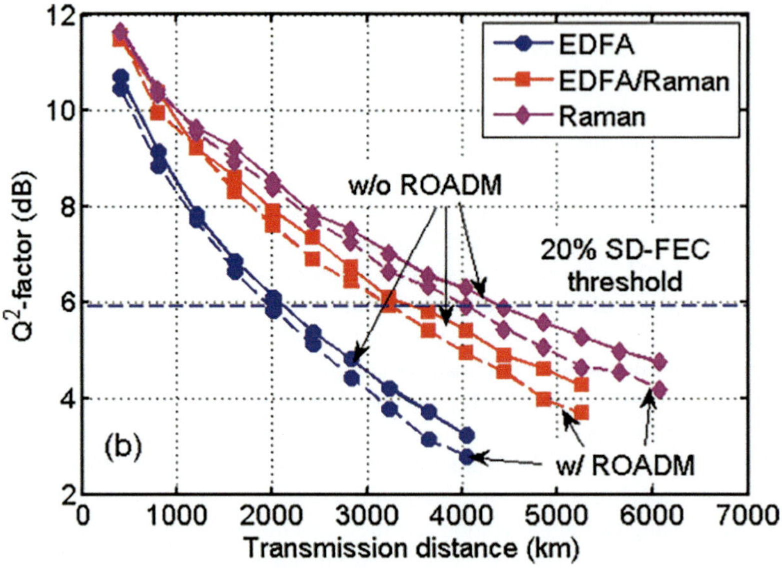

FIGURE 12.23 Q^2 factor versus transmission distance of the ten channel 256 Gb/s PDM 16 QAM system using the three different amplification schemes without (solid curves) and with ROADMs (dashed curves) with 50 GHz channel spacing. Optimum launch powers are used.

Gb/s PDM-16 QAM is a promising modulation format for dual-carrier 400 Gb/s systems due to its relatively simple scheme and the maturity of opto-electronic components. The experiment was conducted with a dispersion uncompensated link featuring a 100 km fiber span length, using three different amplification schemes: EDFA-only, hybrid EDFA/Raman amplifiers, and all backward-pumped Raman amplifiers. It was found that, without a ROADM, ten 256 Gb/s PDM-16 QAM channels were transmitted over 4200, 3500, and 2000 km using TeraWave fibers within a 20% soft-decision FEC (SD-FEC) threshold, respectively, in systems with both 37.5 and 50 GHz channel spacing (see Figure 12.23). The all backward-pumped Raman amplifiers and hybrid EDFA/Raman amplifiers can increase transmission distances by about 100% and 70%, respectively, when compared to EDFA-only amplifiers. It was also found that cascaded ROADMs have a minimal impact on systems with 50 GHz channel spacing. The large-area low-loss fiber has also been studied in other advanced higher-order modulation format systems (for example, time-domain hybrid 32–64 QAM for high SE 400 Gb/s ultra-long-haul transmission systems), and it has been demonstrated that 400 Gb/s-class DWDM signals on the standard 50 GHz ITU-T grid, which is 8.25 b/s/Hz net SE, were transmitted over 4000 km of large-area low-loss fiber with 100 km span lengths for terrestrial optical networks. These results demonstrate that large-area low-loss fibers are indeed beneficial for system reach and margin, regardless of amplifier configuration; they increase system SE and improve the efficiency of utilizing optical network resources for future Flex OTN. However, further enlargement of A_{eff} (e.g., >200 μm²) would inevitably deteriorate bending and cut-off behavior and cause high splicing losses. Hence, fiber design compromises must be made between improved transmission performance and the limitations of careful handling and splicing of new fibers.

12.3 APPENDIX 5: MEMS AND ROADM, EQUALIZER

12.3.1 MEMS and ROADM

Micro-electromechanical switch (MEMS) is an integrated Si-based device built on a silicon substrate and silica insulator in such a way that mechanical force can activate the switch to turn mirrors on and off, routing lightwaves to different output ports as shown in Figure 12.24a for 2D routing in space and panel (b) for 3D routing. The lightwaves are focused in parallel by arrays of microlenses at the input,

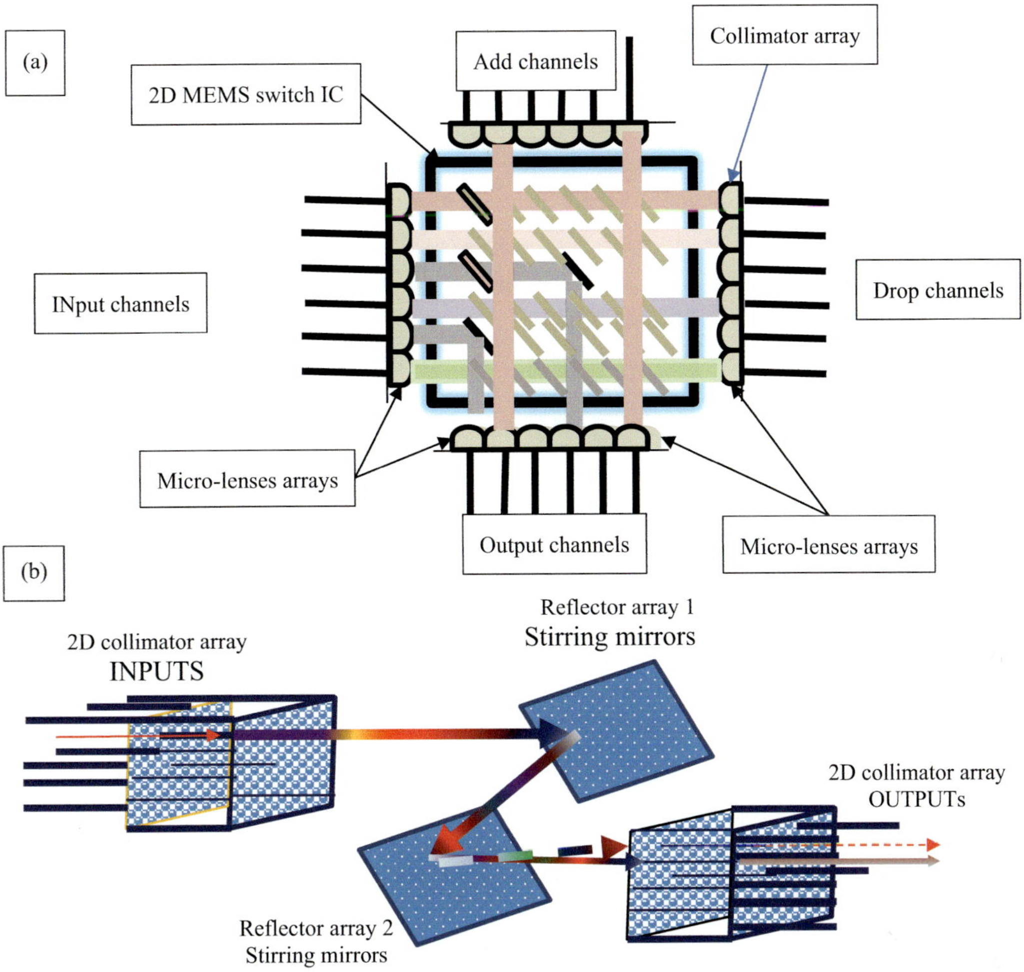

FIGURE 12.24 (a) Two-dimensional optical switch and cross-connect, (b) three-dimensional optical MEMS switch and spatial router.

output, and add/drop ports. Optical collimator lenses are integrated and coupled to fiber at the four ports so that lightwave beams can be injected and coupled to the outputs or add/drop ports.

Lightwave channels are coupled and parallel through the use of micro-lenses, which are quite easy to integrate into SOI or Si_3N_4 interfaces. The MEMS can be controlled to switch up or down to redirect the optical paths, as shown in Figure 12.24a. The MEMS can be integrated with micro-switches using mirrors and cavities for optical filtering to form the ROADM with CDC (colorless, directionless, and contentionless), as shown in Figure 12.24b. The transistors hidden under the grating mirrors act as electrically controlled switches. The massive parallel routing of the channels in free-space MEMS and grating mirrors offers much greater advantages than in integrated structures. These mirrors can be steered to redirect lightwave beams to the appropriate micro-lenses at the output ports. Thus, the resonant cavity and the gratings determine the selected channels and the high quality (Q-) factor in optical filtering.

So, if one of the mirror planes can be fabricated with optical filters or an add/drop multiplexer for different wavelengths then the MEMS can act as a ROADM. If the filtering ports and the directions of the channels can be operated such that both the directions and the wavelengths are independently flexible, then we can have a CDC-ROADM (contention-less, colorless, and directionless reconfigurable optical add/drop multiplexer).

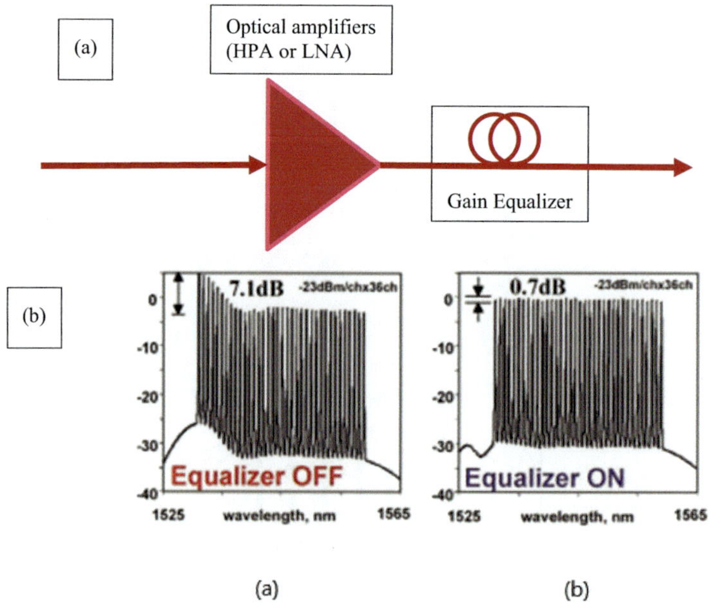

FIGURE 12.25 Gain equalizer for optical amplification and broadband source using integrated broadband filtering (a) Schematic structure; (b) spectra of un-equalized and equalized amplifiers.

Figure 12.24b depicts the schematic of a 3D optical cross-connect and ROADM with arrays of inputs, outputs, and add/drop ports. Two steering mirrors can be used to steer the beams in a similar way as the MEMS. Thus, it is directionless. In addition, if tunable optical filters are integrated into the mirrors of silicon-based MEMS, then the system becomes a CDC-ROADM, or colorless, directionless, and contention-less.

12.3.2 Optical Spectral Gain Equalizer

In SPC, the multi-stage optical link requires the average optical power of all channels to be closely equal so that amplification can provide nearly the same gain and noise factors. Thus, equalization of all channels is essential for the multi-stage system to be realized. Figure 12.25 shows the structure and optical characteristics of the equalizer, which can be achieved by using a special fiber whose optical attenuation is opposite to that of the EDFA amplifier's spectral characteristics.

12.4 APPENDIX 6: MATLAB SIMULINK MODEL OF OPTICAL AMPLIFIERS

This section provides an evaluation of the numerical data generated by our EDFA simulator and EDFA module. Experimental results are compared with those produced by our models. The principal advantage of the Simulink EDFA simulator is its versatility and ease of use. These results are by no means exhaustive.

According to the simulation modeling principle from the previous section, we now focus on implementing EDFA simulation in MATLAB Simulink. This section documents the Simulink simulation design and developments, explaining how each of the blocks within the simulator are integrated. Figures 12.26 and 12.27 illustrate the EDFA simulator, EDFA modules, and associated monitoring equipment. The explorer views of the subsystems within the Simulink model are provided in the following sub-sections.

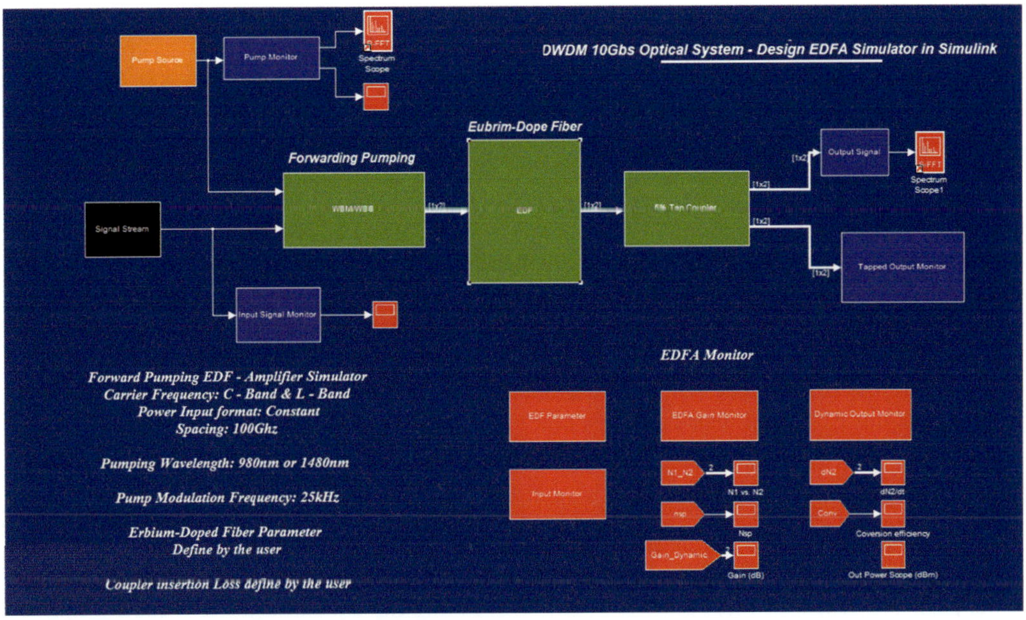

FIGURE 12.26 The EDFA simulator model: amplifier general layout.

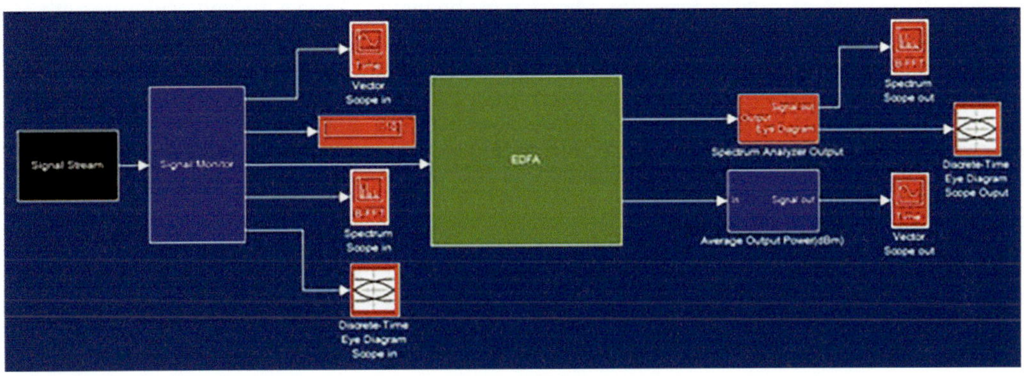

FIGURE 12.27 EDFA module amplifier and monitoring systems. EDFA under forward pumping only, carriers in C- and L-bands, channel bandwidth = 100 GHz and spacing 100 GHz or 0.8 nm at 1550 nm, pump modulation 25 kHz, coupler insertion loss = to be defined by users, parameters of EDFA defined in a MATLAB program. Pump sources wavelength 980 or 1480 nm.

12.4.1 Simulator Design

Simulink can be easier for users unfamiliar with programming methods or general programming techniques, such as operators or system technicians. This reduces the likelihood of human-induced errors when modeling the system of choice, which is important in the rapidly changing field of information transmission.

The Simulink simulation model in this project is basically divided into two parts: (i) EDFA Simulator and (ii) EDFA Module. The EDFA Simulator can simulate the details of amplification along the Erbium-Doped Fiber (Er:doped fiber or EDF) to investigate and scope how the signal can be pumped along the fiber, also observing the effects of gain modulation during amplification. The development of the EDFA Simulator is based on the EDFA dynamic modeling. The EDFA Module is used with the whole optical system simulation for predicting the amplification output and its noise characteristics. The development of the EDFA Module is based on the EDFA steady-state modeling. Thus, the functions of these two simulations are different. The EDFA model can be easily combined with the transmitter simulator and fiber simulator to form an optical communication link.

12.4.2 Simulator Design Process

Intrinsically, MATLAB Simulink is a very powerful design tool. However, this also means many different factors need to be considered to accurately model a system. Little experience with complex system modeling meant that a knowledge base of the system needed to be developed step by step. Therefore, deriving the simulator prototype into a number of small tasks will simplify the design process, depending on time. Thus, initially, the first task of the model setup focuses on how to simulate the pumping schemes of the EDF. This is the primary goal and also the principal part of our simulator that has to be achieved. The simulator is divided into several stages. The following flowchart (Figure 12.28) outlines the design process:

12.4.3 Simulator Requirement

In this section, the requirement relates to basic issues such as software compatibility and simulation efficiency on the MATLAB Simulink[63] software are described. First, the EDFA Simulink simulation is designed based on MATLAB-7 Simulink. Running the simulator in MATLAB Simulink is the first requirement. As there are many versions of MATLAB Simulink available, running an incorrect simulator version may cause bugs, hangs, or jams.

Second, the block sets of the Simulink library must be included and activated. Running without the relevant block sets library initialized on the user's platform may prevent the Simulink simulation from opening. The following specific block sets needed to ensure operation are: Simulink DSP block sets, Simulink Sink, Simulink Communications block sets, and Simulink Extras block sets.

12.4.4 Simulator Design Assumptions

12.4.4.1 On Sampling time

In the design of the simulator, it was decided to employ a discrete rather than continuous sampling method, primarily for practical reasons. In particular, a discrete sampling system would decrease the processing power required to execute simulations and reduce execution time.

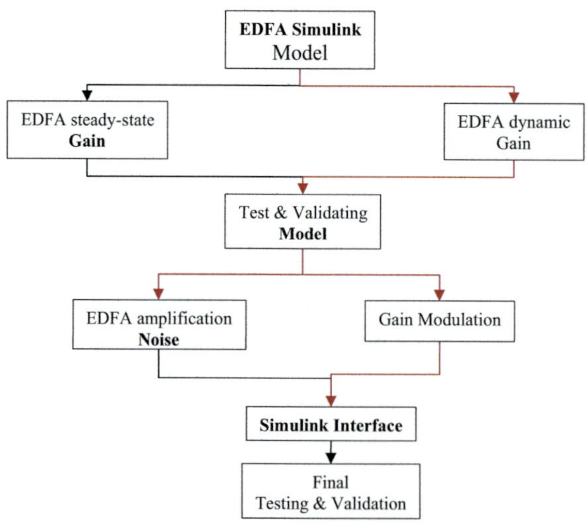

FIGURE 12.28 Simulation design flow chart.

According to the well-known sampling theorem, all data must be sampled at or above the Nyquist frequency. That is, all sampling must be performed at least twice the highest frequency within the simulator (the sampling frequency applied in this project was up to seven times the highest frequency). The use of discrete sampling periods ensures synchronization of various system components as the data signal is propagated, an important consideration in optical systems. Considerations of optical transmission system parameters show that the highest frequency is that of the carrier waveform. For the purpose of this model, the carrier frequency is located in the C-band and L-band, which is around 0.2 PHz (Peta-Hz). Thus, sampling within the system needs to be at seven times this rate, i.e., approximately 1.4 PHz), in order to ensure signal aliasing does not affect the results and to achieve more accurate results. All block sets require a discrete sampling time at which to generate data. Unfortunately, this sampling rate is very high and may cause slowness in execution time.

12.4.4.2 On Signal Streams

Signal streams in the EDFA simulator are not defined during the simulation; the EDFA simulator assumes the average signal power. Thus, the signal stream simulated in the EDFA simulator would be a constant value with superimposed noise on Gaussian pulses. Hence, the determination of the output power is also based on the average signal power.

The signal stream in the EDFA module is developed as a PSK (phase shift key) modulation signal with Gaussian noise. The carrier is defined as a 192.5 THz sinusoidal waveform. The data stream is assumed to be 10 Gb/s (or even higher, up to 40–100 Gb/s with 28 GBaud) binary pulses.

12.4.4.3 Assumptions for EDFA Simulink Model

In the development of the EDFA Dynamic Gain simulator, a number of assumptions have been made.

- Since the pumping wavelengths of 980 and 1480 nm can provide high gain amplification and lower noise, we assume the EDFA can be pumped with either 980 or 1480 nm sources.
- Due to the sufficiently short transient lifetime from the excited state to the metastable state, we assume that the erbium atoms operate in only two levels relevant to the amplification process.
- The effective area of the erbium-doped active region is small compared to that of the optical mode at each wavelength of interest. This assumption for the Er: doped core allows us to treat the overlap factor as a constant value, independent of energy level and optical power.
- Since the amplified spontaneous emission is randomly generated, we assume that the stimulated ions are propagating in one direction to estimate the average mean ASE noise power.
- No reflection occurs during the amplification process.

Assumptions for the modeling of the steady-state EDFA gain are also made:

- As the pumping wavelengths of 980 and 1480 nm can provide high amplification gain and lower noise, it is assumed that the EDFA is pumped with either a 980 or 1480 nm source.
- Due to the short transient lifetime from the excited state to the metastable state, we assume a two-level operation of the doped ions.
- Since excited state absorption only affects the pump absorption along the fiber, it can be neglected.
- The response of each individual ion in the collection is equally broadened, at a picosecond time scale.

- ASE noise is assumed to be small so that it does not saturate the EDFA amplifier.
- The effective area of the erbium-doped active region is small compared to that of the optical mode at each wavelength of interest. This assumption for the Er-doped core allows for a constant overlapping factor that is independent of the energy level.
- As in the steady-state model, the population increment delta N_2 is assumed to be zero.

12.4.5 System Initialization

To correctly operate the simulator an initialization process is required. Written as an m-file, this file loads parameters and variables required for the simulator operation as standard in the MATLAB Simulink platform. The initialization file is required for simulation using MATAB Simulink. User needs to run this file before starting the Simulink simulation. This initiation file contents values of parameters of components and systems. Failing to do so will result in an error warning. It is within the initialization file that the user can modify the system parameters. There are several data points to be defined in the initialization file. Table 12.2 provides the input data that can be defined before the simulation.

Table 12.3 provides the glossary for input signals.

The use of the initialization file simplifies operation for users. All variable parameters are contained here and automatically updated in the Simulink platform upon execution (saved in the MATLAB memory called the "workspace"). Not incorporating the initialization file would require the user to input the same variable in multiple blocks, thus increasing the possibility of errors and complicating the operation of the simulator.

TABLE 12.2
Parameters of the Er:Doped Fiber

Physical Meaning	Symbol Defined in Initialization File
Erbium-Doped Fiber Parameters	
Core radius of Erbium-doped fiber	r (μm)
Overlap factor of EDFA at wavelength (k)	Gamma
Ion density of Erbium-doped fiber	P
Peak absorption on 1530 nm	Peak absorption
Peak absorption on 980 nm	Peak absorption pump
Length of the fiber	L (m)

TABLE 12.3
Input Parameters

EDFA Input Initialize	Symbol Defined in Initialize File
Input signal power	Signal power (dBm)
Input pump power	Pump power (mW)
Input wavelength	Index
Pump wavelength	Pump wavelength
Number of channel input	N
Pump modulation index	Mod
Pump modulation frequency	Mod freq

12.4.6 EDFA Simulator Modelling

The EDFA simulator is combined with various segments as referred to in the prototype of Figure 12.26. Each block represents a different segment of the EDFA simulator. For example, the input signal stream is represented by a block named [Signal Stream], and each block contains more than one subsystem inside. To check the design inside, simply double-click the block. The EDFA in the Simulink simulation is implemented as these subsystems, and they are called from a main program. The following demonstrates the input and output ports in the EDFA simulator and the EDFA Module.

12.4.6.1 Using the EDFA Simulator

Inputs:

- Signal Stream: Power of signal (include C-band and L-band signals)
- Pump Source: Power of the pump (include pump modulation, 980 nm pump source and 1480 pump source)

Outputs:

All the output displayers have been placed into the block named Monitor, and readers can check the output values by opening the Monitor block.

- EDF Parameter (EDF Parameters Monitor): The parameter of the Erbium-Doped fiber such as radius, ions density etc.
- EDFA Gain Monitor: Dynamic gain and steady-state gain scope. All the gain data have been put in order (Ascent).
- EDFA Dynamic Gain Monitor: Output signal power, output pump power, Average ASE power, noise figure, SNR, other noise power.

Also, scopes display the evolution of the signal. Segment by segment of the simulator can be illustrated in the rest of this section.

12.4.6.2 Signal Data Stream Modeling

As illustrated in Figure 12.29, the *signal stream* basically models the C-Band and L-band light wave signals. As we have assumed the system using 100 GHz optical frequency spacing (0.8 nm) between light wave carriers, C-band is also assumed for the wavelength window 1525–1565 nm. Thus, there are around 51 channels that can be used as optical carriers over this band. That we can see from Figure 12.29 that there are five signal power blocks with bold line except the last one. The bold lines are multi-dimensional vector time series associated with ten channels of each signal block. The wavelength section is used for conversion of the signal power in mW unit (or equivalent dBm) into the number of photons (number of photon energy per bit period). The input signal power of each channel has been defined as a constant mean power (Figure 12.30).

The value of the mean signal power is defined in the file [Simulation_initialize.m]. The signal carrier frequency is ultra-high, around 192.5 THz. If we inject a signal generator with a real 192.5 THz signal (e.g., sinusoidal wave at 192.5 THz), the Simulink generates the output based on each sampling point of this signal wave, and it may take a long time (possibly more than 2 days!) to generate the output result, which is naturally very inconvenient. On the other hand, predicting the average output power after the amplifier is sufficient for designing the amplifier. Therefore, an average mean input signal power can be used in our considerations.

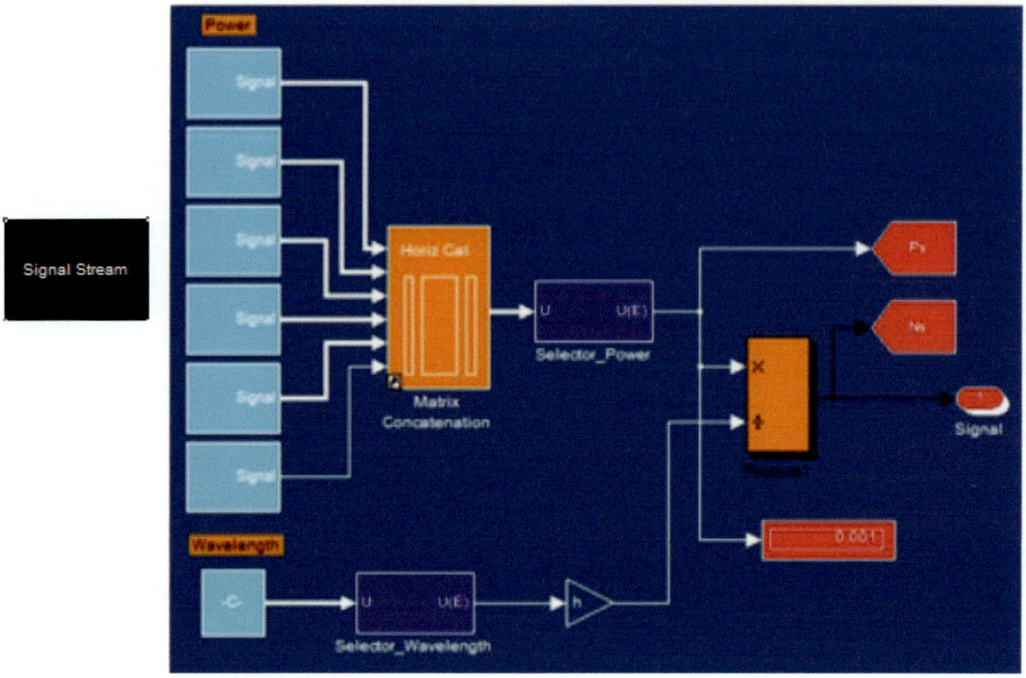

FIGURE 12.29 Signal stream prototype.

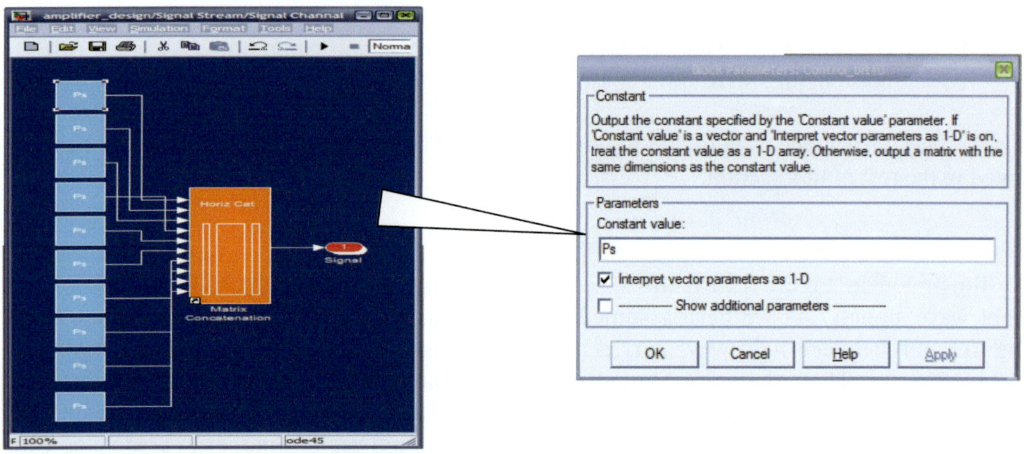

FIGURE 12.30 Signal block prototype.

12.4.6.3 Pump Source

The pump power has also been set as a constant mean power for simulation. In the previous section, we identified three different EDFA configurations: Forward Pumping, Backward Pumping, and Dual (or Bidirectional) Pumping. Only the Forward Pumping is described in this section, as the others can be easily followed. The MATLAB Simulink model for the pump source is shown in Figure 12.31.

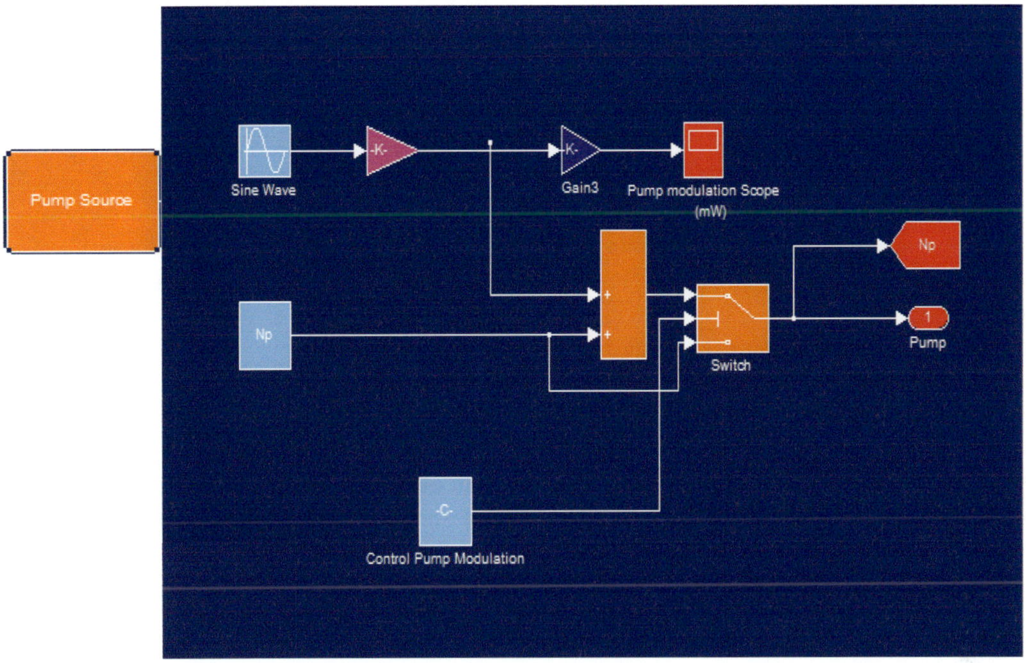

FIGURE 12.31 Pump source prototype.

12.4.6.4 Pumping Wavelength

As described above, two pump wavelengths can offer high amplifier gain. In our simulation, the wavelength modeling is implemented within the initialization file. The following is the instruction code for setting the event when different wavelength channels are propagating through the EDFA.

Algorithm 12.1

According to the absorption and emission characteristics, the absorption and emission properties of the fiber change with different wavelength propagation. The absorption factor is usually provided by the fiber manufacturer or measured. The algorithm steps are given below

If Pump wavelength=980e-9

Absorption factor 980 nm

Emission factor 980 nm

Else if Pump wavelength=1480e-9

Absorption factor 1480 nm

Emission factor 1480 nm

Else

Reset

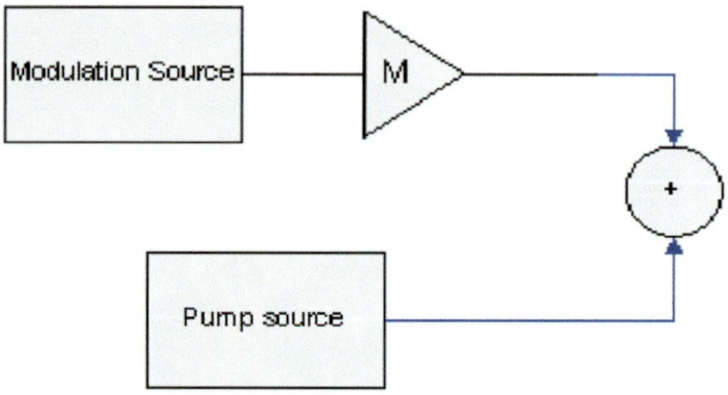

FIGURE 12.32 Pump modulation module.

12.4.6.5 Pump Modulation

Pump modulation can be added, and its effects on gain modulation can be investigated. Freeman and Conradi[2,64] use a dimensionless number called the "modulation index" to measure the power in the modulated signal. Using this definition, the input pump time series has the form $P_p(0,t) = P_p(0) + P_p(0)m_p \cos\omega t$ where $P_p^0(0)$ means the initial pump power (in Watts), m_p is the modulation index. Since the mean pump power is given as a constant value, this is added to the term $m_p\ P_p^0(0)\cos\omega t$ at the summation box as illustrated in Figure 12.32.

$P_p(0)m_p \cos\omega t$ is the output of the sine wave generator in Figure 12.32. The gain labeled with [-M-] is the modulation index m_p. N_p is the constant mean pump power; here the unit of the pump power is photons/seconds. The method for converting Watts to the number of photons/seconds is listed in the previous section.

12.4.7 EDF Modeling

We have now described the theoretical EDF modeling. This section presents the Simulink simulation model design. The Simulink simulation models are based on the rate equations given in Section 12.2. The characteristics of the EDF depend on two factors: the EDFA gain and ASE noise characteristics. The gain characteristic depends on the absorption and emission factors at different wavelengths. Since different erbium-doped fibers provide different absorption and emission characteristics, simulating the absorption and emission spectra would be the first step.

The existing absorption spectra given in Refs.[65,66,67,68] can be used. The absorption spectrum is illustrated in Figure 12.33. Although the data from these references differ, the slopes of these spectra are quite similar. Based on this assumption, normalization of these curves would be the first choice for simulating the spectrum. The following is the normalized spectra simulation.

The EDF simulation can be begin after these spectra. Figure 12.34 illustrates the prototype of EDF simulation.

12.4.7.1 EDFAs Dynamic Gain Model

The equations for solving the dynamic gain model have been described in the previous section. This section describes how all the blocks are integrated. As mentioned in Section 12.4, the gain of the EDFA is based on the equations $G_p = \frac{\Gamma_p(\sigma_{1\text{-}3} + \sigma_{3\text{-}1})}{A}N_2 - \Gamma_p\sigma_{1\text{-}3}\rho L - \gamma_p G_p$ and $G_s = \frac{\Gamma_s(\sigma_{1\text{-}2} + \sigma_{2\text{-}1})}{A}N_2 - \Gamma_s\sigma_{1\text{-}2}\rho L - \gamma_s G_s$. In our simulation, the Simulink model representing these expressions is shown in Figure 12.35.

In Figure 12.35, B_s and B_p are represented by the matrix of $\frac{\Gamma(\sigma + \sigma)}{A}$, and similarly C_s and C_p are represented by the matrix of $\Gamma\sigma\rho L - \gamma G$. Due to the characteristics of absorption and emission,

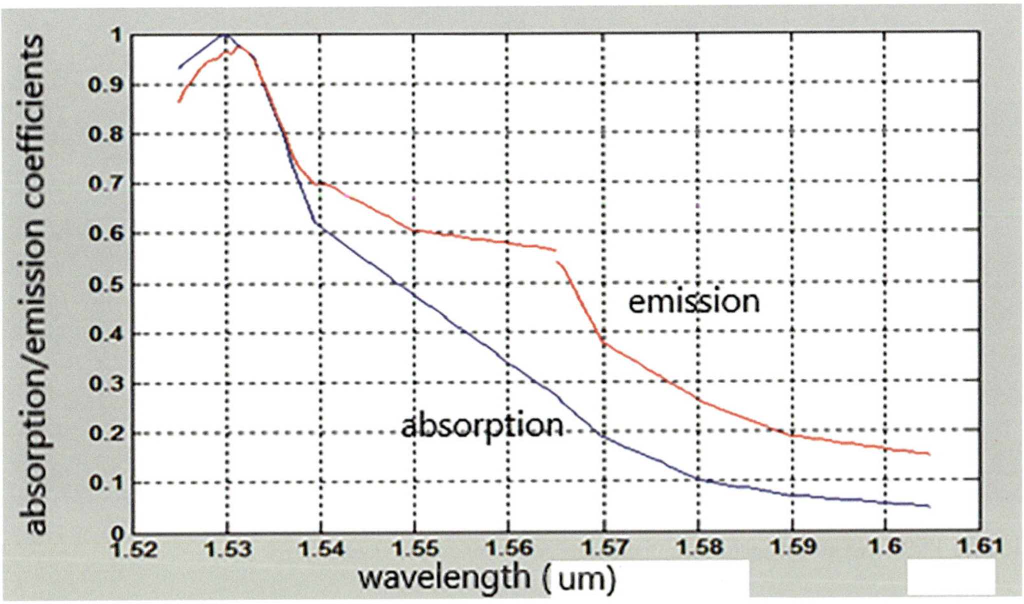

FIGURE 12.33 The absorption (dark line) and emission (light-dark line) factor of the normalized spectra (C-band & L- band). Arbitrary unit versus optical wavelength (in meters).

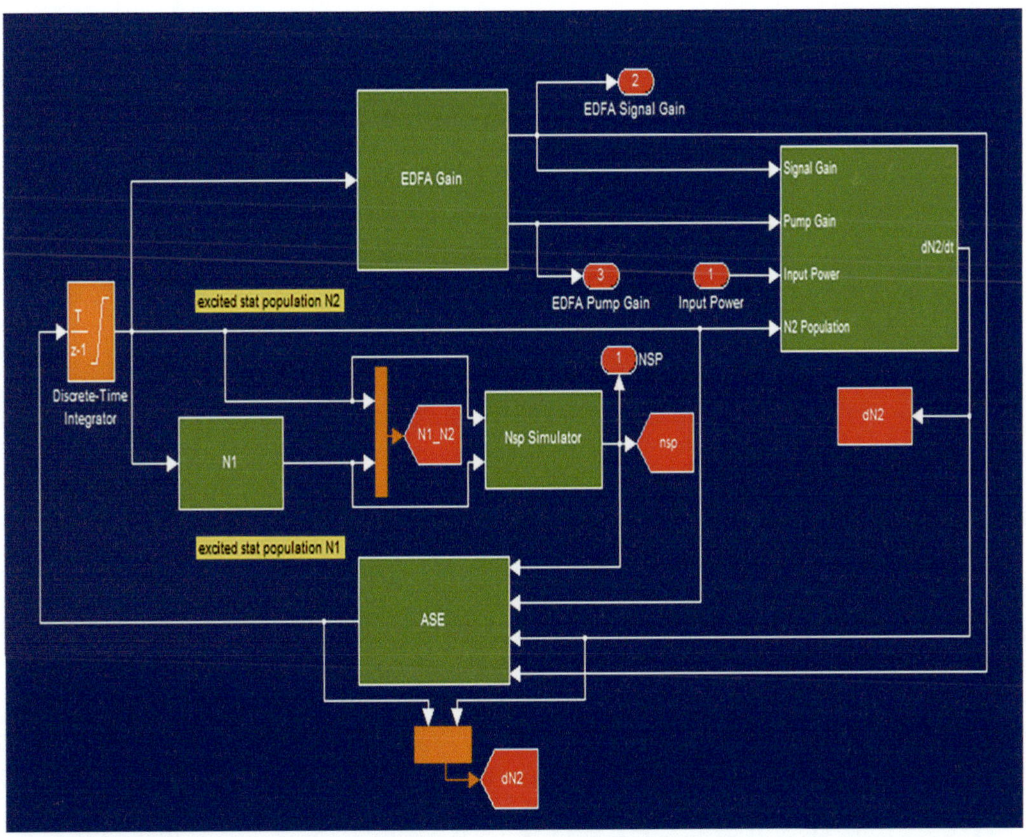

FIGURE 12.34 EDF simulation prototype.

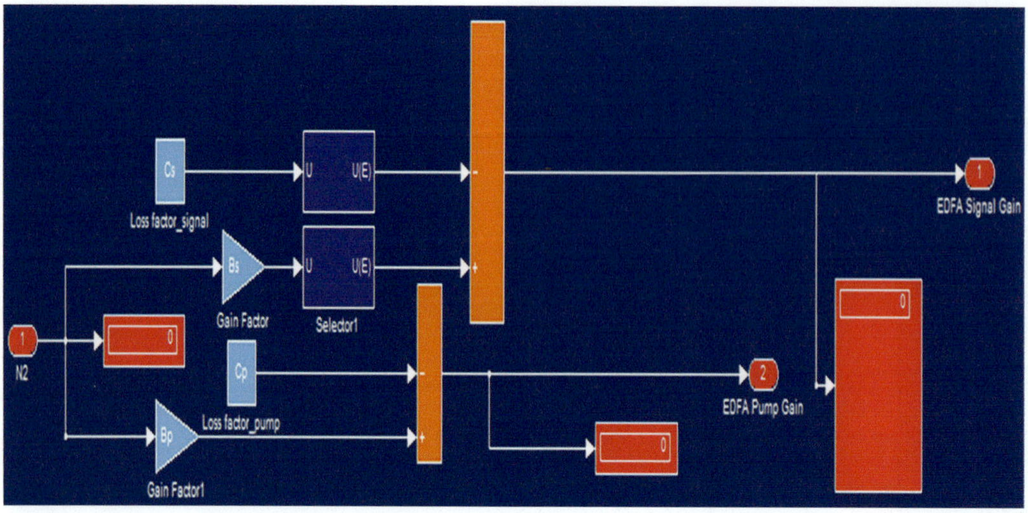

FIGURE 12.35 EDFAs dynamic gain modelling corresponding to Figure 12.34.

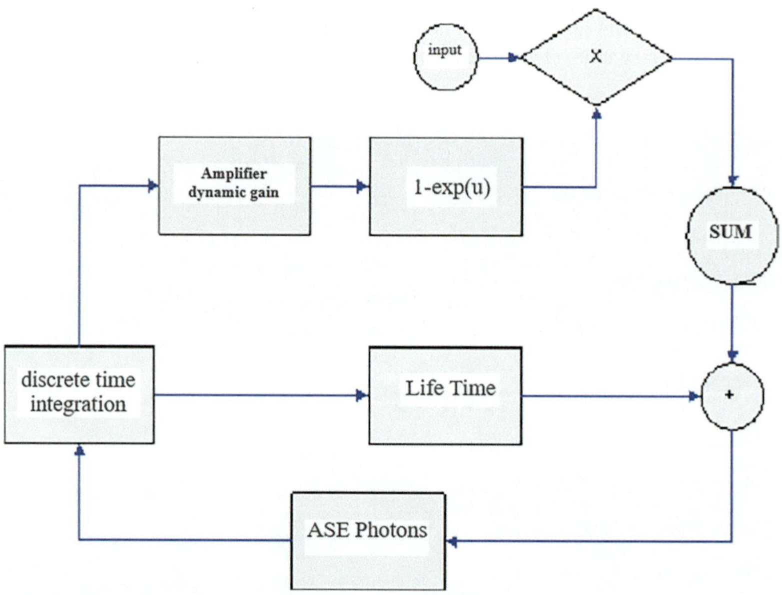

FIGURE 12.36 dN_2/dt simulation block diagram.

different wavelength signals propagating through the EDF cause changes in the absorption and emission factors. So, C_s and B_s *act as* memory locations that store all the absorption and emission values. The following selector is an artificial intelligence block that can select the corresponding values to match the wavelengths. The Selector block works by outputting a selected reference index from the B_s matrix, like $\begin{bmatrix} Bs_1 & Bs_2 & Bs_3 & \cdots & Bs_{51} \end{bmatrix}$ Selected [index 1] → Output – B_{s1}. Accordingly, the absorption and emission factors change with respect to different wavelengths (Figure 12.36).

As shown in the block diagram in Figure 12.35, the loop continues execution until the difference between the input power and output power is the same as that of N_2/τ which means N_2 is saturated.

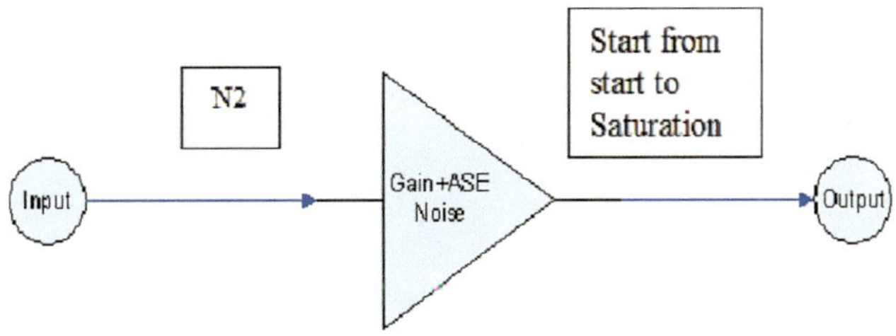

FIGURE 12.37 Schematic model of output algorithm.

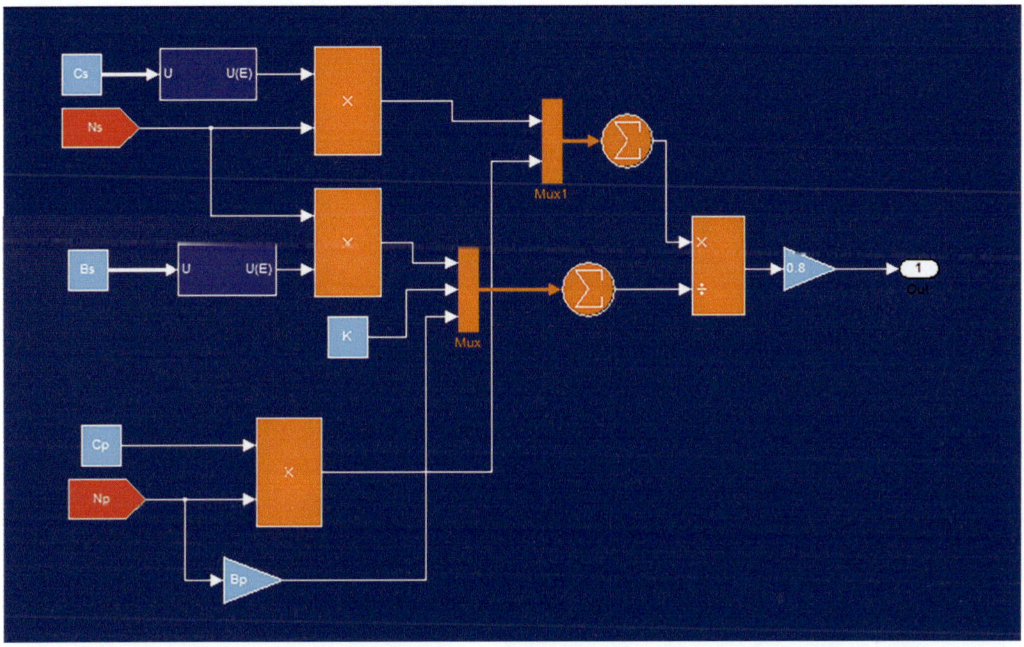

FIGURE 12.38 EDFA steady state N_2 model.

Thus, the gain remains in a steady state. The data inside the discrete-time integrator block sets the step time for each increment; here, we set 3.25e-7 as the sampling step. The output of the model is shown in Figure 12.37.

12.4.7.2 EDFAs Steady State Gain Model

In steady state, $dN_2/dt=0$, the Simulink Model can be illustrated in Figure 12.38.

The signal gain here is equal to B_s * $N2$ (Saturate) – C_s, and the pump gain here is equal to B_p * N_2 (Saturate) – C_p. The Simulink development is similar to the EDFA dynamic model. The output would be the same as given in Figure 12.37.

12.4.7.3 Population Inversion Factor Modeling

- N_1 simulation.

  Recalling the three state populations, and the assumption of N_3 to be zero. The relation between N_1 and N_2 is then $N_1=N_{total}-N_2$ (Figure 12.39).
- N_{sp} simulation (Figure 12.40)

  According to the population inversion factor equation, N_1 would equal N_2. This would lead to the result approaching infinity. Therefore, the [if statement] could limit the result to under 10. Also, the population inversion factor would be greater than 10 in practice.

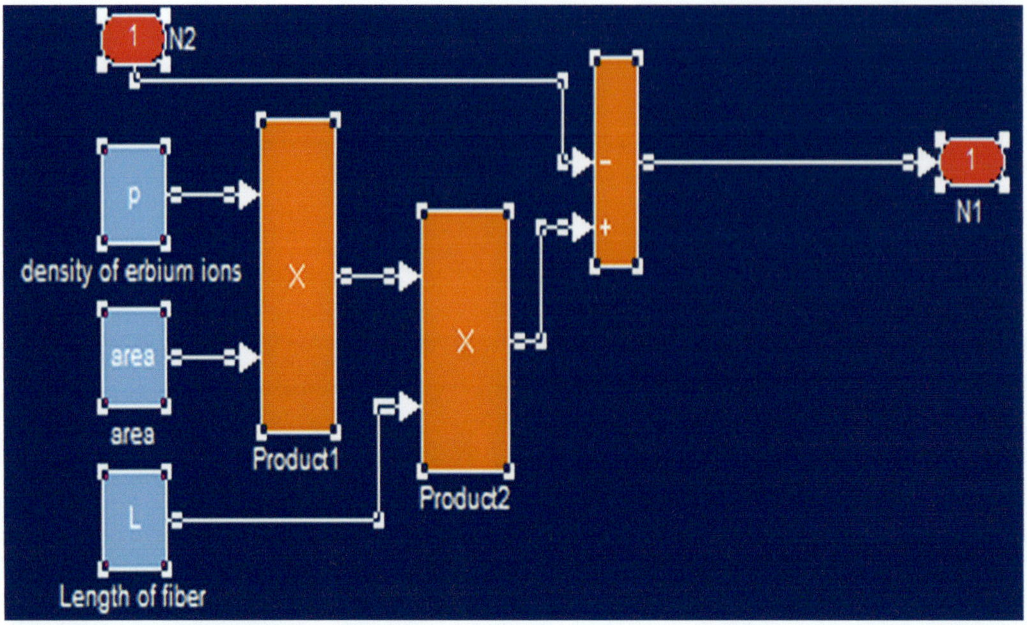

FIGURE 12.39 Representation of N_1 in the simulator.

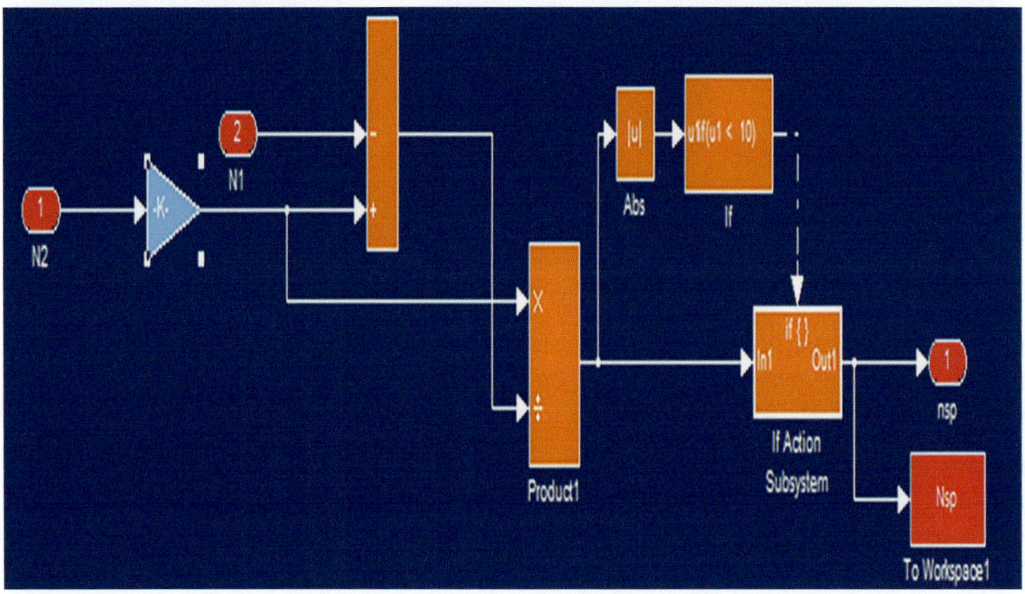

FIGURE 12.40 Representation of n_{sp} in the simulator.

12.4.7.4 Amplifier Noises

In the previous section, we mentioned the simulation method for amplification noise. Therefore, to simulate the ASE noise, we must determine the average mean ASE power in the first step. For the dynamic gain modeling in Simulink, the process can be divided into three sections (Figure 12.41).

Thus, the calculation for determining the ASE is conducted in three different statements. In the simulation, the following algorithm is used to simulate the ASE mean power (Figure 12.42).

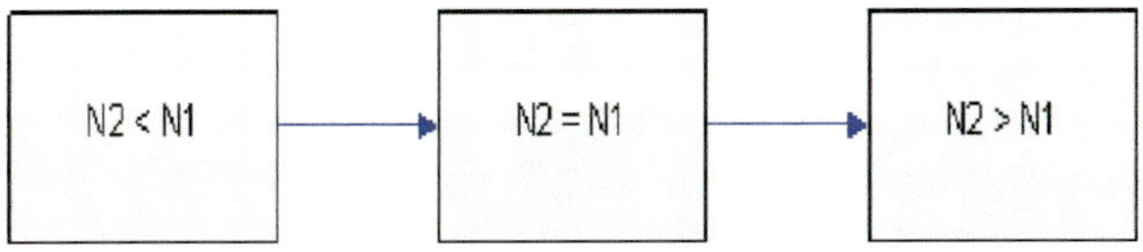

FIGURE 12.41 Amplified population in a sequential process.

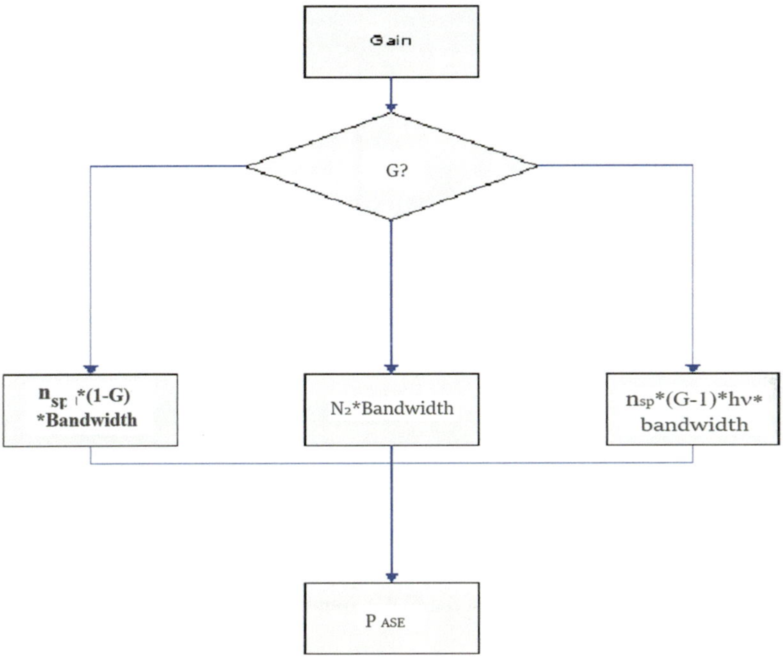

FIGURE 12.42 Modeling flow chart for the ASE noise power and $G<1$.

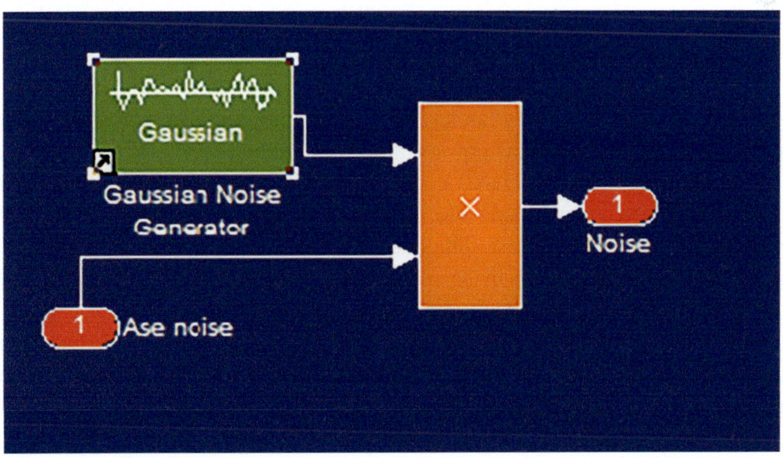

FIGURE 12.43 ASE noise model.

After determining the ASE mean power, the Gaussian assumption of the ASE mean power is used, and the variance is set to model the ASE power and the number of randomly generated photons, as shown in Figures 12.43 and 12.44.

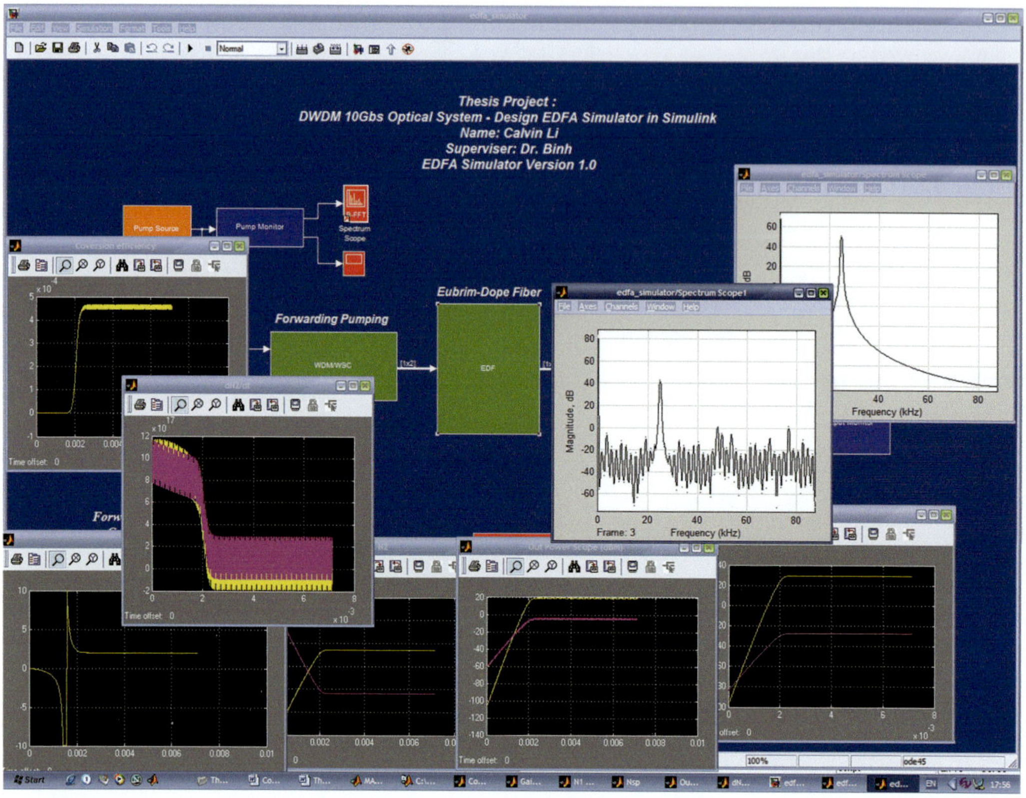

FIGURE 12.44 Samples of the screenshots of the EDFA simulator.

12.4.7.5 Simulink EDFA Simulator: Execution Procedures

The simulator consists of three major parts: the initialization file [EDFA_initialisation.m], the EDFA Simulator [EDFA_simulator.mdl], and the EDFA Module [EDFA_module.mdl].

Steps for executing a simulation:

- Access MATLAB 7.0 R14 or any later version. Open the initialization file from its stored disk, using the file, then open it as per normal.
- The step above brings the MATLAB-coded initialization file to the screen. Refer to the appendices for a hard copy of this file. This file allows the user to define all variables pertaining to the optical simulator. The user is free to modify any value within this file. The file has been commented to instruct the user and future designers on the simulator's operation.
- To run a simulation, enter the required parameter values, then, from the Debug menu, select Save and Run. This will run the file and load the relevant values into both the MATLAB workspace and the Simulink simulator. The simulator should also open in the Simulink workspace. To analyze the initialized system, simply open the Input Monitor block and EDF Parameter, and the initialized input will present there.
- To inspect the result of the EDFA simulator, double-click the Output Monitor block or open the Scope to see the dynamic evolution. Indeed, all results are saved into the workspace for further investigation.
- To inspect the results of the EDFA Module, scopes are popped up automatically, if not, just doodle-click the scope.

Notes: As the EDFA module can combine with other simulation blocks, then the input data stream can be replaced by any other device simulator block. To increase the execution time, the run type in [normal mode], [accelerating mode], [from outside] in the Option menu can be changed.

Typical simulator screenshots of the SIMULINK EDFA simulator are given in Figures 12.44 and 12.45.

The simulated results are based on practical EDFA parameters constructed in our laboratory and some commercial amplifiers operating in the C-band. Table 12.4 provides a set of parameters employed in the simulator. The power generated at the output of the EDFA is plotted versus the input signal power, with the wavelength of the input channel as a parameter, as shown in Figure 12.46. Saturation occurs at around –14 dBm, as expected (Figure 12.47).

The EDFA gain obtained from the simulation agrees with that of the experiment; the only discrepancy occurs during low signal input power. This might be due to the internal settings of the commercial device, which are unknown. All the other factors, such as the absorption and emission spectra in the commercial EDFA, may be slightly different from those in the simulator (Figure 12.48 and Table 12.5).

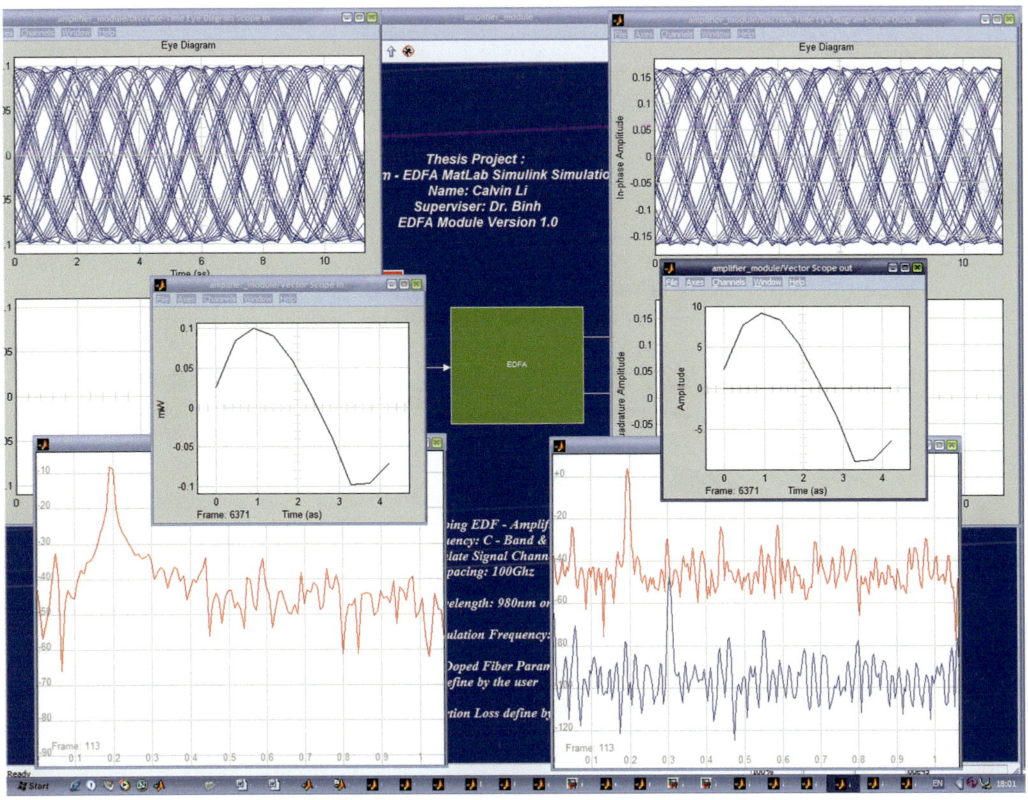

FIGURE 12.45 Samples of the screenshots of the EDFA simulator.

TABLE 12.4
EDFA Simulator Parameters

Single Channel	C-Band	EDF Parameter		
Pump wavelength (nm)	980	MFD=5.9 μm at 1550 nm		
EDF length (m)	16	Cut off wavelength	940 nm	
Pump power (mW)	195	Peak absorption	6.09 dB/m	at 1530 nm
Without gain modulation			5.24 dB/m	at 979 nm

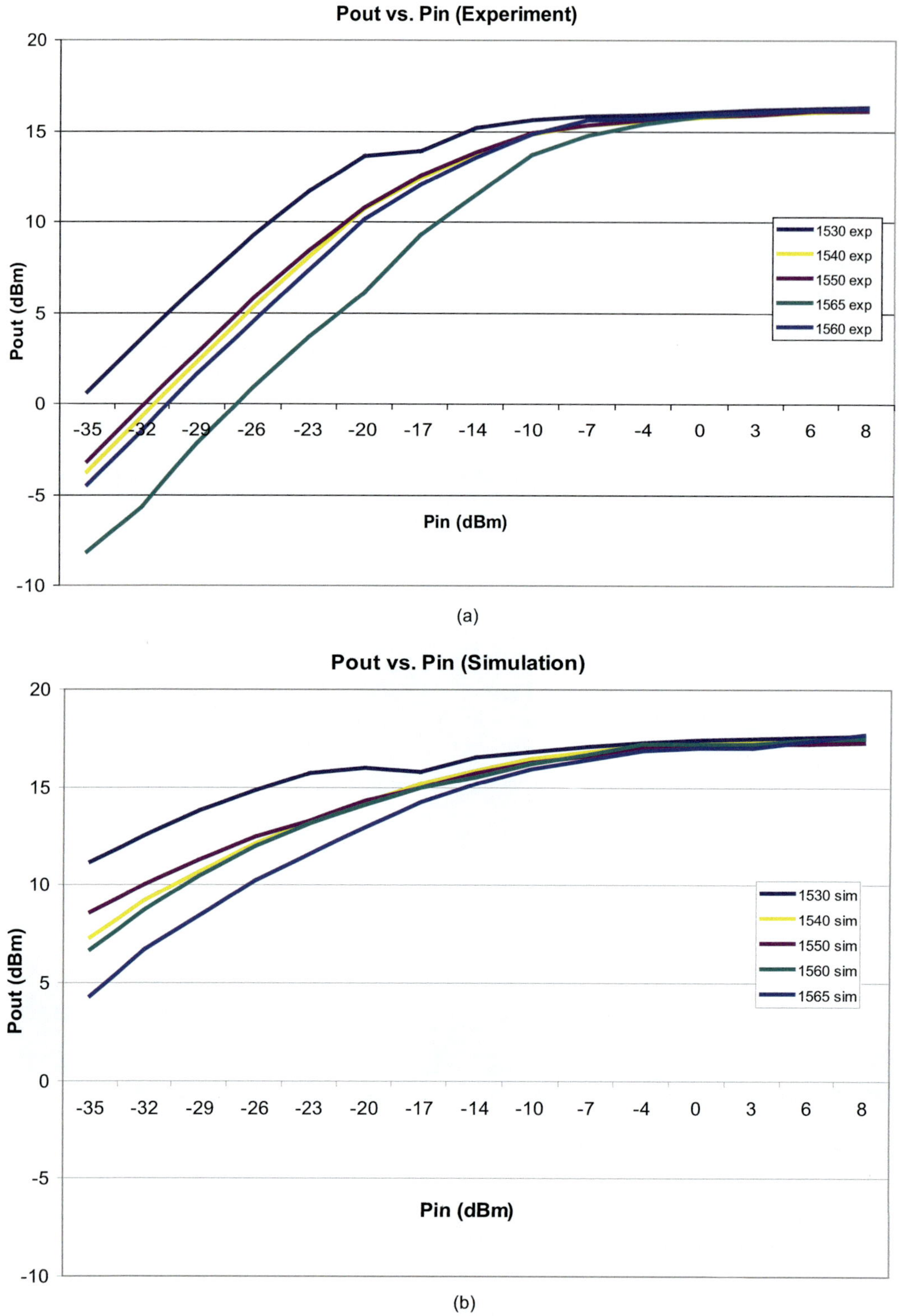

FIGURE 12.46 (a) Experimental results: EDFA output power versus input power level with channel wavelength as a parameter. (b) Simulation results: EDFA output power is plotted against the input signal level with the input channel wavelength as a parameter.

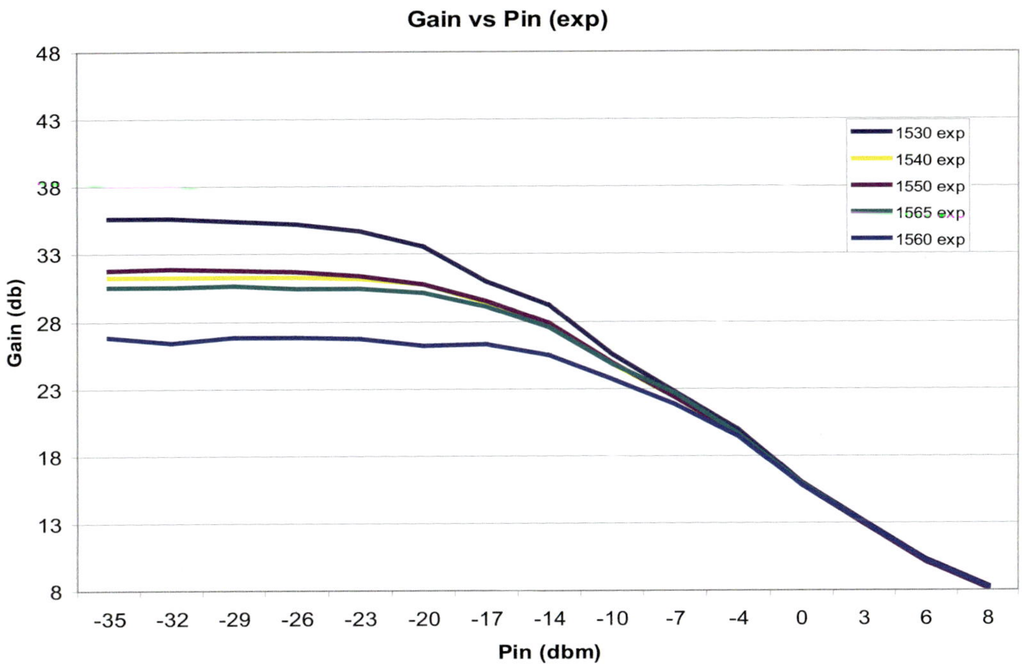

(a)

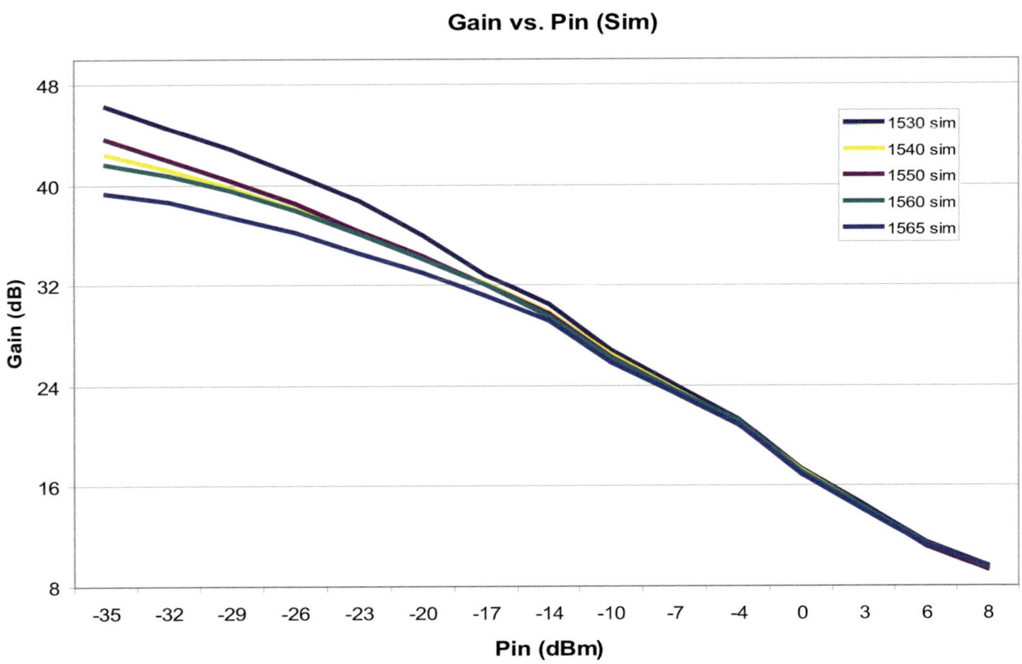

(b)

FIGURE 12.47 (a) Experimental results of the amplification gain. (b) Simulation result of amplification gain.

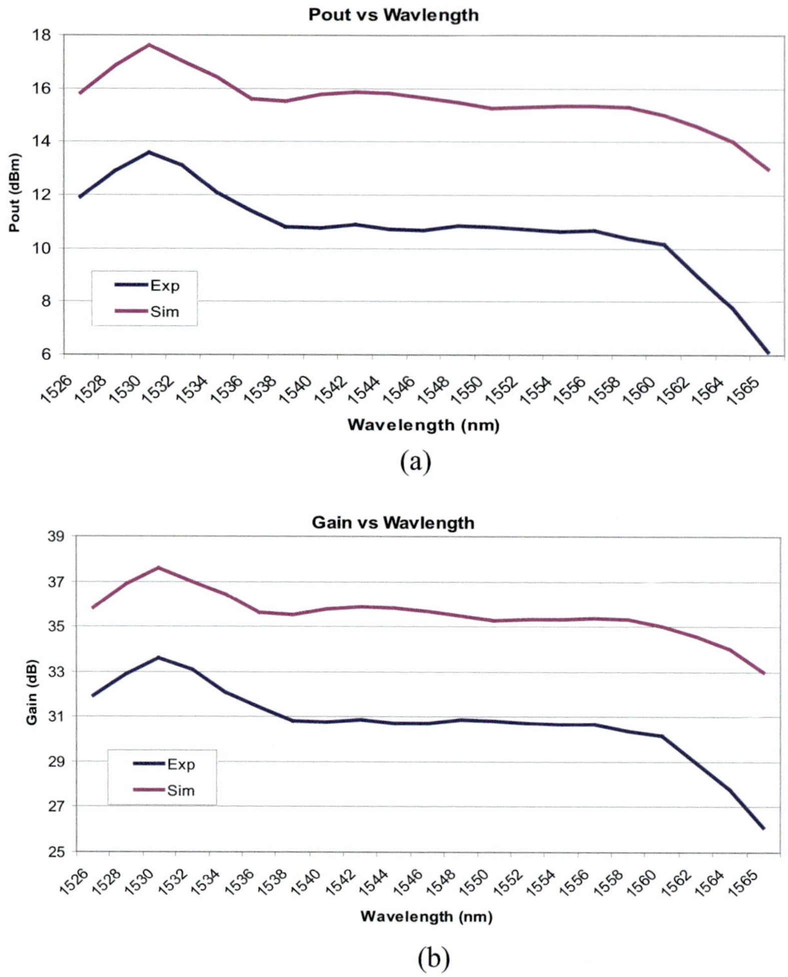

FIGURE 12.48 Spectral properties of (a) output signal power, (b) EDFA gain.

TABLE 12.5
EDFA Simulator Parameters

Single Channel	C-Band	EDF Parameter		
Pump wavelength (nm)	980	MFD=5.9 μm at 1550 nm		
EDF length(m)	16	Cut off wavelength	940 nm	
Pump power (mW)	195	Peak absorption	6.09 dB/m	at 1530 nm
Signal input power (dBm)	–20		5.24 dB/m	at 979 nm

The spectral gain depends on the stimulation emission factor. As the population inversion rate changes with the signal wavelength, the gain peaks at 1528 nm (Figure 12.49 and Table 12.6).

The gain does not increase infinitely with increasing pump power, as the ion density in erbium-doped fiber is constant. Thus, the pump power simply produces energy to pump ions to a higher energy state. Thus, if all the ions have been pumped to the high state, even adding more pump power will not increase the number of simulated photons; hence gain saturation occurs, as illustrated in Figure 12.50 and Table 12.7.

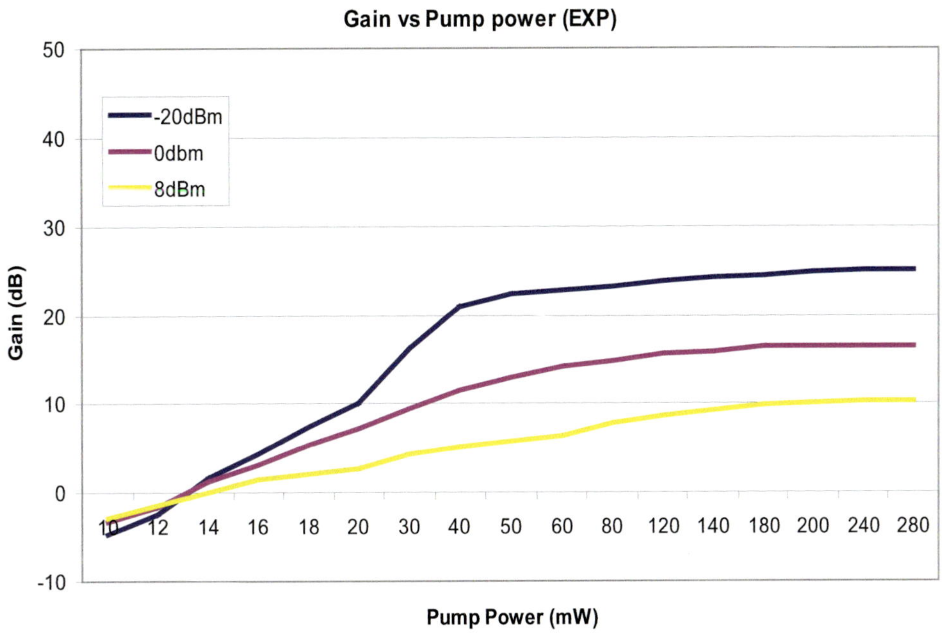

(a)

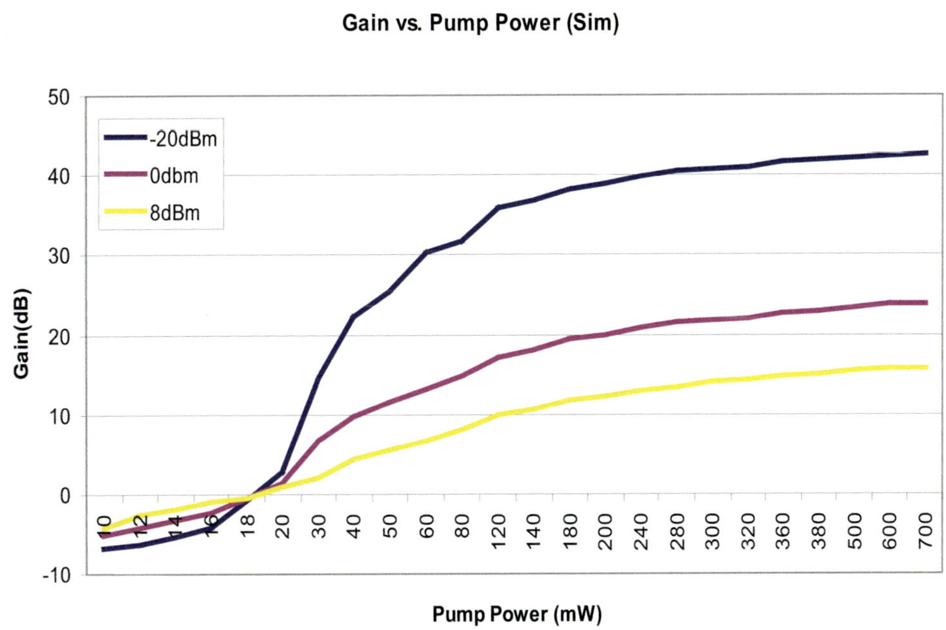

(b)

FIGURE 12.49 Pump power is plotted against the amplification gain with the signal power as a parameter (a) and (b).

TABLE 12.6
EDFA Simulator Testing Parameter

Single Channel	C-Band	EDF Parameter		
Pump wavelength (nm)	980	MFD=5.9 μm at 1550 nm		
EDF length (m)	16	Cut off wavelength	940 nm	
Signal wavelength	1565	Peak absorption	6.09 dB/m	at 1530 nm
Signal input power (dBm)	–20		5.24 dB/m	at 979 nm

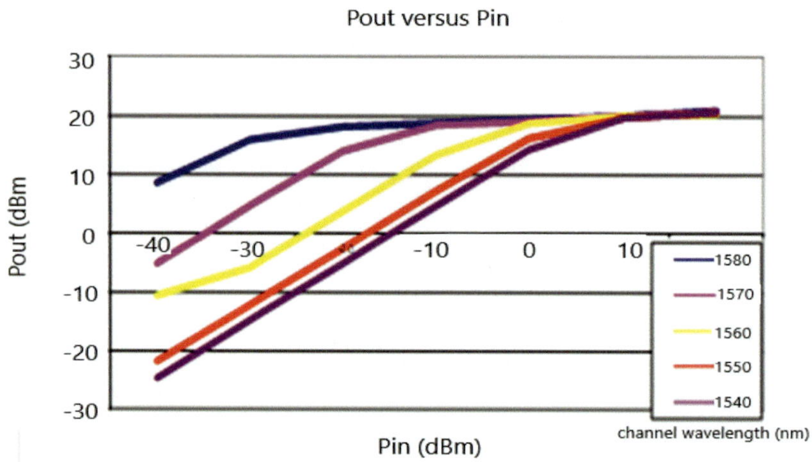

(a)

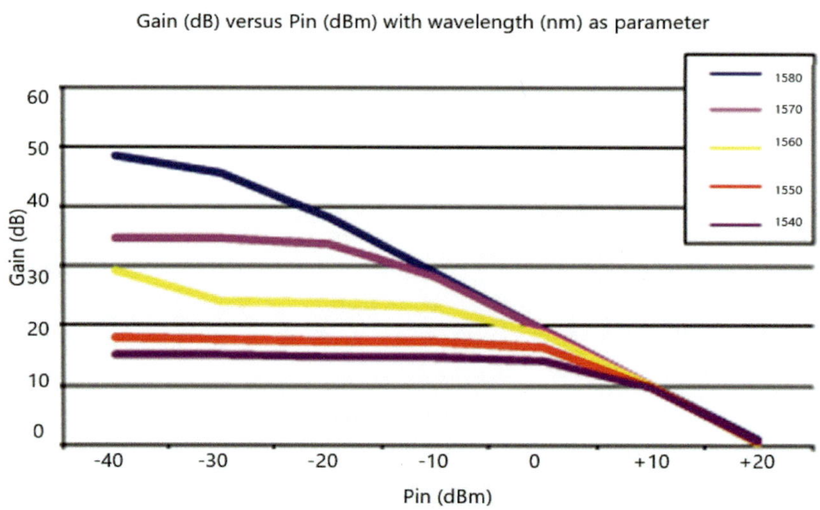

(b)

FIGURE 12.50 EDFA gain versus input signal power level for L-band (a) output power versus input level, (b) gain against input power level with wavelength carriers as a parameter.

TABLE 12.7
EDFA Simulator Testing Parameter

Signal Channel	L-Band	EDF Parameter		
Pump wavelength (nm)	980	MFD=5.9 μm at 1550 nm		
EDF length (m)	16	Cut off wavelength	940 nm	
Pump power (mW)	195 (22.9 dBm)	Peak absorption	6.09 dB/m	at 1530 nm
Without gain modulation			5.24 dB/m	at 979 nm

TABLE 12.8

Signal Level (dBm)/Wavelength (nm)	1530	1540	1550	1560	1565
	16.01	14.22	14.3	–2.709	12.97
Signal Level/Wavelength	**1566**	**1570**	**1580**	**1590**	**1600**
	18.24	13.8	3.838	–2.709	–5.115

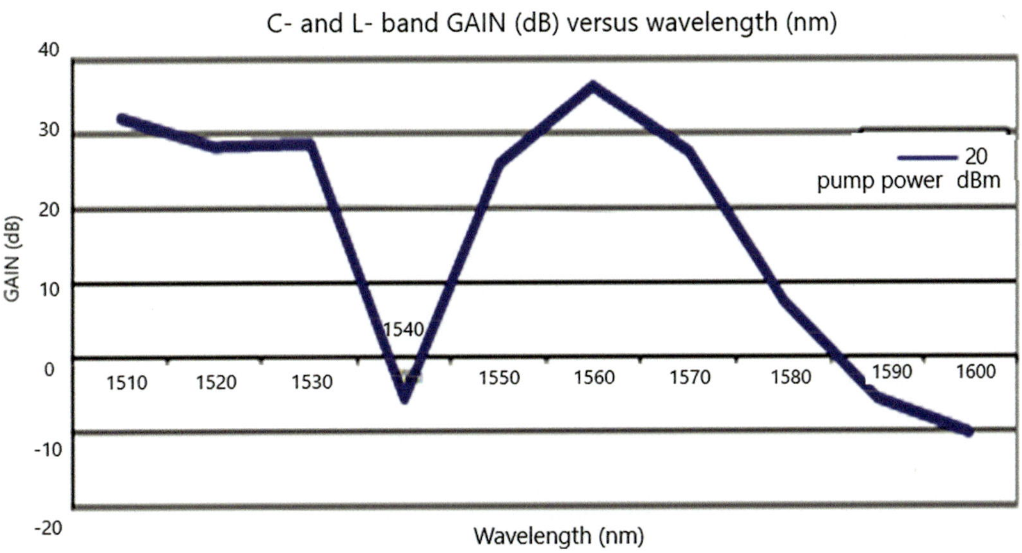

FIGURE 12.51 EDFA gain spectrum in C- and L-band with 20 dBm pump power.

12.4.7.6 Optical Amplification in the L-Band

We can compare the L-Band signal with the C-band signal and tabulate, as presented in Table 12.8, and the gain is plotted over the C- and L-bands, as illustrated in Figure 12.51.

The pump power is varied and the EDFA gain in the L-band is plotted in Figure 12.52 with the parameters of the amplifier given in Table 12.9 (Figures 12.53–15.56 and Tables 12.10–12.12).

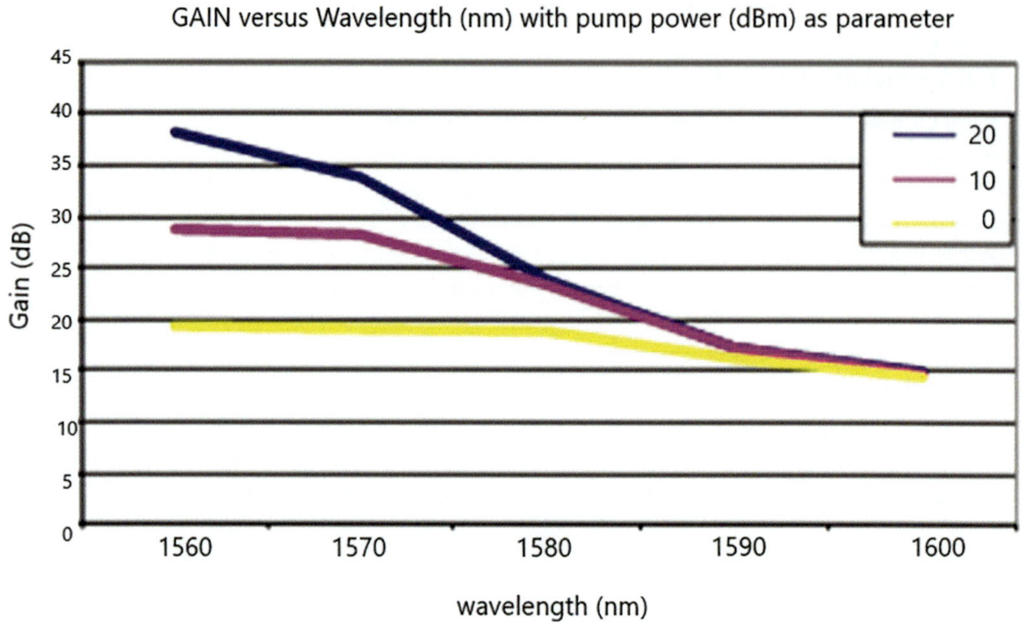

FIGURE 12.52 Optical L-band gain spectrum with pump power as a parameter.

TABLE 12.9
EDFA Simulator Testing Parameter (*Channel Operating Wavelength = 1565 nm*)

Single Channel	C-Band	EDF Parameter		
Pump wavelength (nm)	980	MFD = 5.9 μm at 1550 nm		
EDF length (m)	16	Cut off wavelength	940 nm	
Signal wavelength	1565	Peak absorption	6.09 dB/m	at 1530 nm
Signal input power (dBm)	−20		5.24 dB/m	at 979 nm

12.4.7.7 Samples of the SIMULINK Simulator

This section shows the output scopes observed from the Simulink simulator.[69] They can be divided into two subsections: EDFA simulation scoping and EDFA module scoping. Here, we demonstrate the operation of the Simulink EDFA simulator using simulated results as an example. This provides the user with a guide to follow when using the simulator. The example assumes the system parameters defined in the initialization file. These parameters are loaded using the run command in the debug menu, which allows the MATLAB platform to initialize the simulation values.

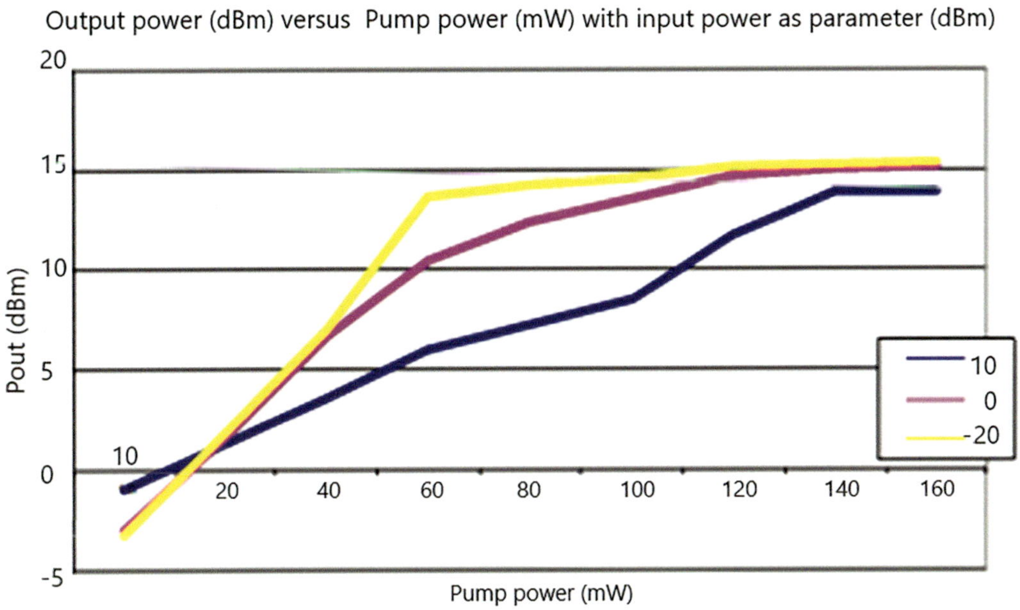

FIGURE 12.53 The output power of the EDFA as a function of the pump power with input power level (in dBm) as a parameter.

12.4.7.7.1 The EDFA Simulator

Figure 12.57 shows a sample of the EDF simulator based on the MATLAB Simulink platform. The simplified model, along with some inputs and outputs of the optical amplifier, is presnted to illustrate the use and outcomes of the simulator. The next subsections describe the simulator inmore detail. Similar models can be developed for HPA and LNA. For HPA, the doping concentration can be changed, and the pump threshold is automatically calculated. In addition, the bidirectional pumping must be included and run in parallel so that the combined gain can be increased to reach the high-power level. For LNA, an optical filter with a sharp roll-off from the 3 dB point to the cut-off region must be inserted in the simulator, the cut-off region level must be low enough to filter out all the noise of the ASE.

12.4.7.7.2 EDFA Simulator Inspection Scopes

The amplifier was tested with the gain plotted against the pump power under critical biasing conditions: one case with biasing in the linear region and the other with saturation bias. The pump power was set at a constant and varied at different levels. Additionally, the pump power was changed in a step function to obtain the transient time of the pump power pulse. These steps are depicted in Figures 12.58 and 12.59a–d.

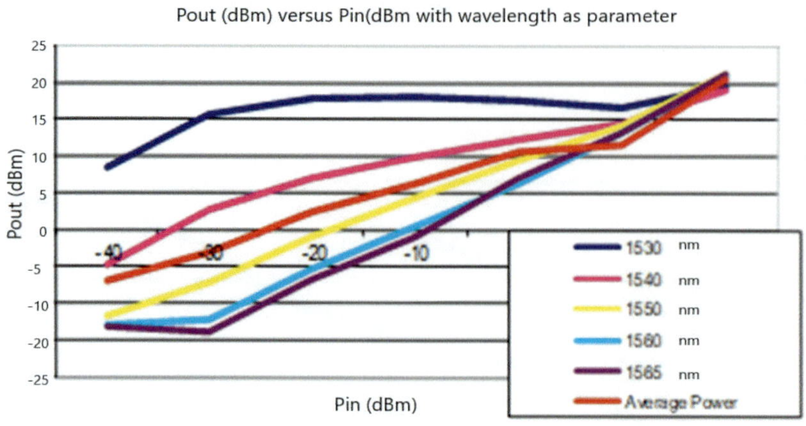

(a)

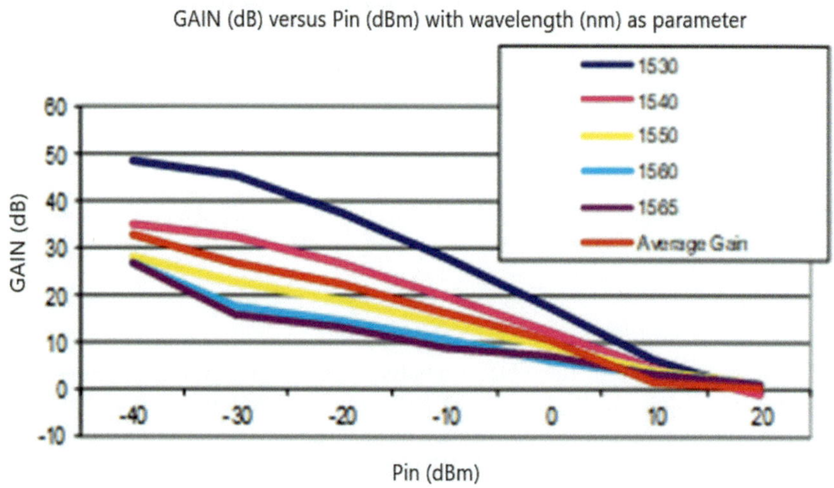

(b)

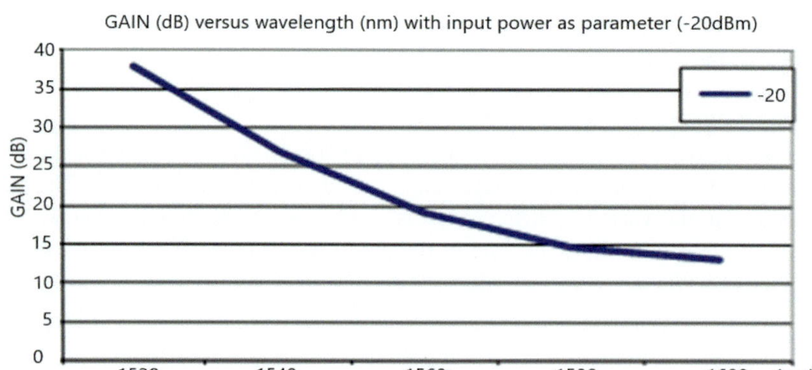

(c)

FIGURE 12.54 EDFA characteristics (a) output power as a function of input signal power level for different channel wavelengths, (b) gain variation as a function of input level, and (c) gain spectra with input average power of –20 dBm.

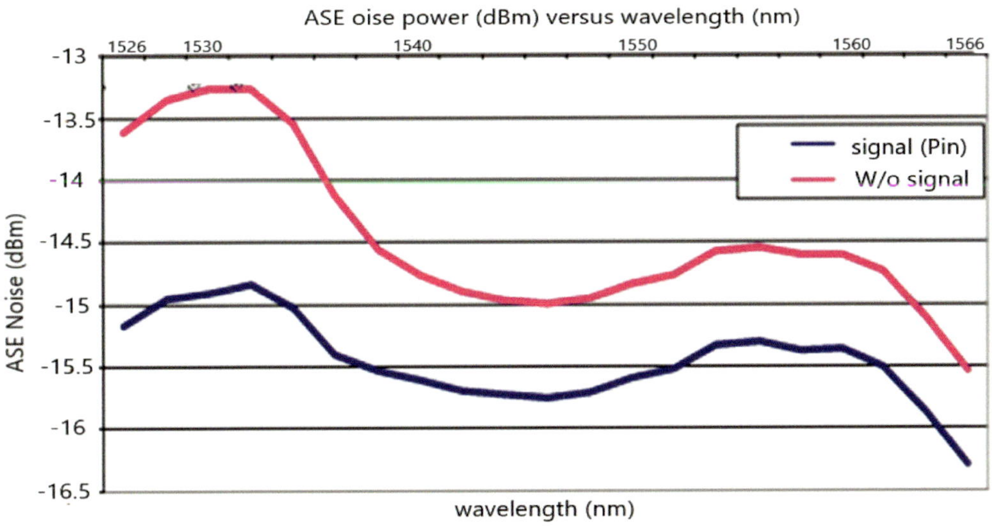

(a)

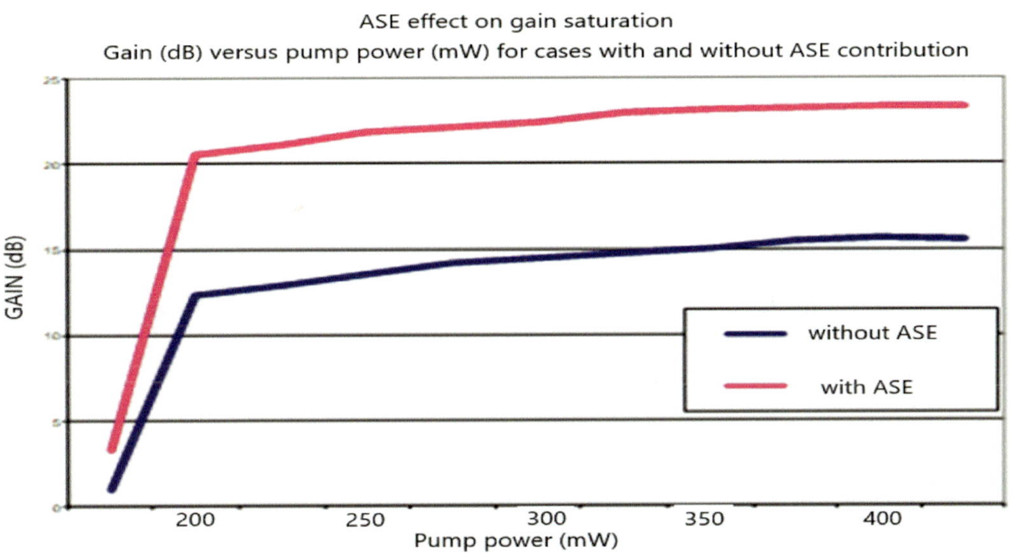

(b)

FIGURE 12.55 ASE noise (a) spectral property and (b) as a function of pump power.

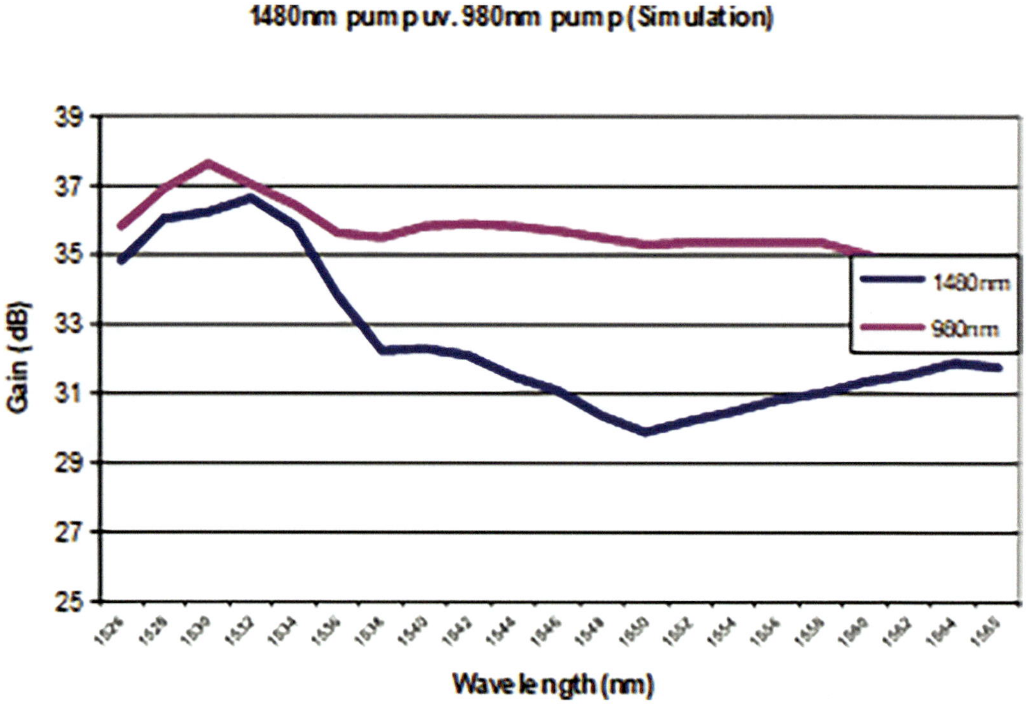

FIGURE 12.56 EDFA gain variation with pump sources of wavelength at 980 and 1480 nm.

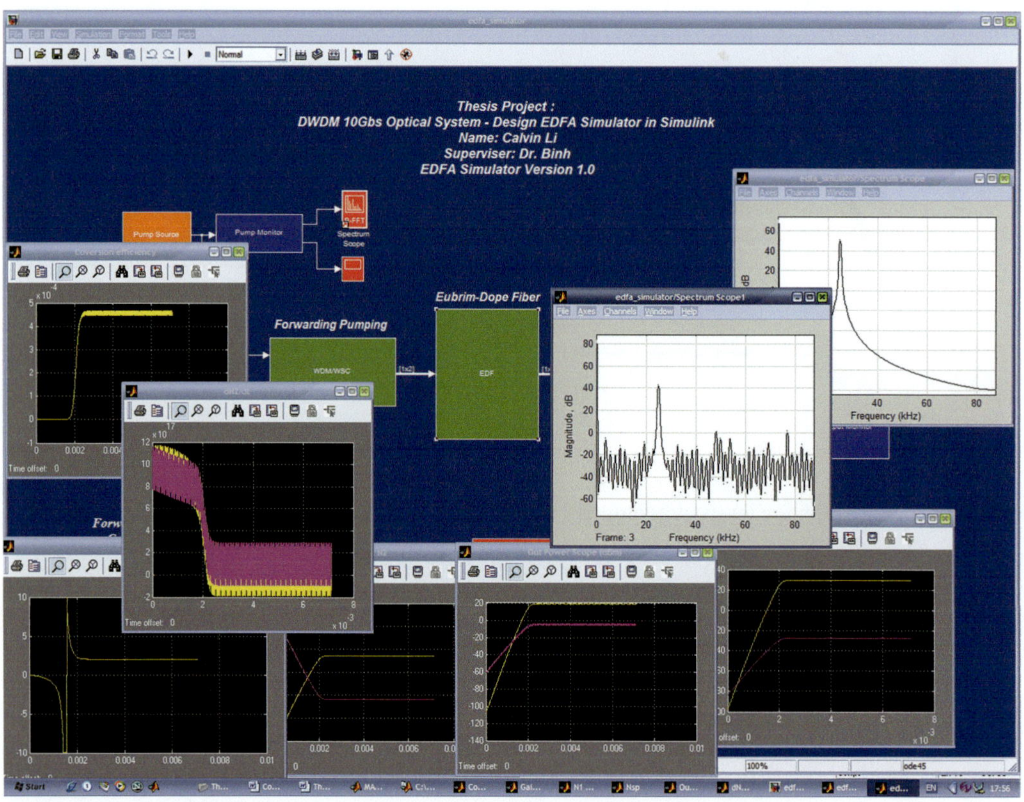

FIGURE 12.57 A sample screenshot on the EDFA simulator based on platform MTLAB Simulink.

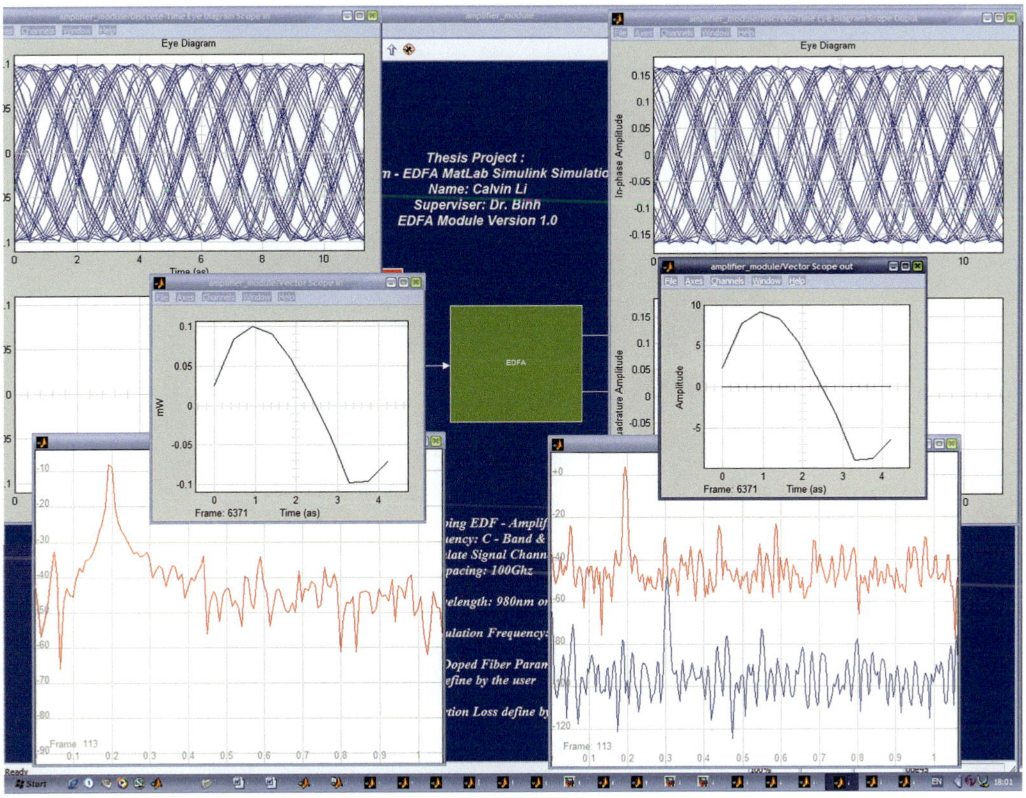

FIGURE 12.58 Sample screen of the EDFA simulator.

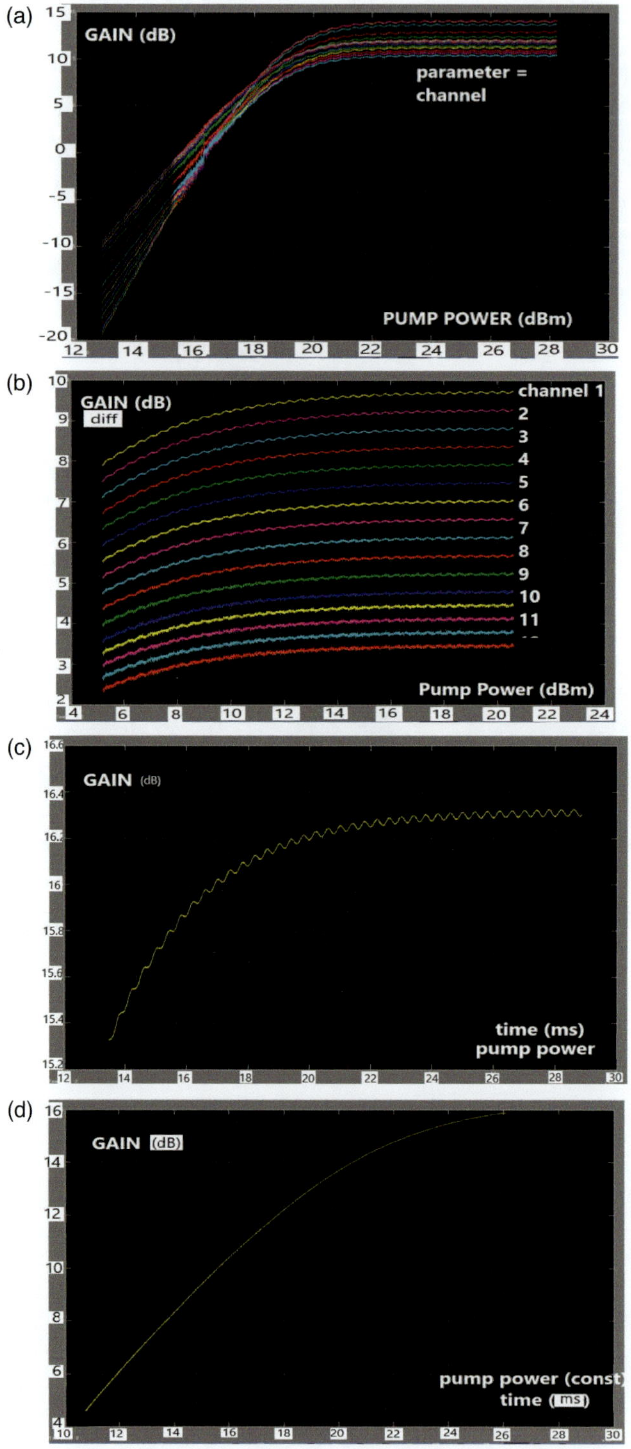

FIGURE 12.59 Sampled outputs of EDFA gain versus pump power (a & b) constant pump power and gain of different wavelength channels (c & d) time-dependent pumps. (a) Amplification gain (dB) versus pump power (dBm). (b) Gain (dB) difference of different channel number (wavelength) against pump power (dBm). (c) Gain (dB) versus pump power time for observation of transient effect from linear to saturation region, with initial bias at saturation. (d) Gain (dB) versus pump power time when biased at linear region, mode.

TABLE 12.10
EDFA Simulator Testing Parameters for Results of Figure 12.54

Multi-Channel	C-Band	EDF Parameter		
Pump wavelength (nm)	980	MFD=5.9 μm at 1550 nm	MFD=	Mode field diameter
EDF length (m)	16	Cut off wavelength	940 nm	
Pump power (mW)	195	Peak absorption	6.09 dB/m	at 1530 nm
Without gain modulation			5.24 dB/m	at 979 nm

TABLE 12.11
EDFA Simulator Parameters for the ASE Simulation Results of Figure 12.55

Signal Channel	C-Band	EDF Parameter		
Pump wavelength (nm)	980 nm	MFD=5.9 μm at 1550 nm	MFD=Mode field diameter	
EDF length (m)	16	Cut-off wavelength	940 nm	
Pump power (mW)	195(22.9 dBm)	Peak absorption	6.09 dB/m	at 1530 nm
Without gain modulation			5.24 dB/m	at 979 nm
Signal power (dBm)	–20			

TABLE 12.12
EDFA Simulator Testing Parameters Used for Figure 12.56

Result of EDFA Amplifier		EDF Parameter		
Pump wavelength (nm)	1480 nm	MFD=5.9 μm at 1550 nm		
EDF length (m)	16	Cut off wavelength	940 nm	
Pump power (mW)	195	Peak absorption	6.09 dB/m	at 1530 nm
Signal input power (dBm)	–20		5.24 dB/m	at 979 nm

12.5 APPENDIX 7: ITU (INTERNATIONAL TELECOMMUNICATION UNION) GRID FOR CHANNEL CARRIER SPECTRAL LOCATIONS

12.5.1 The 100 GHz DWDM Wavelength Channels ITU Grid

ITU[1] Channel Number	Frequency (THz)	Wavelength (nm)	Spectral Band	ITU Channel #	Frequency (THz)	Wavelength (nm)	Spectral Band
1	190.10	1577.03	L-Band	41	194.10	1544.53	C-Band
2	190.20	1576.20	L-Band	42	194.20	1543.73	C-Band
3	190.30	1575.37	L-Band	43	194.30	1542.94	C-Band
4	190.40	1574.54	L-Band	44	194.40	1542.14	C-Band
5	190.50	1573.71	L-Band	45	194.50	1541.35	C-Band
6	190.60	1572.89	L-Band	46	194.60	1540.56	C-Band
7	190.70	1572.06	L-Band	47	194.70	1539.77	C-Band
8	190.80	1571.24	L-Band	48	194.80	1538.98	C-Band
9	190.90	1570.42	L-Band	49	194.90	1538.19	C-Band
10	191.00	1569.59	L-Band	50	195.00	1537.40	C-Band
11	191.10	1568.11	L-Band	51	195.10	1536.61	C-Band
12	191.20	1567.95	L-Band	52	195.20	1535.82	C-Band
13	191.30	1567.13	L-Band	53	195.30	1535.04	C-Band
14	191.40	1566.31	L-Band	54	195.40	1534.25	C-Band
15	191.50	1565.50	L-Band	55	195.50	1533.47	C-Band
16	191.60	1564.68	C-Band	56	195.60	1532.68	C-Band
17	191.70	1563.86	C-Band	57	195.70	1531.90	C-Band
18	191.80	1563.05	C-Band	58	195.80	1531.12	C-Band
19	191.90	1562.23	C-Band	59	195.90	1530.33	C-Band
20	192.00	1561.41	C-Band	60	196.00	1529.55	S-Band
21	192.10	1560.61	C-Band	61	196.10	1528.77	S-Band
22	192.20	1559.79	C-Band	62	196.20	1527.99	S-Band
23	192.30	1558.98	C-Band	63	196.30	1527.22	S-Band
24	192.40	1558.17	C-Band	64	196.40	1526.44	S-Band
25	192.50	1557.36	C-Band	65	196.50	1525.66	S-Band
26	192.60	1556.55	C-Band	66	196.60	1524.89	S-Band
27	192.70	1555.75	C-Band	67	196.70	1524.11	S-Band
28	192.80	1554.94	C-Band	68	196.80	1523.34	S-Band
29	192.90	1554.13	C-Band	69	196.90	1522.56	S-Band
30	193.00	1553.33	C-Band	70	197.00	1521.79	S-Band
31	193.10	1552.52	C-Band	71	197.10	1521.02	S-Band
32	193.20	1551.72	C-Band	72	197.20	1520.25	S-Band
33	193.30	1550.92	C-Band	73	197.30	1519.48	S-Band
34	193.40	1550.12	C-Band	74	197.40	1518.71	S-Band
35	193.50	1549.32	C-Band	75	197.50	1517.94	S-Band
36	193.60	1548.51	C-Band	76	197.60	1517.17	S-Band
37	193.70	1547.72	C-Band	77	197.70	1516.40	S-Band
38	193.80	1546.92	C-Band	78	197.80	1515.63	S-Band
39	193.90	1546.12	C-Band	79	197.90	1514.86	S-Band
40	194.00	1545.32	C-Band	80	198.00	1514.09	S-Band

12.5.2 The 50 GHz DWDM Wavelength Channels ITU Grid

ITU Channel #	Frequency (THz)	Wavelength (nm)	Wavelength Band	ITU Channel #	Frequency (THz)	Wavelength (nm)	Wavelength Band
1	190.10	1577.03	L-Band	40.5	194.05	1544.92	C-Band
1.5	190.15	1576.61	L-Band	41	194.10	1544.53	C-Band
2	190.20	1576.20	L-Band	41.5	194.15	1544.13	C-Band
2.5	190.25	1575.78	L-Band	42	194.20	1543.73	C-Band
3	190.30	1575.37	L-Band	42.5	194.25	1543.33	C-Band
3.5	190.35	1574.95	L-Band	43	194.30	1542.94	C-Band
4	190.40	1574.54	L-Band	43.5	194.35	1542.54	C-Band
4.5	190.45	1574.13	L-Band	44	194.40	1542.14	C-Band
5	190.50	1573.71	L-Band	44.5	194.45	1541.75	C-Band
5.5	190.55	1573.30	L-Band	45	194.50	1541.35	C-Band
6	190.60	1572.89	L-Band	45.5	194.55	1540.95	C-Band
6.5	190.65	1572.48	L-Band	46	194.60	1540.56	C-Band
7	190.70	1572.06	L-Band	46.5	194.65	1540.16	C-Band
7.5	190.75	1571.65	L-Band	47	194.70	1539.77	C-Band
8	190.80	1571.24	L-Band	47.5	194.75	1539.37	C-Band
8.5	190.85	1570.83	L-Band	48	194.80	1538.98	C-Band
9	190.90	1570.42	L-Band	48.5	194.85	1538.58	C-Band
9.5	190.95	1570.01	L-Band	49	194.90	1538.19	C-Band
10	191.00	1569.59	L-Band	49.5	194.95	1537.79	C-Band
10.5	191.05	1569.18	L-Band	50	195.00	1537.40	C-Band
11	191.10	1568.11	L-Band	50.5	195.05	1537.00	C-Band
11.5	191.15	1568.36	L-Band	51	195.10	1536.61	C-Band
12	191.20	1567.95	L-Band	51.5	195.15	1536.22	C-Band
12.5	191.25	1567.54	L-Band	52	195.20	1535.82	C-Band
13	191.30	1567.13	L-Band	52.5	195.25	1535.43	C-Band
13.5	191.35	1566.72	L-Band	53	195.30	1535.04	C-Band
14	191.40	1566.31	L-Band	53.5	195.35	1534.64	C-Band
14.5	191.45	1565.90	L-Band	54	195.40	1534.25	C-Band
15	191.50	1565.50	L-Band	54.5	195.45	1533.86	C-Band
15.5	191.55	1565.09	L-Band	55	195.50	1533.47	C-Band
16	191.60	1564.68	C-Band	55.5	195.55	1533.07	C-Band
16.5	191.65	1564.27	C-Band	56	195.60	1532.68	C-Band
17	191.70	1563.86	C-Band	56.5	195.65	1532.29	C-Band
17.5	191.75	1563.45	C-Band	57	195.70	1531.90	C-Band
18	191.80	1563.05	C-Band	57.5	195.75	1531.51	C-Band
18.5	191.85	1562.64	C-Band	58	195.80	1531.12	C-Band
19	191.90	1562.23	C-Band	58.5	195.85	1530.72	C-Band
19.5	191.95	1561.83	C-Band	59	195.90	1530.33	C-Band
20	192.00	1561.42	C-Band	59.5	195.95	1529.94	S-Band
20.5	192.05	1561.01	C-Band	60	196.00	1529.55	S-Band
21	192.10	1560.61	C-Band	60.5	196.05	1529.16	S-Band
21.5	192.15	1560.20	C-Band	61	196.10	1528.77	S-Band
22	192.20	1559.79	C-Band	61.5	196.15	1528.38	S-Band
22.5	192.25	1559.39	C-Band	62	196.20	1527.99	S-Band
23	192.30	1558.98	C-Band	62.5	196.25	1527.60	S-Band
23.5	192.35	1558.58	C-Band	63	196.30	1527.22	S-Band
24	192.40	1558.17	C-Band	63.5	196.35	1526.83	S-Band
24.5	192.45	1557.77	C-Band	64	196.40	1526.44	S-Band

(Continued)

(Continued)

ITU Channel #	Frequency (THz)	Wavelength (nm)	Wavelength Band	ITU Channel #	Frequency (THz)	Wavelength (nm)	Wavelength Band
25	192.50	1557.36	C-Band	64.5	196.45	1526.05	S-Band
25.5	192.55	1556.96	C-Band	65	196.50	1525.66	S-Band
26	192.60	1556.56	C-Band	65.5	196.55	1525.27	S-Band
26.5	192.65	1556.15	C-Band	66	196.60	1524.89	S-Band
27	192.70	1555.75	C-Band	66.5	196.65	1524.50	S-Band
27.5	192.75	1555.34	C-Band	67	196.70	1524.11	S-Band
28	192.80	1554.94	C-Band	67.5	196.75	1523.72	S-Band
28.5	192.85	1554.54	C-Band	68	196.80	1523.34	S-Band
29	192.90	1554.13	C-Band	68.5	196.85	1522.95	S-Band
29.5	192.95	1553.73	C-Band	69	196.90	1522.56	S-Band
30	193.00	1553.33	C-Band	69.5	196.95	1522.18	S-Band
30.5	193.05	1552.93	C-Band	70	197.00	1521.79	S-Band
31	193.10	1552.52	C-Band	70.5	197.05	1521.40	S-Band
31.5	193.15	1552.12	C-Band	71	197.10	1521.02	S-Band
32	193.20	1551.72	C-Band	71.5	197.15	1520.63	S-Band
32.5	193.25	1551.32	C-Band	72	197.20	1520.25	S-Band
33	193.30	1550.92	C-Band	72.5	197.25	1519.86	S-Band
33.5	193.35	1550.52	C-Band	73	197.30	1519.48	S-Band
34	193.40	1550.12	C-Band	73.5	197.35	1519.09	S-Band
34.5	193.45	1549.72	C-Band	74	197.40	1518.71	S-Band
35	193.50	1549.32	C-Band	74.5	197.45	1518.32	S-Band
35.5	193.55	1548.91	C-Band	75	197.50	1517.94	S-Band
36	193.60	1548.52	C-Band	75.5	197.55	1517.55	S-Band
36.5	193.65	1548.11	C-Band	76	197.60	1517.17	S-Band
37	193.70	1547.72	C-Band	76.5	197.65	1516.78	S-Band
37.5	193.75	1547.32	C-Band	77	197.70	1516.40	S-Band
38	193.80	1546.92	C-Band	77.5	197.75	1516.02	S-Band
38.5	193.85	1546.52	C-Band	78	197.80	1515.63	S-Band
39	193.90	1546.12	C-Band	78.5	197.85	1515.25	S-Band
39.5	193.95	1545.72	C-Band	79	197.90	1514.87	S-Band
40	194.00	1545.32	C-Band	79.5	197.95	1514.49	S-Band

NOTES

1 International Telecommunications Union – Telecommunication, a standard organization, https://www.itu.int/en/ITU-T/Pages/default.aspx

2 S. E. Miller, "Integrated optics: An introduction", *Bell Labs Technical Journal,* 48, 2059–2069, 1969.

3 C. R. Doerr and K. Okamoto, "Advances in silica planar lightwave circuits", *IEEE Journal of Lightwave Technology,* 24(12), 4763–4770, 2006.

4 P. S. Chung, "Waveguide modes, coupling techniques, fabrication and losses in optical integrated optics", *Journal of Electrical and Electronic Engineering,* 5, 201–214, 1985.

5 http://en.wikipedia.org/wiki/Photonic_integrated_circuit, access date December 2009;.

6 R. G. Hunsperger, *Integrated Optics Theory and Technology.* Originally published in the series: Advanced Texts in Physics, 6th Ed. Springer: Berlin, XXVIII, 2009.

7 L. N. Binh and S. V. Chung, "Nonlinear interactions in thin film structures", *Proceedings of 8th Australian Workshop on Optical Communications,* Adelaide, Session VII, 1983.

8 K. Uchiyama and T. Morioka, "All-optical time-division demultiplexing experiment with simultaneous output of all constituent channels from 100Gbit/s OTDM signal", *Electronics Letters,* 37(10), 642–643, May 10, 2001.

9 M. Nakazawa, T. Yamamoto and K. R. Tamura, "1.28Tbit/s–70km OTDM transmission using third- and fourth-order simultaneous dispersion compensation with a phase modulator", *Electronics Letters,* 36(24), 2027–2029, November 23, 2000.
10 C. Schubert, R. H. Derksen, M. Möller, R. Ludwig, C.-J. Weiske, J. Lutz, S. Ferber, A. Kirstädter, G. Lehmann, and C. Schmidt-Langhorst, "Integrated 100-Gb/s ETDM receiver", *IEEE Journal of Lightwave Technology,* 25(1), 122–129, January 2007.
11 H. Suzuki, M. Fujiwara, and K. Iwatsuki, "Application of super-DWDM technologies to terrestrial Terabit transmission systems", *IEEE Journal of Lightwave Technology,* 24(5), 1998–2005, May 2006.
12 J. P. Turkiewicz, E. Tangdiongga, G. Lehmann, H. Rohde, W. Schairer, Y. R. Zhou, E.S.R. Sikora, A. Lord, D. B. Payne, G.-D. Khoe, and H. de Waardt, "160 Gb/s OTDM networking using deployed fibre", *IEEE Journal of Lightwave Technology,* 23(1), 225–234, January 2005.
13 R. Nagarajan, M. Kato, V. G. Dominic, C. H. Joyner, R. P. Schneider, Jr., A. G. Dentai, T. Desikan, P. W. Evans, M. Kauffman, D. J. H. Lambert, S. K. Mathis, A. Mathur, M. L. Mitchell, M. J. Missey, S. Murthy, A. C. Nilsson, F. H. Peters, J. L. Pleumeekers, R. A. Salvatore, R. B. Taylor, M. F. Van Leeuwen, J. Webjorn, M. Ziari, S. G. Grubb, D. Perkins, M. Reffle, D. G. Mehuys, F. A. Kish, and D.F. Welch, "400 Gbit/s (10 channel x40 Gbit/s) DWDM photonic integrated circuits", *Electronics Letters,* 41(6), 347–349, March 17, 200.
14 T. Okoshi, "Recent advances in coherent optical fibre communications", *IEEE Journal of Lightwave Technology,* LT-5, 44–52, 1987.
15 L. G. Karzovky and O. K. Tonguz, "ASK and FSK coherent lightwave systems: A simplified approximate analysis", *IEEE Journal of Lightwave Technology,* 8(3), 338–351, March 1990.
16 S. Tsukamoto, D.-S. Ly-Gagnon, K. Katoh, and K. Kikuchi, "Coherent demodulation of 40-Gbit/s polarization-multiplexed QPSK signals with 16-GHz spacing after 200-km transmission", *Proceedings of CLEOS 2005, San Jose, paper: PDP29, 2005.*
17 D.-Se. Ly-Gagnon, S. Tsukamoto, K. Katoh, and K. Kikuchi, "Coherent detection of optical quadrature phase-shift keying signals with carrier phase estimation", *IEEE Journal of Lightwave Technology,* 24(1), 12–21, January 2006.
18 V. Ramaswamy and R. K. Lagu, "Numerical field solution for an arbitrary asymmetrical gradient index planar waveguide", *IEEE Journal of Lightwave Technology,* LT-1, 408–417, 1983.
19 R. V. Schmidt and I. P. Kaminow, "Metal diffused optical waveguides in LiNbO3", *Applied Physics Letters,* 25, 458–460, 1974.
20 W. Streifer, R. D. Burnham, and D. R. Scifres, "Modal analysis of separate-confinement heterojunction lasers with inhomogeneous cladding layers", *Optics Letters,* 8, 283–285, 1981.
21 D. Marcuse, *Light Transmission Optics.* Van Nostrand: New York, Chapter 8, 1972.
22 E. M. Conwell, "Modes in optical waveguides formed by diffusion", *Applied Physics Letters,* 26, 328–329, 1973.
23 H. Kirchoff, "The solution of Maxwell equations for inhomogeneous dielectric slabs", *AEU,* 26, 537–541, 1972.
24 D. Marcuse, "TE modes of graded index: Slab waveguides", *IEEE Journal of Quantum Electronics,* QE-9, 1000–1006, 1973.
25 A. W. Snyder and J. D. Love, *Optical Waveguide Theory.* Chapman and Hall: London, Chapter 12, 1983.
26 T. R. Chen and Z. L. Yang, "Modes of a planar waveguide with Fermi index profile", *Applied Optics,* 24, 2809–2812, 1985.
27 H. F. Taylor, "Dispersion characteristics of diffused channel waveguides", *IEEE Journal of Quantum Electronics,* QE-12, 748–752, 1976.
28 G. B. Hocker and W. K. Burns, "Mode dispersion in diffused channel waveguides by the effective index method", *Applied Optics,* 16, 113–118, 1975.
29 L. Riviere, A. Yi-Yan, and H. Carru, "Properties of single mode planar with Gaussian index profile", *IEEE Journal of Lightwave Technology,* LT-3, 368–377, 1986.
30 S. K. Korotky, et al., "Mode size and method for estimating the propagation constant of single mode Ti:LiNbO3 strip waveguides", *IEEE Transactions on Microwave Theory and Techniques,* MTT-30, 1784–1789, 1982.
31 V. Ramaswamy and R. K. Lagu, "Numerical field solution for an arbitrary asymmetrical gradient index planar waveguide", *IEEE Journal of Lightwave Technology,* LT-1, 408–417, 1983.
32 A. N. Kaul, S. L. Hossain, and K. Thyagarajan, "A simple numerical method for studying the propagation characteristics of single-mode graded-index planar optical waveguides", *IEEE Transactions on Microwave Theory and Techniques,* MTT_34, 288–292, 1986.

33 L. N. Binh and S. V. Chung, "Nonlinear interactions in thin film structures", *Proceedings of 8th Australian Workshop on Optics Communications,* Adelaide Australia, session VII, 1983.
34 L. N. Binh, et al., "Design considerations for second harmonic generation in thin film grown by ion beam assisted deposition method", *Proceedings of the 3rd International Conference on Lasers,* Melbourne, Session 10A, 1983.
35 J. Killingbeck, "A pocket calculator determination of energy eigenvalues", *Journal of Physics A,* 10, L09-L103, 1977.
36 R. A. Sammut and C. Pask, "Simplified numerical analysis of optical fibres and planar waveguides", *Electronics Letters,* 17, 105–106, 1981.
37 L. N. Binh, *Guided Wave Photonics.* CRC Press, Taylor & Francis: Boca Raton, FL, Chapters 2 and 3, 2012.
38 J. D. Love and A. K. Ghatak, "Exact solutions for TM modes in graded index slab waveguides", *IEEE Journal of Quantum Electronics,* QE 15, 14–16, 1979.
39 D. Marcuse, "The effects of the ∇2n term on the modes of an optical square-law medium", *IEEE Journal of Quantum Electronics,* QE-9, 958–960, 1973.
40 H. Kolgenik and V. Ramaswamy, "Scaling rule for thin film optical waveguides", *Applied Optics,* 13, 1857–1862, 1974.
41 L. N. Binh and S. V. Chung, "Nonlinear interactions in thin film structures", *Proceedings of 8th Australian Workshop on Optics Communications,* Adelaide Australia, session VII, 1983
42 P. K. Tien, "Integrated optics and new wave phenomena", Reviews of Modern Physics, 49, 361–420, 1977.
43 A. Watson, "Physics –where the action is", *New Scientist,* pp. 42–44, January 30, 1986.
44 L. Mammel and L. G. Cohen, "Numerical prediction of fibre transmission characteristics of arbitrary refractive index profiles", *Applied Optics,* 21, 699–703, 1982.
45 K. Petermann, "Constraints for fundamental mode-size for broadband dispersion-compensated single mode fibres", *Electronics Letters,* 19, 712–714, 1983.
46 P. Sansonetti, "Modal dispersion in single mode-fibres: Simple approximation issues of mode spot size spectral behavior", *Electronics Letters,* 18, 647–648, 1982.
47 S.K. Korotsky, et al., "Mode size and method for estimating the propagation constant of single mode Ti:LiNbO3 strip waveguides", *IEEE Transactions on Microwave Theory and Techniques,* MTT-30, 1784–1789, 1982.
48 K. C. Kao and G. A. Hockham, "Dielectric-fibre surface waveguides for optical frequencies", *IEE Proceedings,* 133(3), 1151–1158, June 1986.
49 L. N. Binh, *Advanced Digital Optical Communications Systems.* CRC Press: Boca Raton, FL, 2015.
50 L. N. Binh, *Optical Fibre Communication Systems with MATLAB® and Simulink® Models,* 2nd Ed. CRC Press: Boca Raton, FL, 2016.
51 O. Gerstel, M. Jinno, A. Lord, and S. Yoo, "Time elastic optical networking: A new dawn for the optical layer?" *IEEE Communications Magazine,* 50(2), s12–s20, 2012.
52 G. Wellbrock and T. J. Xia, "How will optical transport deal with future network traffic growth?" In *European Conference on Optical Communication,* Cannes, France, paper Th.1.2.1, 2014.
53 A. Bononi, P. Serena, N. Rossi, and D. Sperti, "Which is the dominant nonlinearity in long-haul PDM-QPSK?" In *European Conference on Optical Communication,* Torino, Italy, paper Th.10.E.1, 2010.
54 G. Charlet, "Impact and mitigation of non-linear effects in coherent transmission," In *Optical Fibre Communication Conference,* San Diego, CA, paper NthB.4, 2009.
55 G. Charlet, "Fibre characteristics for next-generation ultralong-haul transmission systems," In *European Conference on Optical Communication,* Torino, Italy, paper We.8, F.1, 2010.
56 "Characteristics of a cut-off shifted single-mode optical fibre and cable," ITU-T Recommendation G.654, October 2012, Online. Available: https://www.itu.int/rec/T-REC-G.654-202003-S/en
57 B. Zhu, S. Chandrasekhar, X. Liu, and D. W. Peckham, "Transmission performance of a 485-Gb/s CO-OFDM superchannel with PDM-16QAM subcarriers over ULAF and SSMF-based links," *IEEE Photonics Technology Letters,* 23, 1400–1402, 2011.
58 M. Bigot-Astruc and P. Sillard, "Realizing large effective area fibres," In *Optical Fibre Communication Conference,* Los Angeles, CA, paper OTh4I.1, 2012.
59 S. Makovejs, C. C. Roberts, F. Palacios, H. B. Matthews, D. A. Lewis, D. T. Smith, P. G. Diehl, J. J. Johnson, J. D. Patterson, C. R. Towery, and S. Y. Ten, "Record-low (0.1460 dB/km) attenuation ultra-large Aeff optical fibre for submarine applications," In *Optical Fibre Communication Conference,* Los Angeles, CA, paper Th5A.3, 2015.

60 Y. Yamamoto, M. Hirano, K. Kuwahara, and T. Sasaki, "OSNR-enhancing pure-silica-core fibre with large effective area and low attenuation", *Optical Fibre Communication Conference,* San Diego, CA, paper OTuI2, 2010.
61 B. Zhu, "Large-area low loss fibres and advanced amplifiers for high capacity long haul optical network," In *European Conference on Optical Communication,* Valencia, Spain, paper We.2.4.1, 2015.
62 B. Zhu, X. Liu, S. Chandrasekhar, D. W. Peckham, and R. Lingle "Ultra-long-haul transmission of 1.2-Tb/s multicarrier no-guard-interval CO-OFDM super-channel using ultra-large-area fibre", IEEE Photonics Technology Letters, 22(11), 826–828, 2010.
63 MATLAB Simulink is a trade Mark of MATWORKS Inc.
64 S. Novak and R. Gieske 'Simulink model for EDFA dynamics applied to gain modulation', *IEEE Journal of Lightwave Technology*, 20(6), 986–992, June 2002.
65 L. N. Binh, "ECE 4405 lecture notes on optical transmitter and receiver for Gb/s broadband communication system and network", Monash University, 2004.
66 S. Novak and R. Gieske 'Simulink model for EDFA dynamics applied to gain modulation', *IEEE Journal of Lightwave Technology*, 20(6), 986–992, June 2002.
67 Y. Sum, J. L. Zyskind, and A. K. Srivastava, "Average inversion level, modelling, and physics of erbium-doped fibre amplifiers," *IEEE Journal of Selected Topics in Quantum Electronics*, 3(4), 717, August 1997.
68 K. Izuka, "Erbium-doped fibre amplifier", Advanced Labs for Special Topic in Photonics (ECE 1640H), University of Toronto April 28, 2003.
69 MATLAB SimuLink Help Files, refer SimuLink program.

Appendix 1

TABLE 1
Space-Time Duality and Photonic Devices

Governing Equation	Space	Time	Notes
	Space-Time Duality		
Wave propatation equation	$A_z = -\frac{j}{2k}(A_{xx} + A_{yy}) - j\gamma\|A_0^2\|$	$A_z = -\frac{j\beta_2}{2}(A_{tt}) - j\gamma\|A_0^2\|A$	Schrödinger/Helmholtz Spatial spreading and time dispersion
Prop variable	z	z	Same direction as defined in poynting vector
Profile variables	x,y	t	Cross x and y to give z directional components
Fourier variables	k_x, k_y	ω	Frequency optical and RF modulation
Dispersive phase (group velocity) – spectral phase	$\varphi(k_x,k_y) = \frac{(k_x^2+k_y^2)}{2k}z$	$\varphi(\omega) = -\frac{1}{2}\beta_2\omega^2 z$	Frequency-dependent, spatial vector momentum dependence
Lens phase function	$\varphi(x,y) = \frac{k(x^2+y^2)}{2f}z$	$\varphi(t) = -\frac{(\omega_0 t^2)}{2f_t}z$	Quadratic in space and time
Focal lens (time)	$f = \frac{R_1R_2}{(n-1)(R_2-R_1)}$	$f_t = \frac{2V_\pi\omega_0}{\pi V_{pp}\omega_m^2}$	Phase-dependent on modulation amplitude and V_π (applied voltage for π phase change) and modulation frequency
Gaussian profile	$E(x) = E_0 e^{-x^2/w_0^{22}}$	$E(x) = E_0 e^{-t^2/2T_0^2}$	
Gaussian imaging condition	$\frac{1}{d_0 + z_R^2/(d_0 - f)} + \frac{1}{d_j} = \frac{1}{f}$	$\frac{1}{\beta_{2i}d_0 + \frac{T_0^2\gamma^2}{\left(\beta_{2i}d_0 - f_t/\omega_0\right)}} + \frac{1}{\beta_{2i}d_i} = \frac{\omega_0}{f_t}$	Imaging condition
Magnification	$\frac{1}{\sqrt{\left(1-\frac{d_0}{f}\right)^2 + \left(\frac{z_R}{f}\right)^2}}$	$\frac{1}{\sqrt{\left(1-\frac{\beta_{20}d_0}{f_t/\omega_0}\right)^2 + \left(\frac{T_0^2}{f_t/\omega_0}\right)^2}}$	
Prism phase function	$\varphi(x) = \Delta k_x x$	$\varphi(t) = \Delta\omega t$	
Prism (spatial) freq shift	$\Delta k_x = 2k(n-1)\tan\alpha/2$	$\Delta\omega = \frac{\pi V_{pp}\omega_m}{2V_\pi}$	α = prism apex angle
Nonlinear and soliton condition	$L_{diff} = L_{NL}^{\alpha}; L_{NL} = \frac{1}{\gamma\|A_0\|^2}; L_{diff} = \frac{1}{2}k\omega_0^2;$	$L_D = L_{NL}; L_D = \frac{T_0^2}{\|\beta_2\|}$	Nonlinear in silica fibers used as transmission lines and for short connections to optical head in SPC

(*Continued*)

TABLE 1 (*Continued*)
Space-Time Duality and Photonic Devices

Governing Equation	Space-Time Duality: Space	Space-Time Duality: Time	Notes
System Estimation	**Free-Space**	**Time Guided-Wave**	**Comments**
Beam propagation numerical/ analytical	Helmholtz equation and split-step FFT method with phase disturbances at the intermediate step, followed by an IFFT of the beam spreading – over one step size, to ensure no overspreading of the beam the Rykov radius	Schroedinger nonlinear equation (NLSE) and split-step FFT method with nonlinear phase disturbances due to self-phase modulation (SPM), XPM, SRS, SBS, at the intermediate step and then IFFT of the beam spreading – over one step size to ensure no over-dispersion of the nonlinearity. Linear dispersion included in Schroedinger	• Effective using MATLAB Simulink. Helmholtz in free space and no nonlinearity effects, only linear phase effects and turbulence • Schroedinger applied when long distance propagation distance with accumulated nonlinearity, or high-power beam propagation in short-length fiber
Beam alignment	Feedback control of micro-disk and beam feedback – accuracy to 5 μ-rad	No alignment required but coupling and polarization control	Complicated space beam alignment feedback control system. Tunable beam temporal focusing can offer advantages in beam align in free space – integrated optical devices
Insertion losses (IL)	• Free-space devices – IL mainly due to reflection at bulk optic crystal interface and propagation material loss • Friis propagation loss only for purely single frequency lightwaves, different at finite source linewidth, use spatial spreading to estimate insertion and propagation loss between space optics	• Losses over coupling and scattering at core-cladding interfaces • Mode mismatching of guided optic components, but much small compared with free-space counterparts and ease of fabrication	
Power budget	• Modulation and ideal sensitivity of receiver, then prop loss of media, IL of optic heads and receiver sensitivity, optical amplification gain, optical head gain Then design booster amplifier and low noise optical pre-amplifier • Nonlinear cross talk penalty between channels	• Modulation and ideal sensitivity of receiver, then prop loss of media, IL of optical heads and receiver sensitivity, optical amplification gain, optical head gain. Then design booster amplifier and low-noise optical pre-amplifier • Including coupling loss, gain in linear and nonlinear saturation region • Nonlinear penalty	Use dB scale for power budget Amplifier noises estimated by (i) at saturation and noise figure (NF) (ii) at linear gain region (low noise amplifier). Gain and noises are nonlinear so must estimate for saturation region and linear low noise high gain region. Optical filtering with sharp roll-off important for long-distance SPC

(*Continued*)

TABLE 1 *(Continued)*
Space-Time Duality and Photonic Devices

Governing Equation	Space-Time Duality: Space	Time	Notes
Devices	**Space**	**Guided-Wave**	**Notes**
Modulation • **Structure** • **Bandwidth** • **Voltage level**	• Bulk modulator • large separation elect. • π phase change requires kV voltage for 10 mm electrode separation distance	• Integrated structures, micro- separation distance electrode large field interaction. • BW → 220 GHZ $LiNbO_3$ thin-film • 1–5 V for π phase change	No-carrier suppression to ensure coherent carrier and provides coherence of beam in space transmission and carrier phase dispersion
Modulator types	• Acousto-optic, electro-optic, magneto-optic in bulk and integrated guided-wave • Modulation by high-level voltage and low frequency	• Electro-optic mainly in guided wave structures as low driving voltage and broadband RF interaction • Integrated-possible with laser using InGaAsP of InP substrate • EML = external modulation laser – very small assembly and high-speed w/wo polarization multiplexing • Thin-film $LiNbO_3$ can offer BW 240 GHz → multiband RF sensors or modulation • PDM-QAM possible → 400, 800, and 1600 Gbps PDM-16QAM	• Acousto-, electro-, and magneto-optics but electro-optic → largest BW with low V_π. • Thin film $LiNbO_3$ waveguide is ridge-type optical waveguide with refractive index difference w.r.t cladding is large 2.27 to 1.484 (SiO_2) or 1.73 (Si_3N_4), thus the guided optical field is highly confined in the geometrical guide. Thus, shorter electrode with high frequency can be obtained.
Modulation formats	Mainly ASK but ok if integrated mod used • Mary-ASK, Mary-QAM, oOFDM, MSK, MAPM, oOFDM • QPSK (2 bits/symbol)	• *Mary-ASK, Mary-QAM, oOFDM, MSK, MAPM, oOFDM* • QPSK (2 bits/symbol) • PDM-QPSK	Integrated modulators used assumed Capacity increased by constellation points $\log_2(n)$, n = constellation points of modulation format
Polarization	• Pol multiplex	• Two ortho-polarization modes multiplexing	x^2 increase in capacity

(Continued)

TABLE 1 (*Continued*)
Space-Time Duality and Photonic Devices

Governing Equation	Space-Time Duality: Space	Space-Time Duality: Time	Notes
Laser	• Past gas laser but now integrated laser • Spectral windows: 1060 nm with 100 nm spectral window, 1500–1700 nm	• External cavity, tunable laser • Power ~1 to 10 dBm • Narrow linewidth laser (<10 kHz) for coherence transmitter and receiver • In-line EML (external modulation laser) • Spectral windows: 1060 nm with 100 nm spectral window, 1500–1700 nm most suitable and low loss window using integrated optic components	EML for TOSA
Coherence	Linewidth-wide; 10 kHz required for long-distance	• Linewidth ~10 kHz • ~1 to 10 Hz for fiber laser • Important to obtain coherent reception • Minimum relative phase-noise (RIN) < −140 dBm/Hz	Coherence to ensure minimum spatial dispersion and time group delay dispersion
Coupling	• Coupling GRIN SELFOC lenses to couple integrated structure to Bulk optic head • In reverse coupling, OH to fiber with SELFOC lens • Loss reflection and material mismatch Fresnel loss	• Coupling GRIN SELFOC lenses to couple integrated structure to integrated fiber/optics • In reverse coupling, OH to fiber with SELFOC lens • Minimum loss • Integrated interface with matching materials to minimize coupling loss	Fiber-integrated waveguide coupling and integrated to space coupling and launching into space or guided media
Reception	Intensity modulation and direct reception for low data rate, modern employed hybrid optical systems and coherent reception	Both direct and coherent receptions are used, coherent reception is preferred for high bit rate such as 25 Bps as higher sensitivity required	Integrated receiver optics (ICC)

(Continued)

TABLE 1 (*Continued*)
Space-Time Duality and Photonic Devices

Governing Equation	Space-Time Duality: Space	Space-Time Duality: Time	Notes
Processing	• Analog or digital • Adaptive algorithms in DSP if ICO receiver employed	• DSP in both RX and Tx • DSP-TX: pattern generation, pulse shaping and pre-distortion. DSP. • Carrier phase, dispersion, delay difference between polarized channels, skew equalization, distortion equalization, non-linear	Equalization, etc.
Integrated coherent optic Transceiver (ICOTRx)	Adaptable with guided wave ICOTRx, using optical head coupling in and out to optic head	• All integrated with lasers for Tx and LO mixing (coherent det). • Hybrid coupler for coherent mixing and phase-shifter to extract in-phase and quadrature components of QAM. • PD pairs in balanced connections to improve 3 dB sensitivity and coherence mixing to 20–30 dB gain in sensitivity.	ICOTRx offers compact assembly and bit rates to 800 Gbps PDM-16 QAM
Transceiver (OTRx)	• Mainly direct detection or coherent via free-space coupling and photodetection. Coherent ICO can be used with space-integrated structure coupling. • Laser for Tx and LO for coherent Rx	• Optical hybrid coupler (a PDS, PBC, Pi-Shifter), LO laser, PDP, TIA, AGC (Rx), ADC, DSP, FEC • Laser (Tx) narrow linewidth • Hard-FEC, Soft-FEC	Free-space to adapt integrated transceiver to make hybrid OTRx for space transmission
Propagation media **Dispersion** • **Phenomenon** • **Compensation**	Free-space vacuum, free-space turbulent • Spatial dispersion, similar to wireless fading effects over turbulent medium • Adaptive Optics plus fiber (few-mode DGD) compensator with feedback mechanism to minimize under or over compensation • Possible DSP-integrated coherent receiver	Fiber optic guided wave transmission lines • Group delay differential dispersion • Fiber compensator DCF with opposite dispersion factor → higher loss → OA required • Polarization split • DSP algorithms to compensate residual	Helmholtz's equation and Schroedinger's equation
OF(optical filter)	• Bulk OF, thin film coating, high insertion loss	• Fiber-based OF (OADM)	

(*Continued*)

TABLE 1 (*Continued*)
Space-Time Duality and Photonic Devices

	Space-Time Duality		
Governing Equation	**Space**	**Time**	**Notes**
WSS (Wavelength Selective Switch)	• Wavelength selective filtering, add/drop multiplexer	• Wavelength selective filtering, add/drop multiplexer • Mux/Demux sharp roll off	Low loss and sharp roll-off filtering and tunable in both central wavelength and bandwidth
ROADM **Router with add or DROP channels** **CDC-ROADM (colorless, directionless, and contentionless ROADM)**	• Not exit before but ok to adapt ROAM from integrated media • Integrated with WSS and OXC to generate CDC-ROADM	• Reconfigurable OADM with WSS and couplers	Three-port wavelength add/drop and multi-cascade add-drop: high loss But free-space WSS is multi-parallel operation
Coupler	Bulk coupling with PBS/PBC	Fiber or integrated optic-based types, $\pi/2$ phase shift at cross-coupling	Mode-field theory in guided wave, $\pi/2$ phase shift when complete coupling
Isolator	Bulk type with pol rotator and filtering	Bulk fiber-based type	Phase retarder and reflection blocking
Circulator	• Two PBS & PBC with anti-reflection coating and bulk isolator • Reflection from Fresnel effects, isolation 30–40 dB, three ports max due to IL	Fiber-based isolator, three ports-*N*-ports	
Optical amplifiers • **Structure types** • **Nonlinear effects**	• Mid-range and high-level gain optical amplifier, booster, and preamplifier (low noise input −65 dBm • NL FWM, SBS, SRS, SPM, XPS • High power fiber required for additional connections	• EDFA/EYDFA booster amplifier, md-range amplifier, LNA low-noise amplifiers • NL FWM, SBS, SRS, SPM, XPS • High power fiber required for additional connections	Both must employ fiber-based rare-earth doped optical amplifier. For space coupling in and out of amplifiers by SELFOC lenses Note the nonlinearity in space amplifiers when using non-amplified-fibers connect the amplifier output to the optical head, which launches the beam into free-space.

(*Continued*)

TABLE 1 (*Continued*)
Space-Time Duality and Photonic Devices

Governing Equation	Space-Time Duality: Space	Space-Time Duality: Time	Notes
	• Non-equal frequency spacings for DWDM channels to avoid nonlinearity • Raman distributed optical amplifiers possible and required for further gain equalization at the output of optical amplifier	• Non-equal frequency spacings for DWDM channels to avoid nonlinearity • Raman distributed optical amplifiers make it possible to extend long-distance single-span optically amplified transmission	

Notations:

- Coordinate systems: x = horizontal; y = vertical; propagation direction.
- *Waves:* A_{xx} amplitude of wave in x-direction; A_{yy} = amplitude in y direction, A_z = amplitude in the propagation direction z
- Propagation wave number in a medium of refractive index n: $\beta(\lambda)$ = wave number in free-space vacuum; $\beta_1(\lambda)$ = group delay (diff of β w.r.t wavelength; $\beta_2(\lambda)$ = diff of β_1, or second order diff of propagation wave number, i.e. dispersion as a function of group wavelength carrier.
- γ = nonlinear coefficients of transmission media
- k = Propagation vector in vacuum; k_x, k_y = prop vectors in x and y directions; refractive index of air = 1.00
- ω = angular frequency of lightwaves; ω = angular frequency of lightwaves in vacuum
- Δk_x, Δk_y = derivative of variation of the propagation vector n vacuum in spatial x-direction, then y direction.
- L_{NL} = nonlinear length
- L_{diff} = length of spatial dispersion/broadening by group of spectral components; L_D = dispersion length;
- f = focal length in spatial domain; f_t = focal length in time.
- R_1, R_2: radii of spatial propagation, e.g., Rykov' radius

Assumptions:

- Purely single frequency lightwaves – Friis equation not valid if finite linewidth of source or modulated lightwaves with low peak to average power modulation format.
- Finite linewidth laser so Friis equation be modified for spatial spreading to obtain spreading loss due to beam size increase larger than the optical head diameter and any Fraunhofer diffraction loss.
- Turbulence phenomenon similar to multipath fading loss in wireless fading (DGD) of channel → probability of error via probability density function.
- Multi-path fading similar to multi-path ray model of multi-mode fiber, so MMF with opposite group differential delay (DGD) can be used to compensate turbulent multi-path or refractive index of turbulence delay.
- Most commonly spectral window is S-, C-, C+, and L-band also O-band (1300 + −10 nm, 1060 + −100 nm
- Dese wavelength multiplexed (DWDM) channels co-propagation with frequency spacing follows ITU-T.

Appendix 2

TABLE OF ABBREVIATIONS AND TECHNICAL TERMS WITH MEANINGS

Abbreviations/Terms/Notation	Description	Remarks/Meaning/Properties
ADC	Analog-to-digital converter/conversion	Conversion of an analog signal, the received signal at the output of the TIA and AGC stage. Sampling or signal is done here and the digital sequence is fed into DSP for processing
ADD/DROP	Add or drop of channels	Optical devices can be configured as inserting devices of optical channels into a DWDM stream of dropping from such stream to a determined directional device. WSS can be configured to act as an ADD or DROP optical device with filtering characteristics. Such WSS can be integrated with optical amplifiers, especially the LNA (low-noise amplifiers) as an optical preamplifier to enable low-level optical channel input to the level of −65 dBm
ADPM	Amplitude and differential phase photonic modulation	Modulation technique employing both amplitude and phase whose Euclidean distances can be varied
ADS	Advanced design system	
AGC	Automatic gain control	Gain of an electronic amplifier can be controlled so that a higher amplitude input signal can be amplified without reaching saturation of the amplifier, thus the amplifier can operate at low- and high-level input to obtain minimum error and high dynamic range
AM-PSK	Amplitude modulation phase shift keying	equivalent to Duo-binary modulation with carrier phase in two levels (duo-phase)
Anritsu	A commercial company name	
AO	Adaptive optics	Typically, this technique is commonly used in space imaging, especially in astronomy to correct the spatial distortion of images, which are collected by far-distance optical systems. The a system normally consists of a wavefront sensor to measure the tapped signal from an optical splitter, on the phase distribution and hence the beam spreading, then a feedback mechanism is applied to the multi-spatial mirror to compensate for the phase distortion, hence equalization of the spatial image of a beam. In ultra-high-speed transmission, this is not appropriate as the pulse period is only about 40 ps and even much shorter if the operating speed goes up. We propose a DSP-based and optical compensating scheme to perform an adaptive DSP hybrid equalizer
APD	Avalanche photodetector	Photodetector is biased at breakdown voltage so that an electron generated by the photodetection can generate several electrons to obtain avalanche gain. Noises are also higher than normal p-i-n (p-type intrinsic n-type structure) detectors. Typical avalanche gain for wideband APD is ~10 with an applied bias voltage of 12 V. APD provides such a gain in photocurrent with higher current noise

(Continued)

(Continued)

Abbreviations/Terms/Notation	Description	Remarks/Meaning/Properties
ASE	Amplification stimulated emission	Noises are generated during the amplification process, commonly the noises during the spontaneous emission process, which is before the stimulation process. In coherent detection, the ASE beats with the LO to generate noises that are appreciable to signal quality
ASE	Amplification stimulated emission	Noises are generated during the amplification of photonic lightwaves
ASIC	Application-specific integrated circuit	Integrated electronic digital circuit for digital signal processing with a specific application
ASK	Amplitude shift keying	Modulation technique with binary or multiple level amplitudes
ATM	Asynchronous transfer mode	
Att.	Attenuation factor	Attenuation factor unit dB. Ratio in log-based ten of power quantities of output to input
Attenuation by coupling	Coupling loss or insertion loss	Coupling loss of mismatch spot sizes of waveguides and fiber
Attenuation by scattering	Loss by scattering of lightwaves	Scattering to rough surface scattering. Scattered lightwaves are depolarized
Attenuation by scintillation		The ratio in decibels between lightwaves instantaneous amplitude and its average value
B2B	Back to back	Back to back can be a test of system performance by direct connection of transmitter and receiver. For the photodetector pair back-to-back connection indicates the cathode of one PD is connected to the anode of the other photodetector so they operate in differential mode or common mode, that is push-pull operation
BACP	Binary amplitude circular phase	Modulation with two amplitude and phases distributed on circles. Mary-QAM can be organized for optimum Euclidean distance
Bayspec	A commercial company name	
BDPSK	Binary differential phase shift keying	
BER	Bit error rate	A performance measure quantity, the number of error bits over the bit sequence. If 1.0e-3 is required, then there needs to be 1-million-bit sequence to derive a BER of 1e-6 or >1.0 M memory length. This may be the limit of the real-time digital signal analyzer. The BER can also be related to the error-vector-magnitude (EVM) in a constellation, especially useful for QPSK and Mary-QAM
BFR90	Transistor model - brand name	High frequency bipolar transistor with a transition frequency of 5 GHz. Modern transistors can offer 247 and 522 GHz if in SiGe materials
Bit rate 100 G, 200 G	100 or 200 Gbps	Aggregate bits/sec of the modulation scheme. For example, a baud rate of 25 GB and modulation of QPSK, 2 bits/symbol (Baud) plus two orthogonal lightwave channels results in a 100 Gbps aggregate bit rate

(Continued)

(*Continued*)

Abbreviations/Terms/Notation	Description	Remarks/Meaning/Properties
Bit rate and modulation 100 G-DP-QPSK	The 100 Gbps by QPSK modulation per polarization, two orthogonal polarizations	Two polarized lightwave channels, each modulated by QPSK (2 bits/symbol) and a baud rate of 25 GB, thus 25 × 2 × 2 (pol.) gives 100 Gbps
Bit rate and modulation 200 G-DP-QPSK	50 GB × 2 pol. × 2 bits/symbol	GB is Giga baud, two polarizations with 2 bits/symbol by QPSK format to result in 200 Gbps for 50 GBaud. If FEC is used, additional bits are required, for example, 224 Gbps.
BJT	Bipolar junction transistor	
BPF	Band pass filter	In electrical domain with a passband and central frequency with 3 dB passband and cut-off band
BPM	Beam propagation method	Method by propagating optical beam via one-half step by FFT, the operation of linear dispersion or nonlinear effects, and the inverse FFT over the next step to obtain the time domain lightwave signal. Then repeat such steps over the entire transmission length
B_w/B_w	Bandwidth	
C	Common port of WSS	Port of WSS usable as a multiplexed port as input to demultiplex by WSS or as output port of multiplexed channels of a WSS
C	Capacity	The total bit rate of a transmission system
CAPEX	Capital expenditure	
CAZAC	Constant amplitude zero autocorrelation	CAZAC sequence is a sequence that has all components of constant amplitude, but their phases are different and consecutively reduced or increased, i.e., chirp down or chirp up, respectively. Its cross-orrelation is unique and can identify quite several features of the channel. CAZAC is added to the payload data and the FEC in periodic as a training-aided sequence. The cross-correlation products can be used to set the coefficients of the equalization filter, e.g., FDE, the forward decision equalizer
C-band	Central spectral band	Central operational band of optical fiber or space photonic communications systems. Spectral window 1545–1565 nm. Also, C+-band 1545–1570 nm. For 100 GHz spacing, the number of channels that can be multiplexed is about 180 for C- and 110 for C+
CCFE	Coarse carrier frequency estimation	Estimation algorithm for carrier frequency in coarse approximation and thence recovery
CCFE-R	Coarse carrier forward estimation-recovery	
CD	Color dispersion	Dispersion of temporal pulses by different spectral lines of modulated lightwaves, that is lines of double sideband of signals in the frequency domain
CDC-F ROADM	Colorless, directionless, contentionless, and flexible grid ROADM	Advance WSS which integrates additional MEMS (3D switches) so that colorless, directionless, contentionless, and flexible grid (CDC-F) ROADM can be structured. WSS can be configured to operate as a multiplexer or demultiplexer (with filtering and reconfigurable), and switches act as spatial routing devices and tunable optical filtering. Normally, the input port is assigned as a common port, which is either a multiplexed channels port for demultiplexing or a combined port for multiplexing

(*Continued*)

(Continued)

Abbreviations/Terms/Notation	Description	Remarks/Meaning/Properties
CDC-F ROADM	Colorless, directionless, contentionless, and flexible grid ROADM	ROADM with independence on wavelength of channels to be routed (colorless), direction is bidirectional, no contention of channels of the same wavelength. Channel carrier frequencies can be flexibly allocated with spacing from 2.5 to 100 GHz and 200 GHz.
CDC-F WDM muxes and demuxers	CDC-F WDM multiplexers and demultiplexers	Colorless, directionless, and contentionless flexible wavelength division multiplexers and demultiplexers
CDC-F WSS - ROADM	Contentionless, direction-less, and colorless flexible WSS -ROADM	ROADM = reconfigurable optical add/drop multiplexer, which is CDC-F, i.e., independent of color or wavelength, all direction and no contention with flexibility
CFP2-DCO	Compact form package-DCO (digital coherent optics)	Integrated optical circuits and electronic systems for coherent transmitter and receiver of optical channels using coherent reception. Only two components that is only one optical channel, not two polarization multiplexed channels. This is a transceiver module
CFP2-DCO AC1200	Module number	Module as single channel optical transceiver
CH	Channel	Optical channel
CH#	Channel number	Optical channel with a number that matches that defined by ITU-T grids
CHAIS	Charged mode Interleaved Sampler	Digital sampling unit operating by charge mode
Charbonneau relation	Rain attenuation (dB/km)	To obtain the attenuation of lightwaves under rainy conditions
CMA	Constant modulus algorithm	Algorithm for clock recovery and equalization
CMOS	Complementary metal–oxide–semiconductor	
Coh	Coherent	Coherent reception requires a local oscillation laser to co-propagate with signals, then beating in the photodetector to generate a mixing product that preserves the phase difference and the amplitude product of the two lightwaves, resulting in significant gain in amplitude. Homodyne coherence is produced when the frequencies of both lightwaves are the same. DSP assistant phase matching is required
Coh-Bal-ORx	Coherent-balance optical receiver	See above. Balanced Photodetection (PD) by back-to-back connection to produce 3 dB gain compared to the case of only one PD
Coh-ORx	Coherent optical receiver	Mixing of signals and local oscillator laser in a single photodetector, thus, both amplitude and phase of the electronic signals are generated. Shot noises are considered quantum shot noises due to residual high-power LO laser
Constellation	A constellation of satellites that are mounted with optical transceivers and/or optical routers	Constellation of satellites mounted with transceivers is used as optical transmission and receiving nodes, as well as routing in the optical domain, as an optical node

(Continued)

(Continued)

Abbreviations/Terms/Notation	Description	Remarks/Meaning/Properties
CO-OFDM	Coherent optical orthogonal frequency-division multiplexing	
Coupler 3 dB-FC	Fiber coupler with 50% splitting ratio	Fifty percent splitting ratio indicates 3 dB down from input power. In practice, there is coupling loss and a typical value of about 4.2 dB
CPA	Coarse pointing assembly	Laser beam pointing mechanical platform with initial coarse alignment with satellite transceiver platform
CPE	Carrier phase equalizer	Equalization algorithm to recover that carrier phase if temporal distortion or spatial distortion occurs
CPE	Carrier phase estimation	The estimation of the carrier phase by an algorithm so that the frequency difference of the carrier frequency and that of the local oscillator laser can be identified and matched to ensure true homodyne coherent detection
CPFSK	Continuous phase frequency shift keying	Modulation is such that the carrier phase is continuous at the bit boundary. MSK is a technique. With FSK, the carrier phase is changed abruptly at the bit boundary. This continuity makes the passband of the signal like a brick wall
CPR	Carrier phase recovery	DSP algorithm to recover the phase of the carrier, which can then be synchronized with the local oscillator laser frequency to realize homodyne detection
CR	Carrier recovery	Carrier frequency can be recovered using the DSP algorithm
CS-RZ	Carrier suppressed return to zero	Modulation technique in which the carrier is suppressed and the pulse returns to zero within the bit period
CSRZ or CS-RZ	Carrier suppressed return to zero modulation	Modulation such that the lightwave carrier is suppressed and the optical pulse returns to zero within a bit. This is done by biasing the MZIM at the minimum transmission curve point, hence phase difference between the two paths of the interferometer is π, resulting in carrier suppression. The electrical pulse swinging on two sides of this minimum biasing point generates the return to zero pulse sequence
CSRZ_ASK	Carrier suppressed ASK	Modulation format with lightwave carrier suppression and amplitude shift keying, binary, or multilevel
Current noise, $\langle i_{NS}^2(t)\rangle$	Mean square noise current (power) produced by the total optical intensity on the photodetector	Noise spectral density and thence the noise equivalent current power can be derived. Power indicates that the noise current is always positive and bidirectional and always adds to any direction of the current
Current, noise photodetector $\langle i_s^2(t)\rangle$	Mean square current produced by the photodetector by $s(t)$	$\langle i_s^2(t)\rangle$ expected noise current amplitude of photodetector
CV-QKD	Continuous variable-QKD	Quantum key distribution technique with the phase or amplitudes of samples of the whole Baud or bit period and processed by statistical technique. Unlike discrete variable (DV), only one sample at a time is processed. All samples are in quantum states. This is the case of power under Shanon channel capacity.

(Continued)

(Continued)

Abbreviations/Terms/Notation	Description	Remarks/Meaning/Properties
CW	Continuous wave	Laser diode is driven at constant current (bias current), and lightwaves are generated continuously. This is to keep the linewidth of the laser unchanged and stable. To modulate the lightwaves, an external optical modulator is used and signals are fed into the RF port of the modulator
DAC	Digital-to-analog conversion or converter	Similar to ADC but in reverse, converting a digital data sequence to analog form. This is normally implemented after the DSP in which a parallel digital word can be converted to an analog so as to modulate the optical modulator, which is normally outside the laser to preserve the laser linewidth. This is essential for coherent reception and transmitting optical systems; in space, this keeps the beam divergence at a minimum
DAC and ADC	Digital-to-analog converter/conversion	Reverse of ADC is commonly employed to provide analog signals to feed to electrodes of the optical modulator to modulate the lightwaves to generated optically modulated waves. Photonic is used to indicate the coherence of lightwaves
DBM	Duo binary modulation or Duo binary modulation	Note on whether dual-binary in terms of amplitude or in terms of phase, that is dual-phase level of the sub-bits of a baud. If it is in amplitude, then it is equivalent to filtering half the bit rate. If it is phase coding, then the differential phase between the two-level periods can be distinguished
DBR	Distributed BRAGG reflector	
DCF	Dispersion compensating fiber	Dispersion factor is an opposite sign of that of SMF; so, a short length of DCF can be in cascade with SMF to give an equalized transmission fiber span.
DCI	Inter data center networking	
DCN	Data center networking/networks	
DCO	Digital coherent optical module	Optical module for coherent transmission and receptions of polarized channels with integrated digital signal processing units to process signals in the digital domain. At the transmitter, DSP can generate a random pulse sequence or pre-distortion sequence to mitigate any dispersion of the transmission medium. At the receiving section, sampling of analog signals obtained from the optoelectronic detection and processing with specific algorithms to recover carrier phase and thus obtain true homodyne detection, along with further mitigation algorithms of several distortion effects of the transmission media
DD	Direct detection	Detection of Lightwave, which is directly impinged on the absorption surface of the detector, no mixing with any local oscillator laser
DD/PDD	Decision directed/phase decision detection	
Demux	Demultiplexer	Split and filter different frequency lightwaves into separate outputs

(Continued)

(Continued)

Abbreviations/Terms/Notation	Description	Remarks/Meaning/Properties
Demux/Demuxer	Demultiplexing/demultiplexer	In optical or electrical domain, WSSes can be used as either mux or demux with a common port as multiplexed input or output port, respectively
Depolarization	Depolarizing the polarised lights of lightwaves	Optical scattering has depolarized light rays, that is, all rays can have different polarization directions. The depolarization effect creates the problem of unequal amplitudes of the channels in the detection of two orthogonal polarized channels in an integrated coherent optic receiver (ICO-Or). A 2 × 2 MIMO (multiple input-multiple output) algorithm in the DSP can correct this effect
DFB	Distributed feedback	Laser structure using distributed feedback gratings placed at both sides of a gain medium to create a resonant cavity with sharp spectral filtering by the feedback grating which is in a planar structure
DI	Delay interferometer	Interferometers (Mach-Zehnder type) with one path are about one-bit delay longer than the other optical path
DM	Direct modulation	Modulation of lightwaves by alternating the current drive to a laser diode
DMT	Discrete multi-tone	Tones of different frequencies were obtained from the real part samples of FFT. These tones are integrated and converted to analog domain to drive optical modulators and transmission, detection by the non-coherent receiver
DNL	Differential non-linearity	
Doppler shift	Frequency shift by Doppler effect	When two satellites travel at different frequencies, a frequency shift of one oscillator on one satellite with respect to the other is generated. This is the Doppler frequency shift. Such a shift in the receiver can create heterodyne detection rather than homodyne detection. Therefore, the receiver must have passband characteristics to cover this frequency shift. The Doppler shift between GEO and LEO satellites can reach as high as 7.5 GHZ
DP	Dual polarizations	Typically orthogonal dual polarizations of lightwaves
DP04QSFP	Dual polarization with 4- lanes in QSFP	QSFP = quad special form package. It is a coherent reception and transmitter compact optical integrated circuit with 4-input of 2 × polarized channel signals in pairs of in-phase and quadrature components for QPSK modulation. Thus, two polarized channels modulate QPSK at output ports
DP-16 QAM.	Dual polarized lightwave channels modulation format 16 QAM	Two orthogonal optical channels under modulation of 16 QAM, that is 4 bits/baud or 4 bits/symbol, constellation of 16 points on square coordinate system
DPM/PDM	Dual pPolarization multiplexing/ polarization division multiplexing	Two polarized lightwave beams are simultaneously combined or multiplexed to transmit to space channel or received by the receiving optics and split to be detected
DP-QAM	Dual polarization QAM	QAM modulation format of two polarized lightwaves
DRA	Distributed Raman amplification	Optical amplification via Raman scattering effects that are distributed along a guided wave structure

(Continued)

(Continued)

Abbreviations/Terms/Notation	Description	Remarks/Meaning/Properties
DSA	Digital sampling analzser	Oscilloscope using digital sampling and analysis by software processing of sampled data
DSB	Double side band	
DSB-DMT	Double sideband discrete multi-tone	Modulation technique using multi-frequency components and both bands of the signal
DSP	Digital signal processing/processor	Critical subsystem to perform the functions: (i) coding, and modulation format generation to the DAC and then amplification by RF power amplifiers (4× for IQ and polarized channels); (ii) processing and perform decoding, equalization, and recovery of data channels; (iii) providing the monitoring of BER for remote assessment of channel system performance. All operation in digital electronic samples
DSP-ORx	Digital signal processing-based optical receiver	Combining the digital processing and optical received signals by coherent mixing of optical channels and LO laser, small signal amplification of signals from the output of OPreA
DuoB or DB	Duo-binary modulation	Modulation with dual amplitude and/or phases of the carrier
DV FE	Differential (V_{diff}) Viterbi frequency estimation	A digital algorithm
DV-QKD	Discrete variable QKD	Quantum key distribution using discrete samples
DWDM	Dense wavelength division multiplexing	Used for multiplexed multiple channels whose spacing is 50/100 GHz or less
DWDM	Dense wavelength division multiplexing	Dense means the spacing between channels is within 50 or 100 GHz. The 200 GHz can possibly be used by interleaving with other Space NODE channels so 100 GHz spacing can be expected for DWDM
DWDM, WDM	Dense wavelength division multiplexing or wavelength division multiplexing	Dense means the spacing between channels is within 50 GHz. In PEGASUS, 200 GHz and possibly interleaved with other Space NODE so 100 GHz spacing is expected for DWDM. Wider channel spacing, e.g., 200 GHz can be considered WDM
EA	Electro-absorption	Electro-absorption is typically a semiconductor whose absorption coefficient can vary with the intensity level of lightwaves. Thus, the output lightwaves can be of low or high level. EA waveguides can be used as an optical modulator, a photodetector, or optical demultiplexer
EAM	Electro-absorption modulator	
ECL	External cavity laser	A gain medium is placed in the middle whose reflector can be external to the gain cavity medium. This reflector can be tuned to form different cavity lengths and different reflection angles so that the resonant cavity can be changed, thus the wavelength
EDF	Erbium-doped fiber	Fiber doped with rare earth metal impurity to provide an optical gain medium for laser of optical amplifiers

(Continued)

(*Continued*)

Abbreviations/Terms/Notation	Description	Remarks/Meaning/Properties
EDFA	Erbium-doped fiber amplifier	Er:doped fiber amplifier with a high concentration of doped Erbium ions and pumped by laser @980 or 1544 nm via a WDM fiber coupler, in co-direction or contra-direction with respect to the signal input. Gain about 25 dB and ASE noise figure (NF) about 5–6 dB over the whole C-band. Saturated level ~10–16 dBm for single pump. In single pump stage or multiple pump stage amplifiers. Co-doped Er and Ytterbium give higher gain and different spectral windows. Multi-stage optical amplification stages can be used to obtain higher gain and power
Effective refractive index	n_{eff}	Refractive index as seen by guided lightwaves along its propagation direction. It can be derived from the eigenvalue of the propagation constant
E_L	Amplitude of local oscillator optical field	
EM	Engineering model	
EML/DML	External modulation laser	Integrated laser consisting of a DFB laser and inline EA modulator in a semiconductor platform
Enhanced SMF	SMF with attenuation peak suppressed	SMF exhibits a high attenuation loss at about 1300 nm due to water absorption. This inhibits the use of channels operating in this region. Such absorption peak can be suppressed so channels whose carrier wavelength of 1280–1545 nm can be used in coarse spacing, thus increasing the capacity. Corning type SMF-28e
ENOB	Effective number of bits	Quantization errors and noises in the electronic system make the number of bits of quantization become less than the actual design. The distinct levels of such a digital system result in the effective number of bits
E_S	Amplitude of signal optical field at the receiver	
ESA	Electrical spectrum analyzer	Spectrum analysis equipment of electrical RF signals, typical to 63 GHZ commercial units
ESI	Equivalent relative index	Refractive index is equivalent to a non-step index profile of fiber. So, the estimation of the number of guided modes can be estimated using such step equivalent profile as predicted by theory
EVM	Error vector magnitude	An average measurement of sampled points offset from the original error-free magnitude of a constellation point or points of a complex modulated system at the receiver. See also WFE and WFES for determining spatial distortion in spatially distributed plane
EYDFA	Erbium-Ytterbium-doped fiber amplifier	Co-doped Erbium and Ytterbium ions in fiber in multistage to obtain higher power and possibly shift the gain spectrum in the 1557 nm band with low noise, hence low input acceptance
EYDFL	Er-Yb co-doped fiber amplifier	Generation and amplifying in the 1557 nm band, fiber laser using multi-pump can offer high power output
FA	Fixed attenuator	

(*Continued*)

(*Continued*)

Abbreviations/Terms/Notation	Description	Remarks/Meaning/Properties
FASCOD	Fast atmospheric signature code	A radiative transfer code can be employed to simulate the atmospheric transmission and path radiance with customized gas composition (such as CO_2, water vapor, CO, etc.) and concentration to understand the phenomena in a specific wavelength region. Normally, FASCOD is used in combination with HITRAN (High-Resolution Transmission)
FB	Fabry-Perot	Resonant cavity by Fabry and Perot
FBCFE-R	Feedback carrier frequency estimation and recovery	Algorithm with feedback or backpropagation to mitigate unwanted distortion effects
FBG	Fiber Bragg grating	Bragg grating structure for optical filtering and phase shift, which is the phase characteristics of optical filters. Bragg grating can operate in reflection of transmission mode
FC	Fiber coupler	Optical coupler or splitter by fusion of two fibers. The core separation distance determines the coupling ratio, which is the tunneling of one guided mode to the other core region.
fc	Optical frequency	Frequency of lightwave carrier
FDE	Feedback decision equalizer	Algorithm for filtering in the frequency domain and frequency difference between spectral components
FDEQ	Feedback decision equalizer	Digital filter technique in frequency/time domain with feedback to equalize distorted pulse sequence frequency
FDR	Feedback decision recovery	Algorithm for determination decision of frequency of carrier recovery with feedback path
FEC	Forward error coding/coder	Coding with forward error so that at the decoding BER can be as low as 1e-3 is possible for recovery of signal sequences. 15% FEC is defined as open FEC. That is, if 50 GB, then 15% oFEC (Open FEC) required $50 + 50 \times 0.15 = 57.5$ GBaud. Depending on the channel and transmission complexity of the medium 7%, 15%, or 20% additional bits and rates can be required for coding. oFEC= open FEC
FET	Field effect transistor	
FF	Feed forward	
FFCPE	Feed-forward carrier phase estimation	Algorithm to recover the phase of carriers, hence, sequence clock timing
FF-CPE-R	Feed-forward carrier phase estimation and recovery	Algorithm to recover phase of carriers, hence sequence clock timing plus complex estimation technique when under noisy conditions
FFCPR	Feed-forward carrier phase recovery	Algorithm with samples feed-forward for the recovery of carrier phase
FFE	Feed-forward equalization/equalizer	Algorithm for equalization using feed-forward technique
FFE	Feed-forward estimation	Similar to FFE equalization
FFT, IFFT	Fast Fourier transform and inverse FFT	Fast or discrete Fourier transform Accepting serial or parallel binary digital data and transforming into either frequency or time domain signals. Two orthogonal types of functions of the frequency or time signals are Sine and Cosine

(*Continued*)

(*Continued*)

Abbreviations/Terms/Notation	Description	Remarks/Meaning/Properties
FFT/IFFT	Fast Fourier transform/inverse FFT	Fourier transform using sampling points in the discrete domain
Fiber cladding	Outer layer of a circular core/cladding silica glass	Cladding layer of silica fiber, commonly made by collapsing pure silica tube, along with a doped impurity such as Ge_2O as core layer. Radius about 125 µm
Fiber core	Core section of a circular waveguide and continuously cladded by a thicker cladding layer.	Core of circular waveguide (fiber) whose refractive index is higher than that of the cladding layer. For single mode, the difference is ~0.3%, which is very small, thus the guided mode is weakly guided through the fiber
Fiber guided mode	The number of modes supported by a circular waveguide	The number of mode is determined by the core radius, the difference in refractive index of the cladding and core, and the operating wavelength. V-parameter determines the number of mode. If $V < 2.4$, it is single mode but with two polarizations. The number of modes can be approximated by $V^2/2$
Fiber pigtail	A fiber short length connected to an optical planar waveguide	A short length fiber connected to a guided planar or channel optical waveguide, which is commonly known as integrated optical device such as a semiconductor laser, to maximize the coupling efficiency to and from such optical device
Fibre coupler	Fiber device spitting to many fiber outputs or coupling from many inputs to one or more outputs	Coupler couples or splits lightwaves of several input fibers or output fibers, respectively. Commonly made by twisting and fusing the fiber cores so that the optical fields can tunnel from one to the other. The distance between the cores dictates the coupling coefficients.
Fifth generation	5G	Optical network integrating 5th generation band 2–6 GHz RF carrier
FIR	Finite impulse response	Digital filter with a finite response by the sampled points
Fitel	Commercial name of a company	Company manufacturing of special fibers
FM	Flight model	Model of flight or satellite in rotation.
FO	Frequency offset	Offset in spectral or frequency domain from the lightwave carrier frequency
FR	Frequency recovery	Recovery of a carrier frequency for receiving pulse sequence after coherent reception
Friis equation	Equation on space propagation loss	Equation relates to the received intensity and transmitting intensity of radar signals applied to the optical frequency region. This equation assumes an ideal light source, which is only a single frequency laser or zero linewidth that is not real in practice. For semiconductor, laser linewidth of about 10 kHz is typical in modern type. The typical single-longitudinal laser linewidth is 100 kHz. Originally, the idea behind this transmission formula was to dispense with the usage of directivity or gain when describing antenna performance. In their place is the descriptor of antenna capture area as one of two important parts of the transmission formula that characterizes the behavior of a free-space radio circuit; thus, ideal linewidth of the radiation source was assumed. In practice, laser source can never be purely single mode but with finite bandwidth (linewidth in wavelength domain)
FSK	Frequency shift keying	Modulation technique with a frequency of carrier shift between the bit or baud period
FSM	Finite state machine	

(*Continued*)

(Continued)

Abbreviations/Terms/Notation	Description	Remarks/Meaning/Properties
FSO	Free space optics	Non-guided wave optics
FSOSS/FSOCS	Free-space optical communication systems	Communication technique using laser and modulation scheme for long-distance transmission through air vacuum space or turbulent free space media. Alternative term SOCS or SPCS
FTTX	Fiber to the X, X = building, X = transmitter or optical head	X can be home, connections within a satellite transponder or transceiver network, building, OGS, etc.
FWHM	Full width half mark	In power or intensity it is 3 dB points of a passband in log scale, in magnitude it is 0.707 points on both sides of the pulse, if linear intensity then it is at 50% points
FWM	Four-wave mixing	The process of mixing three waves to generate the fourth wave via the nonlinear SPM effects in fiber, single mode type under high-intensity pumping. The mixing of three lightwaves generates the fourth components if the phase matching is satisfied. Alternatively, wave mixing can generate such forward and backward waves. This happens when the nonlinear effect exists in such an interaction medium
GB	Giga baud	1e9 Baud
GBaud/s	Giga-Baud per second	Baud or sSymbol to represent the bits in multi-level modulation. Also equivalent to Gsymbols/sec or Gsy/s
GBaud/s or GSy/s	Giga-Baud per second	Baud or symbol per second to represent the bits in multi-level modulation
GbE	Giga-bit ethernet	
Gbps or Grits/s	Gigabits per second	Line rate in bits/seconds, estimated as modulation number of bits/symbols, the number of polarized channels
Gbps or Grits/s	Giga-bits per second	Line rate in bits/seconds, estimated as modulation number of bits/symbols, the number of polarized channels
GEO	Geosynchronous equatorial orbit	These satellites, sometimes called geostationary satellites, are the most widely recognized satellites. Because they move at the same velocity as the Earth and on a path parallel to the Earth's rotation, they appear to be stationary in the sky. The speed of rotation at the Earth equator is about 1674.4 km/h (1040.4 mph), and it decreases as you get nearer either of the poles. GEO orbit height is about 36,000 km, typically and rotation time is 24 hour one day exactly that of the earth.
GEO	Geostationary earth orbit	Satellite orbit with rotation at the same speed as the earth so it appear stationary
GRIN lenses	Graded index lenses	Lenses whose refractive index is graded. Its structure is a solid cylinder with two flat faces. The graded refractive index acts as an equivalent convex or concave lenses. The length of the cylinder determines the focus spot to couple to other optic devices
GSa/s	Giga-Samples per second	Applicable by ADC, DAC, and DSP system
GSE	Ground support equipment	

(Continued)

(*Continued*)

Abbreviations/Terms/Notation	Description	Remarks/Meaning/Properties
GUI	Graphic unit interface	Interface unit in computer to drive electronic devices for parameters monitoring or setting parameters to drive optical amplifiers, modulators, transceivers, etc.
GVD	Group velocity delay	Group of frequency components such as the band of signals on both sides of the lightwave carrier, they travel or propagate at different velocities due to the index of refractive index varying with respect to wavelength
Hard-FEC	Hardware-based FEC	FEC implemented using hardware is normally less effective than soft FEC
HI amplifier	High-impedance amplifier	Electronic amplifier whose input impedance (as referred to input port) is high as no feedback impedance is used. Bandwidth is low but high impedance allows a higher amplitude voltage signal at the output port
HPFC	**High power fiber coupler**	Fiber coupler using high power resistance fiber. SSMF can operate with a power of 20 W in a core of SSMF, that is 4.1 μm radius or effective area of 50 μm^2
HPF	**High power fiber coupler**	Optical coupler which is resistant to high power
HPFC	High power fiber coupler	Fused optical fibers whose resistance to temperature and optical power is sufficiently high enough to withstand the high power lightwaves. So coupling high power guided waves to other output ports
H-V	Horizontal vertical	Polarization directions
Hybrid optical coupler		An optical integrated circuit placed at the input of an integrated coherent optical receiver or transceiver. It is composed of two optical inputs, one for the dual-polarized optical channel and the other for the local oscillator laser. Two polarizations are used with a combiner and splitter at the input and output respectively. Two optical phase shifters are integrated and tuned to $\pi/2$ so that the I- and Q-components of each polarized channel can be detected. All these functions are implemented in a planar-integrated optic waveguide structure
I DAC	Digital to analog converter for in-phase component	DAC with input digital parallel inputs to analog signals and amplified to drive external optical modulators such as MZM; similar for Q DAC for quadrature components
IC	Integrated circuit	Electronic integrated circuit
ICO	Integrated coherent optics	Optics integrated into an optical structure for photonic coherent transmitter and receiver
ICPTRC	Integrated coherent photonic transceiver	Coherent receiver for coherent reception in integrated optic and integration with photodetector pairs and hybrid optic coupler, integrated DSP units for transmitter and receiver sections
ICPTRx	Integrated coherent photonic transceiver	ICPTRx consists of an optical transmitter section and a receiver section with two integrated DSP systems, one for the transmitting section and one for reception signal processing. In addition, there is a hybrid optical coupler in planar-integrated optics to provide a polarization splitter and combiner as well as optical phase shifters for in-phase and quadrature-phase optical signal paths. Thence, optical polarized channels under QPSK modulation can be generated, as well as the reception of QPSK polarized channels with mixing with local oscillator lasers. Two lasers are integrated, one for lightwave carrier and one acts as a local oscillator for coherent mixing with optical polarized channels

(*Continued*)

(Continued)

Abbreviations/Terms/Notation	Description	Remarks/Meaning/Properties
ICR	Integrated coherent receiver	The integrated receiver of an optical hybrid coupler, PD pair in balanced mode, TIA. Two optical ports: one for LO and one for dual-polarized optical signal channels. Four output electrical signals of in -phase and quadrature components for each polarization. These ports are connected to the ADDC section of the DSP system
IEE	Institute of electrical engineers	Institute of electrical engineers of the United Kingdom.
IF	Intermediate frequency	Used in heterodyne reception. IF is the frequency of down conversion from the frequency-shifted signal to a lower intermediate frequency for ease of detection when limited bandwidth devices are available
IL	Insertion loss, dimensionless unit of dB	In a unit of dB, it is estimated by the ratio of the output with respect to the input power, then converted to logarithm scale. If using a logarithmic scale, add or subtract
IM	Intensity modulation	Modulation by switching lightwaves intensity to "1" or "0"states. Direct modulation by current switching applied to laser diode biased at linear point. Alternatively, by using an external MZIM coupled to an LD operating in CW mode, this external modulation ensures that the LD is stable and its linewidth remains unchanged with TEC (Thermoelectric controller) temperature control
IM/DD	Intensity modulation and direct detection	Direct modulation of the laser current or external modulators in one dimension and no phase information is used. DD = direct detection is one-dimension reception using one photodetector or a balanced photodetector, no phase information of the signal can be recovered. This is a non-coherent transmission and detection system with the intensity of lightwaves being changed in binary or multiple amplitude levels and detection by the variation of photo-diode current and amplifier.
IMEC	Interuniversity microelectronics centre	Interuniversity Microelectronics Centre, funded by EU at its initial stage. Specialized in SOI integrated Silicon photonics
INL	Integral non-linearity	Integral non-linearity (INL) is a measure of how closely the ADC output matches its ideal response.
IOC	Integrated coherent optics	Optics for coherent transmitter and receiver of ultra-high-capacity systems
IoT	Internet of Things	
IP TV	Internet protocol television	
IPTRx	Integrated photonic transceiver	Optical receivers implemented by integrated photonics and eElectronic DSP
IQ	In-phase, quadrature components	Complex plane components for modulation point representation
IQ	In-phase quadrature phase	Optical modulator that can produce in-phase and quadrature optical sequences
IQ	Inphase and quadrature phase	Components in orthogonality of a complex modulation
IQM	IQ modulator	Optical modulator fed by the *I*- and *Q*-component signals to provide the real optical waves and quadrature of $\pi/2$ phase shift lightwaves so that they can propagate or be transmitted over guided or free space medium.

(Continued)

(*Continued*)

Abbreviations/Terms/Notation	Description	Remarks/Meaning/Properties
IQM	I/Q modulator	Modulator with parallel MZMs in real and imaginary directions
IRT	Index of refractive of turbulence	Turbulence generated multipath of lightwaves seen different index due to turbulence of air layer
IRT	Index-of-refraction turbulence	Parameter that quantifies atmospheric turbulence strength, including the randomness and amplitude of changes in the refractive index of the air medium under turbulence, scattering, and multipath propagation, and thus fading effect on signal distortion
ISGL	Inter-satellite ground link	
ISI	Inter-symbol interference	Interference of one bit to the others (adjacent pulses) and hence error between symbols of a data sequence
ISL	Inter-satellite link	Link between satellites in a constellation or cross-constellation of satellites
ITU-T	International telecom union – telecommunication	International union which determines the standard system and parameters such as the spectral grids of optical channels, the spacing between channels so that the bandwidth of a modulated lightwave channel can be defined.
I_x, I_y	In-phase component of X or Y-polarization channel	In-phase components of the polarization X (horizontal) and Y- (vertical)
Keysight	A commercial name of company	Specialized in RF and optical analysis equipment
Kruse relation	Attenuation relation	Attenuation relation with respect to visibility of the air medium, coefficient dependent on the size of the particles, and the operating wavelength
lamda	Wavelength	Wavelength in unit of Nano-meter
Laser, fiber	Fiber laser	Gain medium plus two reflectors, all in fiber form. The gain region is a rare-earth-doped fiber of about 5%–10% and length of 1–2 m. Fiber Bragg gratings (FBG) are formed by a photorefractive effect with UV-eight shining through the fiber core via a phase phase. Linewidth can reach 1–100 Hz for fiber laser and gain about 30 dB. High-power amplifiers also follow this kind of structure, but the back reflector is not 100%. Multiple-stage fiber amplification stages can be cascaded to generate much higher power to about 2–20 dBm output.
Laser, semiconductor	Semiconductor laser in integrated form	Consisting of a gain medium in integrated form with compound semiconductor (InGaAsP for 1550 nm region) in which stimulated emission occurs, along with two reflection mirrors, 30% for front and 100% for the back. Grating feedback structures are commonly used for front mirror, thus reflection and filtering to ensure single longitudinal mode of the resonant cavity. Typical linewidth ~ 1–2 MHz. Higher cavity quality can be created to offer 10–100 kHz laser linewidth for coherent photonic communications
L-band	Long spectral band	The 1568–1682 nm spectral region for fiber optic and equivalent in space air transparent spectral window
LCOS	Liquid crystal on silicon (LCOS)	Switching is done via a transistor fabricated on the silicon substrate in which liquid crystal transmissivity and reflection can be used

(*Continued*)

(Continued)

Abbreviations/Terms/Notation	Description	Remarks/Meaning/Properties
LD	Laser diode	InGaAsP for 1550 nm C-band. Many types such as Fabry Perot (multi-longitudinal modes), DFB laser (single mode linewidth 1–2 MHz), External cavity laser (linewidth 10–100 kHz)
LE	Linear equalizer	Equalization in the digital domain with finite impulse response for first- or second-order filtering, for example, transversal filter
LEAF	Large effective area of fiber	Single-mode fiber whose effective area is larger than that of SMF, that is longer radius of the mode spot size. This larger area would offer a higher level of fiber resistance to nonlinear effects such as SPM, XPM, FWM
LEO	Low earth orbit	LEO satellites occupy the lowest orbit of all satellite types, often between 160 and 20,600 km above the Earth's surface. This proximity to Earth makes them ideal for communications, especially laser/photonic communications at ultra-high capacity. The latency is low as well as shorter distance but has higher Doppler frequency. LEO rotation speed is about twice that of the GEO satellite. Doppler frequency shift between GEO and LEO can reach 7.5 GHz. The LEO satellite constellation requires a smaller number of satellites than the GEO constellation. Rotation speed can be twice of GEO rotation speed delay time is about 50 ms. LEO satellites can complete, with faster rotation speed, an orbit in as short as 40–100 min. Note Doppler frequency shift can be much higher as the speed difference is much higher with respect to the transmitter mounted on Earth's ground station for other satellites of different orbit. Higher hand-shake handover from one satellite to the other
LIDAR	Light imaging, detection, and ranging	LIDAR is a method for determining ranges by targeting an object or a surface with a laser and measuring the time for the reflected light to return to the receiver. This technique can be incorporated with SPC to identify the estimated attenuation of the spectral region where the communication channel can be established
LiNbO3	Lithium niobate	Low loss substrate materials for optical modulators/switches by titanium diffusion and travelling wave electrodes
LLC	Limited liability company	
LLI	Long lead items	
LNA	Low noise amplifier	Optical amplifier whose noise level is very low so that signal power of –65 dBm can be distinct
LO	Local oscillator laser	
LOFO	Local oscillator frequency offset	Different frequency between the lightwave carrier and that of the local oscillator. Doppler shift in SPC creates the main LOFO
LOS	Light of sight	Light of sight of lightwave beam
LPF	Low pass filter	Electrical filter with low pass characteristics. In the optical domain, it is an optical bandpass filter. Digital filter by finite impulse response

(Continued)

(*Continued*)

Abbreviations/Terms/Notation	Description	Remarks/Meaning/Properties
LP_{mn}	Linear polarized mode of order *m* and *n*	Modes of weakly guided optical fiber can be considered linearly polarized due to small difference in the refractive index of core and cladding regions (0.3% only). M indicates the null intensity in the azimuthal direction and n in the radial direction. LP_{01} is the fundamental mode of a single-mode fiber
LP_{mn}	Linearly polarized	Guided lightwave mode in a circular waveguide, the optical fiber. Due to the small refractive index difference of only 0.3% between the core and cladding region. The fundamental mode is guided with only 70% optical field intensity obtained in the core region. Thus, the fundamental mode is weakly guided along the SMF. Their polarization is linearly distributed, thus the name linearly polarization. The fundamental mode is LP_{01} with two orthogonal polarizations
LSB	Least significant bit	Bit of a digital quantizer of lowest levels
MADPSK	Multi-amplitude differential phase shift keying	Modulation scheme with multi-level in amplitude and phase distribution in a circle with optimum Euclidean distance
MAP	Maximum A Priori	Algorithm based on known and experience of data sequence in previous detection/estimation
MAPSK	Multi-level amplitude-phase shift keying	Modulation scheme in which both amplitude and phase are used to code data sequence
Mary QAM/M-M-ary QAM/ mQAM	Multiple level quadrature amplitude modulation	A complex modulation with multilevel. The number of levels is on both real and imaginary axes. Thus, the number of constellation points equals 2^N, with N = number of amplitude levels. The effective number of bits per symbol is N
Mary-ADPSK	M- point constellation ADPSK	Modulation scheme with multi-amplitude level and phase distribution
mASK-MSK	Multi-level amplitude-minimum shift keying	Multi-level amplitude and continuous phase (MSK) modulation schemes
Mbps or Mbits/s	Mega-bits per second	The bit rate in Mega-bits per second and abbreviated as Mbps or Mbits/sec. Higher bit rate can be used such as Gbps (Giga-bits/sec or Tbps (Tera-bits/sec).
MD-ROADM	Multi-dimensional reconfigurable optical add-drop multiplexers	Routing in multi-dimensions using add/drop mux and optical cross-connect with optical amplifiers to compensate insertion losses of photonic devices. Optical filtering of channels can reduce optical noises
MEMS	Micro electro-mechanical switch	
MEO	Medium Earth orbit	MEO satellites orbit at a lower altitude than GEO and higher than LEO, usually occupying space between 5,000 and 36,000 km (3100–7500 miles). Their relative proximity to Earth means they achieve far lower latency than GEO units. Their smaller size and lower orbit mean between 8 and 20 units will be required to provide complete coverage of the Earth. Rotation speed is approximately estimated using the relation given in "orbital period". Hand-shake over other satellites for about one hour via the use of beacons at about 20 kHz in the optical domain

(*Continued*)

(Continued)

Abbreviations/Terms/Notation	Description	Remarks/Meaning/Properties
MEO-MEO	Transmission between MEO-to-MEO satellites	Optical hop between two MEO satellites on which optical transceivers are installed
Mid-OA	Mid-range optical amplifiers	Optical gain is moderate with the input signal level; typically in the range from10 to 10 dBm. Typical single-stage EDFA
Mid-OA	Mid-range optical amplifiers	Optical gain is of moderate level with the input signal level; typically in the range from10 to 10 dBm. Typically, a single stage EDFA and typically gains 20–30 dB
ML	Maximum likelihood	
MLSD	Maximum likelihood sequence detection	Algorithm for equalization based on the optimum sequence point states within the likelihood of all possible sequences. Computational time-consuming and high dynamic storage
MLSE	Maximum likelihood sequence estimation	A signal-processing algorithm in which the best maximum likelihood of match of a sequence is recovered from a combination of many possible sequences
MMI	Multi-mode interference	Planar optical waveguide whose width is much larger than the wavelength of the operating lightwaves. Thus, it can support multimode in a lateral direction. These modes can interfere and maxima and minima interference modes can be created
MOC	Optical modulator component	Optical component that can modulate lightwaves
MOCVD	Metallic organic chemical vapor deposition	Fabrication techniques for planar or circular optical waveguide/circuits
Mod MAPM	Multi amplitude phase modulation	Modulator that can modulate the amplitude of the lightwaves and also the phases of the lightwaves at different levels
Mod Star QAM	Quadrature amplitude modulation with circles of constellation point distribution	QAM constellation distribution with maximizing Euclidean distance between points
Modulation	Technique to manipulate carrier/optical carrier lightwaves so that information can be carried by such lightwaves through space media or guided wave fibers.	Modulation of lightwaves via an optical modulator whose electrodes are fed with electrical signals of data. The electric field developed across the guided lightwave channel can change its refractive index; hence, the phases of the lightwaves and then interference of such waves can be ON or OFF, thus switching "1" to "0"
Modulation complex	Modulation using both the real and imaginary components	QAM is commonly used for complex modulation – QAM = quadrature amplitude modulation Higher order QAM is assigned as Mary-QAM with $M = 2^n$, n = 2, 4, 8, 16 etc. In an optical modulation, the higher the number n, the higher the sensitivity of the receiver is required due to the narrowing of the level of the Euclidean distance. This distance is limited by the modulation amplitude of the modulation such that it must meet the linear region of the modulator

(Continued)

(Continued)

Abbreviations/Terms/Notation	Description	Remarks/Meaning/Properties
Modulation DPSK	Differential phase shift keying modulation	Modulation with the phase difference between two consecutive bits. Thus, doubling the bit rate to obtain such differential coding of phase. This offers an advantage that laser phase fluctuation is minimal at two consecutive bits, hence eliminating the phase fluctuation of a laser
Modulation DuoB	Duo-binary modulation	In this modulation, the duobinary indicates two states of binary signals and returns to zero within one-bit period. Binary can be in both amplitude or phase of the lightwaves. The binary amplitude can be either "1 to 0" or "0 to −1"
Modulation M-ary-QAM	16 QAM, 32 QAM, 64 QAM, 256 QAM	The number of points in the constellation of 2D distribution. The number of constellation points, N, equals the exponential of n which is the number of symbols per bit, thus $N = 2^n$, where n = number of symbols. For example, 16 constellation points requires $n = 4$.
Modulation NRZ-ASK	Non return to zero amplitude shift keying	The data stay ON or OFF for the whole period of a bit or baud period. NRZ_ASK is commonly used when coherent reception is employed. If incoherent, then NRZ-OOK (on-off keying)
Modulation QPSK	Quadrature phase shift keying	Modulation technique using both the in-phase and quadrature components (that is orthogonal components). This is thus a complex modulation
Modulation RZ-ASK	Return to zero amplitude shift keying	The data stay ON and return to OFF within the whole period of a bit or baud period. RZ_ASK is commonly used when coherent reception is employed. If incoherent, then RZ-OOK (on-off keying) or possibly Manchester coded sequence
Modulation statistical	Statistical or probability modulation	Constellation points are modulated such that their position in time can be probability distributed maximizing the Euclidean distance between adjacent points. The symbols per bit may be reduced accordingly. This modulation is to be combined with a complex modulation scheme
MPB	A commercial name of company	Company produces optical amplifiers of high power as boosters and low noise amplifiers used as optical pre-amplifiers
M-QAM or mQAM	M point constellation QAM, M, m = 2, 4, 8, 16, 32, 64, ..., N integer	Number of constellations points in a complex plane, i.e., real and imaginary axes
MQW	Multi-quantum well	Gain medium for optical amplification or laser if two reflectors are integrated
MRR	Micro-ring resonator distributed Bragg reflector	Integrated optic micro-ring can be set to resonant or non-resonant by a signal applied to electrode ion a path of the optical waveguide which changes the phases and pushes the resonator into resonance "1" or non-resonance "0"
MRR_DBR	Micro-ring resonator with distributed Bragg reflector	Resonator whose structure is circular or an optical loop that can be tuned to resonance or out of resonance with feedback frequency component from DBR structure. Thus, it can be a precise autonomous laser or optical filter
MRRM	Micro-ring resonance optical modulator	Optical modulator using micro-ring structure and electrode applied along the path of the ring under applied voltage to tune the ring into resonance or not, thence modulation of the lightwave intensity

(Continued)

(Continued)

Abbreviations/Terms/Notation	Description	Remarks/Meaning/Properties
MSA	Multi-source-agreements	
MSB	Most significant bit	Bits of high levels of a digitizer, e.g., ADC or DAC
MSK	Minimum shift keying	A modulation format with continuous phase at boundaries of the consecutive bits. The phases are linearly distributed so as to obtain the phase continuity. The spectrum of MSK signals is orthogonal and gives sharp and minimum required bandwidth.
MSR	Modified Strehl ratio	A modification of the Strehl ratio such that the phase error is normalized with the reference phase point of the perfect optic system
Mux	Multiplexer/Demultiplexer	O-Mux = Optical multiplexer/O-Demux = Optical demultiplexers
Mux	Multiplexing/Multiplier	Multiplexing of a number of waves, optical or electrical
Mux	Multiplexer	
Mux	Multiplexing/Multiplier	Multiplexing of several waves, optical or electrical, into one multiplexed channels in optical domain
M* × *N WSS	WSS with M input ports and N-output ports	*M* × N WSS (*M* input ports *N* output ports) can also be specified as *N* × *M* WSS if such WSS is operating in the reverse direction. The wavelength of each optical line can be selected and rearranged to propagate to other optical lines and hence the ports. The WSS can operate in the reverse direction as no nonlinear effects or Faraday rotator is used
MZDI	Mach-Zehnder delay interferometer	An optical Mach-Zehnder interferometer whose optical paths are asymmetrical by one path has an optical delay path
MZIM	Mach Zehnder intensity modulator	Optical interferometer of Mach-Zehnder structure and applied signals to the electrode so that the phase of lightwaves in a path to combine with a non-modulated path to result into constructive "1" or destructive "0" optical pulse output
MZM/MZIM	Mach-Zehnder modulator or Mach-Zehnder intensity modulator	Optical modulator whose structure consists of two splitters at the input and output connected then by two optical paths (integrated guided wave structure), thus acts as an interferometer. The applied voltage across these two optical paths can change the refractive indices of the waveguides, hence the lightwave phases, thus combined lghtwaves at the output of the output splitter result in constructive ("1") or destructive ("0") lightwaves
n(x)	Refractive index	Refractive index of a medium can slow down the velocity of lightwaves, hence resulting in an effective index of a guided lightwave in an optical waveguide or optical fiber
N.E.W.S	North east west south	
n+ or p+	Heavy n-type or p-type doped semiconductor region	

(Continued)

(Continued)

Abbreviations/Terms/Notation	Description	Remarks/Meaning/Properties
Nd:YAG	Neodymium doped Ytterbium As Garnet	Laser gain medium, integrated with a reflector to produce laser at 1060 nm line
NEA	Noise equivalent angle	Angle of reception under a boundary for tracking accuracy of the system, which represents the uncertainty of the pointing error estimation due to measurement errors, the error vector magnitude, and phases. Convex or concave lenses can suffer this error and limitation. The use of flat-surface micro-lenses, such as GRIN or SELFOC lensed fiber, can be much more efficient and reduce NEA effects. SELFOC = self focusing
$n_{eff(x)}$	Effective refractive index	Refractive index as seen by a guided wave or along the propagation of a lightwave is the projection of the lightwave propagation vector to the propagation direction. This effective index is always less than the refractive index of the medium, by definition
NEWS_ROADM	ROADM to four directions north-east-west-south	Four directions reconfigurable optical add/drop multiplexer
NF	Noise figure	Ratio of noise contribution to the system under original and system under additional noises, normally in unit of dB but linear noise factor can be derived
NL	Nonlinear	Nonlinear effects over a short fiber length under high power pumping ~1000–2000 mW, effective area = 50 mm^2; SMF-28 type
NLE	Nonlinear equalizer	
NLSE	Nonlinear Schrödinger equation	Wave propagation equation with linear and nonlinear effects on phase
NRZ IM/DD	Non-return-to-zero intensity modulation direct detection	That is ON-OFF modulation of the intensity of lightwaves and detection using a photodetector without mixing with a local oscillator laser (coherent detection)
NRZ-ASK	Non-return to zero amplitude shift keying	Modulation technique with non-return to zero pulses in one bit or baud period and binary level in amplitude
NRZ-PRBS	Non-return to zero pseudo random binary sequence	PRBS of non-return to zero sequence
OA	Optical amplifiers	The amplifying device accepts photons and outputs photons with much higher densities. The typical optical gain is about 30 dB. It can consist of single or multiple stages. It requires external power input in an optical domain such as an external laser pump into the optical gain medium. Optical gain medium can be rare-earth doped silica or semiconductor compound, which can be biased to generate an overpopulation region from where electrons can pump down and generate lightwaves output.
OADM	Optical add/drop multiplexer	Multiplexer in optical domain with thru port and add port or drop port. Different wavelength channels can be added or dropped to appropriate ports
OADMs	Optical add-drop multiplexers	

(Continued)

(Continued)

Abbreviations/Terms/Notation	Description	Remarks/Meaning/Properties
O-band	Zero-dispersion band	Central at 1310 nm and double side band of ten depending on doped impurities.0 or higher nano-meters. Fiber zero dispersion around 1312 nm
OBPF/OLPF	Optical band/low pass filter	
ODL	Optical delay line	
ODQPSK	Offset differential quadrature phase shift keying	Offset Differential Quadrature Phase Shift Keying indicates the QPSK is offset by an angle, normally by $\pi/4$ on the constellation complex plane
OF	Optical filter	Filtering of modulated lightwaves whose bandwidth can be narrower than the passband of the filter. The phase characteristics must be near so that there is no distortion of the modulated lightwaves. Parameters = optical passband (3 dB double side), roll off band (from 3 dB point to −40 dB point). The roll off determines how sharp the filtering band is. 6dB/octave can be used to meaning the roll-off is dB scale of 6dB when the frequency is doubling. This indicates that at this point represents a "pole" existing in the system transfer function in frequency domain.
OFDM	Orthogonal frequency division modulation	A modulation technique in which the bit can be expressed by a combination of orthogonal frequency components. Thus, the higher the number of frequency components, the more accurate but requires higher sampling rate. The spectrum of OFDM is quite sharp and the recovery of such signals can be in frequency domain. In optical modulation, the analog signal requires so the combination of such frequency domain components look like a random noise like signal
OFDR	Optical frequency discrimination receiver	Optical receiver using frequency discriminator, which can be a sharp and narrow band filter tuned to the two frequency components of the modulated lightwaves
O-Front-E	Optical front end	
OGS	Optical ground station	Station on ground in which optical transmitter and receiver are installed for communication between ground and satellite
OGS or OGT	Optical ground station /terminal	Station for laser communication based on ground level. Scintillation effects to transmit signals are expected, i.e., delay broadening of bits.
OH/OS	Optic head/optical system	The optical telescope at the transmitter is used for long-range transmission or at the receiver to receive sharp beam image to focus on the fiber optic line in space laser communication system. Optical gain ~116 dB for 160 mm diameter telescope lenses. Thus, a total of ~210 dB for both telescope lenses (OH) at Tx and Rx
OI	Optical isolator	Isolator to isolate reflected lightwaves. Common component is a combination of a Farady rotator and a phase retarder
OIF	Optical inter-networking forum	

(Continued)

(*Continued*)

Abbreviations/Terms/Notation	Description	Remarks/Meaning/Properties
OISL	Optical inter-satellite link	Optical link in free space between the optical transmitter to optical receiver of optical transceiver in bidirectional transmission
O-MADPSK	Offset MADPSK	
ONF	Optical notch filtering/filter	
ONT	Optical network transporter	
oOFDM	Optical orthogonal frequency division modulation	Optical modulation of scheme in frequency domain whose components are orthogonal via FFT and IFFT
OPA or HPA	Optical power amplifier, or booster amplifier or high power amplifier	Optical amplifiers whose output can reach 40–47 dBm, commonly employing multi-stage pump rare earth doped amplifying stages. Fiber head cooling required
OPC	Optical polarization combiner (integrated optic)	Combining polarization, orthogonal polarized channels into one optical channel with dual polarizations
OPLL	Optical phase locked loop	Optical loop that locks the carrier phase and then regenerates new optical frequency laser of the same frequency
OPM	Optical phase modulator	Phase modulation of lightwaves, normal via an optical waveguide whose refractive index is changed or modulated via electro-optic effect and the applied electrical signal on the electrode
OPM	Optical phase modulator	Consisting of an optical waveguide over which both sides are deposited with two metallic (gold plated o Ti/Al thin layer) so that applied voltage can be inserted to generate electric field that changes the refractive index of the waveguide, thence in turn change the phase of the lightwaves. Thus modulation of the phase of the lightwaves.
OPM	Optical power meter	Meter to measure optical power. For high power lightwaves, optical absorber is used. Important for SPC at the transmitter to indicate the high power launching into space channel
OPreA	Optical pre-amplifier or low-noise amplifier (LNA)	Low-input level and high gain with low noises to receive low level optical signals and minimum noises. Very important for space communication link. Typical OpreA can accept −60, to −65 dBm and gain 50 dB with multiple amplifying stages. Normally, a two-stage amplification with a sharp optical filter inserted in between
Optical noises	Quantum noise spectral density	Shot noises, due to the high-power optical beam to the photodetector, is the product of the electronic charge and power divided by the quantum responsivity (ratio of quantum efficiency multiplied by electronic charge divided by the optical power), then multiplied by the bandwidth of the photodetector and the electronic amplifier. Measurement in terms of A^2/sqrt (Hz) in equivalent electronic current power
optical repeater	O/E/O	Conversion of optical to electrical signals, which are then used to modulate the lightwaves to launch into the next transmission span

(*Continued*)

(Continued)

Abbreviations/Terms/Notation	Description	Remarks/Meaning/Properties
Optical router		
Optical transponder		Optical repeater in optical domain if power meets the next hop transmission. If not, then repeater is to be implemented
Optilab	A commercial company name	Company producing optical amplifiers, active and passive optical components
OptPreAmp	Optical pre-amplifier	Same as LNA low noise amplifier. Input power level acceptable can be down to −655 dBm, amplifier gain of 30–40 dB
Orbital period	Satellite tangential rotation speed around the earth	Satellite orbital periods and speeds can be calculated using the relations $4\pi^2 * R^3 = T^2 * G * M$ and $V^2 * R = G * M$, where R is the radius of orbit in meters; T is the orbital period in seconds; V is the orbital speed in m/s; G is the gravitational constant, approximately 6.673×10^{-11} Nm²/kg²; M is the mass of Earth, approximately 5.98 × 1024 kg (1.318 × 1025 lb)
ORX	Optical receiver	Optical device consisting of a photodetector or a pair of photodetectors connected back-to-back, an electronic pre-amplifier of transimpedance (high speed) or high-impedance (high voltage gain and slow speed). Then possibly connected to an ADC and then a DSP for processing signals in the digital domain. The parameters to evaluate ORX is receiver sensitivity (dBm) under a bit-error-rate specification with or without FEC (Forward error coding)
ORX or ORx	Optical receiver	Coherent receiver which includes balanced PD pair and differential TIA. Used in this book and direct detection receiver is not used as the sensitivity is too low. Coherent detection can achieve an extra 20–30 dB gain in intensity due to the boosting of the amplitude of the local oscillator laser
ORX-DSP or DSP-ORx	DSP based ORX	ORX integrated with DSP for processing signals in digital domain. DSP can generate pre-distortion sequence or processing of sampled in digital domain and processing for mitigation of impairments, clock recovery, phase recovery, frequency shift to mitigate Doppler shift, etc.
OS	Optic system	Optic system for front-end launching beam into space channel; and at front end of receiver so that beam gain can be obtained. OH can also be used
OSA	Optical spectrum analyzer	Analyzing the spectrum of optical channels. Note the vertical reading is power spectral density (dBm/nm). Thus, to obtain the channel power at a particular sampled point, one must multiply this power spectral density with the resolution bandwidth (nm) setting on the OSA. A similar concept is applied to ESA
OSNR	Optical signal to noise ratio	Ratio in dB of the signal average power and that of optical noises
OTDM	Optical time-domain multiplexing	Interleaving lightwave pulses
OTRX	Optical TRX	Optical transceiver consisting of two sections: the optical transmitter (including a narrow linewidth laser) section and an optical receiver section with a local oscillator, along with a DSP sub-system that operates with sift-algorithm for impairment equalization. Coding and pre-compensation can be performed in the transmitting section; ADC and DAC provide signals for modulating the QAM modulator and converting received a signal to digital for DSP

(Continued)

(*Continued*)

Abbreviations/Terms/Notation	Description	Remarks/Meaning/Properties
OTX	Optical transmitter	Optical transmitter consisting of an optical source which can be modulated by modulating the driving current of the laser or externally modulated by an external modulator, commonly an integrated or guided-wave electro-optic modulator (EOM). Typical modulators for 100/200 G PDM-QAM are IQ Mach-Zehnder Modulators based on $LiNbO_3$ and Silicon photonic integrated circuits
OXC	Optical cross-connect	Interconnection between input ports and output ports with no contention. Common OXC is in guided optic structure. OXC, if incorporated with WSSes, can form scalable optical switching matrix and routing of selective wavelength groups of channels of different wavelengths.
P1 … P9	Nine ports of WSS	Nine optical ports of a WSS where channels of different wavelengths can be input for multiplexing or outputs as demultiplexing ports
PBS	(integrated optic) polarization beam splitter	Polarization beam splitter or combiner in guided wave optical platform
PDM	Polarization Division Multiplexing/e roe Demultiplexing	Multiplexing or demultiplexing of polarized beams or guided waves
PDM-IQM	Polarization division multiplexed in-phase quadrature modulator	Optical complex modulator with in-phase and quadrature components and two polarized channels
PDM-QPSK	Polarization division multiplexed QPSK	Same as DP-QPSK if only two orthogonal polarized channels under QPSK modulation scheme
PDP	Photodetector pair	Balanced pair of photodetectors which are connected back-to-back so the 3 dB gain can be achieved, that is operation in push pull mode
PD-QPSK	Dual polarization QPSK	Modulation format QPSK per polarization of lighrwaves, two polarized channels modulation of QPSK whose polarizations are orthogonal to obtain DP-QPSK
PDR	Preliminary design review	
pF	pico-Farad	
PHD	Polarization hybrid coupler	Hybrid coupler with polarization mixing with polarized lightwave channels
Photonics	Used to indicate the art of manipulation of lighwaves at photo levels, that is coherent lightwaves	Photonics term used initially in the mid-1980s when the lightwaves could be coherent and detectedusing both polarizations, the *H*- and *E*-components. This is possible when single-mode optical fibers can be manufactured and the linewidth of the laser is narrow enough to implement coherent detection by mixing the modulated optical signals and the local oscillator laser
PIC	Photonic integrated circuit	Photonic integrated circuit based on Silicon on insulator (SOI) is the most advanced technology for advanced modulation of lightwaves in integrated guided wave structure. Si waveguides are created with pn junction, which operates in reverse bias so that it looks like a dielectric waveguide and transparent at the optical frequency. On the other hand, it can act as an absorber, thus by a modulating voltage signal, the optical field intensity can be changed, hence modulation. The waveguides are designed in such a way that two distinct polarized light can be separated and modulated individually. Optical phase shifter can be built in line to shift the phase of the guided lightwaves to form in-phase and quadrature light beams.
PILAB	A commercial name of company	Company specialized in numerical simulation of laser beam propagation through air media

(*Continued*)

(Continued)

Abbreviations/Terms/Notation	Description	Remarks/Meaning/Properties
PLC	Planar lightwave circuit	Planar integrated circuit is normally used for optical circuits created in SiN_4 (Silicon nitrate) on silica as the refractive index of Si_3N_4 (1.72) is larger than that of silica (1.484) in the C-band. This type of circuit can be used in an integrated coherent receiver as hybrid coupler to provide polarization splitter and combiner and phase shifter of $\pi/2$ for in-phase and quadrature components
PLC	Planar lightwave circuit	
PMD	Polarization mode dispersion	
Pointing loss	Direct point to point loss	Point to point between satellite optical head to optical head. Pointing loss <1.9 dB expected for 5 mrad error. A control feedback mechanism allows such point-to-point alignment implemented to less than 5 μm error in the angle of arc.
Power budget	Budget of loss and power sensitivity of all components of transmission system	Power sensitivity versus transmission distance plus additional loss and attenuation factors of all optical components
PPM	Pulse position modulation	Time domain modulation with pulse at different slot in a bit period
PRBS	Pseudo random binary sequence	Random binary sequence of sufficient length for evaluation of bit error rate or generating eye diagram of channels for evaluating BER by finding eye opening and closure
PS	Phase shifter	Optical phase can be shifted by change of the refractive index of the optical waveguide through which the lightwave channel passes through. Thermal pyro-electric effects are commonly used to shift the phase of the lightwave carrier
P_S, P_L	Optical power of signal and local oscillator laser at the input of the photodetector	Optical signals and optical power of local oscillator
PSD	Power spectral density	
PSK	Phase shift keying	Modulation technique with the phase of the carrier changes at the bit temporal boundary
Q DAC	Digital-to-analogue converter for quadrature phase component	
QAM	Quadrature amplitude modulation	Modulation of waves with in-phase real part) and quadrature or imaginary part, that one is $\pi/2$ phase delayed with respect to the other
QD	Quantum dot	
QD-UTECL	Quantum dot ultra-tune external cavity laser	Same as UTQDECL laser
QKD	Quantum key distribution	Secret keys distribution using quantum level signaling in optical domain in continuous variable, that is the whole bit is used and not sampled amplitudes in discrete mode

(Continued)

(*Continued*)

Abbreviations/Terms/Notation	Description	Remarks/Meaning/Properties
QM	Qualification model	
QPSK	Quadrature phase shift keying	Shifting or binary modulation in x- direction or y-direction ($\pi/2$ phase shift with respect to real component in x-direction)
QPSK	Quadrature phase shift keying	Shifting or binary modulation in x- direction or y-direction ($\pi/2$ phase shift with respect to real component in x-direction)
QPSK25	Qapik modulation with 25 GBaud	The 25 GBaud QPSK modulation, with two orthogonal polarization; the aggregate bit rate is 100 Gbps
Qx, Qy	Quadrature component of X- or Y-polarization channel	Quadrature component of X or Y-polarization channel
rBER	Required BER	Under specific system arrangement, a BER is required and specified
RC	Resistor-capacitor	
Repeater, optical	Optical repeater	Consisting of an optoelectronic receiver and recovering of data band into electrical domain which are used to modulate the transmitter to further transmit to the next optical communication link. If reception in coherence, then use photonic repeater
RF	Radio frequency	Frequency from hundreds of MHZ to a few tens of GHz
RIN	Relative intensity noise	Noises of laser with finite linewidth
RIT	Refractive index of turbulence	Changes of refractive index due to turbulence; normally, this creates multi-optical rays scattering and thus multiple delay paths creating delay distortion in SPC channel signal sequence
RMS	Root mean square	
ROADM	Reconfigurable optical ADD/DROP Multiplexer	Multiplexing and routing of a group of wavelength channels that can be reconfigured, that is shuffling of wavelength channels
ROSA	Receiver optical sub-assembly	
RRC	Root raise cosine	Pulse shaping scheme to obtain the zero-crossing point at the sampling period. This allows sharp roll-off of the signal spectrum and better sampling and lowering of bit error rate
RRM	Ring resonance modulator	
RVP	Runway visual range	Distance that a parallel luminous ray's beam must travel through the atmosphere until its intensity (or luminous flux) drops to 0.05 times its original value
RX or Rx	Receiver	Receiving hardware of signals, amplifying, and followed by an ADC to feed to DSP
Rytov variance	Variance defined Rytov	Variation of refractive index of turbulence, a measured quantity to identify the intensity of turbulence.
RZ33	Return to zero modulation with 33% of bit or baud period	Return to zero pulse with 33% pulse width of the bit period
RZ67	Return to zero modulation (ASK or OOK)	Pulse width of 67% of bit period of return to zero pulse

(*Continued*)

(Continued)

Abbreviations/Terms/Notation	Description	Remarks/Meaning/Properties
RZ-ASK	Return to zero ASK	Modulation technique with amplitude variation and pulses of return to zero within a baud or bit period
s(t)	The modulated pulse	
S/C	Space channel	Channel in space communication consistings of an optical carrier and two sidebands that stored the data
Sa/s	Samples per second	Number of samples per second, usually used for ADC and DAC and the processing in DSP
SOCTN	Space optical communication transport networks or space optical transport networks	Networking and system to transport communicating signals in the optical domain, the communication routes can reach several optical nodes located in satellite and Optical ground stations (OGS)
SAR ADC	Successive approximation register analogue to digital converter	
Satellite constellation	Constellation of satellites in orbits GEO, MO, or LEO	A constellation around the Earth above the distance of a geostationary, medium or low level, GEO, MEO, or LEO. Optical or photonic links of all satellites in the same orbit, that is rotating at the same speed, can form an optical network in space
S-band	Short spectral band	Lower band of central band C-band, spectral window 1500–1545 nm
SBS	Stimulated Brillouin scattering	The process of the generation of frequency shifted waves when fiber under high intensity pumps over a short length of fiber. Initially, acoustic waves or phonons are generated due to the excitation of high intensity beams. Such phonons propagating in the fiber core are also guided as along with the guided optical waves. The moving acoustic grating can match the codirectional propagating optical waves to other optical waves whose frequency is shifted to about 9–11 GHz with respect to its original wave frequency. Also, such SBS generating noises of spectrum in order of few tens of MHz leads to jittering errors on the transported pulse sequence
Scintillation variance		Variation of the index of refraction of turbulence is commonly used for a plane wave (coherent and narrow linewidth laser) under low turbulence and a specific receiver
Scintillations frequency		The speed of the fluctuations of the refractive index of air under turbulence, as well as the amplitude or intensity of such fluctuation
SCOC	Space coherent optical communication	Or SOC communication systems or techniques for space domain or channel using laser beams
SD-FEC or HD-FEC	Software-based/decision FEC (Soft-FEC or Hard-FEC) Soft decision forward error coding and hard decision FEC	FEC using software algorithm embedded in DSP, this is opposite to hard-FEC or FEC using hardware Forward error coding can use either software generated from DSP or by hardware coding circuit. Short terms are normally used as soft-FEC or hard-FEC
SDH	Synchronous digital hierarchy	
SDN	Software defined networks	Networks that can be reconfigured based on programmable software packages
SDRT	Space diversity reception technique	Reception using many photodetectors distributed in space

(Continued)

(Continued)

Abbreviations/Terms/Notation	Description	Remarks/Meaning/Properties
SFFTM	Split-step Fast Fourier method	Numerical technique for propagation of lightwave beam using step by step , each step is split into two-half steps and distortion effects can be inserted in between the half-size steps
SGL	Satellite-to-ground link	Optical link between the satellite and ground station
Simulink	MATLAB Simulink	Simulation model by Matlab with blocks of functions and signal processing tool boxes
SINAD	Signal-to-noise and distortion ratio	
SIT	Spectrum integration technique	Typical spectral integration technique is used to estimate OSNR of lightwave channels
SMF	Single mode fiber	Circular waveguide with a core diameter of about 8.2 μm, cladding diameter 125 μm, refractive index difference of 0.3%, and only one mode of two polarizations (so two polarized modes) can be supported at the operating wavelength of spectral window 1540–1590 nm. Standard SMF or SSMF commonly uses Corning type SMF-28
SMF, SMFe	Single mode optical fiber and enhanced SMF	Circular waveguide which supports only one mode at 1550 nm whose polarizations can be two or two polarized modes. Typical dimension of 4.1 μm radius, refractive index difference of 0.3%, cladding radius about 62.5 μm
SMF-28	Single-mode fiber module SMF-28	Corning Inc. single mode optical fiber
SMF-28e	Enhanced SMF-28	Same as SMF-28 but with water contamination peak suppressed
SMSR	Side mode suppression ratio	Suppression of longitudinal mode in a multi-mode laser spectrum
SNR = signal to noise ratio		Ratio of signal energy (power) to average noise energy over the same interval.
SOC	Space optical communication	The art of communication using lightwaves or laser. The term "Lightwaves" indicates coherent lightwaves
SOCS	Space optical communication systems	System and space channel for communication or routing using lightwaves
SOCTN	Space optical communications transport networks	Networking of optical communication systems in space domain
SOCTN/SOTN	Space optical communication transport network or space optical transport networks	Networking and system to transport communicating signals in optical domain, the communication routes can reach several optical nodes located in satellite and optical ground stations (OGS)
SOCTN/SOTN	Samples per second	Number of samples per second, usually used for ADC and DAC and the processing in DSP
Soft-FEC	FEC using software generated in DSP system and processing, opposite to hard ware FEC	FEC by algorithms embedded in DSP. BER can be at 2e-2 or about a factor of ten better in BER as compared with Hard-FEC. More digital processing load on the DSP
SOI	Silicon on insulator	A platform for photonic integrated circuits (PICs). Substrate is oxide silicon (SiO_2), and then Si films and waveguides are formed to make optical modulators

(Continued)

(*Continued*)

Abbreviations/Terms/Notation	Description	Remarks/Meaning/Properties
Sy./sec or Sy./s	Symbol per second	Number of symbols or baud per second in higher modulation format. For example, QSPK requires 2 bits per symbols so a baud rate or symbol rate of 64 GBaud will contain 64 GSy/s. This will increase the bit rate by a factor of 2 for QPSK
SOI	Silicon on isolator	Integrated optic devices such as modulator by silicon waveguide deposited on an insulator substrate such as Si oxide A platform for photonic integrated circuits (PICs). The substrate is oxide silicon (SiO_2) and then Si films and waveguides are formed to make optical modulators. Alternatively, Si_3N_4 can be used as the insulator so that SI structure can be coupled to Si nitride (Si_3N_4) based optical circuits are used
SON	Space optical NODE	SON is a node configured in a satellite. It is used to route, add, or drop optical channels of different wavelength carriers to the desired OGSes or inter-orbital satellites. All operate in the optical domain; several ports but typically four ports with add/drop capability.
SOTAC	Satellite optical transport access (node)	Optical satellite node to and from which optical channels using one or more 'colors' can be added or dropped to the space optical transport networks
SOTR	Satellite optical transport routing	Routing process over SOCTN employing IP Routing method
SPC	Space photonic communications	Laser modulated beam via using optical manipulation of integrated or guided wave optic devices at photon level or coherent laser lightwaves. Thus, the term photonics. Coherent detection is also the main technique in SPC
SPC	Space photonic communications	Coherent lightwaves are modulated by high-speed signals and transmitted over transparent air (vacuum) or turbulent layer media and then coherent receipted at the receiver with optical amplifications at the transmitter as booster and optical preamplification at the receiver
Spectral bands	Transparent bands over which optical channels can be transmitted with minimum attenuation. These are defined matching to fiber optic bands	C- band 1545–1565 nm C+ band 1545 nm to 1575 nm, L-band = 1565–1580 nm. O-band 1525–1545 nm
SPM	Self-phase modulation	A process of modulation via the changes in the refractive index of the guided medium, the single-mode fiber s pumped with high-intensity beam via the chi-3 coefficient and hence the nonlinear coefficient n_2. The phase modulation leads to in-band crosstalk as the nonlinear interaction process generates waves that fall into the band of other multiplexed channels
SR	The Strehl ratio	The expected value of phase error (complex) over the aperture of the optic head or optical coupling device at the operating wavelength. This quantity can be estimated using WFES (wave front error sensor). SR = 1 = perfect symmetrical optic system and no phase error. Thus, SR is a measure of the magnitude of the square of the vectoral magnitude of the phase front fluctuation caused by random turbulence. SR is a Gaussian distribution of the residual phase variances, thus, it is a real quantity or effective magnitude of the phase error. See also modified SR

(*Continued*)

(Continued)

Abbreviations/Terms/Notation	Description	Remarks/Meaning/Properties
SRS	Stimulated Raman scattering	A process of stimulated emission by the jumping of a lightwaves and electrons of the bonding atoms to other energy levels and then jumping down to other energy levels and hence generating another optical lightwave of different frequencies. The frequency shift is about 8–14 THz, thus very far from the signal bands. This can interfere with other channels if they are in such a frequency shifted region Nonlinear effect via Ramanscattering. The generation of Raman noise carrier term is about few tens of THZ from the main carrier
SSB	Single sideband	Modulation technique with suppression of one sideband but leaving the carrier and one other sideband
SSB-DMT	Single sideband discrete multi-tone	Modulation techniques with one band filtered and discrete frequency component, real components only thus the imaginary components not used. Single sideband in optical domain can be generated by Hilbert transformer
SSMF	Standard single-mode fiber	SMF used throughout fiber optic network with property used as standard for all manufacturers, especially Corning Inc. with lowest loss (0.15 dB/km) and lowest dispersion properties in the C-band (−17 p/nm mm)
STRS	Stimulated thermal Rayleigh scattering	
Sy/s	Symbol per second	Number of symbols or baud per second in higher modulation format. For example, QSPK requires 2 bits per symbols so a baud rate or symbol rate of 64 GBaud will contain 64 GSy/s. This will increase the bit rate by a factor of 2 for QPSK
Tbps	Tera bits per second	Total capacity rate reaching Terabit (1012 bits or 1000 Gigabits/second) level
TCL	Turbulence coherence length	Much less than laser coherence length, which is about few tens of meter for laser of 10 kHz linewidth
TDEQ	Time domain equalizer	
TDM	Time division multiplexing/ multiplexer	Multiplexing of data channels using time delay and combine. Optical delay lines are used for high-speed channel multiplexing. Differential modulation formats and reception in self-coherent reception employ this type of modulation technique
TDR	Time domain reflectometry	
TE/TM	Transverse electric/transverse magnetic	Polarized modes of guided waves in optical fiber or planar channel optical waveguides
TeraWave	Trade name of single mode fiber with wide flat spectral band	
TFF	Dielectric thin-film filter	Optical filter made by thin-film multilayer
TIA	Transimpedance amplifier	Electronic amplifier which accepts input current and transfer to voltage at its output port. TIA is commonly used for wideband amplification of signals in current source form, like photodetector. Classical TIA is a single input port, but advanced TIA is differential form, which is well-suited for balanced detector pairs in coherent receivers as PEGASUS investigated and proposed.

(Continued)

(*Continued*)

Abbreviations/Terms/Notation	Description	Remarks/Meaning/Properties
TOF	Tuneable optical filter	Optical filter whose central band carrier can be tuned to different spectral location
TOF	Tuneable optical filter	Optical filter whose central point and spectral passband can be tuned
TOSA	Transmitter optical sub-assembly	Compact transmitter assembly in pluggable module
Transponder	Optical transponder	Relay of optical channels to next transmission hop
TWE	Travelling wave electrodes	Electrode with long and matched impedance so that RF wave travelling with velocity match theoptical waves for efficient electro-optic interaction
TX or Tx	Transmitter	Transmitting signals with modulation schemes as designed in the phase, amplitude, or complex amplitude-phase.
ULH	Ultra-long haul	Long haul distance in space communication system
UTQDECL	Ultra-tune able quantum dot external cavity laser	External cavity laser with Quantum dot gain waveguide and grating feedback outside the resonant cavity
VE	Vector error	Keysight definition of vector error in a constellation of sampled points
Viavi	A commercial name of company	Specialized in laser, signal generators, and optical modules, analyzer
VOA	Variable optical attenuator	Optical attenuator whose attenuation parameter can be varied, typically from 1 to 50 dB. This is commonly used in adjusting the receiver input power to evaluate the receiver sensitivity. Alternatively, the VOA is used to simulate space propagation loss in a transmission hop
VSA	Vector signal analyzers	Analyzer (instrument) that can analyze complex signals, that is with in-phase and quadrature components
V-V CPE	Viterbi-Viterbi carrier phase estimation	Processing algorithm in digital domain
VVA	Viterbi & Viterbi algorithm	Algorithm in DSP to equalize or compensate distorted signals
WDM	Wavelength division multiplexing	Multiplexing of channels whose wavelengths are in the ITU-grids
WDM	Wavelength division multiplexing	Multiplexing of channels whose wavelengths are in the ITU-grids
WFE	Wave front error	Due to distortion of lens optic system or nonuniform vacuum distribution across the space optical beam. This WFE distorts the signal constellation, e.g., QPSK or M-QAM and creates additional error in vector magnitude of each constellation point. For Space WFE ~1.5 dB as a well-known parameter
WFES	WFE sensor	A matrix of photodetectors in 2D array so that the spatial distribution of the lightwave intensity can be monitored. The spatial matrix can be considered as spatial sampler and hence the complex plane distribution of lightwaves passing through an optical system. This allows the determination of spatial distortion of the lightwaves after passing through such optical system/device. This is equivalent to the EVM of temporal signal distortion. WFES can be used to determine the spatial distortion or broadening of a beam propagating through air space distance due to the finite linewidth of a laser
WFS	Wave front sensor	Same as WFES

(*Continued*)

(Continued)

Abbreviations/Terms/Notation	Description	Remarks/Meaning/Properties
WKB	W.K. Burns	WKB analysis technique
WL	Window length	
WPE	Wall plug efficiency	In optics, wall-plug efficiency or radiant efficiency is the energy conversion efficiency with which the system converts electrical power into optical power at the laser output port. WFE is defined as the ratio of the radiant flux (i.e., the total optical output power) to the input electrical power as measured at the wall plug.
WSC	Wavelength dependent coupler	
WSS	Wavelength selective switch(es)	Optical switching with selection of wavelength channels. Normally via the uses of free space gratings, LOCS switching, and fibers input and output ports. Switching speed is moderate by the speed of transistor behind the liquid crystal. Advance WSS has additional MEMS so that colorless, directionless, contentionless, and flexible grid (CDC-F) ROADM can be structured. WSS can be configured to operate as a multiplexer or demultiplexer. Normally, input port is assigned as common port, that is either as multiplexed channels port for demultiplexing or a combined port for multiplexing. The WSS can operate in diverse direction as no nonlinear effects or Faraday rotator is used. NOTE: the passband of each channel of WSS is tunable from 200 to 12.5 GHz with sharp roll off.
WSS/WSSes	Wavelength selective switch(es)	Optical switching with selection of wavelength channels. Normally, using free space gratings, LOCS switching and fibers input and output ports. Switching speed is at moderate level by the speed of transistor behind the liquid crystal. Advanced WSSes integrate MEMS so that colourless, directionless, contentionless, and flexible grid, CDC-F-ROADM can be structured
XOR	Excusive OR	Logical operation
XPM	Cross-phase modulation	Modulation of phase of different channel due to nonlinear effects via self-phase modulation
XPolM	Cross-polarization modulation	Modulation of different polarized channels due to nonlinear effects
Xtalk	Cross-talk	Crosstalk from one channel to other channel via nonlinear mixing such as SBS, FWM in DWDM channels via nonlinear phase in fiber under high- power launched into fiber from high power amplifier booster
YDFA	Ytterbium – doped fiber amplifiers	Ytterbium-doped fiber core gain spectrum is high and flat over spectral window 1030–1070 nm or 7524 THz active frequency window
YDFL	Yb-Doped fiber lasers	Fiber amplifiers or lasers for operating in a spectral window, 1025–1070 nm or 7524 THz, region. A lasing cavity is normally created by two fiber gratings – tuning of lasing wavelength can be done by tuning the grating length by piezoelectric effects. Yb = Ytterbium
Y-matrix	Admittance matrix	Matrix of admittance of a two-port electrical circuit/network
ZF-NLE	Zero-forcing nonlinear equalization	Algorithm for equalization to mitigate distortion effects

Index

Note: **Bold** page numbers refer to tables; *italic* page numbers refer to figures and page numbers followed by "n" denote endnotes.